EVALUATION OF HUMAN WORK

THIRD EDITION

EVALUATION OF HUMAN WORK

THIRD EDITION

Edited by John R. Wilson
and Nigel Corlett

Boca Raton London New York Singapore

A CRC title, part of the Taylor & Francis imprint, a member of the Taylor & Francis Group, the academic division of T&F Informa plc.

Published in 2005 by
CRC Press
Taylor & Francis Group
6000 Broken Sound Parkway NW, Suite 300
Boca Raton, FL 33487-2742

Printed in the United States of America on acid-free paper
10 9 8 7 6 5 4 3 2 1

International Standard Book Number-10: 0-415-26757-9 (Hardcover)
International Standard Book Number-13: 978-0-415-26757-1 (Hardcover)
Library of Congress Card Number 2004058531

Library of Congress Cataloging-in-Publication Data

Evaluation of human work / edited by John R. Wilson. – 3rd ed.
p. cm.
Includes bibliographical references and index.
ISBN 0-415-26757-9 (alk. paper)
1. Human engineering. 2. Human-machine systems. I. Wilson, John R., 1951- II. Title.

T59.7.E93 2004
620.8'2--dc22

2004058531

Taylor & Francis Group
is the Academic Division of T&F Informa plc.

Visit the Taylor & Francis Web site at
http://www.taylorandfrancis.com

and the CRC Press Web site at
http://www.crcpress.com

Contents

PART VI. Implementation and organisation analysis 917

Preface to the Third Edition

Ergonomics/Human Factors is a fast moving discipline. The domains in which we work and can see demonstrable success from application of our knowledge and skills are constantly expanding. At the same time we are continually improving and enlarging our methodological base, developing new theories, approaches, methods and tools as well as refining those we have used before. This third edition of *Evaluation of Human Work* has been published to try to reflect many of these changes. In every case bar one the chapters from the second edition have been substantially revised, and several new chapters have been added.

In general, the new chapters reflect the growth of interest in a systems viewpoint within ergonomics/human factors, and particularly understanding physical and cognitive work in the context of the social and organisational settings in which it takes place. Thus topics presented in the new chapters include sociotechnical design of work systems (Chapter 29), team design and evaluation (Chapter 30), learning from failures through a joint cognitive systems perspective (Chapter 34), and the analysis of organisational processes (Chapter 38).

In addition to these new chapters, several chapters have been revised not only in terms of updating material but a completely new approach has been taken. Thus the chapter concerned with user trials in the second edition has been expanded into one which addresses many techniques in user-centred design (Chapter 11). The chapter on knowledge elicitation for expert systems has been re-written to reflect the increased interest in understanding the nature of knowledge and of knowledge management in contemporary systems (Chapter 8). The opening chapters on assessment and design of the physical workplace dealing with the environmental factors of climate, visual conditions and noise are now accompanied by a general discussion of environment surveys (Chapter 22). Finally, reflecting what is current practice, the chapter on accident reporting and analysis has been replaced by one on systems for near miss reporting and analysis (Chapter 33).

One thing which has remained unchanged from the first and second editions is that this text is produced NOT as a cookbook of ergonomics methods. Whilst there is a place for handbooks and manuals which describe each of hundreds of different ergonomics/human factors tools and techniques, with some form of look-up table for users to select amongst them, that is not the intention and thrust of this book. Rather the intention is to place ergonomics methodology in context, and each chapter carefully describes the background to method development in that area and to the application of methods and tools. In this way it is intended to make the text suitable for teaching ergonomics and human factors courses beyond those purely concerned with methods of analysis and evaluation, and to try to introduce the topic of ergonomics/human factors from a 'doing it' perspective. We hope that we have succeeded in this aim.

Acknowledgements to the Third Edition

First of all I would like to thank all of the authors for their great efforts in updating chapters and writing new chapters. Particularly for those authors who have been with us since the first edition, I recognise that there is little in it for their careers to publish a revised version of a chapter that has already appeared print, and therefore I am extremely grateful for the diligence and care with which everyone has produced revisions. In addition, I wish to thank the anonymous reviewers who made suggestions on how to improve this third edition and also to all readers and users of the first and second editions who have sent comments and criticisms to the authors and editors.

For this edition the production of the book has moved from Taylor & Francis in the UK to CRC in the United States. It is a measure of the professionalism of all involved that this transition (at the time of writing at least!) appears to have taken place very smoothly. At the publishers we would like to thank Cindy Renee Carelli, Helena Redshaw and Rachel Tamburri-Saunders for all of their help in the production of the book.

A final thank you goes to all my colleagues in the Institute for Occupational Ergonomics at University of Nottingham, and those in other institutions with whom we collaborate. Without the enthusiasm of colleagues for the development of new knowledge in ergonomics/human factors and the application of new and existing methods, I would have little enthusiasm to produce books of this nature. I very much welcome the professionalism and friendship of all my colleagues. Finally I must (yet again) sincerely thank my secretary Lynne Mills for all her hard and insightful work on this book, greatly enhancing its production as near to on time as possible and the quality of the content, as well as acting as a liaison between myself and authors and publishers.

John Wilson
November 2004

Preface to the Second Edition

Since the First Edition of *Evaluation of Human Work* was published, much has happened to change the way we view ergonomics methods and techniques. This has lead to the inclusion of several new chapters in this second edition, and the considerable revision of many others.

Technical, social, political, and legal changes have required continual development and improvement of ergonomics methods. For instance, the ever-increasing power and prevalence of computer systems and the diversity of their user interfaces necessitate parallel improvements in methods of analysis, design and evaluation. We can see this, for instance, in human–computer interfaces generally (revised Chapter 12) and in specialized applications such as control rooms (new Chapter 13). Social and political changes, mirrored by changes in the way industrial work and jobs are organized, have increased recognition of the gains possible from greater involvement of people in what they do and from providing employees with a greater degree of control over their own activities. One manifestation of this is participation, and participative approaches (new Chapter 37) have a long and honourable tradition in ergonomics. Legal developments have had a profound influence on ergonomics in recent years, especially the health and safety regulations governing use of display screen equipment, manual handling work, work equipment and workplaces, which have come into force in Europe, Australia and to an extent in North America. Coupled with costs of compensation claims, such regulations have required structured ergonomics assessments at work (new Chapter 30) within an ergonomics management programme (revised Chapter 1 and new Chapter 35). In addition to the above, this edition includes a new chapter on measurement of physiological functions (Chapter 29), and substantial revision of chapters on task analysis (6), verbal protocol analysis (7), product assessment and user trials (10), knowledge elicitation (14), computer aided methods (20), mental workload (25), and work stress (26).

There are increasing moves towards greater professionalisation in ergonomics, for example the Board of Certification in Professional Ergonomics (BCPE) and the Centre for Registration of Ergonomists in Europe (CREE). Such moves require a recognition that the methods we choose will influence what we find from any investigation, and that methods must produce findings that are valid, reliable and generalizable, meet the objectives of the investigation and be safe and ethical to apply. There can be little excuse for administering questionnaires that make no attempt to use previously validated scales, carrying out experiments without careful piloting, or rigid ill-informed use of assessment checklists.

Any experimentalist should have a good knowledge of statistics. We decided, after long discussion, not to include statistics in this volume — it is heavy enough already! Statistics are necessary not only for experimental design, data compression or testing of results, but for the understanding they bring to the nature of variability and the importance of interactions. Although methods are reported in a 'stand alone' manner, it is rare in ergonomics to find an influence or a cause which has an exclusive relationship with an effect. Hence it is important to retain the ergonomics approach of viewing the whole person within the total environment, an approach which will make it necessary to match a selected group of methods to the requirements perceived. We hope this book will assist ergonomists to do this.

Acknowledgements to the Second Edition

For this second edition we would like to thank all our authors for their considerable efforts in updating their existing chapters or writing new ones; thanks are due also to reviewers of the new chapters. Many readers of the first edition have sent in comments and criticism to authors and editors, and these have been accounted for wherever this improved the book.

Again we must acknowledge the support of Taylor & Francis, and especially Richard Steele and Robert Chaundy, and thanks are due to Chris Stapleton for her professional service on proof reading and the index. The first editor (JW) gratefully acknowledges the Department of Safety Sciences, University of New South Wales for the space and time to carry out most of the editing and writing of new chapters.

Finally, we are very grateful to all our colleagues in the University of Nottingham's Department of Manufacturing Engineering and Operations Management, and especially in the Institute for Occupational Ergonomics, for their collaboration, support and a positive environment. Of these, Lynne Mills has contributed the lion's share in terms of typing, editing, and organization — of the book and of us!

John Wilson and Nigel Corlett
May 1995

Preface to the First Edition

For a long time there existed few books on ergonomics or human factors methodology; Chapanis' *Research Techniques in Human Engineering*, published in 1959, was probably the earliest, as well as the best-known. Lately there has been a slow increase in what is available. For instance one of the contributors to this volume, David Meister, has produced two books dealing with methods (Meister, 1985, 1986) and the present editors have also been involved in two collections of conference proceedings concerned with new methods and techniques (Laboratory of Industrial and Human Automatics, 1987; Wilson *et al.*, 1987).

The books by Meister, excellent in many respects, concentrate upon investigations of large-scale (military) systems design, simulation and evaluation. The two sets of conference proceedings, whilst containing a range of methodological developments and applications, represent what was selected from the papers submitted for presentation, and cannot pretend fully to represent the held. Also produced recently is the authoritative *Handbook of Human Factors*, edited by Salvendy (1987). This does have much to say about methods and techniques, both as separate chapters or as parts of other chapters; nonetheless its intention is to be a comprehensive, general text, with explanation of theories, principles, data and application, as well as of methods.

Our aim with this volume on ergonomics methodology is to produce a text on methods and techniques that is both broad and deep. We intend it to be a companion to the major general textbooks on ergonomics and human factors, particularly and most recently those of Bailey (1982), Grandjean (1988), Kantowitz and Sorkin (1983), Oborne (1987), Salvendy (1987) and Sanders and McCormick (1987). All of these are well known to students, teachers and practitioners of ergonomics, as well as to many of those from other disciplines who take a personal or professional interest in ergonomics. There is, though, little opportunity in such texts to emphasize and make explicit the major part of methodology.

Therefore we have set out to produce a general text on ergonomics methodology. As the book's title implies we are primarily concerned with people at work and with applied rather than basic research. However, the former concern has not ruled out contributions relevant to people's activities at home, leisure or on the road; nor does the latter concern invalidate descriptions of laboratory-based methods — these can have outcomes that are as practically applicable as are those from field investigations.

The contents of the book are intended to be interesting and useful for a wide range of people, including: *students*, to give them a feel for ergonomics investigation and to complement their learning of theory and principles; *industrial and business personnel at all levels*, to allow them to understand better what ergonomics can do for them, why, and how; and *ergonomics practitioners, researchers and teachers*, to give them a compendium of methods and techniques available. For all these groups the contributions here will also point to further sources for more detail on specific topics.

Our text on evaluating human work has brought together experts from many branches of ergonomics theory and practice, and has allowed them the space to introduce and give detail on those methods and techniques of value to then. Since ergonomics is both a science and a technology, these methods can of course be concerned with collecting data or with applying their own or others' data. The primary thrust of each contribution may be general method (e.g. direct observation or protocol analysis), or particular fields of application for several types of method (e.g., mental workload or the climatic environment). Whilst there will no doubt be omissions — of branches of methodology or of techniques within one area — regretted by some readers, we trust that most will find the book to be a comprehensive, readable and useful source of ergonomics knowledge and practice. Certainly we believe that for those students or readers from industry who are relatively new to ergonomics, one of the most interesting and valuable ways to learn about it is through its rich and varied methodology.

John Wilson and Nigel Corlett
University of Nottingham
July 1989

References

Bailey, R.W. (1982). *Human Performance Engineering: A Guide for System Designers.* (London: Prentice Hall), pp. 656 + xxviii.
Chapanis, A. (1959). *Research Techniques in Human Engineering.* (Baltimore: John Hopkins Press), pp. 316 + xii.
Grandjean, E. (1988). *Fitting the Task to the Man: A Textbook of Occupational Ergonomics*, 4th Edition. (London: Taylor & Francis), pp. 363 + ix.
Kantowitz, B.H. and Sorkin, R.D. (1983). *Human Factors: Understanding People-System Relationships.* (New York: John Wiley & Sons), pp. 699 + xii.
Laboratory of Industrial and Human Automatics (1987). *New techniques and ergonomics*: Proceedings of an International Research Symposium. (Paris: Hermes).
Meister, D. (1985). *Behavioural Analysis and Measurement Methods.* (Chichester: John Wiley & Sons), pp. 509 + ix.
Meister, D. (1986). *Human Factors Testing and Evaluation.* (Elsevier Science), pp. 424 + xi.
Oborne, D.J. (1987). Ergonomics at Work, 2nd Edition. (Wiley & Sons), pp. 386 + xvii.
Salvendy, G. (editor) (1987). *Handbook of Human Factors.* (New York: Wiley & Sons), pp. 1874 + xxiv.
Sanders, M.S. and McCormick, E.J. (1987). *Human Factors in Engineering and Design*, 6th Edition. (New York: McGraw-Hill), pp. 664 + viii.
Wilson, J.R., Corlett, E.N. and Manenica, I. (1987). *New Methods in Applied Ergonomics.* (London: Taylor & Francis), pp. 283 + x.

Acknowledgements

Our first debt with this book is to our contributing authors, all of whom have responded to our various requests with great patience, and have produced chapters of high quality within, in some cases, a very limited time. Amongst these authors we must mention those who were with us in the initial discussions about the book at the 2nd International Occupational Ergonomics Symposium at Zadar, Yugoslavia; they were Lisanne Bainbridge, Colin Drury, Ted Megaw, Ken Parsons, Pat Shipley and Rob Stammers. Cohn Drury in particular has contributed much in terms of individual chapters, and the overall content and style of the book.

We would like to thank our colleagues at Nottingham University for contributing to a working environment in which we feel able to embark on and complete this and other publishing ventures. One of us (JW) must also thank the Department of Industrial Engineering and Operations Research, University of California, Berkeley, for allowing him time and facilities to work on this book during periods there as a visitor in 1987 and 1988.

Our editors at Taylor & Francis — David Grist, Sarah Waddell and, for most of the time, Robin Mellors — have been exceedingly supportive, even in the face of a project which seemed to grow exponentially! The style of the book has been enhanced tremendously by the artwork of Tony Aston and cartoons of Moira Tracy. Despite both editors being away from Nottingham for substantial periods of time the production of the book has rolled on relatively smoothly; our colleagues would say this was because we left this and much else in the hands of our excellent secretaries, Lynne Mills and Ilse Browne, to whom we are immensely grateful.

Contributors

John Annett
University of Warwick
Coventry, United Kingdom

Chris Baber
Electronic, Electrical & Computing Engineering
University of Birmingham, Edgbaston
Birmingham, United Kingdom

Lisanne Bainbridge
18 Osler Road
Headington, Oxford OX3 9BJ
United Kingdom

Ann Bisantz
Department of Industrial Engineering
State University of New York at Buffalo
Buffalo, New York

Mark Bullimore
The Ohio State University
Columbus, Ohio

Shawn Burke
Department of Psychology & Institute for Simulation & Training
University of Central Florida
Orlando, Florida

Nigel Corlett
Institute for Occupational Ergonomics
University of Nottingham
Nottingham, United Kingdom

Tom Cox
Institute of Work, Health & Organisations
University of Nottingham
Jubilee Campus
Nottingham, United Kingdom

Patrick Dempsey
Liberty Mutual Research Center for Safety & Health
Hopkinton, Massachusetts

Colin Drury
Department of Industrial Engineering
State University of New York at Buffalo
Buffalo, New York

Ken Eason
Department of Human Sciences
Loughborough University
Loughborough, Leicestershire
United Kingdom

Tim Gallwey
Ergonomics Research Centre/Department of Manufacturing & Operations Engineering
University of Limerick
National Technology Park
Limerick, Ireland

Amanda Griffiths
Institute of Work, Health & Organisations
University of Nottingham
Nottingham, United Kingdom

Helen Haines
School of Mechanical, Materials & Manufacturing Engineering
University of Nottingham
Nottingham, United Kingdom

James Hartley
Department of Psychology
Keele University
Staffordshire, United Kingdom

Christine Haslegrave
School of Mechanical, Materials & Manufacturing Engineering
University of Nottingham
Nottingham, United Kingdom

Sue Hignett
Department of Human Sciences
Loughborough University
Loughborough, Leicestershire
United Kingdom

Erik Hollnagel
CSELAB, Institute of Computer and Information Science
University of Linköping
Linköping, Sweden

Peter Howarth
Visual Ergonomics Research Group (VISERG)
Department of Human Sciences
Loughborough University
Loughborough, Leicestershire
United Kingdom

Åsa Kilboni
Applied Work Physiology Division
Institute of Occupational Health
Stockholm, Sweden

Barry Kirwan
Safety R & D, Eurocontrol Experimental Centre
Centre de Bois des Bordes
Bretigny/Orge, France

Masahara Kumashiro
Department of Ergonomics
University of Occupational & Environmental Health
Kitakyushu, Japan

Ron Laughery
Micro Analysis and Design, Inc.
Boulder, Colorado

Veikko Louhevaara
Finnish Institute of Occupational Health
Kuopio, Finland

Ian McClelland
User Experience Architect, Usage Centred Design
Philips Digital Systems Laboratory
Eindhoven, Netherlands

Ted Megaw
Industrial Ergonomics Group
Electronic, Electrical & Computing Engineering
University of Birmingham
Birmingham, United Kingdom

David Meister
Micro Analysis and Design, Inc.
Boulder, Colorado

Wendy Morris
Land Rover
Coventry, United Kingdom

Ged Morrisroe
Network Rail
Euston, London, United Kingdom

Moira Munro
69 Lomondside Ave
Glasgow 976 7UH, United Kingdom

Leonard O'Sullivan
Ergonomics Research Centre/ Department of Manufacturing & Operations Engineering
University of Limerick
National Technology Park
Limerick, Ireland

Maurice Oxenburgh
Tackjarnsv. 16(tr5)
Bromma 168 68
Sweden

Ken Parsons
Department of Human Sciences
Loughborough University
Loughborough, Leicestershire
United Kingdom

Stephen Pheasant
Department of Household & Consumer Studies
Wageningen University
Wageningen, The Netherlands

Heather Priest
Department of Psychology & Institute for Simulation & Training
University of Central Florida
Orlando, Florida

Jane Rajan
Ergonomiq Ltd.
Markyate
Herts, United Kingdom

Eduardo Salas
Department of Psychology & Institute for Simulation & Training
University of Central Florida
Orlando, Florida

Penelope Sanderson
ARC Key Centre for Human Factors & Applied Cognitive Psychology
University of Queensland
Brisbane, Queensland, Australia

Nigel Shadbolt
Department of Electronics & Computer Science
University of Southampton
Southampton
United Kingdom

Andrew Shepherd
University of Loughborough
Loughborough, Leicestershire
United Kingdom

Carys Siemienuch
Department of Human Sciences
Loughborough University
Loughborough, Leicestershire
United Kingdom

Geoff Simpson
EMS Risk Consulting Ltd.
Lymm, Cheshire, United Kingdom

Murray Sinclair
Department of Human Sciences
Loughborough University
Loughborough, Leicestershire
United Kingdom

Rob Stammers
Centre for Applied Psychology
University of Leicester
Leicester, United Kingdom

Bea Steenbekkers
Department of Household & Consumer Studies
Wageningen University
Wageningen, The Netherlands

Jane Fulton Suri
IDEO
The Embarcadero
San Francisco, California

Tjerk van der Schaaf
Department of Technology Management
Eindhoven University of Technology
Eindhoven, The Netherlands

Pat Waterson
Fraunhofer IESE
Kaiserslautern, Germany

John Wilson
Institute for Occupational Ergonomics
University of Nottingham
Nottingham, United Kingdom

John Wood
CCD Design & Ergonomics Ltd
Golden Cross House
London, United Kingdom

Linda Wright
Postbus 2038
3500 GA Utrecht
The Netherlands

Chapter 1

Methods in the understanding of human factors

John R. Wilson

Introduction

It is common to introduce the idea of ergonomics, and of the importance of the human factor, by referring to one of a number of well publicised disasters in which major loss of life and great commercial costs occurred — from Flixborough to Bhopal, from Three Mile Island to Chernobyl, from Piper Alpha to Ladbroke Grove. Many examples can be found, for instance in Beaty (1995), Bignell *et al.* (1977), Casey (1993), Cushing (1994), Perrow (1984), Petroski (1992) and Reason (1990).

Less seriously, but very indicative of the importance and wide variety of human factors relevant to successful (and unsuccessful) systems, in April 1993, one of the UK's premier horse races, the Grand National, was turned into a debacle when the starting gate stuck. This 70 m strand of cotton tape, 3 cm wide, failed to rise quickly enough and wrapped itself around some of the jockeys and horses. The officials failed to prevent a number of the horses running almost 8 km in two laps of the track, others never started and yet others ran part way before realising something was amiss. Inquiries into the event identified a number of interacting causative factors, all of which had human factors components. These included: poor design and maintenance of equipment; earlier rejection, on grounds of physical environment conditions, of an improved design for this equipment; an inadequate testing programme; blocked lines of sight; impossible auditory communications; confusion caused by an unusual but foreseeable occurrence in the social environment; lack of any thought given to, or rehearsal of, a 'worst case scenario'; errors made by personnel in both omitting to do things (errors of omission) or doing them wrongly (errors of commission); and nobody taking charge of a deteriorating situation by assessing alternatives and using available backup systems to restrict damage. However, at the investigation, attempts were made to place all blame on one low paid employee who was to act as a back-up in cases like this (*The Guardian*, 1993).

In many ways very funny, with no one injured, the incident nonetheless incurred about £6 million in lost tax revenue and other costs. However, it is no laughing matter when events such as those described above occur in situations of potential injury or death, for the workforce or the public, or when the financial consequences of poor design result in closure of a small enterprise, or when operators are blamed for what is really failure in design or management. Yet many examples of inappropriate design of equipment, workplaces, systems, jobs and organisations can be found in large and small companies, in offices and factories, in physical and mental work.

Across a whole range of ICT (information and communications technology) implementation projects we can find examples where end users are not clearly identified and their needs not accounted for, where systems become excessively complex or have far more functionality than really needed, user training and support are not sufficient, interfaces hinder rather than help users, and the development process itself lacks clear objectives, is inflexible, and is too technology driven (e.g., Eason, 1997; UK National Audit Office, 1999; UK Public Accounts Committee, 2000). A very well known example, which broke most of the 'rules' of successful ergonomics, was the failed implementation of a new control system for the London Ambulance Service, with loss of life and great financial costs (e.g., Beynon-Davies, 1999; Wastell and Cooper, 1996).

The common denominator in all cases is that the abilities, needs and limitations of the people working in the system or with the equipment have not been understood and accounted for. On the other hand, successful products or work systems will usually show evidence that the needs of their users have been accounted for during design, implementation and operation.

Taking such account of people — or of the human factor — is the province of *Ergonomics* and of the synonymous *Human Factors* (the interchangeability of the terms is indicated in this chapter through use of the shorthand E/HF). This book is about the *methods* that ergonomists and human factors engineers and scientists use, in analysing, designing and evaluating equipment, tasks, jobs and organisations. Ergonomics has little value unless it is applied and so its practical methodology is of great importance.

The consequences of not applying ergonomics/human factors (E/HF) or of wrongly applying E/HF through inappropriate methodology, can increase risks of ill-health and injury, dissatisfaction, and discomfort for the workforce. For a company the consequences at the least can be a loss of competitiveness, in terms of productivity, quality, flexibility, timeliness and so on. However, this book is even more concerned with the positive side of applying E/HF methodology, with the improvements in well-being that can result for employers, workforces, producers, users, engineers, designers and, indeed, people and society in general.

This application of methodology will be in both of what Kragt (1992) — in the context of manufacturing industry — has distinguished as ***product ergonomics*** and ***production ergonomics***. More widely, that is that we are concerned with both the usability and safety of things people use at work, in the home, in transport and at leisure, and also the viability, effectiveness and safety of people's jobs in making and delivering goods and services. Goods and services are planned and designed by people, produced by people with processes built and managed by people, and are sold, bought, used, maintained and scrapped/recycled by people. Even where automated processes are employed, these interface with people and organisational systems at some point. Therefore understanding of people, and using that understanding wisely, is central to the planning and running of successful organisations, communities and societies.

Whatever the original roots of E/HF, then it is certainly not now limited to the workplace, nor to 'the operator'. E/HF probably cannot properly be applied to all systems involving people, but is relevant to all purposive human behaviour in 'designed' human machine systems. Thus E/HF is highly relevant to, and for, people at home, on the move, in sport, at leisure as well as at work; and for school children, post-retirement populations and those with special needs. Even at work, it is probable that a minority of people are working in jobs where they were traditionally categorised as 'operators' (usually assembly line staff, machine operators, or process control operators). People at work have a vast variety of roles, have multiple goals and means, and require integration of social, cognitive and physical skills.

Some historical background

Clarity of focus for E/HF has been bedevilled by the use of the different terms for the discipline itself. *Ergonomics* and *Human Factors* are both accepted terms worldwide for the theory and practice of learning about human characteristics and capabilities, and then using that understanding to improve people's interaction with the things they use and with the environments in which they do so. Ergonomics tends to be used more in Europe and Human Factors in North America, but even these distinctions are increasingly blurred. One difference is that people from many different disciplines will say they work in 'human factors' but generally only those with a qualification in 'Ergonomics' or, confusingly, 'Human Factors', will say they work in 'ergonomics'. There is a tendency, even amongst ergonomists, to talk about 'human (or people) factors' as being what they study, simply because this is easier grammatically in written and spoken language! The converse is that people working in the profession of human factors find it difficult to know what to call themselves, with 'human factors engineers', 'human factors scientists' and 'human factors professionals' all having been tried. Other efforts to get around the problem of name — for instance using the title of engineering psychology — tend to lose the holistic and total systems flavour of E/HF.

We may regret that the term 'ergonomics' lacks innate meaning or impact for clients or the public who will not be aware of its classical roots, but genies can rarely be put back into bottles; it would now be a futile and damaging exercise to move away from use of the terms ergonomist, ergonomics and human factors. In this book it is what we do that matters, not what we call ourselves, and contributors will use the terms ergonomics and human factors interchangeably.

Although formal consideration of the interactions between people and their working environments can be found in writings of a hundred years ago, for instance from Poland and Germany, the modern history of E/HF emerges from the 1939–1945 World War. As a formal branch of learning, with its own learned societies and scientific journals, E/HF has a history of about 50 years, for instance in Germany, the Netherlands, the United Kingdom and the United States of America. In the UK, the ideas and expertise from different disciplines interested in the effectiveness of human performance (anatomy, physiology, industrial medicine, industrial hygiene, design engineering, architecture and illumination engineering), and an emphasis on theory and methodology, led to the formation of the discipline of ergonomics with two strong sub-groupings of anatomy/physiology and experimental psychology. In parallel, the human factors profession was growing up in the United States, with strong inputs from the disciplines of psychology and engineering. In Germany, the Netherlands and across Scandinavia a basis for ergonomics was growing out of work in medicine and functional anatomy whilst in Eastern Europe growth was largely from the industrial engineering profession. For much more background detail, interested readers are referred to Edholm and Murrell (1973), Singleton (1982), Stockbridge (1989) and Meister (1995) and histories by Waterson and Sell (2005) for the Ergonomics Society and by Kuorinka (2000) for the International Ergonomics Association.

E/HF has drawn from anatomy, physiology and psychology, and has close connections with the applied disciplines of medicine and engineering. Extending this, Chapanis (1996) defines it as a multi-disciplinary field, with psychology (primarily experimental psychology), anthropometry, applied physiology, environmental medicine, engineering, statistics, operations research and industrial design all contributing. For Wickens *et al.* (1998), however, the field of human factors 'originally grew out of a fairly narrow concern for human interaction with physical devices', but has broadened greatly in the last few decades with its various subdomains. They believe that human factors intersects with certain disciplines within psychology and engineering (for instance, experimental psychology, social psychology, industrial engineering and bio-engineering), and that a number of disciplines have overlapped with some aspects of human factors, namely cognitive science, artificial intelligence, industrial design, management and statistics.

Definitions of ergonomics/human factors

It is a feature of the modern world that disciplines of current relevance and value are generally multi-, inter- and trans-disciplinary and therefore less amenable to simple definition. The fact that E/HF was built upon other fundamental disciplines should not in itself be a problem: engineering is built upon mathematics, psychology is built upon biology and economic science upon a number of bases.

E/HF seeks to define itself at regular intervals — for instance, amongst many others, see Chapanis (1976 and 1979), Welford (1976), de Montmollin (1992), Moray (1994) and Meister (1998). A large number of different, if overlapping, definitions of ergonomics and of human factors now exist; Wogalter *et al.* (1998) have considered these. Most definitions stress the view of E/HF as jointly a science — providing fundamental information — and a technology — applying that information to problems of design in their widest sense (e.g., Shackel, 1996). Within this view, the E/HF sphere contains all elements of the total human–environment system, comprising people's interactions with hardware, software and 'firmware' (including space), and with other people ('liveware') both individually and as social groups.

Any acceptable definition of E/HF must emphasise the need for, and the complementarity between, fundamental understanding of people and their interactions and practice in improving those interactions. Meister (1995, p. 9) differentiates between the *theoretical* knowledge within ergonomics, which explains people's interaction with things, and the *instrumental* knowledge, which can be utilised in design. This relationship between theory and practice, between research and application, is under examination continually (e.g., Singleton, 1994; Green and Jordan, 1999). A US National Research Council report (Rouse *et al.*, 1997) makes strong arguments for the value of human factors initiatives in many industries, stressing its multi-disciplinary, systemic, socio-technical and user-centric orientation, but also the difficulties in the current climate of balancing needs for basic research with the high demands for applied activities.

Differences in definitions are probably more to do with where to draw the boundary of the proper domain for E/HF than with fundamental disagreements on approach. To illustrate this contrast, Clark and Corlett (1984, p. 2) saw ergonomics as the study of human abilities and characteristics which affect the design of equipment, systems and jobs, with aims to improve efficiency, safety and well-being. On the other hand Wickens (1984, p. 3) said that human factors is to do with designing machines that accommodate the limits of the user.

The International Standards Organisation, in its various committees on ergonomics standards, has been using as a working definition that:

> Ergonomics produces and integrates knowledge from the human sciences to match jobs, systems, products and environments to the physical and mental abilities and limitations of people. In doing so it seeks to safeguard safety, health and well-being whilst optimising efficiency and performance.

Similarly, the International Ergonomics Association has defined:

> Ergonomics (or human factors) is the scientific discipline concerned with the understanding of interactions among humans and other elements of a system, and the profession that applies theoretical

principles, data and methods to design in order to optimise human well-being and overall system performance.

When pushed at a party for rather more pithy definitions of ergonomics, many of us will respond that it is design for people, or designing fit for human use, or inclusive design, or fitting systems and products to people and not *vice versa* (Figure 1).

A current tendency to define and partition the work of ergonomists into specialisms, whilst understandable, can cause difficulties. Typically this is done into cognitive ergonomics, physical (or musculoskeletal) ergonomics and social (or organisational) ergonomics; other specialisations are also sometimes defined such as rehabilitation ergonomics and forensic ergonomics. There are some good reasons for such partition. It can help explain a potential contribution to clients and funding bodies: for instance, an ergonomist specialising in complex control systems design may define what they practice as cognitive ergonomics; an ergonomist specialising in health and safety at work and re-design of workstations may define this as physical ergonomics. Nonetheless, there are dangers in such a parochial and molecular viewpoint. It is the very systems perspective and holistic nature of E/HF that provides its strength. The breadth of concern to cover all aspects of people's interaction with their environments and the interconnections between these interactions is what allows E/HF to define itself as a unique discipline.

Figure 1 An excellent design from a technical viewpoint but which has not accounted for the needs and limitations of the potential users! © 1984 The Far Side cartoon by Gary Larson is reprinted by permission of Chronicle Features, San Francisco, CA. All rights reserved.

If we do see the value in descriptions of specialisms, these might include:

- Physical ergonomics: fit, clearance, reach, access, tolerance, workload, manual handling, health and safety, workplace layout, displays and controls, product and equipment design, environment, tools
- Cognitive ergonomics: information processing, sensing, perception, decision making, problem solving, reaction, mental workload, fatigue, stress, interface design, reliability, communication, fault diagnosis
- Organisational (social) ergonomics: attitudes, motivation, satisfaction, job and team design, hours and patterns of work, pacing, implementation of change
- Systems ergonomics: most successful ergonomics analysis, design and evaluation integrates the physical, cognitive and social.

Aims and framework for ergonomics

In most definitions of E/HF we will find a list of objectives or criteria that drive its application, for instance jobs, systems or products that are comfortable, safe, effective and satisfying. Aims are often divided into those of gains for the individual (employee or user), and those for the organisation (employer or producer). *These aims, however, are neither mutually exclusive nor independent.* It is not a case of having to implement either a more comfortable workstation or a more productive one for example, nor are the ways of achieving the former necessarily very different from those for the latter. For instance, and at a very simple level, the light intensity, position and colour rendering needed to give best possible performance at a product inspection task are likely to be very similar to those giving least potential risk of visual fatigue for the inspector; the preferred position, size and angle of pedals at an industrial sewing machine to improve work output and quality will be similar to settings to give operator comfort and convenience. Put more strongly, it is difficult to think of cases where work or equipment designed to meet the needs of an employee or user would detract from performance effectiveness; in general user fit will enhance performance. To contribute such 'win–win' outcomes, and to sell them before and to prove them afterwards to managers and engineers, is one of the main tasks facing practicing ergonomists. Generally, if we concentrate upon objectives of ergonomics then we require methods that go some way to helping us 'prove' that we have met certain aims or have achieved a certain level of improvement. Cost benefit analyses are of importance here (see Chapter 37) as is the setting of usability metrics (Chapter 13).

Figure 2 illustrates the twin aims of E/HF in the context of work systems; a similar conceptualisation could be drawn for product (or user) ergonomics. We see that there is a direct connection between design and development criteria for people and organisations, and an indirect or systemic one also.

A traditional view is that, like the epidemiological model of 'host-agent-environment' in disease control or accident prevention, E/HF is concerned with interactions between people and the things they use and the environments in which they use them (Figure 3). Our concern is with the contextually-based (or situated) interfaces between people and the processes with which they interact, whether a toothbrush, training manual, motor car, power plant control room or school for instance. Within this, people's interaction with other people is at the core of an E/HF framework, increasingly so as systems of interest move away from 'one person — one interface' to distributed networks (Wilson, 2000).

The person and the process form a closed loop system (but not a closed system); the output characteristics of each (such as the person's hands, feet or speech, and the process

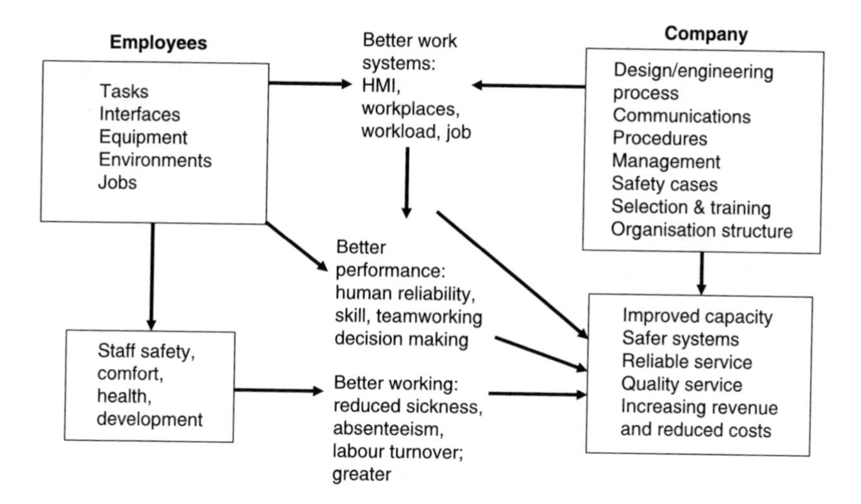

Figure 2 Aims of ergonomics/human factors. E/HF can be seen in the context of its objectives, which are the well-being of people and of organisations. These might be seen as twin aims which are neither independent nor mutually exclusive and which have direct and systemic connections.

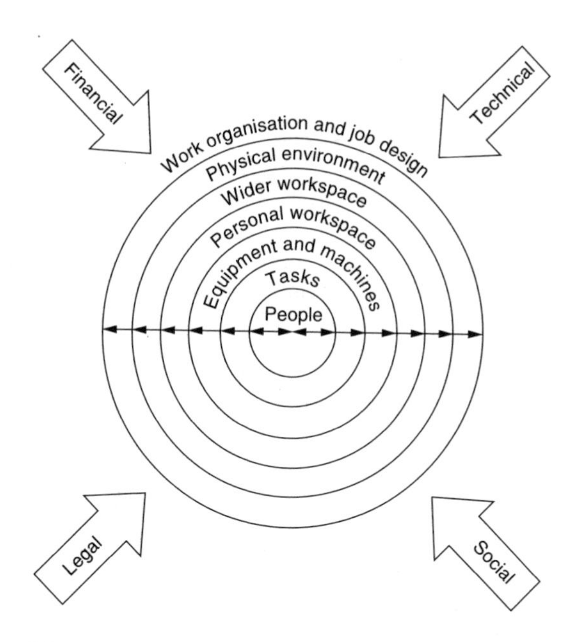

Figure 3 Interactions of factors relevant to the application of E/HF in work design. At the centre is the consideration of people, and of interactions between people. In turn we need to consider the tasks that they do — and physical and mental load as a consequence, the equipment and machines they work with (including software), and the various elements of the environment in which they do so. (Redrawn from Grey *et al.*, 1987.)

displays) must match the input characteristics (process controls and human sensory mechanisms) of the other. If such a match is achieved we talk of a user-system fit or of a successful human–machine interface; this is the subject of many studies and the focus of many methods. Generally the displays and controls themselves are regarded as comprising the interface; however, in systems which are highly automated, where the operator acts as a monitor or supervisor for long periods, the interface may be seen as lying between the person and the displays and controls with the latter considered to be part of the process. Figure 4a shows the interaction between the person and the process (technical or organisational). This might be viewed differently for different types of system. For instance: Figure 4b shows a simplification of Sheridan's (1987) view of the human as supervisory controller; Figure 4c a representation of a participant within a virtual reality (VR) system where they are said to inhabit the interface as well as interacting with various effectors (such as CAVEs or projection walls) and sensors (such as trackers); and Figure 4d the social as well as technical interfaces in the increasingly common and important situations of co-located and distributed teamworking in complex socio-technical systems.

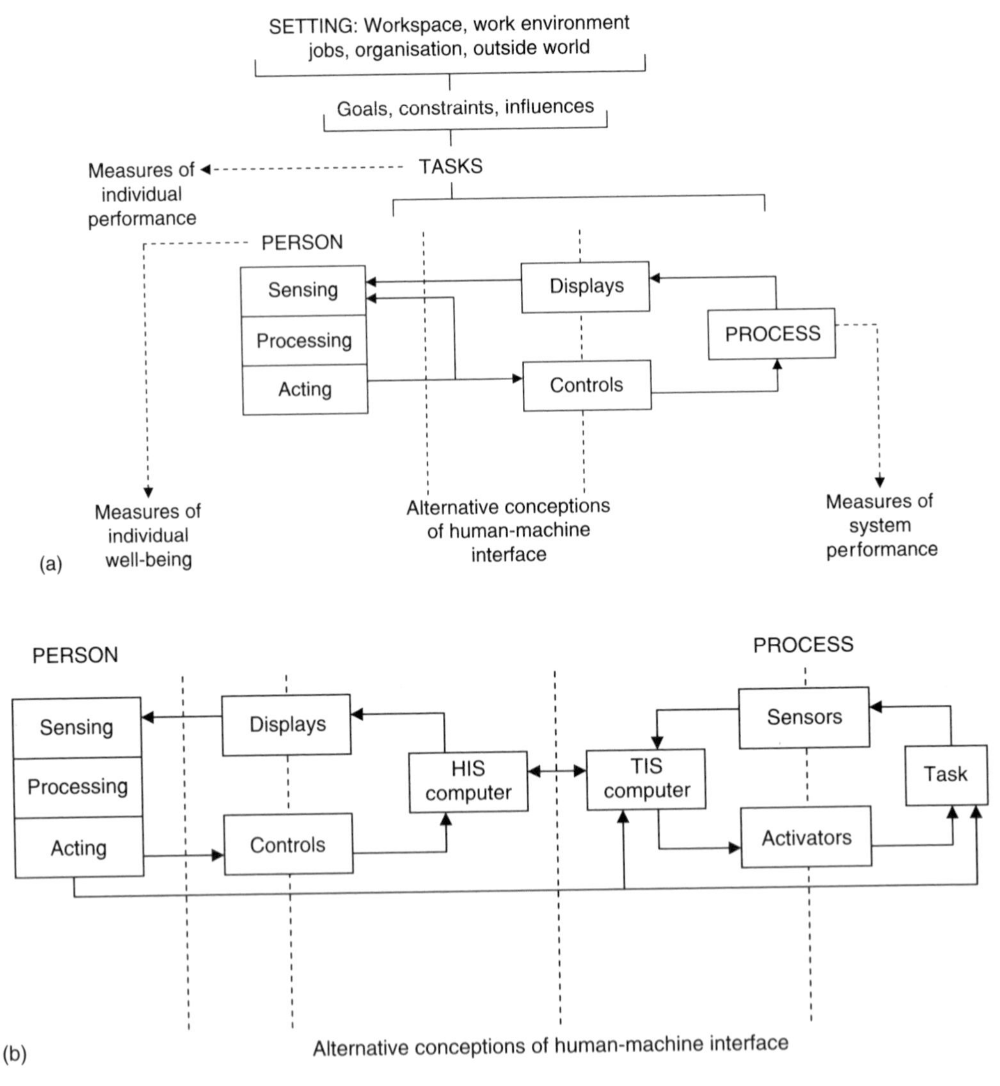

Figure 4 (a) Interaction of the person with the process and with all layers of the setting within which the interaction takes place will define, and be defined by, the tasks completed. The human–machine interface can be conceived as occurring at the displays and controls or between these and the person. (b) Simplified view of interfaces when the human acts as a supervisory controller. (c) Interfaces for participant with a virtual environment within a VR system. (d) Interfaces in teamworking.

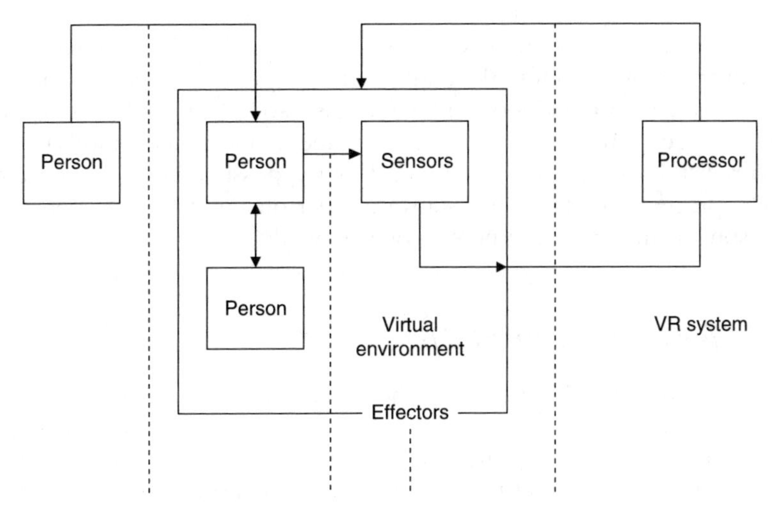

(c) Alternative conceptions of the interfaces in virtual environment participation

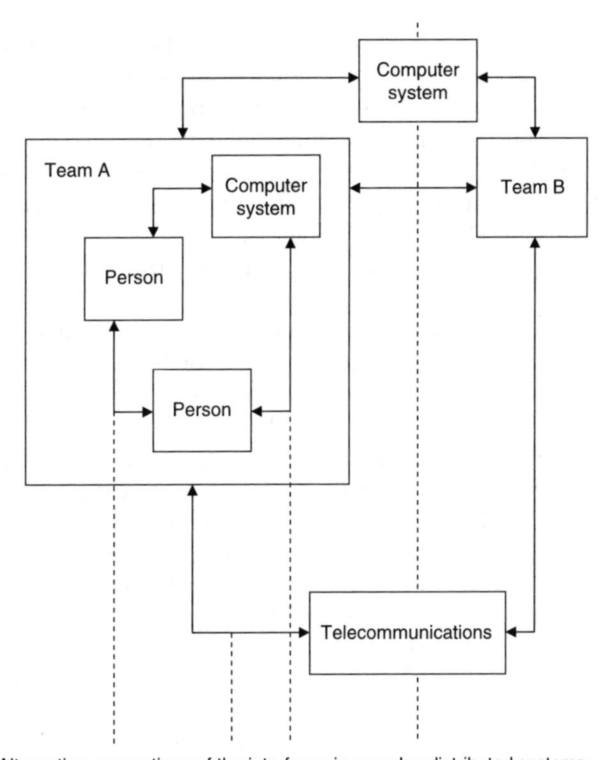

(d) Alternative conceptions of the interfaces in complex distributed systems

Figure 4 Continued.

The tasks that are carried out across the interface which might be studied in ergonomics are many and varied, and include: heavy physical work — shovelling, lifting etc.; moderate physical work — packing, materials handling etc.; psycho-motor skills — assembly, typing etc.; vigilance skills — inspection, monitoring etc.; diagnosis — fault recognition etc.; decision making — in medicine, management etc.; reasoning/problem solving — in engineering, the law etc.; creativity — art, design; and combinations of all these.

Human–machine interaction therefore does not take place in a vacuum; it is directly and indirectly influenced by the physical workplace, the physical work environment and the social environment or organisation of jobs and work. It is also influenced by factors from the world outside. Within such a model we can see E/HF methodology as comprising the techniques needed to predict, investigate or develop each of the possible interactions, between person–task, person–process (hardware or software), person–environment, person–job, person–person, person–organisation, and person–outside world.

Argument for and against E/HF

It was mentioned in passing above that it is not always easy to sell the approach of ergonomics and human factors and the aims to jointly improve work and task performance and also worker and user well-being. We must be prepared for the criticisms or arguments against. Helander (1999) and Pheasant (1986, p. 8) have identified some of these, and the author has directly come across others in his own work. Arguments from engineers, designers and managers can include:

1. 'I can use it and I'm typical' — i.e., I've done my user testing on myself, what's the problem? Some software developers will try their own product out and assume that it can be used by all; whilst such expert walkthrough is a very powerful early assessment technique it is almost always not sufficient. Moreover, there is nothing typical about a designer or engineer — numerate, educated, logical, healthy and generally still male and 25 to 50 — particularly the fact that they should know more about the product than anyone else!
2. 'We designed it for the average' — i.e., I know we should understand the user, but we weren't sure how to do this or even who this was. There are two problems here. First, there is no such thing as an average person; one person might be in a middle range of, say, stature, but generally won't be for leg length, head circumference, shoulder breadth, etc. as well, still less will they be in the middle range of intelligence, strength, vision, hearing, reaction time and many other characteristics as well. Second, where a single characteristic of people is the critical one, and we can identify an average value, it is not usually appropriate to design to it: a doorway for people only of average height, a display that can be read only by people of average eyesight — these are not sensible design choices. Equally we will not always design for the absolute extremes either — we would not want the cost and inconvenience of houses with doorframes 225 cm high nor all instructions to be written in nursery school level language.
3. 'People are too variable to account for' — i.e., I understand about accounting for variability across people for a whole set of critical characteristics, but there are too many of these and too much variability. In fact, very early on in any ergonomics contribution to design we should establish exactly which are the critical characteristics that must be accounted for. Moreover, unless there is good reason not to, we will not design for 100% of people but to best suit 95% or 97.5% of the population, whether through establishing the critical end of the range or through adjustability or by fitting to the average in certain circumstances only.
4. 'People can adapt so why be worried?' — i.e., why should we bother with all the hassle of ergonomics, people will cope. In this, human beings have been their own worst enemy. We are very adaptable — to poor physical design of workstations, to poor information and instructions, to poor design of jobs. Everywhere, workers and consumers are managing with, or working around, the poor designs of equipment, environments and jobs that they have been given, to keep the company working or to

survive in daily life. But such adaptability can come at a cost, to health, satisfaction, spare capacity and reliability if unexpected events or emergencies arise.

5. 'It's the workers that make mistakes' — i.e., let's try to automate everything we can and blame the user if we can't. Yes, human beings are prone to error. But often those errors are made almost inevitable at some time by the planning, organisation, equipment, job and procedures design, training programmes, etc., that have gone before (see Reason's 1990 and 1997 idea of latent failures and organisation pathogens — see also Chapters 32 and 34). Also, to borrow from the legal concept of strict liability, even misuse by individuals is often likely and can often be foreseen by a prudent design team, and should be part of the design decisions that are made. Moreover, it is often the attempts by engineers to design people out of the loop as far as possible that leaves badly designed jobs and circumstances that actually lead to unreliability, and also do not make the most of the capabilities of people — see Bainbridge's (1987) famous ironies of automation.
6. 'Can't they read the instructions?' and 'OK, it's a bad design, but we train them' — i.e., let's increase the adaptability of people to our poor design. Yes, instructions and training are important, but after well thought out human centred design, not instead of it.
7. 'Human factors costs too much'. This is an old argument, and efforts are frequently made to apply cost benefit reasoning to ergonomics (see Chapter 37). Yes, a contribution to take account of human factors will increase the cost of any design project, implementation scheme, or operational overhead — in staff, equipment, study costs, and any extra time needed in the system life cycle. But, as we saw from the aims of ergonomics earlier, these costs will be outweighed by the saving made on the costs of not getting it right and also the financial and less tangible benefits which will accrue (see Table 1).
8. 'Ergonomics is all common sense.' The answer here is yes and no. Some is common sense — but which often seems in short supply without a structured framework within which to identify what needs to be known, apply it and deal with trade offs (see human factors integration plans — Chapter 10). On the other hand, much of ergonomics and human factors is clearly not common sense; some research findings or guidance may even be counter-intuitive. It is use of the approaches and application

Table 1 Summary Costs and Benefits of Ergonomics/Human Factors

Potential costs of ergonomics	Potential costs if *no* ergonomics	Potential benefits from ergonomics
Personnel and administrative resources	Criminal prosecutions	Improved workforce/ customer health and safety
Technical and equipment resources	Compensation claims	Improved workforce performance
Cost of analysis and evaluation process	Insurance premiums increase	Improved customer satisfaction
	Accidents	Legal compliance
Capital costs of redesign features	Absenteeism, labour turnover, cover	Increased job satisfaction
	Recruitment and retraining	Better labour relations
Increased system development time	Lower product sales	Positive company image
Disruption to normal activities	Poor company reputation	Repeat sales

of the methods described in this book that will enable us to address the unusual and opaque issues and to rationally bring together into a coherent process and plan the more obvious and straightforward issues.

Context for application of E/HF methods

Since E/HF must have relevance for real settings, the understanding of findings and their application in practice requires a good grasp of context. This same context will also affect society's view of ergonomics and its value and place in the modern world. It also impacts strongly on choice and application of methods. Relevant contextual factors can be summarised under headings of financial, technological, legal, organisational, social, cultural and political factors.

E/HF works in a world of competing ***financial*** priorities, at both an organisational/ commercial level (e.g., pressure from shareholders, need for rate of return on investments) and a social level (e.g., choices between expenditure on health, education, defence etc). Finance may therefore be a major selection criterion for methods used as well as solutions proposed.

The ***technological*** context is obviously critical, with the rapid rate of change of technology development and the complexity — in some cases unnecessary — of systems, which provide a rich field for E/HF contributions in such areas as control rooms, personal and wearable computers, transport systems and virtual environments. However, we must acknowledge the parallel generations of low and high technology, in so-called developed countries as well as in industrially developing countries, where appropriate technology, first thought about in the 1970s, is still a very relevant concept.

The growth in the market for ergonomists, due to the increases in relevant health and safety regulation across the world and the parallel rise in civil claims for compensation for health and injury problems at work, shows the importance of ***legal*** context. A different, and perhaps less welcome manifestation of our increasingly litigious world, is that E/HF practitioners may become more liable to claims from clients, and clarity and quality in our choice and use of methods is obviously vital to reduce this possibility.

Developments in ***organisational*** systems and structures in part relate to changing technology. Virtual organisations and virtual teams, distributed work, flexible (human and computer) networks, flatter organisations and self directed teams, call centres, participation and people working from home or on the move will impact on development of human factors methods as well as on the lives of the people involved.

At a ***social*** level, ergonomists deal with the interaction of people with and within social and technical systems, and so developments such as an increasingly unstable and fluid job market on the one hand but increased interest to provide a stakeholder society on the other, must affect how (and how well) we go about our work. We also work in many different national, ethnic and professional ***cultures***. By extension, ergonomists work in a ***political*** context, at organisational as well as national and supra-national levels, and we and our methods must be sensitive to this.

Role of E/HF methods

Although we looked at definitions of ergonomics and human factors earlier in this chapter, it is more important that E/HF should be seen as an approach (or as a philosophy) of taking account of people in the way we design and organise; in other words, as designing for people. In this view, E/HF itself is primarily a process, to an extent a meta-method, which makes the clear understanding and correct utilisation of individual methods and techniques even more important. This supports the need for this book and others like it.

Back in 1986, the NRC Committee on Human Factors reported the development of applied methods to be one of the major research needs for human factors (National Research Council, 1986), and nothing has occurred in the years since to change this. Ergonomics is both a science and a technology and thus has need of techniques for both data collection (basic or functional data) and application. The debts we owe to other disciplines are obvious. However, as we gain experience and confidence and as our armoury of knowledge and methodology grows, so the debt to other disciplines is being repaid. Methods developed or adapted within E/HF will be employed by psychologists, engineers, computer scientists, management scientists or health care professionals in turn, just as we are constantly enlarging our own human performance database through results of human–machine systems evaluation.

Bearing in mind the applied nature of ergonomics, if we try to draw parallels with the basic processes of design — *analysis, synthesis and evaluation* — which iterate throughout the design process (e.g., Markus, 1969), and extend these, then we might divide methodology into five types of input: data on people, systems analysis and development, evaluation of system performance, assessment of effects on people, and the organisation of E/HF management programmes.

Methods for the collection of data about people

Our first methodological need is for data about people and this can cover all characteristics: physical size and strength, endurance and physiological capacity, sensory characteristics, mental capacities, psychological responses and so on. Just as important as the collection and reporting of data is the generation of design and evaluation criteria from these. For example, given data on the population range of arm reaches in different directions, what advice can be given on placement of frequently used rotary controls? Or again, given data on working memory limitations, can these be adapted to form design guidance on numbers of different codes that should be used in a coding system? Methods used to produce data about people comprise the scientific base of E/HF and, as pointed out above, methods have often been borrowed directly or adapted from other disciplines.

Methods used in systems analysis and design

The second input of E/HF is its contribution to the analysis and design process. This overlaps with both basic data collection and with evaluation (see below) but here is meant methods to assist in the analysis and design stages of development of equipment, workplaces, software, jobs or buildings. In essence we need methods to *analyse* current or proposed systems (analysis strictly meaning to resolve the system into its constituent elements and critically examine these) and then to *synthesise* data (that is, build up a coherent whole by putting elements back together) into ergonomically sound concepts, prototypes and final designs. Specifications produced out of this process must be capable of transference into design criteria and data, and also have reasoned justifications, in order that ergonomists can work sensibly with engineers and designers.

Methods to evaluate human–machine system performance

In part, analysis at the start of development may involve evaluations of an existing system's performance. Certainly we must evaluate system performance during and at the end of development. Many measures can be defined for this and one challenge today is the search for measures of system performance other than the ubiquitous but often sterile 'time and error' based ones. Manufacturing system performance, for instance, may be assessed by means of production output rates and product quality levels, but we could also use machine utilisation rates, minimisation of finished stocks or work-in-progress, raw material wastage, speed of response to changed schedules, accident rates, sickness or other absence, and job attitudes and job satisfaction measures. Similarly, although a computer interface can be assessed in terms of time taken and errors made in performing a sequence of tasks, more interesting measures might be 'extent of system explored', 'willingness to change direction', 'quality of finished work', etc. These are much more difficult measures to take, undoubtedly, but perhaps they provide a more valid measure of actual system performance.

Any evaluation of subsequent system performance where there is a human factors input into system development is also, in an interesting closing of the loop, an evaluation of just how well E/HF was applied to the design.

Methods to assess demands and effects on people

As we have seen, E/HF has twin aims in its contribution to design and development: improvements for the job-holder or user and improvements for the producer or employer. As a result, any system assessment should be carried out in terms of the demands made on people and of the effects on their well-being, as well as accounting for system performance. There is an argument that the demands made on people should in fact be viewed as an implicit part of system performance and as such their assessment should not be seen as separate; however, on balance it is more persuasive that such a distinction will emphasise the twin, yet interdependent, aims of E/HF. Many methods can be applied to assess the effects that different environments, jobs or equipment have on people. Such impacts might be medical, physical or psychological in nature and methods will vary from direct recording of observable phenomena (e.g., heart rate) to indirect observation of people's affective states (e.g., boredom). In almost all circumstances however, the data collected are not useful by themselves but must be interpreted and any effects inferred, which is again a large part of the ergonomist's input. Moreover, if assessment methodology is developed appropriately then data obtained can be generalised to become part of our first input, the basic data on people.

Methods to develop E/HF management programmes

Our fifth input concerns the management of E/HF programmes. Methods are required here for two situations, although there is not a clear distinction between them. First, there are ergonomists who are working within companies, and they are often doing this in very small groups or even on their own. Second, we have what is termed 'devolving' (Wilson, 1994) or 'giving away' (Corlett, 1991) ergonomics expertise. It is unrealistic to expect all enterprises to employ only ergonomists (either as employees or outside consultants) to handle their ergonomics; it is inappropriate also in many circumstances. In many areas — for instance, health and safety, job redesign, workplace layout — the ergonomics profession must provide methods which allow the development of appropriate strategies and which support the management of programmes which can be run as a part of normal company activities. Nobody is advocating that untrained staff handle all human factors in a company. However, design engineering, production engineering, health and safety, line management and production workers can all make considerable contributions; the methods and support that ergonomists give to them must include enabling them to recognise when specialists must be brought in. This fifth input is certainly the 'messiest' area of E/HF methodology, and it embraces aspects of all other inputs as well as participatory ergonomics, systems implementation and so on (see Chapters 35, 36 and 38). Nonetheless, it is an increasingly important area as E/HF moves more and more into real companies with real problems as well as maintaining its position within research laboratories and universities.

Classification of methods

Classification of ergonomics methods in terms of the parts or stages of any models of E/HF is difficult. Meister (1985) attempts a gross distinction and divides his behavioural methods into the analytic techniques employed during the development of systems and the measurement methods employed to evaluate functioning systems. He does, though, recognise overlaps, particularly that many measurement methods are also used during system development. Similarly, at a first level Megaw in Chapter 18 of this book distinguishes analytical methods or measures — broadly those based on theory, modelling and prediction — from empirical ones — those based on observation or experiments with an actual situation. Crossed with this we have broad categories of measure from Edwards (1973) of direct achievement measurement, operator loading measurement and correlated function measurement (see Table 2).

Starting from Edwards' taxonomy, then reviewing the EIAC (Ergonomics Information Analysis Centre at University of Birmingham) classification, the author attempted an initial taxonomy of ergonomics methodology to inform syllabus development for a human factors MSc course. This was published in the earlier editions of this book, recognising differences in approach, method (or method group), technique, and measure. Review reveals it still to be relevant in general terms to E/HF methodology today, despite almost certainly containing some redundancy, omissions and inconsistencies, and so the general classification scheme or taxonomy is shown in Table 3.

Table 2 Taxonomy of Measures (Edwards, 1973)

Assessment Techniques (measures)	Examples
Direct achievement measurement	
Absolute scores	Distance traveled numbers produced
Speed scores	Time, time per . . ., reaction time
Precision scores (continuous)	Tracking performance
Error scores (discrete)	Faults, missed signals
Information scores	Rate of information handling, redundancy
Multiple scores	Combinations of the above
Operator loading measurement	
Extracted outputs	ECG, EMG, EEG, GSR, O_2 uptake
Secondary tasks	Task loading through secondary activity
Alternative tasks	Decrement assessed on new task
Time function changes	Fatigue, learning
Operator reports	Interviews on, say, fatigue
Correlated function measurement	
Variability	Constitency as measure of skill
Operator reports	Critical incidents
Observer reports	Pilot assessment

Table 3 A Classification of Methods, Techniques and Measures Used with Ergonomics/Human Factors. This Classification has been Split into Categories of 1. General Methods; 2. Collection of Information about People; 3. Analysis and Design; 4. Evaluation of Human–Machine System Performance; 5. Evaluation of Demands on People; 6. Management and Implementation of Ergonomics. It Should be Noted that the General Methods at the Start of the Table can Generally be Used Within Any of the Other Categories

Method			
Group	Subgroup	Technique	Measure/outcome
1. General methods			
Direct observation (laboratory or field)	Unobtrusive, participative, or visible	Human recording: checklists, rating, ranking, critical incident technique, charts (time, spatial, sequence, link)	Event frequency, sequence
			Times, errors, accuracy Overload, underload Descriptive, evaluative, diagnostic measures of performance
		Hardware recording: video, film, tape, event recorder, position/movement recording, computer real time recording	
Indirect observation (laboratory or field)	Psychometrics/scaling	Surveys, questionnaires, rating, ranking, scaling, diaries, critical incidents, checklists, group discussions, interviews	Attitudes, feelings, perceived effort, difficulties, advantages, disadvantages, preferences
Archives and records	Production, medical, personnel, quality	Before/after comparison, trend data, use in cost–benefit calculations	Output, times, quality, etc. Absenteeism, labour turnover, injuries, sickness
'Automated' methods			
Experiments			
Literature and data interpretation		Description, statistical analysis, model building	
Standards and recommendations	Voluntary or enforced		
Multiple methods		Triangulation, validation	

2. Collection of information about people			
Physical measurement	Anthropometry (static or dynamic)	Anthropometer, wall charts, video, photography, CODA, fitting trails, computer modelling	Dimensions, percentiles, other descriptive statistics
	Biomechanics	Dynamometer, strength gauges, goniometer, stadiometer	Values, trends, descriptive statistics
	Performance	Eye/body movements, acuity, CFF, TAF/CMF	Scores, counts, qualities
Physiological measurement		ECG, EEG, EMG, ERP, HRV, O_2 uptake, GSR, pupil diameter	
Perceptual/cognitive assessment	Ability testing	Mental or cognitive tests (e.g., general aptitude test battery); perceptual tests	Prediction of performance
	Psychophysics	Method of limits, method of average error, method of constant stimuli, e.g., aesthesiometer, hearing loss audiometry, CFF, fitting trails	Thresholds and levels of perception, sensitivity
	Visual tests	Acuity, colour, stereopsis, heterophoria	
Knowledge acquisition	Knowledge elicitation (from expert)	Written, audio, video, records, interviews (structured, unstructured), protocol analysis, conceptual mapping, goal decompositions, automatic techniques	'Rules', reasoning, explanations
	Other	Interpretation of records, standards, guidelines, criteria	
Models	Computer mechanical, conceptual, mathematical		
3. Analysis and design			
Task analysis	Data collection methods — see all general methods Description analysis and representation methods	Hierarchical TA, tabular TA, ability requirements analysis, TA for knowledge description, link analysis, cognitive TA, formal mappings (e.g., TAG), job analysis charts, operation sequence diagrams, flow charts, time charts	Requirements for people Consequences of tasks; task sequences, times, probable error rates, criticalities, loading etc.
Expert analysis		Checklist, walkthrough, Delphi technique, method study techniques, expert systems, likelihood matrix, scored assessments	Qualitative reports, critical issues, weightings and priorities, approvals

(Continued)

Table 3 Continued

Method		Technique	Measure/outcome
Group	Subgroup		
Inspection/protocol analysis	Concurrent Retrospective	Written audio, video records; diagrammatic, computational, debriefing, diary, critical incidents, shadowing	Explicit content, implicit content, behaviour transitions, rules, knowledge, models
User models		GOMS, CLG, TAG, etc. mental models	
Statistical analysis		Signal Detection Theory, information theory, reliability assessment	Performance measure, likelihoods
Models		Task network (SAINT, Siegel-Wolf etc.); control theory, deterministic (HOS, etc.); cognitive/information processing, (GOMS, etc.); behaviour (S-R-K)	Performance predictions
Simulation	Computational	Mathematical; computer, including CAD (e.g., man models); virtual reality	
	Physical	Physical mock-up, walkthrough	
Method study		Graphical analysis; process, flow, time charts; filming, micromotion	Movements, times, actions, frequencies
Work measurement		Time study, activity analysis, synthetic analysis, electronic monitoring	Times, standards, task sequence, simultaneity, frequency
Prototyping		Rapid prototyping, storyboarding, mock-ups	
Creative techniques		Brainstorming, decision groups, focus groups	
Participative methods		Design and follow-up groups, user involvements; training and awareness in ergonomics	

4. Evaluation of human–machine system performance			
Work systems analysis	Checklists, walkthrough, expert assessment		
Usability evaluation—bci, products, texts etc.	User trails	Techniques of direct and indirect observation, and physical performance measurement — individuals or groups	Time, reaction time Accuracy, errors Opinions, attitudes, responses Physical fit Workload, stress
	Evaluation environment	CAFE OF EVE	
	Introspection, expert analysis	Protocol analysis, checklists, group decision making	
Measurement by instrumentation		Light (illumination, glare, etc.), climate (temperature, humidity, air, space, etc.), noise (sound, intensity-weighted, frequency), vibration, workplace dimensions	Measurements vs. norms, Comparisons. 'Fit' of the system to people's requirements.
Subjective assessment		Psychophysical techniques, scaling, rating, surveys, questionnaires, etc.	Acceptability Comfort, annoyance, acceptability
Performance measures		Speech intelligibility index, work rate, standard psychomotor and mental tests, etc. Speed/accuracy trade off	'Scores' against norms
Modelling and simulation	Computer, mechanical (manikins), mathematical		
Electronic monitoring		On-line record	Performance measures
Self recording		Gripe button, diary, event recorder	Problems, incidents
Text analysis		Readability formulae (Gunning Fog, Ftesch, etc.); cloze procedures; judgements (rate, rank, etc.); protocol analysis; scan/read tests; checklists	Normative scores, rating
Human reliability analysis	Error analysis and representation	SHERPA, GEMS, PHECA, etc. Fault tree, action tree, etc.	Errors, types, causes Descriptive and predictive charts
	Reliability estimation	THERP, HEART, SLIM, etc.	Human error probabilities

(Continued)

Table 3 Continued

Method			
Group	Subgroup	Technique	Measure/outcome
Accident reporting and analysis		Archive records, reporting system, in-depth follow up interviews, site analysis, statistical analyses	Incidence, severity, epidemiology and aetiology
5. Evaluation of demands on people			
Physical workload		Subjective assessment, perceived exertion (e.g., Borg scale)	Subjective ratings, performance decrement, physical measures
		Performance records, secondary or alternative tasks of psychomotor performance; physical changes (e.g., stadiometer)	
Mental workload		Primary, secondary, alternative task; subjective assessment (e.g., SWAT, Cooper-Harper, TLX); physiological response (e.g., HRV); protocol analysis; CFF, TAF	'Performance' decrement Load (subjective or objective)
Posture anlysis		Biomechanical (mathematical) models; optical methods (CODA, Selspot); notation — paper and pencil (posture target, body part discomfort); video; RULA, OWAS	'Postures' to compare with criteria, postures scores; discomfort ratings
		Psychophysical methods	
Physiological methods		HR, HR variability, O_2 uptake, air analysis, GSR, ECG, EMG, EEG, ERP	Objective data, to be interpreted against norms, criteria

'Fatigue' measurement	Visual fatigue	Direct measurement of ocular function, task performance, asthenopia (reported discomfort)	Eye movements, accommodation, vergence, blink rate, pupil sise, times and accuracy
Environmental response measures	Physiological, perceptual	Sweat rate, body temperature, heart rate; hearing loss; visual acuity, contrast sensitivity; sensation loss; body change	Measurements vs. norms Comparisons Acceptability
Stress assessment	Subjective, physiological	GSR, etc.; indirect observation techniques (SACL, GWBQ, etc.)	
Job and work attitude measurement		Techniques of indirect observation, especially rating scales, e.g., JDS, informal group or individual interviews	Satisfaction, needs, important job characterstics
Guidelines			
6. Management and implementations of ergonomics			
Ergonomics management, organisational analysis			
Project management		Checklists, walkthroughs, project groups, consultation and participation procedures	
Implementation			
Cost-benefit analysis		Investment returns; productivity — life costs, revenue calculation; health and safety valuations	Financial measures
Participative methods		Ergonomics working groups, design decision groups, user representatives, ergonomics programmes	
Ethics in ergonomics			

Choice and combinations of methods

As far as methods are concerned, according to an old expression the proof of the pudding is in the eating. A method which to one researcher or practitioner is an invaluable aid to all their work may to another be vague or insubstantial in concept, difficult to use and variable in its outcomes. More than this, the validity, reliability and sensitivity of methods may well be application specific. We may need examination of utility and generalisability also in order to select or prioritise between methods (see throughout this book).

Once all possible measures are considered, we begin to see some of the problems of classification and selection. Kantowitz (1992) lists 46 possible indicators of nuclear power plant safety, split into seven categories (e.g., operations, quality programmes, management/administration) and reports that 'no single indicator was by itself an adequate measure . . . [nor were any] optimal for predicting plant safety' (p. 391).

It is not surprising that we will often look to use more than one, and often several, methods and measures in any one study. This is particularly so when we are carrying out evaluations in the field. Technically, if this is done formally it is known as triangulation (see Denzin, 1970 or Webb *et al.*, 1972); although the term can encompass data or investigators, triangulation is most often used to denote methodological triangulation — the use of two or more methods to improve the effectiveness of a study or our confidence in any findings; weaknesses in one method can be balanced by strengths in another. To take one simple example, only by questioning and observing operators in complex systems *and* also recording their concurrent verbal reports for subsequent protocol analysis can we begin to understand something about their decision making activities. A multiple-methods study may utilise a mixture of qualitative and quantitative techniques, in field and laboratory settings. Such an approach in the field is often known as contextual inquiry.

Only by use of several of these methods may a full evaluation be possible in any one situation, and thereby effective suggestions be made for redesigning job content, tasks, workstations and environments. However, a word of caution is in order. It is all too easy to fall into the trap, once an investigation has started, of measuring everything possible 'just in case'! This can lead to results that are difficult to interpret or, worse still, an analysis phase that uses large amounts of study resources to no clear or useful purpose.

In selecting an appropriate approach and set of methods for any particular need we will need to weigh up the relative merits of a quantitative or qualitative approach, so-called objective or subjective methods, and studies in the field or laboratory.

Qualitative/quantitative debate for methodology in ergonomics

Qualitative approaches and methodology are being increasingly used in human factors/ergonomics and sister disciplines such as organisational management, product design, psychology and engineering, and chapters throughout this book refer to its much increased acceptability and use. Although there has been considerable debate, even rancour, about its place in disciplines such as psychology, there is good reason to think that E/HF will benefit from the explicit inclusion of qualitative methods and tools in teaching, fundamental research and practice. This, in turn, would help to support the consideration of social, emotional and philosophical factors when conducting both theoretical and practical studies (Hignett and Wilson, 2004a, b). E/HF can be located on the cusp of the sciences and the humanities, and in the centre of the qualitative–quantitative continuum; it has much perhaps in common with anthropology, where the unit of analysis is interaction, in contrast to psychology where the unit of analysis is the individual.

Moray (1994, 2000) has recognised that sociologists, anthropologists, and ethnographers offer a methodology more sensitive to the context of analysis of work, rather than

Table 4 Dimensions of Qualitative and Quantitative Methodologies (adapted from: Hignett and Wilson, 2004b)

Quantitative dimensions	Quantitative dimensions
Words, understanding	Numbers, explanation
Purposive sampling, inductive reasoning	Statistical sampling, deductive reasoning
Social sciences, soft, subjective	Physical sciences, hard, objective
Practitioner as a human being to gather data, personal	Researcher, descriptive, impersonal
Inquiry from the inside	Inquiry from the outside
Data collection and analysis intertwined	Data collection before analysis
Creative, acknowledges of extraneous variables as contributing to the phenomenon	Predefined, operationalised concepts stated as hypotheses, empirical measurement and control of variables
Meanings of behaviours, broad and inclusive focus	Cause and effect relationship
Discovery, gaining knowledge, understanding actions.	Theory/explanation testing and development.

methodologies which result in generalisations based on quantitative analysis, but that there is need to generalise findings of qualitative study to understand other different systems. There should be no claim made for a general superiority or even a preference for either the qualitative or quantitative philosophies, approaches, methods and measures. Each should be used as needs and circumstances dictate, and we need to be clear about differences and similarities in scope, approach and outcomes — see Table 4 from Hignett and Wilson, 2004a.

Objective and subjective methods

The recent debates over the relative merits of quantitative and qualitative approaches interact with debate over so-called objective and subjective measurement (Wilson and Nichols, 2002). This has been an issue from the earliest days of the ergonomics and human factors discipline, with early views or implications that if subjective measurements do not match objective measurements then it is the former which are biased (e.g., Poulton, 1975).

However, what do we mean by objective measurement? Different forms of apparently objective measurement may show many elements of possible observer bias and of subjectivity in the actual selection of measures, criterion levels and methods. This was illustrated for this author in work on mental model identification and representation; the form of mental models reported in the literature often seemed dependent upon the scientists' choices of experimental and measurement method, in turn based on their preconceptions about the form of mental models (Rutherford and Wilson, 1992).

Ergonomics and human factors deal with many concepts which are difficult to define explicitly and therefore have no clear agreement on method of measurement. We work with the complexity of human beings, individually and as social groups. In cases of mental workload, situation awareness, fatigue, mental models and presence for instance, it is quite possible that what we are measuring with objective (performance or physiological) methods may be different — if related — entities to what is measured with subjective methods (participant ratings, interviews, etc.). This means that such concepts should be clearly, often operationally, defined in relation to the measurement method being used, rather than independently of it.

Annett (2002a) in a target paper, and several commentary papers in the same issue of Ergonomics, cover very well the topic of objectivity and subjectivity in ergonomics measurement and assessment.

Field and laboratory study

Within the sporadic debates about the nature of ergonomics and human factors over the past decade there has been much discussion of the merits of field and laboratory study, contrasting formal and informal methodological approaches and setting the advantages of control against those of veridicity (see Wilson, 2000). Moray (1994) believes that ergonomics only makes sense in the full richness of the social settings in which people work. The influence of contextual factors — such as interactions between individuals, the formation and relationships of teams, individual motivations, computer-based support systems and organisational structures — on work in practice must be understood before we can decide on techniques for the measurement and prediction of the outcomes of people's work, or on recommendations for the design of systems.

Laboratory experimentation has long been the dominant approach in, for instance, cognitive research. This is a vital source of information and insight about isolated work variables, but may not be a valid approach to understand work in practice and certainly will not replace the need for field study. The very nature of the controlled laboratory environment means that the complexities and uncertainties of work environments are, by definition, not being simulated. Even excellent — and expensive — simulations probably only capture the information display and individual control aspects of any work. Moreover, there is also the very real danger of a classical problem for experimental ergonomics, whereby the variables that are manipulated and found to have statistically significant effects actually do not matter very much in practice. They may account for little of the variance in the real situation, with all the variance of interest lying within the variables that are actually controlled in the experimental setting — for instance, environmental disturbances, work flow, individual motivation, inter-personal relationships and team relationships (Wilson *et al.*, 2003).

Work is difficult to study in practice, but 'when the unit of analysis is interactions, then field research is arguably the main methodological approach for ergonomics' (de Keyser, 1992). 'Methodologically explicit field research is vital for the core purpose of ergonomics, investigating and improving interactions between people and the world around them.' (Wilson, 2000, p. 563).

This is certainly not an argument against laboratory research — it is important to acknowledge that both laboratory and field studies are necessary — but is an argument for a better balance. Nor is there usually a stark either/or choice to be made. Vicente (2000) presents the idea of a continuum, with prototypical research types identified along it.

We must be aware that field study naturally includes context from the environment, supervision, motivation and disturbances, but the very presence of the investigator will create an additional context which, without care, can invalidate the study. Ethnomethodologists talk of the participant observer, in special cases for when the observer becomes a member of the group being studied and also generally because they also participate as an actor in the workplace by their very presence as a researcher. Well planned, sensitive and long-term studies can minimise any influences (Bernard, 1995; Denzin and Lincoln, 1998; Hammersley and Atkinson, 1995; Webb *et al.*, 1972).

Quality of methods and measures and their choice

In a recent commentary at the end of a special issue on ergonomics models and measurement, Annett (2002b) gently reminded us that ergonomics texts (the second edition of this one included!) do not always treat issues of the quality of ergonomics methods to the extent needed. In particular he suggests that we have not properly considered the reliability and validity of our methods and measurement. In fact, there are a number of ways in which we might judge the adequacy and quality of methods and measures, and therefore we have a number of selection criteria for methods. These are discussed below and the reader should see also any good research methods texts, such as Dane (1990), Frankfort-Nachmias and Nachmias (1996) or Robson (1993). It should be pointed out that we will rarely, if ever, be able to satisfy all of these criteria but equally we will rarely if ever need to for any particular study or investigation.

Validity

The validity of a measure tells us if the findings are trustworthy: are they really about what they appear to be about, does the method actually measure what it purports to measure? It tells us and others about the degree of confidence we can have in the veracity of any interpretations made on the basis of our results.

There are many different forms of validity; some types are embedded within others, some are very similar but are given quite different names. Construct validity represents the extent to which a measure represents the concepts that it should, and does not represent those that it should not. Within this we might look for: convergent validity (the measure is consistent with outcomes from two or more different measures — a form of triangulation); divergent validity (the measure does not correlate closely with measures of different concepts); and face validity — see below. Internal validity is the extent to which a measure provides a plausible demonstration of causal relationships, and external validity relates to its generalisability (to other situations or times, etc.). Content validity asks whether application of a method deals with a representative sample. Criterion validity has two forms: concurrent validity — whether the new measure compares well to outcomes of an existing 'accepted' measure (a limited form of convergent validity); and criterion validity — how well the measure predicts what we eventually see happen in the real world. Finally face validity is, literally, the validity of a measure or method 'on the face of it'. Sometimes seen as a form of construct validity, this is to do with the degree of consensus that a measure actually represents a particular concept. It also has to do with how acceptable a measure or method is to a particular set of stakeholders.

Reliability

It is sometimes said that reliability is a special sort of validity, and certainly it is needed to be sure of validity. Reliability comes in many different forms and manner of assessment as well, but essentially asks whether we would get the same results and interpretations if we repeatedly use a method or measure; it is directly related to the amount of bias or error we may have allowed to creep in. Bias and error may be from the investigator (and her arrangements) or the study participant, and can have conscious or sub-conscious origins. We may look for assessment of reliability through inter-rater comparisons, use of test/retest procedures, application of different forms of measure at different times or on different groups, or through split half techniques whereby the measure is somehow divided into two for each participant (e.g., divide the questions in a scale into two halves), and compare the scores within participants.

Generalisability

Literally this refers to how well a method, measure or measurement will generalise into other domains, situations, settings, populations, etc.

Non-reactivity

It is difficult to think of any ethical measurement of people's performance that does not effect or change the participants or their environment in some way, no matter how trivial. However, we would want to be sure that the reaction to the method and measurement process is not too great for the study purposes, and is known, acknowledged and reported. Rich participant observer accounts are often quite clear about the effect data gathering has had on the focus group, but the value of the data and interpretations frequently outweigh the scientific drawbacks. In an experimental setting, on the other hand, we would expect efforts to minimise any reactivity, to minimise bias and error and enhance reliability.

Sensitivity

We would not want to use a wooden ruler to measure changes in stature due to exposure to vibration — it would just not be sensitive enough to the size of changes we expect. On the other hand we would not want to go to the expense of recruiting 500 participants in an experiment when the effects we are interested in will be shown by almost all people and certainly will show up with 10 participants (some visual functions or performance for instance). Methods and measures should be selected which are sensitive to the right degree to the sorts of effects we are examining, whether to confirm or reject our hypotheses. Related to this, we need to select measures which can provide the level of detail needed for the particular enquiry, but not too much greater than this.

Feasibility of use

Sometimes it is said that methods should be simple to use — a little like product interfaces! This is true of methods which ergonomists have developed for use by others (engineers, managers, health and safety specialists) in practical settings, but is not necessarily required for methods used by human factors and ergonomics specialists in laboratory or field. However, we do want the methods used and the measures taken to be feasible for use in the particular circumstances we define, whether this is in relation to participants, understanding of questions, access points for observation to be set up, or enough data to populate a model meaningfully.

Acceptability and ethics

Strongly related to the idea of face validity and also to feasibility, the methods and measures must be acceptable — to the client, the study population to whom they are applied and to our peers and colleagues in the wider scientific community. Such acceptance may be in terms of resources required (see below), interference with the activities of the population under study, matters raised which the client would prefer not to be (e.g., worker opinions on matters outside the focus of the study), quality of measurement (in terms of validity and reliability in particular) and any potential distress or harm caused to participants (for instance through the very fact of a study being carried out, or else the application of experimental conditions beyond the envelope of acceptability for the population). Very much related is the idea of ethical research and practice, where we should follow codes of conduct and guides to ethical practice such as produced by professional bodies (for instance the Ergonomics Society or the British Psychological Society). Key aspects of these are the way in which we treat our participants, how we represent ourselves and how we report on our findings.

Resources

As with any endeavour, the resources required will be an important determinant of the methods and measures we use. Not only that but it will be the resources needed set against the importance of the study and the potential value of its outcomes, that will be critical. Resources are not just to do with financial cost, although this is a central consideration; we must also assess the resources needed in terms of people (whether as investigators/researchers or as participants), equipment and also time — for the analysis as much as for the data collection, as is seen regularly in observation studies employing video analysis.

Conclusions

Ergonomics/human factors has as its field the interactions of people with all other people and artefacts within an environmental context; its goal is the well being of individuals, organisations and national economies. This requires appropriate theory and practice, models and methods. The thin line which we must tread when we become involved in discussion of methodology is summed up in two views: on the one hand, '[psychology] should not allow itself to be driven by obsession with method to the exclusion of the human problems that are its province' (Barber, 1988, p. 7, reporting Maslow, 1946); on the other hand 'anyone who wishes to reflect on how they practise their particular art or science, and anyone who wishes to teach others to practise, must draw on methodology' (Cross, 1984, p. vii). Certainly ergonomists are driven by human problems, or driven to improve the quality of people's lives, but also we are always concerned to educate others (designers, engineers, politicians, accountants, managers, public, media) in our approach and the necessity for it. If we truly believe that we are, above all, promoters of an approach and of a process, then it behoves us to pay great attention to the roots and current and future state of our methodology. The remainder of this book is but one step in doing this.

Earlier in this chapter we explored frameworks for ergonomics and human factors, and in doing so suggested that the systems with which we should now be concerned cannot be modelled and understood by reference to a single person at the centre of a set of interactions with equipment and environment, but by representations of groups of people acting in social networks. In fact much work and life of the future, and many of the challenges facing ergonomics and human factors, will be found in complex interacting and distributed sociotechnical systems. The activities of interest will no longer just be those of command and control, with individual operators working with single control display interfaces, but will now be collaboration, co-ordination and integration, with constantly shifting and multiple teams of people, co-located and virtual, interacting with each other directly and via various forms of information and communication technologies (see Edwards and Wilson, 2004). Therefore the model of ergonomics and human factors for the future, within which we need to examine our tools and methods, might be represented as in Figure 5. This figure has been adapted from Wilson (2000), where it was presented to illustrate the fact that we are now not so much interested in various artefacts in a person's environment, and how the interfaces for these enable interaction, but more with studying the very interactions themselves.

Because this focus for ergonomics is increasingly important in the future, then we will need to extend our range of methods to deal with it. We certainly will need to draw from fields such as ethnography (e.g., Engström and Middleton, 1996; Hammersley and Atkinson, 1995), social network analysis (Wasserman and Faust, 1994), and computer supported co-operative work (see Bannon, 2003 for instance), and utilise the best new developments in cognitive work analysis and design (e.g., Hollnagel, 2003). This is an exciting time for ergonomics methods in this area, and it will be a considerable challenge to find ways in which we can truly understand peoples' behaviour in such rich sociotechnical

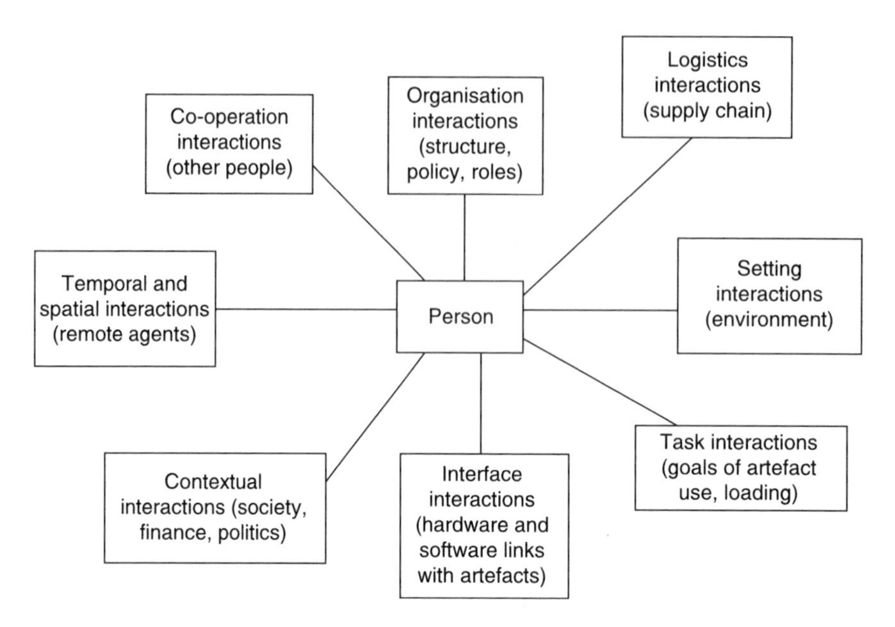

Figure 5 Interacting systems model for ergonomics and human factors.

systems, and at the same time provide data and findings with enough rigour to be applicable in future design and implementation.

This book is about methods in ergonomics and human factors. In introducing the reader to how we study people's interactions with artifacts, environments and other people, how we gather information, form models, develop recommendations and designs, and enhance understanding, we are also introducing them to ergonomics and human factors in general. Many other books do this as well, to different levels of detail, and in doing so introduce human factors methodology. Good sources used by the author, covering ergonomics and human factors generally (as against particular fields such as human–computer interaction) include those from Chapanis (1996), Chengalur *et al.* (2003), Kroemer and Grandjean (1997), Noyes (2001), Salvendy (1997), Sanders and McCormick (1993), and Wickens *et al.* (1998). Other excellent texts on E/HF method are referred to throughout this book also.

References

Annett, J. (2002a). Subjective rating scales: science or art? *Ergonomics*, **45**, 966–987.

Annett, J. (2002b). A note on the validity and reliability of ergonomics methods. *Theoretical Issues in Ergonomics Science*, **3**, 228–232.

Bainbridge, L. (1987). The ironies of automation. In: J. Rasmussen, K. Duncan and J. Leplat (eds) *New Technology and Human Error* (London: J. Wiley).

Bannon, L.J. (2003). Towards a social ergonomics: A perspective from computer supported co-operative work. In: M. McNeese, E. Salas and M. Endsley (eds) *New Trends in Co-operative Activities* (Santa Monica, CA: Human Factors and Ergonomics Society).

Barber, P.J. (1988). *Applied Cognitive Psychology* (London: Methuen).

Beaty, D. (1995). *The Naked Pilot* (Shrewsbury, UK: *Airlife Pub. Co.*).

Bernard, H.R. (ed). (1995). *Research Methods in Anthropology* (2nd Edition) *Qualitative and Quantitative Approaches* (Walnut Creek, CA: Altamira Press).

Beynon-Davies, P. (1999). Human error and information systems failure: The case of the London ambulance service computer-aided dispatch system project. *Interacting with Computers*, **11**, 699–720.

Bignett, V., Peters, G. and Pym, C. (1977). *Catastrophic Failures* (Milton Keynes, UK: *Open University Press*).

Casey, S. (1993). *Set Phasers on Stun* (Santa Barbara, CA: Aegean Pub. Co).

Chapanis, A. (1979). Quo vadis, ergonomia. *Ergonomics*, **22**, 595–605.

Chapanis, A. (1976). Keynote address: Ergonomics in a World of New Values. *Ergonomics*, **10**, 253–268.

Chapanis, A. (1996). *Human Factors in Systems Engineering* (New York: J. Wiley).

Chengalur, S.N., Rodgers, S.H. and Bernard, T.E. (2003). *(Kodak's) Ergonomic Design for People at Work* (New York: J. Wiley).

Clark, T.S. and Corlett, E.N. (1984). *The Ergonomics of Workspaces and Machines: A Design Manual* (London: Taylor & Francis).

Corlett, E.N. (1991). Some future directions for ergonomics. In: M. Kumashiro and E.D. Megaw (eds) *Towards Human Work: Solutions to Problems in Occupational Health and Safety* (London: Taylor & Francis).

Cross, N. (1984). *Developments in Design Methodology* (Chichester: John Wiley).

Cushing, S. (1994). *Fatal Words* (Chicago: University of Chicago Press).

Dane, F.C. (1990). *Research Methods* (Belmont, CA: Wadsworth).

Denzin, N. (1970). *Sociological Methods* (New York: McGraw-Hill).

Denzin, N.K. and Lincoln, Y.S. (eds) (1998). *Strategies of Qualitative Inquiry* (London: *Sage Publications*).

Eason, K.D. (1997). Understanding the organisational ramifications of implementing information technology systems. In: M. Helander, T. Landauer and P. Prabhu (eds) *Handbook of Human–Computer Interaction*, 2nd edition (Amsterdam: Elsevier).

Edwards, E. (1973). Techniques for the evaluation of human performance. In: W.T. Singleton, J.G. Fox and D. Whitfield (eds) *Measurement of Man at Work* (London: Taylor & Francis), pp. 129–133.

Edwards, A. and Wilson, J.R. (2003). *Guide to Implementation of Virtual Teamworking* (London: Gower Press).

Edholm, O.G. and Murrell, K.F.H. (1973). *The Ergonomics Research Society: A History 1949–1970* (Loughborough: The Ergonomics Society).

Engström, Y. and Middleton, P. (eds) (1996). *Cognition and Communication at Work* (Cambridge: Cambridge University Press).

Frankfort-Nachmias, C. and Nachmias, D. (1996). *Research Methods in the Social Sciences* (London: Arnold).

Green, B. and Jordan, P.W. (1999). The future of ergonomics. In: M.A. Hanson, E.J. Lovesey and S.A. Robertson (eds) *Contemporary Ergonomics, '99* (London: Taylor & Francis), pp. 110–114.

Grey, S.M., Norris, B.J. and Wilson, J.R. (1987). *Ergonomics in the Electronic Retail Environment* (Slough, UK: ICL (UK) Ltd).

Hammersley, M. and Atkinson, P. (1995). *Ethnography* (2nd edition) (London: Routledge).

Helander, M.G. (1999). Seven common reasons not to implement ergonomics. *International Journal of Industrial Ergonomics*, **25**, 97–101.

Hignett, S.M. and Wilson, J.R. (2004a). The role for qualitative methodology in ergonomics: A case study to explore theoretical issues. *Theoretical Issues in Ergonomics Science*, **5**, 473–493.

Hignett, S.M. and Wilson, J.R. (2004b). Horses for Courses — but no favourites. *Theoretical Issues in Ergonomics Science*, **5**, 517–525.

Hollnagel, E. (ed). (2003). *Cognitive Task Design* (Mahwah, NJ: Lawrence Erlbaum Assoc.).

Kantowitz, B.H. (1992). Selecting measures for human factors research. *Human Factors*, **34**, 387–398.

de Keyser, V. (1992). Why field studies? In: M. Helander and M. Nagamachi (eds) *Design for Manufacturability: A Systems Approach to Concurrent Engineering and Ergonomics* (London: Taylor & Francis), pp. 305–316.

Kragt, H. (1992). *Enhancing Industrial Performance* (London: Taylor & Francis).

Kroemer, K.H.E. and Grandjean, E. (1997). *Fitting the Task to the Human* (5th edition) (London: Taylor & Francis).

Kuorinka, I. (ed) (2000). *History of the International Ergonomics Association* (IEA Press, International Ergonomics Association).

Leamon, T.B. (1980). The organisation of industrial ergonomics — a human machine model. *Applied Ergonomics*, **11**, 223–226.

Markus, T.A. (1969). The role of building performance measurement and appraisal in design method. In: G. Broadbent and A. Ward (eds) *Design Methods in Architecture* (London: Lund Humphries).

Meister, D. (1985). *Behavioural Analysis and Measurement Methods* (Chichester: John Wiley).

Meister, D. (1995). *Divergent Viewpoints: Essays on Human Factors Questions.* Personal publication.

Meister, D. (1995). Human factors — the early years. *Proceedings of the Human Factors Society 39th Annual Meeting*, San Diego, California, pp. 478–480.

Meister, D. (1998). Twenty-first-century challenges to ergonomics. *Ergonomics in Design,* **April**, 33–34.

de Montmollin, M. (1992). The future of ergonomics: hodge podge or new foundation? *Le Travail Humain*, **55**, 171–181.

Moray, N. (1994). 'De Maximum non Curat Lex' or How context reduces science to art in the practice of human factors. *Proceedings of the Human Factors and Ergonomics Society 38th Annual Meeting*, Nashville, Tennessee, pp. 526–530.

Moray, N. (2000). Culture, politics and ergonomics. *Ergonomics*, **43**, 858–868.

National Research Council (1986). *Research Needs for Human Factors. Report of the Committee on Human Factors* (Washington, D.C.: National Academy Press).

Noyes, J. (2001). *Designing for Humans* (Hove, UK: Psychology Press).

Perrow, C.B. (1984). *Normal Accidents: Living with High Risk Technologies* (New York: Basic Books).

Petroski, H. (1992). *To Engineer is Human: The Role of Failure in Successful Design* (New York: Vintage Books).

Pheasant, S. (1986). *Bodyspace* (London: Taylor & Francis).

Poulton, E.C. (1975). Observer bias. *Applied Ergonomics*, **6.1**, 3–8

Reason, J. (1997). *Managing the Risks of Organisational Accidents* (Aldershot, Hants, UK: Ashgate).

Reason, J. (1990). *Human Error* (Cambridge: Cambridge University Press).

Robson, C. (1993). *Real World Research* (Oxford: Blackwells).

Rouse, W., Kober, N. and Mavor, A. (eds) (1997). *The Case for Human Factors in Industry and Government, Report of a Workshop* (Washington: National Academy Press).

Rutherford, A. and Wilson, J.R. (1992). Searching for the mental model in human-machine systems. In: Y. Rogers, A. Rutherford and P. Bibby (eds) *Models in the Mind: Perspectives, Theory and Application* (London: Academic Press), pp. 195–223.

Salvendy, G. (ed) (1997). *Handbook of Human Factors and Ergonomics* (New York: J. Wiley).

Sanders, M.S. and McCormick, E.J. (1987). *Human Factors in Engineering and Design*, 6th edition (New York: McGraw-Hill).

Sanders, M.S. and McCormick, E.J. (1993). *Human Factors in Engineering and Design*, 7th edition (New York: McGraw Hill).

Shackel, B. (1996). Ergonomics: scope, contribution and future possibilities. *The Psychologist*, **July**, 304–308.

Singleton, W.T. (ed) (1982). *The Body at Work: Biological Ergonomics* (Cambridge: Cambridge University Press).

Singleton, W.T. (1994). From research to practice. *Ergonomics in Design*, **July**, 30–34.

Stockbridge, H.C.W. (1989). *The Ergonomics Society: A History 1971–1989* (Loughborough: The Ergonomics Society).

The Guardian (1993). Criticised tape start could stay; flagman blamed in panto horse race. June 14th.

UK National Audit Office (1999). *The Passport Delays of Summer 1999. Report by the Comptroller and Auditor General, 27th October.* (London: The Stationery Office).

UK Public Accounts Committee (2000). *Improving the Delivery of Government IT Projects. Report of the House of Commons Public Accounts Committee* (London: The Stationery Office).

Vicente, K.J. (1999). *Cognitive Work Analysis* (Mahwah, NJ: L. Erlbaum).

Vicente, K.J. (2000). Towards Jeffersonian research programmes in ergonomics science. *Theoretical Issues in Ergonomics Science*, **1**, 93–112.

Wasserman, S. and Faust, K. (1994). *Social Network Analysis: Methods and Applications* (Cambridge: Cambridge University Press).

Wastell, D.G. and Cooper, C.L. (1996). Stress and technological innovation: A comparative study of design practices and implementation strategies. *European Journal of Work and Organisational Psychology*, **5**, 377–397.

Waterson, P.E. and Sell, R. (2005). Recurrent themes and developments in the history of the Ergonomics Society. *Ergonomics* (in press).

Webb, E.J., Campbell, D.T., Schwartz, R.D. and Sechrest, L. (1972). *Unobtrusive Measures: Non-reactive Research in the Social Sciences* (Chicago: Rand McNally).

Welford, A.T. (1976). Ergonomics: where have we been and where are we going: I. *Ergonomics*, **19**, 275–286.

Wickens, C.D. (1984). *Engineering Psychology and Human Performance*, 1st edition (Columbus, OH: Charles E. Merrill).

Wickens, C.D., Gordon, S.E. and Liu, Y. (1998). *An Introduction to Human Factors Engineering* (New York: Longman).

Wilson, J.R. (1994). Devolving ergonomics: the key to ergonomics management programmes. *Ergonomics*, **37**, 579–594.

Wilson, J.R. (2000). Fundamentals of ergonomics. *Applied Ergonomics*, **31**, 557–567.

Wilson, J.R. and Nichols, S.C. (2002). Measurement in virtual environments: another dimension to the objectivity/subjectivity debate. *Ergonomics*, **45**, 1031–1036.

Wilson, J.R., Jackson, S. and Nichols, S. (2003). Cognitive work investigation and design in practice: the influence of social context and social work artefacts. In: E. Hollnagel (ed) *Cognitive Task Design* (Mahwah, NJ: Lawrence Erlbaum Assoc.).

Wogalter, M.S., Hancock, P.A. and Dempsey, P.G. (1998). On the description and definition of human factors/ergonomics. *Proceedings of the Human Factors and Ergonomics Society 42nd Annual Meeting*, Chicago, Illinois, pp. 671–674.

Part I

General approaches and methods

This book opens with descriptions of general approaches and methods. These approaches might be employed in any area of ergonomics enquiry and investigation, whether in the search for fundamental understanding or in practical application work. They also might be found in any discipline in the human sciences. Broadly we might observe people or ask people about themselves and their behaviour — an empirical approach. Or we might create models and abstractions of people and what they do, and analyse these — an analytical approach. Or we might use already available information, whether produced for human factors purposes or not — an archival approach.

When studying how people behave, perform and think we will employ techniques of *direct observation* of behavioural phenomena, and techniques of *indirect observation* to collect data from people about their interpretations of what they are doing, feeling or thinking. In more detail, direct observation is the collection of information on performance, made directly by the observers themselves or taken from mechanical, electrical, electronic or other artificial recording technologies. Indirect observation involves the provision of verbal or written reports of behaviour, attitudes and opinions by the people themselves, or by other associated individuals such as colleagues or supervisors.

The direct observation group of methods is sometimes seen as 'objective', and the indirect observation methods as 'subjective'. In this view it is assumed that direct observation produces 'true' records, with no biased interpretation by the participant whereas subjective assessment is thought to imply that the participant or other informant provides abstracted and interpreted and possibly biased information. However, this ignores the fact that data collected by direct observation will be re-analysed, summarised and abstracted by the observers, from their memory, notes, tape or video recordings, and then interpreted by the investigators, introducing some possible bias or at least subjectivity. Even the very selection of direct observation is subject to expectations, preferences or other personal choices. Moreover, properly constructed and carried out, subjective assessment or indirect observation can be run so as to maximise as far as possible the data reliability and validity. In any case, much of psychological measurement is possible only through indirect observation; the phenomena involved — what people are thinking or their reasons for certain behaviours — cannot be directly observed or even easily inferred from direct observations.

Both groups of techniques can be used in laboratory or field studies, although only direct observation in the field can be naturalistic and non-interfering with behaviour, and even then only if carried out carefully. The degree of interference will in part be determined by whether the observations are made visibly, participatively or unobtrusively. The last of these should provide a record of behaviour which is unaltered through reaction to an observer's presence, but any attempts to carry out such 'hidden' studies must beware a certain inflexibility of coverage which may result and, most importantly, must first entail careful consideration of the ethics of the particular circumstances. Of course, ethical issues will depend in part on the use to which data might be put but will also be situation specific; justification for unobtrusive observation may be easier for a study of pedestrian behaviour in traffic for instance, than for following an individual industrial worker in their workplace.

Participatory observation, sometimes used in sociological and ethnographic studies, can allow insight into, and non-interference with, participants' behaviour but sometimes leads to a 'worm's eye' view being taken. Also, the observer may change over time, identifying (or not) with those observed, and their presence may well be a catalyst for unusual behaviour.

Generally then, in laboratory or field, the ergonomist relies upon being a visible observer, yet tries not to interfere with, or contaminate, behaviour by word, action or even mere presence.

The decision of whether to collect observable data directly in real-time, using storage in the observers' memory, on paper or event recorder and so on, or by using recording instruments such as video or audio tape, will depend upon the circumstances. Task type, behaviour to be observed, site conditions, study resources in terms of people, equipment and time, and the type of measurements or records needed will all be taken into account. Generally hardware recording will give unlimited data capture rates but be more limited in terms of geographical space covered. It will provide a permanent record to be re-consulted to check accuracy but requires considerable time to analyse down to a written record (generally over ten times the 'real' time for video for instance). Also, although video involves no data reinterpretation or misinterpretation, nor difficulties with the observers' attention, learning and so on in the field, these may be problems which occur later during analysis in the research laboratory.

The formal and traditional scientific way to collect information, especially to confirm (or not) hypotheses about performance or to gather accurate measures of people's physical, mental or other characteristics, is to carry out controlled experiments, usually in a laboratory setting but very occasionally in the field. The ways of doing this, and the factors to be taken into account and procedures to be followed, are laid out by Drury in Chapter 2. Some ergonomists might prefer field studies for their greater realism and thus potentially their validity and meaningfulness, but the production of reliable and usable data will often necessitate use of the controlled laboratory environment. So that such studies and experiments do not end up as costly, sterile exercises, great care must be taken in their design and conduct. The trade-off — being able to isolate the variables of greatest interest but by doing so perhaps losing the influence of setting and context — must be understood and assessed carefully before any choice to carry out experimental research.

In Chapter 3, Bisantz and Drury concentrate upon direct observation and largely upon techniques of observation and measurement which can be applied in field or laboratory. Some of the techniques discussed in this chapter are derived from work study. As ergonomists we have borrowed (and we like to think improved upon) methods used in the industrial engineering world of work measurement and method study, but our philosophy and approach are different — fitting tasks to people instead of vice versa. Also, we have expanded the focus and methods of observation to embrace all kinds of human behaviour and system response, not just those concerned with effective performance.

Before embarking upon any of the extensive and expensive investigations often needed for direct or indirect observation, full use should be made of data that exist already. In Chapter 3 also, Bisantz and Drury identify several sources of archival data usually available within organisations. According to study needs these can be: plans; annual reports; operating records from production, manufacturing engineering, quality control, personnel, medical, and finance departments; maintenance and stores reports; records for absenteeism or labour turnover; or minutes of meetings, for instance worker–management meetings. The beauty of all these archives is that someone else has already taken the time and trouble to produce the data; all we have to do is extract and interpret what we need. The downside is the danger that the records themselves will be incomplete, variable over time or situation, selective or biased, or that we might misinterpret or selectively interpret them. Nonetheless, they are a very useful resource for any field study.

We might also include under the heading of archives our own store of ergonomics principles, methods, data and criteria contained in textbooks, hand-books, journals and reports. Such literature archives are a necessary resource in any study. For instance, in designing the layout of a new car plant assembly line we would much prefer to use good and relevant population anthropometric data than to have to collect our own for all the employees.

In order to truly understand much of human behaviour we must not only observe people but we must question them or obtain verbal and written reports from them also. This permits greater insight into what is being done and why. It is also the method by which we assess affective responses — feelings and attitudes. Indirect observation as an approach is discussed as techniques of subjective assessment by Sinclair in Chapter 4. Such assessment includes use of rating, ranking, questionnaires, interviews and checklists. Psychophysical techniques are also available for such measurement. Ostensibly, indirect observation appears to be a simple approach to data collection, and indeed a questionnaire is often the first choice in a study. Great care is needed though to ensure that subjective reports do indeed reflect accurately and completely the target behaviour, thoughts or knowledge, which means that instrument development, administration and analysis must be treated very carefully. As well as minimising any distortion, omissions or commissions on the part of the participants, we must ensure as far as possible that the investigator and the method and techniques used do not introduce bias themselves and do not unnecessarily constrain the focus of enquiry.

Traditionally there has been a presumption in favour of quantitative measurement wherever possible in human and social sciences. It is possible that this is to do with a traditional desire to appear to be 'scientific', in order to achieve respectability. This is shown in the drive for analytical techniques or models that are computational, in using performance measures derived from objective measurement within direct observation, and in the feeling amongst some ergonomists that we 'must give the engineers numbers to work with'. However, in recent years there has been great interest in use of qualitative approaches, literally approaches that provide information that is not numerical. Thus a questionnaire for instance can provide quantitative data in the form of interval-level scales and statistical analysis, or can provide rich accounts and descriptions from the respondents, that are classified and interpreted by the investigator into qualitative accounts. Hignett, in Chapter 5, reviews and supports a qualitative approach to ergonomics study and measurement, not in opposition to quantitative measurement but as two complementary approaches which should be used by a rich and mature discipline.

There are certain approaches, methods and associated techniques which are seen as particularly a part of ergonomics methodology, or are widely used throughout ergonomics investigations. In the view of many practitioners, an early and vital component of ergonomics investigation is task analysis (Chapter 6, Stammers and Shepherd). General ergonomics opinion, followed in the text here, is that the completion of a task analysis comprises both the description and analysis of tasks; within task analysis, data are collected, represented in an appropriate description form, and are analysed to assess task requirements for the person, their expected behaviour, any constraints and the task and environmental demands on them.

Task analysis is widely used since it can be applied during the analysis of existing systems, the design of new ones, and the evaluation carried out subsequently. Information gained is useful both in development and as criteria against which to assess what is developed. What is represented and analysed is what must be done in order to fulfil certain goals, within constraints from the task environment and from individual or general human limitations. Task analysis and its techniques were originally distinguished from method study both by underlying purpose but also by their concentration upon operator decisions as much as upon actions. Lately this has been extended and we have much interest in cognitive task analysis (CTA), which demands somewhat different methods of data collection, reporting and interpretation (although Stammers and Shepherd are at pains to point out that CTA is not such a departure from the task analyses that have been a staple of ergonomics for three or four decades).

Multiple methods of direct and indirect observation, expert analysis and so on are required in task analysis. What is particular to task analysis beyond general ergonomics study are the description formats or representations selected to best allow requirements analysis; many such formats are described by Stammers and Shepherd. A note of caution is that, contrary to the apparent beliefs of some ergonomists, a task analysis is not a complete investigation and is certainly not an end in itself; it is merely a means to an end, to be of assistance in systems design or redesign.

One methodological approach which can be used within task analysis, but which has a wider utility, is verbal protocol analysis (VPA). As described by Bainbridge and Sanderson in Chapter 7, this is about reports being made by people about what they are doing (and why) whilst they are doing it, sometimes known as 'thinking aloud'. The chapter concentrates upon concurrent VPA, reports made at the same time as task completion, rather than retrospective reports made whilst the worker reviews their performance on video. There is a relationship here with cognitive task analysis in that we are seeking insight into non-observable processes, the thinking which may or may not lead to subsequent action. As with task analysis, stages of data collection, representation and analysis are implied. Data are collected in the form of verbal — often audio tape — reports; reports will be transcribed and placed in the context of a simultaneous record of behaviour (often from video) to produce the actual protocols, divided into phrases or phrase groups; subsequent analysis will be made of explicit and implicit content and of inferred connecting material. Bainbridge and Sanderson quite clearly point out many of the problems of verbal protocol analysis, both in its operation but also in the very concept. However, verbal protocols offer one of the few ways for making inferences about cognitive processes and for understanding complex behaviour in relation to similar past events, present circumstances or predictions of the future.

One particular goal of both task analysis and protocol analysis, not necessarily well addressed, is to understand the nature, manifestation and development of expertise, taken up by Shadbolt in Chapter 8. Human factors has a role to play in addressing the bottleneck in knowledge-based systems — acquiring the knowledge necessary to build the system. Generally *the* major source of such knowledge will be the experts themselves; knowledge elicitation is the name for gathering the relevant information about people's knowledge. In some ways parallels can be drawn between this and chapters in the next section, in that the techniques described there are to elicit information, opinion and insight from users or user 'representatives', to be applied in design.

Elicitating expertise uses a number of methods from other domains but has also seen development of a number of methods explicitly for the purpose of understanding the knowledge and skills that people hold, which is why this chapter appears in this general section. Moreover, the focus is now not so much on the development of expert systems but on the support for knowledge management, knowledge-based systems of all kinds and for design of jobs and organisations that make the most of the skills and knowledge of the staff.

Both task and protocol analysis techniques will be applied when human performance can be observed or reported or, at least in the case of task synthesis, relatively easily predicted. When this is not the case, or when the range of potential behaviours is large, then we must turn to other means. In these circumstances, behaviours or task performance, and related equipment and events, must be modelled or simulated, crudely differentiated as involving physical and symbolic representation respectively. There must be some degree of overlap between the two though, in that models may be used in the simulation of systems and simulators may be built around models. For instance, computer workspace modelling, considered later in Chapter 28, is a modelling technique, which can also produce simulations of activities in the workspace; simulations lead to observations which may give models; models can be superimposed on simulations.

Simulation can involve game or role playing but is especially the physical representation of reality. Hardware plus software simulations with a good deal of fidelity to the actual system are used in system design and evaluation and in training. Expert systems might be seen as a form of simulator, but at a lower level of sophistication simulations can involve expert analysis, including 'walkthrough', of two-dimensional (drawings) or three-dimensional (mock-ups) models of equipment or work places, as described by Laughery and Meister in Chapter 9.

Models of various types will be employed usually when an actual system or even a physical simulation is not available, and thus models are used generally in new system development or in needs analysis. Models of most use to ergonomists are of behaviour in human–machine systems, and are symbolic representations of human performance. The mathematical, usually computer-based, nature of the model allows manipulation of variables to determine system outcomes; they may be used within task analysIs (or synthesIs) in assessing task requirements or consequences.

Developments in computer technology have enabled great improvements in model and simulator power and utility. In the same way, computer-based techniques can make general data collection and analysis much more efficient and comprehensive. However, there is a need for caution. In contrast to the large multi-measure evaluations possible now, the more limited investigations previously possible did have the advantage of producing relatively simple statements on performance or load on the individual, for instance, which colleagues in design, implementation and management would appreciate and utilise. There is a danger of methodology acquiring a momentum of its own, of analysis being performed way beyond the needs of the situation, and of the ergonomist losing sight of the original problem. Methods and techniques are only tools, to be selected and applied with a clear understanding of the design or evaluation objectives. However, although skilful use of methods requires practice we should beware of sticking rigidly to standard methods. Imagination, as well as scientific rigour, is an ingredient of successful ergonomics.

John Wilson

Chapter 2

Designing ergonomics studies

Colin G. Drury

Introduction: what is a study?

A major part of the job description of a human factors engineer or ergonomist is to collect valid and reliable data as part of the design or evaluation of products and processes, or to push forward the boundaries of ergonomics knowledge. Throughout this book there are examples of studies, to illustrate what to measure, how to measure it or how to analyze the data. In the current chapter we tie together many of the concepts to provide an opening guide to conducting studies. Often in science and engineering we consider design of experiments (DoE), but this chapter extends the concepts beyond that special class of studies labelled 'experiments' to consider natural observations, questionnaires and interviews. Each of these has, of course, its own chapter in the book, and indeed a whole library shelf of references and textbooks giving more detail. We will use references liberally in this chapter, both for direct access to specific design ideas, and for examples to illustrate points. The examples will come both from the social science side of ergonomics and the physical side: while the specific theories will be quite different between the two, the study design issues are very similar.

Designing studies is based on accurate goal specification, followed by logical (and/or economic) choice of steps to meet that goal. What is the goal of a study? Typically, those employing the ergonomist would set the goals, e.g., to test different menu design features for computer user acceptance. In many circumstances the goals will be defined in a formal request for proposal (RFP), but in other cases the ergonomist will have some leeway to question the study goals. Perhaps the greatest freedom of goal-choice is in academia, where a student or faculty member can run studies with self-generated goals, subject only to their ability to find funding and satisfy human participants' protection rules. Any good textbook must advise the ergonomist to question the goals of the study, but must also caution that much effort has been wasted on this activity over the years. We will assume that you have a set of goals and proceed as if these were now engraved on stone tablets. The object of this chapter is to provide an orderly way in which these goals can be turned into a designed study, in much the same way as a statistical textbook (e.g., Schiff and D'Agostino, 1996) will demonstrate how to turn a research hypothesis into a statistical hypothesis. As with any scientific activity, the more you know about the mechanism underlying the phenomena of interest, the better you can design the study. While much of human factors may seem rather empirical to the pure scientist, there are underlying bodies of knowledge (and models) without which the design would be ill-informed and perhaps useless. For example, measuring the performance of airport baggage screeners by sending through bags with threats in them and measuring how many are detected would be unthinkable to ergonomists whose knowledge of signal detection theory would cry out for the chance to measure false alarms in the same study. Human factors studies may be empirical, but they are model led.

If a goal is the input then the output is a study design. The specificity of the design is particularly important as minor details can have major consequences, for example see the public debate on whether musculo-skeletal injuries are caused by work (National Academy of Sciences, 2001). A good test of the completeness of the design is that it can be given to a competent technician to carry out, and it will always be done in the way the originator intended. In statistical design of studies, the hypotheses to be tested are defined in terms of the measurements to be taken and the tests to be performed on the data, *before* any data are collected. A well-designed ergonomics study should meet the same criteria; you must know what to do with the data you have collected if you are to avoid the statistical (and economic) error of collecting data *and then* looking for interesting interrelationships. This does not mean that all studies require a formal research hypothesis to justify every measurement. Data collection is expensive, so that it may be possible to augment the study by including some measurements which can be conveniently collected but for which no formal hypotheses are made. In that case, these augmented variables cannot be used to test hypotheses after the fact, but they can be used to formulate new hypotheses for testing in subsequent, independent experiments. While the major goals of any study must always be tested hypotheses, it would be foolish to overlook opportunities to learn more about human interaction with the environment.

Choice of technique

Perhaps the first choice that needs to be made is the choice of study technique. We have a range available, as reading issues of the journals *Ergonomics*, *Human Factors* or *Applied Ergonomics* will demonstrate. The techniques differ on two related dimensions: how much they interfere with the system under study, termed reactivity, and how well they can determine whether effect follows from cause, termed causality. Experiments differ from all other techniques in that they directly change a system and measure the effects of these changes. Because the experimenter determined when and how to change the system, alternative explanations of the results in terms of coincidence are most unlikely. Thus, experiments are able to detect causality with some degree of confidence. In contrast, observational studies, such as those of seat belt use in automobiles (Drury *et al.*, 2002), can only ever show that two variables are related (e.g., seat belt use by driver and by passenger) not that one is causal to the other.

At the extreme low end of reactivity are *task analytic methods*. At the earliest levels of system design, they do not even rely on a system to study. Tasks are described and analyzed for their ergonomic impact by logically deducing the task description from system goals and functions and by deriving task demands and human capabilities from the logic and the literature (see Chapter 6). Later stages of task analysis may well observe the real system, or even call for experiments but by then we are using the techniques that follow.

At the next higher level of system manipulation are the methods of *direct observation* and *archival methods*. Here, a functioning system is studied by reading its records or observing its behaviour (see Chapter 3). We often think that such techniques are non-reactive, but they do involve actively interfering with the system, i.e., no observations would be taken were the study not taking place. The reactivity may only be potential, as in studying ambulance effectiveness from patient records (Barnes and Drury, 1976) or it may be very real, as in a stop-watch time study (Konz and Johnson, 2000, Chapters 27 and 28). As Chapter 3 shows, archival data is also available for human factors analysis in many systems without disturbing the system, although again causality is difficult to prove.

Questionnaires and *rating scales* (see Chapter 4), the next higher level, obviously affect the system but are often used as if they did not. For example, a study of emergency telephone responses (Barnes and Drury, 1976) probably caused many participants to think through their emergency response behavior for the first time as less than 15% had ever used the telephone to call for fire, police or ambulance.

Finally, any *direct manipulation* of the system must be classed as an experiment, whether it is a multifactorial design used in a research laboratory or a case-study (or other pseudo experiment; Kerlinger, 1986) run in a factory. The system is deliberately changed so that the results of the changes can be observed. This can never be done in a non-reactive manner.

Given that manipulating and changing systems must be costly and potentially dangerous, what does this increase in reactivity buy for the ergonomist? In addition to the major advantage of determining causality, the two other advantages of a more highly reactive design are:

1. The ability to be in the right place at the right time to observe. This is particularly important in human factors studies where the system behavior observed is often rare and unexpected, e.g., accidents or breakdowns.
2. The ability to use more obviously invasive, but information-rich, measurement techniques. For example, in inspection research, the response to each individual item inspected can be at considerable detail in an experiment (e.g., Drury and Sinclair, 1983), whereas in the real situation only a simple accept/reject response is often given.

If the human factors engineer gains in experimental control and measurement detail by using highly reactive designs, what is lost? The major loss is in face validity. If we observe a system in its natural state, those associated with the system, and possibly those who commissioned the study, can be convinced that the study is realistic, however that is defined by those to whom it is important. An experiment, particularly one performed in a laboratory with artificial stimuli and non-representative participants, requires much more persuasion on the part of the ergonomist to gain acceptance. The author was once involved in two studies of fork-lift truck control. One (Drury, Cardwell and Easterby, 1974) involved real drivers using real fork-lift trucks in a real warehouse to study lateral control behaviour. The other (Drury, Cardwell and Easterby, 1974) involved real drivers controlling a toy train in a laboratory to study longitudinal control behaviour similar to Fitts law tasks. It is obviously much easier to quote the former study to convince warehouse managers of its design implications.

There are many degrees of realism within the experimental paradigm. This is often seen as a debate or choice between laboratory and field experimentation, although that is a somewhat restricted and sterile view. There is no reason that carefully controlled experiments cannot be conducted outside the laboratory, nor that laboratory experiments should be other than valid predictors of reality. Indeed, the whole approach to the relationship between empirical and knowledge-gathering studies has recently been re-examined (Stokes, 1997). Stokes maintains that research for knowledge and research for empirical results represent two orthogonal dimensions rather than the two end-points of a single dimension. In this light, the choice of level of realism is an entirely empirical one, based on the opportunities for control and face validity in the particular workplace studied. There is no difference *in kind* between running an experiment in a factory or air traffic control center where the participants know that they are being studied (essential for ethical purposes) and running an experiment in a simulator. There may well be practical differences, but you as experimenter have interfered with the normal work conditions in both cases: the only difference is in degree of interference.

Experiments can be conducted using the operational system (e.g., real aircraft), a faithful simulation (e.g., flight simulator) or a task only logically related to the operational system (e.g., tracking or dichotic listening). The issues of fidelity in experimentation are essentially the same as those in simulator design (see Chapter 9). Only a thorough knowledge of the ergonomics literature can guide the experimenter on what it is safe to leave out of an experiment. As an example, there are thousands of references on the component processes of an industrial

Table 1 Summary of Available Study Designs and their Characteristics

	Causality	Reactivity	Face validity	Control	Measurement detail
Task analysis	No	Zero	–	–	–
Archival data	No	Zero	High	Zero	Variable
Observation study	No	Low	High	Zero	Low
Questionnaire/interview	No	Medium	Medium	Low	High
Experiment	Yes	High	Low	High	High

inspection task (visual search, signal detection, vigilance) but only a few dozen that meet the stringent requirements of *experimental* techniques used by *real* inspectors with *real* products. These issues of reactivity, control, measurement detail, and validity are summarised in Table 1.

Choice of independent variables

Independent variables are what you change, while dependent variables are how you measure the effects of the change. Choice of independent variables comes primarily from the hypotheses: if you hypothesise that control-display compatibility affects reaction time, you have chosen compatibility as your independent variable. If life were that simple, this section would not be needed. In order to measure reaction time under different values of the independent variable, you must get down to specifics on other matters: who will be the participants? What will be the instructions? Will the experiment be run sitting or standing, morning or evening, and so on? Thus choice of independent variables is not merely a matter of making obvious deductions from the hypotheses.

It will be simpler to talk of independent variables as if they were manifested in an experiment, implying that the ergonomist actively implements the choices of values of each variable. However, the same principles apply if the technique is observation or questionnaire although choice is more a matter of selecting, from among already existing alternatives, those to be included in the study. In a similar manner, some of the language of experimental design can be applied to observation and questionnaires. Thus an independent variable is a factor and the value which the independent variable takes is the level of that factor. Both naming conventions will be used throughout this section.

The design of any study first consists of choosing levels of all of the independent variables of relevance. Because we must be specific in our study design, then we must specify not only what we will vary, i.e., what levels of which factors, but what we will not vary, i.e., what we will keep constant. One has only to look at the *post hoc* rationalisations of unexpected study results in our journals to see that ergonomists do not always recognise what must be kept constant. Taken with the frequent finding of insignificant effects of a major factor, it recalls the scientist's version of Murphy's law: 'variables won't and constants aren't'. The least that an uncontrolled variable can do is to contribute to random experimental errors: typically it does more and introduces real biases into a study. Systematic techniques are required if we are to bring ergonomics knowledge to bear upon the design of effective and efficient studies. We need a technique for listing all possible factors which can affect the outcome of the experiment, and a technique for deciding what to do with each factor.

To generate a list of all possible factors which could affect the dependent variables, it is simplest to use the categories which ergonomists regularly use: human, machine and environment. To these should rightly be added task, to cover the instructions, restrictions and goals under which the system operates, cover inter-personal relations and management. Thus the five categories are *task, operator, machine, environment* and *social.* Under each of these is listed the factors likely to affect performance. For example, in a study of control/display compatibility and reaction time, obvious factors under operator are: age, experience of similar systems, training on system under test, and national stereotypes for control/display relationships. Less obvious factors, but ones which can certainly affect reaction time, are: intelligence, visual capabilities, risk acceptance/aversion, and motor co-ordination. Some factors *may* affect performance, but our insight tells us that the probability is small: body size, gender, and time since last meal. This list could obviously continue well into the trivial, as could similar lists for task, machine and environment. Note that the lists will be different for different hypotheses; in manual materials handling systems, body size and gender would be of primary importance.

We now have four somewhat orderly lists of possible factors and face the question of what to do with each factor on each list. There are only five alternatives:

1. *Build the factor into the experiment at multiple levels.* This ensures that our hypotheses will be general across the whole range of this factor used in the experiment. Thus if we have five age groups, covering the decades of the working population (i.e., 15–25, 25–35, 35–45, 45–55, 55–65 years), we can be sure that our compatibility conclusions are valid for all working ages, or we will know that they only apply to older workers, depending upon the specific outcome of the study. This is the preferred treatment of each factor, as it maximises the impact of our studies. It is also by far the most expensive option, especially as each factor tends to multiply the size of the experiment by the number of its levels. Typically, only factors specifically named in our hypotheses will be built in at multiple levels.
2. *Treat the factor as a covariate.* Often it is not feasible to choose levels of a major variable in advance, as we may need expensive measurements to determine what the level actually is. For example, although age and body size are simple enough to measure or even to ask for by telephone when scheduling participants, more subtle factors such as perceptual style or visual reaction time will require the participant to undergo tests before being allowed into the experiment. In such cases, the factor of interest can be measured, typically before the main experiment, and used as a covariate in an analysis of covariance design. This means that any correlation between the dependent variable and the covariate is taken out as a specific term in the analysis, typically before the other factors are analyzed, with techniques available in most statistical packages (see later section). The covariate option is usually used for operator variables, although it is not limited to them. In observational and questionnaire studies, many of the variables are treated as covariates, such as number of miles ridden per year in studies of motorcycle accident frequency. At times covariates are the major focus of a design, as for example the classic Fleishman and Rich (1963) study of factors affecting task performance at different stages of learning. At the very least, carefully chosen covariates reduce the waste associated with any study which finds that individual differences account for a large portion of the overall study variance. To give an example, significant correlations between perceptual style and inspection performance (Gallwey, 1982) have enhanced our understanding of the inspection task. Obviously, if multiple participants are to be a part of the design for other reasons, then treating a factor as a covariate is much less expensive than building that factor in at multiple levels.
3. *Fix the factor at a single level.* Here a factor is treated as a constant. We could run our study with a single, narrow, age group; we could choose only third-generation Americans; we could run all experiments in a well-lighted room at constant temperature. By having only a single level of a variable, we do not increase the

size of our study but the price we pay is that the results will not generalise beyond the constant conditions we specify. We may wish to extrapolate the study results, based on other data and other models, but false extrapolation has plagued many disciplines including ergonomics. Human performance has a way of constantly surprising us. Typically, the fixing of a factor at a single level is used for those factors we know are important, but about which we do not need to generalise.

4. *Randomise the effects of a factor.* If a factor may be important, but we cannot control it in one of the ways above, then randomisation will prevent the factor from biasing the study results. By randomly assigning participants to factor level combinations or stimulus presentation orders to participants, any uncontrolled variability is given an equal chance of affecting all levels of the particular factor. Thus the random variability may be increased (resulting in a weaker study) but we will avoid reaching biased conclusions. A well-known example of the use of randomisation is the random assignment of participants to treatment groups, rather than putting the first volunteers in group A, the rest in group B, and so on. People who volunteer early may be different from those who volunteer late and randomisation ensures that this volunteer bias does not result in a biased study.
5. *Ignore the factor.* This final strategy is only included for the sake of completeness. It is never safe to ignore a factor in studies involving entities as complex as human beings. The old adage, 'If in doubt, randomise', applies here. Ignore factors at your peril.

At this point, we have an assignment problem — how to assign each factor to each alternative. In practice, the assignment is usually quite straightforward, because the real work was performed in listing the potential factors in an organised and ordered manner. What has been achieved is that a systematic technique has been used to reduce the chance of the study finding unexpected conclusions for the wrong reasons. The whole procedure may seem laborious, but it forces the ergonomist to think through the issues *before* the study starts and to use ergonomics insights to advantage. A useful exercise to practice these skills is to take a paper from the literature and to use these techniques to see how you would have designed the study. Here you have the advantage of hindsight — you know what the results were. In practice you need to develop foresight if you are ever to get a second chance at designing a study.

Choice of dependent variables

Traditionally, ergonomists have measured the effects of their factors on a single variable (e.g., reaction time, error percentage, heart rate), but advances in the ability to record and analyze multivariate data have to some extent stopped this trend. Now we tend to think in terms of sets of dependent variables, each illuminating a different aspect of human/machine fit. Despite this trend, the current section will concentrate on single measures, as the choice of sets of measures depends to some extent on the branch of ergonomics involved. For example, it is pointless to run a signal detection experiment without measuring (or controlling) both type 1 and type 2 errors.

Before classifying and describing possible measures, it is necessary to have available some criteria by which to judge the adequacy of measures. Social science texts (e.g., Kerlinger, 1986) see three main criteria:

1. *Validity.* Does the measure have a direct relationship to the phenomena described in the hypothesis? Is heart rate a valid measure of metabolic cost? Only under particular circumstances such as lack of heat stress, static muscular tension and emotional stress.

Is reaction time a valid predictor of the relative performance of three computer keyboards? Only when certain major variables such as information content per stimulus are kept constant and only when minimum response time has some relevance to the final criteria of the system evaluation. Technically, validity is the correlation between the variable measured and the phenomenon represented in the hypothesis. A measure may be valid *a priori* (face validity) because it patently measures the phenomenon. Thus accident frequency is a face-valid measure of plant safety. A measure may be valid because it can be logically related to the phenomenon (construct validity). Thus reaction time is a construct-valid measure of control-display compatibility because it is the reciprocal of information processing rate (for constant information per stimulus) and thus is logically related to compatibility. Finally, a measure may be proven valid experimentally. Thus Borg (1982) has validated two scales of rated perceived exertion against heart rate by experimentally correlating the two. Clearly, choosing an invalid measure will ensure that we answer the wrong question.

2. *Reliability*. Does the measure give consistent results when used repeatedly? Is heart rate too variable from pulse to pulse or minute to minute to be a reliable measure of metabolic load? Do we get the same reaction time to the three computer keyboards if we measure them today, tomorrow and next week? If validity asks how well a measure correlates with an external phenomenon, reliability asks how well the measure correlates with itself. While there are many reliability measures, the two most common are test/retest reliability and split-half reliability. Test/retest reliability correlates the measure obtained on a participant in the first (test) period, with that obtained during a subsequent (retest) period. Split-half reliability correlates two halves of the measure obtained during a single time period. Thus in an intelligence test, the score on even-numbered questions should correlate well with that on odd-numbered questions. Similarly, the reaction times measured in the first and second halves of a trial with computer keyboards should correlate highly. Reliability coefficients close to 1.0 are desirable, with values of 0.8 and upwards being used regularly in social science work. Reliability sets a limit to the internal consistency of our studies. Unreliable measures, even if they are valid, cannot prove a hypothesis because we cannot have faith in them to tell how the phenomenon of the hypothesis changes.
3. *Sensitivity*. Does the measure react sufficiently well to changes in the independent variable? It is quite possible that the measure chosen may be valid and reliable, but will not show a large enough effect to be measured easily. Thus, reaction time may not differ between different keyboards because it is not sufficiently sensitive to record the subtleties of keystroke force/distance characteristics. Similarly, heart rate may not react by more than a few beats per minute to differences in the design of handles in manual lifting tasks (Deeb and Drury, 1986), requiring a relatively large design to obtain statistical significance.

Any measure must pass all three tests (validity, reliability and sensitivity) before it can be used, although not every measure will need to perform equally on all three criteria.

Measures themselves fall into two broad classes:

1. **Performance measures** of the effect of the human on the system, typically measured by speed and accuracy. Speed refers to the time for task (or sub-task) completion, to the amount produced in a given time period (e.g., output per shift) or to the speed of movement (e.g., driving). At times it can be a straightforward measure such as task completion time, or it can be normalised with respect to some external criterion, for example speed expressed as a percentage of maximum speed. Accuracy

refers to a comparison between the outcome as measured and the desired outcome. Thus we can have a measure such as probability of missing a threat in an airline checked bag where the desired outcome is that a known threat should be detected. More aggregate measures are error rates over time in operational systems, accident rates of aircraft in a squadron or personal injury rates in a plant.

2. **Well-being measures** of the effect of the system on the human, or cost of the performance to the human (see discussion of evaluating effects on performance and on people in Chapter 1). In general, well-being measures can range from workload (physical or cognitive), through stress ratings and biochemical analyses to long-term measures of human health and freedom from disease/injury.

In general, it is difficult to conceive of measuring one of performance or well-being without measuring (or at least controlling) the other; at times this control is implicit. Thus, when we measure reaction times in a laboratory, it is usually implied that the participant performs at the maximum level, i.e., at a constant level of stress or cost. Similarly, in a manual lifting task, the task is often fixed (box size, lift distance, speed) so that the physiological cost can be estimated using oxygen consumption, heart rate or even subjective ratings. At other times, the joint measurement of performance and stress will be explicit, as in multivariate assessment of jobs where both performance and stress need to be measured separately in a realistic setting to determine how people choose to allocate their processing resources to tasks of variable demand.

A final consideration in choosing a dependent measure is the level of aggregation of the measure. We use data from a variety of levels, from the minute (e.g., time taken for a single cycle of a single component of a task) to the gross (e.g., the annual injury rate of a plant). Measurements can be aggregated over organisational components and over time. Table 2 shows typical levels. Note that we often make the case for change at the higher levels of aggregation, but perform our studies at the lower ones. For example, the recent study of medical errors quoted almost 100,000 deaths per year in the USA due to medical error (IOM, 1999), but error reduction studies are typically at the individual hospital or even physician's office (Weaver *et al.*, 2001).

In choosing a level of aggregation, there are trade-offs to be considered. The higher levels of aggregation, such as the plant's annual quality costs, are more obviously meaningful to the organisation, often referred to as 'system relevant' measures. However, they may be difficult to interpret unambiguously due to the many other factors which can influence these measures. Thus the annual quality costs for the plant may not only reflect ergonomic changes, but those due to industry-wide quality changes demanded by customers, changes in personnel, or even a new accounting system. The discussion of the dangers of archival measures (Chapter 3) is relevant here. Whenever a high level of aggregation is used, special pseudo-experimental designs (Kerlinger, 1986) may be needed to make realistic comparisons against control groups which have not had the intervention, e.g., Cohen and Jensen (1984).

Table 2 Classification of Levels of Aggregation in Dependent Measures

Level	Organisational component	Time period
Lowest	Individual task component	Single cycle
	Individual task	Batch/lot
	Individual job	Shift
	Work group	Day
	Organisation	Week
	Whole industry	Month
Highest	Country	Year

Study designs

While choice of technique, independent and dependent measures have to a large extent fixed the design of the study, there are still more decisions to be taken before the study can be finalised. We must know how many participants are to be used, how many data points for each are to be collected, and how the experimental conditions and participants are to be assigned to each other. These next steps are the province of statistical design of experiments (e.g., Winer *et al.*, 1991), but this section helps bridge the gap between the ergonomics and statistical concepts.

The main issue in study design is human variability and how to obtain reliable results despite this variability. People differ from each other in anthropometry, in physiological performance, in cognitive abilities, and in their reaction to external stressors. This is obvious and any study design must take these facts into account. Equally obvious, but less often considered, is the fact that any individual differs on the same measures from year to year, from day to day, and from minute to minute. Thus we have two sources of variability which any design must take into account, known as: (1) between participant, or inter-participant variability; and (2) within-participant, or intra-participant variability. As discussed previously, both variability sources can be minimised by the correct choice of independent variables. Thus choosing all females for a back packing task would reduce inter-participant variability somewhat in the energy costs measured. Similarly, choosing trained users of spreadsheets in a computer-aided calculation task would minimise trial-to-trial variability in performance and so reduce intra-participant variability. However, within any single, even narrowly-defined, population, both sources of variance will be too large to ignore. In addition, the price of a restricted sample is lack of generality in the study findings.

Human variability is explicitly recognised in how we interpret our results. Statistical tests are used to establish the statistical significance (or otherwise) of a result by estimating the probability of so extreme a result having been obtained through pure chance. The pure chance here is the inter- or intra-participant variability. For example, if use of spreadsheet interface A gives a mean error rate 0.1 higher than the use of spreadsheet interface B, we would interpret the fact differently if we have four participants with an inter-participant standard deviation of error rate of 0.15 than we would if 40 participants with an 0.05 standard deviation had been tested.

A test statistic (such as t or Mann–Whitney's U) is typically calculated as the ratio of the size of an effect to the appropriate variability. The test statistic increases in absolute magnitude as the size of the effect increases and as the variability decreases. Size of an effect is represented by the difference between two means (or the variance between several means for an F test) and is thus a physical fact, although subject to sampling error. For example, given enough participants and measurements, the true difference between spreadsheet interface A and B may be 0.12. We will only conclude that such a difference is significant, however, if the variability is low enough for our test statistic to be in the region of rejection.

The appropriate variability for a test statistic is the standard error of the difference between two means, which in simple cases is:

$$\text{SE} = \frac{\text{Standard deviation}}{\sqrt{(\text{number of data points})}} = \frac{SD}{\sqrt{N}}$$

where SD is the standard deviation of a set of N data points. Thus the test statistic is:

$$\text{Test statistic} = \frac{M_1 - M_2}{SE} = \frac{(M_1 - M_2)}{SD}\sqrt{N}$$

where M_1 and M_2 are the means of the conditions being compared.

It is now obvious that the three ways to increase the size of the test statistic are:

1. Increase $M_1 - M_2$
2. Increase N.
3. Decrease SD.

Each represents a valid option in study design and will be considered in turn.

Increasing difference between means

A large effect will of course be easier to detect than a small effect. But ergonomics is a 'real-world' discipline and the size of effect may be fixed. However, there are a number of steps that can be taken to make the effects larger.

Use a more sensitive measure

As previously outlined, oxygen consumption may be less sensitive than, for example, ratings of perceived body part discomfort in comparing two backpack designs. Part of the ergonomics insight comes from a careful reading of the literature to be able to predict which measures will be sensitive in a new study.

Make the two conditions more extreme

This implies that each mean is physically as different as possible from the other. One tried and tested method is to choose participants who are extreme on a measure likely to be related to the task if the difference in means is between participant characteristics. Thus ageing research often compares 20–30 year olds with 60–70 year olds, rather than comparing the 30–40 with the 40–50 age group. Similarly, rather than split participants at the median in perceptual style for study of an inspection task (Schwabish and Drury, 1984), it would have been possible to choose only people scoring in the upper and lower quartiles on the impulsive/reflexive scale. With task variables, an example would be choosing two weights in a box handling task which would be different enough to show a difference in heart rates. Deeb and Drury (1986) used 7 kg and 13 kg to represent average and difficult tasks based on a large survey of box weights handled in industry. More extreme weights have often been used, e.g., by Fish (1978), who had participants lift barbells weighing either 2 kg or 20 kg to measure disc compressive forces. The danger is that one or the other extreme is unrealistic or dangerous. Physicists often refer to this technique as increasing the magnitude of the forcing function.

Increase task difficulty

This is not just making M_1 more different from M_2 but increasing the difficulty of both tasks so that differences are more likely to show up. Thus if backpacks A and B differ in the amount of padding, then testing both at heavy pack loads will show up the differences more clearly. Another example is in road tests of automobiles where the test drivers negotiate a tight slalom course and measure the performance time. Pushing cars to their limits can reveal differences that may be unimportant in ordinary driving but which become suddenly critical in emergency situations.

Variations on this theme are legion. Instead of changing the task difficulty itself, environmental stressors can be added to push participants nearer to the limit and hence observe subtle performance changes. High noise levels, poor lighting, external pacing, or even competition have all been used in this way.

The dangers, as with making the conditions more extreme, are that the task becomes unrealistic or dangerous. Both apply directly to ergonomics practice and may contravene ethical practice guidelines. A more insidious danger is that of hitting a 'ceiling effect', which occurs when both conditions are so difficult that performance is equally impossible in both. Backpacks weighing 100 kg would represent an obvious ceiling effect. It should be noted that the opposite of a ceiling effect is a 'floor effect' where both conditions are so easy that either 100% performance is achieved whatever the condition or that some other factor limits performance. In a series of studies of self-paced tracking reviewed by Drury (1985), a floor effect is seen whenever the width of the track is so great that speed and errors are limited by vehicle maximum speed or human willingness to go faster.

Increasing number of data points

It is a truism in statistical testing that any difference, no matter how small, could be found significant given a large enough sample size. Ergonomists pride themselves on practicality and hence the need to ask how large a difference must be in order to achieve practical significance. We should thus aim to make the difference which achieves practical significance the same as that which meets the criteria for statistical significance. Doing this we can solve the equation defining the test-statistic for the sample size N and hence have a rational way of limiting our sample. In order to solve for N, we need to know:

1. The practical difference we wish to detect.
2. The critical value of the test statistic for a specific level of type 1 (alpha) error.
3. The appropriate standard deviation.

Unfortunately, while (1) and (2) may be relatively straightforward to specify, no data may exist on the variability to be expected. One can at times obtain it from the literature, but rarely has the same set of conditions been used as you want to test. The recommended method is a pre-test to estimate the variability, but such a recommendation is cold comfort when a proposal is being costed and the prototype is as yet unbuilt. In such a case, many ergonomists guess, although they will rarely admit it in print. There are obviously some broad guidelines. Upper limits on N are provided by cost and lower limits by the face validity. If you showed by calculation that only two participants are needed to give the appropriate level of significance, your sponsor may not feel that such a low number is representative. Such misuse of statistics causes despair among statisticians who rightly point out that if the calculations were correct and the two participants were indeed chosen randomly, there is no need to run more participants. Life is not always kinder to statisticians than it is to ergonomists. Journals, for example in the behavioral sciences, may need much convincing that small numbers of participants are adequate, although in anatomical and physiological studies sample size seems to be less of a controversial issue.

Note that the test statistic formula includes N as a square root term. Thus to double the test statistic means to increase sample size, for instance, from 25 to 100, as $100 = 2^2 \times 25$. Increasing sample size as a strategy is likely to meet with rapidly diminishing returns. However, the cost of a study is usually composed of a fixed cost and a variable cost linear in the number of data points. If the fixed cost is relatively low, then four times as many participants will mean almost four times the cost for the study. However, if the fixed cost is large, a very large increase in the number of data points will only have a small effect on total study costs. For example, the study already quoted which used 30 participants in a manual materials handling task (Deeb and Drury, 1986) had most of its time (and hence cost) spent on the experimental setup, pre-testing and analysis. Decreasing or increasing the sample by 25% would have had less than a 10% effect on total time or total cost. In a more recent extension of that study, each participant took two half days to test and about two weeks to reduce and analyze the data fully. Total study time and cost was closely related to sample size in this case.

Decreasing appropriate standard deviation

In the previous discussion of sample size, the meaning of N was left rather ambiguous. Did it mean the number of participants, the number of replications of a data point on one participant, or both? This ambiguity was deliberate as the arguments used apply to both meanings of sample size, as do arguments concerning the 'appropriate standard deviation'. Not only are there general ways of decreasing variability, which will be considered first, but also ways of altering which is the 'appropriate' standard deviation by varying the experimental design.

First we consider variance reduction techniques. Careful attention to detail is the only way to minimise variability. All of the arguments which apply to the choice of independent variables need to be reviewed to ensure that there are no unexpected sources of variability. All variables which can affect the dependent variable(s) under study should now have been fixed or otherwise controlled. There are, however, some obvious precautions which apply to all studies. All involve standardising the experimental procedure so that it does not differ between or within participants. This means consistent, and consistently presented, instructions. General admonitions to 'do your best' rarely produce the consistency that specific instructions on speeds and errors can provide. Participants must choose some speed-accuracy tradeoff, but if left to their own devices will choose trade-offs which are different from person to person and even from trial to trial for the same person. Specifying a payoff matrix would be ideal, but many experiments cannot be reduced to dollars or pounds or yen. As important as the specification of instructions is their consistency of presentation. Written or tape-recorded instructions ensure that each participant receives the same set, without unwanted variables such as the inflexions of the experimenter's voice. When the experimenter is conducting studies with speakers of another language, he or she must be particularly aware of consistent and clear instructions.

Following consistent instructions are written procedures for running trials. Experimenters learn during an experiment so that the first participant is unlikely to be treated in the same way as the last unless written procedures are adhered to. Obviously these procedures must be perfected in pilot tests. Such pre-testing (whose results are not included in the final data) not only removes bugs from the experiment but gives the experimenter enough practice to prevent adding to the experimental variability. The same concept applies to analysis. When any data reduction must be performed, again the experimenter learns. Analysis of pilot data allows the experimenter to learn analysis without adding to final bias or variance.

Types of design

The choice of appropriate standard deviation is a major factor in experimental design. Between-participants variability almost always exceeds within-participants (or trial to trial) variability and thus any test in which the appropriate standard deviation includes only within-participant variability will be more likely to detect significant differences than one which includes between-participant variability. This section is not meant to be a treatise on experimental design, for comprehensive treatments have wide circulation (e.g., Winer, 1972), but the aim is to point out the major choices available to the ergonomist, with their advantages and pitfalls.

Broadly, the choice is between a within-participants design in which each participant performs in a number of experimental conditions and a between participants design in which different participants are chosen for each condition, although a third design route will also be mentioned.

Between-participants designs

Here each participant is only tested in a single condition. For example, Laughery and Drury (1979) used a between-participants design in a study of optimisation skills because it was suspected that techniques learned during the solution of one type of optimisation problem might transfer in an inconsistent manner to other problems, with an adverse effect on bias and variability. Thus five participants were used in each condition, which meant that any comparison between conditions had to be made against between-participant variability. The groups were kept reasonably homogeneous (engineering students) but this in turn limits the generalisability of the results. Because between-participants variability was large, only large effects could be found with the given sample size.

Within-participants designs

Here each participant receives many experimental conditions or, to put it another way, each participant is their own control. Thus a participant with a slow reaction time is likely to be slow under all conditions when reaction time is measured. Techniques such as analysis of variance or regression can estimate the size of the between conditions effect by using only the deviations from the participant's own mean performance. Hence the slow participant will not add to the standard deviation used to test differences between conditions. As an example, a study of the biomechanics and physiology of handle positions on boxes used ten participants, each performing a box holding task using ten handle positions. The within-participants design allowed small differences to be detected despite the limited sample size. An obvious question is why were the experimenters concerned about inconsistent transfer in optimisation but not in box holding? The answer is equally obvious: no changes to the participant were expected during the box holding experiment, but changes were expected in optimisation. Change occurs in humans in the short-term as they fatigue and in the long-term as they adapt or learn. With appropriate rest periods, no fatigue was expected (or found) in the box holding task and certainly an hour or two of experimentation on a well-practiced task is unlikely to change either a participant's body strength (adaptation) or box holding technique (learning). Here a biomechanical and physiologically limited task is unlikely to exhibit what Poulton's famous (1974) paper called asymmetrical transfer effects.

The same cannot be said for most intellectual skills. What you learn in first solving one calculus or chess problem is quite likely to affect your performance in solving the next. The transfer can be positive, if the same solution techniques are useful in both problems, or negative if the solution to the first problem is inappropriate in solving the second. An optimisation task is *a priori* likely to be closer to an intellectual task than to a biomechanical one, hence the choice of a between-participants design. Any human functions, even anatomical ones, will adapt or change given time, but the key question is not whether or not change will occur but whether enough will occur to bias or desensitise the experimental comparison. This depends upon both the length of the experiment and the resistance to change of a function. We can run short studies to minimise the change, but we are limited by other experimental constraints of how many data need be collected. We can increase resistance to change by either choosing to experiment on systems that are inherently resistant to change (e.g., bone length) or by deliberately ensuring that a performance plateau has been reached. Thus athletes make good participants for physiological tests as their aerobic capacity, for example, is well trained and unlikely to change during even relatively long experiments. Similarly, experienced bus drivers are unlikely to develop new bus driving skills during a few hours of tests on buses they have already driven. Finally, for 'new' skills, we can give participants sufficient practice in all conditions to ensure that plateaus have been reached in each condition. Often the learning/training process required to achieve this level can become of interest in itself (e.g., Bishu and Drury, 1985) so that the time spent achieving stable performance is not time wasted.

Much of the above discussion has centered around creating conditions under which a within-participants design can be used. Such a design is clearly more efficient, but the fact that we need to take special precautions means that this efficiency is bought at the price of potential danger. Transfer between conditions is always possible; what we do is reduce its probability and magnitude. A between-participants design is always safer, and sometimes no other is possible. For example, an experiment comparing training techniques in inspection (e.g., Czaja and Drury, 1981) must always use different participants in different conditions because a fact or skill can only be learnt once. Conversely, some experiments must always use a within-participant's design. If it is desired to derive a functional relationship for each individual participant, then clearly multiple conditions must be given to each participant. Examples are the measurement of the speed-accuracy trade-off or the utility function for each individual participant.

Matched-participants designs

There is a third option for experimental design. If we could somehow clone a participant we could present different conditions to each clone, preserving both the elimination of between-participant variance and the lack of unwanted transfer effects. Obviously this is currently impossible, as well as being morally and legally dubious. We can do the next best thing however, and carefully match participants between conditions. Thus two conditions could be compared using pairs of identical twins, although it would be rather difficult to find sufficient participants who were both, say, airline pilots. Comparing three conditions would lead us to triplets and so on.

A less perfect form of matching is to use unrelated (non-family) participants who are likely to respond in the same way; the art lies in discovering who will be likely to react similarly. If the skill being tested is one of intellectual activity, then one would expect IQ or educational level to determine similarity of response rather than shoe size. On the other hand, shoe size is likely to be correlated with body size and hence strength, making it a likely (although unusual) matching variable for experiments where strength was the limiting human subsystem. Technically, the benefits of matching are determined by the correlation between the matching measure and the experimental dependent variable. If this is denoted by r, the reduction in between-participants variance is directly proportional to r^2. Hence we need to choose matching variables correlating highly with the dependent variable. Again the only guide is the literature, often embodied in models of the human appropriate to the current experiment. Matched participants are used extensively in medical and epidemiological studies. For example, Whitfield (1954) studied accidents in coal miners using participants carefully matched in job type and experience, one of whom was involved in an accident and one who was not. Similarly Saari (1976) matched not participants but situations to study task and environmental factors in accident causation. It should be noted that instead of matching individual participants, groups of participants can be matched. Thus in many studies parameters such as the mean age, height and weight of each group are kept constant to reduce group differences. Group matching is not as powerful statistically as individual matching. Finally, a caution should be raised that the tests used to establish matching criteria should not in themselves produce inadvertent transfer to the experiment proper. Thus a pre-test of a tracking task on a flight simulator to determine RMS tracking error would almost certainly transfer to future simulator tasks for novices.

Conclusion on types of design

What then is the conclusion? Should we go with the costly between-participants design or the more efficient within-participants design, or match our participants? There are no universal answers. If experiments are very costly and participants are limited in number, we may have no alternative to the within-participant design. For example, Drury (1973) measured the speed-accuracy trade-off for inspection using four participants inspecting four batches under four speed conditions with a Graeco–Latin square design. There were only four skilled inspectors in the plant so that this was the only design possible. The complex design used only four trials per participant to measure the effects of participants, trials, batches and speeds. Clearly obtaining four main effects in 16 readings forced the author to make many untested assumptions that he would be unwilling to make in the less demanding environment of a research laboratory. But answers were needed and without the experiment there would be none, so that design decisions would be made without ergonomics input. In that case, it was better to make the assumptions than to abdicate responsibility.

Conversely, if transfer effects are known to exist in a situation, there will be no alternative to the between-participants design; mention has already been made of training experiments in this context. For any between-groups experiment, matching is always good practice, i.e., it never hurts. Whether it is worth the time and effort depends upon the correlation, if known to the experimenter in advance, between the matching variable and the experimental measure. This in turn depends, like so many things in designing ergonomics studies, upon the breadth and depth of the ergonomist's knowledge.

Data reduction and analysis

This chapter has been mainly concerned with how to choose what data to collect and how to organise its collection. The actual collection of the data may be directly into a computer, through a portable instrument that can download data to a computer, or through some non-computer-compatible system such as paper questionnaire or video tape. In all cases, there are now several steps between the data being collected and the final output in terms of whether or not your hypotheses were confirmed. In this section we consider that sequence of steps. Any form of non-computerised data collection may now look rather quaint, but is in many cases inevitable or even desirable. There are times when we need direct observation methods with a human intelligence between the overwhelming complexity of the real world and the cozy simplicity of sustained hypotheses or statistical tests. In an airliner cockpit, or with a medical practitioner helping famine victims, there is little tolerance for interfering data collection equipment. In what follows, we will consider ways in which the data can be prepared for analysis, analyzed and used to refute or support our original research hypotheses.

Data entry

For studies requiring manual data entry, ease and accuracy of entry and checking should be considered at the experimental design stage. An easily read source document for the participant may not be an easily read source document for the data entry person. For observational and questionnaire data it is important to minimise transcription errors, so a well-designed data entry programme is essential. Three alternatives are available. First, a word processor, spreadsheet, or text editor can be used to create a data file, typing in the numerical and text entries in order with suitable delimiters. Although error correction will be easy, there is a considerable chance of forgetting the location of each item in the data file. A second, more compatible method, is to use a database programme (e.g., Access) where each unit of data (a completed questionnaire, a set of workplace dimensions or observation records) is entered onto a custom-designed data input screen. This can be useful in that missing data are very obvious at data entry, and so can be dealt with immediately whilst the original record is still available. Both of these alternatives can then interface easily with the statistical analysis package. The third alternative is to directly enter the data into the statistical package. Packages such as SAS, MINITAB or SPSS allow such data entry, although it may not be as convenient as using an intermediate programme. For example, MINITAB classifies data columns into 'numeric' or 'text' data. If a single entry includes a non-numeric character, the whole column converts to 'text,' which makes sense if the non-numeric entry was intended. However, if a wrong character is entered in error in a column intended to be numeric, the user has to re-convert the whole column, an easy step but annoying if it happens often. There are also programmes to simplify entry of data from videotapes and from records of verbal protocols (e.g., MacShapa, Noldus Observer).

Data checking and verification

The second aspect of data reduction is verification. With full-screen editing and instant print-outs, this is not a difficult problem to tackle, but it is at times neglected, resulting in unreliable data and hence unreliable conclusions. A broader problem in data verification is the logically impossible or implausible data point. In a student sample, any age having other than two digits is suspect. Similarly, sibling birth dates closer together than the gestation period raise questions of data validity. Heart rates above 200 beats/min and negative blood pressures are other obvious examples, but there are less glaring data errors that must be detected. There is always some question of how far the ergonomist should go in editing data. Unusual values will occur legitimately from time to time, indeed there is a whole branch of statistics concerned with extreme values sampled from populations (Gumbel, 1958). Some students are over 60; premature births do occur; very rapid typists exist to cause outlying data points in word processing studies.

The ideal place for data verification is at the time of recording, when the participant is still present and the data point can be repeated if necessary. All subsequent data verification that involves data editing can be called into question. Only logically impossible data can be removed from the database and still withstand serious questioning. Reversed dates for starting and ending employment, mechanical impossibilities, or off scale responses are all 'safe' to remove from the database, but how can they be replaced? If the error is a transcription error, we can re-enter the data, but if it is a recording error, generally it will not be possible to replace the reading at a later time. A participant's response cannot always be re-collected, as the participant may now have changed or disappeared, leaving the problem of missing data, which is discussed later. Database programmes have built-in error checking routines, so that data can be tested as they are input to determine whether they meet certain criteria. In this way logical impossibilities are automatically flagged so that a decision can be taken on their inclusion. The same filtering of data can also be applied on-line to computer recorded data. For example, heart rates based on timing the EKG 'R' wave occasionally count a spurious beat or miss a beat, leading to heart rates double or half the correct value. Data filtering routines can be built into the recording software to detect such anomalies and remove them from the final record. Ease and consistency in such processes can be expected to increase with the more widespread adoption of artificial intelligence techniques for data verification. It should eventually be possible to mimic the decisions of an intelligent and well-informed experimenter to ensure data veracity.

Data reduction

With data collected and verified by a device or computer directly, there is often a data reduction step to produce numbers suitable for statistical analysis. For example, using the NASA TLX scale of workload (Chapter 18), the raw data consist of numerical ratings and paired comparisons. These data need to be combined to produce the five components of TLX, and are often combined further to produce an overall workload score. Other post-processing may be as simple as subtracting the starting date from the leaving date to determine a respondent's period of employment or it may be as complex as calculating the error power spectral density function in a tracking task. Other examples are calculating signal detection theory parameters (A', d', xc, etc.) from hit and false alarm rates, finding disc compressive forces from a biomechanical model of a manual lifting task, finding envelope measures of EMG, or even calculating heart rates from R–R interval times. In all cases, data from various data points are combined to give a measure of greater meaning. The main post-processing decision is whether to perform this post-processing on-line as data are being collected or off-line when data are retrieved from a storage medium. On-line processing obviously requires powerful computational resources, but, as the processed data are usually less extensive than the raw data, storage requirements are minimised. As an example, if four muscles are recorded using EMG techniques at 200 Hz, we need to take $4 \times 200 = 800$ data points each second. If, however, we wish to obtain only fatigue data, then the ratio of high-to-low frequency components (e.g., Kroemer, 1984) can be calculated and recorded much less frequently, perhaps twice per second for each muscle. Our data storage requirements are reduced 100-fold, but high-capacity on-line processing is needed to filter the data by frequency and calculate ratios. Data storage is cheap and widely available, but can still be limiting with large video files.

The main advantage of on-line post-processing is that it can provide useful feedback to participant, experimenter or both. The main danger is that the data cannot be used to answer any questions which require different post processing. To continue our example, level of effort, represented by average or envelope EMG, would not be available if only frequency ratios were recorded. For this reason, recording of the original raw data is always good practice: only record the participant's actual response, particularly in observation or questionnaire data; do not attempt to convert weeks to days just because the questionnaire requires an answer in days and the participant gives the answer in weeks; record original data long-hand and process them later without time pressure.

Data analysis

At this stage, our data are collected, entered into the computer, verified, and the calculations performed so that we are ready for analysis, for which a number of choices are available. Specific analyses may well be beyond the capabilities of the software package: graphs with meaningful but unusual scales may need to be drawn. The typical course of analysis is statistical analysis of data followed by comparison between some non-statistical model and the summarised data. Thus Drury and Hong (2000) ran an experiment on inspection of items containing more than one defect. An ANOVA established significant differences between number of defects, but a model was needed to interpret the particular relationship between performance time and number of defects still remaining undetected. A regression analysis was used to determine how well the model fitted the data.

The handling of missing data is an important topic in its own right and may be different for different packages. In many packages, missing data within one cell of an ANOVA (e.g., four readings instead of five) can be handled easily, although some functions within a package insist on equal numbers in each cell. In other cases there may be different types of missing data. People may have no answer to a question on their age because they do not remember their age, they refuse to provide this information, the experimenter fails to record it, or an obviously wrong value is entered. A good package will allow the analyst to treat these differently, while still denoting all of them as 'missing'. Some missing data may cause other relevant data to be eliminated; e.g., a missing age in a correlation between job satisfaction score and age would cause that participant's job satisfaction score to be omitted from the correlation calculation. If there are more than two attributes and only one is missing the experimenter must choose between omitting all data from the participant with the attribute missing or just omitting data from calculations impossible to perform without that data.

Before considering the statistical packages, it is worth noting that some spreadsheets (e.g., Excel) and many database programmes (e.g., Access) perform rapid data summarisation (means, standard deviations), some statistical functions (regression), and simple graphical plots (line, bar, scatter). There are dangers in using such packages rather than the more appropriate statistical packages for analysis, even though they are well-known to the user and easily available. First, the range of statistical tests available may not match the complexity of the design chosen. A one-way or two-way ANOVA on simple spreadsheets may make quite inappropriate assumptions about choice of error term for tests. Second, the natural tendency of spreadsheet users to find means of all data immediately may be incompatible with appropriate analysis techniques. Typically, we use the raw data (or reduced data) in order to have the correct degrees of freedom for the error term in the ANOVA or regression. Using means eliminates not just variability, but the degrees of freedom required for appropriate testing.

Statistical analysis packages have grown in sophistication over the years so that most now provide a bewildering array of design and analysis features to the analyst. With similar user interfaces for data entry, analysis and graphing, choice of package is often a matter of tradition (which have I always used), access (which does my company support) or specific features (will my usual package perform *post-hoc* testing of means after the ANOVA?). These packages now have design features as well as analysis features. For example, MINITAB will generate design arrays for elaborate partial replicates of factorial experiments, and provide the alias structure so that the ergonomist can see which interactions will be confounded in the final analysis. It will also provide control charts, most used in statistical process control in industry, but also useful in finding when learning has leveled off in a human experiment.

The analysis chosen for the data depends obviously on the hypotheses you started out to prove or disprove. A short chapter in an ergonomics text is no place to provide detailed guidance on probability and statistics, as such is widely available in textbooks on statistics (e.g., Montgomery, 2000) or experimental design and analysis (e.g., Winer *et al.*, 1991). There are even WWW based textbooks on the subject, e.g., the one by the National Institute of Standards and Technology (NIST) at http://www.itl.nist.gov/div898/handbook/. The packages themselves often contain Help files that can be used to check on the statistical background to a particular test, but are not a safe way to learn which tests to choose and when they are appropriate.

Typically, we have extensive data and use parametric statistical techniques to perform sophisticated analyses. Thus we use Analysis of Variance (or Covariance), multiple regression or multi-way comparisons of counting data to test for main effects of factors and interactions between factors. Thus, Drury, *et al.* (2002) analyzed data from extensive observations of seat belt use in automobiles using multiple classifications of the data to show that modelling behavior differed in incidence between different age and gender groups. However, such analyses may not be either required not appropriate in all studies. In many cases, our hypotheses are simpler, or we may be more concerned with exploratory data analysis in a new field. Vincente (2000) provides cogent arguments that ANOVA is not always the appropriate method of analysis. One alternative he suggests is a simple counting of the numbers of participants who performed in different ways. For example, in a recent study of the effects of a human factors training course on aviation maintenance personnel to better investigate adverse incidents (Ma *et al.*, 2003) we used a before-and-after design by having participants investigate a simulated incident before and after training. We also tested an equal number of participants who did not undertake the training. Thus we were able to compare the numbers of participants who improved significantly on the second test as a function of whether they were in the training group or the control group. A chi-square test of the data shown in Figure 1 showed that significantly more of the training group improved on the second test, so that the human factors training did indeed make people better investigators, rather than people merely improving on the second test from practice.

Data presentation

Analysis does not complete the ergonomist's task: no job is complete until the paperwork is done. The data, analysis and conclusions must be presented in a form suitable for the intended user. This is where we must go back to the brief or RFP to remind ourselves of the intended user, and design the data presentation in proper ergonomic fashion to fit the user. Unfortunately, this is often neglected, resulting in poor communication of the findings to the people for whom they could be most useful. Any ergonomics or human factors conference has at least a few examples of inappropriate tables or graphs, that were obviously designed with a message in mind but fail to make that message clear. If human factors practitioners cannot get this right, it does not bode well for those who have less knowledge of human perception and communication.

Fortunately, we have excellent guidelines available, based on cognitive principles, e.g., Gillan *et al.* (1998). The basic choices are between presentation within the text, as a table or as a graph. Small amounts of data, e.g., the values of mean fraction correct before and after an

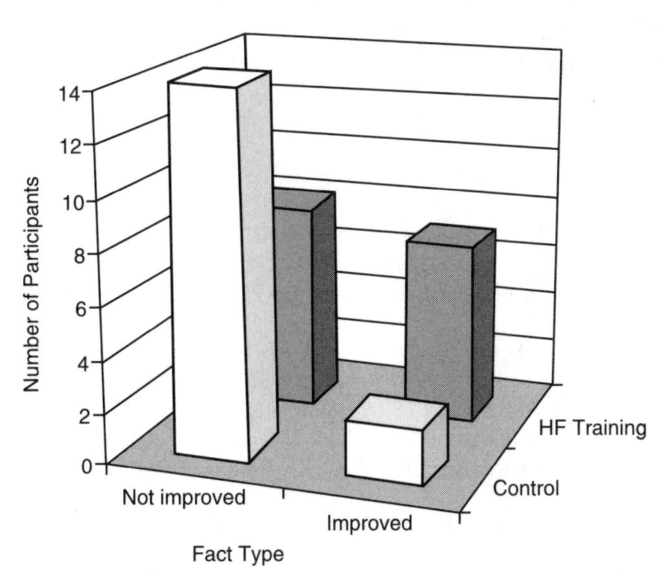

Figure 1 Effect of human factors training on improvement in an incident investigation task (Ma *et al.*, 2003)

ergonomic intervention, should be presented in text, as should single significance levels (e.g., $F(2, 32) = 14.7$, $p < 0.01$). If users need precise values, for example to perform their own calculations or to quote at a later time, then numerical values should be given. This can be in text for small amounts of data or tables for more extensive data sets. If users need to understand relationships, such as how well a model of visual search predicts a participant's speed-accuracy trade-off, then a graph is appropriate. Specific guidelines are given for graphs of different varieties in Gillan *et al.* (1998), but as one general rule, only use line graphs where the abscissa is an ordinal scale or better, otherwise bar charts are appropriate. If an interaction must be shown, the choice is between a series of lines (or bars) representing the different levels of one factor, or a two-dimensional plot of the data as in Figure 1. Note that Gillan *et al.* (1998) would not recommend the format of Figure 1 and would prefer two pairs of bars on a single graphical dimension. I disagree in this case as the change of pattern between training and control groups does not require precise visual estimation of numerical values, a difficult task when the 'floor' of the graph is rendered in perspective.

Conclusions

Designing ergonomic studies represents a choice between many independent variables, dependent variables, study settings and formal designs. Because ergonomics embraces so many traditions from medical, through biomechanics and physiology, to cognitive and social psychology, the choices are particularly broad. This is a strength of the field, but can make the task of choosing a little daunting to the practitioner. However, there is rarely a single correct answer, so that the choice you make may well be quite adequate, even though your company statistician or journal editor disagrees with you at a later date. Without well-designed studies, our field cannot progress, but just because design compromises must be made in any study, this should not stop you from going ahead. Clearly, the first consideration is to 'do no harm,' but after that the amount of positive good that you do depends on your knowledge and skill in study design and analysis. New techniques are always being developed, and every ergonomist realises frequently that their knowledge of statistics and design must always be growing. A critical reading of the literature is essential, and not just to absorb the subject matter of the paper. We need to look carefully at the study design choices our peers have made in papers and reports, to see why they made them and how effective they were. Journal editors and reviewers spend much time doing this, but it is an excellent learning experience for all in the profession.

References

Barnes, R.E. and Drury, C.G. (1976). An application of signal detection theory to EMS decision making. *Ethics in Science and Medicine,* **3**, 187–193.

Baum, S. and Drury, C.G. (1976). Modelling the human process controller. *International Journal of Man-Machine Studies,* **8**, 1–11.

Beggs, W.D.A. and Howarth, I.A. (1972). The accuracy of aiming at a target. *Acta Psychologica,* **36**, 171–177.

Bishu, R.R. and Drury, C.G. (1985). A study of a location task. *Proceedings of the Human Factors Society 19th Annual Meeting*, Santa Monica, CA.

Borg, G.A.V. (1982). Psychological bases of perceived exertion. *Medicine and Science in Sports and Exercise,* **4**, 377–381.

Caplan, R.D., Cobb, S., French, J.R.P., Van Harrison, R. and Pinneau, S.R. (1975). *Job Demands and Worker Health* (Washington, D.C.: US DHEW/NIOSH, Superintendent of Documents).

Cohen, H.H. and Jensen, R.C. (1984). Measuring the effectiveness of an industrial lift truck safety training program. *Journal of Safety Research,* **15**, 125–135.

Corlett, E.N. and Bishop, R.P. (1976). A technique for assessing postural discomfort. *Ergonomics*, **19**, 175–182.

Cox, T. (1978). *Stress* (London: Macmillan).

Czaja, S.J. and Drury, C.G. (1981). Training programs for inspection. *Human Factors*, **23**, 473–484.

Deeb, J.M. and Drury, C.G. (1986). Hand positions and angles in a dynamic lifting task. Part 2. Psychophysical measures and heart rate. *Ergonomics*, **29**, 769–778.

Drury, C.G. (1973). The effect of speed of working on industrial inspection accuracy. *Applied Ergonomics*, **4**, 2–7.

Drury, C.G. (1982). Improving inspection performance. In G. Salvendy (ed) *Handbook of Industrial Engineering* (New York: John Wiley).

Drury, C.G. (1985). Influence of restricted space on manual materials handling. *Ergonomics*, **28**, 167–175.

Drury, C.G. and Brill, M. (1983). Human factors in consumer product accident investigation. *Human Factors*, **25**, 329–342.

Drury, C.G. and Dawson, P. (1974). Human factors limitations in fork-lift truck performance. *Ergonomics*, **17**, 447–456.

Drury, C.G. and Hong, S.-K. (2000). Generalizing from single target search to multiple target search. *Theor. Issues in Ergon. Sci.*, 1(4), 303–314.

Drury, C.G. and Sinclair, M.A. (1983). Human and machine performance in an inspection task. *Human Factors*, **25**, 391–400.

Drury, C.G., Cardwell, M.C. and Easterby, R.S. (1974). Effects of depth perception on performance of simulated materials handling task. *Ergonomics*, **17**, 677–690.

Drury, C.G., Drake, M.L. and Thomas, J.E. (2002). Demographic effects of behavior modeling in seat belt use: analysis of 15,000 observations. *Proceedings of the Human Factors and Ergonomics Society Annual Meeting*, Baltimore, MD, October 2002.

Drury, C.G., Paramore, B., VanCott, H.P., Grey, S.M. and Corlett, E.N. (1987). Task analysis. In: G. Salvendy (ed) *Handbook of Human Factors*, (New York: John Wiley), pp. 370–401.

Fish, D.R. (1978). Practical models of human postures and forces in lifting. In: C.G. Drury (ed) *Safety in Manual Materials Handling*, DHEW (NIOSH) Publication No. 78-185, pp. 72–77.

Fitts, P.M. (1954). The information capacity of the human motor system in controlling amplitude of movement. *Journal of Experimental Psychology*, **47**, 381–391.

Fleishman, E.A. and Rich, S. (1963). Role of kinaesthetic and spatial visual abilities in perceptual-motor learning. *Journal of Experimental Psychology*, **66**, 6–11.

Gallwey, T.G. (1982). Selection tests for visual inspection on a multiple fault type task. *Ergonomics*, **25**, 1077–1092.

Gillan, D.J., Wickens, C.D., Hollands, J.G. and Carswell, C.M. (1998). Guidelines for presenting quantitative data in HFES publications, *Human Factors*, **40**, 28–34.

Hart, S.G. and Staveland, L.E. (1988). Development of a NASA-TLX (Task Load Index): results of empirical and theoretical research. In: P.A. Hancock and N. Meshkati (eds) *Human Mental Workload* (Amsterdam: North-Holland).

Kasprysk, D.M., Drury, C.G. and Bialas, W.F. (1979). Human behavior and performance in calculator use. *Ergonomics*, **22**, 1004–1019.

Kerlinger, F.N. (1986). *Foundations of Behavioural Research*, 3rd edition (London: Holt, Rinehart and Winston).

Kohn, L.T., Corrigan, J.M. and Donaldson, M.S. (2001) *To Err Is Human: Building a Safer Health System Committee on Quality of Health Care in America*, (Washington DC: Institute of Medicine, National Academies Press).

Konz, S. (1983). *Work Design: Industrial Ergonomics* (Columbus, OH: Grid).

Laughery, K.R. and Drury, C.G. (1979). Human performance and strategy in a two-variable optimisation task. *Ergonomics*, **22**, 1325–1336.

Ma, J., Drury, C.G., Richards, I. and Sarac, A. (2003). The effectiveness of human factors training in error investigation. *12th Int. Symposium on Aviation Psychology*, Dayton, OH, April 2003.

Meister, D. (1971). *Human Factors: Theory and Practice* (New York: John Wiley).

Miller, R.B. (1963). Task description and analysis. In: R.M. Gagre (ed), *Psychological Principles in System Development* (New York: Holt, Rinehart and Winston).

Morawski, T., Drury, C.G. and Karwan, M.H. (1980). Predicting search performance for multiple targets. *Human Factors*, **22**, 707–718.

Nie, N.H., Hull, C.H., Jenkins, J.G., Steinbrenner, K. and Bent, D.H. (1975). *Statistical Package for the Social Sciences* (New York: McGraw-Hill).

Poulton, E.C. (1974). *Tracking Skill and Manual Control* (New York: Academic Press).

Rasmussen, J. (1982). Human errors. A taxonomy for describing human malfunction in industrial installations. *Journal of Occupational Accidents*, **4**, 311–333.

Reason, J. (1990). *Human Error* (Cambridge: Cambridge University Press).

Saari, J. (1976). Typical features of tasks in which accidents occur. *Proceedings of 6th Congress of IEA, Human Factors Society*, Santa Monica, CA, pp. 11–16.

Schwabish, S.L. and Drury, C.G. (1984). The influence of the reflective impulsive cognitive style on visual inspection. *Human Factors*, **26**, 641–647.

Senders, J.W. and Moray, N.P. (1991). *Human Error: Cause, Prediction and Reduction* (Hillsdale, NJ: Lawrence Erlbaum Associates).

Sheridan, T.B. and Ferrell, W.F. (1974). *Man-Machine Systems* (Cambridge, MA: MIT Press).

Sperling, G. and Dosher, B.A. (1986). Strategy and optimization in human information processing. In: K.R. Boff, L. Kaufman and J.P. Thomas (eds) *Handbook of Perception and Human Performance* (New York: Wiley).

Swain, A.D. and Guttman, H.E. (1980). *Handbook of Human Reliability Analysis with Emphasis on Nuclear Power Plant Applications* (Washington, D.C.: US Nuclear Regulatory Commission).

Stokes, D.E. (1997). *Pasteur's Quadrant: Basic Science and Technological Innovation*, Brookings Institution, Washington, D.C.

VanCott, H.P. and Kincade, R.G. (1972). *Human Engineering Guide to Equipment Design* (Washington, DC: US Superintendent of Documents).

Vincente, K.J. (2000). The Earth is spherical ($p<0.05$): alternative methods of statistical inference, *Theoretical Issues in Ergonomics Science*, **1.3**, 248–271.

Weaver, J.L., Schell, K. and Grasha, A. (2001). Reducing medical error and improving patient safety: a methodology for studying pharmaceutical error in teams. *Proceedings of the Human Factors and Ergonomics Society 45th Annual Meeting*, Minneapolis/St Paul, Minnesota, The Human Factors and Ergonomics Society, Santa Monica, California, Volume 2, pp. 1758–1761.

Whitfield, J.W. (1954). Individual differences in accident susceptibility among coalminers. *British Journal Industrial Medicine*, **1**, 126–130.

Wickens, C.D. (1992). *Engineering Psychology and Human Performance*, 2nd edition (Columbus, OH: Charles Merrill).

Wierwille, W.W. and Casali (1983). A validated rating scale for global mental workload measurement applications. *Proceedings of the Human Factors Society 27th Annual Meeting*, Norfolk, VA. pp. 129–133.

Winer, B.J., Brown, D. R. and Michels, K.M. (1991). *Statistical Principles in Experimental Design*, 3rd edition (New York: McGraw-Hill).

Chapter 3

Applications of archival and observational data

Ann M. Bisantz and Colin G. Drury

Introduction

Studies in ergonomics and human factors are performed to answer a specific data need or research need (see Chapter 2). As shown in that chapter, studies can range across a continuum of reactivity from merely observing systems, through interacting with participants while they are engaged in system activities, to changing the configuration of systems and measuring the effects of such changes. This chapter covers one end of this continuum of ergonomics study designs: that of the least-reactive measurements. We cover both the use of pre-existing data (archival data) and the collection of reasonably unobtrusive data (observational studies). Two examples will provide a flavor of what is to follow.

First, archival data were used by Drury (1998) who analysed the errors arising from the use of a poorly-designed task document meant to aid a new procedure in inspection of airliners for potential malfunctions. Errors in responses to the document had been recorded by the airline, and found to be excessive in number (about 20% of all records had errors), which led to the study. In the study, these errors were classified by the step number where they occurred, and these steps were then evaluated against published guidelines (Patel *et al.*, 1994). It was found that the errors *all* occurred on those steps where the guidelines were violated, making a powerful argument for ergonomic design of task documentation.

Second, to establish that ergonomics and manufacturing quality were intimately related, Kim *et al.* (2001) made an observational study of two production lines in a factory making disposable cameras. Quality data were collected from archival quality records, while observational techniques were used to collect ergonomic data. Two ergonomic aspects were considered to be of importance on these paced assembly lines: pacing speed and body posture. The first was measured by comparing the time required at each workplace (from observation and timing of videotapes) with the time allowed by the line. Posture was measured from videotapes of operators performing each task, analysed using body angles (Drury, 1987). The finding was again a strong argument for ergonomics: the two ergonomic variables accounted for over 50% of the variance in quality between workstations on each assembly line.

Note that neither of these studies required interference with the on-going operations: they were minimally reactive. The main 'intrusion' was in ensuring that all people involved were informed of our work, and gave their consent to be evaluated, e.g., by videotaping.

Archival and observational data have several uses in human factors. First, they are usually a necessary first step in the study of any system, whether the final step turns out to be a full observational study or a controlled experiment. Second, they are a fruitful source of hypotheses for later studies. Because they are performed in naturalistic settings, they provide interesting questions as well as the 'answers' given in the examples above. Finally, they can answer specific questions of ergonomic interest, as above. Note that in each of the examples presented above, answers were provided using statistical comparisons of different conditions, by careful choice of the units of study (different steps in a document, different workplaces on an assembly line). Observational techniques can test hypotheses rather than generate them, although it is usually impossible to postulate causality without making deliberate changes to the system under study (see Chapter 2). Observational and archival studies are needed where the full complexity of a natural environment is essential, where there is no possibility of direct interference with the system to perform a controlled experiment, and where people cannot be interrupted during they job performance for measures such as questionnaires and interviews. Note however that these non-intrusive studies do not need to be used alone. We often combine observation with other, parallel, measures to obtain a more complete picture of human activities, e.g., by measuring attitudes as well as job content or measuring forces and EMGs as well as body angles.

Use of pre-existing and archival data

A focus of many studies which rely primarily on archival data is to find associations which can help identify causes as to why adverse events (e.g., injuries, accidents) have occurred. Thus, the kinds of archival records often analysed are those which have been generated in response to the specific type of event of interest, such as accident records. However, other types of archival records can be useful, either independently, or in conjunction with event based records.

Incident and event records

A rich source of archival data for use in human factors studies are those records generated in response to particular types of events or incidents. Often, such records describe accidents and incidents which led to injuries, deaths, or other adverse outcomes, or safety critical events which may have caused (but did not) such outcomes. In industrial settings, such records exist in the form of plant generated injury reports, medical logs, and OSHA required records. Aggregate sources of such data also exist. For instance, the Aviation Safety Reporting System (ASRS) maintained by NASA provide a source of aviation safety incident reports, which contain voluntarily submitted, anonymous incident accounts by pilots, air traffic controllers, and other personnel who may have relevant information regarding safety critical incidents. This database has been utilised in a number of human factors research studies (Degani and Wiener, 1993; Jentsch *et al.*, 1999). Kemmlert and Lundholm (2001) utilised a national database of industrial accident records (the Swedish Occupational Injury Information System) and Bentley and Haslam (1998) utilised a database of British postal worker injuries. Other event based records which have been used include those describing motor vehicle accidents (Owens and Sivak, 1996), coroners' reports (Cayless, 2001), and death certificates (Agnew and Suruda, 1993).

Difficulties do exist with these data sources. For instance, the typical enterprise will only record those events which it is legally required to report. In the USA, there are specific criteria which define what must be reported in industry (via OSHA), in aviation (via FAA), in mining (Bureau of Mines) and so on. As many writers (e.g., Hale and Hale, 1972) have pointed out, accidents are not synonymous with injuries, but merely a necessary condition. Indeed, the injury is seen as one event in an accident sequence by Monteau (1977), by Haddon (1973) and by the US Consumer Product Safety Commission (CPSC, 1975).

Recorded injuries are thus the typical archival substitutes for accident data, whether accessed by medical logbook, accident report forms, or in-depth investigation. Because of this, there are many potential biases or artifacts in such data. First, non-injury producing accidents do not get reported, or result in other, less obvious records such as industry scrap reports. Second, not all injuries are reported. The '2000 Accidents' study (Powell *et al.*, 1971) clearly showed that low-severity injuries are remarkably under-reported. So are the less visible injuries, such as back strains and mental health problems. Third, not all facts pertaining to an incident get reported. In a study of airline incident investigators using a novel methodology (Woodcock and Smiley, 1999) in which participants had to ask the experimenter for facts, Drury *et al.* (2002) found that only about a third of available facts were ever discovered. Of those facts discovered, only a fraction survived the investigator's own analysis to make it through to the final summary presented to management. There did not appear to be any deliberate bias in what was discovered or reported, although facts about equipment and environment were less often reported than facts about the mechanics involved in the incidents or the sequence of activities/events. The lesson is that even professional investigators report only a small fraction of what an ergonomist would like to know about the incident.

Given that an injury is reported, it may not be recorded, particularly if it is low severity. OSHA in the USA requires all visits to a medical facility to be recorded in a medical log, but only injuries meeting certain criteria stimulate an accident or injury report. Beyond this only very serious injuries cause an in-depth investigation to be undertaken. Thus, the ergonomist has a progressively filtered system, with broad but shallow coverage in the medical log, more detailed coverage of few events in the accident/injury report, and deep coverage of a minimal number of accidents via in-depth investigations.

From a reasonable accident/injury report comes certain standard information on the victim (age, gender, job category), the task, the equipment in use, and environment (time of day, unusual circumstances). The care with which forms are devised and filled out are both highly variable. Most forms are strong on medical facts and blame-pinning, and weak in details of task, equipment and environment. The archetypal form reports that the victim sustained an (alleged) back sprain due to improper lifting technique and that the remedial action taken was to instruct the victim not to lift that way again. With data sets designed more specifically for human factors interests (e.g., the ASRS), categories of information captured may be more complete; however, there still may be problems regarding the subjectivity and biases of self reports regarding incident progression and cause (Jentsch *et al.*, 1999).

If we are skeptical of the archival incident data, why do we use it at all? First, it is of immediate economic importance. For industrial accident data, each report represents a loss of revenues as well as large direct costs to the company and the employee. Second, accident and incident reports have a high level of face validity. Third, archival reports can be used in a boot-strap procedure to devise better classifications (Drury and Wick, 1984), to see which situations are in most need of human factors interventions (Drury and Wick, 1984) and to derive new and more detailed incident investigation procedures (Monteau, 1977; Wenner and Drury, 2000).

Operating records

Records generated by system operators, as a part of the regular work activities, are another commonly analysed source of archival data. Such data sources may include forms filled out or generated during a production or other work processes, documents at various stages of production (e.g., rough drafts, edited copies), meeting minutes, or any other extant outputs from a working group or process. For instance, Suchman and Trigg (1991) analysed the charts airline operations personnel use to record the transfer of passengers and luggage among incoming and outgoing flights, and numerous studies have analysed the paper flight strips generated during air traffic control operations (Bentley *et al.*, 1992; Mackay, 1999). Salterio (1996) performed an interesting research study regarding information search by analysing records that were generated by accountants during accounting firm audits. Unlike event based records such as accident reports, these records may be less subject to subjective biases, since they are generated as a part of the work task, rather than in response to (and often in explanation of) adverse events. However, they may not include all details of interest to the human factors analyst. The analyst should also be careful to understand, through other techniques such as observation or interviewing, the true role the archived records played during the course of work activities. For instance, Button and Harper (1993) found that invoice and production records resulting from a manufacturing process were better viewed as a post-hoc record of the various operations, which were performed iteratively and in parallel, rather than a strictly sequential description of the process.

Fixed data sources

Another source of archival data is found in fixed information sources. All organisations have sources of data which are relatively invariant over typical project time-scales. Fixed data sources include those records which are generated at a fixed time, often only once, and serve as a source of historic or background information, possibly as a context in which to interpret other archival or observational data sources. For instance, workplaces will usually have a layout drawing, given in more or less detail, showing the physical size of work areas involved, locations of material stores, and fixed equipment. Other fixed data sources include company annual reports and sales data, personnel records, and planning documents. Paquet *et al.* (1999) utilised fixed archival sources such as state highway specifications, contracts, and construction schedules to develop a sampling plan for a further observational study. An obvious danger with the use of fixed data is that they can become outdated without the analyst realising it, or may reflect a normative, rather than descriptive, account of company operations. There is no substitute, for example, for physically checking whether stated equipment still exists at its noted location, or whether the relationships noted in an organisational chart actually function in practice.

Records generated for other analysis purposes

Finally, archival records may exist which were generated as part of other, past analyses. For instance, records concerned with the details of how operators perform their jobs, including detailed task and sub-task breakdowns, can form a starting point for further analyses. In many cases, however, such breakdowns were made for the purpose of estimating times for task completion, and thus there is rarely any information regarding errors, cognitive functions, or performance variability.

Other data considerations

Other considerations regarding archival data sources include the nature of the data as qualitative or quantitative, which can impact methods chosen for analysis. For instance, accident or incident records may contain quantitative (numeric) data such as the age of an individual who was injured; qualitative information may be found in narrative accounts of the accidents. Some textual information (e.g., type of injury) may be easily convertible to categorical variables which are amenable to standard types of statistical analyses (such as those associated with classificatory observational data as discussed below), while other qualitative data sources (e.g., accident descriptions) may require more involved analysis before conclusions can be drawn. A related issue is that of the subjective vs. objective nature of the archival data. In some sense, all data are essentially subjective as they are eventually passed through a human judgment process. For those data sources typically considered 'objective' such as a digital reading of reaction time, or a record of someone's age, there is very low variability regarding the possible beliefs or understandings regarding the value of the data. For those data sources seen as more subjective (e.g., an opinion regarding the cause of an accident), the variability is higher, caused by differences among individuals, or even across time in the same individual. For all sources of archival data, it behooves the investigator to consider whether, and how, the data may have been influenced by such sources of variability.

Advantages and disadvantages

It should be obvious by now that archival data represent a valuable resource for human factors research and analysis, but are full of hidden traps for the unwary.

Measurements from archival data complement the more usual ergonomic surveys and experiments in many way. Primarily, they represent a longer term view of the effects of change over months or years rather than a snapshot of one particular process over a few days. It is possible, for instance, to incorporate archival sources that go beyond the period of even the analyst's lifetime. This long term view, as well as the realistic conditions under which the data were collected, enhance the face validity of the data source and subsequent analysis, since the real system is in operation, with all of its day to day idiosyncrasies as opposed to a 'best behavior' controlled experiment. Additionally, the raw data are already being collected, and so the additional data collection costs look minimal. Inexpensive, face-valid data which reach back into the past and will reach forward into the future are certainly worth considering, but this is not the whole story.

Existing data are not necessarily collected for human factors purposes, and so will not answer human factors questions except by fortunate coincidence. Data that are collected may not be complete, or include all variables relevant to the interpretation of the data. For example, in some cases there is a high level of aggregation over the data, so that accident rates are aggregated across many individuals whose jobs vary to a greater or lesser extent. Recent epidemiological research on the impacts of cell phone use on automotive accidents is hampered by the fact that many archival records of accidents (e.g., police reports) have not captured information regarding cell phone use, or have done so in a superficial or blunt way (e.g., is a cell phone present, rather than if or how it was being used). Archival data sources which include subjective accounts (e.g., accident descriptions) are subject to biases (intentional or not) based on the particular points of view and motivations of the data recorder. For sources generated by many individuals over time, there may be shifts from individual to individual, or time to time, about how or what data to record. Even specialised archival records such as the ASRS, which has been implemented in part to assist in analysis of aviation incident and accident causes (certainly a human factors question), may suffer from incomplete, inaccurate, or biased reports.

Use of observational techniques

Introduction

At a level of intervention somewhat greater than the *post-hoc* analyses of archival data sources, one can consider the application of observation techniques in human factors studies. Within the field of human factors and ergonomics, observational techniques have been utilised in a broad range of studies including those concerned with industrial work analyses, biomechanical and physiological research, cognitive and human–machine systems research studies, and environment, equipment, and computer system design.

One can define an observational measurement as one in which the experimenter or analyst acts as an observer, or the instrument of measurement, with no or minimal interaction with the people and situations being observed. While prototypical applications of observational methods occur in naturalistic settings (e.g., such as the naturalistic inquiry methods described by Lincoln and Guba, 1985), it is important to recognise the difference between studies in real world settings, often considered field or case studies, and the use of observational, or non-intrusive data gathering techniques. It is possible to conduct studies in field settings which do not rely on observation as a measurement technique. For instance, Stal *et al.* (1999) studied wrist positions and movements during milking tasks in a dairy barn (a field setting), but utilised biaxial electrogoniometers (rather than observations) to collect position and movement data. Likewise, it is possible to make observations in settings which are pseudo-realistic, such as task simulators (Bowers *et al.*, 1998) or training exercises (Artman, 2000), or even in laboratory settings, where observations can be made of behavior or activities during the performance of more controlled tasks (Lavender *et al.*, 2000). The apparent contradiction between 'observation' and 'controlled study' brings to light an important distinction regarding methodological reactivity: whether the study design is being considered (e.g., an unaltered, uncontrolled natural setting vs. a controlled laboratory environment) or whether the measurement techniques are being considered (e.g., non-intrusive observations vs. instrumented forms of data collection). For the purposes of this chapter, we will consider studies which have employed observational measurement techniques, primarily within naturalistic or pseudo-naturalistic studies. Some more controlled laboratory studies, where the tasks employed were relatively naturalistic, rather than highly controlled, will also be noted.

Traditions

Observations have played a key role in the methodologies of a variety of science, engineering, and design traditions. Within Industrial Engineering, the analysis of work based on direct observations in the workplace is a almost a century old, beginning at the turn of the 20th century with the work analysis methodologies of Taylor (1911) and the Gilbreths (Gilbreth and Gilbreth, 1919). These methods evolved into more modern task analytic techniques (e.g., Hierarchical Task Analysis; Annett and Duncan, 1967), cognitive work analysis methods (Rasmussen *et al.*, 1994; Vicente, 1999), and ergonomic work analysis techniques (Buchholz *et al.*, 1996) which also rely on observations of work. Methodologies such as process tracing (Woods, 1993) rely on sequential observations of human activities in concert with records and observations of external events and dynamic system states. Observation methodologies such as OWAS (Corlett, 1995; Mattila *et al.*, 1993) and PATH (Buchholz *et al.* 1996) have been developed to assist in assessments of postures and other attributes of physical activities.

Within the social sciences, there is a history of research which has relied on observational methods to provide descriptions of social activities and cultures (e.g., case study methods; Yin, 1989). Some of these methods fall into the category of naturalistic research (Lincoln and Guba, 1985), in which the focus is on understanding activities and tasks in context, from the users' point of view. Recently, there has been cross fertilisation of methods from the social sciences, such as ethnography (Blomberg *et al.*, 1993; Hammersley and Atkinson, 1983) with the more typical work analysis techniques and research and design questions addressed in human factors. This intersection has informed both research studies and designs, particularly regarding studies on the impact and design of modern computer and information technologies (Bentley *et al.*, 1992; Nardi, 1997; Sachs, 1995; Suchman and Trigg, 1991; Zuboff, 1987).

Design oriented methodologies in human-computer interaction, such as Contextual Inquiry (Beyer and Holtzblatt, 1998) also rely heavily on observational methods.

Dimensions of observational methods

Studies utilising observational methods vary across a number of dimensions, including the type of setting, research questions, and the content and style of observations. The nature of the study, in terms of the collection of observations in real time vs. recordings, along with the sampling framework for observations, can also differ. The following sections describe these distinctions in detail.

Setting: field, simulator, laboratory

Observations can be made in a variety of settings and conditions ranging from field settings to task simulators, training situations and more controlled laboratory settings. Human factors studies using observational techniques have been conducted in a wide variety of field settings. For example, observations for human factors and ergonomic analyses have been collected in airplane cockpits (Degani and Wiener, 1993, 1997), corporate settings (Burns and Vicente, 2000; Carletta *et al.*, 2000), manufacturing plants (Bisantz and Ockerman, 2002; Gaudart, 2000), retail environments (Bisantz and Ockerman, 2002), construction sites (Mirka *et al.*, 2000; Paquet *et al.*, 1999), public and home settings (Clark *et al.*, 1990; Connell, 1998), process control rooms (Mumaw *et al.*, 2000; Stanton and Ashleigh, 2000), schools (Knight and Noyes, 1999), and medical environments such as trauma rooms (Mackenzie *et al.*, 1996; Xiao *et al.*, 1996), hospital wards (Kjellberg *et al.*, 2000) and operating theatres (Cook and Woods, 1996; Seagull and Sanderson, 2001).

Additionally, observations have been collected during training exercises (in actual work environments) and in task simulators. For example, Watts *et al.* (1996) collected observations during full scale, simulated training missions at a NASA mission control center, and Artman (2000) made observations during military command and control training exercises. Paquet *et al.* (2001) made postural observations in a simulated construction worksite. Flight simulators (Bowers *et al.*, 1998), and simulations of home environments (Kirvesoja *et al.*, 2000) have also provided settings for observational measurements.

Observations have also been used as a data collection technique in studies that are more similar to controlled laboratory studies, in which participants are directed to perform a variety of realistic, yet simulated tasks. For instance, Lavender *et al.* (2000) had participants carry out tasks typical of paramedic rescue operations, and Obradovich and Woods (1996) made observations of nurses performing specified tasks with a medical device.

Research questions

Observational methods are used in studies addressing a wide variety of research questions. When observational methods are employed within field settings, it is less common to find specific research hypotheses because controlled interventions and comparisons are not conducted. Often, research questions addressed using observational methods in field settings are better characterised as information seeking, rather than hypothesis testing. In this way, they correspond to studies in which the focus is on the 'context of discovery' rather than the 'context of justification' (Lincoln and Guba, 1985).

Some studies focus on providing descriptive accounts of work, particularly in complex or challenging settings. For instance, Mumaw *et al.* (2000) studied how nuclear power plant operators monitor the state of the plant under normal operating conditions, and Artman (2000) described the role of cooperative work in the construction of team situation awareness. Xiao *et al.* (1996) identified factors of complexity in emergency medical care. Information gathered in such studies can be used to set the stage for future research, or to provide input regarding the design of supporting systems. Other studies, while also descriptive in nature, focus more closely on a particular activity or use of technology. For instance, Degani and Wiener (1993) studied the use of procedures in aircraft cockpits, and Carletta *et al.* (2000) described the impact that a newly introduced multi-media collaborative system had on distributed team work. Watts *et al.* (1996) identified and described the role of communication (voice) loops in the coordination of activities in space mission control, and Cook and Woods (1996) described how medical personnel interacted with new, automated technology in operating rooms. Other studies have focused on understanding naturally occurring errors, such as those occurring in the use of a public transportation ticketing machine (Connell, 1998); during prescription drug filling (Flynn *et al.*, 1996), or in emergency medicine (Mackenzie *et al.*, 1996).

Other observational studies have focused on documenting and characterising exposures to physical risks, along with developing structured observational tools for performing such assessments. For instance, Blomkvist and Gard (2000) characterised muscular activities and joint positions during computer work in a variety of cold environments. Mirka *et al.* (2000) and Paquet *et al.* (1999) identified and characterised postures in construction tasks, and Kjellberg *et al.* (2000) developed a tool for motion and posture analysis through observations of patient transfers in hospital wards.

While the research questions and results provided by such studies are quite different than those found in controlled, experimental settings, it is important to note that there is an informative, cyclical relationship between the two types of studies. Observational field studies can form an important basis for developing representative hypotheses and circumstances tested in more controlled environments, and questions raised through controlled experiments can inform the design of further field research (Woods, 1993).

More comparative research questions can be posed in observational studies with characteristics of field experiments. Woods (1993) describes field experiments as studies that occur in field or simulated settings, but where the conditions of observation (the participants, activities, and circumstances) are selected based on a particular question being studied. Because the ability to do active intervention in real world settings is limited, researchers conducting field experiments often capitalise on naturally occurring "experimental conditions." Examples could include comparisons across work shifts, types of jobs, locations, levels of operator experience, or before and after the introduction of new technology or other work oriented intervention. For instance, by collecting observations across different types and phases of surgeries, Seagull and Sanderson (2001) were able to make comparisons in the reactions to anesthesiologic alarms. Stanton and Ashleigh (2000) compared team interactions and activities across teams with different hierarchic structures, and at different points in their compositions. Knight and Noyes (1999) compared classroom behaviors and subjective assessments across classrooms with different types of furniture. Lavender *et al.* (2000) assessed exposure to physical stressors in different paramedic rescue operations, by having participants carry out a range of simulated rescue tasks.

Observation content and style

The content of the observations made is obviously dependent on the research questions of interest, as well as practical factors such as cost and setting constraints. Content captured in many studies focusing on the cognitive and social aspects of work include activities (in more or less detail), conversations, artifacts used, the timing and sequence of activities, locations of individuals, and information sources accessed. Often, these observations are combined with observations regarding the state of the environment or unfolding situation, so that the analyst, in retrospect, can understand the context in which the activities are taking place (Woods, 1993). For instance, Carletta *et al.* (2000) captured observations of meeting activities and verbalisations to assess the impact of new collaborative technology on distributed team meetings. Artman (2000) recorded verbalisations in studying cooperative work and team situation awareness. Stanton and Ashleigh (2000) included observations of people's physical locations and movements in order to understand whether differences in team structured impacted how individuals moved around a process plant control room. Seagull and Sanderson (2001) made sequential observations of the activities, alarms, and the surgical context, to study reactions to those alarms, the use of artifacts and tools, actions, and gestures.

Other studies focus on more detailed observations of postures and work activities, in order to provide basis for documenting and characterising exposure to physical risks, in different occupations. For instance, numerous studies made observations of postures (Kjellberg *et al.*, 2000; Lavender *et al.*, 2000; Mirka *et al.*, 2000; Paquet *et al.*, 1999). Observations in these studies also included task or activity descriptions, the use of tools, and objects being handled. For instance, Paquet *et al.* (1999), in a study which focused on manual material handling activities, recorded the type of activity, weights of loads, trunk postures, types of hand grasps, number of hands used, and the type of tool or material being handled, using a structured coding scheme.

Observation content, along with the researchers questions, can constrain the style of the observations. Observational data can range from the highly structured (i.e., ticks on a form that a particular, predetermined activity occurred), to less structured, qualitative descriptions of activities within predefined categories (e.g., descriptions of cockpit procedure use), to unstructured notes recording any incidents and activities that the observer feels may be relevant. In studies where there are more focused research questions, researchers may be able to develop categories, and classification rating schemes of activities or movements *a priori*, and note the extent to which those activities occur. For instance, the OWAS method for posture classification (Corlett, 1995; Mattila *et al.*, 1993) requires the observer to note if any one of a number of leg, arm, or back postures occur. One could also record the use of particular artifacts (e.g., a written procedure) in a similarly structured fashion. Buchholz *et al.* (1996) gives an example of a computer scannable data entry sheet ('bubble' sheet) giving categories of postures, activities, tools, and hand use which has been customised for data collection in a construction domain. Latko *et al.* (1997) developed an observational tool in which analysts provided a rating of the observed level of repetition and hand activity. In other cases, particularly when the desired output is a descriptive account of a work environment or set of activities, a more expansive, unstructured approach may be used.

The content and structure of the data collected have obvious implications regarding the subsequent analysis required. Qualitative data analysis, which includes methods that can be used to analyse much of the textual data (field notes, unstructured observations, verbalisations) collected through observational methods, is described in Miles and Huberman (1984) and Lincoln and Guba (1985) — and see also Chapter 5. Quantitative data analysis uses statistical tools to test hypotheses about relationships and differences. Note again that observational studies most typically yield relationships (e.g., the error rate on task documents is related to their conformance to guidelines) rather then causality (e.g., lack of conformance to guidelines causes increased error rates).

In general, the type of statistical analysis depends upon the original hypotheses or data needs. If we wish only to estimate the time content of a job on an assembly line, then all we need are descriptive statistics of the observed data set, e.g., mean, variance. If we wish to test particular hypotheses, then, depending upon our measurement level, we can use parametric or non-parametric statistical techniques. These range from chi-square tests of association between two classificatory variables (as in the aviation inspection study at the start of this chapter) to parametric regression techniques (as in the camera assembly line study). In other cases we use the data to develop a classification scheme, and then can perform association tests on such schemes if appropriate. For example, Wenner and Drury (2000) classified ground damage accidents to airliners by the type of event and the presence of antecedent causes, based on archival accident data. They were then able to relate cause and outcome using a chi-square test of association, and derive targeted interventions to reduce accident rates.

Real time vs. video recording

Observations can be made in real time, by analysts in the setting, or from video recordings of activities. Making the choice between real time observations and video tape involves a number of tradeoffs, and also interacts with the content and style of data recorded, and parallel measures used in the study.

Advantages of collecting video data, and making observations from the recordings, are numerous. Unlike observation in person, situations captured in tapes can be reviewed multiple times, in order to fully assess aspects of complex or fast paced activities. Video recordings facilitate measurements of analysis reliability, both across analysts and time. Recorded data can also be kept, in the spirit of archival data, for future research and analysis. If multiple cameras are used, more of a situation than is observable by a single individual can be captured. Video recordings can be slowed down for analysis, even to the point of a frame by frame analysis, when necessary to create highly detailed descriptions of postures or actions, and to capture activity times.

However, there are also disadvantages to the use of video. There are some situations, for legal or practical reasons, where the use of video cameras is not possible. For instance, Seagull and Sanderson (2001) were unable to secure permission to videotape surgeries. For studies involving large areas a single (or even multiple) fixed cameras may be impractical. Setting up, capturing, storing, and converting video data between the variety of available digital and analog formats (if desired) alone can be costly and time consuming, without even considering the time required for analysis. Sanderson and Fisher (1994) suggest that the ratio of time for analysis compared to recording of sequential observational data, such at that stemming from recorded observations, can reach levels as high as 100 to 1, depending on factors such as the granularity of analysis being conducted (i.e., the length or frequency of activities being recorded, such as gestures or motions vs. movement across a room, or topics raised at a meeting), the specificity of research questions being asked, and the complexity of inferences required from the data. Use of multiple cameras requires the consideration of recording synchronisation, as well as increases in analysis time due to the interpretation of events occurring across multiple views. In observing body angles, for example, it is often necessary to use multiple views to ensure that the angular positions of limbs and torso are properly defined in three-dimensional space. With repetitive tasks, a single camera can often be used in different shooting positions across different cycles if multiple cameras are not available (Drury, 1987). Fortunately, body angle data are usually only needed when the issues center around biomechanics and musculo-skeletal injuries, and the tasks where these are most common are highly repetitive. Camera angles and distances from the subject can lead to distorted or obstructed images, making it difficult to identify all activities, or detailed interactions with workplace artifacts, reliably. For instance, it may be possible to determine that a person referred to a piece of paper, but not identify what the paper or its content was. Events which occur out of camera coverage cannot be analysed.

These latter cases, in particular, could be alleviated with the supplement of in-person observation. While a human observer might not be able to capture all elements of sequential activities in a situation, it is possible for documents to be examined, questions to be asked, and activities and equipment to be scrutinised in a more detailed fashion.

A related issue is the level of technological support used in collecting the observations. Methods can range from decidedly low-tech (pen and paper) to audio and video recordings, to computer supported data collection. Handwritten notes can be free form (in the case of unstructured data) or rely on scales and forms developed based on structured categories of data to be collected. Specifically developed software programs combined with portable or handheld computers can be used to allow coding of real time activities. There are also hardware and software systems which facilitate analysis of videotaped data (see Sanderson *et al.*, 1994 for one example). These systems provide functionality such as time based assignment of codes and descriptions to video segments and integrated control of video, as well as support for postural and biomechanical analysis.

Sampling issues

We have already seen that research questions can be answered from archival or observational data by comparing samples from different naturally-occurring conditions. These form a class of 'natural experiments', although unlike actual experiments with deliberately-manipulated factors, causality cannot often be implied. But the issue of how to sample must be answered in any archival or observational study, and how it is answered will affect the conclusions drawn from the study. In the industrial engineering tradition, such considerations have typically been addressed in occurrence sampling studies, although of course they apply to all studies. Occurrence sampling is the systematic sampling by observation of the frequency of occurrence of different system states. While it was originally used to measure non-productive time in industrial work (Tippett, 1934), it has since been a common technique for estimating the probability of occurrence of any system state. For example, we can estimate how much of a pilot's time is taken by communications with Air Traffic Control by making a large number of observations each of which consists of whether the pilot was or was not communicating with ATC. These are very simple observations to make (although quite boring for the observer) but can lead to good estimates of the fraction of time communicating as the ratio of number of times observed communicating to total number of times observed. However, to collect data of value, we need to consider carefully who is observed, how often and when. These are precisely the issues in sampling, and apply to all studies, even though we will use occurrence sampling as the vehicle to present them here.

To obtain data that are representative we must ensure that the conditions sampled are also representative, but of what? If we are studying normal operations, than we should not sample during the introduction of new equipment, at shift turnovers, or the training of a large contingent of new operators. However, if we are studying the change process, these are precisely the conditions we need to sample. In any sampling, we start from a Sampling Frame, the notional or actual enumeration of all of the possible units from which we can sample. This may be the set of all times during a day shift, excluding the first and last 20 minutes. Or it could be all three shifts, or we could limit ourselves to just the shift change periods. The point is that the sampling frame determines the population for which our sample will be used to make inferences. Similar considerations apply to choice of participants. Do we want all operators of machining centers, all hourly workers on a shift, or all personnel on a shift? Again it depends upon whom we wish to make inferences. As an example, to measure the frequency of role modeling behavior in automotive seat belt use, we are normally interested in the population of a particular region. Thus Hong *et al.* (1998) sampled (observed) many automobiles in the Buffalo NY area. In a later study where it was important to match the characteristics of an observational study with those of separate experimental and questionnaire studies, Kritkausky (1999) sampled only from university-based automobiles.

Having determined our sampling frame, we now need to specify how samples will be drawn. Will they be at predetermined times, or times based on what is happening to the system? The former will give inferences about the whole system operation, while the latter will be restricted to the particular system states specified. For example, we could choose to sample only when the system was in overload conditions, or when a statistical process control chart had indicated that a process change had occurred. This would represent event based sampling rather than continuous sampling. Given that we have decided that the samples need to be drawn at particular times, we now have to choose whether these times will be random or systematic. Random sampling, from a computer-generated random listing or even from printed tables, will ensure that there are no systematic biases. Systematic sampling could produce such biases, for example if the sampling frequency coincides with a naturally occurring frequency in the system such as a new batch of product being loaded. In that case all we would ever see was product being loaded, rather than the whole production cycle. If in doubt, the safest way is to randomise. The only downside is that a random schedule is really disruptive to an analyst who has other duties.

Parallel measures used

In most studies which utilise observational techniques, observations are not the only form of data collected. Additional measurement techniques are used to complement the observations, to provide an expanded basis for drawing conclusions about the research questions of interest. The choice of parallel measures varies based on the nature of the study. For instance, observations of postures and work cycles might be combined with direct measurements of muscular activity, forces exerted, or body positions and velocities. Observations of artifact use, team work and collaboration, and information use may be supplemented with interviews (either concurrently, as the work activity progresses, or in a more formal interview setting), focus groups, or through analysis of archival data sources. For example, Degani and Wiener (1993, 1997) supplemented observations of pilot checklist use with interviews with pilots and analysis of archival data (aviation accident and incident records). Questionnaires or rating scales may be used, for example, to collect subjective measurements of physical discomfort, or subjective workload estimates. Analysts may assume roles of participant observers, doing or assisting in the tasks that they are observing, in order to develop a better understanding of the activity. For instance, Burns and Vicente (2000) acted as human-machine design consultants, and thus as participant-observers, during their observations of the interaction between human factors and other design constraints. Sequential observations may be combined with records of system state variables, in order to understand activities within the context of a progressing situation.

Multiple data sources can provide a context from which to interpret the observational data, and likewise, observational data may provide the means for disambiguating more direct measurements. Carletta *et al.* (2000) utilised questionnaires and facilitated group discussions in addition to observations of team meetings, which allowed a characterisation of participants opinions and explanations of use regarding newly implemented distance-meeting technology, in addition to observations of the impact of the technology on meeting communication patterns. Lumbar motion monitors were used to supplement postural assessments made through observations in studies by Lavender *et al.* (2000) and Paquet *et al.* (2001). In the latter case, observations of videotaped activities were used to provide the task context in which to interpret the significance of measurements gathered through accelerometers. A study of the factors affecting role-modeling behavior in automotive seat belt use, Kritkausky (1999) combined observational studies of whether driver and passenger were both wearing seat belts, an experimental study in which subjects rode with a driver who either used or did not use a seat belt, and a questionnaire study to examine attitudes to safety and role-modeling. In this way, the ambiguities in the observational study (e.g., exactly who was modeling whose behavior?) were clarified in the other two studies to produce a more complete description of the behavior.

Process tracing methods (Woods, 1993) also rely on observations combined with other records of a situations, such as events and process variable states, or order to interpret and understand the significance, in terms of cognitive processing, of the activities observed. For instance, Cook and Woods (1996) utilised process tracing methods, collecting data regarding patient states (e.g., physiological parameters) and conducting concurrent interviews in addition to making observations of practitioners using medical devices. Similarly, Seagull and Sanderson (2001) constructed process tracing logs which included surgical events and equipment states, as well observed activities. They also collected and analysed documents related to the cases.

Observational studies may be performed along with validating laboratory studies, or with other types of formal task analyses. For instance, Connell (1998) conducted formal usability analyses of a ticketing machine to predict design problems, in additional to observing customers using the device.

Representations of observational data

If observational data are collected in the form of text (transcriptions of verbalisations, analyst notes), then textual representations such as lists, tables, direct quotes, or narratives are often used to condense and present the data. Examples of such presentations, some of which are consistent with the style of a more ethnographic tradition, can be found in Watts *et al.* (1996), Degani and Wiener (1997), and Zuboff (1987). However, more formal representations of observational data can also be utilised.

Charting methods

When a system of people and machines attains a reasonable level of complexity, it becomes necessary to use some level of abstraction to understand its working. We rarely need full details of every aspect of the system at one time, so we resort to partial representations to aid our dealings with the system. Do we need to be able to operate every machine in a factory when our interest is in how material travels between machines? Do we need the life story of every person in the emergency room to understand how decisions are made under stress? In each case it would be nice to have the complete information, but we recognise that we can gain much understanding with a less-complete picture. In this section we present a selection of charting methods that have been useful to ergonomists. They range from the industrial engineering-based to the socio-technical. Useful references to the former are the *Handbook of Industrial Engineering* (Salvendy, 1998) and the text by Konz and Johnson (2000). For the socio-technical methods, see Taylor and Felton (1993). To understand charting, though, we first need to examine the notion of abstraction.

The original data for any observation study consists of state/time records or event/time records. A state is formally defined as the vector of all the values of all system variables. For air traffic control, this would be a large vector, but at a minimum it would include the characteristics of each aircraft in the relevant airspace (designation, type, speed, altitude, direction) as well as pending transitions, plus weather and the status of any fixed facilities such as airfields. A state/time record would list the states at each time instant when a sample was taken. In the second case, a record only arises when an event happens, where an event is a change of system state. Thus when an air traffic controller communicates to flight BA 756 that it should proceed to Flight Level 36 at a heading of 270, that is an event, as the system has changed the state of BA 756 adjusting its altitude and direction. From an initial state and a complete set of all events, it is possible to collect the complete trajectory of the system over time. With state/time records this would only be possible if the sampling rate was rapid enough to catch all events. We can sometimes make use of such raw records, as when the National Transportation Board releases the transcripts of cockpit and data recorders after an accident. In most cases, though, we abstract such records still further to highlight particular aspects of interest to our study. We will examine a selection of such abstractions.

One of the best known abstractions is the Flow Process Chart or flow chart. It shows how product or information flows between elements of a system, but usually sacrifices any data on when the flows occurred. Because it suppresses temporal information, the flow chart makes the remaining information on the flows along paths between system elements much clearer. This clarity has made the flow chart a staple tool of business improvement efforts such as total quality, business process reengineering or six-sigma. (Just because it is popular does not mean it has little scientific merit!) A related chart is the Link Chart, where the flows between elements are characterised by flow intensity, for example the fact that 1500 kg of raw material goes from the mixer to the extruder in a rubber tire plant rather than just that there is a flow along that link. With the link chart we can focus on either the larger flows (for example to see which elements should be placed close together) or the rarer flows to ensure that all eventualities are covered by a new process layout. Both flow charts and link charts can have abstract symbols and arbitrary locations on a printed page, or can use the actual (or planned) geographic layout of the elements on the ground. These charts can be used to describe movements of physical goods (e.g., product), people, or notional movements such as eye movements on a display. As an example, Figure 1 shows the link chart for hand eye movements needed to restart a metal

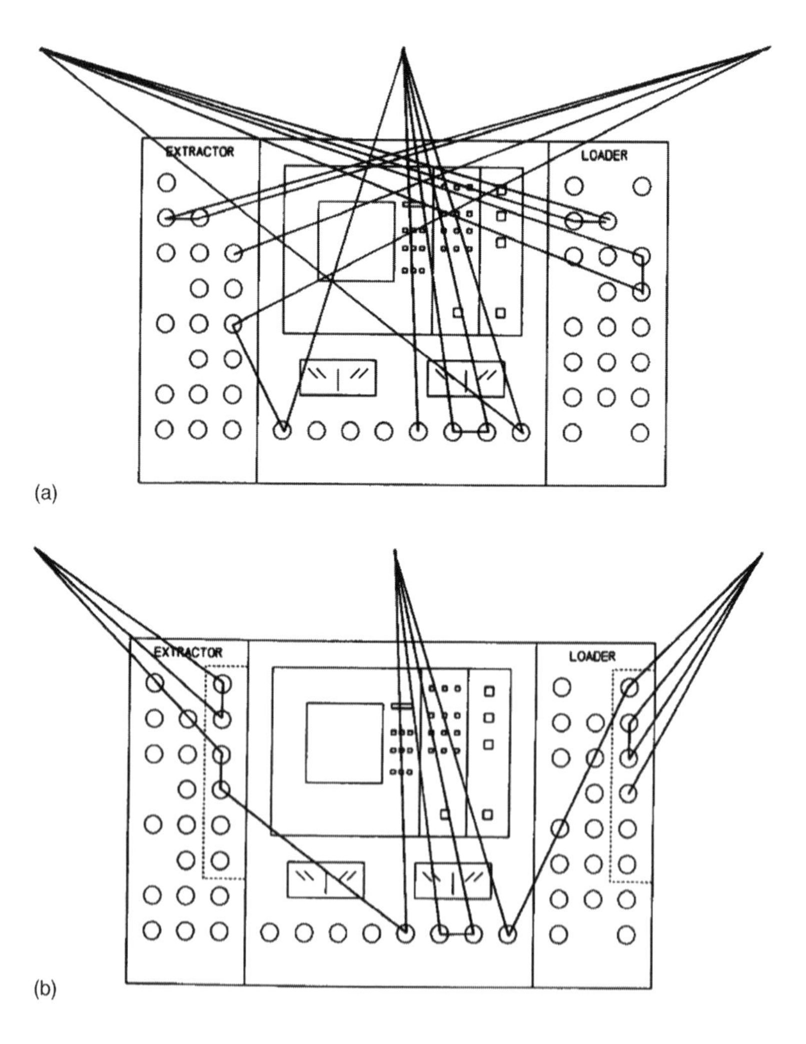

Figure 1 Task sequence required to reset automated system before and after ergonomic changes.

stamping operation after a stoppage. The redesigned panel shows a much more logical movement sequence. The chart was a useful way of presenting the information to management and workforce at the plant, as well as providing a training aid for users.

Charting methods are a useful visual means to communicate the salient points about the working of a system, so it is not surprising that they are used heavily in activities that rely on groups for problem solving. Socio-technical systems design (STS) has developed a number of such methods of which two are presented here as examples. STS bases its methodology on group decision making so that charting techniques are appropriate. The aim of STS is the joint optimisation of the technical system and the social system, rather than giving one preference over the other. Charting methods are needed to collect, analyse and communicate both technical aspects and social aspects of the system (see for example Taylor and Felten, 1993).

For technical aspects of the system, STS uses a Variance Matrix to chart and display how the various sources of system variation interact. Such a chart is heavily influenced by the concept of Unit Operations from STS's legacy in the process industries such as coal mining, chemical plants and oil refineries. A Variance is a source of variation, either from the input to the process (e.g., raw material quality variables) or from the process itself (e.g., machine settings, output variables). The Variance Matrix displays the interactions between these variables by placing X in the appropriate cells. The form of this interaction can be annotated, and the variables themselves can be categorised by importance, for example by identifying Key Variances. These in turn are the basis for control systems, usually based on both human and technical components of the system. Figure 2 shows one such matrix for a distance learning system in an engineering school, developed as part of a 1997 class project by H. Kelly, G. Gattie and K. Markovich. Note how the variances are classified into Unit Operations I to VII following the broad structure of how distance-learning courses are devised, performed and evaluated. Note also the differences in interactions between the variances, with some such as Variance 6: Course Structure influencing many subsequent variances.

Given that the key variances have been identified, the ergonomist now needs to understand the social relationships that impact control of these variables. Who communicates with whom? What do they communicate about? How do they use their communications to control the process? The first step here is a modification of the flow chart where communication patterns are charted from interview data with system personnel. Figure 3 shows the Role Network for participants in the distance-learning system represented in Figure 2 (again from the same source). Each person (or group) is represented by a node in the network, while communications links are the edges. Each link is annotated with a code for what information is transmitted (Taylor and Felten, 1993) as follows:

G = **G**oal Attainment

A = **A**daptation to Environment

I = System **I**ntegration

L = **L**ong-Term Development

These terms are later used with a classification of the contact for communication (e.g., vertical contact between supervisor and subordinate, equal contact between peers in the same work groups, etc.) to form a communications grid, which is finally used as the basis for aligning the control functions with the social organisation.

While these particular charting methods may not be used outside a STS context, they give examples of how observational data can be organised effectively for understanding and design purposes. These examples also show that the basic techniques such as flow charting can be easily extended to new contexts, e.g., the Role Network. Note in addition that the

Stage	No.	Variable	1	2	3	4	5	6	7	8	9	10	11	12	13	14	15	16	17	18	19	20	21	22	23	24	25	26
I. Identify Professors	1	***1. Number of professors willing to participate***																										
	2	2. Number of schools willing to participate																										
	3	3. Number of departments willing to participate																										
II. Course Selection	4	4. Number of courses to be offered	x	x	x																							
	5	5. Classroom availability		x	x	x																						
	6	***6. Course structure***					x																					
III. Advertising	7	7. Funding (from Albany)		x	x	x																						
	8	8. Time available for advertising																										
	9	9. Quality of communication between schools							x																			
	10	10. Marketing strategies							x	x	x																	
IV. Enrollment	11	11. Number of students willing to enroll	x	x	x	x		x		x	x	x																
	12	12. Time required to obtain course number																										
	13	13. Registration timing																										
	14	14. Number of students enrolling not matriculated											x															
	15	15. Lecture organization						x																				
V. Coursework	16	***16. Course content quality***						x	x								x											
	17	***17. Course material quality***						x	x								x	x										
	18	18. Knowledge gained						x										x	x									
	19	19. Student/Professor interaction						x				x				x				x								
	20	20. Number of students attending classes						x			x	x				x					x							
	21	***21. Number of students enrolled***	x	x	x	x						x				x				x		x						
VI. Distribution	22	***22. Course tape quality (mast***						x	x								x	x	x	x								
	23	***23. Course tape quality***						x	x								x	x	x	x				x				
	24	24. Test/Homework						x				x				x	x	x	x				x	x	x			
	25	25. Test/Homework					x	x			x															x		
VII. Evaluation	26	26. Grades on tests, homework, etc.						x				x				x		x	x	x	x	x	x	x	x	x		
	27	27. Instructor evaluation scores						x			x						x	x	x	x	x	x	x	x	x	x		x

Figure 2 Variance matrix for distance learning college courses.

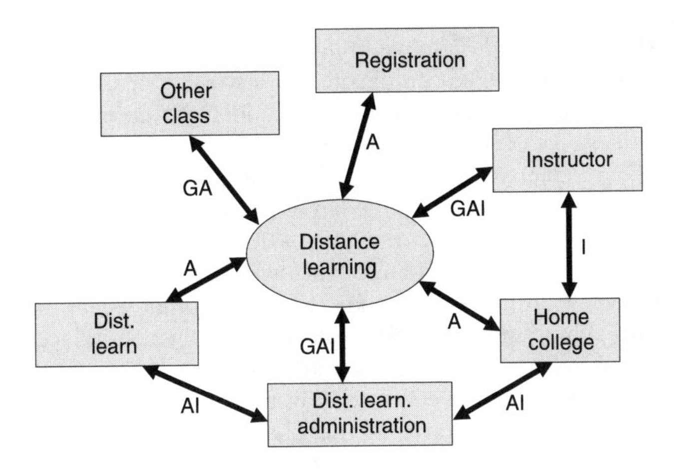

Figure 3 Role network for distance learning college courses.

matrix represented by Figure 2 can also be presented as a flow chart, with only those cells having an X being represented by edges between the nodes. In general, any matrix can be transformed into a network with identical properties.

Issues of reliability and validity

As with any other choice of methodological technique, researchers employing observational methods should be aware of potential problems associated with reliability and validity associated with the techniques, as well as methods to alleviate these concerns.

At their basis, questions of validity try to answer whether the explanations or conclusions drawn from an analysis are truly representative of the state of affairs. An educational test, for example, may be seen as valid if scores on the test represent how much someone knows and can apply the material. Problems with validity can occur in experimental studies for many reasons, including a failure to logically link dependent measures to the concepts they are designed to measure, failures to include appropriate and representative experimental conditions, and failures in meeting the assumptions for the statistical procedures chosen to analyse the data. In observational studies, questions of validity are less tied to those of experimental design and analysis (though certainly, if statistical analyses are conducted, such questions can arise), and are more concerned with the appropriateness of the situations which are sampled for observation, the impact of observational techniques on the activities being observed, and the potential effects of a human observer on data collection and interpretation. Questions of reliability center on whether measurements are repeatable, with low variability — does the measure correlate well with itself. Problems with reliability can come from over-sensitivity of a measurement device to uncontrollable environmental influences, or lack of stability of a dependent measure over time. In observational studies, questions of reliability stem primarily from the variability inherent in human observation. How is an analyst sure that observers are consistent in their choice of things to observe and the language or codes they use in their data recording? Do observations differ across observers, or due to factors such as observer fatigue?

There are a variety of methods for ensuring that the data collected and conclusions drawn using observational methods have high levels of validity and reliability. An important aspect of insuring validity, particularly in observational studies in field settings, is sampling method. The discussion above described methods, such as occurrence sampling and randomisation, which help ensure valid conclusions regarding the proportion of time activities take place. However, if the goal of the study is to be descriptive of all types of activities, even those with very low frequencies, in order to characterise complexities of a work domain, the conclusions may be more valid if a purposive sampling style (Lincoln and Guba, 1985) is adopted. Instead of randomly selecting times for observations, the analyst determines (through continuous, initial observations, or through interviews or discussions with people in the domain) the variety of activities and actors which should be observed in order to capture the variability of activities in the domain. For instance, Gravelle *et al.* (1996) selected observation periods and settings based on known variability in store volume and product mix, to insure the applicability of results over conditions existing within a fast food restaurant chain. Also, in contrast to experimental settings, where the goal is to create situations which induce people to perform particular tasks, making observations in naturally occurring settings is intended to produce data where there has been limited effect of the observation. However, it is unrealistic to think that people, cognisant of the presence of the observer, may not change their behavior. Explaining your role as observer (or the future use of videotaped data), remaining in the setting for an extended period of time, or focusing observations on situations where the focus of the people being observed is most likely to be the work itself (e.g., in high tempo or critical periods of the job) rather than on demonstrating work for the purposes observation, can lessen such concerns.

Miles and Huberman (1984) recommend a variety of methods for ensuring validity in conclusions drawn from qualitative data which can be applied to observational studies, particularly those conducted in field settings. Recommended techniques include developing a sampling plan so that unusual and difficult to access events are observed; confirming the representativeness of observations with other existing records (for example, one might confirm that a representative set of surgical cases was observed, by examining hospital records of surgery types); triangulating, or examining whether conclusions can be supported by observations drawn from different sites or data sets, or are consistent with data collected through other means (e.g., interviews, controlled experiments); considering biases in observations (e.g., due to sampling restrictions, possible impacts of observer on the setting) and discussing those as part of the analysis; looking for differences where they might be expected (if not, why not?); and explicitly considering and reporting 'outlier' observations in drawing conclusions while actively seeking out disconfirming observations. Examples of triangulation include Degani and Wiener (1997) who utilised interviews and archival data, as well as observations, to draw conclusions about cockpit procedure use and Obradovich and Woods (1996) who utilised interviews and benchmark tests in addition to observations of medical device use. The validity of conclusions may also be more formally assessed by performing an initial analysis on part of the data set (e.g., a sample of the observational sites, or part of the videotaped data) and then determining how well conclusions drawn from the first data set apply to the remainder of the data. Finally, and particularly for studies intending to be broadly descriptive of work activities, one can check directly with those observed to see how the observations, as well as the interpretation of events, problems, and situations matches the study conclusions. For instance, Seagull and Sanderson (2001) had anesthesiologists review collected field notes and process descriptions after the surgeries were completed. Mumaw *et al.* (2000) circulated findings among operators, as well as repeating observations with different operators, and at different locations.

With respect to reliability, it may be possible to use multiple observers of the same events, at least for a portion of the observations, and to compare the resultant observations. For structured or coded observations, formal statistical tests of inter-rater reliability can be used. Collection of observations from videotaped data supports a variety of checks on reliability, including comparisons across observers, and comparison within the same observer (making observations of the same tape, at different points in time).

Methods to insure reliability and validity of specific observational techniques or tools being developed are often based on measurement and iterative development of the tools, including comparison of the observational techniques to other measurement methods, and across observers and time. In some cases, such development is supported by the use of videotaped data. For instance, Latko *et al.* (1997) videotaped jobs, and then had multiple observers rate jobs for hand activity using a rating scale under development. Aspects of the scale and associated rating guidelines were iteratively refined after comparisons and discussions among the multiple observers (who were expert ergonomists). The reliability of the method was tested by having a sample of the jobs (which had been recorded on videotape) re-rated one to two years later. Similarly, Kjellberg *et al.* (2000) assessed the validity of a rating scheme for analyzing patient transfer methods by having an expert group view, discuss, and come to a consensus of videotape patient transfer operations. The expert group's ratings were compared to those of individuals trained to use the rating scheme to insure that the scheme produced valid ratings. Buchholtz *et al.* (1996) assessed the validity of a work sampling method for providing estimates of the proportion of time workers spent in particular postures by comparing results collected through the work sampling method at a construction site, to results from a continuous analysis of all postures adopted during several hours of videotaped construction work collected during the field study. Reliability of observations was also assessed, by comparing field observations to those made by the same observers, but of the videotaped work.

Conclusions

Archival and observation data play an important role in human factors and ergonomics research and design. Studies employing these data sources are often less reactive or intrusive than studies which utilise methods in controlled experimentation, and when they are performed in 'real world' settings, can provide the insights and grounding necessary for generalising results to actual situations. The use of observational or archival data may be critical in certain circumstances, such as those which require an understanding of complexities and interactions in a real world environment; cases where it is impossible to interfere with or constrain a system in order to perform controlled comparisons; and safety critical cases where workers cannot be interrupted to collect questionnaire, survey, or verbal protocol data. However, as illustrated in the studies discussed, researchers or practitioners may utilise these data sources for a variety of purposes — both alone, and in conjunction with other data types — in order to provide a basis of understanding for further studies, and to generate as well as answer questions and hypotheses.

References

Agnew, J. and Suruda, A.J. (1993). Age and fatal work-related falls. *Human Factors*, **35**, 731–736.

Annett, J. and Duncan, K.D. (1967). Task analysis and training design. *Occupational Psychology*, **41**, 211–221.

Artman, H. (2000). Team situation assessment and information distribution. *Ergonomics*, **8**, 1111–1128.

Bentley, R., Hughes, J.A., Randall, D., Rodden, T., Sawyer, P., Shapiro, D. and Sommerville, I. (1992). Ethnographically-informed systems design for air traffic control. *1992 Conference on Computer Supported Cooperative Work*, October 31–November 4, Toronto, Ont.

Bentley, T.A. and Haslam, R.A. (1998). Slip, trip, and fall accidents occurring during the delivery of mail. *Ergonomics*, **41**, 1859–1872.

Beyer, H. and Holtzblatt, K. (1998). *Contextual Design: Defining Customer-Centered Systems*. New York: Morgan Kaufmann Publishers.

Bisantz, A.M. and Ockerman, J.J. (2002). Informing the evaluation and design of technology in intentional work environments through a focus on artifacts and implicit theories. *International Journal of Human-Computer Studies*, **56**, 247–265.

Blomberg, J., Giacomi, J., Mosher, A. and Swenton-Wall, P. (1993). Ethnographic field methods and their relationship to design. In D. Shiler and A. Namioka (eds), *Participatory Design: Principles and Practices* (Hillsdale, NJ: Erlbaum), pp. 123–157.

Blomkvist, A. and Gard, G. (2000). Computer use in cold environments. *Applied Ergonomics*, **31**, 239–245.

Bowers, C.A., Jentsch, F., Salas, E. and Braun, C.C. (1998). Analyzing communication sequences for team training needs assessment. *Human Factors*, **40**, 672–679.

Buchholz, B., Paquet, V.L., Punnett, L., Lee, D. and Moir, S. (1996). PATH: A work sampling-based approach to ergonomics job analysis for construction and other non-repetitive work. *Applied Ergonomics*, **27**, 177–187.

Burns, C.M. and Vicente, K.J. (2000). A participant-observer study of ergonomics in engineering design: how constraints drive design process. *Applied Ergonomics*, **31**, 73–82.

Button, G. and Harper, R.H. (1993). Taking the organization into accounts. In G. Button (ed.), *Technology in Working Order: Studies of Work, Interaction, and Technology* (London: Routledge), pp. 98–107.

Carletta, J., Anderson, A.H. and McEwan, R. (2000). The effects of multimedia communication technology on non-collocated teams: a case study. *Ergonomics*, **43**, 1237–1251.

Cayless, S.M. (2001). Slip, trip, and fall accidents: relationship to building factors and use of coroners' reports in ascribing cause. *Applied Ergonomics*, **32**, 149–154.

Clark, M.C., Czaja, S.J. and Weber, R.A. (1990). Older adults and daily living profiles. *Human Factors*, **32**, 537–550.

Connell, I.W. (1998). Error analysis of ticket vending machines: comparing analytic and empirical data. *Ergonomics*, **41**, 927–962.

Cook, R.I. and Woods, D.D. (1996). Adapting to new technology in the operating room. *Human Factors*, **38**, 593–613.

Corlett, E.N. (1995). The evaluation of posture and its effects. In J. Wilson and E.N. Corlett (eds), *Evaluation of Human Work* (London: Taylor & Francis), 2nd ed., pp. 662–714.

CPSC (1975). *In-Depth-Investigations* (Washington, D. C.: Consumer Product Safety Commission).

Degani, A. and Wiener, E. (1993). Cockpit checklists: Concepts, design, and use. *Human Factors*, **35**, 345–360.

Degani, A. and Wiener, E. (1997). Procedures in complex systems: The airline cockpit. *IEEE Transactions on Systems, Man, and Cybernetics — Part A: Systems and Humans*, **27**, 302–312.

Drury, C.G. (1987). A biomechanical evaluation of the repetitive motion injury potential of Industrial jobs. *Seminar in Occupational Medicine*, **2**, 41–49.

Drury, C.G. (1998). Case study: error rates and paperwork design. *Applied Ergonomics*, **29**, 213–216.

Drury, C.G., Ma, J. and Woodcock, K. (2002). *Measuring the Effectiveness of Error Investigation and Human Factors Training*, FAA Office of Aviation Medicine.

Drury, C.G. and Wick, J. (1984). Ergonomic applications in the shoe industry. Paper presented at the *Proceedings of the 1984 International Conference on Occupational Ergonomics*, Toronto.

Flynn, E.A., Barker, K.N., Gibson, J.T., Pearson, R.E., Smith, L.A. and Berger, B.A. (1996). Relationships between ambient sounds and the accuracy of pharmacists' prescription-filling performance. *Human Factors*, **38**, 614–622.

Gaudart, C. (2000). Conditions for maintaining aging operators at work—a case study conducted at an automobile manufacturing plant. *Applied Ergonomics*, **31**, 453–462.

Gilbreth, F. and Gilbreth, L. (1919). *Applied Motion Study* (London: Sturgis and Walton).

Gravelle, M., Cohen, S.M., Wilson, K.S. and Bisantz, A.M. (1996). May I take your order? Human factors field studies in the fast food environment. *Proceedings of the 40th Annual Meeting of the Human Factors and Ergonomics Society*.

Haddon, W. (1973). Energy, damage and the ten countermeasure strategies. *Human Factors*, **15**, 355–366.

Hale, A.R. and Hale, M. (1972). *A Review of Industrial Accident Research* (London: National Institute of Industrial Psychology, Committee on Safety and Health at Work).

Hammersley, M. and Atkinson, P. (1983). *Ethnography Principles in Practice* (London: Tavistock Publications).

Hong, S., Kim, D., Kritkausky, K. and Rashid, R. (1998). Effects of imitative behavior on seat belt usage: Three field observational studies. Paper presented at the *Proceedings of the Human Factors Society 42nd Annual Meeting.*

Jentsch, F., Barnett, J., Bowers, C.A. and Salas, E. (1999). Who is flying this plane, anyway? What mishaps tell us about crew member role assignment and air crew situation awareness. *Human Factors,* **41**, 1–14.

Kemmlert, K. and Lundholm, L. (2001). Slips, trips and falls in different work groups — with reference to age and from a preventative perspective. *Applied Ergonomics,* **32**, 149–153.

Kim, S., Drury, C.G. and Lin, L. (2001). Ergonomics and quality in paced assembly lines. *Human Factors in Ergonomics and Manufacturing,* **11**, 1–6.

Kirvesoja, H., Vayrynen, S. and Haikio, A. (2000). Three evaluations of task-surface heights in elderly people's homes. *Applied Ergonomics,* **31**, 109–120.

Kjellberg, K., Johnsson, C., Proper, K., Olsson, E. and Hagberg, M. (2000). An observation instrument for assessment of work technique in patient transfer tasks. *Applied Ergonomics,* **31**, 139–150.

Knight, G. and Noyes, J. (1999). Children's behavior and the design of school furniture. *Ergonomics,* **42**, 747–760.

Konz, S. and Johnson, S. (2000). *Work Design: Industrial Ergonomics* (Scottsdale, AZ: Holcomb Hathaway).

Kritkausky, K. (1999). *The Effect of Imitative Behavior and Gender on Seat Belt Use* (Buffalo, NY: University at Buffalo).

Latko, W.A., Armstrong, T.J., Foulke, J.A., Herrin, G.D., Rabourn, R.A. and Ulin, S.S. (1997). Development and evaluation of an observational method for assessing repetition in hand tasks. *American Industrial Hygiene Association Journal,* **58**, 278–285.

Lavender, S., Conrad, K.M., Reichelt, P.A., Meyer, F.T. and Johnson, P.W. (2000). Postural analysis of paramedics simulating frequently performed strenuous work tasks. *Applied Ergonomics,* **31**, 45–58.

Lincoln, Y.S. and Guba, E.G. (1985). *Naturalistic Inquiry* (Newbury Park, CA: Sage Publications).

Mackay, W.E. (1999). Is paper safer? The role of paper flight strips in air traffic control. *ACM Transactions on Computer–Human Interaction,* **6**, 311–340.

Mackenzie, C.F., Jefferies, N.J., Hunter, W.A., Bernhard, W.N., Xiao, Y., LOTAS-Group, and Horst, R.L. (1996). Comparison of self-reporting of deficiencies in airway management with video analyses of actual performance. *Human Factors,* **38**, 623–635.

Mattila, M., Karwowski, W. and Vilkki, M. (1993). Analysis of working postures in hammering tasks on building construction sites using the computerized OWAS method. *Applied Ergonomics,* **22**, 43–48.

Miles, M.B. and Huberman, A.M. (1984). *Qualitative Data Analysis* (Beverly Hills, CA: Sage Publications).

Mirka, G.A., Kelaher, D.P., Nay, D.T. and Lawrence, B.M. (2000). Continuous Assessment of Back Stress (CABS): A new method to quantify low-back stress in jobs with variable biomechanical demands. *Human Factors,* **42**, 209–225.

Monteau, M. (1977). *A Practical Method for Investigating Accident Factors* (Luxembourg: Commission of the European Communities).

Mumaw, R.J., Roth, E.M., Vicente, K.J. and Burns, C.M. (2000). There is more to monitoring a nuclear power plant than meets the eye. *Human Factors,* **42**, 36–55.

Nardi, B.A. (1997). The use of ethnographic methods in design and evaluation. In M. Helander, T.K. Landauer and P.V. Prabhu (eds), *Handbook of Human Computer Interaction* (Amsterdam: Elsevier Science-North Holland), pp. 361–367.

Obradovich, J.H. and Woods, D.D. (1996). Users as designers: how people cope with poor HCI design in computer-based medical devices. *Human Factors,* **38**, 574–592.

Owens, D.A. and Sivak, M. (1996). Differentiation of visibility and alcohol as contributors to twilight road fatalities. *Human Factors,* **38**, 680–689.

Paquet, V.L., Punnett, L. and Buchholz, B. (2001). Validity of fixed-interval observations for postural assessment in construction work. *Applied Ergonomics,* **32**, 215–224.

Paquet, V.L., Punnett, L. and Buchholz, B. (1999). An evaluation of manual materials handling in highway construction work. *International Journal of Industrial Ergonomics*, **24**, 431–444.

Patel, S., Drury, C.G. and Lofgren, J. (1994). Design of workcards for aircraft inspection. *Applied Ergonomics*, **25**, 283–293.

Powell, P.I., Hale, M., Martin, J. and Simon, M. (1971). *Two Thousand Accidents*. (London: National Institute of Industrial Psychology).

Rasmussen, J., Pejtersen, A.M. and Goodstein, L.P. (1994). *Cognitive Systems Engineering* (New York: Wiley and Sons).

Sachs, P. (1995). Transforming work: Collaboration, learning, and design. *Communications of the ACM*, **38**, 37–44.

Salterio, S. (1996). Decision support and information search in a complex environment: Evidence from archival data in auditing. *Human Factors*, **38**, 495–505.

Salvendy, G. (1998). *Handbook of Human Factors* (New York: John Wiley & Sons).

Sanderson, P. and Fisher, C. (1994). Exploratory sequential data analysis. *Human Computer Interaction*, **9**, 251–317.

Sanderson, P., Scott, J., Johnston, T., Mainzer, J., Watanabe, L. and James, J. (1994). MacSHAPA and the enterprise of exploratory sequential data analysis (ESDA). *International Journal of Human-computer Studies*, **41**, 633–681.

Seagull, F.J. and Sanderson, P.M. (2001). Anesthesia alarms in context: an observational study. *Human Factors*, **43**, 66–78.

Stal, M., Hansson, G.A. and Moritz, U. (1999). Wrist positions and movements as possible risk factors during machine milking. *Applied Ergonomics*, **30**, 527–533.

Stanton, N.A. and Ashleigh, M.J. (2000). A field study of team working in a new human supervisory control system. *Ergonomics*, **43**, 1190–1209.

Suchman, L. and Trigg, R.H. (1991). Understanding practice: Video as a medium for reflection and design. In J. Greenbaum and M. Kyng (eds), *Design at Work: Cooperative Design of Computer Systems* (Hillsdale, NJ: Lawrence Erlbaum), pp. 65–89.

Taylor, F.W. (1911). *The Principles of Scientific Management* (New York: Norton and Company).

Taylor, J.C. and Felten, D.F. (1993). *Performance by Design* (Prentice Hall).

Tippett, L.C.H. (1934). Statistical Methods in Textile Research. *Shirley Institute Memoires*, **13**, 35–93.

Vicente, K.J. (1999). *Cognitive Work Analysis*. (Mahwah, NJ: Erlbaum).

Watts, J.C., Woods, D.D., Corban, J.M., Patterson, E.S., Kerr, R.L. and LaDessa, C.H. (1996). Voice loops as cooperative aids in space shuttle mission control. Paper presented at the *Computer Supported Cooperative Work '96*, Cambridge, MA.

Wenner, C. and Drury, C.G. (2000). Analyzing human error in aircraft ground damage incidents. *International Journal of Industrial Engineering*, **26**, 177–199.

Woodcock, K. and Smiley, A. (1999). *Developing Simulated Investigations for Occupational Accident Investigation Studies*. Toronto, CA.

Woods, D.D. (1993). Process tracing methods for the study of cognition outside of the experimental psychology laboratory. In G.A. Klein, J. Orasanu, R. Calderwood and C.E. Zsambok (eds), *Decision-Making in Action: Models and Methods* (Norwood, NJ: Ablex Publishers), pp. 228–251.

Xiao, Y., Hunter, W.A., Mackenzie, C.F., Jefferies, N.J., Horst, R.L. and LOTAS-Group (1996). Task complexity in emergency medical care and its implications for team coordination. *Human Factors*, **38**, 636–645.

Yin, R. (1989). *Case Study Research: Design and Methods* (Newbury Park, CA: Sage Publications).

Zuboff, S. (1987). *In the Age of the Smart Machine* (New York: Basic Books).

Chapter 4

Participative assessment

Murray A. Sinclair

Characterising the ergonomist's tasks

Assume you have to carry out some design project. Considering only the information and knowledge that you need to do this, we can picture it as in Figure 1, as an 'egg' of knowledge, divided as shown.

The important thing about the representation is that it emphasises the need to gain knowledge as the project continues, and after it has entered service; almost never are you in a position to say, 'I know enough'. Of course, this applies not only to design projects, but to almost all projects.

Participative methods are typically used in the centre and right-hand divisions in the egg. They are used whenever it is necessary to ask people for their opinions; for example, 'What should this product do for you?', or 'Why is this version better than that one?'. Since most product design is iterative, these questions and other similar ones will recur often.

The next general point to make is to characterise the activities of ergonomists. Let us take an example familiar to us all — the educational process. Imagine you have been asked to improve the efficiency with which statistics knowledge is transferred to students on a given course. A simple model of this might be Figure 2.

Given that this represents the process, you as ergonomist might have three roles to play:

- *Explorer*: your first task is to show that this model does in fact represent what happens. One way in which you might wish to gain relevant information is by talking to people involved in the current system.
- *Optimiser*: knowing the goals of the system, your task is to optimise the flow down the various channels to ensure both more efficient learning and control of the learning process. For evaluation purposes, you might wish to talk to those involved about their perceptions of the changes you have made.
- *Innovator*: you introduce new channels of flow to improve even more the efficiency of the system — see Figure 3. Again, you might wish to evaluate the changes by talking to people involved.

For each of these activities you will have to gain knowledge. This chapter is about a class of methods for accomplishing this, known as 'Participative Methods'.

What are participative methods, and why use them?

The common characteristic among the methods is the environment in which they operate. This is shown in Figure 4. The generic problem faced by anyone in obtaining the relevant

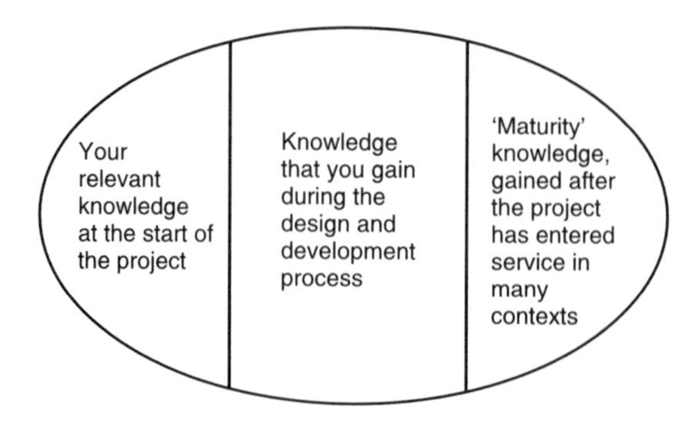

Figure 1 Representation of knowledge required for a design project.

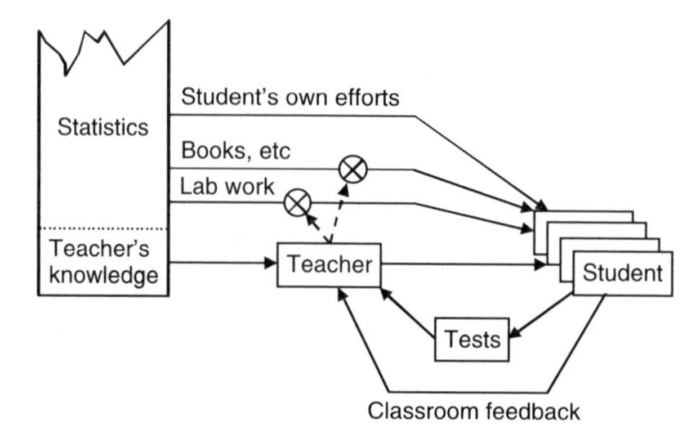

Figure 2 A simple model of a system for transferring knowledge to a student. The dotted lines represent the teacher's ability to control these information flows.

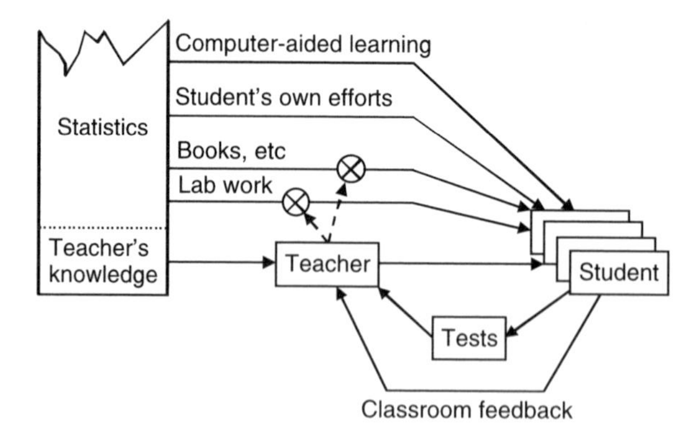

Figure 3 The learning model with innovation: computer-aided learning has been added.

information is that the query must be transmitted to the respondent in such a manner that the respondent can understand what is required. Then the respondent must delve into long-term memory to retrieve the information trace required. This may not be a textual memory; it could be a sensation of comfort or of anger, or an image. The respondent must then reconstruct and analyse the situation from the memory trace to find the relevant

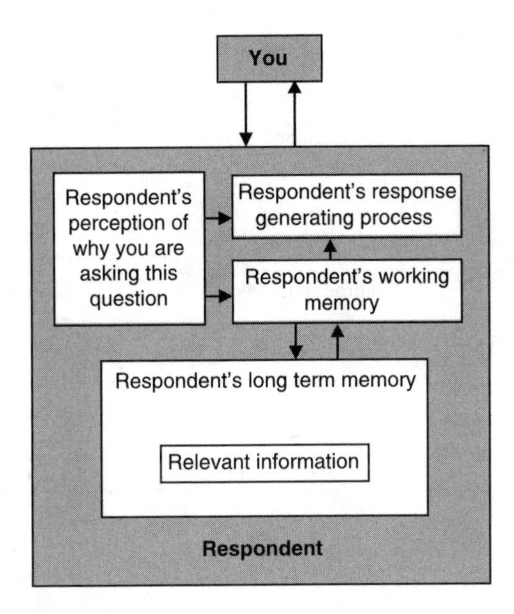

Figure 4 Representation of the information-seeker's basic problem. In most cases information sought requires the respondent to access long-term memory, retrieve the information and then report it accurately to the information-seeker.

information, and then create a suitable form of words and utter them. In each of these steps it is possible for errors or bias to creep in.

There are many methods for gleaning participative information, and there are several classifications of them by ergonomists. Edwards (1973) offers one based on the intrinsic nature of the measurement, another is given by Alluisi (1975) based on the purpose of the measurement exercise, and a third is by Meister and Rabideau (1965) based on the source of the data. An edited version of the last is given in Figure 5.

The first class, 'Observational methods', includes most of the methods that are derived from those used in the so-called 'hard' sciences, such as physics. The thing in common among these methods is that some degree of formal 'objective' measurement is involved; the idea is that you are reaching out to touch 'reality' as directly as possible. It should be noted however, that while you may measure accurately what you decide to measure, what you decide to measure may not be a good measure of reality.

The second class constitutes the historical class, where records of one sort or another are created and stored. The main problems here are firstly, that in many cases the information is not collected for the kinds of purpose that ergonomists have in mind and therefore important types of data are either not collected, or are lost during data conflation. Second, where you are consulting expertise, it is often the case that because the expertise is intended to be generic for a whole class of problems, it is good on rules and standard procedures, but poor on specific facts and specific situations. Third, especially where magazines, newspapers and the Internet are concerned, their immediacy and the lack of moderators means that scepticism is a useful approach when considering these sources.

The third class constitutes the participative methods. The common thread here is to use as measuring instruments the people involved in the system that you wish to study. In effect, you rely on the system stakeholders to come to some sort of conclusion about the system and its behaviour, and then access that conclusion as a measurement of the system. This class contains any method that draws its data from the psychological contents of people's heads.

Observational methods	Experimental methods	Laboratory experiments
		Simulations & games
		Field experiments
	Observational studies	Continuous observation
		Time-sampled observation
		Motion studies
		Checklists
		Unobtrusive methods
Database methods	Internet sources	Quality-checked
		Others
	Paper sources	Journals, books
		System records
		Investigative reports
		Newspapers, magazines
Subjective methods	Question methods	Questionnaires
		Interviews
		Focus groups
		Panels
	Protocol analysis	Think-aloud
		Recorded conversation
		Retrospective
	Scaling methods	Magnitude estimation
		Pair comparisons
		Rankings & ratings
		Others
	Others	(lots)

Figure 5 Classification of data-gathering methods, derived from Meister and Rabideau (1965). It should be noted that class boundaries in this scheme are fuzzy.

This chapter is concerned with the last of these classes. This particular class of participative measurements is important for a number of reasons:

- There are size limits to what can be measured 'objectively'. Objective measurement usually requires the use of expensive equipment, with people to maintain and use it. Very rapidly, one runs out of resources as the scope of the measurement problem increases.
- There are type limits as well. Many types of information can be measured objectively, as we all know, but there are some that cannot be — at least, not easily. For example, how would one measure the skills involved in figure-skating without using human judges? Rather more germane to our problems, we often wish to assess human decision-making skills. For this, one needs to know what the decision-maker was thinking, not just what data were presented and what the resulting actions were. This kind of information is not accessible without participative methods.
- As a sage once remarked, 'Seeing is believing, but experience gives you an original truth'. The point of this is that people who are part of a system (for example, the 'teacher' and the 'student' in Figure 2) will have different views of the system, both from each other and from an external observer. Furthermore, their perceptions will almost certainly be deeper, more complex and more subtle than those of the external observer as far as their local area of the system is concerned, and therefore will constitute a valuable source of information.

- There are accuracy limits. In many cases, instruments can be made more sensitive, more reliable, more precise and more accurate than humans, but this is not always so, especially where qualitative judgement is concerned, or where a particular measurement is multivariate. Taste is a good example; it is not for nothing that most food and beverage companies allocate considerable resources to taste panels. Certainly, there are problems of human bias, but the participative methods are intended to reduce, if not nullify, such sources of error.
- Finally, validity. Irrespective of the methods used, we would like to have valid data. The most common approach to try to ensure this is to use the notion of 'Convergent Validity'. The principle is that if you use two different, independent methods to get data about the same topic, and both produce the same (or nearly the same) results, then it is likely that the data are valid. Participative methods comprise an independent group of measures, to complement objective measures or historical records.

The arguments in favour of participative measurement thus lie in the independence of the measures and the ability to acquire data that cannot be obtained easily by other methods. The two main arguments against the use of such methods are firstly the inherent biases in human judgement, and secondly the resource requirements necessary to get reliable data. These two arguments will be considered later, after a discussion of some of the basics of participative assessment.

Discussion of the methods

Theoretical details regarding the various classes of method listed above are not discussed here; whole books are devoted to these methods and for such details you might wish to consult some of the texts listed in the references, such as Birnbaum (1998) and Nunnally and Bernstein (1994).

Instead, there is an outline of the practical principles of the method and a brief discussion of the characteristics of the techniques. In case there is some difficulty with the jargon in this section, some definitions follow:

Entity	The objects that are going to be scaled. For instance, if you wished to assess cookers for ease of use, each cooker that is assessed would be an entity. Similarly, if you assess people for intelligence, each person would be an entity. If you are assessing statements typed on cards, each statement is an entity.
Attribute	The property of the entity that you are scaling. For ease of use of cookers, you might be scaling the 'ease of controlling temperature'. This is the attribute that you are assessing, and it is the extent to which each cooker possesses that attribute that you are measuring with your scale.
Participant	The people that you use to do the scaling. In the cooker example, you might recruit people off the street as 'lay persons' to try the cookers and register their opinions. These people are the participants. If these people are assessing other people (for instance politicians), the latter would be entities, from the definition above.
Respondent	Usually used in connection with questionnaires. Again, they are the people who give their opinions which are recorded on the questionnaire, and thus are participants, from the definition above.
Judge	A participant used for special purposes, usually for the creation of the scales in the first place.

Magnitude estimation

This is a common method among those investigating sensory information, for example in design projects. It is important where the intention is to provide 'delight' and 'emotional benefit' from the customer's experience with the product (in other words, in situations where measurement of the senses and emotions are important). A typical scenario might be where the designer wants to know what surface texture would be good for a given product. We assume that there are six candidate textures, and six prototypes are available, each with a different texture. One of these is deemed the 'standard', and is allocated, say, 50 points. Each person from a sample of likely customers is then presented with the standard, and is told it is worth 50 points. Then each of the other prototypes is presented in turn, for comparison with the standard. The person responds by giving a ratio of how much better, or worse, the texture is. If, for example one of them is thought to be twice as good, the response would be '100 points'; if only a third as good, then the response would be '17 points'. An average value (using the Geometric Mean) can be obtained for each of the textures from the points allocated to it by the whole sample. A number of checks can then be carried out to ensure that the assumptions of the method have not been compromised. Good sources of information will be found in Lodge (1981) and Gescheider (1997); however, both of these authors are more interested in academic explanation than in usage. You will have to do some process design work to make the methods work.

Characteristics

- The theory underlying this method embraces wholeheartedly the Power Law, which makes no allowances for social aspects, nor of the context in which the exercise takes place.
- Lay people who act as participants need some training in using the method; it is not an assessment which they are used to making, and need a while before they feel comfortable with it.
- The method is said to be very good at separating entities which tend to be at one extreme or the other of a given scale, such a texture. However, it does not tell you whether the 'best' texture is actually 'delightful'.

Ranking methods

These methods are concerned with questions of the type:

> 'Given four typefaces, [Courier, Helvetica, Palatino, Times], rank them from first to last for ease of reading.'

What you obtain are data that distinguish between the examples quite well, and are easily handled by statistical methods, but do not tell you for example whether the best is actually easy to read.

Details of ranking methods will be found in Birnbaum (1998) and Nunnally and Bernstein (1994). The analysis of the rankings produced is considered in many texts of non-parametric statistics; one of the best in the author's opinion is Siegel and Castellan (1995). A single question as in the example about typefaces earlier is rare; usually, there is a series of questions forming a 'battery', probing a number of different attributes of the situation in question. These batteries are usually used in experimental situations, where the participants have a number of different alternative entities to rank. Entities are usually physical objects, but there is no reason why people, software products and suchlike should not be included. The essentials of the method involve presenting the entities to the respondent in such a way that the attribute in which you are interested can be assessed. This may involve user trials, etc. The entities should be presented in a random order for each participant involved, for experience. The respondent should then be presented with all the entities together, and be asked to rank them. If there is a large number of entities to be ranked, it is sometimes better to ask the respondent to rank the best ones, then the worst, and then the ones in the middle.

Characteristics

- For each attribute to be ranked in the battery, each entity should possess the attribute in some degree. While this is an obvious statement, it can be difficult to establish in practice.
- It may well be the case that for several of the entities, a participant may not have any real preferences. The ranking process disguises this for an individual participant, and its occurrence can only be detected by using Coombs' Unfolding Technique (or some derivative — see Nunnally and Bernstein, 1994) or by using many participants and examining the results statistically. It is commonly accepted that ranking methods only record real preferences for the first two or three ranks, and the last two or three ranks. In between, the rankings are possibly unreliable; hence it is unwise to give too many objects to a participant for ranking; about nine is usually taken to be the upper limit.

Rating methods

These methods deal with questions of the type:

> Given the same four typefaces, rate each one in turn for ease of reading on the 5-point scale in Figure 6.

These methods will provide information on the perceived ease of reading of the examples, but are less sensitive to differences between them.

Rating methods have been studied extensively since the early 1900s. The literature is replete with texts describing exhaustively the details of the various methods for generating scales. The most commonly-used methods are:

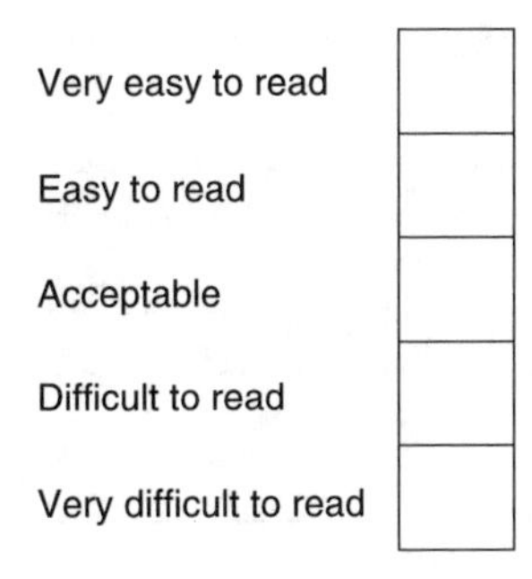

Figure 6 Example of a simple rating scale, in vertical format. The scale is shown as a series of boxes, each one with a label. See also Figure 7; another example.

- Simple rating scales (as above)
- Thurstone's Paired Comparisons Technique
- Thurstone's Equal-appearing Intervals Method
- Likert's Summated Ratings Method

A good introductory text to these techniques is Oppenheim (1992). A more technical text is Nunnally and Bernstein (1994).

Simple rating scales

Simple rating scales are usually depicted as shown in Figure 7. The 'parking bays' representation is typical.

The method requires appropriate questions to be generated about the attributes of the entities involved, typically in the form of a battery of questions, as for rankings. Scales are then created for these questions. The important points here are that the researcher must decide whether it is necessary to have a genuine 'neutral' region in the middle of the scale (the example does not have one), and how the two ends of the scale shall be 'anchored' (i.e., given unambiguous labels that will be interpreted the same way by the majority of users of the scale). The rating scale itself is usually shown as a 100 mm line; subdivisions are optional, as are labels for the subdivisions. As for the ranking methods, the respondents should be presented with the entities in a random order, and should be given an opportunity to experience the attribute of interest. The respondents should then place a mark on the scale to indicate their opinions.

Characteristics of rating scales

- This is the most common technique used by ergonomists. This is because of its ease of use, particularly for respondents.
- You must take care about the meanings of the scale anchor points and the labels used along it. It is not easy to establish clear anchor points, nor is it easy to get good labels; if in doubt, leave them off. It appears that almost all respondents are good at dividing the line into equal participative intervals, and in any case, if you use a 100 mm line, the usual way to get the rating converted to a numeric value is to measure to the nearest cm. This unit of measure will encompass most of the variability shown by respondents.
- There are a number of biases that can affect rating scales. A list will be found in Spector (1991). The worst of these is the 'leniency' effect, where respondents are unwilling to be critical. A second problem is the 'halo' effect, where the respondent has already decided that one of the entities is better than the rest, and subconsciously 'adjusts' his or her ratings to demonstrate this clearly.

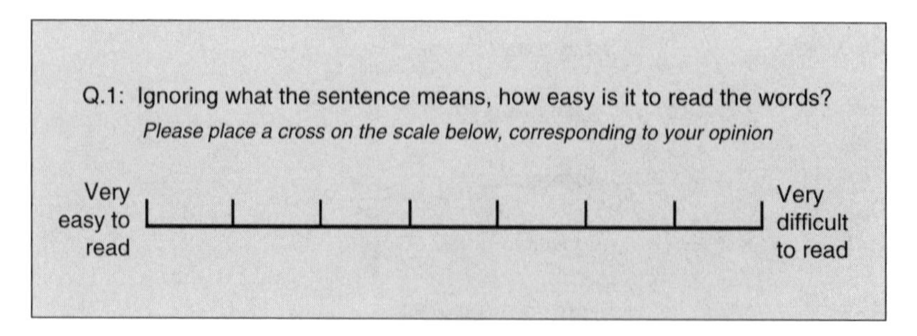

***Figure* 7** Another example of a simple rating scale. It features the question to the respondent, instructions to the respondent, and the scale, as the standard 100 mm line; in this case with tick marks. Scale 'anchors' are also given at each end, but there are no intermediate labels.

Paired comparisons technique

The Paired Comparisons technique, described in Nunnally and Bernstein (1994), is the 'gold standard' technique, against which other scaling methods are compared. This is because more data are collected than necessary to create the scale, and the 'extra' data can then be used for internal checks of validity. In this technique, participants are typically asked to compare two entities, A and B, and make a decision whether A is 'better than' or 'worse than' B. The entities are taken two at a time from a range of entities provided by the researcher, and the judgements are recorded for each possible pair. By making some statistical assumptions about the nature of human judgements, and using a pool of participants, it is possible to derive a quantitative scale from these simple judgements, with the positions of the entities marked along it. For example, using the four typefaces judged by each of ten judges individually for ease of reading, we might generate the frequency matrix of Figure 8.

These frequencies are then transformed into probablilities, from which Normal deviates (*z*-values) are calculated, to produce Figure 9.

Averages for each column are obtained, in Figure 10. These represent the best estimate of where these typefaces fall on an 'ease of reading' scale. Note that this scale is now at an Interval level of measurement, where we started with just ordinal judgements by the participants — a handy property of this technique. However, all we know is the relative readability of these typefaces, not their absolute readability. However, this can be achieved (i.e., 0.00 on the scale really does mean 0.00) by carrying out a regression analysis after asking people 'do you like to read this typeface?'

If instead of objects we use verbal statements, we can create a scale for measuring other objects, to be used by participants in a field situation, where each statement has a numerical value. For example, we may have created a scale for sitting comfort. A participant would then sit at a workstation, for example, and assess how comfortable he or she feels, then select a statement from the scale set which best represents that assessment. The numerical value for this statement is then allocated to the workstation seat.

	Courier	Helvetica	Palatino	Times
Courier		4	3	4
Helvetica	6		5	0
Palatino	7	5		2
Times	8	10	8	

Figure 8 A 'Frequency matrix', capturing the results of the two-by-two trials for the 10 subjects. Main diagonal is empty; no point in comparing a typeface with itself. In the first column, the 6 in the 'Courier/ Helvetica' cell means 6 of the 10 people judged Courier more easy to read than Helvetica.

	Courier	Helvetica	Palatino	Times
Courier		−0.25	−0.52	−0.25
Helvetica	0.25		0.00	−3.00
Palatino	0.52	0.00		−0.84
Times	0.25	3.00	0.84	

Figure 9 A 'Z matrix', in which the frequencies have been transformed into Normal distribution z-values. Note that this transformation fails when everybody agrees that one typeface (e.g., Helvetica) is better than the other (e.g., Times). One fix, among several, is to insert the value 3.00, as above.

	Courier	Helvetica	Palatino	Times
Courier		−0.25	−0.52	−0.25
Helvetica	0.25		0.00	−3.00
Palatino	0.52	0.00		−0.84
Times	0.25	3.00	0.84	
Average	0.26	0.69	0.08	−1.02

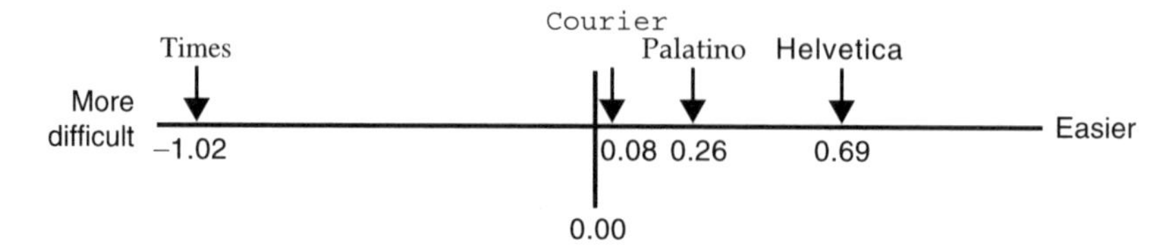

Figure 10 Column averages have been calculated, and these have been plotted on an 'Ease of reading' scale, at an interval level of measurement.

Characteristics of paired comparisons

- During creation of the scale, the judgement required of participants is a relative judgement, and does not require the participant to assess 'by how much' one is greater than the other. This is a simple judgement, which most participants find fairly easy to make about almost anything.
- Under normal circumstances, more data than strictly necessary are collected. This 'overdetermination' allows a number of internal checks for consistency to be applied to the resulting scale.
- The number of judgements required per participant rises by a factor of $[n(n-1)/2]$ as n, the number of entities, rises. This can require. There are penalties in this; a bored participant towards the end of the session is a very different person from the keen, slightly apprehensive, participant who started the session. There are ways of reducing the number of judgements required; see, for example, Trawinski *et al.* (1985).
- The scale that results is usually taken to be uni-dimensional, but there is no guarantee of this (for example, the comfort scale might be made up of two scales, one that measures ease and well-being, and other which measures localised pain in the rump). Small departures from uni-dimensionality may not be detected by the internal checks.
- Because of the resource problems it is commonly accepted that there should not be more than nine entities to be compared.
- The success of the scale hinges on the selection of entities. If the attributes being assessed are not commensurable or if the entities are too clearly dissimilar to each other, the scale will be invalid. In the case of scales of statements, the choice of statements is critical. One requires a set of statements that cover the range of the scale, each of which has a relatively fixed point on the scale. Most of the work should be allocated to this initial choice of statements.
- Care must be taken to ensure that the group of participants used to create the scale is equivalent to the group of participants who will use the scale subsequently, otherwise the scale values allocated may not be accurate.

Thurstone's equal-appearing interval technique

The method of equal-appearing intervals produces the same scales as for Paired Comparisons, but is much quicker. However, there is less opportunity for internal checks. It was developed by Thurstone to provide a quick means of obtaining scales for use in the field, typically in questionnaires. The method emphasises the selection of statements to comprise the final scale; it assumes that a pool of statements can be created by the researcher, and provides a means of reducing this pool to the final selection. Once created, each scale is represented as a randomised list of statements, with no scale values showing. The participant using the scale is asked to select that statement which best represents his or her judgement, and the scale value for that statemnent is then used in subsequent analyses.

Characteristics of Thurstone's technique

- The scales that are produced by this technique are almost comparable with the Paired Comparisons approach for validity and reliability.
- The data produced by this technique are very easy to analyse.
- Participants can experience problems if their feelings do not quite match the statements given. This can create the impression that words are being put in their mouths, which can have fatal effects on their motivation.

Likert's summated ratings method

The Likert scaling technique is based on a very different approach, but is taken to be as powerful as the Thurstone techniques. Whereas in the Thurstone techniques the participant is asked to select a statement from a set of alternatives with which he or she most agrees, in the Likert technique the participant must respond to every statement in the set, showing his or her disagreement with it. Typically, each statement will have a simple 5-point scale associated with it as shown in Figure 11 and Figure 12.

Each 'bay' on the scale has a value (for example 1 to 5), and, using the appropriate value depending on which bay is selected, the scale values are summated and the result represents the participant's opinion. Clearly, there is a premium in the selection of statements and in ensuring that the points allocation is matched to the statement. What one seeks is that if a participant's response to one statement obtains a score of 5, then a similar score will be obtained from another statement in the set. For example, if a participant strongly agreed with the statement in Figure 11, scoring 5 points, one might expect the participant to disagree to some extent with the statement in Figure 12, again scoring high points. This enables individuals with differing viewpoints to be discriminated from each other by accumulation of these scores. Some 10 or 20 statements are used to create the scale.

Characteristics of Likert's method

- It is said that participants prefer the Likert scaling technique, because it is more 'natural' to fill in and because it maintains the participants' direct involvement.
- The Likert approach is said to require less effort to generate scales compared to the Thurstone techniques.
- There appears to be no great difference between the Likert and Thurstone techniques in validity and reliability.
- Thurstone techniques are said to be better at measuring 'snapshot' views, whereas Likert techniques are better at measuring changes over time.

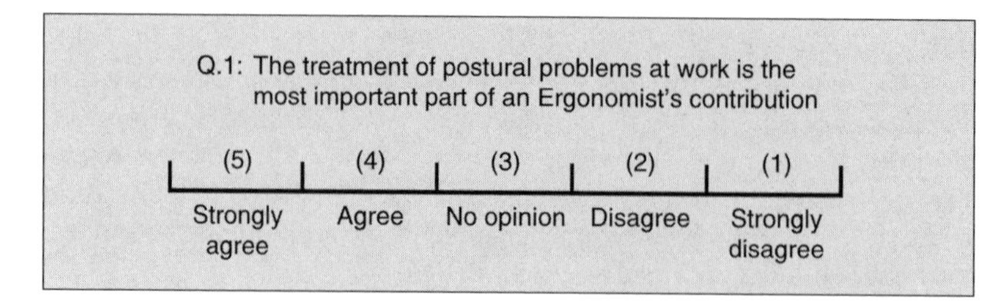

Figure 11 Example of a Likert scale item. It contains a statement with which the respondent can either agree or disagree, and indicate this opinion on the scale. Sometimes, the three scale labels in the centre are not included. The points values in brackets are never shown. They have been included here for illustration purposes only.

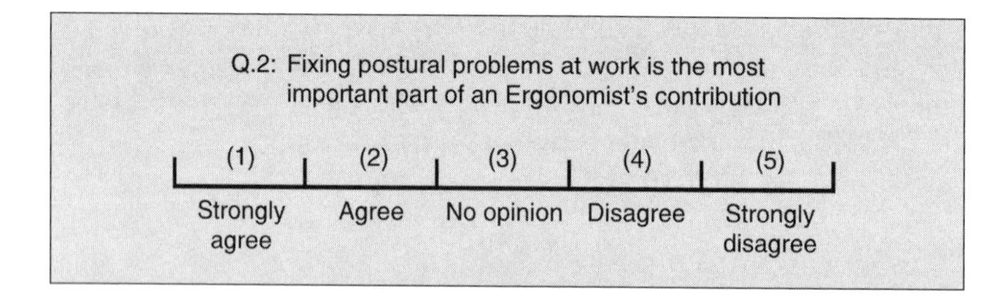

Figure 12 A second example of a Likert scale. Since the statement conflicts with the example in Figure 11, one would expect the (invisible) scoring for this item to be reversed, as shown.

More complex methods using ratings and rankings

The methods outlined above are used as described. However, it is more common to find several scales bundled together to make a battery' of scales, as mentioned earlier; it is seldom that you will wish to assess any group of entities just for one attribute alone.

Such batteries are often found in questionnaires, as discussed below. At this point, two other techniques that use batteries of scales will be mentioned.

Semantic differential technique

This technique is described in detail by its originators, Osgood *et al.* (1990), and shorter descriptions will be found in most textbooks on attitude measurement (e.g., Moser and Kalton, 1985; Oppenheim 1992) or Rea *et al.* (1997). Their original intention was to discover how groups interpreted the meaning of words; since then the technique has been used for all sorts of purposes, including evaluation of household goods.

A series of rating scales is generated which describes the class of entities to be studied. By convention, the scales are usually seven categories long. The end points of the scales are given anchors which are single word adjectives, and are 'polar opposites' (e.g., good–bad; strong–weak; fast–slow). The participant then rates each entity according to these adjectival scales. Generally, the data resulting from a group of participants carrying out such an exercise are participants to a Factor Analysis (described below), to determine what 'latent dimensions' are being used. The assumption in this is that the scales are a superficial representation of deeper underlying evaluation dimensions that can be ascertained from the adjectival scales. These dimensions may not be apparent beforehand, nor within the normal vocabulary of the participants involved, whereas the adjectival scales are fairly commonplace. The differences between this technique and the Likert technique are twofold. First, superficially in the Likert technique there is a range of statements but only one version of the scale, whereas in the Semantic Differential there are different versions of the scales but only one entity to be evaluated at a time. Rather more deeply, in the Likert technique the statements are all representations of a single underlying issue, whereas in the Semantic Differential an individual scale may be examining a unique aspect of the entity unrelated to any of the other scales.

The strength of the Semantic Differential lies in its explanatory power, in elucidating the underlying dimensions and, once these are obtained, in showing the relationships between the entities in the *n*-dimensional space constructed from these dimensions. These are not necessarily apparent before the exercise, though you, the investigator, may have some shrewd suspicions (and wrong ones), and it is this power to reveal what you might otherwise miss that is the appeal. Of course, the power depends critically on the quality of the data collected. Statistical techniques by themselves do not provide explanatory power; they deal with numbers, not with the meanings of the numbers.

Repertory grids

This technique was developed by Kelly (Fransella *et al.*, 2003). Kelly's aim was to provide a psychiatric tool to help understand an individual's representation of his or her environment. The tool has been widened in its use since the early days, and has become popular in knowledge engineering as a means of exploring experts' conceptions of their skills and knowledge; it is discussed in this context elsewhere in this book (see Chapter 8).

As with the Semantic Differential, there is an assumption that there are underlying dimensions, in this case called constructs, which are to be established. Because this is a single-participant method, the typical approach is slightly different; using the 'Method of triads'. Once some entities have been selected (let us assume we are dealing with different telephones), they are presented three at a time to the respondent (e.g., telephones A, B and F). The respondent is asked to arrange them so that two are similar and the third is different, and to state what the difference is (e.g., 'these two are small (B and F), and that one is bulky (A)'). The participant repeats this with the same triad (e.g., 'these two are modern (A and B) and that one (F) is old-fashioned'), until no further differences are found. Another triad is then presented (e.g., A, B and G) and the process is repeated, until no new differences are recorded. The more common, or most interesting, of these differences are then selected, and for each of these differences (e.g., 'Small–Bulky') the respondent ranks the entities along the implied continuum. A standard statistical technique (usually a Factor Analysis using correlations between the rankings) is used to elucidate the underlying dimensions.

Factor analysis for the exploration of latent variables

This is one of a suite of methods for analysing multivariate data — for example, a set of anthropometric measures to do with body size. The underlying assumption here is that, while each measure may be interesting in itself, there are deeper, underlying dimensions ('factors') from which the scales have arisen by linear combination. If this is so, then Factor Analysis might be able to reveal them. More detailed descriptions will be found in many statistics texts; two of them are Harris (2001) and Kim and Muller (2001).

Assume we are interested in the body size of 4-year-old males. We identify six anthropometric variables which we believe will assess this, and gather data from 1000 children (multivariate analyses require large numbers of participants). The variables are as shown in Figure 13.

If there is a hidden structure underlying these variables, then we would expect some degree of correlation among those variables which embody each of the underlying dimensions. So that's where the analysis starts, with a correlation matrix of the variables, prepared in a statistical package such as SPSS (this analysis is not something to attempt by hand).

If you look along the rows, or down the columns, you will see that some correlations are relatively high, and some are near zero. This implies that there might be an underlying structure. The usual first step is the performance of a Principal Components Analysis, to establish that there is some kind of meaningful structure of factors (or principal components). This will present you with a table such as Figure 14.

There will be as many factors as there were scales, each with an Eigenvalue (roughly speaking, this measures the strength of the underlying factor), and an associated 'per cent variance explained' value. The correlation matrix is a measure of the variance and covariance in the data, and this value indicates how much of the variance and covariance is due to this factor. Note also that these factors are 'orthogonal' to each other; there's no correlation between them.

At this point, the second stage of the analysis takes place. The identified factors are 'rotated to get a better fit', usually using the Varimax method (though other methods, for other purposes, are available). The criterion here is parsimony of explanation; the idea is to maintain orthogonality of the factors, but to try to increase the large loadings of each factor, and decrease the small loadings of each factor. The net result is that it is easier to allocate a

	Stature	Hip height	Biacromial breadth	Waist girth	Thigh girth	Chest girth
Stature		0.60	0.50	0.19	0.20	0.35
Hip height	0.60		0.47	0.22	0.19	0.30
Biacromial breadth	0.50	0.47		0.21	0.20	0.24
Waist girth	0.19	0.22	0.21		0.43	0.32
Thigh girth	0.20	0.19	0.20	0.43		0.42
Chest girth	0.35	0.30	0.24	0.32	0.42	

Figure 13 A correlation matrix, using values from Table 1 in Kim and Muller (2001). The empty cells should be filled with 'communality' values; in the absence of these, 1.00 is usually inserted.

	Factor 1	Factor 2	Communality
Stature	0.75	-0.30	0.65
Hip height	0.70	-0.27	0.56
Biacromial breadth	0.60	-0.18	0.39
Waist girth	0.43	0.36	0.31
Thigh girth	0.51	0.61	0.62
Chest girth	0.53	0.25	0.37
Eigenvalues	2.13	0.75	

Figure 14 A factor Loading matrix, from Figure 13. This table shows the 'loading' of each scale on the individual factors. Only the two most important factors are shown; there will be four more (corresponding to the number of scales), but these are 'unimportant', mapping up spare variance in the data.

	Factor 1	Factor 2
Stature	0.79	0.17
Hip height	0.73	0.17
Biacromial breadth	0.60	0.19
Waist girth	0.15	0.54
Thigh girth	0.08	0.78
Chest girth	0.31	0.50

Figure 15 Factor loadings, after Variance rotation. Compare with Figure 14; one might now label Factor 1 as a 'lengthiness' factor, and Factor 2 as a 'blobbiness' factor.

meaning to the factor. This might result in Figure 15. Now, we identify the factors, participatively, by looking at the factors and their 'loadings' on the original scales.

Questionnaires, interviews and focus groups

Methods of questionnaire, interview and focus group form part of the stock-in-trade of ergonomists; almost every project on which you work will involve one or more of these methods.

Their purposes are different; while all of them involve the extraction of information from people, it is purpose and context which will determine which method is selected. With reference to the explorer/optimiser/innovator classification discussed earlier; in the exploration phase, the problem is to create a map of the problem area. Interviews are probably the best way to commence this activity, because they allow the interviewee (who, presumably, understands the system better than you do) to assist you in guiding the interview into interesting directions. Once you have enough information about the system and its behaviour, it then becomes possible to ask questions such as 'what percentage of people have problems in controlling this thing?'. Where numerical answers are sought, and it is necessary to access large numbers of people, a fixed-format questionnaire becomes a sensible vehicle for data-gathering. For innovations, the big problem is imagination. These days, we all take mobile phones for granted. However, had you asked someone back in the 1970s to tell you what functionality he/she would like in a mobile communications device, it is highly unlikely that the person would describe the modern mobile phone. Focus groups are one of the best methods for obtaining useful information; as the name suggests, the interviewer addresses a group of people, who discuss the issues and in doing so generate ideas which none of them would have had individually.

All three techniques can be used in scenarios other than those above. However, as a general rule, you should heed the following constraints:

- Where the participant matter may arouse strong, opposing emotions it is usually best to avoid focus groups; too easily, they can degenerate into shouting matches.
- Questionnaires, while superficially appealing and able to reach large numbers of people, are not good at probing complicated issues; nor are they good where long answers are expected from respondents.
- Interviews are a very flexible medium for information gathering; they have the added advantage of producing memorable quotations that throw a completely new light on the issues. However, they are an expensive data capture form.

Of these methods, fully structured questionnaires have the most requirements in terms of preparation. For this reason, the discussion that follows deals with the preparation of paper-based, mass-distribution questionnaires; although the medium is different, they also apply to email-and web-based questionnaires (in fact, the participant is so well-understood that websites exist to create an e-questionnaire from templates; an example will be found at: www.perfectsurveys.com/home/howitworks.html).

Many of the considerations are also relevant both to focus groups and to interviews; these considerations will be made apparent in the discussion.

The appeal of questionnaires lies in their ability to obtain large amounts of information from large numbers of people, at relatively low cost, and relatively quickly. There are many texts discussing in detail the preparation, administration and analysis of questionnaires, two good texts are Oppenheim (1992) and Rea *et al.* (1997).

However, there are problems with questionnaires. Unless the information required is factual and fairly easily checked, their reliability and validity can be quite low, so considerable care must be taken. Furthermore, they do make a substantial demand on resources. Designing and administering a high quality questionnaire is a skilled task; a specialist in one of the behavioural sciences, a statistician, a computer expert and a graphic designer working as a team may be needed. Then there is the fieldwork; interviewing is a time-consuming and skilled task, requiring fairly large numbers of people. Finally, there are the analysis issues; those of correct data entry, data cleansing, and, often, trying to make sense of data from what are now seen to be poor questions (nobody has ever created the perfect questionnaire). Many questionnaires are administered to respondents by interviewers, but this is omitted from the discussion here because the topic is discussed in the Chapter.

Based on many peoples' experience, the points below indicate the necessity of careful planning of your work. These are applicable to interviews and focus groups as well:

- Data-gathering presupposes a fairly high level of inter-personal understanding, a common culture, and a common language. Your questionnaire must stay within the boundaries imposed by this.
- Clumsy presentation may lead respondents to 'close down', for example by giving minimal answers, devious answers, or by outright refusal to continue.
- The opportunities for misinterpretation are much greater than you might suppose. This is considered in more detail under the heading, 'Questionnaire construction', and is illustrated in Figure 16.
- Your respondent's knowledge may not be organised usefully, it may be limited to a small range of circumstances, and it may be wrong. It is extremely difficult to guard against these problems.
- Respondents often have only a partial knowledge of the extent of their own knowledge; this is usually the case when opinion is passed off as truth. Careful question design can ameliorate this problem, typically by the use of closed questions (discussed later).
- The most common characteristic of respondents' knowledge is that they have a good grasp of generalities and rationales, but a poor grasp of particular, necessary facts. There are no solutions to this.
- Particularly for non-verbal knowledge (e.g., driving skills), a skilled respondent may not know what the skill is, let alone be able to describe it.

Planning the questionnaire

Typically, when you administer a questionnaire to a respondent you are implicitly asking him or her to reconstruct from memory the environment or the problem area that you are interested in, and then to pluck from it the information or opinions that you require. This is not

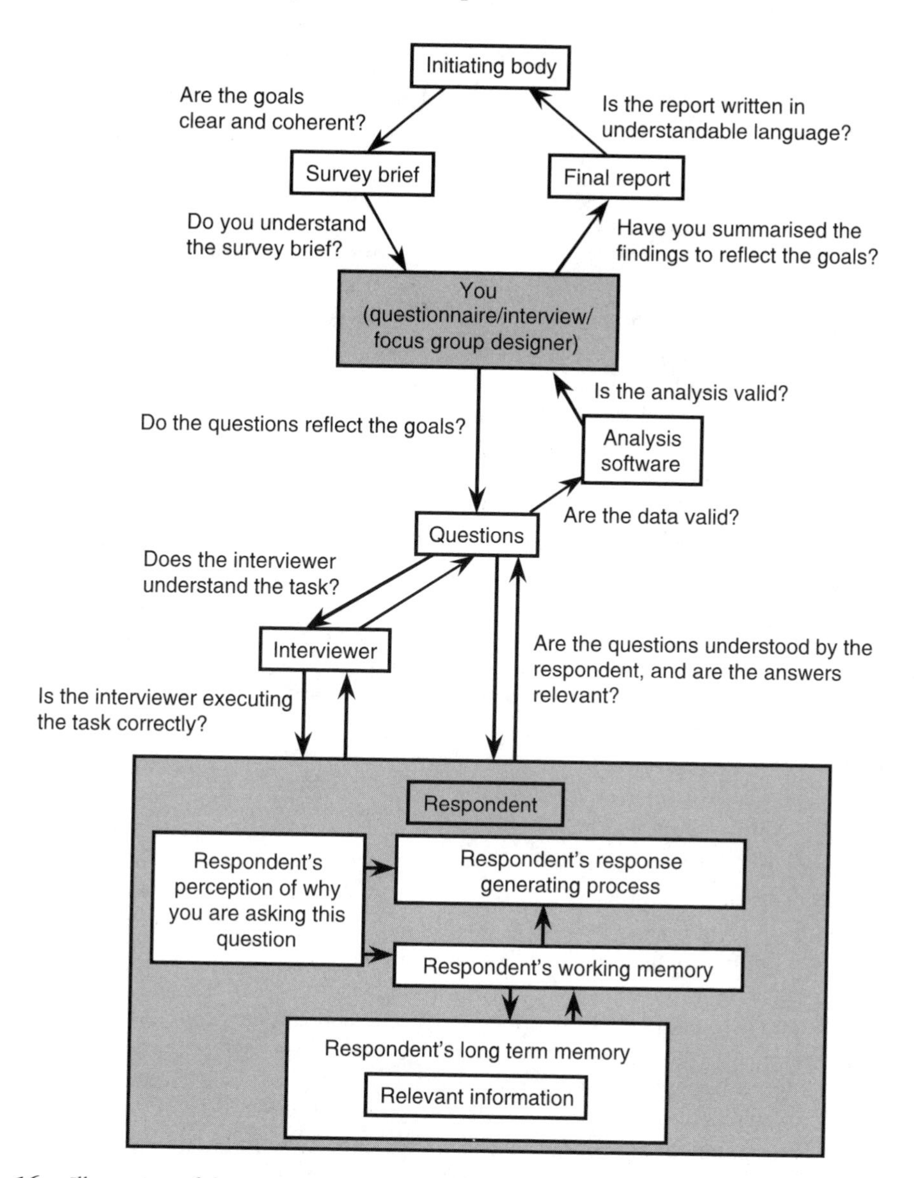

Figure 16 Illustration of the communication problems in questionnaire design. The arrows represent communication process, and the questions associated with them indicate some of the problems that can occur. It should be noted that once you have designed your questionnaire, any errors are then 'frozen in' and cannot be retrieved. Redrawn from Sinclair (1975).

necessarily an easy task (how much detail can you recall of the first pedestrian you encountered on your journey to work today?) and it requires a careful, methodical approach to make your efforts worthwhile. The steps are discussed briefly below.

Definition of objectives and resources

This is the most essential and hardest step. First, you must define your objectives *in detail.* For example, it is not sufficient to have as your objective: 'Find out about accidents on grinding machines'. You should be able to define precisely what you mean by 'accident' (does it include damage to clothing? Does it include near-accidents? What about minor accidents such as bruised fingers, which are almost never reported?), and what range of machines is covered by 'grinding machines' (all of them, or just the 'common' ones? What does 'common' mean?). At this point, there are three important questions to be answered:

- What are the results supposed to show?
- What level of accuracy is required?
- What additional data will be required to link this survey with other data sets?

This stage must be done thoroughly; down to deciding the form of the questions you will be asking. As a rule of thumb, you will have spent enough time and effort at this stage when the questions virtually write themselves, and the format of the final report is clear. Another rule of thumb is that between a third and one-half of the total time available should be spent on this first stage.

There is no substitute for this part of the design work. No matter how sophisticated you are from here on, if you start with fuzzy thinking you will continue with fuzzy questions, and however ingenious the analyses, at best you will finish with sophisticated fuzzy answers, and a fairly clear perception of the futility of your efforts.

Where interviews and focus groups are concerned, the same rules apply. However, it may not be possible to be detailed about the questions, because of the exploratory nature of the work. But at least you should be clear about the topics to be covered, and a few starting questions.

Sampling

This step deals with whom you will survey. First, you must define the population you wish to investigate (for example, which organisations your respondents work for, and what skills and/or grades they must possess), and then decide on the sample size (which could be 100%). This is determined by the resources available and the accuracy required — see any statistics text.

The next step is to generate a 'sampling frame', which is a list of all the members of your population (e.g., constructed from personnel records), from which you then draw your sample. You should be aware that sampling frames are not always accurate, usually because the data sources are not up to date.

The aim is to eliminate any systematic bias (of course, if you have decided upon a 100% sample, this is not a problem). Bias can arise from three sources:

- Non-random sampling. For example, if the personnel records have been arranged in order of seniority, a decision to sample every tenth name in a small workforce will result in a sample with a bias towards lower levels of management.
- Bias in the sampling frame. This is usually due to the sources not being up to date.
- Non-response. In any population there are people who refuse to respond, as a matter of principle. There are also those who are never available, such as salesmen and drivers. Exclusion of these groups inevitably means introducing bias.

The chief means of overcoming bias is by random sampling. This allows you to alleviate the effects of the first two sources, but only perseverance on your part will deal with the third.

'Stratification' is often used with random sampling, to improve the representativeness of the sample. The population is arranged into strata on the basis of age groups, gender, income levels, etc. You then sample at random within the strata, but note that to do this you need this specific information about each individual in your sampling frame right at the beginning, in order to sort them into strata.

Another approach used is quota sampling. This is particularly useful where you cannot generate a sampling frame. In this case each interviewer is given a quota of people to interview, participant to them falling into certain age groups, income levels, and so on. The interviewer is then left to find the people to fill the quota. This method is open to interviewer bias, but produces about as accurate results as the stratification method.

There are also panels, usually groups of people to whom you make repeated reference over a period of time, to measure changes in opinions or judgements over time. The approach has its own problems of attrition or conditioning of panel members.

Having selected the sample, there remains the problem of ensuring maximum participation in the survey. This is important because many experiments have shown that non-responders are not typical of the total population. The main causes of this together with suggested remedies are as follows.

- Units outside the sampling frame — someone who should be in the sampling frame but is not available (e.g., dead). In this case, select more people from the sampling frame.
- Units unsuitable for interview — too old, too deaf. No relevant knowledge, or language barrier (there are generally fewer instances of this than field workers claim). In this case, select more people from the sampling frame.
- Movers. In this case the people who have moved in to replace those who moved out should be used.
- Refusals to answer. This is an important source of bias. A 'good' response rate would be between 60–80% for mail questionnaires, but to obtain this figure it is generally necessary to do follow-ups of one sort or another. E-questionnaires generally have much lower response rates, of the order of 20% or less. Hence these questionnaires are particularly open to charges of bias.
- People on holiday. In this case up to two follow-up calls should be made. After this, the return from further follow-ups is usually not justified in terms of expense.
- People who are away at the time of call. Accepted practice is to make two follow-up calls, at different times, to ensure that shift-workers and the like are included.

Questionnaire construction, and question wording

At this stage you should know specifically what questions are going to be asked. Question wording has to do with how you ask for the information. The problem is discussed below and in Oppenheim (1992) and Rea *et al.* (1997), and is exemplified in Figure 16. This applies to e-questionnaires and to focus groups as well.

If each arrow in the diagram is regarded as a process of communication, it is obvious that there is ample scope for error. It is important to realise that your only chance to control most of these errors is in the wording of the questions. Once the questionnaire is published your control over the information is almost nil. What complicates the matter further is that the wording of questions is an art, rather than a science, and consequently there are no rules, just a few guidelines. Quite apart from the problems in Figure 16, you have to ensure that at the same time:

- The respondent is motivated to respond.
- The respondent has the particular knowledge required.
- The questionnaire takes into account the respondent's limitations and personal frame of reference, so that he or she will understand easily the aim and meaning of the questions.
- The respondent has produced an adequate answer, from his or her own knowledge.

Questionnaire sequencing

Typically, a questionnaire is made up of four parts:

- A prologue, which introduces the topic to the respondent, provides any information he or she will need, and tries to motivate the respondent to answer the questions. It may also include examples of difficult questions and instructions for answering them.
- An information section, in which you ask your questions (e.g., 'How long have you worked on this machine?' 'How many of your fingers has it chopped off?').
- A classification section, in which you obtain personal data and any other background information (e.g., 'How old are you?' 'How long have you been doing this job?').
- An epilogue, which thanks the respondent for his or her efforts, and includes any further instructions, if necessary.

Within the information section, the questions should be arranged in consistent groups which follow each other in a reasonably logical way. This should be from the respondent's point of view, not from yours; it is important that the respondent should feel comfortable.

You should also consider the use of 'filter' questions. These serve two purposes: firstly, they determine whether your respondent has relevant knowledge or not (e.g., 'Have you taken a training course for . . .?'), and secondly they can guide your respondent past sections that are not applicable in his or her case (e.g., 'If you have had back pain, answer this section; if not, go to section . . .').

Note that in the case of e-questionnaires, the flexibility to do this is much greater, and more control can be exerted over progress through the questionnaire (for example, in preventing respondents from looking back at the answers to earlier questions). There is even more flexibility for interviews and focus groups; however, this may in turn produce problems of coherence for the respondent and comparability in the analysis.

Degree of structure in questions

After a while it becomes apparent that questions, irrespective of content, can be classified into certain structured classes. Two of the classes have already been discussed earlier in the chapter, ratings and rankings, and some of the sub-classes within these have been outlined. Others are: factual questions, for example asking for birthdates, gender, etc. (Question 7 in Figure 17 is an illustration); questions with mutually exclusive answers (Question 5 in Figure 17); questions with multiple non-exclusive answers (Question 6 in Figure 17); matrix questions (Question 4 in Figure 17); open questions (Question 2 in Figure 17); and closed questions (Question 1 in Figure 17). There are many more. Your questionnaires will most likely be composed of a variety of these, as in the example. This structured classification is of use subsequently in organising the analytical routines, and is the reason why web sites are able to offer a simple, cheap service for e-questionnaires.

Irrespective of the class of question, it must be structured correctly. For ratings and rankings the methods discussed above will have ensured this. It is with the other classes of questions that we are now concerned. Belson (1968) and Kalton *et al.* (1978) have a number of important, interesting and practical comments to make about the problems. The first decision that must be taken is whether the questions should be open (i.e., participants compose their own answers) or closed (participants choose an answer from a given set).

PART 1 – General

1. What are the risks in your brigade area? *Please show the distribution of risks by putting a percentage in the box. (e.g. if your brigade is a county with B-D risk, the B risk might represent 60% of the area, C 25% and D 20%*

High risk	%
A risk	%
B risk	%
C risk	%
D risk	%
E risk	%
Remote rural risk	%

2. Do you consider that there is a suitable production commercial vehicle chassis currently on the market that will meet the requirements of the fire service?

Yes ☐
No ☐

If Yes, who manufactures it?

☐

If No, what features render them unacceptable for your use?

____
____
____

3. Do your drivers receive training similar to the programmes given to police drivers?

Yes ☐
No ☐

If Yes, please explain the training procedure

____
____
____

4. How frequently are your vehicles involved in accidents?

	Total calls	Total accidents		
		On road	Off road	Fire area
1970				
1971				

5. In fighting fires in recent years, how many times have you used open water, with/without a main or portable pump?

More than 100 times per year ☐
50-100 times per year ☐
11-50 times per year ☐
0-10 times per year ☐

6. Have you had any vehicle mechanical failures in the last year?

Yes ☐
No ☐

If yes, please tick relevant boxes below

Mechanical
Engine ☐
Transmission ☐
Axles ☐

Electrical
Batteries ☐
Generators ☐
Wiring ☐

7. As brigade procedures, do crew members don BA sets on the way to a call?

Yes ☐
No ☐

Figure 17 An extract from a well-designed questionnaire (Gray 1975), illustrating different question types, the use of different typefaces for different purposes a layout to optimise both the use of space and the clarity of the questions, and the pleasing overall appearance of the document.

The advantages of closed questions are:

- They clarify the alternatives for the respondent and avoid snap responses being given.
- They reduce keying errors in analysis and eliminate the need for extra people to code the answers.

- They eliminate the useless answer (e.g., 'How long have you done this job?' 'Since I moved here').

The disadvantages are:

- It is difficult to make the alternatives mutually exclusive.
- They must cover the total response range (which presupposes that you know what answers are likely to appear — hence the importance of pilot studies).
- They create a forced-choice situation which rules out marginal or unexpected answers.
- All the alternative answers must seem equally logical or attractive, for fear of biasing the results.
- In complex or difficult questions, participants may dive for the safety and ease of the 'don't know' alternative.

The wording of questions

In framing the questions, whether for focus groups, interviews or questionnaires, the following important points must be considered:

- *Question specificity.* Your questions should be precise and unambiguous. Fortunately, if you have carried out the planning stages thoroughly, this is not likely to be a problem. If you are in difficulties, go back to planning and piloting.
- *Language.* You must use language that your population understands. You should also be aware of localised interpretations; for example, consider the word 'tea' — which in Britain may mean a 'pot of tea', 'high tea', or 'the main evening meal', depending on where in Britain you are. It is important to use short words rather than long ones, and to avoid scientific or professional jargon. In this context, look at some of the editorials in the so-called 'popular press'. As a method of communication, the style is superb, whatever you might think of the content.
- *Clarity.* The rule is, 'keep the questions short'. This rule has two useful consequences; firstly, it makes you clarify your thinking and remove unnecessarv words; secondly, it reduces the chance of overloading the respondent with too much information to digest. Avoid vague phrases such as 'on the whole', 'generally', 'normally' and 'frequently' (what do they mean?). Double-barrelled questions should be avoided, such as 'Do you suffer from headaches or stomach pains?' (what would the answer 'yes' mean?).
- *Leading questions.* Clearly these must be avoided. Obvious examples such as 'Do you agree that the policies of the present government are unfair?' (which invites the answer 'yes') are quite easy to detect. More insidious examples are questions that contain such loaded words or phrases as 'get involved in', 'student' and the like. You should be aware of questions that become leading questions because of the nature of the questionnaire. As an example of this, the question 'How many cigarettes do you smoke in a day?' may be innocuous in a questionnaire about household expenditure, but may produce different answers in a questionnaire concerned with medical matters.
- *Prestige bias.* This is a bias that can arise in questions that involve socially desirable behaviour. Thus, a question such as 'Which magazines have you looked at recently?' is likely to reveal that such journals as *The Economist* and the *Literary Review* have a considerably larger readership than is actually the case. Care is required to overcome this sort of bias; filter questions and careful wording should be used, so that low prestige answers appear equally as acceptable as high-prestige ones. In the

example above, you might introduce a filter question such as 'In the past seven days, have you had any time to read magazines?' to identify those who have not opened a magazine, but are not prepared to say so.

- *Embarrassing questions.* There is seldom an easy way to obtain correct information of a personal nature (e.g., 'do you molest children?'). If such questions must be asked, they are better asked in interviews by skilled interviewers, making sure that the whole tone of the interview is relaxed and permissive. This requires considerable skill and possibly the use of euphemisms.
- *Hypothetical questions.* These are the 'What would you do if . . .' type of question. They almost never yield reliable results, and should be avoided. There is usually a noticeable difference between people's self image in a particular set of circumstances and their actual behaviour.
- *Impersonal questions.* These questions tend to produce spurious answers because the respondent becomes disengaged from the participant-matter, and consequently can lose interest in the questionnaire, sometimes to the extent of refusing to answer any more questions.

Layout of the questionnaire

By this we mean the arrangement of words on the page. This aspect is almost always overlooked by questionnaire designers and yet it can make a difference of about 20% in the response rates. Figure 17 illustrates a well laid-out questionnaire, and you should note the two-column layout, the use of different weights of type, the use of italics for instructions, the use of boxes to restrict the size of written answers, and the pleasing aesthetic appearance (for its date) of the page. Slightly different rules apply for e-questionnaires, whether email or web-based; this is a skilled task, and you are advised to find an expert to help you in this.

Piloting the questionnaire

This stage is vital. It is here that the last chance occurs to discover the fallacies and hidden assumptions in your thinking. It is here that the respondent's understanding of the questions and the problems of analysis are revealed. It is also the last occasion on which remedial action can be taken. You should test all aspects of the questionnaire at this stage; the introductory passage, the questions, the alternative answers (or coding frames for interpreting open questions), and the form of the analysis. If you are running focus groups or interviews, this is the time to practice your skills at putting people at ease, and keeping unobtrusive control of proceedings.

This is best done in three stages:

1. Individual criticism: the questionnaire should be handed to a colleague or several colleagues who have experience of questionnaires (but not of this particular one) for comment.
2. Depth interviewing: once the criticisms generated above have been examined and any appropriate changes made, the questionnaire be given to a small sample of respondents (about ten) for their reaction. On completion of the questionnaire, each respondent should be questioned in detail about the answers to the questions, to ascertain what the respondent understood the question to be asking, and the exact meaning of the responses given.

3. Finally, the questionnaire should be given to a larger sample of respondents to investigate the implications of the desired analysis, and to detect whether any invalid or meaningless patterns of answers are occurring. This also enables estimates to be made of the reliability and validity of the questionnaire, the reliability of the sampling frame, and the likely non-response rate. This stage should be repeated until the questionnaire appears to be error-free (but you will never get rid of them all — nobody has ever created the perfect questionnaire).

Fieldwork

For many surveys, where the number of respondents required exceeds 100 and the survey is to be interviewer-admininistered, the most appropriate course of action may be to obtain the services of one of the commercial market research companies. In view of the expertise and service that is provided, they are good value for money. As was stated earlier, interviewing is a job requiring training and experience; co-opting students or other 'odd-job people' is unlikely to produce reliable or accurate results. Hence, if the time, trouble and cost of obtaining a field force, and then training and maintaining it is taken into account, there is seldom a cheaper alternative to the commercial organisations. Where the numbers are below 100, it is feasible (and crunchingly instructive) for the designer to do the interviewing personally.

Where you intend to use a mail questionnaire or an e-questionnaire, the considerations are different; the problems become those of layout and design of the questionnaire, and retrieval of the data. If the layout and the introduction to the questionnaire are good and look professional, the problems of retrieval are usually reduced. Nevertheless, it is usually necessary to consider some means of improving the response rate. First, there are encouraging reminders. It is advisable to send these within one week after the expected return date, together with a duplicate questionnaire, in case the first one was tossed away or deleted. Second, a shortened version of the questionnaire may be sent, to ask for the most important information. Third, the use of raffles or prizes to encourage the return of questionnaires might be considered. However, the cost of these can offset the chief advantage of mailed questionnaires, which is their cheapness. The final method of follow-up is to contact the non-respondents by telephone. This method has received increasing use in recent years.

You should note that for any of these follow-up methods to work, it is necessary to know who has responded and who has not, which necessitates removing the cloak of anonymity from respondents. It appears, however, that this is not generally a serious cause for non-response provided the questionnaire is well designed.

Analysis

Having completed the fieldwork and captured the data, there is still the problem of analysis. There are two major areas of interest here: the first is data cleansing, to identify any inconsistencies in the responses to questions and to take appropriate action (for example, a man in the household has seemingly given birth to a baby), and the second is data tabulation, where the tables and statistics required for the report are generated.

Once the data sets are clean the analysis can be performed. If you consider the data to be in matrix form, with data items across the top and respondents down the side, you can then conveniently use the manipulative power of spreadsheets to do most of the simple analyses and tabulations. For more complicated analyses, there is a wide range of statistical packages available to accomplish this; Minitab and SPSS are two of these which are widely available and for which it is usually easy to find expert help. However the analysis is carried out, it is essential that the arithmetical and statistical operations performed on the data are valid, as it is very easy to arrive at false conclusions by the use of inappropriate analyses.

General characteristics of questionnaires

- Most of the characteristics have been discussed in the paragraphs above; they are best means for gathering data from a large group of people, especially when they are geographically dispersed.
- As the participant matter becomes more complicated and/or difficult to verbalise, the usefulness of questionnaires decreases fairly rapidly to the point where interviews or focus groups are a better option, even though it is more difficult and resource-intensive to get acceptable numbers of people.
- Questionnaires should not be used until the structure of the participant matter is well-understood. They are not good for exploratory projects.
- The key issue in questionnaire work is to obtain a good response rate (more than 50%). This will only be achieved by you if you plan carefully and thoroughly. Piloting your survey until you are sure it will work is a vital step in this.
- E-questionnaires are a cheap option, but most times, so are the results. They are an easy way to get global access, but the response rate tends to be so low that the data is almost certainly biassed, and therefore of low quality.
- The physical presentation of the questionnaire is very important, whether it is mailed, interviewer-based, or web-based. For each of these, it is initial impressions that count most for a potential respondent to decide whether to become a respondent or to find a better thing to do. The endemically low response rate for e-questionnaires means that this is a key issue; you are strongly advised to make use of a web design professional.

Interviews

Interviews can be considered to be questionnaires carried out face to face by an interviewer, and the comments above regarding the need for planning, sampling, and so on apply here as well. However, because there is an interviewer present, the typically rigid structure of the questionnaire can be relaxed; this of course brings its own dangers unless the interviewer is skilled and experienced, and the interview is carefully planned beforehand. A good text for describing different kinds of interviews and discussing the practicalities of doing interviews is Oppenheim (1992).

While interviews have always been part of the repertoire of ergonomists, the emergence of knowledge engineering has lent a new importance to this technique (see Chapter 8).

The big danger in deciding to carry out interviews is to use the rationale that because the interviewer is an intelligent person, he or she can adapt to the needs of the situation, and do the right thing, thereby reducing the need for careful planning at the outset. Unless you have an extremely knowledgeable and skilled interviewer, this is unlikely to happen. All too often, this argument is used as an excuse to ease the mental pain at the outset. It almost never works well; the old sporting cliché, 'No pain, no gain' still applies. You will have to plan your interviews with the same dedication as for questionnaires.

However, it is because there is an intelligence in the interviewer that certain benefits do accrue, provided care is taken. In some circumstances, interviews are the best way to capture information. If any of the following apply, then use an interview:

- matters are very personal to the respondent,
- complex information is involved,
- you think respondents might need to have some help in giving their answers,
- different people may have markedly different views about reality,
- you do not know what is involved and are exploring the problem space.

There is a continuum of interviewing styles, ranging from directed interviews to non-directed interviews. Directed interviews are those where the questions to be asked, and the order of questions are specified beforehand. Interviewer-administered questionnaires are an example. Non-directed interviews are those where essentially the respondent controls the interview, and the interviewer is there to help the respondent express himself or herself, hopefully eliciting the important information during the process. This latter style can get close to that used in psychiatry. The questions are not formulated in any detail beforehand (though the topic areas should always be well-defined), and the main role of the interviewer is to help the conversation along, with interjections such as 'Really?' and so on. If the respondent stops, the interviewer may start the process again by using a non-directive question, for example, by repeating the respondent's last phrase with a questioning tone of voice.

The former technique has been criticised for its rigidity and its intolerance of unanticipated individual differences, whereas the latter has been criticised for the time required, and, to quote a memorable phrase, for 'leaving behind a posse of cured souls, but not necessarily producing much worthwhile information'. Most interview techniques fall between these two extremes, where some direction of the respondent occurs, if only to direct the respondent towards the areas that you want discussed. In some cases, the initial questions might be defined, and so on.

General characteristics of interviews

- The use of an interviewer can serve to direct and accelerate the information flow.
- The interviewer can explore unexpected information, or unexpected occurrences.
- A well-trained interviewer will be sensitive to the individual needs of the respondents, and will adjust his or her behaviour accordingly, thereby improving the quality of the information flow.
- Interviewers can help to motivate respondents to give more information about the topic during the interview.
- For the advantages above to occur. the interviewer must be well-trained in interview technique, should have at least some knowledge of the topic areas (the more the better), and must be sensitive to people. Collectively, these criteria are not easy to meet.
- It can be difficult to find and schedule people for the interview session.
- Sessions should not last for more than 10 minutes for sreet-corner interviews, or 20 minutes in someone's home or 90 minutes in an office situation. You should never let the interview run for more than 90 minutes; if you do, you will discover that your brains will stop working.
- Consider having two people (an interviewer and a scribe, swapping roles as necessary) for complicated topics. The scribe has more time to think in the interview, and can spot gaps, 'read between the lines', and detect errors.
- Always take notes, even when taping the interview. It is amazing how often the tape fails to capture the vital piece of information.
- Interviewer's bias may creep in; this might be due to the interviewer's own knowledge of the topics, interpersonal relationships between the interviewer and the respondent, or to more mundane things such as fatigue, and so forth. This constitutes an extra source of error.
- Systematic recording of data is difficult, and in some cases impossible. In certain instances (an example is knowledge elicitation from experts), it may take up to ten times as long to sort and assimilate the data as it took to obtain it.

Focus groups

The focus group may be construed as an interview with a group, rather than an individual. However, in this case it is group interaction that provides the 'extra', beyond what you would expect from interviews. The assumption that underlies this is that ideas or comments from one individual will remind another of a forgotten fact, or that the conversational interaction will sharpen a viewpoint, or even develop a new opinion. An example might be in a discussion about slips, trips and falls in the home, where one person describing a particular incident might prompt the memories of similar incidents in other people. From these, it might be possible for the group organiser to infer a generic cause, with a little further careful prompting. A good discussion of focus groups will be found in Morgan (1997).

In focus groups, the role of the interviewer becomes that of group organiser and prompter. Typically, the role involves gathering the respondents together in appropriate accommodation, getting everybody introduced and (especially in the United Kingdom) comfortable with each other and ready to talk. This usually requires a few easy, non-directive questions that are guaranteed to generate paragraph-length replies, such as 'Did you find it easy to get here?' It is important to ensure that everyone has their say, even the shy ones. Once the conversation has started, the topic of interest can be introduced. It is often easier to get this part of the discussion started if there is something concrete on which the group can focus; a range of products, or a photograph, for example, which can be passed around. Props such as these are very useful for ensuring that the participants are thinking and talking about the same issues.

Once the information starts appearing, the role of the group organiser is firstly to encourage people to keep participating. This may mean deftly stopping the dominant people, and encouraging the quieter ones. Second, as best as can be managed, it is to stay on the topics and programme that was planned beforehand — and, as for questionnaires, planning is essential.

Because the workload on the organiser will at times be excessive (especially if the group turns out to be a loud, argumentative one), it is usually necessary to have a scribe there as well, to help the organiser and to take notes of the significant issues.

General characteristics of focus groups

- The group should not exceed seven people; more than this and some people are unlikely to contribute much.
- The social skills of the organiser are extremely important in making the group and the discussion fruitful. The organiser should have ready-prepared control comments for most situations; for example "That's interesting. I'd like to hear if [person X] agrees" and, when arguments arise, "Do you think you might be talking about two different situations?," and so on.
- Care is necessary in choosing a suitable location for the discussion; easy for all participants to get there, coffee/tea is available, surroundings and furniture are comfortable, and not too noisy.
- Props should be used whenever possible.
- The organiser should have an agenda of topics to discuss, and good introductory questions for each of these.

The wider process — audit trails

Whichever participative method you choose to answer the particular problem facing you, most likely it will be part of a larger process dealing with a higher-level project. For example, you might be considering the introduction of a computer-aided learning package, as outlined earlier in this chapter. Or you might be testing a particular characteristic of a product on its way to market. Therefore, there are wider issues to be faced as well as the direct methodological ones outlined above. Some of these are:

- The cultures of Western society are moving steadily towards greater transparency in processes.
- The codes of conduct published by the professional societies are more and more particular about professional practice.
- There is a concomitant trend towards litigation for any perceived adverse effects arising from products, processes or professional advice. The main protection against this is to be able to show that you acted in a thoroughly professional manner in whatever you did, using the best available knowledge at that time.

Consequently, it will be in your own interest to maintain 'audit trails' of the work you have undertaken, from when you started on the project to its final completion and signing-off. This includes the planning of the project, its administration, and the final activities of project closure. The central issue is to be able to show for any stage of the project, suitable answers both to the usual six questions: what, why, how, where, where, and who, and to the seventh question, what was the outcome?

Most texts on 'qualitative assessment' give a lot of detail on audit trails, albeit in education settings.

Conclusion

A number of participative assessment methods have been explored briefly in this chapter. If you wish to make use of any of the methods, the texts given in each section should be sufficient to enable you not only to get started with the method, but also to achieve reasonable results. As in most areas, using these methods skilfully requires lots of practice, and you will make the same blunders that all the experts have made in their time (including the editors and authors of this book), from which you will learn the practical lessons about the methods. As Rasmussen (1985) has said rather more cogently, it is only by making errors that you learn skills.

You should be warned: a trap into which you might fall is to stick rigidly to the methods outlined in the texts, because there is a fairly strong scientific basis to them, and a wide acceptance of them. While these are laudable attributes of the methods, what counts in the end is the quality of the data that are gathered, and what is revealed about the participant of your interest. In this regard, it is suggested that you should read the text by Webb *et al.* (1966) if you can find it; in its own way it is a highly entertaining book for the originality of its thoughts and the subtlety of its methods.

Finally, it should be remembered that even though the methods may seem laborious, and the statistical complexities daunting, the data that result are usually worth all the trouble that was caused, and the warm feeling engendered by this will certainly outweigh the pain you experienced at the time.

References

Alluisi, E. A. (1975). Optimum uses of psychobiological, sensorimotor, and performance measurement strategies. *Human Factors* **17**: 309–320.

Belson, W. (1968). Respondents' understanding of survey questions. *Polls* **3**: 52–70.

Birnbaum, M. H. Ed. (1998). *Measurement, judgement and decision-making*, Boston, MA: Academic Press.
Edwards, E. (1973). Techniques for the evaluation of human performance. *Measurement of man at work*. W. T. Singleton, J. G. Fox and D. Whitfield. London, Taylor & Francis: 129–134.
Fransella, F., Bell, R. and Bannister, D. (2003). *A manual for repertory grid technique*, Chichester: John Wiley & Sons.
Gescheider, G. A. (1975). *Psychophysics: the fundamentals, Mahwah, NJ:* Lawrence Erlbaum Associates.
Gray, M. (1975). Questionaire typography and production. *Applied Ergonomics* **6**: 81–89.
Harris, R. J. (2001). *A primer of multivariate statistics*. Mahwah, NJ, Lawrence Erlbaum Associates.
Kalton, G., M. Collins, et al. (1978). Experiments in wording of opinion questions, *Applied Statistics* **27**: 149–161.
Kim, J.-O. and C. W. Muller (2001). *Factor Analysis: statistical methods and practical issues*, Thousand Oaks, CA: SAGE Publications.
Lodge, M. (1981). *Magnitude scaling*. London, SAGE.
Meister, D. and G. F. Rabideau (1965). *Human factors evaluation in system development*. New York, John Wiley & Sons.
Morgan, D. L. (1997). *Focus groups as qualititative research*. London, SAGE University Papers.
Moser, C. A. and G. Kalton (1985). *Survey methods in social investigation*. London, Ashgate Publishing.
Nunnally, J. C. and I. H. Bernstein (1994). *Psychometric methods*, New York: McGraw-Hill.
Oppenheim, A. N. (1992). *Questionnarie design, interviewing and attitude measurement*. London, Pitman.
Osgood, C. E., P. H. Tannenbaum, et al. (1990). *The measurement of meaning*, Urbana, IL: University of Illinois press.
Rasmussen, J. (1985). Trends in human reliability analysis. *Ergonomics* **28**: 1185–1196.
Rea, L. M., P. A. Shrader, et al. (1997). *Designing and conducting survey research: a comprehensive guide*, San Francisco, CA: Jossey-Bass.
Siegel, S. and N. J. Castellan (1995). *Nonparametric statistics*, London: McGraw-Hill.
Sinclair, M. A. (1975). Questionnaire design. *Applied Ergonomics* **6**: 73–80.
Spector, P. E. (1991). *Summated rating scale construction*, Thousand Oaks, CA: SAGE Publications.
Trawinski, B. J., R. Bechhofer, et al. (1985). *Three sets of tables: expected sizes of a selected subset in Paired Comparison experiments: Tables of admissible and optimal Balanced Treatment Incomplete Block designs for comparing treatments with a control: expected values and variances*, Providence, RI: American Mathematical Society.
Webb, E. J., D. T. Campbell, et al. (1966). *Unobtrusive Measures*. Skokie, IL, Rand McNally.

Chapter 5

Qualitative methodology

Sue Hignett

Introduction

This chapter will:

- Take a broad look at the difference between qualitative and quantitative methodologies
- Give an introduction to the philosophy of qualitative methodology
- Provide guidance on some of the basic steps needed for carrying out a qualitative project by looking at what people say (using a thematic-based analytical approach) rather than how they say it (linguistic approach)

The first question often asked is what is the difference between qualitative and quantitative methodology? This is not a simple question and although one-line answers can be given, for example words versus numbers, this does not do justice to the mature debate.

It is easier to start by looking at the difference on two levels: philosophical and operational (Bryman, 1988: 102). Qualitative methodology represents the philosophical position of a broadly interpretavist explanation of the social world (Mason, 1996: 4), whereas quantitative has a positivist position.

Dimensions in qualitative and quantitative methodologies

A summary of the dichotomy of qualitative compared with quantitative is shown in Table 1 (Hignett, 1999, 2001). It draws on, and summarises, contributions from Burrell and Morgan (1979), Crabtree and Miller (1992), Evered and Louis (1981), Guba and Lincoln (1981), Hammersley (1992), Leininger (1985), Lincoln and Guba (1985), Marshall (1994), Miles and Huberman (1984, 1994), Patton (1990), Perry (1996), Robson (1993) and Webb (1992).

The key points in defining qualitative methodology are:

- A consistency in representing the world in terms of words or pictures rather than numbers.
- The scale: with qualitative studies focussing on a few cases with many variables and the opposite in quantitative studies (many cases, few variables).
- The sampling strategy: pre-assigned for projects using quantitative methodology, whereas the sampling strategy for qualitative research develops during the study.
- The iterative nature of data collection and analysis, which also drives the sampling strategy

Table 1 Dimensions of Qualitative and Quantitative Methodologies

Qualitative dimensions	Quantitative dimensions
Words, understanding	Numbers, explanation
Purposive sampling, inductive reasoning	Statistical sampling, deductive reasoning
Social sciences, soft, subjective	Physical sciences, hard, objective
Practitioner as a human instrument to gather data, prescriptive, personal	Researcher, descriptive, impersonal
Inquiry from the inside	Inquiry from the outside
Data collection and analysis intertwined	Data collection before analysis
Creative, acknowledgement of extraneous variables as contributing to the phenomenon	Predefined, operationalised concepts stated as hypotheses, empirical measurement and control of variables
Meanings of behaviours, broad and inclusive focus	Cause and effect relationship
Discovery, gaining knowledge, understanding actions	Theory/explanation testing and development
Micro-sociology	Macro-sociology

- The emphasis on identifying the influence of the researcher. This is achieved by the researcher reflecting on their interaction before and during the project.

The wide range of qualitative methodologies available can be confusing so a framework devised by Tesch (1990) has been used and developed to give brief definitions for twelve methodologies (Table 2). It shows the disciplines in which the methodology originated, and is mostly used, and gives a brief description of the background philosophical thought. A simplified summary question is also given for some categories to indicate the fundamental area of concern.

These methodologies are represented on an epistemology continuum (Figure 1) to show the theoretical poles of qualitative and quantitative methodologies with the types, and levels,

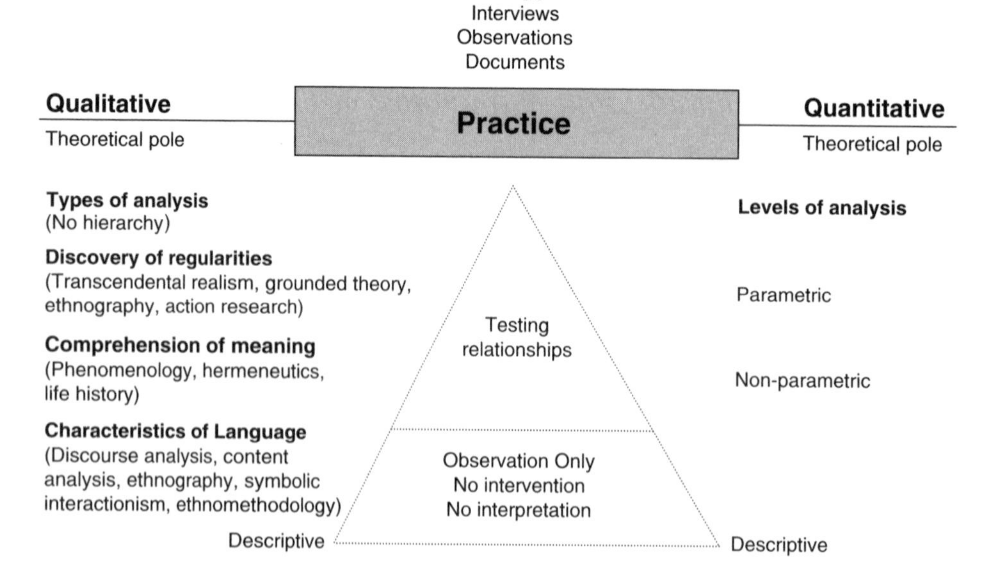

Figure 1 Qualitative/quantitative continuum.

Table 2 Twelve 20th Century Qualitative Methodologies (Chenitz and Swanson, 1986; Cresswell, 1998; Miles and Huberman, 1994; Patton, 1990; Schwandt, 1997; Strauss and Corbin, 1990)

Methodology and central questions (Q)	Disciplinary roots	Development
Transcendental realism Q = Is there both a causal explanation and the evidence to show that each entity or event is an instance of that explanation?	Sociology Philosophy	Harré and Secord (1972). Social phenomena exist not only in the mind but also in the objective world and that some lawful and reasonably stable relationships are to be found among them. Human relationships and societies have peculiarities that make a realistic approach to understanding them more complex but not impossible. Calls for both a causal explanation and for evidence to show that each entity or event is an instance of that explanation (explanatory structure and care descriptive account).
Grounded theory Generate and develop propositions, processes and substantive theory	Sociology	Highly systematic research approach for the collection and analysis of qualitative data for the purpose of generating explanatory theory that furthers the understanding of social and psychological phenomena. Developed by Glaser and Strauss (1967). Closely linked to symbolic interactionism.
Ethnography Q = What is the culture of this group of people?	Anthropology Sociology	An ethnograph is a description and interpretation of a cultural or social group or system. There is disagreement in the meaning of 'culture' between the two disciplines.
Action research Process with the researcher as change agent	Social Psychology	Derived from the work of Kurt Lewin (1890–1947) to describe the uniting of the experimental approach to social science with programmes of social actions to address social problems.
Phenomenology Q = What is the structure and essence of the experience of this phenomenon for these people?	Philosophy Sociology Psychology	Description of the experience of every day life as it is internalised in the subjective consciousness of the individual. This has a multiple philosophical derivation, including the transcendental phenomenology of Husserl (1859–1938), the existential views of Merleau-Ponty (1980–1961) and Sartre (1905–1980) and the hermeneutic phenomenology of Heidegger (1889–1976). Schultz (1899–1955) popularised existential phenomenology and influenced the development of ethnomethodology.

(Continued)

Table 2 Continued

Methodology and central questions (Q)	Disciplinary roots	Development
Hermeneutics Q = What are the conditions under which a human act took place or a product was produced that makes it possible to interpret its meanings?	Theology Philosophy Literary Criticism	Refers to the art, theory and philosophy of the interpretation of meaning of an object (text, work of art, human action etc.). Draws mostly on the work of Dilthey (1833–1911) and Heidegger (1889–1976) leading to a split between the theory of interpretation as a methodology (Dilthey) and an ontological position of existentialism (Heidegger). The hermeneutic circle refers to an ontological position of understanding as an inescapable condition of what it means to be human. The interpretation of each part depends on the interpretation of the whole and vice versa.
Life History Variety of approaches that focus on the generation, analysis and presentation of the data of a life history.	Sociology	Developed by the Chicago school, initially a biographical account, more recently developed into an interpretive process of developing the biographical account. Assumes that human action can best be understood from the accounts and perspectives of the people involved.
Discourse analysis Concerned with the analysis of the process of communication.	Sociology Psychology Communication Studies	Study of language, structure, function and patterns. De Saussure (1857–1913), Foucault (1926–1984). The linguistic categories into which people or things are sorted do not reflect any natural, objective order in reality.
Content analysis	Sociology Psychology	Textual analysis to compare, contrast and categorise data. Originally using word counts and semantic concepts, it has developed further by exploring stylistic characteristics and themes.
Symbolic interactionism Q = What common set of symbols and understandings have emerged to give meanings to people's interactions?	Social Psychology	Developed by George Herbert Mead (1863–1931) and Herbert Blumer (1900–1987), but also derived from Weber's (1864–1920) ideas that sociologists should proceed to understand those they have studied. Introduces the concept that human action is different to human behaviour, where action depends on an individual's ability to plan their actions and reflect on surrounding objects and past experience. So it is appropriate to study social life as it occurs in its natural social setting.
Ethnomethodology Q = How do people make sense of their everyday activities so as to behave in socially acceptable ways?	Sociology	The study of everyday practical reasoning, and the study of the processes whereby rules that cover interactional settings are constructed. Originated principally from Garfinkel (1917–), drawing on phenomenology and the work by Schultz and Husserl.
Conversation analysis	Sociology Communication Studies	Detailed analysis of audio and audio-visual recordings of naturally occurring social interaction to identify the interactional practices used by speakers to produce their own conduct and to interpret and deal with the conduct of others.

of possible analysis. In practice data types are used often across the continuum so a middle ground philosophical position will be suggested for ergonomics.

For quantitative methodology levels of analysis are given, as defined by the strength of statistical analysis, going from parametric (interval/ratio) down to non-parametric and descriptive. In contrast, types of analysis are given for qualitative methodology to differentiate between methodologies used for testing relationships and for observation-only. The methodologies that mostly provide a description of the situation or language characteristics have been placed in the observation-only tier. Those seeking comprehension of meaning or looking for regularities are categorised as testing relationships. Some of the methodologies appear in more than one group and anyway are changing in use.

Philosophy

Before getting into the philosophy of qualitative methodology, some working definitions are given for the following terms: epistemology, methodology, paradigm, objectivity. Validity and reliability are discussed later in the chapter.

Ontology is the fundamental study of being (Hollis, 1996: 8). It has also been defined as the study of reality and as a branch of metaphysics which provides a certain view of the social world. It is concerned with beliefs and understanding the kinds of things that constitute the world (Schwandt, 1997: 90). Simply put, ontology can be used to mean a world view or the fundamental way in which a person or society sees the world. This could, for example, be the concept of religion, setting out a defined structure to a world view and would include: how the world came into being (creationist theory rather than evolution); how societies function with respect to codes of behaviour (morals and ethics); as well as justice (legal systems).

The next level is *epistemology*. Grayling (1996: 38) gave the definition as the theory of knowledge. How can things be known? Do we learn things or do we innately know them? This is a fascinating area with a rich debate broadly split between the rationalists and the empiricists, and can be traced from Ancient Greek history as shown in Table 3.

Methodology is used as an umbrella term to indicate the theory and account of how research is, or should be, carried out in contrast to method, which is used to describe a technique for gathering evidence (Harding, 1987). Silverman (2000: 300) uses methodology to mean a general approach to studying research topics which relates theories to methods. So for his definition, a theory could be the Foucaultian discourse analytic approach, with the world viewed as discourses (forms of knowledge which work like languages; Jones, 1993: 103), and the method could be the analysis of language, to identify the discourses being used.

Paradigm is often used to describe a school of thought. Kuhn (1962) popularised the word when he used it to describe the changes in fashions in scientific knowledge as paradigm shifts. His thesis was based on the notion that all scientific knowledge was produced from within a particular tradition or paradigm which then determined what research was done and how it was carried out (Jones, 1993: 156). This applies equally to qualitative and quantitative methodologies.

Qualitative data types have tended to be classified as subjective rather than objective. This takes us into another debate with respect to whether it is possible for any research to be objective if a human has been involved in choosing the area of research/question/equipment, etc. *Objectivity* in qualitative research is generally not an issue as the goal is not usually to seek neutrality but to recognise the researcher's effect on the research, making explicit how this may affect the interpretation and findings.

In philosophy there are two extreme positions (rationalism and empiricism) with respect to ontology, epistemology and methodology. This tradition of a dichotomy can be traced

Table 3 Dichotomy of Philosophical Thought

Period	Reason	Senses
Ancient Greek	Socrates, Plato	Aristotle, Democritus
Medieval	St. Augustine	St. Thomas Aquinas*
Renaissance	Descartes, Spinoza	Copernicus, Bacon, Hobbes, Locke, Hume, Kant*
Enlightenment	Kierkegaard, Hegel, Nietzsche, Dilthey, Weber*	Comte, Mill, Marx, Darwin

*Philosophers who explored a middle ground.

back to Ancient Greek philosophy between mind and matter, or reason versus senses (Murphy *et al.*, 1998: 15).

A rationalist has the belief that reason is the primary source of knowledge, with certain innate ideas that exist in the mind prior to all experience. The rationalists are roughly grouped as taking a world view, or ontology, with human reason as the central tenet. In contrast an empiricist believes that there is absolutely nothing in the mind that is not experienced through the senses. Table 3 shows some of the key thinkers over the last 200 years who classified themselves into one of these two schools of thought. The summary is drawn from the following authors: Bond and Bond (1994); Bunnin and Tsui-James (1996); Chalmers (1982); Gaardner (1995); Hughes and Sharrock (1997); Jones (1993); Marshall (1994); Richards (1983); Williams and May (1996). Even within this limited number of texts there were differences of opinion, so only one classification has been used here: of reason versus senses rather than other alternative classifications, for example conflict, consensus and action theory (Jones, 1993: 15) or structuralism versus functionalism (Bond and Bond, 1994: 18).

In the twentieth century it is harder to follow such a simple dichotomy. The philosophy of science is discussed by philosophers such as Popper (1902–1994) who developed his theory of falsification for a hypothesis rather than the previous justification approach. More recently Lakatos and Feyerabend have added to the discussion. Lakatos suggested a model for the epistemology of physical science as a hard core of basic assumptions (e.g., the Earth moves around the Sun), which could not be rejected or modified and was protected from falsification by a belt of auxiliary hypotheses, initial conditions, etc. Feyerabend is the post-modernist of the philosophy of science, presenting an anarchistic theory of knowledge. The physical sciences, and scientific approach, are now being increasingly critically appraised.

A middle ground for ergonomics?

A challenge to all researchers (academic and/or practice based) is to define their philosophical position with respect to each and every research question (qualitative and/or quantitative) they investigate. This sets the scene for the subsequent choice of methodology, method, analysis and interpretation of findings.

One philosophical position relevant to ergonomics is transcendental realism, developed by Bhaskar (1989) from the historical tradition of realism. Williams and May (1996: 81) summed up this philosophy with one phrase 'the world has an existence independent of our perception of it . . . a common sense position'. So concepts such as causality, explanation and prediction are just as appropriate in the social sciences as physical sciences (Williams and May, 1996: 82). This is an ontological realism in which the dichotomy of qualitative and quantitative methodologies can co-exist, it requires the 'recognition of the existence of a real, independent world which operates according to natural necessity with a corresponding position of epistemological relativism' (Bhaskar, 1975: 250). So there is a difference between our descriptions of reality and the reality itself. This means that the kinds of descriptions given in science are themselves historically and socially formed products, the result of the work by previous researchers. This offers a very pragmatic viewpoint with the possibility of multiple realities such that the description of a thing in one way, rather than another, does not change the nature of the thing.

Doing a qualitative project

This section looks at the steps of sampling, data collection, data management and data analysis and shows that there are common processes across qualitative methodologies for the first three. Data analysis is the exception, remaining within a philosophical sphere where each methodological school applies their own body of literature to the research. A parallel can be drawn with quantitative research, where different versions of the same statistical tests (e.g., chi-squared, ANOVA) are preferred in different academic disciplines. When planning a qualitative project there are basic design decisions to be made (Table 4). The framework (point 5) is a key qualitative defining dimension. One way of achieving this is to use a conceptual framework to identify implicit and explicit theories and relationships. In a qualitative project this will be on-going so a project diary can be useful to record memos, mind maps, reflective thoughts and iterations throughout the project. This diary can also be used as part of the audit trail in establishing reliability.

Data sources

Although qualitative data sources are many and varied, there are three basic types: interviews, observations and documents. Table 5 gives a summary ontology and epistemology for each. The rest of this chapter refers to thematic-type analysis, looking at the discovery of regularities and comprehension of meaning (Figure 1), in contrast to looking at the characteristics of language as the 'linguistic turn or rhetorical approach' (Gomm *et al.*, 2000: 262). So rather than collecting spoken data to find out about what people think or feel, their words are interpreted in terms of what they are doing with the language at that time. For example asking or answering (or not) questions, issuing compliments or complaints, humouring the person they are speaking to and so on. This can be interpreted in terms of relationships with respect to power, sympathy, empathy etc., and there are specific transcript annotations (Silverman, 2000: 298; Jordan and Henderson, 1995). Some of the qualitative methodologies in Table 2 tend to use linguistic analysis more than thematic analysis, e.g., conversation analysis, ethnomethodology.

Table 4 Design Decisions (Janesick, 1998)

1. What is studied?	Intellectual question, site, participants (data sources)
2. Under what circumstances?	Access and entry to site and participants. Ethics.
3. For what duration?	Time frame.
4. Research strategy?	Methodology? Methods?
5. Framework?	Personal position, viewpoint with respect to the research question, site and participants.

Table 5 Qualitative Data Types

Miles and Huberman (1994: 9)	Wolcott (1992)	Dingwall (1997)	Mason (1996)
Interviews	Asking	Asking questions	Ontology = people's knowledge, views, understandings, interpretations, experiences and interactions are meaningful properties of the social reality which the research questions are designed to explore.
		Researcher: researched relationship	Epistemology = a legitimate way to generate data on these ontological properties is to interact with people, to talk to them, to listen to them, and to gain access to their accounts and articulations.
			Knowledge and evidence are contextual, situational and interactional so each interview will be different, reflexive and responsive to the situation, context and interaction.
Observation	Watching	'Hanging out'	Ontology = a data collection method which sees interactions, action and behaviours, and the way people interpret these and act on them, as central.
		Transactions between members	Epistemology = the knowledge, or evidence or the social world can be generated by observing, participating in, or experiencing natural or real life settings, interaction situations and so on, based on the premise that these kinds of settings, situations and interactions reveal data, and that it is possible for the researcher to be an interpreter, or knower of such data as well as an experiencer, observer or participant observer.
Documents	Examining	Reading the papers	Ontology = (1) the written word, texts, documents, records, visual or spatial phenomena or aspects of social organisation are meaningful constituents of the social world themselves, (2) interest in the processes by which they are produced or consumed; (3) belief that they act as some form of expression or representation of relevant elements of the social world; or (4) that aspect of the social world can be traced or read through them.
			Epistemology = Words, texts, visual documents, visual records, visual artefacts and phenomena can provide or count as evidence of these ontological properties.

Sampling

The sampling strategy for any research project should be defensible with respect to the appropriate relationship (or logic) of the sample and the intellectual question.

Sampling is one of the key dimensions in defining qualitative methodology. Inductive reasoning is used in qualitative analysis, to interact with the data and drive the sampling, whereas quantitative methodology uses deductive reasoning to test (or falsify) a pre-existing theory (Mason, 1996: 99).

The range of sampling strategies is symptomatic of the range of qualitative methodologies. Table 6 is developed from a table by Patton (1990: 182) with additional contributions

Table 6 Sampling Strategies

Type	Purpose
	Spreading the net
Purposive sampling	Maximum variation/open sampling. Picking a wide range of variation on dimensions of interest (time, location, events, people) to provide the greatest opportunity to gather the most relevant data about the phenomenon under study.
Mixed purposeful sampling	Triangulation, flexibility, meets multiple interests and needs.
Convenience sampling	Save time, money, effort. Poorest rationale, lowest credibility. Yields information-poor cases.
	Following up leads
Theoretical sampling	Analyst jointly collects, codes and analyses the data and then decides which data to collect next and where to find them in order to develop the theory as it emerges. Central tenet of Grounded Theory (secondary or analysis sampling strategy).
Snowball or chain sampling	Identifies cases of interest from people who know people, who know what cases are information-rich, i.e., good examples for study.
Opportunistic sampling	Following new leads during field work, taking advantage of the unexpected flexibility.
	Focusing
Homogenous sampling	Focuses, reduces variation, simplifies analysis.
Typical case sampling	Illustrates what is typical, normal, average, trying to find more than one case.
Intensity sampling	Information-rich cases that manifest the phenomenon intensively but not extremely such as above/below average.
Stratified purposeful sampling	Illustrates characteristics of a particular subgroup of interest, facilitates comparisons.
	Analysis sampling (inductive analysis)
Extreme or deviant case sampling	Learning from highly unusual manifestations of the phenomenon of interest. Qualify findings and specify variations or contingencies in the main patterns observed.
Confirming and disconfirming cases	Elaborating and deepening initial analysis, seeking exceptions, looking for variations. Disconfirming cases limit conclusions and indicate points of greatest variation.
Criterion sampling	Picking cases that need some criterion, such as children abused in a treatment facility. Quality Assurance.
Multiple case sampling	Grounding a finding using replication strategy.
Indiscriminate sampling	Choosing sites, persons and documents that will maximise opportunities for verifying the storyline, relationships between categories and for filling in poorly developed categories.

from: Miles and Huberman, 1994; Kuzel, 1992: Coyne, 1997; Strauss and Corbin, 1990, Glaser, 1978; and Yin, 1991.

The strategies are grouped into similar logics for:

- Spreading the net.
- Following up leads.
- Focusing.
- Analysis.

Table 7 Alternative Terms for Validity and Reliability

Internal validity	External validity	Reliability
Credibility	Fittingness	Auditability
Truth Value	Applicability	Consistency
Trustworthiness	Transferability	Dependability
Authenticity	Generalisability	

It is usually necessary to use more than one sampling logic during a qualitative project (Sandelowski *et al.*, 1992). The sampling strategy will develop during the project, so one might start by spreading the net, and then, possibly, go on to following up leads or focussing on a specific characteristic. Finally, some form of analysis sampling is generally evident in most qualitative studies.

Validity and reliability

The words validity and reliability are taken from quantitative methodology and need interpretation in a qualitative context. Table 7 pulls together some of the terms from Guba and Lincoln (1981), Lincoln and Guba (1985), Miles and Huberman (1994) and Robson (1993).

Internal validity addresses issues of credibility and authenticity in the research. At an operational level this is established through the audit trail and the analytic induction process of testing theory. External validity looks at issues of generalisability and transferability. The detail given with respect to the context, researcher bias, sampling strategy and history of the research question can all help to establish the conditions whereby the findings could be transferred to another setting.

Dingwall (1997: 62) gave three tests for general validity

1. Distinguish clearly between data and analysis.
2. See how the study has looked for contradictory or negative evidence and set out to test statements proposed on theoretical grounds or reported from previous studies.
3. See how it reflects the interactive character of social life and deals even-handedly with the people being studied.

Respondent validation, also known as member checking, is when the interpretation of the researcher is presented back to the subjects as part of the conclusion drawing and verification (Walker, 1989). This is a different process to accuracy checking of data where an interview transcript is returned to the interviewee. Mays and Pope (1995) suggested that member checking could be used to add to both the internal (authenticity check) and external validity (transferability of findings).

Triangulation is another method which can be used to establish validity, both internal and external. It refers to the use of more than one data source, method, or investigator and the convergence of these to add credibility to a study. There are concerns that if the philosophical (ontological) positions have not been defined the combination of different analyses would look at if they had been stuck 'together like children's building blocks in order to create a single edifice' (Coffey and Atkinson, 1996: 14).

Reliability addresses the issues of auditability or quality control. This could be the consistency by which instances are assigned to the same code in analysis, or on a broader level could relate to the process itself. There is a problem with replicability for both qualitative and quantitative research as, for any study looking at human actions within a social context, there will be change for example of the people involved or the social situation.

At a fundamental level one aim of all research should be to convince the reader. Whether this is achieved using large sample sizes and statistical tests, or by detailed descriptions of a situation or point of view, depends on the design of the investigation or exploration. If the reader is able to use the research by incorporating the findings in their own work then boundaries have been extended and knowledge has been generated, and robust scientific research has been done.

Data management, display and analysis

It is particularly important to have a good audit trail showing how data were collected and then managed with respect to the analysis due to the iteration of data collection and analysis in qualitative research.

Table 8 shows a three stage process for data management as proposed by Miles and Huberman (1994: 10), Dey (1993) and Marshall and Rossman (1989). All start by organising, reducing or describing the data. Step two starts the analysis by classifying, either visually or as text. Step three mostly involves interpretation or conclusion drawing. As the qualitative process is iterative these steps are intertwined cyclically, rather than in a linear relationship.

Data reduction

Initial tools for reducing data can include the use of:

- *Contact summary sheets.* This method pulls together the data immediately after collection and makes them available for further reflection and analysis. It acts as a quality assurance or reflective mechanism to review data collected, any key points raised, and any areas which need further exploration.
- *Memoing during coding.* This is a continuation of reflection which started when formulating the conceptual framework. Memos are the ideas about codes and their relationships as they strike the researcher during coding (Glaser, 1978).

Primary analysis starts by trying out coding categories before moving to identifying themes and trends. Codes are labels for assigning units of meaning to 'chunks' of varying size (words, phrases, sentences or paragraphs). Codes drive the retrieval and organisation of data for analysis. The use of CADQAS (Computer Assisted Qualitative Data Analysis Software) can support this process.

Coding at this first stage can be considered as just data reduction or simplification. However it can also be used as an analytical strategy by noticing relevant phenomena; collecting examples of those phenomena; and analysing the phenomena in order to find commonalities, differences, patterns and structures. Coding is the technical operation of a more subtle process of having ideas and using concepts to describe the data.

Pattern coding is the second level, where coding is used to expand, transform and reconceptualise the data (Coffey and Atkinson, 1996). Miles and Huberman (1994) suggested ways of using pattern coding:

- To map the codes by network displays to show how components interact.
- To check out the codes in the next wave of data collection.

Data display

Step two in Table 8 is data display. Miles and Huberman (1994) give a number of suggestions for data display including:

1. Context chart, where the inter-relationships between roles and groups is mapped in graphic form.

Table 8 Data Management Steps

Author	Step One	Step Two	Step Three
Miles and Huberman (1994)	**Data Reduction** Managing field notes, transcripts etc. Data are reduced in anticipatory ways as conceptual frameworks are chosen and cases and questions are refined. Data are summarised, coded and broken down into themes, clusters or categories.	**Data Display** For example, matrices, charts, graphs, networks. Data display describes diagrammatically pictorial or visual forms in order to show what those data imply to give an organised compressed assembly of information that permits conclusion drawing and/or action taking.	**Conclusion Drawing/Verification** Regularities, patterns, explanations, causal flows. Conclusion drawing and verification using different tactics, e.g., analytic induction.
Marshall and Rossman (1989)	**Summarising and packaging the data** Organising the data.	**Repackaging and aggregating the data** Generalising categories, themes and patterns.	**Developing and testing propositions to construct an explanatory framework** Testing emergent hypotheses against the data, looking for contradictions. Searching for explanations for the data. Writing the report.
Dey (1993)	Describing, including context of action, intentions and process of social action.	Classifying, as themes and codes, to give meanings.	Connecting concepts.

2. Checklist matrix to tabulate the data in terms of a specific question.
3. Time ordered display to show the flow and sequence of events. This is similar to an activity record or critical incident chart.
4. Conceptually ordered display to show well-defined themes and their interactions.

Data display can also be used as part of the analysis, to try and find relationships and then test them against the data. This creates the framework for the next process of analytic induction as part of the conclusion drawing.

Conclusion drawing

Conclusion drawing is listed as step three but there is a fuzzy boundary between it and data display in step two. Analytic induction is the process whereby negative or extreme cases are sought to (1) test, (2) extend the scope, and (3) determine the limits of the proposed theory. Basically the theory is revised until all the exceptions are eliminated by inclusion (Silverman, 1993: 160; Mason, 1996: 94; Fielding and Fielding, 1986: 89).

At this stage it is important to bring all the reflective strands together. Testing the interpretation will include checking against one's own biases as acknowledged at the start of the project in the conceptual framework, as well as the influences of the epistemology of the methodology if one has chosen to carry out the study within a particular school of thought.

Table 9 shows a question set used to critically appraise qualitative methodology studies for a systematic review project (Hignett *et al.*, 2003). This could also be used to internally review a project prior to writing the final report to identify the strengths and limitations of individual project designs.

Table 9 Question set to critical appraise qualitative studies

1.	Is the hypothesis/aim/objective of the study clearly described?
2.	Are the research methods appropriate to the question being asked?
3.	Was the qualitative method that was used made clear in the aims of the study?
4.	Is there a clear connection to an existing body of knowledge/wider theoretical framework?
5.	Is the context for the research adequately described and accounted for?
6.	Are the criteria for, and approach to, sample selection, data collection and analysis described clearly and systematically applied?
7.	Does the paper describe the sample in terms of gender, ethnicity, social class etc. (if appropriate)?
8.	Was the sample appropriate?
9.	Were the processes of fieldwork and the means of data collection described adequately?
10.	Is the relationship between the researcher and the researched considered and have the latter been fully informed?
11.	Is sufficient consideration given to how findings are derived from the data and how the validity of the findings were tested (negative examples, member checking)?
12.	Has evidence for and against the researchers interpretation been considered?
13.	Are the findings systematically reported and is sufficient original evidence reports to justify relationships between evidence and conclusions?

CADQAS (Computer Assisted Qualitative Data Analysis Software)

There are a number of data management software packages available to assist with the handling of qualitative data. It is important to reinforce the point which is often made when discussing the use of computers in analysis: the computer cannot do the thinking, interpreting or relationship exploring, this must come from the researcher.

Tesch (1990) and Fielding and Lee (1998) give good discussions of the range and applicability of most of the currently available software packages.

Conclusion

This chapter sets the scene for using qualitative methodologies in ergonomics research philosophically by establishing that there has been an on-going debate between two poles (represented here as qualitative-quantitative) for over two thousand years. However it is possible to take up a middle ground philosophical position and combine both qualitative and quantitative methodological approaches for the same research question.

In taking this pragmatic position a generic process for doing qualitative research has been described using a thematic analysis approach with three steps of data reduction, data display and conclusion drawing.

At the moment it seems (from conference proceedings and journal publications) that ergonomics is currently more at the quantitative end but there are influences tipping the balance back towards the qualitative side.

References

Bhaskar, R. (1975). *A Realist Theory of Science* (Leeds: Leeds Books).

Bhaskar, R. (1989). *Reclaiming Reality — a Critical Introduction to Contemporary Philosophy* (London: Verso).

Bond, J. and Bond, S. (1994). *Sociology and Health Care: An Introduction for Nurses and Other Health Care Professionals.* (2nd ed.) (Edinburgh: Churchill Livingstone).

Bryman, A. (1988). *Quality and Quantity in Social Research* (London: Unwin Hyman).

Bunnin, N. and Tsui-James, E.P. (1996). *The Blackwell Companion to Philosophy* (Oxford: Blackwell Publishers).

Burrell, G. and Morgan, G. (1979). *Sociological Paradigms and Organisational Analysis* (London: Heinemann).

Chalmers, A.F. (1982). *What is this thing called Science?* (Milton Keynes: Open University Press).

Chenitz, W.C. and Swanson, J.M. (1986). *From Practice to Grounded Theory — Qualitative Theory in Nursing* (Menlo Park: Addison-Welsey Publishing Company).

Coffey, A. and Atkinson, P. (1996). *Making Sense of Qualitative Data* (Thousand Oaks, CA: Sage Publications).

Coyne, I.T. (1997). Sampling in qualitative research. Purposeful and theoretical sampling: merging or clear boundaries? *Journal of Advanced Nursing*, **26**, 623–630.

Crabtree, B.F. and Miller, W.L. (1992). *Doing Qualitative Research* (Newbury Park: Sage Publications).

Creswell, J.W. (1998). *Qualitative Inquiry and Research Design: Choosing among Five Traditions* (Thousand Oaks, CA: Sage Publications).

Dey, I. (1993). *Qualitative Data Analysis: A User Friendly Guide for Social Scientists* (London: Routledge).

Dingwall, R. (1997). Accounts, Interviews and Observations. In: G. Miller and R. Dingwall (eds) *Context and Method in Qualitative Research* (London: Sage Publications).

Evered, R. and Louis, M.R. (1981). Alternative perspectives in the organisational sciences. *Academy of Management Review.* **6**, 385–395.

Fielding, N.G. and Fielding, J.L. (1986). *Linking Data. Qualitative Research Method Series 4* (London: Sage Publications).
Fielding, N.G. and Lee, R.M. (1998). *Computer Analysis and Qualitative Research* (London: Sage Publications).
Gaardner, J. (1995). *Sophie's World* (London: Phoenix, Orion Books).
Glaser, B. and Strauss, A. (1967). *The Discovery of Grounded Theory: Strategies for Qualitative Research* (Chicago: Aldine).
Glaser, B. (1978). *Theoretical Sensitivity. Advances in the Methodology of Grounded Theory* (Mill Valley, CA: Sociology Press).
Gomm, R., Needham, G. and Bullman, A. (2000). *Evaluating Research in Health and Social Care.* (London: The Open University/Sage Publications).
Grayling, A.C. (1996). Epistemology. Chapter One. In: N. Bunnin and E.P. Tsui-James (eds) *The Blackwell Companion to Philosophy* (Oxford: Blackwell Publishers).
Guba, E.G. and Lincoln, Y.S. (1981). *Effective Evaluation* (San Francisco: Jossey-Bass Pub.).
Hammersley, M. (1992). Ethnograph and realism. In: M. Hammersley (ed) *What's Wrong with Ethnography: Methodological Explorations* (London: Routledge).
Harding, S. (ed) (1987). *Feminism and Methodology* (Milton Keynes: Open University Press).
Harré, R. and Secord, P.F. (1972). *The Explanation of Social Behaviour* (Oxford: Blackwell).
Hignett, S. (1999). Qualitative methodology in ergonomics. In: M.A. Hanson, E.J. Lovesey, and S.A. Robertson (eds) *Contemporary Ergonomics, 1999.* (London: Taylor & Francis), pp. 105–109.
Hignett, S. (2001). *Using Qualitative Methodology in Ergonomics: theoretical background and practical examples.* Ph.D. thesis, University of Nottingham.
Hignett, S., Crumpton, E., Ruszala, S., Alexander, P., Fray, M. and Fletcher, B. (2003). *Evidence-based Patient Handling: Tasks, Equipment and Interventions* (London: Routledge).
Hollis, M. (1996). The philosophy of social science. Chapter 11. In: N. Bunnin and E.P. Tsui-James. *The Blackwell Companion to Philosophy* (Oxford: Blackwell Publishers).
Hughes, J.A. and Sharrock, W.W. (1997). *The Philosophy of Social Research*, 3rd ed. (Harlow: Addison-Wesley Longman Limited).
Janesick, V.J. (1998). The dance of qualitative research design: metaphor, methodolatry and meaning. Chapter 2. In: N.K. Denzin and Y.S. Lincoln (eds) *Strategies of Qualitative Inquiry* (Thousand Oaks, CA: Sage Publications), pp. 35–55.
Jones, P. (1993). *Studying Society. Sociological Theories and Research Practices* (London: Collins Educational).
Jordan, B. and Henderson, A. (1995). Interaction analysis: foundations and practice. *The Journal of the Learning Sciences,* 39–103
Kuhn, T.S. (1962). *The Structure of Scientific Revolutions* (Chicago: University of Chicago Press).
Kuzel, A.J. (1992). Sampling in qualitative inquiry. In: B.F. Crabtree and W.L. Miller (eds) *Doing Qualitative Research* (Newbury Park, CA: Sage Publications).
Leininger, M.M. (1985). *Qualitative Research Methods in Nursing* (Philadelphia: W.B. Saunder).
Lincoln, Y.S. and Guba, E.G. (1985). *Naturalistic Inquiry* (Beverly Hills, CA: Sage Publications).
Marshall, C. and Rossman, G.B. (1989). *Designing Qualitative Research* (Thousand Oaks, CA: Sage Publications).
Marshall, G. (1994). *Oxford Concise Dictionary of Sociology* (Oxford: Oxford University Press).
Mason, J. (1996). *Qualitative Researching* (London: Sage Publications).
Mays, N. and Pope, C. (1995). Reaching the parts other methods cannot reach: an introduction to qualitative methods in health and health sciences research. *BMJ* **11**, 42–45.
Miles, M.B. and Huberman, A.M. (1984). *Qualitative Data Analysis: A Source Book of New Methods* (Beverly Hills, CA: Sage Publication).
Miles, M.B. and Huberman, A.M. (1994). *Qualitative Data Analysis: An Expanded Source Book.* (2nd ed.) (Thousand Oaks, CA: Sage Publications).
Murphy, E., Dingwall, R., Greatbatch, D., Parker, S. and Watson, P. (1998). Qualitative research methods in health technology assessment: a review of the literature. *Health Technol Assessment.* **2**, 16.
Patton, M.Q. (1990). *Qualitative Evaluation and Research Methods.* (2nd ed.) (Newbury Park: Sage Publications).
Perry, A. (ed) (1996). *Sociology — Insights in Health Care* (London: Arnold, Hodder Headline Group).

Richards, S. (1983). *Philosophy and Sociology of Science: An Introduction* (Oxford: Blackwell Publications Ltd.).

Robson, C. (1993). *Real World Research. A Resource for Social Scientists and Practioner-Researchers.* (Oxford: Blackwell Publications Ltd.).

Sandeslowski, M., Holditch-Davis, D. and Harris, B.G. (1992). Using qualitative and quantitative methods: the transition or parent-hood of infertile couples. In: J.F. Gilgun, K. Daly and G. Handel (eds) *Qualitative Methods in Family Research* (Newbury Park, CA: Sage Publications) pp. 301–323.

Schwandt, T.A. (1997). *Qualitative Inquiry: A Dictionary of Terms* (Thousand Oaks, CA: Sage Publications).

Silverman, D. (1993). *Interpreting Qualitative Data: Methods for Analysing Talk, Text and Interaction* (London: Sage Publications).

Silverman, D. (2000). *Doing Qualitative Research. A Practical Handbook* (London: Sage Publications).

Strauss, A. and Corbin, J. (1990). *Basics of Qualitative Research: Grounded Theory Procedures and Techniques* (Newbury Park, CA: Sage Publications).

Tesch, R. (1990). *Qualitative Research: Analysis Types and Software Tools* (Bristol, PA: Falmer Longman).

Walker, M. (1989). Analysing qualitative data: ethnograph and the evaluation of medical education. *Medical Education,* **23**, 498–503.

Webb, C. (1992). The use of the first person in academic writing: objectives, language and gate keeping. *Journal of Advanced Nursing,* **17**, 747–752.

Williams, M. and May, T. (1996). *Introduction to the Philosophy of Social Research* (London: UCL Press).

Wolcott, H.F. (1992). Posturing in qualitative inquiry. In: M.D. LeCompte, W.L. Milroy and J. Preissie (eds) *The Handbook of Qualitative Research in Education* (New York: Academic Press), pp. 3–52.

Yin, K. (1991). *Applications of Case Study Research* (Washington, D.C.: Cosmos Corp.).

Chapter 6

Task analysis

Andrew Shepherd and Rob B. Stammers

Summary

This chapter describes the background to task analysis, setting out the important influences of industrial and technological changes in the 20th century and the developments in task analysis prompted by these changes. The *processes* involved in carrying out task analysis are discussed in order to demonstrate how the analyst needs to focus properly on the requirements of a client and keep abreast of the progress of the project. The *product* of the analysis is described in terms of organising information to best represent the task and make decisions concerning performance weaknesses and design solutions. Hierarchical Task Analysis is presented as a methodology in which the various lessons learnt from other approaches are incorporated. The chapter concludes with an examination of developments in cognitive task analysis (CTA). Concern is expressed that many writers on CTA have overstated the unique features of computer related work in dismissing the relevance of more traditional approaches to task analysis. The chapter argues that CTA methods are usefully treated as specialist techniques for use within a wider task analysis framework.

Introduction

Task analysis is a term used to describe activities concerned with the examination of human performance in systems, both from the perspective of the behaviour of the human and the factors that shape that behaviour. This includes the manner in which information and artifacts are presented and the environment in which people work.

Task analysis is used by ergonomists to gain an understanding of what people do in the jobs they carry out. This is done either to identify weaknesses in human performance or to develop solutions to improve productivity, safety, worker satisfaction, or the long-term health and wellbeing of workers. When using task analysis to identify problems, the analyst needs to understand how to judge what is critical in the work situation in terms of events that can arise, including operating error, and the consequences and costs of these events. When task analysis is used to generate solutions to problems, it is necessary to understand the costs and benefits of potential solutions before sensible recommendations can be made, as well as appreciating the constraints on how things should be done, for example safety or cost constraints (see Chapter 37). Task analysis also provides an interface with the client to ensure that the ergonomist is working according to the client's requirements and in accordance with the client's values.

Often ergonomists do not use formal approaches to task analysis in their work, preferring intuitive judgement. This may be satisfactory if the ergonomist is very familiar with a domain, but it can lead to mistakes in attributing problems or designing solutions. A process of systematic analysis is generally beneficial. It leads to better insights and enables solutions to be justified more effectively than expert judgement on its own, especially where tasks are complicated. Task analysis is an activity that everybody agrees is vitally important and should be done systematically, yet which people often ignore in practice. It is important to understand what task analysis is and what the issues are. There are many different techniques that people call 'task analysis' and there are controversies about which is best or which should be applied for which purpose. It is most important to understand what information any task analysis method is providing for the analyst and whether this is sufficient for the decisions and judgments that the analyst needs to make.

Some background to the ideas

Work study

Task analysis methods share many ideas with 'work study'. Work study or method study techniques are concerned with establishing norms of performance to measure how quickly people can do the things they are paid to do and to find ways to reorganise work in order to obtain improvements in performance. This has given rise to the familiar issues of 'time and motion study', 'payment by results', 'experienced worker standard', etc. Work study is concerned with measuring the observable outcomes of work, rather than the psychological processes that underpin work. For some critics, this is sufficient reason to dismiss work study as a precursor to task analysis, but there are too many similarities that cannot be ignored.

Early ideas of work study as a formal discipline were established through the work of Taylor, an American mechanical engineer working around the latter half of the 19th and the early part of the 20th century in the management of the steel industry. Taylor set out to determine, through precise measurement, the most efficient ways of doing things and to establish what constituted a fair rate of work to ask people to do. He broke tasks down into elements, timed these elements, worked out their best organisation and used this information to organise workers and set rates of work and rates of pay. This led to the coining of the term 'scientific management'. A major contribution to scientific management of this kind came from Frank and Lillian Gilbreth. They developed the idea of the 'one best way of doing a job'; people entering an organisation were assigned to jobs according to how their capabilities matched the best way for particular jobs. They established formal methods of presenting work elements and rules for establishing which elements were crucial to success.

Work study and its application developed substantially between the two world wars and continued after the Second World War. Reliance on these approaches, however, has declined as the dependence on manual work in industry has declined. Now, industrial work tends to be machine-paced or concerned with the supervision of automated and partly automated systems. In these cases work is more *cognitive*, involving planning and decision-making to a far greater extent than before. We also have far less industrial and far more service work.

Work study and *scientific management* are often used as derogatory terms. Work study techniques are often used as a basis for limiting payments. Like many things, such ideas are good or bad depending on how they are utilised. The Gilbreths, for example, saw work study methods as a means of paying people more, to enable them to share in the increased profits of the company. They also did pioneering work in the employment of disabled workers (Gotcher, 1992).

A more technical objection to these approaches is that many jobs that people are required to do entail a high cognitive component involving, for example, reasoning, judgment, discretion and flexibility. While there may be a best way of organising actions, there is no best way of organising cognition because it is covert. It is extremely difficult to establish how cognition is organised for real jobs and particularly difficult to argue that all satisfactory performance derives from the same or similar strategy. Different people will achieve the same standards and reach the same decisions in different ways. Moreover, individuals may modify their strategies as the context and their experience dictates. There are individual differences in the way in which individuals most comfortably handle information; even if it were possible reliably to coerce people to adhere to a particular cognitive strategy it may impose a method that some people find difficulty in working with.

Further problems relate to what work study methods do not deal with, for example, human error and its consequences (see Chapters 32 and 34), and issues of motivation and well-being. Task analysis methods in ergonomics are applied more broadly than work study and seek to explore psychological and work physiology aspects as well as working methods.

The military influence

One major impetus in the evolution of task analysis methods was from the military context. The Second World War in particular raised a number of human resource concerns. One was to help soldiers and civilian staff cope with new forms of equipment. Using advanced technology systems, or controlling new sorts of vehicle prompted a need to understand better the manner in which human beings processed information. This stimulus led to developments in skill psychology and also prompted a need to address the psychological processes entailed in what people had to do. There was a requirement to deploy large numbers of people in these jobs and this directed attention towards developing principles of task and job design to enable tasks to be carried out more easily and improve instruction. The years following the war saw the publication of important papers on task analysis, most notably by Miller (1962, 1967).

Miller's work introduced many of the ideas in task analysis that remain important today. He advocated two main stages, *task description* and *task analysis.* A *task description* is a statement of what an operator has to do in his or her job, expressed in *systems* or *operational* terms, i.e., describing what must be done by the operator to change the state of the system. One way of expressing this is in a table. Table 1 illustrates part of a job description concerning maintenance of a radar system.

The left-hand column in Table 1 contains one of several component statements for the task, namely *adjust radar receiver.* The general heading 'control' lists component tasks — *power on, AC voltage adjustment, POS regulated voltage,* etc. The 'activity' column states what to do to maintain the particular control; 'indication' lists cues for carrying out the task. Finally 'remarks' is used to record any other information thought to be relevant and which does not have a home elsewhere.

Implicit in this table is the theme, common within task analysis methods, of dealing with tasks hierarchically — *turn power on* is a part of *adjust radar receiver* which is a part of *maintain radar system.* Any hierarchical system of task description, where lower descriptors are subsumed within higher descriptors, must have some method of conveying how the lower descriptors are organised to meet the requirements of the higher descriptor. For example, being able to make the various control actions in this task is hardly satisfactory if they are made in the wrong order. This issue of *when* to do things is dealt with in Table 1 within the 'indication' column. This contains cues for action to prompt the execution of the different control actions in adjusting the radar receiver.

In Miller's scheme, this phase of *task description* is followed by a phase called *task analysis* where the analyst explores the behavioural structure of tasks. For this, Miller

Table 1 A Task Description Table (adapted from Miller, 1962), Showing the Anlaysis of a Radar System Adjustment Task

JOB ELEMENT FORM

Position Line mechanic — Radar system
Duty 1. Adjust system

TASK	TIME minutes		ELEMENTS			REMARKS Alternatives and/or precautions
	In seq	Out seq	CONTROL	ACTIVITY	INDICATION (include when to do task and frequency of task)	
1.1 Adjust radar receiver	40	40	⟶		Adjust every 25 hours of a/c time. See a/c log.	
			1.1.1 POWER ON button	Press	Inverter starts and makes audible hum, pilot light comes on, range indication lights come one, tilmeter pointer comes on scale.	Avoid starting system with covers removed from high voltage units: personal hazard.
			1.1.2 AC voltage adjustment (screwdriver)	Turn	AC voltmeter aigns to 117 ± 4 volts.	
			1.1.3 POS regulated voltage adjustment (screwdriver)	Turn	POS regulated voltage meter indicates 300 ± 5 volts.	
			1.1.4 BRIGHT control knob	Turn clockwise	Sweep trace becomes visible on CRT.	
			1.1.5 FOCUS control knob	Turn as required	Sweep trace becomes sharper (focused).	

proposes an *information processing model.* This is presented graphically in Figure 1 and involves stages through which signals pass from the task environment through *reception of task information, interpretation and problem solving,* which deals with the manner in which a response selected, to *motor response mechanisms.* Supporting this performance is *retention of task information,* which provides a store for current task information, such as the current status of specific parameters, as well as learned rules and principles governing decision-making. Providing direction to all activities is *goal orientation and set.* The representation in Figure 1 is consistent with many such human–system interaction models presented in ergonomics.

Miller's approach is pragmatic. His purpose was not to try to defend a specific view of human information processing, but to present a practical framework for considering different influences on performance. Miller suggests that an analyst should consider execution of a task in terms of the information processing stages in Figure 1 and judge where difficulties are likely to arise. Identifying a potential weakness prompts the analyst to seek a solution. For example, if it is judged that short-term retention of information will be affected in a task where the operator has to remember a number of system variables, then an external memory aid, such as an instrument or a note-pad, is advised. Other models of the human–task interaction, or even models of human behaviour, may serve exactly the same purpose (e.g., Wickens and Hollands, 2000).

Industrial changes

As well as military applications, the second half of the 20th century saw major changes to the technology, scale and organisation of industry and commerce. It also saw major changes to the legislative context in which work was carried out affecting safety, training and employment conditions. As late as the 1950s, the world of work was dominated by labour intensive industry, where workers were employed in large numbers to carry out short cycle repetitive work, such as assembling bicycles, cars or domestic appliances. Such work was subject to the scrutiny of work study engineers, as described above. The 1950s and 1960s also saw a major effort in supporting skills development for instance in the UK through work sponsored by various government or quasi government agencies: the Department of Employment, the

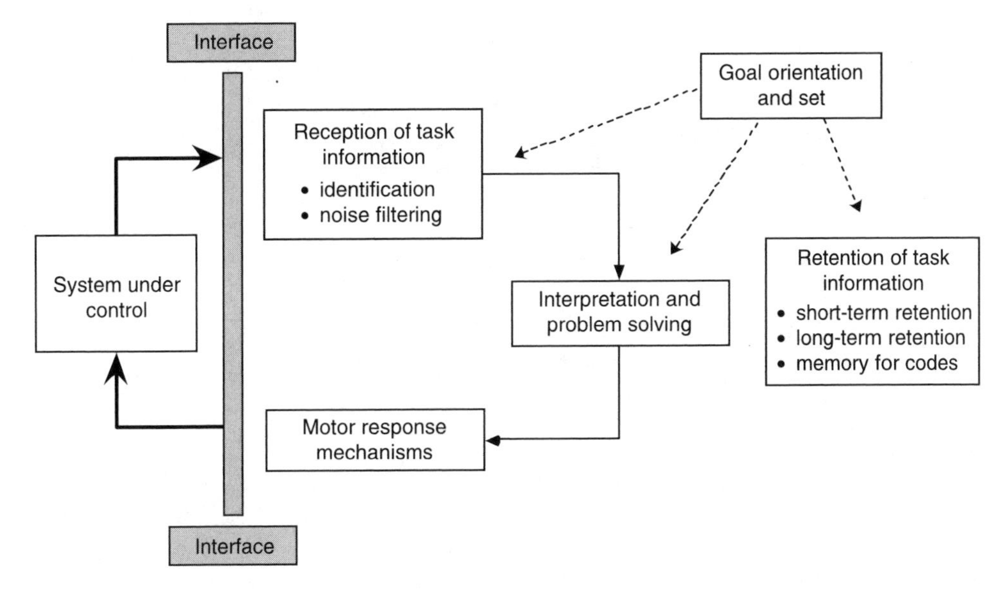

Figure 1 Schematic representation of Miller's scheme for task analysis.

Manpower Services Commission, Industrial Training Boards and the Armed Forces. This provided a funding base for many developments.

Influential at that time was the work of W.D. Seymour (1966; also Salvendy and Seymour, 1973). He used the term 'job analysis' to describe the different activities a worker carries out in performing a task and the term 'skills analysis' to describe how task activities are carried out. Thus, *job analysis* accounts for what the worker was required to do, while skills analysis shows how these things are done with a view to organising them more efficiently and developing training. Seymour developed a set of basic descriptions of different types of action, for example:

R	Reach	G	Grasp
M	Move	P	Position
PIU	Pickup	P/A	Place aside
RL	Release	D	Disengage

Seymour's dichotomy of *job analysis* and *skills analysis* is much the same as the distinction Miller draws between *task description* and *task analysis.* This is just one of many confusions in terminology that abound in this area (Stammers, 1995). It is most important, therefore, to pay attention to what people mean when they use terms, rather than make assumptions based purely on the terminology they use.

The approach adopted by Seymour's scheme has many variants. For example, Crossman (1956) brought together elements of work-study with a greater acknowledgement of human information processing. He produced the 'Sensorimotor' process chart, which recorded cycles of skill in terms of the relationship between observable actions, as with the earlier work-study approaches, and psychological processes such as perception, planning and memory.

One of the major engineering developments over the last 50 years has been the increase in automation. Many hitherto manual activities have been automated by the development of sensors that automatically detect when process conditions should be adjusted and then send signals to machinery to affect changes (for example, by adjusting automated valves or activating switches and motors to change operating conditions). Automation ensures that, in principal, complex processes can be controlled more reliably than by the human operator. Automation is also used to enable larger and more hazardous systems for manufacturing to be developed in industries such as power generation, chemicals and petrochemicals, food production and transportation. Where the operator once maintained moment-to-moment contact with raw materials, automation and process control require operators to assume a *system supervisory* role. Supervising an automated system entails *monitoring, problem detection, diagnosis* and *dealing with problems.* Dealing with problems includes *managing safety* and *planning recovery* in terms of *problem compensation, rectification* and *recovery.* A major characteristic of these tasks is that the decision-making involved is covert and not available for observation in the manner of manual operations. These are, in essence, cognitive tasks and require different types of approach to those adopted hitherto.

Hierarchical Task Analysis (HTA) was developed explicitly to deal with such issues (e.g., Annett and Duncan, 1967; Annett *et al.*, 1971; Duncan, 1972, 1974). The ideas and methods of HTA will be described later in this chapter. Briefly, however, HTA entails examining hierarchies of goals in terms of *subordinate operations* and *plans* that state when subordinate operations should be carried out. This method results in a hierarchy of goals and subgoals and enables the analyst to identify performance weaknesses and suggest improvements. As such, HTA combined many of the methods developed by earlier practitioners (see Duncan, 1972).

HTA has proven to be a versatile and effective method for task analysis and can be applied to a wide range of tasks and its results applied to a wide range of benefits (see Shepherd, 2001). Despite its obvious benefits, HTA is mainly *functional* in the sense that it is concerned with identifying what people do and identifying the skills they need to decide what to do next. But it does not explicitly describe the psychological processes entailed in doing these things. However, the analyst is able to use other methods to explore aspects of the task identified through HTA if further examination is justified.

One influential approach, used by some people analyzing supervisory control tasks, is Rasmussen's model of performance in terms of skill-, rule- and knowledge-based behaviour (Rasmussen, 1986; Sanderson and Harwood, 1988).

Skill-based behaviour is concerned with well-practiced and stereotyped actions.

Rule-based behaviour is where operators resort to a strategy where they match features of a current situation to a class of responses and can, therefore, deal with unfamiliar situations.

Knowledge-based behaviour is where operators must try to construct a response from scratch to cope with an unfamiliar set of circumstances.

The basic idea of this approach, is that, confronted with a set of circumstances requiring a response, the operator may try a proven, well-practiced, response before adopting a more novel approach. The different types of behaviour implied by this framework may be best supported by different types of information display, they may also give rise to different patterns of errors.

The different types of behaviour implied by this framework will different types of display in different ways; they may give rise to different patterns of errors. An application of Rasmussen's ideas which has been used particularly in the area of human error identification is the GEMS framework (Reason, 1990). GEMS (Generic Error Modelling System) can be applied to determine the type of behaviour that has led to, or could lead to, an error. Specifically, it helps the analyst distinguish between 'slips' and 'mistakes'. *Slips* are seen as departures of action from intention — that is, the operator knows what to do, but does the wrong thing. *Mistakes* are where the operator acts according to plan, but selects an inadequate plan to achieve the desired outcome. The importance of this distinction is that slips and mistakes are caused by different processes of behaviour and, therefore, need to be eradicated by different solutions. For example, slips can be made by experts, possibly due to lack of attention when carrying out familiar routines; solutions such as training, which might be applied to eradicate problems due to inadequate skill, would be unsuitable in dealing with this type of problem. Often HTA and GEMS are used in tandem to investigate human error (see also Chapters 32 and 34).

Many of the perspectives discussed so far are *systems oriented* in that they seek to establish how the human fits into a system of work; the aim is to explore ways of modifying the user's behaviour or the equipment or some other facet of the task.

There are other task or job analysis methods that could be regarded more as *personnel psychology* approaches. These are less concerned with work redesign than with investigating how people carry out designated forms of work. One approach is that of Fleishmann (e.g., Fleishmann and Quintaince, 1984). Fleishmann and co-workers have sought to explore performance by using a factor analytic approach to identifying constituent skills deployed in tasks. By factor analysing numerous activities, they have sought to explain actual performance of tasks in terms commonly occurring skills.

Another personnel approach is characterised by the Position Analysis Questionnaire (PAQ) developed in the USA by McCormick and co-workers (McCormick, 1979). This approach gathers job information by questionnaire in order to identify general job characteristics. It has been used primarily for personnel purposes, such as selection, job comparison/grading and career development, rather than for operational purposes, such as work design and training.

Applications in information and communications technology

A principle impetus for developments in task analysis has been information and communications technology. Developments in information technology have affected the ways in which work can be organised. Computing applications, such as word processors, spreadsheets and databases have created new ways of doing familiar tasks and place emphasis on different types of work organisation. For example, we are all familiar now with the power of databases and how they have changed shopping methods and other services. This has spawned new types of jobs in tele-sales, for example. We are also familiar with the power of communications systems, including the internet. These developments mean that people can now work together in novel ways. Because many commercial systems have become automated, much of the effort of the human worker is directed towards managing or supervising systems of people and computers. As with other supervisory systems already discussed, this entails the processes of monitoring system performance and dealing with problems when they arise.

It is assumed that these developments have created new challenges to task analysis methods, but the novelty of such tasks is often overstated by researchers. In the design of novel computing environments, where computer users are able to view the world in novel ways, there may be a case for investing in new task analysis methods. But in many tasks and work organisations in situations where IT has been introduced, existing task analysis methods are likely to be satisfactory and helpful.

During the 1980s, when human computer interaction research became prominent, it was assumed that new methods of *cognitive task analysis* were needed to deal with the predominance of cognitive skills required by computer users. This topic will be dealt with in later.

Process and product in task analysis

Much of the writing about task analysis gives the impression that task analysis is always a tidy affair. The methods that researchers present are generally orderly and the completed task analyses that are reported in the literature usually look neat and tidy. Yet when analysts first attempt to translate what they have read, especially in a new context, the whole activity can look very messy indeed. They may abandon their efforts, or they may improvise an alternative approach and publish it as a new method of task analysis or they might persevere with a published account and derive new insights and new analytical skills.

Published accounts of task analysis methods and projects often report the *product* of the work and fail to give insight into the *processes* used in gathering and organising information. Task analysis is a problem solving activity that entails the analyst using a range of approaches to elicit and organise information as the analysis unfolds.

This section will illustrate the types of issue that analysts must consider as they progress. It will describe the phases that are implicit in conducting task analysis. In doing so, it will refer to some of the numerous methods that have been developed.

The main activities that the analyst is engaged in at any time in a task analysis project include: *appraising the problem; collecting information; organising information and representing the task; modelling behaviour; suggesting solutions.* By looking at these activities, we can see how different methods of task analysis relate to each other. The next section will provide an account of Hierarchical Task Analysis (HTA) because this method provides a general and systematic approach to task analysis that can be used by practitioners to organise their task analysis intervention in order to accommodate these five phases discussed here in this section.

Appraising the problem

At the outset of any task analysis activity it is advisable to explore the problem informally at a general level. One important aspect is to make a general appraisal of the task context and how the main activities to be considered in the task analysis relate to wider activities in the organisation. It is useful to try to identify key *actors* whose views might be sought, both in terms of obtaining their perspective and in helping to manage the political aspects of the intervention. It is important to ask questions, sketch information and product flows, learn some terminology and understand something of how the work and technology is organised. Apart from providing essential language and concepts to help later on in the more formal stages of the task analysis, this helps to develop a good relationship with the client or the client's agents.

While the task context is being explored, it will also be possible to obtain an indication of the nature of the task, an indication of the resources available for dealing with problems arising from the analysis and the various constraints that will need to be overcome in dealing with these problems. It will also often be possible to infer the client's attitude and level of commitment to dealing with any issues that emerge — sometimes a client will be reluctant to acknowledge certain outcomes or countenance certain solutions, because of costs, personal biases, current company initiatives or fashions, project time-pressures or damage to reputations. This understanding will serve to anticipate how best to deal with issues that emerge from the task analysis. The sift will also indicate what support may be expected when carrying out the task analysis — what collaboration will be available in providing information, what assistance will be given in carrying out the stages of task analysis and, generally, how supportive people in the organisation will be. Whilst it is always advisable to sift through the problem at the outset, the analyst should also repeat this type of activity from time to time as the analysis progresses. The task analysis will continually refocus the analyst's view of the task, raising new issues that need to be understood.

Knowing what aspects of the organisation to consider, and how best to locate the information, emerges with experience. Experience is gained as the analyst recognises how organisations and industrial and commercial processes function. There are no formal methods for doing these things routinely during the task analysis, although there are formal approaches to *organisational analysis* that can help a person new to task analysis appreciate the bigger picture of how organisations work (see also Chapter 38).

Collecting information

An important aspect of any task analysis is the collection of information, including information related to managerial preferences, information needs, operator knowledge requirements, task cycle times, performance errors etc. The methods for doing these things include *interviewing, verbal protocol analysis, questionnaires, observation, simulation, examining production and examining safety records*, etc. (See especially Chapters 3, 4, 7 and 9). Some people refer to these data collection methods as task analysis methods; this adds further to the plethora of methods and techniques with which the reader has to contend. Competence at appropriate techniques of information collection is essential. However, of prime importance is the appropriate identification of situations where the costs of using a specific technique are really justified. In the majority of situations, *interviewing* domain experts is quite sufficient. Where this proves less than satisfactory, and the task element currently under scrutiny is seen as sufficiently important, then applying a more costly or time-consuming technique is justified.

The following information collection methods are reasonably common and accounts of many are given in methods courses and texts, including Kirwan and Ainsworth (1992), Robson (2002), Stanton and Young (1999) and elsewhere in this book:

Interviews	Masking/blanking off	Time line analysis
Observation	Witholding information	Walk-through/talk-through
Verbal protocol analysis	Link analysis	Simulation

Organising and evaluating information and representing the task

Information collected must be organised so that it can be evaluated systematically and is properly represented. Clear representation serves three main purposes.

1. It must be used to present findings to a client so that the task analysis can be validated and approved.
2. Representation itself helps clarify many of the complexities of the task for the analyst — by clear representation it can be shown which parts of the task can be treated independently whilst retaining their relationship to the task as a whole.
3. The manner in which the task is represented will assist in developing solutions. For example, interface design, job design training programmes and operating manuals can all be derived from the way the task is organised.

Typically, diagrams are a favoured method for organising and representing tasks. Hierarchical diagrams show the relationship between tasks and sub-tasks. Flow diagrams and time-lines illustrate the temporal relationships between task elements, including different contingencies for action. Influence diagrams are often helpful to clarify how different parts of the task environment influence each other. Diagrams are popular, because they can help analysts and their clients think constructively about the task and share their understanding. However, diagrams are by no means essential or necessarily the best method of communicating all task information.

Often, straightforward text is both concise and explicit. Task descriptions should always be clear and without ambiguity. They should indicate the things that operators do in terms of the actions they must carry out, on which object and for which purpose. Many cognitive task analysis methods use *formal* languages to describe units of action and their logical interrelationships. There is a problem here for practical task analysis projects whereby the client is excluded from a proper appreciation of the task analysis as it emerges, because specialist knowledge is needed in order to understand the method of task representation. This could result in the analyst pursuing an elegant method of task representation but failing to represent the task that the client wants to be carried out.

Tables are another common solution for organisating and examining task information. Tables can be used to record task steps in successive rows, with comment on respective task steps set out in adjacent columns. Thus, tables enable the analyst to expand thoughts about task elements, either informally or according to a formal procedure. Spreadsheet packages are particularly effective for this purpose and can be adapted in a number of useful ways. Another use of spreadsheets is to create matrices of task elements with a view to recording how these different elements relate to each other in terms of frequency or information flow, for example.

No more will be said about the issue of representation for the present, because many of the issues concerning representation of tasks are dealt with by Hierarchical Task Analysis, and will be described more fully in the next section.

Modelling behaviour

The reason for carrying out any task analysis should be established at the start of the project, because this will influence where effort should be directed and what information should be obtained. Typical reasons are to identify factors that adversely affect performance or to assist in the design of training or a user interface. To meet these requirements fully, it may be necessary to explore the behaviour which underlies aspects of the task.

Psychological demands of tasks may be modelled by, for example, the information processing model of Miller (1967) described earlier, or a variant of this such as Wickens' information processing model (Wickens and Hollands, 2000, p. 11). Another system of modelling is Reason's Generic Error Modelling System (GEMS), described earlier. Problems identified through these approaches may be resolved by methods such as job redesign, interface design and training, or they may simply be used to highlight the risk involved in such tasks.

Psychological constraints are not all that should concern the analyst. Many tasks entail physical effort that requires the operator to have suitable access to controls, sufficient space to adopt a safe and effective working posture and sufficient space to manipulate the controls to exert appropriate forces. Thus, aspects of anthropometry may need to be modelled (e.g., Pheasant, 1996). Often factors affect the visual, auditory or mechanical environments in which people work. Psychological, physical and physiological demands should be dealt with in an even-handed manner by the analyst. The underlying assumption in any approach to work design is that a range of human characteristics need to be matched by appropriate features of the physical work space and working environment. These topics are more fully dealt with in other chapters in this volume.

Suggesting solutions

Finally, when a task element is shown to warrant attention, the analyst must hypothesise a solution or a set of possible solutions. Generating hypotheses will depend on the analyst's and client's knowledge of the types of things that can be done to improve matters in a particular domain. Often *guidelines* or *checklists* devised by specialists can help suggest how aspects of tasks can be improved, although suggestions may also emerge from the analyst's or the client's experience. There are usually several ways of dealing with a particular problem identified through a task analysis and an evaluation is necessary to choose which hypotheses are worth exploring in greater depth. This initial evaluation of hypotheses may be done by discussion or by simulating the task to explore the efficacy of respective suggestions. There are cost implications in choosing between design hypotheses. These should take into account the costs and adverse consequences of the innovation and any benefits that are likely to accrue.

Whether task analysis is seen to stop at the point where *performance* has been modelled or continues into hypothesis generation is a point for debate. Often a task analysis is presented as something neutral — *the* categorical model of what operators do to carry out a task. This perspective may be of limited practical use. One problem is that psychological models generally assume a task interface and a working environment. Design to improve performance entails suggestions to change the task interface, the environment or the operator's skills. To do this constructively generally requires that the task be further analysed in some way. Should a design hypothesis prove unacceptable for any reason, further analysis may be warranted.

The activity of suggesting solutions is clearly on a fuzzy boundary between *task analysis* and *development of design*. It is important to emphasise that the task analysis is likely to be revisited at various stages of system development — to consider fresh design hypotheses, in representing an interface, for example, or to take design further in developing the training necessary to provide the skills that operators will need in order to use the interface.

One practical issue that the analyst must be aware of is that the costs of an innovation may themselves change as the analysis develops. Early on the suggestion to install a local emergency panel, for example, may have been rejected and an alternative solution preferred. However, later, the emergency panel might become justifiable for some other purpose and so the earlier decision can now be revisited. This happens frequently when considering simulation as a training solution. Its cost may initially appear prohibitive to management, but as more and more uses for a simulation facility emerge its costs may be justified. If the early decision is reversed, the various parts of the analysis should be revisited to see how best use can be made of this new training facility.

These five general task analysis activities are only listed in order for convenience. It is rarely appropriate to start at the first activity, then move systematically through till the last is completed. It is more common to switch between the various activities as the task analysis progresses. This is an important practical point that must be emphasised. It is common for people to assume that task analysis is a progression from initial consideration of a system or system specification through successive stages of data collection and analysis. Indeed the present authors conveyed this view in their chapter in the previous edition of this volume (Stammers and Shepherd, 1995). The diagram presented as Figure 6.2 in that volume to illustrate the progression of task analysis activities has been modified in Figure 2 to illustrate a more accurate representation of the task analysis process. The present version is not as tidy as that presented previously, but it does reflect a more accurate picture of how the analyst must switch the focus of activity as circumstances dictate within a task analysis. This is particularly important when considering *Hierarchical Task Analysis*, which will be considered next.

Hierarchical task analysis — a practical framework for task analysis

The discussion above showed how the analyst moves through different activities in conducting his or her work. We shall now consider Hierarchical Task Analysis (HTA) in greater detail because it provides the analyst with a practical approach that encompasses many of these considerations within a coherent scheme. Moreover, it can be utilised in a variety of domains and can be used to support a wide range of investigations and design activities.

Hierarchical task analysis (HTA)

HTA was developed by John Annett and Keith Duncan (Annett and Duncan, 1967; Annett *et al.*, 1971; Duncan, 1972, 1974) to provide an approach to analyzing tasks that would cope with, among other things, the increasing number of system supervision tasks. When operating a continuous process plant, for example, the operator has to do a number of simple things, such as *switching equipment on and off, opening, closing* and *adjusting valves*. These things are often done from a control room with no significant physical effort involved. The complexities of these tasks are concerned with collecting and processing information to judge deviance from operating targets and then planning and coordinating actions to re-establish these targets. In other words, it is futile simply to scrutinise actions because the principle skills are concerned with planning and decision-making.

In HTA, a task is first expressed in terms of a *goal* that the operator is required to attain, then redescribed into *sub-goals* and a *plan* governing when each sub-goal should be carried

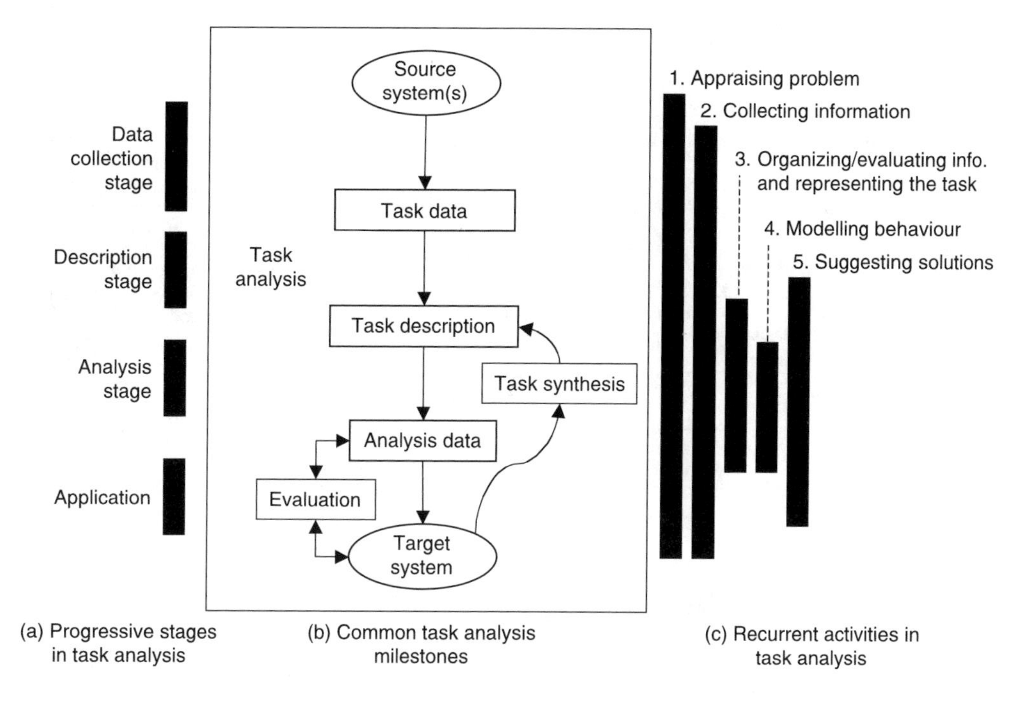

Figure 2 Processes employed in task analysis. The central box (b) represents the main deliverables in a task analysis project, moving from consideration of the source of information through to the specification of a target system. The left-hand column (a) represents the task analysis activity as a progression of stages from 'data collection' through to 'application'. These parts of the figure were presented as Figure 6.2 in Stammers and Shepherd (1995). However, the current chapter emphasises the recurrent nature of activities within the task analysis process — as represented in column (c) — as the analyst focuses on different aspects of dealing with the problem.

out. Each sub-goal is then considered to decide whether a solution is forthcoming or whether further redescription is warranted. The result of this process is illustrated in Figure 3.

It is useful to point out that the earlier writing on HTA by Annett and Duncan and their co-workers sought to discuss the *principles* for analyzing tasks, without explicitly 'branding' HTA as a method — this acronym was coined later. As such, HTA was, and is, a reasonably catholic framework to guide the analyst's decision process, calling on various techniques as necessary in order to explore tasks systematically with a view to suggesting how improvements could be brought about. HTA relies on common notions of hierarchical organisation, information processing, action sequences and flow-charting, together with ideas of cost benefit analysis and risk analysis to help the analyst communicate task detail and focus attention. It uses what has been shown to be helpful in a practical framework that can guide the analyst through a principled set of procedures.

Analysing goals and operations

In HTA, tasks are analysed by considering the operation that has to be carried out to meet a given system goal. A *goal* is often a business requirement along with the constraints that must be observed in meeting such targets (i.e., goals here are not psychological constructs but they are something to aim at). An *operation* is what people carrying out the task do to a system in order to move it towards its goal state. In HTA a goal or its operation is conveyed as an imperative — a verb-object instruction (e.g., 'make cake', 'serve customer', 'repair photocopier'). In recreational, domestic and creative situations, people set their own goals

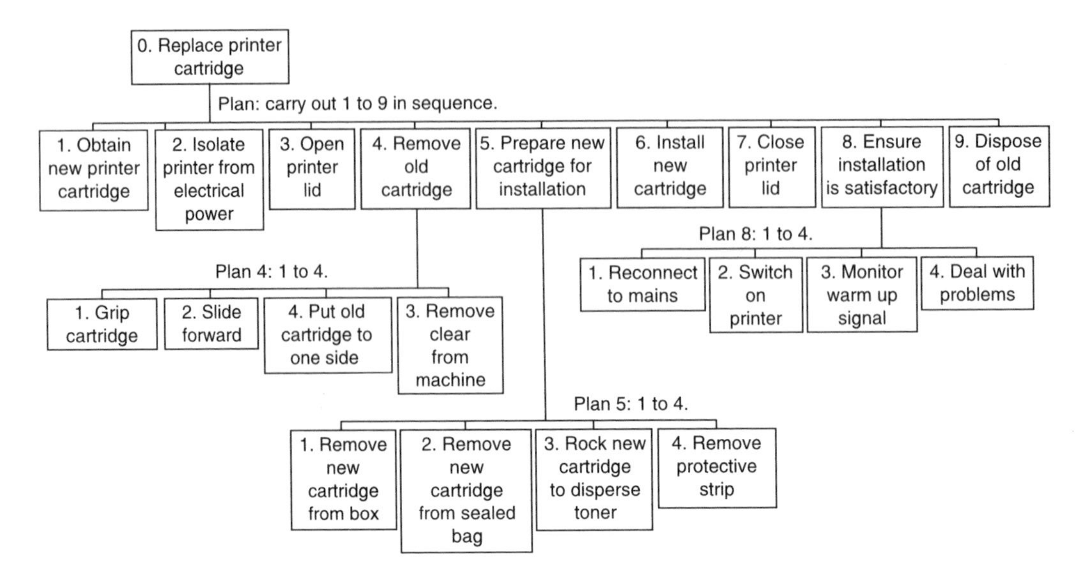

Figure 3 A simple HTA to illustrate hierarchical organization of goals and plans.

for the part of the world they are trying to control, then use what skills they have to operate on their environment to achieve these goals.

Examining human–task interactions

Goals are examined by two main methods: (i) by examining the human–task interaction and (ii) by redescription. To examine the human–task interaction Annett and Duncan suggest that an operation (i.e., what the operator does) should be thought of in terms of *input, action* and *feedback* (Annett and Duncan, 1967; Annett *et al.*, 1971). That is, competence at an operation implies an ability to collect information in order to decide what, if anything needs to be done (input); an ability to carry out the action selected in order to move towards the goal state (action); and an ability to monitor appropriate information to determine whether the action is being executed correctly, whether it is appropriate for dealing with the goal in question, and when the action has been completed satisfactorily in terms of attaining the goal state (feedback). Examining the human–task interaction in this way is similar to applying Miller's and other's information-processing approaches to task analysis. In practice, the I-A-F classification is rarely used formally by the analyst but is used intuitively and routinely to remind the analyst of the components of performance that should be present.

Examining goals by redescription

If making a judgment about the human–task interaction proves difficult, unsatisfactory in generating useful hypotheses, or the analyst feels that he or she has not yet gained a suitable grasp of the issues then further insights may be gained by *redescription*. A goal is redescribed in terms of a set of sub-ordinate goals and an organising component known as the *plan*, that specifies the conditions under which sub-ordinate goals have to be carried out to meet the system goal in question. Repeating the process of redescription results in a hierarchy of operations and plans. This is seen in the simple example in Figure 3 and in the more complex example in Figure 4. Figure 4 is a familiar task to most people. It is the analysis of a supermarket checkout operator's task — so the reader can visit a local supermarket to determine how their method of working conforms to or varies from the supermarket where

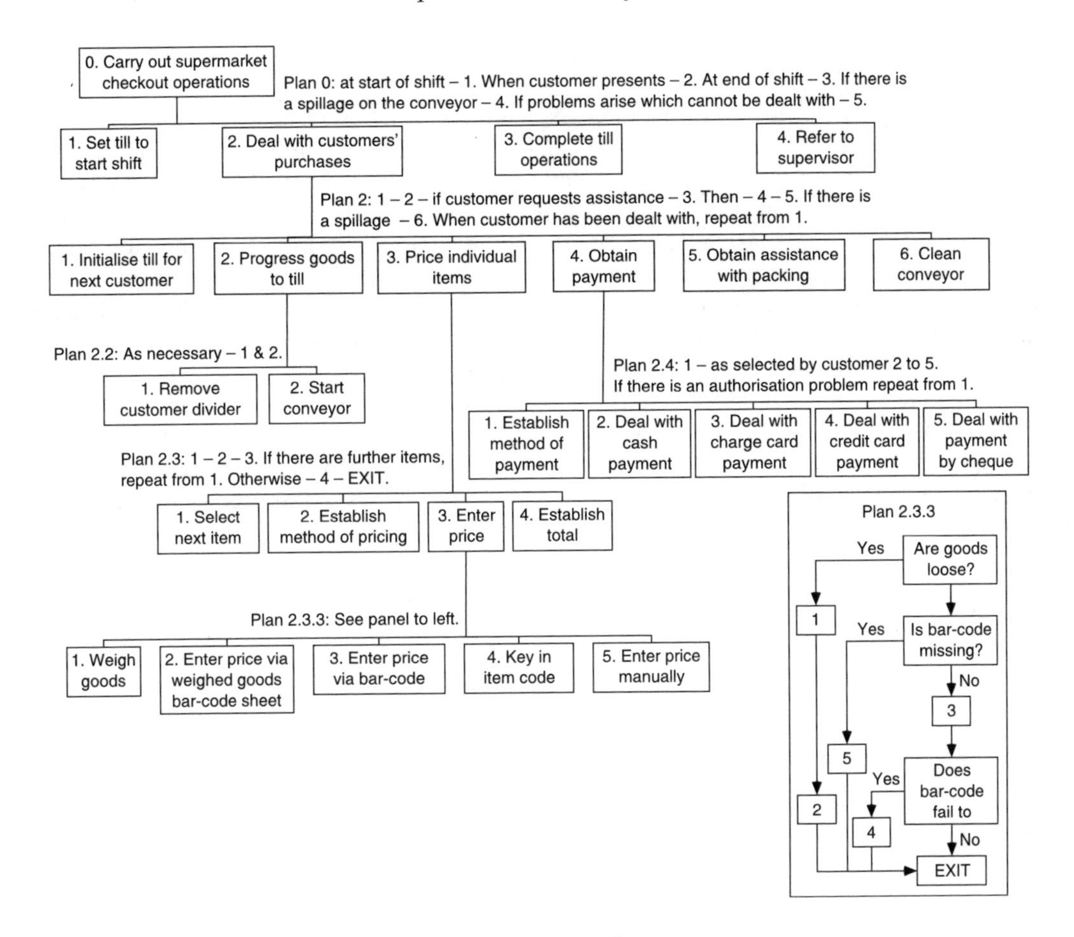

Figure 4 HTA of a supermarket checkout operator's task.

this example was taken. A particular feature of this task is how the cycle of activity described by plan 2.3.3 is nested within plan 2.3, and how the cycles of activity described by plans 2.3 and 2.4 are nested with plan 2. The resultant behaviour can look quite complicated until the benefits of the hierarchical organisation of operating goals are appreciated.

Redescription always entails stating a plan. Herein lies a problem because it is not always possible to state a plan and an exhaustive set of subordinate goals explicitly such that the redescription provides an accurate account of the task. In diagnostic tasks, for example, the set of possible faults in a system is not always known, nor is the set of information to be used to make a diagnosis always clear. Many systems require staff to be versatile, meaning that they must be able to deal flexibly with problems that arise. To do this, they must understand the principles that govern system behaviour and know how to apply these in solving problems. Equally, many planning activities defy the process of redescription that would yield an explicit plan with an HTA. The analyst cannot, therefore, gain further insight into the task by exploring it with a greater grain of analysis and must try to deal with the problem as it stands. One solution is to research classes of operation in order to identify ways in which instances, encountered later, may be treated. As an example, several useful fault diagnosis training methods have been devised which entail using system knowledge and diagnostic rules-of-thumb, together with simulation of the task (see Patrick, 1992 for an explanation and review).

Types of plan

A key feature in organising a task hierarchy is the representation of appropriate plans. There are several different types of plan and they can be stated in many different ways. It is tempting to simply state difficult plans as merely reiterating a super-ordinate goal — that is by stating that the sub-operations must be carried out in order to attain the overall goal. For example, plan 0 in Figure 3 could be stated as 'carry out sub-goals 1 to 9 in order to change the printer cartridge'. This logically completes the redescription, but it is not much use as a plan because the cues for action are not explicit and the analyst cannot be confident that the set of sub-operations exhaust the different activities that are implied by their super-ordinate goal. The most useful plans specify the conditions that indicate when sub-goals must be attained.

Generally, there is little benefit in trying to categorise plans in HTA. There are many ways in which sub-operations can be carried out to attain different goals and, for most practical purposes, we gain little by trying to standardise them. It is useful, however, to convey some of the ways in which plans might vary. Real plans often combine aspects of the following characteristics.

Fixed sequence	Contingent fixed sequence	Cycles
Cued actions	Choices	Time-sharing
Discretionary plans		

Fixed sequence

A fixed-sequence component in a plan is where a specified second operation is carried out when the first goal has been successfully attained. For example, to use a toaster, a slice of bread is inserted then the lever is pressed down and so on. To save text in a word-processor, the user may need to move the cursor to the 'File menu' *then* hold the mouse button down to 'pull down' the file menu *then* move the highlight down to the word 'save' *then* release the mouse button. The behaviour necessary to conduct a small task such as this includes the capability to carry out each of the suboperations and the ability to do them in the right sequence. The *cue* to carry out the next operation is the *feedback* that shows that the previous goal has been attained.

Cued actions

When actions are cued, an appropriate operation from a set of possible operations is selected. This is a very common type of plan element. It occurs very frequently in *monitoring* tasks for example, when a monitored signal becomes a cue for action. Cued actions can entail straightforward behaviours where the person merely has to link an action to a cue. Such cues are often time-based, for example a diary appointment or an instruction to do everything at least once every fifteen minutes. Another common form of cue is an instruction from, say, a supervisor — if the supervisor says 'start the pumps', the operator must start the pumps. It is important to record this type of instructional cue in an analysis, because it defines the manner in which authority and control is distributed throughout a team and, ultimately, who is responsible for invoking certain actions. Then it is easier to ensure that the person with the authority is provided with the appropriate information with which to make the decision. Cued actions can be represented in the form of *production rules.*

Contingent fixed sequence

The contingent fixed sequence is a combination of the *fixed sequence* and the *cued action.* In the contingent fixed sequence, following successful completion of a first goal (i.e., confirmatory feedback), a second specified goal is carried out when a specified system condition develops. Plan elements such as this occur in tasks concerned with controlling large systems where there is a slow-response to system changes due to inertia of some kind, such as process control — a first action may change the system state to cause the temperature in a vessel to rise; a second may have to wait until a required vessel temperature is reached.

Just as with the *choice plan* element, an important class of cues is the instruction from a manager, supervisor or other colleague. This is particularly important in complex systems where several people have to collaborate for a successful outcome. Often a person knows *what* to do next, but must wait for someone else with authority to say *when* it should be done. Authority is used here in a broad sense and not simply in the sense of managerial control. A sales manager may need to wait for the word from a production supervisor to say when fresh product will be available before customers can be told when they may expect it. Other examples can be given in hazardous environments, where one operator overseeing one crucial area may provide the cues for others to carry out operations to ensure that everything comes together at the right time. An important facet of genuine teamwork is that individual team members supply specific cues for the plans of colleagues; effective teamwork depends on how well instructions are transmitted and received.

Choices

Choices are logically similar to *cued* actions in that one of a set of subordinate operations is carried out according to prevailing conditions. However, the basis for choice is a decision or multiple discrimination that the operator has to make (e.g., in a fault diagnosis situation). This means that the HTA should identify a decision-making operation to indicate that the operator must first make the choice on which to then act.

Cycles

In some tasks, operators have to repeat a sequence of operations until conditions arise when they must stop. A cycle is a combination of fixed sequences and a cue to indicate when stopping is required — see the cycle plans in Figure 4. In a tele-sales procedure, for example, if there are no customers, the operator will monitor a screen to respond when a customer calls. When there is a call, a common cycle of activities is followed, from identifying the caller and the service required through to arranging delivery and payment. On termination of the call, the operator returns to monitor the next call. Sometimes the cue to stop is given by another person e.g., a supervisor. This is typically the case in repetitive work situations. This type of cycle can be called a *procedural* cycle.

Sometimes the cue to stop is provided by a test result. This is typical in situations where targets must be attained and action sequences must continue until this criterion is reached. Process control and maintenance tasks are work contexts where such *remedial* cycles are extremely common. For example, a process operator may need to carry out a cycle of activity entailing successive adjustment and measurement until a target is reached.

Time-sharing or concurrent plans

Sometimes it is necessary for two (or more) operations to be carried out together. This must be stated quite explicitly in the plan. Time-sharing elements are very common. Sometimes they signal serious problems for the person carrying out the operation in terms of task loading and demands on attention, but this is not necessarily the case and the respective sub-goals would need to be explored further. Often in HTA time-sharing elements are encountered reasonably high up in the hierarchy. It is only when the respective goals are analysed further that the full implications for task loading can be appreciated.

Time-sharing problems are particularly important where two or more of the constituent task elements require prompt responses, such as emergency situations. In many cases, time pressure means that even the most experienced operator will be challenged. In other cases skill and experience enables the operator to balance two or more activities, with experience indicating when attention can safely be switched between two or more goals. Alternatively, management can use the task analysis to plan off-line how two sets of activities should be accomplished concurrently — either by specifying a procedure to be followed or redesigning jobs to reduce the burden.

Discretionary plans

In a discretionary plan the person is at liberty to do any one of a set of subordinate operations without constraint. This may seem a strange type of plan, but it is a logical alternative and does sometimes occur in practice — sometimes it does not really matter which of several things is done. It is good to look busy when the boss comes round, for example, so the shop assistant may check the stock, or tidy the shelves or familiarise themselves with different product lines. These are all useful things to do, but it does not matter which is done and which is left — and the customers might be impressed. It may be important for nurses to be seen to be doing something to provide a pretext to monitor the patient's condition.

Mixed plans

The plan elements listed above have been outlined simply to show features of plans that might be encountered. Actual plans will be a mixture of these, as will be seen shortly, and there is often little practical benefit in trying to turn them into one of these specific types. Additionally, plans have a multiplicative effect in a hierarchy. If a plan entails a 'choice' element resulting in at least two possible ways of carrying out subordinate goals and these sub-goals in turn involve choices, then the range of variations possible soon builds up. Thus, hierarchies of simple plans can account for very complex sequences of action, as Figure 4 illustrates.

Stopping analysis

Tasks, goals or operations should not be examined unless performance is judged to be unacceptable — otherwise it is a waste of resources. To deal with this, Annett and Duncan proposed the $P \times C$ rule, where P refers to the *probability of inadequate performance* and C the *cost of inadequate performance* (Annett and Duncan, 1967; Annett *et al.*, 1971). If this product is acceptable, then there is no need to progress with the analysis. The $P \times C$ rule is an important rule-of-thumb, reminding the analyst to weigh both of these factors. It is usually applied informally as a judgment process.

If an operation is likely to be carried out badly (high P), yet resultant error is of no consequence (insignificant C), then $P \times C$ will be insignificant and the operation will not be worth examining. If an operation is likely to be carried out reasonably well (low, but finite, P), but error will result in a catastrophe with loss of life (unacceptably high C), then $P \times C$ implies an unacceptable risk and examination and suitable treatment must follow. If without any training, and within an existing interface design, $P \times C$ is judged unacceptable, the analyst can propose modifying the interface or can propose the training needed, then test this solution by re-applying the $P \times C$ rule. In each of these cases, the decision rule should be extended to anticipate the costs of the innovation, such that costs do not exceed benefits. Issues associated with these stopping rules are discussed by Shepherd (2001).

HTA continues until the analyst has reason to stop — because current performance is judged satisfactory (via $P \times C$) or a hypothesis for improving matters has been proposed and recorded. When this has been achieved it means that the analysis has been taken as far as is necessary to consider all constituent problems. This has always been one of the most useful, economical and yet little appreciated features of HTA. It reflects on the discussion above of where task analysis ends and design development begins. In Annett and Duncan's view HTA is complete when the whole task has been considered to a level of detail where each operation is either shown to be benign in terms of the $P \times C$ rule or where operations of concern have been dealt with by providing a feasible and acceptable hypothesis for improvement. Of course, the ultimate test of the analysis is whether support for these hypotheses is evident in practice, by successful trials of prototype designs, successful trials of the complete design or successful system operation. There will always be scope for dusting off the HTA to have a fresh look at something should one or more elements prove unsatisfactory at a later stage or if the operating context changes.

HTA as a process

We have already discussed task analysis as a process. HTA can be treated in the same way to show how the analyst moves through a cycle of considerations until the analysis is complete. This process is represented in Figure 5.

The flowchart captures the flow of decision-making typically involved in HTA. There may be concern where an accurate redescription of a cognitive operation is not forthcoming because it is impossible to state categorically an exhaustive set of subordinate goals, how these goals would be 'planned' to achieve their superordinate, or both of these things. It was suggested that the analyst's course of action was to review the task in various ways and look again for a solution. This amounts to revisiting stage 5 and re-examining the human task interaction to seek a solution. Sometimes the task that cannot be redescribed could be dealt with if management conceded that further resource be provided to secure a solution. Sometimes, it is necessary to refer the problems to someone with expert knowledge of such activities, possibly involving different forms of analysis. Sometimes it is necessary to research possibilities. This is justified according to the criticality of the operation in question. It is often the case that such problematic operations are sufficiently critical to warrant this expense, since they are often concerned with operating decisions in high risk, safety critical operations.

Representing and recording HTA

Clear recording of the task is essential to ensure that a consistent record is maintained. This allows the analyst to remember the stage he or she has reached. Clear recording is also vital for communicating information to other people, in order to justify any recommendations made.

Generally, it is a good idea to produce both tables and diagrams. Diagrams are easy to understand and help people see the structure of the task most clearly. Tables enable far more detail to be included. Moreover, if the hierarchical description is to be extended to incorporate one of the other forms of analysis, which has entailed some form of task classification, then development of a table is essential.

Hierarchical diagrams

Figure 6 shows the hierarchical diagram of operating the radar system (the task illustrating Miller's radar adjustment task in Table 1). Diagrams are very useful as a means of conveying the structure of the hierarchy. Diagrams are limited if the analyst wishes to make copious design notes and comments.

Numbering the analysis

It is important to adopt a rational numbering system to keep track of the analysis. Any numbering system that works may be adopted. The system used in Figure 6 has proven to be effective. The overall goal is numbered 0. Its sub-goals are numbered from 1 to whatever is

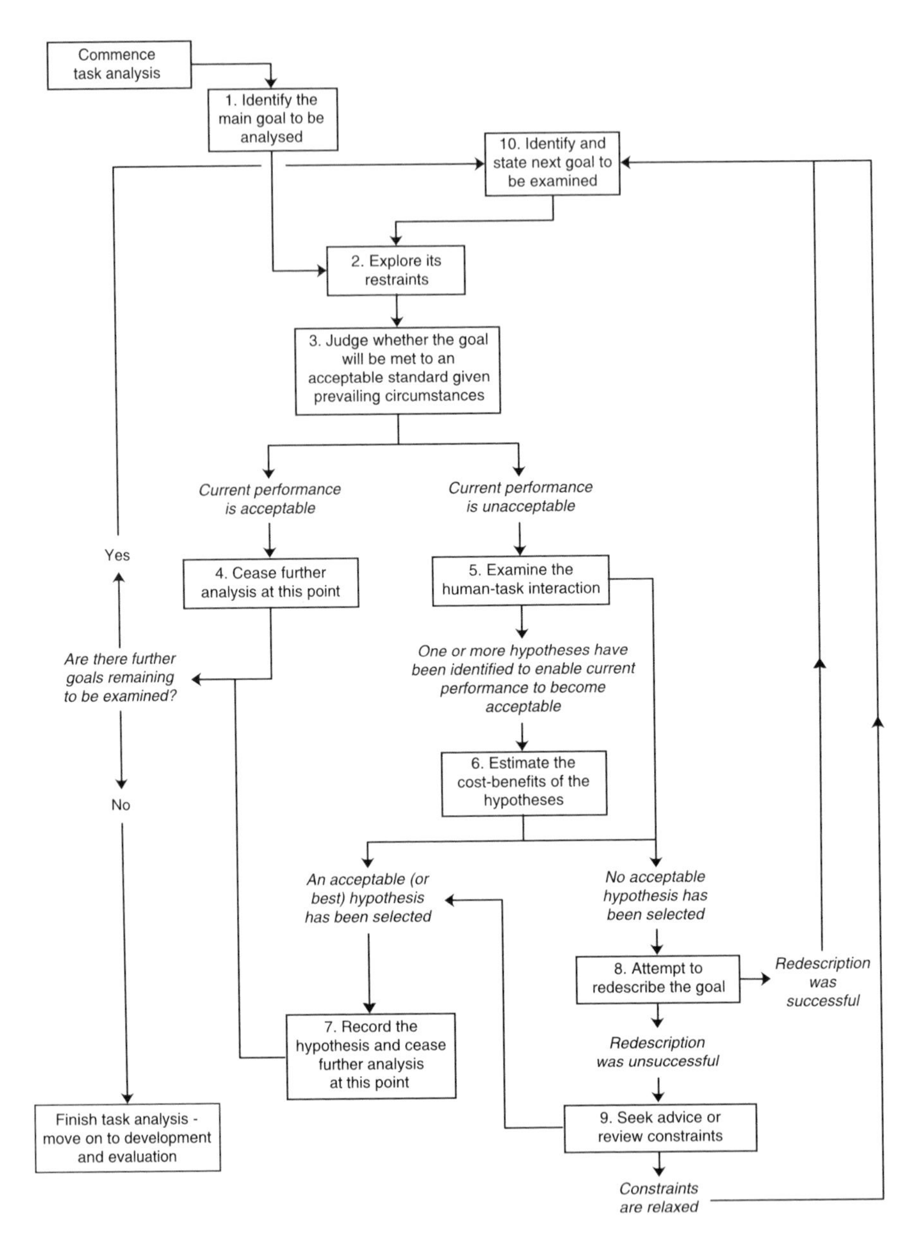

Figure 5 The cycle of task analysis decisions. Working through this framework enables the analyst to generate the hierarchical task structure characteristic of HTA and follow each of the stages typical of task analysis.

necessary (9 in this case). The plan governing its sub-goals is given the same number as its super-ordinate goal (in this case 'plan 0') and can then refer to the sub-goals in terms of their numbers. Operation 3 is redescribed. Its sub-goals are numbered from 1 to 5; its plan is labelled 'plan 3'. Its operation 3 is further redescribed; its sub-goals are numbered 1 to 2; and its plan is given the label 'plan 3.3'. People may find it clearer to write down a full number for each operation, rather than just a single digit. An objection to this is that numbers can become unwieldy if the analysis is taken to several levels. Another advantage of numbering operations

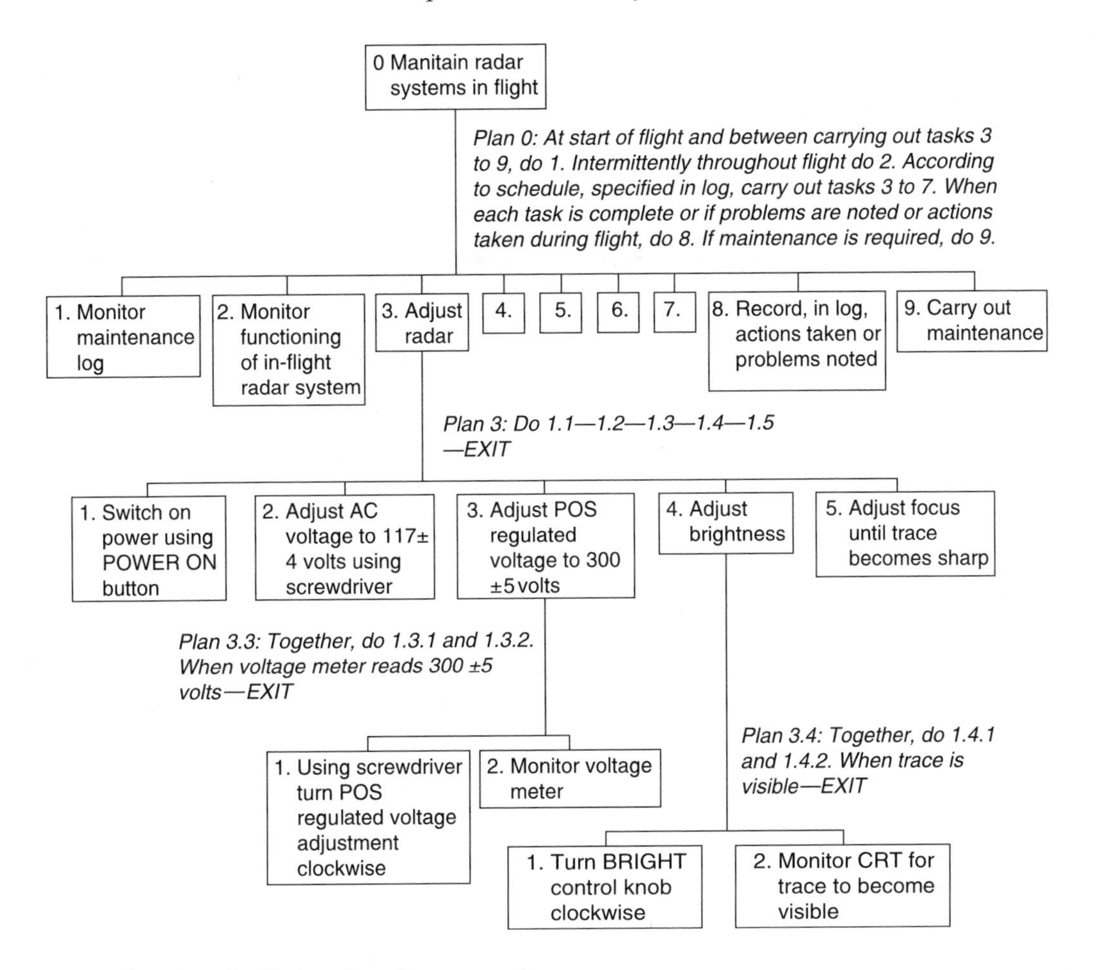

Figure 6 HTA of Miller's radar adjustment task.

with single digits is that it is easy to edit an analysis numbered in this way — modifying the numbers of higher level operations will not affect the numbering of lower level operations. But it is, after all, simply a clerical tool and other numbering methods could be adopted. It is most important that the analyst clearly states how the adopted numbering system works for each HTA recorded so the reader is able to find their way around.

Tabular formats

There are two main aspects in representing task analysis in a table. First there is the issue of how to represent the hierarchical structure in a linear sequence down the page. Second is the issue of what additional information should be provided in the analysis record. This entails extending the columns across the page to include some of the category information in the task analysis methods described above. These two issues will be treated separately.

Table 2 shows one way of sequencing task analysis in a table. The task is the radar adjustment task, represented graphically in Figure 6. The same numbering system is adopted. In this version, the left-hand column contains the absolute number of the goal or sub-goal currently being described and the operations and plans that form the task analysis. The next column indicates whether there is further redescription of each operation — this is particularly important when working through large tables. The next column indicates

Table 2 Illustration of a Tabular Format Using the Task Represented in Figure 6

Task analysis	scribe d	I-A-F	Notes
0 Maintain radar systems in flight			
plan 0: At start of flight and between tasks do 1. Intermittently throught flight do 2. According to schedule specified in log carry out tasks 3 to 7. When each task is complete or if problems are noted or actions taking during fight, do 8. If maintanence is required, do 9.			
1 Monitor maintenance log	no	I	Ensure a format for the log which prompts technician to carry out routine tasks and reminders to check-up out on unscheduled contingenices are clear. Review ceasing description.
2 Monitor functioning of in-flight radar system	no	I	How are key parameters displayed for monitoring? Review ceasing redescription.
3 Adjust radar receiver	yes		This will ormally be required very 25 hours.
4 *for illustration only*	no		
5 *for illustration only*	no		
6 *for illustration only*	no		
7 *for illustration only*	no		
8 Record, in log, actions taken or problems noted	no	I	Ensure layout of log is clearly designed to record action taken. Devise means of highlighting and bringing forth problems.
9 Carry out maintenance	no		Requires redescription.

Task / plan			Comments
3 Adjust radar receiver			
plan 3: *Do 1.1 — 1.2 — 1.3 — 1.4 — 1.5 — Exit*			
1 Switch on power using POWER ON button	no	F	Avoid starting system with covers removed from high voltage units: personal hazard. Redescription would focus on the feedback to indicates proper functioning of the system — hum, pilot light, range indication lights, tiltmeter pointer.
2 Adjust AC voltage to 117 ± 4 volts using screwdriver	no	F	
3 Adjust POS regulated voltage to 300 ± 5 volts	yes		
4 Adjust brightness	yes		
5 Adjust focus until trace becomes sharp	no		
3 Adjust POS regulated voltage to 300 ± 5 volts			
plan 3.3: *Together, do 1.3.1 and 1.3.2. When voltage meter reads 300 ± 5 volts — Exit*			Siting of adjustment screw given need to monitor voltage may be problematic.
1 Using screwdriver turn POS regulated voltage adjustment	no	A	Review equipment design or training recomment or plan 3.3
2 Monitor voltage meter	no	F	Review equipment design or training recomment or plan 3.3
3 Adjust brightness			
plan 3.4: *Together do 1.4.1 and 1.4.2. What trace is visible — EXIT.*			
1 Turn BRIGHT control know clockwise	no		
2 Monitor CRT for trace to become visible	no	F	Criterion for visibility appears too subjective. Check that subjective criteria satisfy the requirements of operational staff. Modify training or job design as appropriate.

whether there are any problems that can be attributed to input, action or feedback difficulties. This I-A-F column is often not necessary if the task does not involve any significant perceptual/motor operations. The right hand column contains any notes that are felt worth including. The 'notes' column provides one of the main benefits of the tabular format, since these can be allowed to extend as far as required. It is the notes column that tends to be expanded to develop a 'tabular task analysis'. Diagrams may be more easily understood by people, but tables are more thorough. It is often best to use both to record and communicate the analysis.

It is a common misconception that HTA is always presented as a hierarchical diagram. The hierarchical nature of HTA refers to the relationship between goals and sub-goals and not the drawing. Thus, Table 2 is the same HTA as Figure 6, albeit in a tabular format. A detailed account of HTA, including examples of HTA from different contexts, is provided by Shepherd (2001).

Cognitive task analysis

The term 'cognitive task analysis' (CTA) is broad, like task analysis itself, and refers to a wide variety of topics dealing with methods to analyse the processes and products of cognition. CTA is predicated on the view that work has changed in a qualitative fashion since the advent of the computer and that more traditional approaches to task analysis cannot be expected to cope. The justification for CTA methods stems from the view that 'traditional' (i.e., non-CTA) methods of task analysis rely on observing actions, and are, therefore, unsuitable for analyzing tasks where cognition plays a major role. This charge is simply wrong and fails to take into account many of the challenges that task analysis researchers have been dealing with for years. Task analysts confronted with developments in technology and work organisation have not shied away from the cognitive aspects of tasks and many task analysis approaches have been shown to be entirely satisfactory with respect to dealing with industrial and commercial requirements in a practical way. It is true that methods such as HTA do not provide a categorical account of human cognition, but this is not its purpose. There is no *a priori* reason why task analysis should represent the psychology of people carrying out work. Indeed, it is explicitly not trying to do this, because its purpose is to reflect and examine the *demands* that a system places on the human operator and not to provide a defensible account of human behaviour. Moreover, focusing on the cognitive processes that a worker is adopting merely provides opportunity to comment on that strategy for working. It limits opportunities for the analyst and the client to propose alternatives that might be more satisfactory in serving the requirements of the system or more straightforward for the worker to accomplish. CTA should be viewed with some caution to ensure that it is not pursued on the basis of inadequate arguments; it should be used only where its use is justified.

Common justification for CTA

The need for a separate field of cognitive task analysis has often been justified because increasing attention must be paid to the cognitive aspects of tasks (Hoffman *et al.*, 1998). This view is challenged by Neerincx and Griffioen (1996) who point out that justification for cognitive forms of analysis seems to ignore the fact that existing forms of task analysis are well able to handle many aspects of the cognitive content.

Preece *et al.* (1994) draws a distinction between CTA and HTA, representing HTA as, '... concerned with establishing an accurate description of the steps that are required in order to complete a task ...,' whereas the focus in CTA is to, '... capture some representation of the knowledge that people have, or need to have in order to complete the task' (p. 417). They point out that some actions are physical, such as pressing buttons, and some are mental or cognitive, such as deciding which button to press or when to press it. This, of course, is obvious, but it does not mean that HTA cannot handle cognition in the context of a task and relate it to the actions that must take place. The importance of cognition in performance is fundamental to experimental psychology and it has not been overlooked by innovators of task analysis methods; successful performance in all tasks — even simple actions — depends upon the interaction between physical and cognitive elements.

Hoffman and Woods (2000) have outlined what they see it as a number of *threads* of activity that have led to an increasing focus on the cognitive content of work.

> Among the threads that began to use or draw a link to the term CTA were
>
> 1. the study of expertise for the development of different forms of computer aided training ... — what might be labeled its *cognitive learning* thread;
> 2. the need to aid human cognition and performance in complex, high-consequence settings ... — what might be labeled a *functionalist cognitive engineering thread*;
> 3. the need to understand work cultures as a result of the consequences of technology change ... — what might be called an *ethnography* of work thread;
> 4. the need to understand work that is conducted using computers, and to support the design of human computer interfaces ... — what might be called a *cognitive simulation* thread; and
> 5. the inability of formal or normative methods of decision-making to capture observations from field studies of experts in complex settings ... — what might be called a *naturalistic decision-making* thread.
>
> Hoffman and Woods (2000, p. 1)

This list overstates changes in work practices; all of these are threads that have affected how work is organised and analysed for as long as task analysis methods have been considered. Granted, the types of work dealt with by early forms of work-study, for example, did focus on that which could be observed, but this suited a type of work problem that needed to be addressed at the time for which work-study methods fitted the bill. This did not mean that other forms of work did not exist, or were not recognised by people concerned with analyzing work, or were not addressed successfully by other task analysis methods. Interest in cognitive tasks has been prominent for many years in the context of military, nuclear power, steel, transportation and petrochemical industries, for example. Indeed, the challenges of some of these industrial tasks are greater than those of the tasks that have emerged with the development of computers because technologies in these other industries do not always provide the same freedom to design tasks with people in mind that is available in computer systems.

It is to be regretted that much of the debate on task analysis methods tends to be confrontational, with one group of protagonists arguing how other methods fail to deal with a set of phenomena. Such claims are usually rhetorical, serving to clear the path for a new set of ideas to take their place. They miss the point that task analysis methods are tools to aid the understanding of work, sufficient to allow hypotheses to be suggested in areas that are deemed to be critical to the success of a system; they are not scientific constructs. CTA methods, with their focus on the processes and products of cognition, often fail to appreciate the work context in which their focus resides. It is equally the case that many task analysis methods, including HTA, that focus on the work context, do not provide a detailed account of cognition. HTA is able to pin-point where cognition, in the form of planning, decision-making and monitoring, is required in a task, but is then often forced to stop and look elsewhere to illuminate the problem, as in the case of fault-diagnosis, discussed above. This issue will be developed further below.

Variants of cognitive task analysis

CTA is a generic title for a range of methods that seek insights into applied cognition. Several CTA methods use task decomposition, similar in structure to HTA, but entailing assertions about how parts of the task description are represented. Thus a *GOMS* analysis (*G*oals, *O*perations, *Methods* and *S*election rules — Card *et al.*, 1983) requires the analyst to identify the rules for selecting methods for organising operators to achieve goals. TAG (Task Action Grammar — Payne and Green, 1986) entails coding actions according to rules of syntax, to facilitate making comparisons between human–computer interfaces for example.

TAKD (Task Analysis for Knowledge Description — Diaper, 1989) entailed utilising rules for knowledge elicitation against task descriptions. However, Diaper *et al.* (1998) have now developed an approach that integrates elements of TAKD with a technique commonly used in software engineering, *data flow analysis*, and now no longer advocate the use of TAKD (Diaper, 2001).

More recently, approaches to cognitive modelling in support of CTA have attempted to develop what Kjaer-Hansen (1995) describes as 'unitary theories of cognitive architectures'. These are more global theories which aim to understand cognition in terms of the role of knowledge within the processes of learning, inference and decision-making in producing behaviour, for example SOAR (Laird *et al.*, 1987) and ACT* (Anderson, 1983). From a performance perspective, the unitary theories would account for the manner in which information is acquired, stored, organised and retrieved to account for decision-making and performance across a range of task elements.

Reconciling CTA methods with HTA

One concern with making comparisons between CTA and more traditional task analysis methods is that the choice is often represented as exclusive. If a task contains elements of cognition that cannot be fully dealt with by a particular form of task analysis, that should not be a reason for abandoning that method, but it is a reason for seeking ways to accommodate other methods. HTA is itself catholic, taking on board a range of task analysis ideas aimed at enabling the analyst to focus attention where it is required. In any event, there is a flaw in the argument that says the CTA methods are useful for some tasks while more traditional methods are useful for others. A better solution is to reconcile both approaches so that the judgment of where CTA methods should be adopted can be made according to the merits of the task.

By starting off with a more general and neutral method of task analysis, such as the HTA framework, the analyst can appraise the overall task, identify which aspects are critical and deal with those aspects that are amenable to the method. As past of the process of conducting the HTA, the analyst will identify which aspects of the task are critical and be able to infer where *monitoring*, *judgment* and *decision-making* is required. The analysis will show how these cognitive elements relate to one another. For example, it is common in task analysis to relate the processes of *monitoring* to *problem detection*, *diagnosis*, *compensation*, *rectification* and *recovery*. These relationships will be recorded in the analysis. If these elements need further investigation the CTA methods or ethnographic methods can be deployed as appropriate to provide necessary insights about sources of operator weakness or how to proceed in developing a solution.

This combination of methods can be very effective. Much of the task will be dealt with economically by the HTA framework. The specialist effort required to carry out CTA methods will be focused where it is justified. The specialist CTA will also be better informed in appreciating how that element fits into the broader task. For example, it will be clearer whether the operator has to timeshare between two or more operations. Air Traffic Control provides an interesting example. Part of this task is to plan and oversee changes of routes by aircraft in a sector to avoid possible collisions. This should be done to minimise inconvenience and cost to respective aircraft companies as well as to maintain safety. To optimise such savings as well as maintain safety would entail extensive detailed analysis by the ATC officer which would take time and detract attention from other aircraft and events in the sector. The view often taken by ATC managers is that too much attention on optimising route changes can risk a reduction of vigilance of other aircraft with possible disastrous consequences. They, therefore, tolerate suboptimal solutions from the viewpoint of efficiency in order to ensure that safety standards remain paramount. It is important fully to appreciate any concurrent operations that must be maintained, rather than focusing too closely on a particular area of concern.

In other contexts, where decision-making is required, the HTA can indicate the nature of the decision to be taken and the conditions that will affect it. HTA should make clearer the conditions that the operator experiences prior to carrying out the operation in question, how other task elements must compete for cognitive resources and what options or consequences will follow the task element in question. For example, process control tasks require operators to carry out fault-diagnosis. It is important to know whether the operator has been monitoring and controlling the plant prior to the disturbance, since this will affect the working knowledge available to carry out the task. The operator will know, for example, which actions preceded the onset of the problem and should have up-to-date knowledge of the state of the process and the plant. It is important to know whether *diagnosis* has to compete for mental resources with other tasks, such as *planning safety contingencies* or *monitoring other safety systems*, or whether there is excessive time pressure. It is important to know the range of options available — deciding whether to *continue production* or *shut down* is a simpler choice than having to select from a range of remedial actions as a consequence of diagnosis. It is also important to be aware of the stressors on the operator in terms of the consequences of making a mistake. Unless these contextual factors are understood, cognitive task analysis will be of little value because the behaviours investigated may have no bearing on what actually happens and the contextual constraints affecting behaviour cannot be known.

Rather than distinguishing between cognitive and non-cognitive task analysis we might more usefully consider how a general task analysis strategy identifies tasks where performance is driven substantially by cognitive factors. It is then that CTA methods will be justified.

Concluding remarks

This chapter has provided a brief history of the development of task analysis methods outlining the influences that shaped the task analysis methods that we use currently. There is a substantial literature in this area and much of it can be confusing because writers about task analysis often have different experiences and different perspectives from each other. One of the main problems with task analysis literature is that it tends to provide accounts of completed analyses without showing the reader how the analysis was conducted and what factors were taken into consideration. For this reason the activites involved in five common stages of a task analysis have been outlined. The first of these, *appraising the problem*, is usually overlooked in published accounts of task analysis because it lacks the straightforward prescriptiveness that people want when they are seeking guidance on a new technique.

Without an initial appraisal of the problem, however, the analyst may misdirect efforts and fail to appreciate what is expedient in moving the analysis forward. In particular they may fail to establish an effective working relationship with the client. Other stages — *collecting information, organising information, modelling behaviour*, and *suggesting solutions* — relate more closely to the activities that have become known as task analysis.

A description of Hierarchical Task Analysis is provided because it offers a broad methodology for analysis that enables the analyst to take into account many of the lessons learned from its precursors. It is also effective in dealing with a wide range of industrial and commercial tasks. The analyst is able to use the analysis to identify operating concerns. Moreover, the form of task description used is transparent to a client, so the needs of the client can be kept in focus and the analyst can avoid straying from what is required in the task. HTA is a catholic approach that allows the analyst to accommodate a range of associated techniques in the process of examining the task. This includes cognitive task analysis (CTA) methods. CTA is discussed in order to demonstrate how it relates to the more traditional approaches to task analysis. The view taken in the chapter is that CTA methods should be used within the framework of a wider HTA so that its use is properly justified and focused by appreciating how cognitive task elements fit into the broader picture.

References

Anderson, J.R. (1983). *The Architecture of Cognition*. (Cambridge, MA: Harvard University Press).

Annett, J. and Duncan, K.D. (1967) Task analysis and training design. *Occupational Psychology*, **41**, 211–221.

Annett, J., Duncan, K.D., Stammers, R.B. and Gray, M.J. (1971). *Task Analysis*. Training Information Paper No. 6. (London: HMSO).

Card, S.K., Moran, T.P. and Newell, A. (1983). *The Psychology of Human–Computer Interaction*. (Hillsdale, NJ: Erlbaum).

Crossman, E.R.F.W. (1956). Perceptual activities in manual work. *Research*, **9**, 42–49.

Diaper, D. (1989). Task analysis for knowledge descriptions (TAKD): the method and an example. In: D. Diaper (ed.) *Task Analysis for Human–Computer Interaction*. (Chichester: Ellis Horwood), pp. 108–159.

Diaper, D. (2001). Task analysis for knowledge descriptions (TAKD): a requiem for a method. *Behaviour and Information Technology*, **20**, 199–212.

Diaper, D., McKearney, S. and Hurne, J. (1998). Integrating task and data flow analyses using the pentananalysis technique. *Ergonomics*, **41**, 1553–1582.

Duncan, K.D. (1972). Strategies for the analysis of the task. In: J. Hartley (ed.) *Programmed Instruction: An Education Technology*. (London: Butterworths), pp. 19–81.

Duncan, K.D. (1974). Analytical techniques in training design. In: E. Edwards and F.P. Leeds (eds) *The Human Operator and Process Control*. (London: Taylor & Francis), pp. 283–319.

Fleishmann, E.A. and Quintaince, M.K. (1984). *Taxonomies of Human Performance*. (New York: Academic Press).

Gotcher, J.M. (1992). Assisting the handicapped: the pioneering efforts of Frank and Lillian Gilbreth. *Journal of Management*, **18**, 5–13.

Hoffman, R.R. and Woods, D.D. (2000). Studying cognitive systems in context: Preface to the special section. *Human Factors*, **42**, 1–7.

Hoffman, R.R., Crandall, B. and Shadbolt, N. (1998). Use of the critical decision method to elicit expert knowledge: a case study in the methodology of cognitive task analysis. *Human Factors*, **4**, 254–276.

Kirwan, B. and Ainsworth, L.K. (eds) (1992). *The Task Analysis Guide*. (London: Taylor & Francis).

Kjaer-Hansen, J. (1995). Unitary theories of cognitive architectures. In: J-M Hoc, P.C. Cacciabue and E. Hollnagel (eds) *Expertise and Technology: Cognition and Human–Computer Cooperation*. (Hillsdale, NJ: Erlbaum), pp. 43–54.

Laird, J.E., Newell, A. and Rosenbloom, P. (1987). SOAR: an architecture for general intelligence. *Artificial Intelligence*, **33**, 1–64.

McCormick, E.J. (1979). *Job Analysis: Methods and Applications.* (New York: American Management Associations).

Miller, R.B. (1962). Task description and analysis. In: R.M. Gagné (ed.) *Psychological Principles in Systems Development.* (New York: Holt, Rinehart & Winston), pp. 187–228.

Miller, R.B. (1967). Task taxonomy: science or technology? In: W.T. Singleton, R.S. Easterby and D. Whitfield (eds) *The Human Operator in Complex Systems.* (London: Taylor & Francis), pp. 167–176. (also, *Ergonomics*, 1967, **10**, 167–176).

Neerincx, M.A. and Griffioen, E. (1996). Cognitive task analysis: harmonizing tasks to human capacities. *Ergonomics*, **39**, 543–561.

Patrick, J. (1992). *Training: Research and Practice.* (London: Academic Press).

Payne, S.J. and Green, T.R.G. (1986). Task-action grammars: a model of the mental representation of task languages. *Human–Computer Interaction*, **2**, 93–133.

Pheasant, S. (1996). *Bodyspace: Anthropometry, Ergonomics and the Design of Work.* (London: Taylor & Francis).

Preece, J., Rogers, Y., Sharp, H., Benyon, D., Holland, S. and Carey, T. (1994). *Human–Computer Interaction.* (Harlow: Addison Wesley).

Rasmussen, J. (1986). *Information Processing and Human–Machine Interaction: An Approach to Cognitive Engineering.* (New York: North-Holland).

Reason, J. (1990). *Human Error.* (Cambridge: Cambridge University Press).

Robson, C. (2002). *Real World Research* (2nd edn). (Oxford: Blackwell).

Salvendy, G. and Seymour, W.D. (1973). *Prediction and Development of Industrial Work Performance.* (Chichester: Wiley).

Sanderson, P.M. and Harwood, K. (1988). The skills, rules knowledge classification: a discussion of its emergence and nature. In: L.P. Goodstein, H.B. Anderson and S.E. Olsen (eds) *Tasks, Errors and Mental Models.* (London: Taylor & Francis), pp. 21–34.

Seymour, W.D. (1966). *Industrial Training for Manual Operations.* (London: Pitman).

Shepherd, A. (2001). *Hierarchical Task Analysis.* (London: Taylor & Francis).

Stammers, R.B. (1995). Factors limiting the development of task analysis. *Ergonomics*, **38**, 588–594.

Stammers, R.B. and Shepherd, A. (1995). Task analysis. In: J.R. Wilson and E.N. Corlett (eds) *Evaluation of Human Work.* (London: Taylor & Francis), pp. 144–168.

Stanton, N. and Young, M.S. (1999). *A Guide to Methodology in Ergonomics.* (London: Taylor & Francis).

Wickens, C. and Hollands, J.G. (2000). *Engineering Psychology and Human Performance* (3rd edn). (Upper Saddle River, NJ: Prentice Hall).

Chapter 7

Verbal protocol analysis

Lisanne Bainbridge and Penelope Sanderson

Editor's note

This chapter has been reprinted in its original form from the second edition. Although the chapter has not been revised and updated, it is felt that the basic description and understanding that it gives of verbal protocol analysis is still relevant 10 years on. The reader will be able to understand how to conduct a verbal protocol analysis, and what they might and might not gain from such study, by reading the chapter.

Introduction

There are many complex jobs in which the outcome of thinking does not emerge in observable action. For example, one can think out a plan of action, assess it, and decide it is inadequate for the purpose, or one can work out the implications of a situation, and memorise the decision for use later. If we want to train and support these types of work, then we need information about these mental processes. One apparently obvious way of getting this information is to ask people to 'think aloud' while they are doing the task. These verbal reports are called 'verbal protocols'. To distinguish them from other knowledge elicitation techniques (discussed by Shadbolt and Burton in Chapter 14 of this book), such protocols essentially report mental processes used during a task, rather than being the answers to questions about the task which are given while the person is not actually doing it. This chapter will focus on the analysis of naturalistic, 'on-line' (concurrent) verbal protocols, rather than retrospective, walkthrough, or prompted verbal reports.

This review of verbal protocol methodology will be in four sections:

1. The status of verbal protocols as evidence; their validity as data, and what types of data can be obtained.
2. Technique; practical aspects of collecting verbal protocols, and devising task situations.
3. Analysis; how to extract valid and useful information from the verbal protocols.
4. Tools; different hardware and software media that can help researchers perform verbal protocol analysis.

From the outset it should be emphasised that there is as yet no simple, quick method of collecting and analysing protocols, and the classic techniques of Newell and Simon (1972) and Ericsson and Simon (1984/1993) are likely to remain practical only for research. However verbal protocol analysis can be performed at different levels of detail, and there are signs that some simpler versions of the technique can be useful in well-chosen applied settings, especially if appropriate supporting software tools are used.

Verbal protocols as evidence

As there is no way of observing someone's mental behaviour directly, it is not possible to test whether there is a correlation between what someone actually thinks and what they say they think. Consequently, when verbal reports do not fit a theory of mental behaviour one does not have to reject that theory. Because the main test for the scientific value of data is that they can be used to reject theories, some academic psychologists feel strongly that verbal data are useless.

In more practical contexts, there are two main uses for verbal protocol data. First, they can be a source of hypotheses about cognitive processes, and so of predictions about non-verbal behaviour. The predictions can be tested without involving verbal reports. Suppose one note from verbal reports that someone doing a task uses prediction and complex working memory. One can then compare performance using equipment designed on the assumption that prediction is used, and equipment designed assuming that prediction is not used, and test, which is best.

Secondly, if someone says something, evidently they do have this knowledge somewhere in their heads. The question then becomes: Is this reported activity and knowledge actually what is being done and used at the time that it is reported, or is it epiphenomenal? We need to determine when verbal reports are likely to correlate sufficiently highly with cognitive processes for them to be useful for practical purposes, and when they are not. Several studies give indirect evidence of the circumstances under which people do give valid reports of their thought processes.

In a highly influential paper, Nisbett and Wilson (1977) reported studies which should be taken seriously by anyone involved with knowledge elicitation, and which have developed into a major research area in social psychology. Their studies are of people's self-perception' of the influences on their attitudes and judgements. Nisbett and Wilson carried out experiments in which people were observed acting in various circumstances. They asked the people, in questionnaires, what they would do if faced with these situations, and what they thought was influencing their behaviour. Nisbett and Wilson concluded that people can give the same verbal report for two situations in which they actually behave quite differently. People can also claim that factors influenced their behaviour which actually did not, and can deny that factors influenced their behaviour which actually did. They concede that an individual does know more than an observer about private facts, not only the results of their thinking but also their focus of attention at a given time, current sensations, emotions, evaluations, plans, intentions and personal history. Nisbett and Wilson concluded that people are conscious of the results of their thinking, such as their attitudes or judgements, but have no special access to the process of thinking, such as how they made those judgements. Overall, Nisbett and Wilson's research strongly suggests that people are not very, good at reporting the factors which influence their decisions.

This suggests that knowledge elicitation techniques, which concentrate on the content of thinking, are more likely to be valid than reports, which claim to be observations of the processes underlying thinking. Verbal protocols mainly contain factual statements, especially when people are under time pressure in a real task. The contents of verbal protocols and interviews differ in syntax and type. This is illustrated in these report fragments:

1. 'I shall have to cut (furnace) E off, it was the last to come on, what is it making by the way? E makes stainless, oh that's a bit dicey, I shall not have to interfere with E then.'
2. 'If a furnace is making stainless, it's in the reducing period, obviously it's silly, when the metal temperature and the furnace itself is at peak temperature, it's silly to cut that furnace off.'

These two examples come from the same furnace operator, the first while he was making decisions about the task, the second during a lull in activity a few minutes later. The verbal protocol contains mainly observed facts and results of decisions. These are the data from which an analyst infers the underlying cognitive processes and knowledge used, rather than getting direct information about them. Cuny (1979), who explicitly compared verbal protocols and interview data from the same process operators, came to the conclusion that interviews give more general information about strategies but omit the details of a particular working context which can mean that behaviour differs.

The research by Nisbett and Wilson (1977) did not go unchallenged. The most important replies to it were from Ericsson and Simon (1980, 1984/1993) who, while acknowledging that verbal reports could misrepresent mental processing, argued that in certain circumstances verbal reports should be expected to represent the products of mental processing rather accurately. They detailed different types of distortion, and suggested a simple human information processing model of how verbal reports are generated. They proposed that if the results of mental processing are normally in short-term memory (even if the person is not thinking aloud), if these short-term memory contents are in an easily verbalisable code (e.g., not spatial or motor code) and if the short-term memory contents are not changing so fast that they cannot be articulated fast enough, then verbal protocols have a much better chance of being valid (see Simon and Kaplan, 1989, for a recent summary). However, valid verbal protocols depend upon other factors as well. In the next sections we will outline the types of distortion that can affect verbal protocols, which a researcher should take into account before undertaking a verbal protocol study.

We might add — though this is a hypothesis based on experience rather than being based on careful experiments — that the information people report about their behaviour in familiar working situations may be more accurate than the information they report about their behaviour in general situations. More detailed and accurate information is available to them about the state of the working environment and the results of their behaviour. In addition, in many working environments people are explicitly — and usually verbally — trained to respond to particular aspects of the environment. Experienced workplace instructors, in particular, are very fluent at expressing their behaviour and thinking as they carry out their work.

Types of distortion, and types of information obtainable

In practice, we can consider the ways in which a verbal report is likely to be distorted, with the aim of minimising these effects. Note that many types of distortion could occur in all verbal reporting situations, not just with verbal protocols.

First, having to give a verbal protocol changes the task and may change the way the task is done. When there are alternative methods of doing the primary task, the need to give a verbal protocol may influence the person to use a method which is more easily described. Someone could do a task in a mechanical rather than an abstract way, as the former is better fitted to the pace of speech, or do things in sequence rather than doing several things at the same time. They may use a version of the task which is more verbal in form, for example, a beginner's method which has been explained in words, or they may follow the official regulations more closely than usual because these are in verbal form.

Second, there are time constraints when giving a verbal protocol. In problem solving situations, many things may quickly pass through people's minds and be forgotten before there is time to report them. If people are working under time pressures, there may not be time to mention everything that is relevant. Moreover, they may not mention information, which they collect while reporting other activities, which may lead to unexplained behaviour later.

Third, giving a verbal protocol is a social situation involving self-presentation, and so can be influenced by social biases. People can select what they think it is appropriate to say. People in experiments are often overly co-operative, and try to say what they think the experimenter wants to hear. If the person reporting feels that the listener has a superior status, and perhaps is powerful, then there may be pressures to present a non-habitual approach to the work, such as appearing rational, knowledgeable and correct or, conversely, being uncooperative and unforthcoming. An experienced worker will talk much more freely and fully if they think that the listener will understand and value what they are talking about. On the other hand, the worker may overestimate the listener's understanding and may not mention points which seem to them to be obvious.

Fourth, distortions arise when the person usually does the job in a nonverbal way, but is asked to give a verbal report. People are not consciously aware of how they carry out perceptual-motor skills, for instance when skating or swimming. Parts of cognitive skills are automatic, and there is evidence that words, pictures and the feel of movements are each dealt with by different parts of the brain. Thus, images and movements may not be accurately represented when they have to be reported in words. Moreover, if the person's vocabulary is limited, they may not be able to find an expression for all that they know. The knowledge expressed may therefore appear limited, whereas actually it is the language, which is limited, not the skill.

Fifth, knowledge about the components, mechanisms, functions and causal relations in a machine, memories of specific past events, and helpful categories will be mentioned explicitly only if the task involves some problem solving that requires the person to review this sort of evidence. Otherwise the researchers will only be able to infer such knowledge indirectly from the fact that the person would not have been able to act in a certain way if they had not had particular knowledge. This underlines the general point that verbal protocol evidence may provide only a limited sample of the total knowledge available to the person being studied. Even if the task is well-controlled, it will not be possible to test a large enough number of situations to explore all of the person's knowledge. When collecting protocols in a real task situation, the researcher must be aware that the evidence collected will be limited to what happened during the recording.

In summary, there are many factors that can compromise the validity of verbal protocols. Rather than reject the technique entirely, researchers should critically consider each new situation in light of the above factors to see if valid data might be expected from verbal protocols. The same critical scrutiny should be applied to the results of the protocol analysis. If there is reason to suspect that verbal protocols may be subject to distortion, then this suggests that verbal protocols should be combined with more conventional behaviour observation techniques (see Chapters 2 and 3) to check on these distortions and to provide potentially corroborating data. If protocols are likely to be distorted because of the task or social situation, this has implications for the technique used in collecting the protocols, which will be mentioned below.

Getting the most out of verbal protocol analysis

Clearly, verbal protocols are more suited to obtaining some types of information about cognitive processes than others. We have already seen that they may give data on the outcomes but not the processes of skill. As the protocols are made while performing a task, they can give explicit information about the sequence of items considered. From this the strategy being used may be inferred, and also the working memory contents. As we will see, if there are enough data it is possible to identify decision-making choice points, and what determines how the decisions are made. Even leaving aside all the problems outlined in the previous section, verbal protocol analysis is anyway a very difficult and time-consuming technique. There is no one accepted way of performing verbal protocol analysis — no 'canon' — and researchers must adapt existing methods to new situations, and maybe even develop new methods of their own. First, when embarking on a new study the researcher should seriously examine whether verbal protocol analysis is needed at all. The answer is often 'yes', but there are often situations which are served equally well by more structured knowledge elicitation techniques. Second, the researcher needs to decide how to sample the data: how many subjects will be used, how much data will be collected, under what conditions, etc. Third, the researcher needs to have some a priori idea of the general kind of approach he or she will take to the analysis of the data. Some approaches are considerably more time-consuming than others. On the one hand, the researcher may wish to document, and loosely describe, certain types of problematic situations. This is less time-consuming because a detailed analysis of every phrase may not be needed to identify and analyse the situation, and only certain parts of the protocol need to be analysed. On the other hand, if the researcher wishes to build a complete computational model of how the subject performs the task, a much more microscopic approach is needed. Sanderson and Fisher (1993) have discussed the 'analysis time : sequence time' (AT : ST) ratio, which indicates how much analysis time will be required for every unit of recorded time (sequence time). Estimates range from 10 : 1 to 1000 : 1, and transcription alone can account for 5 : 1 or more. Thus, researchers should keep approximate estimates of the AT : ST ratios they experience for different types of analysis, and set realistic goals for new research projects.

Verbal protocol analysis process

Figure 1 diagrams the typical verbal protocol analysis process, showing how protocols are used to move from research or design questions through to statements or conclusions that answer those questions. The figure divides verbal protocol analysis into the three stages of (1) collecting and storing protocol data, (2) preparing protocol data for analysis, and (3) analysing the explicit and implicit content of the protocol. We will use this structure to present protocol analysis in the sections that follow.

Technique of collecting the protocols

In order to collect protocols, the person is asked to 'think aloud' whilst performing some task of interest, and their comments are recorded and usually transcribed. Where possible it is useful to make short recordings which can then be transcribed and checked with the speaker during the same session; the speaker can read through the transcripts to correct mistakes and clarify vague passages (though this can involve distortions of the interview situation).

It is possible to make a much more detailed analysis of the data if a simultaneous recording is made of the information the person receives and the actions they make.

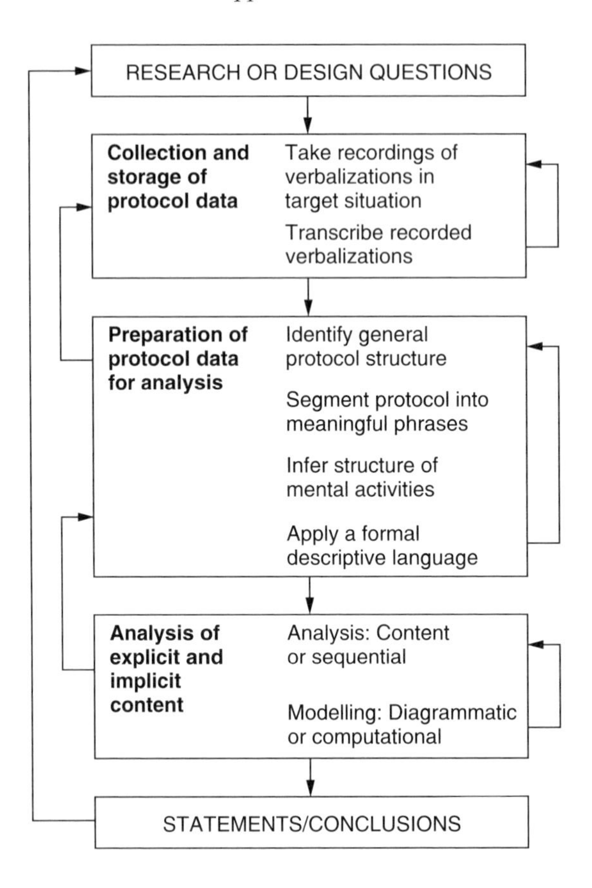

Figure 1 Flow chart of the verbal protocol analysis process.

The more context present, the easier it is for the researcher to interpret ambiguities. This is useful in at least two ways:

1. People talking while doing a job frequently use pronouns, e.g., '*it's* at 55'. The objective record shows what is 'at 55', and therefore what is being referred to. Taking video rather than audio recordings can also help when people use general anaphoric references supplemented by pointing, such as '*that's* too high so I'll lower *this* until it's between *these*'.
2. It is useful to have information about the total task situation which may later be used to see if people's behaviour is influenced by the situation without being mentioned in the protocol (see above on the completeness of protocol data and below on analysis).

This observation record can be made in several ways, but does require considerably more work to collect and analyse than a simple audio record. An observer can log the events (for example, see Figure 2). Alternatively, if the investigation is in a high-technology environment, the data can be captured and stored in a data file that can later be printed or imported into a software protocol analysis environment (see also Chapter 9). Moreover, video recordings of the workplace and the operators' movements often substitute for audio recordings entirely. Video preserves the audio, but reveals aspects of the task that may not be verbalised. Video greatly increases the amount of data that a researcher can analyse if he or she so wishes, and the researcher must exercise discipline in choosing what to analyse.

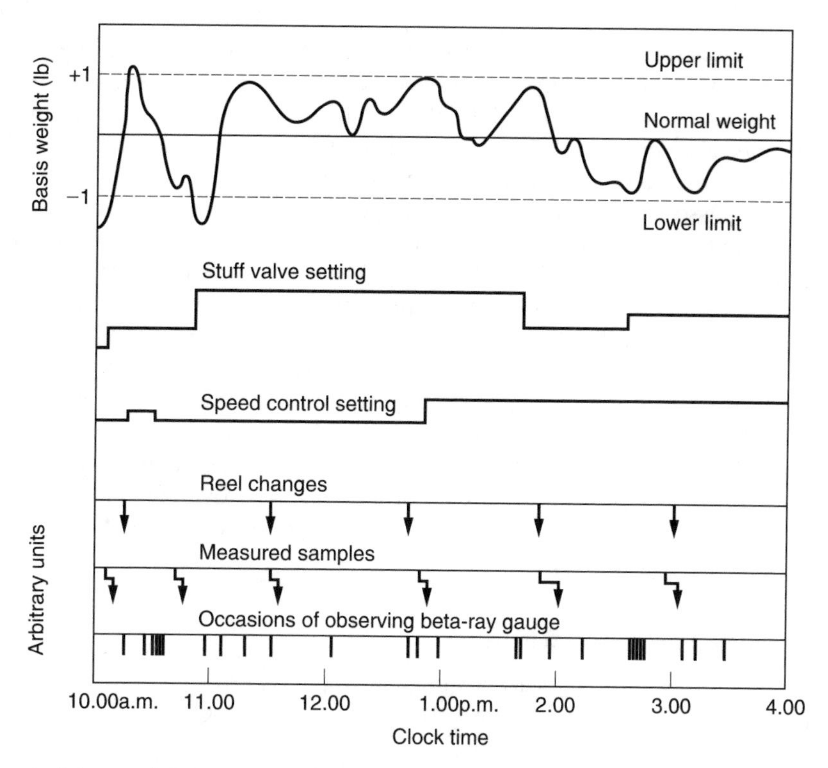

Figure 2 Activity graph for paper machineman (from Beishon, 1967).

The task and social situations

Asking people to 'think aloud' is asking them to do something unfamiliar and awkward. Some people find it a very uncomfortable requirement. It helps if the researcher first demonstrates what is meant by giving an on-the-spot example — for instance, a verbal protocol of doing complex multiplication. The subject can then try the example for themselves and this usually breaks the ice.

Another technique that helps promote more fluent verbalisation is to ask two or more people to work together, particularly where there is a problem-solving situation, or to ask one person to guide another in how to do the task. Communication of thoughts and knowledge, or admissions of lack of knowledge, can then emerge naturally from the people's interaction. However, the verbalisations will now be influenced by interpersonal communicative purposes, rather than being solely a report of mental contents. This will affect how the knowledge is expressed and how its contents might be inferred by the researcher (for discussions of social factors, see research on conversation analysis and interaction analysis: Suchman, 1987; Jordan and Henderson, in press).

There are two principal activities in protocol analysis once data have been collected, which are deeply interwoven. The first activity is the preparation and description of the protocols, which includes developing and applying a system of summarisation to them. The second activity is the analysis of the explicit and implicit content and structure of the protocols, always with respect to the research or design question at hand.

Preparing the protocol for analysis

As many of its practitioners have commented, verbal protocol analysis is a highly iterative process. Although it is usually presented as a linear process, in practice researchers usually go through several reworkings of how the data are characterised and analysed before they find the most cogent and useful descriptions and formalisations of the results. Throughout the verbal protocol analysis process, it is helpful if a researcher can remain as closely in contact as possible with the raw verbal and behavioural data. This serves as a 'reality check' for the higher-order transformations being imposed on the data.

We have discussed limitations to the validity of verbal protocols in their raw form as evidence of thought processes. There are also problems with the validity of the analyses that are performed on the protocols. There are no objective independent techniques for doing these analyses. This means that investigators have to use both their own natural language understanding processes, and their knowledge of the task, in order to make sense of what is going on, to infer missing passages, and to interpret the results of summary analyses. This worries many investigators, who wonder whether all they are doing is rediscovering their own knowledge. There are several replies to this. First, researchers do find activities, strategies and structure which had not been anticipated. Secondly, the analysis can give a description of the behaviour which is consistent with other data or theories, so providing converging evidence. Ideally, the analysis should be made by several investigators who work independently and then come together to agree on a final versions.

Identifying general protocol structure

The first stage of protocol analysis is to identify the structure of the recorded material. This can be done at several levels:

1. finding the most general stages of activity over the period recorded, such as 'taxi out', 'take off', 'fly to cruising altitude', 'straight and level flight', 'diversion around bad weather', 'descent', and 'landing'.
2. finding units of activity within each stage of activity, such as 'request clearance', 'configure aircraft for landing', 'decide to make diversion'.
3. dividing the material in each unit of activity into a sequence of elemental phrases, each of which represents a separate thought or activity.

The structure can be noted with pencil marks, or if working with a work processor by the use of spaces, indenting, or use of outlining capabilities.

Segmenting material into phrases

The next stage is to identify meaningful units within the protocol that may suggest discrete mental processes. The following piece of report will be used in the examples below (the dots indicate pauses):

> C is on oxidation now that's something you can make an estimate for it's a quality so I must leave it alone . . . oxidation average length is one hour 30 minutes for C and started at time zero no it didn't it started at time 33 minutes how confusing of it so it's got nearly one and a half hours to run . . . I'd better check that oxidation for C one hour 30 minutes started 50 minutes ago so it's got 37 minutes to go . . .

Segmented into phrases this becomes:

1. C is on oxidation
2. now that's something you can make an estimate for
3. it's a quality

4. so I must leave it alone
5. oxidation average length is one hour 30 minutes on C
6. and started at time zero
7. no it didn't
8. it started a time 33 minutes
9. how confusing of it
10. so it's got nearly one and a half hours to run
11. I'd better check that
12. oxidation for C one hour 30 minutes
13. started 50 minutes ago
14. so it's got 37 minutes to go.

In this section of protocol a steel-works operator is talking about a furnace called 'C'. Phrases 1–4 are concerned with general properties of the 'oxidation' stage through which the furnace is going, and phrases 5–14 make an estimate of the time at which C will finish oxidising. The average stage length is given in a job aid booklet, and the time the stage started is displayed on the operator's panel. The segmentation of the text into phrases (which might loosely be described as minimum grammatical units, though the language in such reports is often not at all well formed) is done by natural language understanding. This segmentation can often be done by people who have no knowledge of the specialist content of the material.

Inferring structure of mental activities

The structure of mental activities can be inferred by combining phrases into groups. There are two methods for doing this, both of which make use of the semantic content. The first approach is to identify the pronominal referents, since these indicate cross-references between phrases. There are three ways of identifying the pronoun referents. The most reliable requires an independent record (e.g., video) of the states of the environment during the task, from which what started at 33 minutes' time can be identified. Another method involves going through the report immediately afterwards with the speaker to check on the meaning. The third method is to use judges' semantic knowledge of the task.

The second approach to combining phrases into groups is possible after the links between phrases made by the pronouns have been identified. At this point further groupings can be made on the basis of judges' knowledge of what items go together in the task. The result of doing this for the material above is shown in Figure 3. One can also take advantage of the necessity to use judges' knowledge, by making an explicit record of one's own knowledge which was used in the analysis, and using this as a record of the knowledge one is attributing to the speaker. For example, lines 1–4 list properties of C furnace, and lines 5–14 recount a method of calculation. These points are also what one needs to mention to a person unfamiliar with the task before they can understand the report, as in the task explanation given earlier.

The semantics of an individual phrase may imply that the speaker is referring to wider background knowledge. Consider the phrases (from a process operator):

> 'I'll try to run the temperature down to about 400 degrees ... no trouble, it's far away now.'

The first phrase implies the operator knows the required future process state (400 degrees). The second phrase appears to assess the ease of making the change, which implies that he knows how to make both this change and this assessment. A full protocol analysis could detail the types of more general knowledge which can be inferred in this way. Unfortunately, although one can infer from the protocol that such knowledge exists, there may be very little

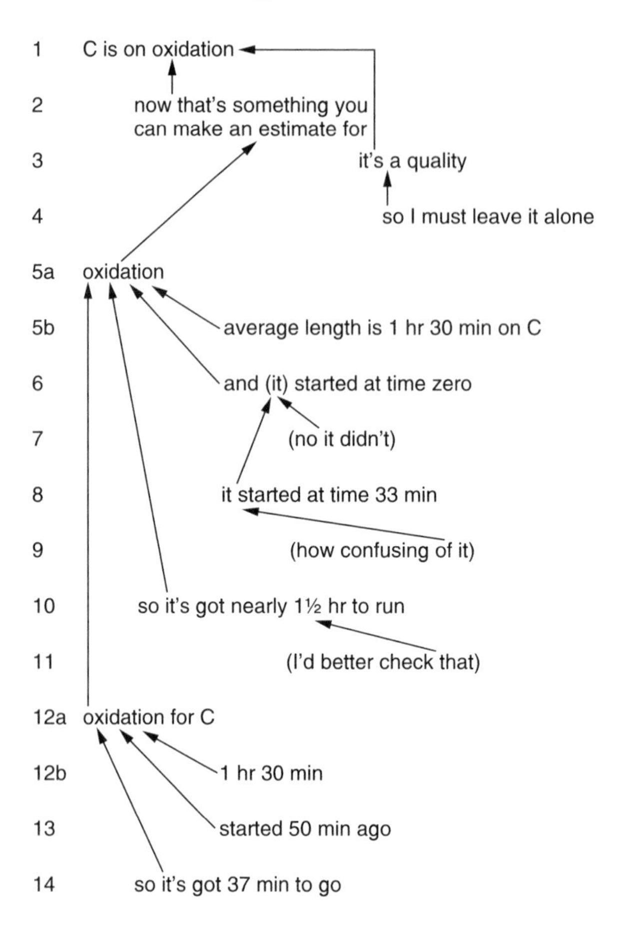

Figure 3 Cross-references between phrases in protocol analysis.

direct evidence in the protocol about it structure and use. Information about this might be obtained by following the protocol study with interviews on these points.

Applying a formal descriptive language

Having reached this point, many researchers wish to use a formal descriptive language to summarise and systematise, in the context of a chosen research question or framework, what they are seeing in the protocol. Examples of research frameworks are 'Knowledge about system relations', 'Interpersonal warmth', or 'Strategies for control'. The research framework points to the kinds of distinctions the researcher wants to make, such as 'friendly', 'cold', 'helpful', and 'authoritarian', or 'knows X' and 'knows Y', where X and Y represent specific classes of fact a subject has mentioned. These distinctions form a set of standardised categories that can be applied to elemental phrases or units of activity in the protocol. This process is known as 'encoding'.

Encoding is not a necessary step for extracting meaning from protocols, but it is frequently used to systematise the researcher's understanding of the protocol. Moreover, a formal encoding lends itself to symbolic and statistical manipulations which can bring out features of the situation recorded that would otherwise be missed. The categories developed, to be useful, must be ones that encourage further inferences about the material. For examples, the frequencies with which certain categories appear may suggest emphases in the way the speaker thinks about the topic. Alternatively the categories can be used simply as a preliminary sort before further analyses. For example, the analyst could look at all instances of the phrases that have been categorised as 'comment on own behaviour' to see if the phrases have any common properties. If the categories are based on semantic aspects of the reports, one might want to count only the occurrence of overall concepts (e.g., birds) or to count the frequency of individual instances (e.g., robins vs. blackbirds) (content analysis will be discussed further below in the section on analysis). Different aspects of syntactic structure could be differentiated. For example, one might wish to distinguish between active and passive voice as reflecting different emphases by the speaker. Alternatively, one could differentiate between different types of conditional statement, perhaps comparing the incidence of 'A therefore B' with 'B because A'. Whether one does this depends not only on whether one is interested in this level of analysis but also on whether the concepts occur sufficiently often to make a frequency count something more informative than a simple listing. The categories may also imply inferences about the types of cognitive process underlying them, for example, 'statement of fact' and 'comment on own behaviour' may imply different types of underlying cognitive activity.

The categories are therefore always a function of particular empirical questions. If there is a set of categories that can be applied in many different circumstances, they are likely to be so general that the results of using them will not be very rich. Indeed, one of the main analytic problems is the development of categories which are both relevant to the investigation and reliably usable by the judges who assign the material to them. Rasmussen and Jensen (1974) describe the iterative method that must be used. First, several independent judges each develop a set of categories. Then they attempt to use each other's categorisation schemes. This both combines the inferences the judges have made about the important distinctions to be made, and also tests whether different people can make the same allocation of material to categories. If not, then analysis using these categories cannot give reliable data, and the categories must be revised, again with the judges working independently during development, and coming together for assessment. The categories arc repeatedly defined and revised by trying to use them to analyse the protocol, until the category definitions and phrase assignments stabilise. This procedure is arduous and repetitive but must be done with precision. The degree of success can be measured with reliability statistics, such as Cohen's kappa (1960). Suen and Ary (1989) provide a comprehensives account of how reliability can be promoted and measured in behavioural encoding.

To take a further look at categorising, examples of identifying characteristic structures within phrases at two levels will be examine: the types of referent word that it is useful to identify within the phrases, and the characteristic patterns in which these occur. For example, the phrase:

'average length is 1 hour 30 minutes'

can be interpreted as a statement of the general form:

'VARIABLE has VALUE'.

The phrase 'steam pressure as 101' can then be categorised as a statement of the same type. Having identified all statements of this type, one could then look at the category members further, e.g., identifying how many different VARIABLEs occur, and with what frequency, or concentrating on the VALUEs and seeing how accurately they are specified. One would have to consider whether to categorise 'there's steam temperature — it's rising' as a statement of the same type, or whether this would lose some of the information in the report, so this phrase should be identified as 'VARIABLE has CATEGORISED VALUE'. This also shows how categorising can retain information considered important for one type of analysis, but can lose other aspects. This phrase is syntactically different from the previous instances, but it has been assumed that this is not relevant to the question of how VARIABLE VALUEs are thought about by the speaker. The syntactic change rather than the semantic one might, of course, be the emphasis of an analysis being made for other purposes. Simple phrases also occur as a component in more complex ones, for example:

'I'll try to run the temperature down to about 400 degrees'

can be interpreted as:

'ACTION causes VARIABLE has VALUE'.

Statements of this type show the speaker's knowledge about how changes in the outside world can be made. Speaker may also give information about their knowledge of the conditions in which certain effects occur, for instance.

'We have to have 50 degrees superheat before we can run it up'

could be described as:

'when VARIABLE has VALUE then ACTION causes VARIABLE has VALUE'.

Or more simply, if one is not interested in distinguishing between different ways of expressing conditional knowledge (e.g., to test whether different expressions occur in different task contexts) the same phrase can be described as:

'ACTIONS given VARIABLE VALUE'.

These examples have categorised material within phrases. Related methods can be used to study categories of phrase type, or of groups of phrases. As an example, the phrases in Figure 3 might be categorised as follows:

1. Fact
2. Strategy
3. Fact
4. Strategy
5. Fact
6. Fact
7. Comment
8. Fact
9. Comment
10. Prediction
11. Strategy
12. Fact
13. Fact
14. Prediction

These categories are more general, and they use a simple one-word notation. In this case there may be an even more distant relation between the actual material in the report and the way in which it is described. Therefore more care may be needed to ensure that the categories are unambiguous to judges. Again, it may be useful or interesting to distinguish sub-categories, for example, between statements of fact about past and present, or between comments expressing general strategy compared with rules for behaviour at this particular time.

In some tasks, it is not very useful to categorise individual phrases, because similar task situations do not recur at this level of detail. In this case, analysis is done at the level of groups of phrases. As an example, the explanatory notes given for Figure 3 are equivalent to a categorisation of types of activity in that piece of report. Rasumssen and Jensen (1974) and Umbers (1981) given examples of this type of analysis in maintenance and process control tasks, respectively.

The above example have shown two different types of encoding notation: the first was a semantic net notation, and the second a single word notation. There are other variants, such as coding each phrase or phrase group with multiple unconnected words, or using the kind of relational notation advanced by Ericsson and Simon (1984/1993). Choosing an encoding notation can be as difficult as choosing the actual encoding categories. The choice of notation will depend upon the meanings that must be captured, and the types of analysis the researcher wishes to perform with the encoding. As encoding categories are developed in the process of analysing the data, neither the data nor the categories can be justified in terms of each other, and some external tests of objectivity are necessary. One test is that the categories can be applied to the data, and with the same results, by judges who were not involved in developing the categories. Another test is that is should be possible to apply the categories meaningfully to other data. The latter test is difficult to carry out because there are too few case studies and they tend to be for purposes that are too different for comparisons to be useful. One solution to this problem is to develop the categories using one half of the data, and test their adequacy on the other half.

Inferring what is not spoken

As mentioned earlier, there are many reasons why verbal protocols may be incomplete reports. First, material may be glossed over because events are proceeding faster than they can be verbalised. Second, certain material may be encoded spatially, acoustically, or haptically, and thus not be in a form amenable to verbalisation. Third, significant activity may be verbalisable, but it may be more effective, appropriate or conventional to communicate it through silence, pauses, tonality, or indirect speech acts (utterances whose communicative significance is different from their literal content). It can be possible, however, to reconstruct some types of intervening material, when the speaker says something that must be the result of some thinking — that they have not mentioned. For example, in lines 10 and 14 of the steelmaking example, the controller states the result of a calculation, so he must have made this calculation in some way. The report does not necessarily indicate how the intervening processes were carried out, only that they must have been done. A question can arise about how deeply these inferences can be taken and still be useful. One way is to concentrate on describing behaviour at a level at which there are some explicit clues to constrain the range of possible underlying mechanisms suggested, but some people would not agree with this. In tasks in which the same situation recurs frequently, it may be possible to combine what is said on different occasions to obtain a fuller account of what is happening. In the example, lines 5–10 can be interpreted as:

5. read STAGE LENGTH
6–8. read STAGE START TIME
10. TIME TO GO = X.

(Lowercase indicates operations that have been inferred, upper case items that are explicitly mentioned. This version in terms of basic operation has already involved considerable inference about underlying processes, which will be discussed more fully later.)

When lines 5–10 are combined with lines 12–14, a complete picture of the processes can be inferred:

(5) 12. read STAGE LENGTH
(6–8) read STAGE START TIME
read time now
13. TIME SO FAR = time now − start time
(10) 14. TIME TO GO = stage length − time so far.

Together with the results from the identification and grouping of phrases, such inferences provide the material that is used in further analyses. In other cases, inferring the unspoken material involves developing an interpretation of the communicative significance of what was said, rather than filling in its surface content.

Analysis of explicit and implicit content

Analysis and encoding are not the same, although they overlap considerably. In order to determine the categories to use in encoding, the researcher must work within some analytical framework, deciding what the most important distinctions are that should be represented if the protocol is to speak to the research issue at hand. Encoding is the application of categories to the raw protocol material, whereas analysis is the process of inferring the ultimate significance of the protocols for the research issue. Sometimes a description of the protocol data provides this, but very often the significance of the protocols is not fully appreciated until qualitative and quantitative manipulations have been performed on the encoding. We shall now look at analysis techniques, all of which can be used for non-verbal just as much as verbal protocol analysis.

Content analysis

There are many styles of analysis called content analysis. Content analysis typically involves counting frequencies of occurrence of chosen words or encoding categories. For example, in the steelmaking example a researcher could count the number of times certain VARIABLEs such as 'steam pressure' or 'average length' were mentioned, or the number of phrase groups encoded as 'Prediction'. Further examples have been given in the previous section. Many of the software tools listed in the final section of this chapter can help researchers to perform content analysis. Moreover, if the segments and phrases of the protocol (and encoding) have been associated with timestamps, then duration analysis can be performed as well. For example, the researcher could calculate the total amount of time, or total duration, that the subject was performing activities categorised as 'Prediction'.

The researcher can analyse the frequency of occurrence of particular categories of phrase types, or types of phrase groups, or within phrases. These categories, their instances, and the frequency counts can be used as the basis for further analyses, such as comparing the protocols of subjects who have received different training, or with different information management systems.

Sequential analysis

One of the most important features of on-line verbal protocols is that — if elicited appropriately — they represent the normal sequential flow of working and thinking. The person, however, does not talk about this sequential flow while it is unfolding: instead, it emerges from his or her talk. Thus, we can characterise the analysis of sequential aspects of protocols as the analysis of structure which is implicitly rather than explicitly expressed by the subject. Looking for sequences in the material can be done either at the level of individual phrases or of groups of phrases: the choice depends on the material. For example, in the fairly constrained 'world' of controlling a simple industrial process, similar sequences of phrases may occur sufficiently frequently to allow the researcher to identify a standard sequence phrase by phrase. However, sequential structure does not always entail repetitions of the same literal sequence. For example, in Rasmussen and Jensen's (1974) verbal protocol study of fault diagnosis, each technician was working on a different piece of equipment with a different fault. Consequently, the fault finding behaviour was not repeated at the level of individual phrases. To search for the common properties of the behaviour in these disparate situations, Rasmussen and Jensen had to look at the data more globally and think about functioning at a more abstract level.

Sequences of phrases

The sequences in which individual items are mentioned in the report can indicate the standard 'routines' or 'programmes' with which the speaker thinks about a particular topic. This is the main approach of workers who use verbal protocols as the basis for simulations of cognitive processes. There is extensive literature on this type of theory development, of which Newell and Simon (1972) is a classic example. It is only possible to reach strong conclusions about these activities, however, if there are several examples of each type of behaviour. The reports are always incomplete at some level, and one may have several hypotheses about the processes intervening between two phrases. Unless one has other examples of the same behaviour, the analysis remains very speculative and unwieldy. For example, in Figure 3 it would be possible for phrases 8 and 10 to be linked by the following serial calculations:

$$\text{Stage end time} = \text{stage start time} + \text{stage length};$$
$$\text{Time to go} = \text{stage end time} - \text{time now}.$$

However, we have another instance of the same behaviour, in which phrase 13 shows which of the two possible linking calculations is being used.

Sequences of groups of phrases

Identifying the sequences of groups of phrases is more difficult because the researcher wants to infer what influences the speaker to move from one topic to another. One cannot take the specaker's work for it, because Nisbett and Wilson (1977) have shown that speakers do not necessarily have good access to this type of information about their behaviour. To identify sequences of groups of phrases, the researcher needs to take into account earlier behaviour by the speaker, because previous items can affect the choice of later behaviour. Also a record of the environment is almost essential in studying dynamic tasks (one which change over time), because it is often these changes which influence what is the most appropriate thing for the speaker to think about next. A first step is to identify transitions from one type of behaviour to others. For example, behaviour A may be followed by behaviour B or by behaviour C. One then looks at the whole context of the speaker's behaviour and the environment to see whether any dimension of the task or previous thinking consistently has one value when A is followed by B and another when A is followed by C. If so, then one infers that the value of this dimension determines the choice of behaviour at this point.

An example comes from a process control task. The operator frequently made remarks such as 'it's above now', 'it's below', and the problem was to identify what dimension of the process he was using to make this judgement. There were two main candidates; the total power being used at the time, or the discrepancy between present power usage and target power usage. Table 1 shows the distribution of judgements at different readings of the total power display, and Table 2 shows the distribution of judgements at different readings of the discrepancy meter. It is clear that the judgements correlate with the discrepancy meter reading and not with the total power display, so the speaker is assumed to be using the discrepancy meter when making these judgements. (Note that his example is not drawn from an analysis of phrase sequences, as no compact example is available. Here the technique has been used to identify pronominal referents. This illustrates that most of the techniques used in analysing verbal reports can be used to study any level of organisation in the material.)

Table 1 Frequency of Assessments at Different Total Power Values

Total power	Above	Alright	Below
31–40		1	3
41–50	2	2	2
51–60	2	3	1
61–70	2		
71–80			1

Table 2 Frequency of Assessments of Discrepancy Meter Readings

Total power	Above	Alright	Below
– 41 to – 50			1
– 31 to – 40			
– 21 to – 30			
– 11 to – 20			3
– 6 to – 10		3	2
– 1 to – 5		2	
1 to 5	1	1	
6 to 10	1		
11 to 20	2		
21 to 30			
31 to 40	1		
41 to 50	1		

Sequential statistics and exploratory techniques

There are certain statistical techniques that can sometimes help the researcher extract informative sequential structure in protocol data (see van Hooff, 1982; Bakeman and Gottman, 1986; Gottman and Roy, 1990 for details). These techniques need to be used with care in verbal protocol analysis. An important problem is that for full statistical validity a great deal of data usually needs to be available, and this is often not the case in protocol studies.

A commonly used procedure for analysing sequences is to perform a Markov analysis, i.e., to find the probability of transition from one item to another, or from a strict sequence of items to the next (Kemeny an Snell, 1960; Howared, 1971). An example was given in the previous section. Markov analysis can be a useful preliminary technique to determine which transitions are most frequent, and therefore most likely to be rewarding to study further. However, this method gives a very limited description of the properties of a sequence, hiding some properties completely. First, the statistical structure of a person's sequential behaviour can change during a course of protocol, and this will not be picked up unless special techniques are used. Second, a person's behaviour often includes considerable amounts of interpolated material, or noise, that makes it difficult to see sequential structure with a technique like Morkov analysis which requires exact instances of sequences, rather than approximate instances.

Other techniques make weaker assumptions about sequences, and can be used for exploratory analyses just as much as confirmatory ones. Lag sequential analysis is a technique

Table 3 General Classes of Tools for Protocol Analysis

Pencil and paper
General software applications: • Word processors • Spreadsheets • General multimedia software
Specialised protocol analysis software

for finding dependencies between events that are separated in time by one, two or more steps (Bakeman and Gottman, 1986). Lag sequential analysis does not require events to fall into strict sequences, as Markov analysis does. For example, a researcher may find, more often than by chance alone, that subjects make a judgment three or four phases after reading a discrepancy meter, but not one or two phases after reading it. Such a pattern of results would suggest that some other activity is taking place before the judgement can be made. This other activity can be sought, and its meaning for the sequence inferred.

Further sequential techniques are even more exploratory, helping the researcher find potentially informative patterns in the data. Some researchers who perform protocol analysis in UNIX environments have successfully used regular expressions to define and seek loose patterns in protocol data. Regular expression notation allows the researcher to describe a pattern, such as 'find all sequences that start with "read STAGE LENGTH", followed by one or more statements of indeterminate nature, and conclude with "TIME TO GO=" '. The computer finds all such patterns, and prints them out or performs some other operation on them. The researcher can now see all the different ways people handle the calculation of 'time to go' in a stage, and start to see commonalities and differences.

Another exploratory computer-based technique involves finding 'maximal repeating patterns' in a protocol. The computer scans the protocol for all sequences of categories that repeat, such as 'ABC' in the sequence 'FDABCGABC' and tries to find whether these repeating patterns form parts of even larger repeating patterns. In this way both local and global structure can be induced, although ways have to be found for handling noise if this technique is to be completely useful.

Analysis through modelling

Many researchers use symbolic modelling techniques to analyse protocols and to convey the results of those analyses. There are two principal approaches; diagrammatic modelling and computational modelling.

Diagrammatic modelling is a qualitative technique used to capture the content, structure, and/or sequential flow of a protocol. Figure 4 represents a verbal protocol from Roth, Bennett and Woods (1988) in a study in which technicians tried to diagnose a faulty electronmechanical devices with the help of an expert system. The boxes represent groups of phrases, the lines represent connections between different parts of the protocol, and other visual enhancements have been used to emphasise the structure that the researchers have inferred. Figure 5 moves further away from the physical form of protocols themselves. It is a subjective 'influence diagram' that shows the results of a study by Sanderson, Verhage and Fuld (1989). Subjects were asked to think aloud while controlling a simulation of a thermal hydraulic system. Their protocols were combed for statements about the effect of variables on other variables in the system. The combined results were superimposed on a diagram of the

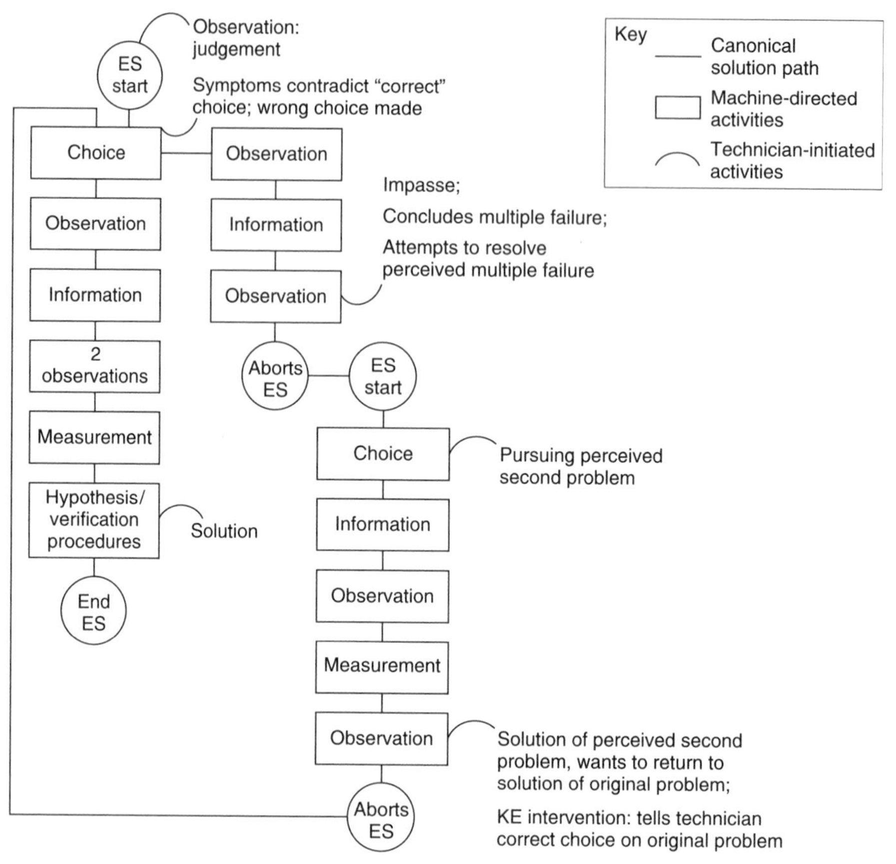

Figure 4 Summarised protocol of aided troubleshooting session (From Roth, Bennett and Woods, 1988).

correct cause–effect relations, so that the most common patterns of verbally-expressed understanding could readily be seen.

Computational modelling based on protocol analysis is a vast topic that cannot be addressed here. Briefly, however, the idea is to analyse protocols in sufficient detail that a computer programme can be written of how the subject performed the task (Simon and Kaplan, 1989; Ritter, 1992). The computer programme is then actually run, and it generates a 'protocol' of how it performs the task. This protocol can include actions and also an indication of what the current mental contents should be, which the researcher would expect to see reflected in a verbal protocol. The computer's 'protocol' can then be compared with the protocols of the original subjects or, even better, with the protocols of subjects whose data were not used to build the computer model, to see how well they fit. To date, the best examples of this type of modelling have been performed in the context of basic, comprehensive cognitive models such as ACT (Anderson, 1983) and SOAR (Newell, 1990), but the principle holds for any model that might generate traces of behaviour and assumed mental activity.

Tools for aiding protocol analysis

As pointed out at the beginning, verbal protocol analysis is very time-consuming, with AT : ST ratios of anywhere from 10 : 1 to 1000 : 1, depending on the type of data, the research question, and the level of detail of the analysis.

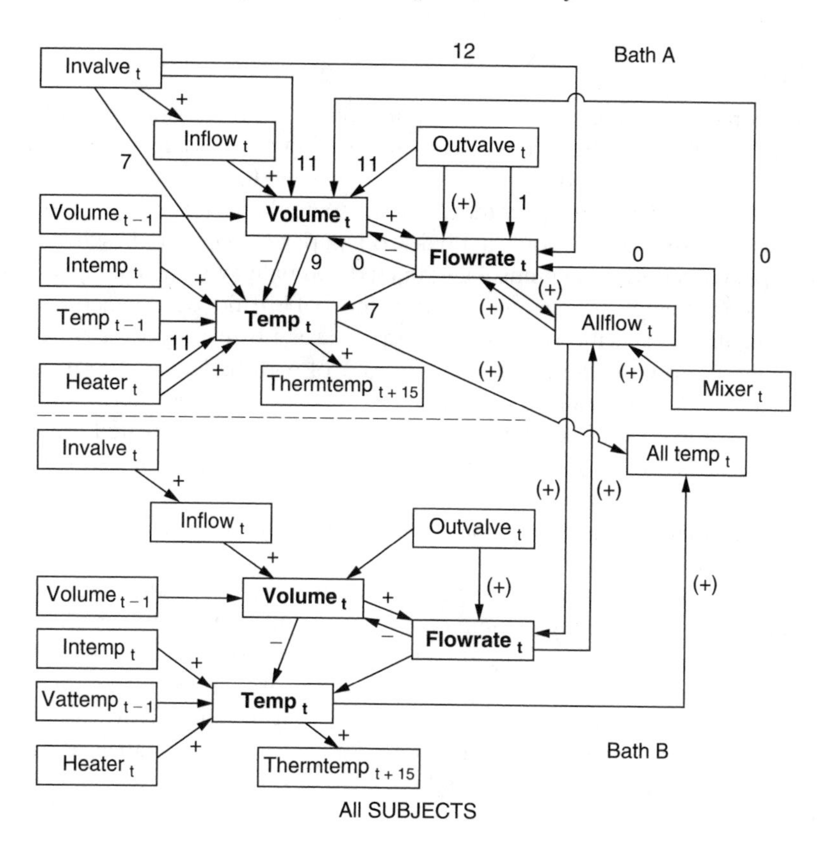

Figure 5 Influence diagram combining twelve subjects' knowledge about system relations in a thermal hydraulic process (from Sanderson, Verhage and Fuld, 1989).

Traditionally it has been performed with paper and pencil and very often still is. However, many researchers believe that AT:ST ratios can sometimes be reduced by the use of appropriate software (Sanderson and Fisher, 1993). Because software can potentially turn verbal protocol analysis — a typically unwieldy research technique — into a fully viable one, some space will be devoted to this topic. Of the software tools mentioned here, some have been developed for verbal and nonverbal protocol analysis and some for broader types of observational analysis. The remainder are general software applications which were not designed for any kind of behavioural data analysis, but which happen to have features that can be successfully exploited in protocol analysis. The majority of the tools mentioned below are for the Macintosh™. However, tools for many other machines and operating systems are available as well. The treatment below offers comments about the strengths and weaknesses of the different approaches. It cannot be emphasised too strongly that researchers should exercise critical juldgement when contemplating the use of software to support protocol analysis. Each researcher should find a tool — or combination of tools — that supports how he or she wants to analyse protocols, rather than adopt a tool because someone else has found it useful, or because it purports to be designed for protocol analysis. The software designer's intentions and the tool's success elsewhere do not guarantee that it will be suitable for the needs of a given research programme. If a tool is not helpful this does not mean that the researcher has failed to follow some canon for performing protocol analysis, but instead simply that the tool was not helpful.

Pencil and paper

Traditionally, verbal protocols have been transcribed, printed out, and then analysed with paper and pencil. There are distinct disadvantages but also equally distinct advantages of this approach. On the one hand, if the researcher wishes to change his or her categories during encoding, or to change the way they have been applied to protocol segments or units, then these changes are very tedious to make on paper. Moreover, if statistics are to be performed then the necessary data must be extracted manually, which is highly unreliable and extremely time-consuming. On the other hand, the paper and pencil medium provides a flexible, high-quality visualisation of the transcript that is unsurpassed for basic annotation, such as chunking the phrases and units of the protocol with hand-drawn boxes, connecting spatially separate but thematically related parts of the protocol with lines and arrows, and making informal comments in the margins. Moreover, the whole annotated transcription can be spread out on a desktop so that its overall structure is apprehended in a way that cannot be achieved on even the largest computer screens.

It is widely felt that anyone contemplating using verbal protocol analysis should first train themselves with paper and pencil analysis so that they fully understand the types of manipulation and types of thinking that go into protocol analysis. Moreover, at least some paper and pencil analysis is highly recommended at the outset of a new research programme. Only then can the full data manipulation requirements be appreciated, and an appropriate software environment chosen from the many that the available. Paper and pencil coding may be all that is needed for an entire research programme. This is particularly true if a researcher has a stable set of categories, does not plan to do any statistical analysis or qualitative modelling, and if he or she will be using a style of analysis that relies heavily upon connections between widely disparate parts of the protocol.

General software applications

Researchers turn to various types of general software application for help with specific aspects of protocol analysis. These applications include work processors, spreadsheet programmes, and multimedia applications. Their most general advantages are proven reliability, good documentation, and the availability of many expert users who can help researchers find the best way to use the application for protocol analysis.

Word processors

Word processors have been used by some researchers for certain aspects of protocol analysis. After transcription, some of the features of advanced word processors such as Microsoft's Word™ can be put to good use. Formatting features, such as indentations, line feeds, font and font size, allow researchers to give meaningful visual structure to a protocol. Annotations and categories can be added if tabular or columnar formats are used. Word processors designed for collaborative writing and multi-author editing add the ability to mark and link certain parts of the text visually and to make marginal annotations (e.g., PREP; Chandhok, 1993). Furthermore, categories can be stored as glossary items to save time typing them fully every time they are used and to promote standardisation. Glossary changes or find and replace capabilities can be used to change the categories, if global changes are needed. Finally, document outlining capabilities can also be used to good effect. For example, encoding categories can be handled as headings at one or more levels, and inserted in the raw protocol. Then the document can be displayed using outlining to show just the top level headings, the top *n* level headings, or all headings plus raw protocol for all or some of the whole document. However, there are considerable disadvantages. Word processors do not allow statistical analysis, modelling or certain types of querying to be performed, and they do not offer any connection through software to the raw data in audio or video form. Moreover, their ability to conditionalise word counts and textual changes on other features of the headings or protocol is limited. Finally, they do not make it easy to juxtapose and align multiple streams of information, such as verbalisations, actions, environmental states, researchers' comments, and formal encoding categories. In summary, although word processors provide some of the benefits of paper and pencil and case certain kinds of manipulations, they are useful principally for formatting and encoding.

Spreadsheets

Spreadsheet programmes such as Microsoft's Excel™ overcome some of the limitations of word processors, but add other constraints. Spreadsheets provide a row/column array of cells into which various kinds of data can be entered. Thus spreadsheets make it easy to handle multiple streams of information and to keep them properly aligned, relieving the researcher of this concern. The researcher can define selected parts of the spreadsheet as containing data do certain types, such as timestamp, text, integer, or float data, so that different kinds of calculation can be performed on them. Moreover, because the cells of a spreadsheet can contain formulae that use information in other cells, the spreadsheet can store summary information about itself. Spreadsheets often incorporate many of the features of word processors, such as find and replace, the use of font and font size, and outlining capabilities. However, they have added advantages because cells conforming to certain descriptions can be selected so that further operations such as counting, printing, and formatting can be performed on them alone. Finally, some spreadsheets allow the researcher to perform queries, such as 'every time the subject mentions neutron flux, print out the corresponding cell in the "activity" column'.

The limitations of spreadsheets arise from the fact that they were not designed to handle events occurring at different points in time. For example, because of the row/column grid, it is difficult to enter, display, and perform manipulations on events in different columns that occur out of synchrony with each other, as things will in the real world. Moreover, although some spreadsheets offer time series analysis, they do not offer the types of sequential data analysis that a researcher more usually wants to use for protocol analysis. Finally, like word processors they do not offer any connection through software to the raw data in audio or video form. In summary, though, spreadsheets can be very effectively used for the more straightforward kinds of protocol analysis (Hoeim and Sullivan, in press), and a competent user can quickly adapt them to this purpose.

As a final note, it is possible to annotate word processing or spreadsheet documents with digitised sound or video inserts. This tends to serve presentation purposes better than analysis purposes because the insert is placed at a fixed location rather than being associated continuously with a timeline, and because only small inserts can usually be handled. However it is a useful capability if analysis is being shared among investigators who wish to bring certain episodes to each other's attention during the analysis process.

General multimedia software: digitised audio and video

Many researchers are currently exploring the feasibility of digitising their audio or video, and accessing and annotating it using audio and video editing programmes. In principle this can cut down on the need for transcription if there is good support for the type of thinking required in protocol analysis. Editing tools for digitised audio such as SoundEdit™ or SoundDesigner™ often provide timeline visualisations of a sound signal, showing the 'on–off' pattern of verbalisation in a way that makes it easy to navigate through the protocol. Some tools allow sections of the signal to be marked, annotated, and played in any order, and the annotation associated with any part of the timeline easily retrieved. Such fluent contact with the raw data can be of enormous help to the researcher doing protocol analysis. If relying upon audio, it is important to find software that stores in quickly accessible form the amount of sound recording required — some will only handle a few seconds or minutes at a time. Moreover, audio and video editing programmes are seldom designed to support large amounts of transcription and annotation, and so it is often best to run them alongside a word processing or spreadsheet programme. This means there are no logical links between points in the sound signal and points in the programme document, and the researcher must handle the coordination. Finally, such programmes do not include querying and statistical capabilities of the kinds usually needed for protocol analysis.

The growth of multimedia has brought many video digitising and editing programmes onto the market, building on products such as Apple Macintosh's Quicktime™ movies. Because many of these products are hypermedia based, or designed for end-user authoring, such as Hypercard™ and MacroMind Director™ they are easily moulded to a variety of uses. Many of these video editing tools have the same limitations as the audio editing and sound analysis programmes discussed above, and the storage requirements for digitised video are far greater — usually quite prohibitive if anything more than a few minutes are to be recorded. CD storage circumvents the problem, but only the most well funded research institutions can afford one-time pressings of many hours of observational data.

An alternative to digitisation that removes the storage problem at the cost of increasing access time, is to control external analogue devices such as audio cassettes and VCRs. Analogue video is easily timestamped, which goes part of the way towards handling coordination between document and medium. Many tools linking VCRs and computers have developed for the video production industry. At present, software tools for video production that might be adapted to protocol analysis are usually very costly and often require expensive supporting equipment. However, many of the specialised tools for protocol analysis now being developed use this approach and will be discussed in the next section. Fewer software tools link computers to audio cassettes, although some have been developed in the context of office automation, as stenographic aids.

Specialised protocol analysis software

For over 20 years there have been efforts to automate protocol analysis, or at least to facilitate it through software (see Sanderson, James and Seidler, 1989; Ritter, 1992; Harrison and Baecker, 1992; Roschelle and Goldman, 1991 for details of these and related efforts). The ideal tool would seemingly combine the advantageous features of pencil and paper, word processors, spreadsheets, database programmes with good querying and browsing features, multimedia editing tools, hypermedia, and statistical programmes, but of course it is no simple matter to achieve all this. Nonetheless, promising specialised tools are in various stages of development, and there is space to mention only some of them here. Many are research tools rather than commercially available products, but some authors are willing to share their software or to distribute it for a small fee. Sources for further information are given at the end, of the chapter.

The simplest tool is probably CVideo (Roschelle, 1992) which coordinates a customised text editing tool with an external VCR. The VCR is controlled from the computer, and tape locations can be found either by moving a marker up and down a scroll bar on the computer screen or by selecting some text in the editor and commanding the VCR to find the corresponding time. Timestamps can be captured from the videotape, inserted into the word processor, and any subsequent text will be associated with that timestamp. No support for sophisticated formatting, encoding, multiple streams of activity, or data analysis is provided. However, the sort of arrangement found in CVideo can help protocol analysis a great deal.

The logical connection between the VCR and the document seen in CVideo is at the heart of many of the more sophisticated tools.

Another class of tool provides support for the encoding and analysis phase of protocol analysis, assuming that transcription has been done elsewhere and can be either viewed in another application, or imported as a text file. PAW (Fisher, 1988, 1991) and SHAPA (James and Sanderson, 1991; Sanderson *et al.*, 1989) are software environments for performing verbal protocol analysis very much in the style outlined in Ericsson and Simon (1984/1993). Both tools allow complex vocabularies of encoding categories to be established, altered, and applied to the data. To differing degrees, both allow the data — or more precisely the encoding — to be filtered, queried, and submitted to some of the sequential data analysis techniques listed in an earlier section of this chapter. However, neither offers connections with external recording devices, and neither naturally allows multiple streams of information to be represented.

A further class of tool combines the capabilities of the above two classes and adds further capabilities. Tools such as EVA (MacKay, 1989), VideoNoter (Roschelle and Goldman, 1991), MacSHAPA (Sanderson, 1993), VANNA/Timelines (Harrison and Baecker, 1992), and the Observer/Tracker/Reviewer suite (Hoeim and Sullivan, in press) support text entry and editing, connection with external video devices, multiple streams of information, and the establishment and use of vocabularies of encoding categories. These tools differ in other respects. EVA, VideoNoter and VANNA/Timelines offer particularly sophisticated video manipulation, MacSHAPA offers a broad set of sequential and non-sequential data analysis routines, EVA and VideoNoter offer powerful ways to link data in different forms, and both the Observer/Tracker/Reviewer suite and MacSHAPA offer particularly sophisticated querying and filtering capabilities.

Certain tools for protocol analysis have been developed to perform the kind of cognitive modelling that often accompanied protocol analysis. Very early, Waterman and Newell (1971, 1972) developed software tools (PAS I and II) to perform the types of verbal protocol analysis reported in Newell and Simon (1972). Bhaskar and Simon (1977) later developed a protocol analysis support tool called SAPA that adjusted its model of human cognitive processes as it learned more about how the human performed a task. Some current protocol analysis tools form part of a suite of programmes centred around a strong model of cognition (such as SOAR, grammar-based, or production system models of human performance) which support model development and manipulation as well as the comparison of behavioural results with a model's predictions. For example, Ritter's (1992) SMT provides researchers with a suite of tools for aligning and measuring the degree of fit between two 'traces' (protocols): one generated by a human and the other by a computational model representing the researcher's idea of how the task might be done and what the human should be doing and saying at each point.

There is a considerable amount of related software than can also be used for certain types of protocol analysis. First, hypermedia-based tools have been developed for ethnographic studies (Fielding and Lee, 1992; also see special issue of *SIGCHI Bulletin*, 1989, on the use of video for HCI). AI-oriented tools have been developed to support knowledge elicitation in the context of knowledge engineering, and some handle continuous verbalisation (see special issues of *International Journal of Man-Machine Stuides*, 1987, on knowledge elicitation). For studying computer-supported cooperative work, a system named Conversation Editor has been developed that digitises and separates the speech signal from multiple speakers, and represents the signal in timeline form so it can be edited, explored and analysed (Sellen, 1992) and a related system has been developed by van der Velden (1992). Finally, an extensive and powerful set of tools called Childes-Clan (MacWhinney, 1991) has been developed for studying child language which, although it takes considerable training to learn because of its complexity, could be equally well used for verbal protocol analysis.

In conclusion, researchers interested in using software to aid protocol analysis should keep abreast of developments in such journals as *Behavior Research, Methods, Instruments and Computers, Behaviour and Information Technology, Human-Computer Interaction,* and particularly in the proceedings of annual conferences of the Association of Computing Machinery's Special Interest Group on Computer-Human Interaction (ACM SIGCHI) and the Human Factors and Ergonomics Society (HFES) in the United States.

Conclusion

This methodological review has taken an optimistic approach to the difficulties of collecting and analysing verbal protocols. Analysing verbal reports is not easy, and even with the best software support it can still be time consuming and difficult. Ironically, a thorough verbal protocol analysis study of a task may remove the need to use verbal protocol studies in future investigations with the same task. If the task becomes well known, and the range of strategies used for it identified, then in future studies the influence of different variables may be detected by more focused probes of subjects' knowledge, such as questionnaires, interviews, or walkthroughs of situations that are certain to be sensitive to the manipulation, rather than through the more tortuous route of verbal protocol analysis. Sometimes the process of verbal protocol analysis can be sufficient training for the researcher to understand behaviour with less dense data. However verbal protocol analysis may be needed again when different classes of question, or different systems, are being used.

A person's choice of complex behaviour is influenced not only by immediate circumstances but also by planning in relation to the predicted future or by reference to similar past events. It is difficult or impossible to get sufficient evidence from observed nonverbal behaviour to suggest or constrain hypotheses about such cognitive activities. As a consequence, we know very little about the processes underlying complex behaviour. Verbal protocol analysis is currently one of the richest ways of investigating the nature of behaviour that is a function of past, present, or future. The methods described here illustrate the flexible analytic techniques that can be used, and the software tools outlined provide a wide range of options as to how the researcher might proceed.

References

Anderson, J.R. (1983). *The Architecture of Cognition* (Cambridge, MA: Harvard University Press).

Baecker, R. (1993). Timelines software. Dynamic Graphics Project, University of Toronto, Toronto, Ontario, Canada.

Bainbridge, L. (1979). Verbal reports as evidence of the process operator's knowledge. *International Journal of Man-Machine Studies*, **11**, 411–436.

Bainbridge, L. (1985). Inferring from verbal reports to cognitive processes. In *The Research Interview*, edited by M. Brenner, J. Brown and D. Canter (London: Academic Press), pp. 201–215.

Bainbridge, L. (1986). Asking questions and accessing knowledge. *Future Computing Systems*, **1**, 143–149.

Bakeman, R. and Gottman, J. (1986). *Observing Interaction: An Introduction to Sequential Analysis* (Cambridge: Cambridge University Press).

Beishon, R.J. (1967). Problems of task description in process control. *Ergonomics*, **10**, 177.

Bhaskar, R. and Smimon, H.A. (1977). Problem-solving in semantically rich domains: an example from engineering thermodynamics. *Cognitive Science*, **1**, 193–215.

Chandhok, R. (1993). The PREP editor. Department of Computer Science, Carnegie-Mellon University, Pittsburgh, PA.

Cohern, J. (1960). A coefficient of agreement for nominal scales. *Educational and Psychological Measurement*, **20**, 37–46.

Cuny, X. (1979). Different levels of analyzing process control tasks. *Ergonomics*, **22**, 415–425.

Duncan, K.D. and Shepherd, A. (1975). A simulator and training technique for diagnosing plant failures from control panels. *Ergonomics*, **18**, 627–641.

Ericsson, K.A. and Simon, H.A. (1980). Verbal reports as data. *Psychological Review*, **87**, 215–251.

Ericsson, K.A. and Simon, H.A. (1984/1993). *Protocol Analysis: Verbal Reports as Data* (Cambridge, MA: MIT Press).

Fielding, N.G. and Lee, R.M. (1992). *Using Computers in Qualitative Research* (London: Sage Publications).

Fisher, C. (1988). Advancing the study of programming with computer-aided protocol analysis. In *Empirical Studies of Programmers, 1987 Workshop*, edited by G. Olson, E. Soloway and S. Sheppard (Norwood, NJ: Ablex Publishing Corporation).

Fisher, C. (1991). Protocol Analyst's Workbench: Design and evaluation of computer-aided protocol analysis. Unpublished Ph.D. thesis. Department of Psychology, Carnegie-Mellon University, Pittsburgh, PA.

Fisher, C. and Sanderson, P.M. (1993). Exploratory sequential data analysis: traditions, techniques and tools. Report of the CHI '92 workshop. *SIGCHI Bulletin*, **25**, 31–40.

Gottman, J.M. and Roy, A.K. (1990). *Sequential Analysis: A Guide for Behavioral Researchers* (Cambridge: Cambridge University Press).

Harrison, B.L. and Baecker, R.M. (1992). Designing video annotation and analysis systems. *Proceedings of the Graphics Interface '92 Conferences*, Vancouver, BC., May 11–15.

Howard, R. (1971). *Dynamic Probabilistic Systems* (New York: Wiley).

Hoeim, D. and Sullivan, K. (in press). Data collection and analysis tools: Challenges and considerations. Manuscript accepted for publication in *Behaviour and Information Technology*.

James, J.M. and Sanderson, P.M. (1991). Heuristic and statistical support for protocol analysis with SHAPA version 2.01. *Behavior Research Methods, Instruments and Computers*, **23**, 449–460.

Jordan, B. and Henderson, A. (in press). Interaction analysis: foundations and practice. *Journal of the Learning Sciences*.

Kemeny, J.G. and Snell, J. (1960). *Finite Markov Chains* (New York: Van Nostrand).

Leplat, J. and Bisseret, A. (1965). Analyse des processus de traitement de l'information chez le controleur de la navigation aerienne. *Bulletin du CERP*, 1–2, 51.

MacKay, W.E. (1989). EVA: An experimental video annotator for symbolic analysis of video data. *SIGCHI Bulletin*, **21**, 68–71.

MacWhinney, B. (1991). *The CHILDES Project: Tools for Analyzing Talle* (Hillsdale, NJ: LEA).

Newell, A. (1990). *Unified Theories of Cognition* (Cambridge, MA: Harvard University Press).

Newell, A. and Simon, H.A. (1972). *Human Problem Solving* (Englewood Cliffs, NJ: Prentice Hall).

Nisbett, R.E. and Wilson, T.D. (1977). Telling more than we can know: verbal reports on mental processes. *Psychological Review*, **84**, 231–259.

Rasmussen, J. and Jensen, A. (1974). Mental procedures in real-life tasks: a case study of electronic trouble shooting. *Ergonomics*, **17**, 293–307.

Ritter, F.E. (1992). A methodology and software environment for testing process models' sequential predictions with protocols. Unpublished Ph.D. thesis. Department of Psychology, Carnegie-Mellon University, Pittsburgh, PA.

Roschelle, J. (1992). *CVideo Manual* (Palo Alto, CA: Institute for Research on Learning).

Roschelle, J. and Goldman, S. (1991). VideoNoter: A productivity tool for video data analysis. *Behavior Research Methods, Instruments, & Computers*, **23**, 219–224.

Roth, E., Bennett, K.B. and Woods, D.D. (1988). Human interaction with an "intelligent" machine. In *Cognitive Engineering in Complex Dynamic Worlds*, edited by E. Hollnagel, G. Mancini and D.D. Woods (New York: Academic Press).

Sanderson, P.M. (1993). Designing for simplicity of inference in observational studies of process control: ESDA and MacSHAPA. *Proceedings of the Fourth European Conference on Cognitive Science Approaches to Process Control (CSAPC '93): Designing for Simplicity*, Frederiksborg, Denmark, August 25–27.

Sanderson, P.M. and Fisher, C. (1993). Exploratory sequential data analysis: Foundations. Manuscript accepted for publication in *Human-Computer Interaction*.

Sanderson, P.M., James, J.M. and Seidler, K.S. (1989). SHAPA: An Interactive Software Environment for Protocol Analysis. *Ergonomics*, **32**, 1271–1302.

Sanderson, P.M., Verhage, A.G. and Fuld, R.B. (1989). State space and verbal protocol methods for studying the human operator in process control. *Ergonomics*, **32**, 1343–1372.

Sellen, A. (1992). Speech patterns in video-mediated conversation. *Proceedings of the ACM Conference on Human Factors in Computing Systems.* New Orleans, 27 April–2 May. (New York: ACM Press).

Shadbolt, N.R. and Burton, A.M. (1989). Empirical studies in knowledge elicitation. *ACM-SIGART Special Issue on Knowledge Acquisition*, p. 108.

Simon, H.A. and Kaplan, C. (1989). Foundations of cognitive science. In *Foundations of Cognitive Science*, edited by M. Posner (Cambridge, MA: MIT Press).

Suchman, L. (1987). *Plans and Situated Actions: The Problem of Human-Machine Communication* (New York: Cambridge University Press).

Suen, H.K. and Ary, D. (1989). *Analyzing Quantitative Behavioral Observation Data* (Hillsdale, NJ: LEA).

Umbers, I.G. (1981). A study of control skills in an industrial task, and in simulation, using the verbal protocol technique. *Ergonomics*, **24**, 275–293.

van der Velden, J. (1992). *Delft-WIT Lab: Research issues and methods for behavioral analysis.* CSCW '92 Technical Video Program, Toronto: ACM SIGGRAPH Video Review.

van Hooff, J.A.R.A.M. (1982). Categories and sequences of behavior: methods of description and analysis. In *Handbook of Methods in Nonverbal Behavior Research*, edited by K.R. Scherer and P. Ekman (Cambridge: Cambrdige University Press).

Waterman, D.A. and Newell, A. (1992). Protocol analysis as a task for artificial intelligence. *Artificial Intelligence*, **2**, 285–318.

Waterman, D.A. and Newell, A. (1973). *PAS-II: An Interactive Task-Free Version of an Automatic Protocol Analysis System* (Pittsburgh, PA: Department of Computer Science, Carnegie-Mellon University).

Chapter 8

Eliciting expertise

Nigel Shadbolt

Introduction

Since the last edition of this book there have been rapid developments in the use and exploitation of formally elicited knowledge. Previously (Shadbolt and Burton, 1995), the emphasis was on eliciting knowledge for the purpose of building expert or knowledge-based systems. These systems are computer programs intended to solve real-world problems, achieving the same level of accuracy as human experts. Knowledge engineering is the discipline that has evolved to support the whole process of specifying, developing and deploying knowledge-based systems (Schreiber *et al.*, 2000).

Now there is a much wider interest in capturing and modelling knowledge and expertise. This has arisen because the importance of Knowledge Management (KM) is universally recognised by organisations large and small. There are many different characterizations of KM but the central assumption is that knowledge is a valuable asset that must be managed (Nonaka and Takeuchi, 1995; Stewart, 1997). What we are looking for in KM is a means to get the right knowledge to the right people at the right time and in the right form. These are difficult challenges, many of them identical to those encountered when building early knowledge-based systems (Hayes-Roth *et al.*, 1983). Acquiring, documenting, distributing, reusing and maintaining knowledge are all difficult and time-consuming tasks. We have argued elsewhere that the tools and techniques, methods and approaches of knowledge engineering are well suited to the KM enterprise (Milton *et al.*, 1999).

This chapter will discuss the problem of knowledge elicitation for knowledge intensive systems in general. These systems range from classical knowledge-based systems through to structured intranets, from workflow support tools through to best practice guidelines. The content elicited from experts does not have to exist in an electronic format at all. However, increasingly the results of hard won knowledge elicitation and expertise modelling find their way into some form of digital system.

Knowledge elicitation comprises a set of techniques and methods that attempt to elicit an expert's knowledge through some form of direct interaction with that expert. The first section will review the nature and characteristics of this 'bottleneck' in system construction. We will then look at a range of methods and techniques for elicitation. Where appropriate we will describe their implementation in software. Methodologies for expertise modelling will be described and we will illustrate the kinds of knowledge that will be present in expert behaviour. We will consider the different types of expert that may be encountered and the attendant consequences for elicitation. Finally, we will consider the extent to which the burgeoning amount of content on the web is changing the way might think about aspects of the knowledge acquisition problem.

There is still no comprehensive theory of knowledge acquisition available. It remains an art as much as a science. It is not the purpose of this chapter to investigate the theoretical shortcomings of knowledge acquisition but to deliver practical advice and guidance on performing the process.

Knowledge intensive systems

In the early days of Artificial Intelligence much effort went into attempts to discover general principles of intelligent behaviour. Newell and Simon's (1963) General Problem Solver exemplified this approach. They were interested in uncovering a general problem solving strategy that could be used for any human task.

In the early 1970s this position was challenged. A new slogan came to prominence — 'in the knowledge lies the power'. A leading exponent of this view was Edward Feigenbaum of SRI. He observed that experts are experts by virtue of domain specific problem solving strategies together with a great deal of domain specific knowledge. Programs that attempted to implement detailed knowledge about tasks and the subjects to which they applied resulted in the class of programs called Expert or Knowledge-Based Systems. These are now widely used and often quite invisible to the end user. The spelling and grammar checker that is being used to write this chapter owes its origins to knowledge-based systems technology. There are systems that look for patterns to detect credit card fraud, classify radar tracks, interpret patient vital signs and support in the design of aero engines.

There are a variety of ways in which expertise is encoded in run time systems. Many systems will incorporate some type of rule-based or object-orientated representation. For a review of the major types of knowledge-based system architecture and of different knowledge representation formalisms, see Stefik (1993).

The problem of elicitation

The people who build knowledge intensive systems are typically not people with a deep knowledge of the application domain. However, it is they who must gather the domain knowledge and then implement it in a form that the machine can use.

In the simplest case, one may be able to gather information from a variety of non-human resources: textbooks, technical manuals, case studies and so on. However, in most cases one needs actually to consult a practising expert. This may be because there isn't the documentation available, or because real expertise derives from practical experience in the domain, rather than from a reading of standard texts. The task of gathering information generally, from whatever source, is called *knowledge acquisition* (KA). The sub-task of gathering information from the expert is called *knowledge elicitation* (KE). In this chapter we will be concentrating on KE. Few knowledge intensive systems are ever built without recourse to experts at some stage. Those systems not informed by actual expert understanding and practice are often the poorer for it.

Many problems arise before elicitation of the detailed domain knowledge. We need to understand the purpose and requirements for any knowledge intensive system. Sometimes the failure is in formulating the role of the system, on other occasions it is a failure to appreciate what it is realistic to build. Systems can fail because no one has thought of the social and organisational problems that must be resolved in deploying a system. Very often the effort and resources required to build systems are underestimated: this occurs in both the development and maintenance of systems. A particularly nasty situation arises when one is expected to conjure up knowledge for areas in which no evidence of systematic practice exists at all. Here one is expected to provide theories for domains where there is no theory.

Providing we can resolve these issues then we get down to KE in the expectation that it will be time well spent.

Two questions dominate in KE. How do we get experts to tell us, or else show us, what they do? How do we determine what constitutes their problem solving competence? This is a hard enough problem in itself but there are a variety of circumstances that contrive to make the problem even harder. Much of the power of human expertise lies in laid-down experience, gathered over a number of years, and represented as heuristics. (An heuristic is defined as a rule of thumb or generally proven method to obtain a result given particular information.) Often the expertise has become so routinised that experts no longer know what they do or why. In many cases the knowledge required to build a system is distributed across an organisation and in the heads of a number of experts. Experts do not always agree so there is the problem of reconciling conflicting or differing views.

There are obviously clear commercial reasons to try to make KE an effective process. We would like to be able to use techniques that will minimise the effort spent in gathering, transcribing and analysing an expert's knowledge. We would like to minimise the time spent with expensive and scarce experts. And, of course, we would like to maximise the yield of usable knowledge.

There are also sound engineering reasons why we would like to make KE a systematic process. We would like the procedures of KE to become common practice and conform to clear standards. This will help ensure that the results are robust. Robust methods are ones that can be used on various experts in a wide range of contexts by any competent knowledge engineer or KE practitioner. We also hope to make our techniques reliable. This will mean that different practitioners can apply them with the same expected utility. Placing elicitation on such a systematic footing will also be important in the development of methodologies that direct the process of specifying, constructing and maintaining systems.

We will begin by describing, in sufficient detail for the reader to apply them, examples of major KE methods. We will mention other techniques and where the reader can find out more about them. We will then review aspects of expertise and human information processing that are likely to directly affect the KE process. We will also indicate various software tools that implement or support some of these KE methods. We will describe the methodologies for acquisition and modelling expertise that are beginning to emerge. Finally, we will discuss the effect that the presence of evermore content on the web is having on the knowledge acquisition problem.

Elicitation techniques

The techniques we will describe are methods that we have found in our previous work to be both useful and complementary to one another. We can subdivide them into *natural* and *contrived* methods. The distinction is a simple one. A method is described as natural if it is one an expert might informally adopt when expressing or displaying expertise. Such techniques include interviews or observing actual problem solving. There are other methods we will describe in which the expert undertakes a contrived task. The task elicits expertise in ways that are not usually familiar to an expert. The first two categories of elicitation method are both natural under this definition and are varieties of interview and protocol analysis.

The structured interview

Almost everyone starts in KE by determining to use an interview. The interview is the most commonly used knowledge elicitation technique and takes many forms, from the completely *unstructured* interview to the formally-planned, *structured* interview.

The structured interview is a formal version of the interview in which the person eliciting the knowledge plans and directs the session. A significant benefit of the structured interview is that it provides structured transcripts that are easier to analyse than unstructured conversation. In reality the structured interview is a class of techniques (Hoffman *et al.*, 1995).

The formal interview specified here constrains the expert-elicitor dialogue to the general principles of the domain. Experts do not work through a particular scenario extracted from the domain by the elicitor; rather experts generate their own scenarios as the interview progresses.

A template for such an interview is as follows.

1. Ask the expert to give a brief (10 minute) outline of the target task, including the following information:
   An outline of the task, including a description of the possible solutions or outcomes of the task;
   A description of the variables that affect the choice of solutions or outcomes;
   A list of major rules or procedures that connect the variables elicited to the solutions or outcomes.
2. Take each rule or procedure elicited in Stage 1, ask when it is appropriate and when it is not and if it is a procedure how it is preformed. The aim is to reveal the scope (generality and specificity) of each existing rule, and hopefully generate some new rules.
3. Repeat Stage 2 until it is clear that the expert will not produce any additional information.

A useful way of obtaining a domain overview (stage 1 of the structured interview) is to ask probe questions that relate to an individual's specific experience. It is also important in this technique to be specific about how to perform stage 2. We have found that it is helpful to constrain the elicitor's interventions to a specific set of *probes*, each with a specific function. Here is a list of probes (P) and functions (F) that can help in stages 1 and 2.

P1.1	Could you tell me about a typical case?
F1.1	Provides an overview of the domain tasks and concepts.
P1.2	Can you tell me about the last case you encountered?
F1.2	Provides an instance based overview of the domain tasks and concepts.
P2.1	Why would you do that?
F2.1	Converts an assertion into a rule.
P2.2	How would you do that?
F2.2	Generates *lower order* rules.
P2.3	When would you do that? Is <the rule> always the case?
F2.3	Reveals the generality of the rule and may generate other rules.
P2.4	What alternatives to <the prescribed action/decision> are there?
F2.4	Generates more rules.
P2.5	What if it were not the case that <currently true condition>?
F2.5	Generates rules for when current condition does not apply.
P2.6	Can you tell me more about <any subject already mentioned>
F2.6	Used to generate further dialogue if expert dries up.
P2.7	Can you tell me about an unusual case you encountered/heard about from some other expert?
F2.7	Refines the knowledge to include rare cases and special procedures.

The idea here is that the elicitor engages in a type of slot/filler dialogue. The provision of template questions about concepts, relations, attributes and values makes the elicitor's job very much easier. It also provides sharply focused transcripts that facilitate the process of extracting usable knowledge. Of course, there will be instances when none of the above probes are appropriate (such as the case when the elicitor wants the expert to clarify something). However, you should try to keep these interjections to a minimum. The point of specifying such a fixed set of linguistic probes is to constrain the expert to giving you all, and only, the information you want.

The sample of dialogue below is taken from a real interview of this kind. It is the transcript of an interview by a knowledge engineer (KE) with an expert (EX) in the domain of geological analysis. (In the transcripts we use the symbol + to represent a pause in the dialogue.)

KE What would you do at this stage?
EX I would look at the grain size of the hand specimen and see how fine it was.
KE Why would you look at the grain size?
EX That will tell me if the rock has been formed near to the surface or deep inside the earth. The finer the grain size the faster it cooled. Coarse crystals indicate that the rock was cooling slowly + forming deeper down + we say its emplacement is plutonic + if it cooled near the surface its emplacement is volcanic.
KE Are there any alternatives to coarse and fine grain size?
EX There are glasses + you can't see any structure here because the rock cooled so fast.
KE What would you look at next?
EX Colour is important + the lighter the rock the more acidic it is.
KE Why is a lighter rock more acidic?
EX Acidic rocks are higher in quartz and colour is a good indicator of quartz content — leucocratic or light things have a lot of quartz — melanocratic that is darker rocks have olivines and pyroxines.

This is quite a rich piece of dialogue. From this section of the interview alone we can extract numerous rules such as

IF grain size is large
THEN rock is plutonic
IF rock is leucocratic
THEN rock has high quartz content

Of course these rules may need refining in later elicitation sessions, but the text of the dialogue shows how the use of the specific probes has revealed a well-structured response from the expert. (In fact, a possible second-phase elicitation technique would be to present these rules back to the expert and ask about their truthfulness, scope and so forth.)

Semi-structured interviews

Techniques exist to impose a lesser amount of structure on an interview. We mention two examples here. One of these is the Knowledge Acquisition Grid (LaFrance, 1987). This is a matrix of knowledge types and forms: examples of knowledge forms are *layouts* and *stories*; examples of question types are *grand tour* and *cross-checking*. A grand tour involves such things as distinguishing domain boundaries and the overall organisation of goals; cross-checking involves the engineer attempting to validate the acquired knowledge by, for example, playing devil's advocate.

Second, there is the teachback technique (Johnson and Johnson, 1987). In this technique the KE formulates a representation of the knowledge that has been acquired in an interview. This is then 'taught back' to the expert, who can then check or, when necessary, amend the information.

Unstructured interviews

Unstructured interviews have no agenda (or, at least, no *detailed* agenda) set either by the knowledge elicitor or by the expert. Of course, this does not mean that the elicitor has no goals for the interview, but it does mean that she has considerable scope for proceeding; there are few constraints and herein lie its advantages. First, the approach can be used whenever one of the goals of the interview is to establish a rapport between the expert and the knowledge elicitor. There are no formal barriers to the discussion covering whatever material either participant sees fit. Second, one can get a broad view of the topic easily; the knowledge elicitor can 'fill in the gaps' in her own perceived knowledge of the domain. Third, the expert can describe the domain in a way with which he is familiar, discussing topics that he considers important and ignoring those he considers uninteresting.

The disadvantages are clear enough. The lack of structure can lead to inefficiency. The expert may be unnecessarily verbose. He may concentrate on topics whose importance he exaggerates. The coverage of the domain may be patchy. The data acquired may be difficult to integrate, either because it does not form a coherent body of content, or because there are inconsistencies. This last will be an even more likely occurrence if the information provided by *several* experts is to be collated.

In all of the interview techniques mentioned so far (and in some of the other generic techniques as well) there exist a number of dangers that have become familiar to practitioners of knowledge elicitation.

One problem is that in an interview experts will only produce what they can verbalise. If there are non-verbalisable aspects to the domain, the interview will not recover them. It may be that the knowledge was never explicitly represented or articulated in terms of language (consider, for example, pattern recognition expertise). Then there is the situation where the knowledge was originally learnt explicitly in a propositional or language-like form. However, in the course of experience such knowledge can become routinised or automatised. (We often use a computing analogy to refer to this situation and speak of the expert as having *compiled* the knowledge.) This can happen to such an extent that experts may regard the complex decisions they make as based only on hunches or intuitions. In actual fact, these decisions are based upon large amounts of remembered data and experience, and the continual application of that knowledge. In this situation they tend to give *black box* replies 'I don't know how I do that . . .', 'It is obviously the right thing to do . . .'.

Another problem arises from the observation that people (and experts in particular) often seek to justify their decisions in any way they can. It is a common experience of the knowledge elicitor to get a perfectly valid decision from an expert, and then to be given a spurious justification as to why it was made and how it originated.

For these and other reasons one should always supplement interviews with additional elicitation methods. Elicitation ought always to consist of a programme of techniques and methods. This brings us on to consider another family of techniques much favoured by knowledge engineers.

Protocol analysis

Protocol Analysis (PA) is a generic term for a number of different ways of performing some form of analysis of the expert(s) actually solving problems in the domain (see also Chapter 7).

In all cases the elicitor takes a record of what the expert does — preferably by video or audio tape — or at least by written notes. Transcripts or protocols are then made from these records and the elicitor tries to extract meaningful structure, rules and processes from the protocols.

We can distinguish two general types of PA — *on-line* and *off-line*. In on-line PA the expert is being recorded solving a problem, and concurrently a commentary is made. The nature of this commentary specifies two sub-types of the on-line method. The expert performing the task may be describing what they are doing as problem solving proceeds. This is called *self-report*. A variant on this is to have another expert provide a running commentary on what the expert performing the task is doing. This is called *shadowing*.

Off-line PA allows the expert(s) to comment retrospectively on the problem solving session — usually by being shown an audio-visual record of it. This may take the form of retrospective self-report by the expert who actually solved the problem. It could also be a critical retrospective report by other experts, or there could be group discussion of the protocol by a number of experts including its originator. In the case in which only a behavioural protocol is obtained then obviously some form of retrospective verbalisation of the problem-solving episode is required.

Before PA sessions can be held, a number of pre-conditions should be satisfied. The first of these is that the elicitor is sufficiently acquainted with the domain to understand the expert's tasks. Without this the elicitor may completely fail to record or take note of important parts of the expert's behaviour.

A second requirement is the careful selection of problems for PA. This sampling of problems is crucial. PA sessions may take a relatively long time, only a few problems can be addressed in any programme of acquisition (Shadbolt and Burton, 1989). Therefore, the selection of problems should be guided by how representative they are. Asking experts to sort problems into some form of order (Chi *et al.*, 1981, 1982) may give an insight into the classification of types of problems and help in the selection of suitable problems for PA (see also the next two sections on concept sorts and laddering for methods that can be used to help structure a classification of types of problem).

A further condition for effective PA is that the expert(s) should not feel embarrassed about describing their expertise in detail. It is preferable for them to have experience in thinking aloud. Uninhibited thinking aloud has to be learned in the same way as talking to an audience. One or two short training sessions may be useful. In these training sessions a simple task can be used as an example. This puts the expert at ease and familiarises them with the task of talking about their problem solving.

Where a verbal or behavioural transcript has been obtained we next have to undertake its analysis. Analysis might include the encoding of the transcript into 'chunks' of knowledge (actions, assertions, propositions, key words, etc.), and should result in a rich domain representation with many elicited domain features together with a number of specified links between those features. The example below is from a self-report of an expert geologist. It is immediately apparent that protocols can be extremely dense sources of information. A very significant amount of work is required to analyse and structure the content in this very small fragment of a self report on one specimen.

> To start off with it's obviously a fairly coarse-grained rock ... and you've got some nice big orthoclase crystals in here — this is actually SHAP GRANITE — I know it just because everybody's seen SHAP GRANITE — or it's a very strong possibility that it's SHAP GRANITE ... it's a typical teaching specimen — as I say the obvious things are these very big orthoclase crystals pink colouration and you can certainly see some cleavage in some of them — you can certainly make out there are feldspar cleavages in there — it's a coarse-grained rock anyway, you

> can see the crystals nice and coarsely — these large porphyritic crystals — you can see, in the ground mass, you can see quartz — get some light on it (HOLDS SPECIMAN UP TO WINDOW) quartz, which is this fairly clear mineral you can actually look into it and see through it as opposed to calcite or feldspars where it's more cloudy — you can't actually see any good crystal faces on these cut sections — small flakes of biotite, black micacious looking — small plates, you can certainly see some on this specimen even without a hand lens.

There are a number of principles that can guide the protocol analysis. For example, analysis of the verbalisation resulting in the protocol can distinguish between information that is attended to during problem-solving, and that which is used implicitly. A distinction can be made between information brought out of memory (such as a recollection of a similar problem solved in the past), and information that is produced 'on the spot' by inference. The knowledge chunks referred to above can be analysed by examining the expert's syntax, or the pauses he takes, or other linguistic cues. Syntactical categories (e.g., use of nouns, verbs, etc.) can help distinguish between domain features and problem-solving actions, etc.

In trying to decide when it is appropriate to use PA bear in mind that it is alleged that different KE techniques differentially elicit certain kinds of information (Hoffman *et al.*, 1995). With PA it is claimed that the sorts of knowledge elicited include the 'when' and 'how' of using specific knowledge. It can reveal the problem solving and reasoning strategies, evaluation procedures and evaluation criteria used by the expert, and procedural knowledge about how tasks and sub-tasks are decomposed. A PA gives you a complete episode of problem solving. It can be useful as a verification method to check that what people say is what they do. It can take you deep into a particular problem. However, it is intrinsically a narrow method since usually one can only run a relatively small number of problems from the domain.

When actually conducting a PA the following are a useful set of tips to help enhance its effectiveness. Present the problems and data in a realistic way. The way problems and data are presented should be as close as possible to a real situation. Transcribe the protocols as soon as possible, the meaning of many expressions is soon lost, particularly if the protocols are not recorded. In almost all cases an audio recording is sufficient, but video recordings have the advantage of containing additional and disambiguating information. Avoid long self-report sessions. Because of the need to perform a double task the process of thinking aloud is significantly more tiring for the expert than being interviewed. This is one reason why shadowing is sometimes preferred. In general, the presence of the elicitor is required in a PA session. Although the elicitor adopts a background role, her very presence suggests a listener to the interviewee, and lends meaning to the talking aloud process. Therefore, comments on audibility, or even silence by the elicitor, are quite acceptable.

Protocol analyses share with the unstructured interview the problem that they may deliver unstructured transcripts that are hard to analyse. Moreover, they focus on particular problem cases and so the scope of the knowledge produced may be very restricted. It is difficult to derive general domain principles from a limited number of protocols. These are some of the practical disadvantages of protocol analysis. However, there are more subtle problems.

Two actions, which look exactly the same to the knowledge elicitor, may be very different in their extent and intent. For example, our geologist who performs a particular test to a specimen may apply that same test to another but with a quite different purpose. The knowledge elicitor simply does not know enough to discriminate the actions. The obverse to this problem can arise in shadowing and the retrospective analyses of protocols by experts. Here the expert(s) may simply wrongly attribute a set of considerations to an action after the event. This is analogous to the problems of misattribution in interviewing.

A particular problem with self-report, apart from being tiring, is the possibility that verbalisation may interfere with performance. The classic demonstration of this is for a driver to attend to all the actions involved in driving a car. If one consciously monitors such parameters as engine revs, current gear, speed, visibility, steering wheel position and so forth, the driving invariably gets worse. Such skill is shown to its best effect when performed automatically. This is also the case with certain types of decision making expertise. By asking the expert to verbalise, one is in some sense destroying the point of doing protocol analysis — to access procedural, real-world knowledge.

Having pointed to these disadvantages, it is also worth remembering that context is oftentimes important for memory — and hence for problem solving. For most non-verbalisable knowledge, and even for some verbalisable knowledge, it may be essential to observe the expert performing the task. For it may be that this is the only situation in which the expert is actually able to perform it.

Finally, when performing PA it is useful to have a set of conventions for the actual interpretation and analysis of the resultant data. Ericsson and Simon (1993) provide the classic exposition of protocol analysis although it is oriented towards cognitive psychology. Useful additional references are Kuipers and Kassirer (1983), Belkin *et al.* (1987), McGraw and Harbison-Briggs (1989), and Firlej and Hellens (1991), as well as a fuller account in Chapter 7 of this book.

Critical decision method

This method contains elements of both interviewing and protocol analysis but in a context that stresses the examination of problem solving in natural decision making contexts (Zsambok and Klein, 1997). Klein and his colleagues developed a set of opening queries to stimulate recall of salient cases — cases that involved critical decisions (Klein *et al.*, 1986). A set of probe questions were designed to elicit specific, detailed information about the important cues, choice points, options, actions plans and the role of experience in decision making. A distinctive feature of this approach was that it seemed well suited to eliciting knowledge relating to highly dynamic situations where the requirement was to rapidly assess a situation and identify an effective and feasible course of action (Klein, 1993). Domains examined using the approach included acute clinical care, military planning, fire fighting and industrial process control.

A CDM session is organised around an account of a specific incident from the expert's own experience. The expert is guided in the recall and recounting of the incident and its context. There then follow three information-gathering passes back through the incident. First a time line is built that verifies the points at which decisions are made. Second there is a phase of deepening that produces a more comprehensive and contextually rich account of the incident — focusing, for example, on the cues used to recognise salient features of the incident. A final information sweep uses a 'what if' approach to identify potential errors, alternative decision points, and expert/novice differences.

Table 1 contains a range of probe question types with exemplars that we have found to be particularly useful in various phases of the CDM. There is no reason to use these questions exclusively for an individual phase although it is clear that the *options* and *choice* probe types are likely to feature substantially in the 'what if' phase of information gathering.

A typical CDM session can last around 2 hours; depending on the domain more or less time might be spent on recollecting a rich complex incident whilst in another setting the majority of the effort is devoted to examining counterfactual situations. The CDM does have limitations. In distributed problem solving no one individual may handle more than

Table 1 Sample CDM Probe Questions

Probe Type	Probe Examples
Cues	What were you seeing, hearing, smelling?
Knowledge	What information did you use in making this decision? How was it obtained?
Analogues	Were you reminded of any previous incidents?
Scenarios	Does this case fit a standard or typical scenario? Does it fit a scenario you were trained to deal with?
Goals	What were your specific goals and objectives at the time?
Options	What other courses of action were considered or available?
Choice	How was this option selected/other options rejected? What rule was being followed?
Anticipation	Did you imagine the possible consequences of this action? Did you imagine the events that would unfold?
Experience	What specific training or experience was necessary or helpful in this decision? What more would have helped?
Decision making	How much time pressure was involved in making the decision? How long did it take to make the decision?
Aiding	What training, knowledge or information could have helped?
Situation assessment	If you were asked to describe the situation to a colleague at this point, have would you summarise the situation?
Errors	What mistakes are likely at this point? How might a novice have behaved differently?
Hypotheticals	If a key feature of the situation had been different, what differences would it have made in your decision?

one element of a task. They would never know whether their judgements or assessments were correct. In high workload environments we have observed that incidents and events can become merged. When responding to an opening query one sometimes sees an expert recount an incident but then become confused when asked for a time line or other details. Despite these shortcomings the style of interview and problem solving reflection provides a rich output from which the elicitor can extract important task relevant knowledge — a more detailed account of the method can be found in Hoffman *et al.* (1998).

The techniques discussed so far are *natural* and intuitively easy to understand. Experts are used to expressing their knowledge in these sorts of ways. The techniques that follow are what we what we have termed *contrived* and permit the expression of knowledge in ways that are likely to be unfamiliar to the expert.

Concept sorting

Concept sorting is a technique that is useful when we wish to uncover the different ways an expert sees relationships between a fixed set of concepts.

In the version we will present an expert is presented with a number of cards on each of which a concept word is printed. The cards are shuffled and the expert is asked to sort the cards into either a fixed number of piles of else to sort them into any number of piles the expert finds appropriate. This process is repeated many times.

Using this task one attempts to get multiple views of the structural organisation of knowledge by asking the expert to do the same task over and over again. Each time the expert sorts the cards he should create at least one pile that differs in some way from previous sorts. The expert should also provide a name or category label for each pile on each different sort.

Performing a card sort requires the elicitor to have some basic conception of the domain. Cards have to be made with the appropriate labels before the session. However, no great familiarity is required as the expert provides all the substantial knowledge in the process of the sort. We now provide an example from our geology domain to show the detailed mechanics of a sort.

The concepts printed on the cards were the names of igneous rocks drawn from a structured interview with the expert. He had described 18 rock types:

1 adamellite
2 andesite
3 basalt
4 dacite
5 diorite
6 dolerite
7 dunite
8 gabbro
9 granodiorite
10 granite
11 lherzolite
12 microgranite
13 peridotite
14 picrite basalt
15 rhyodacite
16 rhyolite
17 syenite
18 trachyte

The expert was shown possible ways of sorting cards in a *toy* domain, as part of the briefing session, and then asked to sort the real elements in the same way.

The dimensions/piles which the expert used for the various sorts were as follows:

Sort 1: grain size	Piles 1=coarse, 2=medium, 3=fine
Sort 2: colour	Piles 1=melanocratic, 2=mesocratic, 3=leucocratic
Sort 3: emplacement	Piles 1=intrusive, 2=extrusive
Sort 4: presence of olivine	Piles 1=always, 2=possibly, 3=never
Sort 5: presence of quartz	Piles 1=always, 2=possibly, 3=never
Sort 6: % of silica	Piles 1=>68%, 2=<68%, 3=about 68%
Sort 7: density	Piles 1=v.light, 2=light, 3=medium, 4=dense, 5=v.dense

Table 2 shows the pile of each sort for each element. You will see that many of the elements are distinguishable from one another — even with these few sorts.

Using this information we can attempt to extract decision rules directly. An example of a rule extracted from the sorting is:

IF	the grain size is fine	(sort 1/pile 3)
AND	the colour is mesocratic	(sort 2/pile 2)
AND	its emplacement is extrusive	(sort 3/pile 2)
AND	it does NOT contain olivine	(sort 4/pile 3)
AND	may possibly contain quartz	(sort 5/pile 2)
AND	it contains less than 68% silica	(sort 6/pile 2)
AND	its density is medium	(sort 7/pile 3)
THEN	the rock is andesite	(outcome 2)

As you can see from the example such sorts produce long and cumbersome rules. In fact many of the clauses may be redundant — once you have established that the grain size is small, then it is going to be an extrusive rock.

Table 2 Tabulated Results from the Card Sort

	SORT						
ROCK	1	2	3	4	5	6	7
1	1	3	1	1	1	2	1
2	3	2	2	3	2	2	3
3	3	2	2	2	2	2	4
4	3	2	2	3	2	3	2
5	1	2	1	3	2	2	3
6	2	1	1	2	2	2	4
7	1	1	1	1	3	2	5
8	1	2	1	2	2	2	4
9	1	3	1	3	1	3	1
10	1	3	1	3	1	1	1
11	1	1	1	1	3	2	5
12	2	3	1	3	1	1	1
13	1	1	1	1	3	2	5
14	3	1	2	1	3	2	4
15	3	3	2	3	1	1	2
16	1	3	2	3	1	1	1
17	1	3	1	3	1	2	3
18	3	3	2	3	2	2	2

However, the utility of this technique does not reside solely in the production of decision rules. We can use it, as we have said, to explore the general inter-relationships between concepts in the domain. We are trying to make explicit the implicit structure that experts impose on their expertise.

When using any of these KE methods knowledge elicitors should beware a type of semantic mindset whereby the expert or elicitor focuses on only one type of knowledge element. To derive the full benefit of a KE method one should play many variations on the theme. For example, in concept sorting the cards can name knowledge elements of any type not just objects in a domain. The cards might name tasks, goals, actions, resources, etc. The restriction is that in any sorting session the cards should be of the same knowledge type.

Variants of the simple sort are different forms of *hierarchical* sort. One such version is to ask the expert to proceed by producing first two piles, on the second sort three, then four and so on. Finally we ask if any two piles have anything in common. If so you have isolated a higher order concept that can be used as a basis for future elicitation.

The advantages of concept sorting can be characterised as follows. It is fast to apply and easy to analyse. It forces an explicit format on the constructs that underlie an expert's understanding. In fact it is often instructive to the expert. A sort can lead the expert to see structure that he himself has not consciously articulated before. Finally, in domains where the concepts are perceptual in nature (i.e., x-rays, layouts and pictures of various kinds) then the cards can be used as a means of presenting these images and attempting to elicit names for the categories and relationships that might link them.

There are, of course, features to be wary of with this sort of technique. Experts can often confound dimensions by not consistently applying the same semantic distinctions throughout an elicitation session. Alternatively, they may over simplify the categorisation of elements, missing out important caveats.

An important tip with all of the contrived techniques we are reviewing is to always audiotape these sessions. An expert makes many asides, comments and qualifications in the course of sorting ranking and so on. In fact one may choose to use the contrived methods as a means to carry out auxiliary structured interviews, with the structure this time centred on the activity of the technique.

It is worth noting that we have found (Schweikert *et al.*, 1987) an expert's own opinion of the worth of a technique to be no guide to its real value. In methods such as sorting we have a situation in which the expert is trying to demonstrate expertise in a non-natural or contrived manner. He might be quite used to chatting about his field of expertise, but sorting is different and experts may be suspicious of it. Experts may in fact feel they are performing badly with such methods. However, on analysis one finds that the yield of knowledge is as good and sometimes better than for non-contrived techniques (Shadbolt and Burton, 1995).

Laddered grids

This is another somewhat contrived technique that you will need to explain carefully to the expert before starting. The expert and elicitor construct a graphical representation of the domain in terms of the relations between domain or problem solving elements. The result is a qualitative, two-dimensional graph where nodes are connected by labelled arcs. No extra elicitation method is used here, expert and elicitor construct the graph together by negotiation.

In using the technique the elicitor enters the conceptual map at some point and then she attempts to move around it with the expert. A formal specification of how we use the technique is shown below together with an example of its use.

Start the expert off with a seed item
Move around the domain map using the following prompts

To move DOWN the expert's domain knowledge:

Can you give examples of <ITEM>?

To move ACROSS the expert's domain knowledge:

What alternative examples of <CLASS> are there to <ITEM>?

To move UP the expert's domain knowledge:

What have <SAME LEVEL ITEMS> got in common?

What are <SAME LEVEL ITEMS> examples of?

To elicit essential properties of an item:

How can you tell it is <ITEM> ?

To discriminate items:

What is the key difference between <ITEM 1> and <ITEM 2>?

The elicitor may move around the knowledge map in any order which seems appropriate or convenient. As the session progresses, the elicitor keeps track of the elicited knowledge by drawing up a network on a large piece of paper or if computer supported via some other graphical characterisation. This representation allows the elicitor to make decisions (or ask questions) about what constitutes higher or lower order elements in the domain, what differences exist between elements in the network. In order to give the reader the flavour of the technique, there follows an extract from a laddered grid elicitation session. Once again, the knowledge domain is geology.

KE So how could you tell something was dacite?
EX Well + examine the fresh surface and the weathered surfaces first + looking at grainsize, the relationship between the grains.
KE Can I just stop you there? What type of grain size is it?
EX Coarse, medium, fine grain, oh, you want me to actually say what dacite is?
KE The grain, in dacite what would it be?
EX Er + medium grained.
KE Medium grained, right. So can you give me other examples of medium grained rocks?
EX Medium grained rocks + dolerite.... Granodiorite as well.... And we'll stay with that.
KE Right, erm, what alternative is there to a medium grained rock?
EX Well, you can have a coarse grained one or a fine grained one, those are sort of the three major ones.
KE Right, can you give me examples of coarse grained rocks?
EX Er, gabbro, granite ... hmm, yeah, those two.
KE And any examples of fine-grained rocks?
EX Er, basalt ... er andesite, trachyte ... microgranite as well.
KE Right, erm so. What about others?
EX Some of these are sort of a metamorphic ones where you're going to get large grains in a fine-grained matrix. There are phenocrysts in them, that's what we call the large grains.
KE Is, is there a word for that kind of texture or?
EX Porphyritic mixture.
KE Can you give me the examples of the porphyritics ...
EX Nepheline-syenite, oh and Kentallenite.
KE How would you go about telling the difference between dolerite and granodiorite? What is the key difference?
EX Whether it's got quartz or hasn't got quartz or the percentage of quartz present will define whether it's an acidic rock or a basic rock, basic not having any quartz in it at all, and then er if there's a low amount, that's going to be an intermediate rock.
KE Which, which are the intermediate?
EX Dacite + you've got high quartz are granite, microgranite, and andesite, and no quartz gabbro, basalt, dolerite and trachyte, intermediate dacite.

In the course of this laddered grid interview the elicitor drew up a hierarchical representation of the domain as shown in Figure 1. This is only one of a number of representations that could have been made. In this case the concepts of fine, medium and coarse grained rocks have been understood to be classes of rock type. Similarly the concept of an acidic, intermediate or basic rock has been treated as a class of rock type. However, the grain size and acidity (amount of quartz) could have been represented as properties of the particular rock types. These sorts of representational decisions abound in any knowledge elicitation exercise.

This hierarchy gives rise to the following set of rules that could be included in the knowledge base of a knowledge intensive system for geological rock classification.

IF the rock is of medium grain size
AND the rock is intermediate
THEN the rock may be dacite

IF the rock is of coarse grain size

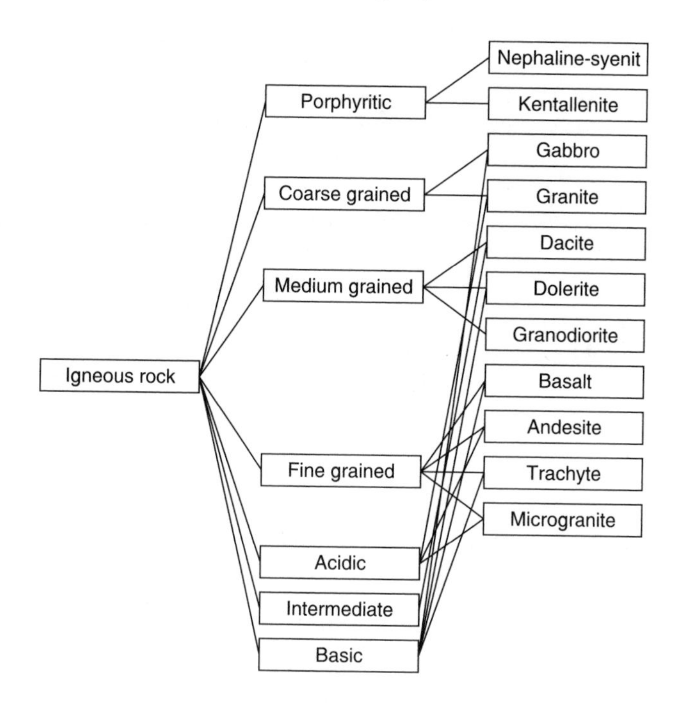

Figure 1 Laddered grid in the geology domain.

AND the rock is acidic
THEN the rock may be granite

IF the rock is of coarse grain size
AND the rock is basic
THEN the rock may be gabbro

As with the previous contrived method it is important to keep an audio record of the session for future review or transcription. Laddering is an excellent way of carrying out a structured interview. Also it is a technique that can be used on a variety of knowledge types — objects, actions, tasks, goals, etc.

We have found that this form of knowledge elicitation is very powerful for structured domains. As with other contrived techniques we have found that whilst an expert may think this technique is revealing little of interest, subsequent analysis provides good quality content.

The limited information task

A technique which can prove an excellent complement to the methods already outlined does not provide a spatial representation of the domain, but rather a set of hints or suggestions. This technique, which may prove useful in the construction of knowledge intensive systems, is called the *limited information task* (Hoffman *et al.*, 1995) or *20 questions* (Grover, 1983). The expert is provided with little or no information about a particular problem to be solved. The expert must then ask the elicitor for specific information that will be required to solve the problem. The information that is requested, along with the order in which it is requested, provides the elicitor with an insight into the expert's problem solving strategy. One difficulty with this method is that the elicitor needs a good understanding of the domain in order to make sense of the experts' questions, and to provide meaningful responses. The elicitor should have forearmed themselves with a problem from the domain together with a *crib sheet* of appropriate responses to the questions.

In one of the versions of the limited information task that we use we tell the expert that the elicitor has a scenario in mind and the expert must determine what it is. The scenario might represent a problem, a solution or a problem context. The expert is told that they may ask the elicitor for more information, though what the elicitor gives back is terse and does not go much beyond what was asked for in the question. The expert may be asked to explain why each of the questions was asked.

An example of the kind of interaction produced by this technique is shown below. Here the problem domain is in the construction of lighting systems for the inspection of industrial products and processes.

EX Is this in the manufacturing industry?
KE Yes
EX So we've ruled out things like fruit, vegetables, cows?
KE Yes
EX Is it the metal industry?
KE The material is wood
EX So we could be dealing with a large object here like a chair or table
KE The object is large
EX It's likely to be a 3-D object, you've got to pick it up and turn it over
KE That's right
EX So what I need now are the dimensions of this object in terms of the cube that will enclose it
KE It would have similar dimensions to the table top
EX Do I inspect one surface or all the surfaces?
KE All of them
EX Is the inspector looking for one or many faults?
KE One particular fault
EX Can you describe it for me?
KE It's pencil marks about half an inch long
EX What colour is the wood?
KE Dark unfinished wood
EX We've got a contrast problem here. At this point I'd go and look at the job + to see if the graphite pencil marks reflect light + sometimes it does, but it depends on the wood + if it does you can select the light to increase the contrast between the fault and the background

:

:

:

EX I'd be doing this in three phases: first a general lighting, then specific for surface lighting, and then some directional light [expert then gives technical specifications for these types of light]

This interview gives us an interesting insight into the natural line of enquiry of an expert in this domain. Often traditional knowledge-based systems gather the right data but the order in which it is gathered and used can be very different from how an expert works. This can decrease the acceptability of any implemented system if other experts are to use it, and it also has consequences for the intelligibility of any explanations the system offers in terms of a retrace of its steps to a solution.

It will be seen that we can once again extract decision rules directly from the dialogue:

IF fault colour is black
AND object colour is dark
THEN contrast is a problem

The drawbacks to this technique are that the elicitor needs to have constructed plausible scenarios and the elicitor has to be able to cope with questions asked of her. The experts themselves are sometimes uncomfortable with this technique; this may well have to do with the fact that, as with other contrived techniques, it is not a natural means of manifesting expertise. Whilst a few scenarios may reveal some of the general rules in a domain the elicitation is very case specific. In order to get a broad range of knowledge for a sweep of situations many scenarios would need to be constructed and used.

An interesting variation on this method is a form of telephone consultancy. Here we take two domain experts and place them at opposite ends of a table and ask them to imagine that one is a 'client' who is ringing up the other, a 'consultant', to ask for advice concerning a particular problem. They then engage in a conversation in which the consultant tries to elicit the nature and context of the problem, and finally attempts to offer appropriate advice. In this variation of the limited information task you can rely on one of the experts to generate interesting cases. In addition, the expert 'role playing' as the client can provide appropriate responses to the consultant's enquiries. The only drawback is that sometimes the experts construct extremely difficult cases for each other in order to test each other's mettle!

A taxonomy of KE techniques

We have sampled some of the major approaches to elicitation and, where appropriate, given a detailed description of techniques that are likely to be of use. There are many variants on the methods we have described. Below we have provided a taxonomy of methods with which we are familiar together with a primary reference for each one.

- Non-contrived
  - Interviews
    - Structured
      - Fixed Probe (Shadbolt and Burton, this volume)
      - Focused Interviews (Hart, 1986, Clayton *et al.*, 1991)
      - Forward Scenario Simulation (Grover, 1983)
      - Critical Decision Method (Hoffman *et al.*, 1998)
    - Semi-Structured
      - Knowledge Acquisition Grid (LaFrance, 1987)
      - Teach Back (Johnson and Johnson, 1987)
    - Unstructured (Weis and Kulikowski, 1984)
  - Protocol Analysis
    - Verbal (Sanderson and Bainbridge, this volume)

On line (Johnson *et al.*, 1987)

Off line (Elstein *et al.*, 1978)

Shadowing (Clarke, 1987)

Behavioural (Ericsson and Simon, 1984)

Contrived

Conceptual Mapping

Sorting and Rating (Gammack, 1987)

Repertory Grid (Shaw and Gaines 1987)

Pathfinder (Schvaneveldt *et al.*, 1985)

Goal Decomposition

Laddered Grid (Hinkle, 1965)

Limited-Information Task (Grover, 1983, Hoffman, 1987)

Having discussed the principal methods of elicitation we should spend a little time reflecting on the nature of two other major components of the KE enterprise, namely the experts and the expertise they possess.

On experts

Experts come in all shapes and sizes. Ignoring the nature of your expert is another potential pitfall in KE. A coarse guide to a typology of experts might make the issues clearer. Let us take three categories we shall refer to as *academics, practitioners* and *samurai* (in practice experts may embody elements of all three types). Each of these types of expert differs along a number of dimensions. These include; the outcome of their expert deliberations, the problem solving environment they work in, the state of the knowledge they possess (both its internal structure and its external manifestation), their status and responsibilities, their source of information, the nature of their training.

How are we to tell these different types of expert apart when we encounter them? The academic type regards their domain as having a logically organised structure. Generalisations over the laws and behaviour of the domain are important to them. Theoretical understanding is prized. Part of the function of such experts may be to explicate, clarify and teach others. Thus they talk a lot about their domains. They may feel an obligation to present a consistent story both for pedagogic and professional reasons. Their knowledge is likely to be well structured and accessible. These experts may suppose that the outcome of their deliberations should be the correct solution of a problem. They believe that the problem can be solved by the appropriate application of theory. They may, however, be remote from every day problem solving.

The practitioner class on the other hand are engaged in constant day-to-day problem solving in their domain. For them specific problems and events are the reality. Their practice may often be implicit and what they desire as an outcome is a decision that works within the constraints and resource limitations in which they are working. It may be that the generalised theory of the academic is poorly represented and articulated by the practitioner. For the practitioner heuristics may dominate and theory is sometimes thin on the ground.

The samurai is a pure performance expert — their only reality is the performance of action to secure an optimal performance. Practice is often the only training and responses are often automatic.

One can see this sort of division in any complex domain. Consider for example medical domains where we have professors of the subject, busy doctors working the wards, and medical ancillary staff performing many important but repetitive clinical activities.

The knowledge elicitor must be alert to these differences because the various types of expert will perform very differently in KE situations. The academic will be concerned to demonstrate mastery of the theory. They will devote much effort to characterising the scope and limitations of the domain theory. Practitioners, on the other hand, are driven by the cases they are solving from day to day. They have often *compiled* or *routinised* any declarative descriptions of the theory that supposedly underlies their problem solving. The performance samurai will more often than not turn any KE interaction into a concrete performance of the task — simply exhibiting their skill.

But there is more to say about the nature of experts and this is rooted in general principles of human information processing. (An excellent review of the psychology of expertise is Chi *et al.* (1988) and a fascinating glimpse into the constituents of some aspects of expertise can be found in Ericsson (1996).) Psychology has demonstrated the limitations, biases and prejudices that pervade all human decision-making — expert or novice. To illustrate this consider the following facts, all potentially crucial to the enterprise of KE.

It has been shown repeatedly that the context in which one encodes information is the best one for recall. It is possible then, that experts may not have access to the same information when in a KE interview, as they do when actually performing the task. So there are good psychological reasons to use techniques that involve observing the expert actually solving problems in the context in which they normally work. In short, protocol analysis techniques may be necessary, but will not be sufficient for effective knowledge elicitation.

Consider now the issue of biases in human cognition. One well-known problem is that humans are poor at manipulating uncertain or probabilistic evidence. This may be important in KE for those domains that require a representation of uncertainty. Consider the rule:

IF the engine will not turn over
AND the lights do not come on
THEN the battery is flat with probability X

This seems like a reasonable rule, but what is the value of X, should it be 0.9, 0.95, 0.79? The value that is finally decided upon could have important consequences for the working of any knowledge intensive system, but it is very difficult to decide upon it in the first place. Medical diagnosis is a domain full of such probabilistic rules. However, even expert physicians cannot accurately assess probability values in their own domains of expertise.

In fact there are a number of documented biases in human cognition that lie at the heart of this problem (see for example the classic work of Kahneman, *et al.*, 1982). People are known to undervalue prior probabilities, to use the ends and middle of the probability scale rather than the full range, and to *anchor* their responses around an initial guess. Cleaves (1987) lists a number of cognitive biases likely to be found in knowledge elicitation, and makes suggestions about how to avoid them. Faced with these difficulties many knowledge elicitors prefer to avoid the use of uncertainty wherever possible.

Cognitive bias is not limited to the manipulation of probability. A series of experiments has shown that systematic patterns of error occur across a number of apparently simple logical operations. For example, *Modus Tollens* states that if 'A implies B' is true, and 'not B' is true, then 'not A' must be true. However people, whether expert in a domain or not, make errors on this rule. This is in part due to an inability to reason with contrapositive statements.

Also in part it depends on what A and B actually represent. In other words, they are affected by the content. This means that one cannot rely on the veracity of experts' (or indeed anyone's) reasoning.

All this evidence suggests that human reasoning, memory and the representation of knowledge is rather more subtle than might be thought at first sight. The knowledge engineer should be alert to some of the basic findings emanating from cognitive psychology. Whilst no text is perfect as a review of bias in problem solving the book by Meyer and Booker (1991) is reasonably comprehensive.

On expertise

Clearly the expertise embodied by experts is not of a homogeneous type (Feltovich *et al.*, 1997). In constructing any knowledge intensive system it is likely that very different types of knowledge will be uncovered which will have very different roles in the system.

There are a number of analyses available of the epistemology of expertise. Our analysis is based to a large extent on that of CommonKADS (Schreiber *et al.*, 2000).

First, we can distinguish what is called domain level knowledge. This term is being used in the narrow sense of knowledge that describes the concepts and elements in the domain and relations between them. This sort of knowledge is sometimes called declarative, it describes what is known about things in the domain. The propositions below can both be seen as domain level knowledge in this sense.

Extract 1 Domain Knowledge from an analysis of a laddered grid obtained from an expert geologist

Granite is a coarse grained rock

Andesite has a high quartz content

There is also knowledge and expertise that has to do with what we might call the inference level. This is knowledge about how the components of expertise are to be organised and used in the overall system. It tells us the type of inferences that will be made and what role knowledge will play in those inferences. This is quite a high level description of expert behaviour and may often be implicit in expert practice. The following is a description of knowledge about part of an inference level structure called systematic diagnosis.

Extract 2 Analysis of verbal and behavioural protocols obtained from an expert in abdominal medicine

> To perform systematic diagnosis we will have knowledge about a complaint, and knowledge about observables from the patient or object. We select some aspect of the complaint and using a model of how the system should be performing normally we look to see if a particular parameter of the system is within normal bounds.

Another type of expert knowledge is the task level. This is sometimes called procedural knowledge. This is knowledge to do with how goals and sub-goals, tasks and sub-tasks should be performed. Thus in a classification task there may exist a number of tasks to perform in a particular order so as to utilise the domain level knowledge appropriately. This type of knowledge is present in the following extract.

Extract 3 Analysis of a verbal protocol obtained from an expert geologist

> First of all perform a general inspection of the object. Next examine the sample with a hand lens. Next use a prepared thin-section and examine that under a cross-polarising microscope.

Finally, there is a level of expert knowledge referred to as strategic knowledge. This is information that monitors and controls the overall problem solving. This can have to do with the way resources are used. What to do if the proposed solution fails or is found to be inappropriate in some way. What to do when faced with incomplete or insufficient data. Such information is contained in the following extract from an interview.

Extract 4 Part of a structured interview transcript obtained from an expert computer technician

> If I had time I would always check the disc head alignments. If its a BRAND X machine I'd always check that because they are notorious for going wrong.

Any field of expertise is likely to contain these various sorts of knowledge to greater or lesser extents. At any particular knowledge level the information may be explicit or implicit in an expert's behaviour. Thus in some domains the experts may have no real notion of the strategic knowledge they are following whilst in others this knowledge is very much in the forefront of their deliberations. Also, of course, the requirements on any knowledge intensive system about how far it needs to implement these various levels will vary. It is almost universally acknowledged that significant reasoning about problem domains requires more than just modelling simple relationships between concepts in the domains. It may require causal models of how objects influence and affect one another, models of the processes in which objects participate. This is a hard problem. And often the limitations of implementation technologies means that sophisticated domain models cannot be supported.

This brings us to a final important feature of KE. Since knowledge elicitation is such a time consuming and expensive business, not all of whose results can be immediately used, there is an increasing interest in developing ways of storing, archiving and retrieving knowledge that makes the best use of the elicitation investment (Neches *et al.*, 1991). The key to this lies in a change in our way of thinking about the content of knowledge-based systems. This has already been outlined earlier in this section. It is called the knowledge level view and was originally conceived by Allen Newell (1982).

Managing knowledge in this way requires using an expressive and unambiguous intermediate representation of the knowledge to be stored. A number of candidates exist for this including graphical and language oriented representations — many of these draw inspiration from knowledge representation languages developed in AI (Young and Gammack, 1987; Rich and Knight, 1991; Sowa, 1999; Norvig and Russell, 2003). One of the major deliverables in many projects we have worked on has not been any implemented system but a set of knowledge documents that describe in a structured way the knowledge in a particular domain. In these cases we use CommonKADS as a modelling and representational standard (Schreiber *et al.*, 2000).

Common KADs embodies the knowledge level thesis put forward by Bill Clancey in his classic 1985 paper on heuristic classification. The structure, shown in Figure 2, was the

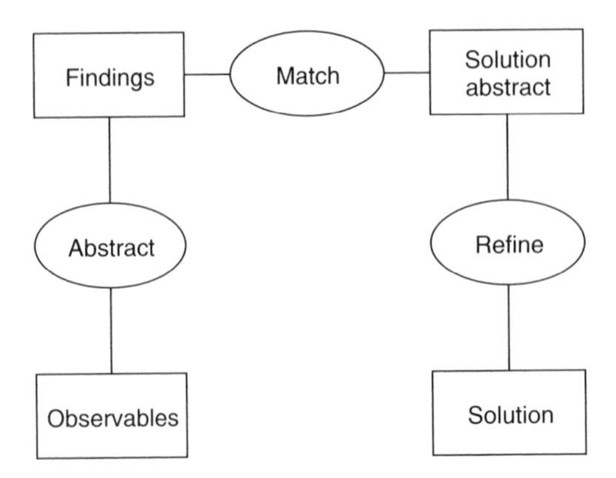

Figure 2 Heuristic classification.

result of a rational reconstruction of a number of existing knowledge intensive systems. His claim was that the knowledge bases of many systems were for the most part undifferentiated. The knowledge bases of these systems had been built with little regard as to how the knowledge was used. His analysis uncovered what he saw as an important type of problem solving system. Not all systems would contain and use knowledge in this way. Not all systems would be examples of heuristic classification, but some important ones were.

One system which Clancey characterised as heuristic classification was the classic MYCIN medical knowledge-based system (Shortliffe, 1979). We shall illustrate his approach using MYCIN-like knowledge. Figure 2 is to be understood as a structure at what we have termed earlier the inference level. It tells us what kinds of inferences are performed in this domain and the type of knowledge used by these inferences. The rectangles should be seen as types of data and the ellipses as types of inferences.

Let us take the left hand side of this structure that contains a process called abstraction. This is the process by which observations or data are transformed into abstract observations or findings. The process of abstraction can be realised using a number of methods. One of these is qualitative abstraction. Examples are shown below — in this example we move from quantitative observations to qualitative findings.

if patient has white blood cell count <2500
then patient has low white blood cell count

if patient has temperature >101
then patient has fever

What we have provided above is the actual domain level knowledge that plays the inference layer role of abstraction. Much of the knowledge in MYCIN's knowledge base is to do with this type of knowledge level processing — the process of abstraction, moving from a quantitative description of data to a qualitative one.

The top part of Figure 2 involves inferences from findings to abstract solutions. This type of inference is called match — and is understood as a type of association knowledge. An example of such knowledge might be a rule such as:

if patient has fever
and patient has low white blood cell count
then patient has gram negative infection

The concept of gram negative infection might be viewed as a diagnosis or solution but it is not a very specific one. What sort of gram negative infection is it? The right hand side of the heuristic classification structure deals with inference types that refine general to specific solutions. Such knowledge might consist of hierarchical typologies containing knowledge such as

gram negative infection

has sub types

e. coli infection

:

:

To establish a particular infection the system is likely to use knowledge that discriminates the sub-types.

Notice that in this account although we have talked about the inference structure from left to right no explicit control knowledge has been given. We might have stipulated that the system start with patient data and reason forward to a possible solution. We might have stated that the system should hypothesise a solution and see if there was evidence from the patient's condition to support the hypothesis. Or else a mixture of these ways of moving around the structure of Figure 2 might have been adopted. This additional knowledge is, of course, the task layer we mentioned earlier. Sometimes the standard method of moving around the inference structure is modified. This might arise if when a particular disease is suspected then one immediately looks for a particular piece of supporting observational data. Such knowledge comprises the strategic layer.

What we have described was exemplified via a MYCIN-like example. But heuristic classification as a type of problem solving might apply to many domains; financial assessment of an individual's creditworthiness, the likelihood of finding a mineral resource at a particular location, classifying a particular workplace setting as conforming to a particular health and safety standard, etc. It is this generality that is the power of these knowledge level approaches. If we know what kind of application we are building we can use the models to indicate the type of knowledge we need to acquire, how we might structure the knowledge base, how we can archive and index knowledge for future use.

These knowledge level models have formed the basis for a number of important methodologies that aim to support the knowledge engineering process (Shadbolt and O'Hara, 1997). A similar attempt to exploit structured templates can be found in modern approaches to designing software — for example the work on Design Patterns (Gamma *et al.*, 1995) and Object Oriented Design (Booch, 1993). The principles from OOD are ones that can be usefully adopted in any knowledge modelling exercise, in particular the concentration on acquiring hierarchical descriptions of a domain in the form of class hierarchies of the sort we see in Figure 1. OOD stresses the importance of associating with each class the necessary properties to distinguish it from other objects. It also holds that the classes should represent the most general levels of abstraction consistent with discriminating between objects.

There is a growing need to standardise, share, and exchange knowledge descriptions in all application areas and across a wide range of individuals and organisations. For example, efforts are underway to build 'knowledge-rich' thesauri to define the relevant terms in diverse fields such as medicine (Humphreys *et al.* 1998), genetics (Gene Ontology Consortium, 2000) and art (Petersen, 1994) — but there are also attempts to provide such structured resources for general terms in language, for example WordNet (Fellbaum, 1998). An organising principle in these thesauri is the *subsumption* or *is-a* relation — but there are others such as *part-of*. Recently, computer scientists seeking to promote the exchange of knowledge between machines and humans have promoted the use of ontologies. These contain an explicit description of the semantics ('meaning') of the types introduced. Tools and methods are now becoming available to support the modelling of ontologies (Noy *et al.*, 2001). The construction of ontologies will be an important new application context for knowledge elicitation techniques.

Methodologies and programmes of KE

We turn next to the question as to how KE techniques should be assembled to form a programme of acquisition and when we should use the various techniques. The choice may depend on the characteristics of the domain, of the expert, and of the required system. Furthermore, it is clear that some techniques are going to be more costly than others in terms of time with the expert, or else the effort required for subsequent analysis of transcripts.

There are a number of articles and books available on 'how to do knowledge elicitation'. These often contain advice of the most general kind, and emphasise the pragmatic considerations of knowledge intensive system development. General reviews can be found in Welbank (1983), Hoffman (1987), Kidd (1987), Hart (1986) McGraw and Harbison-Briggs (1989), Firlej and Hellens (1991) and Clayton *et al.* (1991). While these reviews are based on experience of the general kind, there have also been a number of attempts to make formal recommendations.

Knowledge engineers have developed a number of principles that form the basis for the techniques and tools used for knowledge acquisition and modeling. Moreover, there are a number of assumptions in much of this area that are worth making explicit.

> *Broad repertoire of techniques*: There is much evidence to suggest that different techniques can be more or less efficient in the types of knowledge they can elicit (Burton *et al.*, 1987; 1988), the so-called differential access hypothesis (Hoffman *et al.*, 1995). Hence, to efficiently acquire the knowledge in a domain often requires a range of techniques.
>
> *Acquisition as modelling*: Traditionally, knowledge engineering was viewed as a process of 'extracting' or 'mining from the expert's head' and transporting it in computational form to a machine. This has turned out to be a crude and rather naive view. Today, knowledge engineering is approached as a modelling activity. *A model is a purposeful abstraction of some part of reality.*
>
> *The knowledge-level principle*: In knowledge modelling, first concentrate on the conceptual structure of knowledge, and leave the programming details for later. Many software developers have an understandable tendency to take the computer system as the dominant reference point in their analysis and design activities. But there are two important reference points: the computational artefact to be built but, most importantly, there is the human side.
>
> *Knowledge structures*: Knowledge has a stable internal structure that is analysable by distinguishing specific knowledge types and roles. It goes without saying that knowledge, reasoning, and problem-solving are extremely rich phenomena. Knowledge may be complex, but it is not chaotic: knowledge appears to have a rather stable internal structure.
>
> *Evolutionary development*: A knowledge project must be managed by learning from your experiences in a controlled 'spiral' way. The

> development of simple or very well-known types of information systems usually proceeds along a fixed management route. This is especially clear in the so-called waterfall model of systems development.

As we have already noted the most thorough attempt to integrate KE procedure is provided by CommonKADS — Knowledge Acquisition and Domain Structuring (see Schrieber *et al.* 2000 for an overview of CommonKADS; Wielinga *et al.*, 1992 for its roots). KADS embodies a number of principles for the elicitation of knowledge and construction of a system. These principles are:

1. The knowledge and expertise should be analysed before the design and implementation starts.
2. The analysis should be model-driven as early as possible.
3. The content of the model should be expressed at the knowledge level.
4. The analysis should include the functionality of the prospective system.
5. The analysis should proceed in an incremental way.
6. New data should be elicited only when previously collected data have been analysed.
7. Collected data and interpretations should be documented.

Principle 1 is quite straightforward and requires no further explanation. Principle 2 requires that one should being to bear a model of how the knowledge is structured early on in the process, and use it to interpret subsequent data. Principle 3 means that one should use an appropriate intermediate level knowledge representation device, and try to characterise knowledge in terms of its use and functioning. Principle 4 is a reminder that a complete analysis includes an understanding of how the system is to work — e.g., who will use it, and in what situation. One cannot gain a full understanding of the problem simply by trying to map out an expert's knowledge without regard to how it will be used. Principle 5 emphasises the fact that there is a wide variety of related topics within a domain. This means that construction of a model should be 'breadth-first', embodying all aspects at once, rather than attempting fully to represent one sub-part after another. Principles 6 and 7 are once again straightforward. Like many of the best recommendations the utility of these statements is most apparent when they are not adhered to.

The identification of an appropriate model for an application becomes an important knowledge elicitation exercise when adopting a CommonKADS approach. The models themselves can be organised into tree like structures, see Figure 3. A series of questions about the domain attempt to establish which node in the tree best corresponds to the features of the application (O'Hara *et al.*, 1998).

Currently the most complete model set is that of CommonKADS (Schreiber *et al.*, 2000). However, these models introduce a new level of complexity in the acquisition process and whilst they may by extremely helpful when trying to understand the implementation of expertise models, whether the adoption of these various methodologies makes for more efficient and effective KE is a moot point as these claims have not been formally evaluated (Shadbolt *et al.*, 1999).

Automatic knowledge acquisition

As KE is acknowledged to be a time consuming and difficult process, the idea of automated elicitation is most attractive. A number of programs have been developed towards this goal, and we will briefly consider some of them in this penultimate section.

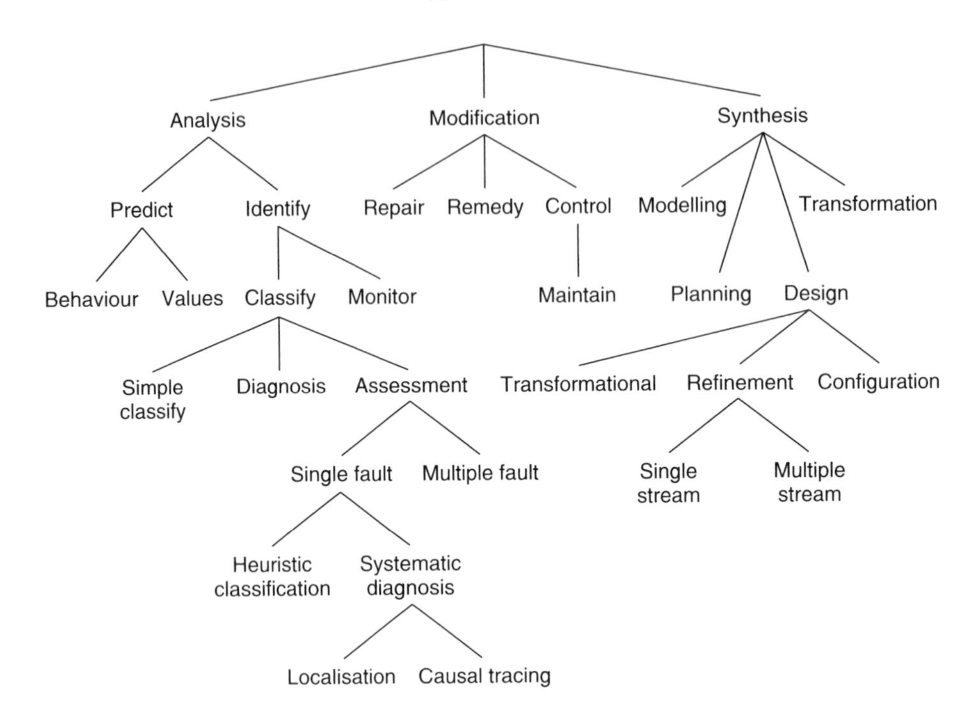

Figure 3 A taxonomic hierarchy of problem solving models.

Software tools for KE can be split into three categories: (1) *domain dependent* tools which have been developed for specific task domains; (2) *domain independent* tools which are computer implementations of one particular KE technique; and (3) *integrated systems* for acquisition and elicitation support.

Domain specific tools are tailored to elicit the types of knowledge known to be important in a particular application. An early example of this sort of system was SALT (Marcus, 1989). This had built in knowledge of the problem solving strategies used when configuring electro-mechanical systems. The expert interacts through an automated interview which allows the selection by menu of elements in the domain. The interview results are automatically converted into rules, and the expert has an opportunity to edit the resultant rules. In this task-oriented approach, the complexity of the knowledge acquisition process is reduced through the use of a model of the required knowledge as a template for customising otherwise general KA techniques and tools. Similar approaches can be found in medical domains such as therapy and treatment planning (Tu *et al.*, 1995). These systems are by definition restricted to particular tasks and domains. Moreover, most of these domain oriented tools have remained research prototypes.

The second approach to automation of the KA process is to focus on one particular technique and support its use. Of those systems which implement individual standard techniques, many of the most successful are based on the *repertory grid* (see Chapter 4). This technique has its roots in the psychology of personality (Kelly, 1955; Jankowitz 2003) and is designed to reveal a conceptual map of a domain, in a similar fashion to the card sort as discussed above. The work of Shaw and Gaines was particularly influential in promoting its use (Shaw and Gaines, 1987). The technique as developed in the 50s was very time-consuming to administer and analyse by hand. This naturally suggested that an implemented version would be useful.

One of the earliest and best known programs was ETS (Boose, 1985) although this was developed primarily as a research tool. KSS0 (Gaines, 1990; Gaines and Shaw, 1989) which we illustrate below formed the basis for a number of commercial products. More recently a web enabled freely accessible version of the software has become available (Gaines and Shaw, 1997, http://tiger.cpsc.ucalgary.ca:1500/WebGrid/WebGrid.html) that provides an excellent means of experimenting with the approach and indeed undertaking machine supported elicitation sessions.

Briefly, subjects are presented with a range of domain elements and asked to choose three, such that two are similar, and different from the third. Suppose we were trying to uncover an astronomer's understanding of the planets. We might present him with a set of planets, and he might choose Mercury and Venus as the two similar elements, and Jupiter as different from the other two. The subject is then asked for their reason for differentiating these elements, and this dimension is known as a construct. In our example 'size' would be a suitable construct. The remaining domain elements are then rated on this construct.

This process continues with different triads of elements until the expert can think of no further discriminating constructs. The result is a matrix of similarity ratings, relating elements and constructs. This is analysed using a statistical technique called *cluster analysis*. In KE, as in clinical psychology, the technique can reveal clusters of concepts and elements which the expert may not have articulated in an interview.

The automated versions are run in such a way that the repertory grid is built-up interactively, and the expert is shown the resultant knowledge. Experts have the opportunity to refine this knowledge during the elicitation process. In Figure 4 we can see that the expert has so far generated seven constructs along which the planets vary. In this case a seven point rating scale has been used and in the case of the construct size the smallest planet, Mercury, has been given a rating 1 and the largest, Jupiter, a rating of 7. The other planets have been rated in a comparative manner along the size construct. (In Figure 4 shading in the matrix is also used to highlight ratings. Heavy shading designates a high value for an element of a construct.) The analysis has already revealed clusters of both constructs and elements. Thus Jupiter and Saturn are clustered together at around 82% similarity, Neptune and Uranus at around 88%, and these two pairs are clustered at around 80%. An astronomer might well observe that this group of four planets constitute the gas giants. A new concept has been

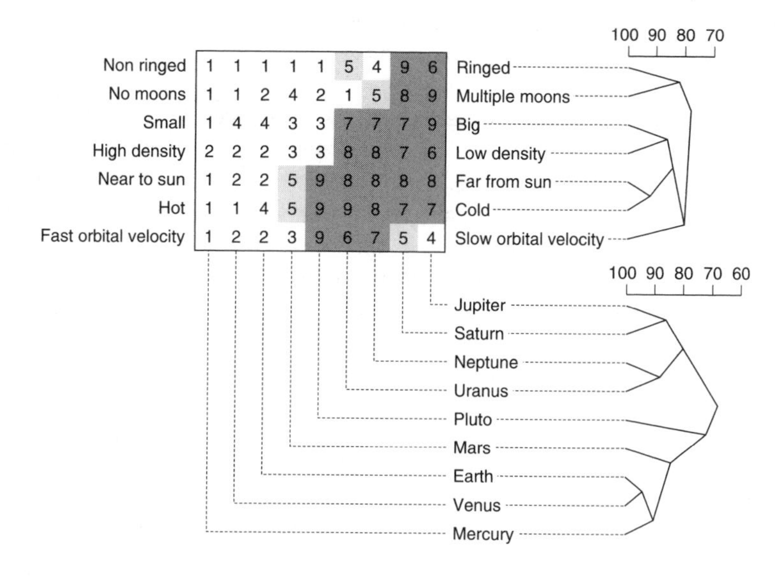

Figure 4 Knowledge elicited using WebGrid-III.

uncovered. Similarly, constructs can be clustered. We see that the constructs relating to temperature and distance from the sun are clustered. Such associations can reveal causal or other law-like relations in the domain.

We are also able to draw inferences from these structures in terms of implications between the constructs and elements. Examples of the sorts of implications we can draw are shown below.

```
high density -> hot (3)
low density -> cold (5)

Overall Evaluation
  Correct 8/8  100.00%

fast orbital velocity -> high density (4)
slow orbital velocity -> low density (5)

Overall Evaluation
  Correct 9/9  100.00%

high density -> near to sun (4)
low density -> far from sun (5)

Overall Evaluation
  Correct 9/9  100.00%

big -> ringed (2 E1)
small -> non ringed (5)

Overall Evaluation
  Correct  7/8  87.50%  Errors  +:1 -:0  Total Errors  1/8  12.50%
```

Variants on this technique allow you to run sociograms so that one can compare one individual's view of a domain with another's — highlighting areas of consensus and difference. These systems can be found a place in any programme of elicitation.

Another widely used machine supported method is concept mapping. The maps were developed by Joseph D. Novak in educational contexts to help students express and share their knowledge. In this technique the expert and knowledge elicitor construct a graphical network of nodes and relations representing knowledge about a domain. A concept map is a two-dimensional representation of a set of concepts and their relationships, shown as concept names connected by directed arcs encoding propositions in the form of simplified sentences.

A number of computer supported versions of this tool have been developed but one of the most accessible can be obtained from the Institute for Human Machine Cognition (Cañas *et al.*, 1999) at the following URL http://cmap.ihmc.us. In a browser, concept-map links can lead to diagrams, digital video, text, and arbitrary remote resources. Using these tools, domain experts can easily construct, navigate, share, criticize, and collaboratively refine knowledge models.

A related technology allows one to elicit and construct graphical networks that represent the rationale behind actions and decisions (Selvin and Buckingham Shum, 2000). Research has also been carried out trying to understand how to build KA tools for domains in which graphical representations are the most important (Cheng *et al.*, 2001).

Although programs continue to be written to support single KA techniques there is an increasing trend towards the third type of approach mentioned at the beginning of this section. This approach is to integrate several KA tools. The idea is that the whole is greater than the sum of the parts (Shadbolt *et al.*, 1993; Motta *et al.*, 1993). One such system, PCPACK (Schreiber *et al.*, 2000, Chapter 8, includes protocol editing (enabling text and documents to be annotated and analysed), concept and process laddering, card sorts and various other rapid knowledge formation methods. The results of elicitation are stored in a persistent object-oriented database. All the tools are able to access this database providing a means of transferring knowledge between the various tools. The entire package is interfaced through an intuitive direct manipulation interface. A flexible web publication hypertext system is able to annotate objects in the database together with concise documentation and tutorial material. A demonstration version of PC PACK can be downloaded from www.epistemics.co.uk and evaluated. More recent versions have begun to incorporate templates and problem solving models that guide the user through the elicitation process so as to populate a knowledge repository.

We have not included individual or collections of tools that originate from the disciplines of data mining (Witten and Frank, 1999) and machine learning (Mitchell, 1997). Such a discussion is beyond the scope of this chapter and as such they are not techniques that are used in sustained face to face knowledge acquisition sessions with experts. However, good references to the state of the art are contained in the sources given above.

Expertise and the web

The most significant difference between the world of knowledge we now inhabit as against that of a decade ago is the extraordinary rise of the importance of the internet. Any figures provided become rapidly out of date but as of 2002 estimates are that the indexed web — the web the search engines can get at — comprises some 10 billion indexed pages and that this is dwarfed by the so-called deep web. This deep web consists of huge numbers of databases, innumerable excel spreadsheets, a deluge of other content that is potentially available as an information resource but is as yet either not included or not fully indexed.

What is undeniable is that this provides access to a huge potential resource for the construction of any prospective knowledge intensive system. Recently systems have been built with significant recourse to knowledge on the web. The MIAKT project (Hu *et al.*, 2003) has used information from the web to help build an ontology for the domain of breast disease. It has also located on the web an extensive library of images that are indexed with a range of information about patients and symptoms. Similarly if one takes the domain used throughout this chapter — classification of rocks and minerals — there are substantial online resources. These range from dictionaries and definitions of terms, succinct summaries of the process of rock formation, compact representations of diagnostic heuristics, and extensive online databases.

The ability to search out content of this sort offers a new and powerful way to build initial knowledge structures. However, one should be aware of the very considerable problems that attach to this sort of content. These include issues of provenance, context and interpretation.

When we locate an apparently relevant piece of content on the web how are we to judge its provenance? Who asserted the content, how long ago, what sources did the author use, what qualifications are associated with the content? At the moment very little if any of this information is associated as meta-data with the content we are interested. In the absence of such information one tends to resort to the tried and trusted notion of looking at the brand associated with the content. One is likely to take the content hosted on an IEEE standards web site or a governmental or medical agency on trust rather more than that offered in more informal networks. However, it is now apparent that individuals and organisations will go to real lengths to appear to be the trusted sites of reputable organisations when in fact they are seeking to misinform. This whole issue of trusting digital content is attracting considerable interest and attention.

A more general problem is that of context — we may be able to download significant amounts of content but how are we to recognise the context in which it is appropriate to apply that knowledge. Human experts are extremely sensitive to the conditions under which knowledge should be applied. They are also often very aware of the context in which a piece of knowledge should not be applied or else relied upon.

The third problem — interpretation — is one that to solve would require a general solution to providing machines with a full understanding of the meaning of language. Take

1	The orbit of a planet/comet about the Sun is an ellipse with the Sun's centre of mass at one focus	Planets move in orbits that are ellipses
2	A line joining a planet/comet and the Sun sweeps out equal areas in equal intervals of time	The planets move such that the line between the Sun and the Planet sweeps out the same area in the same time no matter where in the orbit
3	The squares of the periods of the planets are proportional to the cubes of their semimajor axes	The square of the period of the orbit of a planet is proportional to the mean distance from the Sun cubed

quite straightforward scientific knowledge such as the above, taken from two web sites dealing with Kepler's Laws of planetary motion. Consider the differences and ask how it is that we see them as essentially the same. We as humans seldom notice because we are so adept at understand the equivalences between statement, the nuances of different phrases, and the conditions under which more or less precise statements are required.

There is no doubt that the web as an extended knowledge repository is set to figure even larger in years to come. This is precisely the aim of research underway in bringing about the so-called Semantic Web (Berners-Lee *et al.*, 2001). The aim is to build content that is much more richly described and enriched with information about the content itself — its provenance, its context of acquisition and so on. This attempt to bring about a range of knowledge services that can populate the web with such content and then exploit it in a number of ways is a focus of current research for the author (www.aktors.org).

With a new generation of knowledge technologies, with techniques from data mining and machine learning we can expect to see larger amounts of knowledge intensive processing driven by knowledge that has been acquired in an automatic or semi-automatic fashion (Crow and Shadbolt, 2001). These topics go well beyond the scope of this chapter but they will set the agenda for research in knowledge intensive systems for the next decade and beyond.

Conclusions

Notwithstanding the developments alluded to in the last section it is folly to imagine that face to face knowledge elicitation, one human with another, will cease to be important and necessary. In fact it is essential for all of the reasons outlined above — to know the provenance of the content, to understand the conditions under which it is to be applied, to know how to interpret a problem solving context. Human experts are able to demonstrate a mastery of when, where and how their knowledge applies and, more particularly, when, where and why it might not be. Human experts have accumulated their expertise over thousands of hours. This experience enables them to recognise situations and contexts that fall inside and outside of their competence. It enables them to make subtle judgements about the quality of the information they are presented with and the decisions they make. The fact that expertise and knowledge is ultimately grounded in human practice means that we need to understand the methods and techniques, problems and opportunities afforded by knowledge elicitation.

The problem of knowledge elicitation remains a subtle and complex one. This chapter has described some of the methods and techniques that are used in this enterprise. We have also sought to provide an indication of the difficulties inherent in doing this kind of work. Knowledge elicitation is itself a form of complex expertise. Experienced knowledge engineers come to recognise the characteristics of expert thinking. They develop skills that allow them to capture an expert's knowledge despite the many obstacles they face. Recently, methodologies have begun to emerge that seek to structure and manage the acquisition process.

As knowledge intensive systems become more widely deployed more people will face the challenges of knowledge elicitation. Whether it is in the form of a corporate intranet or a clinical decision support system knowledge is still the power driving these applications. Knowledge elicitation remains an important area of research and practical application.

References

Belkin, N.J., Brooks, H.M. and Daniels, P.J. (1987). Knowledge elicitation using discourse analysis. *International Journal of Man-Machine Studies*, **27**, 127–144.

Berners-Lee, T., Hendler, J. and Lassila, O. (2001). The Semantic Web. *Scientific American, 204*/5, 34–43.

Booch, G. (1993). *Object-Oriented Design with Applications*, 2nd edition. (Reading, MA: Addison-Wesley).

Booch, G., Rumbaugh, J. and Jacobson, I. (1998). *The Unified Modelling Language User Guide* (Reading, MA: Addison-Wesley).

Boose, J.H. (1985). A knowledge acquisition program for expert systems based on personal construct psychology. *International Journal of Man-Machine Studies*, **23**, 495–525.

Burton, A.M., Shadbolt, N.R., Hedgecock, A.P. and Rugg, G. (1987). A formal evaluation of knowledge elicitation techniques for expert systems: domain 1. In: D.S. Moralee (ed.) *Research and Development in Expert Systems IV* (Cambridge: Cambridge University Press).

Burton, A.M., Shadbolt, N.R., Rugg, G. and Hedgecock, A.P. (1988). Knowledge elicitation techniques in classification domains. ECAI-88: *Proceedings of the 8th European Conference on Artificial Intelligence*. pp. 136–145.

Cañas, A.J., Leake, D.B. and Wilson, D.C. (1999). *Managing, Mapping and Manipulating Conceptual Knowledge*, AAAI Workshop Technical Report WS-99-10: Exploring the Synergies of Knowledge Management and Case-Based Reasoning. AAAI Press, Menlo, Calif. (July 1999).

Cheng, P., Cupit, J. and Shadbolt, N.R. (2001). Supporting diagrammatic knowledge acquisition: an ontological analysis of Cartesian graphs. *International Journal of Human Computer Studies*, 54, 457–494.

Chi, M., Feltovitch, P. and Glaser, R. (1981). Categorization and representation of physics problems by experts and novices. *Cognitive Science*, **5**, 121–152.

Chi, M.T.H., Glaser, R. and Rees, E. (1982). Expertise in problem solving. In: R.J. Sternberg (ed.) *Advances in the Psychology of Human Intelligence: 1* (Hillsdale NJ: Lawrence Erlbaum), 7–75.

Chi, M.T.H., Glaser, R. and Farr, M.J. (1988). *The Nature of Expertise* (Hillsdale, NJ: Lawrence Erlbaum Associates).

Clancey, W.J. (1985). Heuristic classification. *Artificial Intelligence*, **27**, 289–350.

Clayton, J., Gibson, E. and Scott, C. (1991). *A Practical Guide to Knowledge Acquisition* (Boston: Addison-Wesley).

Clarke, B. (1987). Knowledge acquisition for real time knowledge based systems. *Proceedings of the first European Workshop on Knowledge Acquisition for Knowledge Based Systems.* Reading University, UK C2 1–7.

Cleaves, D.A. (1987). Cognitive biases and corrective techniques: proposals for improving elicitation procedures for knowledge based systems. *International Journal of Man-Machine Studies*, **27**, 155–166.

Crow, L. and Shadbolt, N.R. (2001). Extracting focused knowledge from the semantic web. *International Journal of Human Computer Studies*, **54**, 155–184.

Elstein, A.S., Shulman, L.S. and Sprafka, S.A. (1978). Medical problem solving: an analysis of clinical reasoning (Cambridge, Mass.: Harvard University Press).

Ericsson, K.A. (1996). *The Road to Excellence* (Mahwah, NJ: Lawrence Erlbaum).

Ericsson, K.A. and Simon, H.A. (1984). *Protocol Analysis: Verbal Reports as Data* (Cambridge, Mass.: MIT Press).

Ericsson, K.A. and Simon, H.A. (1993). *Protocol Analysis: Verbal Reports as Data*, revised edition (Cambridge, MA: MIT Press).

Fellbaum, C. (1998). *WordNet: An Electronic Lexical Database* (Cambridge, MA: Bradford Books).

Feltovich, P., Ford, K. and Hoffman, R. (1997). *Expertise in Context — Human and Machine* (Cambridge, MA: MIT Press).

Firlej, M. and Hellens, D. (1991). *Knowledge Elicitation: A Practical Handbook* (New York: Prentice Hall International).

Gaines, B.R. (1988). Knowledge acquisition systems for rapid prototyping of expert systems. *INFOR*, **26**, 256–285.

Gaines, B. R. (1990). An architecture for integrated knowledge acquisition systems. *Proceedings of 5th AAAI Knowledge Acquisition for Knowledge-Based Systems Workshop*, Banff, Canada.

Gaines, B.R. and Linster, M. (1990). Development of second generation knowledge acquisition systems. In: B. Wielinga, J. Boose, B. Gaines, G. Schreiber and M. van Someren (eds) *Current Trends in Knowledge Acquisition* (Amsterdam: IOS Press), 143–160.

Gaines, B.R. and Shaw, M. (1989). Comparing the conceptual systems of experts. *Proceedings of 11th International Joint Conference on Artificial Intelligence*, 633–638 (San Mateo, CA: Morgan Kaufmann).

Gaines, B.R. and Shaw, M. (1997). Knowledge acquisition, modeling and inference through the World Wide Web. *International Journal of Human-Computer Studies* **46**, 729–759.

Gamma, E., Helm, R., Johnson, R. and Vlissides, J. (1995). *Design Patterns: Elements of Reusable Object-Oriented Software* (Reading, MA: Addison-Wesley).

Gammack, J. G. (1987). Different techniques, and different aspects of declarative knowledge. In: A.L. Kidd (ed.) *Knowledge Acquisition for Expert Systems: A Practical Handbook* (New York: Plenum Press).

Gammack, J.G. and Young, R.M. (1985). Psychological techniques for eliciting expert knowledge. In: M. Bramer (ed.) *Research and Development in Expert Systems* (Cambridge: Cambridge University Press), 137–163.

Gene Ontology Consortium (2000). Gene Ontology: tool for the unification of biology. *Nature Genet.* **25**, 25–29.

Grover, M.D. (1983). A pragmatic knowledge acquisition methodology. *IJCAI-83: Proceedings 8th International Joint Conference on Artificial Intelligence*, 436–438.

Hart, A. (1986). *Knowledge Acquisition for Expert Systems* (London: Kogan Page).

Hayes-Roth, F., Waterman, D.A. and Lenat, D.B. (1983). *Building Expert Systems* (Reading, Mass.: Addison-Wesley).

Hinkle, D.N. (1965). The Change of Personal Constructs from the Viewpoint of a Theory of Implications. Unpublished Ph.D. thesis, University of Ohio.

Hoffman, R.R. (1987). The problem of extracting the knowledge of experts from the perspective of experimental psychology. *AI Magazine*, **8**, 53–66.

Hoffman, R.R., Crandall, B. and Shadbolt, N.R. (1998). Use of the critical decision method to elicit expert knowledge: a case study in the methodology of cognitive task analysis. *Human Factors*, **40**, 254–276.

Hoffman, R., Shadbolt, N.R., Burton, A.M. and Klein, G. (1995). Eliciting knowledge from experts: a methodological analysis. *Organizational Behavior and Decision Processes*, **62**, 129–158.

Hu, B., Dasmahapatra, S. and Shadbolt, N. (2003). From lexicon to mammographic ontology: experiences and lessons. In: D. Calvanese, G. De Giacomo and E. Franconi (eds) *Proceedings International Workshop of Description Logics — DL'03*, pp. 229–233, Rome, Italy.

Humphreys, B.L., Lindberg, D.A., Schoolman, H.M. and Barnett, G.O. (1998). The unified medical language system: an informatics research collaboration. *J. Am. Med. Inform. Assoc.* **5**, 1–11.

Janowitz, D. (2003) The Easy Guide to Repertory Grids. (London: Wiley).

Johnson, L. and Johnson, N. (1987) Knowledge elicitation involving teachback interviewing. In: A. Kidd (ed.) *Knowledge Elicitation for Expert Systems: A Practical Handbook* (New York: Plenum Press).

Johnson, P.E., Zualkernan, I. and Garber, S. (1987) Specification of expertise. *International Journal of Man-Machine Studies*, **26**, 161–181.

Kahneman, D., Slovic, P. and Tversky, A. (eds) (1982). *Judgement under Uncertainty: Heuristics and Biases* (New York: Cambridge University Press).

Kelly, G.A. (1955). *The Psychology of Personal Constructs* (New York: Norton).

Kidd, A.L. (ed.) (1987). *Knowledge Acquisition for Expert Systems: A Practical Handbook* (New York: Plenum Press).

Klein, G.A. (1993). A recognition-primed decision (RPD) model of rapid decision making. In: G.A. Klein, J. Orasanu, R. Calderwood and C.E. Zsambok (eds.) Decision Making in Action: Models and Methods, 138–147 (Norfolk, NJ: Ablex).

Kuipers, B. and Kassirer, J.P. (1983). How to discover a knowledge representation for causal reasoning by studying an expert physician. *IJCAI-83: Proceedings of the 8th International Conference on Artificial Intelligence*, 49–57.

LaFrance, M. (1987). The knowledge acquisition grid: a method for training knowledge engineers. *International Journal of Man-Machine Studies*, **26**, 245–255.

Marcus, S. and McDermott, J. (1989). SALT: a knowledge acquisition language for propose-and-revise systems. *Artificial Intelligence*, **39**, 1–38.

McGraw, K.L. and Harbison-Briggs, K. (1989). *Knowledge Acquisition: Principles and Guidelines* (Englewood Cliffs, NJ: Prentice-Hall International).

Meyer, M. and Booker, J. (1991). *Eliciting and Analysing Expert Judgement: A Practical Guide* (London: Academic Press).

Milton, N., Shadbolt, N., Cottam, H. and Hammersley, M. (1999). Towards a knowledge technology for knowledge management. *International Journal of Human-Computer Studies*, **51**, 615–664.

Mitchell, T. (1997). *Machine Learning*, (New York: McGraw-Hill).

Motta, E.(1999). *Reusable Components for Knowledge Modelling: Principles and Case Studies in Parametric Design* (Amsterdam: IOS Press).

Motta, E., O'Hara, K. and Shadbolt, N. (1993). Grounding GDMs: a structured case study. *Journal of Knowledge Acquisition*, **5**(4), 315–347.

Neches, R., Fikes, R., Finin, T., Gruber, T., Patil, R., Senator, T. and Swartout, W. (1991). Enabling technology for knowledge sharing. *AI Magazine*, **12**, 37–56.

Newell, A. (1982). The knowledge level. *Artificial Intelligence*, **18**, 87–127.

Newell, A. and Simon, H.A. (1963). GPS, A program that simulates human thought. In: E. Feigenbaum and J. Feldman (eds) *Computers and Thought* (New York: McGraw-Hill), 249–293.

Nonaka, I. and Takeuchi, H. (1995). *The Knowledge-Creating Company* (Oxford, UK: Oxford University Press).

Norvig, P. and Russell, S. (2003). *Artificial Intelligence: A Modern Approach* (Prentice Hall).

Noy, N.F., Sintek, M., Decker, S. Crubezy, M., Fergerson, R.W. and Musen, M.A. (2001). Creating semantic web contents with protege-2000. *IEEE Intelligent Systems*, **16**, 60–71.

O'Hara, K. and Shadbolt, N.R. (1997). Interpreting generic structures: expert systems, expertise and context. In: Paul Feltovich, Ken Ford and Robert Hoffman (eds) *Expertise in Context* (AAAI and MIT Press), 449–472.

O'Hara, K., Shadbolt, N.R. and Van Heijst (1998). Generalised directive models: integrating model development and knowledge acquisition. *International Journal of Human-Computer Studies*, **49**, 497–522.

Petersen, T. (1994). *Art and Architecture Thesaurus* (New York: OUP).
Rich, E. and Knight, K. (1991). *Artificial Intelligence.* 2nd Edition (New York: McGraw-Hill).
Schreiber G., Akkermans, H., Anjewierden, A., de Hoog, R., Shadbolt, N.R, Van de Velde, W. and Wielinga, B. (2000). *Knowledge Engineering and Management* (Cambridge, MA: MIT Press).
Schvaneveldt, R.W., Durso, F.T., Goldsmith, T.E., Breen, T.J., Cooke, N.M., Tucker, R.G. and De Maio, J.C. (1985). Measuring the structure of expertise. *International Journal of Man-Machine Studies,* **23**, 699–728.
Selvin, A.M. and Buckingham Shum, S.J. (2000). Rapid knowledge construction: a case study in corporate contingency planning using collaborative hypermedia. *Proc. KMaC 2000: Knowledge Management Beyond the Hype.* Aston Business School, Aston University, Birmingham, UK, 16–19 July 2000, Operation Research Society, 48–58.
Shadbolt, N.R. (1989). Knowledge representation in man and machine. In: R. Forsyth (ed.) *Expert Systems: Principles and Case Studies,* 2nd Edition (London: Chapman & Hall), C7 142–173.
Shadbolt, N.R. and Burton, A.M. (1989). Empirical studies in knowledge elicitation. *ACM-SIGART Special Issue on Knowledge Acquisition,* p. 108.
Shadbolt, N.R. and Burton, M. (1995). Knowledge elicitation: a systematic approach. In: J.R. Wilson and E.N. Corlett (eds.) *Evaluation of Human Work: A Practical Ergonomics Methodology* (London: Taylor & Francis), pp. 406–440.
Shadbolt, N., Motta, E. and Rouge, A. (1993). Constructing knowledge-based systems. *IEEE Software,* **November**, pp. 34–39.
Shadbolt, N.R and O'Hara, K. (1997). Model-based expert systems and the explanation of expertise. In: P.J. Feltovich, K.M. Ford and R.R. Hoffman (eds.) *Expertise in Context* (Cambridge, MA: AAAI and MIT Press), 315–337.
Shadbolt, N.R., O'Hara, K. and Crow, L. (1999) The experimental evaluation of knowledge acquisition techniques and methods: history, problems and new directions. *International Journal of Human-Computer Studies,* **51**, 729–755.
Shaw, M.L.G. and Gaines, B.R. (1987). An interactive knowledge elicitation technique using personal construct technology. In: A.L. Kidd (ed.) *Knowledge Acquisition for Expert Systems: A Practical Handbook* (New York: Plenum Press), 109–136.
Shortliffe, E.H. (1979). *Computer-Based Medical Consultations: Mycin* (New York: American-Elsevier).
Sowa, J.F. (1999). *Knowledge Representation: Logical, Philosophical, and Computational* (Pacific Grove, CA: Brooks Cole Publishing Co.).
Stefik, M. (1993). *Introduction to Knowledge Systems* (Los Altos, CA: Morgan Kaufmann).
Stewart, T. (1997). *Intellectual Capital — The New Wealth of Organizations* (London: Nicholas Brealey).
Tu, S.W., Eriksson, H., Gennari, J., Shahar, Y. and Musen, M.A. (1995). Ontology-based configuration of problem-solving methods and generation of knowledge-acquisition tools: Application of PROTEGE-II to protocol-based decision support. *Artificial Intelligence in Medicine,* **7**, 257–289.
Van Heist, G., Schreiber, A.Th. and Weilinga, B.J. (1997). Using explicit ontologies in KBS development. *International Journal of Human-Computer Studies,* **45**(2), 183–292.
Weis, S.M. and Kulikowski, G.A. (1984). A Practical Guide to Designing Expert Systems (Totowa, NJ: Rowman and Littlefield).
Welbank, M.A. (1983). *A Review of Knowledge Acquisition Techniques for Expert Systems* (Martlesham Heath: British Telecom Research).
Wielinga, B.J., Scheiber, A.Th. and Brevker, J.A. (1992). KADS: a modelling approach to knowledge engineering. *Knowledge Acquisition Journal,* **4**, 5–53.
Witten and Frank (1999). *Data Mining: Practical Machine Learning Tools and Techniques with Java Implementations* (San Fransisco: Elsevier).
Young, R.M. and Gammack, J. (1987). Role of psychological techniques and intermediate representations in knowledge elicitation. *Proceedings of the first European Workshop on Knowledge Acquisition for Knowledge Based Systems.* Reading University, UK, D7 1–5.
Zsambok, C.E. and Klein, G. (1997) *Naturalistic Decision Making,* (Mahwah, NJ: Erlbaum).

Chapter 9

Simulation and modelling as a tool for analysing human performance

Ron Laughery

Introduction

The need for *quantitative estimates of human performance within a system early in the design process* has become even more important over the past few years. We've all seen the curves relating how system problems are much more expensive to fix the later in the development process that we find them. Variations of the curve shown in Figure 1 abound in textbooks and presentations everywhere. It's better to get it right from the beginning than to have to *make* it right later. To meet this challenge, many human factors and ergonomics tools and technologies have evolved over the years to support early analysis and design. Two specific types of technologies are design guidance (e.g., O'Hara *et al.*, 2002; Boff *et al.*, 1986) and high fidelity rapid prototyping of user interfaces (e.g., Dahl *et al.*, 1995); readers should also refer to Chapters 10, 13 and 38. Design guidance technologies, either in the form of handbooks or computerised decision support systems, put selected portions of the human factors and ergonomics knowledge base at the fingertips of the designer, often in a form tailored to a particular problem such as nuclear power plant design or UNIX computer interface design. However, design guides have the shortcoming that they do not often provide methods for making quantitative tradeoffs in *system* performance as a function of design. For example, design guides may tell us that a high-resolution color display will be better than a black and white display, and they may even tell us the value in terms of increased response time and reduced error rates. However, this type of guidance will rarely provide good insight into the value of this improved element of the human's performance to the *overall system's* performance. As such, design guidance has limited value for providing concrete input to *system level performance prediction.*

Rapid prototyping, on the other hand, supports analysis of how a specific design and task allocation will affect human and system level performance. In many respects, the use of rapid prototyping is one kind of simulation that has become a mainstay of human factors analysis. Essentially, by developing a prototype that allows the humans to interact with the simulated system, designers can conduct experiments and test design concepts with actual human participants. The disadvantage of prototyping, as with all human participant experimentation, is that it can be costly and can only be done relatively late in the design process. In particular, prototypes of hardware-based systems, such as aircraft and machinery, are very expensive to develop, particularly at early design stages when there are many widely divergent design concepts. In spite of the expense, rapid prototyping is an important tool for the human factors practitioner, and its use is growing in virtually every application area. While we encourage and support their use and recognise that rapid prototyping is a valid form of simulation, this chapter does not focus on that aspect of the world of simulation.

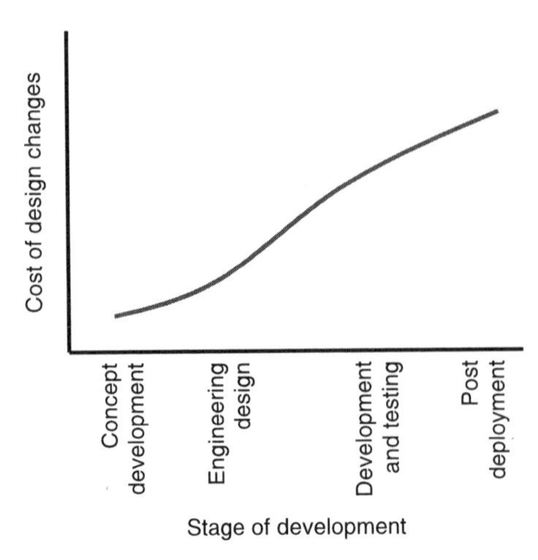

Figure 1 Cost of system changes over the life cycle of a system.

What is often needed is an integrating methodology that can extrapolate from the base of human factors and ergonomics data, as reflected in design guides and the literature, in order to support system level performance predictions as a function of design alternatives. In other words, we need methods and tools that allow us to answer the ever present question about our design suggestions, *So what?* These methods and tools should also bind with rapid prototyping and experimentation in a mutually supportive and iterative way.

As has become the case in many engineering disciplines, a prime candidate for this integrating methodology for considering human performance is computer modelling and simulation. In fact, it is not just the human factors practitioner that is being called upon to use computer modelling and simulation — the entire engineering community is increasingly embracing the use of modelling and simulation as the basis for design. For example, the U.S. Department of Defense has determined that all new systems be designed through the extensive use of simulation. Policies regarding simulation-based acquisition dictate that simulation must be used to prove the necessity of a system, to test system concepts, and to support the design, development, and testing process. Other examples of the engineering communities' increased reliance on simulation include other agencies of government who are embracing this approach, such as NASA and the FAA, and Boeing Commercial Aircraft who have demonstrated that you really can design an aircraft relying almost exclusively on computer simulation. Computer modelling and simulation is the way of the future. It reduces the cost of prototyping as well as cutting engineering cycle time — the time it takes to evaluate a design or operational concept — thereby allowing more concepts to be explored and, ultimately, a better design found earlier and at a lower engineering cost.

If human factors engineers are going to continue to influence the systems engineering and design processes, then we must be able to provide quantitative estimates of human performance early in design just like the other engineering disciplines must. To be sure, our estimates will involve imperfect models populated heavily with data based upon expert guesses. The good news is that this isn't all that different from many of the rest of the engineering models that are used early in concept exploration and design. These models require that analysts often must 'make things up' too. Human factors practitioners, like others in the engineering process, must also be willing to select a reasonable approach to modelling human performance in the context of the system design, populate the model with the best available data, and run the model to come up with estimates of human performance that other engineers can use and understand. As the design process moves forward, we will have the chance, like other engineers, to build prototypes and conduct

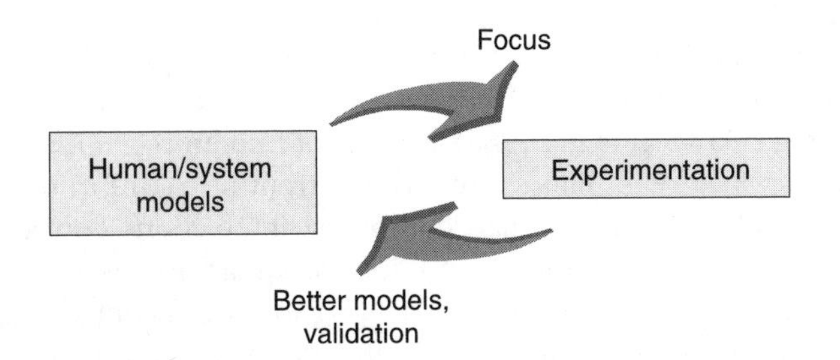

Figure 2 Human performance models and experimentation working together in systems design.

testing with real people. These experiments can, in turn, provide better models to guide further testing. For many complex systems, we will have to rely on a model of design and testing as shown in Figure 2 rather than the traditional 'experimentation with humans' model that we have relied on for too long. To continue with this 'experimentation only' model, we will be relegated to offering too little too late in the way of engineering solutions. By embracing the concept presented in Figure 2, we will have stepped up to the table with the rest of the engineering community.

Computer modelling of human behavior and performance is not a new endeavor. Computer models of complex cognitive behavior have been around for over 20 years (e.g., Newell and Simon, 1972) and tools for computer modelling of task level performance have been available since the 1970s (e.g., Wortman *et al.*, 1978). However, two things have changed appreciably in the past decade that promote the use of computer modelling and simulation of human performance as a standard tool for the practitioner. First is the rapid increase in computer power and the associated development of easier to use modelling tools. Individuals with an interest in predicting human performance using modeling and simulation can select from a variety of computer-based tools. Second is the increased focus by the research community on the development of *predictive* models of human performance rather than simply descriptive models. For example, the GOMS model (Gray *et al.*, 1993) represents the integration of research into a model for making predictions of how humans will perform in a realistic task environment. Another example is the research in cognitive workload that has been represented as computer algorithms (e.g., McCracken and Aldrich, 1984; Farmer *et al.*, 1995). Given a description of the tasks and equipment with which humans are engaged, these algorithms support assessment of when workload-related performance problems are likely to occur, and often include identification of the quantitative impact of those problems on overall system performance. Lots of progress has been made in the past twenty years and there are enough modelling technologies and success stories to support their use. The human factors practitioner has the means to model and simulate human performance in a systems context and the purpose of this chapter is to help lead individuals and teams interested in this technology down the path to successful application.

A taxonomy of human performance models

In any study of modelling, it is important to determine what aspects of the phenomena of interest need to be included in the model. Models are, by definition, abstractions of reality and in building a model, the decisions about what to include and what to ignore are critical. For example, the type of model that you would build to simulate human chess playing behavior would focus on very different aspects of the human organism than a model of a human flying an airplane. Therefore, we must begin our study of human performance modelling with a view of how to abstract human behavior in complex systems.

Of course, human performance can be highly complex and involve many types of processes and behavior. Over the years many models have been developed that predict sensory processes (e.g., Gawron *et al.*, 1983), aspects of human cognition (e.g., Newell, 1990), and human motor response (e.g., Fitts's law). The current literature in the areas of cognitive engineering, error analysis, and human/computer interaction contains many models, descriptions, methodologies, metaphors, and functional analogies (e.g., see Chapters 13 and 32). These models are all valuable and can provide significant component models for these elemental human behaviors. However in this chapter, we are not focusing on these models of specific aspects of human behavior but, rather, on models that can be used to describe human performance in systems. These human/system performance models we discuss will typically include some of these elemental behavioral models as components, but, perhaps more importantly, they provide a structural framework that allows them to be put in the context of human performance at tasks in systems. Therefore, our taxonomy of human performance models focuses not on what aspects of human perception, cognition, and action are required but rather on how the individual models of human behaviors are organised and interfaced to provide a higher level prediction of human/system performance.

We have found a reasonable way to characterise the world of human performance models is to separate them into two general categories: (1) *reductionist* models and (2) *first principle* models. *Reductionist* models use human/system task sequences as the primary organising structure as shown in Figure 3. The individual models of human behavior for each task or task element are connected to this task sequencing structure. We refer to it as reductionist because the process of modelling human behavior involves taking the larger aspects of human/system behavior (e.g., 'perform the mission') and then successively reducing them to smaller elements of behavior (e.g., 'perform the function,' 'perform the tasks'). This continues until a level of decomposition is reached at which reasonable estimates of human performance for the task elements can be made. One can also think of this as a top-down approach to modelling human/system performance. The example of this type of modelling that we will use in this chapter will be *task network modelling* where the basis of the human/system model is a task analysis (see Chapter 6).

First principle models of human behavior are structured around the underlying goals and principles that dictate human performance. Tools that support first principle modelling of

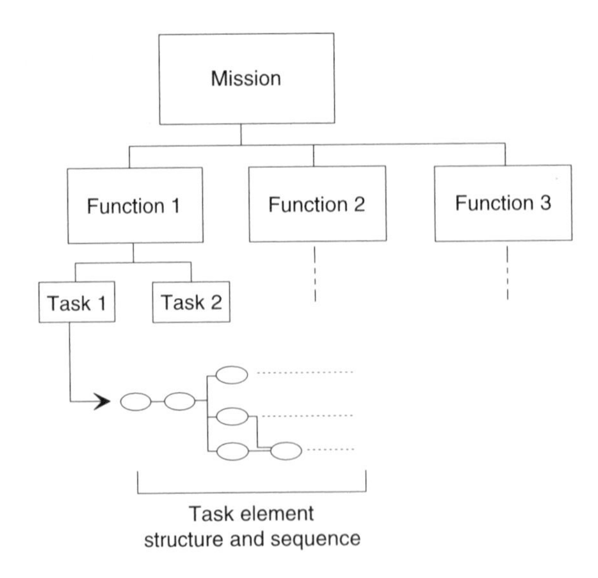

Figure 3 The concept of reductionist models of human performance.

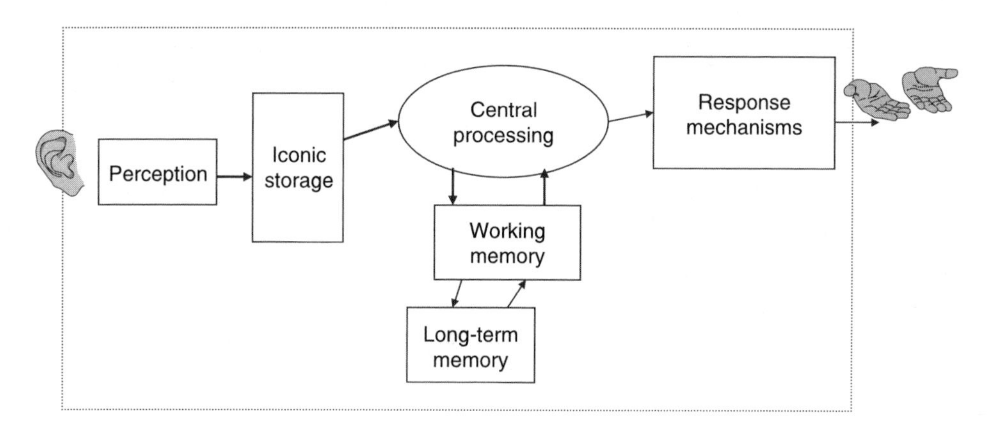

Figure 4 The concept of a model of human performance based upon first-principles of human behavior.

human behavior have structures embedded in them that represent elemental aspects of the human. For example, these models might directly represent processes such as goal seeking behavior, task scheduling, sensation and perception, cognition, and motor output. To use tools that support first principle modelling, one must describe how the system and environment interact with the modelled human processes. An example of a very simple structure that supports this type of modelling environment is presented in Figure 4.

It is worth noting that these two modelling strategies are by no means mutually exclusive and, more often, are mutually supportive in any given modelling project. When one is modelling using a reductionist approach, one needs models of basic human behavior to accurately represent specific behavioral phenomena (e.g., cognitive workload) and, therefore, must draw on first-principle models. Alternately, when one is modelling human/system performance using a first-principled approach, some aspects of human/system performance and interrelationships between tasks may be more easily defined using a reductionist approach. Either class of model has been used to model individual and team performance. It is also worth noting that recent advances in human performance modelling tool development are blurring the distinctions between these two classes (e.g., Hoagland *et al.*, 2001; LaVine, 2000). Increased emphasis on interoperability between models has caused researchers and developers to focus on integrating reductionist and first principle models.

Next, we will provide a more detailed example of reductionist modelling and how it can be used to model aspects of human/system performance. Then, we will demonstrate how the incorporation of first principles into a reductionist model can enhance the capability of human performance models.

We would also like to note that the utility of human performance modelling depends upon the availability of tools to support the modeler. Building the types of models we discuss here in the absence of good modelling tools would be tantamount to doing spreadsheet analysis without programs designed to support spreadsheets (e.g., Excel). There are many good tools available and at the end of this chapter, we will provide a list of modelling tools available as of the publication of this book.

Modelling human performance 'from the top down' using a reductionist approach

One reductionist-based technology that has proven useful for predicting human-system performance is *task network modelling*. In a task network model, human performance is

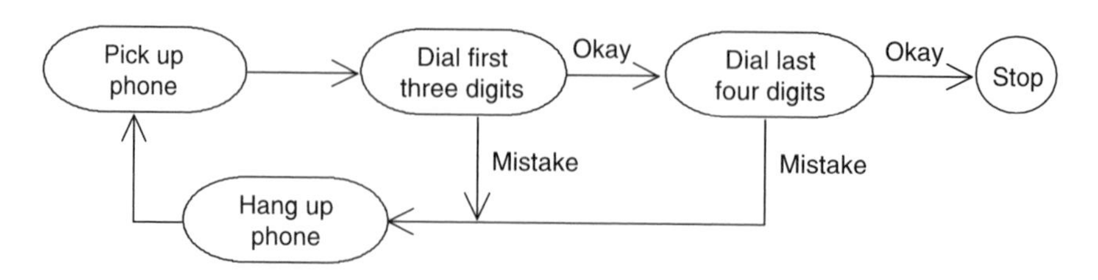

Figure 5 An example of a task network model representing a human dialing a telephone.

decomposed into tasks. The fidelity of this decomposition can be selective, with some functions being decomposed several levels, and others just one or two, just as would be the case in any human engineering task analysis. The general sequencing of task performance is defined by constructing a *task network*. This concept is illustrated in Figure 5, which presents a sample task network for dialing a telephone. To create an executable computer model, additional information is required such as task timing, interdependencies among tasks, and branching logic, all of which will be discussed below in some detail.

Task network modelling is an approach to modelling human performance in complex systems that has evolved for several reasons. First, it is a reasonable means for extending the human factors staple — the task analysis. Task analyses organised by task sequence are the basis for the task network model. Second, task network modelling is relatively easy to use and understand. Our experience has been that individuals with no training in computers or modelling can still look at a task network model and understand the basic flow of the human process being modelled. Finally, task network modelling has been demonstrated to provide efficient, valid and useful input to many types of issues. With a task network model, the human factors engineer can examine a design (e.g., control panel redesign) and address questions such as 'How much longer will it take to perform this procedure?' and 'Will there be an increase in the error rate?'

Task network models of human performance have been subjected to validation studies with favorable results (e.g., Lawless *et al.*, 1995; Allender *et al.*, 1995). While validation issues must be considered with respect to a particular model and not a modelling approach, these studies have demonstrated that a reductionist strategy for modelling human performance, if properly applied, can be a productive one that provides meaningful and useful predictions of human performance in complex systems.

What goes into a task network model?

The basic ingredient of a task network model is the task analysis and sequencing as represented by a network or series of networks. However, to represent complex, dynamic human/system behavior, many aspects of the system in addition to simply task lists and sequence may need to be modelled. Obviously, task time distributions and error rates might be important. Also, some of the interdependencies among the tasks may need to be represented (e.g., through shared variables affected by task performance). For example, when an operator manipulates a control in a power plant, this may initiate an 'open valve' task in a part of the model representing the plant. This valve opening could ripple through to a network representing other operators and subsystems and their response to the open valve. Therefore, while the essential behavior of the human is readily defined through the sequencing of tasks implied in the task network, other aspects of task and system behavior must be represented to capture and predict complex dynamic system behavior. For example, each task in a task network could include information on the following:

Time distribution — Often, it is not average human performance that is of interest, but the variability. To get a sense of the cost of variability, distributions and their parameters are defined so that Monte Carlo simulations can be run with task performance times sampled from a distribution as defined by this option (e.g., normal, beta, exponential).

Mean time — This parameter defines average task performance time for this task. This might be something as simple as a number, or some type of algorithm representing how mean performance is affected by other attributes of the human or system (e.g., the change in performance induced by fatigue under sustained operations).

Standard deviation — This parameter would be the standard deviation of task performance time or other parameters defining the statistical distribution. Again, this can be a number or algorithm.

Release condition — A release condition determine conditions that have to be met before a task can begin. For example, a condition stating that this task will not start before an operator is available (e.g., they are performing another task) might be represented by a release condition such as the following:

operator >= 1.

In other words, there must be at least one operator available for the task to commence. If all operators were busy, the value of the variable 'operator' would equal zero until a task is completed at which time an operator becomes available. This task would wait until the condition was true before beginning execution, which would probably occur as a result of the operator completing the task he or she is currently performing.

Beginning effects — Beginning effects permit the user to define how the system will change as a result of the commencement of this task. For example, if this task used an operator that other tasks might need, we could set the following condition to show that the operator is unavailable while he performed this task:

operator := operator − 1.

Assignment and modification of variables in beginning effects are one key way in which tasks are interrelated.

Ending effects — Ending effects permit the user to define how the system will change as a result of the completion of this task. From the previous example, when this task was complete and the operator became available, we could set the ending effect as follows

operator := operator + 1;

at which point another task waiting for an operator to become available could begin. Ending effects could also be used to represent control inputs made by the human through changes in variables. Ending effects are another key way in which tasks can be interrelated through the assignment and modification of variables.

Another notable aspect of task network models are the logic that determines task sequencing. Any time more than one path out of a task is defined, the logic that controls the flow of these tasks must also be defined. Often, this is used to represent the rules and logic used as part of the human decision making process, although task sequencing logic can be used to represent other non-human aspects of the system included in the simulation. Since this task sequencing logic represents, from the model's perspective, *decisions* that the model must make at these branching point, that is how we will refer to them. There are only three general types of decisions to model:

- *Probabilistic* — In probabilistic decisions, the human will begin one of several tasks based on a random draw weighted by the probabilistic branch value. These weightings can be dynamically calculated to represent the current context of the decision. For example, this decision type might be used to represent human error probability and would be connected to the subsequent tasks that would be performed.
- *Tactical* — In tactical decisions, the human will begin one of several tasks based on the branch with the highest 'value'. This could be used to model the many types of rule-based decisions that humans make that can consider many aspects of system operation and control. More complex 'fuzzy' decisions can also be represented using this type of decision model through the introduction of uncertainty and random variation.
- *Multiple* — This would be used to begin several tasks at the completion of this task, such as when one human issues a command that begins other crewmembers' activities.

There are other aspects of task network model development. Some items are defining a simulation scenario, defining continuous processes within the model, and defining queues in front of tasks. An example of how these features work and contribute to modelling complex human performance can be obtained from the Micro Saint User's Guide (Micro Analysis and Design, 2001).

Again, there are a number of tools to support task network model construction. What is important here is not the tool, but the concept of task network modelling and how it fits into the human and systems engineering processes. Task network modelling is a relatively straightforward concept that is a logical extension of function and task analysis. Task network modelling is an evolution, not a revolution to the human factors practitioner. Much of the information needed to build a task network model is generally gathered as part of the task analysis. Task network modelling, however, greatly increases the power of task analysis since the ability to simulate a task network with a computer permits *prediction* of human performance rather than simply the *description* of human performance that a task analysis provides. What may not be as apparent, however, is the power of task network modelling as a means for modelling human performance in systems. Simply by describing the systems activities in this step-by-step manner, complex models of the system can be developed where the human's interaction with the system can be represented in a closed-loop manner. Since task network modelling is a derivative of discrete-event modelling, which is widely used in the systems engineering process, there is often a clean fit between task network models of the humans and other discrete-event models of other system components or processes.

Example — Using task network modelling to determine the value of speech input vs. a keyboard and mouse

To illustrate how task network modelling can be used to support a human factors trade-off study, we are going to look at the use of task network modelling to evaluate user–computer interface designs. Here, what we will do is compare predictions of human performance on a simple text search and replace task, using two different user–computer interaction methods:

1. User input using a mouse and keyboard in a windows-style interface, and
2. User input using a speech recognition device.

Specifically, we want to determine combinations of acceptable speech rates and word recognition probabilities that result in overall equal performance to standard mouse/keyboard entry on a simple text search and replace task. The measure of performance here is simply time — how long it takes to do it with speech technology vs. standard mouse and keyboard user-interface technology. (*Note*: the goal is not to definitively answer this question but, rather, to demonstrate how task network modelling can be used to evaluate human–computer interface design concepts and technologies.) In practice, of course, there may be many more criteria for any choice than simply task completion time.

Since this is a comparative study, the first step is to model the baseline for comparison — performance of the task with a mouse and keyboard. While one could model this task at great levels of detail, to address the above question a model that includes the following tasks is sufficiently detailed:

1. Press 'file open' button
2. Select file from list
3. Press 'find text string' button
4. Enter text string
5. Press 'OK'
6. Wait for computer (not dependent on human performance, per se, but an important parameter that may be affected by the technology being evaluated)
7. Enter new text string

Again, there are a variety of additional details and error paths that could be built into this model. However, for the purpose of gaining some insight as to how good a speech recognition system needs to be to be equal to current technology, this model will certainly get us into the ballpark with a minimum of complex model development time and effort. An axiom of computer modelling is that one should never underestimate the value of model simplicity in reducing model development effort and in explaining it to others, while still answering the basic questions the model was built to address.

Since these tasks must be performed in series, the task network is as shown in Figure 6. The first task in this network is a 'system' (i.e., non-human) task that sets the number of words. The time to complete the task with a mouse/keyboard as well as speech technology will depend upon the number of words to be found and replaced. Therefore, we have introduced this variability in this model based experiment. The model simulates text search and replacement of between one and four words.

To set task time parameters, one can either collect data for the tasks or use existing data sets or detailed models such as GOMS. However, a factor that will vary and that will make the use of speech more or less effective will be the typing speed of the user. Obviously, people who can type faster will need a faster and more accurate speech recognition system to see a benefit. Towards this end, we defined a variable in the model, *words_per_second,* that represents the typing rate of the human and we define the words to be searched and replaced to be a variable named *words*. Based upon these parameters, the mean times for all of the above tasks are shown in Table 1.

Since task network modelling allows the time to be represented as a probability distribution and, as we stated earlier, it is often the distribution of human performance that is as interesting as the mean, all times in this model are represented as normally distributed with

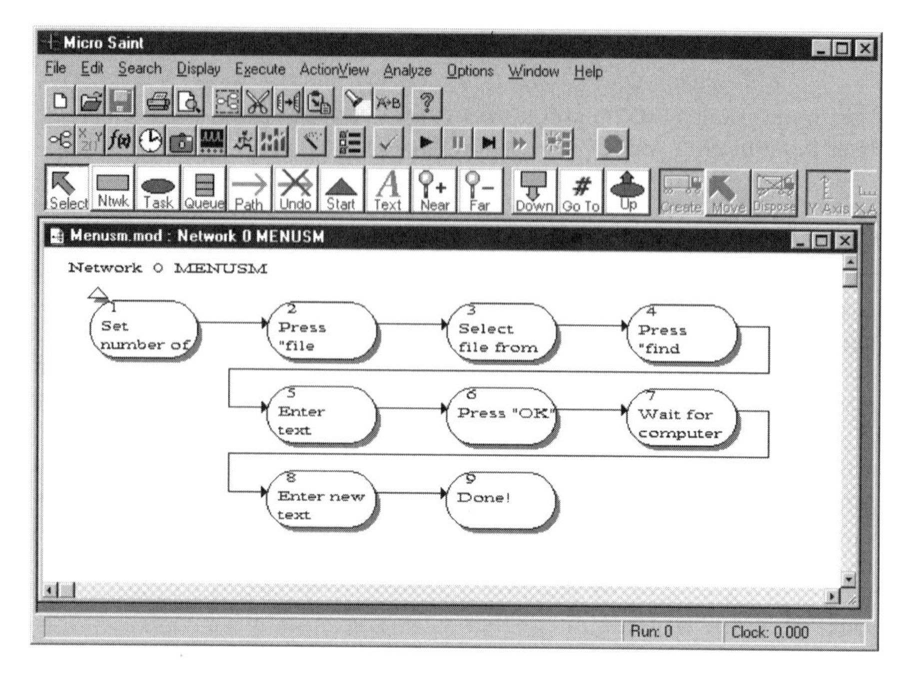

Figure 6 Task network for mouse and keyboard entry.

Table 1 Times for the Tasks for Keyboard and Mouse Entry Model

Task	Estimated mean task time
Press 'file open' button	1.0 seconds
Select file from list	1.5 seconds
Press 'find text string' button	1.0 seconds
Enter text string	Words/words_per_second
Press 'OK'	1.0 seconds
Wait for computer (computer delay)	1.5 seconds
Enter new text string	Words/words_per_second

mean times equal to the numbers in the above table and standard deviations equal to 30% of the mean. Laughery and Gawron (1984) conducted a study of the literature in a decade of research reported in the journal Human Factors and found that, on average, the standard deviation was equal to approximately 30% of the mean of performance. Based upon this finding, we have found that, in the absence of a better basis for estimating task time variability, a standard deviation set to 30% of the mean is a reasonable estimate.

Using the above parameters (i.e., one to four words to be replaced and task times as indicated in the above table) and with a typing rate of 0.5 words per second, we ran the model shown in Figure 9, 1000 times and obtained the following data:

Mean time — 19.6 seconds

Standard deviation — 7.7 seconds

Minimum time — 6.0 seconds

Maximum time — 40.4 seconds

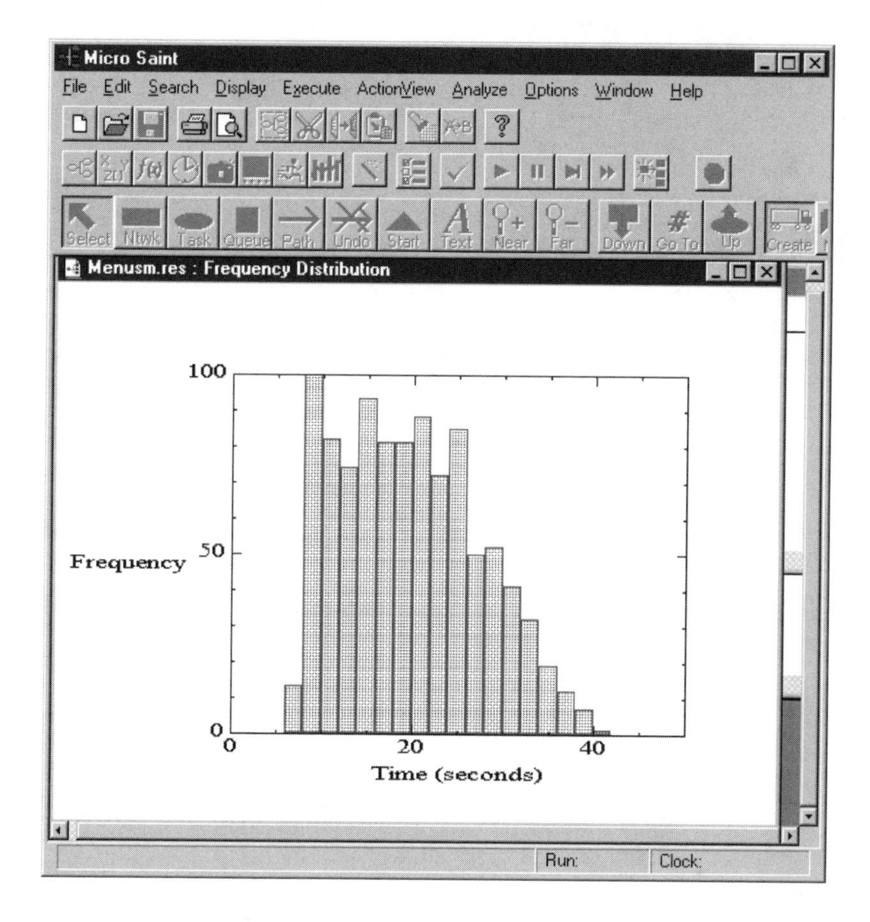

Figure 7 Frequency distribution for keyboard and mouse inputs.

The frequency distribution of these 1,000 runs is shown in Figure 7. This provides the baseline data against which a speech recognition system will be compared.

When running a model-based comparison study, it is important to maintain an approximately equal level of model detail and granularity between the models representing the phenomena being compared. Accordingly, we developed a model of the speech recognition system that includes the following tasks:

1. Say 'OPEN filename'
2. Wait for system (not dependent on human performance, per se, but an important parameter that may be affected by the technology being evaluated)
3. Say 'FIND word string'
4. Wait for system (not dependent on human performance)
5. Say 'REPLACE WITH new word string'
6. Wait for computer (not dependent on human performance)

The task network for this is shown in Figure 8. What is immediately obvious is that this task network involves some branching. The purpose of the branching is to represent the ability of the speech recognition system to recognise the spoken words. The higher the likelihood of this, the less likely that the system will need to ask the user to repeat the words because it did not understand the human. This repeating is what the loops of task 2 back to task 1, task 4

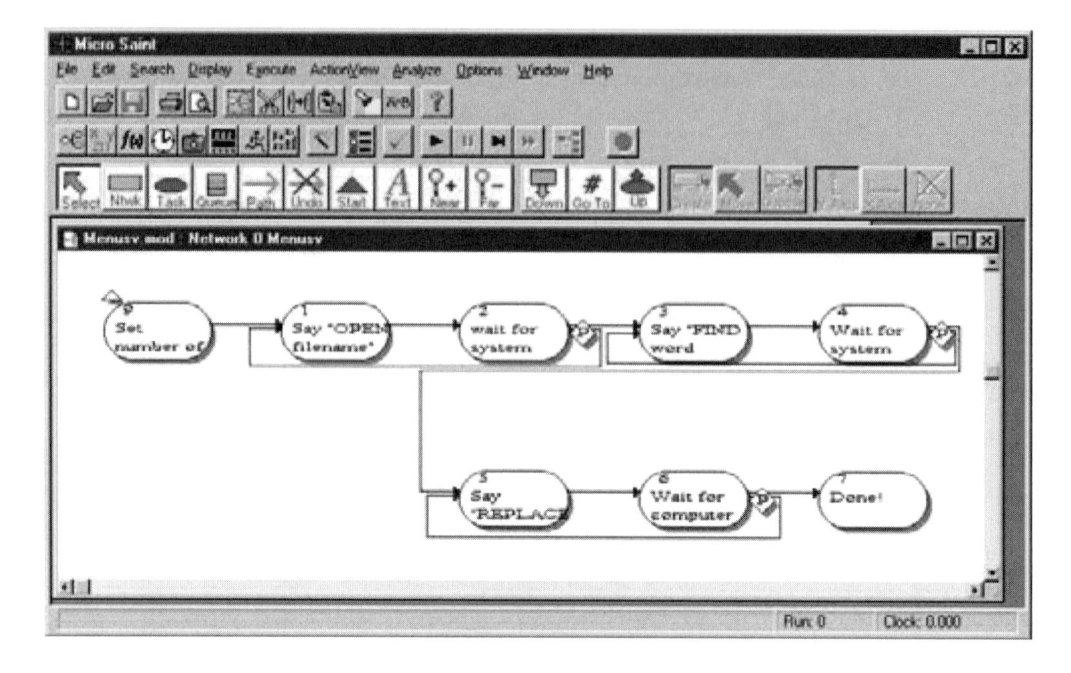

Figure 8 Task network for speech recognition system.

Table 2 Times for Tasks for Speech Entry Model

Task	Estimated mean task time
Say 'OPEN filename'	seconds_per_word*2
Wait for system (not human performance)	1.5
Say 'FIND word string'	seconds_per_word*(words + 1)
Wait for system (not human performance)	1.5
Say 'REPLACE WITH new word string'	seconds_per_word*(words + 2)
Wait for computer (not human performance)	1.5

back to task 3, and task 6 back to task 5 represent — that the system did not understand what the user said and he or she needed to say it again. In speech recognition systems, the likelihood of correct recognition depends upon the technology and the allowable speech rate. So, to represent these two fundamental aspects of speech technology, we have defined two variables, (1) *word_recognition_probability* which represents the probability that a single word will be correctly recognised and (2) *seconds_per_word* which represents the rate of speech that the speech recognition system can manage while maintaining the *word_recognition_probability*.

Based upon these parameters, the times for each of the tasks are presented in Table 2. Again, we have set the standard deviations equal to 30% of the mean.

However, what will influence the effectiveness of speech technology is the word recognition rate and the ensuing likelihood that phrases in tasks 2, 4, and 6 will need to be repeated. Given an individual *word_recognition_probability,* the probability of recognising n words in a row is equal to $word\_recognition\_probability^n$. Therefore, the probabilities of correct recognition for tasks 2, 4, and 6 are as indicated in Table 3.

Table 3 Probabilities of Successful Recognition for Speech Recognition Tasks

Task	Estimated probability of correct recognition
Say 'OPEN filename'	$\text{word\_recognition\_probability}^2$
Say 'FIND word string'	$\text{word\_recognition\_probability}^{(\text{words}+1)}$
Say 'REPLACE WITH new word string'	$\text{word\_recognition\_probability}^{(\text{words}+2)}$

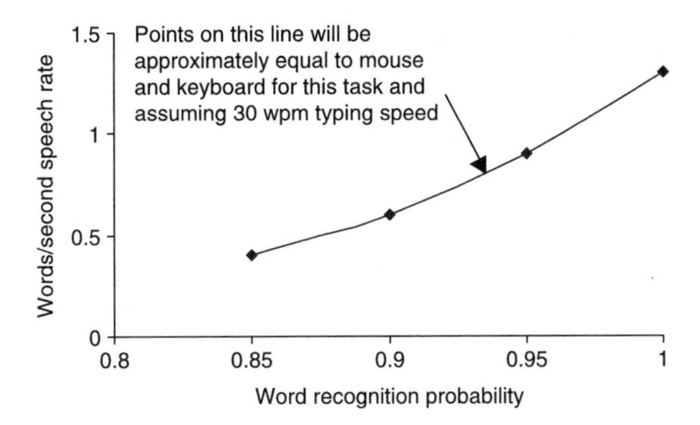

Figure 9 Combinations of speech rate and word recognition probability that are equivalent to mouse and keyboard model.

Using the above model and parameters, we ran the model 1,000 times with various combinations of speech recognition rates (by adjusting the value of *seconds_per_word* and word recognition probabilities (by adjusting the parameter *word_recognition_probability*) to find combinations that yielded a mean performance time of roughly 19.6 seconds (i.e., what was predicted with the mouse and keyboard task). The first test was to determine what the speech rate could be for a system with perfect recognition (i.e., *word_recognition_probability* = 1). It turned out that the speech rate of about 1.3 seconds per word resulted in roughly equal performance to the mouse and keyboard interface. With *word_recognition_probability* = 1 and *seconds_per_word* = 1.3, we obtained the following summary statistics:

Mean time — 20.1 seconds

Standard deviation — 5.2 seconds

Minimum time — 8.4 seconds

Maximum time — 35.5 seconds

However, when the word recognition probability dropped to 95%, to maintain a level of mean performance of approximately 20 seconds, we had to increase the speech rate to 0.9 seconds per word to achieve roughly equal performance. In fact, we ran several combinations and developed the graph that presents the combinations of speech rate and word recognition probability that are equivalent to the performance using a standard mouse and keyboard assuming the 30 wpm typing speed. These data are presented in Figure 9.

If this were a real design problem, this analysis would have defined, from a human performance perspective, where the break-even points were for going to be for a speech recognition system. Of course, in a realistic analysis, we would probably want to explore, at a minimum, other typing rates, errors, and more complex combinations of realistic tasks. However, this example should illustrate how these very basic concepts of task network modelling and simulation can be used to study aspects of the human-computer interface for realistic tasks and to make quantitative estimates of how new user interface strategies will affect human and system performance.

Other types of problems addressed through task network modelling

In the next section in this chapter, we will discuss how the power of task network models can be greatly enhanced through the incorporation of first principles. However, even with this purely reductionist approach, we can study and predict aspects of human performance. For example:

- *Human control of systems* — Many aspects of closed-loop human behavior can be predicted through task networks. For example, in a study reported by Hoagland *et al.* (2001), a task network model was integrated with an actual flight simulator and used to study control strategies and tactics in a fighter aircraft. These task network models representing pilot behavior actually responded to aircraft state changes and presented the flight simulator with simulated human control inputs. Through the task network models, significant flaws in proposed aircraft tactics were uncovered that were later confirmed with man-in-the-loop simulation.
- *Team composition and behavior* — Through the development of multiple task networks operating simultaneously, team composition and behavior can be studied. For example, in a study reported by Laughery, Scott-Nash, Wetteland, and Dahn (2000) attempting to define optimal crew size and task allocation on a ship, many different strategies of allocating tasks to jobs and the number of people qualified to perform each job were explored. By trying out different task allocation and staffing concepts with respect to their ability to perform, the many specific instances of tasks that were required over the course of a sustained cruise could be explored. When tasks were unable to be performed because of unavailable staff, insights could be gained as to how the task allocation and staff size could be adjusted to optimise performance.

Essentially, task network modelling simply brings the power of computer simulation to the analysis of human tasks. However, humans cannot always be studied as automatons. We have many unique characteristics that set us apart from other system components. The next section will discuss how we can bring some of these *first principles* that are uniquely human into the analysis.

Incorporating first principles of human behavior into models of human performance

The second fundamental approach to modelling human performance is to build models of human behavior from the mechanisms that underlie and cause human behavior. We refer to this as modelling using the *first principles* of human performance. Over the years, the human engineering, psychology, and physiology communities have established many first principles of human performance. Quantification of these first principles has been ongoing for years with models such as Fitt's Law and Signal Detection Theory being developed in some of the earliest days of the human engineering discipline. More recently, as the need for model-based analysis has grown, the quantification of theoretical human factors principles has continued. Some examples of and references for these emerging models are:

- Cognitive workload and human response to workload (Wickens, 1984)
- Human error and system response to error (Archer and Adkins, 1999)
- Micro models of human time and accuracy (e.g., Newell, 1990)
- Human performance shaping factor effects (LaVine *et al.*, 1994)
- Linkage to first-principle based anthropometric, biomechanical models (Dahl *et al.*, 1991)
- Goal-driven task scheduling (Hoagland *et al.*, 2001)

There is much to be gained in combining these approaches in the future. Both reductionism and first principles have their strengths and they can complement one another very well as has already been demonstrated. Of course, an ideal world for the human performance modeler would be one where the power of first-principle models was combined with the power of reductionist models to create a true *combined first-principles and reductionist* modelling environment. In effect, that is what has been happening in the world of human performance modelling tool development for the past decade. Those modelling from the reductionist perspective have found that first principles are needed to populate the models with data and algorithms to represent complex human behavior. At the same time, first principle modelers have come to appreciate the power and ease of use of a reductionist approach for many aspects of human behavior. The net result is that a number of model applications have taken a hybrid approach to representing human-system behavior. Of course, there are an infinite number of ways that hybrid models can be developed and used. Below is one example that illustrates this concept.

An example of combining a first principled and reductionist approach to predicting human workload

Cognitive workload has been a central issue in designing systems, particularly when a goal is to replace humans in the control loop with automation and/or decision aiding. Promising theories of human workload that were based on first principles of human cognition were developed in the 1980s. However, the question from system designers was, 'How can this theory help me determine when the human is going to get overloaded and start to fail?'

To address this question, several methods and tools were developed around the concept of modelling the humans' activities using a task network approach while laying on top of the task network model a first principled model to predict combined cognitive workload.

The basis of the first principle model of cognitive workload prediction we will use in our example here is an assumption that excessive human workload is not usually caused by one particular task required of the operator. Rather, it is the human having to perform several tasks simultaneously that lead to overload, such as driving while reading information from an in-vehicle display. Since the factors that cause this type of workload are intricately linked to these dynamic aspects of the human's task requirements, computer modelling with task networks provides a basis for studying how task allocation and sequencing can affect operator workload.

However, task network models are not inherently a model of human workload. The only relevant output that necessarily falls out of a task network model is the time required to perform a set of tasks and the sequence in which the tasks are performed. Time information alone would suffice for some workload evaluation techniques such as Siegel and Wolf (1969) whereby workload is estimated by comparing the time available to perform a group of tasks to the time required to perform the group of tasks. However, it has long been recognised that this simplistic analysis misses many aspects of the human's tasks that influence both perceived workload and performance. At the very least, this approach misses the fact that some pairs of tasks can be performed in combination better than other pairs of tasks.

The cognitive workload model that we use here is based upon the multiple resource theory of attention and human performance proposed by Wickens (e.g., Wickens *et al.*, 1983). Simply stated, the multiple resource theory suggests that humans have not one information processing resource that can only be tapped singly but several different resources that can be tapped simultaneously. Depending upon the nature of the information processing tasks required of a human, these resources would either (a) have to process information sequentially (if different tasks require the same types of resources) or (b) be able to process information in parallel (if different tasks required different types of resources).

Using this approach in its simplest form, each operator task in a task network can be characterised as requiring some amount of each of the attentional demands, as represented by a value between one and seven. All operator tasks can be analysed with respect to these demands and values assigned accordingly.

In performing a set of tasks pursuant to a common goal (e.g., driving a car), an operator frequently must perform several tasks simultaneously, or at least nearly so. For example, a driver may be required to monitor position on the road while receiving a communication. Given this, the workload literature indicates that the operator may either accept the increased workload (with some risk of performance degrading) or begin dumping tasks perceived as less important.

During a task network simulation, the model of the operator may indicate that he or she is required to perform, or at least be responsible for, several tasks simultaneously. In the simple implementation of the multiple resource theory, this would involve the task network model evaluating total attentional demands for each of the channels (visual, auditory, psychomotor, speech and cognitive) by simply summing the attentional demands across all tasks that are being performed simultaneously. For example, let us assume that at some point in driving a vehicle, the operator is simultaneously maintaining the location of the vehicle on the road while looking at an in-vehicle display to get navigational information.

Let us assume that the attentional demands of these tasks are as follows:

Channel	*Maintain location*	*Navigate with in-vehicle display*	*Combined tasks*
Visual	3	4	7
Auditory	1	0	1
Cognitive	2	5	7
Psychomotor	5	0	5

The last column above indicates what the combined single task attentional demands would be for each of the four channels.

An example of the type of data that can be generated is presented in Figure 10.

This technique was further refined by Plott (1995) to account for more complex representations of human behavior under conditions of high workload. For example, specific conflicts between specific control/display pairs can be modelled. Plott's models allow us to consider that, even though the visual workload for each of two tasks was low, if the displays presenting this information required the operator looking in two distinctly different directions at the same time, the tasks could not be performed simultaneously. Also, Plott conducted a

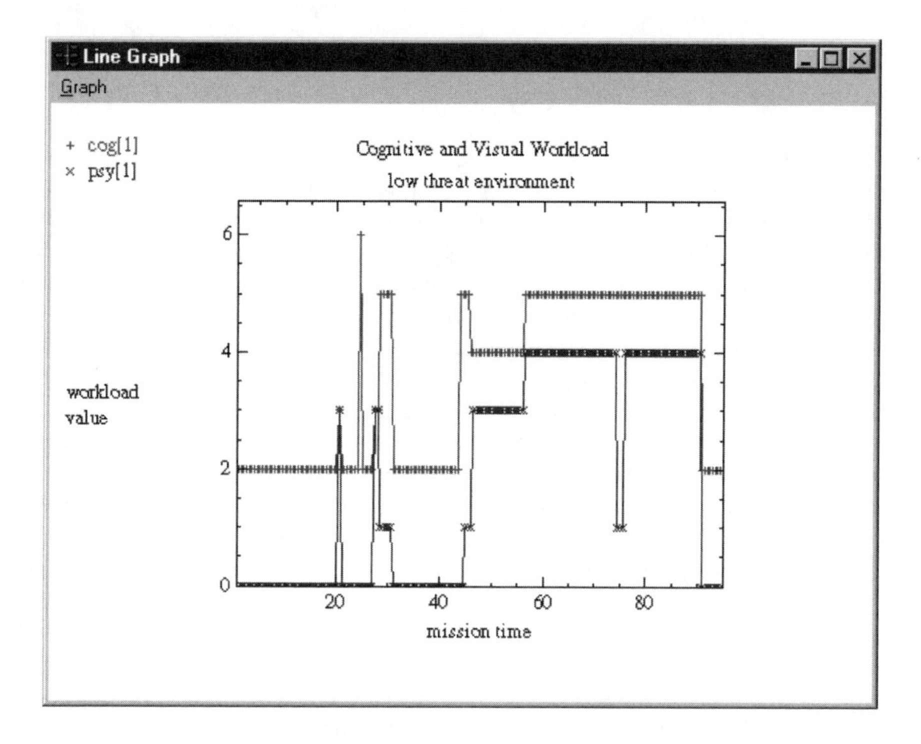

Figure 10 Sample Predicted Workload Profile for a Vehicle.

'first principles' analysis of how humans respond when a cognitive workload threshold is exceeded. The following workload management strategies were identified when multiple tasks demanded more workload than the human could deliver, commonly thought of as 'overload':

- perform tasks concurrently despite overload, but absorb a time and accuracy penalty;
- serialise the tasks;
- drop the task of lesser priority and resume it later; or
- reallocate the task to another qualified operator.

Models can be built that include dynamic modification of the tasks being performed by an operator when the model of workload indicates an overload condition has been reached. In this way, reductionist models of basic task activity can be effectively combined with first principle models of cognitive workload to create hybrid models that are far more powerful predictors of human response to workload than each individually.

Again, the purpose of the above example is to show how the relative ease of reductionist modelling can be combined with predictive power of first principle models of human behavior to create a hybrid that takes advantage of both approaches.

The steps for modelling and simulating human performance in complex systems

How does one go about building human performance models to address system design issues? Perhaps the most important note is that there is really no preset set of steps. Recently (AGARD, 1998), a NATO study group focusing on human performance modelling, offered the following as a list of activities that someone planning on using human performance modelling should consider:

1. *Define the goals of the modelling project* — There is probably no step that is more important than this. Models are, by definition, abstractions of reality and what the analyst chooses to model at higher or lower levels of fidelity should be driven by the issues the model is being built to address. Modelling human performance should always begin with a clear definition of why you are building this model and what you hope to gain from it.
2. *Define the constraints of the modelling project* — Given a set of modelling goals, the analyst must also consider how the modelling project will be constrained. Constraints can take many forms — time, people, money, and available data being the most salient. By clearly and explicitly defining these constraints, the remainder of the modelling process can be properly focused.
3. *Determine the human performance modelling tool(s) that are most appropriate for the project* — There are a variety of human performance modelling tools available. Each tool has its merits relative to the human performance issues under consideration. At the end of this chapter, we provide a list of some of the tools, but more are appearing on a regular basis.
4. *Develop the design reference scenarios for analysis* — In using models to address human performance issues, the scenarios that are selected for study will be critical. This is, in fact, no different than in experimentation with human subjects although the costs of running more and different scenarios should be far less with models. As with human subjects, scenarios should be selected to sample a range of the problem space to ensure that model results and conclusions are representative of the range of scenarios that humans would experience in the real system.
5. *Design the model structure and scope* — This step represents the first step of actual model design and, it should be noted, is presented after a large amount of groundwork has been laid in the above four steps. Again, it is important to focus model structure on the problems at hand and limit model scope as much as possible. In general, less scope is better at this point. If it becomes apparent later that parts of the model must be made more complex, then that complexity can always be added.
6. *Conduct a task analysis* — The specific nature of the task analysis and data required will depend upon the modelling tool. Generally, more will be required than with a typical task analysis or cognitive task analysis. Often, the sources for data will be limited and the analyst will be forced to rely on subject-matter estimates for many of the data that go in to model construction. This is not unusual nor unique to modelling human performance. Remember, it is better to guess how long it will take a human to perform a small task, for example, than it is to guess how well the human will perform the overall mission. By getting the model structure correct and relying on guesswork to fill in some of the data holes, we have reduced the likelihood that large overall errors in net human performance estimation will occur.
7. *Build the model* — Depending upon the tool and model complexity, this can be a relatively straightforward problem that can be done by a human factors specialist, or this can be something that requires someone who has made a career out of human performance modelling with a specific type of tool. While the emphasis of the human performance modelling tool development community has been on making tools ever easier to use, many are still quite complex. Again, we in the human performance community are not unique in this regard — complex problems require complex analyses, and the analysis process itself is usually educational and

almost always better than building the system and finding out it doesn't work as expected.

8. *Calibrate and validate the model* — To the extent possible, models should be calibrated with existing human performance data and validated through an independent review of the model's structure and data. However, there is no 'gold standard' here as to what is the proper course for model calibration/validation nor for when a model is 'good enough.' As a rule, if a model is transparent enough for experts to understand and believe, and modelling is the best tool available to address complex design issues, that is sufficient.
9. *Run experiments with the model* — Finally, the model should be used to run the experiments under the selected scenarios to address the issues for which it was designed. In many ways, this can be seen as essentially the same thing as running experiments with human subjects. However, the analyst should be cautious about relying on statistical significance in interpreting the results. Different experimental conditions reflected in the different models are different — the only question is whether the sample size is sufficient to prove it statistically, and sample size can always be increased with a few more runs on the computer. Experimentation with human performance modelling is one place where the difference between *statistical* significance and practical significance is paramount to consider. What the analyst should assure, however, is that enough model data have been collected to get stable and reliable results.

Summary

Modelling and simulating human performance is a sufficiently mature methodology and technology that it belongs in the toolkit of the human factors practitioner. To be sure, we do not yet know how to model the 'complete' human, and it is not always the most efficient way of studying human behavior and performance. However, in many instances, particularly early in the design process, it provides a practical means for identifying and studying performance issues that would otherwise be ignored. As the design process proceeds, it allows the human factors team to make better use of the experimental data that they collect by allowing them to be interpolated and extrapolated to study other design issues. Human performance modelling technology will undoubtedly continue to mature over the foreseeable future, largely based upon the experience we gain in using it to solve real design problems.

Current web sites describing human performance modelling software

As discussed earlier, there are a number of tools available that make the process of human performance modelling much more efficient and manageable. Below are a number of web sites for tools that are current as of the publication of this book:

ACT-R — http://act-r.psy.cmu.edu/
iGen — http://www.chisystems.com/
IMPRINT — http://www.maad.com/
Integrated Performance Modelling Environment (IPME) — http://www.maad.com/
Jack — http://www.eds.com/products/plm/efactory/jack/classic_jack.shtml
Micro Saint — http://www.maad.com/
Safework — http://www.safework.com/
SOAR — http://www.eecs.umich.edu/~soar/

References

Allender, L., Kelley, T., Salvi, L., Headley, D.B., Promisel, D., Mitchell, D., Richer, C. and Feng, T. (1995). Verification, validation, and accreditation of a soldier-system modeling tool. In: *Proceedings of the 39th Human Factors and Ergonomics Society Meeting*, October 9–13, San Diego, CA. Available from the Human Factors and Ergonomics Society, Santa Monica, CA.

Archer, S.G. and Adkins, R. (1999). *Improved Performance Research Integration Tool (IMPRINT) Analysis Guide*. Army Research Laboratory Technical Report, Aberdeen Proving Ground, MD.

Boff, K.R., Kaufman, L., and Thomas, J.P. (1986). *Handbook of Perception and Cognition* (New York: Wiley and Sons).

Dahl, S., Laughery, K.R., and Hood, L. (1991). Integrating task network and anthropometric models. *Contemporary Ergonomics — Proceedings of the Ergonomics Society's Annual Conference*, Southampton, England.

Dahl, S.G., Allender, L., Kelley, T., and Adkins, R. (1995). Transitioning software to the windows environment — challenges and innovations. Published in the *Proceedings of the 1995 Human Factors and Ergonomics Society Meeting, Human Factors and Ergonomics Society*, Santa Monica, CA.

Farmer, E.W., Belyavin, A.J., Jordan, C.S., Bunting, A.J., Tattersall, A.J. and Jones, D.M. (1995). *Predictive Workload Assessment: Final Report*, DRA/AS/MMI/CR95100/1.

Gawron, V.J., Laughery, K.R., Jorgensen, C.C., and Polito, J. (1983). A computer simulation of visual detection performance derived from published data. *Proceedings of the Ohio State University Aviation Psychology Symposium*, Columbus, Ohio.

Gray, W.D., John, B.E. and Atwood, M.E. (1993). Project Ernestine: Validating a GOMS analysis for predicting and explaining real-world task performance. *Human–Computer Interactions*, **8**, 237–309.

Hoagland, D.G., Martin, E.A., Anesgart, M., Brett, B.S., Lavine, N., and Archer, S.G. (2001). Representing goal-oriented human performance in constructive simulations: validation of a model performing complex time-critical-target missions. *Proceedings of the Simulation Interoperability Workshop*, Orlando, FL.

Laughery, K.R., Scott-Nash, S., Wetteland, C., and Dahn, D. (2000). Task network modeling as the basis for crew optimization on ships. *Proceedings of the Meeting on Human Factors in Ship Design on Automation*, sponsored by the Royal Institute of Naval Architects, London, England.

LaVine, N.D., Peters, S.D. and Laughery, K.R. (1995). *A Methodology for Predicting and Applying Human Response to Environmental Stressors*, Micro Analysis and Design, Inc., Boulder, CO.

Lawless, M.L., Laughery, K.R., and Persensky, J.J. (1995). *Micro Saint to Predict Performance in a Nuclear Power Plant Control Room: A Test of Validity and Feasibility*, NUREG/CR-65, U.S. Nuclear Regulatory Commission, Washington, DC.

McCracken, J.H. and Aldrich, T.B. (1984). *Analysis of Selected LHX Mission Functions: Implications for Operator Workload and System Automation Goals*. Technical Note ASI 479-024-84(B) prepared by Anacapa Sciences, Inc., June.

Micro Analysis and Design (2001). *Micro Saint User's Guide*. Micro Analysis and Design, Boulder, Colorado.

Newell, A. (1990). *Unified Theories of Cognition*. (Cambridge, MA: Harvard University Press).

Newell, A. and Simon, H.A. (1972). *Human Problem Solving*. (Englewood Cliffs, NJ: Prentice-Hall).

O'Hara, Brown, Stubler, Wachtel, and Persensky (2002). *Human-System Interface Design Review Guidelines* (NUREG-0700, Rev. 2). U.S. Nuclear Regulatory Commission, Washington, DC.

Plott, B. (1995). *Software User's Manual for WinCrew, the Windows-Based Workload and Task Analysis Tool*. U.S. Army Research Laboratory, Aberdeen Proving Ground, MD.

Siegel, I.L. and Wolf, I.J. (1969). *Man-Machine Simulation Models*. (New York: Wiley Interscience).

Wickens, C.D. (1984). The multiple resource model of human performance: implications for display design. In the *AGARD/NATO Proceedings*, Williamsburg, VA.

Wortman, D.B., Duket, S.D., Seifert, D.J., Hann, R.l., and Chubb, A.P. (1978). *Simulation Using SAINT: A User-Oriented Instruction Manual*. Aerospace Medical Research Laboratory, AMRL-TR-77-61. Wright-Patterson AFB, Ohio.

Part II

Techniques in design

The first part of this book contained chapters describing fundamental groups of methods and techniques, and also implicitly proposed certain approaches to ergonomics study. This second part has within it descriptions of how the basic methods and approaches may be modified and combined in particular examples of application in design: this is for systems, products, text, human–computer interfaces (HCI), and control rooms and control facilities. In some cases the techniques generally available are not sufficient and so special ones have been developed. Although the focus is on design, in order to contribute to this the methods described in this part have relevance also for the analysis and evaluation of equipment and systems, within the whole systems life cycle.

It will be apparent that there is overlap between the chapters in this section in terms of the types and application of appropriate methods, and that much of the methodology delineated has potential value in many other domains. However, these five have been selected so as to provide an interrelated set of application domains, and as likely areas of importance into the foreseeable future. Within this whole part we will find all the methods and techniques of part I — direct and indirect observation, archives, experiments, task analysis, protocol analysis, modelling, simulation and computerised data collection — with explanation of their strengths and weaknesses in the *particular* circumstances.

Chapter 10 from Wilson and Morrisroe takes as its focus human machine systems as whole, and concentrates on the systems analysis techniques needed at the outset of the systems design process. Some of this has been covered earlier in Chapter 6 on task analysis, but Chapter 10 extends this into function and user analysis as well. Also, the increasing prevalence of human factors integration plans and standards — to provide a more formal inclusion of ergonomics in design — is described.

In Chapter 11 McClelland and Fulton Suri describe a large range of methods and techniques from the perspective of user centred design. This is particularly in terms of product design, an application area that has been of interest to ergonomists for over 30 or 40 years. The dominant methodological approach has been user trials, an approach with wide applicability and particularly valuable in the development and assessment of products and equipment, and the main focus of the equivalent chapter in the second edition of this book. Linkage with human computer interaction evaluation is obvious and current concern with usability engineering in the whole area of information technology implies amongst other things that user trials must be an integral part of an iterative design process. User trials are a way of bringing basic measurement principles into the product development process, and their use provides both quantitative and qualitative information on product effectiveness. However, McClelland and Fulton Suri make clear that user trials are but one approach to user centred design, and that there is a vast array of analytical and empirical methods available, each with their own strengths and weaknesses.

In a product development process described in Chapter 10 the parallel development of product and manuals and instructions was proposed. This might seem a little optimistic or excessive, but the usability and even market success of a product can be impaired by inadequate instructions for use and maintenance. Nowhere is this more so than in the world of modem consumer products — videorecorders and DVD players, digital cameras, dishwashers, mobile phones etc. — where manuals have historically been the subject of mirth or frustration or both. Hartley in Chapter 12 summarises the evaluation of text, describing the sometimes specialised, sometimes general, methods to do this.

Much human factors effort is put into the understanding of human computer interaction (HCI) and development of human–computer interfaces that match the needs and limitations of users and tasks. After a period spent producing guidelines and criteria, the HCI community is now much taxed by the problems of evaluation (related of course to relevant criteria) and of what should be evaluated. In doing so they are confronting problems, compromises and gaps in knowledge that have faced ergonomists in other domains for years. What is interesting is that the enormous concentration of human factors resources in HCI has allowed exploration of the use of virtually any method or technique — from psychological through to physiological measurement — and the combination of some of these into broad-based approaches to HCI evaluation. In turn, results of interest and significance for all ergonomists and new methodological approaches are emerging. Baber in Chapter 13 gives a comprehensive account of the area. He also makes clear that HCI is not just about PC interfaces and use; HCI now embraces many types of Information and Communication Technology (ICT), from personal devices and wearable computers to large collaborative virtual environments.

For 50 years, ergonomics textbooks have included chapters on 'controls and displays'. Of late there has often been distinction between human-computer interfaces and so-called 'traditional' interfaces, with the latter including dials and scaled displays, status and warning information, switches, levers, knobs and valves, for instance. Such distinction is not really valid any longer since increasingly all interfaces are computer-based and the computer interfaces are not just VDU-based. Nowhere is the blurring of generations of display and control technology more apparent than in control rooms and control facilities, and Chapter 14 from Wilson, Rajan and Wood looks generally at human–machine interfaces — their analysis, design and evaluation — but with specific reference to systems and process control applications.

Common to all methodological approaches in the chapters in this section is the need to understand and allow for who the users or participants will be and what type of tasks they will carry out. Most of the chapters address explicitly the question of 'what users?' Between them they define attributes that need to be identified to specify and characterise the target user group: personal characteristics; physical, cognitive and attitudinal characteristics; specific skills, training, previous experience; the characteristics of their jobs, and roles; and personal preferences. Perhaps most critical is to know whether the users will be beginners, novices, regular users or experts, whether use will be frequent or not, and the degree of discretion they have in using the product or system.

A second main underlying need within design and is to understand what it is the users will have to do, the tasks they must undertake. In most cases task analysis (Chapter 6) will have allowed such identification, and the user characterisation in terms of experience, frequency and discretion of use will also be relevant. The number, complexity, sequence, simultaneity, loading and criticality of the tasks will, *inter alia*, be important to the making of design and development decisions, in determining how each of the products or systems must be evaluated, and in selecting the appropriate methods and techniques.

John Wilson

Chapter 10

Systems analysis and design

John Wilson and Ged Morrisroe

Introduction

Imagine that you are a software engineer, design engineer or architect. You have been given a leading role in a major systems design project — say a new airport cargo handling terminal or a new railway signalling centre. You have enough problems already, of course, but also you are aware enough to know that getting the human factors right will be vital and that you need to incorporate ergonomics thinking from the start. But, how should you do this? How will you integrate concerns for physical, cognitive and organisational ergonomics, and for the wide variety of stakeholders who will use the system? How will you bring development of the humanware into line with development of hardware, software and organisational structures?

Now, closer to home, imagine you are an ergonomist who is contacted by a company, and asked to lead the human factors effort in a complete redesign of a factory facility. Ergonomists often complain that they have not been brought in early enough, that people were considered too late in the design process and that as a result the ergonomics would be costly or the design too 'locked in' to remove all its flaws. But, coming in at the start with a blank piece of paper can be worrying, even frightening. Where do we start — with which of physical layout, task design, team selection and role design, training and awareness programmes, motivation and reward schemes, equipment and interface design . . .? Each of these, and the design decisions made about it, will have implications for the others. Also, how do we know where we are heading — do we have a framework for the process, or a roadmap?

The authors have been in situations and development projects similar to those described above. These questions for ergonomics — where do we start?, what are the directions or options?, how do we set priorities?, how do we integrate human with technical, organisational and economic factors?, when do we know if a design element is adequate? — are not easy to answer. They are important questions however, both in any particular project and also in general. To begin to address such questions we need to understand the relevance of human factors to the whole system life cycle, and especially to the systems development stages of specification, design and testing. In parallel, we need to understand the value and role of techniques of systems analysis and design in the application of ergonomics.

This chapter opens by describing a number of the better known specifications of the systems development process. We then describe in more detail the early, more analytical, stages in product and systems design, especially stages such as information gathering, requirements specification, function analysis and task analysis.

In the last few years all of this has in part been managed through what are known as human factors integration plans (HFIPs), strategies for real integration of human factors concerns and insights throughout the system life cycle, integrated with the other development stages rather than being an add-on.

For much more information about systems analysis and design, and especially the role and contribution of human factors, the reader should consult Chapanis (1996), Hollnagel (2003), Meister (1985), Meister and Enderwiel (2002), O'Brien and Charlton (1996), Vicente (1998) and Wickens *et al.* (1998).

System development processes

The literature is full of processes for system development. In the context of this chapter, 'system' is being used as a description of any artefacts which may be designed, including social and organisational artefacts, so that something like job design can be included also. In fact the different levels of focus and different domains for the development activity, from graphics design to product design to human machine systems design to software design, bring different families of development processes. The last of these, software design, has a whole set of methods devoted to it alone, and within that, of great interest to ergonomics, user interface design has sets of methods also (see Chapters 11 and 13).

Early proposed processes for man–machine (nowadays human–machine) systems design were similar to that of Singleton (1974) whereby human and hardware subsystems are developed in parallel, followed by subsequent integration both in terms of the interface and then the operational system itself (Figure 1). It is probably no accident that models of this kind were proposed at a time of great interest in socio-technical systems design (see Cherns, 1987, Clegg, 2000, and also Chapter 29), with its emphasis on joint development of social and technical systems, although it has to be said that practice did not always match theory.

Subsequently, representations of the systems design process handled the parallel hardware and human design needs by placing ergonomics considerations within all stages of the total systems design process. Thus Bailey (1989) and Sanders and McCormick (1987, p. 522) talked of stages — which overlap and are iterative — as: Stage 1 — Objectives and performance specifications; Stage 2 — Definition of system; Stage 3 — Basic design; Stage 4 — Interface design; Stage 5 — Facilitator design; Stage 6 — Testing. In other views, Kragt (1992, p. 15) and Kirwan and Ainsworth (1992, p. 28) define multiple overlapping human factors activities, within a 'parent' system design process flow of 'concept–specification/definition–design–production/commissioning operation'; Kragt (1992) still makes explicit the parallel development of machine, jobs and interfaces (Figure 2). Wickens *et al.* (1998) associate a series of human factors activities with a classical 'staged' model of the product life cycle (see Figure 3).

In two further representations of the design process we can see a philosophical contrast. For product development a process has been suggested that defines ergonomics interventions within the total product design process (Figure 4), whereas the systems design process shown in Figure 5 is itself an ergonomics process, developed as part of a programme to embed ergonomics within a company's system design activities (Aikin *et al.*, 1994). It was constructed to parallel the existing system of design and approvals in the company (for example, for safety testing, environmental impact, etc.).

One of the simplest models of the design process comes from software engineering and this has become known as a waterfall model, in which each activity leads naturally into the next (Figure 6). In more recent contributions the specialists in software engineering, and particularly those involved in interface design, have expanded upon this waterfall model, to define: a V model (in which the two stages of the cycle are structured into two processes, downward for specification design and upward for validation and testing); a spiral model (e.g., Figure 7), in which we start to see representation of the iterations that should take place in development, where needs are formulated progressively, risks analysed and resolved as and when they are encountered; and an incremental model, which also explicitly recognises the

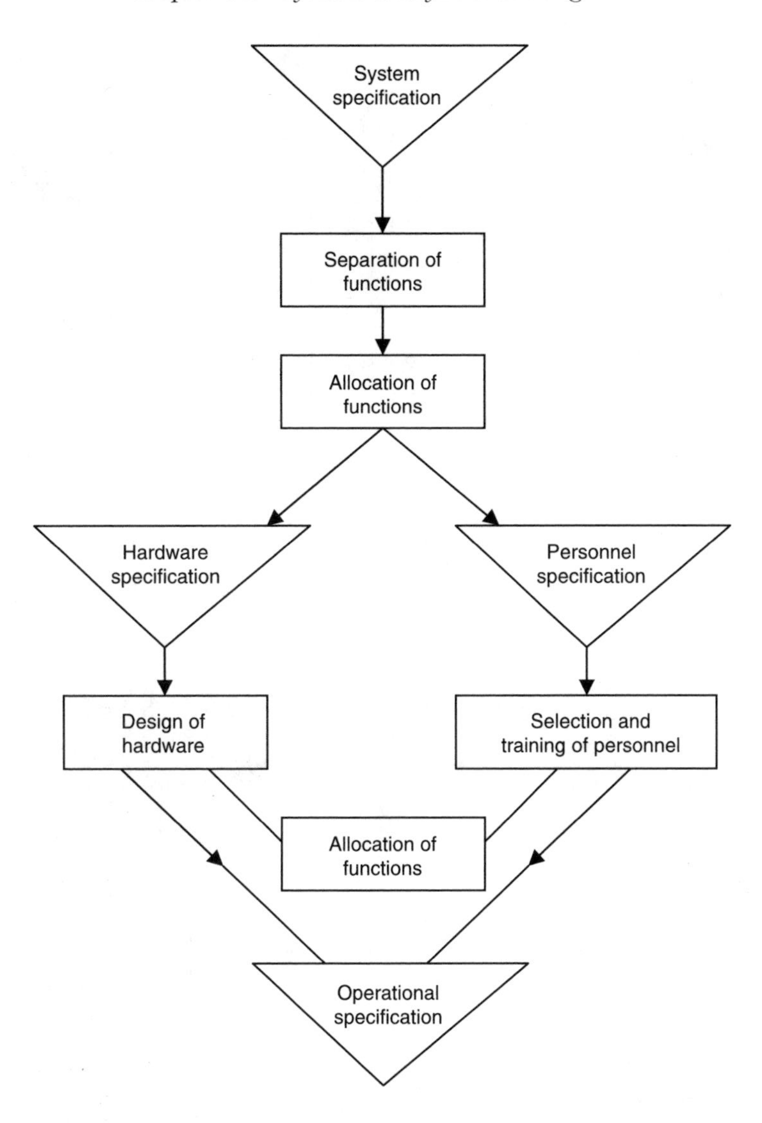

Figure 1 The systems design process. From Singleton (1974), reprinted by permission of the author.

iterative development of a design (see Lepreux *et al.*, 2003). In a simplified form the iterative model can be represented in the well known two-dimensional version which distinguishes design morphology from design process (see Figure 8a and b).

Virtual environments (using VR technology) are an interesting case whereby a mixture of different design process stages and elements are required, to analyse the scenario or work system that the virtual environment (VE) will be applied to, and also to analyse the work of participants within the VE. Figure 9 shows a virtual environment development system in current use.

Design process in the real world

It is somewhat ironic that so much effort has gone into so many descriptions and specifications of the design process, since design in the real world so rarely goes neatly to plan. Anyone who has been involved in a design, whether large or small, will know that it rarely if ever follows

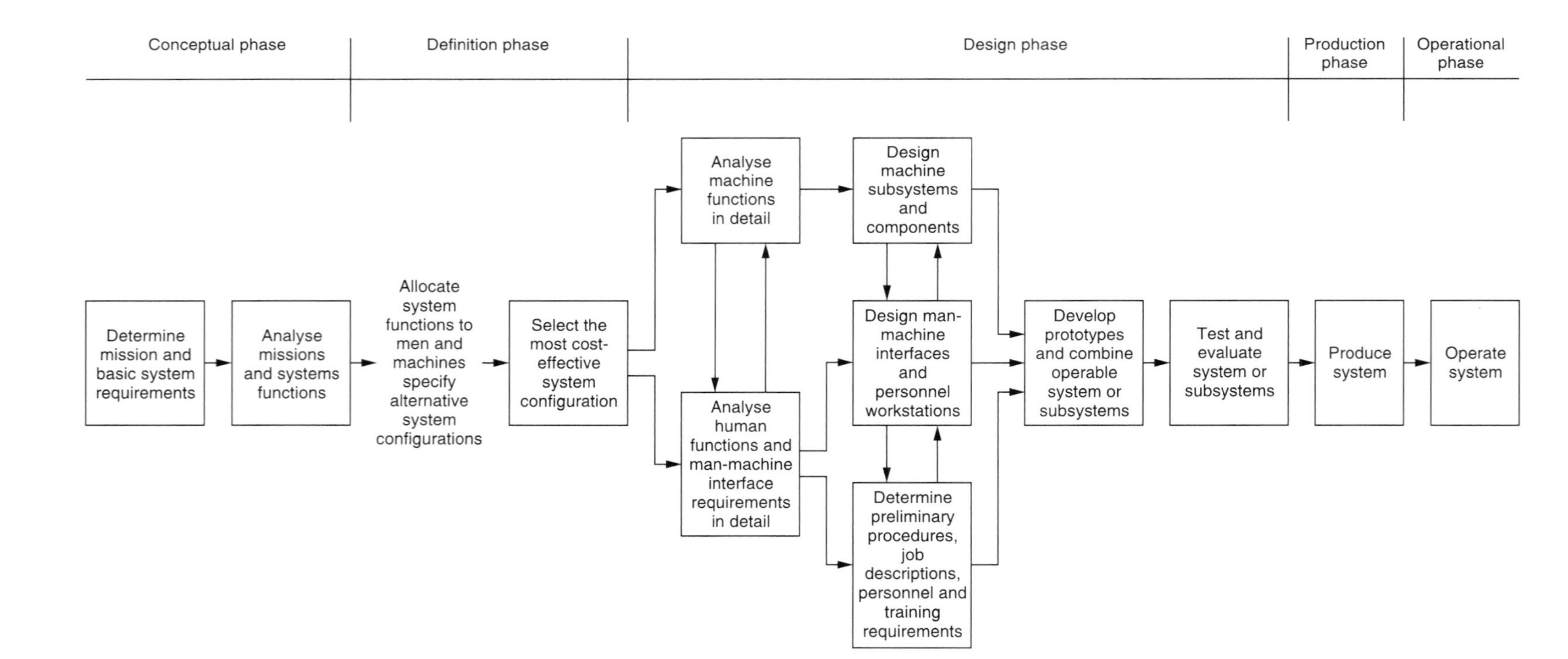

Figure 2 Design activities occurring during phases of the system development process (adapted from Kragt, 1992).

Stages in product life cycle	Human factors activities
Front-end analysis	User analysis Function analysis Preliminary task analysis Environment analysis User preferences and requirements Input for system specifications, including ergonomic criteria
Conceptual design	Functional allocation Support conceptual design
Iterative design and testing	Task analysis Interface design Develop prototype(s) Heuristic evaluation (design review) Additional evaluative studies/analyses: Cost-benefit, trade-off, workload; safety analyses; stimulations of modelling Usability testing
Design of support materials	Develop materials, such as manuals
System production	
Implementation and evaluation	Evaluate system
System operation and maintenance	Monitor system performance
System disposal	

Figure 3 System development life cycle and associated human factors activities: adapted from Wickens *et al.* (1988, p. 46).

the pre-defined path and that development will actually include the adaptation of planned stages, iterations around loops in the process, frequently redefined objectives or means, and ever changing priorities.

In real developments, rather than a neat agreement on goals and then setting of requirements and constraints early, on we will usually find several competing requirements and goals. Classically, marketing may require one thing — something very eye catching for the customer with a large number of variants to allow targeting of many markets — whereas manufacturing would prefer something which requires minimal re-settings of machines, no difficult machining processes, and with as few variants as possible. In another example, some parts of a service company may think it best that their new computer system be merely an adaptation and improvement of the one they already have and which is on the market; others in the company, perhaps more adventurous, may be pushing for a radical re-think and an entirely new form of system.

In complete contrast, in other development processes the team may be bereft of early ideas. They may know that they need to have a new product for the market but may not know where to start or may lack inspiration.

Whether we have many competing requirements or a paucity of ideas, we are not helped when there are considerable gaps in the information and market intelligence available to the company, or when the information available is contradictory and confusing. Much design in fact, does not start from a clean new set of goals and ideas for completely new design, but consists of development teams begging, borrowing or stealing ideas from others. One large UK domestic products manufacturer known to the author in fact admitted, in confidence, that the majority of their development started from visiting trade fairs overseas and seeing what other companies were putting on the market.

Once the design process actually starts, again the process does not necessarily match the classic or idealised system, of a neat move from conceptual design to various levels of detail design, albeit with iterative steps. In fact there may be a great deal of developing and trying out various designs of different levels of sophistication, prioritising these in a crude fashion as the team goes along. When this is working well, as the models or prototypes themselves

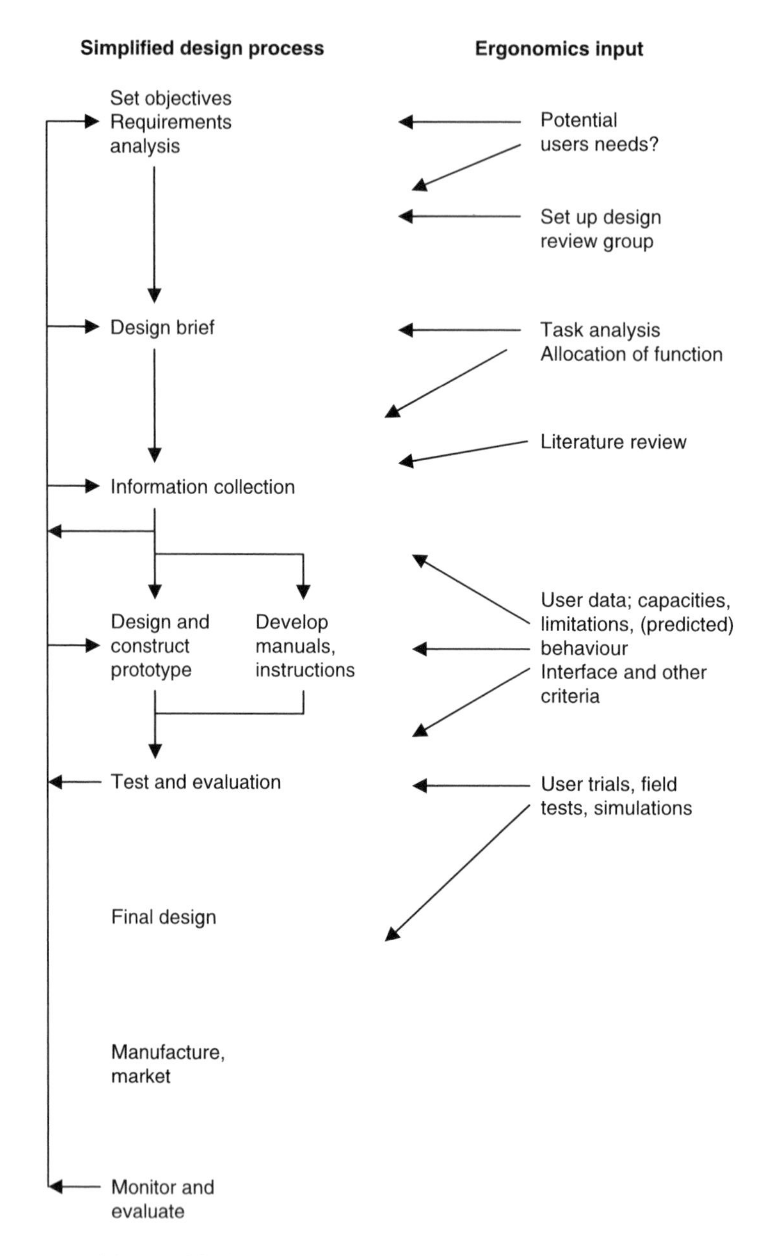

Figure 4 One possible simplified product design process showing types of ergonomics input (files of the author).

increase in sophistication so can the test and evaluation methods, including those for human factors, become more detailed and extensive.

In practice, as the design process goes on, and for anything which has any substantial amount of resource invested, there will be pressure to short cut. This pressure is usually from the finance department, but may also be from marketing, who want to get the product to the customer as soon as possible, and perhaps also from production who want to see their facilities being fully employed. Also of critical importance to ergonomics, there may often be a late realisation that the real user needs and requirements have been forgotten, leading eventually to system misuse or disuse, or at the least expensive re-working and modification at a late stage in design.

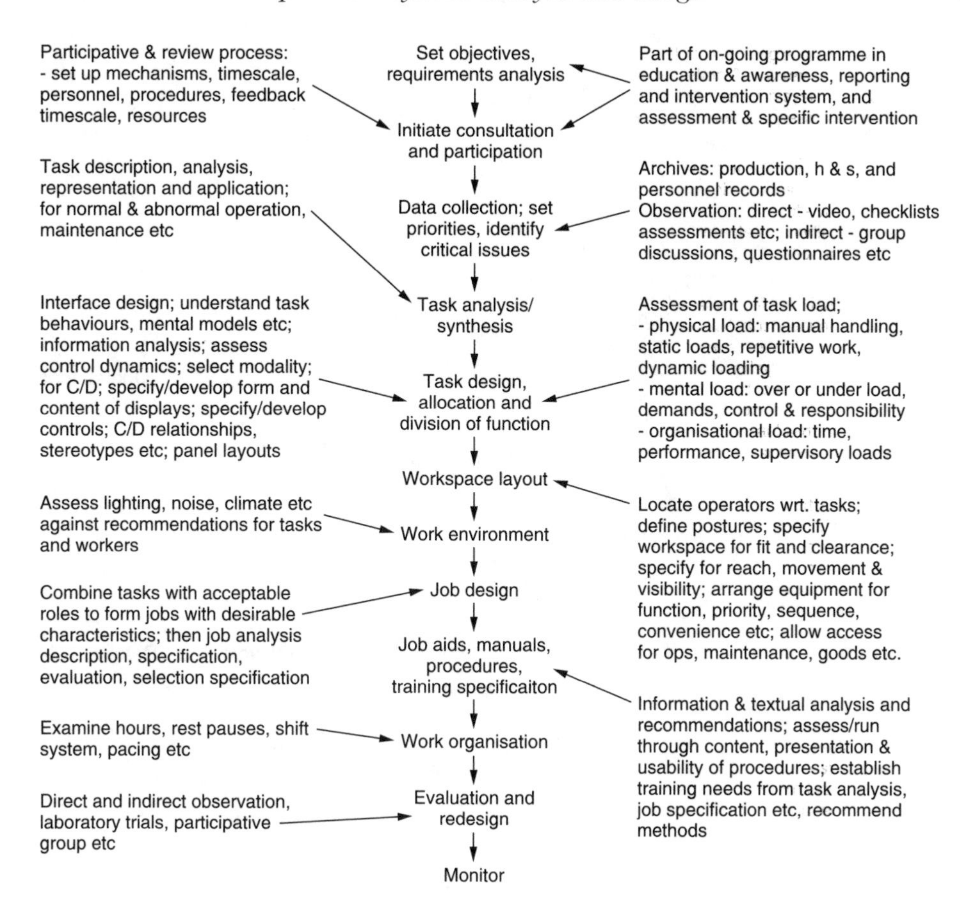

Figure 5 Ergonomics design process. Each of the stages of ergonomics input has itself got particular input requirements, which imply use of a large variety of methods in analysis, design and evaluation (files of the author).

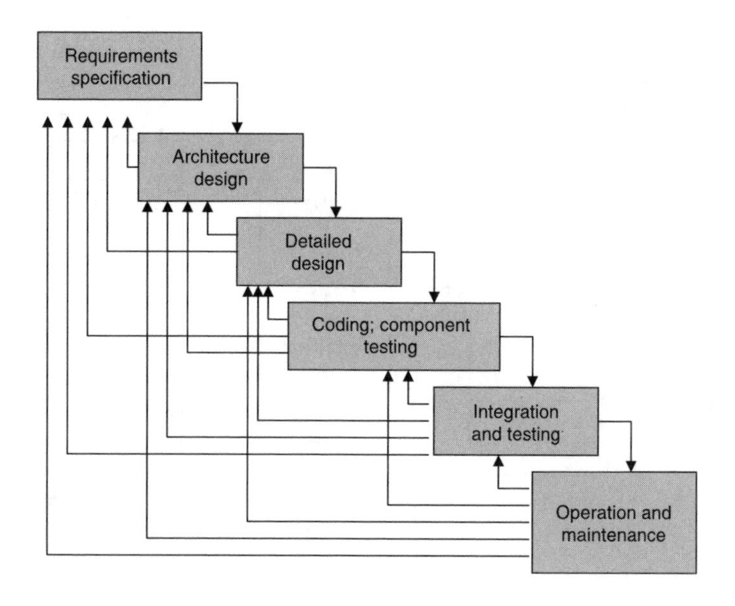

Figure 6 The waterfall model of the software life cycle.

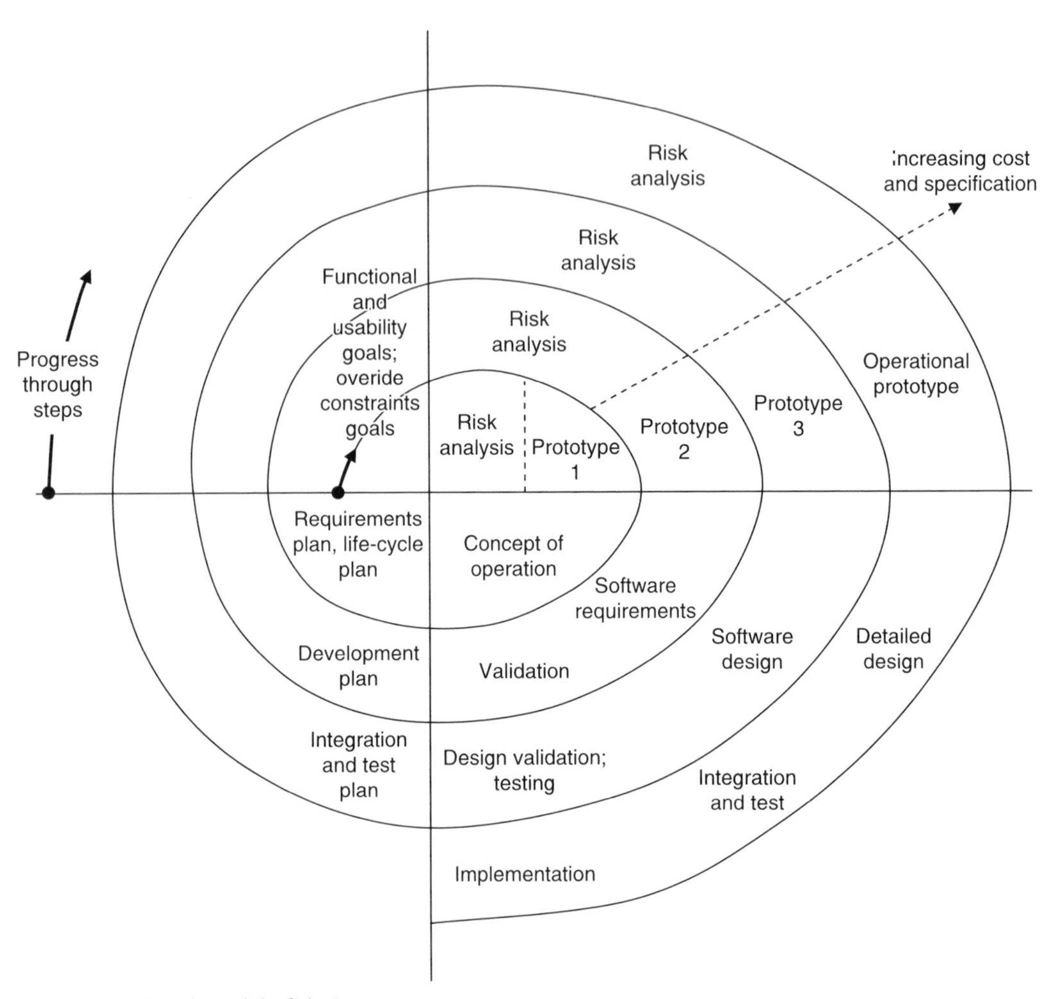

Figure 7 Spiral model of design.

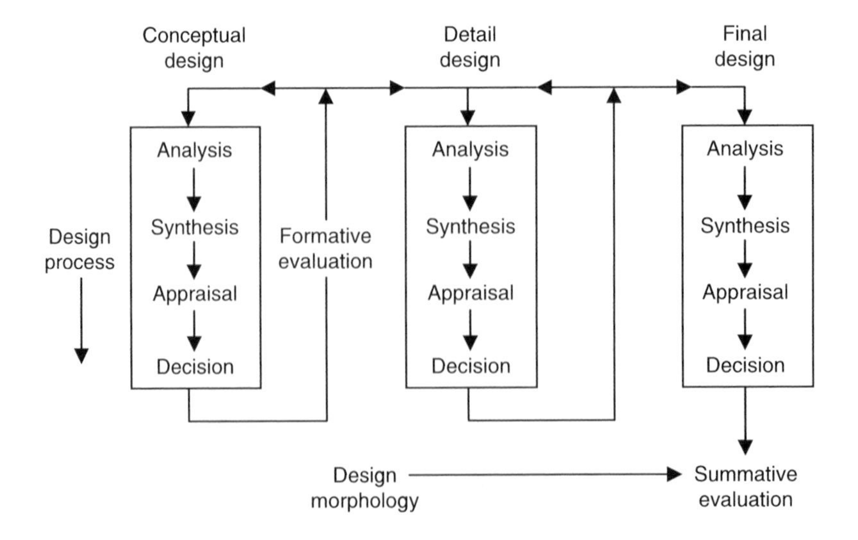

Figure 8a Two dimensional model of the design process. Adapted from Markus (1969).

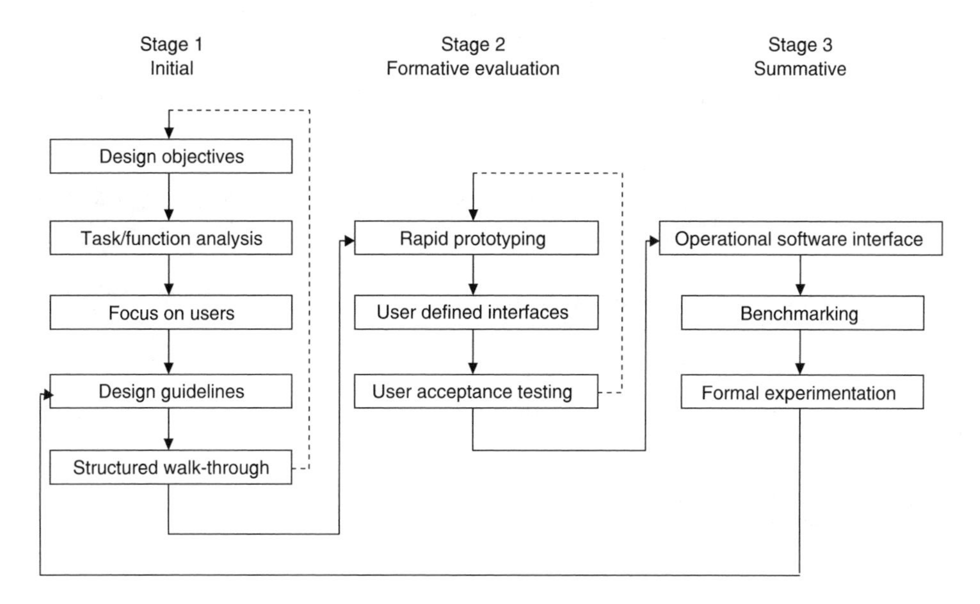

Figure 8b Flow diagram of the three stages in the design of human–computer software interfaces. (Source: Williges *et al.*, 1987, reprinted by permission of the publisher John Wiley & Sons.)

Ergonomics in a major construction project

Some time ago one author was involved, together with several members of his own group, in a major UK systems development and construction project. A massive new facility was being built in a project costing hundreds of millions of pounds. Many of the usual sorts of partners involved in this project were collected together in a loose form of supply chain, connected together through quality procedures and formal reporting systems. There were two main clients, the owner of the total facility where the new development was to be built and the eventual operator of the actual development, working together in a partnering relationship. There were a number of different firms of design engineers, architects, systems engineers and then suppliers of different sub-systems. At various times we, as the ergonomics team, reported directly to the design engineers who were about number 3 or 4 in the chain, and at other times to sub-system suppliers who were numbers 9, 10 and 11 in the chain. Despite, or perhaps because of, all the very tight quality procedures on document generation, distribution lists, approval, processes, etc., we often found as ergonomists that we were communicating directly with several other partners in the supply chain, contributing to their own particular developments. Whilst this was very good from the point of view of making good relationships, and establishing ergonomics criteria and design contributions at an early stage, it often meant that we were working on generations of drawings and design concepts that were different to those of our direct clients. Subsequently we discovered that this was the norm across the whole project, and that different aspects (buildings, services, logistics etc.) were all at different generations at different times.

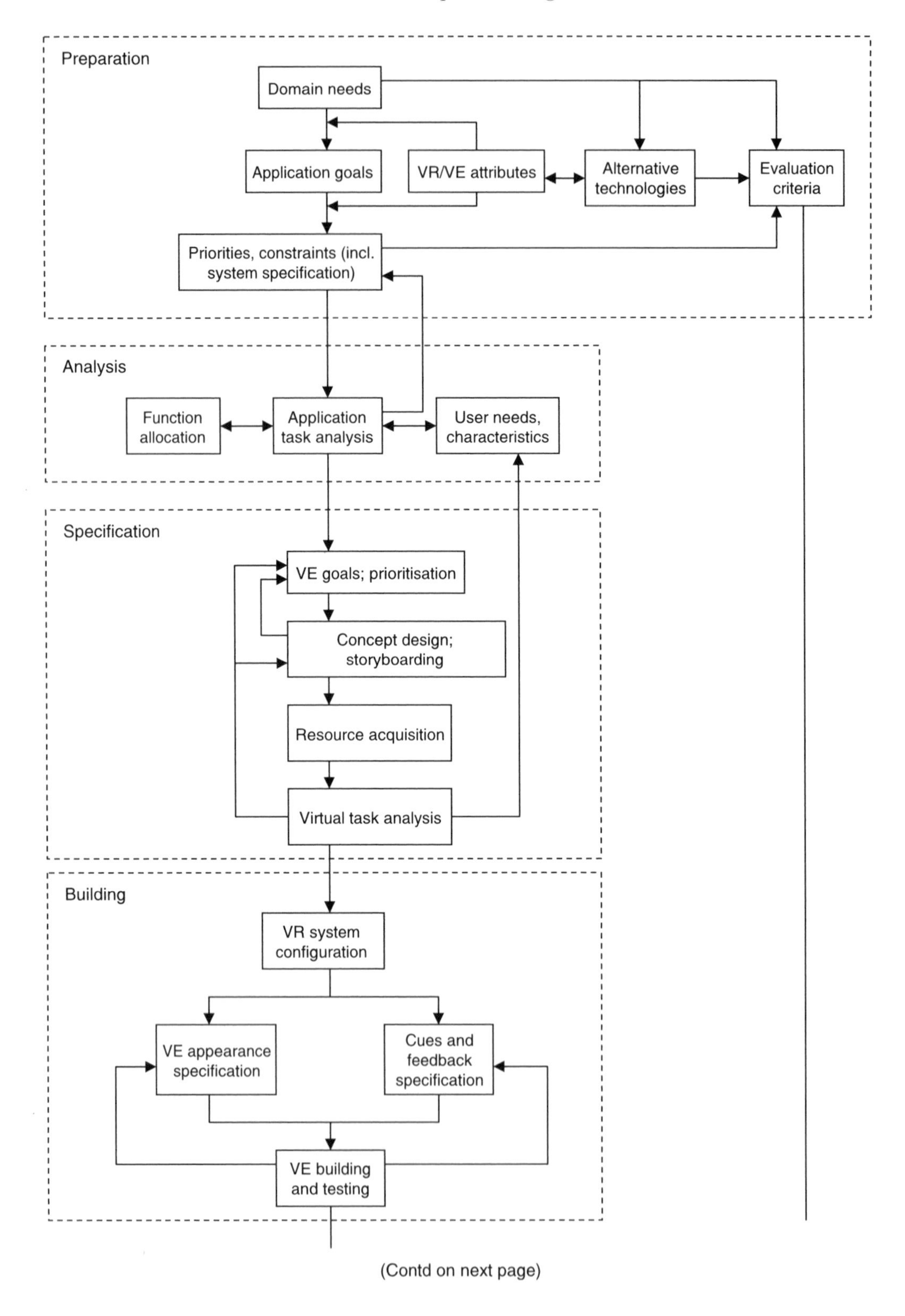

(Contd on next page)

Figure 9 VEDS: Virtual Environments Development Structure (Wilson *et al.*, 2002).
Note: Only major feedback loops are shown, for clarity, in fact, the whole process is an iterative one.

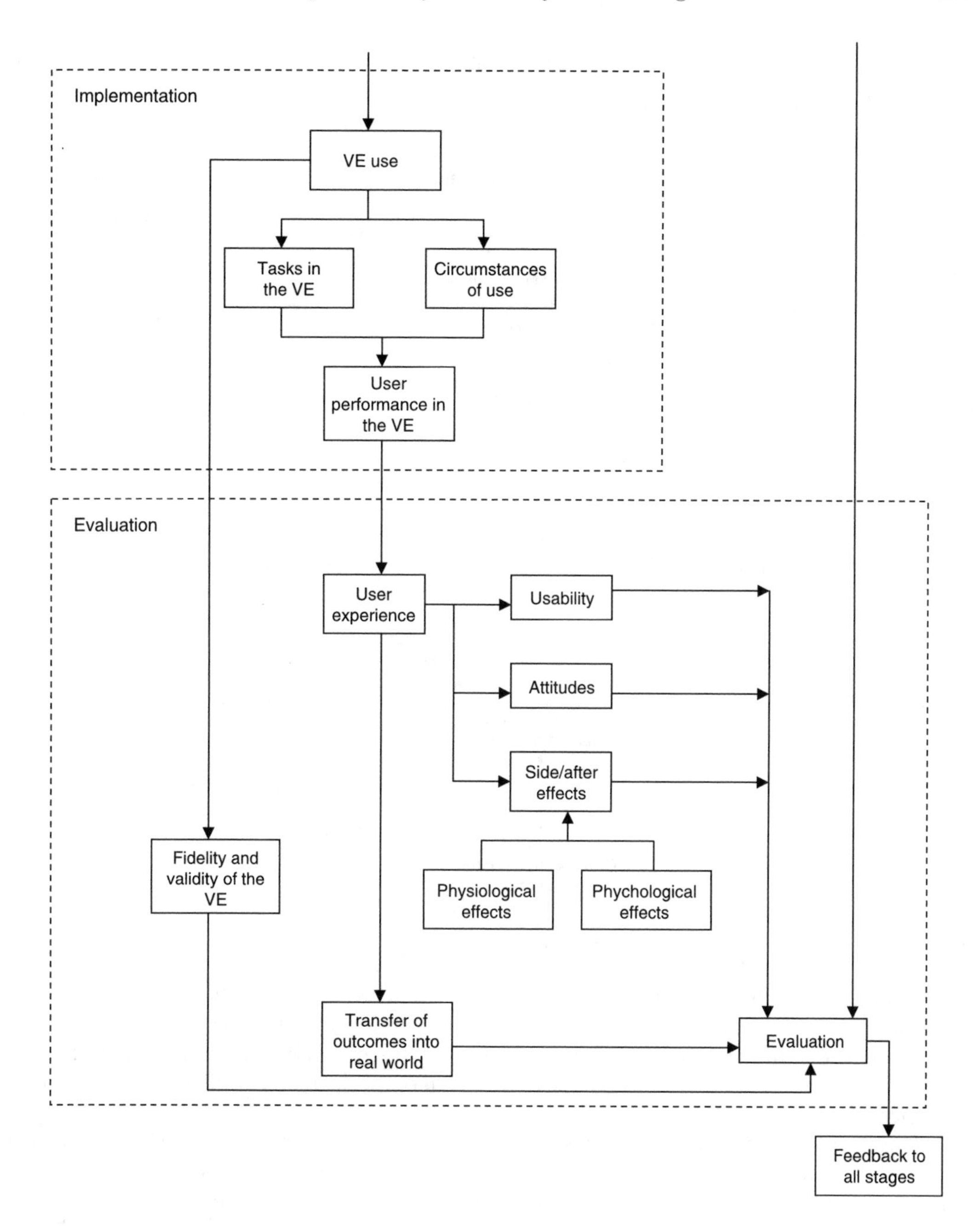

Figure 9 Continued.

Early human factors stages in the development process

The ergonomics/human factors contribution in the early stages of system or product design is crucial. The questions that must be asked by the ergonomist in the design team include: Who is going to be affected by the design and implementation?: What is really going on in the activity carried out within or with the system?: How can we study the activity or problem or concept, as it is now or in some mock-up version of how it will be?: Why are we finding particular information or results in early data collection or experimentation?: What should the

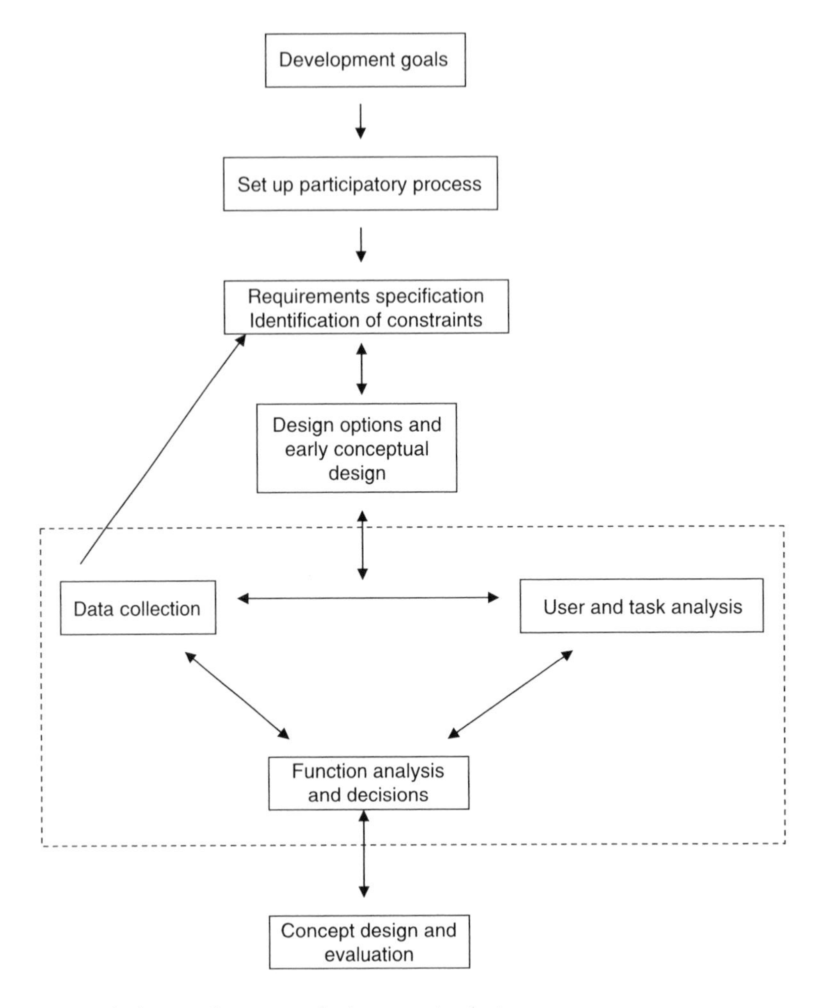

Figure 10 Key early human factors analysis stages in design.

designers and developers do about it? The input of the ergonomist will be in terms of data collection, analysis of tasks or systems, provision of criteria for design, generating participation and involvement amongst stakeholders or representatives of stakeholders, and carrying out early evaluations of potential effects of the new system or product, on people involved and on their performance.

In figures 10 and 11 appear two somewhat different representations of where human factors is accounted for, and how, in the early stages of the system life cycle. In the first representation (Figure 10) we see that stages of setting up the development process (in which the establishment of goals, requirements, constraints and criteria as well as enabling participation are included) feeds into early conceptual design and analysis contributions. This set of processes is drawn deliberately with no set sequence, since in anything more than a trivial development process the decisions that are made in one stage will redirect attention and contribution from the others. For instance, as decisions are made about functions, there may be a decision that more analysis of user needs is required, which requires greater amounts of data to be collected; in turn, these data may change the conceptualisation of the functions which must be carried out within the system. Each of the relevant stages will be discussed later on in this chapter.

In the other representation (Figure 11), produced more explicitly for the development process in human computer interaction, we can see particularly the input of information from

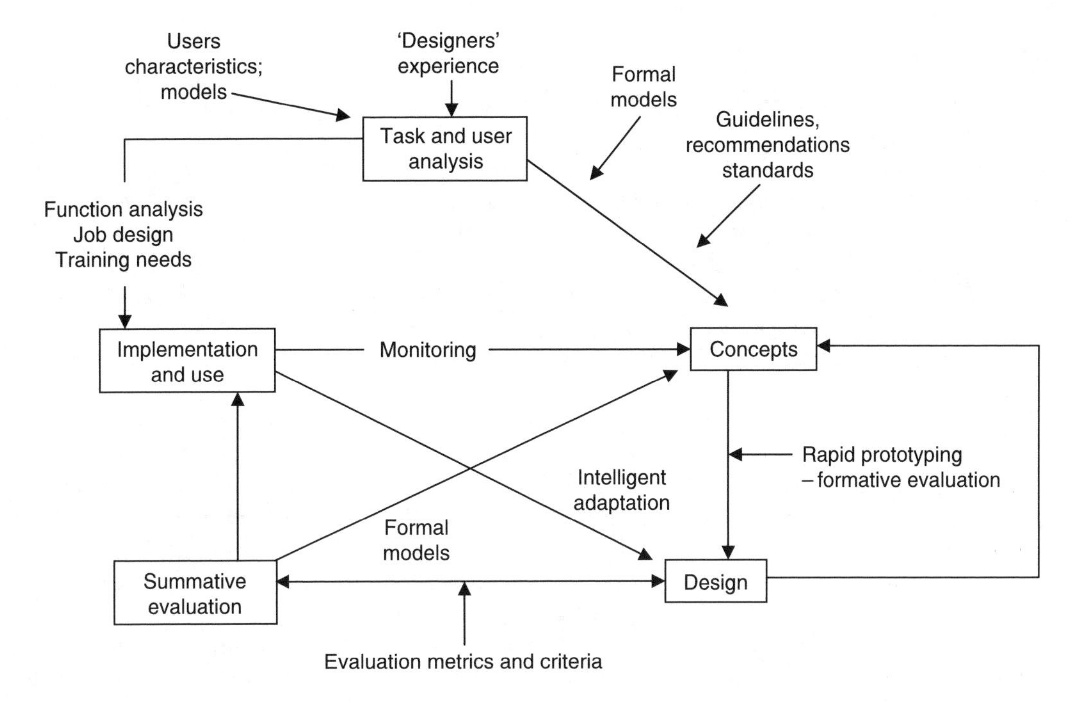

Figure 11 HCI design cycle.

user models and formal models, and guidelines and recommendations into the design process. This system representation also goes so far as to show the roles of formative and summative evaluation in the design process.

Requirement and constraints

At the outset of the development process, goals and constraints should have been identified. Also, the involvement of stakeholders should be ensured, using participatory processes and methods (see Chapters 35 and 36). Following this, the systems requirements should be specified (see Viller *et al.*, 1999 for a review of approaches).

Take a case of a new computer system for maintenance in a manufacturing plant, with associated new job designs. The requirements here might include: the company increases its rate of prevention rather than repair; there is a 20% decrease in major breakdowns month on month; and the maintenance function not only responds 30% more quickly but costs less, by a factor of 10% annually. Constraints identified might be: that there is limited space actually at the site for any information terminal; that the workforce need to be mobile, travelling out to particular maintenance jobs, and therefore any information support should also be portable or at least mobile; and that the company wants to work with the same or fewer staff in maintenance than previously.

From these requirements and constraints we can identify a number of initial design options before moving into functional analysis. In the maintenance case, these options might be: a fully automated system of diagnosis and repair; diagnosis being automated but the repair being carried out by people; the information display showing the status of the machine which is interpreted by the maintenance engineers to decide on whether to carry out repair; everything being carried out by the maintenance engineers from inspection through to final repair; or the role of interpreting the information and deciding on level of repair being given to the machine operators rather than maintenance engineers.

Data collection

In an ideal world, the ergonomics team will continue to collect relevant data throughout the system design life cycle, to support decisions about the form of design and about the user and task criteria that need to be incorporated into design considerations. Any or all of the methods and techniques which appear throughout this book may be used as appropriate, including those of direct and indirect observation, simulation and modelling and examining archives and other sources of data in the literature. Broad examples are listed in Table 1. Task analysis is regarded by some as a method in its own right (Chapter 6), but also employs many other methods of data collection. In fact making decisions on data collection for task analysis provides an excellent illustration of the problems of method choice in ergonomics generally.

A particularly important form of data collection used at the outset of any design project is to establish what guidelines, recommendations and standards already exist. Where there is guidance this may be explicit in the form of data sheets or similar, or may purely be guidance that can be gained from literature on similar design projects carried out elsewhere. A major difficulty of using human factors guidelines comes from the well known generality/specificity paradox. On the one hand, if guidance is general enough that it can be applied for the particular system or product that you are designing, then often the constraints or detail in the guidance are not sufficient for the particular product you are concerned with. Much guidance in human factors, unfortunately, remains at the very general level of 'know thy user' or 'design to match user abilities' or 'allow the user to understand where they are in the system'; whilst very worthy, these sort of guidelines can be very frustrating for designers, since they offer nothing to directly support the development process. On the other hand, guidance which is produced exactly for a particular type of system, to a high level of detail, will rarely if ever be capable of being transferred elsewhere, because of differences in context and setting if not in the system itself.

Function analysis

Whilst some ergonomists would hold that the critical stage in any design project is task analysis, this can only take place if a proper function analysis is also carried out. Functions in any human–machine system are the major elements of what must be done, related to the goals. Therefore a function level description of going to work in the morning might consist of: get out of bed and get dressed; make and eat breakfast; leave house and walk to train station; buy ticket and ride on train to work; walk into building, open office door and enter office. In other words these are large groupings of tasks, each containing many activities and decisions. The identification of functions, their analysis and subsequent allocation will take place on an iterative basis throughout the design process. The initial specification of requirements and constraints will give a performance specification, and this will allow identification of gross level functions. As ideas on functions change, so will the task and user analyses be modified, and vice versa.

A number of different means might be adopted to represent and analyse the functions involved in the new system or use of the new product. Examples are an input/output diagram, a system level flow chart or a rich picture diagram (see the case later in this chapter for examples, and also Chapter 38 and Tudor and Tudor, 1995).

At this stage in development it might be useful to make simple representations of the operational view (how the system is or will be used), physical view (what the system actually contains or will contain) as well as a functional view (what the system does or will do) — see Chapanis (1996, p. 41).

Table 1 Data Collection for Analysis and Evaluation

Documentation
Archives/logs — production, quality, maintenance, sickness, absence System or job specifications, operating procedures

Direct observation
Notes, audio, photos, slides
Event recorders
Video
Electronic recording
Eye tracker
Body position or posture recorder
Checklists — aide memoire, items, scored, graphical etc.
Critical Incident Technique
Concurrent verbal protocol analysis

Simulation and modelling
Mock-ups — drawing, scale, physical
Computer — workspace, flows, biomechanics, cognitive ...
Rapid prototyping

Physical measurement
State of site, traces or evidence of activity
Physical workplace measurement — dimensions
Environment recording

Performance measurement
Times
Errors
Behaviour (personal) — fidget, blink, yawn, sweat ...
Behaviour (social) — talk/avoid, positioning ...

Operation loading
Physiological
Biomechanical
Psychological

'*Expert*' *assessment*
Walkthroughs — 'angels', 'demons' etc.
Scenario diagnosis
Estimations — likelihoods, ratings etc.
Delphi method

Indirect observation
Questionnaire, rating scales
Interviews — unstructured, semi-structured
Group discussions
Diaries, self-reports
Retrospective verbal protocols

New library issuing system — function analysis

If we have been brought in to implement the human factors for a new library book issuing system, a function analysis would start from the top level system goals. These are to allow users to borrow books according to their status, and to associate the user with the book and the date due for return; a related goal would be to maintain security. Therefore we might identify the main functions as: a book is found and brought to registration system; eligibility and current status of user is checked (i.e., how many books can be borrowed, for how long, no fines outstanding etc); a record is made of user name against book details (author, title, classification number) and date of borrowing; a record is made for the user of the date due for return; user leaves library with book through security screening system.

At some point early on in development (although Fuld (1993) questions the whole concept of function allocation in preliminary design, suggesting its use may be of greater value at a test stage) some initial decisions need to be made about which functions and tasks will be the responsibility of people and which should be partially or fully automated; this is known as allocation of function (see Price, 1985; Sharit, 1997; Sheridan, 2000). Also, we may wish to decide which people have responsibility for different functions — the service clerk or the customer in a service system, or the operator or the engineer in manufacturing. This is division of function.

Allocation of functions was historically often seen from an engineering (or military) perspective as comprising only a checklist method, selecting functions to be carried out by people or machines according to capability. Engineers may still use variations of a 'Fitts List' of comparative abilities, the PABA–MABA (people are best at — machines are best at) approach, despite early warnings against rigid application of such an approach (e.g., Jordan, 1963).

In fact in system design there can be many criteria for making allocation decisions, to decide whether people or computers will carry out functions and also to decide between different people (e.g., operator, supervisor, maintainer etc). Comparative capability of people and machines is relevant of course, but some of the other criteria shown in Table 2 will be very relevant in certain situations.

To pick up on our running example, for the maintenance system, a part of the functional analysis flow chart and the first allocation and division of function might be represented as follows.

Functional flow chart (part)

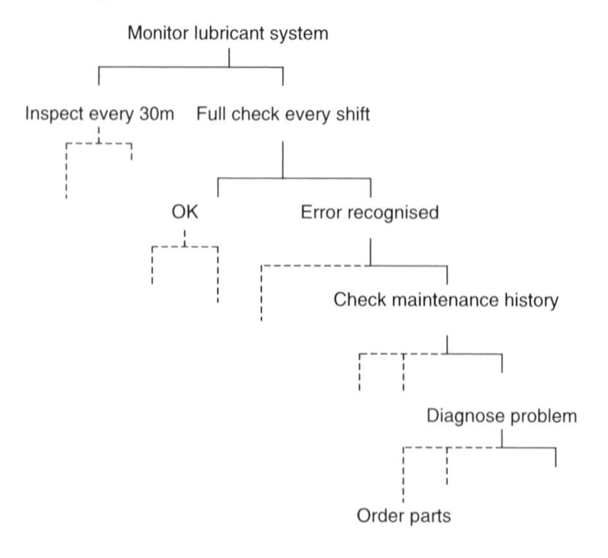

First allocation and division of functions decisions

	Human		Computer	Criterion
	Operator	Engineer		
Monitor	•			Cost
Inspect			•	Reliability
Check	•		•	Flexibility
Recognise	•			Job design
History			•	Performance
Diagnose		•		Performance
Order			•	Record keeping

Table 2 Possible Criteria for Function Allocation Decisions

Performance capability
Reliability and quality
Flexibility
Costs in implementation
Technological feasibility
Time to implement
Maintainability
Record keeping
Power requirements
Number of personnel
Safety
Good job and team design
Support requirements
Political considerations

The main problems with a PABA–MABA (or PABA–CABA for computers) approach are: the lack of use that may be made of human capabilities (because capabilities of the machine are generally more quantifiable and tangible and therefore appealing to a certain sort of engineer); the 'odd job' or 'left over' designs which may emerge (giving people only an unrelated and incoherent set of tasks; and the very unpredictability of all circumstances of system use which can make a machine/computer centred approach inflexible and maybe dangerous or expensive. See Older *et al.* (1997) for a critical review.

Interesting developments are flexible, complementary or dynamic allocation of function, whereby the extent of the role taken by the person can change according to personnel and operational conditions (e.g., Clegg *et al.*, 1989; Grote *et al.*, 1995; Hancock and Scallon, 1995). In modern systems we will find less rigidity in design and much more flexible or dynamic allocation of function. This means that responsibility for tasks or functions will depend on the goals currently in operation, the tasks being carried out by individuals or teams and the particular situation requirements. Therefore the computer or machine support can be on call, offered by the computer when it feels it is needed (using some form of learning or intelligence), or can be automatically available in all circumstances. In particular computer support will be offered or brought in when there are high loads on the operators or there are emergencies, when detailed calculations need to be carried out, when display of a checklist or a procedure would provide useful support, when providing multiple or alternative ways of displaying information would be helpful, in prediction and in support of decision making by presenting alternative courses of action. Such systems are known as adaptive systems.

A flexible and sociotechnical systems oriented method is proposed by Waterson *et al.* (2002). This differentiates allocations made on mandatory grounds (legislation, standards or technical feasibility), provisional allocations between people and between people and machines, provisional dynamic allocations, a global examination of allocations and then reporting on the proposed allocations.

Task and user analysis

For some ergonomists, task analysis is the central core of their activity. In design of some large systems, the task analysis may be the complete specification or blueprint of systems design, especially in highly proceduralised industries. In such circumstances, enormous efforts may go into delivering a very formal, usually hierarchical, task analysis. On the other hand, for smaller (product) design projects the task analysis may be a much less formal set of representations, covering only the most salient product functions (see Hahn *et al.*, 1995). The methods used to gather information to produce such different types and level of task analysis may be any covered in this book, but will at the least usually involve some form of interviewing and some form of observation.

In its original form, task analysis was principally used to help determine training programme content (e.g., Annett and Duncan 1965). Therefore one view of task analysis is that it is strictly a representation of what must be done to achieve system goals, no matter how (that is, current performance is, at most, only marginally relevant). In more recent times, task analysis has also represented how work, or human-machine systems interactions, are performed in practice. Vicente (1998), using parallels with decision making research, has very usefully distinguished normative work analysis (what must be done) and descriptive work analysis (how work is actually done); he also proposes formative work analysis as a means to move us from rich accounts of what is going on to valuable contributions in design.

Whichever task analysis approach is taken — and far more information is provided in Chapter 6 of this book and in Kirwan and Ainsworth (1994) — it is this author's contention that to be of value the task analysis must also include representation of the constraints on, and requirements of, the people involved. In other words, what are the characteristics of the system or product users (and others effected), what performance is expected of them, what limitations might they bring to total systems performance? Here we are providing a user analysis.

In the maintenance system example described above, we might now identify three main groups of users — maintenance engineers, operators and managers. Requirements from the three groups might include, respectively: enabling detailed predictions to schedule future preventive maintenance; providing information support to carry out minor repairs; and giving down-time information. They will also have many common requirements in addition. Special system needs might be to overcome scepticism and give confidence in the knowledge base amongst the engineers, to allay suspicions about grading and rosters and to give confidence in using the system amongst operators, and to retain clarity in responsibilities and provide timely information amongst managers. Relevant physical and cognitive characteristics of the users must be identified — for current and likely future stakeholders — to feed into design of any physical and information interfaces. Also, special needs — for instance for training or team design — must be identified and planning begun during user analysis.

Art gallery information system — user analysis

We have been asked to carry out the initial analysis for a new computer-based Art Gallery Information System. We might begin to gather our initial information to help with user analysis from observations carried out at the art gallery itself — what do people appear to be looking for and at, what questions are they asking of staff etc? Also we can carry out interviews of existing gallery visitors and staff, and perhaps by carrying out observations at other types of public information system, for example at a transport terminus, in a shopping centre or at a special exhibition event.

We need to identify the different groups of potential users and their use requirements. There will be professionals — art historians, artists, curators and the staff themselves, there will be the art knowledgeable general public, and there will also be the casual visitors. Therefore the type of information and the level of detail which might be wanted by different users could vary from why an artist uses certain sorts of brush strokes, to outline descriptions of paintings in a special exhibition, to the location of toilets and restaurants. We need to ask about the general characteristics of our user populations; there will probably be a wide age range, from little children (including school parties) through to the elderly, there will be both genders (although in this case this characteristic may not be relevant to design), there will be visitors with a wide range of different native languages and degrees of competence in the 'home language' (in this case English), and we certainly must design for visitors with special needs, for instance the visually impaired or wheelchair users. Given the range of users we predict, we can then determine ranges of design-related characteristics, such as physical dimensions (for instance stature for display heights or finger size for control buttons) and perceptual and cognitive capabilities for using different forms of computer interface (whether these should be mobile/personal or static, and whether touch screen, keyboard etc).

Our user and use analysis will also identify a number of special factors relevant both to constraints on the design and to requirements placed on the users. The robustness of the technology, given the multiplicity of different types of users including children, will be vital; the author has been involved in a project making virtual environment technology available to the public whereby the real problem has been the degradation — gradual and catastrophic — of the peripheral devices! Security is also an issue, whether stand alone terminals or personal technology is provided. Maintenance and cleaning will be relevant, particularly if the technology is worn personally or if touch screens are used. Finally there is the interesting question of how both to motivate users to make use of the system but at the same time ensure that they do not block the terminals for long periods and prevent use by other users.

The art gallery example should identify for the reader one other aspect of analysis required in the task and user analysis stage. This is to predict and examine the likely impact of the settings and environments of use of the new system. What will be the influence of environmental conditions of lighting, climate, noise etc? Will there be effects of pressures — of time, of the presence of other people or of task stressors such as needs for safe or effective performance. For instance, in the art gallery we need to understand the physical space available and lighting conditions if terminals are to be placed in a foyer area or at the entrance to different galleries. In the earlier maintenance example then there may be impact on use of the system from all the environmental conditions of lighting, noise, climate and space; there will also be performance and time pressures to consider and also any group or team effects.

A case example

To illustrate some of the stages in the early parts of the development process, especially those to do with function analysis, take the case of a new airline check-in desk. To begin with, in turn we would want to set requirements, identify constraints, and then identify some top level options for the operational concepts of the system. These should be prioritised in the light of technical, organisational, performance, economic and human factors goals.

First of all we might identify systems goals and requirements as:

- Associate passenger, ticket, flight, seat, baggage
- Security — 100% is the target
- Accuracy — 1% error is acceptable
- Speed — process 18 passengers per hour
- Provide a better job — in terms of measures of satisfaction, health, staff retention
- Show increase in customer satisfaction of 5% per annum

We have to review these requirements, and assess overall design options, knowing all possible constraints. These constraints may include:

- Security needs must be paramount.
- Passengers moving around the terminal may be in conflict with passengers queuing to check in.
- Check-in desks will be adjacent to each other, and the work of different agents or needs of different groups of passengers might conflict.
- There will be a proportion of passengers who arrive with particular problems, including original booking errors, missed flights, no identity papers, overweight baggage, etc.
- Baggage as well as passengers must be processed at the same point and time.
- Passengers will be of all ages and cultures, with different and often special needs, and may be under stress.

A number of options will be defined for design of the system and for its operational concept. At a very top level, the operational options as regards allocation of function might be:

- An agent carries out all customer service, using a computer to produce the passenger — seat — baggage — flight links.
- Two stage check-in, with baggage checked by agent and self-seating allocated by passenger with public terminal.
- Entirely automated check-in, baggage and passenger seating processed by computer with original entry by passenger.

Assuming that the first option is chosen, and so the system is to be a more effective version of the current operation, we then need to identify and analyse all users. The key stakeholders are check-in agents and passengers, for whom we need to know demographic profiles, range of physical and psychological characteristics, and to understand the likely incidence of special needs and special problems. Interviewing will establish staff concerns and ideas for improvements, and also passenger perspectives. Other stakeholders whose requirements and limitations must be known about are supervisors, baggage handlers (for access across conveyors) and maintenance engineers (for work on hard equipment and computer systems software). A participatory process should be established, with structural input from main stakeholder groups informing decisions on physical layout, task design and sequencing, interface designs etc.

A function flow chart can be produced to help understand what may happen in the system and what human factors are relevant. Part of such a flow chart is:

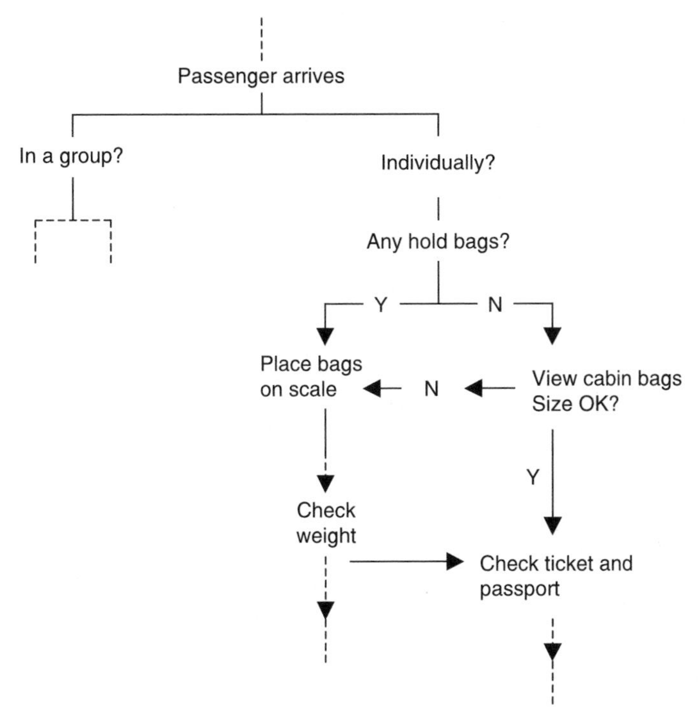

From this small fragment we will identify design factors of relevance as including sightlines for the agent to see cabin bags, horizontal and vertical lift distances to place hold bags on the scale, differential counter heights for passenger and agent to both feel at ease, usability of security software to help passport check, and so on.

The allocation and division of function will start now. Given the gross functions identified, for each we make a judgement of whether it is carried out or controlled by an individual or computer or both, using criteria similar to those of Table 1. Part of this structured allocation might look like:

Function	Agent	Passenger	Computer	Criteria for choice
Place bags on scale		•		Cost, safety
Record weight			•	Performance
Security questions	•	•		Communications
Check ticket	•			Flexibility
Check validity			•	Performance
Decide seating etc.	•	•	•	Good job, good service

At this stage it might be appropriate to produce an input/output diagram, to start to define communication flows, and system inter-connections. This will be particularly useful for the system designers, and especially the software team, to be able to start to define processing requirements as well as inputs and outputs. Again, part of an example is:

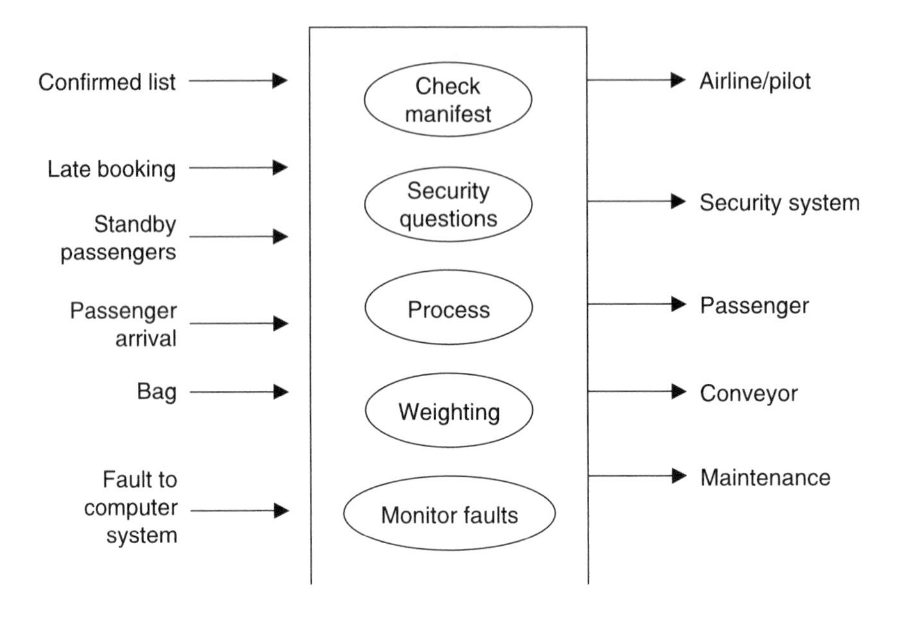

Another way of describing the system and system concepts, especially of value to ensure the needs of all stakeholders are accounted for, is to use a rich picture representation, developed as part of soft systems methodology (see Checkland, 1981, Checkland and Scholes, 1990, and Chapter 38). Part of an example is:

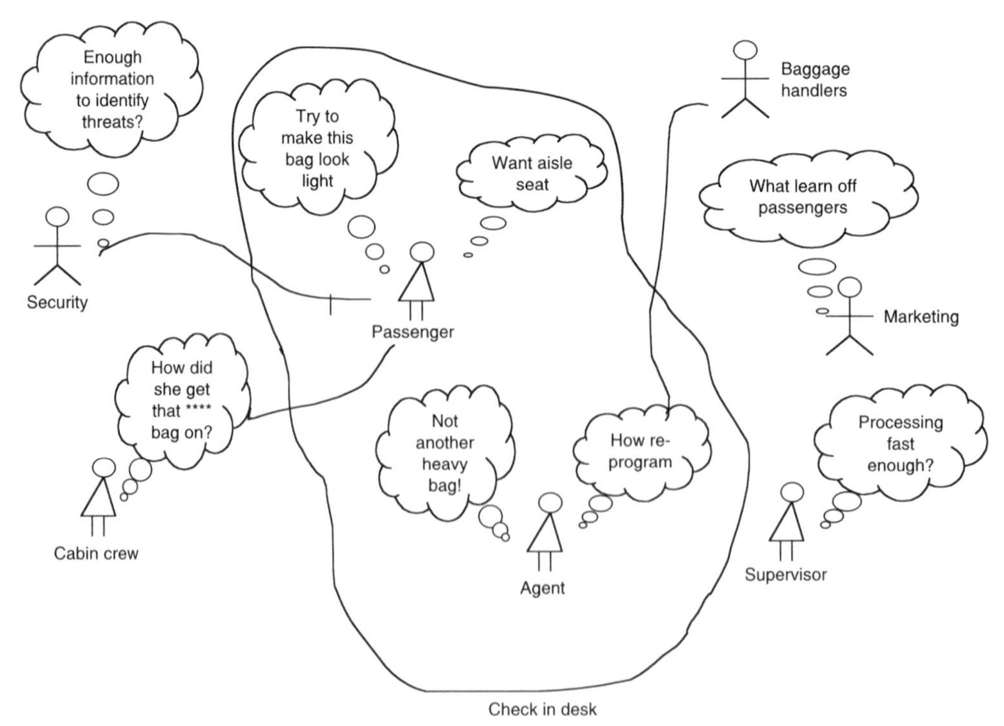

From all this preparatory work on function analysis — sometimes afterwards, sometimes in parallel as we saw above — the task analysis will proceed, making the human factors issues and choices more explicit. Flow chart forms and tabular forms are used most often. The former will provide an idea of task sequencing and inter-dependency, and the hierarchical version most often used will make priorities, causal influences and plans explicit. The tabular version will permit annotation of criticalities, frequencies, failure probabilities, settings, and human factors requirements and constraints (e.g., for information processing) to each task.

Ergonomics and systems engineering

There are a number of challenges that face the deployment of ergonomics in development projects. The first risk is that it is not carried out at all, a risk reduced by company standards requiring human factors integration. The second risk is that even when a human factors integration plan (HFIP) is agreed and resourced, the products of the work fail to make a real impact on the system. To be specific, the failure is in not shaping and influencing the design decisions that have been made for the system.

The role and activities of the systems engineering function and those of the ergonomists bear a striking resemblance. This, and the proposition that the ergonomics work is concerned with shaping and influencing design decisions, leads to a compelling need for the ergonomics work to be undertaken within or in close collaboration with, systems engineering.

If ergonomics is part of the systems engineering function, then it needs to be able to communicate successfully with the other members of the team. Ergonomists, in the main, use techniques and methods whose outputs are not mainstream and not always easily assimilated by non-ergonomists.

For many years, systems engineering has been very vigorous in defining and developing tools and methods for modelling systems, and there is a good level of maturity and standardisation in the approaches. In particular, there is widespread adoption of UML (Unified Modelling Language) developed by Booch, Rumbaugh and Jacobson (Booch *et al.*, 1998; Fowler, 2003).

Models are created to be a simplification or shorthand representation of the existing (or future real) world. There is a danger in 'modelling for modelling's sake', i.e., a speculative approach to 'define everything and see what comes out'. It is wishful thinking to assume that someone else's model of the system will address the specific questions of others. If we as ergonomists are to use UML models, we must choose and set up the models to investigate the properties of the system in which we are interested. That is, the information and evidence that we require determines the objective, scope and type of modelling activity that we undertake; this ensures that what is modelled will help us formulate and choose between design solutions.

There are broadly two types of models: static models and dynamic models. From an ergonomics viewpoint, static models can be used to represent the conceptual model that users will need to learn, think and operate the system, and dynamic models are used to describe the logical and temporal elements of the system such as work flow, user–system interaction, system state transitions, navigation models etc.

Static models

Conceptual modelling is an important approach used by ergonomists. This involves the definition of the conceptual objects and the actions that can be performed on them. For instance, a simple railway conceptual model may contain the following conceptual objects: train, path, route, signal, point, track, territory, signaller. Further for each conceptual

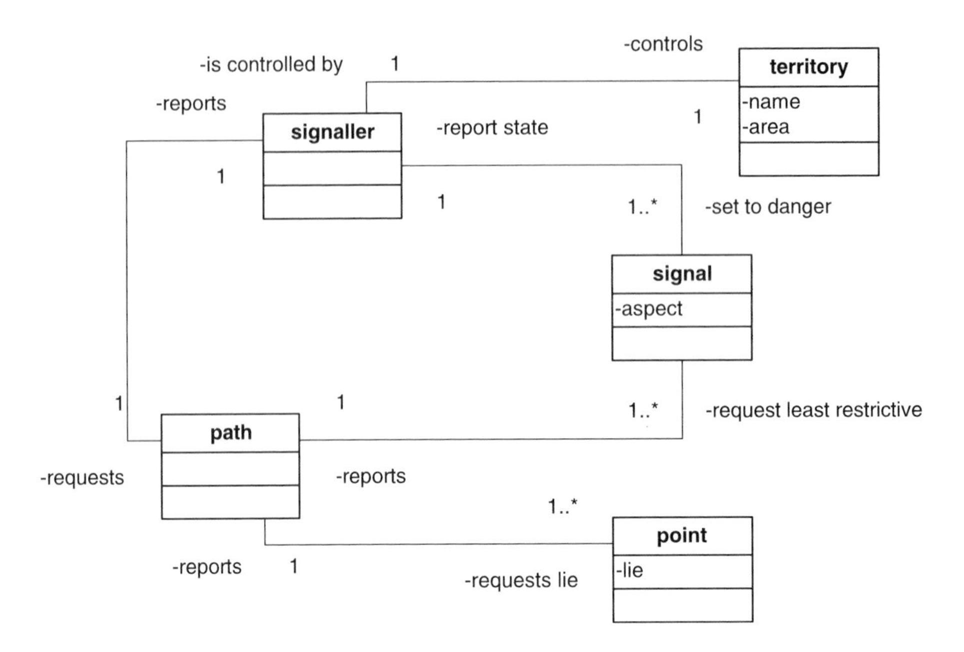

Figure 12 Example class diagram.

object, we can describe the actions that can be preformed, for instance, set point. In a similar way we could use UML Class diagrams to represent the same key entities in the system.

Figure 12 is a Class diagram that represents five entities. The diagram also describes the relationships or associations between the signaller, path, point, signal and territory entities. The object (actor) signaller controls an area (territory), can request a path to be set and set signal to danger. The path object has been modelled to request the relevant signal and point to be set to achieve the path specified by the signaller. Finally, signal and point can either implement or reject the path setting request. There is feedback to the signaller regarding the success of the path setting request.

Also of interest is the multiplicity (or cardinality) of the relationships. In the figure there is a rule stated that one territory has one signaller and vice versa. If this relationship was to be changed so that a signaller could control more than one territory the cardinality will change to 'one or more' (denoted as 1..*).

If these Class diagrams are produced, and if ergonomists are unable to interpret and understand the implications of these condensed representations, the status and relevance of the ergonomist's conceptual model (objects and actions) prepared earlier must be in doubt. Therefore it is important the conceptual modelling (if needed) agrees with the static class diagrams.

Finally, the example model above is extremely conceptual. Systems and software engineers will want to produce much more computationally detailed models. If the top level models and definitions are agreed, there is a good chance that the more detailed implementation level modelling will also be in alignment. This is why involvement in the early, top level modelling really pays off.

Dynamic models

In the area of dynamic models, two techniques are particularly suited for task definition: sequence diagrams and state transition diagrams.

Table 3 Candidate Scenarios for Development

1000	Signaller makes point-to-point voice call to Train Driver
0110	Driver makes call to signaller
0120	Driver and signaller communicate
0130	Signaller ends call to driver
0140	Driver ends call to signaller
0150	Start up, registration, change mode, de-register and close down operations
0200	Emergency calls
0300	Priority and pre-emption calls
0400	Radio functions (e.g. hold, transfer)
0500	Formulation and receipt of text messages
0600	Targeting signallers
0700	Light duty working and combination and splitting territories
0800	SMS
0900	Group, multi party and multi-driver calls
1000	Failure modes and fallback behaviours
1100	Shunting operations

The starting point for sequence diagrams is the definition of scenarios that are typical and/or critical to the operation of the system. Table 3 lists a candidate scenario for development.

Once a set of operational scenarios that cover the range of operational conditions that may exist in the system have been determined and agreed (e.g., normal, disturbed/perturbed, emergency and migration), each can be described in the form of sequence diagrams. Once the actors and system entities have been defined, a time based activity line can be created as shown in Figure 13.

In the example above there are two human actors at the left and right hand extremes. In the central area there are system objects: the signaller radio and the driver radio (including the user interface for both — there is a note to indicate the multiple methods for initiating a call), and the object in the centre (called RT Manager) provides the infrastructure for the call.

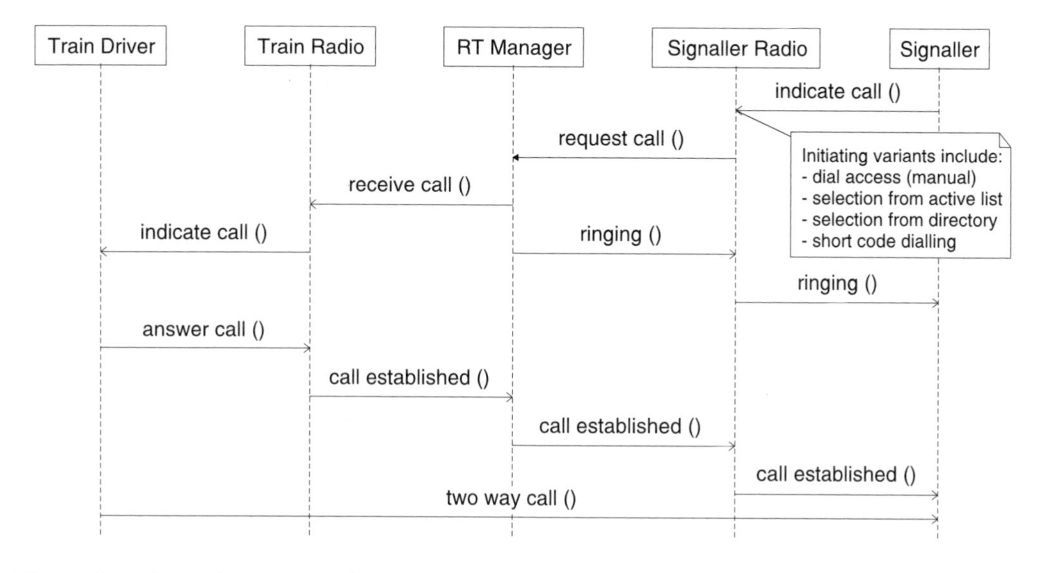

Figure 13 Example sequence diagram.

These sequence diagrams can be constructed with operational users and technical experts. Both are needed to make sure that the system is meeting the operational requirements and that the requirements are technically achievable.

Once the sequence diagram has been constructed, there are number of insights possible. Taking the example above, we can ask or observe:

1. What is heard and seen by the signaller as the call progresses through the various systems? Are these indications consistent with user expectations?
2. The ringing indication for the signaller: The ringing tone does not indicate that the call is ringing at the driver's radio, just that the call has been requested by the RT Manager. There is a likely failure case where the radio does not respond to the request (new scenario).
3. Once the driver has answered the call there are three operations ('call established') prior to the channel for two way communication being opened. Does this appear as instantaneous to both parties? Are there failure modes that can occur and how would the user detect them?

In all systems there are system states that will affect the user. Depending on the reason for the state changes, it is likely that the user will be exposed to different methods of operation in the various states. Figure 14 shows an example state transition diagram of how cab radio modes are moved between by the user.

There is often a high degree of complexity encountered in innocuous user operations such as logging on and off, selecting user roles and user account management. In the example above, the transitions available to a user are modelled. The states are shown in

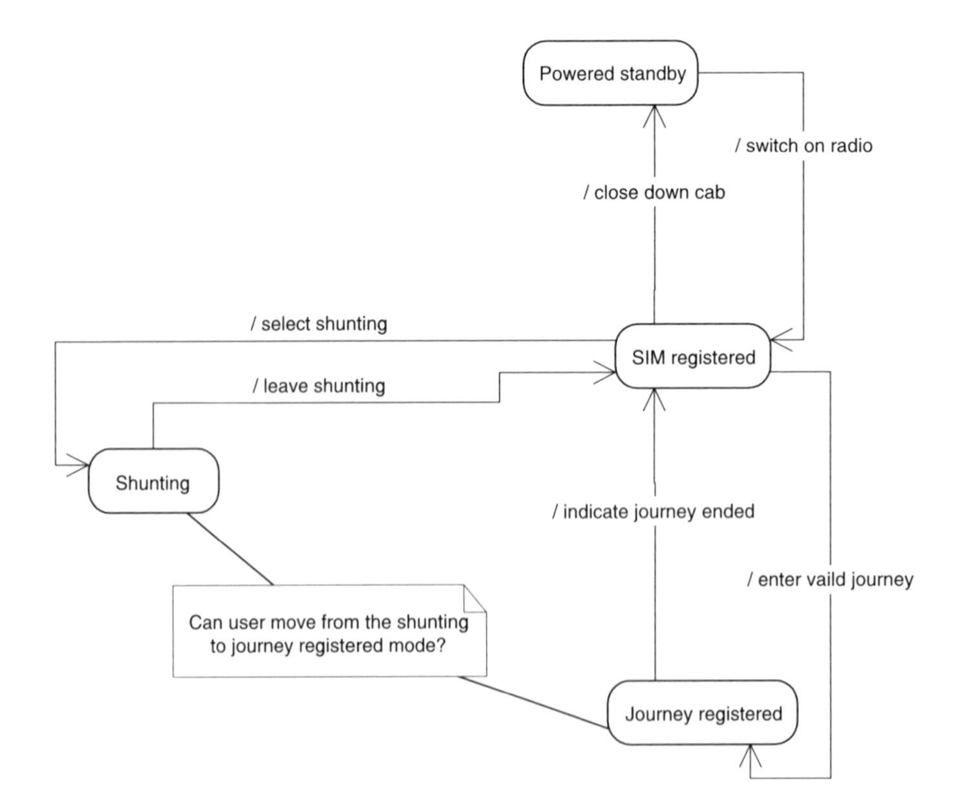

Figure 14 Example state transition diagram.

the round cornered boxes and the operation (by the user in this case) is shown on the arrowed line.

The radio in its 'off' state is actually active even though the HMI is not powered up. When the driver turns the radio on, the radio HMI is powered up and the radio declares itself active to the network. On completion of this transition the radio has a subset of the full radio functionality available. Similarly when the driver selects the Shunting or Journey Registered states, the appearance and functionality will change. The note reflects the debate about entering an active Journey Registered state from Shunting.

Human factors integration and plans

Because people are involved in almost every aspect of any new systems, and therefore human factors is relevant to all stages and parts of the systems design, in recent years there has been development of understanding and good practice in human factors integration. In projects this is carried through in terms of Human Factors Integration Plans, sometimes supported by a standard. The rationale is that for human factors to be adequately addressed it is essential that it is managed as integrated as part of the whole project rather than as something separate, oppositional and alien. As a consequence, human factors integration plans are aimed as much at project managers, design and construction engineers as at human factors experts.

A human factors integration plan (HFIP) defines the integration of human factors into system development, and also defines relevant assurance procedures to monitor and sign off all the activities needed to support such integration.

Human factors integration and HFIPs may be developed and applied just to particular projects or system designs, but more usually will be a generic document produced by an organisation or group of organisations, which can be operationalised for any particular project or system.

The content of a HFIP will usually include:

1. Identification of those responsible for particular aspects of the human factors systems development, and the arrangements made organisationally including lines of reporting.
2. Providing for end user representation, how stakeholders will be consulted and preferably enabled to participate in the project or development.
3. Showing how there will be continuation of availability of expertise, how the target company and suppliers will have human factors expertise throughout the system development life cycle.
4. Co-ordination of human factors delivery across all the organisations involved in the project, including sub-contractors, and how decisions will be made, priorities set and unresolved issues settled.
5. Understanding relevant regulations for the project and how human factors considerations will be incorporated into the consultation process with the regulators.
6. Identification of operational concepts, their content, and how and when these will be produced.
7. The analyses required of any existing or related system (technical or organisational) in order to identify and assess key aspects of people, processes, equipment, performance and tasks.
8. Definitions of measurement methods and criteria in order that outcomes of performance and well being can be measured.

9. Listing of relevant human factors standards and guidelines, from the particular domain and from elsewhere.
10. The management of contractual arrangements so that these fit with the requirements of the overall project.
11. Definition of detailed requirements for human performance, including reliability, consequent loading on people (including workload, situation awareness etc.), and risk of injury or ill health.
12. Detailed criteria and methods for the evaluation of the operability and performance of the total system and its design.
13. A time line plan for formative and summative evaluation of the whole system and any coherent sub-systems.
14. Support requirements for documentation, procedures and training throughout the project.
15. Initial plans for human factors testing of the system prior to and during the commissioning process.
16. Provision for monitoring of the performance of the system once it has been commissioned and is in full operation.
17. Full description of terms and abbreviations used.
18. Reference to further sources useful to the different parties using and drawing from the HFIP.

One of the difficulties of Human Factors Integration Plans and Standards is, to paraphrase the words of a number of design engineers and operational staff that the author has worked with, that it may be a very worthy document but the document itself, and particularly putting it into practice, can be akin to 'pouring concrete into the veins of creative design'. In words from elsewhere, from the head of an ergonomics group at a UK organisation, the last thing that the human factors team, the engineers they work with or the project managers want to do is have a 600 page search for one or two key aspects.

In order to illustrate, at a simple level, a Human Factors Integration Plan, one is included with permission of Network Rail — in the appendix to this chapter. It is an example of a simple and short HFIP, proposed to ensure its acceptance and use.

Final thoughts

Of course, if we just mechanically follow the advice in this chapter, or indeed in this book as a whole, we will not necessarily produce a good systems design. Nor will the guidance even ensure that all human factors are accounted for during development. The ergonomist and the ergonomics team will often use the craft of ergonomics as much as its science in supporting, sometimes leading, systems development projects. But good ergonomics design will require well-informed choices to be made — about functions, the role of people, and the requirements and consequences for them, and for these choices to be made in a systematic manner. Carrying out requirements specification, intelligence gathering, function analysis, task analysis and user analysis, and where warranted implementing a human factors integration plan, will improve the human factors in systems development and in the design of human-machine systems.

Systems engineering goals and products are very similar to those of the project ergonomist. In addition, systems engineering have developed and standardised on a core set of tools. There are good reasons for ergonomists to use these tools.

However, it should be remembered that analysis and modelling is not ***design***. Any analysis method or modelling tool selected must provide evidence that is useful in making design decisions.

References

Aikin, C., Rollings, M. and Wilson, J.R. (1994). Providing a foundation for ergonomics: Systematic Ergonomics in Engineering Design (SEED). *Proceedings of the 12th Congress of the International Ergonomics Association, Toronto*, **5**, 276–278.

Annet, J. and Dunan, K.D. (1967). Task analysis and training design. *Occupational Psychology*, **41**, 211–221.

Bailey, R.W. (1989). *Human Performance Engineering: Using Human Factors/Ergonomics to achieve Computer System Usability*, 2nd edition (Englewood Cliffs, NJ: Prentice Hall).

Booch, G., Jacobson, I. and Rumbaugh, J. (1998). *Unified Modeling Language User Guide.* (Addison-Wesley).

Chapanis, A. (1996). *Human Factors in Systems Engineering.* (New York: J. Wiley).

Checkland, P.B. (1981). *Systems Thinking, Systems Practice.* (Chichester: J. Wiley).

Checkland, P.B. and Scholes, J. (1990). *Soft Systems Methodology in Action.* (Chichester: J. Wiley).

Cherns, A. (1987). Principles of sociotechnical design revisited. *Human Relations*, **40**, 153–162.

Clegg, C., Ravden, S., Corbett, M. and Johnson, G. (1989). Allocating functions in computer integrated manufacturing: a review and a new method. *Behaviour and Information Technology*, **8**, 175–190.

Clegg, C.W. (2000). Sociotechnical principles for systems design. *Applied Ergonomics*, **40**, 463–478.

Fowler, M. (2003). *UML Distilled: A Brief Guide to the Standard Object Modeling Language*, 3rd Edition. (Addison-Wesley).

Fuld, R.B. (1993). The fiction of function allocation. *Ergonomics in Design*, **January**, 20–24.

Grote, G., Weik, S., Wäfler, T. and Zölch, M. (1995). Criteria for the complementary allocation of functions in automated work systems and their use in simultaneous engineering projects. *International Journal of Industrial Ergonomics*, **16**, 367–382.

Hahn, H.A., Houghton, F.K. and Youngblood, A. (1995). Job-task analysis: Which way? *Ergonomics in Design*, **October**, 22–28.

Hancock, P.A. and Scallen, S.F. (1996). The future of function allocation. *Ergonomics in Design*, **October**, 24–29.

Hollnagel, E. (ed.) (2003). *Handbook of Cognitive Task Design.* (Mahwah, New Jersey: L. Erlbaum).

Jordan, N. (1963). Allocation of functions between man and machines in automated systems. *Journal of Applied Psychology*, **47**, 161–165.

Kirwan, B. and Ainsworth, L. (1992). *A Guide to Task Analysis.* (London: Taylor & Francis).

Kragt, H. (1992). *Enhancing Industrial Performance* (London: Taylor & Francis).

Lepreux, S., Abed, M. and Kolski, C. (2003). A human-centred methodology applied to design support systems design and evaluation in a railway network context. *Cognition, Technology and Work*, **5**, 248–271.

Markus, T.A. (1969). The role of building performance measurement and appraisal in design method. In: G. Broadbent and A. Ward (eds). *Design Methods in Architecture.* (London: Lund Humphries).

Meister, D. (1985). *Behavioural Analysis and Measurement Methods.* (New York: John Wiley).

Meister, D. and Enderwick, T.P. (2002). *Human Factors in System Design, Development and Testing.* (London: L. Erlbaum).

O'Brien, T.G. and Charlton, S.G. (1996). *Handbook of Human Factors Testing and Evaluation.* (Mahwah, New Jersey: L. Erlbaum).

Older, M.T., Waterson, P.E. and Clegg, C.W. (1997). A critical assessment of task allocation methods and their applicability. *Ergonomics*, **40**, 151–171.

Price, H. (1985). The allocation of functions in systems. *Human Factors*, **27**, 33–45.

Sanders, M.S. and McCormick, E.J. (1987). *Human Factors in Engineering and Design*, 6th edition (New York: McGraw-Hill).

Sharit, J. (1997). Allocation of functions. In: G. Salvendy (ed.) *Handbook of Human Factors and Ergonomics.* (New York: J. Wiley), pp. 301–339.

Sheridan, T.B. (2000). Function allocation: Algorithm, alchemy or apostasy. *International Journal of Human-Computer Studies*, **52**, 203–216.

Sheridan, T.B. (2002). *Humans and Automation: Systems Design and Research Issues.* (New York: J. Wiley).
Singleton, W.T. (1974). *Man-Machine Systems.* (Harmondsworth: Penguin).
Tudor, D.J. and Tudor, I.J. (1995). *Systems Analysis and Design: A Comparison of Structured Methods.* (Oxford: NCC Blackwell).
Vicente, K. (1998). *Cognitive Work Analysis.* (Mahwah, New Jersey: L. Erlbaum).
Viller, S., Bowers, J. and Rodden, T. (1999). Human factors in requirements engineering: A survey of human sciences literature relevant to the improvement of dependable systems development processes. *Interacting with Computers*, **11**, 665–698.
Waterson, P.E., Older Gray, M.T. and Clegg, C.W. (2002). A sociotechnical method for designing work systems. *Human Factors*, **44**, 376–391.
Wickens, C.D., Gordon, S.E. and Liu, Y. (1998). *An Introduction to Human Factors Engineering.* (New York: Longman).
Wilson, J.R., Eastgate, R.M. and D'Cruz, M.D. (2002). Structured development of virtual environments. In: K. Stanney (ed.), *Handbook of Virtual Environments.* (London: Lawrence Erlbaum Associates), pp. 353–378.

Appendix Ergonomics Integration Plan for the Telecommunications Engineering Centre

Network Rail, London, UK

Version/issue record

Filename	Version/Issue	Date	Comments
ErgsTECPlanV01a.doc	0.1a	04/07/2003	Version for pre-issue review
ErgsTECPlanV10.doc	1.0	11/09/2003	First Issue

Executive Summary

The TEC (Telecommunications Engineering Control) is a new facility and will perform the task of control of the major telecommunications systems used by Network Rail. The major elements of the TEC will be the control of the new FTN and GSM-R telecommunications systems, and the existing Ops. Comms network. In addition there may be a number of legacy systems that will be moved to and controlled by the TEC.

This Ergonomics Integration Plan (EIP) addresses the Ergonomics or Human Factors activities that are needed to ensure that the new TEC performs its intended role in an effective manner.

Strategy for Implementing the TEC

The strategy for the implementation of the TEC has many dynamic elements, including:

1. Phased approach to the control tasks being undertaken
2. Phased approach to the people carrying out the work
3. Flexibility of the operational staff to carry out a range of tasks for a range of systems
4. Flexibility of the control room layout to accommodate changing roles and systems

Phased approach to the control tasks being undertaken in the TEC

Tasks carried out by the operators in the TEC will change over time. The initial tasks that are likely to be carried out in the TEC are the commissioning activities as part of the networks being brought under control.

Phased approach to the people carrying out the work in the TEC

The initial staff will be specialists in building and expanding FTN and GSM-R networks. The responsibilities of the TEC will develop over time to include a wider range of skill sets, involving maintenance, support, and the Help Desk.

Flexibility of the operational staff to carry out a range of tasks for a range of systems in the TEC

The initial tasks will be undertaken primarily during the day. The tasks performed in the fully implemented TEC will involve comprehensive operations, administration, and maintenance (O, A & M) of the networks, technical support, and customer facing functions. Many of these activities will be carried out during continuous (24/7) shift patterns.

Flexibility of the control room layout to accommodate changing roles and systems in the TEC

The control room will be developed from a basic (interim) facility, to one that will be similar to the control rooms of the major telecommunications operators. This modern accommodation will provide a pleasant and comfortable working environment for the staff.

Overview of ergonomics activities

The wide scope of the TEC requires that the scope of the ergonomics activities be wide also. The diagram below illustrates the three main categories of ergonomics activities.

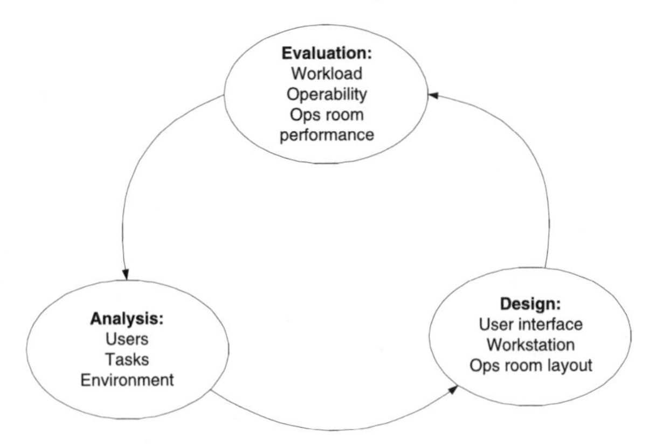

In the course of this programme a range of activities will need to be undertaken and these are described in the following sections.

Analysis

The key ergonomics activities comprise the analysis of:

1. The users of the system and their roles and responsibilities, and their characteristics and competencies
2. Tasks that the users carry out

3. The environment in which the tasks are carried out, including the social, organisation and physical context

The requirements that are derived from the ergonomics analysis are used to influence the project design decisions.

Design

The key areas of design are:

A. Workstation design
B. Job design and workflow
C. User interface design
D. Control room layout
E. Tools and user support

Evaluation

A number of key ergonomics evaluations will be required including:

I. Workload assessment and manning levels
II. Operability review and assessment
III. Control room performance assessment

Ergonomics Analysis

In order to make informed decisions about the design of the TEC, a number of analysis activities are required to be carried out. These analysis data are not an end in themselves and serve to enlighten decision makers about the implications of design options on the operability of the system.

The ergonomics analysis relies heavily on analysis and assumptions made elsewhere in the project and the outputs of the main project activities will be required as input to the ergonomics activities. For instance, a key early input to the ergonomics activities is a Concept of Operations for the TEC and the strategy for its implementation and operation.

There follows in the table a list of analysis activities that will be carried out in the course of this programme.

Activity	Description	Inputs and outputs
A1. Define functional goals of the system	Definition of goals is undertaken based on the Concept of Operations. The whole of the system goals are defined at a high level. The important aspect here is to understand the system boundary: what is in scope of the system and what is not. The definition of the in-scope goals need to be comprehensive and the out-of-scope goals need to be complementary (i.e. allow in-scope activities to take place and avoid duplication which may lead to ambiguity and conflict)	Input: — Project Concept of Operation Output: — Documented Scope of System and what it offers to provide in terms of goal and responsibilities. — Risk register of known project risks and tests of assumptions that need to be validated.

Activity	Description	Inputs and outputs
A2. Define user tasks and role responsibilities	Once functions have been allocated the user task descriptions can be defined in more detail. An overall task analysis and lists of typical and critical operational scenarios that involve the user are defined.	Input: — Project Concept of Operation and Functional Goals documents — Partition of user goals and operations Output: — documented overall task analysis — lists of typical and critical operational scenarios.
A3. Define user characteristics and competencies required	The list of user types needs to made explicit. The allocation of the tasks to individual users requires definition and their roles and responsibilities defined. The characteristics of these users that are relevant to tasks being carried out should be described.	Input: — all products above Output: — list of user types — definition of roles and responsibilities — definition of user characteristics and competencies
A4. Define interactions within and between roles	Analyse and understand the interactions between the various user groups and define the key operations that are critical for the success of the overall system.	Input: — all products above Output: — Information flows between users — information types and medium — link analysis
A5. Define information requirements for the user and necessary tools/support	Using the task descriptions and design options the information requirements of the type of user can be defined. These information requirements may need to be system based or paper based tools or they may be addressed as training requirements for the users.	Input: — all products above Output: — System support — Paper based support — Communications support — Training Requirements
A6. Define Workstation Assets	There needs to be a definition of all the equipment and materials that are required by users carrying out a particular role.	Input: — all products above Output: — List of assets required at a workstation
A7. Define Operability Targets	This activity considers the performance objectives, usability and operability of the system used by the various user roles. The targets are identified as throughput, speed and accuracy (error) of key operations. Acquisition of learning tasks may also be used as operability criteria.	Input: — all products above Output: — list of key operability tasks — error likelihood estimation — specification of performance targets

(*Continued*)

Activity	Description	Inputs and outputs
A8. Define workload assessment and derive staffing complement	User tasks will be defined and workload will be predicted in order to identify the areas and operations that may be susceptible to operator overload. If deemed necessary, the ATLAS workload prediction tools will be used to confirm the planned staff complement.	Input: — task description and analysis — user and role definitions — operability target descriptions Output: — detailed task description — results of ATLAS workload assessment tool — workload prediction report and staff complement definiton
A9. Training Needs Analysis	Arising from the task analyses, gaps in skills and knowledge for the user roles are identified and become training requirements. The TNA lists the degree of user support and training needed so that it acan be addressed in the Training and Transition Plans.	Input: — task descriptions — user definitions. Output: — training needs

Ergonomics Design

Ergonomics design activities will concentrate on the aspects of the TEC that are open to design specification. Whereas the computer based COTS products are not expected to require design input, the workstation design and control room layout will be a major design activity.

Ergonomics Evaluation

Ergonomics evaluation by review of requirements and design will be undertaken. In addition, opportunities to evaluate designs by trials with user representatives will be sought as the programme develops.

Because the TEC commences with a small group of staff and incrementally increases, there will be opportunities to evaluate the operation at a number of points using the real and actual user population. This will allow the monitoring and correction of design that is deemed to be deficient.

Opportunities for ergonomics evaluation using trials and prototypes will be sought to address areas of ergonomics risk.

Other Activities

The ergonomics activities are not carried out in isolation. For the ergonomics activities to progress there needs to be input from key sources. Similarly, the ergonomics findings and recommendations need to be acted upon by other parties within the project. In the table below, related activities are described.

Technical approach

In the preceding sections the range of ergonomics activities that are needed to be addressed in the course of the design, implementation and commissioning of the TEC are described *in toto*.

In this section the activities that are planned for the next 6 months and the technical approach being taken to achieving the initial phase of the ergonomics work is described.

Activity	Description	Inputs and outputs
D1. Allocate functions between system and user role	Here is the first design decision to be considered – who does what. The main inputs to this are the statement of system goals and the user definition; it considers how these goals are to be allocated to user roles. In reality, we expect that the division of work between users (network operators and supporting staff) is a common enough model in the telecommunications area, therefore we will adopt and review the industry models of operation rather than creating a novel arrangement.	Input: — Project Concept of Operation and Functional Goals documents — User definitions — Understanding of shift patterns and light duty working — Understanding of good telecom network management operations in the industry Output: — Documented partition of the goals of the system to relevant user roles.
D2. Job design	The preferred model of operation need to be evaluated from a ergonomics perspective to understand the implications for job design. In particular, this design activity will identify informationrequirements of users and identify the need system support and job aids. Preliminary high level procedures can be identified at this stage.	Input: — Task Analysis — User definition — Work flow — Information requirements analysis — Communication and link analysis Output: — Documented roles and responsibilities — Interactions between roles — Interactions between individuals holding the same role
D3. Define Workstation Design options	The workstation assets used by each user role will be taken to determine the number of workstation types. There will be a constraint to keep the workstation or work position as generic as possible and this activity will indicate the degree to which this can be achieved. From the outset there is a requirement for workstations to be modular and able to be flexibly grouped. Other considerations will be the interaction with the room and the need for task lighting at each workstation.	Input: — all products above — site plans and areas — lighting plan and specifications Output: — Workstation design options and trade-offs

(*Continued*)

Activity	Description	Inputs and outputs
D4. Define Control Room(s) layout	In collaboration with the project building refurbishment consultants, options for the layout of the workstations and accessories will be reviewed. The analysis of user roles, workflow, and communications requirements will be employed to evaluate the trade-offs for layout options. Finally, the implications for lighting, HVAC, access, cabling and services will be reviewed.	Input: — Requirements for co-location of users/roles — Workflow — Communications and link analysis — Site plans and areas — HVAC plans — Lighting plans — Cable and services run plans Output: — Review of control room layout
D5. User Interface Design	It is not expected the COTS products will be modifiable and subject to change. The areas that will need to be designed are the data driven displays specific to this network. In addition, the way in which the alarms and attention getting devices are implemented in the control room will be considered. These will be reviewed and inspected from a usability perspective, using user representative input.	Input: — data determined display designs Output: — review of user interface design — view on acceptability of design — review of alarm and alerts policy
D6. User Support	User requirements for support and information will have been addressed in the analysis activities. These requirements will be met by the design of job aids, aides memoires, working instructions and information sources.	Input: — all products above Output: — Design of job aids

In the course of development of the TEC there will be a series of operational states that are followed by transitions to a new state (shown as Interim states X/Y). In this initial plan, the approach proposed is to target and address the 'Initial State' and the end state (TEC goes Operational) are in the course of the next six months of work. Only where necessary is the Interim State X addressed. This is shown in the following tables.

Definitions of the terms in the table are:

- Final — the analysis, design and evaluation tasks are complete.
- Draft — the analysis, design and evaluation tasks have been commenced but are not complete.
- Initial Draft — analysis, design and evaluation has progressed to an early stage but is 'work in progress' and may have incomplete or missing sections.

In summary, the objective of the initial phase is to commence the ergonomics activities that will:

Activity	Description	Inputs and outputs
E1. Reviewing activities	Specification and design documents will be reviewed from an ergonomics perspective. Areas that contain ergonomics risk will be documented in the risk register.	Input: — all relevant specification and design documents Output: — Review comments — Risk register
E2. Evaluation of steady state operations	During the phases when the TEC is in a steady state (i.e. between transitions) staff and the operation will be surveyed and inspected for ergonomics issues that may need corrective action in the following phase.	Input: — steady state operation Output: — issues and proposed remedy or solutions
E3. Trials and prototypes	In order to reduce areas of risk that are identified in the course of the TEC programme, we will conduct trials to assess the source and extent of the risk. These might include workstation trials, lighting trials, paper handling trials, communication trials and walk-throughs, transfer of responsibility walk-throughs, failure and incident trials, recovery trials, etc.	Input: — detection of programme risk Output: — trials results and recommendations
E4. Formal Management and DAG approval	The products of this plan will be offered for management and DAG approval. Deficiencies and review comments will be considered and actions for their disposition carried out.	Input: — products of the ergonomics programme Output: — comments and their disposition — updated documents

- Capture the ergonomics aspects of the 'Initial State' system
- Provide early feedback on the operation of the 'Initial State' system
- Select a modular workstation design
- Evaluate control room layout design
- Understand the lighting, heat, ventilation and services requirements of the control room layout
- Construct an preliminary ergonomics view of the operational TEC in terms of users and tasks

Schedule

The objective of the initial phase is to commence the ergonomics activities that will:

- Capture the ergonomics aspects of the 'Initial State' system
- Provide early feedback on the operation of the 'Initial State' system
- Select a modular workstation design

Activity	Description	Inputs and outputs
O1. Identify Hazards and essential PRAMS requirements	Early examination of hazards that may arise with the use of these systems. It is expected that the programme as a whole will be designated as being safety related or not in accordance with the ESM process. Similarly, the PRAMs requirements impact on the user needs to be examined. It is assumed that this will be planned as an overall project activity with ergonomics input as required.	Input: — all products above Output: — initial list of hazards (if any identified) — user related PRAMs requirements — PRAMs constraints on the user
O2. Review of relevant standards	There are a number of general and industry based standards that need to be considered and applied to the design of the operation. The relevant standards need to be identified and reviewed. Mandatory Requirements that apply to this area will be flagged.	Input: — all relevant international, railway group and company standards Output: — selected standards and precedence — non-applicable sections — derogations and waivers.
O3. Liaise with key functional areas, departmental and user representatives	These include Telecom and Systems Engineers, Systems Designers, User representatives, Subject Matter Experts, Installation, HR and Training.	Input: — Liaison meetings — Issues and risks log Output: — Liaison meeting agreements

- Evaluate control room layout design
- Understand the lighting, heat, ventilation and services requirements of the control room layout
- Construct an preliminary ergonomics view of the operational TEC in terms of users and tasks

The activities of this plan are iterative over the duration of this programme.

Resources

The staff resources for this plan are as follows:

- HFIP manager
- Ergonomics Consultants — as required to carry out part of the plan.

Analysis tasks	Initial state	Interim state X	Interim state Y	TEC goes operational
A1. Define functional goals of the system	Final			Initial Draft
A2. Define user tasks and role responsibilities	Final			Initial Draft
A3. Define user characteristics and competencies required	Final			Initial Draft
A4. Define interactions within and between roles	Final			Initial Draft
A5. Define information requirements for the user and necessary tools/support				Initial Draft
A6. Define Workstation Assets	Final			Initial Draft
A8. Define workload assessment and derive staffing complement				Initial Draft

Design tasks	Initial state	Interim state X	Interim state Y	TEC goes operational
D1. Allocate functions between system and user role				Initial Draft
D2. Job design				Initial Draft
D3. Define Workstation Design options		Draft		Initial Draft
D4. Define Control Room(s) layout		Draft		Initial Draft

Evaluation tasks	Initial state	Interim state X	Interim state Y	TEC goes operational
E1. Reviewing activities	Final			
E2. Evaluation of steady state operations	Final			
E3. Trials and prototypes	Final			
E4. Management and DAG review	Final			

Other task	Initial state	Interim state X	Interim state Y	TEC goes operational
O3. Liaise with key functional areas, departmental and user representatives	Final			Initial Draft

Summary

The plan for the TEC implementation is an incremental build up of functions and control responsibility. Tasks carried out by the operators in the TEC will change over time. Initially, the tasks that are carried out in the TEC will be commissioning activities as parts of the network are brought under control.

It follows that the ergonomics activities need to be scheduled in a way that allows the steady states between transitions from intermediate states up until the end-state needs to be addressed. This Ergonomics Integration Plan is the first iteration and will be updated for each major phase of the programme.

Chapter 11

Involving people in design

Ian McClelland and Jane Fulton Suri

Introduction

Purpose of the chapter

This chapter is primarily about creation. It is about how the people we design for can participate in, and otherwise contribute to, the design of things that are useful, usable and enjoyable to use. It is about methods that enable people's needs, desires and values to be effectively embodied and integrated into design activities so that they are embraced by design and development teams.

This book is aimed primarily at practicing ergonomists (= human factors engineers). Ergonomists certainly have the skill base to make use of the methods covered in this chapter, but there are also many other professional practitioners who play the role of 'user advocate' in design and are equally well equipped to use them. Indeed, many of the methods have been developed and practiced by professionals who would never describe themselves as ergonomists. This reflects the reality that the practice of creating satisfactory designs from the end-user perspective requires information and skills that go beyond those that ergonomists have traditionally offered. Specifically, the practice goes beyond concern for a design's functional suitability and usability — its physical and cognitive 'fit' — to include consideration of emotional, social and cultural dimensions.

Our focus is on the design of devices, services and systems that are intended for use by large numbers of people in the course of their everyday lives, at home, work or play. Consumer products, communication tools, vehicles and public transit systems all fall within the scope of this chapter. We will focus on the kind of questions that crop up at the start of a development programme with particular emphasis on innovation rather than simple face-lifts for existing products. By 'innovation' we mean the introduction of new-to-the-world products and services or radical new design within an existing category. Where a development organisation wants to create a 'ground breaking' design there are greater challenges and opportunities for user-orientated investigations — hence our focus. Of course, even a request to 'face lift' a mature product can be overturned by raising a few critical questions about how and where the design will be used, who will use it, etc. in favour of a more fundamental approach. Developers may well be persuaded to reconsider their design in more fundamental terms — 'designing the right thing' vs. 'designing the thing right' — such that many of the methods here would apply.

Chapter structure

The chapter continues next with our perspective on some issues as orientation for the reader. This includes outlining the way we use the terms 'design', 'designing' and 'designers', notes on participants, and notes on the investigator and the skills they require. This is followed by an overview of the Human Centred Design (HCD) process, and the main section on the Methods. We conclude with some observations on directions for the future.

Our perspective

On design, designing and designer

We first want to clarify these terms and how we will use them throughout this chapter.

'Design' as noun — the thing designed. We will refer to the design as the 'thing' that will be used by people in some context to achieve some particular end. The design might in practice be a product such as a mobile phone, it might be a service that is mediated through some machine such as an automated railway ticketing system, or it might be a system such as airline check-in.

'Design' as verb — the activity of designing. The professional practice of designing is increasingly recogned as a multidisciplinary activity and typically involves people of different professional orientations working in design teams of one sort or another (see for example discussions by Bannon, 1991, Frascara, 2002, Rosson, 2002). Organisational parameters will determine how, in practice, design teams are set up. Design teams usually involve people working in close collaboration with several other professionals, some of whom will often work together in a formal team, and/or sometimes collaborate in a more *ad hoc* and informal way.

'Designer' — a person making design decisions. A person working in a professional context whose decisions directly influence the way a thing is designed. This person may, in a professional context, describe themselves explicitly as a Designer, or they may be one of a number of several complementary professional communities such as Software Engineer, Human Computer Interaction Specialist, Human Factors Engineer, Usability Engineer, Applied Psychologist, Information Architect, Interaction Designer, amongst others. In this context we regard ergonomists as designers in that they directly use their expertise to make decisions that contribute to the creation of designs.

Participants

Traditionally, investigations carried out by ergonomists (and comparable professions) have been modeled to a large extent on formal 'experiments' that recruit subjects. Subjects are typically introduced into a context where they carry out some task that is used as a means of 'measuring' their response. Referring to people as subjects tends to depersonale the individual, implies that they are just sources of data, and that they contribute in rather narrow and specific ways to an 'experiment'. In many contexts designing experiments using such parameters may still be appropriate. But in helping us to understand and accommodate what matters to people in the context of design, this type of perspective is overly limiting.

In this chapter we use the term participant to denote a more holistic, and realistic, perspective on the contribution that people can make to design creation. The term participant conveys the notion that people can collaborate with design teams in positive and creative ways to identify what makes for an appropriate design. In recent years there has been a growing advocacy of considering people as partners in the design process. The strongest movement has come from the advocates of participatory design (see for example Greenbaum *et al.* (1991) and Chapter 36 in this book) where the main focus has been on the design of multi-user computer systems. But in recent years there has been a growth of interest in developing the same principles to a much broader scope of artifacts and user communities; see for example Bodker (2000), Gaver *et al.* (1999), Druin (1999, 2002), Ferris *et al.* (2002), Poggenpohl (2002), Sanders (2002), Suchman (1987) for discussions on the role of participants in design. See also several papers in the following collections; Bekker *et al.* (2002), Frascara (2002), Greenbaum *et al.* (1991), and Scrivner *et al.* (2000).

In addition to their basic capacities and limitations as human beings, participants bring their own experience, expertise and expectations about the products, services and systems they use in their everyday lives. And to an extent people often design for themselves, a feature of our lives that is often overlooked — see Moran (2002) for a stimulating commentary on people as designers. This chapter is about the methods that can be used to tap into the explicit and tacit knowledge, and feelings that people have, and facilitate the application of their creative energy to the design issue in question.

We acknowledge that there are many developer organisations that are skeptical about the value of including people in the design process. Common negative prejudices are reflected in statements like 'people lack the appropriate insights', 'they lack imagination and cannot be expected to "envision the future" ', 'people cannot tell you what they want', and so on. It is not easy for people to imagine the future in the abstract and, because of a lack of know-how and expertise, people often find it difficult to be precise about what they would like or expect. But, on a daily basis people expend a great deal of energy and imagination designing aspects of the world they inhabit, whether it be the way they dress or the configuration of environments in which they live and work. And they continue to adapt and change things as their lives and needs evolve, often exhibiting remarkable levels of imagination in the process.

Experience shows, time and again, that skilled investigators are able to harness this creativity and provide opportunities for people to participate in professional design activities in productive ways. The goal of the investigator is to create 'dialogue' between the participant and the design team, and through this dialogue enable the participant to illuminate the design issue in question. The dialogue should enable the knowledge, creativity and energy of the participant to be applied in a focused and productive way. Investigators can, using the type of methods described in this chapter, create circumstances which provoke the imagination of participants and help them to reflect upon their behaviour and experience, to articulate their own ideas about what they expect, what is likely to please them, interest them, and what they might buy. Tapping into this kind of information gives investigators the opportunity to identify and interpret the critical human focussed and contextual focused variables that need to be taken into account. See Dunne *et al.* (2001), Gaver *et al.* (1999), and Hutchinson *et al.* (2003) for examples of how artifacts can be used to provoke the imagination and creativity of participants.

Finally, participant involvement is not a 'panacea'. The involvement of participants does not of itself guarantee that designs will be beyond criticism and without faults. Nor does it guarantee startling breakthroughs. Significant innovations in design come from many sources and so the methods described in this chapter will benefit from other complementary forms of enquiry. But participant involvement can provide a critical and valuable stimulus. Participants can help make the difference providing that (1) their attention is focused by the investigator on the questions participants can answer and on activities that reveal useful insights, and (2) the investigators (and the design team) use their professional expertise to carefully analyse, understand and interpret the information that participants provide. Participants provide a critical point of reference for the design team because they are external to the team and the developer organisation, and because their point of view is fundamentally usage centric.

Investigator

We are using the term investigator (sometimes referred to as user-advocate) to refer to the individual (or people) on a design team who have primary responsibility to ensure that a design is usable, useful and enjoyable for those who will ultimately encounter it. They will take the lead in employing the methods we describe, act as facilitator between a design team and the user community, and be the primary interpreter of the information investigations yield.

Skills and attitudes required

The skills and attitudes needed by the investigator fall into three categories.

Technical skills. All the methods we describe will require, to varying degrees, technical skills in conducting investigations effectively. These include such skills as formal design of investigations, planning and logistics, observation, use of data gathering tools such as audio and video recording, interview and questionnaire design, data analysis and interpretation.

Social skills in relation to managing and involving participants in investigations. Success in drawing useful insights and involving potential end-users is not simply a matter of technical competence. The detail, enthusiasm and creativity that people are willing to contribute is dependent in large measure on an investigator's ability to establish mutual trust and integrity. Generally there is only a very short space of time in which to develop a positive relationship and to engender co-operation. A good investigator needs to be able to rapidly engage people. Participants need to understand what to expect and how they will contribute. The investigator needs to convey a genuine interest in the personal opinions, descriptions and ideas of participants. It is helpful to rehearse briefing statements and develop courteous opening remarks and warm-up topics/activities to relax participants and put them at their ease. But nothing beats genuine curiosity, interest, and enjoyment of people in these kinds of encounters. Many of the methods invite the design team to use their empathic skills to identify with potential users' experience of designs, rather than to remain as objective observers and investigators.

Social skills in relation to being a persuasive team player. The design team must embrace the discoveries and integrate them effectively into the developing design. To be effective as a 'user advocate' it is important to be a team player. There are many ways to contribute to a particular team. Some teams are best persuaded by delivering objective factual information and clear specific direction. But other teams respond best when they are personally involved in discovering and empathing with user issues. Which is the best way must be considered case by case, though there are some basics that generally apply. These include an ability and willingness to:

- to "see another's point of view",
- involve others in your work process,
- understand the technical, organisational and political constraints the team must confront,
- go beyond your own specific assignment, identify with the team's objectives, work collaboratively on the synthesis of results, presentations, report writing and project planning.

Several of the methods in this chapter have the effect of 'co-discovering' and communicating a shared vision of user issues within the design team. In many cases the investigator can take on the role of facilitator and use such methods to help focus the efforts of the design team, and make the human issues visible, tangible, believable, manageable and measurable.

The methods — introduction

In this section we describe HCD in the context of the design and development process, discuss a number of general issues concerning setting up an investigation, and finally overview the methods.

The design and development process

All organisations that develop designs have some form of 'process model' that enables them to coordinate efforts and achieve a desired result in a managed way. Typically the processes adopted by industrial and commercial organisations are determined by commercial objectives (and these are not entirely financial), technical requirements and resource management considerations.

Human centred design (HCD)

The goal of HCD is to focus attention on the critical human issues throughout the design and development process so that the inevitable trade-offs between human, commercial and technical issues can be made in a balanced way. If the human issues are not articulated in a clear and manageable form they are likely to get very easily pushed aside when the pressure of costs, schedules, and technical issues come to bear. To be effective the activities within a HCD process need to be aligned with the 'host process'. The key to success is to 'ask the right questions at the right time' and 'deliver the right information to the right people when they need it'. Ideally HCD activities will be fully integrated into an organisation's 'process model'. An excellent example of how one large corporation approaches HCD can be found on the IBM Ease of Use web site (http:/www-3.ibm.com/ibm/easy). However, for many user advocates working within organisations with well-established technical and commercial operations, such integration remains a long-term goal. For example, in the case of IBM the current 'state of the art' builds on many years of excellent work. For those working outside an organisation, even as an occasional consultant, it is advisable to understand the receiving organisation's process, and at what points human-centered contributions will best complement and influence outcomes. In any case, knowledge of the host process in general, and the specific phase(s) where an investigator plans to contribute, often has a big influence over the choice of method(s) for a particular investigation.

HCD is NOT just fantasing about usage. HCD practitioners need to *gather* evidence of what is appropriate in usage centric terms, to *interpret* this evidence in terms of the decisions that confront the design team, and to *communicate* the evidence in a form which the design team can use. See for example Curtis *et al.* (1999) who report on a project aimed at communicating customer focused design data within a large industrial organisation.

The historical origins of HCD

The idea that we can measure interactions between people and designs is a fundamental principle that underpins ergonomics, and, equally important, that we can do so in ways that can guide design development (see for example Bailey, 1982). The body of knowledge that now exists about the methods available is considerable and continues to develop. And not least because there is a need for more effective methods (Meister *et al.*, 1992).* Over the years it also became apparent that a corresponding need existed for a

*For a more extensive discussion of the history, principles and practice of test and measurement in human factors work see Meister (1986). The application of ergonomics to design testing is also discussed by Bailey (1982), Chapanis (1959), Cushman and Rosenberg (1991), Gould *et al.* (1997), Holleran (1991), Kirk and Ridgeway (1970, 1971), Macleod (1992), Rennie (1981), Rubinstein *et al.* (1984) and Whiteside *et al.* (1988).

process perspective within which the methods available could be positioned. In answer to this need the most significant development in recent years has been the emergence of 'usability engineering'. This concept was developed primarily as a way of applying human factors to 'information technology' products and systems (Gould *et al.*, 1997; Nielsen, 1993, 1994; Shackel, 1986, 1987; Whiteside *et al.*, 1988, ISO 9241 Part 11), ISO TR 16982 (2002) see also Chapter 13 in this book. The concept of 'usability engineering' provides a process framework for designing IT systems from a usage point of view. As the principles of 'usability engineering' evolved so the notion of HCD as a more 'holistic' approach to the design of interactive systems has evolved. HCD is now being driven by the wide dispersion of computational technologies into all sectors of life in advanced industrialised societies. A broad range of professional communities beyond the traditional HCI or usability professionals now subscribe to the HCD perspective. HCD now includes issues such as life-styles, aesthetic considerations, emotional value, as well as the traditional concerns of usability engineering.

Benefits of HCD

In broad terms it:

- helps inspire new ideas and design directions
- helps create new paradigms and value for existing product or service offerings
- increases real and perceived value of a design
- provides better experiences for people using a design
- helps avoid developing a bad design idea
- helps make a weak idea much better
- helps turn a good idea into an outstanding one
- reduces exposure to product liability issues
- reduces customer complaints and product recalls
- helps to create a more efficient design process in relation to:
  a faster and more precise definition of functionality,
  a faster choice of appropriate interaction technologies
  an earlier determination of design performance targets
  a more cost effective way of assigning design team effort

For those who wish to read more about HCD and its development there are several authors such as Cushman (1991), Damodaran (1996), Hackos *et al.* (1998), Hix *et al.* (1993), Macleod (1992), Mayhew (1999), Norman *et al.* (1986), Preece *et al.* (1990), Rosson *et al.* (2002), Torres (2001) and Vredenburg *et al.* (2001, 2002). Wixon (1996) discusses HCD and the associated processes and methods as applied to interactive system design. Norman (1999) provides for a wider interpretation of the application of HCD than traditional IT systems, and Vredenburg *et al.* (2002) has a recent survey on HCD practices. Increasingly, HCD is being recogned as providing value from a business perspective as a source of innovation — see Leonard (1991), Norman (1999) and Kelley (2001).

The phases of the HCD process

There are many ways of describing the HCD process. For the sake of simplicity, we have chosen to present it as four key phases in evolving a design:

1. Explore and understand needs and opportunities
2. Identify design options and directions
3. Create and refine specific solutions
4. Evaluate designs in the real world

In reality, the HCD process is iterative and the phases are not strictly sequential and self-contained as this description implies. But this will serve as a useful structure to explore the kinds of questions that a design team may have to deal with as they make progress through a design programme.

Phase 1. Explore and understand needs and opportunities

Objectives. The purpose of this phase is to discover what matters to people that could be better supported through design, and what might motivate or inhibit the appeal and use of a future design. The questions, implicitly or explicitly, may be:

- what are the key characteristics of expected users and contexts?
- what are their unmet needs in these contexts?
- what are the appealing aspects of an activity or existing design? what are the frustrations?
- what would be desirable characteristics of a design that addressed these aspects?
- what are people's expressed and latent needs? what are their relative importance?

> *Example: Your team is asked to come up with design concepts for an in-car information system*:
> What different kinds of information might people need or want in a vehicle?
> How do the needs of a busy parent, a travelling sales-person, a holiday-maker differ?
> What are the delights and frustrations of people's current experience?
> How could existing methods or tools be improved?
> What are the characteristics of a product/service that would enhance people's experience?

Although the design team probably already has some type of design in mind the focus of an initial investigation should not be directed upon this specifically. Rather, the focus should be on understanding which group of people are expected to use the design, the context in which it will be used, key motivators that create the perceived need, and the criteria that people will likely use to judge the value of the design.

How people can contribute. By examining their current experiences, people can almost always provide valuable insights into benefits they will seek in future designs. Exploring their current experience can also reveal why certain designs are, in one way or another, not appropriate. A useful approach is to explore, together with their users, weaknesses or inadequacies of current designs and identify what remedies would be appropriate. Such exercises act as catalysts in revealing underlying motivators and sought-after benefits. People can also contribute to an understanding of the attributes of a design or the usage context that will ensure user satisfaction.

Understanding of potential users, their needs and the use context enables the design team to begin to identify a hierarchy of needs, and characterise key design challenges and opportunities as well as establish initial design criteria. To do all this the design team needs to interpret the information in terms of user requirements:

- those requirements that the design aims to meet
- those requirements that will not be met
- those requirements that may be better supported by some complementary device or activity

The outcomes of this phase should enable the design team to profile the people who are expected to use the design, and the circumstances under which the design will be used. This should also lead to the generation of preliminary usage scenarios based on the evidence that the design team has been able to gather. The usage scenarios encapsulate descriptions of typical users, typical usage situations, and the range of tasks that users will be expected to undertake. In short usage scenarios characterise how the design team expects the design to be used in practice. Usage scenarios are the first steps in generating particular solutions.

Phase 2. Identify design options and directions

Objectives. The goal of this phase is to respond to the discovered needs and opportunities by generating and representing design options. And, from these options select directions for further development. Questions that are commonly addressed in this phase 2 will include:

- what new design ideas emerge from understanding people's needs and desires?
- how do the proposed ideas map onto the anticipated usage situations?
- what functional and technical attributes could enhance usage?
- how will the design fit within contextual constraints: physical, socio-cultural and socio-technical?
- what attributes should the design have to meet the critical acceptability factors?

> *Example: Your team is tasked with exploring ideas for a new health-drink-making machine for use in cafes and restaurants.*
> How might the configuration of a machine enhance the operator's workflow and interactions with colleagues and customers?
> How might the machine be positioned in relation to other equipment?
> Is it more appropriate to conceal the operation of the machine or to feature it?
> How might this influence machine form-factor to be as efficient, safe and comfortable as possible in use?

In practice there is substantial interplay between phase 1 and phase 2. Often the design team already has a good idea about possible design directions before setting out to explore needs. Nevertheless it is also usually the case that, whatever the early ideas, the design team will want to explore other possibilities in response to the new evidence available to them.

How people can contribute. People can easily highlight features of a design that are attractive as well as those that are not. People are often able to identify the characteristics that are desirable and those to be avoided. But to achieve this kind of contribution design concepts or new technologies need to be experienced by people in tangible ways that clearly relate to their own context. Given this participants can provide relevant detail and texture about the usage context and help identify the strengths and weaknesses of particular design proposals. Combined with the skill and knowledge of the design team, they help in generating and selecting a range of design directions that are well-suited to the target user population and context of use.

Phase 3. Create and refine specific solutions

Objectives. The goal of phase 3 is to further develop and detail the attributes of selected options to the point that a specific solution can be implemented as a final design. Questions that are commonly addressed in this phase 3 will include:

- is the functionality, usability and appeal of the design compatible with what we know of people's needs and desires?
- how rapidly do people understand the potential of the design and ways to interact with it?
- what are the detailed appearance and technical performance requirements that will meet usage needs?
- how can we resolve users' desire for ease of use with conflicting legislated or industry standards?

> *Example: Your team has established a couple of possible design directions for a medical monitor that people will need to wear 24 hours a day. You plan to try out some weighted configuration models.*
> How will this feel when you have to wear it 24 hours per day?
> How can people cope with it through daily activities like sleeping, dressing, bathing, getting in and out of a car?
> What do we need to change to make it as secure and comfortable as possible?

Phase 3 clearly builds on the outcome of phase 2. Phase 2 tends to be a divergent phase in which the design team pushes the boundaries to explore many design possibilities. Phase 3 can be characterised as one of convergence and consolidation. It involves detailing the specific attributes that will eventually be incorporated into the design. This will often involve the need to compare different solutions with each other, or to check solutions against specific performance requirements that have been set as targets for the design.

How people can contribute. In phase 3 people can contribute to the creation of designs as well as evaluate them. Participants can be involved in co-development with the design team in varying stages of the development, especially in exercises aimed at resolving high level system or product configurations.

A key issue is to provide participants with sufficient information to enable them to provide useful insight. There is almost always a compromise between the ease with which options and contexts can be represented and the level of fidelity of those representations. This compromise is schematically shown in Figure 1. The challenge for the investigator is to match the type of questions that can be investigated to the level of design representation available. In the early phases, when design concepts are fairly loose, it is generally easy and inexpensive to represent them in some form of low-fidelity simulation that people can respond to. As concepts become more robust and detailed, accurate representations become more time-consuming and often more costly to realise. In most cases, participants are quite able to appreciate the nature of a rough prototype and are tolerant of simulations 'breaking down'. Furthermore, participants can often be more motivated to help a design team seeing that there is clearly a chance to shape the solution before it is finalised.

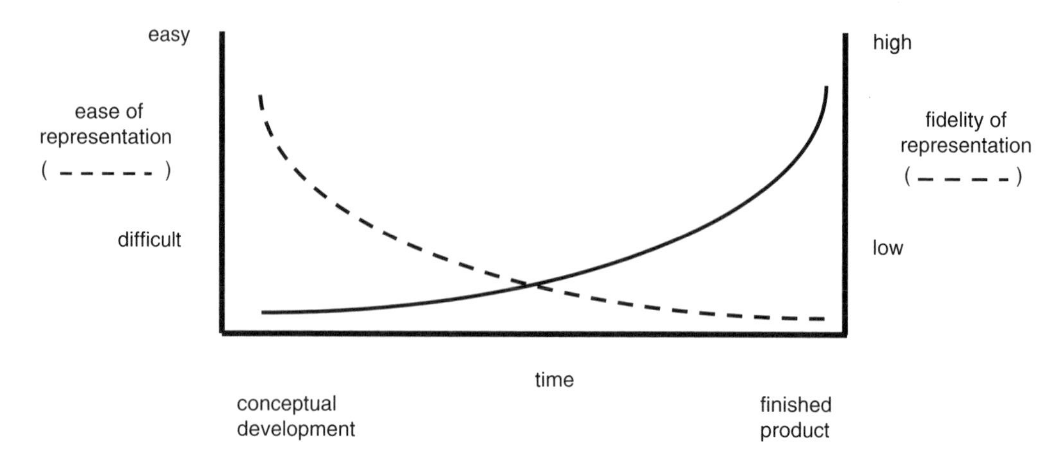

Figure 1 Ease and fidelity of design representation

The value people provide here is less in deriving details of specific solutions as much as their ability to highlight requirements, prioritise functionality, and reflect users' own priorities in terms of what is of interest in the solution. Their involvement also helps to remind the design team about aspects concerning the reality of usage and context that might easily be overlooked as commercial and technical factors influence the final implementation of the design.

Phase 4. Evaluate implemented designs in the real world

Objectives. The focus of phase 4 is on investigating implemented designs in use. The purpose is to discover whether the design is being used as was anticipated during phases 2 and 3, and to check on the assumptions made during the course of development.

Questions that are commonly addressed in phase 4 will include:

- is the design being used by the kinds of people and for the purposes expected?
- how valid were the assumptions about when, where and how the design would be used?
- is there sufficient support for the design in use, in terms of, e.g., training, servicing?
- what considerations and applications were overlooked?
- how is the design changing people's behaviour and experiences

> *Example: Your company has recently launched a new mobile phone with digital image capture and communication capabilities?*
> Are people able to use the device as easily as they use a regular mobile phone?
> In what circumstances, both in personal and work life, is the imaging capability most used?
> What are the most frequently used functions?
> Did we provide the easiest access to these?
> Are there any functions that are hardly ever used?

How people can contribute. Evaluating designs in use usually involves monitoring actual performance and logging usage data, preferably over an extended period of time. People contribute by volunteering information about their behaviour and experience, often in response to pro-active requests to them to participate. They also provide information to investigators via incidental data they generate, such as records of complaints, service and product returns held by the manufacturer or supplier. Information can also be gathered from existing on-line communities and discussion groups related to specific kinds of products and services.

By revealing specifics about the use of a new design in context, this phase provides valuable input into phase 1 of any subsequent development of the design and related products and services. It also provides the opportunity to examine how the solution impacts on and is impacted by the actual context of use. This is of particular value in the case of new-to-the-world products and services where inevitably new patterns of behavior and expectations develop. Consider, for example, how the availability of mobile phones and text messaging services has changed the way people, especially the young, plan and conduct social events from the days that telecommunications were essentially land-line based. Such evolving usage patterns create new challenges and opportunities for design, at both the detailed and conceptual level. Awareness of these opportunities and data about their prevalence is a key competitive advantage, and offers great possibilities for user advocacy to drive commercial innovation.

Setting up an investigation — general considerations

Later in this chapter we discuss, in some detail, a range of different methods for providing human-centered input to the four main design phases. But there are several basic issues that need to be considered when planning the involvement of people, irrespective of the particular methods that may be employed.

Ethical issues

Participants share their time, energy, information and sometimes even intimacies that are of commercial value to developers. It is important to conduct all investigations with the highest degree of respect and consideration for people's time, health, safety and privacy. Participants should be informed about the purpose of their involvement, what is being recorded and what will be done with the data. Sometimes it is necessary that participants are NOT made fully aware of an investigator's specific focus ahead of time, for example if the knowledge will alter their behaviour, but in these cases it is generally courteous and tends to build trust to 'let them in on the secret' when a session is completed.

Objectives

Investigations can usually be designed to provide useful information within the practical constraints of costs and time-scales. The basic aim of an investigation is to improve on estimates based on opinions by obtaining evidence that can guide the design team. Any investigation needs to start with a clear set of objectives. These will depend upon the current phase of the design process, the questions that have arisen and who on the team needs the information. Here are some typical examples of different information needs within a development team and at different phases:

- Product planner; a sporting goods company is considering its strategy for integrating communications technology into products. What sporting situations and kinds of products would benefit from this kind of integration?
- Interaction designer; the interface elements and behaviours for a new digital radio system are being developed. What user considerations should we be aware of in migrating a familiar system from analogue to digital format?
- Industrial designer; a new remote-control device is being designed for museum visitors to use in engaging with interactive exhibits. What form, factors and materials will be pleasant to handle, easy to keep clean and enable it be stowed, carried and used most easily?
- Marketing specialist; a new range of innovative personal entertainment products is being considered. What are the key features that will appeal to the second and third generation users?
- Manager of a development department; a choice has to be made between two technologies. Will it matter to the user which we choose?
- Design engineer; what are the critical dimensions in the layout of a particular workstation?
- Software engineer; a particular user group is accustomed to a particular style of interaction. Will a new style be acceptable and are there any negative transfer effects?
- Documentation specialist; how will customers learn or be trained most effectively and where might particular difficulties with the product occur?
- Maintenance engineers; a specific time limit has to be met for the exchange of particular machine components. Will the design meet this requirement?

Many investigations can, and need to, provide answers to more than one question of this kind. In designing an investigation it is a useful discipline to articulate and prioritise each of the questions to establish appropriate focus.

Participant goals, tasks and activities

The investigator needs to work out what activities they need to observe, and be clear about how much they need to influence them in order to satisfy the objectives of the investigation. In general the earlier in a programme the broader the investigation. It also depends on how much the investigator wishes to diverge and invent, and how much there is a need to converge on a solution that is already fairly well defined. In some cases, such as on-site observations, the investigator may have no direct control over what participants do but may need to be selective about what aspects they need to observe. In situations where the investigator is necessarily absent, for example in the case of 'diary keeping studies', the investigator has little control and can only describe their requirements in general terms. In others, such as user trials, the investigator is in the position to instruct participants to carry out specific tasks. The investigator also needs to consider the relationship between the tasks and activities that are directly served by the design, and those that are complementary in some way. For example, users of Automatic Teller Machines often have to deal with bags and other types of luggage when making cash withdrawals. This can add to anxieties about personal security when using ATMs in airports, railway stations and similar crowded areas. The complementary aspects may not be the subject of the investigation but they may be very important in ensuring that the local context is adequately taken into account when designing the investigation.

Location

The appropriate location for an investigation depends upon the nature of the information sought and the phase of the design process. Exploring needs and opportunities, phase 1, often involves at least some activity in the context where the design will ultimately be used. Investigating design options, and refining specific design solutions, phase 2, may also be carried out 'in context' but is often carried out within artificial or simulated situations, such as a usability laboratory or workshop space, where circumstances can be controlled by the investigator. But in these controlled situations it is important to be aware of the extent to which there may be significant contextual factors that are relevant to the design, and to accommodate these as far as possible using props and models to simulate elements where desirable. Investigators can often make reasonable assumptions based on their own professional experience, experience of real contexts of use, and information provided by participants. Whiteside *et al.* (1988) provide an interesting discussion on this topic in relation to human–computer interface development. There is a growing conviction amongst practitioners that studies conducted in a real-life context provide opportunities to gain important insights that would not be revealed in the laboratory (see Wolf *et al.*, 1989). It is nearly always beneficial to ensure that at least some parts of phases 2 and 3 involve studies in-context. But is it is also worth noting that given the widespread use of network systems, and not least the development of the Web, conducting studies in-context might also mean dealing with participants on a remote basis; see for example Hill *et al.* (1997) who involved remote users in the development of Web site content.

Participant selection

An obvious point, but nevertheless essential, is that people selected to participate should have relevant experience or expertise to be able to provide insightful contributions. The investigator needs to profile a range of people who will provide useful perspectives on the design. At a high level there are several key dimensions to consider in deciding whom to involve:

- the person as user; a future operator, passenger, visitor who will interact with the design and may represent 'typical' or 'extreme' use, be novice, experienced, or expert, etc.
- the person as an individual human being; their physical and psychological capacities and limitations, such as age, strength, literacy level, demographic and/or psychographic segment etc.
- the person as a social being; their role in relation to the context of use, e.g., student, teacher, doctor, and their domain of expertise and ultimate goal.
- the person as a cultural being; their world-view, life-style and values, habits, rituals and expectations based upon national, ethnic, generational, religious or other adopted group membership.

These dimensions overlap, and of course within any one person, they interplay at various levels. The participant profile should identify the dimensions that are most relevant to the design and type of investigation that needs to be carried out. In some cases there may be data to assist in deciding what is relevant (for example, if a marketing plan has been defined there is likely to be demographic and/or psychographic data available). But, more often than not, the investigator has to make an informed estimate based on experience and other useful sources of information. Apart from the professional ergonomics literature, information sources could include market research, service, sales, application specialists, training departments, occupational health specialists, etc. All may be able to offer useful guidance. Again, assumptions will have to be made which should be agreed with those who requested the investigation.

Once profiles are developed, participants can then be recruited to match them. Often these profiles will serve as a guide rather than strict criteria for selection. However, if the validity of an investigation relies on participants complying with specific requirements, then they should be appropriately screened before being recruited.

Selected participants do not always need to be 'typical' of the eventual use population. Sometimes it is useful to seek out people who characterise the 'fringes' in order to help the design team establish the boundary conditions in terms of their needs and requirements (for example see Gilmore (2002)). However, whenever possible or practicable it is good practice to collect descriptive data from the participants that enable comparison with, and extrapolation to, the user population as a whole. Of course, such comparisons can only be made if the corresponding data describing the user population are available in an appropriate form.

Participant numbers

It is difficult to provide general advice on how many participants to involve in an investigation. There are methods available for estimating sample sizes in formal experiments, where statistical probabilities are required to examine the role of key variables in an outcome (Winer, 1971; Collins, 1986; see also Chapter 2 of this book). Practitioners often find that the number of variables relevant to a design-related investigation is so large that such experimental methods are impracticable and unwieldy. It is also often the case that statistical information about the variance of human characteristics within a population is not available and best estimates have to be made. In short, this usually results in estimates of participant numbers being very large, sometimes alarmingly so, to achieve even moderate levels of accuracy.

In practice, involvement of very small numbers of, for example 4–6, participants selected to represent a good range of characteristics and contexts, can provide significant information and evidence of issues that were not apparent beforehand. Many investigators aim to involve small numbers of participants representing as broad a range of perspectives as possible on an incremental basis. Thus the investigator stops recruiting new participants or running new sessions when they are confident that no significant new knowledge is likely to be generated.

For user trials specifically, experience shows that dependable results can be obtained from as few as five users in a single trial (Rubinstein *et al.*, 1984). More than five users will generally always be beneficial but how many more is an open question. Virzi (1992) concludes that 80% of usability problems can be detected by 4–5 participants, and fewer new insights are revealed as the numbers increase, with the first one or two participants detecting the most severe problems. These conclusions reflect a widely-held view amongst practitioners that small numbers are often sufficient to guide the design team. However there are many examples where investigations have involved more than 30 users and, in some cases, samples in excess of 100 are quoted. However always bear in mind that, for example, a visit to even a single surgical procedure and subsequent discussion with the one attending physician and anesthetist will not necessarily provide a repeatable or representative 'result'. But it beats relying solely on a discussion around the interaction designer's computer about how the blood-gas analyser interface should be laid out.

Representing the design

It is often necessary to represent the design in some way in order to undertake an investigation. The appropriate level of fidelity will influence the type of information required and vice versa — see Figure 1. Many types of representations are commonly used and include:

- *'Paper-based' descriptions of product concepts*
  These are sketches, narratives, annotated drawings, screen graphics or other concept descriptions that can enable initial explorations of ideas on product functionality to be made, important usability characteristics to be identified, or 'walkthrough' studies of protocols for product control systems to be conducted.
- *Part prototypes or simulations*
  Part prototypes are used to simulate specific functional attributes of a design. They might be mock-ups of physical form, scale or mass, mechanical models, static or animated screen graphics that enable people to interact with them. The prototype may look nothing like the final design but will accurately represent those aspects under investigation. Part prototyping of software through the use of rapid prototyping tools is one area where this type of testing is now very common.
- *'Experience' prototypes*
  These are representations in any medium, that help people to appreciate experiential issues beyond the purely functional attributes of a design. They are designed to include contextual and affective qualities conveyed through a relevant subjective experience (see Buchenau *et al.*, 2000).
- *Full prototypes*
  Full prototypes perform as the final product is intended to perform and incorporate the complete functionality and appearance of the product.
- *Complete products*
  Complete products enable the complete user-interface to be examined. This opens up the possibility of carrying out field investigations, comparative studies with other products, in-service studies etc. For further readings about design representations particularly in HCI see Houde *et al.* (1997) and Wong (1992).

Measures, observations and subjective data

Some investigations will reveal simple objective data, such as error rates or times to complete a transaction. Others will enable the investigator to make direct objective observations of behaviour and events. Still others will provide subjective information in the form of participant's expressions and self-reports of behaviour, opinions, attitudes, concerns, and habits.

Objective measurement has traditionally been a central tenet of Human Factors research. Kantowitz (1992) summarises its role: 'Measures are the gears and cogs that make empirical research efforts run. No empirical study can be better than its selected measures'. In the design context, the standard empirical approach is time-consuming and appropriate for only some of the issues that need to be addressed. Nevertheless, it is wise to obtain objective data whenever it is possible and practicable. Such data are readily accepted by many individuals and organisations. It can complement subjective and qualitative data, and insights gathered from other more exploratory methods with participants. There are three main kinds of objective data that are frequently collected:

- direct measurements taken from a person. Examples include body dimensions and physiological measurements such as heart rate, oxygen intake, body temperature, etc.

- data resulting from user actions recorded by the investigator or by some remote means such as video or automatic event recording such as keystroke capture. Examples include time-based measures (e.g., task duration, event duration, response time, reaction time); error or accuracy scores (e.g., mistakes in procedures, incorrect responses to stimuli, error rates in relation to time or events).
- data taken directly from product or prototype as the result of adjustments or modifications by participants. Examples include positions for seats, shelves, or controls, levels for lighting, sound, colour, quality, brightness, contrast settings on displays.

Observations of user actions (see also Chapter 3) also fall into three main categories (adapted from Meister (1986)):

- descriptive techniques, where the observer simply records events as they take place (e.g., time based, frequency based, event sequence, postures adopted, controls used, etc);
- evaluative techniques, where the observer evaluates the outcome or consequence of events that have taken place (e.g., degrees of difficulty, incidence of hazardous events, errors of judgment, etc.); and
- diagnostic techniques where the observer identifies the causes that give rise to the observed events (e.g., positioning of controls, inadequacy of displays, poor user instructions, etc.).

Descriptive observations are generally the easiest to set up and conduct. The degree of difficulty increases with evaluative observations and is greatest with diagnostic observations. This is primarily because of the type of on-the-spot judgments the investigator is required to make during the observation. Ability to conduct evaluative or diagnostic observations depends greatly on the background knowledge of the investigator. To make effective evaluative or diagnostic observations the investigator needs extensive background knowledge of tasks and objectives. It is essential also to decide what to look at and what to record beforehand. The greatest problems come from the speed and number of events occurring concurrently. It often surprises people who are new to observing the behaviour of people just how quickly the accuracy and reliability of an investigator can break down when there are too many events to monitor. Even moderately experienced investigators often find themselves being far too ambitious in the variety of events they wish to record. Consequently it is most important that the method of observation is carefully piloted and video-recording used as back-up whenever possible.

Subjective data such as participant's self-reported behaviour, opinions, and habits are often difficult to quantify in any meaningful way. However, they are frequently the basis for rich insights that can be applied to design and should be carefully documented and represented. These data often find their way into *stories* told about specific people observed, their behaviour and expressions. Such stories have an important role in carrying information from investigations to design teams. Investigators bringing 'tales from the field' — narrative descriptions and anecdotes highlighting specific examples derived from both objective data and observations — help to bring data to life, make them actionable and make connections for other team members who may find charts and tables of data irrelevant. Stories are inherently not objective, but are selectively edited and integrated descriptions of phenomena. Care must be taken in the choice and telling of such tales that they maintain the integrity of their source and reflect significant design issues in appropriate ways.

Gathering behavioural data

There are many techniques for gathering behavioural data. The one technique we want to highlight here is video. In recent years video technology has become the most prolific method for recording participant behaviour. Video has not only become standard equipment in any usability laboratory (Nielsen, 1994) but, due to the portability of modern cameras, and the excellence of the recorded images, video is recorded much more frequently in less formal on-site studies. The great advantage of video is that it readily captures concurrent events and of course it enables one to repeatedly revisit the data. A particular sequence of events can be analysed several times over from different viewpoints, enabling an investigator to overcome many of the problems associated with analysing events which occur in parallel or very quickly. An example of this was in a study on warning systems for railway trackside workers (McClelland *et al.*, 1983). Strommen *et al.* (1992) used this approach in a study of three year old children using a computer games controller. However, a few cautionary words are in order. First, detailed analysis of video recordings can be extremely time-consuming. Ratios at least of between 1:5 and 1:10 ('recorded time: analysis time') are commonly quoted (Mackay *et al.*, 1988); Bainbridge and Sanderson (1995) report 'analysis time: sequence time' ratios of 10:1 to 1000:1. Second, the need for detailed preparatory work is not reduced because an extensive record of events is available. The use of video requires at least as much care as paper based approaches in deciding beforehand what events are to be observed and the form in which the data are to be analysed.

Interpretation of results

No matter what kind of information emerges from investigations involving users, it requires careful interpretation in order to pull out applicable insights. Rarely will an investigation in itself provide information that can be literally applied to a design. Rather, the investigation will first provide broad insight into people's current perceptions and sensitivities and help to establish a frame of reference from which the design team can think about future offerings. Second, the investigation will illuminate issues that can be turned into specific design criteria and design principles that need to be met if the design is to be successful. These criteria will guide design efforts and serve as a reference for the evaluation of proposals.

Design criteria

Deciding on which criteria to use to guide the design process and evaluate a design is key. In some cases the appropriate criteria will be self-evident, but defining them is often the specific focus of investigations with users, particularly when the technologies are new and usage patterns are still evolving (see, for example, Dillon, 1992, Egan *et al.*, 1989, and Francik *et al.*, 1989).

The ultimate success of a design may be related to diverse issues including social, commercial, organisational, subjective, emotional, value for money concerns in addition to those physical, cognitive, safety, reliability and usability criteria that are more usually considered. The goal of defining design criteria is to define those dimensions that will correlate with a successful design, be measurable, and produce information that facilitates design decisions.

Criteria may be phrased as, for example, 'the haircutting tool should be quiet enough to allow stylist and client to hold a conversation', or 'the voting transaction should feel reassuring.' Even for these it would ultimately be possible to develop some objective criteria (e.g., decibel and speech interference levels, system response and speed). In practice though, these criteria may be more effectively judged subjectively by participants and design team members as a design evolves (Fulton Suri, 2001). In any case, each criterion needs to be expressed in such a way that the design, or a participant's performance with it, can be assessed, whether subjectively or objectively. It is usual practice to use several measures rather than to rely on one, and to include both objective and subjective measures.

The methods

We have clustered the methods for involving people in design into a number of categories. Within each category we discuss the general approach and refer to specific examples used by different investigators. We do not pretend to provide a comprehensive inventory of all examples of each method. That would require a book in itself. As said in the introduction, the purpose of this chapter is to raise awareness of the scope of methods available.

Many of the methods outlined are about how the investigator can create the right kind of circumstances that help participants contribute, and help the design team to create the appropriate design. On the one hand, good methods should help participants to articulate their needs, their wishes, their experiences, their expectations, and provoke their imagination. And on the other hand, good methods should enable the investigator (and the design team) to generate the most useful information that helps the design team make the best decisions.

More in depth overviews of HCD methods can be found in Beyer *et al.* (1998), Muller *et al.* (1997), Torres (2001) and Vredenburg *et al.* (2001). Overviews of methods can also be found at the Usability Net (2002)* web site (http:/www.hostserver150.com/usabilit/home.htm). The International Standards Organisation has also issued a Technical Report outlining usability methods supporting human-centred design, ISOTR 16982 (2002). See also Jordan *et al.* (1996) for a collection of papers on usability evaluation in industry.

We have rated each method according to its ability to deliver the type of information required at each design phase. The rating is meant as only a guide, and is based on a combination of our professional experience and comments by investigators on the methods in case study examples. The legend is:

strong = •••
neutral = ••
weak = •
NA = not applicable

Contextual observations

Overview

Contextual observations, sometimes termed ethnographic methods, include a range of methods that involve learning about people's behaviour and activities as they occur in a real-

*At the time of writing (Dec 2003) the Usability Net web site was still available. Usability Net was funded by the European Union to provide resources and networking for usability practitioners, managers and EU projects. The project was completed in July 2003.

Contextual observations

Methods	References	Explore and understand needs	Identify design options and directions	Create and refine specific solutions	Evaluate designs in real world situations
Behavioural archeology	IDEO (2003)	•••	••	NA	••
Behaviour mapping	IDEO (2003)	•••	••	•	•••
Blueprint mapping	Muller *et al.* (1997)	•••	••	•	NA
Cultural probes	Gaver *et al.* (1999)	•••	••	NA	NA
Day in a life	IDEO (2003)	•••	••	NA	••
Contextual inquiry	Beyer *et al.* (1998); Holtzblatt *et al.* (1996)	•••	••	NA	••
Ethnography; rapid/video	Ford *et al.* (1996); Nardi (1997); Hughes *et al.* (1997); O'Brien *et al.* (1997)	•••	•	NA	•••
Guided tours	IDEO (2003)	•••	••	•	•••
Narration	IDEO (2003)	•••	•••	•••	•••
Shadowing	IDEO (2003)	•••	•••	NA	•••

Diary-keeping

Methods	References	Explore and understand needs	Identify design options and directions	Create and refine specific solutions	Evaluate designs in real world situations
Audio/video/ written-diaries	Sellen *et al.* (2002); Palen *et al.* (2002); Robinson (1996)	•••	•	NA	•••
Narration	IDEO (2003)	•••	•••	•••	•••
User photo-surveys	IDEO (2003)	•••	•	NA	•••

Framework development

Methods	References	Explore and understand needs	Identify design options and directions	Create and refine specific solutions	Evaluate designs in real world situations
Brain draw	Muller *et al.* (1997)	•••	•••	NA	NA
Cluster and network-diagrams	Tufte (1990)	•••	•••	•	NA
User-modeling	Hasdogan (1996)	•••	•••	••	••
Journey-mapping		•••	••	NA	NA
Task-flow schematics		•	•••	•••	••

Interviews

Methods	References	Explore and understand needs	Identify design options and directions	Create and refine specific solutions	Evaluate designs in real world situations
Chat-rooms		••	••	NA	•
Focus groups	Krueger *et al.* (2000)	•••	•••	•	•••
Interviews (one on one)	Oppenheim (1999), Rea *et al.* (1997)	•••	•••	•••	•••
Interviews (small group)	Krueger *et al.* (2000)	•••	•••	•••	•••

Questionnaires

Methods	References	Explore and understand needs	Identify design options and directions	Create and refine specific solutions	Evaluate designs in real world situations
Questionnaires (in person)	Oppenheim (1999), Rea *et al.* (1997)	•••	•••	•••	•••
Questionnaires (On-line)		•••	••	•	•••
Questionnaires (survey)	Moser *et al.* (1971), Morton-Williams (1986); Sinclair Ch. 4	•••	••	•	•••

Projective techniques

Methods	References	Explore and understand needs	Identify design options and directions	Create and refine specific solutions	Evaluate designs in real world situations
Brain draw	Muller *et al.* (1997)	•••	•••	•	NA
Collage-making	Sanders *et al.* (2001), Serpiello (2001)	•••	••	NA	NA
Draw the experience	IDEO 2003	•••	••	NA	NA
Word/image associations	IDEO 2003	•••	•••	•	•

Role-playing

Methods	References	Explore and understand needs	Identify design options and directions	Create and refine specific solutions	Evaluate designs in real world situations
Artefact walkthrough	Muller *et al.* (1997)	•••	•••	•••	•••
Cognitive walkthrough	Polson *et al.* (1992)	NA	••	•••	•••
Do-it-yourself	Moore (1985)	•••	•••	•••	•••
Body-storming	Buchenau *et al.* (2000)	•••	•••	••	NA

Scenario building

Methods	References	Explore and understand needs	Identify design options and directions	Create and refine specific solutions	Evaluate designs in real world situations
Journey maps		•••	••	NA	•••
Informance	Burns *et al.* (1995, 1997)	•••	•••	••	NA
Storyboards	Verplank *et al.* (1993)	•	•••	•••	NA
Story telling	Moggridge (1993), Joe (1997)	•••	•••	•	••
Task-flow schematics	Holtzblatt *et al.* (1996).	•	••	•••	•••
Usage scenarios	Carroll (1995), Fulton Suri *et al.* (1999), Nardi (1992), Rosson *et al.* (2002), Welker *et al* (1997)	•••	•••	•••	•
User profiles	Calde *et al.* (2002), Hasdogan (1996)	•	•••	•••	•
Work mapping	Holtzblatt *et al.* (1996)	•••	•••	•	•••

User trials

Methods	References	Explore and understand needs	Identify design options and directions	Create and refine specific solutions	Evaluate designs in real world situations
Acceptance tests	Cantwell *et al.* (1985).	•	••	•••	•••
Co-discovery	Kemp and Gelderen (1996)	NA	••	•••	••
User performance trials	Fulton Suri (1993)	•	••	•••	•••
Usability tests	Benel *et al.* (1985); Nielsen (1994); Rubin (1994);	NA	••	•••	••
Verbal and think aloud protocols		••	•••	•••	•••
Wizard of oz		•	•	•••	NA

User workshops

Methods	References	Explore and understand needs	Identify design options and directions	Create and refine specific solutions	Evaluate designs in real world situations
Activity groups	Druin (2002)	•••	••	NA	•
CARD	Muller *et al.* (1995), Tschudy *et al.* (1996)	•	•••	•••	NA
CISP	Muller *et al.* (1997)	•	•••	•••	NA
Collaborative design workshops	Sanders (2000); Bekker *et al.* (2002); Pedersen *et al.* (2000)	•	•••	•••	NA
Cooperative evaluation		•••	•••	•••	•••
Cooperative Requirements capture	Muller *et al.* (1995)	•••	•••	•	•
Future workshop		•••	••	•	•
PICTIVE	Muller (1992)	NA	•••	••	NA
Storyboards	Muller *et al.* (1995)	•	•••	•••	NA
Storytelling	Moggridge (1993)	•••	••	•	NA
TOD	Muller *et al.* (1997)	NA	•••	•••	NA

world setting, rather than in a controlled environment. They are an excellent way to see how technologies, products and artefacts are used in practice, and how the real conditions of everyday interaction impact on the way they are used. Applications vary in breadth of focus, in the degree to which the investigator influences and interacts while collecting information, and in the extent to which current, rather than historical events, are of interest. The focus may be upon a very limited set of behaviours, e.g., how people empty vacuum cleaners, or upon a broader domain, e.g., how are house-cleaning activities performed. Observations may involve minimal disruption to natural and real-time behaviour, for example, watching without interrupting or video-recording using a discreetly positioned time-lapse camera. Alternatively they may rely heavily on a participant's involvement with the investigator, as in shadowing an individual over a period of time. Others depend to a higher degree upon recall of past or usual ways of behaving. For example, participants may be asked to demonstrate, explain and/or reflect upon how they perform a particular task, or to act as a tour-guide and interpreter for the investigator, pointing out and commenting on the significance of artifacts and design elements in an environment. The investigator may be studying the use of technologies in general as they exist, evaluating specific existing designs, or advanced design prototypes in context. Two variations on this theme are behaviour sampling and shadowing.

Behaviour sampling

Behaviour sampling involves making a series of brief, sometimes momentary, observations of people's activities, often over an extended period of time and in a variety of locations. It is helpful in situations where variation over time of day, week and month are of interest but it is not practical to maintain continuous observation. Behaviour sampling can be achieved by remote monitoring through time-lapse video-recording or by an investigator making personal visits, for example to identify the locations within a hospital that are congested or underused and where staff and patients tend to congregate or seek private moments. Behaviour sampling is also sometimes achieved by enlisting participants' help remotely. An example is described in Fulton Suri (2001) in which people were given pagers and called intermittently over a one-week period and asked to record where they were, what they were doing, and with whom. The goal was to anticipate the needs and fears of future patients with implanted defibrillators that might shock them at any time as they go about their daily lives.

Shadowing

Shadowing is a form of contextual observation that involves following selected individuals as they conduct everyday activities through space and over time. This is a useful method for examining, for example, how specific work-roles play out in practice, as individuals interact with other people, tools, technologies and processes in the conduct of their working day. It is also a helpful way to understand the relative delights and frustrations that people experience as they interact with products and systems over time. In the context of designing hospital systems, products and services, for example, it may be appropriate to shadow a sample of care-givers, including doctors and nurses, as well as a sample of patients and their loved-ones to elucidate the range of different behaviours, perceptions and needs.

When to use

Contextual observations can be used at all stages of the design process but they are often most valuable in the early stages when a design team needs to develop an understanding of user needs, requirements and preferences. This may be particularly important when the design team is not familiar with the usage context. Contextual observations can reveal:

- the physical and social environmental factors that need to be taken into account
- the realities of people's practice in contrast with formal processes
- insights concerning user preferences and attitudes, and the impact of social-cultural issues
- work-arounds developed to cope with design inadequacies and/or people's own physical/cognitive abilities
- the functional and emotional significance of specific artefacts or rituals
- opportunities for design improvements in terms of people's processes and experience
- unusual physical or cognitive demands where dynamic behaviour in space and temporal factors may need to be accommodated.

Contextual observations can provide realistic contextual texture for the design team to use as reference material throughout the design process.

Participant involvement

Participants may be passively or actively involved in collecting information through these methods. Where participants are only passively involved, it is helpful to have at least one or two participant representatives to assist with interpreting observed data, or to triangulate using other methods such as debrief interviews. Active involvement of an investigator with participants can create an empathic relationship which can lead to unexpected insights, but also to undue influence over participant behaviour. The appropriate relationship will depend in part on whether you are really seeking insight and inspiration, or factual data. It is generally best not to maintain a formal tone. Aim for a conversation rather than an interview to maintain naturalism. Long periods of immersion are often better, especially in situations where time is needed to settle down into the 'culture'.

Designer and developer involvement

It is recommended that as far as possible the design team is directly involved in field-work and subsequent analysis. Designers and developers can make observations themselves, or at least review raw recorded materials, so that all design team members get first-hand exposure to the realities of how the context might affect the design.

How to use

Process involved

Decide on the contexts and vantage points (static or dynamic) that are needed. Decide how participants will be involved and to what extent the investigator needs to influence or participate with them. The investigator needs to take care that they get to see what is relevant to their enquiry, and that enough time is taken to ensure that the required information is collected. Develop a checklist of things the design team needs to know about and decide on the kind of data collection to be used. The investigator also needs to take care to get any 'permissions' to observe and/or get access to the context of interest. Prepare the way in which the data will be analysed and reported. As with all investigations pilot your data collection techniques and your analysis methods first.

Expertise needed

If active collaboration with participants is required the investigator needs to quickly embed themselves in the local culture and establish an effective rapport with the participants. Parameters to consider may include language, age, experience and so on depending on the nature of the enquiry.

Outcomes

The reporting of discoveries derived from contextual observations is most valuable when it preserves the rich audio/visual data that is in itself informative and inspiring to the design team. Raw data are frequently in the form of audio/visual records, whether still-images, video, sound recordings, maps, plans and sketches. The goal of contextual observations is often to get a holistic view of the usage circumstance of interest. Consequently contextual observations can often deliver large quantities of very rich and diverse data which can be difficult to analyse. The process of design requires the observations to be interpreted in terms of how the design needs to accommodate the 'lessons learnt'. This challenges the investigator to draw out useful information for the design team quickly and efficiently. The investigator will need to apply structure to get at high level issues and more generally applicable principles. So it is often best to report back in debrief presentations or workshops that incorporate video-clips and photographic imagery to illustrate the general principles.

Examples of use

Dray *et al.* (1996) report on an ethnographic study of family life and the use of information technology.

Eggen *et al.* (2003) report on a study into the home in which contextual observations featured as a method for eliciting peoples' perceptions about what made a 'house' a 'home'.

Gaver *et al.* (1999) report on a study into the perceptions of elderly people to the use of technologies in a large housing estate.

Gilmore (2002) discusses the value of home visits in development of an internet appliance.

Juhl (1996) reports on how interviewing and observing people at home can provide key insights into the requirements for future products for a major developer of software.

O'Brien *et al.* (1997) discuss the role of ethnographic studies in investigating the design of interactive systems in domestic environments.

Spicer *et al.* (1984) report on a study that involved direct observation of techniques that people used to get into and out of passenger cars.

Tomlie *et al.* (2002) report on the use of ethnographic methods to investigate domestic routines as part of a European Union project on the application of ubiquitous computing.

Fulton *et al.* (1981) describe the use of fixed cameras and time lapse recording of passengers on the London Underground and in other locations to capture the incidence of key behaviours.

Mäkelä *et al.* (2001) describe a design case-study involving the invention of computer-based toys for young children.

Diary-keeping

Overview

Diary-keeping methods include a range of self-documentary techniques in which participants are asked to record events, thoughts and feelings through the course of specific activities or time periods within their natural context. The key feature of diary-keeping is that a participant is not directly supervised by an investigator. The methods involve asking participants to make written, web-based, photographic, video or audio recordings that can be later reviewed by the investigator, with or without further involvement of the participant. Depending on exactly what you ask people to do and record, you can gather information about contexts of use, activity though time and space. Diaries can also be used to collect data on preferences, frustrations, needs, motivations and attitudes.

When to use

Diary-keeping is valuable at all stages of the design process, though most useful during initial explorations and where prototype solutions are available for evaluation in context. Diary-keeping also allows examination of how people are really using and reacting to the design once it is implemented. Diary-keeping methods allow you to gather information from the context of use without having to be there physically. This is useful when you need to know about behaviour taking place over time and in places that are difficult to observe directly. It also means that you can gather information from geographically/culturally diverse contexts fairly inexpensively. Diary-keeping can be useful as a starting point for follow-up interviews or contextual observations. For example to learn more about specific events or interactions of interest. In particular, methods that involve visual records can serve as a useful focus for interviews with people where there might be initial reluctance or difficulty in just talking as, for example, with children or people from another culture. Dairy-keeping can also be used as a form of behaviour-sampling technique. For example, participants can be prompted via phone or pager (as described by Robinson, 1996) at random times of day to capture thoughts, activities and context instantaneously. This allows you to gather more objective data as it is rather less dependent upon self-editing. It some cases it works well to ask participants to be involved in diary–keeping as a follow-up to interviews or other activities. They may be more motivated to participate after they have made a connection to the subject matter and the investigator.

Participant involvement

Participants are obviously actively involved in collecting the information required. These methods put particular emphasis on ensuring that instructions are clear and that the data can be easily collected. It may be advisable to give participants some training in the type of data the enquiry requires as part of the briefing session.

Designer and developer involvement

By definition designers and developers have no involvement in the collection of the data from participants. However, it can be useful to involve designers and developers in the diary-keeping exercise as respondents themselves; their participation can heighten their general awareness and sensitivity to user issues, and whet their appetite for information obtained from users with experiences similar and dissimilar to their own. As with other methods close involvement with the preparation of the questions, the design of the data gathering tools and reviewing the results is advised.

How to use

Process involved

To get the most value out of these methods you need to provide participants with equipment and instructions that make it very easy to do what you want them to. It is best to design and create self-contained 'kits' that include everything they will need. A kit might include instructions about what and when you want them to record, a log-book with defined fields to check or fill in, prompts to remind them to think of specific issues, labels/stickers to attach to photographs they take, camera/video/tape recorder, instructions about how to use the equipment, contact information, and instructions about returning their contributions.

Expertise needed

The key to designing and implementing good diary-keeping methods is to make it simple and enjoyable for participants to provide the information you need. Like questionnaires, diary-keeping tools usually benefit from careful design of graphic layout so that experienced graphic designers can be a great help. Data are generally qualitative in nature and so benefits from being analysed sensitively by someone, or a team, experienced in this kind of data analysis.

Outcomes

As with all self-reporting methods, what you learn depends upon people's motivation, ability and willingness to record information. It can be difficult to control the quality of the data recorded and to ensure that the investigator receives a complete record. Investigators can often be faced with big differences in the level of detail and the accuracy of the data recorded. In some cases this might be the information the enquiry requires but if not then the only defence is to carefully prepare the participants. It is often valuable to explore issues using complementary methods such as debrief interviews, focus groups and user workshops.

Examples of use

Palen *et al.* (2002) review many different dairy-keeping techniques in the context of studying mobile work.
Robinson (1996) describes 'beeper studies' in which participants are prompted, via a pager, at random times to capture thoughts, activities and context instantaneously. This is essentially a behavior sampling technique that allows the gathering of more objective data than many diary-keeping methods, as it is rather less dependent upon self-editing.
Sellen *et al.* (2002) report on a diary study of how knowledge workers use the Web.

Framework development

Overview

Frameworks are interpretive tools for a design team. They help build consensus and make sense of complex information around user issues. They comprise simple graphical models and diagrams that enable the team to structure information and provide a shareable view of key issues and inter-relationships. Their purpose is to make these issues and relationships explicit in a way which helps the design team make their design decisions. Framework development involves many different ways of representing ideas, discoveries or knowledge about people's behaviours, activities, thoughts and feelings in simplified summary form. The focus may be upon the relationship of behaviours through time or space, e.g., a sequence of events or actions involved in a specific task or experience — these are sometimes referred to as task synthesis or journey maps. Alternatively the focus may be on more abstract similarities and differences between types of activities, motivations, people or contexts. For example, in designing a new airline check-in system one simple relevant framework might be a diagram sequencing information flow between a user and the system; another might be a matrix differentiating users as experienced vs. inexperienced and resident vs. foreign visitors. In many cases, framework development is a matter of making explicit and agreeing upon what are actually implicit conceptual models among members of the design team or user community. Frameworks may be formalised descriptions as in traditional flow charts describing task sequences, or looser more informal visual descriptions that suggest relationships, such as Venn-type diagrams, or network maps.

When to use

Framework development is a useful synthesis activity that is applicable to all kinds of investigative methods. The activity of creating frameworks enables the design team, with or without the involvement of participants, to discuss and agree upon useful patterns that capture otherwise complex details. The greatest value is in creating an explicit and shared conceptual model for the team to reference. So, in other words, frameworks can be powerful stimulants of communication primarily within the design team, but also between design team and participants. Framework development is valuable for different reasons in all phases of the design process identified earlier.

- In Phase 1 it is useful to develop 'frameworks for looking' to focus attention and help decide on specific activities/issues/people to involve in explorations. It is useful, following explorations, to develop 'frameworks for explaining' that summarise the discoveries made and areas where more information might be necessary.

- In Phase 2 the same frameworks can be used to structure opportunity areas and provide a basis for the generation of design ideas — 'frameworks for idea generation'. These frameworks also serve as simple tool to communicate why a team has focussed on a specific set of design aspects.
- In Phase 3 frameworks serve as a reference for planning evaluation studies and highlighting where problems arise.
- In Phase 4 focus the enquiry on the key issues that need to be evaluated.

Participant involvement

Users may be passively or actively involved in the development of frameworks. Users are passively involved when they implicitly provide the data that underpins the framework. In such cases, it is important to check the validity of the framework by reference to specific individuals who have been, or will be, involved in first-hand enquiries, e.g., contextual observations and/or interviews. When participants are actively involved in the development of a framework they might be asked to contribute their own conceptual maps and then together with the design team and a facilitator to develop a consensus view.

Designer and developer involvement

As far as possible, the design team should be involved directly in developing frameworks. All team members should be encouraged to surface their own preconceptions and together reach understanding and take ownership of the framework.

How to use

Process involved

As described above, frameworks can be used for different purposes. Although sometimes the same basic framework can be used at any phase, it is often useful to consider alternative models. Developing useful frameworks is a creative activity. The way frameworks evolve is highly influenced by the nature of a design project, the phase the design team is in, and the type of material the design team wants to describe. In essence it is about finding and representing graphically the relationships that exist between elements. Elements might be ideas, individual people, types of people, activities, tasks, goals, physical things, spaces, etc. One good way to start is simply to list these elements and look for clusters, sequences, connections and patterns that are interesting. There is no standard way to create a useful/relevant framework although there are a couple of failsafe ways to proceed. One is to use the intrinsic spatial and temporal dimensions associated with interactions, such as the steps involved in a process (e.g., sending a text message or being the victim of a medical emergency), or the links between people and spatial elements (e.g., types of communications between people in a work setting). Another is to look for polar opposites that describe important dimensions, e.g., personal–shared, specific–generic, technophobic–technophilic, high investment–low investment etc.

Expertise needed

Framework development benefits from the involvement of a combination of analytical and visual thinkers who are also able to help others interpret their own ideas or points of view in a visual way. People skilled in visualisation of information can also help maintain the pace of the process by quickly articulating structure and relationships between elements.

Outcomes

Frameworks and diagrams are helpful in providing an easy way of understanding often complex information. They can be very useful in not only clarifying issues for a design team but also for presenting the demands and challenges of a project to clients and senior management. They very often reveal the relative strengths of relationships, gaps in knowledge, opportunities for improvement or innovation and so on. Frameworks such as task-flow diagrams for example, can reveal where delays or excessive mental work load and hence errors might occur. Network diagrams will show up key formal/informal communication channels and social structures that the design must support. Matrices of people types, activities and/or contexts can show up areas or parts of the process that have been neglected.

Interviews

Overview

Interviews are an inherent part of involving participants in the design process. Interviews are an essential and valuable method of data collection for all phases of the design process. Interviews are basically a guided conversation between a respondent and an investigator about his or her perspective on some issue of mutual interest from which the investigator wants to collate certain information. Interviews may be highly structured and focussed on specific questions, they may be open free ranging discussions, or they may be any one of several combinations. Unfortunately there has been little discussion in the literature on the strengths and weaknesses of interview techniques in design development work, and more particularly any systematic evaluation of the approaches (McClelland, 1984; Meister, 1986). However, most experience with interview techniques shows them to be extremely important.

When to use

Using a design is a dynamic process that has many facets concerning the physical, perceptual and cognitive aspects; how information is absorbed from the design, how it is interpreted, and what actions follow as a result. Interviews are an excellent method for gathering such information. Interviews are good for revealing participant opinions and subjective judgments. Useful for gathering insights into complex cause and effect relations.

Type of participant involvement

The interviewer should aim at giving the interview the flavour of a conversation rather than an inquisition. Achieving this is largely a question of professional expertise and the ease with which the interviewer can establish a rapport with the participant(s). Ensure that the participant is put at ease. Investigations involve participants in unfamiliar circumstances and, however well they are briefed, participants tend to be uncertain about what to expect. The way an interview is handled is important if the co-operation and interest of the participant is to be obtained and maintained. One way to achieve this is to use 'props' as part of the interview. 'Props' can be the catalyst that provokes responses and helps to engage participants in a dialogue. Some representation of the design(s) of interest are often used in this way. The presence of a design can be an invaluable aid to explaining points.

Interview formats

Interviews may be one on one, or involve several respondents and several investigators simultaneously. One of the main stimulants for the evolution of these approaches has been the desire to explore more productive formats for obtaining information from users. The format of one investigator and two users has been employed, sometimes referred to as 'co-discovery' or 'dyads'. In one example it was used to investigate the problems of users installing a small computer (Comstock, 1983). More generally this format has been advocated as an aid to protocol analysis and in expert knowledge elicitation (see Chapters 1 and 8) where users are required to verbalise their interpretation of a product as a trial progresses (Rubinstein *et al.*, 1984). Ericsson *et al.* (1980), Lewis (1982), and Olson *et al.* (1984) also discuss the use of this approach as part of using verbal reports in user interface evaluation. In recent years group interviews, often referred to as focus groups, have become much more common in design development work. Focus groups involve one or two investigators and several participants. The value often claimed for this approach is that participants are stimulated by the observations of their peers and additional insights emerge which would not otherwise be the case. On the other hand concerns are that particular individuals, or issues, can dominate a discussion with the effect that important issues are obscured or lost. Good facilitation can counter these tendencies — see Krueger *et al.* (2000).

Type of designer/developer involvement

Interviews can be excellent formats for enabling design and developer team members to experience first hand participant opinions. But an important caution is not to overwhelm participants with several interviewers.

How to use

Process involved

At an early stage the investigator needs to decide on the type of interview that an investigation requires. Whether an open style or closed style of interview is used the investigator needs to be clear about the information that the interview needs to deliver. Correspondingly, the investigator needs to work out the questions that need to be asked, their sequence, and the way in which responses will be recorded. The investigator also needs to consider whether any particular devices, artefacts or other 'props' will be used to help the participant(s) answer the questions.

Expertise needed

The interviewer needs to be able to put participants at ease and cultivate a rapport with them quickly. In multi-participant formats the interviewer also needs to be able to achieve a balanced discussion between participants.

Outcomes

Sometimes just talking is at times an inadequate form of expression. So consider the use of different media for recording the outcome of the interview such as writing, drawing, videoed explanations etc. Whichever approach is used the purpose for the investigator is to gain insight into the opinions of the respondent(s), and generate useful data for the design team.

Examples of use

Belloti *et al.* (2000) report on a study into the use of an office information management system that featured on-site interviews as a key part of the study.

Spicer (1987) carried out a study on solid fuel space heaters for domestic use. Users individually carried out a set of tasks with the products under investigation and gave their own assessments. Subsequently they were brought together in small groups to compare their experiences.

O'Brien (1982, 1987) used group interviews in the context of 'participatory design' exercises for the identification of user requirements and subsequent evaluations using product simulations for the design of control rooms. Similar approaches to the design of workstations involving groups have been adopted by Davies *et al.* (1986), Murphy *et al.* (1986), Pikaar *et al.* (1985), Stubler *et al.* (1986) and Wilson (1991, 1995). Although these have involved well defined and smaller user groups the principles could be equally well applied to user groups representing larger populations (see also Chapter 36).

Questionnaires

Overview

Questionnaires are traditionally paper based tools that incorporate a series of questions to be answered in a predefined order. Their main purpose is to collect people's responses to questions of fact or opinion in a systematic way so that the data can be easily analysed. They can vary greatly in complexity. In recent years telephone, PC based or on-line presentation techniques have become more common ways to administer questionnaires.

When to use

Questionnaires are suitable for any type of question that can be answered 'on paper' — either by a verbal or written response. The types of questions asked are either open ended or closed. Open-ended questions do not limit the respondent to a pre-defined type of answer. Open-ended questions have the advantage that they are likely to generate a wide variety of responses and provide more detailed information but they require post-hoc coding during analysis. For large surveys this type of question can be very revealing but time-consuming to analyse. In closed questions the respondent chooses between a set of predefined categories. This is an efficient way of collecting basic factual information, and collecting rating and ranking data for scaling preferences and attitudes. In general they are very easy to set up and administer providing they are well designed (Oppenheim (1999); Rea *et al.* (1997) and Chapter 4 of this book). Questionnaires are not good for getting participants to describe complex cause and effect relations.

Participant involvement

Questionnaires can be completed either by the respondent or by the investigator. Often a combination is used. In either case questionnaires often form part of a structured interview. Questionnaires may also be used in surveys where participants complete the questionnaire without direct supervision of the investigator.

Designer and developer involvement

Designers or developers tend not to participate in administering questionnaires unless they form part of an interview. But it is often useful to involve them in deciding on the topics that need to be addressed and the kind of information that the questionnaire needs to deliver.

How to use

Process involved

The key issue for the investigator is to decide on the questions that need to be asked, and the way they should be answered. The investigator also needs to decide how to administer (paper based, screen based or on-line), in what context will participants be completing the questionnaire, and will it be administered by the investigator or will it be self completion, It is always good practice to draft up the questionnaire and run a pilot. This should include analysing some hypothetical data as a means of checking whether the questions are likely to reveal the information that the design team needs. It is also important to check the layout of the questionnaire, especially if it is to be completed by the respondents at some remote location.

Expertise needed

Basic questionnaires are relatively straightforward to design, but if the array of questions is large and large quantities of data will be generated then a specialist in questionnaire design may be required. Questionnaires usually benefit from carefully designing the graphic layout. Experienced graphic designers can be a great help. Finally, the way the data is to be analysed needs to be considered and planned by someone experienced in the statistical analysis of data.

Outcomes

Questionnaires can be used to generate the following type of data.

Factual statements where simple yes/no responses or specific items of information are required.

Multiple category questions where categories are specified and the user chooses which of the categories apply. They can be answered on a yes/no basis or using some form of rating scale.

Rating scales used to assess the attitudes of users to specified product attributes. Commonly used to assess comfort, degrees of convenience, ease of use, perceived degrees of difficulty and so on.

Ranking scales used to indicate the relative order of a set of conditions or attributes according to a specified criterion.

Projective techniques

Overview

Projective techniques are methods that invite people to express ideas, thoughts and feelings in forms that rely less upon verbal expression and more upon making things, creating or reacting to imagery. Imaged-based methods — such as collage-making, word/image association, and model-making — offer a level of enquiry that taps people's non-verbal cognitive and emotional experience.

When to use

Projective methods are suitable primarily for exploring emotional aspects of design or a domain and the meaning that people attach to it. The methods are valuable in situations where people may find it difficult to articulate or reveal attitudes and thought-processes verbally. This might include occasions when: the topic is inherently non-verbal (e.g., describing spatial or cognitive perceptions); there are social taboos involved (e.g., discussing finance or personal hygiene); the topic of interest is an abstract or complex one (e.g., entertainment or parenting).

Projective techniques are useful primarily at the exploration phase as a way of uncovering latent feelings and needs and in revealing key qualities that can later be interpreted as early design ideas.

Participant involvement

These methods challenge participants to do things that may be unfamiliar to them or that they are nervous about. Participants need to be put at ease, invited to engage in a spirit of fun and reassured that the exercise is about a process of expressing their ideas and perceptions' not to produce art. The role of the investigator here is to ensure a non-threatening environment and convey supportive curiosity about what and why the participant has produced what they have. Participants themselves are encouraged to explain the meanings and associations of the images. Projective techniques can be applied one-on-one with participants, or in group sessions.

Designer and developer involvement

In group sessions it can be useful to engage design team members, even clients, in the creative process. Workshop settings provide more scope for their involvement in the process to encourage personal insights and also a sense of shared discovery. When conducted one-on-one, design team members may be remote from the creative event itself. In reporting discoveries there is great value in exposing selected examples of the actual products themselves — collages, maps or models with commentary to provide interpretation — to illustrate specific points of a more general nature. Of collages for example, Serpiello (2001) writes: 'A project report — text upon paper, is easily ignored and hardly passionate, but a three-foot poster plastered with images of what is most important to the client's customer packs a punch.'

How to use

Process involved

Projective methods can be used in association with one-on-one interviews, or in activity/workshop settings. The methods all need to be introduced to participants after they have become comfortable and feel safe and reassured that you are genuinely curious about their internal mental world and reactions. This would usually be at least 20 minutes into an open-ended interview session, or after group warm-up activities at a workshop. All these activities can be done in a wide-open exploratory manner, but it is helpful to provide participants with a clear request and a fairly short time limit so that their responses are spontaneous rather than rationalised, which would defeat the point. It is best to take the lead in deciding how much discussion goes on during their creation phase. Some people like to talk aloud while they create and enjoy explaining and being asked questions. Others like to focus exclusively upon what they are making and engage in conversation when they have finished. In either case, the participant is invited to explain the elements and layout of their creation, why they selected specific images, arranged elements as they did or chose a particular shape. This inquiry requires sensitivity. Some decisions will have been made at a less conscious level than others (and these may be the most interesting) so it is important not to force the participant to make up explanations. The discussion should rather take the form of a mutual discovery process where the investigator is the encouraging audience. Participants need to be provided with whatever tools they need. Examples are clay, foam board, paper or card, drawing instruments, stickers, images culled from magazines or photocopied from books, photographs, word labels on cards, glue, scissors and modelling materials.

Expertise needed

People with excellent social skills who are able to put participants at ease, provide clear instructions, and, perhaps of most importance, are skilled interviewers. The value of these methods often lies in the participants' explanations of the significance of what they have created. It can also, in group sessions, be very helpful to have skilled model makers and illustrators on hand to help participants express their ideas.

Outcomes

The methods provide an alternative form of expression for thoughts and feelings. The process itself enables participants to unearth associations and make discoveries about their own perceptions and use the design as a starting point for discussion and personal storytelling. The methods provide insight into: the role and meaning of personal and culturally relevant rituals; mental models and cognitive structures; emotional valence of specific activities and design elements; significance, associations and cause and effect relating to experience.

Examples of use

Lynch (1960) shows examples of people's maps and drawings to explore how they understand and perceive local and world geography.

Serpiello (2001) describes how collage-making was used in a design programme involving the development of new kinds of water-delivery mechanisms and plumbing fixtures.

Role-playing

Overview

Role-playing methods include a range of immersive techniques in which members of the design/development team are asked to personally take part in using products, environments, events or services from the perspective of a user. The methods involve the design team, sometimes with props to assist them, taking on a specific set of characteristics — motivations, abilities, limitations — and performing or participating in activities in the real world or contrived environments. Role-playing is used mainly by design teams alone but it is an approach that also provides an opportunity for participants and a design team to work together to explore a particular design issue.

When to use

Role-playing is valuable at all stages of the design process. In the early exploratory phase people can learn about design requirements to support various users and roles. It is beneficial as a direct method of generating ideas in situ as role-players are engaged in life-like activities and contexts. As prototypes and simulations are developed, role-playing enables the players to evaluate their strengths and weaknesses. And finally, once a design is implemented, role-playing provides an opportunity to gain insight into the kinds of experiences that people have in using and reacting to the design.

Participant involvement

Participant involvement in role playing techniques tends towards acting out the roles they would have in reality.

Designer and developer involvement

Role-playing methods allow you to engage the design team in first-hand explorations and discoveries about the design domain from a range of perspectives different from their own. It is useful when you want team members to deeply understand the issues facing different users in different contexts. This personal understanding is valuable as inspiration in generating, evaluating and refining design ideas to take diverse users' issues into account. In contrived settings particularly, when it is not their turn to 'play', other team members may be available to observe the action of their colleagues. This provides an opportunity for another source of insight about issues and solutions.

How to use

Process involved

Role-playing methods require a balance of sufficient structure and direction for team members to feel comfortable participating, and flexibility for them to improvise and make their own discoveries through the process. You need to define the role you are asking the players to play, and provide some basic props that will help with that. For example, each player can be given a reference card that describes the personal characteristics of the abilities, motivations, goals and specific activities for their role. These require careful planning to ensure that you are exploring an appropriately diverse set of conditions. Simple props, modelling materials or prototypes may be used to enable particular contexts for the action to be created. The players may also be provided with contrivances designed to limit specific abilitiese, e.g., earplugs to limit hearing, latex gloves to reduce sensation, bandaged joints to restrict movement.

During or immediately following enactment, it is important to facilitate reflection about individual experiences and what's been learned. This could be by asking players to keep diaries, for example, or to participate in a group session where individuals and observers share their discoveries in a semi-structured way, e.g., first listing problems and issues and then brainstorming solutions. Role-playing methods allow a design team to engage in first-hand explorations and discoveries about the design domain from a range of perspectives different from their own and to understand deeply the issues facing different users in different contexts. This personal understanding is valuable as inspiration in generating, evaluating and refining design ideas to take diverse users' issues into account.

It is important to recognise that role-playing, while it has the beneficial aspects of creating personal insights into the experience, needs and perceptions of other people, is to a large extent based upon imagination and fiction. It should never be used as a substitute for information gathering from the real world and attempts to understand the nature of other people's reality thorough direct interaction with them.

Expertise needed

Running a successful role-playing session benefits from good social and group facilitation skills but it relies also on a team's willingness to participate. It can help to give team members confidence in overcoming natural shyness to work with skilled teachers of improvisation methods, often actors themselves.

Outcomes

Designers' own experiences while role-playing will engender empathy for users, enabling them to appreciate issues including physical difficulties, cognitive confusion, and contextual issues in use. By encouraging members to enact the role of potential users you can facilitate a common team understanding of the critical user issues. Information becomes more vivid and engaging when it resonates with personal experience. If designers and clients can have informative personal experiences, it is easier for them to grasp the issues and feel greater empathy with both the people who will be affected by their decisions, and the experiences users may face.

Examples of use

Moore (1985) describes the author's experiences while she was disguised as an elderly person and is a classic example of learning through role-playing about the needs and challenges of people different from ourselves.

Buchenau *et al.* (2000) describe several applications of role-playing methods including body-storming.

Scenario building

Overview

Scenarios mean different things to different people (Campbell, 1992). But the common thread of scenario building, in a design context, is about creating one or more fictional portrayals involving specific characters, events, products and environments, to explore and envision design ideas, technologies or issues in the context of a realistic future. These stories may take many forms: text narrative; annotated sketches; cartoons; photographs; video; or live enactment. They may vary in scope and scale, involving entire processes or events over a period of time, storyboarding of a product's interaction behaviour, or vignettes that portray a brief moment or single event. Scenarios should not be just fantasies, but carefully constructed projections of anticipated usage situations that are based on available evidence.

When to use

Scenario building is a powerful exploration, prototyping and communication tool, particularly useful early on in the design process, before committing substantial resources to detailed design and development. Campbell (1992) identified four distinct purposes for future-focused scenarios: to illustrate the use of a system; to evaluate system functions; to design attributes or features; and, though of less value in a design context, to test theory. In phases 1 and 2 of design scenarios are useful for exploration and idea generation, to learn and respond to usability and lifestyle issues that will be relevant to a particular group of users, and how contextual and procedural factors might be accommodated. In phase 3 they are useful as an evaluative and diagnostic tool, to discover faults that exist in a design concept, and how they might be rectified or how alternative or competing solutions, interaction protocols might play out in practice.

Participant involvement

Participants in some cases may be directly involved in developing scenarios about their current or future activities. In others they may themselves not be directly involved but will provide the base material about real people in real contexts which will inform the development of a small set of fictional characters, personas, and settings that feature in the stories. The value of abstracting lessons from a fictional character set is threefold: it allows the team to know intimately and talk publicly about them without being invasive of real people's lives, it allows the team to easily grasp and relate to a huge range of important human differences embodied in a memorably small group of characters, and it allows the team to project people into future situations.

Designer and developer involvement

Scenario-building may be the responsibility of an individual on a team, often the user advocate, but often it may also involve participation of all members of the design team. Since it benefits from complementary skills of analysis and synthesis, verbal and visual fluency, there are roles for diverse talents. Moreover, one of the main benefits of scenario-building is that it provides a shared view of user issues that the entire design and development team can refer to. For this reason it is valuable to develop a sense of shared-ownership through a high level of participation by the design and developer teams in creating scenarios and interpreting the lessons. Scenario building offers the opportunity to involve the developer organisation in appreciating the user perspective on the design in question.

How to use

Process involved

Scenario building has its roots in the more traditional techniques of user profiling, task analysis and system ergonomics (see Chapters 6 and 10). The process starts by analysing the relevant human characteristics, motivations, tasks, social and technology trends and contextual issues that need to be considered. The elements represented in scenarios are (1) a set of individual users, detailed with respect to abilities, lifestyle and circumstances, (2) issues, goals, tasks and situations, and (3) the design itself. The design might be a well-defined proposition, or a loosely defined idea to be developed through the process of building a scenario. Next, the elements are woven together to create a coherent and believable story or vignette. Like all creative endeavours, scenario building is itself an iterative process. Gaps may appear in knowledge about the people, contexts or technology which prompt a further round of analysis and restructuring.

Expertise needed

The key skill in scenario building is the ability to synthesise and weave together diverse information concerning people, their characteristics and motivations with contextual issues and technology, product and service ideas to create a believable story. Narrative development skills are essential and, depending upon the medium of expression, technical skills in video, illustration or photography can be useful. Frequently it is beneficial to have an expert in the appropriate technologies available.

Outcomes

Scenario building has value both as process and product. As a process, creating scenarios has the effect of forcing the team to think through usage issues early. With very low investment it provides an easy way for users, team members and clients to explore, discuss and grasp how a design might work in practice. It encourages the formulation and exploration of multiple questions about usage and context over time as questions arise such as 'what if . . . he doesn't notice the blinking light? . . . someone interrupts him now? . . . he has brought a child?'

As a product, once built, scenarios promote understanding and identification among a design team and clients with the users portrayed as specific individuals. They provide a succinct way of representing and realising the temporal and spatial consequences of design decisions as they may play out in reality and so allow qualitative evaluation of dynamic aspects of interactions with a design. Scenarios have the effect of bringing to life a detailed task analysis and embedding it in context so that it becomes an easily shared portrayal of positive and negative impacts of context upon interaction with a design. Its value is in providing easily shared evidence of human factors issues for design teams. A further advantage is that the resulting scenarios provide stimulus material for use in evaluation by users, in interviews or in workshop or focus group settings.

Fulton Suri *et al.* (1999) discuss some dangers to be avoided in scenario creation. Since they are inherently fictional there may be temptations to avoid or gloss over difficult situations, rely on stereotyped descriptions of people, or offer only single solutions and justify weak ideas.

Examples of use

Carroll *et al.* (1997) report on the development of requirements for a virtual physics laboratory based on creating scenarios together with staff and students.

Fulton Suri *et al.* (1999) give examples of scenarios used in various consumer product design projects.

Rosson *et al.* (2002) provide an extensive discussion of the use of scenarios in the development of computer systems.

User trials

Overview

A user trial is primarily about creating an environment, often in a usability laboratory, that enables the interaction between a design and a user to be systematically examined and measured under controlled conditions. User trials enable an investigator to measure the effectiveness of designs both from quantitative and qualitative points of view. User trials involve applying the basic principles of experimental design, measurement techniques and data analysis that apply to formal 'experiments'. So, to adapt the description Chapanis (1959) gave for the experimental method, a user trial is 'a series of controlled observations undertaken in an artificial situation with the deliberate manipulation of some variables in order to answer one or more specific questions about the effectiveness of the design.' In practice, when user trials are used as part of a design development programme, they tend to examine the initial learning phase and not habituated use.

When to use

A user trial is the classic method to use when a design team needs to evaluate a design under controlled conditions. The design is usually in the form of a simulation or prototype but the method lends itself equally well to the evaluation of implemented designs in actual use. A user trial is an excellent method when a design representation can be evaluated using usability (and similar) metrics. Metrics may include specific formal or informal performance standards that the design should meet. So the method can be used in phases 2, 3 and 4. In phase 2 the method can be used to investigate user needs and requirements. For investigations of this sort the investigator needs be aware of different type of objectives and adapt accordingly. More traditionally the method is often used in phase 3 from very early stages of concept development up to evaluation of complete prototypes. The method can be adapted to suit design evaluation under real world conditions. In this context the investigator needs to be aware that attempting to evaluate a design under 'controlled conditions' in real life circumstances can create artificial situations. User trials can be adapted to:

- exploratory studies; what usability issues are relevant to a particular group of users? What expectations do users have of early design concepts?
- diagnostic studies; what faults exist in the design, and how might they be rectified? Design faults can include operating procedures, physical configurations, instructions for use, and labelling.
- measuring performance; does a particular design meet specific performance requirements? They can be used for benchmarking a design both 'within' and 'between'; within itself as design versions evolve to check on progress towards the design objectives, and between the design and competing solutions. Competing solutions may be comparable designs from other manufacturers, or other designs from within the developers own organisation.
- investigating contextual factors; how does the design interrelate with the context in which it will be used?

Participant involvement

User trials are excellent for demonstrating user perceptions and experiences of a design to the design and developer teams. User trials can use different formats in terms of how to involve participants. Two or more participants can be involved in one user trial. Examples include the co-discovery technique (Kemp *et al.*, 1996). Monk *et al.* (1993) have also developed a 'co-operative evaluation' technique specifically aimed at evaluating early design proposals.

Designer and developer involvement

A trial enables the design team, developer teams, and clients to remotely observe the trials in real time. Real time or post hoc review of videos can be extremely persuasive when design review discussions get underway. There is potential for a high level of participation by the design and developer teams in observing, analysing, and interpreting the outcomes. There is also great potential for involvement by others in the developer organisation in appreciating the user perspective on the design in question.

How to use

Process involved

Probably the three most important issues to consider when setting up a user a trial are the key aspects about the design that need investigating, the form in which the design will be represented, and the location of the user trial. The investigator then needs to consider the tasks the participants will carry out, the performance criteria that will be used and their associated data collection method(s). The tasks need to follow a pattern that make sense to the participants in terms of how the design would be used in practice. The tasks should be also designed so that the investigation can get at the aspect(s) of the design that concern the design team. How the data is to be collected will also be heavily influenced by the choice of location. Often user trials incorporate interviews and questionnaires. It is very common these days to run user trials in a usability laboratory in which audio and video data can be recorded.

Expertise needed

Running user trials requires expertise in experimental design, statistical analysis of data and interview techniques. Expertise is also required in audio and video data analysis. In terms of the user trial itself the investigator also needs staff skilled in creating a rapport with participants.

Outcomes

User trials typically generate quantitative and qualitative data, and often in large amounts. The specific form of the data obviously depends on the metrics selected. Typical examples are task times, error scores, physical fit, comfort scores, attitudinal measures, general impressions, etc.

Examples of use

Fulton Suri (1993) reports on the use of user trials to evaluate information graphics.

User workshops

Overview

User workshops are events where a group of participants work on a design issue facilitated by the investigator. A critical characteristic is that such workshops generate tangible evidence, such as drawings or models for example, of the participants perspective on design(s) for a particular purpose. In this respect they differ significantly from focus groups which typically rely upon discussion only. The design issue chosen for user workshops needs to be one which participants have the knowledge and expertise to tackle. User workshops can be used to explore such issues as functional needs, workplace layout, general configuration of a user interface, etc. The location for user workshops can be various. It is a question of what suits the purpose of the investigation and what is convenient. In general user workshops are best supported by flexible informal spaces in which participants feel free to move about, make things (if needed), change things, without fear of 'making a mess'. The great value of such workshops is that participants often get to express their ideas through media other than simply verbal or written. While participants may be working on particular details of a design they can be also stimulated to address more fundamental concerns in terms of 'needs and requirements', characteristics that make a design acceptable or not, and so on. It is this kind of output that is what makes user workshops so valuable for design teams.

When to use

The idea of user workshops grew out of the simple idea that participants and designers could work productively together to resolve design questions. The development of the concept has evolved to encompass a wide variety of specific approaches that cover all phases of the design process. In phase 1 they can be a useful complement to ethnographic methods as a way of participants elaborating on perceived needs and requirements, and contextual issues. In phase 2 they can support the definition of the design space in terms of requirements, contextual issues and the features of a design that will make it suitable for the purpose intended. In phase 3 they can be used to support the definition and evaluation of specific design proposals, or, in the case of phase 4, fully implemented designs.

Participant involvement

Participant involvement revolves round particular activities set out by the investigator. The activities in themselves usually produce particular tangible outcomes. But one of the key objectives of any user workshop is typically to get the participants to interact with each other (as well as designers and/or developers if they participate) and thereby benefit from the cross fertilisation of their individual experiences.

Designer and developer involvement

If designers and/or developers also take part they can often collaborate with participants where the designers facilitate the articulation of participant ideas. This can be an excellent way to foster interaction between the design team and participants. The result is often that the designers not only get to understand particular ideas but also develop a more in depth appreciation of the people who are expected to use their design.

How to use

Process involved

As with all forms of workshops the investigator needs to work out a clear activity agenda that successfully facilitates the achievement of the chosen objectives. All the participants need clarity about the type of outcome expected from them. The activities and the outcome expected need to be carefully selected so that the participants do not have to climb a steep and difficult learning curve. The participants should be able to become productive quickly and thereby focus on the subject of the workshop and not be hindered by a lack of skill or expertise. If members of the design team are involved it is important that the investigator takes care to brief them before hand on the role that they should play.

Expertise needed

However carefully the agenda is designed workshops require an excellent facilitator to ensure that the objectives of the workshop are achieved.

Outcomes

The great value of such workshops is that participants get to express their ideas through media other than simply verbal or written. This can include representations of how a design might appear, visual descriptions of a system should behave, video walkthroughs showing how a user might interact with a system, and lifesize models of workplace layouts and hardware configurations. While participants may be working on particular details of a design they can be also be stimulated to address more fundamental concerns in terms of 'needs and requirements', characteristics that make a design acceptable or not, and so on. User workshops can be excellent for helping a design team to identify critical design features and rank design issues in importance.

Examples of use

Dayton *et al.* (1996) report on the use participatory design for the development of Graphical User Interfaces.

Eggen *et al.* (2003) report on a study into the home in which user workshops featured as a method for developing new application concepts.

Karasti (1997) discusses the use of workshops involving participants and designers in the development of an experimental tele-radiology project.

Mackay *et al.* (2000) report on a study into a GUI design for a 'Petri Net' application using participatory design techniques and video based tools.

Muller (2001) reviews three recent case studies that used the CARD (Collaborative Analysis of Requirements and Design) technique.

Future directions

In this chapter we have introduced a broad range of different ways of involving people in design. We now discuss some future directions and focus on a number of issues that will influence the development of methods and the professional practice of designing.

Organisational and technical barriers

As designs become more complex, technology more powerful, commercial pressures more severe and resources more expensive, there is greater need to tackle the human impact of designs.

Developers, designers and providers of products, services, environments and media are increasingly interested in methods which enable them to better anticipate people's needs and desires so that they can provide more successful designs in both commercial and human terms. Implicit in this interest is the drive to base design decisions as far as possible upon 'evidence' rather than on 'opinions'. This is precisely where the effective use of methods, of the kinds we have described here, can play a vital part. But the relevance of such 'methods' to mainstream commercial design development will depend upon their effective adoption and use. They need to demonstrate value in creating and inspiring successful design directions and in efficiently guiding implementation.

Design as a strategic business advantage

Many companies in many business sectors have long valued design as a central function in their business that complements marketing, manufacturing or engineering. In recent years design has been recognised by more companies — beyond its obvious function in generating products — as an important way that businesses can favourably differentiate themselves from their competitors. See Peters (1998, 2000) about design from a business perspective. With this key business role, it is even more important that designers successfully anticipate and provide solutions which elicit positive responses from people. Involving people in the design process has much to offer in the form of a range of ways to learn about people's behaviour and motivations.

In the longer term implementation of human centred design (HCD) processes in industry will depend on managers recognising that ensuring user satisfaction with their designs needs to be embedded in the way their organisation manages product development and product implementation. This requires a more systematic approach to design quality and usage issues than is typically the case. In other words design quality and usage issues need to be incorporated into the specific design objectives that product developers have to meet. On this basis it is then possible to select the appropriate methods and maximise their benefit by focusing on the design issues in question. The principles are easily stated but, in practice, are very difficult to implement. Gould (1988) and, more recently, Vredenburg *et al.* (2001) present valuable discussions on this topic (see also Grudin, 1990).

One of the major benefits of using the methods outlined in this chapter is that they help to make usage issues tangible and manageable. The usage issues get expressed in terms and in ways which allow design teams to manage in a balanced way the inevitable comprises and trade-offs with the technical, commercial and project planning issues. The methods also confront design teams with the 'real questions' concerning the purpose of a design and the value it has for peoples lives, in contrast to just checking on device performance as an artifact in itself.

Proving the value and relevance of methods

There is considerable scope for research work aimed at assessing the most effective methods for ensuring that human issues are taken into account within design development. In terms of the more traditional human factors and usability methods, Meister, as long ago as 1986, noted the lack of attention given to evaluating methods that are used to guide design and called for further efforts to improve our understanding of their value — see also Anderson *et al.* (1985) and Karat *et al.* (1992). It is a critique that is still relevant today. Many of the technical issues discussed are of long-term and of academic interest that lie beyond the pragmatic concerns of most investigators within industry.

Investigators within an industrial design-development environment, of necessity, have to take a short to medium term view. Nevertheless, it is of great practical importance to designers and developers to review critically where and when specific methods are most effective, reliable and valid. This kind of critical appraisal will undoubtedly be a focus for the future. However, the real and changing world of designs-in-use — where an enormous number of variables determine makes a design successful — will be a very challenging arena in which to investigate the effectiveness of methods with high levels of confidence. The successful implementation of methods is likely to depend on demonstrating their effectiveness in terms of:

- how far is it possible to predict how a design will perform in practice?
- can the performance of designs in use be improved by evaluating their performance during development?
- the cost-benefit to development organisations of improving the quality of designs, ensuring that people's latent needs are met, and that significant design deficiencies do not get to market
- reducing the time it takes for investigations to reveal valuable insights to the design team. This means reducing the time it takes to execute investigations, to interpret the results, and incorporate the results into design work.
- how people's involvement in the design process positively changes peoples perceptions of the organisation and the designs it sells.

Earlier involvement and faster response times

As mentioned above a key issue is the demand for earlier and more timely input concerning human issues, to ensure that efforts are directed appropriately at all stages in the process. Tools and methods are required which reduce the time taken to carry out investigations, including set up, collection, interpretation, reporting and application of discoveries. New technologies — e.g., digital audio and image capture, networked communications, the Internet — offer the investigator many opportunities to gather and disseminate a wealth of data rapidly, but also bring with them the potential handicap of gathering much more data than are useful or necessary and adding time for collation and analysis. In many cases, alert observation and thoughtful use of traditional tools — paper and glue, face-to-face communication — still prove to be very effective. But there is clearly room for improvement in tools and methods now available to help reduce turn-around times and speed up the information transfer process.

As HCD continues to gain ground in both organisations that develop designs, and in the professional practice of designing, the scope and boundaries of HCD concerns will evolve. And as these changes occur the methods at our disposal must also evolve. We have not the space to discuss the issues in detail but nevertheless we want to highlight some major interrelated trends that we expect to have a significant influence on the development of our methods in the coming decades.

'Ecological validity'

How well do the methods used help design teams accurately anticipate the way designs will be used in practice? In the final analysis the value of HCD is in the extent to which organisations can both increase their rate of success in introducing products and services that positively delight their users and customers, as well as reducing the risk of design failures and errors. Errors in design may have no more serious consequences than mild inconvenience. But on the other hand design errors may at times have lethal consequences for a few or, worse still, for many people. At the heart of our methods lies an explicit or implicit usage model, and it is the validity of the usage model on which rests the value of HCD. The methods we use are either concerned about understanding and model ling use, or the methods assume a certain model of use to either guide design decisions or evaluate the suitability of solutions. So a major issue for the future is the accuracy and reliability of the usage models design teams use. The scope and level of detail to which any usage model must go obviously depends on the type of design required and the organisational context of the design team. There are clearly, in principle, many factors to consider. In practice a design team has to take a pragmatic approach and make its own best estimate of the usage model required, the critical factors to consider, and the extent to which they can rely on assumptions rather than explicit evidence.

Design criteria

HCD as a concept relies on the use of explicit criteria to both guide design decisions and to evaluate the outcomes. These criteria should reflect the concerns and interests of the people who will use the design. In recent years an important and growing debate has emerged about how methods will need to adapt to cope with the variety of human issues that affect the success of a design. Design practitioners of all professional persuasions have begun to look for methods and frameworks that embrace the greater variety of human issues. The challenge is to discover what is 'simultaneously useful (needed), usable (understandable) and desirable (wanted)' (Sanders 2002). This is particularly true in the realm of designs for consumers, where emotional, experiential aspects relating to appeal, aesthetics, image, and lifestyle-fit are crucial to a product's success. A central theme in this debate is the way people *experience the design*. Correspondingly the debate also centers on the ways and means by which the *experience* can be both anticipated when designing, and measured when solutions are evaluated. Affective, social and cultural issues all contribute and influence people's experience of designs but have been largely neglected by the human factors profession (Jordan, 2000). But other professional groups concerned with psychology, market research, sociology, anthropology (e.g., Csikszentmihalyi *et al.*, 1981; Csikszentmihalyi, 1991) have made attempts to address these issues and discover tools and methods that enable these issues to be managed. One celebrated example of the kind of work now emerging is by Reeves *et al.* (1996). They report on a number of studies that suggest that people use the same kind of social constructs in their relationship with designs as they do in handling their relationships with other people. In a similar vein there is increasing interest in the affective aspects of designs, their aesthetical impact on the users emotions, and the significance of the design in terms of the social context of use. The central question is which are the critical criteria that design teams need to design to, and have we the corresponding methods that enable the criteria to be used effectively to guide design decisions and evaluate solutions?

Aesthetics and Preferences

Even today there is little established theory regarding the human aesthetic response, although practitioners are beginning to embrace the issue, particularly in relation to product design, for example: Desmet (2002), Fulton Suri (2002), Jordan (2000), Jordan *et al.* (1996), Macdonald (1998), Segal *et al.* (1997), and Sinclair *et al.* (2002). Crozier (1994), provides an overview of contributions to questions of aesthetics and preference including Gustav Fechner's experimental psychophysics, explorations of the 'golden section', the Gestalt theories of human perception, Berlyne's theory linking aesthetic preferences with physiological arousal, and Arnheim (1995) making connections between art and psychology. None of these theoretical approaches is able to fully explain or predict the way people will react to designs, not least because aesthetic preferences represent only a part of the significance of objects for people (Desmet, 2002). But our tools and methods must continue to address these matters even when we lack predictive theory and principles.

Personal and socio-cultural

Social and cultural factors are intrinsic to some kinds of design, such as shared spaces, communication and collaborative systems. But, there are also subtle and complex social and cultural implications for all kinds of designs. Designers, marketers, anthropologists and consumers know, as Helga Dittmar (1992) says '... material possessions have a profound symbolic significance for their owners, as well as for other people ... they influence the ways in which we think about ourselves and about others.' Miller (1997) suggests that we need a more formal discipline of 'social ergonomics' and notes several ways in which designs have social significance for people:

- as tools of non-verbal communication, Goffman (1971)
- establishing meaning about ourselves and our lives, Csikszentmihalyi (1991)
- stimulating certain kinds of social behavior — e.g., mobile-phones demand a response wherever we are, or the way we talk too loudly when listening to a personal stereo
- affecting our agency — e.g., when the computer is down or busy, we can't communicate with our family
- mediating social interactions — e.g., gathering-places like doors and thresholds can be complicated by automatic doors and door closures
- ascribing personality to designs — e.g., when a device or service seems 'stupid', 'friendly' or 'demanding'

These considerations will continue to be important aspects to be addressed by HCD methods.

The sustainability frontier

Since its origin in a concern with fitting tasks and environments to human need, ergonomics in design has expanded its scope. Indeed, the notion of 'human fit' has evolved over time from physical and perceptual to cognitive to social to cultural to ecological fit. We are right now at the brink of this ecological frontier. Sustainability issues loom in the face of the human creative and commercial enterprise of design and production of products, services and environments. Customers and users, especially in the leading industrially developed nations, are becoming increasingly aware of the ways that our individual behaviours and the products and services we consume are negatively impacting quality of life in a holistic sense. Businesses and consumers are beginning to grapple with systemic issues relating to human use of the earth's resources and subsequent destructive interference with natural cycles of regeneration and replenishment. If the underlying motivation for design is to enhance human existence and our 'quality of life' collectively and individually then our design practice needs to encompass consideration of:

- ways to design products, systems and services that support and encourage ecologically responsible behaviour [*what correct/incorrect mental models of ecologically responsible behaviour exist in the minds of designers, users etc.? how are these reflected in design? how can desirable behaviours can be encouraged and simplified by design?*]
- a regeneration of concern with fundamentals of human health and well-being [*how can design support and encourage health and wealth-production for the under-privileged? how can design ensure clean air and water as a widespread biological and economic necessity?*]
- establishing greater understanding of cultures and value systems other than our own [*what methods are appropriate and useful in working with people from different educational, economic and cultural experience from our own and in learning about their needs, desires and perceptions?*]

Creative and integrated methods

The overview by Stanton *et al.* (1998) of ergonomics methods used in product design reveals a clear emphasis on evaluative and analytical tools, whether quantitative or qualitative in nature. No longer is it appropriate simply to evaluate what others design and produce, nor to work as isolated purist specialists on the sidelines of design. Designers and developers have been frustrated sometimes by the results of human factors analyses and evaluative studies. Such methods provide information about people's capabilities, problems that arise and people's reactions to specific design variables. But, as we have seen, people's response to design is more complex than the traditional scope of this work and information does not, by itself, lead to design solutions. Design requires synthesis and creativity — integration of human and technology capabilities to create coherent and workable designs. To make useful contributions, we need to emphasise and explore methods that bridge this gap between analysis and synthesis; to help translate human factors information into a form which stimulates well-conceived, human-centered design ideas (Hasdogan, 1996). Our energies must be directed not simply to providing information but in engendering a human-centered and people-inspired approach to tackling design issues. Many of the methods explored in this chapter are designed to do just this, through activity, reflection and imagination by participants, designers and developers together exploring and creating designs that will be beneficial for people. We hope that this will inspire the development of more new and effective methods, and that their use will become more widespread and well-established in HCD.

References

Anderson, N.S. and Olson, J.R. (1985). Methods for designing software to fit human needs and capabilities. In *Proceedings of the Workshop on Software Human factors* (Washington, DC: National Academic Press).

Arnheim, R. (1974). *Art and Visual Perception: A Psychology of the Creative Eye.* (Berkeley, CA: University of California Press), (New version; expanded and revised edition of the 1954 original).

Bailey, R.W. (1982). *Human Performance Engineering* (Englewood Cliffs, NJ: Prentice Hall Inc.)

Bainbridge, L. and Sanderson, P. (1995). Verbal protocol analysis. In: J. Wilson and N. Corlett (eds) *Evaluation of Human Work*, 2nd edition. (London: Taylor & Francis).

Bannon, L. (1991). From human factors to human actors: the role of psychology and human–computer interaction studies in system design, In: J. Greenbaum and M. Kyng (eds) *Design at Work: Cooperative Design of Computer Systems* (Hillsdale, NJ: Lawrence Erlbaum Asociates).

Bekker, M.M., Markopoulos, P. and Kersten-Tsikalkina, M. (eds) (2002). *Interaction Design and Children*, Proceedings of the International Workshop. (Maastricht: Shaker).

Belloti, V. and Smith, I. (2000). Informing the design of an information management system with iterative fieldwork. In *DIS 2000, Proceedings of Designing Interactive Systems*, ACM Press, pp. 227–237.

Benel, D.C.R. and Pain, R.F. (1985). The human factors usability laboratory in product evaluation, *Proceedings of Human Factors Society, 29th Annual Meeting* (Santa Monica, CA: Human Factors Society).

Beyer, H. and Holtzblatt, K. (1998). *Contextual Design; Defining Customer-Centered Systems* (San Francisco, USA: Morgan Kaufmann Publishers Inc).

Bodker, S., Nielsen, C., and Petersen, M.G. (2000). Creativity, cooperation and interactive design. In *DIS 2000, Proceedings of Designing Interactive Systems*, ACM Press, pp. 252–261.

Buchenau, M. and Fulton Suri, J. (2000). Experience prototyping. In *DIS 2000, Proceedings of Designing Interactive Systems*, ACM Press, pp. 424–433.

Burns, C., Dishman, E., Johnson, B. and Verplank, B. (1995). 'Informance': Min(d)ing future contexts for scenario-based interaction design. Presented at BayCHI (Palo Alto, August 1995). Abstract available at http:/www.baychi.org/meetings/archive/0895.html.

Burns, C., Dishman, E., Verplank, B. and Lassiter, B. (1997). Actors, hair-dos and videotape: Informance design. Presented at Presence Forum (Royal College of Art, London, November 1997). Paper available at http:/www.presenceweb.org/papers.

Calde, S. Goodwin, K. and Reimann, R. (2002). SHS Orcas: The first integrated information system for long-term healthcare facility management. Co-published with permission in the Association for Computing Machinery Digital Library. ©2002 American Institute of Graphic Arts Experience Design Case Study Archive.

Campbell, R.L. (1992). Will the real scenario please stand up? *SIGCHI Bulletin*, **24**, 6–8.

Cantwell, D. and Stajano, A. (1985). Certification of software usability in IBM Europe, *Ergonomics International 85 Proceedings of the Ninth Congress of the IEA*, edited by I.D. Brown.

Carroll, J.M. (ed) (1995). *Scenario-Based Design: Envisioning Work and Technology in System Development*. (New York: John Wiley & Sons).

Carroll, J.M. Rosson, M.B., Chin, G., and Koeneman, J. (1997). Requirements development: stages of opportunity for collaborative needs discovery. *DIS 97 Designing Interactive Systems Conference Proceedings* (New York: Association for Computing Machinery), pp. 55–64.

Chapanis, A. (1959). *Research Techniques in Human Engineering* (Baltimore, MD: The John Hopkins Press).

Collins, M. (1986). Sampling. In: R. M. Worcester and J. Downham (eds) *Consumer Market Research Handbook* (Amsterdam: North-Holland).

Comstock, E. (1983). Customer installability of computer systems. *Proceedings of the Human Factors Society, 27th Annual Meeting* (Santa Monica, CA: Human Factors Society).

Crozier, R. (1994). *Manufactured Pleasures: Psychological Responses to Design*. (Manchester, U.K.: Manchester University Press).

Csikszentmihalyi, M. (1991). Design and order in everyday life. *Design Issues*, 8, 26–34

Csikszentmihalyi, M. and Rochberg-Halton, E. (1981). *The Meaning of Things: Domestic Symbols and the Self* (Cambridge, U.K.: Cambridge University Press).

Curtis, P., Heiserman, T., Jobusch, D., Notess, M. and Webb, J. (1999). Customer-focused design data in a large, multi-site organisation. *CHI 99 Conference Proceedings, Association for Computing Machinery* (New York: Addison-Wesley), pp. 608–615.

Cushman, W.H. and Rosenberg, D.J. (1991). *Human Factors in Product Design* (Amsterdam: Elsevier).

Damodaran, L. (1996). User involvement in the systems design process — a practical guide for users, *Behaviour and Information Technology*. **15**, 363–377.

Davies, D.K. and Phillips, M.D. (1986). Assessing user acceptance of next generation air traffic controller workstations. *Proceedings of the Human Factors Society, 30th Annual Meeting* (Santa Monica, CA: Human Factors Society).

Dayton, T., Kramer, J., McFarland, A. and Heidelberg, M. (1996). Participatory GUI design from task models. *CHI '96 Conference Proceedings, Association for Computing Machinery* (New York: Addison-Wesley).

Desmet, P.M.A. and Hekkert, P.P.M. (2002). The basis of product emotions. In: W.S. Green and P.W. Jordan (eds) *Pleasure with Products: Beyond Usability* (London: Taylor & Francis).

Dillon, A. (1992). Reading from paper versus screens: a critical review of the empirical literature. *Ergonomics*, 35, 1297–1326.

Dittmar, H. (1992). *The Social Psychology of Material Possessions: To Have Is To Be.* (Hemel Hempstead, U.K.: Harvester-Wheatsheaf).

Dray, S.M. and Mrazek, D. (1996). A day in the life of a family: an international ethnographic study. In: D. Wixon and J. Ramey (eds) *Field Methods Casebook for Software Design* (New York: J. Wiley & Sons).

Druin, A. (1999). Cooperative inquiry: developing new technologies for children with children. *CHI 99 Conference Proceedings, Association for Computing Machinery* (New York: Addison-Wesley), pp. 592–599.

Druin, A. (2002). The role of children in the design of new technology. *Behaviour and Information Technology*, **21**, 1–25.

Dunne, A. and Raby, F. (2001). *Design Noir: The Secret Life of Electronic Objects.* (Basel, Switzerland: Birkhauser).

Egan, D.E., Remde, J.R., Landauer, T.K., Lochbaum, C.C. and Gomex, J.M. (1989). *CHI '89 Conference Proceedings, Association for Computing Machinery* (New York: Addison-Wesley), pp. 205–210.

Eggen, B., Hollemans, G. and van de Sluis, R. (2003). Exploring and enhancing the home experience. *Cognition Technology and Work,* **5**, 44–54.

Ericsson, K.A. and Simon, H.A. (1980). Verbal reports as data. *Psychological Review*, **3**.

Ferris, K. and Bannon, L. (2002). ... a load of ould Boxology. *DIS 2002, Designing Interactive Systems Conference Proceedings.* (New York: Association for Computing Machinery Press), pp. 41–49.

Ford, J.M. and Wood, L.E. (1996). An overview of ethnography in system design. In: D. Wixon and J. Ramey *Field Methods Casebook for Software Design* (New York: J. Wiley & Sons).

Francik, E. and Akagi, K. (1989). Designing a computer pencil and tablet for handwriting, *Proceedings of the 33rd Annual Meeting of the Human Factors Society*, pp. 445–449.

Frascara, J. (ed) (2002). *Design and the Social Sciences; Making Connections.* (London: Taylor & Francis).

Fulton, J. and Stroud, P. G. (1981). Ergonomic design of automatic ticket barriers for use by the Travelling Public. *Applied Ergonomics*, December.

Fulton Suri, J. (1993). User trials for information graphics: replacing designers' assumptions with feedback from users. *Information Design Journal*, **7.**

Fulton Suri, J. and Marsh, M. (1999). Scenario building as an ergonomics method in consumer product design. *Applied Ergonomics*, **31**, 151–157

Fulton Suri, J. (2001). The Ergonomics Society — the Society Lectures 1999: The next 50 years: future challenges and opportunities for empathy in our science. *Ergonomics*, **44**, 1278–1289.

Fulton Suri, J. (2002). Whether to measure pleasure or just tune in. In: W.S. Green and P.W. Jordan (eds) *Pleasure with Products: Beyond Usability* (London: Taylor & Francis).

Gaver, W. and Dunne, A. (1999). Projected realities: conceptual design for cultural effect. *CHI 99 Conference Proceedings, Association for Computing Machinery* (New York: Addison-Wesley), pp. 600–607.

Gaver, W., Dunne, A. and Pacenti, E. (1999). Cultural Probes. *Interactions*, **January–February**, pp. 21–29 (New York, NY: Association for Computing Machinery).

Gilmore D.G. (2002). Understanding and overcoming resistance to ethnographic research. *Interactions*, **May–June**, pp. 29–35.

Goffman, E. (1971). *The Presentation of Self in Everyday Life.* (Harmondsworth, U.K.: Pelican).

Gould, J.D. (1988). Designing for usability: the next iteration is to reduce organizational barriers, *Proceedings of the 32nd Annual meeting of the Human Factors Society*, pp. 1–9.

Gould, J.D., Boies, S.J. and Ukelson, J. (1997). How to design usable systems. In: M. Helander, T. K. Landauer and P. Prabhu (eds) *Handbook of Human-Computer Interaction* (Amsterdam: Elsevier).

Greenbaum, J. and Kyng, M. (eds) (1991). *Design at Work: Cooperative Design of Computer Systems.* (Hillsdale, NJ: Lawrence Erlbaum Asociates).

Grudin, J. (1990). The computer reaches out: the historical continuity of interface design. *CHI 90 Conference Proceedings*, Association for Computing Machinery Press (New York: Addison Wesley), pp. 261–268.

Hackos, J.T. and Redish, J.C. (1998). *User and Task Analysis for Interface Design* (New York: J. Wiley & Sons).

Hasdogan, G. (1996). The role of user models in product design for the assessment of user needs. *Design Studies*, **17**, 19–33.

Hill, W.C. and Terveen, L.G. (1997). Involving remote users in continuous design of web content. In: G. van der Veer, A. Henderson and S. Coles (eds) *DIS 97, Designing Interactive Systems Conference Proceedings* (New York: Association for Computing Machinery Press), pp. 137–145.

Hix, D. and Hartson, H.R. (1993). *Developing User Interfaces; Ensuring Usability Through Product and Process* (New York: John Wiley & Sons).

Holtzblatt, K. and Beyer, H. (1996). Contextual design: principles and practice. In: D. Wixon and J. Ramey (eds) *Field Methods Casebook for Software Design* (New York: J. Wiley & Sons).

Houde, S. and Hill, C. (1997). What do prototypes prototype? In: M. Helander, T. Landauer and P. Prabhu (eds), *Handbook of Human-Computer Interaction* (2nd edn.) (Amsterdam: Elsevier Science B.V).

Hughes, J.A., O'Brien, J., Rodden, T., Rouncefield, M. and Blythin, S. (1997). Designing with ethnography: a presentation framework for design. In: G. van der Veer, A. Henderson and S. Coles (eds), *DIS 97, Designing Interactive Systems Conference Proceedings* (New York: Association for Computing Machinery Press), pp. 147–158.

Hutchinson, H., Mackay, W., Westerlund, B., Bederson, B., Druin, A., Plaisant, C., Beaudouin-Lafond, M., Conversy, S., Evans, H., Hansen, H., Roussel, N., Eiderback, B., Lindquist, S. and Sundblad, Y. (2003). Technology probes: inspiring design for and with families. *CHI 2003 Conference Proceedings* (New York: Association for Computing Machinery), pp. 17–24.

IDEO (2003). *IDEO Method Cards* (Palo Alto: IDEO).

ISO TR 16982. (2002). *Ergonomics of Human–System Interaction — Usability Methods Supporting Human-Centred Design.* International Standards Organisation TR 16982:2002(E).

ISO 9241. *Ergonomic requirements for office work with visual display terminals.* International Organization for Standardization.

Joe, P. (1997). Scenarios as an essential tool; stories for success. *Innovation Quarterly Journal of the Industrial Designers Society of America*, **Fall**, 20–23.

Jordan, P.W. (2000). *Designing Pleasurable Products* (London: Taylor & Francis).

Jordan, P.W., Thomas, B., Weerdmeester, B.A. and McClelland, I.L. (1996). *Usability Evaluation in Industry* (London: Taylor & Francis).

Juhl, D. (1996). Using field-orientated design techniques to develop consumer software products. In: D. Wixon and J. Ramey (eds) *Field Methods Casebook for Software Design* (New York: J. Wiley & Sons).

Kantowitz, B.H. (1992). Selecting measures for Human Factors research. *Human Factors*, **34**, 387–398.

Karasti, H. (1997). Bridging the analysis of work practice and system redesign in cooperative workshops. In: G. van der Veer, A. Henderson and S. Coles (eds) *DIS 97, Designing Interactive Systems Conference Proceedings* (New York: Association for Computing Machinery Press), pp. 185–195.

Karat, C.M., Campbell, R. and Fiegel, T. (1992). Comparison of empirical testing and walkthrough methods in user interface evaluation. *CHI 92 Conference Proceedings*, Association for Computing Machinery Press (New York: Addison-Wesley), pp. 397–404.

Kelley, T. (2001). *The Art of Innovation.* (London: Harper Collins).

Kemp, J.A.M. and Gelderen, T. van. (1996), Co-discovery exploration: an informal method for the iterative design of consumer products. In: P.W. Jordan, B. Thomas, B.A. Weerdmeester and I. L. McClelland (eds) *Usability Evaluation in Industry* (London: Taylor & Francis).

Krueger, R.A. and Casey, M.A. (2000). *Focus Groups: A Practical Guide for Applied Research* (London: Sage Publishing).

Leonard, D. and Rayport, J.F. (1997). Spark innovation through empathic design. *Harvard Business Review*, November–December, pp. 102–113.

Lewis, C. (1982). Using the 'thinking allowed' method in cognitive interface design. *IBM Research Report* RC 9265.

Lynch, K. (1960). *The Image of the City* (Cambridge, MA: MIT Press).

Macdonald, A. (1998). Developing a qualitiative sense. In: N. Stanton (ed.) *Human Factors in Consumer Products* (London: Taylor & Francis Ltd).

Mackay, W.E., Guindon, R., Mantel, M.M., Suchman, L. and Tatar, D.G. (1988). Panel session: Video: Data for studying human-computer interaction, *CHI '88 Conference Proceedings* (Reading, MA: Addison-Wesley).

Mackay, W.E., Ratzer, A.V. and Janecek, P. (2000). Video artifacts for design: bridging the gap between abstraction and detail. In *DIS 2000, Proceedings of Designing Interactive Systems* (ACM Press), pp. 72–82.

Macleod, M. (1992). *An Introduction to Usability Evaluation*, National Physical Laboratory, Department of Trade and Industry.

Mäkelä, A and Fulton Suri, J. (2001). Supporting users' creativity: design to induce pleasurable experiences. In: M. Helander, H.M. Khalid and M. Tham (eds) *Proceedings of The International Conference on Affective Human Factors Design* (London: Asean Academic Press).

Mayhew, D.J. (1999), *The Usability Engineering Lifecycle: a Practitioners Handbook for User Interface Design* (San Francisco: Morgan Kaufmann).

McClelland, I.L. (1984). Evaluation trials and the use of subjects. In: E.D. Megaw (ed.) *Contemporary Ergonomics 1984 Proceedings of the Ergonomics Society Conference* (London: Taylor & Francis).

McClelland, I.L., Simpson, C.T. and Starbuck, A. (1983). An audible train warning for track maintenance personnel. *Applied Ergonomics*, **14**, 2–10.

Meister, D. (1986). *Human Factors Testing and Evaluation* (Amsterdam: North-Holland).

Meister, D. and Enderwick, T.P. (1992). Measurement in human factors; special issue preface. *Human Factors*, **34**, 383–385.

Miller, H. (1997). The Social Psychology of Objects. *Understanding the Social World Conference*, The Nottingham Trent University, U.K.

Moggridge, B. (1993). Design by story-telling. *Applied Ergonomics*, **24**, 15–18.

Monk, A., Wright, P., Haber, J. and Davenport, L. (1993). *Improving your Human Computer Interface: A practical technique* (New York: Prentice Hall).

Moore, Patricia A. (1985). *Disguised!* (Waco, TX: World Books).

Moran, T. (2002). Everyday adaptive design. *DIS 2002 Designing Interactive Systems Conference Proceedings*. (New York: Association for Computing Machinery Press), pp. 13–14.

Morton-Williams, J. (1986). Questionnaire design. In: R.M. Worcester and J. Downham (eds) *Consumer Market Research Handbook* (Amsterdam: North-Holland).

Moser, C.A. and Kalton, G. (1971). *Survey Methods in Social Investigation* (London: Heinemann Educational Books).

Muller, M.J. (1992). Retrospective on a year of participatory design using the PICTIVE technique. *CHI 92 Conference Proceedings*, Association for Computing Machinery Press (New York: Addison-Wesley).

Muller, M.J. (2001). Layered Participatory Analysis: New Developments in the CARD Technique. *CHI 01 Conference Proceedings*, Association for Computing Machinery Press (New York: Addison-Wesley), pp. 90–97.

Muller, M.J., Tudor, L.G., Wildman, D.M., White, E.A., Root, R.W., Dayton, T., Carr, R., Diekmann, B. and Dykstra-Erickson E.A. (1995). Bifocal tools for scenarios and representations in participatory activities with users. In: J. Carrol (ed.) *Scenario-Based Design for Human Computer Interaction* (New York: Wiley).

Muller, M.J., Halswanter, J H. and Dayton T. (1997). Participatory practices in the software lifecycle. In: M. Helander, T.K. Landauer and P. Prabhu (eds) *Handbook of Human-Computer Interaction*, (Amsterdam: Elsevier).

Murphy, E.D., Coleman, W.D., Stewart, L.J. and Sheppard, S.B. (1986). Case study: Developing an operations concept for future air traffic control. *Proceedings of the Human Factors Society, 30th Annual Meeting* (Santa Monica, CA: Human Factors Society).

Nardi, B.A. (1992). The use of scenarios in design. *SIGCHI Bulletin*, **24**, 13–14.

Nardi, B.A. (1997). The use of ethnographic methods in design and evaluation, In: M. Helander, T.K. Landauer and P. Prabhu (eds) *Handbook of Human-Computer Interaction* (Amsterdam: Elsevier).

Nielsen, J. (1992). Finding usability problems through heuristic evaluation. *CHI 92 Conference Proceedings*, Association for Computing Machinery Press (New York: Addison-Wesley), pp. 373–380.

Nielsen J. (1994). Ed Usability Laboratories, a special issue of *Behaviour and Information Technology*, **13**, 1&2.

Nielsen, J. and Molich, R. (1990). Heuristic evaluation of user interfaces. *CHI 90 Conference Proceedings*, Association for Computing Machinery Press (New York: Addison-Wesley), pp. 249–256.

Nielsen, J. (1993). *Usability Engineering* (Boston, MA: Academic Press).

Nielsen, J. (1994). Usability Laboratories, Special Issue, *Behaviour and Information Technology*, **13**, 1&2.

Norman, D.A. (1999). *The Invisible Computer* (Cambridge, MA: The MIT Press).

Norman, D.A. and Draper, S.W. (eds) (1986). *User Centered System Design*. (Hillsdale, NJ: Lawrence Erlbaum Associates).

O'Brien, D.D. (ed.) (1982). Design methods. *Seminar on Control Room Design*, Lancashire Constabulary HQ.

O'Brien, D.D. (1987). Personal communication from the U.K. Government Home Office, London.

O'Brien, J. and Rodden, T. (1997). Interactive systems in domestic environments. In: G. van der Veer, A. Henderson and S. Coles (eds.) *DIS 97, Designing Interactive Systems Conference Proceedings* (New York: Association for Computing Machinery Press), pp. 247–259.

Olson, G.M., Duffy, S.A. and Mack, R.L. (1984). Thinking-out-loud as a method of studying real-time comprehension processes, In: D.E. Keiras and M.A. Just (eds) *New Methods in Reading Comprehension* (Hillsdale, NJ: Lawrence Erlbaum).

Oppenheim, A.N. (1999). *Questionnaire Design, Interviewing and Attitude Measurement* (London: Pinter Pub.)

Palen, L. and Salzman, M. (2002). Voice-mail diary studies for naturalistic data capture under mobile conditions. *Computer Supported Cooperative Work Conference 2002* (CSCW 02). New Orleans, LA.

Pedersen, J. and Buur, J. (2000). Games and movies: towards innovative co-design with users. In: S.A.R. Scrivner, L.J. Ball and A. Woodcock (eds) *Collaborative Design* (London: Springer).

Peters, T. (1998). Design is IT! *Design Management Institute Journal*, **9**, No. 3, Summer.

Peters, T. (2000). Design as Advantage No. 1: The Design + Identity 50 Design Management. *Institute Journal*, **11**, No. 1, Winter.

Pikaar, R.N., Lenior, T.M.J. and Rijnsdorp, J.E. (1985). Control room design from situational analysis to final layout; operator contributions and the role of ergonomists. In: G. Mancini, G. Johannsen and L. Martensson (eds) *2nd IFA C/IFIP/IFORS/IEA Conference; Analysis, Design and Evaluation of Man-Machine Systems* (Oxford: Pergamon Press).

Poggenpohl, S.H. (2002). Design moves: approximating a desired future with users. In: J. Frascara (ed.) *Design and the Social Sciences* (London: Taylor & Francis).

Polson, P.G., Lewis, C., Rieman, J. and Wharton, C. (1992). Cognitive walkthroughs: A method for theory-based evaluation of user interfaces. *International Journal of Man-Machine Studies*, **36**, 741–773.

Preece, J., Davis, G. and Keller, L. (eds) (1990). *A Guide to Usability* (Milton Keynes: The Open University in association with the Department of Trade and Industry (UK)).

Rea, L.M. and Parker, R.A. (1997). *Designing and Conducting Survey Research: A Comprehensive Guide* (San Francisco: Jossey-Bass).

Reeves, B., and Nass, C. (1996). *The Media Equation* (Stanford, CA: Center for the Study of Language and Information, and Cambridge, U.K.: Cambridge University Press).

Robinson, R. (1996). Panel the participatory design of work space. In: Harrison, S. (Panel Organizer) *Proceedings of the 1996 Conference on Participatory Design*, Cambridge, MA.

Rosson, M.B. and Carroll, J.M. (2002). *Usability Engineering; Scenario Based Development of Human–Computer Interaction* (San Francisco: Morgan Kaufman Publishers).

Rubin, J. (1994). *Handbook of Usability Testing* (New York: John Wiley & Sons).

Rubinstein, R. and Hersh, H.M. (1984). *The Human Factor* (Maynard, MA: Digital Press).

Scrivner, S.A.R., Ball L.J. and Woodcock, A. (2000). *Editors of Collaborative Design: Proceedings of CoDesigining 2000* (London: Springer).

Sanders, E.B.N. (2000). Generative tools for codesigning, In: S.A.R. Scrivner L.J. Ball and A. Woodcock (eds) *Collaborative Design.* (London: Springer).

Sanders, E.B.N. (2002). From user-centered to participatory design approaches. In: J. Frascara (ed.) *Design and the Social Sciences* (London: Taylor & Francis).

Sanders, E.B.N. and William, C.T. (2001). Harnessing people's creativity: ideation and expression through visual communication. In: J. Langford and D. McDonagh-Philip (eds), *Focus Groups: Supporting Effective Product Development* (London: Taylor & Francis).

Segal, L.D. and Fulton Suri, J. (1997). The empathic practitioner: measurement and interpretation of user experience. In: *Proceedings of the Human Factors and Ergonomics Society 41st Annual Meeting Albuquerque*, HFES Santa Monica, CA.

Sellen, A.J., Murphy, R., and Shaw, K.L. (2002). How knowledge workers use the web. *CHI 02 Conference Proceedings* (New York: Association for Computing Machinery Press), pp. 227–234.

Serpiello, N. (2001). Picture This: Collage as a Human Centered Research Method for Product Design. HFES Consumer Product Technical Group.

Shackel, B. (1986). Usability–context, framework, definition, design and evaluation, *Proceedings of the SERC CREST Course; Human Factors for Informatics Usability*, HUSAT Research Centre, University of Technology, Loughborough.

Shackel, B. (1987). Human Factors for Usability Engineering, ESPRIT '87; Achievements and impact, *Proceedings of the 4th Annual ESPRIT Conference*, Brussels, edited by Commission of the European Communities (Amsterdam: North-Holland).

Sinclair, R.C., Moore, S.E., Lavis, C.A. and Soldat, A.S. (2002). The influence of affect on cognitive processes. In: J. Frascara (ed.) *Design and the Social Sciences* (London: Taylor & Francis).

Spicer, J. (1987). Personal communication from the Institute for Consumer Ergonomics, Loughborough.

Spicer, J., Wilkinson, S. and McClelland, I.L. (1984). Access to cars by disabled and elderly people; Report No.3, Car trials. Working Paper WP/VED/84/10, Transport and Road Research Laboratory, Department of Transport.

Stanton, N.A. and Young, M. (1998). Ergonomics methods in consumer product design and evaluation. In: N. Stanton (ed.) *Human Factors in Product Design* (London: Taylor & Francis Ltd.), pp. 21–52.

Strommen, E.F., Razavi, S. and Medoff, L.M. (1992). This button makes you go up: three-year-olds and the Nintendo controller. *Applied Ergonomics*, **23**, 409–413.

Stubler, W.F. and Bernard, T.E. (1986). Office ergonomics: design methodology and evaluation. *Proceedings of the Human Factors Society, 30th Annual Meeting* (Santa Monica, CA: Human Factors Society).

Suchman, L. (1987). *Plans and Situated Actions: The Problem of Human Computer Communication* (Cambridge, U.K.: Cambridge University Press).

Tomlie, P., Pycock, J., Diggins, T., MacLean, A. and Karsenty, A. (2002). Unremarkable computing. *CHI 02 Conference Proceedings* (New York: Association for Computing Machinery Press), pp. 399–406.

Torres, R.J. (2001). *Practitioner's Handbook for User Interface Design and Development* (New York: Prentice Hall).

Tschudy, M.W., Dykstra-Erickson, E.A. and Holloway, M.S. (1996). PictureCARD: a storytelling tool for task analysis. In: *PDC 96 Proceedings of the Participatory Design Conference.*

Tufte, E.R. (1990). *Envisioning Information* (Cheshire, CT: Graphics Press).

Usability Net (2002). A project funded by the EU Framework V IST Programme (1999-29067). (http://www.hostserver150.com/usabilit/home.htm).

Verplank, B., Fulton, J., Black, A., and Moggridge, B. (1993). Observation and invention: The use of scenarios in interaction design. *Tutorial Notes 18 CHI 93 and Interact 93*, 24–29 April 1993, Amsterdam, The Netherlands.

Virzi, R. A. (1992). Refining the test phase of usability evaluation: How many subjects is enough? *Human Factors*, **34**, 457–468.

Vredenburg, K., Isensee, S. and Righi, C. (2001). *User-Centered Design: An Integrated Approach* (New York: Prentice Hall).

Vredenburg, K., Mao, J-Y., Smith, P.W. and Carey, T. (2002). A survey of user-centered design practice. *CHI 02 Conference Proceedings* (New York: Association for Computing Machinery Press), pp. 471–478.

Welker, K., Sanders, E.B.-N. and Couch, J.S. (1997). Design scenarios to understand the user. In: *Innovation Quarterly Journal of the Industrial Designers Society of America*, **Fall**, 24–27.

Whiteside, J., Bennett, J. and Holtzblatt, K. (1988). Usability engineering: our experience and evolution. In: M. Helander (ed.) *Handbook of Human–Computer Interaction* (Amsterdam: North-Holland).

Wilson, J.R. (1991). Design decision groups — a participative process for developing workplaces. In: K. Noro and A. Imada (eds) *Participatory Ergonomics* (London: Taylor & Francis), pp. 81–96.

Wilson, J.R. (1995). Solution ownership in participative work redesign: The case of a crane control room. *International Journal of Industrial Ergonomics*, **15**, 329–344.

Winer, B.J. (1971). *Statistical Principles in Experimental Design* (New York: McGraw-Hill).

Wixon, D. and Ramey, J. (eds) (1996). *Field Methods Casebook for Software Design* (New York: J. Wiley & Sons).

Wolf, C.G., Carroll, J.M., Landauer, T.J., John, B.E. and Whiteside, J. (1989). The role of laboratory experiments in HCI: Help, hindrance, or ho-hum?, Panel discussion *CHI 89 Conference Proceedings*, Association for Computing Machinery Press, (New York: Addison-Wesley), pp. 265–268.

Wong, Y.Y. (1992). Rough and ready prototypes: Lessons from graphic design. In: *Proceedings of CHI '92 Posters and Short Talks* (New York: ACM Press), pp. 83–84.

Chapter 12

Is this chapter any use? Methods for evaluating text

James Hartley

Summary

How can we evaluate a piece of text? What questions should we ask, and how can we answer them? The literature in this area reveals a concern with at least four issues. We can ask questions about the content of a text; the way it is presented; the way it is supported by illustrative materials; and about its suitability for its intended purpose. These four inter-related issues form the basis of this chapter.

Evaluating content

We can approach the task of assessing the content of textual materials from many different points of view. When the material is a procedural document, a form to be completed, or commercial website, then the procedures used to evaluate its content will be rather different from the ones used to evaluate a chapter in a textbook. Nonetheless, our main concern in all of these cases will be to decide whether or not the content is fit for its purpose.

Personal opinions

One common way of doing this (in addition to making one's own judgement) is to ask other people — experts, users and colleagues — for their opinions. In the case of chapters in textbooks such as this one, people typically skim through the materials, study the section headings, examine any illustrative materials, and perhaps read the introductory section and/or concluding summary. They then decide whether or not the coverage is sufficiently detailed and sufficiently relevant for them to pursue it in more detail. And, since they may be evaluating text for others to use rather than themselves, they will look to see if there are outdated materials, errors of facts or important omissions, biases of any kind — national, racial and sexual — and whether or not the text may need to be supplemented by additional materials.

This much may seem obvious, but what is not obvious is how we can be sure that the content presented meets the required specifications. Not only are the above descriptions hard to specify but also at times they might be misleading. Users of technical documents often complain, for instance, that such documents contain *too much* information — that is that they contain more than is needed for the task in hand. The same kind of problem can also arise with textbooks. How much background information, of interest historically but now out of date or even inaccurate by today's standards, should go into, say, an

Please rate the book in the spaces provided on each of the items given, using a scale of 0 (very poor) to 5 (very good).

_____ general appearance	_____ relevance of content
_____ practicality of size	_____ ease of reading
_____ durability of binding	_____ use of chapter sub-headings
_____ quality of paper	_____ use of illustrative materials
_____ appeal of page layout	(tables, figures, graphs)
_____ legibility of typefaces	_____ degree of challenge for able students
_____ details of index	_____ suitability for less able students

Figure 1 An excerpt from a typical checklist for judging the quality of a school text book.

ergonomics textbook? And how far can we assume that a reference to a past event or author will be understood by younger readers?

Checklists and rating scales

Evaluating content is at best a difficult activity, and it is usually a subjective one. One way of increasing its objectivity is to increase the number of judges involved and to provide some sort of checklist or rating scale to help ensure that they all evaluate the same concerns. Figure 1 provides an example of part of such a checklist.

This kind of approach is commonly used for evaluating school textbooks in countries that have state-controlled school systems (but it can also be used by colleagues for evaluating course textbooks such as this one). Although such checklists are useful in making the judges' ratings more systematic and consistent, no standardised tools for doing this have been developed. Different investigators have created their own instruments. In one early study, for instance, Farr and Tulley (1985) reported that the number of items on checklists for evaluating school textbooks in the USA ranged from 42 to 180, with an average number of 73.

Such checklists and rating scales are usually completed *before* recommending a particular text for use. However, this information can also be collected *after* texts have been used by teachers and students. Information gained in this way is helpful in deciding whether or not to use the text again, and for authors who are planning subsequent editions.

Content analysis

Content analysis provides a more detailed approach to assessing the content of a text. This term in fact describes a multitude of approaches (e.g., see Weber, 1990) but, in general, these involve analysing phrases, clauses and sentences in the text and making thematic analyses. Abraham *et al.* (2002) for instance, used this technique to isolate 45 themes that occurred in health promotion leaflets on condom use, and to compare the frequency with which sentences or phrases referring to these themes appeared in British and German leaflets. (Thus the sentence, 'Unprotected anal sex carries a particularly high risk of contracting HIV', for example, would be categorised as an exemplar of the theme: 'risk information'; and 'Always carry your own condoms' as an exemplar of the theme: 'availability of condoms'.) Content analysis allowed these authors not only to discriminate between more and less effective leaflets but it also enabled them to see more generally what key information was being transmitted and what omitted.

Discourse analysis

Discourse analysis is somewhat related to content analysis, but here the emphasis is more on determining the — sometimes hidden — meanings of the phrases, clauses and sentences in the text, and in assessing how this meaning might be interpreted by the readers. The argument here is that text is never neutral (see Banister, Burman *et al.*, 1994; Gill, 1996). Thus the sentence, 'It is dangerous to exceed the stated dose', for example, may mean different things to different people in different contexts.

Furthermore, different sentences within the same texts may serve different functions. Cook and O'Halloran (1999), for instance, in an analysis of labels on babyfood products, found a variety of sentence types with different functions. They wrote:

> A typical label will contain instances of persuasion (referring to taste, health environmental protection, value for money), warning (consumption date, allergenic properties), advice (for storage, cooking) exposition (ingredients, nutrients, guarantee) and so on. (p. 140).

They went on to show that readers with different backgrounds and levels of education found it difficult to disentangle some of these aspects of the discourse. Banister *et al.* (1994) provide a similar detailed analysis of the instructions for using children's toothpaste.

Evaluating the layout of text

So much for the content: but how is it presented in the text? I now turn to examining issues of layout.

Print versus screen

Print and screen-based text share several common features, but differ in many other respects. Perhaps the most obvious difference, apart from the use of colour, is that printed texts width) and screen-based text is usually (but not always) oriented in a *landscape* style (where the width is greater than the height). Furthermore, screen 'pages' usually contain much less actual text than do typically printed ones. Average computer screens, for instance, have about 5,000 characters whereas the average page from a phone book has approximately 35,000. These three factors — colour, orientation and size — determine the choices that designers can use in setting the text in these different media, and each has consequences for presenting, understanding and finding information (Wright, 2000). These tasks also vary with different types of readers, and different types of text (e.g., see Morrell *et al.*, 2001; Walker and Reynolds, 2000).

Colour

One of the most important features of screen-based text is the availability of multi-coloured formats on multi-coloured backgrounds. In some systems many colours are used without any seeming coherence, and text is printed without any apparent reference to notions of contrast. The research suggests that green, white, yellow and cyan are the most useful colours to use on a dark background, and that legibility is impaired when black (or dark) text is printed on a strongly coloured background. As far as reading coloured text is concerned it is also perhaps important to remember that about 8.5% of males and 0.5% of females are colour blind to some extent (Post, 1997).

Coloured cueing can be used to emphasise particular words or phrases on screen (e.g., red for DO NOT. . .) in much the same way that italic or bold print can be used in printed text (Post, 1997). Colour can also be used to convey categories of importance, as in the News on *Ceefax,* where the main paragraphs appear in white, and sub-paragraphs in blue. One difficulty arises here, however, in that there does not appear to be an intuitive range of colours that suggests a hierarchy of importance (unlike the use of space). If colours are to be used to convey particular functions across several pages (in text as well as on screen) then it is important that they should be used consistently, and that they should be clearly differentiated from other colours used in the text.

Colour can be used in place of — but perhaps not as successfully as — space to convey organisation and structure. Thus rows or columns in tables (or groups of them) may be presented in alternating colours to aid search and retrieval, but the resulting 'striped' effect can cause difficulties if readers are expected to read both across and down the tables. Further technical problems arise from the fact that warm colours (red and yellow) usually appear larger than cooler colours (green and blue) on screen and thus, for example, a bar presented in one colour in a bar chart may seem larger or smaller, depending upon the background colour on which it is presented (Tedford *et al.*, 1977).

In much informational text presented on screen it is hard not to consider that the primary function of all this colour seems to be to attract and excite the readers, rather than to inform them. Similar kinds of arguments have been used, too, with respect to many children's textbooks that appear to be designed to sell to adults rather than to reflect the reading requirements of the children (Hartley, 1994).

The use of space

In printed text the initial choice of page-size (and orientation) determines what decisions are made about key factors such as column widths, interline spacing, choice of typeface(s), and the positioning of illustrative materials, such as tables and graphs (Hartley, 1994; Schriver, 1997). In the early days of screen-based text there was much less variety possible in these respects than there was for printed text. Today, however, with the advent of the world wide web (WWW) there seems to be an almost equal richness of choice available for both methods.

So, in both printed and electronic text, space can be used systematically to convey the underlying structure of the text. Figures 2 and 3 provide an illustration. In Figure 2 the text is presented in a centred format, using what is called 'justified' text (i.e., it has a straight left and right-hand edge. (This is achieved by varying the space between the words and, indeed, sometimes between the letters, or by using hyphenation.) Furthermore, each paragraph is present in a 'block' format. In Figure 3 the text is ranged from the left, and uses 'unjustified' text. Here there is equal word-spacing, no hyphenation, and (in this example) each new sentence within a paragraph begins on a new line. Research suggests that, in general, people prefer this more 'open' approach for complex texts (Grabinger, 1993; Hartley, 1994). And, in particular, experts widely recommend using unjustified text for screen based presentations (e.g., Nielsen, 2000).

It is common in designing printed text to use an underlying grid to help specify the spacing and positioning of all the elements in the text. Although some people consider this constricting, it is almost essential to use a grid when planning complex text that has many subcomponents that need to be displayed in a consistent format. Hartley (1994), Schriver (1997) and Roberts and Thrift (2002) provide illustrations. However, using such a grid does not

INSULATING GLOVES

1. GENERAL

1.01. This section covers the description, care and maintenance of insulating gloves provided for the protection of workmen against electric shock, and the precautions to be followed in their use.

1.02. This section has been re-issued to include the D and the E insulating gloves.

2. TYPES OF INSULATING GLOVES

2.01. All types of insulating glove are of the gauntlet type and are made in four sizes: 9-1/2, 10, 11 and 12. The size indicates the approximate number of inches around the glove, measured midway between the thumb and finger crotches. The length of each glove, measured from the tip of the second finger to the outer edge of the gauntlet, is approximately 14 inches.

2.02. There are various kinds of insulating gloves. The original ones were just called Insulating Gloves. After that the B, C, D and E Insulating Gloves were developed. As described below, the D Gloves replaced the original Insulating Gloves and the E glove replaced the B and C Gloves.

2.03. Insulating Gloves are thick enough to eliminate the need for protective gloves and are intended for use without them. These gloves have been superseded by the D Insulating Gloves

Figure 2 Extract from the first page of a piece of technical text.

INSULATING GLOVES

1. GENERAL

1.1. This section covers the description, care and maintenance of insulating gloves provided for the protection of workmen against electric shock, and the precautions to be followed in their use.

1.2. This section has been re-issued to include the D and the E insulating gloves.

2. TYPES OF INSULATING GLOVES

2.1. All types of insulating glove are of the gauntlet type and are made in four sizes: 9-1/2, 10, 11 and 12.
The size indicates the approximate number of inches around the glove, measured midway between the thumb and finger crotches.
The length of each glove, measured from the tip of the second finger to the outer edge of the gauntlet, is approximately 14 inches.

2.2. There are various kinds of insulating gloves.
The original ones were just called Insulating Gloves.
After that the B, C, D and E Insulating Gloves were developed.
As described below, the D Gloves replaced the original Insulating Gloves and the E glove replaced the B and C Gloves.

2.3. Insulating Gloves are thick enough to eliminate the need for protective gloves and are intended for use without them.
These gloves have been superseded by the D Insulating Gloves

Figure 3 The same page with a revised typographical layout.

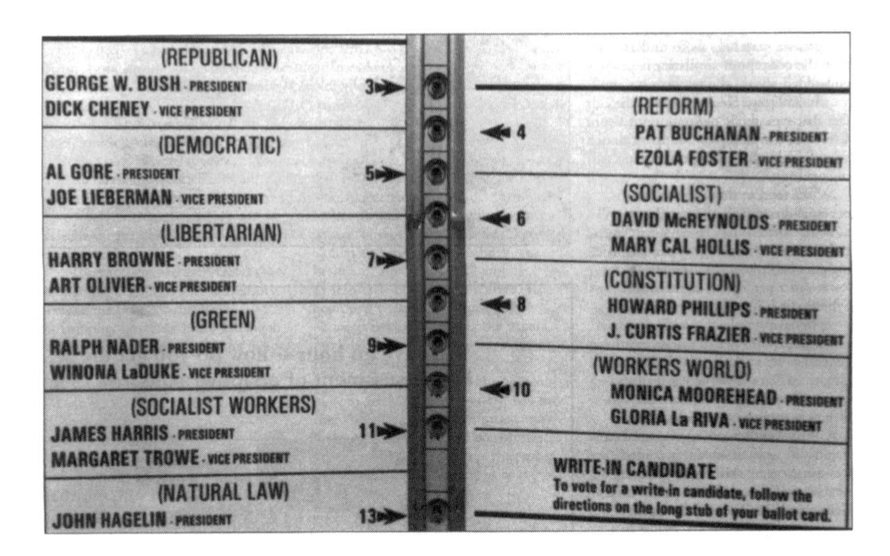

Figure 4 Which hole do you have to punch to vote for George Bush? Using a central axis for candidates on both the left and the right is bad practice (see text). Many people, it seems, managed to vote for Pat Buchanan (a Republican) when they thought that they were voting for Al Gore (a Democrat).

necessarily guarantee that a good fit will always be achieved between the text presentation and the concerns of the reader. We are all familiar with tables or graphs that appear at the top or the bottom of the page (or column) irrespective of where they are mentioned in the text (see next section). However, a grid does ensure that all the elements in the text are presented in a *consistent* manner.

One good example of the difficulties that can arise when spacing is not considered properly arose in the US presidential election in 2000. Here many people in a crucial vote in Florida found themselves voting inadvertently for the wrong candidate because (as shown in Figure 4 and discussed in Clay, 2001) the punch holes for each candidate were not systematically aligned with the candidates' names. If the punch holes had been placed systematically to the right of each candidate's name then this mistake would not have occurred, and possibly a different president elected.

It seems obvious to the current writer that many WWW and printed pages are not planned in a systematic way but rather on a 'let's put this here' basis. Different parts of the text are printed in one, two, three or even more columns. Different sections are 'blocked off' in different ways. Different pieces of the text are printed in differently coloured type-faces and type-sizes. One problem with such inconsistency is that the resulting text is difficult to follow — and this contravenes all that we know about screen design. As Tullis (1997) put it:

> The main point to remember about the placement and sequence of elements on the screen is that the user should be able to develop very clear expectations about what information will fall where (p. 517).

Gaining access to the text

Readers come to text with many different purposes. They need to be able to find the right page, to skim, to search, to look ahead, to return to previous information, and to read parts of the text in detail. In a word, readers need to be able to 'navigate' around a text easily.

'Text navigation' is aided by good typographic practice and by the use of what Waller (1979) called *access structures* — devices that facilitate readers' access to the text. Such access

Table 1 Recent Research on Various 'Access Structures'

Abstracts	Hartley (2000)
Captions	Hartley (1991a)
Footnotes	Jansen *et al.* (2001)
Headings	Lorch and Lorch (1996)
Indexes	Van der Meij (2002)
Links	Price and Price (2002)
Lists	Seki (2000)
Outlines	Hofman and Van Oostendorp (1999)
Questions	Van den Broek *et al.* (2001)
Signals	Meyer and Poon (2001)
Summaries	Dupont and Bestgen (2002)
Titles	Sadoski, Goetz and Rodriguez (2002)

structures can be found at the beginning and end of texts (e.g., contents pages, glossaries, indexes, summaries) and embedded in the texts themselves (e.g., chapter indicators and summary titles at the top of the page, page numbers, numbered headings and sub-headings, figure and table numbers, etc.). Hartley (1994) and Misanchuk (1992) summarise research on access structures, and Table 1 lists some more recent materials in this respect.

Screen-based text has some advantages and limitations compared with print when it comes to access structures. One of the advantages is that screen-based text is not constrained by its length. With conventional text the paperwork can simply become too cumbersome. One study of a technical manual, for example, found that in order to diagnose and repair one aircraft radar malfunction, a technician had to refer to 16 pages in eight documents and to look in 41 different places in these documents. One of the main benefits of electronic text is that one can accelerate access to such information. Furthermore, what the readers see is much less than what they do not see. Readers do not need to handle a whole encyclopedia in order to look up one entry.

Another advantage of screen-based text is that different levels of text can be provided on screen whereas this is hard to do in print. In writing this chapter, for instance, I debated with myself about whether or not to provide detailed references to the research literature that would support or offer different viewpoints on what I was writing about. The scholar in me wanted to document every point: the writer in me said that the resulting product would be difficult to read. But I could achieve both aims if I were writing an electronic text. I could, for instance, place a link, or 'electronic button', behind any researcher's name, or any sentence that starts, 'The research shows . . .' or 'See also . . .'. Interested readers could then click on to those parts of the text to reveal the supporting references, or even more detailed discussion.

However, compared with printed text, it is difficult, if not impossible, to flip through, or skim an electronic text and make side by side comparisons of different pages. You cannot put your finger in an electronic text to keep your place whilst you check an earlier point. Thus particular attention has to be paid to the design of indexes and contents pages to help users find their way around the text. Such pages list topics and sub-topics and readers have to select appropriately from them. It is conventional to structure such 'menus' in a hierarchical or 'tree-like' manner, with the basic or primary choices first and the lower-level, more detailed choices later — but a simpler way might be to list all the entries alphabetically. The difficulty with the former arrangement is that the choices deemed appropriate by the author may not match what the reader has in mind, making search difficult. Furthermore, the author has to choose between producing a few detailed (and crowded) menus, or a larger number of less detailed (but more spacious) ones that will take the reader time to work through. And preferences may change with time. Readers may well like the simpler menus to start with but, as they become more knowledgeable, prefer the more detailed ones because these will involve less page finding (Paap and Cooke, 1997).

Finally, we should note that book chapters, such as this one, do not translate readily to screen-based presentations. The line-lengths and the paragraphs are too long, and the inter-line spacing and the type-size too small. All of this suggests that writing for the screen is different. Basically, materials have to be simplified, and written in shorter 'chunks' which can then be displayed more clearly by appropriate typography and spacing (Nielsen, 2000). Price and Price (2002), for instance — following Nielsen — advise, 'Write 50% less text." (p. 86), and they provide interesting 'before' and 'after' illustrations of how this might be done.

Evaluating web-page design

In a paper published in 2000 De Jong and Van der Geest wrote:

> A quick Internet search using the keyword 'Web design' produces 754,904 hits. A search within those hits using the keywords 'design guidelines,' 'design principles,' 'design rules,' 'design criteria,' or 'heuristics' produces a total of 384 hits. These 384 hits form a broad collection of guidelines on the technology and tools that make the Web run (e.g., Javascript or HTML), galleries of images and sounds, user interface design principles, page and site design guidelines, and style guides for Web pages to specific organisations. Imagine the poor Web designer who wants to make sensible choices from this jumble of guidelines. Which of these to choose? (p. 311).

Which indeed! Guidelines vary from detailed specific checklists, to broad statements of principle, and to compendiums of illustrative practices. De Jong and Van der Geest (2000) point out that checklists may focus on particular kinds of web sites (such as university home pages) or they may be feature specific, focusing on certain site characteristics such as navigation, graphics and layout, or accessibility for people with disabilities. Nielsen (2000) offers useful advice and Nielsen and Tahir (2002) present 113 guidelines for improving homepages, and use them to analyse 50 examples of well-known sites (each of which is illustrated in full colour).

Readers interesting in evaluating web sites will find much useful information in these sources. De Jong and Van der Geest indicate which sets of guidelines are research based, and their strengths and limitations in this respect. (More specific examples and discussion can also be found in Ant Ozok and Salvendy, 2001; Morrell *et al.*, 2001; Sears, 2000; Upchurch *et al.*, 2001; and Van der Geest, 2001.)

Evaluating illustrative materials

Within passages of text it is common to find a variety of what I shall call here 'illustrative materials'. These include tables, charts, graphs, diagrams and illustrations. The effectiveness of such materials can be evaluated in their own right but, perhaps more important, is the problem of how well such different materials are integrated with the text.

Tables

Tables vary enormously in their complexity, detail and size. Wright (1982) suggested that in order to use a table successfully, readers have (1) to understand how the table has been organised; (2) to know where to look to find the answers to their questions; and (3) to be able to interpret the answers to their questions — a process that may involve comparing figures within and across the same or different tables. Wright argues that as a table becomes more complex, each of these elements of the process becomes more difficult.

A large body of research has been assembled on the merits of different ways of presenting tables (see Hartley, 1994; Tufte, 1983; Warren, 1999). Putting it rather simply, there are two main ways of making tables easier to use. First, one can *simplify their content* — and this is especially important for electronic text. Ehrenberg (1997) gives some particularly useful suggestions in this regard (which are summarised and illustrated in Hartley, 1994). Second, one can consider their *typographic layout*. Here again I stress the importance of systematically using the 'white-space' both vertically and horisontally to separate out and group related sections (see Hartley, 1994).

Graphs

Just as there are reviews of the research literature on the presentation of tabular materials, so there are parallel reviews on the design and presentation of graphic and illustrative materials in printed and electronic text (e.g., see Kosslyn, 1994; Lohse, 1997; Westendorp, 1999; Wilkinson, 1999). Again, the points I take from all of these reviews are that graphical materials need to be simple, clear and consistently presented.

Simple bar charts and line graphs are probably the easiest kinds of graphical representation to understand. Line graphs are probably better in most circumstances than bar charts or tables for showing trends, and tables probably better than line graphs for providing exact quantities. Pie charts are said to be easy to understand but, at times they can be misleading. When the segments are small it is difficult to judge the proportions accurately, and to label them clearly. It is also difficult to compare the segments of two or more pie charts if the charts have different diameters. One possible reason for this is that in order to make one circle (or a square) look twice the size of another, it has to be drawn four times as large.

Another consideration here is that new technology allows us to do things more easily than before. It is now much simpler, for example, to produce graphs and charts. However, this has lead to a proliferation in more recent texts of three-dimensional charts as opposed to two-dimensional ones. Such 3-D graphics may be appropriate in some circumstances (e.g., showing a change over surface area that differs along different axes) but, on the whole, clarity is being sacrificed here for aesthetic considerations (Nielsen, 2000). Research has shown that although readers find 3-D displays attractive, they find it easier to extract information from 2-D ones (Hartley and Yates, 2001; Hicks *et al.*, 2003). Figure 5 shows a 2-D and a 3-D version of a simple bar chart to illustrate this point.

Illustrations and diagrams

Words are good for conveying abstract ideas and concepts that have already been learned (for example, the word 'mammal'). Illustrations, on the other hand, are good for conveying concrete images (e.g., a picture of an elephant). And diagrams are good for conveying structures and processes.

If the information to be presented can be readily described in words, then there may be no need for pictures. Conversely, if the information can be readily conveyed in a picture or a diagram, then there may be no need for words. Illustrations and diagrams may be good ways of avoiding complex wording and technical jargon. Thus one can argue that Figure 6 is a more effective way of communicating the instruction, 'See that the sliding dog associated with the reverse drive bevel is rotating freely before tightening the log differential casing'.

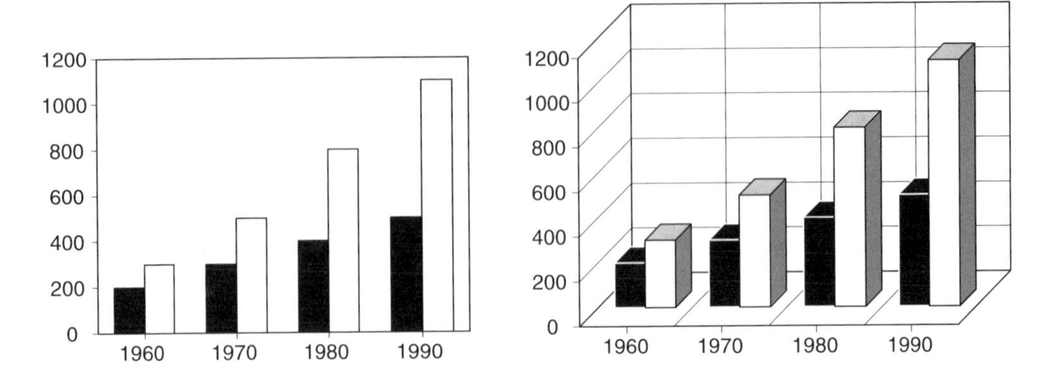

Figure 5 A 2-D and a 3-D bar chart of the same data. Although readers may prefer the 3-D version, it is probably easier to extract information form the 2-D one.

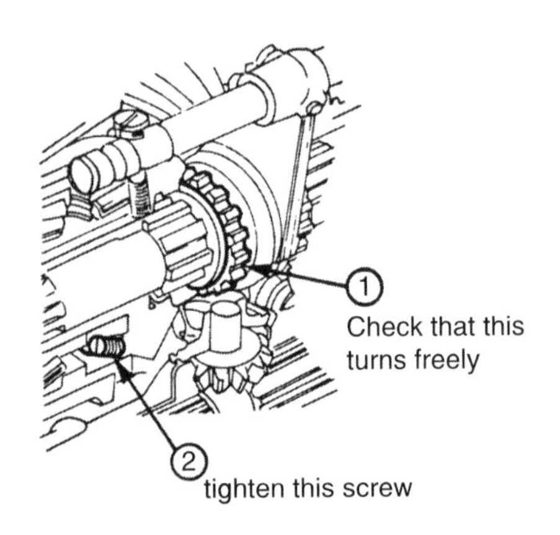

Figure 6 A picture is sometimes worth a 1,000 words.

One approach to assessing the effectiveness of illustrations and diagrams has been to consider them in terms of their functions. Some authors have suggested, for example, that illustrations and diagrams fulfil one or more of the following roles in informational text:

- An affective role — enhancing interest and motivation
- An attentional role — attracting and directing attention
- A didactic role — facilitating learning by explaining or showing something that it is difficult to convey solely in words
- A supportive role — enhancing the learning (of, say, less-able readers)
- A retentional role — facilitating long term recall

These roles are clearly important as far as informational text is concerned, and many researchers have attempted to assess the effectiveness of the didactic role in particular. Clearly in such studies there are a large number of variables to consider and, at first sight, it would seem difficult to draw any general conclusions. Researchers (such as Dwyer, 1987; Houghton and Willows, 1987; Schriver, 1997; Willows and Houghton, 1987) have worked with different kinds of texts (from children's books to technical manuals), used different kinds of illustrative materials (from line-drawings to coloured photographs), studied different groups of readers (from young to old, with low and high ability) and measured different things with different kinds of measuring techniques (from factual recall to drawing).

The statistical technique of meta-analysis, however, allows researchers to pool all of the studies on one particular issue and to look for overall average effects. Levie and Lentz (1982) were probably the first to use this technique to consider the evidence for the didactic role of illustrations. In examining over 40 studies on the topic they found first that they had to distinguish between three sources of information in illustrated texts. These were:

1. Information provided only in the text.
2. Information provided only in the illustration(s).
3. Information provided in both the text and the illustration(s).

Levie and Lentz then asked (from the results of the pooled studies) whether or not the illustrations aided the recall of information from all three sources. The results were surprisingly clear. There were marked effects for condition 3 (that is recall of *text that was illustrated*), but text that was not illustrated (condition 1) and information that was illustrated but not discussed (condition 2) did not fare so well. Thus it appears that pictures have an additive function in informational text: they aid the recall of the textual material that they illustrate. But they do not really help the recall of the non-illustrated text.

Lowe (1993) suggests that there are a number of general and specific questions that investigators need to ask when evaluating diagrams. His general questions include:

- Does the learner consider the diagram to be useful?
- Does the learner process the diagram sufficiently deeply overall?
- Does the learner understand and remember the material presented?
- Can the learner apply the material presented?

Questions such as these can be used to assess the value and effectiveness of both illustrations and diagrams. This can be done, somewhat crudely, by asking potential readers either singly, or in small groups, to comment on, explain or discuss the proposed illustrative material with the investigator. It is helpful to start investigating the effectiveness of a particular diagram in this way before moving on to full-scale trials (Nielsen, 2000).

Integrating illustrative materials with the text

Deciding where to place illustrative materials in the text is a matter of concern, both for printed and for electronic text. On screens and in printed texts, tables, graphs, illustrations and diagrams are typically put at the top or bottom of the page without reference to where they are mentioned in the text. Often, too, because of their size, they may be positioned on the following page. Clearly it is not always possible to put such materials immediately after their first textual reference especially when, for example, there are several tables in a row, or the illustration is particularly large. Nonetheless, some thought needs to go into these matters in order to avoid confusing the reader.

A good example of the kind of problem I have in mind occurs in Cleveland's (1985) otherwise excellent book on graphs. Here, at the beginning of Chapter 1, each figure occurs not in its own section but in the following one. Thus all the figures are out of step.

Empirical research has in fact supported the suggestion that it is important to consider the *integration* of illustrative materials and their related texts. Hartley (1991b), Schreiber *et al.* (2002) and Stylianidou *et al.* (2002) have all shown that it is helpful if the text is written in a way that is consistent with how the readers process such materials. Indeed, precise measures of such integration can be made using eye-movement recorders (e.g., see Walley and Fleming, 1975; Rayner *et al.*, 2001).

A special case involving the integration of pictures and text occurs in the context of procedural text or instructions which use pictorial sequences (e.g., see Ganier, 2001; Novick and Morse, 2000; Spinillo and Dyson, 2001). It is tempting, in an international context, to develop pictorial instructions without words, but it may be wiser to have minimal text in addition. Horn's (1998) text, *Visual Language*, argues that texts and pictures have become so interdependent today that they have become a language in their own right.

Finally, if readers will forgive the pun, I should point out that there is more to reading illustrative materials than meets the eye. As in reading text, understanding, experience and prior knowledge of what to expect are important (e.g., see Roth, 2002). So when choosing participants for evaluation trials it is useful to include participants with wide ability and age-ranges.

Evaluating the suitability of the text for the reader

Perhaps the most common question asked about a piece of text concerns its suitability for its intended readers. Sometimes, when the readership is well known, it is easy to arrive at an answer. However, if the text is going to be used by multiple readers for a variety of different reasons, then the task becomes more difficult.

Several methods can be used to evaluate the suitability of a text for its intended audience and they can be grouped, following Schriver (1989, 1997) under three main headings: 'expert-based', 'reader-based' and 'text-based'. Many of the measures within each group can be applied during the 'construction' of the text, or afterwards with the finished product (or on both occasions).

Expert-based measures

Experts in this context are people who have a high level of knowledge about (a) a particular subject matter, (b) the potential readership of a text, and (c) the skills of writing. Such people typically use their judgement to assess texts. As noted earlier in first section above, the tools that experts use to evaluate text can include checklists and rating scales.

Reader-based measures

Reader-based tools for evaluating text require readers to carry out some activities. Such activities can be many and varied (e.g., see Gould *et al.*, 1997). Schriver (1989) distinguishes between those which are *concurrent* with reading the text, and those which are *retrospective,* or come after this. Table 2 lists examples of different reader-based measures under these two headings. Here I will consider two of them in more detail.

Cloze tests

The cloze test was originally developed by Taylor (1953) to measure people's understanding of text. Here samples of a passage are presented to readers with, say, every sixth word missing. The readers are then required to fill in the missing words.

Table 2 Some Examples of Concurrent and Retrospective Reader-based Text Evaluation Measures

Concurrent measures	*Retrospective measures*
Eye-movement pattems	Comprehension tests (including cloze)
Verbal commentaries	Readers' judgements of difficulty
Oral reading errors	Readers'preferences
Search tasks	Readers' feedback sheets
Reading times	Usability tests
Cloze tests	
Usability tests	

Technically speaking, if every sixth word is deleted, then six versions should be prepared with the gaps each starting from a different point. However, it is more common ——— prepare one version and, perhaps ——— to focus the gaps on ——— words. Whatever the procedure, the ——— are scored either (a) by ——— accepting as correct those responses ——— directly match what the original ——— actually said, or (b) by ——— these together with acceptable synonyms. Since the two scoring methods (a) and (b) correlate highly, it is more objective to use the tougher measure of matching exact words. (In this case: 'to', 'even', 'important', 'passages', 'only', 'which' 'author' and 'accepting'.)

The scores obtained can be improved by having the gaps more widely dispersed (say every tenth word); by varying the lengths of the gaps to match the lengths of the missing words; by providing dashes to indicate the number of letters missing in each word; by providing the first of the missing letters; by providing multiple-choice answers; or even by having readers work in pairs or small groups. These minor variations, however, do not affect the main purpose of the cloze procedure which is to assess readers' comprehension of the text and, by inference, its difficulty.

The cloze test can be used by readers both concurrently and retrospectively. It can be presented concurrently (as in the paragraph above) as a test of comprehension, and readers asked to complete it, or it can be presented retrospectively and readers asked to complete it after they have read the original text. In this latter case the test can serve as a measure of recall as well as comprehension. The cloze test is particularly useful in that it can also be used to assess the effects of different textual organisation, readers' prior knowledge and other textual features, such as illustrations, tables and graphs, on the readers' understanding of the actual text (e.g., see Couloubaritsis *et al.*, 1994; Greene, 2001; Reid *et al.*, 1983).

Readers' judgements and preferences

A rather different but useful measure of text difficulty is to ask readers to judge it for themselves. One simple procedure here is to ask readers to circle on the text those areas, sentences, or words, that they think *readers less able than themselves* will find difficult. In my experience, if you ask readers to point out difficulties *for others* they will be much more forthcoming than if you ask them to point out their own difficulties.

An elaboration of this technique is to ask readers to give a running commentary (or verbal protocol) on the difficulties they experience as they are reading or using a text (see Chapter 7). This technique has proved extremely valuable in evaluating complex text such as that provided in instructional manuals where there is often a rich interplay between text and diagrams (see Shriver, 1997). Some critics of this approach suggest that talking about a task whilst trying to do it can cause difficulties, and this does seem to be a reasonable objection. However, such problems can be partly overcome by using retrospective protocols or, for example in this case, by videotaping readers using manuals to complete a particular task, and then asking them to talk through the resulting tape. The tape can be stopped at any point to allow them to make an extended commentary.

Readers can also be asked to state their preferences for different kinds of texts, and for different layouts of a specific text (e.g., see Ant Ozok and Salvendy, 2001; Grabinger, 1993), and to predict the difficulties that other readers might have (e.g., see Young and Wogalter, 2001). Some experts dismiss preference judgments by readers because they think that such preferences might be based on inappropriate considerations (such as a lavish use of colour rather than the clarity of the wording). However, most readers have clear views about what they like in texts, and how they expect texts to perform. So, first impressions might colour attitudes to a text. A book that looks dense and turgid is not going to encourage one to read it — no matter how important the content.

A common method of measuring preferences is to ask people to rate (or mark out of 10) original and revised texts. The results can tell you whether a revised text is preferred to the original, whether people see no difference, or whether people prefer the original version. However, one has to be careful here. For some reason or other, when people rate any *two* things out of ten they often rate one of them 5 or 6, and the other one 8 (Hartley and Ganier, 2000). So it is useful to have a baseline text for comparison. The same text might be rated 5 or 8 depending on what it is being compared with.

Another useful tool to use here, if you want preference judgements for a number of texts that vary in different ways, is the method of *paired comparisons* (see Chapter 4). Suppose, for example, you have 15 designs for a website. You could ask potential readers to judge them (overall, or on some specific aspect) and to make paired comparisons. Essentially this involves each judge comparing design 1 with design 2 and recording the preference, then 1 with 3, 1 with 4, 1 with 5 and so on, until 1 with 15 is reached. Then the judge starts again, this time comparing 2 with 3, 2 with 4, 2 with 5 and so on until 2 with 15. This procedure is repeated again, starting with 3 with 4, 5, 6, etc., 4 with 5, 6, 7, etc., until all the designs have been systematically compared. Finally, you total the number of preferences recorded for each design to see which one has been preferred the most often. Reliability can be assessed by asking participants to do the task again backwards. An example illustrating this approach can be found in Hartley, Trueman and Burnhill (1979), and the issues of rating and ranking generally are discussed in Chapter 4.

Finally, in this section, we should note again the advisability of using a wide range of readers in making reader-based measures.

Text-based measures

Text-based tools for evaluating text can be used without recourse to readers. These measures, too, can be applied concurrently — whilst one is writing the text, and retrospectively, once it has been written — either by the author(s) or by others who might be thinking of using it. In this chapter I shall describe two computer-based tools for evaluating written text.

Computer-based measures of text difficulty

Measures of text difficulty — or readability formulae — were originally developed in order to predict the age at which children, on average, would have the necessary reading skills and abilities to understand a particular text. And this is still their main aim today, although the scope of application has widened (Klare, 1963).

Most readability formulae are in fact not as accurate at predicting this age as one might wish (and different formulae produce slightly different results), but the figures they provide do give a rough guide. Furthermore, if you use the same formula to compare two different texts, or to compare an original with a revised version, then you do get a good idea of relative difficulty.

Readability formulae typically combine two main measures to predict the difficulty of text. These are (i) the average sentence length of samples of the text, and (ii) the average word length in these samples. One simple formula — the Gunning Fog Index (Gunning, 1952) — is as follows:

- Take a sample of 100 words.
- Calculate the average number of words per sentence in the sample.
- Count the number of words with three or more syllables in the sample.
- Add the average number of words per sentence to the total number of words with three or more syllables.
- Multiply the result by 0.4.

The result is the 'reading grade level' as used in US schools. (Grade 1 = 6 years old; Grade 2 = 7 years old, etc.). You can add 5 to the answer to obtain an equivalent British reading age (if you think that British and US children are similar despite the fact that British children start school one year earlier).

Most readability formulae, however, are much more complex to calculate than is the Gunning Fog Index — hence the interest in computer-based methods. One better known formula, but one which is harder to calculate by hand, is the Flesch Reading Ease (RE) formula (Flesch, 1948). This is:

$RE = 206.835 - 0.846w - 1.015s$ where w is the number of syllables per 100 words and s is the average number ofwords per sentence.

In this case, the higher the RE score, the easier the text. Table 3 shows the relationship between RE scores, difficulty and suggested reading ages.

When I ran a readability programme on my word-processor on the first section of this chapter (with the headings deleted) I obtained a Flesch score of 39. Different formulae would no doubt produce slightly different results. Furthermore, an additional difficulty has arisen with computer-based readability formulae because different programmers have worked out different ways of computerising what are ostensibly the same formulae. Thus you might find that, for example, if you use the *Word for Windows* version of the Flesch formula you will get a slightly different RE result from using that provided by, say, *Grammatik 5* or MicroSoft's *Office 97*. This problem is not too serious with simple texts, but it can become more of an issue when working with complex ones (Mailloux *et al.*, 1995; Sydes and Hartley, 1997). So, if you want to be consistent, the moral is to always use the *same* computer programme when evaluating *different* texts.

Table 3 The Relationships Between Flesch Reading Ease Scores, Suggested Reading Ages and Difficulty Levels

Flesch socre	Reading age	Difficulty level
90–100	10–11 years	Very easy
80–89	11–12 years	Easy
70–79	12–13 years	Fairly easy
60–69	14–15 years	Average
50–59	16–17 years	Fairly difficult
30–49	18–20 years	Difficult
0–29	Graduate	Very difficult

Scientists divide the different forms of life into two main groups. There are animals called *vertebrates* that have backbones, and there are animals called *invertebrates* that do not. *Vertebrates* can be divided into several sub-groups. There are *reptiles* such as snakes and crocodiles; *amphibians*, such as frogs and toads; *fish*, such as salmon and sharks; *birds*, such as sparrows and eagles; and *mammals*, such as dogs, horses and people.

Scientists divide the different forms of life into two main groups. There are animals called *vertebrates* that have backbones. There are animals called *invertebrates* that do not. *Vertebrates* can be divided into several sub-groups. There are *reptiles* - such as snakes and crocodiles. There are *amphibians* - such as frogs and toads. There are *fish* - such as salmon and sharks. There are *birds* - such as sparrows and eagles. And there are *mammals* - such as dogs, horses and people.

Figure 7 How one can make marked changes in the results from a readability formula simply by shortening sentences. The top passage has a Flesch reading age of 15–17 years. The bottom passage has one of 13–14 years. However, the top passage seems easier to read.

The basic idea underlying readability formulae is that the longer the sentences and the more complex the vocabulary in these sentences, then the more difficult the text will be. Clearly such a notion, whilst generally sensible, has its limitations. Some technical abbreviations are short (e.g., 'DNA') but difficult for people who have not heard of them. Some words are long but, because of their frequent use, become quite familiar (e.g., 'ergonomist' in the present context). With readability formulae, the order of the words and the sentences is not taken into account and nor are the effects of other devices used to aid comprehension (e.g., typographical layout, tables, graphs and illustrations). And, most importantly (unlike reader-based tools) the readers' motivation, abilities and prior knowledge are not assessed (Davison and Green, 1987). Clearly there is more to text than just sentence and word lengths — otherwise it would be easy to make text simple by just shortening the words and the sentences. But, as Figure 7 shows, text that has short, choppy sentences is not always easier to read.

Computer-based style and grammar checkers

Most readers who use word-processors will be familiar with 'spelling-checkers': tools that enable us to check the spelling in our documents. Style and grammar checkers, as their name suggests, are but an extension of this idea — they aim to help us with our style and grammar. Essentially the procedure is to run these checkers over the text once it is completed (but it can be done concurrently if this is desired). The checker stops at every point where the programme detects a possible stylistic or grammatical error.

Early studies of style and grammar checkers focused on assessing how useful they were to writers. This research suggested that many people found them rather tedious to use, but that they did find them helpful (Hartley, 1994). More recent research has focused on making comparison studies between different programmes to see which is the most effective. Typically what one does here is to assemble a set of ungrammatical, or poorly written sentences or passages, and then try out different grammar checkers on them to see which errors are detected and what sort of advice is given (e.g., see Kohut and Gorman, 1995; Pedler, 2001). Other, more theoretical research in this area concerns itself with developing more sophisticated programmes than the ones currently available (e.g., see Dale and Douglas, 1996; Harrison and Bakker, 1998; Woolls and Coulthard, 1998).

Grammar checkers are very good at spotting the minutiae of errors in punctuation and grammar but, naturally, they cannot help with matters of content. In my experience it is best to use both computer-based and human editors (experts and readers) to evaluate the effectiveness of style.

Combining different measures

Experiments have been carried out to see if the information provided from expert-, reader- or text-based measures is equally effective or not in improving texts. De Jong and Lentz (1996) for instance, compared the usefulness of expert versus reader feedback in assessing the effectiveness of a public information brochure about rent subsidies. Here the criticisms of fifteen expert technical writers were compared with those of fifteen members of the public. The main conclusions of this study were that criticisms of the two groups were very different. The readers pointed out significantly more problems associated with the typographic design of the brochure and with their understanding of it. The technical writers pointed out significantly more problems with the use of appropriate expressions and conventions, and with matters of writing style.

In another study, Weston *et al.* (1997) gave to a new set of instructional designers the suggestions from previous experts, readers, and instructional designers for re-writing a six-page instructional unit on diet and cancer. The new designers most frequently used the suggestions from the readers and the previous instructional designers rather than the experts in making their revisions. Subsequent comprehension tests showed that the most important information for improving the comprehension of the passage came from the readers' earlier comments.

In a third study, Wilson *et al.* (1998) reported, amongst other things, the responses of General Practitioners (GPs) and patients to questions concerning the content and usefulness of patient information leaflets. Both the GPs and the patients thought that the leaflets were useful, but they had widely disparate views about the content. Thus, for example, 80% of the GPs responded 'No' and 75% of the patients responded 'Yes' to the question, 'Is there anything you feel is essential to include but is omitted?' Similarly, 86% of the GPs responded 'No' and 46% of the patients responded 'Yes' to the question, 'Is there anything you feel should be left out that is included'. Finally, 86% of the GPs responded 'No' and 50% of the patients 'Yes' to the question, 'Is there anywhere where you feel the style of the language is not appropriate (e.g., patronising/confusing)? Berry *et al.* (1997) reported similar results.

A different kind of study (Hartley and Benjamin, 1998) showed how making multiple measures could be more informative than making single ones. Here comparisons were made between the traditional abstracts from journal articles and what are called *structured* abstracts. (These contain sub-headings, such as 'background to the study', 'aims', 'methods', 'results' and 'conclusions'.) In this investigation the effects of these changes were assessed in five ways. The results showed that:

- In terms of *length* the structured abstracts were significantly longer.
- In terms of *information content* the structured abstracts were significantly more informative — as assessed by readers.
- In terms of *readability* the structured abstracts were significantly more readable — as assessed by computer-based readability formulae.
- In terms of *searchability* the readers were able to find information more quickly with the structured abstracts.
- In terms of *preferences* the authors of the abstracts were almost unanimous in their preferences for the structured versions.

These overall results suggested that the structured abstracts were more effective than the traditional ones — but that they took up more journal space to achieve this. The use of several evaluation methods — rather than just one — strengthened this conclusion.

Studies such as these (and others reported by Schriver, 1997) point to the usefulness of combining different sources of evaluation and using different — complementary — tools. This is an important consideration that applies to almost all evaluation studies.

Cyclical testing and revision

However, we may not always be interested in testing to see whether, say, one text is easier to use than another. We may be concerned with using a selection from the tools described above to *improve* a text. One of the most useful approaches here involves cyclical testing and revision. This approach requires testing the first version of a text with appropriate tools, revising it on the basis of the results obtained, testing this revised version again, revising it, re-testing and so on, until the text obtains its objectives.

Examples of this approach are described by England (1989), Schriver (1997) and Waller (1984). Waller described how he and his colleagues used this iterative or cyclical approach to improve a form published by the then Department of Health and Social Security (DHSS). This form, used by unemployed people claiming supplementary benefits in the UK, was attractive to look at but difficult to complete. In fact only 25% of the forms were completed satisfactorily. Such a high error-rate had enormous costs for the DHSS — forms had to be returned for correction, or respondents had to be followed up in some way before an assessment of benefit could be made.

In order to improve the form a revised prototype was first tested with small groups of appropriate people. Part of this assessment included the use of an eye-movement recorder to assess which pieces of text were read and in what order. The aim of this first assessment was to isolate the main causes of difficulty, and to collect data against which the final version of the form could be compared. The investigators concluded from this first test that the form was not asking the right questions to obtain the information that was needed, and that many of the questions were ambiguous.

Thus a re-designed form was prepared. The emphasis now was on making clearer the language of the form and its sequencing. This version was again tested with small groups of appropriate people. It was clear that improvements had been made, but that more could be done. So a third version was prepared. Headings were added to make clearer the different sections of the form, and the routing instructions were further clarified.

The testing of this third version showed that this had solved most of the problems. So a fourth and final version was prepared. This version introduced colour-coding for the main headings (the earlier versions had been in black and white), a larger page-size, and yet another re-sequenced order. This final version was tested with larger groups of appropriate people. Now about 75% of the forms were completed satisfactorily. These results, although not perfect, ensured massive cost-savings for the DHSS. Indeed, *cost-benefit analysis* is yet another important tool for evaluating instructional text (see, e.g., Karat, 1997 and also Chapter 37).

Summary

This chapter has discussed the evaluation of text from four different viewpoints:

1. evaluating the content,
2. evaluating the layout,
3. evaluating structural devices and illustrative materials, and
4. evaluating the suitability of the text for its readers.

In each section a variety of methods for assessing texts have been mentioned. Content can best be assessed by using survey techniques or more detailed analyses. Typographical layout, structural devices and illustrative materials can be evaluated both by survey techniques, laboratory and field experiments. The suitability of the text for its readers can be evaluated by applying readability formulae, but such an approach would seem unduly restrictive when there are other more useful survey, field and laboratory approaches. Approaches that emphasise combining different methods in an iterative approach are likely to be the most helpful for ergonomists.

References

Abraham, C., Krahe, B., Dominic, R. and Fritsche, I. (2002). Do health promotion messages target cognitive and behavioural correlates of condom use? A content analysis of safer-sex promotion leaflets in two countries. *British Journal of Health Psychology*, **7**, 227–246.

Ant Ozok, A. and Salvendry, G. (2001). How consistent is your web design? *Behaviour and Information Technology*, **20**, 433–447.

Banister, P., Burman, E., Taylor, M. and Tindall, C. (1994). *Qualitative Methods in Psychology* (Chapter 6) (Buckingham: Open University Press).

Berry, D.C., Michas, I.C., Gillie, T. and Forster, M. (1997). What do patients want to know about their medicine, and what do doctors want to tell them? A comparative study. *Psychology and Health*, **12**, 467–480.

Clay, R.A. (2001). It was bad design, not dumb voters. *Monitor on Psychology*, **32**, 30–31.

Cleveland, W.S. (1985). *The Elements of Graphing Data* (Monterey, CA: Wadsworth).

Cook, G. and O'Halloran, K. (1999). Label literacy: Factors affecting the understanding and assessment of baby food labels. In T. O'Brien (ed.), *Language and Literacies* (Clevedon: British Association for Applied Linguistics).

Couloubaritsis, A., Moss, G.D. and Abouserie, R. (1994). Evaluating curriculum materials: Do Greek children understand their history textbook? *Education Training and Technology International*, **31**, 268–275.

Dale, R. and Douglas, S. (1996). Two investigations into intelligent text processing, In M. Sharples and T. van der Geest (eds), *The New Writing Environment: Writers at Work in a World of Technology* (London: Springer), pp. 123–145.

Davison, A. and Green, G. (eds) (1987). *Linguistic Complexity and Text Comprehension: A Re-examination of Readability with Alternative Views* (Hillsdale, N.J.: Erlbaum).

De Jong, M. and Van der Geest, T. (2000). Characterizing web heuristics. *Journal of Technical Communication*, **47**, 311–326.

De Jong, M.D.T. and Lentz, L.R. (1996). Expert judgements versus reader feedback: A comparison of text evaluation techniques. *Journal of Technical Writing*, **26**, 507–519.

Dupont, V. and Bestgen, Y. (2002). Structure and topic information in expository text overviews. *Document Design*, **3**, 2–12.

Dwyer, F.M. (1987). *Enhancing Visualized Instruction — Recommendations for Practitioners*. Learning Services, Box 784, State College, Pennsylvania, PA 16801.

Ehrenberg, A.S.C. (1977). Rudiments of numeracy. *Journal of the Royal Statistical Society*, **140**, 227–297.

England, E. (1989). Case study: Iterative screen design — errors as the basis of learning. *Educational Training and Technology International*, **26**, 149–155.

Farr, R. and Tulley, M.A. (1985). Do adoption committees perpetuate mediocre textbooks? *Phi Delta Kappan*, **66**, 467–471.

Flesch, R. (1948). A new readability yardstick. *Journal of Applied Psychology*, **32**, 221–223.

Ganier, F. (2001). Processing text and pictures in procedural instructions. *Information Design Journal*, **10**, 146–153.

Gill, R. (1996). Discourse analysis: Practical implementation. In J.T.E. Richardson (ed.). *Handbook of Qualitative Research Methods for Psychology and the Social Sciences* (London: British Psychological Society Books), pp. 141–156.

Gould, J.D., Boies, S.J. and Ukelson, J. (1997). How to design usable systems. In: M.G. Helander, T.K. Landauer and P.V. Prabhu (eds), *Handbook of Human-Computer Interaction* (2nd edn) (Amsterdam: Elsevier), pp. 231–254.

Grabinger, R.S. (1993). Computer screen designs: Viewer judgements. *Educational Technology and Research Development*, **41**, 35–73.

Greene, B. (2001). Testing reading comprehension of theoretical discourse with cloze. *Journal of Research in Reading,* **24**, 82–98.

Gunning, R. (1952). *The Technique of Clear Writing* (New York: McGraw-Hill).

Harrison, S. and Bakker, P. (1998). Two new readability predictors for the professional writer. *Journal of Research in Reading,* **21**, 121–138.

Hartley, J. (1991a). Captions for tables and figures. *British Journal of Educational Technology,* **22**, 149–150.

Hartley, J. (1991b). Tabling information. *American Psychologist,* **46**, 655–656.

Hartley, J. (1994). *Designing Instructional Text* (3rd edn) (London: Kogan Page).

Hartley, J. (2000). Clarifying the abstracts of systematic literature reviews. *Bulletin of the Medical Library Association,* **88**, 332–337.

Hartley, J. and Benjamin, M. (1998). An evaluation of structured abstracts in journals published by the British Psychological Society. *British Journal of Educational Psychology,* **68**, 443–456.

Hartley, J. and Ganier, F. (2000). Which do you prefer? Some observations on preference measures in studies of structured abstracts. *European Science Editing,* **26**, 4–7.

Hartley, J., Trueman, M. and Burnhill, P. (1979). The role of spatial and typographic cues in the layout of journal references. *Applied Ergonomics,* **10**, 165–169.

Hartley, J. and Yates, P. (2001). Referees are not always right! The case of the 3-D graph. *British Journal of Educational Technology,* **32**, 623–626.

Hicks, M., O'Malley, C., Nichols, S. and Anderson, B. (2003). Comparison of 2D and 3D representations for visualizing telecommunication usage. *Behaviour and Information Technology,* **22**, 185–201.

Hofman, R. and Van Oostendorp, H. (1999). Cognitive effects of a structural overview in a hypertext. *British Journal of Educational Technology,* **30**, 129–140.

Horn, R.E. (1998). *Visual Language: Global Communication for the 21st century.* Bainbridge Island, Washington 98110: MacroVu Press.

Houghton, H.A. and Willows, D.M. (eds) (1987). *The Psychology of Illustration,* Vol. 2 (New York: Springer Verlag).

Jansen, F., Van Lijf, A. and Toussaint, E. (2001). A note on the evaluation of footnotes and other devices for background information in popular scientific texts. *IEEE Transactions on Professional Communication,* **44**, 195–201.

Karat, C.-M. (1997). Cost-justifying usability engineering in the dotware life cycle. In: M.G. Helander, T.K. Landauer and P.V. Prabhu (eds), *Handbook of Human-Computer Interaction* (2nd edn) (Amsterdam: Elsevier), pp. 767–778.

Klare, G.R. (1963). *The Measurement of Readability* (Ames, Iowa: Iowa State University Press).

Kohut, G.F. and Gorman, K.J. (1995). The effectiveness of leading grammar/style software packages in analyzing business students' writing. *Journal of Business and Technical Communication,* **9**, 341–361.

Kosslyn, S.M. (1994). *Elements of Graph Design.* (New York: Freeman).

Levie, W.H. and Lentz, R. (1982). Effects of text illustrations: A review of the research. *Educational Communication and Technology Journal,* **30**, 195–232.

Lohse, G.L. (1997). Models of graphical perception. In: M.G. Helander, T.K. Landauer and P.V. Prabhu (eds) *Handbook of Human-Computer Interaction* (2nd edn) (Amsterdam: Elsevier), pp. 107–135.

Lorch, R.F. and Lorch, E.P. (1996). Effects of headings on text recall and summarization. *Contemporary Educational Psychology,* **21**, 261–278.

Lowe, R. (1993). *Successful Instructional Diagrams* (London: Kogan Page).

Mailloux, S.L., Johnson, M.E., Fisher, D.G. and Pettibone, J. (1995). How reliable is computerized assessment of readability? *Computers in Nursing,* **13**, 221–225.

Meyer, B.J.F. and Poon, L.W. (2001). Effects of structure strategy training and signalling on recall of text. *Journal of Educational Psychology,* **93**, 141–159.

Misanchuk, E.R. (1992). *Preparing Instructional Text: Document Design using Desk-top Publishing* (Englewood Cliffs, NJ: Educational Technology Publications).

Morrell, R.W., Dailey, S.R., Feldman, C., Mayhorn, C.B. and Echt, K.V. (2001). *Older Adults and Information Technology: A Compendium of Scientific Research and Web Site Accessibility Guidelines* (Washington, D.C.: National Institute on Aging).

Nielsen, J. (2000). *Designing Web Usability: The Practice of Simplicity* (Indianapolis, IN: New Riders).
Nielsen, J. and Tahir, M. (2002). *Homepage Usability: 50 Websites Deconstructed* (Indianapolis, IN: New Riders).
Novick, L.R. and Morse, D.L. (2000). Folding a fish, making a mushroom: The role of diagrams in executing assembly procedures. *Memory and Cognition*, **28**, 1241–1256.
Paap, K.R. and Cooke, N.J. (1997). Design of menus. In: M.G. Helander, T.K. Landauer and P.V. Prabhu (eds) *Handbook of Human-Computer Interaction* (2nd edn) (Amsterdam: Elsevier), pp. 533–572.
Pedler, J. (2001). Computer spellcheckers and dyslexics — a performance survey. *British Journal of Educational Technology*, **32**, 23–37.
Post, D.L. (1997). Color and human-computer interaction. In M.G. Helander, T.K. Landauer and P.V. Prabhu (eds), *Handbook of Human–Computer Interaction* (2nd edn) (Amsterdam: Elsevier), pp. 573–615.
Price, J. and Price, L. (2002). *Hot text: Writing That Works* (Indianapolis, IN: New Riders).
Rayner, K., Rotello, C.M., Stewart, A.J., Keir, J. and Duffy, S.A. (2001). Integrating text and pictorial information: Eye movements when looking at print advertisements. *Journal of Experimental Psychology: Applied*, **7**, 219–226.
Reid, D.J., Briggs, N. and Beveridge, M. (1983). The effects of pictures upon the readability of a school science topic. *British Journal of Educational Psychology*, **53**, 327–335.
Roberts, L. and Thrift, J. (2002). *The Designer and the Grid* (Hove: Rotovision).
Roth, W.-M. (2002). Reading graphs: contributions to an integrative concept of literacy. *Journal of Curriculum Studies*, **34**, 1–24.
Sadowski, M., Goetz, E.T. and Rodriguez, M. (2000). Engaging texts: Effects of concreteness on comprehensibility, interest, and recall in four text types. *Journal of Educational Psychology*, **92**, 85–95.
Sears, A. (ed.) (2000). Special issue: WWW usability. *International Journal of Human–Computer Interaction*, **12**, 167–277.
Schreiber, J.B., Verdi, M.P., Patock-Peckham, J., Johnson, J.T. and Kealy, W.A. (2002). Differing map construction and text organization and their effects on retention. *Journal of Experimental Education*, **70**, 114–130.
Schriver, K.A. (1989). Evaluating text quality: The continuum from text-focused to reader-focused methods. *IEEE Transactions on Professional Communication*, **32**, 4, 238–255.
Schriver, K.A. (1997). *Dynamics in Document Design: Creating Text for Readers* (New York: Wiley).
Seki, Y. (2000). Using lists to improve access to text: The role of layout in reading. *Visible Language*, **34**, 280–295.
Spinillo, C.G. and Dyson, M.C. (2001). An exploratory study of reading procedural pictorial sequences. *Information Design Journal*, **10**, 154–168.
Stylianidou, F., Ormerod, F. and Ogborn, J. (2002). Analysis of science textbook pictures about energy and pupils' reading of them. *International Journal of Science Education*, **24**, 3, 257–283.
Sydes, M. and Hartley, J. (1997). A thorn in the Flesch: Observations on the unreliability of computer-based readability formulae. *British Journal of Educational Technology*, **28**, 143–145.
Taylor, W.L. (1953). Cloze procedure: A new tool for measuring readability. *Journalism Quarterly*, **30**, 415–433.
Tedford, W.H., Berquist, S.L. and Flynn, W.E. (1977). The size-color illusion. *Journal of General Psychology*, **97**, 145–149.
Tufte, E.R. (1983). *The Visual Display of Quantitative Information.* Graphics Press, PO Box 430, Cheshire, CT 06410.
Tullis, T.S. (1997). Screen design. In M.G. Helander, T.K. Landauer and P.V. Prabhu (eds) *Handbook of Human-Computer Interaction* (2nd edn) (Amsterdam: Elsevier), pp. 503–531.
Upchurch, L., Rugg, G. and Kitchenbaum, B. (2001). Using card sorts to elicit web page quality attributes. *IEEE Software*, July/August, 84–89.
Van den Broek, P., Tzeng, Y., Risden, K., Trabasso, T. and Basche, O. (2001). Inferential questioning: Effects on narrative texts as a function of grade and timing. *Journal of Educational Psychology*, **93**, 521–529.
Van der Geest, T. (2001). *Website Design is Communication Design.* (Amsterdam: John Benjamins).

Van der Meij, H. (2002). Styling the index: Is it time for a change? *Journal of Information Science*, **28**, 243–251.

Walker, S. and Reynolds, L. (2000). Screen design for children's reading: Some key issues. *Journal of Research in Reading*, **23**, 224–234.

Waller, R. (1979). Typographic access structures for instructional text. In P.A. Kolers, M.E. Wrolstad and H. Bouma (eds) *Processing of Visible Language* (New York: Plenum), pp. 175–187.

Waller, R. (1984). Designing a government form: A case history. *Information Design Journal*, **4**, 36–57.

Whalley, P. and Fleming, R. (1975). An experiment with a simple reading recorder. *Programmed Learning and Educational Technology*, **12**, 120–123.

Warren, T.L. (1999). Prolegomena for a theory of table design. In H.J.G. Zwaga, T. Boersema, and H.C.M. Hoonhout (eds) *Visual Information for Everyday Use*. (London: Taylor & Francis), pp. 203–208.

Weber, R.P. (1990). *Basic Content Analysis* (2nd edn). Newbury Park, CA: Sage.

Westendorp, P. (1999). Text, pictures or flowcharts for specific types of information. In H.J.G. Zwaga, T. Boersema, and H.C.M. Hoonhout (eds). *Visual Information for Everyday Use* (London: Taylor & Francis), pp. 83–89.

Weston, C., Le Maistre, C., McAlpine, L. and Bordonaro, T. (1997). The influence of participants in formative evaluation on the learning from written instructional materials. *Instructional Science*, **25**, 368–386.

Wilkinson, L. (1999). *The Grammar of Graphics* (New York: Springer).

Willows, D.M. and Houghton H.A. (eds) (1987). *The Psychology of Illustration* (Vol. 1) (New York: Springer Verlag).

Wilson, R., Kenny, T., Clark, J., Moseley, D., Newton, L., Newton, D. and Purves, I. (1998). Ensuring the readability and understandability and efficacy of patient information leaflets. *Prodigy Publication, No. 30*, Sowerby Centre for Health Informatics, Newcastle University.

Woolls, D. and Coulthard, R.M. (1998). Tools for the trade. *Forensic Linguistics: The International Journal of Speech, Language and Law*, **5**, 33–57.

Wright, P. (1982). A user-oriented approach to the design of tables and flowcharts. In D.H. Jonassen (ed.). *The Technology of Text*. (Englewood Cliffs, NJ: Educational Technology Publications), pp. 317–340.

Wright, P. (2000). The psychology of layout: Consequences of the visual structure of documents. *Technical Report FS-99-04*. American Association for Artificial Intelligence, Menlo Park, CA: AAAI Press, pp. 1–9.

Young, S. and Wogalter, M.S. (2001). Predictors of pictorial symbol comprehension. *Information Design Journal*, **10**, 124–132.

Chapter 13

Evaluation in human–computer interaction

Chris Baber

This chapter presents an overview of evaluation in human–computer interaction (HCI). The chapter begins with a discussion of similarities and differences between ergonomics and HCI, in order to illustrate different notions of evaluation. This theme is developed further in the discussion of four main approaches to HCI: craft, applied science, engineering and sociological. It is argued that each approach has different interpretations of evaluation. Furthermore, the chapter proposes that evaluation is conducted for four distinct reasons: evaluation against competing products, evaluation against design targets, evaluation against requirements and evaluation against Standards. Each reason requires different approaches and different types of data. Consequently, there is a need to consider multiple methods to support multiple reasons for conducting evaluation that reflect multiple approaches to HCI. However, it is proposed that appropriate HCI evaluation requires, as a minimum, the clear statement of a referent model against which the evaluation takes place.

Ergonomics and HCI

To some extent, research into human–computer interaction (HCI) could be seen as an extension of ergonomics; after all, the focus of both disciplines is on the relationship between people, their technology and their working environments. An alternative view of HCI, sees the discipline being (largely) developed from the efforts of cognitive psychologists and software engineers and their search for new ways to study computer systems that could be used by humans, with roots in work conducted in the 1970s. In recent years, the range of disciplines having something to say about HCI has been steadily growing, such that one can find research drawing on disciplines from sociology to architecture. Given the broad church that HCI represents, it strikes me that evaluation in HCI is not simply a matter of aping other ergonomic (or psychological or software engineering) evaluation methods. Rather, there is a two-way traffic between HCI and ergonomics when it comes to designing, developing and employing evaluation methods. Indeed, whilst it is the HCI community that has been speaking of, measuring and proselytising the concept of 'usability', the International Standards that relate to HCI are being developed by an Ergonomics Committee. My point, in this opening paragraph, is to suggest that HCI and ergonomics are complementary areas of research study (it is fruitless to ask whether one is subordinate to the other), and that both are concerned with the fundamental questions of what makes technology hard to use, and how can we design *better* technology. In this chapter, I will consider the role that evaluation plays in HCI. In order to do this, it is probably useful to begin with some basic assertions (I will explain and elaborate on these assertions during the chapter but ask the reader to take them on trust at the moment):

- Evaluation requires comparison with a referent model;
- Evaluation measures more than a product's features;
- Evaluation is integral to the process of design;
- Evaluation implies critique.

Approaches to HCI

Long and Dowell (1989) proposed that HCI could be considered in terms of three distinct approaches. A 'craft' approach seeks to employ designers' experience and common-sense through heuristics and through design practice. Such an approach might have little need for formal evaluation tools or for rigorously defined data. An 'engineering' approach would seek to generalise scientific principles into laws and guidelines and would seek to produce specifications of the products/systems to be built. Such an approach might view evaluation as a means of testing products against benchmarks and targets. An 'applied science' approach would seek to develop fundamental principles of human behaviour with technology. Such an approach would, perhaps, view evaluation as an extension of the experimental methodology used in psychology. This tripartite classification gives a good reflection of the state-of-play in HCI as it was in the 1990s, with three main interest groups represented: the software engineers who were developing user interfaces tended to take a craft approach; the ergonomists tended to take an engineering approach; and the cognitive psychologists tended to take an 'applied science' approach (I accept the gross over-simplification of this demarcation but feel it gives a reasonable feel for what was happening). In recent years, sociology has made a significant impact on HCI, primarily through the critique of applied science approaches (in the field of Computer Supported Cooperative Work), but also partly through the rise of research methods derived from ethnography, and partly through the increasing trend to employ 'Scandinavian' approaches to user participation (see Chapter 36). Therefore, to the three classes offered by Long and Dowell (1989), I will add a fourth class that I will term 'sociological'. This approach seeks to understand the responses that people have to their work and to technology. In this context, 'evaluation' becomes a matter of reflecting on such responses.

Each approach would view the 'data' produced from evaluations (and the manner in which these data should be collected) in different ways. Thus, the applied science approach might require proof of reliability of the data, e.g., through appropriate experimental design and through statistical significance testing. This would typically require a reasonable sample size of participants, e.g., one might suggest in the region of 10–30+, depending on the statistical test employed, and adequate control over confounding variables. The sociology approach might argue that the control of variables serves to curtail the relevance of any findings, and that one needs to reflect real practice in real environments, perhaps seeking to present typical instances of user activity in the form of case studies or scenarios. Thus, data might be collected over a long period of time, in order to allow the analyst to become fully acquainted with the ways in which the organisation operates. The engineering approach would seek to derive quantifiable metrics of user performance in order to compare with targets. Thus, data might be collected from relatively small samples of users (or artificially generated through models). Finally, the craft approach might require immediately usable information, e.g., through design critiques and user comment. This might require a very small sample size of participants, e.g., only one or two people might provide sufficient information for the designer to rethink some of the design concepts. Clearly, given such a diversity of requirements for the output of evaluation, there is not going to be a single 'best' way of conducting evaluation.

In order to develop these points further, the chapter will begin with brief review of the concept of usability, followed by an overview of reasons why people conduct HCI evaluation. This will be followed by a brief (but not exhaustive) review of evaluation methods. Following this, consideration is given to how evaluation fits into the design lifecycle (and why it is necessary to consider evaluation in the context of design). The paper concludes with a discussion of the ways in which future computing technologies could be evaluated.

Usability: metric or concept?

'Usability' is central to much of the discussion of evaluation in HCI. The term usability is generally agreed to have been defined for HCI/ergonomics purposes by Shackel (1981), and originally referred to 'the capability in human functional terms [of a product] to be used easily and effectively by the range of users, given specified training and user support, to fulfill the specified range of tasks, within the specified range of environmental scenarios' (Shackel, 1984). This definition has been developed into several International Standards. ISO9241-11 defines usability as '. . . the extent to which a product can be used by specified users to achieve specified goals with effectiveness, efficiency and satisfaction in a specified context of use'. According to ISO 9216, usability comprises '. . . a set of attributes that bear on the effort needed for use, and on the individual assessment of such use, by stated or implied set of users' and '. . . the capability of the software product to be understood, learned, used and attractive to the user, when used under specified conditions'. ISO9216 covers the apparent discrepancy between these definitions (i.e., is usability an attribute of the product or a consequence of using the product?) by defining a further concept, 'Quality in use: the capability of the software product to enable specified users to achieve specified goals with effectiveness, productivity, safety and satisfaction in specified contexts of use.' Figure 1 illustrates the main components of ISO9126's concept of 'quality in use'.

There are two points of note from this brief consideration: (i) each definition, implicitly or explicitly, emphasises the notion of 'context of use' as a key component of usability; (ii) usability can be measured in terms of at least 34 factors (efficiency, effectiveness, satisfaction, effort, ease of understanding, ease of learning, ease of use, attractiveness, safety, productivity, operability, time-behaviour, resource, utilisation, analysability, changeability, stability, testability, adaptability, installability, co-existence, replaceability, maintainability, portability, functionality, accuracy, suitability, interoperability, security, reliability, maturity, fault-tolerance, recoverability, availability). This suggests that we are dealing with a concept that has multiple attributes, and that using a single term ('usability') may only serve to mask to complexity of the inter-relationships at work.

There is more than a passing resemblance between the debates surrounding 'workload' (see Chapter 18) and usability. The question for concepts such as workload and usability is what do they add to analysis that other concepts (such as the 34 taken from the ISO Standards) do not? My proposal is that usability ought to read as a shorthand description of the complex inter-relationship between people and technology, and most definitely *not* as an attribute of a product. This raises significant and, as yet unresolved, questions about evaluation in HCI. To make matters worse, 'context of use' seems to imply every aspect that might have a bearing on user performance. In other words, we have an dependent variable (usability) that varies along more than a single dimension and for which there is not a single measure, and an independent variable (context of use) that seems to include all possible contributing variables (in experimental design, such an independent variable would be treated with caution as it

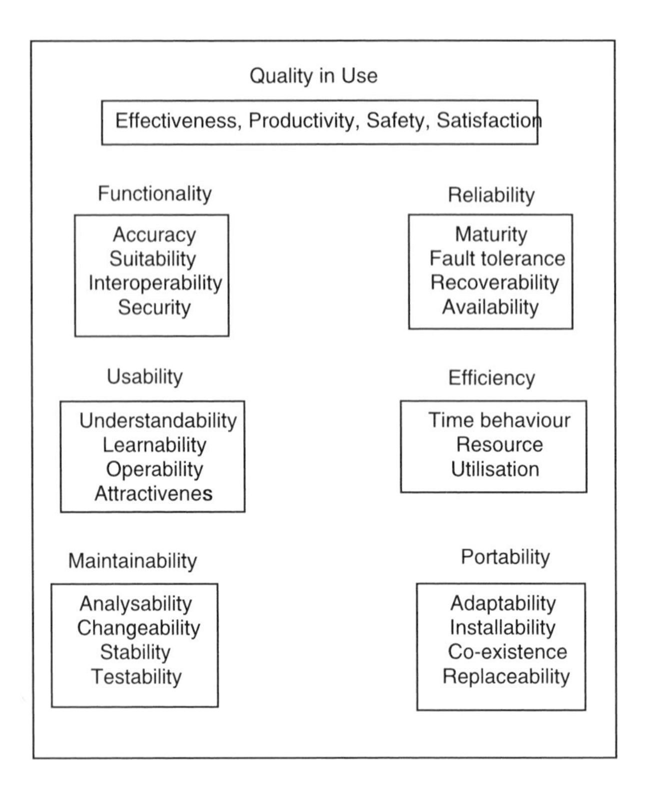

Figure 1 Quality in use [ISO 9126].

would appear to comprise of myriad confounding variables). As Stanton and Baber (1992) point out, 'While there are structured techniques that can be used to define hardware and software considerations, usability is not so easy to operationalise. This means that it is often used at a micro level, e.g., in terms of keyboard layout or font size, that can be defined in terms of specifiable criteria. Obviously, these factors can be measured and judged against benchmarks and Standards .... However, some factors will only be loosely specified ..., and other factors, which may have equally detrimental effects on usability, may only be defined at a ... macro level...' [p. 152]

The International Standards specify the need to measure aspects of usability, and provide an indication of what to measure, but they leave the precise definition of usability to the evaluator. This is primarily because, unlike physical measurements such as length or weight or voltage, the application and interpretation of the measure will vary according to context of use. This inevitably makes the idea of a standard measure of 'usability' highly problematic. Whilst one can define the measurement test and report results, it is not clear how one ought to compare the outcome of one test with another (particularly when slight variations in test might be used). For example, can one say that system *X* (for remote monitoring of asthma patients) which scores '3' is, therefore, inferior to system *Y* (for computer-aided design) which scores '7'? Having said this, if one cannot report an agreed set of benchmark measures, how can one speak of measurement? It is like using the notion of a 'hand' to measure height, and allowing both children and adults to use their own hands as reference. However, this is not simply a problem for usability evaluation; the majority of studies in HCI rely on measurements and tests that are devised by the individual practitioners, with little recourse to standard testing and (often) with insufficient sample size to provide (statistically) valid results. Thus, for instance, in the application of Fitts' Law to cursor control, there is a large variation in even measures as simple as time to move the cursor (Baber, 1997). Such variation could be related to variation in experimental, participant, device and environmental conditions (MacKenzie, 1992).

A concept with multiple attributes (usability) can be interpreted according to different purposes. Thus, a 'craft' practitioner might wish to take the list of 34 components of usability listed above (or a subset of these) and simply provide definitions of the interaction between user and product that demonstrates how these components are applicable. In this sense, the measurement would barely qualify as ordinal and would, more likely, simply be a verbal description summarising the opinion of the evaluator (or a small number of evaluators). Obviously such an approach would be quick to undertake, but is clearly open to accusations of bias and lack of control. On the other hand, an applied scientist might seek to control the components of context of use (whatever these might be) in order to determine the effects on different combinations of user performance. In this approach, standard laboratory measures, such as time and error, would be taken in order to quantify performance. This has the advantage of experimental control but might lack generalisability (e.g., how would the constrained combinations of context of use reflect the actual situation in which a product would be used?), and would be far more time-consuming than the craft approach. The 'sociology' approach might take a more grounded theory approach, deriving components of usability from the observations made in the field. But this means that the components are, by definition, very much subject to the observations being made and would be difficult to generalise across different conditions. Finally, an engineering approach would seek to operationalise the components of usability, and to define benchmarks against which the product could be tested. This is, to a great extent, the approach that appears to be preferred by the ISO Standards (although these Standards do leave the definition of many of the components, measures and tests fairly open-ended). In many respects, this is similar to the manner in which concepts such as 'Quality' are handled — the evaluator sets out a set of terms of reference, providing specific definitions of the terms, and then seeks to measure against this set.

This brief discussion indicates the general tone of the chapter: if there are four main groups of protagonists in HCI, then each group is seeking something different from the processes of evaluation. While one of the groups, say the applied scientists, might lay claim to moral high-ground in a debate on methods, e.g., because it can claim a tradition of experimental rigour, this does not guarantee that it is correct or that the other approaches are wrong. The general sense the reader will get from this chapter is that HCI is still in the process of deciding what it is doing when it comes to evaluation, and the various internecine squabbles only hide the general lack of focus. Having said this, the chapter will also argue that (paradoxically perhaps) evaluation is central to the practice of HCI research.

Reasons for conducting evaluation

There are many reasons why one would want to conduct evaluation of computer-based products. The primary reasons for conducting evaluation in HCI would seem to be to influence the design of a product (ideally, of course, to *improve* the product). The brief discussion of the ISO Standards above suggested a fairly lengthy set of characteristics of HCI that could be studied, measured or otherwise considered in evaluation. Clearly not all characteristics will be measured in all forms of evaluation, and it is useful to consider the reasons why someone would need, or want, to conduct HCI evaluation. In this section, I shall consider four main reasons for conducting HCI:

- Evaluation against other products;
- Evaluation against Design Targets;

- Evaluation against Requirements (User and Organisational);
- Evaluation against Standards.

The reader will notice that each type is considered as 'evaluation against' something. This develops one of the points made earlier; evaluation requires a referent model. It is naïve to believe that one can 'evaluate' something in a vacuum, i.e., to think that one can take a single product and evaluate it only in terms of itself. Put bluntly, such activity is not evaluation (I am not sure what it is, but would be very skeptical of any conclusions arising from such activity).

Evaluation against other products

If one has developed a new product, then it makes sense to compare it with competitor products, i.e., products that offer similar functionality. Clearly, one might think, the reference model in this situation will be the other products. The evaluation will then simply involve lining up the various products and conducting some sort of measurement, with a view to determining which product is the best or deciding how well the new product fares against such competition. From the perspective of HCI, this view is wrong for several reasons.

One of the problems with the view that 'evaluation against other products' ought to focus on products *per se* is that one rarely finds a set of products that can be matched on all functions, e.g., one might find that product *X* is excellent for performing activity *j* and very poor for activity *i*, whereas the converse is true for product *Y*, and that product *Z* does not even support activity *j*. To make matters worse, one might not be able to find any products that are comparable with the new product (you might have such a radically new design that there is nothing like it in existence). The question for 'evaluation against other products', therefore, hinges on the issues of what functions to address, what aspects of performance to measure, and how to make sensible comparisons. I would suggest that these are, probably, far more important than the issue of selecting the products for comparison (although the selection of products is not necessarily trivial).

In terms of selecting the functions to address, one could, for instance, consider three spreadsheet packages. If one reviews the packages, listing the functions that they can perform, one will find that not all functions are available in all packages. Consequently, selecting a function that can only be performed by one package would unfairly bias any evaluation. Having said this, producing an initial table summarising whether the packages offer specific functions can be informative and can highlight initial differences between them. Such a listing of functions offered by packages represents a simple form of comparison, and one that can be used to discuss the pros and cons of the packages in general terms. Indeed, the *UK Consumers Association* produces a very informative magazine (*Which?*) that contains many such comparisons of products in terms of their features (see Figure 2).

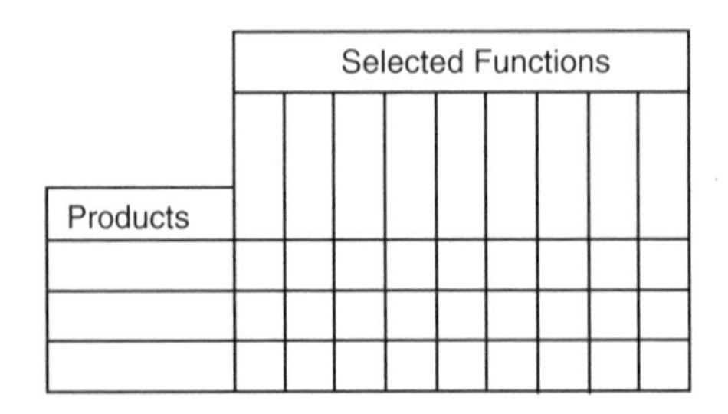

Figure 2 Product comparison table.

Whilst a table showing similarities and differences across products (or packages) can be a useful starting point, it also highlights one of the significant (and I feel unresolved) problems surrounding *all* forms of evaluation in HCI; the items that one wishes to compare are not identical. This might appear a truism. However, obvious or not, the point that one is typically comparing different things in HCI evaluation is often overlooked. For instance, if I reported a study comparing cursor control using a mouse and a trackball in terms of movement time, would it be sufficient simply to report average time? After all, if one device was faster, surely this is the main result to report. However, if the mouse supports movement controlled from the wrist, while the trackball supports movement controlled by the fingertips, one is not comparing similar user activity. I will accept that the outcome of the task (move the cursor from A to B) remains the same, but would argue that the processes involved are so different that one is bound to find differences in crude measures of performance (such as total movement time). I would further say that, given that such differences are inevitable, it is almost pointless to conduct the evaluation because comparing chalk with cheese only leads to the conclusion that it is very difficult to write on a blackboard using cheese (or that chalk does not cook well on toast).

So, how can the problem of comparing unlike things be resolved? In order to consider this, we need to revisit the problem raised earlier: how can you compare a radically new design with anything (when nothing like it exists)? Let us assume that we have recently developed a tour-guide that, knowing where the user is and what the user is doing, can provide up-to-the-minute information to help the user. Such systems are a common research vehicle for the study of context-aware and mobile computing. The product, perhaps a handheld device using Global Positioning System (GPS) to determine location and a wireless link to websites to collect information, differs from commercial products, which makes evaluation against other products tricky. Having said this, tourists have a variety of strategies and devices that they currently use and that need not feature a 'hi-tech' solution. One could ground an initial evaluation of the product in a comparison with existing practices. Thus, for example, set the goal of 'Find out about "Old Joe"' — participants might be given the handheld system, a guidebook to The University of Birmingham, an A–Z of Birmingham or allowed to walk on The University of Birmingham campus and talk to people. If they were successful, the participants would discover that "Old Joe" is the nickname of the clock tower that stands in the centre of the campus (named after Joseph Chamberlain).

What does this activity-based comparison of products tell us (and, more importantly, why is this different from the 'chalk and cheese' comparison dismissed above)? It is my proposal that conducting evaluation against other products in terms of a set of activities offers the analyst the following benefits:

- The evaluation will cover a range of functions on the products. It is important to ensure that the comparison provides a fair and accurate view of the product. After all, it is not really the point of evaluation to just demonstrate that product *X* is better than product *Y* — partly because there are bound to be occasions when products *X* and *Y* are similar, or where product *Y* is better than product *X*, and partly because simply knowing that $X > Y$ tells us very little about how to improve *X* (or *Y*) or why *X* is superior.
- The focus of the evaluation is less on product functioning than on user activity. This might appear, at first glance, to be tautological — surely product evaluation is about evaluating the product? This is, of course, true in a technical sense. However, HCI is about human–computer interaction, and the defining feature of this relationship is the interaction (rather than either human or computer). If one is concerned with technical evaluation then, perhaps some of the features to be included in a comparison table

(like the one shown in Figure 2) would be some of the technical features, e.g., processor speed, RAM, memory etc.

- As the evaluation is concerned with user activity (as opposed to product functioning), the type of metrics that could be applied may well change. While 'metrics' are discussed in more detail later in this chapter, it is worth considering their role here. When comparing user activity on two or more products, it is important to decide what information is really being sought. Do we want to know only that $X > Y$? (in which case, why not simply refer to the product summary table discussed above). Or do we want to know that using product *X* or *Y* has differing effects on user activity? Returning to the mouse vs. trackball example above: is it sufficient to know that the trackball will (probably) yield slightly slower (albeit not always statistically significant) performance? Or is it more useful to know that the mouse tends to produce long ballistic movements (arising from a sweeping movement about the wrist), with few control corrections, and a short homing time, in comparison with the trackball that requires shorter ballistic movements and more control movements (arising from the need to reposition the fingers on the trackball), and similar homing times? Of course, it depends what one thinks one needs to know: however, I would suggest that the latter conclusions are more useful in that they are not merely dependent on that task being performed by those participants with those devices under those conditions (as is the case with movement time), but allow the designer to reason about possible variations in performance and to generalise to new situations.

Evaluation against design targets

The notion of evaluating against design targets is central to the potential role of evaluation as a significant component of the human-centred design lifecycle. There are several reasons why design targets are beneficial. First, the design team can specify the type of targets that they believe are important, e.g., if the technology is intended to support training, the design team will need to specify a target relating to training effectiveness. Second, the design team can set increasingly specific targets as a means of supporting rapid prototyping, i.e., when a version of the prototype meets one set of targets, the team can decide whether to change the target. Third, the design team will be able to demonstrate performance improvements through showing either how the targets have changed during the design process or showing how the product has been modified to meet the targets. Fourth, the International Standards (discussed below) can be easily related to design targets, e.g., under the ISO9241 notion of usability one could begin with a target of '66% of the specified users would be able to use the 10 main functions of product *X* after a 30 minute introduction.' Once this target has been met, the design team might want to increase one of the variables, e.g., 85% of the specified users, or 20 main functions, or 15-minute introduction.

The notion of specifying targets lends itself to the development of usability specifications. In an early discussion of usability engineering, Good *et al.* (1986) propose that it is important to define both usability goals and metrics that relate to these goals. Given these notions, it is then possible to determine planned or acceptable levels of performance on each metric. For example, in a study of conferencing systems Whiteside *et al.* (1988) identified 10 attributes that they felt reflected the use of the conferencing system, e.g., ranging from a fear of feeling foolish to user preference to number of errors made during task performance. For each attribute, Whiteside *et al.* (1988) defined a method for collecting data about that attribute, e.g., questionnaires, observation, etc. and then set performance limits relating to best, worst and planned levels. A recent study of a wearable computer for paramedics (Baber *et al.*, 1999) used this concept to produce the information in Table 1. Three measures of performance

Table 1 Defining Design Targets (adapted from Baber *et al.*, 1999)

Factors	Method	Metrics	Worst	Target	Best	Current
Performance task						
	CPA	Time	−15%	0	+5%	−2%
	User trials	Time	−15%	0	+5%	−10%
Practice	1st vs. 3rd trial	% change	1st > 3rd	3rd > 1st	0	3rd > 1st
Usability user evaluation						
	SUS	Scale: 0–100	50	60	70	65
	SUMI	Scale: 0–100	50	60	70	60
	Heuristics	Scale: 0–10	<6	6	>6	8

were undertaken, i.e., predictive modelling (using critical path analysis), user trials and performance improvement arising from practice. In addition, three usability evaluation self-reporting methods were used. Table 1 shows that the current version of the system meets (or exceeds) some of the target criteria, but is still falling below the target for time (although is not within the 'worst' case range). One benefit of such a technique is simply to allow the design team to decide whether there is a need for more effort to put into refining the product, or whether, having met (some or all of) the requirements, the product design process can be closed.

It is not always easy to quantify design targets, so it might not be possible to produce tables such as Table 1 in all cases. However, it is possible to advance a set of qualitative measures that could be considered in evaluation. For instance, Holcomb and Tharp (1991) propose a 'model' of interface usability that contains seven components: Functional; Consistent; Natural and Intuitive; Minimal memorisation; Feedback; User help; User control. In their paper, they provide weightings for these components, and illustrate the components with specific terms. Table 2 lists the components and their terms proposed by Holcomb and Tharp (1991).

Over the past decade or so, a number of researchers have been interested in techniques that support design rationale. A design rationale technique, such as Questions-Options-Criteria QOC (MacLean and McKerlie, 1995), could be used to compare current concepts against a set of criteria that the design team feel are appropriate for the project. In this manner, informal evaluation of designs against targets can be conducted. For example, Figure 3 shows a QOC diagram for the Question 'How to identify a person as a valid user of an automated-teller machine (cashpoint)'? In this example, the design team has posed a specific question and then sought to generate possible design Options. Throughout the design project, the team will be developing a set of Criteria. These Criteria will be initially developed at the start of the project, and then added to as the project progresses. The design Options are then simply considered in terms of the Criteria, with an indication of whether the Option meets a given Criterion. It should be noted that Figure 3 is intended only as an illustration and not as a real set of solutions. However, the approach does allow the design team to consider designs against specific targets (or criteria).

Evaluation against user and organisational requirements

HCI has long recognised that a core feature of design is the capturing of user requirements (see also Chapter 10), and the derivation of specifications from such requirements. There are three problems with this view: (i) Users do not know what they want, and have great difficulty in expressing these wants (Norman, 1988). This is not to say that users cannot contribute

Table 2 Holcomb and Tharp's (1991) 'Model' of Interface Usability (after Holcomb and Tharp, 1991)

Component	Term
Functional	Able to accomplish tasks for which software is intended
	Perform tasks reliably and without errors
Consistent	Consistent key definitions
	Show similarly information at same place on screens
	Uniform command syntax
Natural and Intuitive	Learnable through natural conceptual model
	Familiar terms and natural language
Minimal memorisation	Provide status information
	Don't require information entered once to be re-entered
	Provide lists of choices and allow picking from the lists
	Provide default values for input fields
Feedback	Prompt before destructive operations like DELETE
	Show icons and other visual indicators
	Immediate problem and error notification
	Messages that provide specific instructions for action
User help	On-line help system available
	Informative, written documentation
User control	Ability to undo results of prior commands
	Ability to re-order or cancel tasks
	Allow operating system actions to be performed within the interface

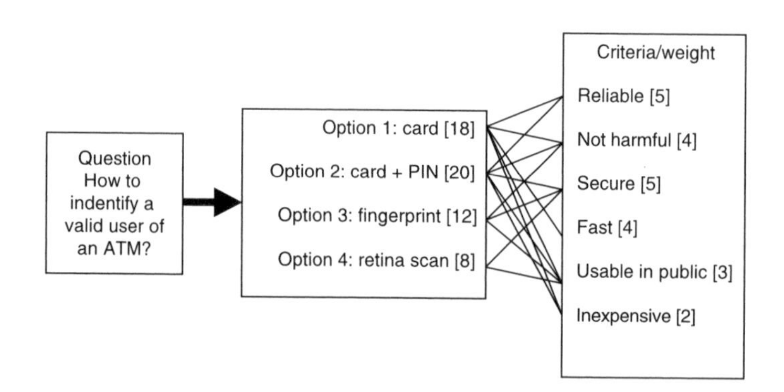

Figure 3 Example of a QOC diagram.

enormously to the design process. Indeed, some of the participative design and evaluation methods (see Chapter 11) demonstrate that user involvement can lead to many benefits to both the design team and to the users themselves; (ii) User requirements are typically incomplete, grounded in everyday experience and do not reflect the future system. This means that any specification taken solely from user requirements will be flawed. This is not to say that users should be left out of the requirements process, but that simply asking them is not enough; (iii) Any specification (typically) represents a 'model' of the future system, i.e., a formal description of what the system will look like and how it will function. Any model, by definition, is a simplification of a complex process, and (probably) omits significant aspects, either to make the model tractable (i.e., easy to develop) or to ensure coherence (i.e., to avoid contradictions in requirements — even when these exist) or to fit the modelling language (i.e., a dataflow diagram does not easily capture the inter-personal politics between members of a management team, but such 'politics' might affect how the system is used).

There have been (and will continue to be) many approaches that seek to capture user requirements. There have been some approaches that seek to relate requirements to specification. For instance, HCI is currently experimenting with the use of scenarios as a means of representing requirements through case studies (Carroll, 1996), and can then use scenarios to generate more formal descriptions, perhaps through object-oriented programming techniques. A recent development in structural user requirements capture is the Volere template (http://www.volere.co.uk).

There has been less work on how one could define organisational requirements. Furthermore, the requirements arising from an organisation might relate to factors that lie outside the experience of individual users, e.g., how well does the system allow transfer of data with existing software? Or how will the system influence current work practices? Kyng (1991) speaks of scenarios in the narrow (to describe individual experience) vs. scenarios in the wide (to describe organisational impact of new technology). Cooper *et al.* (1999) demonstrated that the use of simple scenarios, describing how a speech-controlled call handling product would affect an organisation, led to clearer specification of product requirements than simply asking the organisation which features it might need.

Evaluation against standards

When one thinks about Standards in engineering, one probably assumes that these documents define ranges of performance that products or devices should meet in order comply with the Standards, or define procedures that one needs to follow in order to demonstrate compliance. While there are some Standards that cover physical aspects, such as key size or displacement, the more recent Standards for HCI are somewhat different in that they tend not to report quantitative limits that products need to meet. There is a simple reason for this; as mentioned above, HCI is not merely about studying products (or users) but the interaction between users and computers. This means that it is meaningless to define absolute limits of performance that all devices, products or systems would need to reach in order to meet Standards. For example, would it be sensible to require that a computer keyboard should always result in typing speeds of 100 words per minute, i.e., irrespective of user capabilities, task demands etc.?

As Bevan (2001) points out, HCI Standards are generally concerned with the following areas:

(1) The use of the product.
(2) The user interface and interaction.
(3) The process used to develop the product.
(4) The capability of an organisation to apply user-centred design. [Bevan, 2001, p. 534]

The focus of the Standards being developed by the International Standards Organisation's (ISO) Technical Committee ISO/TC 159/SC4 *Ergonomics of human–system interaction* has been to develop documentation, procedures and support to enable people developing interactive systems to design, develop products that are demonstrably able to support effective, efficient and satisfying performance by specified users in specified contexts. An interesting conclusion from this objective is that the notion of designing for 'everyone' is of dubious validity; rather one is designing for specified user groups performing specified collections of activities in specified contexts. (This does not, of course, mean that design is in any way exclusive; rather it means that one needs to clearly identify groups of users, to outline their requirements and to use these specifications as a basic component in the design process.) The Standards can be roughly divided into those that focus on the user–product interaction, and those that focus on attributes of the product. Table 3 distinguishes between the focus of

Table 3 A Collection of HCI Standards (adapted from Earthey *et al.*, 2001)

Focus of evaluation	Attribute	Standard	Evaluation against...
Attributes of the product	Assessing whether some features of a product conforms to ergonomic guidelines	ISO9241; ISO10741; ISO11581; ISO14958; ISO20282; ISO18021; ISO13406; ISO18789	Standard products
User–product interaction	Measure usability against definition or requirements	ISO9241; IS013407; ISO9126; ISO14915; ISO TR 16982; ISO TR 18529 ISO16071	Standard target

evaluation for the various Standards and provides an indication of the referent against which the evaluation will be performed.

ISO 9241 (1998)

ISO 9241 (1998) is a 17 part Standard which is primarily concerned with the use of computers for office work, but which has a wide remit, e.g., from defining the types of tasks and working postures that should be considered, to the design of visual display terminals (VDTs) and keyboards, to the concept of usability. As far as Evaluation is concerned, this Standard offers procedures on conducting evaluations, e.g., for measuring glare and reflection from VDTs (ISO9241-7), for assessing interaction devices (ISO9241-9), as well as guidance on specific aspects of HCI, e.g., use of colour on displays (ISO9241-8) or dialogue design (ISO9241-14, ISO9241-15, ISO9241-16, ISO9241-17). In ISO9241-11, the Standard consider usability and provides a definition: 'Usability: the extent to which a product can be used by specified users to achieve specified goals with effectiveness, efficiency and satisfaction in a specified context of use.' [ISO9241-11 (1998)]. The main focus of this document is to provide advice on how to define measures and on how to perform evaluation trials, rather than on providing targets to aim for.

ISO 13407 (1999)

ISO 13407 (1999) is a single part Standard which is concerned with encouraging the involvement of human-centered design processes and activities throughout the lifecycle of interactive systems. The primary characteristics of the human-centered approach advocated by ISO 13407 are:

(a) the active involvement of users and a clear understanding of user and task requirements;
(b) an appropriate allocation of function between users and technology;
(c) the iteration of design solutions;
(d) multi-disciplinary design. [ISO13407 (1999) 5.1]

ISO13407 advises the use of an evaluation plan (7.5.2) and suggests that the evaluation should provide feedback into the design process (7.5.3), using either experts or users for this process. It also provides a template for writing an evaluation report. It would appear that the main focus of evaluation in this Standard is to demonstrate compliance with requirements, at both an organisational and user level. The Standard does not detail how the requirements are captured, nor how compliance is demonstrated.

ISO TR 18529 (2000)

The current activity of ISO/TC 159/SC4 appears to be focused on providing practitioners with the means to demonstrate compliance with the Standards notions of usability (and other aspects of the design process that can be evaluated). ISO TR 18529 (2000) describes the human-centred lifecycle, and suggests that evaluation is concerned with socio-technical aspects of the system, i.e., demonstrating that the organisation benefits from changes in its processes.

Summary of HCI related standards

The previous discussion highlights the developing range of Standards that will bring HCI into product and software design. The earlier Standards tended to focus on product attributes, but more recent Standards have recognised the multifaceted nature of the concept of usability and have sought to encourage an approach that is similar to Quality assessment. This development, to allow evaluation criteria to be developed for specific purposes, on the one hand encourages analysts to focus their attention on those aspects of context of use and HCI that will have a major bearing on the person-product performance, but, on the other hand, removes the notion of 'standard' measures of performance. It is my feeling that the current batch of International Standards represent a significant step forward in HCI evaluation, but that there now needs to be some means of demonstrating 'standard' measurement. At the moment, demonstrating compliance with the Standard would require the analysts to prepare a document describing their evaluation and demonstrating how the evaluation meets the objectives of the Standard (thus supporting evaluation against Standards and, I feel, evaluation against targets). However, if another team produced an equally appropriate evaluation but used different methods, then comparison of products remains problematic. In conclusion, evaluating against Standards requires the design/evaluation team to demonstrate that it has been able to measure aspects of HCI that adequately reflect the notions of effectiveness, efficiency and satisfaction, and to provide documentary support for the evaluation process.

Methods for conducting HCI evaluation

There may well be as many methods for evaluating usability as there are practitioners undertaking usability evaluation. While this might be a slight exaggeration, it does indicate that the field of usability evaluation methods is very large and growing each year. There are two main reasons why there are so many methods for evaluating HCI: (i) given the distinct approaches in HCI research (craft, engineering, applied science, sociology), there have arisen sets of methods that have been developed to support each of the approaches; (ii) each evaluation method tends focus on a specific aspect of the design lifecycle or for a specific form of evaluation. This is not to deny that some methods have been adopted by more than one form of HCI research or can be used at more than one stage in the design lifecycle or can be used for more than one form of evaluation. The following section will review a sample of HCI evaluation methods, and will consider the approaches to HCI (Craft, Applied Science, Engineering, Sociology) and form of evaluation (Evaluation against other Products, Evaluation against Standards, Evaluation against Targets, Evaluation against Requirements) to which a method is most suited, the type of expertise required to conduct the evaluation (Task expert, HCI expert, Product expert), and the approximate duration required to conduct the evaluation (hours, days, weeks).

A common approach to classifying evaluation methods employs four classes:

1. Analytic Methods — concerned with user models to predict performance
2. Usability Inspection Methods — 'experts' evaluation relating to HCI, product or domain
3. User Reports — subjective responses from potential or actual users
4. Observation — field or laboratory-based studies of user activity

To this list, I would add a fifth class:

5. Participative

Examples of each class of method will be presented a little later in this chapter. The reader could also usefully turn to the many product evaluation methods discussed in Chapter 11.

Comparing evaluation methods

Prior to discussing some of the methods in detail, it is worth pausing to consider how one might consider the relative performance of the various HCI evaluation methods. Throughout the 1990s, there were attempts to compare the performance of methods. For example, Jeffries *et al.* (1991) compared heuristic evaluation, cognitive walkthrough, guidelines review and usability laboratory testing. Recently, a number of papers have criticised much of this work and suggested that we are not in a position to report on whether the methods are actually of any use.

Gray and Salzman (1998) presented a damning indictment of five studies comparing evaluation methods (including the Jeffries *et al.* study mentioned above). The studies typically compared 'quick and dirty' techniques, such as cognitive walkthrough and heuristic evaluation, with other methods. However, Gray and Salzman criticise the small sample sizes used in the comparisons, the lack of control of variables and inappropriate reporting and manipulation of data. Andre *et al.* (1999) consider 17 usability evaluation studies and note a general lack of consistency between findings. They also criticise the methodology followed in the comparison studies. Thus, the question of which methods are 'best' remains open. This is a common problem in ergonomics research (Stanton and Young, 1999), and there remains a tendency to employ methods that 'feel' appropriate rather than investigating their appropriateness.

A brief survey of usability evaluation methods

In order to select a representative sample of evaluation methods to discuss in this section, I have chosen HCI textbooks from my bookshelf and checked through them for reports of methods. Obviously, such a 'straw-poll' has problems of coverage, reliability, validity etc. However, the summary of methods reported (shown in Table 4) presents what I consider to be a fair reflection of the range of methods that feature in HCI research, and also allows identification of a set of popular methods.

In broad terms, one can relate these types of evaluation method to the reasons for conducting evaluation (see Table 5). As mentioned above, most of the methods tend to be more suited to some types of evaluations than others. The reader might take issue with, for

Table 4 Survey of Evaluation Methods Covered in 10 HCI Textbooks

Type of method	Method	Books	Count
Analytic	GOMS, inc. Keystroke Level Model	1, 2, 3, 4, 5, 10	6
	Command Language Grammar	3	1
	Task Analysis for Knowledge Description	3	1
Specialist report	Metrics	1	1
	Heuristic	1, 2, 4, 5, 6	5
	Discount usability	1	1
	Walkthrough	1, 2, 3, 4, 5	5
User report	Verbal protocol	1, 3, 5	2
	Interviews	1, 2, 4, 5, 6, 8	6
	Questionnaires/Surveys	1, 4, 5, 6, 7, 8	6
	Focus groups	1, 2	2
Observation	Direct observation in the field	1, 3, 4, 5, 7, 8	6
	Video logging	1, 4	2
	Software logging	1	1
	Experiment	1, 3, 4, 5, 6, 7, 9	7
	Usability laboratory/Benchmark tests	1, 8, 9	3
Participative	Contextual inquiry	1	1
	Participative evaluation	1	1
	Cooperative evaluation	1, 8	2

Key: 1. Preece *et al.*, 1994; 2. Newman and Lamming, 1995; 3. Johnson, 1992; 4. Dix *et al.*, 1998; 5. Noyes and Baber, 2000; 6. Shneiderman, 1986; 7. Maddix, 1990; 8. Jordan *et al.*, 1996; 9. Downton, 1991; 10. Eberts, 1994.

Table 5 Comparing Types of Method with Type of Evaluation (after Preece *et al.*, 1992)

Evaluation against. . . .	Observation	User opinion	Usability inspection	Analytic	Participative
Other products standards	X	X	X	X	X
Design targets requirements	X		X	X	
		X	X		X

instance, the omission of 'user opinion' and 'participative' from Design Targets. This is because the use of Design Targets (in this chapter) is considered to be an activity that is managed by the design team. While users might be invited to participate or their opinions might be canvassed, they are not the final arbiters of the decisions as to whether the product meets (or fails to meet) design targets. (I have been in meetings where user opinion more or less negated the actual design taget and the product, and also have heard these opinions glossed over and both taget and product continued with.)

Analytic methods for HCI evaluation

By and large, analytic methods have tended to concentrate on the production of predictive models of user performance, e.g., by considering time (Card *et al.*, 1983; Olson and Olson, 1990), error (Young and Whittington, 1990; Baber and Stanton, 2001) or information processing (Johnson, 1992). As such the methods are mainly used in 'engineering' approaches to HCI. Table 6 shows an overview of Keystroke Level Models (KLM), which Table 4 shows to be one of the commonly referenced of the engineering approaches.

Table 6 Brief Description of KLM

Name of method	Keystroke Level Model (KLM)		
Evaluation type:	*Analytic*	Approach to HCI:	*Engineering*
Background knowledge:	HCI		
Brief description:	The method employs a form of task analysis, in which a goal is decomposed into subgoals and tasks. The subgoals and tasks can then be related to standard description, using either Production Rules or using standard times, such as:		
	Keypress	0.12 s (90 words per minute typist); 0.5 s (random letters); 0.75 s (complex codes)	
	Pointing	1.1 s	
	Homing	0.4 s	
	Mental operation	1.35 s	
Source references:	1. Card *et al.* (1986) 2. Olson and Olson (1990) 3. Stanton and Young (2001)		
Performance:	Number of evaluators:	1[1,2]	
	Duration:	Minutes to hours[depends on analyst]	
	Agreement with user data:	80–90%[1]	
	Inter-rater reliability:	d = 0.754[3]	
Overall:	Reducing human activity to standard tasks. Can be useful to develop predictions of performance. However, the reductionist approach might omit key features of performance or ignore the influence of contextual features.		

Note: The superscript numbers refer to the source references.

Analytic methods can be used to determine the likely performance of users with a particular device. One use of such models would be to define performance benchmarks, e.g., one could predict that the design of device *X* ought to support a performance time of 3 seconds to complete a specific task. One could then use this target performance time as a criterion for performance testing, e.g., stating that you expect, say 66% of your user group to complete the task in 3 seconds (or less). A second use of these models is to conduct 'evaluation' of products prior to building. For example, Gray *et al.* (1993), using critical path analysis, were able to demonstrate that a proposed computer workstation to support call handling (for a major telephone company) would not effect overall call handling time as the use of the computer did not lie on the critical path (most of the operator's time was spent speaking to the caller, rather than using the computer). Consequently, the study showed that investment in the technology would have resulted in an undue expense and not net gain in productivity.

Referent model

The various analytic techniques are aimed at developing their own referent models, employing standard descriptions or performance data. The validity of the models can be open to question because it is not straightforward to compare the prediction against other data (although see Baber and Mellor, 2001 for a discussion of this issue). Also, any model stands or falls by the simplifying assumptions that are inevitably required for its construction.

Relationship with perspectives on HCI

From an engineering perspective, these techniques offer the design team a number of benefits. For instance, having a set of standard times to describe performance can allow the development of fairly reliable predictive models. The efficiency of such models can be further enhanced through the use of critical path modelling (Gray *et al.*, 1993; Baber and Mellor, 2001). From a craft perspective, the approach can appear quite daunting as it requires some proficiency in developing the models, and can be time-consuming to develop from first principles. Furthermore, the lack of user involvement might, at first glance, suggest that these approaches are not human centred. In response to these criticisms, it should be noted that developing a model that can predict (with a reasonable degree of accuracy) performance that is similar to the average results obtained from a user trial (Baber and Mellor, 2001). From both the sociological and applied science perspectives, the reductionism of the approach leads to some tenuous assumptions concerning the manner in which people process information. There also remains a problem of incorporating external and contextual information into the model (although some of the critical path analysis work has addressed some aspects of this problem).

Conclusions

Analytic methods can often provide surprisingly good predictions of performance, and have been shown to have quite high inter-rater reliability. However, the methods are most appropriate for activities that can be described in terms of a set of unit-tasks that can have quantifiable attributes, e.g., times or probabilities of failure. This means that, providing one can describe a sequence of tasks and that these tasks are primarily observable activity, then the methods work well. When one turns to more cognitive activity, then the methods become a little less reliable. The models (like any model) exclude certain properties of the object being modeled — in this instance, the models exclude pretty much the entire range of external influences on performance, i.e., they do not capture context. Having said this, there are ways in which some contextual factors could be used to modify the data used in the model, e.g., if one calculated the time it took to move a cursor using a mouse using a version of Fitts Law [time $= a + b \log 2\ (2A/W)$] and then assumed that vibration, say due to a moving armoured vehicle, would lead to an increase in '*a*', then a modified calculation can be produced.

Usability inspection methods

Usability inspection concerns the evaluation of products through the examination of their features. Usability inspection methods are most closely associated with craft approaches to HCI. Typically the evaluation considers the 'static' aspects of the features, i.e., the inspection will be performed on the features that are shown on the interface. Having said this, most HCI practitioners who employ inspection methods will ask the evaluator to perform a set of selected tasks in order to provide them with sufficient context in which to evaluate the product. Tables 7 and 8 show two commonly employed usability inspection methods: heuristic evaluation and cognitive walkthrough.

Referent model

In a provocative statement at CHI'95, Gray *et al.* (1995) argued that usability inspection methods are of dubious utility. They noted that HCI has turned away from the idea of evaluating against Guidelines. I suppose that the era of HCI guidelines was the mid to late 1980s, e.g., the publication of the Smith and Mosier (1986) guidelines and versions of style-guides being produced by Apple, IBM and Microsoft and the MOD. The primary reasons why

***Table* 7** Brief Description of Heuristic Evaluation

Name of method	Heuristic		
Evaluation type:	*Inspection*	Approach to HCI:	*Craft*
Background knowledge:	Product, task or HCI		
Brief description:	The evaluator is asked to consider the product in terms of a set of simple rules (see below). For each rule, the evaluator determines how the product meets (or fails to meet) the rule, perhaps assigning a severity score when the product fails to meet the rule.		
	Heuristics: Use simple and natural language; Speak the user's language; Minimise user memory load; Be consistent; Provide feedback	Provide clearly marked exits; Provide short cuts; Good error messages; Prevent errors;	
Source references:	1. Nielsen, (1993) 2. Andre *et al.* (1999)		
Performance:	Number of evaluators: Duration: Percent of problems found: Agreement with 'expert':	3–5[1] Minutes 36 (±16)% [range: 17–46%][2] 51 (±25)% [range: 20–86%][2]	
Overall:	Quick method to apply and so can be useful for rapid evaluation of designs. However, the method is highly subjective and depends on the experience of the evaluator and interpretation of the rules, so open to bias. Furthermore, the method is dependent on the 'detectability' of the problem.		

HCI guidelines fell out of favour was that they were tremendously tedious to apply, and often failed to capture the right level of detail, i.e., the guidelines were either too prescriptive (and so could not easily be applied to all designs) or too vague (and so could not be applied with any rigour). However, at least the Guidelines represented an agreed referent model, i.e., a 'good' interface was one that complied with the Guidelines. Gray *et al.* (1995) suggested that reducing the vast corpus of Guidelines down to a handful of heuristics is an insult to HCI; they ask whether the entire knowledge that the HCI community has been accumulating over the past 40 years can be easily distilled into a potpourri of 'rules of thumb'. A further problem with heuristics is that they carry many of the problems that guidelines possessed, i.e., the difficulty of striking an appropriate level of detail. However, while one could debate the content of an entry in a set of guidelines (e.g., in terms of its specific content), the heuristics are sufficiently open-ended to limit such debate. It is probably the open-ended nature of heuristics that makes them attractive, i.e., they appear plausible and can be fitted to most evaluation activities. Ironically, it is this open-endedness that reduces their inter-rater reliability and makes them so context (and rater) dependent.

Monk *et al.* (1986) presented a detailed set of guidelines on how to use walkthrough approaches to evaluation in their Cooperative Evaluation method. The main aim of the approach is to capture problems that users experience when using a product. The approach requires recruitment of users from a specific target population, and ask them to perform tasks that reflect 'real' use of the product. The users are asked to think-aloud during task performance, and to ask for help (from the analyst) when any problems are encountered.

Table 8 Brief Description of Walkthrough

Name of method	Walkthrough		
Evaluation type:	*Inspection*	Approach to HCI:	*Craft*
Background knowledge:	Product, task or HCl		
Brief description:	The evaluator is asked to perform specific tasks, or to achieve specific goals, using the product and to provide a simultaneous commentary. The commentary should contain what the evaluator is thinking, doing or looking at. The technique is a form of verbal protocol (see Chapter 7). The resulting transcript can be considered in terms of the following questions: i. what does the user want to achieve? ii. Are the correct actions sufficiently clear for the user? iii. Will the user connect the correct action with the system image? iv. Will the user correctly interpret the system response?		
Source references:	1. Newman and Lamming (1995) 2. Andre *et al.* (1999)		
Performance:	Number of evaluators: Duration: Percent of problems found: Agreement with 'expert':	2+[1,2] Minutes [depends on task and detail required] 13 (±5)% [range: 7–17%][2] 41 (±30)% [range: 5–73%][2]	
Overall:	Method is relatively quick to apply and can provide useful insight into the misunderstandings that users might make when using the product. However, the results are highly dependent on the type of users employed and the type of tasks considered. Relatively quick method to apply, but can take time to transcribe or analyse results. However, the method depends on the evaluator's ability to generate 'problems'.		

The process is recorded (on video or audio tape), and the tape-count is used to index any reports of problems.

Relationship with approaches to HCI

From an 'applied science' perspective, the usability inspection methods fare quite poorly. They have relatively low accuracy/reliability scores, which means that one ought to treat the outcome of these methods with some caution. On the other hand, from a craft perspective, the methods have proved to be very popular in HCI as they are quick to apply and often provide informative results. The obvious question, therefore, is which particular criterion ought one to apply in order to judge the suitability of these methods?

Before exploring this question, it is probably worth noting Gray and Salzman's (1998) critique of the various studies that have compared heuristic/walkthrough with other evaluation methods (see above), and to consider the meaning of the 'reliability' measures in Tables 7 and 8. Most studies comparing evaluation methods report 'validity' in terms of the proportion of 'usability problems' identified using a given method against the total number of usability problems. Presumably, this 'total number of usability problems' is either derived *post-hoc* (i.e., simply totaling the number of problems that all participants identified) or is produced by the experimenters (acting as experts) prior to the study. However, neither approach ensures that

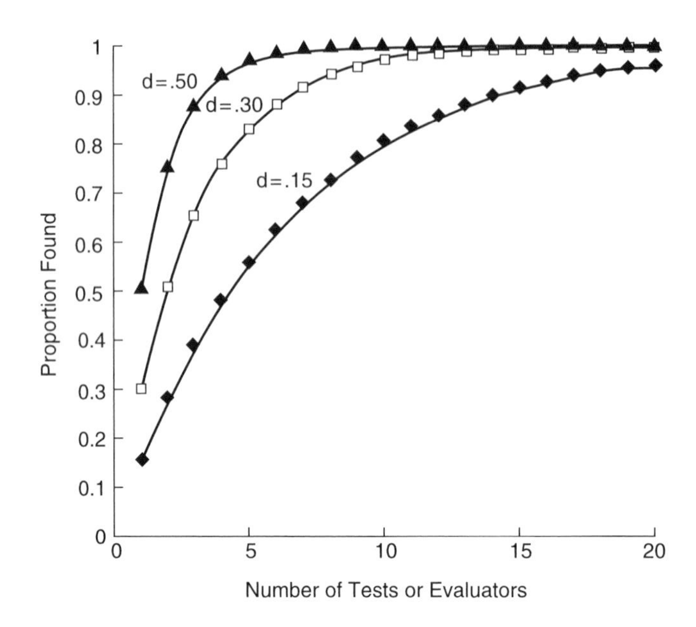

Figure 4 Detectability of usability problems related to number of evaluators (from Landauer, 1995).

one has produced a comprehensive set of such problems, and so any manipulation of data relative to this term seems a little arbitrary. Furthermore, reporting the number of problems identified provides no indication of the importance (or indeed, relevance) of the problems. Finally, as a number of authors have noted, usability problems typically vary in their level of detectability, e.g., the 'easy to spot' problems will be found by the majority of participants and the 'hard to spot' problems will be found by a minority (see Figure 4). While this last point may strike the reader as obvious, it also highlights a significant problem with the reliability of the methods and potential for bias.

Conclusions

In the 1990s, there was a call for usability inspection and evaluation methods to be 'quick and dirty'. This led to various forms of 'discount usability' techniques, of which Nielsen's (1993) various accounts of heuristic evaluation are, perhaps, the best known. These methods could be performed with minimal outlay, experience, effort or users, and could produce some useful data. Some comparative studies suggested that heuristic evaluation was superior to user trials, while others showed the opposite effect. The point to note is that the methods are acknowledged to provide little in the way of consistent, reliable, replicable data, i.e., they are 'dirty' in the sense that they do not provide 'clean' data. While such approaches have undoubtedly had an impact on HCI, it has been proposed that future usability evaluation methods need to be 'Quick and Clean', i.e., 'Quick studies, which are highly efficient... [-[when]... developing products that evolve as rapidly as 3 months. Clean studies are necessary... to provide valid and reliable data for correct decisions to be made.' [Wichansky, 2000, p. 1004].

Heuristic evaluation and walkthrough are (and will remain) popular usability inspection methods. Part of their attraction lies in their adaptability (they can be easily modified to fit any evaluation), and in the fact that one is apparently collecting 'data' (regardless of the pedigree of these data). It is not apparent how one can resolve any conflict in the results from such evaluations, e.g., if one evaluator says feature *X* is good and meets heuristic *Y*, and another evaluator say feature *X* is bad and fails to meet heuristic *Y*, how could one decide which evaluator is 'correct' (particularly when one is using a small set of respondents).

User reports for HCI evaluation

User reports are employed by all three perspectives, although a craft or engineering approach might use less structured methods than the applied science approach. It is a truism to state that '... ultimately it is the users of a software system [or any product] who decide how easy its user interface is to manipulate ...' [Holcomb and Tharp, 1991, p. 49]. One might feel that asking people about the product would the obvious and most useful approach to take. At a craft level, this might simply mean using a fairly open and unstructured interview to elicit user opinion (see Chapter 4 on interview techniques). While such an approach might yield some useful insight, it can lack the potential to provide generalisable data. One of the problems that is worth noting before discussing some of these techniques is that there are few off-the-shelf methods that can be employed. Consequently, evaluators will need to develop their own methods and instruments. This can be time-consuming (particularly if one is planning to use a questionnaire). However, it is possible that companies simply re-use existing instruments. For instance, product evaluation in industry typically involves the re-use of prepared checklists (Baber and Mirza, 1998). While this might save time, it could also lead to problems in the reliability of the data being collected.

There are many surveys that have been designed to evaluate usability. Some, like CUSI — Computer User Satisfaction Inventory (Kirakowski and Corbett, 1988) and QUIS — Questionnaire for User Interface Satisfaction (Chin *et al.*, 1988), are designed to capture user response to an interface, particularly in terms of affective components (such as satisfaction). This approach can be seen as a simplification of the use of Guidelines; the evaluator is provided with a set of characteristics of a 'good' interface, and asked to state how well the product under evaluation compares. Some surveys, like Ravden and Johnson (1989), have been designed to cover both aspects of the interface and characteristics of usability. This checklist can also be used as source material for developing reduced evaluation sets. Other surveys, like SUS (Brooke, 1996), have been designed to support 'quick and dirty' evaluation. While the approach can be applied quickly and the items are relatively easy for the lay evaluator to interpret and apply, the use of 10 items could suggest that the approach is open to the same criticisms as heuristics.

I suppose that any survey technique ought to meet some of the criteria laid out by occupational psychologists when they consider psychometric techniques. At the very least, one ought to be able to demonstrate the validity and reliability of the technique and the appropriateness of its findings. SUMI (Kirakowski, 1996) was developed using a rigourous approach to defining appropriate components of usability and presents results in terms of a comparison with a database of previous evaluations.

Referent model

The surveys considered in this section appear to employ a variety of reference models. Thus, some of the approaches are based on the assumption that one can evaluate a product against Guidelines, while others focus on usability heuristics. SUMI employs two referent models: (i) A multidimensional concept of usability; (ii) A database of previously evaluated products. Thus, the results from SUMI reflect both a set of measures concerning characteristics of usability, and a comparison of the evaluated product against the 'average' score for these characteristics (Table 9).

Table 9 Brief Description of SUMI

Name of method	Software Usability Metric Inventory (SUMI)		
Evaluation type:	*User report*	Approach to HCI:	*Applied Science*
Background knowledge:	Product or task		
Brief description:	Evaluators are asked to consider the product in terms of a 50 item survey. Each survey item is rated in terms of 3 responses (x, x, x). The evaluation requires a minimum of 15 evaluators to complete the survey. The results are fed into a computer and compared against a database of 'standard' scores. The analysis produces scores relating to: Global Usability, Efficiency, Affect, Helpfulness, Control, Learnability.		
Source references:	1. Kirakowski and Corbett (1996)		
Performance:	Number of Evaluators:	15[1]	
	Duration:	Minutes for analysis, under one hour for evaluation	
	Agreement with other method:	Produces similar findings to other user reports; internal verification, based on comparison with previous studies.	
Overall:	The results can be compared with a database of 'standard' responses (derived from some 200 previous analyses). This provides an indication of how well the software relates to other products and, as such, is useful in benchmarking exercises.		

Relationship with approaches to HCI

User surveys (and the other approaches associated with eliciting user opinion) can cover all of the approaches to HCI. For an 'applied science' approach, the collection of user opinion can be considered both in terms of additional data to support laboratory data and in terms of intersubject reliability of response. For a 'sociology' approach, user opinion represents a significant source of data, particularly when obtained through open-ended interview and discussion. For a 'craft' approach, user opinion can be collected easily and fed back into the design process as quickly as possible. For an 'engineering' approach, user opinion can provide useful indication of pros and cons of a design.

The differences in approach also highlight the differences in how the resulting data from user surveys are treated, e.g., in terms of how the data are represented. Thus, the engineering and applied science approaches might seek to derive quantifiable data, e.g., through reporting average scores or percentage responses, where the sociology approach might seek to relate key stories or comments, and the craft approach might summarise the gist of the users' comments.

Observational methods for HCI evaluation

The notion that studying human interaction with technology is best conducted through observing people actually using the technology is central to much of HCI research. However, a little reflection on this apparently obvious approach introduces several problems, particularly when it comes to evaluation. For instance, an immediate question concerns what is being evaluated: previously the chapter has proposed that the focus of HCI evaluation is the interaction between person and technology. However, when one is observing people at work, their interaction with technology is often intertwined with a host of other activities, often in pursuit of multiple goals. This raises two possibilities: (i) To employ a methodology which is able to capture and reflect the richness and diversity of this intertwined activity. Broadly speaking, this is what the sociology approaches seek; (ii) To effectively control or eliminate as many activities as possible, in order to focus on core activities. This is effectively what the applied science approach seeks. Each approach could criticise the other for being too broad or too narrow, but the point is that the outcome of the study depends on the approach and the objectives of the researchers. As mentioned in the introduction, it is fatuous to argue whether one approach is 'better' than another.

In terms of evaluation, the goal of observation is often to see how technology affects user performance. An engineering approach might employ task analysis (see Chapter 6) to define descriptions of user activity before and after any change in technology. An applied science approach might run an experiment (or user trial) in which participants are asked to perform specific tasks under two conditions (e.g., with and without the technology). The sociology approach would examine working conditions and context of use under both conditions. It is a little more difficult to imagine 'craft' approaches to observation, short of some form of unstructured and cursory examination of what people are doing with the technology.

Referent model

The referent model used in observation is, typically, the activity itself. Thus, many observation studies rely on comparison of conditions. The 'control' condition ought to represent a neutral state of affairs against which any changes brought about by the technology can be assessed. Unfortunately, it is not easy to define a 'neutral' condition for HCI; this is partly due to the preconceptions and experiences that people might bring to the study, and partly due to the influences of context of use on performance. Thus, one might want to compare performance on two designs of keyboard, say a standard Qwerty and a Dvorak layout. While the Qwerty represents the control condition in this study, there are additional factors relating to experience of this keyboard layout that mitigate against adequate comparison: anyone taking part in the trial will, probably, have at least seen, if not actually used, a Qwerty keyboard, and so the baseline for performance in this condition is different for that in the Dvorak condition.

Relationships with approaches to HCI

Traditionally, the 'observational' approach was the province of applied science, and a great deal has been written advising researchers on how to design experiments. A competing tradition has concerned field study, using ergonomics methods of observation (see Chapter 3) and task analysis (see Chapter 6). More recently, researchers have been applying principles of ethnography to HCI (Thomas, 1995). Thus, when considering observational methods, one needs to distinguish between place of study, i.e., field or laboratory, and research tradition, i.e., experimental psychology, ergonomics or ethnography.

Participative HCI evaluation

One of the seminal works in the 'sociology' approach to HCI is Suchman (1988), in which pairs of users of a complex photocopier were asked to perform a series of tasks. The conversations between the users were transcribed and, through conversation analysis, some of their misunderstandings and difficulties were illustrated and discussed. The idea of having pairs of users work together to examine products has grown in HCI and represents one of the common forms of participative evaluation.

From a 'craft' perspective, a simple participative evaluation technique, Co-Discovery, has been successfully employed at Philips. In this approach, a pair of users is given a product and asked to explain how it works. Their explanation is video-taped, and the video is then edited to extract key comments and problems to relay to the design team. Such an approach can yield user problems relatively quickly, although it is not clear how comprehensive or reliable such an approach might be.

Referent model

The referent model for participative evaluation is, possibly, the shared experiences of other products that the participants bring to their discussion. Thus, there might not be a simple, single focus of evaluation. Rather the evaluators might introduces ideas and concepts in order to explain their decisions. In a sense, therefore, the process of talking about the product will lead to the development of a referent model, with analogies, metaphors and experiences with other products all being drawn upon by the participants. It is not clear whether participative evaluation methods make much use of these developing referent models (as Suchman, 1988 did), or whether the approaches tend to simply look to extract problems and comments.

Relationships with approaches to HCI

The notion of participative evaluation, as indicated above, tends to find most synergy with the craft and sociology approaches. It is less easy to see how it can be used in an applied science or engineering approach. Consequently, it remains a moot point as to how one can determine whether the outcomes of the methods can be generalised across participants, or how well participants agree with each other, or whether the problems would persist over trials.

Reporting usability evaluation

The American National Standards Institute (ANSI) has produced a document that sets out a standard reporting format for usability evaluation: ANSI NCITS 354-2001 'Common Industry Format for Usability Test Reports' (copies can be purchased from http://www.nsit.gov/iusr/). The standard, based on the ISO 9241 definition of usability, covers a report format that should help to improve the consistency with which usability evaluation is conducted and reported. The basic content of the report covers the following points: indication of who conducted the test, when and for whom; clear definition of the product under test; specification of test objectives; clear description of the participants in user trials and respective demographic data; detailed description of procedure followed, including tasks performed and where the tasks were performed; definition and explanation of usability metrics and methods used; clear presentation, analysis and interpretation of the resulting data from the evaluation. I assume that recommendations for redesign, points of specific interest and results that point to particular problems will be included under the presentation of results. In broad terms, the report format follows what might be termed a conventional experiment write-up, the sort of report that one might write for high-school or undergraduate laboratory classes. The seems to me to be a reasonable approach to the writing-up of usability evaluation. It does, however, tend to support the 'applied science' and 'engineering' models of evaluation (see above) rather than the other approaches; this is not to say that the craft approach cannot be accommodated within the format, but it does mitigate against slapdash, ill-conceived and poorly conducted attempts at evaluation.

Evaluation and the design lifecycle

In order to ensure adequate coverage, it is important for evaluation to be located explicitly against a system design framework or lifecycle (Lim and Long, 1994). The majority of books on software engineering and user interface design introduce the notion of a design lifecycle. This is presented as a series of landmarks past which a project will pass during its development life. It is generally accepted that the notion of lifecycle is a useful planning tool, but that projects rarely follow the landmarks in an orderly and linear fashion. Noyes and Baber (1999) have summarised the generic design lifecycle as comprising six main stages: requirements specification, architectural design, detailed design, implementation and unit testing, integration and unit testing, operation and maintenance.

While there is some consensus as to the lifecycle followed in engineering design, there appears to be less agreement as to how one might define the ergonomics lifecycle. In this chapter, we follow the lead of Bertaggia *et al.* (1992), whose Human Factors (Ergonomics) Lifecycle is defined by five key stages: needs analysis, requirements specification, conceptual design, prototype development and product evaluation. Table 10 contrasts this Ergonomics lifecycle with the Design lifecycles (for a number of other representations of design life cycles see Chapter 10).

A significant point to note from this notion of stages in design or ergonomics lifecycles is that evaluation plays different roles at each stage. This means, first and foremost, that evaluation is not a one-off activity to be conducted at the end of the design lifecycle in order to allow a design to be signed-off. Rather it means the following:

1. Evaluation is a recursive activity that cuts across the entire design lifecycle;
2. Evaluation should be incorporated into as many stages of design as possible;
3. Evaluation should be designed to maximise the impact of the evaluation of the design stage in which it is used
4. Evaluation should guide and inform design activity

Table 11 indicates how the types of evaluation methods considered previously can relate to notional lifecycle stages.

Table 10 Ergonomics Lifecycle Compared with Generic Design Lifecycle

Generic design lifecycle	Ergonomics lifecycle
	Needs analysis
Requirements specification	Requirements specification
Architectural design	Conceptual design
Detailed design	Prototype development
	Prototype testing
Implementation and Unit testing	Product evaluation
Integration and testing	
Operation and maintenance	

Table 11 Comparing Types of Method with Lifecycle Stage

Lifecycle stage	Observation	User opinion	Usability inspection	Analytic	Participative
Needs analysis	X	X			X
Requirements specification		X			X
Conceptual design		X		X	X
Prototype development			X	X	
Prototype testing	X	X	X		
Product evaluation	X	X	X		X

Evaluating future HCI

At one level, the evaluation of future HCI calls for the application of current evaluation techniques. When I say future HCI I mean mobile and ubiquitous computing (including wearable computers), virtual and augmented reality and computer gaming. Thus, one can imagine the four reasons for evaluating HCI mentioned previously to be relevant to future HCI, that the four approaches to HCI will have a bearing on future HCI and that the five types of method will be employed. However, there are other aspects of future HCI that call for rethinking of evaluation. In other words, it might not be entirely appropriate to take methods that have proven useful for evaluating desktop HCI and apply these to future HCI. Indeed, it is an open question as to whether the concept of 'usability' is entirely appropriate to future HCI, e.g., how would one define the usability of a Virtual Reality experience? It seems to me that a problem in evaluating future HCI lies in defining adequate referent models. As Wilson and Nichols (2002) point out, 'There are only a limited number of ways in which we can assess people's performance generally and in VEs. We can measure the outcome of what they have done, we can observe them doing it, we can measure the effects on them of doing it or we can ask them about either the behaviour or its consequences.' (p. 1032) At one level, this is simply because future HCI is attempting to develop approaches to interaction with technology for which there are no existing models. As mentioned earlier, one way to deal with this problem is to focus on activities that people are performing using a variety of products. However, this will only cope with part of the problem. For instance, the electronic tour-guide given above, could be evaluated in comparison with other ways of performing activities, but this does not tell us whether any differences between the electronic tour-guide and the other products are due to the concept or to the realisation of the concept. In other words, if we find that the electronic tour-guide performs less well than speaking to someone, is this because the tour-guide lacks information, or because it lacks clear presentation of information, or because it lacks speedy access to the information, or because of some other reason (our evaluation would only point to all of these, not to specific reasons).

At another level, evaluation of future HCI means reconsidering and redefining context of use. Thus, in a virtual environment one might need to consider technical factors such as rendering, frame rate, image quality etc., and subjective responses such as comfort, presence or credibility. In virtual reality, there has been a great deal of research into evaluating the comfort (or more appropriately the discomfort) associated with the use of VR equipment and users' responses to the simulated environments, particularly in terms of nausea. Much of this work has resulted in user survey, e.g., into subjective responses to virtual reality from the perspective of 'simulator sickness' or 'fidelity'. For example, Cobb *et al.* (1999) report a subjective assessment survey called VRISE (Virtual Reality Induced Symptoms and Effects). This survey is given to people following immersion in a virtual environment and is used to score the experience, particularly in terms of the potentially negative aspects such as nausea, disorientation and visual fatigue. A major focus of research in the VR community, certainly in the 1990s, was the definition and measurement of 'presence'. Again, much of this work has led to the development of self-report questionnaires (e.g., Slater *et al.*, 1994; Witmer and Singer, 1998). The issue of whether presence (or any other phenomenon that can be associated with the experience of engaging with a virtual reality) can be measured in a fashion that can use more objective measures is a matter of some debate. Thus, one can examine performance measures by manipulating some aspect of the virtual reality and examining how this leads to a change in user action (Moody *et al.* 2001) or one can begin to catalogue possible objective correlates with subjective impressions (Sheridan, 1996).

In the field of mobile and wearable computers, much of the evaluation research has focused on comparing performance on a wearable computer with performance using other media. Thus, studies might compare performance using a wearable computer, say to perform a task involve following instructions and recording data, and find that sometimes performance is superior in the paper condition (Siegel and Bauer, 1997; Baber *et al.*, 1999) and sometimes it is superior in the wearable computer condition (Bass *et al.*, 1995, 1997; Baber *et al.*, 1998). This highlights the potential problem (mentioned earlier) of comparing disparate technologies in an evaluation; it is not clear that any differences in performance are due to the experiment favouring one technology over another or whether there are other factors in play here. For example, a common observation is that people using the wearable computer tend to follow the instructions laid out on the display, whereas people using paper tend to adopt a more flexible approach (Siegel and Bauer, 1997; Baber *et al.*, 1999). Of course, the notion that technology influences the ways in which people work is often taken as 'common-sense' by ergonomists. However, the question of how and why such changes arise ought to have a far deeper impact on evaluation than is currently the case. In addition to evaluating performance, it is possible to consider a wearer's response to the technology. For example, Knight *et al.* (2002) report a comfort rating scale, that can be used to gauge the subjective response to wearing and using such technology.

One of the current dilemmas in evaluating future HCI lies in establishing appropriate benchmarks for comparison. After all, the point of these technologies is to move beyond the conventional desktop-bound personal computers and to create new forms of interaction. However, the move to very different technologies makes it hard to establish a sensible basis for evaluation. Recall that it was proposed that evaluation requires a comparison with a referent model. So, what is the referent model for future HCI? As we have just seen, wearable computers often take paper as a referent model; the idea is that paper represents the 'current' way of working. Rather than solving the problem of a referent model, this, I think, only serves to bring the problem into stronger relief. Providing the two (or more) conditions of an experiment require similar actions and use similar technologies, then one can have some faith in the results. As soon as the actions or technologies begin to change, then care must be taken to ensure that these changes are catered for so as not to induce artefactual results. For instance, much of the research comparing reading from screen and paper (see Dillon, 1992 for a review of this work) indicated that aspects of the screen, such as refresh rate, font and layout, significantly affected performance. This meant that the 'paper is better than screen' argument is not true; rather some aspects of a visual display unit seem to influence some of the psychophysical aspects of the reading task. Furthermore, paper is clearly manipulable while a 17' monitor is not and these aspects of document manipulation create quite different actions for the experimental participants. This is not to say that we cannot produce well-grounded evaluations of future HCI, but that care needs to be taken to ensure that we make use of a referent model, i.e., we do not simply ask people who have used our new 'ghee-whizz' device 'do you like it?' and expect to get sensible, useful, and valid data.

Discussion

This chapter has proposed that HCI is currently being pulled in four directions (craft, applied science, engineering, sociology) and that each direction has its own research traditions and methods. It is suggested that any debate on the primacy of the various approaches is pointless and would readily degenerate into personal opinion and bias. However, it is equally clear that each approach is making useful contributions to the study of HCI and that the future of HCI lies in greater amalgamation and assimilation of approaches. Ideally, of course, one would like to see HCI (or ergonomics) stand as a coherent, complete discipline in its own right, with its own research methods and its own set of evaluation techniques. The key concept in HCI evaluation, 'usability', is seen as a multidimensional concept, and each dimension has different measurement requirements. Indeed, Frøkjær *et al.* (2000) demonstrate that effectiveness, efficiency and satisfaction need independent measurement in order to produce satisfactory results. The International Standards have been working towards such an approach, although they tend to give priority to an engineering approach at the expense of other approaches. Having said this, the International Standards offers a reasonable route out of the current *impasse* of HCI evaluation: evaluators should clearly state what they are evaluating, how they are evaluating and why they are evaluating things in this manner. To this set of simple requirements, I would argue that HCI needs to clearly identify the referent models that it is using in its evaluation, and to think more clearly about the level of reliability and validity that it wants from its evaluations.

What is important to note from reading this chapter is that evaluation involves some notion of 'critique' — the aim is not to 'prove' that the design is OK, but to ask how it could be improved. In other words, the rationale for undertaking evaluation should not be simply conformance testing but should focus on improving the design, or asking how the design fails to support specific tasks or to meet specific design targets. This is tantamount to viewing evaluation as a form of hypothesis-testing, albeit with less rigour than usually found in experiments. Simply demonstrating an 'OK design', is thus a form of proving the null-hypothesis (which, statistically, is usually meaningless, invalid and of little use).

References

Andre, T.S., Williges, R.C. and Hartson, H.R. (1999). The effectiveness of usability evaluation methods: determining the appropriate criteria. *Proceedings of the Human Factors and Ergonomics Society 43rd Annual Meeting* (Santa Monica, CA: Human Factors and Ergonomics Society), pp. 1090–1094.

Baber, C., Haniff, D.J., Knight, J., Cooper, L. and Mellor, B.A. (1998). Preliminary investigations into the use of wearable computers, In: R. Winder (ed.) *People and Computers XIII* (Berlin: Springer-Verlag), pp. 313–326.

Baber, C., Arvanitis, T.N., Haniff, D.J. and Buckley, R. (1999). A wearable computer for paramedics: studies in model-based, user-centred and industrial design, In: M.A. Sasse and C. Johnson (eds), *Interact'99* (Amsterdam: IOS Press), pp. 126–132.

Baber, C., Haniff, D.J. and Woolley, S.I. (1999). Contrasting paradigms for the development of wearable computers. *IBM Systems Journal*, **38**, 551–565.

Baber, C. and Mellor, B.A. (2001). Modelling multimodal human–computer interaction using critical path analysis. *International Journal of Human Computer Studies*, **54**, 613–636.

Baber, C. and Mirza, M. (1998). Ergonomics and the evaluation of consumer products: surveys of evaluation practices, In: N.A. Stanton (ed.) *Human Factors in Consumer Products* (London: Taylor & Francis), pp. 91–104.

Baber, C. and Stanton, N. (2001). Analytical prototyping of personal technologies: using predictions of time and error to evaluate user interfaces, In: M. Hirose (ed.) *Interact'01* (Amsterdam: IOS Press), pp. 585–592.

Bass, L., Siewiorek, D., Smailagic, A. and Stivoric, J. (1995). On site wearable computer system. *CHI'95* (New York: ACM), pp. 83–88.

Bass, L., Kasabach, C., Martin, R., Siewiorek, D., Smailagic, A. and Stivoric, J. (1997). The design of a wearable computer. *CHI'97* (New York: ACM), pp. 139–146.

Bertaggia, N., Montagnini, G., Novara, F. and Parlangeli, O. (1992). Product usability, In: M. Galer, S. Harker and J. Ziegler (eds) *Methods and Tools in User-Centred Design for Information Technology* (Amsterdam: North-Holland), pp. 127–176.

Bevan, N. (2001). International Standards for HCI and usability. *International Journal of Human Computer Interaction*. **55**, 533–552.

Brooke, J. (1996). SUS: a quick and dirty usability scale. In: P.W. Jordan, B. Weerdmeester, B.A. Thomas and I.L. McLelland (eds) *Usability Evaluation in Industry* (London: Taylor & Francis), pp. 189–194.

Card, S.K., Moran, T.P. and Newell, A. (1983). *The Psychology of Human-Computer Interaction* (Hillsdale, NJ: LEA).
Carroll, J.M. (1999). *Scenario-based Design* (New York: Wiley).
Chin, J.P., Diehl, V.A. and Norman, K.L. (1988). Development of an instrument measuring user satisfaction of the human-computer interface. *CHI'88* (New York: ACM), pp. 213–218.
Cobb, S.V.G., Nichols, S.C., Ramsey, A.D. and Wilson, J.R. (1999). Virtual reality induced symptoms and effects (VRISE). *Presence: Teleoperators and Virtual Environments*. **8**, 169–186.
Cooper, L., Williams, D. and Baber, C. (1999). A user-centred deployment methodology, In: D. Harris (ed.) *Engineering Psychology and Cognitive Ergonomics*, volume 4 (Aldershot: Ashgate), pp. 433–438.
Dillon, A. (1992). *Designing Usable Electronic Text* (London: Taylor & Francis).
Dix, A., Finlay, J., Abowd, G. and Beale, R. (1993). *Human Computer Interaction* (Hemel Hempstead: Prentice-Hall).
Downton, A. (1991). *Engineering the Human-Computer Interface* (London: McGraw-Hill).
Earthey, J., Sherwood Jones, B., and Bevan, N. (2001). The improvement of human-centred processes — facing the challenges and reaping the benefit of ISO 13407. *International Journal of Human Computer Studies*, **55**, 553–585.
Eberts, R.E. (1994). *User Interface Design* (Englewood Cliffs, NJ: Prentice-Hall).
Frøkjær, E., Hertzum, M. and Hornbæk, K. (2000). Measuring usability: are effectiveness, efficiency and satisfaction really correlated? *CHI'2000* (New York: ACM), pp. 345–352.
Good, M., Spine, T.M., Whiteside, J. and George, P. (1986). User-derived impact analysis as a tool for usability engineering. *CHI'86* (New York: ACM), pp. 241–246.
Gray, W.D., Atwood, M.E., Fisher, C., Nielsen, J., Carroll, J.M. and Long, J. (1995). Discount or disservice? Discount usability analysis — evaluation at a bargain price or simply damaged merchandise? *CHI'95* vol. 2 (New York: ACM), pp. 176–177.
Gray, W.D., John, B.E. and Atwood, M.E. (1993). Project Ernestine: validating a GOMS analysis for predicting and explaining real-world performance. *Human–Computer Interaction*, **8**, 237–309.
Gray, W.D. and Salzman, M.C. (1998). Damaged merchandise? A review of experiments that compare usability evaluation methods. *Human–Computer Interaction*, **13**, 203–261.
Holcomb, R. and Tharp, A.L. (1991). What users say about software usability. *International Journal of Human–Computer Interaction*, **3**, 49–78.
ISO 10741 (1995) *Dialogue Interaction — Cursor Control for Text Editing* (Geneva: International Standards Office).
ISO 9241 (1998) *Ergonomics of Office Work with VDTs-guidance on Usability* (Geneva: International Standards Office).
ISO 13407 (1999) *Human-centred Design Processes for Interactive Systems* (Geneva: International Standards Office).
ISO 18789 (1999) *Ergonomic Requirements and Measurement Techniques for Electronic Visual Displays* (Geneva: International Standards Office).
ISO 13406 (1999) *Ergonomic Requirements for Work with Visual Displays Based on Flat Panels* (Geneva: International Standards Office).
ISO 9126 (2000) *Software Engineering — Product Quality* (Geneva: International Standards Office).
ISO DTS 16071 (2000) *Guidance of Accessibility For Human-Computer Interfaces*, (Geneva: International Standards Office).
ISO TR 18529 (2000) *Ergonomics of Human System Interaction-Human-centred Lifecycle Process Descriptions* (Geneva: International Standards Office).
ISO 11581 (2000) *Icon Symbols and Functions* (Geneva: International Standards Office).
ISO 20282 (2001) *Usability of Everyday Products* (Geneva: International Standards Office).
ISO 14915 (2000) *Software Ergonomics for Multimedia user Interfaces* (Geneva: International Standards Office).
ISO 14958 (2000) *Information Technology — Evaluation of Software Products* (Geneva: International Standards Office).
ISO 18021 (2001) *Information Technology — User Interface for Mobile Tools* (Geneva: International Standards Office).

ISO 16982 (2001) *Usability Methods Supporting Human-centred Design* (Geneva: International Standards Office).
Jeffries, R., Miller, J.R., Wharton, C. and Uyeda, K.M. (1991). User interface evaluation in the real world: a comparison of four techniques. *CHI'91* (New York: ACM), pp. 119–124.
Johnson, P. (1992). *Human Computer Interaction* (London: McGraw-Hill).
Jordan, P.W., Weerdmeester, B., Thomas, B.A. and McLelland, I.L. (1996). *Usability Evaluation in Industry* (London: Taylor & Francis).
Kirakowski, J. and Corbett, M. (1988). Measuring user satisfaction, In: D.M. Jones and R. Winder (eds) *People and Computers IV* (Cambridge: Cambridge University Press), pp. 329–430.
Kirakowski, J. and Corbett, M. (1993). SUMI: the software usability measurement inventory. *British Journal of Educational Technology*, **24**, 210–214.
Knight, J.F., Baber, C., Schwirtz, A. and Bristow, H. (2002). The comfort assessment of wearable computers. *Digest of Papers of the 6th International Symposium on Wearable Computing* (Los Alamitos, CA: IEEE Computer Society), pp. 65–74.
Kyng, M. (1991). Designing for cooperation: cooperating in design. *Communications of the ACM*, **34**, 65–73.
Landauer, T.K. (1995). *The Trouble with Computers* (Cambridge, MA: MIT Press).
Lim, K.Y. and Long, J. (1994). *The MUSE Method for Usability Engineering* (Cambridge: Cambridge University Press).
Long, J. and Dowell, A. (1989). Conceptions of HCI as a discipline: craft, applied science and engineering, In: A. Sutcliffe and L. Macauley (eds) *People and Computers V* (Cambridge: Cambridge University Press), pp. 9–34.
MacKenzie, I.S. (1992). Fitts' Law as a research and design tool in human–computer interaction. *Human-Computer Interaction*, **7**, 91–139.
MacLean, A. and McKerlie, D. (1995). Design space analysis and use representation. In: J.M. Carroll (ed.) *Scenario-based Design: Envisioning Work and Technology in System Development* (New York: Wiley), pp. 183–208.
Maddix, F. (1990). *Human–Computer Interaction* (Chichester: Ellis Horwood).
Monk, A.F., Wright, P.C., Haber, J. and Davenport, L. (1986). *Improving your Human-Computer Interface: a Practical Technique* (London: Prentice-Hall).
Moody, C.L., Baber, C., Arvanitis, T.N. and Elliott, M. (2003). Objective metrics for the evaluation of simple surgical skills in real and virtual domains. *Presence: Teleoperators and Virtual Environments*, **12**, 207–221.
Newman, W. M. and Lamming, M. (1995). *Interactive System Design* (Reading, MA: Addison-Wesley).
Nielsen, J. (1993). *Usability Engineering* (London: Academic Press).
Norman, D.A. (1988). *The Psychology of Everyday Things* (New York: Basic Books).
Noyes, J.M. and Baber, C. (1999). *User-Centred Design of Systems* (Berlin: Springer).
Olson, J.R. and Olson, G.M. (1990). The growth of cognitive modelling in human–computer interaction since GOMS. *Human–Computer Interaction*, **3**, 309–350.
Preece, J., Rogers, Y., Sharp, H., Benyon, D., Holland, S. and Carey, T. (1994). *Human–Computer Interaction* (Reading, MA: Addison-Wesley).
Ravden, S.J. and Johnson, G.I. (1989). *Evaluating Usability of Human–Computer Interfaces* (Chichester: Ellis Horwood).
Shackel, B. (1981). The concept of usability. *Proceedings of the IBM Software and Information Usability Symposium* (Poughkeepsie, NY: IBM).
Shackel, B. (1984). The concept of usability. In: J. Bennett, D. Case, J. Sandelin and M. Smith (eds) *Visual Display Terminals: Usability Issues and Health Concerns* (Englewood Cliffs, NJ: Prentice-Hall), pp. 45–88.
Sheridan. T.B. (1996). Further musings on the psychophysics of presence. *Presence: Teleoperators and Virtual Environments*, **5**, 241–246.
Shneiderman, B. (1986). *Designing the User Interface* (Reading, MA: Addison-Wesley).
Siegel, J. and Bauer, M. (1997) On site maintenance using a wearable computer system. *CHI'97* (New York: ACM), pp. 119–120.
Slater, M., Usoh, M. and Steed, A. (1994). Depth of presence in virtual environments. *Presence: Teleoperators and Virtual Environments*, **3**, 130–144.

Smith, S.L. and Mosier, J.N. (1986). *Guidelines for Designing User Interface Software* (Bedford, MA: The Mitre Corporation), Report ESD-TR-86-278.

Stanton, N.A. and Baber, C. (1992). Usability and EC 90/270. *Displays*, **13**, 151–160.

Stanton, N.A. and Young, M.S. (1999). *A Guide to Methodology in Ergonomics* (London: Taylor & Francis).

Suchman, L.A. (1988). *Plans and Situated Actions* (Cambridge: Cambridge University Press).

Thomas, P.J. (1995). *The Social and Interactional Dimensions of Human-Computer Interfaces* (Cambridge: Cambridge University Press).

Virzi, R.A. (1992). Refining the test phase of usability evaluation. *Human Factors*, **34**, 457–468.

Whiteside, J. Bennett, J. and Holtzblatt, K. (1988). Usability engineering: our experience and evolution. In: M. Helander (ed.) *Handbook of Human–Computer Interaction* (Amsterdam: Elsevier), pp. 791–817.

Wichansky, A.M. (2000). Usability testing in 2000 and beyond. *Ergonomics* **43**, 998–1006.

Wilson, J.R. and Nichols, S.C. (2002). Measurement in virtual environments: another dimension to the objectivity/subjecitivity debate. *Ergonomics*, **45**, 1031–1036.

Witmer, B.G. and Singer, M.J. (1998). Measuring presence in virtual environments: a presence questionnaire. *Presence: Teleoperators and Virtual Environments*, **7**, 225–240.

Young, R.M. and Whittington, J. (1990). Using a knowledge analysis to predict conceptual errors in text-editor usage. *CHI'90* (New York: ACM), pp. 91–96.

Chapter 14

Control facilities design

Jane A. Rajan, John R. Wilson and John Wood

Introduction

In 1983 a British Airways helicopter crashed into the sea off the Isles of Scilly, with the loss of 20 lives out of 26 on board. At the inquest, the pilot agreed that he had not checked his altimeter properly, but also claimed that he did not see a warning light on the altimeter that should have come on automatically when the helicopter was within 200 ft of the surface. He thought that the light might have been partly obscured by the helicopter's control column; 'The stick can cover part of the instruments. It depends on your height and build. In my case it does.' (*The Guardian*, Feb. 24th, 1984).

During maintenance at the BP Oil Refinery at Grangewood in 1987, an explosion and fire occurred 6 hours after the plant had been put on standby because a temperature cut-out had tripped the plant. One operator was killed. Over-pressurisation in a low pressure separator brought about the explosion but the underlying root causes were found to lie in a series of multiple failures in operation, design, procedures and equipment. Amongst these were instrument and display failings which included: inaccurate gauge readings due to the (foreseeable) physical effects of wax and cold; a display which read higher than true because it was offset without the operator's knowledge; displays going off-scale; operators having to estimate flow rates due to lack of displays; controls for different functions being of similar appearance, allowing for inadvertent adjustment; and a breakdown of expectations in the relationship between direction of movement of controls and valve actions (Health and Safety Executive, 1989).

These issues are not limited to older or existing systems. The design of the UK's new air traffic control centre at Swanwick had what appeared to be display inadequacies reported due to the poor legibility of computer screens. One issue highlighted was the legibility of numbers and letters which were critical to identify altitude levels, with possible confusions between 6, 8 and 0 amongst others. This was coupled with a further increase in overload reports, in which controllers considered that they have too many planes to handle in too short a period of time, indicating a situation in which human error is more likely to occur. Although the degree of and reasons for any such problems has caused controversy in the ergonomics community, there have been considerable concerns about the safety and effectiveness of the system: 'The worrying thing is that the number of situations in which it is increasingly likely that . . . near misses occur has gone up.' (*Daily Mail*, June 14, 2002).

In these cases and countless other examples we find inadequate design of the displays and controls making up the human–machine interface, even at a very basic level of size and positioning, never mind any issues involving information support for problem solving and reasoning. They are particularly inadequate in the way they can lead operators to make errors or lead to the need to violate safety procedures. Poor interface design and implementation can manifest itself in physical problems (e.g., the display cannot be seen), cognitive problems (e.g., a coding is not understandable) or organisational problems (e.g., two team members have a different understanding about what is really happening).

At the heart of human–machine systems is the exchange of information between people and the system through displays and controls. Indeed, the early years of ergonomics are often characterised by the expression 'knobs and dials ergonomics'. The interaction of person and process is at the centre of a total systems view, and that displays and controls provide the media for that interaction and for carrying out tasks. According to the type of system and the view of the analyst, the human–machine interface might physically or conceptually lay between the person and the displays and controls, or between the controls/displays and the machine, or might comprise the controls and displays themselves. In the first of these views the interface is seen as a conceptual entity, in the third as a physical entity, and in the second as either or both. In virtual reality and virtual environments the interface can be seen as being ubiquitous, comprising the interaction experience itself.

How we define the interface depends on where we draw the system boundary. For instance, a book can be seen as a 'read only' display system, or we can see reading a book or a set of instructions as a multiple-person system with a temporal dimension: one person uses controls — pen, typewriter, wordprocessor, etc. — to produce a display of information which is seen later, on one or several occasions, by themselves or by other people. Controls and displays that make up the human–machine interface take a wide variety of forms and this in large part is the reason for the awkward way in which authors often refer to controls and displays of a 'process' or 'system' in order to avoid having to refer explicitly to tools, equipment, machines or products.

'It is unfortunate that there is no obvious word which suggests all types of method for communicating between the operator and the plant, or one operator and another, because many human factors recommendations are the same whatever the technology used. For example, most recommendations about the shape of numbers which can be read most quickly and accurately apply whether those numbers appear on conventional instruments, VDUs, controls, or printed reference materials. Questions about how easy controls are to use, apply whether the operators are indicating what action they want to make by turning a knob, pressing a switch, using a list of alternatives on a VDU, or making a telephone call. In the following ... the word '*display*' refers to all methods of giving information to the operator, the word '*control*' applies to all means by which the operator gives instructions.' (Ball, 1991)

Amongst the variety of interface forms we might find:

- industrial machines which have hand or foot controls (pedals, levers, buttons) and displays of numerical quantities or qualities (states);
- computer systems, which can utilise input devices operated by hand, foot or even head (for the disabled) or where input may be achieved 'directly' through speech, gestures, eye movements, etc., but where information presented to the user is largely by visual display with some auditory feedback; personal devices, with bio-metric input and output, extend on these channels.
- simulation and virtual environment systems which aim to give the user some feeling of being in the world which is modelled, where they may have no control, or control via a veridical interface (driving or aviation simulator), or control via novel input devices (dataglove or spaceball in virtual reality for instance);
- essentially display-only systems, usually with visually displayed information (e.g., signs, books) but which also can be auditory (warnings) or tactile (e.g., braille);
- products where the 'control' is built into or is part of the product, and information is displayed back in part through visual confirmation of the control action but also

tactilely, kinesthetically or proprioceptively (e.g., toothbrush, razor, screwdriver, hammer, manual gear change);
- transportation where, although traditional type controls and displays exist, the greatest part of information is received by directly seeing the course or track (road or river) and registering the vehicle's state along it.

As technology develops these systems become increasingly multi-modal and the operator is able to select the preferred mode of interaction, for example in groupware conferencing systems offering video, whiteboard, messaging, group calendars and tools and voice communications.

We will use the convention in this chapter of defining the interface as being the medium through which the two-way exchange of information takes place, between people and computers and between people and other people via computers and telecommunications. The chapter will not address all types of interface but focus on the control and display issues generally associated with control panels and control rooms in process and transport control. Evaluation of human–computer interfaces is dealt with in detail in Chapter 13. Here we will deal with the sorts of control and display issues which include conventional instruments — analogue (quantitative and qualitative) and digital, annunciators, alarms, computer generated displays, information from written or verbal sources, information from the 'state' of a machine, process or environment, and all forms of controlling a process or system. Within our review we will introduce the need for a considered approach, both when analysing existing human–machine interfaces and when designing and evaluating new ones. Following a brief overview of the human–machine interface and control room issues we discuss at length a set of activities that must be carried out in any analysis and design of interfaces for control rooms. Then some detailed discussion of interface specification, especially for displays, covers choice of display mode, display formats, coding and navigation and structuring of displays. (The increasing role that auditory display of information plays in control rooms is recognised, although not explicitly here.) Finally, although we cannot cover all issues of evaluation we provide a basic control room interface evaluation checklist. As such the operator is an integral part of the system and thus for the system to be both safe and perform effectively this human component must be designed as part of the system. Whilst an integrated approach to HMI design is becoming increasingly common:

> Many design engineers accept the arguments (of human based design) but when safety by design becomes difficult they relapse into saying 'we shall have to rely on the operator'. This is the essential issue in ergonomics. If reliance on the operator is necessary in order to assure that the plant design is safe and reliable one has to first understand how the human behaviours of that operator that provides the critical link in plant functioning (which in such cases is often safety critical) (Kletz, 1991).

The scope of HMI issues specific to control rooms is addressed in ISO standard 11064 Ergonomic Design of Control Centres: This set of standards provides information on principles pertinent to the ergonomics design of control rooms, with part 3 on ergonomic design of control centres, part 4 on layout and dimensions of control centres and part 5 providing a specific standard for displays, controls and interactions in a control room context. This standard also points the reader to other ISO standards relevant to the design of controls and displays (for instance ISO 9241 and 13406). To consider control-display interfaces and control room ergonomics, other useful sources of information, some a little dated but still with useful information and guidance, include: Ball, 1991; Goodstein *et al.* 1988; International Instrument Users Association, 1998; Ivergård, 1989; Kirkade and Anderson, 1984; Kirwan, 1994; Moray,

1997; Noyes and Bransby, 2001; NUREG-0700, 1996; NUREG-5908, 1994; Sheridan, 2002; Stanton, 1994; US Department of Defence, 1999.

The human–machine interface

A classic view of the interface has been to understand the operator (person controlling the interface) as a passive and limited capacity processor of information. In this view, the operator and 'machine' are in a closed loop (although comprising an open system), connected by displays and controls. 'Machine information' is converted into 'operator information' via displays, and controls act as transducers to allow the operator to change a system state. Feedback to the operator comes via the displays and via interaction with the controls.

In more complex situations, with increased development and use of computer generated information systems, the operator is seen as needing higher level cognitive skills in both normal and abnormal conditions. Skill requirements in perceptual judgement, decision making, problem solving and diagnosis have led to more sophisticated models of an operator. From the work of authors in the collections by Edwards and Lees (1974), we can see an expansion of the original simple operator-process loop. The way people interact with systems is modelled as including attributes of the operator such as their mental model, experience, etc., and includes representation of their interaction through formal and informal procedures (Figure 1).

The model is frequently conceived of as the 'human as supervisory controller' (Sheridan, 1987). In this view, computer systems mediate between the operators plus their displays and controls on the one hand and the task or process and its sensors and actuators on the other. Sheridan identifies ten cause-effect loops in supervisory control and defines possible supervisory roles for the operator as planning, teaching, monitoring, intervening and learning, so we can see the need for a structured comprehensive approach to the design of display–control interfaces. If system control can occur in so many ways, and the needs and roles of operators can be so varied, then interface analysis and design must be based on an appreciation of much more than checkpoints for selecting individual instruments. In addition, the interaction with any given element of the interface may be fluid in its nature and interfaces may adapt or be adapted to suit the needs of each individual operator.

The control room context

With the increasing use of computers to control processes and systems the tendency for the control of such systems to be centralised has continued, usually from one primary location, to a central control room. Control rooms are found in many different domains — transport systems (e.g., air traffic control, railways, metros, etc.), emergency services (police, fire, ambulance), industrial processes (steel, chemical, food, etc.) and power plants (nuclear, electricity, etc.), security applications (banks, prisons, public or private buildings), etc. Importantly, although the term used generally and in this chapter is 'control room', these may in fact be a series or suite of rooms. 'A control room or control centre is the place where one or more people, sitting at control desks, conduct control activities. A control suite . . . is a group of co-located, functionally related rooms, including a control room . . . [and] control complex is . . . a series of functionally related rooms which are on different sites . . .' (Wood, 1994).

The complexity of technological systems has continued to grow so the demands for the safety and efficiency of these systems has grown apace. Increasingly the overall reliability of such systems as limited by the reliability of the hardware components has increased (although perhaps not as much as engineers might like to think!), so human operators can become the most unreliable components of a system, unless appropriate ergonomics input is made. It is imperative that the interfaces which provide the means of interaction between the system and

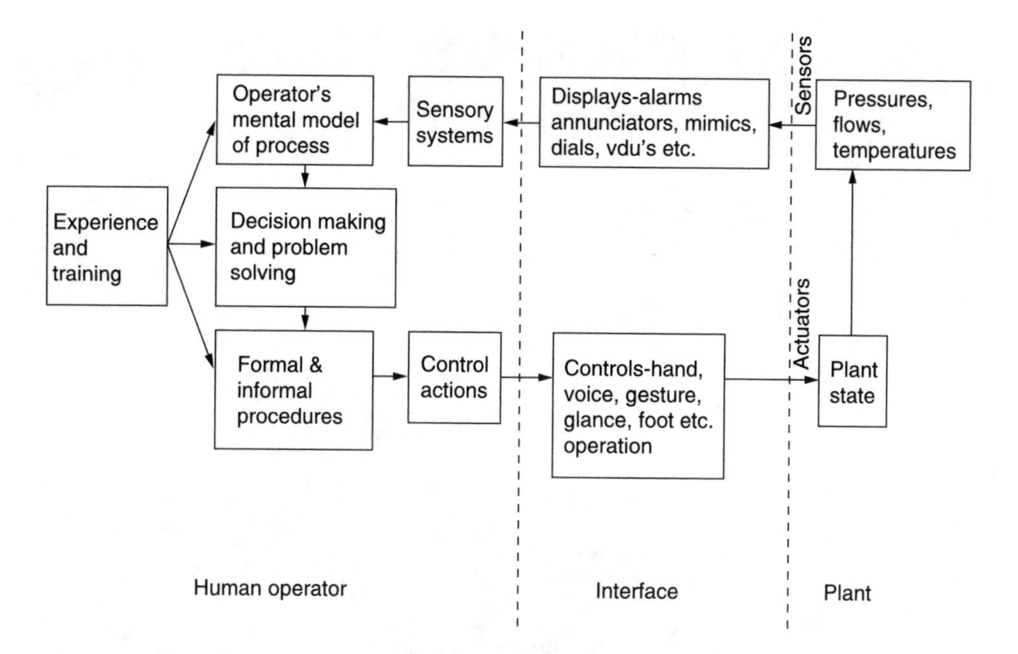

Figure 1 Model of human operator in process control.

the people who operate them, allow the operation of the system to be as efficient and error free as possible. As systems become more complex and increasingly automated, the requirements placed on the system operators are evolving. Each group of operators, or in some cases each operator, will have choices about the configuration of the human–machine interface they interact with the system through. The allocation of function and level of automation may not be fixed but flexible. This flexibility may in its turn be automated (e.g., if an emergency scenario occurs, the operator may not be allowed to have any direct control over the system for a defined period of time, to allow him or her to gain a more complete understanding of the current situation). Alternatively the operator may select the use of automation, for example, switching autopilot on, for navigation of an aeroplane or a ship. This may facilitate periods of high workload or allow other tasks to be given more attention. This chapter addresses the issues to be considered in the design of interfaces to support the operation of such complex systems.

In line with the increase in system complexity has come an increase in the amount of information available to the operator to support system operation, and the capability to display this information in a variety of formats. Control room interfaces are increasingly based around several VDU (Visual Display Unit) displays — rather than wall or console mounted panel displays which use dials and similar display types to present the information to the operator (Figure 2). This has had a significant impact on the design of control room interfaces. No longer is all the information concerning system parameters available to operator at the same time. VDUs give far greater flexibility in the presentation of the information, the same data can be presented in a variety of different formats, but there is a limited amount of information that can be presented on one display screen. To overcome this one operator may have several displays from which the system can be operated. 'How many VDUs do the operators need?', is one of the questions most often asked by control room designers (Figure 3) (see also Part 4 of ISO 11064). It must be remembered that, as with any computer interface, the operator has to deal with the complexity of both the plant/process and also the control and surveillance

Figure 2 A conventional rail power panel signalbox showing panel based control room and an IECC rail signalbox showing a VDU based control.

Figure 3 A BBC operator working from a number of VDUs displaying different information in different formats.

systems. For example, in a chemical manufacturing plant an operator may be required to control and monitor the process parameters such as flow rate, pressure and temperature on VDU-based process displays whilst also monitoring CCTV pictures of the plant that show visually the equipment and any personnel on the plant.

Process for analysis and design of human–machine interfaces

We make no attempt to define a full process for control-display system design or control room design, which would have integral staged evaluations and iterations embedded within it. Nonetheless, the chapter is built around a set of activities which, taken in order, outline an analysis and design process. It is suggested that the procedures and activities shown in Table 1 comprise a logically ordered listing of what must be done to develop control/display interfaces for control rooms and human–machine systems generally, which has been used in practical control room human factors (although not all stages are necessarily included in each study or development). Any ergonomics approach to design will follow the same basic steps; for instance, current discussions about new versions and parts of ISO 11064 include consideration of a flow chart to inform and guide on the overall process, which matches very well the one we use here (see Figure 4). It is important to note that the selection of specific interface types and their development or their specification do not take place until very late in the process. The implication is that development cannot merely be built around design guidelines and equipment selection rules; before any such guidance can be used the designer must gain a thorough understanding of user, task, system and environment requirements in a detailed analysis.

Table 1 Human–Machine Interface Design — Factors and Choices

Intial Analysis
1. User and task analysis
   - Establish tasks and users (operators and others)
   - Allocation of function
   - User needs and constraints: skills analysis
   - Task analysis/synthesis
2. Mental model and behaviour assessment
   - Consider potential operator mental models
   - Assess expected task behaviours
3. Environmental influences and circumstances of use

Outline Design
4. Information analysis
   - Consider on-line and off-line information
   - Establish information content, sequence (and form)
5. Analysis of potential operator overload/underload
6. Determination of control dynamics
   - Analyse or model operator control behaviour
   - Assess needs for aiding, quickening, prediction etc.
7. Selection of modality
   - Displays
   - Controls
8. Prototyping and formative evaluation

Detail Design
9. Display instrument specification
   - Meet criteria for form then content
   - Integrate displays
10. Control specification
11. Control–display integration
12. Detailed evaluation
13. Integration into control room

Evaluation
14. Evaluation and modification

Source: Author's (JRW) own records.

The remainder of this chapter takes up discussion of the activities defined in Table 1. First we will examine the initial analysis activities — establishing user needs, constraints on design and so on. Second, there is a more cursory look at initial design decisions in terms of the information content and modalities in display and control. Finally, some of the available guidance on interface specification is reviewed, concentrating upon display systems with particular application in control rooms.

User needs and initial task analysis

As with all ergonomics analysis and design, we need first to establish the requirements for the system, to identify and describe tasks and users, which will involve carrying out a task analysis (see Chapter 6 and below). It also involves making some first level decisions on the functions that must be performed and these are described at a gross level as activities that are needed to meet the system objectives (and usually collections of tasks). We need to consider the balance of responsibilities for functions between people and computers or other equipment (allocation

Task elements	Information requirements	Task type	Display format options
Carry out pre-start up checks	Valve positions Pump in a ready state Isolations in place Current maintenance activity	Procedural Checking	Mimic display Sequential Checklist
Initiate start up sequence	Pre-start up checks complete Feedback that sequence has initiated	Operational input Checking	Sequence display
Monitor start up for faults and to check correct sequence	Progress through start up sequence Faults occuring	Monitoring Fault Detection	Alarm display Sequence display

Figure 4 Extract from a task analysis to identify information requirements for display design for a pump start up task (*Source*: Rajan, 1993).

of function) and between different people (division of function), including potential needs for automation. The outcomes and decisions made on the basis of such analysis can be returned to, reviewed and revised, in each subsequent stage, on the basis of new or amended information, ideas and opinions. Moreover, the task and user analysis themselves are not once-and-for-all exercises; they will be revised as appropriate in later stages on the basis of new information and decisions. However, it should be noted that task analysis should also be focused as there may be a danger of 'overdoing' analysis and producing substantial amounts of information, which may not have direct application in the design process.

Where an existing interface is being assessed and analysed many of the methods, measures and techniques described in this book will be appropriate, particularly verbal protocol analysis (Chapter 7) since the work carried out from using information interfaces is often not directly observable. At this time it is important to identify any likely significant constraints on the interface from the potential users or their tasks. Factors such as user experience, training and support must be predicted or assessed. Certain user attributes may determine a major decision at the outset. Disabled users are an obvious case but other needs may be to design for a variety of cultures and languages, or to understand that military and civilian operators may well behave very differently from each other in circumstances where a technology is being transferred from defence to industrial application.

Often there will be one operator or a team of operators who are highly skilled and knowledgeable about the system, but there may also be a variety of other users of the system, with a wide range of activities, who need to be considered in the design of the interface. Such users include those responsible for maintenance, managers or supervisors who require information on system performance, and systems engineers who may require information on the way in which the control system itself is performing. For supervisors, an early decision will be required on the degree to which they are monitoring systems or operators, which will determine whether emphasis should be an equipment position for access, or on workstation configuration for communication. We must appreciate the different levels of operator knowledge involved in working with any particular system, knowledge we may wish to provide, support, enhance or merely to be aware of in designing interfaces. Wirstad (1988) provides a listing of such potential knowledge in process control, including categories of:

- plant layout
- function, construction and capacity of components
- 'manoeuvring' (location and operation of controls and displays)
- system function, construction and flows
- process theory and practice
- identification of disturbances and consequence prediction
- measures to take for disturbances
- procedures for serious incidents
- organisation structure
- administrative arrangements
- safety regulations
- supervision

As well as knowledge required of operators we must also be aware of all the tasks that must be supported by the interfaces (e.g., Table 2).

For interface design there is a variety of task analysis methods that can be applied to analyse the information requirements of the users; these need to show who will use the information, when it will be used, what it will be used for and how it will be used. Such an analysis may also provide a basis for estimation of the potential errors that could occur in performing each task element and their probability of occurrence (see Chapter 32 and Kirwan, 1994). We can also use some analyses to determine what the information requirements of each of the tasks are, thus identifying the content of the displays that should he provided to support performance, and first decisions can be made on the display format for this information (see Figure 4, for instance). Of course the design of interfaces is dependent on the information available, which in turn is dependent on the system hardware and software. Whilst technology may mean that there is a large pool of potential data for display, it is often the case that the display format options are limited, by the control system software supplied by the manufacturer of the control system.

The selection of a task analysis method will depend on the interface design application it is to he used for. For example, it may be used to identify the frequency of use of the different displays and controls, the content of the information required for VDU-based displays, or the optimal format for displaying task information. Methods which may be used to identify information requirements include Job Process Charts (Figure 5) and a modified version of

Table 2 Types of Control Room Task To Be Supported

Procedural	Tasks that involve following a pre-determined sequence of events
Sensory-motor	For example the physical manipulation of input devices
Communication	The transmission of information between operators without the information being translated into another format (i.e., verbal communication or logs)
Monitoring	The surveillance of the system to identify any change in system status
Fault detection	The identification of an abnormal or unexpected system status
Decision making	The selection between alternative options/actions
Problem solving	The process of resolving uncertainty about system states. A particular kind of problem solving that is especially relevant to this context is fault diagnosis
Prediction	Judgement of likely future system states.

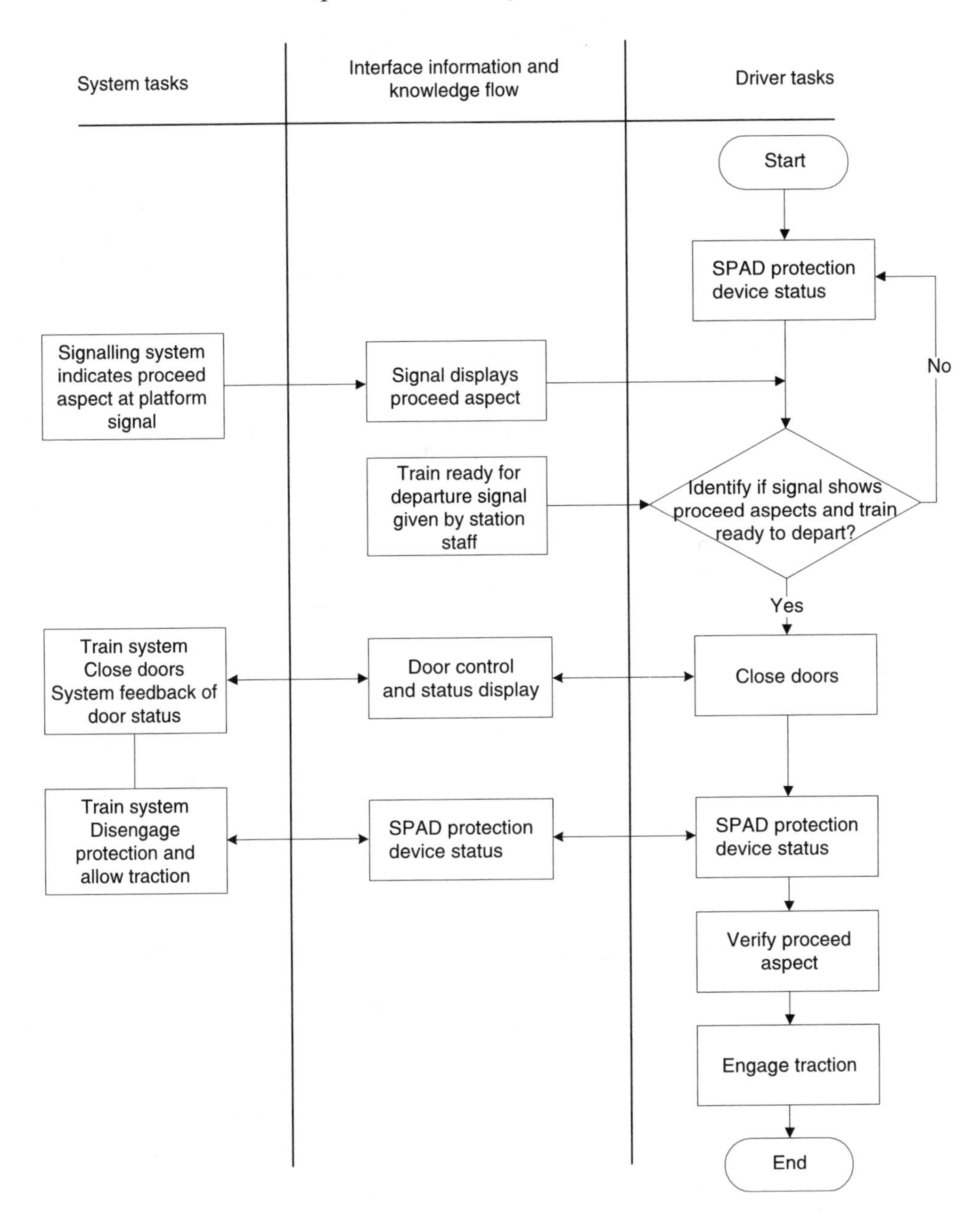

Figure 5 Simplified Job Process Chart of train drivers task (excerpt). *Source*: author's own records.

Hierarchical Task Analysis (see Figure 6 and Chapter 6). Other methods which may be useful in this context include Task Analysis for Knowledge Description (Diaper, 1989) to identify the knowledge requirements of the operator and Operational Sequence Diagrams (Johnson, 1993) to provide information on the sequencing of control and display use. In addition the layout of panel displays can also be analysed using simple methods such as link analysis or frequency counting (see Chapter 3). Where there is no existing interface for comparison, a task synthesis can he carried out using some of the techniques of task analysis.

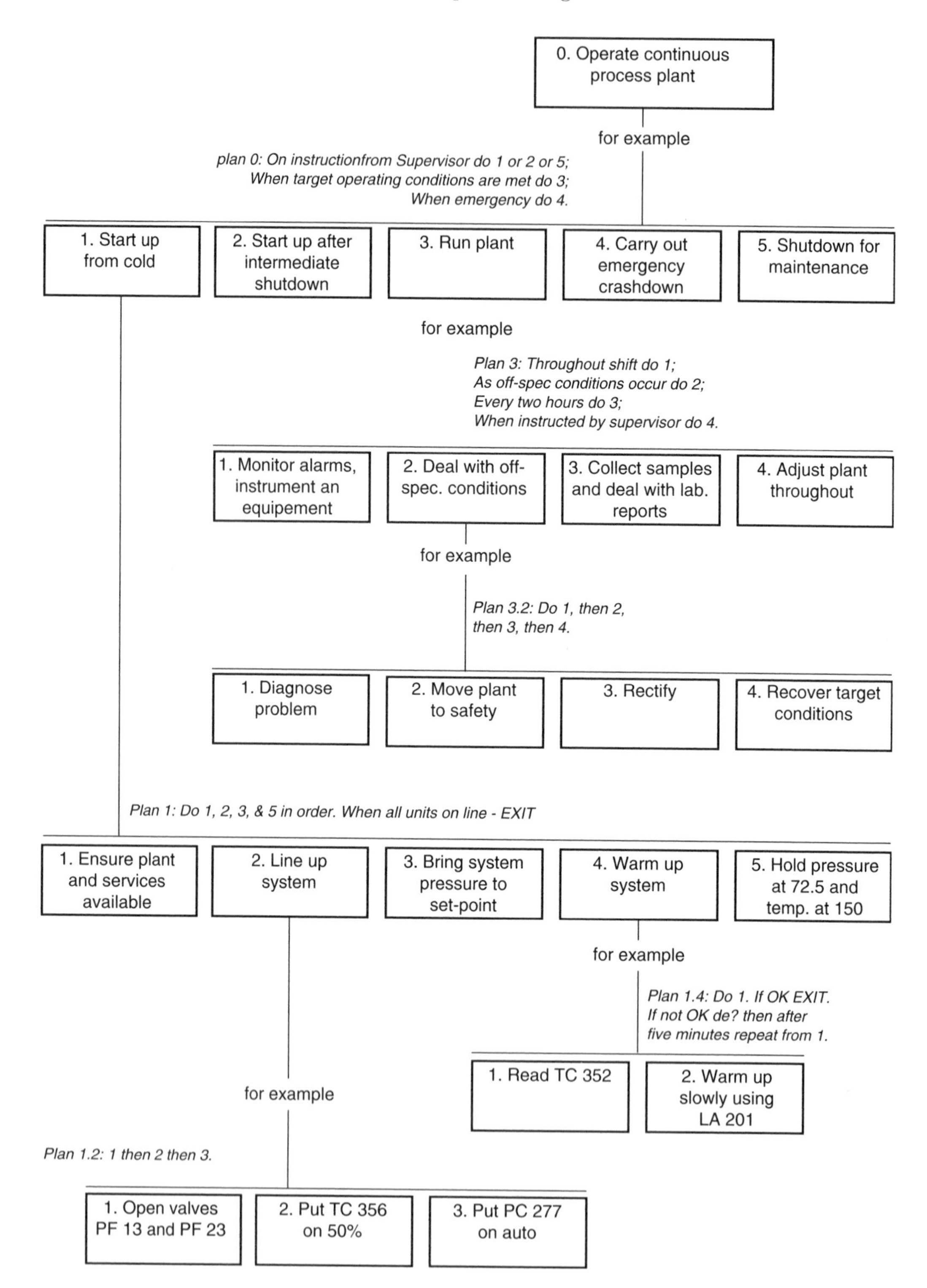

Figure 6 Hierarchical task analysis for a continuous process control task (*Source*: Kirwan and Ainsworth, 1992).

As complex control tasks are primarily cognitive, methods of collecting information for task analysis must reflect this, including verbal protocol analysis, interviews, direct observation, archives, scenario diagnosis and walk-throughs (see Kirwan and Ainsworth, 1992 and throughout this book).

Situation awareness, mental models and task behaviours

Situation awareness

We need to capture user requirements to define the baseline from which design and selection of displays and the control interface can commence. In complex systems we must consider the context in which the displays are viewed and other information that may contribute to the operating decisions which govern plant safety and performance.

The concept of situation awareness, that is, the awareness of the current operating context, and anticipation of how the system is likely to change and perform, has developed as a key construct in the understanding of how people perform in complex and information rich environments. It has been first proposed for understanding performance of pilots, and is now discussed in relation to a wider spread of jobs. As Endsley (1993) notes: 'focussing on situation awareness as a major design goal, the emphasis shifts from a 'knobs and dials' approach to a focus on the integrated system.' It has to be recognised that the notion is not without controversy, since opinion varies as to whether it should be seen as a personal characteristic (e.g., she shows high SA) or as a description of the situation or scenario (e.g., that display improves situation awareness).

Situation awareness is viewed as a state of knowledge, which is achieved by a variety of processes one of which is the interaction with the system by the operator. It can be simply defined as 'knowing what is going on' (Endsley, 1995). Generally it is conceived as comprising three levels: awareness of the current situation and the entities in it; comprehension of what this means; and assessment of what is likely to happen into the future. Therefore, the elements of situation awareness are situation specific but include factors such as current facts about system status, understanding of the system goals and objectives, and associated timings and projected system events and schedules. The implications for the operation of, and interaction with, complex systems are that the interface must be considered in the context of the whole system and of the information environment. The available system information and its use will be influenced by the knowledge of the operator leading to the actions taken by the operator to meet the demands of system performance.

This is important for display and control design as the displays will significantly influence situation awareness. The performance of the operator will be influenced not only by the current situation and particulars of the given context, but also their internal representation or mental model that they hold of the system. Tools developed for the measurement of situation awareness include such as the Situation Awareness Global Assessment Technique (SAGAT) and the Situation Awareness Rating Tool (SART) (see Endsley, 1996). It is often difficult to use such measures in real time environments and their use is most helpful in simulation trials.

User mental models

Imagine an operator at a remote control task who must carry out a job from a vantage point not used before, trying to get orientated and guide the system through critical operations in the most reliable, safe, effective manner; or the process worker who must try to understand why problems are occurring, and what combination of the many input parameters available would solve the problem, and must do this from process displays as well as by directly viewing the operations; or a maintenance engineer on a continuous process plant trying to relate information from system diagnostics to that on the control panel and to that on a giant mimic display. In all these cases, it is reasonable to think of the people involved as having or forming a mental model of the system. This mental model (or conceptual model) may be accurate or inaccurate, usable or worthless, but nonetheless the notion of mental models is attractive and useful for ergonomics.

In fact, in many cases the worker will construct and use several mental models: perhaps a symbolic one of the interaction of the variables in the systems — electrical, or chemical, or mechanical; a pictorial one of the form of the system being worked with; and a model of rules governing the operation (correct or faulty) of the system. These models can he relatively concrete or abstract. The representations may be formed from the system itself, from operating, emergency or maintenance procedures, from instructions and training, or from other systems worked on in the past.

As regards methods to identify and represent mental models, much debate has focused on the degree of formality needed. Within cognitive psychology there is the understanding that the notion has utility only if mental models can be described in computational form (e.g., Johnson-Laird, 1983). On the other hand, within human factors we are usually willing to postpone questions of how people represent and use knowledge in favour of understanding what knowledge is represented and how it is used to make inferences in specific domains (Payne, 1988). The majority of human factors literature, and especially in human–machine systems, appears to refer to conceptual (or non-computational) mental models, that constitutes their topographical, structural and functional understanding of a physical system and which allows them to describe, explain, understand and predict system behaviour. It is easiest to conceive such a model as comprising a system simulation that can be reconstituted and run in order to derive or confirm understanding.

There is some agreement about mental models that they:

- are internal representations of objects, actions, situations, people, etc.;
- build on experience and observation, of the world and of any particular system;
- are simulations run to produce qualitative inferences;
- constitute topography, structure, function, or operation of the system;
- may contain spatial, causal, contingency relations;
- allow people to describe, predict and explain behaviour;
- underpin people's understanding and behaviour;
- are instantiated each time they are required, and are parsimonious, and therefore are incomplete, unstable and often multiple.

The last point, made by Norman (1983) and others, has high relevance for control/display interface design. For example, the models instantiated by, say, a central control room operator to assist in fault diagnosis at a flexible manufacturing cell will vary each time in type and content. If the problem is product quality related then the cell may be conceptualised in functional form, modelled in terms of the series of transformation processes and the tooling needed to do this. Alternatively, if the problem is to do with hold-ups in components delivery then the cell may be conceptualised physically and spatially in terms of element flows and bottlenecks. In many cases more than one mental model may be formed, which will nonetheless overlap in their content and how they are employed, and which will have gaps or 'inaccuracies' according to operator experience and training.

There are many serious questions about the notion of mental models, though, and little agreement on definition, identification, representation and utilisation or even adequacy in methodology for their identification (Rutherford and Wilson, 1991). But, if we can predict or understand even in some fashion what mental models a new operator or user might hold about a system and its relevant domain, and what model they might build through subsequent interaction with the system, then we can improve interface design, training, operating procedures and so on. By understanding the potential users' mental models, and by adapting their own conceptual model accordingly, designers might develop a 'system image' that better matches, sustains and helps develop an appropriate user mental model (Wilson and Rutherford, 1989).

Task behaviours

Extending from the notion of mental models is the consideration of task behaviours found in the work of Rasmussen and his associates (e.g., Goodstein *et al.*, 1988, Rasmussen, 1986, Vicente, 1999). We can see elsewhere in this book the pervasiveness of Rasmussen's model of skill-, rule- and knowledge-based (S-R-K) behaviour, with its relevance to human reliability and to accident causation (Chapters 32, 33 and 34). There is also an intimate relationship between expected task behaviours and the information displayed to support them; in a basic description of how operators perceive information within the S-R-K model, Rasmussen uses a simple display/control system as illustration (Rasmussen, 1986, p. 107).

Skill-based behaviour is what is shown in tasks such as 'in-the-loop' controlling or steering, adjusting and calibrating instrument settings, or assembly tasks. The behaviour is akin to sensory-motor performance — where there is a fairly direct connection between sensory input and motor output with little mediation in cognition. In essence, skill-based behaviour is shown in familiar situations and is where the 'operator' recognises a 'signal' from the system and understands that this requires a normal routine, then executes a well-learned skilled act more or less automatically. Generally this type of work is undertaken with simple feedback control (comparison of the actual and intended states) that provides error information and thus defines the motor output response. However, there are suggestions that skill-based performance may also be based on feed-forward control and knowledge of the environment, for instance in riding a bicycle.

Rule-based behaviour is said to require a more conscious effort than does skill-based behaviour, and involves following a set of stored rules — in the mind or written down. Here, performance is goal oriented and takes place in familiar but non-routine situations; the operator perceives a 'sign' indicating environmental state(s), and then he or she uses learned rules and procedures once they have recognised certain cues. Control is fed forward through the stored rules. These rules will be derived empirically through experience, or communicated to the operator by colleagues, institutions or training.

The boundary between skill- and rule-based behaviour is indistinct, and may depend in the same situation on training or attention levels. A person working in skill-based mode may not be able to describe how they work, because the behaviour required may have become so automatic to them. The result of working in rule-based behaviour is frequently to co-ordinate and control a sequence of skill-based acts.

At the highest level is knowledge-based behaviour. Here the operator is in an unfamiliar situation with no, few or partial rules available from past experience. A goal is formulated by the operator, based on their perception of the state of the world and on some overall or global aims. Perceiving these 'symbols' allows them to develop a plan, using knowledge, reasoning and experience, and the goal they themselves have formulated helps them work to that plan. The plan itself may be selected and tested through a process of conceptual or physical trial-and-error.

Rasmussen's model has been criticised, and subsequent authors have not always distinguished what they mean by 'skill', 'rules' and 'knowledge' clearly and consistently. On the other hand, a large number of important human factors studies have made use of the model and approach. Vicente (1999) gives a very useful account of its value in making cognitive task design decisions.

Within consideration of control-display interfaces, the least that might be said is that any displays must, in general, support all kinds of behaviour. For instance, the same information display might be needed to guide normal operations, routine maintenance, and fault diagnosis in abnormal pressurised situations, and we have already seen that what starts out as a knowledge-based task may subsequently decompose into a rule- and then a skill-based task. As an example, an in-car navigation and diagnostics system might be used (1) to feed back a continual check on the state of the car subsystems or position on the road for the driver, (2) to support a routine check by the driver of the brake, suspension, tyre and engine sub-systems, (3) to allow substantial replanning of a route whilst on the move in heavy traffic, or (4) to be used by a garage mechanic in diagnosing fuel injection problems. Not only should the display support all three types of behaviour, it should also allow an operator to move 'up and down' between levels, in terms of information detail and degree of abstraction. Furthermore, in many circumstances it will he valuable for the system itself to support the operator by providing guidance on appropriate behaviour for different situations.

Environment and circumstances of use

Although the person, process, controls and displays constitute — to an extent — a closed-loop subsystem at the heart of human–machine systems, this sub-system is itself open to the environment. Interaction will occur with all physical, psycho-social and organisational environment factors.

Basic physical constraints are imposed by the location of the interface (Figure 7a, b). If such a location is a control room, then the most basic of these are the architectural constraints of the building in which the control room is located. For example, there may be architectural constraints on the positioning of consoles (e.g., blast proof walls or channelling under raised floors) which reduce the options concerning the number and type of interfaces that can be housed in the control room. There may be environmental constraints (for example the temperature and humidity of the room) affecting suitability of the equipment. The size and shape of the control room will have an impact on the workspace design and consequently on the number and positioning of the VDUs. This will in turn determine how operators will interact together and work as part of an operating team. Social contact should be catered for by grouping operators so that conversation is possible without compromising efficiency, especially important in larger facilities during quieter periods when staffing levels are lower. The interfaces provided in a centralised location such as a control room may be supplemented by interfaces actually on plant or in other locations, to provide information on system status. These may require ruggedisation to allow them to conform to safety requirements for equipment to be located on the plant or factory floor. Other specialised requirements are for seismic protection in certain geographical regions or for protection against shock or vibration on a ship or aircraft.

Normal and emergency operation, and planned maintenance

The interface design effort must take into account the consequences of events and look at different scenarios and modes of operation and predict the information needs of the operators under these different scenarios (Figure 8). This will include normal operational scenarios and also abnormal conditions, emergencies and maintenance. In some emergency situations the information for maintaining safe and effective system status may be provided through existing display formats; in other situations special formats may be necessary.

In the case of an abnormal event, VDU-based displays are vulnerable to factors such as loss of power. For reasons of reliability and safety, panel-based displays may be provided as

(a)

(b)

***Figure** 7* (a, b) Location of displays and controls placing physical load on work of operators (Courtesy of Tom Mayfield, Rolls Royce and Associates).

back-up to, or even in place of VDUs. The requirement of an overview of plant state is sometimes difficult to meet with VDUs alone, and state boards or wall-mounted hard wired mimic displays may be needed here. Such displays are also useful for shift changeovers to allow a reference point for rapid review of system state.

It is important to ensure that information concerning critical system states is available, to allow the operator to monitor the system and to bring it to safe status, perhaps using an emergency shut-down panel remote from the control room. In an emergency, automatic safety systems may operate and intervention by personnel prevented or restricted. Information provided to the operators is critical to ensure that they know what is happening, that

Figure 8 Interface design and layout of workstations should allow for all tasks, including ones not directly using interface elements.

appropriate action is taken and that personnel are able to anticipate future system states. In emergencies it is imperative that the interface does not add to the workload of staff.

Designers of some systems have adopted a strategy to reduce the tendency of people to react in certain inappropriate ways under stress, by ensuring that user intervention in the system in the event of an emergency is prevented for a set period of time to allow the operators to familiarise themselves with the conditions of the system and with information concerning the failures. This avoids the tendency of operators to try and fit the information they have into familiar diagnoses and solutions, rather than to explore a range of solutions to the problem in hand (this is shown in availability bias and cognitive lock-up — see Hogarth, 1980, pp. 204–234). The famous example is the Three Mile Island incident (Reason, 1990, pp. 189–191, 251) where operators considered that they had correctly diagnosed the origin of failure and took corrective actions based on this. Additional information, which would have indicated that their diagnosis was incorrect, was fitted into the original diagnosis or largely ignored, leading to a delay before the correct cause was identified and mitigating action could be taken.

There are a number of factors which impact on interface design under abnormal operating conditions. These include differences in the tasks to be performed, differences in personnel (including engineers or managers who may be unfamiliar with the displays), differences in workload and thus increased stress from unfamiliar operational conditions, and the fact that during an abnormal scenario the control room is often used as an emergency control centre, which may change needs for access to consoles or for numbers of people in the room and will certainly increase background noise and distractions.

In the design of VDU interfaces for complex systems, maintenance is critical and often overlooked. In many cases, errors which bring about system failures actually occur during the maintenance phase of the system life cycle. Maintenance personnel frequently use the same VDU display formats (or at least the same system) as operating personnel. Even if the information presentation is not common between these two groups of users, information concerning which equipment is undergoing maintenance (both emergency and planned maintenance) is essential to operating personnel, to assure the safety of personnel and equipment during the operational phases of the system. Although it is usual good practice physically to lock-off controls at the panels for equipment undergoing maintenance or repair, it is more difficult to do this reliably working from a VDU. Also, in many situations a shut-down may not be possible, for instance in air traffic control or the emergency services. A further potential problem in safety critical systems, where instrumentation must be checked at intervals as short as 24 hours, is that mistakes in checking can bring about an unforced shut-down of the plant.

Information analysis

The purpose of any display at work is to deliver information to someone to allow them to perform a task, whether this be passive (e.g., monitoring) or active (e.g., calibrating or fault diagnosis). In doing this, display design must eliminate or reduce errors of sensing, recognition, perception and decision; in other words, people should be able to recognise the relevant information against its background, distinguish its meaning to understand what is required and use the information to make decisions and perform tasks. The right information must be communicated in the right form to the right person at the right time.

One useful technique to employ in early screening of display designs can be borrowed from the world of Work Study — the systematic questioning process, used in examining a method and process analysis. This can be applied mainly to the information content of displays (but to an extent also to their form), and is a systematic look at what information is to be contained in displays before any consideration of how such information is to be presented. The questioning procedure shown in Table 3 allows a systematic and rigorous look at all parts of the system, whether in analysing existing displays or assessing needs for new ones. (see also ISO 11064 part 5 section 5).

Operator underload or overload

Increasingly people at work are interacting with machines or processes almost entirely through an interface rather than by direct sensing and physical actions. As a consequence, the interface will be a major determinant of the load on the operator. Broadly speaking we can talk of physical workload and of mental workload. In general we should seek to minimise static or dynamic physical workload imposed on operators by modes of control (e.g., heavy cranks, multiple valve opening, etc.) or by equipment layout (e.g., the position of a display giving a worker neck ache to see it, or controls requiring reach to awkward positions).

For mental workload (MWL) of process control operators the position is less clear, mirrored by much debate over the whole notion of MWL and its measurement (see Chapter 18). Despite the theoretical arguments against it, the notion of the arousal curve, indicating lower performance or higher errors when people are at high or low arousal levels has great face validity for designers and allows general guidance to be given. Stressors from the work

Table 3 Questioning procedure for information analysis

What information is to be displayed?	Why is it necessary?	What else could be displayed?	What should be displayed?
Where is the information to be displayed?	Why there?	Where else could it be displayed?	Where should it be displayed?
When is the information to be accessed/communicated?	Why then?	When else could it be communicated?	When should it be communicated?
Who is to have/use the information?	Why then?	Who else could have/use it?	Who should have/use it?
How is the information to be presented?	Why that way?	How else could it be presented?	How should it be presented?

Source: Author's (JRW) own records.

(e.g., time pressures), environment (e.g., intrusive noise), and personal factors (e.g., lack of sleep) should be maintained at intermediate levels, certainly not at extremes. Consequences of loads on operators which are too high or too low may be lapses in attention, cognitive lock-up, less co-ordination and timeliness in performance, shedding of tasks in random fashion, impatience, irritability and so on.

If we take driving as an example almost all of us will have experienced anxious or angry reactions when roads are very busy, we are late, direction signs are confusing and other drivers are behaving erratically, with consequent effect on our driving performance. On the other hand, motorway driving at night can produce very low arousal levels with consequent small and large errors of judgement, and driver fatigue being shown by their fidgeting and being unable to control speed and course simultaneously.

Control dynamics

Although this is in theory a logical place to consider control dynamics in an idealised process for human–machine interface design, in real life most of the considerations discussed above would have been in the light of knowing at least whether the task involved discrete or continuous control. Examples of the former are switch and valve operation to start-up a plant; the latter is when an operator has to keep a plant working with several parameters (e.g., temperatures, flow rates, pressures) held at specified values, and when fluctuations or transients mean that continual adjustment of the variables is needed. Continuous control is also exhibited in most kinds of transport, whether a car, boat, helicopter or submarine. Many modern work tasks require a mixture of continuous and discrete control, and interface design must allow for this. PC use will involve both discrete control — e.g., the sequence of keyboard initiated steps required to edit text, and also continuous control — e.g., use of mouse or space-ball or graphics tablet to input graphics or to 'walkthrough' or around the display. Handwriting is a fundamental form of continuous control.

Amongst the factors to be accounted for in a continuous control interface are:

- What order of control is required? Is the operator making simple step (or zero order) inputs to control distance and position, or is the control of rate (1st order), acceleration (2nd order) or of the much higher orders found in complex systems?
- If the operator is tracking a course, is this pursuit tracking (where target and cursor both move) or compensatory tracking (where change is shown in the discrepancy between target and cursor)? The former mode has control advantages but the latter can

make for more effective panel layouts by saving space. Whatever the choice, it will have considerable implications for display design.
- Would any form of assistance for the operator be of value? We may apply control aiding (e.g., control can be of cursor position or rate of movement according to whether large distances must be tracked), or display quickening, preview or prediction (all techniques to help when operators might want advance warning of what is to come or of what effects their actions might have).

For a full discussion of human factors issues to do with continuous control, the reader is referred to Kantowitz and Sorkin (1983), Sanders and McCormick (1992) and Wickens (1992).

Selection of control and display modalities

Although controls are most often hand operated and displays are most often visual ones, other modalities are possible. Control may be affected by feet or legs, especially when power must be transmitted, when the hands may be fully occupied and when the control task is one of discrete actuation. Examples are an on–off button for a power press, pedals in most forms of land transport, a foot mouse for a computer or a pedal/shuttle mechanism as on a sewing machine. Novel forms of control may be found with computers, including use of eye or head movements (and see Chapter 13). Finally, one of the most common forms of input, albeit more for human–human systems, is speech. Although technical and human factors difficulties remain, speech is increasingly used for system input.

Selection of mode of input, and especially any decision of whether to use other than hand controls will be on the basis of the task analysis and earlier design decisions. This is true also for display modality selection. Displays may be tactile (e.g., braille or shape coding of control knobs), proprioceptive (e.g., sensing of correct speed and track for cornering in a car), but most will be visual or auditory with emphasis on the former. In general, an auditory display may be chosen when:

- messages are simple and short (even for speech synthesised displays);
- immediate action is required;
- no later referral to the message is needed;
- the information is continually changing;
- the message refers to events in time;
- work is in poor viewing conditions — for instance low illumination or high vibration;
- the recipient is moving around;
- the recipient is receiving a large amount of other visual information;
- the task environment does not have a high degree of background noise.

From the above, it is clear that a principal use of auditory displays is in warnings or other short signals for action. However, the vast majority of interaction with systems though will be via visual displays and the remainder of this chapter concentrates on this. The reader should, though, bear in mind the role that auditory signals may be a very important source of information in communication centres. Guidance on the selection and use of input devices in a control room context is again given in ISO 11604 Part 5.

Interface specification — display of information

We cannot cover all relevant aspects of interface specification in this chapter. The available research and guidance on, e.g., computer input devices or screen coding is legion. We will have to be selective. There is good coverage in many of the general texts on control rooms and process control identified at the start of the chapter.

The interface provides the basis for communication between the system and the user. It provides information to the user on current status and performance of the system. In turn the user can manipulate and monitor the system via the interface. In the design of an interface the consideration of the information needs of the user(s) is probably the single most important element, and it is the display of information we concentrate upon here. Within this, we will first of all compare the two display formats of most applicability to control room design — panel displays and VDUs — before looking at some methods of information coding on displays.

Panel displays and VDUs

One of the principal choices facing the control room interface designer has historically been that between panel and/or visual display unit (VDU) displays (or increasingly flat panel LCD displays). To an extent panels have been replaced by VDU interfaces — so much so that the authors debated whether or not to include this section in the chapter. However, there are good operational (and operator support reasons behind having panels) and many control facility designs still have a mixture of displays, so an idea of the advantages of each and preferred use is still useful.

Panel-based displays are usually wall mounted panels in modern control rooms, although they can be found as floor-based consoles. These may be mainly static mimic type displays, which include dynamic information through LCDs or indicator lights, or they may be projected displays. They are usually hardwired and will display the information in one format only, e.g., an analogue dial or light emitting diode (LED), unless information is repeated in other formats to provide redundancy for safety or improved performance (see Figure 2a). For such panel interfaces the critical decisions are how the information will be displayed (modality) and where it will be located on the control panel. Controls are also located on the panel amongst the displays and will usually be allocated to individual displays, rather than having the one keyboard with a mouse or joystick that is often the case with VDUs. This is because the different displays require individual controls (e.g., to allocate a set point) and because the size of a panel often means that it would be impractical for several displays to share one control without causing problems of display-control compatibility.

One of the most critical factors in the design of panel-based displays is the layout and grouping of information. Unlike VDU-based displays this formatting cannot be flexible and so must accommodate the full range of tasks that the operator is required to perform. It must also accommodate tasks where two or more operators are required to work side by side. For example, information that is required by more than one operator to carry out different functions concurrently will necessitate special consideration for display placement, or information redundancy through display duplication elsewhere on the panel. Alternatively, VDU overview displays can be projected onto a screen on the wall of the control room, to give a large dynamic overview of key parameters, accessible to all personnel. Such shared overview displays can provide support for team working in a control room environment. However, they raise additional issues for consideration in the design of information, for instance:

- Is an overview display needed?
- What information should be available on the display?
- How should the information relate to desk-based or individual operator displays?
- Who will operate the configuration of the display?
- What back-up capabilities are available?

The configuration of shared displays has an impact on the design of control room and individual workspace layout.

The other important factor in consideration of the design of panel displays is the relationship between controls and displays located on the interface. The location of the control associated with a particular display or range of displays must clearly indicate which display will provide feedback of the effects of that control action. The movement of the control must also conform to the expectations and stereotypes of the user in terms of the effect it will have on the display (see Wickens, 1992, pp. 324–334).

Control-display compatibility, in terms of the expectations of users and design which fits with common population stereotypes will depend on displays indicating a response that is expected as a result of the control action taken. Stereotypes may be related to the expectations of a significant proportion of the population — for instance that a movement of a control in one direction would be expected to change the associated display in a particular way. They may be culturally based — national or professional — for instance in terms of the meaning ascribed to certain colours. Placement or activation of controls and displays that do not conform to majority expectations or which do not meet principles of compatibility can result in greater training requirements, increased errors or slower reaction or operation times. See Sanders and McCormick (1992) for more information.

Increasing technological complexity has meant that far more demands are placed on the design of the interface to convey relevant and timely information to the operator. The role of the operator is not only one of operation, but through that operation he or she has a major role to assure the safety of personnel and maintain the integrity of the equipment and the environment. A vast number of data points could potentially be displayed to the operator and the importance of displaying the right information at the right time is critical. The use of VDU-based interface technology means that many different formats and combinations of information are possible, so the interfaces can be configured to meet the requirements of a range of different users. However, the move from panel to VDU-technology is not without its disadvantages. The user of a VDU-based system will typically have several screens to access, but even so these screens can only display a limited sample from the range of data available. In addition the operation of the interface to access the information required to operate the system is an additional task, which forms part of the operator's workload.

VDU displays are now the most typically used medium for the display of process parameters. The number and location of VDUs is dependent on who the users will be and what they will be used for. The number of VDUs required for a given control room application can be calculated by envisaging a realistic worst case operational scenario, for example a situation where two unrelated failures occur concurrently. In such a situation a minimum of two screens would be needed to display information relating to each failure and a third screen would be needed to display alarm information. In general, most systems require a dedicated screen to display alarm or emergency information, allowing this critical information always to be available to the operator regardless of the current system status information that is being displayed at the interface (Bainbridge, 1991). The number of VDUs actually installed should be considered in the light of anticipated normal and worst case scenarios, number and types of users, number of data points for display, and the control room configuration (see also ISO11604 parts 4 and 5).

In control rooms, primary operation is often carried out from a VDU console arrangement, with panel displays providing an emergency back up, usually to convey information that for safety reasons must be on hardwired displays. Panels may also provide redundancy of information to supplement the VDUs. There may be several parameters critical to operation or safety which would tie up one or more of the display screens if the operators were to display them continuously under normal operational circumstances; one alternative is to present this information on a panel-based display (usually wall mounted) — for example, the fire and gas sensors on an offshore oil rig. Such 'hybrid interfaces', where VDUs display some of the system information and are located in a panel with other display media surrounding them, provide a means of achieving a better balance between the strengths of VDU based process displays (in terms of flexibility, alternative display formats for data etc) with the advantages of the panel based format (e.g., all parameters displayed simultaneously, rapid detection of specific alarms etc). Such hybrid interfaces can be problematic to design from the environmental point of view however; for example, lighting should be adjusted to suit both viewing of the VDUs and also the panel displays. Also, a systematic approach must be taken to co-ordinating information format between, say, screen-based and mimic-panel displays; particular attention must be paid to commonality of coding (e.g., colour) between the two media.

For the display of alarm information, annunciator panels provide a means of instantly recognising an alarm by its location on the panel, whilst VDU-based alarm lists can be more difficult to read and make it difficult to assimilate information at a glance.

Display formats

Whatever the display type, there are a number of possible formats for the display of process information: mimic, sequence, alphanumeric, graphical/trend, deviation, object oriented, pictorial, fault and alarm, etc. VDU technologies mean that the information can be displayed in almost any format limited only by the screen size and resolution. Increasingly displays are developed specific to the system under control, and operators may be allowed their own individual configurations of the workstation. However, whilst some displays still replicate the format of the local hardwired equipment display (e.g., a 3 point controller), formats such as mimics and graphs are more commonly used.

Mimic displays

Mimic displays offer a graphical representation of the system (Figure 9). This representation can either reflect the functional relationships between elements of the system (for example, a schematic representation of process flows) or the geographical/topographical layout of the elements of the system. Whilst research (Vermeulen, 1987) has indicated that neither is more effective when evaluated in terms of operational performance, many displays that are to be used for more than one task are based on a functional presentation (Figure 10). Such a presentation allows the sequence of system processes to be followed and can help the operator to visualise functional relationships between system elements, thus assisting with problem solving tasks. These display types may have a significant impact on the development of the operators' mental models of system function and so need to be designed carefully to ensure that any inferences an operator may make concerning system functioning are accurate and do not compromise system safety.

Sequence displays

The use of operating procedures to support a wide range of tasks is common practice in many complex systems (but see, for instance, Dekker, 2003). Such procedures may be in the form of a written manual or may be computer based, although procedural tasks which follow a predetermined sequence are increasingly automated, leaving the operator free to perform other operational tasks. (It is well recognised, though, that the trend to automate task elements wherever possible does not always enhance the safety or reliability of the system.) For an automated process sequence, the operator will need to know the stage in the process that has been reached, to allow monitoring and early detection of any abnormal events,

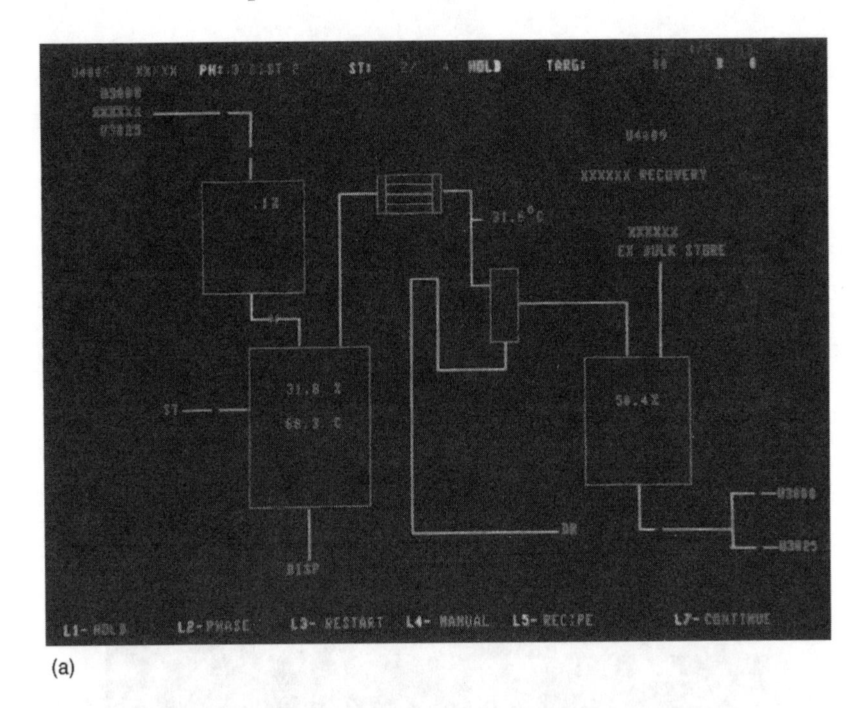

(a)

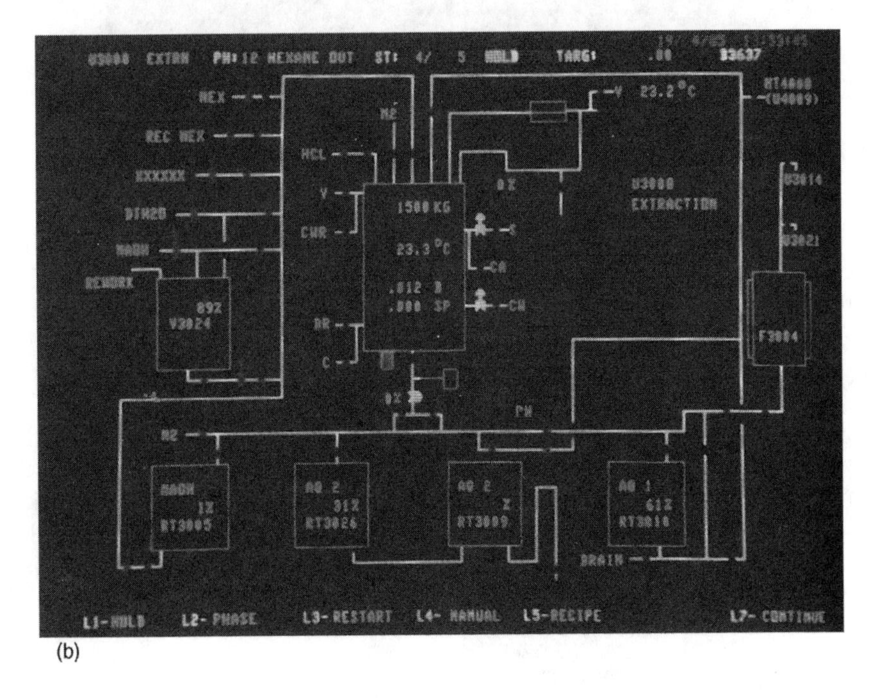

(b)

Figure 9 (a, b) Examples of mimic displays (*Source*: The Boots Company plc).

anticipation of requirements for manual inputs or operator intervention, and, in the event of a fault, whereabouts in the automated sequence it occurred. Sequence displays can make this information explicit. Where procedural tasks are still the operators' responsibility then display requirements are to act as a reminder of, and guide through, the sequence of tasks. Sequence displays may take the form of textual lists (see also alpha-numeric formats), however, a graphical form such as a flow chart or network diagram may be preferable

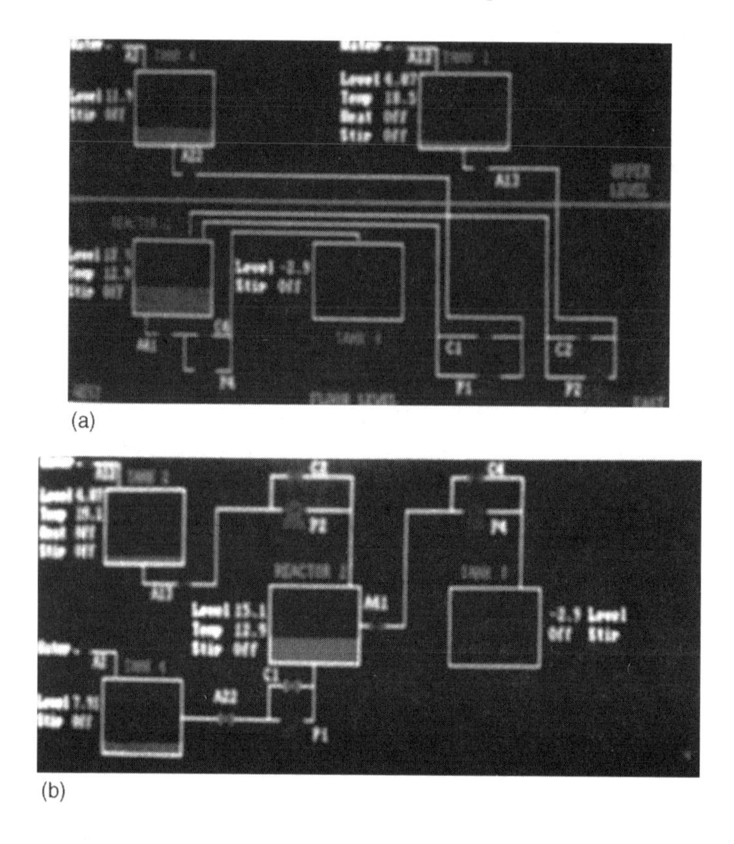

Figure 10 Illustrations of (a) geographical mimic, (b) topographical mimic (*Source*: ErgonomiQ).

particularly where the sequences are complex. The displays should provide an overview 'map' of the sequence, indicating operator inputs and giving a positive indication of progress through the sequence.

Deviation displays

These are used primarily for monitoring and fault detection tasks. The displays are used to indicate when a variable deviates outside given threshold values and to present the degree of deviation. The display usually takes the form of a horizontal bar chart, with each variable being represented by a bar and the height of the bar indicating the magnitude of deviation and threshold limits (Figure 11). If the bars are placed centrally on the display both positive and negative deviations can be indicated. Deviation displays are good for monitoring and fault detection to show movement towards abnormal conditions and to allow a rapid response if required.

Alphanumeric displays

At the most simple level, alphanumeric displays can either be static (e.g., printed instructions on a display) or dynamic (i.e., the values or the written information change over time). On panel-based displays information is usually static with the exception of digital numeric displays (e.g., an LED indicating quantitative information). On VDUs the formats for alphanumeric displays are varied, from lists to codes to natural language. The literature offers a range of guidelines on formatting and coding (e.g., Helander *et al.*, 1997; Shneiderman, 1998)

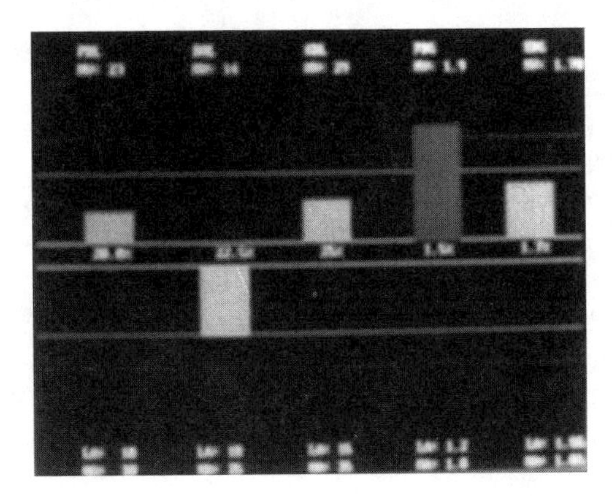

Figure 11 Illustration of a deviation display (*Source*: ErgonomiQ).

covering issues such as layout, density, and use of abbreviations. In control room contexts alphanumeric displays most commonly take the form of lists, static (e.g., process steps for a batch process) or dynamic (e.g., alarm lists) where the values of the variables or information displayed change according to system state. In addition, there are also dynamic displays in which the data change as a result of operator actions rather than process changes initiated by the system, for example, entering numerals to adjust a set point or entering default values prior to plant start up. In practice there are many hybrid formats where sections of a display (e.g., overlays or windows) are alphanumeric in style.

Alarms and warnings

Well designed alarm displays are critical to support correct decision making and problem solving in the event of system failure or abnormal conditions. There is a wide body of relevant research and guidance, in terms of information format and content, and interaction (e.g., Stanton, 1994). A range of display formats are found in alarms, often hybrid interfaces with hardwired LED displays or annunciators complemented by dedicated VDU alert lists or highlighted mimics.

Graphical/trend displays

In a primarily panel-based control room, trend and other graphically based information will usually be presented on chart recorders, because such information usually requires the provision of an historical record of plant operation and so records are often archived for future reference. VDU-based graphic formats are also used for the display of trend information. They offer the advantage that a large amount of historical data can be stored and readily recalled and that the time axis can be readily manipulated. Such displays can be used as a diagnostic aid and as a prediction tool to anticipate future plant states. The display of graphs on a VDU is limited by the size and resolution of the display although several graphs can be overlaid for comparison (a maximum of 4–6 is recommended). Line graphs are recommended for ease of use for most tasks (Figure 12).

Polar co-ordinate/object oriented displays

Polar co-ordinate or object oriented displays (also called integrated or shape displays) aim to provide rapid recognition of overall process or system status and can be used as a VDU-based overview display. They commonly take the form of a geometric shape, with the variables to be

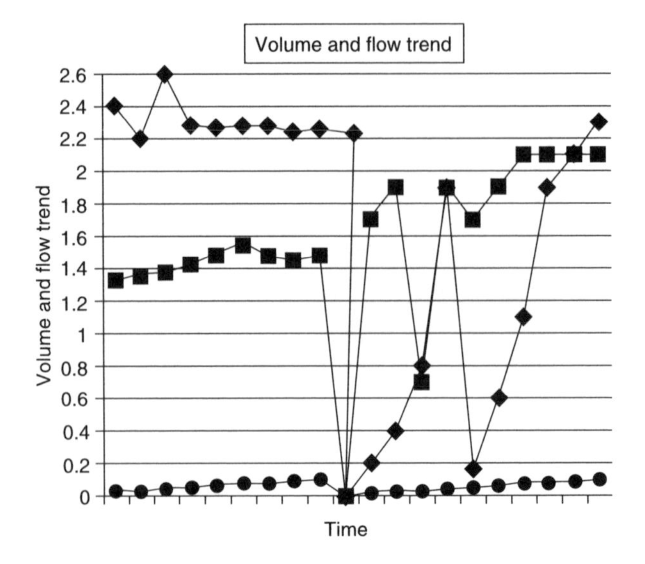

Figure 12 Example illustration of a trend display.

displayed indicated as points on the geometric object (e.g., at each angle of a hexagon or on the circumference of a circle). The variables are normalised so that when the system is functioning normally the figure will appear in the expected format and deviations in the system lead to deviations in the shape. Operators are able to recognise the shapes formed and so will be able to identify quickly any unusual or unexpected change in system status (Figure 13).

The displays are based on the principles of Gestalt psychology, that people will look for form or shape in information and that processing such information takes up less cognitive capacity than processing the individual data elements. The idea of object oriented displays was originally proposed by Coekin (1969) and developed by Goodstein (1981) and Wickens (1986). Whilst any geometric figure could be used, studies have proposed polygons, triangles and rectangles.

Pictorial display

VDUs can offer options for graphics displays that are not possible with panel-based formats, for example high fidelity simulations. This flexibility means that pictorial displays and other

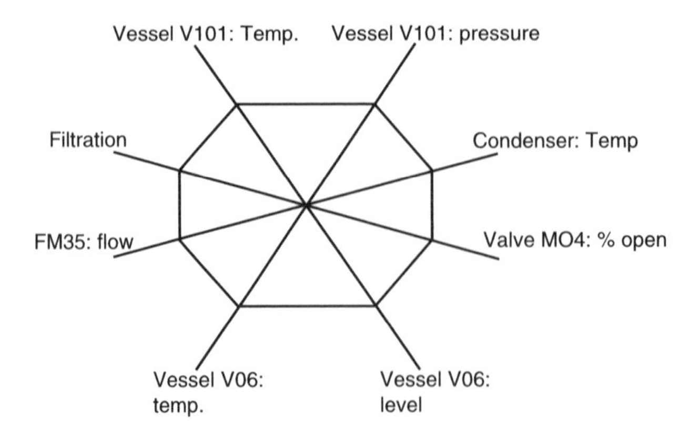

Figure 13 Detail from object oriented display format.

high resolution graphics can be used to present information in a way that is easily assimilated and used by the operator. The main advantage of such displays is that they are not constrained by convention and can be tailored to suit the task. As an example, in a paper production process plant a three-dimensional graphic was introduced to show process flows; operators found it easier to see the impact of problems on the overall continuous process than by using the more conventional two-dimensional mimic displays. Other applications where such displays might be useful include provision of a general qualitative overview of a process which can be backed up by detailed quantitative information, a pictorial representation of a particular item of process equipment in a complex plant, which assists the operator to relate operational problems and maintenance issues to the equipment on plant, and generally any information which cannot be easily represented using pre-defined display formats.

We are increasingly seeing use of close circuit television (CCTV) for many monitoring applications, monitoring valves or vessels for steam or gas leaks, observation of personnel working in potentially hazardous areas or in an emergency when access is risky, identifying the location of system problems, or aiding communication in noisy environments or any other situation where there are advantages in giving body signals. Increased use of CCTV will present a new set of ergonomics questions to do with detection, diagnosis, display system navigation, optimum positioning and panning, numbers which can be monitored, operator workload and control room environments.

Coding

One of the advantages of the use of VDUs is their flexibility to display data in a variety of formats. A range of coding options are available to the display designer including colour, shape, brightness, flash and spatial coding, which permit information density to be increased without increasing the perceived display density. Careful use of coding can help to break a complex display down into comprehensible but integrated elements (and see van Laar, 2002; Umbers and Collier, 1990). The important factors to consider in the application of coding techniques to the presentation of information are:

- Consistency should be maintained in what codes are used across all displays. To minimise the likelihood of error this includes consistency between all VDU formats and the use of the same code on panel displays and on plant (where appropriate).
- Consistent spatial coding of information fields should be provided (e.g., the menu bar always located at the top of the screen).
- Any code should be easily learnt and follow any conventions and stereotypes of the user population.
- The code should be unambiguous — and items in the code should be easily discriminated from one another.
- Alphanumeric codes should not be too long.
- The code should be unique and not confusable with any other form of coding used within the system.
- Low saturation colours can be used to code 'background' information, which needs to be present but is not immediately important or critical to operation.

Guidelines for visual display coding types are given in Table 4. Other forms of coding of information received by different sensory channels have an important role in signal reinforcement and distinction; this includes auditory coding (especially for alarms) and tactile coding of manual controls. There are many standards and reference documents which offer guidance on the design and use of symbols. These are often industry specific (e.g., symbols for

Table 4 Coding Guidelines for Visual Displays in control Rooms

Coding type	Description of code	Application of code
Size	Use of different sizes of same display element to convey changes in magnitude etc.	Useful way of representing magnitude for levels, temperature, etc. When it is the only code only 3 different sizes are recommended in the code for easy discrimination. Non-linearity of CRTs can make comparative judgements difficult if variations in magnitude of size are small.
Shape	Use of different (usually geometric) shapes to represent categories of items displayed, e.g., different classes of marine vessel on a display of a sea channel.	A relatively large number can be used and be readily distinguishable (up to 10). To be effective good resolution (e.g., to ensure a hexagon is not mistaken for a circle) is required and effective contrast, especially if colour is used. It is a useful means of coding for search, counting, and comparative type tasks.
Alphanumeric	Use of letters and numbers to form codes to (uniquely) identify elements within a system — often used for labelling purposes, e.g., to label valves with a common code on plant and on the display.	Avoid confusability between similar letters, numbers — e.g., X and K, S and S. They should form a natural category and stand out from other items on the display.
Brightness	The use of different luminance levels to highlight certain items of information or to indicate degrees of magnitude, e.g., temperature.	No more than 2 levels should be used to be distinguishable and effective.
Spatial	Use of the location on a display to give a meaning to an item of information, e.g., to identify it as a menu item, etc.	Should be used consistently throughout the system to indicate: —Title pages —Information fields —Alarms —Active and static display areas
Colour	Use of different displayed colours to provide differentiation between items on a display or to impart inherent meaning (colour should only be used as a redundant code).	Colour has a variety of uses which are task dependent: —To identify and classify information —As a formatting aid —To collate related information across different displays —Reduction of clutter —Visual display structure —Aid to visual search —To connotate danger, etc. —To highlight items or status change
Flash	Used to attract attention to items of information on the display.	Up to 4 blink rates are distinguishable — but less is preferable. Rate should be between 1–4 Hz. It is useful for redundant coding. It should be used sparingly. Operators should be able to cancel the blinking.

Figure 14 Control room operator utilizing six different types of information, controlled via virtual keyboards (courtesy of Honeywell Control Systems Ltd. and Tom Mayfield of Rolls Royce and Associates).

valves in the process industries, and for signals in the railways industry). The reader is referred in particular to the following helpful resources; ISO 3461: General principles for the creation of graphical symbols, ISO 7000: Graphical symbols for use on equipment and ISO 3511: Process measurement control functions and instrumentation — symbolic representation.

Navigation, structuring and layout of control room displays

As already pointed out, the use of VDUs in control applications means that, whilst the operators can access a vast amount of system data, all these data are not displayed simultaneously and continuously. Therefore, they have the additional task of navigating around the display structure and accessing the displays to find the required information. It is important that this structure is clear to the user and accessible in ways which assist, rather than detract from, task performance (Figure 14). Users may be offered a variety of options for accessing the different displays within the system, including indirect manipulation of a cursor or other pointing device via a trackerball or joystick, direct manipulation via a touchscreen, or the use of a keyboard with dedicated keys or a coded input. Often more than one option is offered, and for expert users shortcuts to access displays (e.g., to avoid having to progress through the levels of a hierarchy) should be made available. The design of control room VDU formats differs from those of, for example, office automation packages, as the users will be trained to a defined level and will be practised in the use of the system for normal operation. However, the training of new operators will often take place with them sitting by and observing experienced operators, and the structuring of the displays forms an important part of how their mental model of plant or system functioning is developed (see earlier).

The main issues to be addressed in display system structure and navigation are: what is to be presented on a page (the number of data items and the options for accessing other pages, e.g., menu items), the number of levels of pages in a system, the naming and grouping of items on a page, and the method of display selection.

Grouping of items into display pages and the means of navigating around those display pages are interrelated design issues, dependent on functional relationships and task needs; for example, if the operator controls the system on plant as well as in the control room then topographical layout may be of importance. The naming of groupings is also important but can be problematic when groupings do not have one clear and unambiguous collective description.

Once the potential user population has been defined and the information content and format of the displays determined, then the division of information into display pages for a VDU-based system is the next stage, along with decisions concerning how those pages will be structured and accessed. Hierarchical menus are the most commonly used means of accessing different display formats. There has been a range of research aiming to establish the theoretically optimum structure; the optimal number of menu items on a page is dependent on the way in which the items are grouped, keying and system response times and the number of levels of selections. A broad and shallow hierarchy may often be better for experienced operators; increases in hierarchy depth can produce slower data access times and lead to an increase in errors in performance. Although support of navigation between pages is commonly achieved by the use of menus, there are alternative mechanisms including: direct interaction (screen item selection), direct selection, tag code access, indexed access, keyword or command access.

The structure of the displays within a control system context may not be a straightforward application as other influencing factors may apply. Functional or geographical groupings of items into displays may not give even sized and consistent groupings. A larger 'top' level may be required for overview displays. The detail of control information required for the different 'levels' of display may be fixed by organisational constraints.

Where possible the number of pages in a system should be reduced to a minimum, and the information presented on each maximised by the careful use of coding, structuring and labelling. The grouping of items into display 'pages' is critical to effective and safe performance. As information can only be accessed in a limited way, the grouping of items impacts on how individual displays and groups of displays can be used in operation. In grouping (and naming) items for allocation to a display page, important factors include:

- Groupings should be consistent with the functional and causal groupings within the actual system.
- Groupings should be compatible with the operators' mental models of system functionality. Operators should be able to anticipate which formats will contain which information and the groupings and structure should be meaningful.
- Do not separate directly interacting variables across pages — or where this is unavoidable provide an overview schematic of interacting functionally-related variables.
- Avoid operators having to carry information in memory from one display page to another.
- Displays which may be commonly used together should be checked both for consistency, and the division of information between them.

As a last point, the careful design of window systems is crucial to good control room interface design. They have enormous potential to assist the operator in navigation around the information structure, to underpin intelligent decision support sub-systems and to make the simultaneous display of different data structures or formats more effective. Multimedia interfaces, certainly require good windowed top-level screen displays. Virtual environments also are seen as having a role in control rooms of the future, and usability of these — including navigation — will he just one of a number of consequent ergonomics and technical issues (Wilson *et al.* 1995; Wilson, 1999).

Evaluation of interfaces

Finally, having carried out our initial analysis and conceptual and detailed design, we must evaluate the resulting interface, or maybe even the whole control room. Evaluations will also be carried out independent of design, for instance as an input to safety audits. Many of the techniques discussed in other chapters of this book must be applied here. We present a set of guidelines written as a checklist, to guide any first level expert evaluation (Table 5). These guidelines are particular to displays in control rooms, and largely VDUs, but many parts would be relevant in other contexts also. The evaluation checklist is sectioned by topic and is merely

Table 5 Checklist for Evaluation of Control Room Interfaces

Displays structure

- Maintain a consistent relationship in the way in which all displays are structured throughout the system.
- Avoid the user needing to maintain anything other than simple items in memory when moving from one display to another.
- Ensure that the display structure is transparent to the user, i.e., can the user locate any given item of information within the structure and is the interrelationship between different display pages evident?
- The operator should be able to apply a consistent set of rules for navigation throughout the system.
- Navigation between pages should be simple and the user should be able to enter and exit the structure at any point.
- Ensure the structure of any hierarchical system is appropriate to the user e.g., broad and shallow for expert and frequent users, deeper for novices.
- Ensure the paging structure corresponds to the operator's mental model of the system (or that a coherent mental model can be established through use).
- Provide a page to give an overview of the paging structure.

Information structure

Navigation

- The organisation of the display structure should be transparent.
- Moving between screens frequently used in conjunction with each other should be as simple as possible.
- Frequently used displays should be directly accessible (e.g., by dedicated keys).
- Each display format should be labelled with a unique identifier which also shows its place in the navigational structure.

Division and partitioning of information pages

- The operator should be able to perform tasks without having to carry significant information from one display to another in memory.
- Information concerning variables which interact with one another should not be divided across display pages.
- Fields for the presentation of particular types of information (e.g., menus, display titles) should be consistent throughout all displays.
- Simple variable relationships can be divided across pages, but there should be a repeat of information to ensure the relationship is clear.
- Preferred display density is dependent on a range of issues such as the format of information, types of coding, number of dynamic data points, display screen resolution and frequency of use.

Formatting information on the screen

- Displayed information should be standardised in location of particular kinds of information.
- Fields or screen positions for certain types of information, e.g., titles, menus, commands and input fields, should be consistent between displays.

(*Continued*)

Table 5 Continued

- Where possible displays should be symmetrically balanced.
- Important information should be positioned in the upper left/central/upper right areas of the screen.
- Position and layout of information should provide the user with extra coding about the nature of the information in different parts of the display.
- Data should be grouped to assist with the tasks the user has to perform. This may mean that groupings are on a task-based or functional level.
- Items may be grouped or arranged in a variety of ways dependent on user requirements. For example, Criticality — important items are placed prominently and together; Frequency — frequently used information is placed prominently on the screen, data frequently used together are displayed together; Function — items can be grouped based on function when sequence and frequency are not important; Sequence — items are displayed in the order in which they occur e.g., process flow.
- The use of too many windows or partitions should be avoided.

Display formats

- The formats used should take into full account the task(s) they are used to perform.
- Formats should be matched to user attributes, e.g., polar coordinate displays make use of operators' pattern recognition capabilities, for rapid detection of system deviations.
- Information should be presented in a format which is consistent with the task requirements of the user (e.g., if qualitative state information is required then a digital readout may not assist the user and may actually detract from task performance).

Coding

General coding

- Codes should be used to make information more easily assimilated and to structure displays. In general codes should be: Consistent — both within the display system and with other codes used on plant or in the process; Unambiguous — items in the code should not be confusable; Unique — any coding scheme should be clearly distinguishable from other coding schemes.

Colour

- The use of colour is generally subjectively preferred over monochrome, although it does not always give improvements in performance. Colour can be used to enhance the appearance of a display or as a code. Care must be taken in the use of colour as it can increase the potential for human error.
- Colour should only be used as a redundant code, i.e., all items should be distinguishable without colour, using colour only to enhance the coding.
- Consistent colour coding should be used over all interfaces, in the control room and on plant.
- For accurate discrimination of colours in a code a maximum of seven should be used.

Colour can be used to relate items that are similar but separated spatially.

Highlighting

- Highlighting should not be overused but used selectively for emphasis and to give feedback.
- Blinking is good for attention getting.
- High brightness has attention getting properties, but less urgency.
- Reverse video should be used in moderation.
- Underlining to highlight text should only be used if it will not contribute to display clutter and if spatial layout permits.
- Do not overuse different fonts and upper case.

Labelling

- Consistency is essential and should be observed in size, font, use of abbreviations and positioning of labels.
- Information content of displays.
- Ensure that a task analysis and systematic identification of information needs has been applied within system design.

(*Continued*)

Table 5 Continued

- Check that all potential users of the system have been considered in display design, including operators, management, control system engineers, and maintenance personnel.
- Ensure that the information content of displays matches the tasks it will be used for (i.e., information required for a particular task should be accessed easily and structured in a way that facilitates task performance).

Ensure that the consistency in the presentation of information is maintained wherever possible between plant and control room.

Information content of displays
- Ensure that a task analysis and systematic identification of information needs has been applied within system design.
- Check that all potential users of the system have been considered in display design, including operators, management, control system engineers, and maintenance personnel.
- Ensure that the information content of displays matches the tasks it will be used for (i.e., information required for a particular task should be accessed easily and structured in a way that facilitates task performance).
- Ensure that consistency in the presentation of information is maintained wherever possible between plant and control room.

a simple guide to the issues involved. There is inevitably some overlap between sections since the whole checklist may not always need to be used.

References

Bainbridge, E.A. (1991). Multiplexed VDT display systems: a framework for good practice. In: G. Weir and J. Alty (eds) *Human Computer Interaction and Complex Systems* (London: Academic Press), pp. 189–210.

Ball, P.W. (ed.) (1991). *The Guide to Reducing Human Error in Process Operations.* Report SRDA-R3 of The Human Factors in Reliability Group (Warrington: The SRD Association).

Coekin, J. (1969). A versatile presentation of parameters for rapid recognition of total state. In: J. Moraal and K.-F. Kraiss (eds) *Manned System Design* (New York: Plenum), pp. 153–179.

Dekker, S. (2003). Failure to adapt or adaptations that fail: contrasting models on procedures and safety. *Applied Ergonomics*, **34**, 233–238.

Diaper, D. (ed.) (1989). Task analysis for knowledge description; the method and an example. *Task Analysis for Human Computer Interaction* (London: Ellis Horwood).

Edwards, E. and Lees, F.P. (eds) (1974). *The Human Operator in Process Control.* (London: Taylor & Francis).

Endsley, M.R. (1993). Predictive utility of an objective measure of situational Awareness. *Proceedings of the Human Factors Society 34th Annual Meeting* (Santa Monica California: Human factors and Ergonomics Society), pp. 41–45.

Endsley, M.R. (1995). Measurement of situation awareness in dynamic systems. *Human Factors*, **37**, 65–84.

Endsley, M.R. (1996). Situation awareness measurement in test and evaluation. In: T.G. O'Brien and S.G. Charlton (eds), *Handbook of Human Factors Testing and Evaluation* (Mahwah, New Jersey: Lawrence Erlbaum).

Goodstein, L.P. (1981). Discriminative display support for process operators. In: J. Rasmussen and W.B. Rouse (eds) *Human Detection and Diagnosis of System Failures* (New York: Plenum), pp. 433–449.

Goodstein, L.P., Andersen, H.B. and Olsen, S.E. (1988). *Tasks, Errors and Mental Models* (London: Taylor & Francis).

Health and Safety Executive (1989). *The Fire and Explosion at BP Oil (Grangemouth) Refinery Ltd* (London: HMSO), pp. 15–35.
Helander, M.G., Landauer, T.K. and Prabhu, P.V. (1997). *Handbook of Human–Computer Interaction*, 2nd edition. (Amsterdam: Elsevier).
Hogarth, R. (1980). *Judgement and Choice* (Chichester: John Wiley & Sons).
International Instrument Users Association (1998). SIREP-WIB-EXERA Report M2656X98. Ergonomics in Process Control Rooms. Part 2: Design Guidelines.
ISO 11064, various parts and dates, Ergonomic Design of Control Centres. International Standards Organisation.
ISO 9241, various parts and dates, Ergonomic requirements for office work with visual display terminals (VDTs). International Standards Organisation.
ISO 13406 (BS ENISO 13406), various parts and dates, Ergonomic requirements for work with visual displays based on flat panels.
Ivergård, T. (1989). *Handbook of Control Room Design and Ergonomics* (London: Taylor & Francis).
Johnson, G.I. (1993). Spatial operational sequence diagrams in usability investigations. In: E.J. Lovesey (ed.) *Contemporary Ergonomics 1993* (London: Taylor & Francis).
Johnson-Laird, P.N. (1983). *Mental Models* (Cambridge: Cambridge University Press).
Kantowitz, B.H. and Sorkin, R.D. (1983). *Human Factors: Understanding People–System Relationships* (New York: John Wiley).
Kinkade, R.G. and Anderson, J. (eds) (1984). *Human Factors Guide for Nuclear Power Plant Control Room Development*, EPRI Report NP-3659 (Palo Alto, CA: Electric Power Research Institute).
Kirwan, B. (1994). *A Guide to Practical Human Reliability Assessment* (London: Taylor & Francis).
Kirwan, B. and Ainsworth, L.K. (1992). *A Guide to Task Analysis* (London: Taylor & Francis).
Kletz, T. (1991) *An Engineers View of Human Error* (Institution of Chemical Engineers).
Moray, N. (1997). Human factors in process control. In: G. Salvendy (ed.) *Handbook of Human Factors* (New York: J. Wiley), pp. 1944–1969.
Norman, D.A. (1983). Some observations on mental models. In: D. Gentner and A. Stevens (eds) *Mental Models* (Hillsdale, NJ: Erlbaum), pp. 7–14.
Noyes, J. and Bransby, M. (2001). *People in Control: Human Factors in Control Room Design.* (Stevenage, UK: The Institution of Electrical Engineers).
NUREG-0700 (1996). *Human–System Interface Design Review Guidelines.* Revision 1. US Nuclear Regulatory Commission, Washington DC.
NUREG/CR-5908 BNL-NUREG-52333 (O'Hara, J.M., Brown, W.S., Baker, C.C., Welsh, D.L., Granda, T.M. and Vingelis, P.J.) (1994). *Advanced Human System Interface Design Review Guidance.* US Nuclear Regulatory Commission, Washington DC.
Payne, S.J. (1988). Methods and mental models in theories of cognitive skill. In: J. Self (ed.) *Artificial Intelligence and Human Learning* (London: Chapman & Hall).
Pethick, A.J. and Wood, J. (1989). Closed circuit television and user needs. In: E.D. Megaw (ed.) *Contemporary Ergonomics 1989* (London: Taylor & Francis), pp. 450–455.
Rajan, J.A. (1993). Human factors in control room design: reducing risk and maximising safety. *Loss Control Newsletter*, **4**, 20–22 (London: Sedgwick Energy).
Rasmussen, J. (1986). *Information Processing and Human–Machine Interaction* (Amsterdam: North Holland).
Reason, J. (1990). *Human Error* (Cambridge: Cambridge University Press).
Rutherford, A. and Wilson, J.R. Searching for the mental model in human–machine systems. In: Y. Rogers, A. Rutherford and P. Bibby (eds) *Models in the Mind: Perspectives, Theory and Application* (London: Academic Press), pp. 195–223.
Salvendy, G. (ed.) (1987). *Handbook of Human Factors* (New York: John Wiley).
Sanders, M.S. and McCormick, E.J. (1992). *Human Factors in Engineering and Design*, 7th edition (New York: McGraw-Hill).
Sheridan, T.B. (1987). Supervisory control. In: G. Salvendy (ed.) *Handbook of Human Factors*, (New York: John Wiley & Sons), pp. 1243–1268.

Sheridan, T.B. (2002). *Humans and Automation: System Design and Research Issues.* (Santa Monica, CA: J. Wiley and the HFES).

Shirley, R.S. (1992). *Computer Graphics for Industrial Applications.* (Englewood Cliffs, N.J.: Prentice Hall).

Shneiderman, B. (1998). *Designing the User Interface: Strategies for Effective Human–Computer Interaction*, 3rd edition (New York: Addison-Wesley).

Stanton, N. (ed.) (1994). *Human Factors in Alarm Design* (London: Taylor & Francis).

The Guardian (1984). Helicopter pilot admits piloting error. Feb. 24th.

Umbers, I.G. and Collier, G.D. (1990). Coding techniques for process plant VDU formats. *Applied Ergonomics*, **21**, 187–198.

US Department of Defense (1999). Design Criteria Standard — Human Engineering MIL-STD 1472F.

van Laar, D.L. (2002). Psychological and cartographic principles for the production of visual layering effects in computer displays. *Displays* **22**, 125–135.

Vermeulen, J. (1987). Effects of functional or topographically presented process schemes on operator performance. *Human Factors*, **29**, 383–395.

Vicente, K.J. (1999). *Cognitive Work Analysis* (London: Lawrence Erlbaum).

Wickens, C.D. (1986). *'The Object Display': Principles and a Review of Experimental Findings*, Technical Report No. CPL 86-6 (Champaign-Urbana, IL: University of Illinois).

Wickens, C.D. (1992). *Engineering Psychology and Human Performance*, 2nd edition (New York: Harper Collins).

Wilson, J.R., Brown, D.J., Cobb, S.V., D'Cruz, M.D. and Eastgate, R.M. (1995). Manufacturing operations in virtual environments. *PRESENCE*: Teleoperators and Virtual Environments, 4/3, 306–317.

Wilson, JR. and Rutherford, A. (1989). Mental models: theory and application in human factors. *Human Factors*, **31**, 617–634.

Wilson, J.R. (1999). Virtual environments applications and applied ergonomics. *Applied Ergonomics*, **30**, 3–9.

Wirstad, J. (1988). On knowledge structures for process operators. In: L. Goodstein, H. Andersen and S. Olsen (eds.) *Tasks, Errors and Mental Models* (London: Taylor & Francis), Ch. 3.

Wood, J. (1994). The developing international standard on control room ergonomics, ISO 11064. *Proceedings of the 12th Congress of the International Ergonomics Association*, Toronto, 15–19 August.

Part III

Consequences of work activities

Having examined and explained the methods and techniques basic to all areas of ergonomics investigation, looked in more depth at techniques developed or adapted to study some particular design and evaluation applications, and before devoting a number of chapters to the broad area of physical environment assessment, we turn now to analysis techniques for work activities. Each of the chapters in this section takes either a specific, if broad, area of work activity — such as visual performance — or a specific consequence of certain types of work — as in stress. The individual contributions draw upon any number of the methods discussed earlier in the book in more basic and general form. In general the chapters are concerned with assessment of the consequences of carrying out a wide range of work activities.

Work involves both physical and mental activities and this section considers both of them. The physiological and biomechanical costs of work are a reality to many, and are not necessarily alleviated by modern technology. This emphasises the importance of knowing about, and knowing how to correctly apply, both field and laboratory methods for evaluating physical effort, to recognise when this effort is high and particularly when it is excessive. In most circumstances this is not a simple matter.

Although dynamic work (Louhevaara and Kilbom, Chapter 15) is obvious when it is performed it can be manifest in so many ways that it is the selection of the appropriate approach and then assessment method that requires human factors skill. In contrast, static and postural load (Corlett, Chapter 16) is often overlooked. It is easy for managers and engineers to assume that if someone is sitting down there is no load of any significance, and forget that muscles are still active holding a posture. If the job restricts the opportunity for change between muscles, then the situation is exacerbated. People will often discount the discomfort of postural loading, expecting to 'feel tired after work'. However these manifestations of discomfort often give guidance to the ergonomist on where to look for the sources of problems, and methods to assess the effects on the person are important in very many situations. This means that self report measures of various kinds will be as useful and as powerful as physical measurement and direct observation. (In the chapter on seating later in this section, subjective methods and people's responses are said to be vital assessment approaches.)

When risks and boundaries for levels of exposure, for example, have to be assessed, we often need the support of numbers. Chapter 17 by Dempsey and Munro on biomechanics is cautious about the acceptance of figures but strongly in favour of biomechanical analyses for 'before and after' comparisons. If figures are to be used as absolute measures and matched against criteria, the availability of a range, or measure of the likely spread of results, together with the proposals for using 'worst-case' values, contributes to the assurance that the decisions taken will be conservative. Since the second edition of this book methodology in this area has increasingly embraced the dynamic as well as the static loading, and this revision reflects this.

If the assessment of physical workload is difficult and contentious, then the situation for mental workload (Megaw, Chapter 18); is possibly more so. The methods used do parallel some of those for physical workload — especially performance (or primary task measures), subjective assessment and physiological measurement. Even the group of techniques known as secondary or alternate task measures can be used in somewhat different form for physical workload too, for instance checking the effects of extended keyboard use by testing subsequent performance on one of the many tests of manipulative ability. Mental workload is very instructive as a notion in human factors; as with notions such as mental models and situation awareness it is a very attractive one to our clients, seemingly very meaningful as well as important, but we struggle when we have to explain that in fact this is a multi-factorial notion, there are differences of opinion as to what it reflects or defines, and routes to its measurement are, to say the least, controversial.

Of course, one way in which mental workload assessment differs from physical workload assessment is in the less overt and visible nature of the behaviour involved. Direct observation of task performance or worker response can often at least confirm or illuminate investigations of physical work. This is less easy for mental work, and techniques such as protocol analysis (see Bainbridge and Sanderson; Chapter 7) may be needed. A similar problem, i.e., that the causative factors are not easily or directly observable, is also found in the assessment of stress. In Chapter 19, Cox and Griffiths discuss the various models or definitions of stress, as with many topics in this book the model or definition chosen will affect our selection of methods as well as of preventive strategies. They then address the identification of stress potential, and recognition and measurement of stress in individuals.

The next chapter in this section covers the assessment and consequences of performing visual work. Such special attention is appropriate since for over 90% of work the task-related input is visual. Howarth, Bullimore and Megaw (Chapter 20) are concerned with the performance of visual tasks and upon possible outcomes of visual performance, such as visual fatigue. As well as the types of method that may be adapted for use in almost any situation, visual processes and their consequences may be assessed by use of some particular techniques also, including specific performance tests (e.g., for colour vision) and observed behavioural measures such as eye movements.

Visual task performance is a very rich topic as far as methods and techniques development is concerned. This is enlarged upon in Chapter 21 in which Kumashiro gives a view on psychophysiological measurement of workload which concentrates upon a measure of visual function (critical flicker frequency), as well as measures of attention and performance (e.g., concentration maintenance function) and of physiological function (e.g., heart rate variability).

In conclusion, if there were one theme which connects all the work activity analysis in this section, it is that the measurement and assessment of consequences of work activity — in terms of the effects on people — is made doubly difficult because of problems in defining, in unambiguous terms, the phenomena concerned. For instance, workload, physical or mental, is made harder to understand and evaluate if we are unsure of its causes, their interaction and how they are made manifest; disagreement about what visual fatigue is does not make the task of assessing it any easier; and so on.

Chapter 15

Dynamic work assessment

Veikko Louhevaara and Åsa Kilbom

Introduction

The purpose of physical muscle work is to produce muscle contractions resulting in the output of dynamic or static strength. Muscle contractions require the complex and harmonised co-operation of musculoskeletal, cardiorespiratory and nervous systems. Muscle contractions require energy, and therefore, the supply of nutrients and oxygen as well as the removal of waste- or endproducts and carbon dioxide are necessary. The brain, via the kinetic nervous pathways, controls voluntary muscle contractions. The autonomic nervous system with chemical and humoral mechanisms controls the involuntary organic functions (Figure 1).

In industrialised countries about 10–20% of workers are still employed in jobs requiring physical muscle work, in spite of rapid technological developments. In less industrialised countries, all kind of muscle work is very common. Heavy dynamic and static muscle work with large muscle groups is most frequently needed in jobs in forestry, agriculture, building, installation, transportation and warehouse, manual sorting, cleaning, health and home care, and in some special occupations, such as fire-fighters, police officers and soldiers.

Muscle work can be considered according to the concept where the two main elements are work and a worker (Figure 2). External load (stress, burden, demand, exertion, effort, cost) factors at work may be physical, psychological and social in nature. Physical factors include muscle work and environmental exposures. During a job performance load factors are transmitted various cardiorespiratory, musculoskeletal or neural loads through internal loading process. Organic load results in psychophysiological strain responses and strain outcomes, which depend on individual characteristics and organic tolerances of a worker. The most important individual characteristic relating to muscle work are gender, age, anthropometrics and functional capacity (Figure 2).

Muscle work in occupational activities

Muscle work in occupational activities can be roughly categorised as heavy dynamic work, manual materials handling, static postural work and repetitive work (Figure 3). Heavy dynamic work with large muscle groups consists mainly of activities requiring the moving of a worker's own body mass, and his or her physiological strain responses and strain outcomes are mostly cardiorespiratory (overall) in nature. The load of heavy dynamic work increases in relation to moving speed, distance, the degree of ascent at the covered distance, and the amount of a worker's own body mass as well as the additional mass of personal protective equipment which must be worn in many heavy physical tasks. Manual materials handling involves mixed dynamic and static muscle work with large muscle groups. The ordinary activities of manual materials handling are lifting, carrying, pulling and pushing of external loads of various weights and sizes. Muscle work of manual materials handling equally affects the cardiorespiratory and musculoskeletal system. Static postural and repetitive types of muscle work predominantly produce musculoskeletal (local) strain responses and strain outcomes (Figure 3).

The type of muscle contraction (dynamic or static) and the amount of muscle mass are very important factors as regards physical workload as well as physiological strain responses and strain outcomes. In addition, physical workload is greatly affected by the output of strength, the frequency of sudden peak load efforts, and work-rest regimens, as well as environmental exposures such as basic thermal parameters (ambient temperature, relative humidity, and air velocity), and work rate or the intensity of work.

Physiology of muscle work

Dynamic muscle work

In voluntary dynamic work, active skeletal muscles contract (shorten) and relax (lengthen) rhythmically and simultaneously. The shortening phase is called concentric contraction, while that of lengthening is called eccentric contraction. The blood flow to the muscles is increased to match metabolic needs as active muscles consume more oxygen and nutrients. The increased blood flow is achieved through increased pumping of the heart (cardiac output), decreased blood flow to inactive areas such as kidneys and liver, and an increased number of open blood vessels in the active muscles. Heart rate (HR), blood pressure, and oxygen consumption (VO_2) in the muscles increase linearly in relation to the amount of active muscle

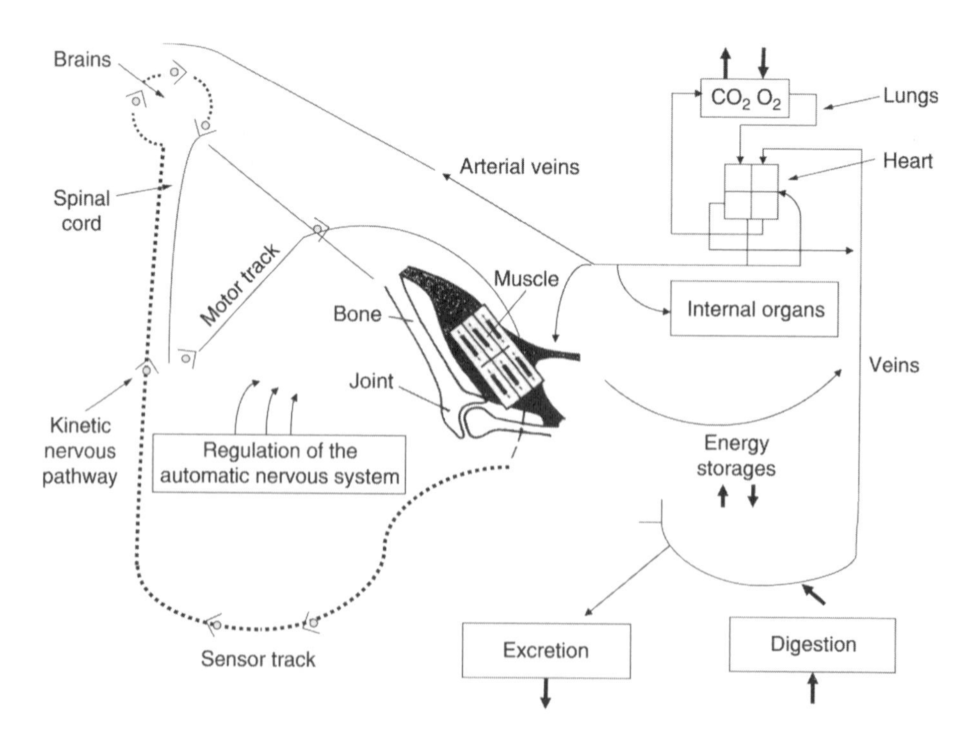

Figure 1 The simplified schematic description of the musculoskeletal, cardiorespiratory and neural systems needed for voluntary muscle contractions and output of strength, and their supportative and regulative functions.

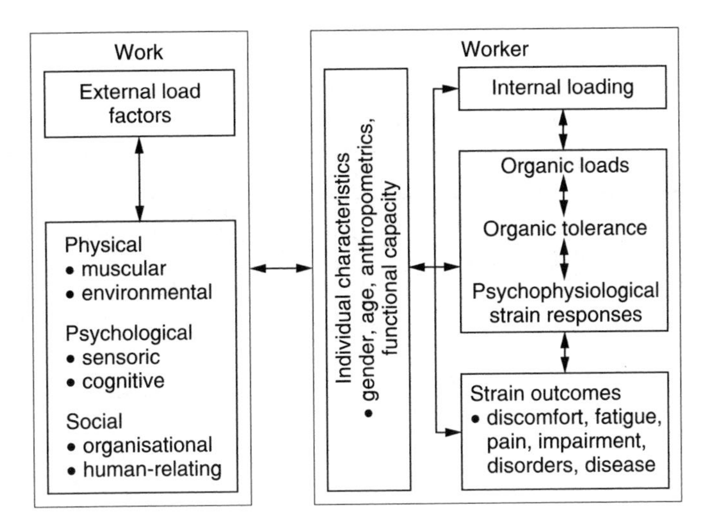

Figure 2 Conceptual model for the relationships of workload, and individual strain responses and outcomes. Individual characteristics and organic tolerances can be considered intervening factors that modify strain responses and outcomes due to various workload factors. Modified from the stress-strain concept by Rutenfranz (1981) and the model by CoBaSSaE (2001).

mass and working intensity. Also pulmonary ventilation is heightened owing to larger tidal volume in form of deeper breathing and increased breathing frequency. The purpose of activating the whole cardiorespiratory system is to enhance oxygen delivery to the active muscles (Figure 1). The level of VO_2 measured during dynamic muscle work indicates the intensity of work. The maximal VO_2 indicates a worker's individual cardiorespiratory capacity for dynamic muscle work (Figure 4).

In the case of dynamic work, when the active muscle mass is smaller (as in the arms and hands) the maximal VO_2 is smaller than in dynamic work with large muscle masses. At the same external work output, dynamic work with small muscles elicits higher cardiorespiratory responses (e.g., HR, blood pressure) than work with large muscles. The high-speed repetitive type of muscle work with small muscles such as fingers during handgrip exercise produced almost similar cardiorespiratory responses as the static muscle work with same small muscle groups.

Static muscle work

In static work, the muscle contraction produces no visible movement, for example in a limb. In the static (isometric) contraction, the muscle length is unchanged whereas the strength (power output) may be unchanged or change over a period of time varying from a few seconds to several hours. The static contraction may be called isotonic when muscle strength also remains unchanged. In intermittent work, dynamic or static muscle contractions are maintained for a period of a few seconds up to several minutes, interrupted by rest, and continued again.

Static work increases the pressure inside the muscle, which together with the mechanical compression occludes blood circulation partially or totally. The delivery of nutrients and oxygen to the muscle and the removal of metabolic endproducts from the muscle are hampered. Thus, in static work the strain outcome of fatigue is more easily met than in dynamic work.

The most prominent circulatory feature of static work is a rise in blood pressure. HR and cardiac output do not change much, particularly, if the static contraction is submaximal in

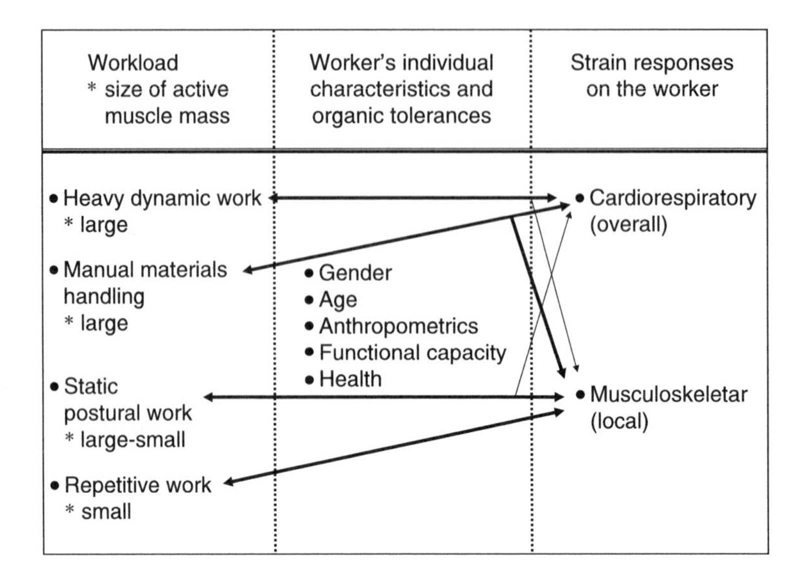

Figure 3 The effects of different types of physical workload on the cardiorespiratory and musculoskeltal strain responses. The individual characteristics and organic tolerances are considered as intervening factors influencing strain due to physical workload. Adapted from the model by Louhevaara (1992).

nature. Above a certain intensity of effort, blood pressure increases in direct relation to the intensity and duration of the effort. Furthermore, at the same relative intensity of effort, static work with large muscle groups produces a greater blood pressure response than does work with smaller muscle groups.

In principle, the regulation of pulmonary ventilation and circulation in static work is similar to that of dynamic work, but the metabolic signals from the muscles are stronger, and induce a different response pattern.

In occupational work, purely static contractions hardly ever occur, the most common being a low intensity contraction with small variations both in muscle length and strength.

The size of active muscle mass

The physiological responses to muscle work also depend on the size of the active muscle mass. The maximal VO_2 is higher in work tasks requiring large muscle groups than in those requiring smaller muscle masses. In young individuals, the maximal VO_2 during two-arm cranking exercise is about 70% of the maximal VO_2 during two-leg exercise. The maximal VO_2 for two-leg exercise declines progressively with age, whereas during two-arm exercise, and generally during exercises with small muscle masses, the differences between individuals due to age are smaller. At a given submaximal VO_2, HR, systolic blood pressure and pulmonary ventilation are higher during armwork than during legwork or combined arm- and legwork.

The maximal VO_2 values are usually based on dynamic leg exercise with large active muscle mass such as pedalling on a cycle-ergometer, walking or running on a treadmill or stepping up and down on a bench. The relative workloads are often expressed as percentage of the individual's maximal VO_2 attained during dynamic exercise with large muscle masses. The maximal VO_2 for arm work, however, cannot be estimated

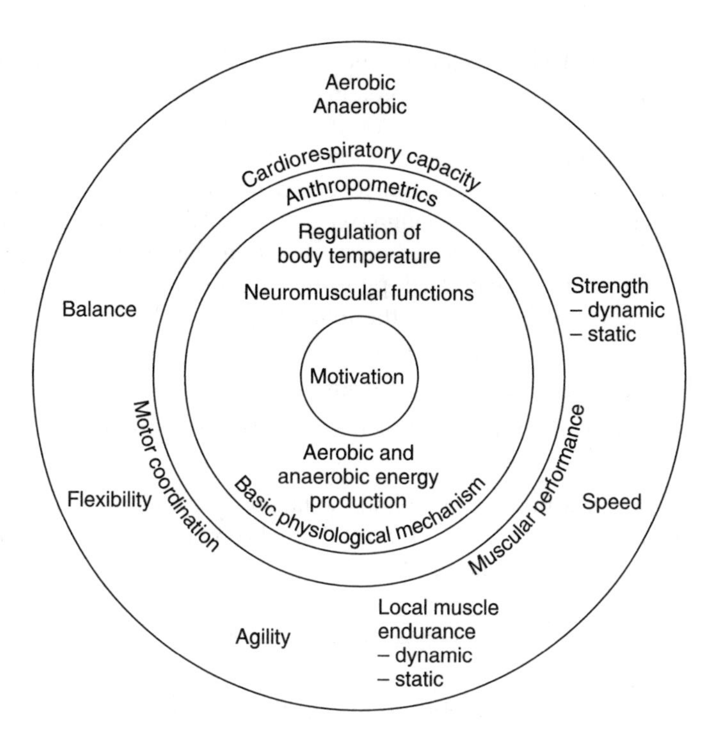

Figure 4 The main dimensions of physical functional capacity. The quality and quantity of output variables within cardiorespiratory capacity, muscle performance, and motor coordination depend on the basic physiological mechanisms affected by individual characteristics such as anthropometrics (Louhevaara, 1998).

from legwork, and vice versa. The maximal VO_2 varies greatly between arm and leg work. The external workload that is quite easy for legwork may be quite exhausting for arm work.

Physiological responses are specific in nature, and depend on the type of muscle work, the size of active muscle mass and the ratio of dynamic and static work. Therefore, the assessment of the maximal VO_2 should be done during muscular exercise that resembles those activities required during actual work. When the workload is related to the maximal VO_2 attained with the equal amount of active muscle mass, the differences in physiological responses become smaller, but are still higher, for instance, during prolonged armwork than during prolonged legwork.

Muscle metabolism during work

The brain controls voluntary skeletal muscles and sends neurosignals via kinetic pathways to produce muscle contractions and strength (Figure 1). Finally, a discharge of a nerve impulse on the motor endplate of the muscle is the signal for a fast conversion of chemically bound energy from the form of ATP (adenosine triphosphate), to adenosine diphosphate (ADP) and mechanical energy allowing voluntary muscle contractions as follows:

$$\text{ATP} \leftrightarrow \text{ADP} + \text{phosphate} + \text{energy}$$

Available stores of ATP are very limited, and spent in heavy dynamic muscle contractions or static contraction of a few seconds. Therefore, they have to be built up continuously from energy obtained by the oxidation of glucose and fatty acids. Protein is also metabolised but at a much lower rate. The main function of protein is to provide material for tissue repair and growth.

The metabolism leading to the release of energy can take place either aerobically (with oxygen) or anaerobically (without oxygen). Oxygen is transported to the muscles by the cardiorespiratory system (Figure 1). If enough oxygen is available the aerobic pathway is used because it is more efficient as it leads to a more complete metabolism of energy-rich nutrients and allows longer and more efficient work performance due to less fatigue. The form of the aerobic energy production with oxygen is the following:

$$\text{Glucose and free fatty acids} + \text{oxygen } (O_2)$$
$$\Rightarrow \text{carbondioxide } (CO_2) + \text{water } (H_2O) + \text{energy}$$

In anaerobic muscle work without oxygen, muscle fatigue occurs rapidly, probably due to the lowering of the acidity (i.e., pH-value of blood) that takes place when lactate is produced in anaerobic energy production as follows:

$$\text{Glucose} \Rightarrow \text{lactate} + \text{energy}$$

The breakdown of ATP is also anaerobic, but no lactate is produced.

Muscle work with the help of anaerobic energy production leads to the accumulation of lactate and muscular fatigue. The muscle work is always anaerobic in the following situations:

1. At the onset of dynamic work;
2. During heavy dynamic work when the energetic demands exceed 50% of the individual maximal VO_2;
3. During static work exceeding about 10% of the muscle strength during a maximal voluntary static contraction.

Some of the lactate produced during muscle work can be metabolised in the muscles and removed. However, during very heavy dynamic work or during static contraction, blood circulation cannot keep up with the demands on oxygen supply and lactate removal, and this leads to the accumulation of lactate, lowered pH, perception of fatigue and the reduction of maximal strength output and local muscle endurance. The endurance time for maintaining a static contraction decreases as the output strength of the contraction increases. The endurance time is relatively unlimited without perceived fatigue when the intensity of static contraction is no more than 5% of muscle strength during a maximal voluntary static contraction.

Cardiorespiratory system at work

The cardiorespiratory system, which allows the energy supply to the voluntary contracting muscles includes the heart, lungs, blood, blood veins, and nutrient storages (Figure 1). During muscle work the main functions of the cardiorespiratory system are to transport:

1. Oxygen (O_2) from lungs to muscles with blood flow,
2. Glucose and fatty acids from storage in the liver and fatty tissue to active muscles,
3. Carbon dioxide (CO_2), water (H_2O) and lactate from muscles to the lungs, liver and kidneys for excretion and metabolism, and
4. Heat from active muscles to the body surface.

In order to meet the demands of increased muscle activity the cardiorespiratory system can increase the transporting capacity by a factor of 100 within a few minutes after the onset of muscle work. This is mainly achieved by the increase of blood flow to active muscles, the increase of pulomonary ventilation and VO_2 and the exertion of carbon dioxide, and the increase of the total blood flow by increasing the cardiac output with larger stroke volume and higher HR.

Recovery after fatiguing dynamic muscle work depends on the intensity and duration of exercise, and the cardiorespiratory capacity of the worker (Figure 5). The rate of recovery depends to a large extent on the capacity of the blood circulation to supply oxygen to tissues (repayment of oxygen debt). Therefore, recovery has usually been studied by measuring HR before, during and after work reflecting relative workload intensity (Figure 5).

The recovery of muscle function may not he complete even though all cardiorespiratory variables have returned to initial levels. After static contractions HR may normalise, but on repeated contractions muscle strength and endurance may still be reduced. After prolonged static or eccentric muscle contractions to exhaustion full recovery may take several hours or even days. The most likely cause of the reduced performance capacity is damage in the structures of the muscles (including ruptures and oedema formation) and maybe also the depletion of energy-rich nutrients in the muscles.

Capacity for dynamic work

An individual worker's capacity to perform dynamic muscle work varies within very wide limits, and depends on his or her physical functional capacity, and particularly on cardiorespiratory capacity (Figure 4). Physical functional capacity in terms of cardiorespiratory (aerobic) capacity determined by the maximal VO_2, muscular performance (strength and endurance), and motor co-ordination (body control) is based on physiological mechanism: aerobic and anaerobic energy production, neuromuscular functions, and the regulation of body temperature. Anthropometric characteristics can be regarded as intervening factors in association with the output parameters of functional capacity. The utilisation of different functional capacities is done by means of voluntary muscle contractions, which are impossible without an adequate level of motivation (Figure 4).

Physical work capacity is based on functional capacity, If the worker's profile of functional capacity and his or her anthropometry are in harmony with the external physical workload factors, the situation is optimal or at least acceptable in terms of strain responses and outcomes (Figure 6). The balanced situation is neither static nor permanent, but highly dynamic, and depends on changes in workloads and individual capacities. Disturbances in the production process or the breakdown of automatic data processing systems may change physical workloads completely within a few seconds. Similarly, sudden illness or injury may decrease a worker's level of functional and work capacity dramatically. With the exception of these quick and often unexpected changes, the implementation of advanced technology and the unavoidable ageing of workers will affect the sensitive balance between physical workloads and physical work capacity.

In a dynamic muscle performance the principal individual determinant of work capacity is the maximal VO_2. It has a large individual variation mainly due to genetic factors, which influence functional capacity, anthropometrics, and basic physiological mechanisms. Directly measured maximal VO_2 is the highest consumption of oxygen per minute that can he achieved during dynamic exercise with large muscle groups. Physical inactivity, such as spending a few days in a bed, leads to a fast reduction of the maximal VO_2, i.e., detraining effects (Figure 6). Regular physical training gives nearly as fast an increase of the maximal VO_2. Thus the individual cardiorespiratory capacity is the effect of an adaptation process. The variations in average values for men and women in different age groups are presented in Table 1.

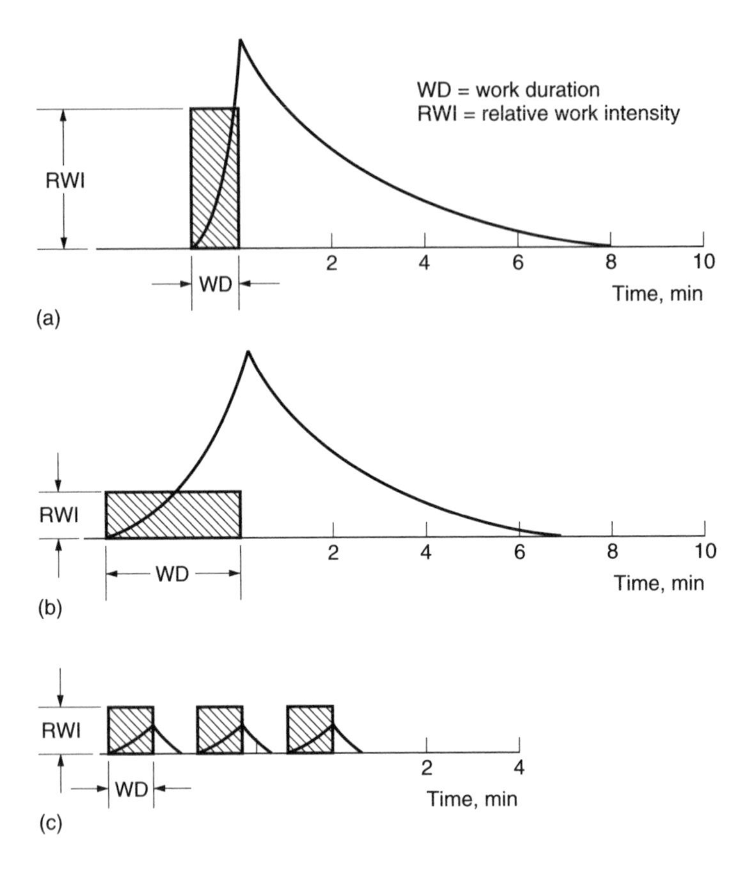

Figure 5 Schematic representation of the build-up of fatigue and recovery (solid line) in three types of work: (a) very high intensity work of one minute, followed by the recovery which in this case took around nine minutes, (b) work at the 33% level of the previous intensity, and performed for 3 minutes, followed by the recovery which took for around seven minutes, and (c) intermittent work consisting of three work periods at the 33% level of the initial intensity. The duration of each work period was for one minute, and followed by a rest period.That work-rest regimen resulted in a rapid recovery.

Strain outcomes due to muscular overload and underload

The physiological strain responses to muscle work depend on the type of muscle work individual characteristics and the tolerances of loaded organs (Figure 2). When the worker's physiological capacities are not exceeded by the muscle workload, the body will adapt to the load, and recovery is quick when the work is stopped. If the muscle load is too high, fatigue ensues, work capacity is reduced, and recovery slows down (Figure 5). Severe loads or prolonged overload may result in strain outcomes in terms of organ damages, injuries, work-related disorders and symptoms or even occupational diseases (Figure 2). On the other hand, dynamic muscle work of certain intensity, frequency, and duration may also result in training effects, just as underload in terms of excessive, low and prolonged dynamic muscle activity causes detraining effects (Figure 6).

Heavy muscle work is a risk factor for decreased work capacity, and work-related disorders and diseases. When poor health, impaired work capacity and subjective overload at work converge the risk is higher, and especially with older workers. Older women who are working in basic service occupations performing cleaning, home care, nursing, kitchen and

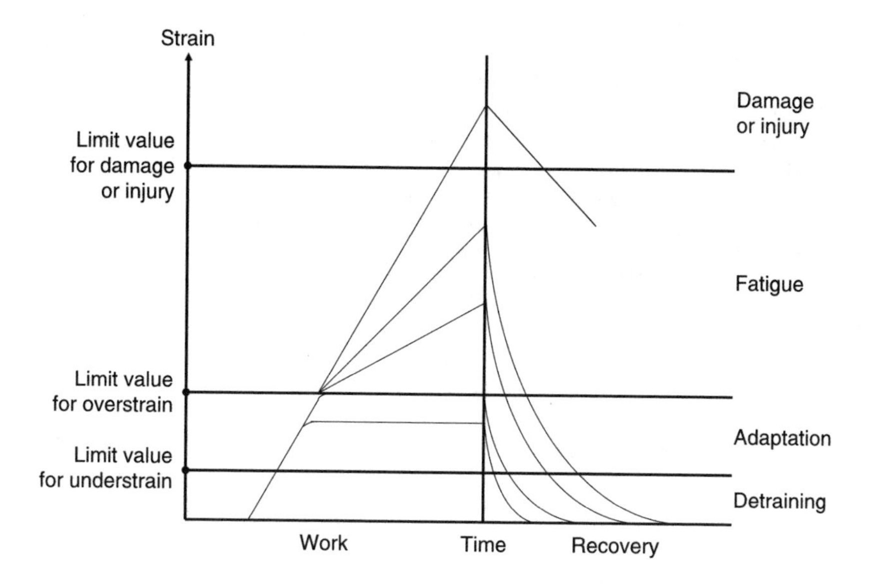

Figure 6 Schematic desription of the different levels of strain at work, and their possible strain outcomes such as organic damages due to overstrain or detraining effects due to understrain. Adapted from the model by Louhevaara (1992).

laundry work need specific attention. Many risk factors for work-related musculoskeletal disorders and diseases are connected to different features of muscle work. The negative strain outcomes may cause, particularly, sudden peak loads, heavy lifting, high exertion of strength, poor work postures, and repetitive muscle contractions at high-speed. Usually the risk will increase if more than one risk factor occurs at the same time at work.

Assessment of dynamic muscle work

Estimation of energy expenditure

Energy expenditure during dynamic muscle work is usually expressed power qualities such as Joules per second (J/s = Watt, W), kilojoules per minute (kJ/min) or VO_2 in litres per minute (l/min) or related to body mass (ml/min/kg). Also the energy expenditure at work can be expressed as multiples of basal metabolic rate which is 100% or one MET unit. For instance, during walking the energy expenditure may be about four times higher than that at complete rest. Then, the energy expenditure is about 350 W, 400% MET, four MET units or 20 kJ/min (Tables 1 and 2). That means that VO_2 is about 1.0 l/min or 14.3 ml/min/kg with an individual weighting 70 kg.

Energy expenditure due to muscle activity can also be estimated with scales based on average values of energy expenditure in different muscle activities of the whole body (Table 3) (Edholm, 1966) or body segments (Table 4) (Spitzer and Hettinger, 1969).

Assessment of VO_2

The rationale for assessing VO_2 is that the amount of oxygen consumed during aerobic muscle work is directly proportional to the amount of energy produced within the body. Thus, VO_2 directly measures dynamic workload in a job. During heavy and very heavy dynamic

Table 1 The Classification of Cardiorespiratory Workload According to the Average Values of Energy Expenditure (Watts, W) at Work for Men and Women, and their Maximal Oxygen Consumption (VO_2max) in Different Age Categories Expressed as Litres per Minute (1/min) and Watts (W) in Maximum (VO_2max and Wmax). The Cardiorespiratory Workload is Classified as Follows: Light <25% VO_2max, Moderate 25–50%VO_2max, Heavy 51–75%VO_2max, and Very Heavy >75% VO_2max (Andersen *et al.*, 1978). The Average Energy Expenditure is about 85 W at Rest Corresponding to the Oxygen Consumption (VO_2) of about 0.25 1/min

Men Age (years)	Average energy expenditure (W)				Wmax (W)	VO_2max (l/min)
	Light	Moderate	Heavy	Very heavy		
20–29	–229	300–579	580–872	873–	1163	3.44
30–39	–272	273–544	545–816	817–	1088	3.22
40–49	–258	259–495	496–747	748–	996	2.95
50–59	–223	224–440	441–663	664–	884	2.62
60–69	–174	175–349	350–523	524–	697	2.06
Women Age (years)	**Average energy expenditure (W)**				**Wmax (W)**	**VO_2max (l/min)**
	Light	Moderate	Heavy	Very heavy		
20–29	–223	224–356	357–488	489–	651	1.93
30–39	–202	203–293	294–454	455–	605	1.79
40–49	–188	189–279	280–419	420–	559	1.65
50–59	–154	155–265	266–384	385–	512	1.51
60–69	–133	134–244	245–349	350–	465	1.38

(energetic) work, energy expenditure and VO_2 are, on the average, about six-seven and nine-ten times higher that at rest (Tables 1, 2 and 3).

For most purposes, including manual work, VO_2 is a good indicator of physical workload, and the various factors which affect physical workload (Figure 7). Important exceptions are work tasks which include a heavy exposure of heat, much static muscle work, or other activities which demand the use of anaerobic metabolism. For such tasks VO_2 gives valuable information about the aerobic component, but the assessment of workload must be supplied with other measurements. For example, a short and very demanding athletic event such as a run of 400 meters cannot be evaluated only on the basis of VO_2. However, if the VO_2 is measured over the entire run and during recovery period the total requirement of energy for the run can be estimated by calculations (Table 2).

Because occupational work activities usually are maintained for several hours, it is very uncommon that they contain an appreciable degree of anaerobic exercise. Exceptions may be various emergency operations or extreme peak load situations met occasionally, for instance, by fire-fighters, police officers and soldiers. These operations are usually brief and time is given afterwards for recovery. However, such operations may he potentially hazardous, especially, when combined with high heat exposure and the use of heavy protective equipment. Therefore, individuals required to perform such operations should have a good physical work capacity, and the time for operations may have to be controlled in order to avoid heat stroke or accident caused by fatigue.

The assessment of VO_2 during work aims at evaluation of the dynamic physical workload in different tasks or during an entire workshift. The level of VO_2 is used to identify the most demanding tasks in a job. Often weaker members of the work team cannot do such tasks. The most strenuous tasks can be redesigned to make them less loading. For instance, job rotation

Table 2 Average Energy Expenditures and Oxygen Consumption (VO_2) in Different Occupational and Sport Activities

Activity	Energy expenditure (W)	Energy expenditure (kJ/min)	Energy expenditure (MET, %)	VO_2 (l/min)
Sleeping, rest	85	5	110	0.25
Light energetic work	−249	−14	−299	−0.7
Video display unit work				
Laboratory work				
Shop assistant work				
Installation of electronics				
Sorting of mail				
Slow walking with speed > 4 m/h				
Pedalling with load < 50 W				
Moderate heavy energetic work	250–399	15–24	300–499	0.7–1.1
Cleaning				
Home care				
Laundry work				
Reinforce work				
Installation of tiles				
Sorting of postal parcels				
Brisk walking with speed > 4 km/h				
Pedalling with load < 50 w				
Heavy energetic work	400–699	25–42	500–899	1.2–2.1
Loading of aircraft				
Pushing wheelbarrow with concrete				
Mail delivery by cycling				
Slow jogging with sped of 8 km/h				
Slow skiing with speed of 8–10 km/h				
Cycling with speed of 15–20 km/h				
Pedalling with load <150 W				
Very heavy energetic work	700–	43–	900–	2.2–
Intensive shovelling of snow				
High speed logging work				
Smoke diving in firefighting				
Running or skiing with speed of 12 km/h				
Cycling with speed of 25 km/h				
Most ball games				
Pedalling with load > 150 W				
Maximal energetic work	2000	120	2500	5.7
Maximal short-distance running* (400 m or 800 m)				
Pedalling with load of 400 W				

*At least half of the energy production is anaerobic.

or pauses can be introduced or the work pace can be reduced. With the help of VO_2 time demands of the alternative ways of the work performance can be compared. For example, tools and/or work routines can be improved. VO_2 can be used for the evaluation of the component of dynamic muscle work in a complex task. Many jobs have components of static and dynamic muscle work as well as heat exposure.

Table 3 The Edholm Scale for the Evaluation of Energy Expenditure at Work Expressed as Watts (W), Kilojoules per Minute (kJ/min), and Percentages from Basal Metabolic Rate (MET, %) (Edholm, 1996)

Scale	Type of muscle work	Energy expenditure		
		(W)	(kJ/min)	(MET, %)
1	Sleeping	85	5	110
2	At rest sitting or lying	95	6	120
3	Light sedendary work, standing without movement	120	7	150
4	Light work standing	235	14	300
5	Moderate physical work activity such as brisk walking	395	24	500
6	Heavy physical work activity such as climbing stairs	635	38	800
7	Very heavy physical work activity such as brisk running	790	48	1000

Table 4 The Evaluation of Average Energy Expenditure in Watts (W) at According to Work Movements and Postures (Spitzer and Hettinger, 1969). For Instance, the Average Energy Expenditure for the Standing Work with the Back Bent Forwards and Light Dynamic Work with Both Arms is $40+15+125+85/\text{W}=265\ \text{W}$

Muscle movements and postures	Average energy expenditure (W)
Legs	
Sitting	20
Standing	40
Squatting, kneeling	35
Walking	165
Upper body	
Elevation of shoulders	15
Elevation of arms	15
Back bent forwards	15
Upper body bending or twisting	85
Arms	
Dynamic work with hands and fingers	
Light	30
Heavy	65
Dynamic work with one arm	
Light	70
Heavy	125
Dynamic work with both arms	
Light	125
Heavy	180
Basal metabolic rate	85

Mechanical efficiency

Of the energy produced a large part is in the form of heat. The remaining energy production during work is used for performing the external workload such as lifting a box. The lifting requires the output of muscle strength and moving of arms and legs.

$$\text{Mechanical efficiency} = \frac{\text{Energy for external work}}{\text{Total production of energy}}$$

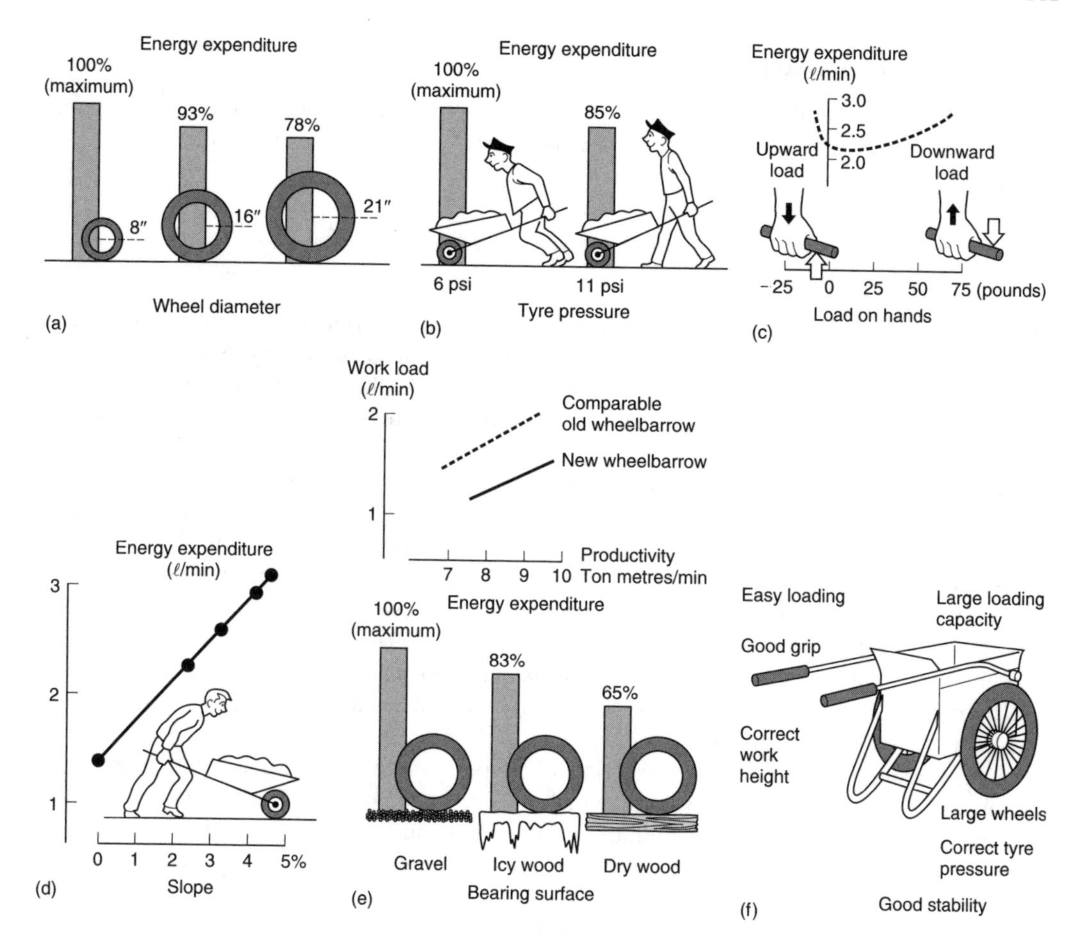

Figure 7 Energy expenditure in percentages of the maximum (100%) or in litres per minute (l/min), and the productivity of work during wheelbarrowing. Large differences in the energy expenditure were obtained with changes in (a) wheel diameter, (b) tyre pressure, (c) positioning and grip of handles, (d) slope, and (e) structure of the terrain. The optimal design and good work conditions (f) improved the productivity of wheelbarrowing by 40% without increasing energy expenditure (Hanson, 1970).

The mechanical efficiency during different kinds of muscle activities varies from 0 to 40%. During static work, for example, no external work is performed and no movements take place, and therefore the mechanical efficiency is 0%. In most everyday and occupational activities mechanical efficiency varies between 0 and 20%. In the basic modes of dynamic physical work such as cycling, walking, cranking, and carrying, VO_2 varies very little between individuals who perform the same type of dynamic work. In work tasks where an individual's own body mass is carried, as in walking, there will be some differences between individuals due to variations in body mass. If these differences are not extreme, measurements obtained in a small group of individuals can he used in the assessment of the energy, which describes dynamic physical workload in that situation. The more complex and skill demanding a physical performance is, the more VO_2 varies between individuals. Consider, for example, the difference in VO_2 between horse riders. VO_2 of a skilled rider is less than that of an expert in horse riding. The unskilled rider uses additional muscle contractions for balancing on horseback and for preparation for unexpected actions from the horse.

Measuring and analysing VO_2

VO_2 is calculated after the measurement of the volume of expired or inspired air in a certain time period. Then, expired is analysed for its content of oxygen and carbon dioxide. The equation is the following:

$$VO_2 = VE(CO_2 - CO_2e)$$

VO_2 is oxygen consumption in litres per minute, VE is expired pulmonary ventilation in litres per minute, CO_2i is concentration of carbon dioxide in inspired air, and CO_2e is concentration of carbon dioxide in expired air. VO_2 is usually expressed in litres per minute in standardised STPD conditions when temperature is 0°C, barometric pressure 760 mmHg, and relative humidity is 0, i.e., the sample is dry.

The volume of expired air can be either measured or calculated from the volume of inspired air. The volumes of inspired and expired air are not the same, as expired air volume is slightly expanded through heating and the addition of carbon dioxide and humidity. Therefore, the volume of inspired air must be corrected according to differences in temperature, humidity, and the concentration of carbon dioxide. Thus, a complete calculation of VO_2 includes the analysis of the volume of expired air (pulmonary ventilation), and the concentrations of oxygen and carbon dioxide, and the STPD standardisation. The concentration of oxygen and carbon dioxide in ambient air at sea level is constant being 20.94% and 0.04%, respectively.

The measurement of VO_2 requires relatively expensive and complex instrumentation. For the laboratory, fully automated systems which measure the volume of pulmonary ventilation directly or via airflow, and concentrations of oxygen and carbon dioxide in expired air, ambient temperature, current barometric pressure and humidity are available. The recordings can be done with the intervals of 60, 30, 15 or 5 seconds or breath by breath. Large errors in the calculation of VO_2 can be introduced if the volume of expired air is not correctly measured, or if the air sample for oxygen analysis is mixed with ambient air. Therefore the system for collection and analysis must fit tightly to the individual's mouth via a mouthpiece while a nose clip is used to eliminate leakage from the nose. A full-face mask can also be used but it is less reliable due to risk of leakage and a larger dead space inside the mask.

For field studies automated equipment is also available. The simple devices only analyse the volume of expired or inspired air and the concentration of oxygen, making assumptions for other factors. Thus, a small error is introduced, but within the most common range of VO_2 this error is usually less than 5%. The modern devices measure VO_2 in the field as reliably as those in the laboratory. In the field, however, the work situation and/or environment may be very demanding in terms of the reliability and feasibility of the direct VO_2 assessments.

Sometimes, a simple measurement of pulmonary ventilation can be used to predict VO_2 instead of the complex procedure required for the measurement of VO_2. During a light to moderately heavy dynamic muscle work the volume of pulmonary ventilation is closely related to VO_2. During heavy work when VO_2 is more than 2.0 l/min and pulmonary ventilation above 40–50 l/min, pulmonary ventilation is an unreliable predictor of VO_2.

Assessment of physical work capacity

The maximal VO_2 can be accurately measured during dynamic exercise with large muscle groups such as during cycling, uphill walking, running and skiing. Other physical activities which may produce maximal or near maximal VO_2 are rowing and swimming, but such activities are not often used in the tests of the maximal VO_2 for practical reasons.

Usually the maximal VO_2 is assessed during exercise on a cycle-ergometer or on a treadmill. There are some advantages for the use of the cycle-ergometer instead of the treadmill. The test person is more stationary than on the treadmill and therefore all measurements are performed more easily and accurately. The mechanical efficiency during cycling varies little between individuals. At a given external workload on the cycle-ergometer it can be expected to yield very similar values for VO_2 in a group of individuals with different size, cardiorespiratory capacity, age or gender (Figure 8).

On the treadmill, however, uphill walking or running at a certain speed will give a larger range of VO_2 for a group of individuals because of their varying body masses. Older individuals usually experience maximal effort on the cycle-ergometer to be less frightening than on the treadmill.

The advantage of treadmill exercise is that maximal VO_2 is up to 10% higher than on the cycle-ergometer. Moreover, young and fit individuals usually prefer tests on the treadmill.

Protocols for cycle-ergometer tests

The maximal VO_2 can be assessed according to direct or submaximal test protocols on the cycle-ergometer. The conventional way to test the maximal VO_2 is to use incremental submaximal workloads of 3–5 minutes and one maximal workload without pauses between the workloads. The submaximal workloads are chosen to correspond to 35–40%, 50–60%, and 70–80% of the individual's predicted maximal VO_2. The prediction may be done according to gender and age (Table 1). In practice, this often coincides with external workloads of 25–50 W, 75–100 W and 100–125 W for women, and 50–75, 100–125

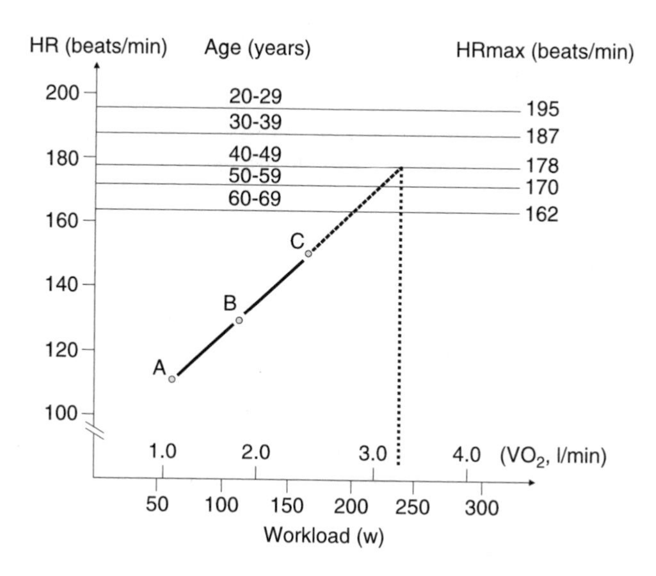

Figure 8 The prediction of maximal oxygen consumption (VO_2max) according to the submaximal exercise test on a cycle-ergometer. In this case, the prediction is based on the three-point (A, B, and C) extrapolation method. Each point represents the measured value of heart rate (HR) at a given external workload which can be expressed as the energy expenditure in Watts (W) or as the oxygen consumption in litres per minute (l/min). The age-specific maximal heart rate (HRmax) can be predicted according to age. In the present case the estimated VO_2max is about 3.1 l/min.

and 150–175 W for men. For small or old as well as well-trained individuals, the test protocols must be adjusted.

The use of submaximal workloads serves several purposes. The investigator can observe the individual's responses to exercise, register pulmonary ventilation, HR, blood pressure, and subjective reactions and experiences, and, thereby, more easily estimate which maximal workload should be used. Contraindications to increased workloads can also be identified, i.e., symptoms and physiological responses indicating cardiovascular, pulmonary or musculoskeletal problems or diseases. Furthermore, submaximal workloads serve as a warm-up exercise, and they increase muscle temperature. Thereby, the exchange of oxygen and carbon dioxide is facilitated, and the risk of sprains and injuries is decreased through improved muscle co-ordination.

At each submaximal workload VO_2 should not he measured until during the last 1–2 minutes as it takes for 1–2 minutes from the beginning every workload before VO_2 to stabilises at a steady state. With the use of the conventional test protocol the maximal workload is chosen for exhausting the test person completely within 2–8 minutes as the total duration of the test may be about 20 minutes. Usually VO_2 is measured from two to three samples after the onset of maximal exercise, and the highest value obtained is used as the maximal VO_2.

There are numerous different test protocols for measuring the maximal VO_2 on the cycle-ergometer. Today the physiological responses can be continuously monitored and registered using the on-line techniques and fast computers. The test protocol may be quite short, and the increase of external workload can be done, for instance, every or every second minute. The main criteria are that the true maximal VO_2 can be reached before the local fatigue of legs or the onset of other problems related to the use of a mouthpiece or mask.

Protocols for treadmill tests

On the treadmill the workload is usually gradually increased through increments of slope and speed of the treadmill. Usually a test person is tested for 2–3 min on each workload without pauses between, and VO_2 is measured during the last minute of each workload. As the test person approaches exhaustion, continuous measurements are made for the assessment of the maximal VO_2. A number of different protocols for treadmill tests are available. If VO_2 is not measured in the test, it can be estimated according to the slope and speed of the treadmill. VO_2 is related to the body mass because it greatly affects VO_2 values during exercise on the treadmill.

Alternative test protocols

An alternative to cycle-ergometer and treadmill exercise tests is the so-called step test, which may be suitable in field studies, where sometimes no other equipment for testing are available. For instance, in the Harvard step test, the physical work capacity is estimated by letting the subject step up and down a bench at a given pace. In modified step tests the height of the bench may be varied to suit to the estimated maximal cardiorespiratory capacity of an individual in the test. Incremental test protocols, using submaximal and maximal work rates, have also been described. Step tests, however, are difficult to standardise, measurements of HR are often difficult to perform, and the tests often lead to muscle soreness. In summary, step tests using any profiles tend to give biased results.

Submaximal exercise testing based on HR

During submaximal dynamic exercise at a constant external workload, HR increases during the first 1–3 minutes and then reaches a steady state level. Steady-state HR values are linearly related to external workload and VO_2 (Figure 8). An individual with low maximal VO_2 will have his or her HR–VO_2 relationships shifted to the left, and a well-trained individual with a high maximal VO_2 will have it shifted to the right. This means that the HR of unfit individuals is much higher than that of well-trained individuals at the same workload and VO_2.

The relationship between HR and VO_2 is the basis for the method to predict maximal VO_2 from submaximal HR values. With the measurement of HR, for instance, at three to five incremental submaximal workloads, the maximal VO_2 can be predicted by the extrapolation to the maximal HR (Figure 8). The extrapolation method requires knowledge of the maximal HR, which gradually decreases with age, from around 220 beats/min in young children to around 160 beats/min at the age of 60 years. Also the continuous monitoring of HR is needed so that the increase of workloads would correspond the cardiorespiratory capacity of a test person. It is essential that HR would be at the highest submaximal workload — 80–90% — of the age-related maximal HR. However, there is also a relatively large variation in maximal HR between individuals, even of the same age. Thereby an error of about 10% is introduced in the prediction of maximal VO_2, compared with direct measurements. In order to avoid some other factors which influence HR, submaximal tests should not be performed within one hour after smoking or eating a large meal, and the air temperature should be around 18°C. A test participant should be in light sports wear with sneakers.

Contra-indications to exercise testing

Before both submaximal and maximal exercise tests the test participants must be asked about cardiorespiratory disease, musculoskeletal disorders, acute infections and medications, and physical activity. Acute infections and chest pain are absolute contraindications, whereas previous cardiac insufficiency, angina pectoris, myocardial infarctions and hypertension are relative contraindications. In such cases exercise testing must only be done under the monitoring of an electrocardiograph (ECG) done by trained medical and laboratory personnel, and with resuscitation equipment available. Symptoms like chest pain, excessive breathlessness and certain arrhythmia that occur during exercise are indications to stop the test immediately. In individuals above 40 years of age it is advisable to register ECG before, during and after the exercise test. With the above precautions the health risks of exercise testing are small.

HR during work

HR increases during both dynamic and static exercise, during heat exposure, and as an effect of psychological stress. Besides muscle work HR is greatly affected by individual cardiorespiratory capacity i.e., the maximal VO_2. Thus a HR increase is a unspecific cardiorespiratory strain response, and the interpretation of HR recordings must always be made against a background knowledge of the circumstances of the recording and individual characteristics. Other factors that can influence HR are tobacco smoking, certain types of medication, infections, drugs, and the lack of sleep.

HR as a strain measure during occupational work

The continuous measurement of HR during work is a common method to evaluate cardiorespiratory strain. The measurements are relatively simple to perform, and the results are usually reliable. The most commonly used system is telemetry where the ECG signal is registered via chest electrodes and transmitted to a receiver. The receiver identifies the R-waves of the ECG signal and stores them in a microprocessor, where the number of beats is counted for given time periods (60, 30, 15 or 5 seconds). The receiver can be positioned either on the wrist of the subject or in an external station, where, for instance, the signals from several transmitters, using different frequencies, can be stored. The circuits of the receiver can be constructed to identify the R-wave with good accuracy, so that no artefacts are recorded. Another method is to record the ECG signal continuously using miniaturised tape recorders. This also permits analysis of ECG such as arrhythmia and so is often used for clinical purposes.

The long-term recordings of ECG also allow the analysis of HR variability, which refers to the amount of HR fluctuation around the mean HR. HR variability can be assessed by calculating indices based on statistical operations on RR (beat to beat) intervals using so called time domain analysis. The sympathetic and parasympathetic nervous systems are the primary mediators of HR modulations during exercise or set physical work. HR variability seems to describe several physiological parameters such the level of cardiorespiratory capacity, the maximal HR, and the degree of systemic fatigue (e.g., Tulppo, 1998).

HR during work or exercise can be recorded manually with reasonably good accuracy if no specialist measuring devices are available. Usually the work tasks to be studied are interrupted regularly at intervals of 1–5 minutes and the time for 10–15 beats is recorded with a stopwatch.

HR is a very unspecific strain response. Therefore, during occupational work measurements of HR must be supplemented with activity observations of the worker. The observations should indicate the type of physical activity, psychologically stressful situations, simultaneous heat exposure, and other factors that might affect HR. Often the investigation is supplemented with measurements of VO_2, skin and core temperatures of the body, the ratings of perceived reactions and experiences and environmental measurements.

Analyses of HR recordings during work

The occupational HR recordings should be supplemented with a submaximal or maximal exercise test. The purpose of this testing is to expose the subject to a standardised external workload or activity for reference, and also to measure or estimate the maximal VO_2. At least some of the exercise workloads should be designed so that they resemble the target type of occupational work. For example, if the occupational tasks studied are mainly performed when walking, the exercise test should preferably include testing at one or two walking speeds on the treadmill. If the upper body is mainly used in occupational tasks, arm crank exercise might be most appropriate in the test.

The analyses of HR usually include the calculation of average HR over the entire workshift or for different work tasks. This average value is related to VO_2 or external workload at the same HR level attained in the standardised exercise test. This will permit an estimation of the proportion of the maximal VO_2 used at work. Identification of the most demanding tasks or peak load situations can be done according to HR levels. Further analyses comprise comparisons between the distribution of HR values for the studied job or work task and those previously reported in other jobs.

The HR analyses can be used to estimate the severity or the amount of dynamic muscle load of the work tasks in terms of calculated proportion of maximal VO_2 and cardiorespiratory strain.

Acceptable physical work load according to VO_2 and HR measurements

The percentage relation of VO_2 measured during work to the maximal VO_2 of the worker measured in the laboratory is defined as the relative aerobic strain (RAS) as follows:

$$\text{RAS} = \%\text{ of the maximal } VO_2 = \%VO_2\text{max}$$

$$= \frac{(VO_2 \text{ at work})}{(VO_2 \text{ max})} \times 100 \quad \text{or} \quad \frac{(VO_2 \text{ at work} - VO_2 \text{ at rest})}{(VO_2 \text{ max} - VO_2 \text{ at rest})} \times 100$$

Usually the VO_2 at rest is considered so small that it is left out from the equation.

If only HR measurements are available, two indices can be calculated as follows:

$$\%\text{ of the maximal HR} = \%\text{ HRmax} = \frac{(\text{HR at work})}{(\text{HRmax})} \times 100$$

or

$$\%\text{ of the maximal HR range} = \%\text{HRR} = \frac{(\text{HR at work} - \text{HR at rest})}{(\text{HRmax} - \text{HR at rest})} \times 100$$

The maximal HR can be measured directly in an exercise test or taken from the age-specific tables. The following equations are commonly used:

$$220 \text{ beats/min} - \text{age in years} = \text{age-specific maximal HR} \quad \text{or}$$
$$205 \text{ beats/min} - 1/2 \times \text{age in years} = \text{age specific maximal HR} \quad \text{or}$$
$$210 \text{ beats/min} - 0.7 \times \text{age in years} = \text{age specific maximal HR.}$$

At the age of 30 years all these equations give almost equal predictions of the maximal HR as 189–190 beats/min.

The percentage of the HR range (%HRR) is closely related to $\%VO_2$max (Table 5).

$\%VO_2$max should not exceed 50% during an 8-hour workshift. In the experiments at $50\%VO_2$max level, body mass decreased, HR did not reach steady state, and subjective discomfort increased during prolonged exercise periods. Therefore, the $50\%VO_2$max level is considered the ultimate upper level for an 8-hour workshift supplemented with regular pauses

Table 5 The Physiological Responses to the Maximal Exercise Test on a Cycle-Ergometer with a Male Test Person Aged 30 years. % VO_2max= =RAS=Relative Anerobic Strain, %HRmax=Percentage of the Maximal Heart Rate, %HRR=Percentage of the Functional Range of Heart Rate

Workload (W)	VO_2 (l/min)	HR (beats/min)	$\%VO_2$ max*%	$\%VO_2$ max†	%HRmax%	%HRR
50	1.0	90	29	23	45	21
100	1.5	110	43	38	55	36
150	2.1	150	60	57	75	64
200	2.8	180	80	78	90	86
250, max	3.5	200	100	100	100	100

*Without VO_2 values at rest.

† With VO_2 values at rest.

of 10–15 minutes during each hour. Most often the 30–40%VO_2 max levels are recommended for dynamic muscle work of 8 hours. The lowest level should be implemented when the frequency and duration of rest pauses are inadequate.

The accepted %VO_2max levels were developed for purely dynamic muscle work, which rarely occurs in actual work situations. It may happen that acceptable %VO_2max levels are not exceeded, for instance in a lifting task, but the local load on the back may greatly exceed acceptable levels.

Usually the maximal VO_2 is measured on a cycle-ergometer or on a treadmill, in which the mechanical efficiency is high (20–25%). When the active muscle mass is smaller the static component is higher. Then the maximal VO_2 and the mechanical efficiency are smaller than in exercise with large muscle groups. For example, in sorting of postal parcels, the maximal VO_2 was only 65% of the maximal values measured on a cycle-ergometer, and the mechanical efficiency was less than 1%. When guidelines for acceptable levels of dynamic load are based on VO_2, the test mode in the maximal test should be as close as possible to the actual work task. This goal, however, is difficult to achieve. Despite its limitations, the determination of %VO_2max has been widely used in assessing acceptable dynamic workloads in different jobs.

In Table 5 the %VO_2max, %HRmax and %HRR values are compared in a test person aged 30 years. He has maximal VO_2 of 3.5 l/min, and maximal HR of 200 beats/min. His HR at rest is 60 beats/min. The test person performed a maximal exercise test on a cycle-ergometer. The comparison of the values shows that the %VO_2max values and %HRR values are quite similar at different load levels. The %HRmax values differ from others particularly at the submaximal workloads because the scale is from 0 to 100, and HR at rest covers 30% of the maximal HR. The corresponding proportion using VO_2 values is only 7%.

Analysis of HR during recovery

If no automatic system for HR measurement is available, and if the work cannot be interrupted for manual recordings, the cardiorespiratory strain can be estimated using recovery HR measurements done according to the Brouha method (Brouha, 1960). The method has been claimed to be even more reliable than HR measurements during exercise. In spite of that, HR should always be measured during actual work, supplemented with observations of work activites whenever it is possible.

According to the Brouha method a worker is seated immediately after the termination of a work task, HR is measured during the recovery period of 3 minutes. Recovery HR measurements can be performed repeatedly during a workshift in order to evaluate whether the workload is too high, and whether recovery is complete between the different tasks. The practical procedure can be used for evaluating work and recovery. HR during the first, second and third minute after exercise is obtained by counting the beats during the last 30 s of each minute, and doubling the values to express HR for each minute (beats/min). The average HR is the mean of these three values (HR_1, HR_2, and HR_3), and is highly correlated with total cardiorespiratory strain as follows:

1. If HR_1–$HR_3 \geq 10$ beats/min, or if HR_1, HR_2, and HR_3 are all below 90 beats/min, then recovery is normal;
2. If the average HR_1 obtained from the number of recordings is ≤ 110 beats/min, and HR_1–$HR_3 \geq 10$ beats/min, the workload is not excessive; and
3. If HR_1–$HR_3 < 10$ beats/min, and if $HR_3 > 90$ beats/min, then the recovery is inadequate with respect to workload.

The absolute level of HR at work, the work-rest regimens, and the cardiorespiratory capacity of the worker influence the rate of the recovery HR. In addition, heat exposure influences the rate of recovery. If recovery is unsatisfactory, the work must be redesigned in such a way as to reduce the physical load. The intensity of work is usually difficult to influence, especially in industrial tasks where machines set the work pace. The cardiorespiratory capacity of the worker can be affected but often the behavioural change in the form of increased physical activity is temporary. With well-designed, intensive and regular training, improvements of 10–20% for the maximal VO_2 can be achieved. The best measure for securing sufficient recovery is usually to limit the duration of each task or, if this is not possible, to increase the duration of the pauses between tasks. Short recovery pauses at high and regular frequency is recommended instead of a few long breaks (Figure 5).

Rating of perceived exertion

There is a curvilinear relationship between the intensity of a range of physical stimuli and the perception of their intensity. A positively accelerating relationship has been found between physical workload and perceived exertion. Thus, there is a highly reproducible individual relation between e.g., workload on a cycle-ergometer and perceived exertion. In the rating of perceived exertion (RPE) scale, the scale steps have been adjusted so that the ratings from 6 to 20 are linearly related to HR divided by ten (Table 6). The scale is presented to the test person before the start of the exercise test and the endpoints of the scale (6 and 20) are thoroughly defined. The scale is then shown to the test person at the end of each exercise load and he or she is asked to rate the perceived exertion. The verbal explanations are used as support information. Also a non-linear scale from 1 to 10 with ratio properties is available for assessments in the laboratory or field.

These scales of perceived exertion can be used to supplement physiological measurements during exercise and occupational work. They often provide valuable additional information about subjective responses, especially in cases where HR values are unreliable for some reason, where HR responds to external muscle load too weakly or strongly. Similar subjective scales have also been developed to quantify the intensity of pain.

RPE scales can be used to evaluate perceived exertion in actual work situations. The idea of substituting physiological measurements by perceived evaluations is attractive, as ratings do

Table 6 The Rating of Perceived Exertion with the Scale From 6 to 20 (Borg, 1970)

Rating	Intensity of exertion
6	
7	VERY, VERY LIGHT
8	
9	VERY LIGHT
10	
11	LIGHT
12	
13	FAIRLY HARD
14	
15	HARD
16	
17	VERY HARD
18	
19	VERY, VERY HARD
20	

not require any instrumentation. However, findings in industry and in industrial tasks show that ratings are not only influenced by the overall perception of exertion, but also by previous experiences and motivation of the workers. Thus highly motivated workers may tend to underestimate their exertion and their physical workload.

Body temperature during heat exposure and exercise

Body temperature increases during exercise, since a large proportion of the energy produced is converted to heat. To some extent the exercise performance benefits, since the metabolic processes work faster in higher temperatures. Thus, the core or deep temperature of the body is adjusted according to relative workload. The adjustment is accurate although it is slow and takes at least 30 minutes. For example, during prolonged exercise corresponding to 30% of maximal VO_2, the core temperature of the body is about 38.0°C at temperate conditions.

Human tissues have a limited tolerance to high or low temperatures. The core temperature should be close to 37.0°C, and the skin temperatures should not be over about 40°C or under about 15°C. The production of heat may be too high due to intensive dynamic muscle work, heat exposure or the use of heavy clothing. Then, the additional heat must be dissipated. In low environmental temperatures, radiation and convection are the main means of heat dissipation, whereas sweating is the only possibility in high environmental temperatures. With an ambient air temperature above 37°C, or in intense radiant heat, external heat is transferred to the body and must also be dissipated through sweating. Since sweating demands increased blood circulation to the skin, heat exposure puts large demands on the cardiorespiratory system. Hence muscle work in hot environments induces increments in HR far above those obtained at the same work in a cold environment (see Chapter 23).

Prolonged heavy sweating also leads to loss of fluid, which can severely reduce the blood volume and thereby further increase the cardiovascular strain. A loss of 1% of body mass through sweating leads to deterioration of physical work capacity and reduced orthostatic tolerance. The World Health Organisation recommends that sweat production should not exceed 4 liters in an 8-h workshift. Thus, heavy physical work in hot environments imposes comprehensive stress on the worker. If the production of heat cannot be balanced with the dissipation of heat, the core temperature of the body increases with a concomitant risk of heat exhaustion and heat stroke. The core temperatures above 38.0°C indicate that this balance may be upset, and the termination temperature during exercise testing is usually 39°C.

Through environmental measurements of wet bulb temperature, radiant heat, and air velocity, a commonly used wet bulb globe temperature (WBGT) index can be calculated. The WBGT can be implemented to predict the risk of heat exhaustion during work or exercise at different intensities. A more accurate prediction can only be obtained through physiological measurements of HR, core and skin temperatures, and fluid balance.

Conclusions

Work and thermal physiology measurements have a long history of contributions to ergonomics. They provide evidence-based indications of the dynamic muscle loads of workers. Well-documented and reliable limits can be used to ensure that the physical load and strain of workers is not too high. On the other hand, few limits are available for the consideration of very low workloads, which may hamper physical work capacity in the long run. In actual work situations, there are many individual and environmental confounding factors, which may bias physiological responses to dynamic muscle work and hamper their interpretation.

Perceived ratings give a rough comparison with the physical measurements, and allow the recordings of subjective experiences at work. In work situations where a single extreme workload is acting, one valid measurement could be appropriate. Most work situations are complex, and several measurements of different aspects of muscle work are needed to understand how to re-design work and work environments with ergonomics interventions. Furthermore, the psychosocial stress factors of work may be more relevant and important to investigate than physical load factors.

References

Andersen, K.L., Rutenfranz, J., Masironi, R. and Seliger, V. (1978). *Habitual Physical Activity and Health.* (Copenhagen: World Health Organisation).

Andersen, K.L., Shephard, R.J., Denolin, H., Varnauskas, E. and Masironi, R. (1971). *Fundamentals of Exercise Testing.* (Geneva: World Health Organisation).

Åstrand, I. (1960). Aerobic work capacity of men and women with special reference to age. *Acta Physiologica Scandinavia (Suppl. 169)* 49.

Åstrand, P.O. and Rodahl, K. (1986). *Textbook of Work Physiology* (New York: McGraw-Hill).

Borg, G. (1970). Perceived exertion as an indicator of somatic stress. *Scandinavian Journal of Rehabilitation and Medicine* **2**, 92–98.

Borg, G., Ljunggren, G. and Cegi, R. (1985). The increase of perceived exertion, aches and pain on legs, heart rate and blood lactate during exercise on a bicycle ergometer. *European Journal of Applied Physiology and Occupational Physiology* **54**, 343–349.

Brouha, L. (1960). *Physiology in Industry* (Oxford: Pergamon).

CoBaSSaE (2001). *Panel on Musculoskeletal Disorders and Workplace.* National Research Council and the Insitute of Medicine. *Musculoskeletal Disorders and the Workplace: Low Back and Upper Extremities.* (Washington, D.C: National Academy Press).

Edholm, O.G. (1966). The assessment of habitual activity. In: Evang, K. and Andersen, K.L. (eds.) *Physical Activity in Health and Disease*, (Oslo: Universitetbolaget).

Hansson, J.F. (1970). *Ergonomics in the Building Industry.* Research Report No. 8. Byggforskningen, State Council of the Building Industry.

Louhevaara, V. (1992). Cardiorespiratory and muscle strain during manual sorting of postal parcels. *Journal of Occupational Medicine—Singapore* **4**, 9–17.

Louhevaara, V. (1995). Assesment of physical load at work sites: A Finnish–German concept. *Internal Journal of Occupational Safety and Health* **1**, 144–152.

Louhevaara, V. (1999). Job demands and physical fitness. In: W. Karwowski and W.S. Marras (eds.) *The Occupational Ergonomics Handbook.* (Boca Raton: CRC Press), pp. 261–273.

Louhevaara, V., Smolander, J., Aminoff, T. and Ilmarinen, J. (1998). Assessing physical work load. In: W. Karwowski and E. Salvendy (eds.) *Ergonomics in Manufacturing. Raising Productivity through Workplace Improvement.* Society of Manufacturing Engineers. Engineering & Mangement Press, Norcross, pp. 121–133.

Punakallio, A., Lusa-Moser, S. and Louhevaara, V. (2001). Fire-fighting and rescue work in emergency situations and ergonomics. In: Karwowski, W. (ed.) *International Encyclopaedia of Ergonomics and Human Factors*, Volume I. (London: Taylor & Francis) pp. 449–453.

Rutenfranz, J. (1981). Arbeitsmedizinische Aspekte des Stressproblems. In: J.R. Nitsch (ed.), *Stress: Theorien, Untersuchungen, Massnahmen.* (Bern-Stuttgard-Wein, Verlag Hans Huber) pp. 379–390.

Spitzer, H. and Hettinger, Th. (1969). Tafeln fur den Kalorienumsatz bei körperlicher Arbeit. (Berlin, Köln & Franfurt am Main: Beuth-Vertrieb-GmbH).

Smolander, J. and Louhevaara, V. (1998). Muscular work. In: J.M. Stellman (ed.) *Encyclopaedia of Occupational Health and Safety*, 4th ed. (Geneva, Internationak Labout Office), pp. 1:29.28-29.31.

Tulppo, M. (1998). Heart rate dynamics during physical exercise and during pharmacological modulation of autonomic tone. *Acta Universitatis Ouluensis.* D Medica 503. Oulun Yliopisto, Oulu (doctoral dissertation).

Chapter 16

Static muscle loading and the evaluation of posture

E. Nigel Corlett

Introduction

The maintenance of postures and the support of loads are particular examples of the performance of static work. Although these are both quite common, what can be overlooked and present particular difficulties for the analyst are those cases where postures and other physical activities intermingle. If one or the other situation predominates it may be sufficient to assess the whole situation on the basis of its worst aspect, but if serious levels of effort are required whilst posture is maintained, this procedure is unsatisfactory.

In general we can say that, whilst the limitations on dynamic physical activities felt by a person would be high heart rate and a shortage of breath, the limits to static work will be the experience of muscular pain. As described in Chapter 15 this arises particularly from the anaerobic metabolic activity of the muscles, whose blood supply is restricted due to the increased intramuscular pressure. A consequence of this pressure is that the heart rate does not represent the static effort involved, although post-effort heart rate can be an important indicator of the existence of static load (see Brouha, 1960).

If we are to evaluate posture and static work, therefore, we must have some overview of the major contributors to static workloads. Five dimensions relevant to the definition of a posture are given below, from which, by suitable measures, we can deduce key components which contribute to the loads experienced.

1. the angular relationship between body parts.
2. the distribution of the masses of the body parts.
3. the forces exerted on the environment during the posture.
4. the length of time that the posture is held.
5. the effects on the person of maintaining the posture.

Measurement of some, or all, of the above five dimensions helps us to define the following:

1. the stability of the person in that posture.
2. estimates of muscle loads and joint torques.
3. estimates of fatigue levels and recovery times (which may require some additional measures).
4. comparisons of conditions against criteria, e.g., to assess hazards in load handling.

Although this chapter recognises the above structure, as with so much else in this book it cannot be too strongly emphasised that what is chosen for measurement from the first list depends on what it is we want to know from the second list. Thus for those wishing to use the information given here, a review of the whole chapter is desirable before selecting any particular methods. It should also not need emphasis that the physical and psychophysical measurements to be described are not all that contribute to people's experience of workloads.

Methods for the direct measurement of the effort involved in posture, as well as its effects, are less common than for dynamic work. The range of methods embraces estimation techniques (biomechanics and estimates from maximum voluntary contraction tests); measures of muscular activity (analysis of the electromyographic-EMG-signal); measures of the resultant effects (e.g., spinal shrinkage); subjective measures (e.g., discomfort recordings) and a wide range of interpretive methods. These last range from epidemiological studies to estimates of likely effects from posture recordings (using such techniques as OWAS, RULA, the NIOSH formula, the Nordic questionnaire or posture targetting) and posture measurements (e.g., goniometers, SELSPOT or CODA). The measurement of posture and its effects is more extensively discussed in Corlett *et al.* (1986).

Posture recording

As a first thought, it may be proposed that photographs or videos, perhaps in pairs of views orthogonal to each other, would be enough to record postures. It is true that for some situations this is adequate, if for example just a visual record is needed or some particular angles are to be measured. In other cases, however, it will be realised that, although the posture may have been recorded using a video or by still photography, the data for analysis still have to be retrieved from the recorded images. Where accuracy is needed, problems of parallax in a plane transverse to the optical axis of the lens can be overcome, but inaccuracies along the optical axis can be considerable.

However, rarely is a record of posture of use on its own; it is necessary also to have data on task activities, the loads moved, conditions of the people concerned, and the workplace. It is fitting, therefore, before continuing with a number of methods for recording postures, to comment on the wider aspects of postural loading. It is also appropriate to note that postural data are also frequently used in biomechanical analysis, and Chapter 17 on biomechanics should be consulted to assess what data are required, so that the method most appropriate to the need is selected.

The longer term effects of posture and work activities, whilst under less control by the investigator, are of great importance. Epidemiological methods are outside the scope of this text but the ergonomist has need of data from epidemiologists and occupational health professionals and is sometimes in a position to contribute data to them.

Direct observation methods

The measurement of the angles between body parts, or their angles in relation to the environment, are frequently required. For biomechanical analysis accuracy should, in general, be high. There are several procedures available which assess postures in relation to their contributions to discomfort, strain, stability or force exertions and for these a wider margin of error can often be acceptable.

OWAS

Although work study (methods engineering) offers several charts and symbol systems for recording work activities, these do not embrace the whole posture. One of the earliest whole posture coding systems for industrial use was developed in Finland to investigate the working postures in a steelworks. The company, Ovako Oy, in conjunction with the Finnish Institute of Occupational Health (1992) developed the OWAS method. Postures are observed, and recorded as shown in Figure 1(a). An accompanying assessment sheet, similar to Figure 1(b), enables each posture to be assessed for acceptability, or else for appropriate remedial action.

The OWAS code for a posture comprises a record of the posture itself in the first three figures, the load or force used is indicated in the fourth figure and a record of the stage in the cycle or task is recorded in the fifth figure. The procedure is to glance at the work to take in the posture, force and work phase, then to look away and record it. Thus the work activities can be sampled, and from these samples estimates can be made of the proportions of time during which forces are exerted or postures held.

The assessment sheet of Figure 1(b) allows the assessment of the likely musculo-skeletal load even for a single posture combination of back, arms and legs, with the action categories. Where the activity is frequent, though the load is light, the sampling procedure permits the estimation of the proportions of time the limbs or back are spent in the various working postures. These postures may then be evaluated for adequacy by using the table of Figure 1(c), where action categories for the various postures in relation to their various times of use during the working day are given.

It is evident that these measures could be taken from video recordings and it is recommended that video should always support direct observations. The usual precautions when using activity sampling should be followed, such as not using a sampling period which is a factor of the task cycle time, although for non-cyclic work equal intervals of thirty or sixty seconds can be suitable. Observers require training to a standard which ensures that their recordings are consistent and that their observation errors are at an agreed level; under 10% is desirable. Regular testing of users is also desirable if confidence in the system is to be maintained.

It will be evident that OWAS seeks to identify postures which put the body in positions where force exertions can be dangerous. Balanced and symmetrical postures are the ones which achieve the lowest scores, and are seen as acceptable. Pushing, pulling or moving loads when the person is twisted, or the body is in other ways asymmetrically loaded give the highest scores and are recommended for change.

Procedures such as OWAS have some limitations. They take limited account of external loads, or time pressures, for example, and were devised for heavy industrial work, with broad categories in their posture and force estimations. Their great strength is the facility they provide for the rapid identification of most of the major inadequate postures. Furthermore, as they are easy to learn and use, they can be used by a wide spectrum of the workforce, and alert people to those aspects of activity which may be hazardous. The presentation of results in diagrams, which show the proportion of the harmful categories in each part of a person's job, on a time base, can illustrate forcefully the conditions of the job. Measurements after changes have taken place, and subsequently, will demonstrate the benefits and monitoring the job over years is facilitated by the relative simplicity of the process.

REBA

Whilst OWAS is valuable in other than heavy industries and has been widely used, workers in health care found the need for a tool which would deal with the additional factors. Examples of these are the use of parts of the body other than the hands to take some of the load when assisting people and to recognise the likelihood of instability in the handler. These are all factors which arise in health care, as well as in work activities in other industries.

Using the proven utility of the OWAS format and the experience of RULA (see later), Hignett and McAtamney (2000), developed REBA, Rapid Entire Body Assessment, to

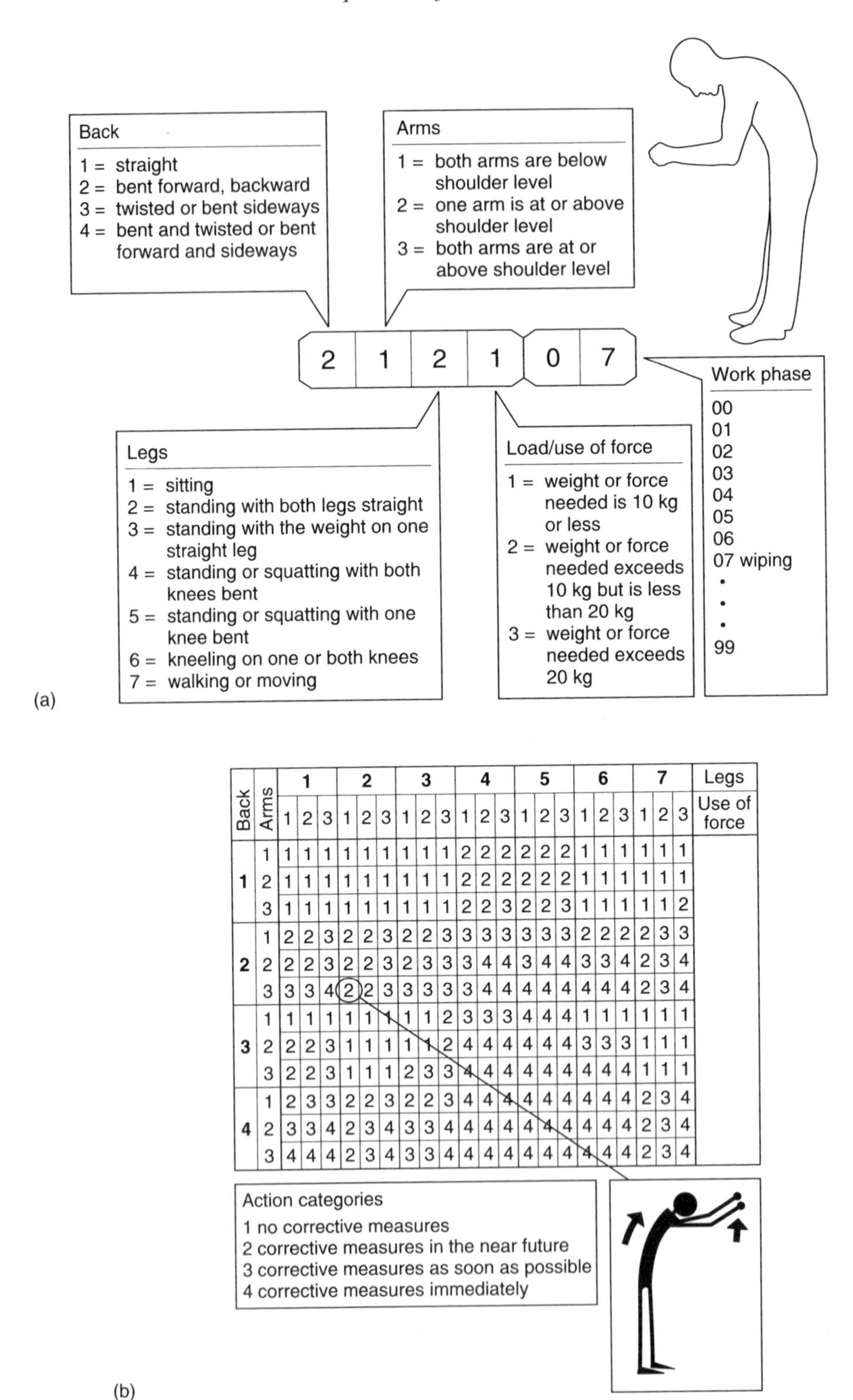

Back	Arms	1			2			3			4			5			6			7			Legs
		1	2	3	1	2	3	1	2	3	1	2	3	1	2	3	1	2	3	1	2	3	Use of force
1	1	1	1	1	1	1	1	1	1	1	2	2	2	2	2	2	1	1	1	1	1	1	
	2	1	1	1	1	1	1	1	1	1	2	2	2	2	2	2	1	1	1	1	1	1	
	3	1	1	1	1	1	1	1	1	1	2	2	3	2	2	3	1	1	1	1	1	2	
2	1	2	2	3	2	2	3	2	2	3	3	3	3	3	3	3	2	2	2	2	3	3	
	2	2	2	3	2	2	3	2	3	3	3	4	4	3	4	4	3	3	4	2	3	4	
	3	3	3	4	2	2	3	3	3	3	3	4	4	4	4	4	4	4	4	2	3	4	
3	1	1	1	1	1	1	1	1	1	2	3	3	3	4	4	4	1	1	1	1	1	1	
	2	2	2	3	1	1	1	1	1	2	4	4	4	4	4	4	3	3	3	1	1	1	
	3	2	2	3	1	1	1	2	3	3	4	4	4	4	4	4	4	4	4	1	1	1	
4	1	2	3	3	2	2	3	2	2	3	4	4	4	4	4	4	4	4	4	2	3	4	
	2	3	3	4	2	3	4	3	3	4	4	4	4	4	4	4	4	4	4	2	3	4	
	3	4	4	4	2	3	4	3	3	4	4	4	4	4	4	4	4	4	4	2	3	4	

Figure 1 (a) Items of the OWAS method and an example code for a particular task. (b) Action categories in the OWAS method for work posture combinations. (c) Action categories in the OWAS method for work postures according to percentage of use over the work period.

Back	1 straight	1	1	1	1	1	1	1	1	1	1
	2 bent forward	1	1	1	2	2	2	2	2	3	3
	3 twisted	1	1	2	2	2	3	3	3	3	3
	4 bent and twisted	1/2	2	2	3	3	3	3	4	4	4
Arms	1 both arms below shoulder level	1	1	1	1	1	1	1	1	1	1
	2 one arm at or above shoulder level	1	1	1	2	2	2	2	2	3	3
	3 both arms at or above shoulder level	1	1	2	2	2	2	2	3	3	3
Legs	1 sitting	1	1	1	1	1	1	1	1	1	2
	2 standing with both legs straight	1	1	1	1	1	1	1	1	2	2
	3 standing with one leg straight	1	1	1	2	2	2	2	2	3	3
	4 both knees bent	1/2	2	2	3	3	3	3	4	4	4
	5 one knee bent	1/2	2	2	3	3	3	3	4	4	4
	6 kneeling	1	1	2	2	2	3	3	3	3	3
	7 walking	1	1	1	1	1	1	1	1	2	2
	% of working time	0	20		40		60		80		100

Action categories
1 no corrective measures
2 corrective measures in the near future
3 corrective measures as soon as possible
4 corrective measures immediately

(c)

Figure 1 Continued.

deal with these problems. For recording and analysis the body is divided into segments, which are dealt with individually and then combined into two groups, the arms, and the rest of the body. Muscle activity is included in the assessment. The position of the arms, as well as the posture of the head, arms and legs, are recorded according to the diagrams in (Figure 2). The appropriate tables, Table A for the body or Table B for the arms (see Figure 3), allow postures and loads to be combined, including the particular types of load as experienced in health care, to give a score for each group.

These scores are then combined (see score sheet in Figure 4) to give a single score which, when linked with a score dealing with the activity involved (Table C in Figure 3) results in a final score.

This score is assessed for its indicated severity by considering the proposed action levels in Figure 5. These levels are the result of decisions by a number of health care professionals. Applications of REBA are given in Ferreira and Hignett (2003), where the postures of paramedics in ambulances were studied. Workplace changes were sufficiently effective that they were adopted by ambulance services across Britain. A review (McAtamney and Hignett, 2004) covers both the method and some applications.

RULA

This procedure was developed by McAtamney and Corlett (1993) in a form analogous to OWAS, to assess the exposure of people to postures, forces and muscle activities known to contribute to upper limb disorders (ULD). This Rapid Upper Limb Assessment (RULA) technique uses observations of postures adopted by the upper limbs, the neck, back and legs, classifying them according to the values given on charts A and B of Figure 6. After recording the values representing the observed postures in the first column of the score sheet, Figure 7, Tables A and B (Figure 8) are used to obtain a posture score for the A and B body groups. Values for muscle use and loads are then extracted from the tables of Figure 9 and also entered into their appropriate spaces on the score sheet.

Trunk

Movement	Score	Change score:
Upright	1	+1 if twisting or side flexed
0°–20° flexion 0°–20° extension	2	
20°–60° flexion >20° extension	3	
>60° flexion	4	

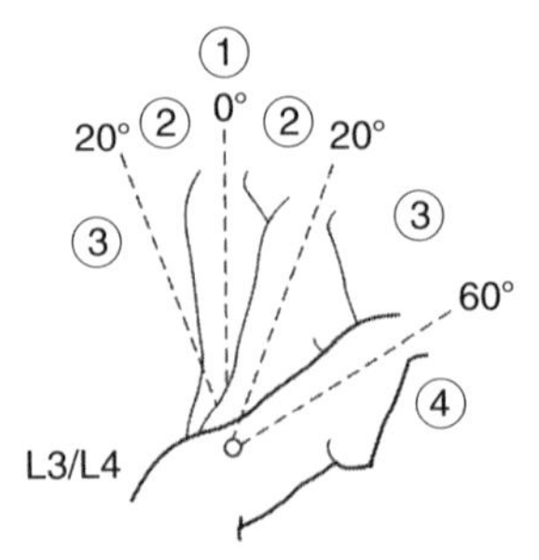

Neck

Movement	Score	Change score:
0°–20° flexion	1	+1 if twisting or side flexed
>20° flexion or in extension	2	

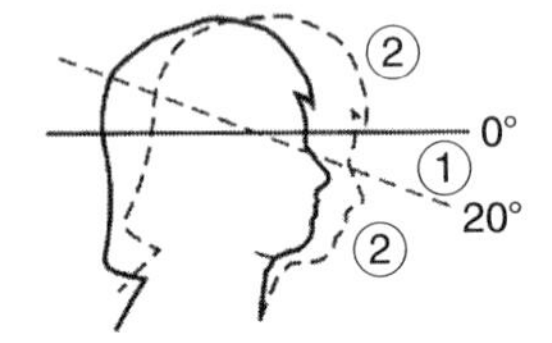

Legs

Position	Score	Change score:
Bilateral weight bearing, walking or sitting	1	+1 if knee(s) between 30° and 60° flexion
Unilateral weight bearing Feather weight bearing or an unstable posture	2	+2 if knee(s) are >60° flexion (n.b. not for sitting)

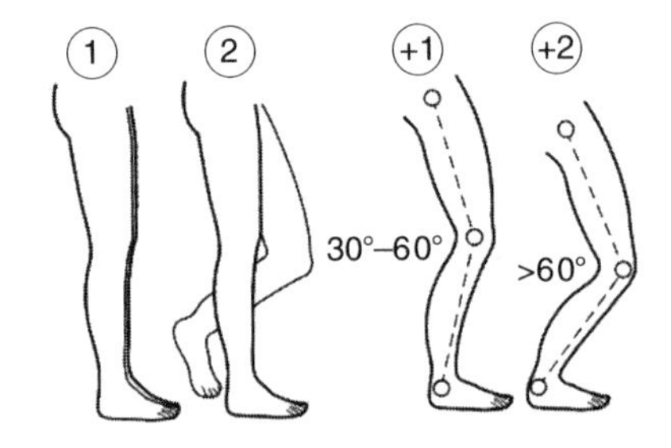

Upper arms

Position	Score	Change score:
20° extension to 20° flexion	1	+1 if arm is: • abducted • rotated
>20° extension 20°–45° flexion	2	+1 if shoulder is raised
45°–90° flexion	3	–1 if leaning, supporting weight of arm or if posture is gravity assisted
>90° flexion	4	

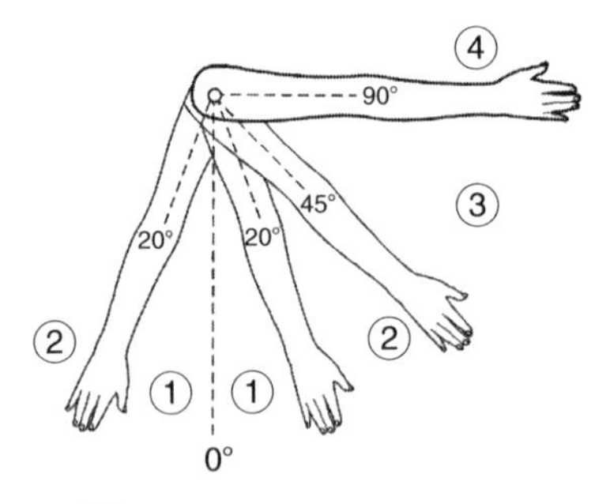

Lower arms

Movement	Score
60°–100° flexion	1
<60° flexion or >100° flexion	2

Wrists

Movement	Score	Change score:
0°–15° flexion/ extension	1	+1 if wrist is deviated or twisted
>15° flexion/ extension	2	

Figure 2 Group A and B body part diagrams.

Table A

Trunk		Neck 1				Neck 2				Neck 3			
	Legs	**1**	**2**	**3**	**4**	**1**	**2**	**3**	**4**	**1**	**2**	**3**	**4**
1		1	2	3	4	1	2	3	4	3	3	5	6
2		2	3	4	5	3	4	5	6	4	5	6	7
3		2	4	5	6	4	5	6	7	5	6	7	8
4		3	5	6	7	5	6	7	8	6	7	8	9
5		4	6	7	8	6	7	8	9	7	8	9	9

Load/force

0	1	2	+1
<5 kg	5 – 10 kg	>10 kg	Shock or rapid build-up of force

Table B

Upper arm		Lower arm 1			Lower arm 2		
	Wrist	**1**	**2**	**3**	**1**	**2**	**3**
1		1	2	2	1	2	3
2		1	2	3	2	3	4
3		3	4	5	4	5	5
4		4	5	5	5	6	7
5		6	7	8	7	8	8
6		7	8	8	8	9	9

Coupling

0 Good	1 Fair	2 Poor	3 Unacceptable
Well fitting handle and a mid-range power grip	Hand hold acceptable but not ideal or coupling is acceptable via another part of the body	Hand hold not acceptable although possible	Awkward, unsafe grip, no handles Coupling is unacceptable using other parts of the body

Table C

Score A	Score B 1	2	3	4	5	6	7	8	9	10	11	12
1	1	1	1	2	3	3	4	5	6	7	7	7
2	1	2	2	3	4	4	5	6	6	7	7	8
3	2	3	3	3	4	5	6	7	7	8	8	8
4	3	4	4	4	5	6	7	8	8	9	9	9
5	4	4	4	5	6	7	8	8	9	9	9	9
6	6	6	6	7	8	8	9	9	10	10	10	10
7	7	7	7	8	9	9	9	10	10	11	11	11
8	8	8	8	9	10	10	10	10	10	11	11	11
9	9	9	9	10	10	10	11	11	11	12	12	12
10	10	10	10	11	11	11	11	12	12	12	12	12
11	11	11	11	11	12	12	12	12	12	12	12	12
12	12	12	12	12	12	12	12	12	12	12	12	12

Activity score

• +1	• 1 or more body parts are static e.g. held for longer than 1 minute
• +1	• Repeated small range actions e.g. repeated more than 4 times per minute (not including walking)
• +1	• Action causes rapid large range changes in posture or an unstable base

Figure 3 Tables for the body (A) and arm (B) for REBA, and combined table (C).

The scores C and D are then found by adding the separate scores as shown in Figure 6. From these the Grand Score is found from table C of Figure 10. Enter scores C and D into the boundaries of the diagram and note the value where row and column intersect. Appropriate action requirements for the different scores are given at the bottom of Figure 10.

In RULA, as in OWAS and REBA, the higher the code number, at any stage of the analysis, the further does the part concerned depart from a desirable posture. So changes concentrate on reducing the magnitude of the individual numbers, which in turn reduce the total score. Thus the analyst has some guidance regarding where to introduce changes.

OWAS, RULA and REBA are easily learned and give consistent and reasonable accuracy. Evidently their effectiveness depends on the expertise of their users in understanding the ergonomics of the working situations under study and the technical possibilities for changes. As with other methods given later in this chapter, their effectiveness can be increased by sound work analysis (see Drury, 1987) and also by incorporating individual techniques into a more comprehensive intervention programme (McAtamney and Corlett, 1992).

For investigating upper-limb activities, Armstrong *et al.* (1982) illustrate a procedure using data from a single camera. The upper-limb posture is coded, according to the diagram of Figure 11(a), at equal, short, intervals through a number of work cycles. Hand grip forces should be obtained where these are noted as forceful. The observations can be checked off in columns, one for each posture, and then plotted on a time base, as in Figure 11(b). Right and

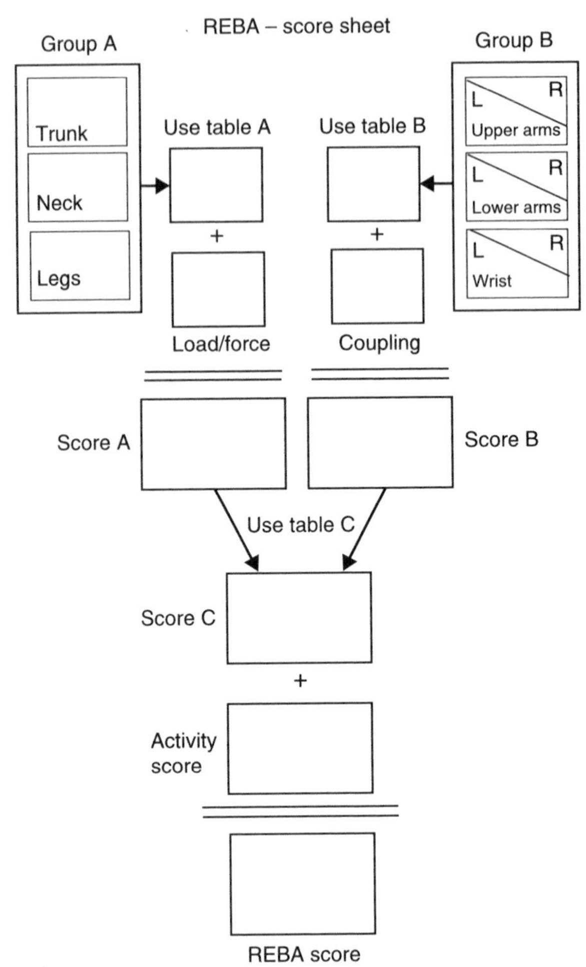

Figure 4 REBA scoresheet.

Action levels for REBA

Action level	REBA score	Risk level	Action (including further assessment)
0	1	Negligible	None necessary
1	2 – 3	Low	May be necessary
2	4 – 7	Medium	Necessary
3	8 – 10	High	Necessary soon
4	11 – 15	Very high	Necessary NOW

Figure 5 Action levels for REBA.

left arms, or values before and after a change, can be shown on the one plot, thus providing comparisons in a convenient form. Such a detailed analysis will demonstrate where adverse postures and forces are maintained or frequently repeated. An analysis after changes have been introduced will then show the improvements to postural loadings and any other benefits from the intervention.

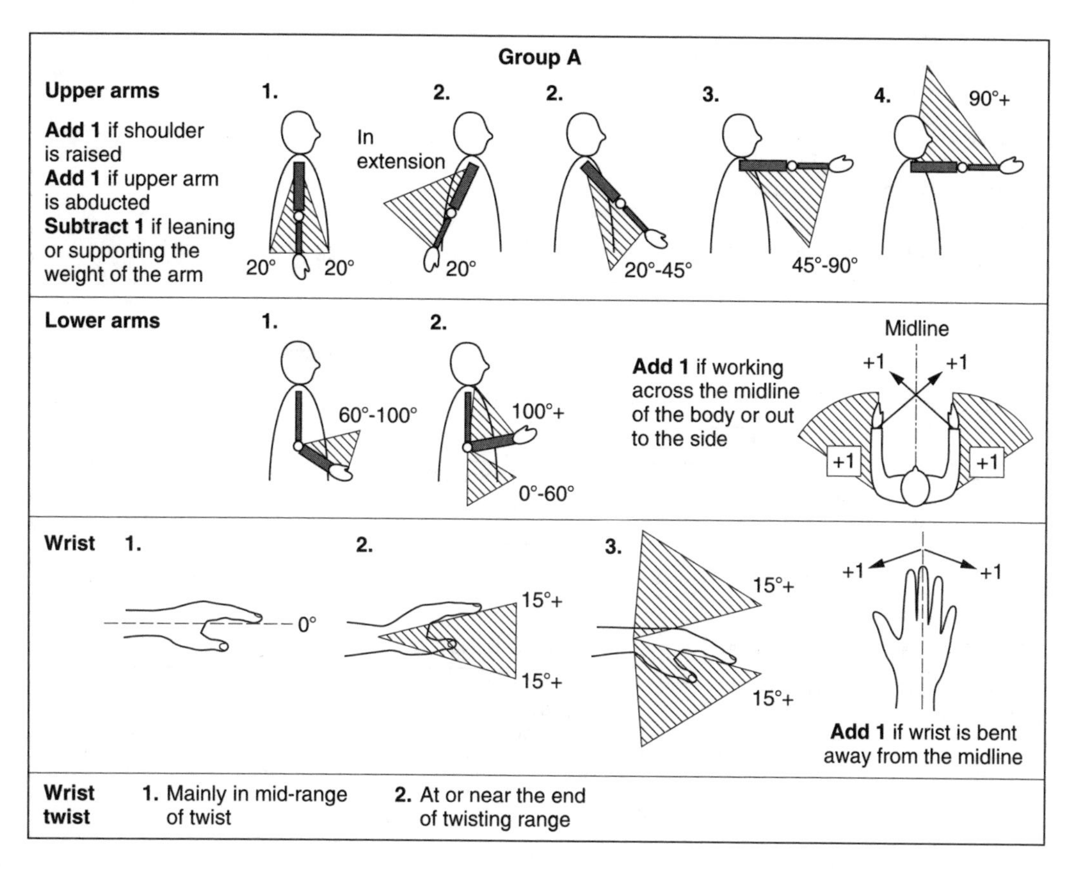

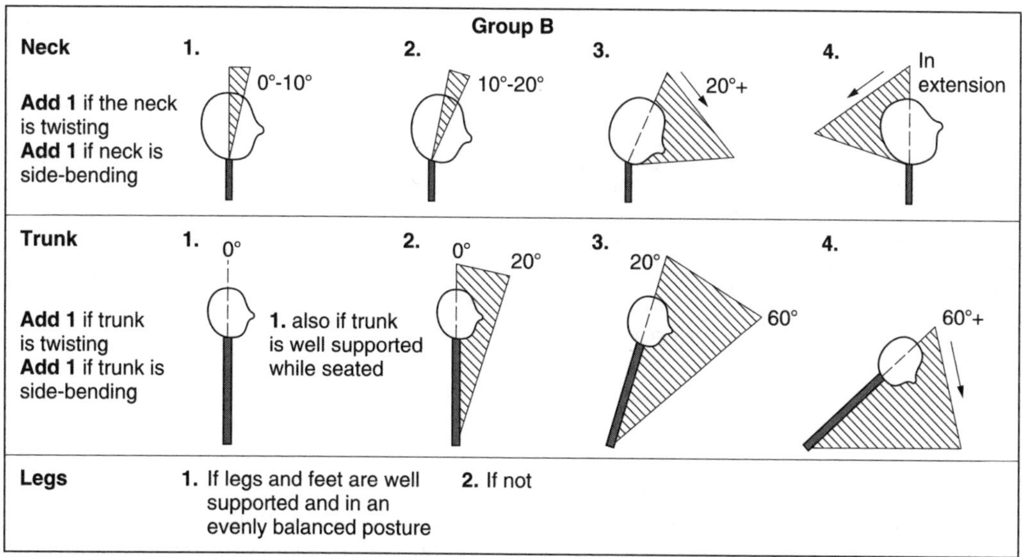

Figure 6 Items for assessment when using the RULA method.

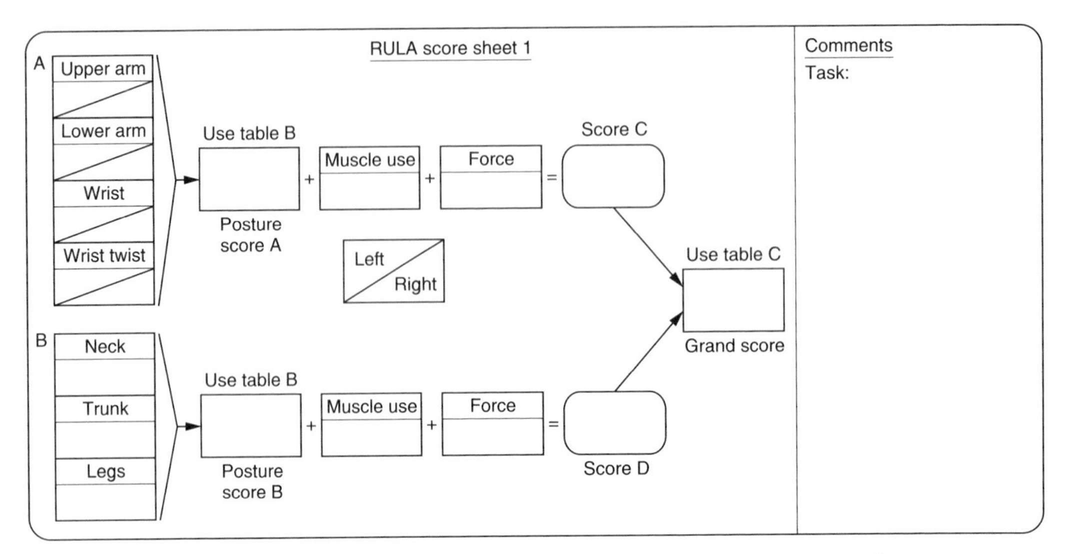

Figure 7 Score sheet for recording observations, using diagrams shown in Figure 6.

Table A Upper limb posture score

Upper arm	Lower arm	Wrist posture score 1		2		3		4	
		Twist 1	Twist 2	Twist 1	Twist 2	Twist 1	Twist 2	Twist 1	Twist 2
1	1	1	2	2	2	2	3	3	3
	2	2	2	2	2	3	3	3	3
	3	2	3	3	3	3	3	4	4
2	1	2	3	3	3	3	4	4	4
	2	3	3	3	3	3	4	4	4
	3	3	4	4	4	4	4	5	5
3	1	3	3	4	4	4	4	5	5
	2	3	4	4	4	4	4	5	5
	3	4	4	4	4	4	5	5	5
4	1	4	4	4	4	4	5	5	5
	2	4	4	4	4	4	5	5	5
	3	4	4	4	5	5	5	6	6
5	1	5	5	5	5	5	6	6	7
	2	5	6	6	6	6	6	7	7
	3	6	6	6	7	7	7	7	8
6	1	7	7	7	7	7	8	8	9
	2	8	8	8	8	8	9	9	9
	3	9	9	9	9	9	9	9	9

Table B Neck, trunk, legs posture score

Neck posture score	Trunk posture score 1		2		3		4		5		6	
	Legs 1	Legs 2	Legs 1	Legs 2	Legs 1	Legs 2	Legs 1	Legs 2	Legs 1	Legs 2	Legs 1	Legs 2
1	1	3	2	3	3	4	5	5	6	6	7	7
2	2	3	2	3	4	5	5	5	6	7	7	7
3	3	3	3	4	4	5	5	6	6	7	7	7
4	5	5	5	6	6	7	7	7	7	7	8	8
5	7	7	7	7	7	8	8	8	8	8	8	8
6	8	8	8	8	8	8	8	9	9	9	9	9

Figure 8 Tables for evaluating posture scores A and B, to be entered on score sheet of Figure 7.

As with the other methods for sampling postures, sampling at appropriate intervals can build up a picture of joint activities and forceful exertions from which interpretations of body loadings can be derived. Inspection of the video and discussions with the job holders in relation to identified extreme or frequent loadings can lead to reductions of overall work load and levels of hazard.

Notation

Detailed records of postures have been used to record ballet for some two centuries. Two popular methods, in wide use, are the Benesh Notation, and Labanotation (Hutchinson,

Muscle use score

Give a score of 1 if the posture is;
mainly static, e.g. held for longer than 1 minute repeated more than 4 times/minute

Forces or load score

0.	1.	2.	3.
No resistance or less than 2 kg intermittent load or force	2-10 kg intermittent load or force	2-10 kg static load 2-10 kg repeated load or force	10 kg or more static load 10 kg or more repeated loads or forces. Shock or forces with a rapid buildup

Figure 9 Muscle and load score tables; values to be entered into score sheet of Figure 7.

Score D (neck, trunk, legs)

Score C (upper limb)	1	2	3	4	5	6	7+
1	1	2	3	3	4	5	5
2	2	2	3	4	4	5	5
3	3	3	3	4	4	5	6
4	3	3	3	4	5	6	6
5	4	4	4	5	6	7	7
6	4	4	5	6	6	7	7
7	5	5	6	6	7	7	7
8	5	5	6	7	7	7	7

Table C Grand score table

Action level 1 A score of one or two indicates that posture is acceptable if it is not maintained or repeated for long periods.

Action level 2 A score of three or four indicates further investigation is needed and changes may be required.

Action level 3 A score of five or six indicates investigation and changes are required soon.

Action level 4 A score of seven or more indicates investigation and changes are required immediately.

Figure 10 Table to determine action levels, using scores C and D from score sheet of Figure 7.

1970). An example of the use of the former is given by Kember (1976). Both methods incorporate a record of timing, which is a major advantage, but each requires about three months of training in order to become even minimally proficient.

A trained notator can record in real time, and in more detail than the methods described in the previous section. This detail can extend to the positions of the fingers, for example, as would be needed if a ballet was being recorded. Although proposals have been put forward to include effort in the recording, this has been based on the appearance of the performer,

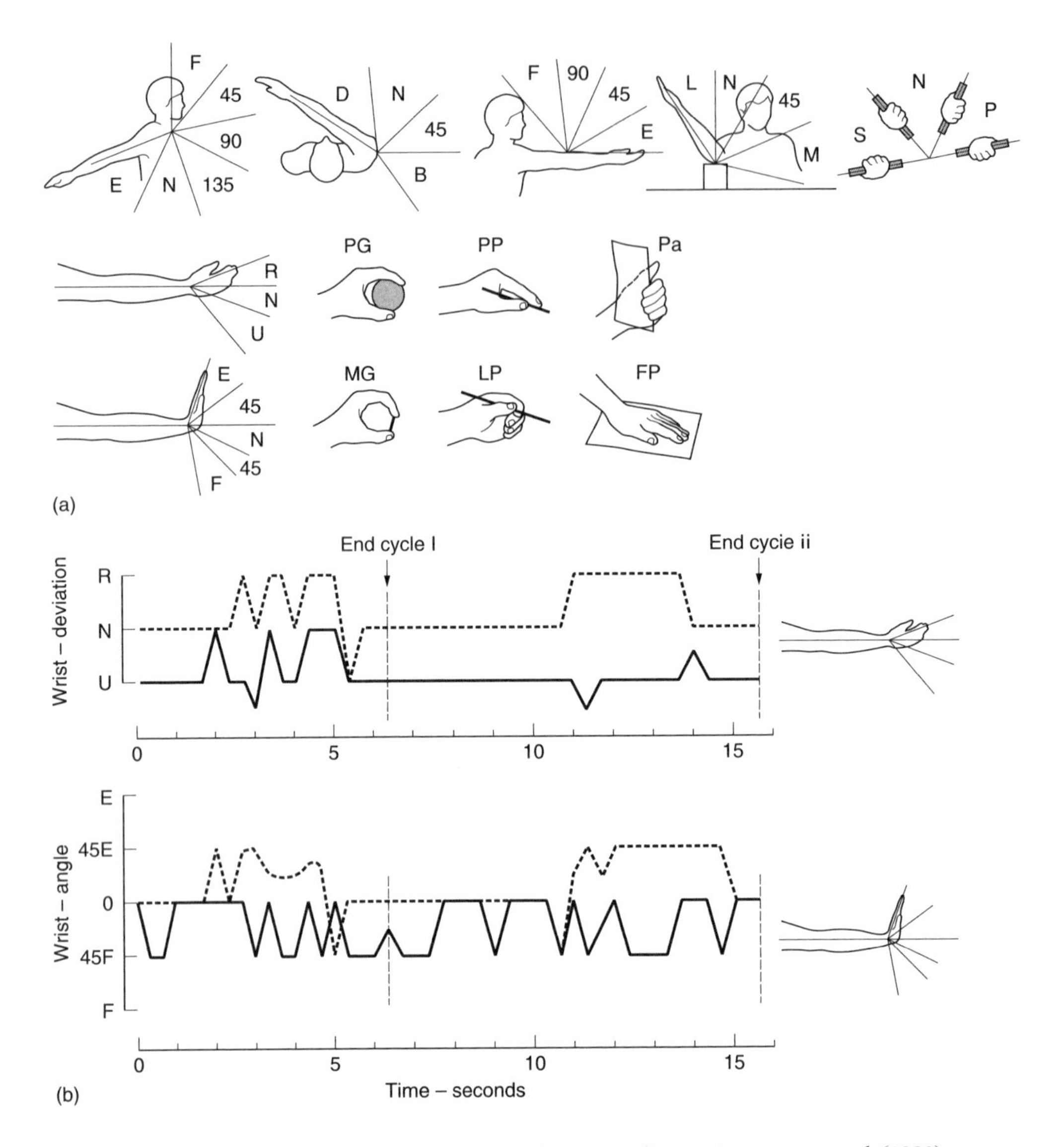

Figure 11 Recording of investigation, using coding according to Armstrong *et al.* (1982).

appropriate in recording a piece of choreography but not based on any measures which would be suitable for work analysis.

A less comprehensive recording procedure which does not incorporate timing data is posture targeting (Corlett *et al.*, 1979). This makes use of a diagram (Figure 12) which has 'targets' located alongside each of the major limb segments, and on head and trunk. As shown in the posture target record of Figure 13, the same position of the targets, but without the human shape, can be used, saving space on the recording form.

Marking the form requires the user to take the posture of Figure 12 as the normal, or zero position. For a person in this position a mark would be made on the centre of each target. Movements forward by limbs, trunk or head (in the sagittal plane), would require a mark along the vertical axis of the target (up for forward) and positioned on that axis according to the estimated angle of displacement. Each concentric circle marks off a 45 degree step. Displacements to the side of the body would be marked on the horisontal axes, again in a position according to the estimated angle. Directions in a horisontal plane are marked along

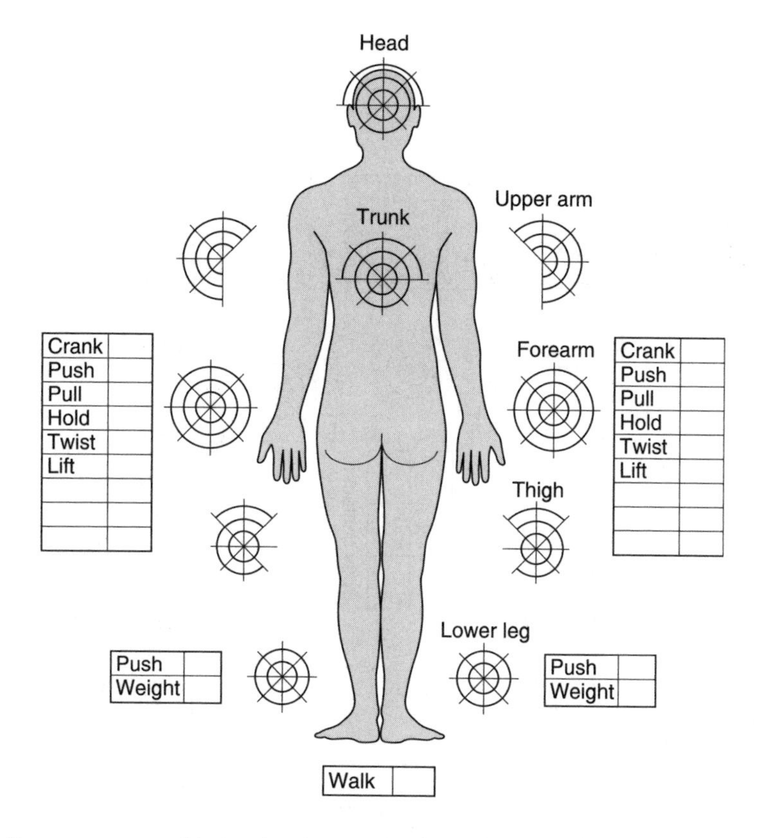

Figure 12 Posture targets, with the datum position shown by the dotted figure.

the appropriate radius or between appropriate radii. An example of the use of the system is shown in Figure 13.

There are cases where the trunk may be twisted rather than bent, or where a bend and a twist occur together. In most of these cases the recording of the positions of arms and legs, together with the trunk position on its target (but without the "trunk twist" marking), will be sufficient. Where it is felt desirable to record the angle of twist of the trunk with respect to the hips, the arc between the head and trunk targets — marked 'trunk twist' on Figure 13 — may be marked at the appropriate point, using the same angular scale as for the targets.

The example of Figure 13, which shows this 'trunk-twist' arc, also shows, just beneath it a point for recording where the person is contacting the environment. The left or right side of the body, in the x plane, has spaces, and front (anterior) or back (posterior) in the z plane are also provided for. Other, self explanatory additions are shown, and it will be evident that the diagram can be modified to suit the user.

With only a modest amount of practice this procedure can be learnt and its accuracy can be tested with goniometers (see below) during the learning period. The postures can be reconstructed from the diagrams, and with simple mathematics the angles of all segments can be related to each other, to permit the whole posture to be redrawn, or input to a computer. With measures of body weight, body sizes and exerted forces, a biomechanics programme can then be used to calculate the required torques and loads.

A biomechanics programme has been modified to use posture targeting as an input (Tracy and Corlett, 1991). The transfer of the targeted points directly from the recording form to the computer screen, which displays a similar target to that on the form, avoids errors in measurement or interpretation of the record. The computer converts the target representation

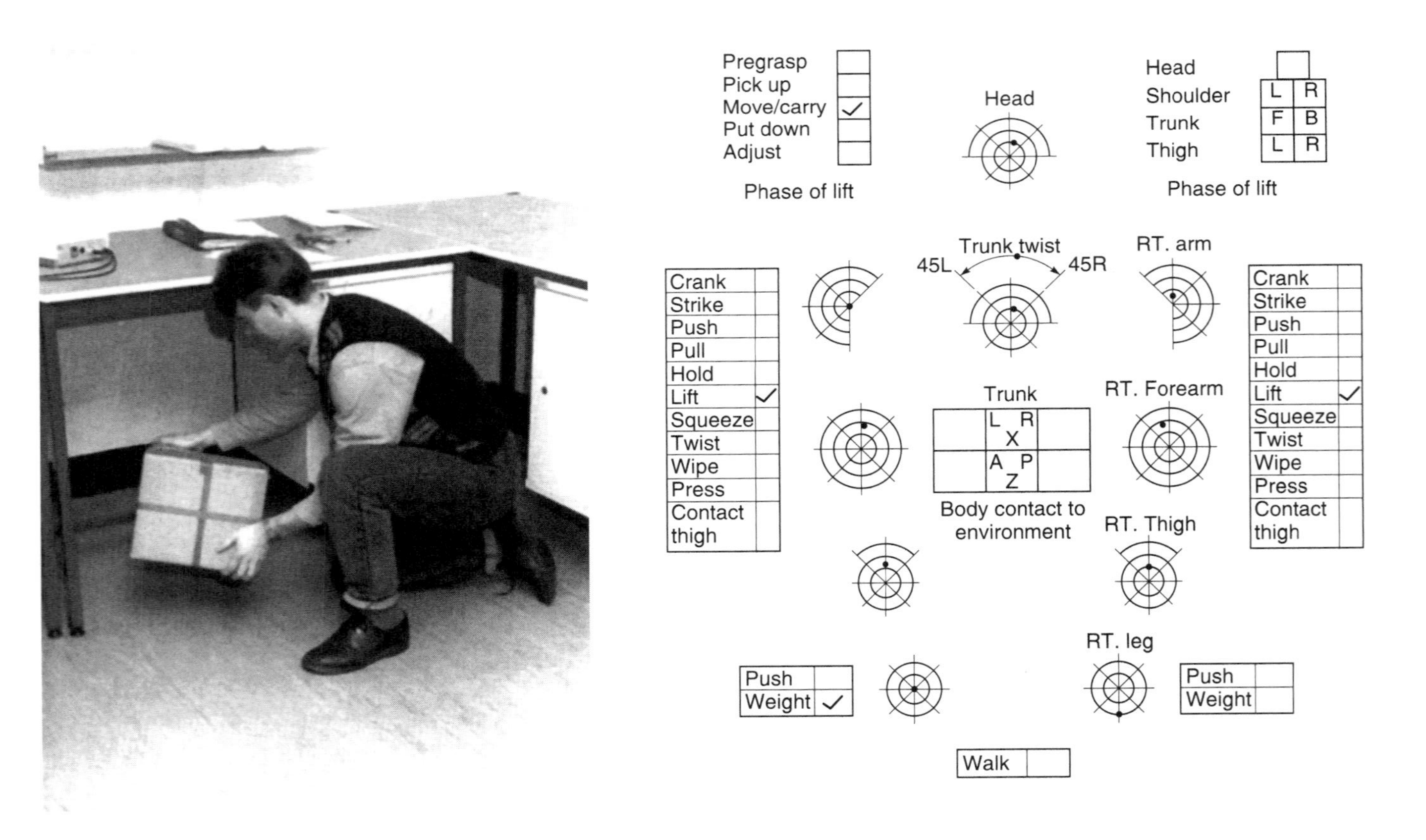

Figure 13 Posture target records taken from a photograph.

to angles, builds up and displays the posture on the screen and then pursues the analysis, giving results more quickly than if measurement or interpretation had to be used. With the current power of lap-top technology it is feasible to undertake direct input at the work site.

A comment on observational methods

All observational methods have a deceptive appearance of simplicity, giving the potential user the impression that their use is easy and their results simple to determine and conclusive. Unfortunately, this is not so, and potential users should be aware of the need for training in the method, monitoring of its use and supporting knowledge for the effective application of the results.

Direct measurement

Goniometers

Individual angles can be measured using a simple goniometer, which will give angles to the vertical or the angle between adjacent body segments. Accuracy is perfectly adequate for most purposes. The pendulum goniometer is quick and simple to use. Where spinal angles are being measured, care is needed since flexion of the spine introduces a pronounced curvature. The goniometer can be used at the level of L3 to L5 to get an estimate of the angle at the lumbar–sacrum junction, and on the lower part of the thoracic spine for an approximate estimate of the spinal angle.

A combination of instruments in conjunction with a flexible rod, to record spinal shape, can be used for a closer estimate (Burton 1986). The changes in spinal flexibility have been well documented by use of such a system, and it could be used to demonstrate the difficulties in performance faced by some workers with limited mobility, due either to their condition or to workplace restrictions.

Recordings for segment or for whole body postures are possible. The ease of recording from small electronic pendulum potentiometers can allow several to be mounted at selected points on a subject, sufficient to reconstruct from the recordings the postures adopted. Some of these instruments record only in one plane and are of limited use in practical cases. However, with a careful choice of instrumentation, recordings can cover the whole waking period, gathering the data on magnetic tape so that computerised analysis is possible.

Instrumentation for the recording of spinal motion over a period of time has been developed by several workers; see for example Marras (1990). These devices strap to the subject's body around the chest and around the hips. The space between, along the line of the spine, is linked by an instrumented flexible unit which will sense the movement of the spine in any direction and is recorded on-line by a computer. Software allows all deflections, as well as velocities and accelerations, to be calculated by the computer. It has been demonstrated by Marras *et al.* (1990) that the difference between conditions, and investigation of the level of hazard, in materials handling jobs, can be better distinguished by including the velocity and acceleration data in the analysis of lumbar movements than by relying on displacement alone.

A different approach has been to use flexible lightweight tubes containing strain-gauged strips, which are strapped to the joint, or along the spine. Each allows joint movement to be measured in two planes simultaneously and the results can be stored on unobtrusive recorders for direct reading or later transfer. The procedure gives consistent results, and is commercially available. It has been used in clinical studies as well as in ergonomic investigations, e.g., spinal flexion of the neck (Parsons and Thomson, 1990) and lumbar movement by motor mechanics (Boocock *et al.*, 1994).

Indirect observations

There are several commercial methods which use three or more video cameras, together with software, to track markers on a subject moving in the cameras' field of view. Advances in computer software are making it increasingly possible for laboratories to develop their own systems, using e.g., 'frame-grabbing' techniques and taking data from video recordings.

The use of a video camera has increased the ease with which work activities may be recorded for subsequent analysis. As yet, however, the analysis is still a long and tedious process. Unless care is taken to record truly representative samples, to do a proper work analysis and to record from several positions so that joint angles can be estimated accurately, the analyst is probably better off using simpler, direct measurements at the workplace itself.

An example of the use of a single video camera, combined with work analyses and other techniques, has been described by Drury (1987). The problem was to evaluate the hazard of upper limb disorders due to posture and repetition. Drury recorded at least two representative cycles from five different camera positions: (1) a front three-quarter view from above the operator's head, (2) a direct left side view, (3) a direct right side view, (4) a direct front view, and (5) a direct rear view.

From the task analysis the sub-tasks were decided upon, so that a sequence of single motions was identified. Each was defined, using the video record, onto an analysis form (Figure 14). For the purpose of the study the joint angle values were then coded, using the classification of Figure 15, and recording the class number on another copy of the same form. The coding gives higher numbers to more serious deviations from the neutral position and the resultant sheet of coded values demonstrates where serious problems are to be expected. By further calculations, using performance data or work standards, estimates of the repetitions per day were made, to give information on the potential for upper limb disorders.

From this, and other data relevant to the case described in the paper, a set of requirements to be addressed was drawn up, so that the working group responsible for change had a succinct summary of the relative importance of the various factors. A repetition of the procedure after change gave clear evidence of improvement.

It is possible to calculate true joint angles using a single camera and one exposure, but the position of the camera and the subject must be controlled. Computation methods for the analysis of images from a single camera were given by Jian Li *et al.* (1990). They demonstrated calculated angles with very small errors, and considered that where the plane of the movement can be controlled, the procedure is not difficult. This last constraint is often a natural feature of many industrial tasks, such as assembly work, and could make the method more practically useful than at first might appear.

Video methods often use markers on the body to provide more precise measurement points. These can be stripes placed, for example along the axes of the long bones, reflective spots on appropriate body parts (e.g., over the joints of the limbs), or small lights, similarly mounted.

A dedicated motion analysis system which uses LED markers on the subject is the CODA motion system. Three infra-red sensors in a solid state unit incorporating its own PC can scan up to 56 markers at rates up to 800 Hz with a resolution of up to 0.1 mm over a field of view of up to 6 metres. The system, which is illustrated in Figure 16, can be used with one unit or with more than one suitably placed around the subject. This overcomes a common difficulty with many optical systems, that they can only see one side of a subject.

With the introduction of whole body posture to the computer, the introduction of force data, from strain-gauged instruments or a force plate, or EMGs with CODA's own optical telemetering system, on-line biomechanic calculations can be done in real time, with stick figures displayed on a screen together with other data from the subject or environment.

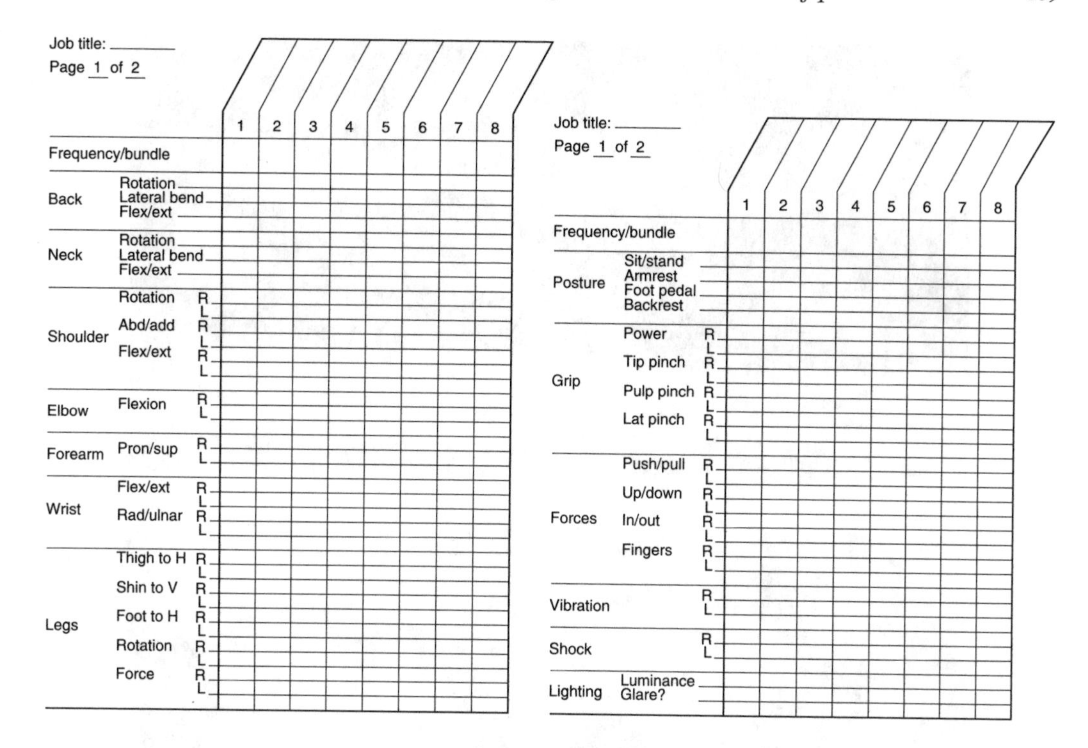
Job title: ______
Page 1 of 2
1 2 3 4 5 6 7 8
Frequency/bundle
Back: Rotation; Lateral bend; Flex/ext
Neck: Rotation; Lateral bend; Flex/ext
Shoulder: Rotation R L; Abd/add R L; Flex/ext R L
Elbow: Flexion R L
Forearm: Pron/sup R L
Wrist: Flex/ext R L; Rad/ulnar R L
Legs: Thigh to H R L; Shin to V R L; Foot to H R L; Rotation R L; Force R L

Job title: ______
Page 1 of 2
1 2 3 4 5 6 7 8
Frequency/bundle
Posture: Sit/stand; Armrest; Foot pedal; Backrest
Grip: Power R L; Tip pinch R L; Pulp pinch R L; Lat pinch R L
Forces: Push/pull R L; Up/down R L; In/out R L; Fingers R L
Vibration R L
Shock R L
Lighting: Luminance; Glare?

Figure 14 Posture analysis form, using data from video recordings (Drury, 1987).

Zones and joint angles

	Zone, degrees			
	0	1	2	3
Neck				
Rotation	0–8	8–20	20–40	40 +
Lateral bend	0–5	5–12	12–24	24 +
Flexion	0–6	6–15	15–30	30 +
Extension	0–9	9–22	22–45	45 +
Back				
Rotation	0–10	10–25	25–45	45 +
Lateral bend	0–5	5–10	10–20	20 +
Flexion	0–10	10–25	25–45	45 +
Extension	0–5	5–10	10–20	20 +
Shoulder				
Outward rotation	0–3	3–9	9–17	17 +
Inward rotation	0–10	10–24	24–49	49 +
Abduction	0–13	13–34	34–67	67 +
Adduction	0–5	5–12	12–24	24 +
Flexion	0–19	19–47	47–94	94 +
Extension	0–6	6–15	15–31	31 +
Elbow				
Flexion	0–14	14–36	36–71	71 +
Forearm				
Pronation	0–8	8–19	19–39	39 +
Supination	0–11	11–28	28–57	57 +
Wrist				
Flexion	0–9	9–23	23–45	45 +
Extension	0–10	10–25	25–50	50 +
Radial deviation	0–3	3–7	7–14	14 +
Ulnar deviation	0–5	5–12	12–24	24 +

Figure 15 Joint angle codes for posture analysis (Drury, 1987).

(a)

(b)

Figure 16 The CODA mpx 30 optical posture recording instrument (a), with examples of the retroreflective pyramidal markers (b).

Effects on the person of maintaining a posture

Maximum voluntary contractions

Perhaps the earliest scientific approach to estimating the appropriateness of static loads was to evaluate, experimentally, the holding times for various loads and express the results as the times for which a person could hold various proportions of the maximum load. The force required to achieve maximum load is referred to as maximum voluntary contraction (MVC).

Whilst MVC may be measured quite simply (in many cases using a spring balance) there are some essential controls. An impulse force is not required for the measurement; the instruction to a subject is usually of the form, 'build up your maximum force gradually, over a

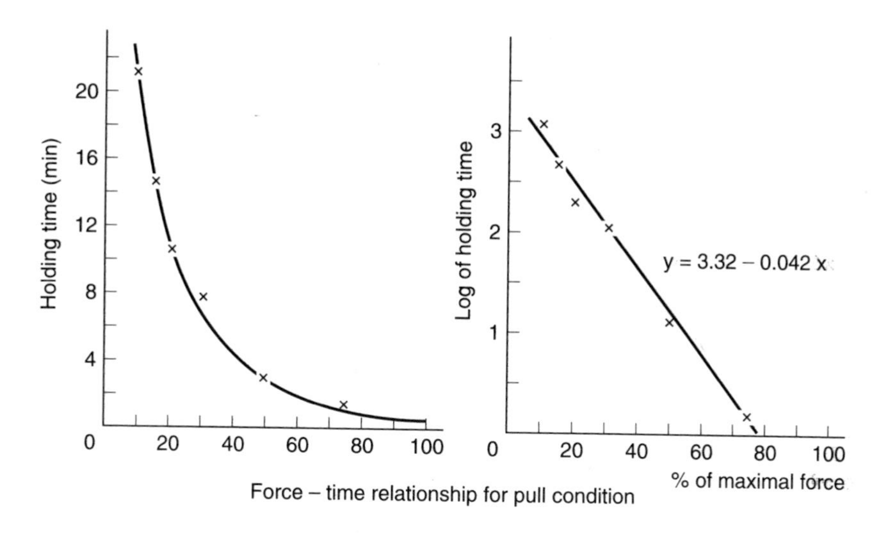

Figure 17 Subjects held various percentages of their maximum force, exerted by pulling at shoulder level, for as long as they could. The log-normal relationship gives a means for calculation of intermediate values.

period of 2–3 seconds, and hold it for 3 seconds'. The value used is the mean force over the last, relatively constant period.

The relationship between force and holding time, demonstrated by Monod and Scherrer (1965) and by Rohmert (e.g., 1973a, b) is a logarithmic one (Figure 17). Today it is accepted that a long term *constant* static effort greater than 2–3% of MVC is unacceptable, although at one time 15% was thought to be possible. Knowledge of the force-holding time relationship, which appears to hold for most skeletal muscles, does allow us to estimate the effects of some postures and provide guidance as to their appropriateness.

The maximum holding time for a posture is not, by itself, a very useful measure, since we usually wish to know the frequency with which the position may be held and the consequent likelihood of damage. Hence recovery from static workloads is of interest. In experimental work, evidence of recovery has been taken as being when the same posture can be held again for the same maximum time. In gathering such data we must seek to achieve the same level of motivation for each test, and treat subjects as their own controls. Thus we calculate the forces as a percentage of the subject's own MVC, and provide rest periods as a proportion of the subject's own holding time. A typical recovery curve is shown in Figure 18 (Milner *et al.* 1986). It arises from a number of different forward bending postures held for as long as possible, (T1), after which the subjects had a rest interval equal to twelve times their own T1, and then repeated the posture again, (T2). An important consequence of the relationship shown in this curve is that if the maximum holding time is considerable, e.g., the posture is such that the discomfort builds up over a long period, then recovery also takes a very long time.

If we are seeking MVC for a particular situation it is important that the posture adopted, including foot positions and any constraints due to the workplace, are repeated during the tests, so that as nearly as possible the same muscle groups are used. Few *practical* force exertions are undertaken by just a single group of muscles, so, if a number of different muscles is recruited to do the task, unless the posture is identical, the groups of muscles could be operating differently, or else other groups of muscles may be called into play. It is more usual to find static work interspersed with rest pauses or other opportunities to relax the muscles,

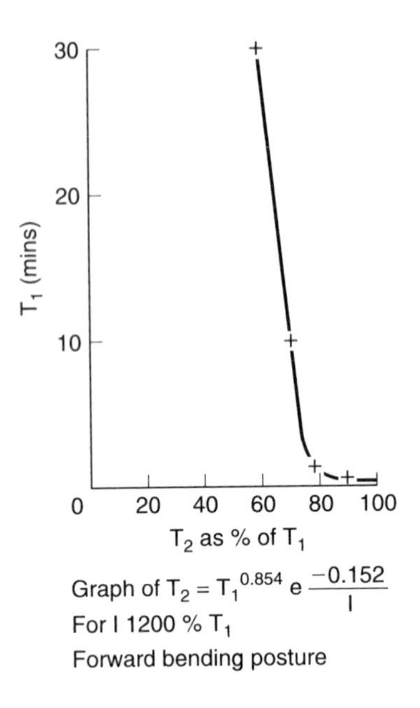

Figure 18 The graph shows the recovery (T_2) after a rest pause 1 equal to twelve times the first holding time (T_1). Data for the formula came from 42 subjects experiencing five different postures and five rest intervals.

and in these practical situations the requirement is often to estimate the feasibility of a given work-rest cycle.

Subjective methods

Of the available subjective methods there are two which can make a considerable contribution to the evaluation of static work. Borg's scale has already been described, together with its rationale, in Chapter 15, on dynamic work. Although its use in static work is not valid in terms of the relationship of the numbers to a person's heart rate, the judgements of severity do give important information. An example of this arose from a demonstration of the difficulties involved in butchery. A rig was available which permitted subjects to exert single handed forces on a set of knife handles, pulling them in directions across and down the body in the vertical plane, and exerting as much force as possible. Although differences in the forces were found when working at different heights, they were modest. However, when subjects used Borg's scale to judge the difficulty of pulling in each of the directions, large differences were identified in the ease of the knife movements in each of the directions, which separated them more clearly than the analysis of the imposed forces could do.

The second subjective method uses muscular pain as a measure. Because 'pain' is sometimes seen as a specific and localised experience, the term 'discomfort' is used. Experiments (Corlett and Bishop, 1976) demonstrated that, if a force was exerted for as long as possible, until the discomfort was unbearable, and estimates of the discomfort levels made on a scale, 5 or 7 points, at intervals during the holding time, the growth of discomfort was linearly related to the holding time, regardless of the level of the force being exerted (Figure 19). So, discomfort itself can be used as a linear scale. There are deviations from linearity at the top end of the scale, where subjects may reach close to the end of the scale before reaching their own discomfort limits. A magnitude estimation technique can be

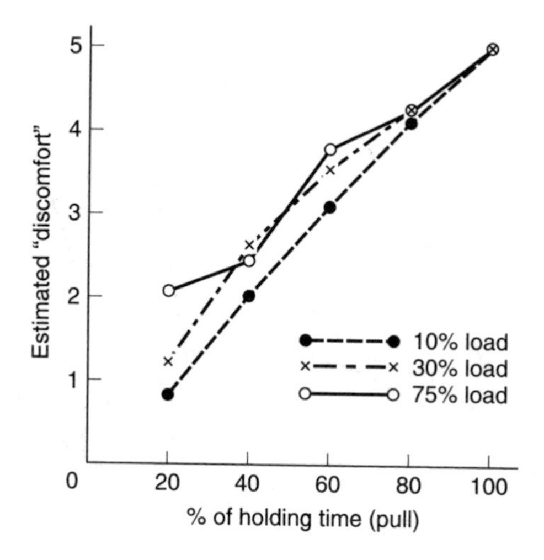

Figure 19 Mean values for overall discomfort ratings when different percentages of MVC are exerted for as long as possible.

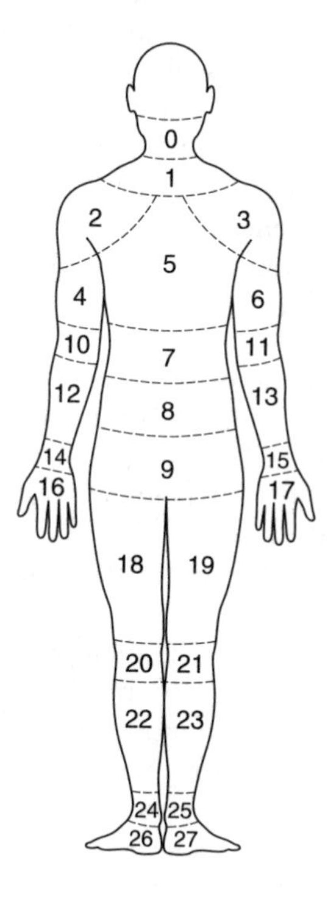

Figure 20 The body map for evaluating body part discomfort, either by rating or ranking.

adopted if the extreme end of the scale is important (see Stevens, 1975; Kvalseth, 1983, pp. 246–250).

To specify the sites of discomfort a body map is used, divided into segments (Figure 20), depending on the sites of discomfort experienced by those engaged in the tasks being investigated. This information is found from preliminary trials or enquiries. The procedure for mapping the development of discomfort can proceed in one of two ways.

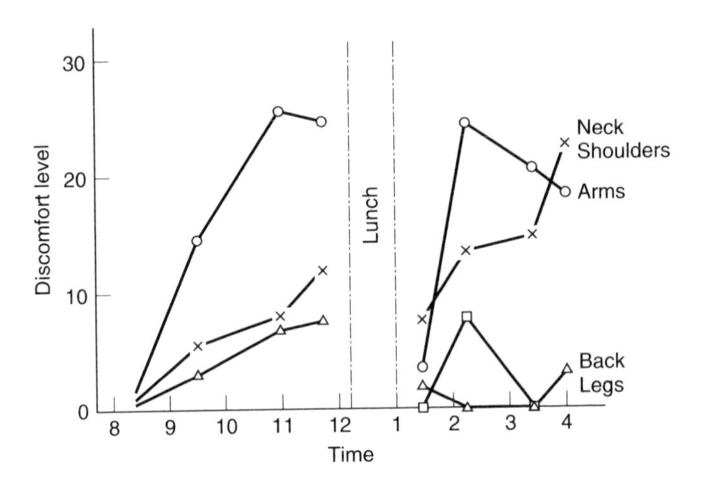

Figure 21 Discomfort scores, rating on a 7-point scale by four workers. The relative continuity of the neck and arms graphs over the lunch period will be noted.

(1) At intervals during the whole working day, people are asked to point to the site(s) of current discomfort on the body map. Then they are asked to rate the intensity of discomfort at each of the identified sites on a 5 or 7 point scale, preferably by marking a paper scale which is 'anchored' at the 0 and the 5th or 7th points by 'no discomfort' and 'extreme discomfort' respectively. These scores are plotted against time of day for each body site, (or related groups of body sites), dividing the scale during analysis at the half points to give an effective 10 (or 14) point scale. Since it is likely that differences in body size or in person-equipment relationships will cause changes in the differences in the distribution of discomfort around the body, the effect of adding the scores from several subjects who are engaged on the same tasks should be considered carefully; it is usually unwise.

The reason for urging the collection of data throughout the working day is that recovery from static load is slow, see above, and a lunch break is often not sufficient time to achieve full recovery. A study of engravers (Figure 21) where four workers were studied and the results averaged, illustrates the point forcibly. It is evident from the high discomfort levels reached that the posture was extreme, and the curves representing the most heavily loaded body parts appear as one curve spread across the whole day, rather than the morning curve being repeated in the afternoon.

(2) The second way in which the body mapping procedure can be used is recommended when the use of rating scales will take up too much of the subject's time. The person can be asked to point out the site(s) which are most uncomfortable, which are noted. Then the sites which are judged 'next most uncomfortable' are pointed out, and so on until no more sites are reported. We have asked the person to identify a sequence of just noticeable differences in discomfort, so it is unlikely that more than five or six will be differentiated, as any text on psychophysics will confirm. Again these results are plotted against time of day, but a numerical value is obtained by counting back the number of levels of discomfort reported and numbering them, using the 'no discomfort' sites as zero, the last reported sites as 1, and so on. Although less detailed than the previous method, it is much quicker, involves less explanation to the subjects and reveals the most heavily loaded body parts equally well. Of course, all the other aspects of experimental control apply for this method as well.

As described in the original paper (Corlett and Bishop, *op cit*), the interpretation of the results in order to suggest changes can be aided by information relating posture and discomfort such as that given by van Wely (1970). It will often be evident, from an inspection of the workplace in conjunction with the results, where the problems lie.

Postural load data

As a major contribution to the faster accumulation of data on musculo-skeletal problems, the Institutes of Occupational Health in the Nordic countries have designed the Nordic Questionnaire (Kuorinka *et al.*, 1987). This provides a standard format for gathering data on musculo-skeletal problems. Increased information about the incidence and epidemiology of these complaints is very necessary. Where data are needed for a particular investigation, such a questionnaire can be supplemented by additional questions, but its use will enable data from different studies to be compared, and the large data pool arising from its use in the Nordic countries can also be used for comparison purposes.

The NMQ was very attractive to the UK Health and Safety Executive and in consequence it was evaluated for widespread use by that authority. Small modifications were made, to make sure that it was unambiguous for native UK English speakers (Dickinson *et al.*, 1992). This modified questionnaire is included as an appendix at the end of the chapter, by kind permission of the Health and Safety Executive. It follows a format as given in Chapter 4 on questionnaires, with a prologue, the core questions, some classification questions and an epilogue. The musculoskeletal disorders survey begins, after the personal details section, with a general survey to give estimates of prevalence and disability. Following this are four sections seeking more specific information for each of four body areas. These seek to establish the severity of any disorder. The final sections include a general health questionnaire and an opportunity to give more details of the work of the respondent.

The example shown was for a survey of supermarket checkout operators (Mackay *et al.* 1998). Ten supermarkets and 481 operators were used in the initial evaluations. Their results in terms of reliability of responses were similar to those of Kuorinka *et al.* (1987) but they emphasised that a minimum 80% return is necessary if prevalence rates are to be realistic. The HSE experience was that to bring respondents together in groups to answer the questionnaire provided a better return rate than just issuing the forms and asking for their return. Numbers suffering some disability appeared to be less affected by the way the form was administered, those suffering presumably having more reason to see it to be to their advantage to reply.

Another tool developed at the Swedish National Board of Occupational Safety and Health, is the single sheet analysis for identifying musculoskeletal stress factors (Figure 22). This is self-explanatory, and uses the sites of discomfort or injury to focus attention on a number of possible workplace faults which could be their causes. The list of possible causes is equally applicable to the body mapping procedure described earlier, enabling a direct link to be made to the sources of the problems. After changes have been made it will be clear that the same methods can be used to demonstrate any improvements which have been achieved.

Electromyography

Electromyography (EMG), the recording of myoelectric signals which occur when a muscle is in use, can be used to assess the level of activity occurring over a period of time. It can also be used to show the presence of muscle fatigue, a state when a skeletal muscle is unable to maintain the required force of contraction (Hagberg, 1981). A high correlation has been shown between EMG activity and muscular force, for both static and dynamic activities

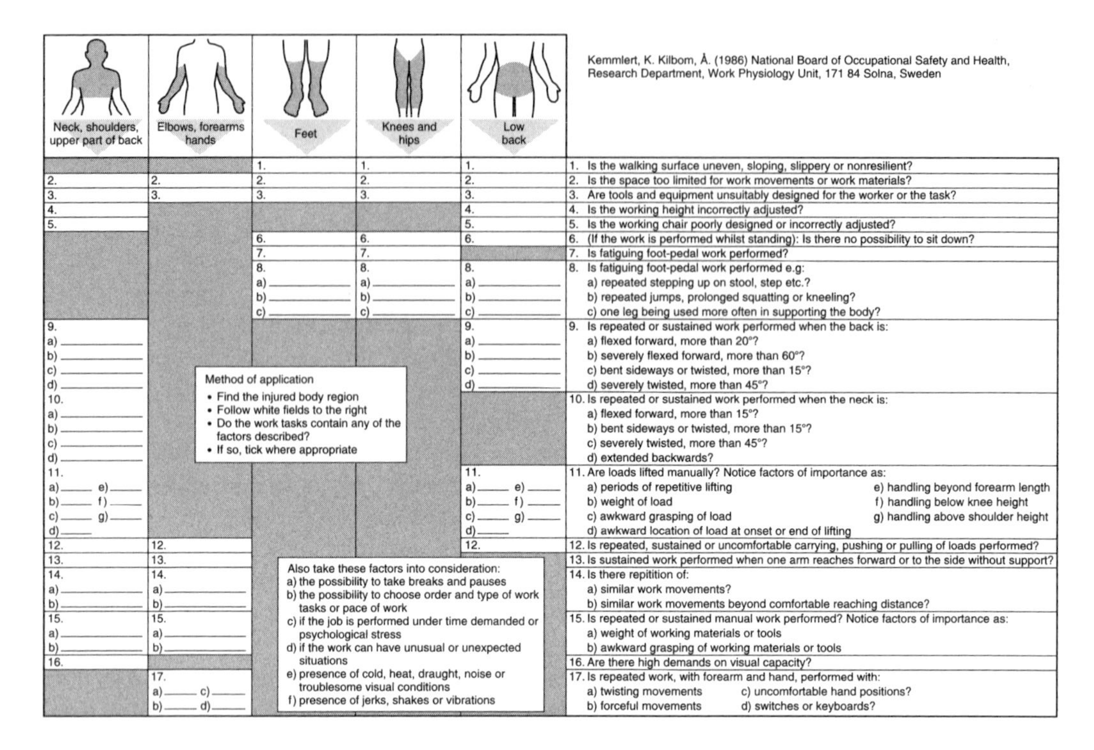

Neck, shoulders, upper part of back | Elbows, forearms hands | Feet | Knees and hips | Low back

Kemmlert, K. Kilbom, Å. (1986) National Board of Occupational Safety and Health, Research Department, Work Physiology Unit, 171 84 Solna, Sweden

1. Is the walking surface uneven, sloping, slippery or nonresilient?
2. Is the space too limited for work movements or work materials?
3. Are tools and equipment unsuitably designed for the worker or the task?
4. Is the working height incorrectly adjusted?
5. Is the working chair poorly designed or incorrectly adjusted?
6. (If the work is performed whilst standing): Is there no possibility to sit down?
7. Is fatiguing foot-pedal work performed?
8. Is fatiguing foot-pedal work performed e.g:
   a) repeated stepping up on stool, step etc.?
   b) repeated jumps, prolonged squatting or kneeling?
   c) one leg being used more often in supporting the body?
9. Is repeated or sustained work performed when the back is:
   a) flexed forward, more than 20°?
   b) severely flexed forward, more than 60°?
   c) bent sideways or twisted, more than 15°?
   d) severely twisted, more than 45°?
10. Is repeated or sustained work performed when the neck is:
   a) flexed forward, more than 15°?
   b) bent sideways or twisted, more than 15°?
   c) severely twisted, more than 45°?
   d) extended backwards?
11. Are loads lifted manually? Notice factors of importance as:
   a) periods of repetitive lifting
   b) weight of load
   c) awkward grasping of load
   d) awkward location of load at onset or end of lifting
   e) handling beyond forearm length
   f) handling below knee height
   g) handling above shoulder height
12. Is repeated, sustained or uncomfortable carrying, pushing or pulling of loads performed?
13. Is sustained work performed when one arm reaches forward or to the side without support?
14. Is there repitition of:
   a) similar work movements?
   b) similar work movements beyond comfortable reaching distance?
15. Is repeated or sustained manual work performed? Notice factors of importance as:
   a) weight of working materials or tools
   b) awkward grasping of working materials or tools
16. Are there high demands on visual capacity?
17. Is repeated work, with forearm and hand, performed with:
   a) twisting movements
   b) forceful movements
   c) uncomfortable hand positions?
   d) switches or keyboards?

Method of application
- Find the injured body region
- Follow white fields to the right
- Do the work tasks contain any of the factors described?
- If so, tick where appropriate

Also take these factors into consideration:
a) the possibility to take breaks and pauses
b) the possibility to choose order and type of work tasks or pace of work
c) if the job is performed under time demanded or psychological stress
d) if the work can have unusual or unexpected situations
e) presence of cold, heat, draught, noise or troublesome visual conditions
f) presence of jerks, shakes or vibrations

Figure 22 Method for the identification of musculoskeletal stress factors which may have injurious effects.

(Hagberg, 1981). This relationship was once thought to be linear, but is now proposed as exponential (Lind and Petrofsky, 1979; Hagberg, 1981).

When a muscle begins to fatigue there is an increase in the amplitude in the low frequency range and a reduction in the amplitude in the high frequency range of EMG activity (Petrofsky *et al.*, 1982). There is also a shift in the frequency sprectrum towards the lower end as fatigue occurs.

Although needle electrodes, entering specific muscles, are used for medical research, occupational EMG records are usually taken from surface electrodes. These are stuck over the central part of the muscle and leads taken, via preamplifiers, to amplification and recording equipment. Telemetering can be done, but the existence of miniaturised circuitry and recorders has rendered this less necessary.

Skin preparation is necessary for reliable recordings. This involves the use of fine sandpaper to remove layers of dead skin prior to fixing the electrodes. Many electrodes, usually small silver discs, have a central hole into which electrode jelly, a saline grease, may be inserted with a syringe after the electrode is in place. Otherwise electrode jelly is used on the electrode itself prior to attachment to the skin. The placement of the electrode is generally over the central part of the muscle because that is where most of the active fibres will lie; and the electrodes will be from 3–5 cm apart. An 'earthing' electrode is sometimes included, placed away from the muscle being recorded and where other muscle activity is unlikely to be picked up. Signals from the electrodes should be preamplified as close to the electrodes as possible, to increase the signal-to-noise ratio, before passing them via a low pass filter to a recorder or analyser.

As opposed to clinical EMG, where the quality of the signal is of importance, in occupational EMG the quantity of the signal is usually the important factor. The signal is analysed with respect to its frequency or amplitude, and the amplitude is usually analysed to find what percentage it represents of that achieved when a test MVC was undertaken prior to the experiment. This conversion permits comparison across different tasks and between different people.

There are three major methods for EMG analysis: the integrated EMG (IEMG), Fourier analysis and the *amplitude probability distribution function* (APDF) analysis.

IEMG

The integrated EMG gives a measure of the power of the signal. Integrating circuits accumulate the root mean square (rms) values of the signal, recording them until a certain selected total value has been reached and then starting the addition again. The visual record shows a series of triangular waves, with equal peaks but spaced more closely where the EMG signal is greater. Counting the peaks per unit time or calculating the rms value, again per unit time, provides values representative of the muscular activity.

Fourier analysis

The speed of response of muscle fibres varies, and they are conventionally divided into fast and slow twitch fibres. As a muscle is used, slow twitch activity becomes more evident, and is taken to be a sign of increased muscular fatigue. The frequencies bound up in a raw EMG signal are identified by Fourier analysis, a procedure which breaks down the signal into its component sine waves. Usually the analysis is done by taking successive short samples of the EMG, analysing them for their component sine waves and presenting the results as a frequency spectrum, showing the distribution of the frequencies and amplitudes of these component waves.

Amplitude probability distribution function

Work by Jonsson (1976) and Hagberg (1979) demonstrated the utility of analysing the EMG in terms of the amplitudes present in the signal. Each excursion of the signal represents the innervation of muscle fibre(s) to exert a force. Large excursions are related to the exertion of external force or rapid movement; small and frequent excursions can be interpreted as indicating the maintenance of static work, e.g., for holding a position.

The analysis is relatively simple in concept. Again short, successive samples of the signal are taken and the amplitudes of all the peaks counted and grouped. They are plotted as an amplitude spectrum or as a cumulative amplitude distribution function. If the latter plot is adopted, the 10th, 50th and 90th deciles can be identified. The 10th decile is proposed as the level which demonstrates the static work load.

Although close correlations have been reported between EMG analyses and force, Hagberg (1981) has noted some sources of variance in experiments. The relationship will change with temperature, such as might arise from high levels of work activity, with fatigue, with whether the contraction is concentric or eccentric, and with changes in the velocity of contraction. For much occupational work these factors may not be serious influences on results, but should be considered in relation to the quoted literature for any extensive studies. Where defined test contractions are easy to apply and the muscle action substantially isometric, the APDF can be a useful measure of muscle performance.

Comment on EMG

Each method has its uses, depending on the problem being studied. As will be evident, all methods may be used from the same EMG recordings. The amount of data collected in occupational EMG is usually very large, as long periods of recording are required. This may create difficulties in storage and analysis. Also, in field studies artefacts can be more common, such as electrode movement or changes in temperature.

Occupational EMG is a good technique for assessing which muscles are used in a task but is of more limited use in accurately assessing the fatigue process. It is, however, of value to establish the changes in the amplitude and frequency domains, with respect to time, to gain some understanding of how a muscle is operating. It is a tool which should be used in conjunction with other assessment techniques for a good understanding of a work situation.

Spinal loading

Although in the investigation of postural loading, where loads are of short duration, a biomechanical analysis may be suitable, exposure over a longer period can introduce changes which alter the spinal characteristics. The effect of exposures over longer periods, or loads which are frequently repeated can be assessed by measuring changes in total stature (Eklund and Corlett 1984).

The forces imposed on the lumbar spine during the day are the gravity loadings from thorax, head and arms, together with the components of forces exerted by the arms which have to be transmitted to the pelvis. These forces cause a reduction in stature over the day of 15 mm or more due to a slow decrease in spinal disc height; this will vary with age. Such shrinkage is recovered when lying down. To compare

Figure 23 Precision stadiometer.

the effects of different workplaces, postures or work regimes on spinal loading, the use of a precision stadiometer is required (Figure 23). By close control of the experiment and measurement protocol, changes in stature of about 0.5 mm can be identified. As with many biological measures, it is preferable to use subjects as their own controls, since averaging across subjects, due to differences in responding, increases the variance considerably.

There are several points to note when using the technique, which has been used in the workplace as well as the laboratory. As with all precision measurements, tight experimental control is required. If repeated tests are to be made, the time of day and prior activities of the subjects should be consistent. As recovery of height loss is quite rapid, the periods between exposure to load and subsequent measurement should be kept short and, again, consistent. Rest pauses between any sequences of measurements should also be controlled so that no extra increase or major decrease of load arises, say from a major change in posture. Thus, if at one point in an experimental sequence, a subject should lie down, the resultant change in disc condition would make the effects of a subsequent test condition very different from earlier trials (Abu Amin *et al.*, 1988).

Some instruction to subjects is needed on how to position themselves on the stadiometer, to help in maintenance of a consistent posture. The experimenter must also check and control weight distribution between heels and soles, location of the spine on the micro-switch pads set into the backboard of the instrument, head position (which is assessed by the nose marker), and ensure that the subject has folded arms. About five recordings over 2 or 3 seconds are taken. The subject then steps off, and back onto the stadiometer immediately, is repositioned and five more readings taken. The repetition gives measures for the estimation of error as well as evidence for the consistency of the measuring posture.

For any study it is desirable to differentiate stature changes due to load from those which would normally arise under gravity loadings in the postures associated with the situation under investigation. Thus it will often be advisable to run a number of preliminary measures to establish the rate of shrinkage prior to loading, following with the trials. Extrapolation of the initial measures to the time of the final measures under load will allow assessment of a loading effect which will be independent of the expected rate of shrinkage at that time of the day. Work by Althoff *et al.* (1992) has demonstrated the importance of this point, as well as the difficulty of establishing it in some cases.

A study by Foreman (1989) noted that compressibility of the heel is a major influence on total stature. He showed changes ranging between 2 and 6 mm in a group of 20 subjects over a period of about 1.5 min. Hence it is desirable that, where experimental conditions reduce the load on the heel, some control is exercised to take this change into account. This might be done by recording initial calibration curves for heel compression and correcting results in relation to the time of standing on the stadiometer. The work quoted earlier, by Althoff *et al.* (1992), describes techniques to exercise control over heel pad shrinkage and regular diurnal variations, as well as noting the effects on shrinkage of lumbar disc diameter. They note that, where close control is exercised, an accurate and reliable measurement of stature change, and hence an estimation of spinal stress, is possible in practical situations.

Some results from the use of the method are shown in Figure 24. Comparisons are made between seat back angles for drivers of a mock-up of a car. Subjects 'drove' a video driving game for 1-hour periods. The changes in stature for seat geometries support the work of Andersson and Ortengren (1974), who showed that spinal load was least for a seat-back angle of about 110 degrees.

The technique has been used to study nursing activities, (Foreman and Troup, 1987), design of working seats, (Corlett and Eklund, 1983), the effects of circuit weight training and running (Leatt *et al.*, 1986) the effects of equipment arrangement

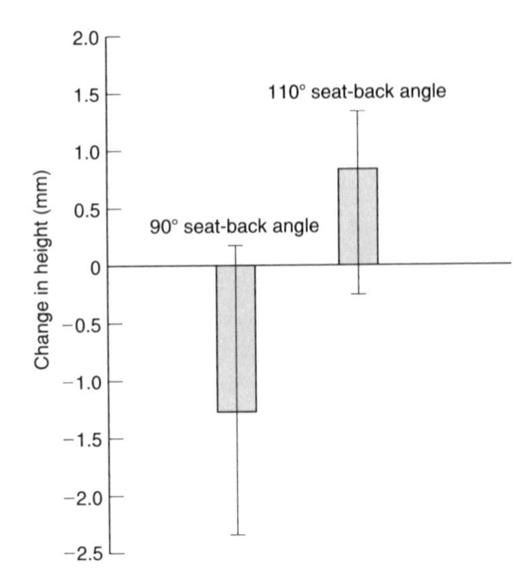

Figure 24 Changes in height during simulated driving for two different seat-back angles.

and vibration during vehicle driving (Bonney, 1988; Bonney and Corlett 2002, 2003) and the effects of overhead working (Burton and Tillotson, 1991).

Research by Jaffry (1993) used a stadiometer designed for a seated subject. She found differences in the measures taken by this stadiometer and one where the subjects stood on the platform, surmising that when moving from sitting to standing the change in spinal posture may allow a relatively rapid change in the disc height which could confound the effects of the experimental loadings.

Advances in posture evaluation

In the introduction we set out five dimensions relevant to the investigation of postural loads and static work. However, consideration of some of the measures discussed will show that more than these factors have been brought into the analysis. Although research techniques — especially in the laboratory — can be relatively specific and focus on one or two measures, the field worker must look more widely. This wider view is one of the features of field investigations which make causation such a problem to define in a court case, but at the same time helps to elucidate apparently conflicting field results.

Approaches will be sought which may not pin down in a quantitative manner the contributing share from each of a number of variables in an investigation, yet still a logical and satisfactory explanation for the observations resulting from a study can be obtained. This is a natural order of things; the complexities of nature do not reveal themselves immediately. We gradually recognise the ways in which apparently disparate components interact. Only when they are recognised as relevant and contributory can we begin to measure their contributions. A qualitative understanding is where we start.

Even in the rather restricted — yet still complicated — field of posture evaluation it is inevitable that the list in the introduction will be extended. This is clear from the four key problem areas which succeed our five components; what we know about them already tells us that there is, for example, more to fatigue than just muscle metabolism and biomechanics. Ways of quantifying these other influences will arise, be built into the toolkit and become normal practice. This is one challenging area for methods development. It is also one of the ways in which practical studies, illuminated by the broad ergonomics base of physiological, psychological, environmental and social scientific knowledge, contribute to the more reliable understanding of the real world. The iterative interplay between science and practice, conducted by a wondering mind, is the process through which reliable knowledge is consolidated into a more comprehensive understanding. Readers will realise that the research literature is where they must look for such developments in methods. That is where they are tested and their contributions assessed, after which their utility can be explored in a wide range of field studies. We can expect, therefore, and should welcome, many changes to ergonomics methods, including those for posture evaluation, over time.

References

Abu Amin, A., Corlett, E.N. and Bonney, R.A. (1988). Does wearing a seat belt alter the load on the back whilst driving. In *Contemporary Ergonomics*, edited by E.D. Megaw (London: Taylor & Francis).

Althoff, I., Brinkman, P., Frobmin, W., Sandover, J. and Burton, K. (1992). An improved method of stature measurement for quantitative determination of spinal loading. *Spine*, **17**, 682–693.

Andersson, B.J.G. and Örtengren, R. (1974). Lumbar disc pressure and myoelectric back muscle activity during sitting. IL Studies on an office chair. *Scandinavian Journal of Rehabilitation Medicine*, **6**, 115–121.

Armstrong, T.J., Foulke, J.A., Joseph, B.S. and Goldstein, S.A. (1982). Investigation of cumulative trauma disorders in a poultry processing plant. *Journal of the American Industrial Hygiene Association*, **43**, 103–115.

Bonney, R.A. (1988). Some effects on the spine from driving. *Clinical Biomechanics*, **3**, 236–240.

Bonney, R.A. and Corlett, E.N. (2002). Head posture and loading of the cervical spine. *Applied Ergonomics*, **33**, pp. 415–417.

Bonney, R.A. and Corlett, E.N. (2003). Vibration and spinal lengthening in simulated vehicle driving. *Applied Ergonomics*, **34**, pp. 195–200.

Boocock, M.G., Jackson, J.A., Burton, A. and Tillotson, K.M. (1994). Continuous measurement of lumbar posture using flexible electrogoniometers. *Ergonomics*, **37**, 175–185.

Brouha, L. (1960). *Physiology in Industry* (Oxford: Pergamon).

Burton, A.K. (1986). Measurement of regional lumbar sagittal mobility and posture by means of a flexible curve. In *The Ergonomics of Working Postures*, edited by E.N. Corlett, J.R. Wilson and I. Manenica (London: Taylor & Francis).

Burton, A.K. and Tillotson, K.M. (1991). Measurement of spinal strain to estimate loads on the spine in overhead working postures. Report of the Spinal Research Unit, School of Human and Health Sciences, Huddersfield University, Huddersfield, UK.

Corlett, E.N. and Bishop, R.P. (1976). A technique for assessing postural discomfort. *Ergonomics*, **19**, 175–182.

Corlett, E.N. and Eklund, J.A.E. (1983). The measurement of spinal load arising from work seats. *Proceedings of the Human Factors Society 27th Annual Meeting*, Santa Monica, CA.

Corlett, E.N., Madeley, S. and Manenica, I. (1979). Posture targetting: a technique for recording working postures. *Ergonomics*, **22**, 357–366.

Corlett, E.N., Wilson, J.R. and Manenica, I. (1986). *The Ergonomcis of Working Postures* (London: Taylor & Francis).

Dickinson, C.E., Campion, K., Foster, A.F., Newman, S.J., O'Rourke, A.M.T. and Thomas, P.G. (1992). Questionnaire development: an examination of the Nordic Musculoskeletal Questionnaire. *Applied Ergonomics*, **23**, 197–201.

Drury, C.G. (1987). A biomechanical evaluation of the repetitive motion injury potential of industrial jobs. *Seminars in Occupational Medicine*, **2**, 1, 41–50 (New York: Thieme Medical Publishers Inc.).

Eklund, J.A.E. and Corlett, E.N. (1984). Shrinkage as a measure of the effect of loads on the spine. *Spine*, **9**, 189–194.

Ferreira, J. and Hignett, S. (2003). Improving posture within the ambulance rear patient compartment. *Ambulance Today*, **2**, 17–20.

Finnish Institute of Occupational Health (1992). OWAS, a method for the evaluation of postural load during work. Publication office, Topeliuksenkatu 41 aA, SF 00250 Helsinki, Finland.

Foreman, T.K. (1989). Low back pain prevalence, work activity analyses and spinal shrinkage. Unpublished Ph.D. thesis, University of Liverpool.

Foreman, T.K. and Troup, J.D.G. (1987). Diurnal variations in spinal loading and the effects on stature. *Clinical Biomechanics*, **2**, 48–54.

Hagberg, M. (1979). The amplitude distribution of surface EMG in static and intermittent static muscular exercise. *European Journal of Applied Physiology*, **40**, 265–272.

Hagberg, M. (1981). An evaluation of local muscular load and fatigue by electromyography. *Arbete och Hälsa*, **24**, Solna, Sweden.

Hignett, S. and McAtamney, L. (2000). Rapid entire body assessment (REBA). *Applied Ergonomics*, **31**, 201–205.

Hutchinson, A. (1970). *Labanotation* (London: Oxford University Press).

Jafry, T. (1993). The effects of vibration, posture and operating foot pedals on spinal loading. PhD thesis, University of Nottingham.

Jian Li. Bryant, J.T. and Stevenson, J.M. (1990). Single camera photogrammetric technique for restricted 3D motion analysis. *Journal of Biomedical Engineering*, **12**, 69–74.

Jonsson, B. (1976). Evaluation of the myoelectric signal in long-term vocational electromyography. In *Biomechanics V*, edited by A.P.V. Komi (Baltimore: University Park Press), pp. 509–514.

Kember, P.A. (1976). The Benesh movement notation used to study sitting behaviour. *Applied Ergonomics*, **7**, 133–136.

Kuorinka, I., Jonsson, B., Kilbom, Å., Vinterberg, H., Biering-Sørenson, F., Andersson, G. and Jorgensen, K. (1987). Standardized Nordic questionnaires for the analysis of musculaskeletal symptoms. *Applied Ergonomics*, **18**, 233–237.

Kvalseth, T.O. (1983). *Ergonomics of Workstation Design* (London: Butterworths), pp. 246–250.

Leatt, P., Reilly, T. and Troup, J.D.G. (1986). Spinal load during circuit weight training and running. *British Journal of Sports Medicine*, **20**, 119–124.

Lind, A. and Petrofsky, J.S. (1979). Amplitude of the surface EMG in fatiguing isometric contractions. *Muscle and Nerve*, **2**, 257–264.

Mackay, C. and others. (1998). Musculoskeletal disorders in supermarket cashiers. HSE Books, Sudbury, C010 2WA. UK.

McAtamney, L. and Corlett, E.N. (1992). Reducing the risks of work related upper limb disorders: a guide and methods. Institute for Occupational Ergonomics, University of Nottingham.

McAtamney, L. and Corlett, E.N. (1993). RULA: a survey method for the investigation of work-related upper limb disorders. *Applied Ergonomics*, **24**, 91–99.

Marras, W.S., Ferguson, S.A. and Simon, S.R. (1990). Three dimensional dynamic motor performance of the normal trunk. *International Journal of Industrial Ergonomics*, **6**, 211–214.

McAtamney, L. and Hignett, S. (2004). Rapid entire body assessment. In: Stanton N, Hendrick, H.W., Salas, E. and Brookhurst, K (eds) *Handbook of Human Factors and Ergonomic Methods* (London: Taylor & Francis).

Milner, N.P., Corlett, E.N. and O'Brien, C.O. (1986). A model to predict recovery from maximal and submaximal isometric exercise. ch. 13 in: *The Ergonomics of Working Postures*, edited by E.N. Corlett, J.R. Wilson and I. Manenica (London: Taylor & Francis Ltd).

Monod, H. and Scherrer, J. (1965). The work capacity of a synergic muscle group. *Ergonomics*, **8**, 329–338.

Parsons, C.A. and Thompson, D. (1990). Comparison of cervical flexion in shop assistants and data input VDT operators. In *Contemporary Ergonomics* (London: Taylor & Francis).

Petrofsky, J.S., Glaser, R.M. and Phillips, C.A. (1982). Evaluation of the amplitude and frequency components of the surface EMG as an index of muscle fatigue. *Ergonomics*, **25**, 213–223.

Rohmert, W. (1973a). Problems in determining rest allowances. 1. Use of modern methods to evaluate stress and strain in static work. *Applied Ergonomics*, **4**, 91–95.

Rohmert, W. (1973b). Problems in determining rest allowances. 2. Determining rest allowances in different tasks. *Applied Ergonomics*, **4**, 158–162.

Stevens, S.S. (1975). *Psychophysics* (Chichester: John Wiley & Sons Ltd).

Tracy, M.F. and Corlett, E.N. (1991). Loads on the body during static tasks: software including the posture targetting method. *Applied Ergonomics*, **6**, 362–366.

van Wely, P. (1970). Design and disease. *Applied Ergonomics*, **1**, 262–269.

*Appendix**

	2050
Surv. No.	☐☐☐☐ 0 1

Musculoskeletal disorders survey of supermarket checkout operators

Dear Sir/Madam,

With the co-operation of your employer and trade unions we are conducting a survey to find out the extent to which muscle and joint aches and pains are experienced by employees in retail occupations.
We are interested in mild and severe problems affecting muscles, ligaments, nerves, tendons, joints and bones suffered both at work and away from work. This could mean sprains, strains, inflammations, irritations, and dislocation. For the purpose of this survey we are not interested in any injuries to the skin.
We would like you to complete this questionnaire about your health. All answers will be treated as strictly confidential and individual answers will not be made known to anyone other than the survey team.
The more questionnaires that are completed, the greater will be the accuracy and usefulness of the findings, the better to help us improve health and safety at work.

Thank you for your help.

Claire Dickinson

HOW TO ANSWER THE QUESTIONNAIRE

Please complete this questionnaire by answering ALL questions as fully as possible. Some of the questions require a written answer, for others need only tick a box. ☑

Please do not write in the margin.

PERSONAL DETAILS

1. Today's date — Day Month Year ☐☐☐☐☐☐
2. Sex — Male 1☐ Female 2☐
3. Date of birth — Day Month Year ☐☐☐☐☐☐
4. What is your weight?
   pounds ☐ ounces ☐ or kg ☐
5. What is your height?
   feet ☐ inches ☐ or cm ☐
6. Are you right or left handed?
   right 1☐ left 2☐ able to use both hands equally 3☐

*The Nordic Musculoskeletal Questionnaire (NMQ) as modified by the UK Health and Safety Executive, and applied to a study of supermarket checkout operators.

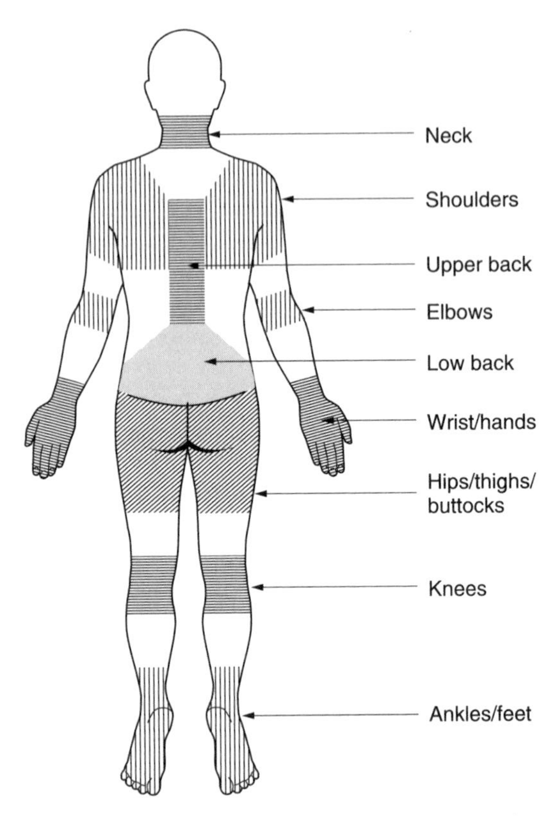

This picture shows how the body has been divided. Please answer the three questions shown on next page for each body area.
Body sections are not sharply defined and certain parts overlap. You should decide for yourself which part (if any) is or has been affected.

Musculoskeletal disorders

Rec [0|2]

Please answer by using the tick boxes ☑ - *one tick for each question*
Please note that this part of the questionnaire should be answered, even if you have never had trouble in any parts of our body.

Have you at any time during the fast 12 months had trouble (such as ache, pain, discomfort, numbness) in:	Have you had trouble during the last 7 days:	During the last 12 months have you been prevented from carrying out normal activities (e.g. job, housework, hobbies) because of this trouble.
1 *Neck* No Yes 1☐ 2☐	2 *Neck* No Yes 1☐ 2☐	3 *Neck* No Yes 1☐ 2☐
4 *Shoulders* No Yes 1☐ 2☐ in the right shoulder 3☐ in the left shoulder 4☐ in both shoulders	5 *Shoulders* No Yes 1☐ 2☐ in the right shoulder 3☐ in the left shoulder 4☐ in both shoulders	6 *Shoulders (both/either)* No Yes 1☐ 2☐
7 *Elbows* No Yes 1☐ 2☐ in the right elbow 3☐ in the left elbow 4☐ in both elbows	8 *Elbows* No Yes 1☐ 2☐ in the right elbow 3☐ in the left elbow 4☐ in both elbows	9 *Elbows (both/either)* No Yes 1☐ 2☐
10 *Wrists/hands* No Yes 1☐ 2☐ in the right wrist/hand 3☐ in the left wrist/hand 4☐ in both wrists/hands	11 *Wrists/hands* No Yes 1☐ 2☐ in the right wrist/hand 3☐ in the left wrist/hand 4☐ in both wrists/hands	12 *Wrists/hands (both/either)* No Yes 1☐ 2☐
13 *Upper back* No Yes 1☐ 2☐	14 *Upper back* No Yes 1☐ 2☐	15 *Upper back* No Yes 1☐ 2☐
16 *Lower back (smell of the back)* No Yes 1☐ 2☐	17 *Lower back* No Yes 1☐ 2☐	18 *Lower back* No Yes 1☐ 2☐
19 *One or both hips/thighs/buttocks* No Yes 1☐ 2☐	20 *Hips/thighs/buttocks* No Yes 1☐ 2☐	21 *Hips/thighs/buttocks* No Yes 1☐ 2☐
22 *One or both knees* No Yes 1☐ 2☐	23 *Knees* No Yes 1☐ 2☐	24 *Knees* No Yes 1☐ 2☐
25 *One or both ankles/feet* No Yes 1☐ 2☐	26 *Ankles/feet* No Yes 1☐ 2☐	27 *Ankles/feet* No Yes 1☐ 2☐

Neck trouble

Rec [0|3]

How to answer the questionnaire:

By neck trouble we mean pain, ache or discomfort in the shaded area only.

Please answer by using the tick boxes ☑ -one tick for each answer.

1 Have you ever had any neck trouble (ache, pain, numbness or discomfort)?

Yes 1☐ No 2☐

If you have answered NO to this question, do not answer questions 2–12 but please go to the section on shoulder trouble page 6.

---

2 Have you ever hurt your neck in an accident?

Yes 1☐ No 2☐

If the answer is NO, please go on to Question 3.

If YES:

2a Was the accident at work?

Yes 1☐ No 2☐

2b What was the approximate date of the accident? Month Year ☐☐☐☐

---

3 Have you ever had to change duties or jobs because of neck trouble?

Yes 1☐ No 2☐

---

4 What do you think brought on this problem with your neck?

1 Accident☐ 2 Sporting Activity☐ 3.Activity at Home ☐

4 Activity at work ☐ 5 Other☐ (please specify) ☐

5a What year did you first have neck trouble? 19☐

5b What year was your worst neck trouble? 19☐

---

6 How bad was the pain during the worst episode?

Mild 1☐ Severe 2☐ Very. Very Severe 3☐

---

7 Have you ever been absent from work because of neck trouble?

Yes 1☐ No 2☐

If the answer is NO, please go on to Question 8.

If YES:

How many times?

7a ☐☐

How many days have you been absent from work with neck trouble in total?

7b ☐☐ days

How many days have you been absent from work with neck trouble in the last 12 months?

7c ☐☐ days

8 How often do you get or have you had neck trouble?

daily	1☐
one or more times a week	2☐
one or more times a month	3☐
one or more times a year	4☐
one or more times every few years	5☐
one episode of trouble only	6☐

---

9 What is the total length of time that you had neck trouble during the last 12 months?

0 days	1☐
1–7 days	2☐
8–30 days	3☐
More than 30 day, but not every day	4☐
Every day	5☐

---

10 Has neck trouble caused you to reduce your activity during the last 12 months?

10a Work activity (at home or away from home)

Yes No
1☐ 2☐

10b Leisure activity

Yes No
1☐ 2☐

---

11 What is the total length of time that neck trouble has prevented you from doing your normal work (at home or away from home) during the last 12 months?

0 days	1☐
1–7 days	2☐
8–30 days	3☐
More than 30 days	4☐

---

12 Have you been seen by a doctor, physiotherapist, chiropractor, or other such person because of neck trouble during the last 12 months?

Yes No
1☐ 2☐

If the answer is NO, please go on to the next section.

If YES:

12a Where? (more than one box can be ticked)

Medical centre at work	1☐
GP	2☐
Hospital	3☐
Private doctor	4☐
Osteopathor chiropractor	5☐
Other*	6☐

*If you have ticked *Other* please give details ☐

Shoulder trouble

Rec 0 4

How to answer the questionnaire:

By shoulder trouble we mean pain, ache or discomfort in the shaded area only.
Please answer by using the tick boxes ☑ -one tick for each answer.

1 Have you ever had any shoulder trouble (ache, pain, numbness or discomfort)?

Yes 1☐ No 2☐

If you have answered NO to this question, do not answer questions 2–12 but please go to the section on low back trouble page 8.

---

2 Have you ever hurt your shoulder in an accident?

No 1☐

Yes 2☐ my right shoulder

3☐ my left shoulder

4☐ both shoulders

If the answer is NO, please go on to Question 3.

If YES:

2a Was the accident at work?

Yes 1☐ No 2☐

2b What was the approximate date of the accident? Month Year ☐☐☐☐

---

3 Have you ever had to change duties or jobs because of shoulder trouble?

Yes 1☐ No 2☐

---

4 What do you think brought on this problem with your shoulder?

1 Accident☐ 2 Sporting Activity☐ 3 Activity at Home☐

4 Activity at work☐ 5 Other☐ (please specify) ☐

5a What year did you first have shoulder trouble? 19

5b What year was your worst shoulder trouble? 19

---

6 How bad was the pain during the worst episode?

Mild 1☐ Severe 2☐ Very. Very Severe 3☐

---

7 Have you ever been absent from work because of shoulder trouble?

Yes 1☐ No 2☐

If the answer is NO, please go on to Question 8.

If YES:

How many times?

7a ☐☐

How many days have you been absent from work with shoulder trouble in total?

7b ☐☐ days

How many days have you been absent from work with shoulder trouble in the last 12 months?

7c ☐☐ days

8 How often do you get or have you had shoulder trouble?

daily 1☐
one or more times a week 2☐
one or more times a month 3☐
one or more times a year 4☐
one or more times every few years 5☐
one episode of trouble only 6☐

---

9 What is the total length of time that you had shoulder trouble during the last 12 months?

0 days 1☐
1–7 days 2☐
8–30 days 3☐
More than 30 day, but not every day 4☐
Every day 5☐

---

10 Has shoulder trouble caused you to reduce your activity during the last 12 months?

10a Work activity (at home or away from home)

Yes No
1☐ 2☐

10b Leisure activity

Yes No
1☐ 2☐

---

11 What is the total length of time that shoulder trouble has prevented you from doing your normal work (at home or away from home) during the last 12 months?

0 days 1☐
1–7 days 2☐
8–30 days 3☐
More than 30 days 4☐

---

12 Have you been seen by a doctor, physiotherapist, chiropractor, or other such person because of shoulder trouble during the last 12 months?

Yes No
1☐ 2☐

If the answer is NO, please go on to the next section.
If YES:

12a Where? (more than one box can be ticked)

Medical centre at work 1☐
GP 2☐
Hospital 3☐
Private doctor 4☐
Osteopath or chiropractor 5☐
Other* 6☐
*If you have ticked *Other* please give details ☐

Low back trouble

Rec [0|5]

How to answer the questionnaire:

By low back trouble we mean pain, ache or discomfort in the shaded area whether or not it extends from there to one or both legs (sciatica).

Please answer by using the tick boxes [√] -one tick for each answer.

1 Have you ever had any low back trouble (ache, pain, numbness or discomfort)?

Yes 1☐ No 2☐ If you have answered NO to this question, do not answer questions 2–12 but please go to the section on wrist/hand trouble page 10.

2 Have you ever hurt your low back in an accident?

Yes 1☐ No 2☐

If the answer is NO, please go on to Question 3.

If YES:

2a Was the accident at work?

Yes 1☐ No 2☐

2b What was the approximate date of the accident? Month Year ☐☐☐☐

---

3 Have you ever had to change duties or jobs because of low back trouble?

Yes 1☐ No 2☐

---

4 What do you think brought on this problem with your back?

1 Accident☐ 2 Sporting Activity☐ 3.Activity at Home☐

4 Activity at work☐ 5 Other☐ (please specify) ☐

5a What year did you first have low back trouble? [19]

5b What year was your worst low back trouble? [19]

---

6 How bad was the pain during the worst episode? Mild 1☐ Severe 2☐ Very.Very Severe 3☐

---

7 Have you ever been absent from work because of low back trouble?

Yes 1☐ No 2☐

If the answer is NO, please go on to Question 8.

If YES:

How many times?

7a ☐☐

How many days have you been absent from work with low back trouble in total?

7b ☐☐ days

How many days have you been absent from work with low back trouble in the last 12 months?

7c ☐☐ days

8 How often do you get or have you had low back trouble?

daily 1☐
one or more times a week 2☐
one or more times a month 3☐
one or more times a year 4☐
one or more times every few years 5☐
one episode of trouble only 6☐

---

9 What is the total length of time that you had low back trouble during the last 12 months?

0 days 1☐
1–7 days 2☐
8–30 days 3☐
More than 30 day, but not every day 4☐
Every day 5☐

---

10 Has low back trouble caused you to reduce your activity during the last 12 months?

10a Work activity (at home or away from home)

Yes No
1☐ 2☐

10b Leisure activity

Yes No
1☐ 2☐

---

11 What is the total length of time that low back trouble has prevented you from doing your normal work (at home or away from home) during the last 12 months?

0 days 1☐
1–7 days 2☐
8–30 days 3☐
More than 30 days 4☐

---

12 Have you been seen by a doctor, physiotherapist, chiropractor, or other such person because of low back trouble during the last 12 months?

Yes No
1☐ 2☐

If the answer is NO, please go on to the next section.
If YES:

12a Where? (more than one box can be ticked)

Medical centre at work 1☐
GP 2☐
Hospital 3☐
Private doctor 4☐
Osteopath or chiropractor 5☐
Other* 6☐
*If you have ticked *Other* please give details ☐

Wrist or hand trouble

Rec 06

How to answer the questionnaire:

By wrist or hand trouble we mean pain, ache or discomfort in the shaded area only.

Please answer by using the tick boxes ☑ -one tick for each answer.

1 Have you ever had any wrist or hand trouble (ache, pain, numbness or discomfort)?

Yes 1☐ No 2☐

If you have answered NO to this question, do not answer questions 2–12 but please go to General health questionnaire on page 12.

---

2 Have you ever hurt your wrist or hand in an accident?

No 1☐

Yes 2☐ my right wrist or hand
3☐ my left wrist or hand
4☐ both wrists or hands

If the answer is NO, please go on to Question 3.

If YES:

2a Was the accident at work?

Yes 1☐ No 2☐

2b What was the approximate date of the accident? Month Year ☐☐☐☐

---

3 Have you ever had to change duties or jobs because of wrist or hand trouble?

Yes 1☐ No 2☐

---

4 What do you think brought on this problem with your wrists or hands?

1 Accident ☐ 2 Sporting Activity ☐ 3 Activity at Home ☐

4 Activity at work ☐ 5 Other ☐ (please specify) ☐

5a What year did you first have wrist or hand trouble? 19☐

5b What year was your worst wrist or hand trouble? 19☐

---

6 How bad was the pain during the worst episode? Mild 1☐ Severe 2☐ Very.Very Severe 3☐

---

7 Have you ever been absent from work because of wrist or hand trouble?

Yes 1☐ No 2☐

If the answer is NO, please go on to Question 8.

If YES:

How many times?

7a ☐☐

How many days have you been absent from work with wrist or hand trouble in total?

7b ☐☐ days

How many days have you been absent from work with wrist or hand trouble in the last 12 months?

7c ☐☐ days

8 How often do you get or have you had wrist or hand trouble?

daily 1☐
one or more times a week 2☐
one or more times a month 3☐
one or more times a year 4☐
one or more times every few years 5☐
one episode of trouble only 6☐

9 What is the total length of time that you had wrist or hand trouble during the last 12 months?

0 days 1☐
1–7 days 2☐
8–30 days 3☐
More than 30 day, but not every day 4☐
Every day 5☐

10 Has wrist or hand trouble caused you to reduce your activity during the last 12 months?

10a Work activity (at home or away from home)
Yes No
1☐ 2☐

10b Leisure activity
Yes No
1☐ 2☐

11 What is the total length of time that wrist or hand trouble has prevented you from doing your normal work (at home or away from home) during the last 12 months?

0 days 1☐
1–7 days 2☐
8–30 days 3☐
More than 30 days 4☐

12 Have you been seen by a doctor, physiotherapist, chiropractor, or other such person because of wrist or hand trouble during the last 12 months?

Yes No
1☐ 2☐

If the answer is NO, please go on to the next section.
If YES:

12a Where? (more than one box can be ticked)

Medical centre at work 1☐
GP 2☐
Hospital 3☐
Private doctor 4☐
Osteopath or chiropractor 5☐
Other* 6☐
*If you have ticked *Other* please give details ☐

General Health Questionnaire Rec 07

We should like to know how your health has been in general, OVER THE PAST FEW WEEKS. Please circle the answer which you think most nearly applies to you.

HAVE YOU RECENTLY:

1	been able to concentrate on whatever you're doing?	Better than usual	Same as usual	Less than usual	Much less than usual
2	lost much sleep over worry?	Not at all	No more than usual	Rather more than usual	Much more than usual
3	felt that you are playing a useful part in things?	More so than usual	Same as usual	Less useful than usual	Much less useful
4	felt capable of making decisions about things?	More so than usual	Same as usual	Less useful than usual	Much less useful
5	felt constantly under strain?	Not at all	No more than usual	Rather more than usual	Much more than usual
6	felt that you couldn't overcome your difficulties?	Not at all	No more than usual	Rather more than usual	Much more than usual
7	been able to enjoy your normal day-to-day activities?	More so than usual	Same as usual	Less so than usual	Much less than usual
8	been able to face up to your problems?	More so than usual	Same as usual	Less able than usual	Much less able
9	been feeling unhappy and depressed?	Not at all	No more than usual	Rather more than usual	Much more than usual
10	been losing confidence in yourself?	Not at all	No more than usual	Rather more than usual	Much more than usual
11	been thinking of yourself as a worthless person?	Not at all	No more than usual	Rather more than usual	Much more than usual
12	been feeling reasonably happy, all things considered?	More so than usual	About same as usual	Less so than usual	Much less than usual

SCORE 0011

13 How often do you experience any of the following symptoms during or after work? For each symptom, put a tick in the appropriate box.

	Frequently	Sometimes	Rarely	Never
Fatique	1☐	2☐	3☐	4☐
Headaches	1☐	2☐	3☐	4☐
Distributed vision	1☐	2☐	3☐	4☐

14 Do you wear spectacles or contact lenses whilst working a check-out?

Yes	No
1☐	2☐

Information about your job Rec 08

1 How many years and months have you been doing your present type of work at this supermarket?

Years Months If less than one month, how many weeks?

2 Have you worked in other supermarkets?

No Yes
1☐ 2☐

2.1 If yes, what is the total length of time you worked on checkouts elsewhere, before working at this supermarket?

Years Months If less than one month, how many weeks?

3 Do you have any other paid job other than at this supermarket?

Yes No
1☐ 2☐

4 On average, how many hours a week do you work at this supermarket? (incuding overtime but excluding the main meal break)

Hours

5 How many of these hours are spent working on a check-out?

Hours

6 Do you rotate or change your duties regularly during the day?

Yes No
1☐ 2☐

If YES,

6a How often?

Changing once every hour 1☐
Changing once every 2 hours 2☐
Changing once every 2–4 hours 3☐
Other 4☐

If you have ticked *Other* please say how often

7 On average how many breaks do you have each working day?

---

8 Ignoring your lunch-break, how long is each of your breaks on average?

minutes

---

9 Do you experience any difficulty in operating the following equipment?

	Yes 1	No 2	Don't use it 3
Laser scanning			
Electronic cash register			
Weighing scales			

		Yes 1	No 2
10a	Do you adjust the backrest of your seat?		
10b	Do you adjust the footrest to your seat?		
10c	Do you adjust the height of the seat?		
10d	Do you move the seat to or from the desk?		

Chapter 17

Biomechanical methods for task analysis

Patrick G. Dempsey and Moira Munro

Introduction

Biomechanics is the study of forces acting on the human body. It is used in the ergonomics field most often to assess manual handling tasks. This chapter outlines the types of task for which a biomechanical analysis is or is not an appropriate method, details an overview of the tools required to carry out an analysis, and describes various methods and models used in the field, with their advantages and disadvantages. This description focuses on the fundamental concepts and calculations required, as biomechanical models and the requisite equipment are constantly evolving.

When interpreting results of a biomechanical analysis, we come to the questions of how valid are the simplifying assumptions and approximations made in the analysis, and how robust are the results of the analysis. And finally, what are our criteria for 'safe' levels of force and how reliable are they? Just as the approach used to estimate forces on the body is important for a reasonably valid estimate, the interpretation of these results is also critical.

Relevant tasks

Biomechanics is a useful tool in most manual handling situations, whether people are lifting, lowering, pushing, pulling, or even when no load is handled but the body's own weight is creating postural stress. The human body is complex and biomechanics cannot at this stage give very fine detail. For instance it is, to our knowledge, not possible to show through calculations which of two backrest shapes would be superior: the posture or force from the backrest would need to change quite noticeably for calculated results of body loadings to show differences. In general, applications are often focused on high-force activities requiring gross body motions.

Biomechanics is best used as a comparative method because, as will be shown throughout this chapter, all biomechanical modelling results rest on simplifications and approximations, since direct measurement of forces acting on and within living body tissues is not possible, except in rare experimental settings. Because of these factors, biomechanics is particularly successful at demonstrating possible improvements obtained from redesigning a task, or helping the ergonomist choose between alternate task or workplace designs. It can also be used to identify the most stressful tasks or elements of a job so that resources can be directed at the most harmful aspects first.

Interpretation of results

Having calculated forces at joints such as elbows or shoulders, and forces on the spine or within trunk muscles, results for various tasks or designs can be compared. It is also possible to assess the feasibility and estimate the safety of tasks even at the design stage, by comparing these forces with recommended limits. However, as will be discussed in this chapter, these limits are not absolute guarantees of safety, so results of a biomechanical analysis should not be used in isolation, but combined with other assessment methods.

Direct observation, discomfort charts or questionnaires (see Chapters 3 and 4) may identify sources of discomfort which force calculations cannot. For repetitive tasks, a physiological assessment is often necessary as well. Injury statistics will help to identify problems for which biomechanics can give relevant answers and help towards an assessment of the cost-effectiveness of redesign.

Drury *et al.*'s (1983) evaluation of a palletising aid provides a good example of the use of biomechanics with other methods. Very simple calculations based on video recordings showed to what extent the load on the lumbar spine was reduced when a palletising aid was used. The heart rate was also shown to go down, and the authors used these factors and injury statistics to evaluate the cost effectiveness of introducing the aids. If this study had been based only on accepted 'safe' spinal load limits, this particular biomechanical model would have indicated that they were quite acceptable. The use of several methods to evaluate the task ensured that problems were not-overlooked and biomechanics demonstrated the benefits of introducing an aid.

How detailed should the analysis be?

The simplest biomechanical calculations are those relating to a static posture and force in the sagittal plane (no twisting or lateral bending). The problem is purely two-dimensional (2D), and there is no consideration of forces caused by accelerations and inertia. It is easy to record the posture in that one plane and all the calculations can be done with a calculator.

If the task is characterised by much lateral bending or twisting, problems or possible improvements would be overlooked with a 2D analysis. Recording posture in three dimensions (3D) requires special methods outlined later. Calculations are quite lengthy and a computer programme is usually necessary. A 3D analysis is therefore quite time-consuming unless the posture-recording method and computer programme have already been set up.

If the task is not static, the forces resulting from the accelerations need to be included in the calculations. This requires sampling the posture at a reasonably frequency (e.g., 50 Hz or higher), along with the value and direction of the acceleration of each limb. Dynamic analyses were once restricted to the laboratory, although technological advances are increasing the potential for conducting dynamic analyses in the workplace. In 2D it is possible sometimes to avoid recording posture and acceleration continuously by using simulations. This has been done for lifting, the accuracy of the results depending on how similar the lifting technique is to the one used as a database by the computer programme. Finally it is quite common to analyse dynamic tasks statically, freezing at the beginning, middle, and end of the task, for instance. If the task was done slowly and smoothly, this is a good approximation. Otherwise this may underestimate forces during the acceleration phase up to two or three times (Garg *et al.*, 1982). These approaches will be discussed in more detail later.

Equipment required

A biomechanical analysis requires measurements of posture, hand-force (and any other external forces acting on the body) and, in the case of dynamic tasks, accelerations. Hand-forces can be measured with equipment as simple as a spring balance, but in some cases a tool equipped with strain gauges is required. These tools can be interfaced to a computer for on-line recordings. For pushing and pulling, analog and digital scales are available commercially, including load cells designed for generic applications. These load cells can be connected to personal computers or even small, portable "Personal Digital Assistants" (McGorry *et al.*, 2002).

The choice of a posture recording method will depend on whether the analysis is 2D or 3D, in the field or in the laboratory. It may not be necessary to record posture when using some 2D computer programmes: the posture is input via a stickman on the screen or can be estimated from a photograph. It is best to base this on a photograph in order to input an accurate posture. Videos can be digitised for 2D or 3D posture recordings. This is time consuming in 3D but often used as a simple field method in 2D. Another simple method, although more suited to the laboratory since work would be interrupted, is to take measurements from the subject with a tape measure and plumb bob (Schultz *et al.*, 1983).

For 3D analyses or dynamic tasks, a dedicated acquisition system is often required. A comprehensive review goes beyond the scope of this chapter, but there are numerous systems available on the market that collect postural data at rates sufficient for dynamic analyses. These systems use various approaches including retrieving the postural data from standard videotape, tracking active and passive motion markers, and tracking postures using electro-magnetic fields.

Human variability

Biomechanics is sometimes used with the aim of determining safe limits for most of the population. At this stage, human variability must be borne in mind. Two critical variances are the large ranges in the population in muscle strength and in the strength of the spinal column. Predicting the percentage of the population capable of performing a task requiring strength, or developing criteria for safe limits based on compression tolerance of the spine, are two of the most commonly applied applications of biomechanics in ergonomics.

Human variability in body weight and stature also affects biomechanical calculations. The posture people adopt for a given task affects loading, but given a particular posture, the loads on the body are greater for larger body weights and limb lengths. It is therefore recommended to use 95th percentile body weight and stature in order to err on the side of safety.

Finally, when results of biomechanical calculations are evaluated against 'safe' limits, it must be borne in mind that people differ in their susceptibility to back pain or injury. Hutton and Adams (1982) have shown that as a rule female spines are more susceptible to fracture than male, and age also is a weakening factor. Troup *et al.* (1987) found that people who had experienced back pain chose to lift lower loads. The mechanism of back pain is still uncertain in many cases: tests on cadaveric spines do not necessarily produce the same effects as those observed *in vivo* (Brinckmann, 1986). The best predictor of susceptibility to back pain seems, at the moment, to be a history of previous low-back trouble (MacDonald, 1984).

Principles of biomechanical calculations

This section demonstrates how loads at body joints are estimated. These results can be used in a comparative way, or they can be compared with population data on maximum strength capabilities, in order to assess how strenuous the task is. (This last aspect is discussed in the next main section. Readers who do not need to know the details of calcuations can turn directly to that section after reading the paragraph on moments and lever arms.) The present section will also demonstrate how forces within the low back can be evaluated with a simple 2D model.

The calculation of forces rests on the principle that all forces must balance each other if the body is to be in equilibrium. If there is a resultant force in any direction, the body will move in that direction. The calculation of loads at body joints which follows is in fact the

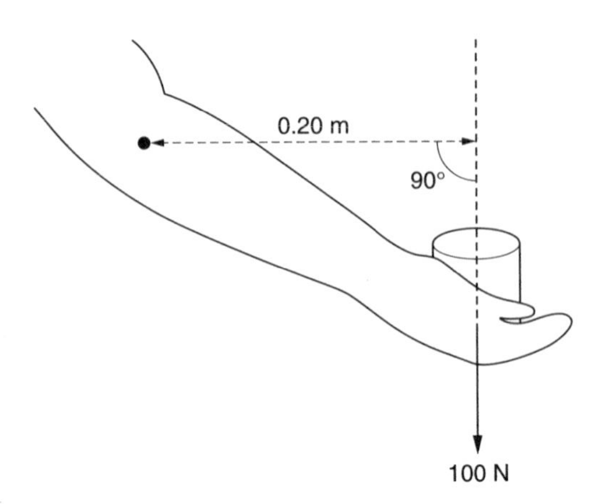

Figure 1 The moment at the elbow due to the 100 N weight held at the hand is the product of the force (100 N) and the perpendicular distance (0.20 m) through which it acts.

calculation of moments, or turning forces around a point. Moments must also balance, so that the sum of moments around any point is zero, if there is to be no rotation.

Moments and lever arms

The moment, or torque, of a force around a point, is a measure of the turning force round the point. For instance holding a weight in the hand creates a moment around the elbow, tending to make it extend. Muscles spanning the elbow provide the opposite moment by contracting, so that the elbow is able to support the weight. The greater the weight, the larger the elbow moment.

The moment of a force about a point is the product of the force and the perpendicular distance between the point and the line of action of the force (see Figure 1). A weight of 100 N held in the hand creates a moment of $100 \times 0.20 = 20$ Nm (Newton metres) for the position shown. (Newtons (N) are units of force of weight. A 1 kg mass weighs 9.81 N. More on this and other units appears at the end of this section.) If the weight was held with the arm hanging down, the lever arm of the force would be zero, and so the moment about the elbow would be zero. There would, of course, still be a force at the elbow, resisting the downward pull of the 100 N weight.

Another force exerting a moment round the elbow is the weight of the hand and forearm. The location of the centre of gravity of the hand and forearm, and their weight, can be estimated from tables, presented later in the chapter (see section 'Inputs to biomechanical calculations').

A simple example with a 2D low-back model

To evaluate forces in the lumbar region, the moment around a point of the low back is calculated in the same way as has been demonstrated for the elbow. Then a model of the muscular layout is used to calculate how much force the back muscles exert to counteract this moment. This enables the compression force on the spine to be estimated, which is a commonly-used criterion.

What follows is a simple example in 2D (see Figure 2): the posture and hand-force are in the sagittal plane. Calculations in 3D, for lateral bending and twisting, will follow after that.

A participant is depicted holding a 100 N weight and moments are calculated around the point indicated by a star, situated on a point of the lumbar spine. The weight of the body

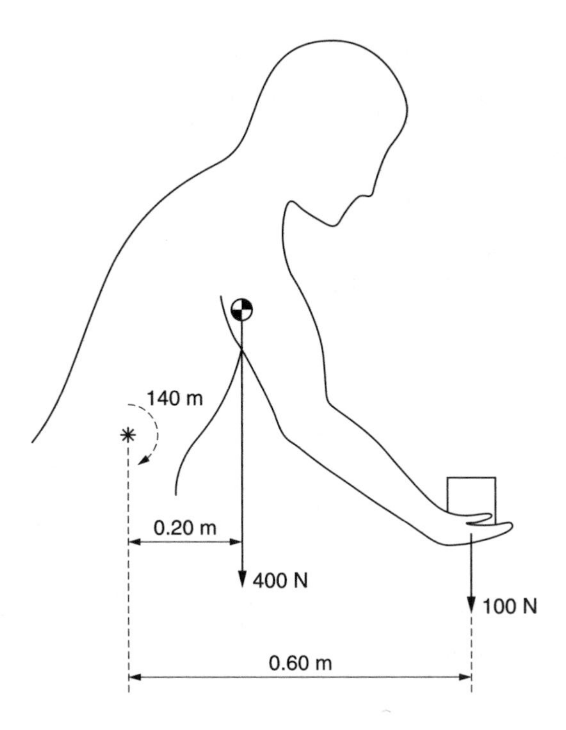

Figure 2 Forces and low-back moment 100 N at the hands.

above this point is 400 N in this example, acting with a lever arm of 0.20 m; creating an 80 Nm moment. Add to this the effect of the weight at the hands, acting with a 0.60 m lever arm, and the total moment created around the starred point is 140 Nm. Thus the weight at the hands and the subject's own body weight tend to flex the trunk: trunk muscles and ligaments must counteract this so that the posture is held.

Forces within the trunk can be evaluated at this stage, using a model of the low back. Such models vary in detail and complexity, as will be seen later, but the principle can be shown here. In a simple 2D model, the 140 Nm trunk flexion moment is resisted by back muscles alone. The greater the leverage those muscles have from the spine, the smaller the force needed from them. Let us suppose the line of action of the back muscles is 5.8 cm posterior to the spine (this choice of values will be discussed later). The force these muscles need to exert is $140/0.058 = 2414$ N. This is much larger than the body weight or the hand-force, because the muscles are balancing the moments through a very small lever arm.

As the back muscles pull to counteract moments, they compress the lumbar spine. The weight of the body above the lumbar spine and the weight at the hands also compress it, so finally the total compression force is the sum of all these components: $100 + 400 + 2414 = 2914$ N. This again may seem large, but compression tests have shown that the spine can, in general, withstand this type of force. Further discussion on this is given in the section 'Forces on the low back'.

Figure 3 shows all the forces involved in this simple problem: the sum of all these forces is zero, and so is the sum of all moments around any point. The point indicated by a star on the lumbar spine was chosen only to simplify calculations, as the compression force does not create a moment about this point. In this example the centre of mass of the upper part of the body is shown as being 0.20 m from the lumbar spine. In fact this would not be known in an ordinary problem; what is estimated is the location of the centre of mass of each body segment, and so the moments created by each segment can be summed. (The section, 'Moments in 3D space', shows in detail how this is done. Calculations are simplified because in

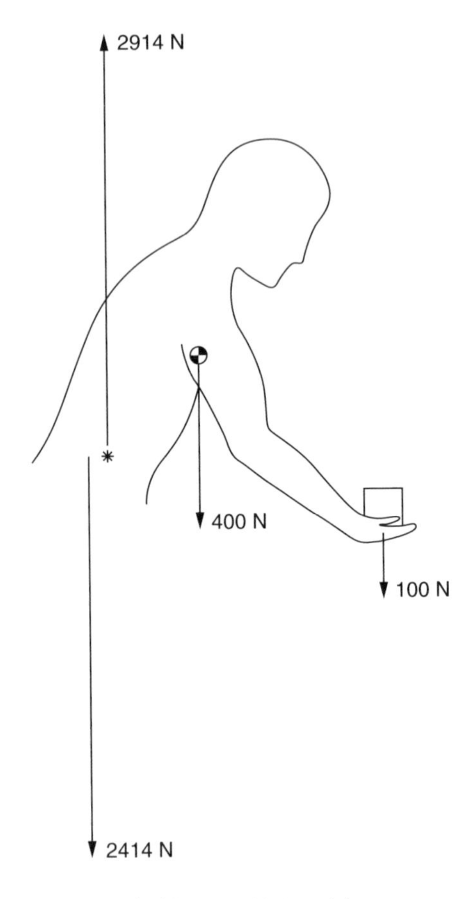

Figure 3 Equilibrium of forces when holding a 100 N weight.

2D situations, $r_y=0$, $F_y=0$, $M_x=0$, $M_z=0$.) We will now go on to calculations in 3D static or dynamic tasks.

Moments in 3D space

The following, on moments in 3D space, contains details which need not be read by users who will not actually be performing calculations.

Previous examples were restricted to postures and forces in the sagittal plane. For asymmetric postures, or forces not contained in this plane, the calculation of the moments can be made by the following method. Figure 4 shows a force **F** (bold type indicates a vector) acting at the hand, with **r** the vector running from the elbow to the hand. The simplest way to determine the moment, **M**, around the elbow is to record the *x, y, z* components of vectors **r** and **F**, along a set of perpendicular axes. Most measuring equipment will allow this. Thus the components of **r** are r_x, r_y, r_z along the *x, y, z* axes, and those of **F** are F_x, F_y, F_z. The moment **M** around the elbow is also a vector and it is simple to calculate its components. M_x is the turning force in the (*y*, z) plane, M_y is the turning force in the (*z, x*) plane, and M_z is the turning force in the (*x, y*) plane.

The resultant of M_x, M_y, M_z is the size of the vector **M**. From Pythagoras' theorem:

$$M^2 = M_x^2 + M_y^2 + M_z^2.$$

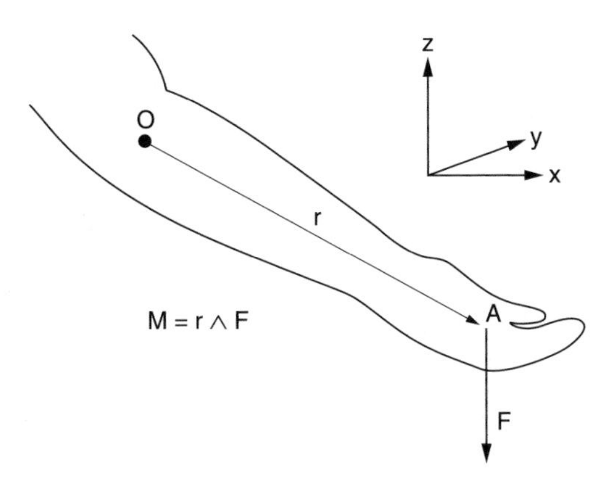

Figure 4 The moment of a force about *O* is the vector product (noted ∧) of the vetcor **r** running from *O* to *A*, with the force vector **F**.

M_x, M_y, M_z are obtained through the following equations:

$$M_x = r_y F_z - r_z F_y$$
$$M_y = r_z F_x - r_x F_z$$
$$M_z = r_x F_y - r_y F_x$$

A short-hand notation for these is:

$$\mathbf{M} = \mathbf{r} \wedge \mathbf{F}$$

M is described as the vector product (noted Λ) of **r** and **F** (the order is important).

For calculations of moments in a 2D situation we have seen that the moment was the product of the force and its perpendicular distance (lever arm). This gives the same result as the equations above, so the method chosen depends on which is easiest to record: the lever arm, or components along the *x*, *y* and *z* axes.

Each component of the moments, M_x, M_y, M_y, represents the turning force about the *x*, *y*, or *z* axis. For instance, in Figure 4 the *y* axis is directed into the paper, and M_y is the moment round that axis, and represents the flexion/extension moment about the elbow.

This is the only moment in the case of Figure 4: the reader can verify from the above equations that M_x and M_z are zero. (The only component of **F** is along the z axis and **r** and **F** are in one plane: $r_y = 0$, $F_x = 0$, $F_y = 0$.) This means that there are no twisting or abduction/adduction requirements on the elbow.

The sign (positive or negative) of a moment indicates the direction of the turning force. In the example of Figure 4, F_z is negative, so from the equations, M_y is positive. This represents the extension effect force **F** has on the elbow. A negative M_y would represent a flexion effect.

One method of working out the meaning of a positive or negative moment is as follows: stick out your right-hand thumb in the direction of the selected axis, *y* in this case. Your other fingers naturally curl round in the direction of the rotation corresponding to a positive M_y, in this case, elbow extension.

For this system, a 'right-handed' set of axes is needed, i.e., *y* should go into the paper, as in Figure 4, not out of it. An easy trick to ensure the axes are right-handed is to point the right-hand thumb along the *z* axis. The fingers curl round to indicate the direction from *x* to *y* (Figure 5).

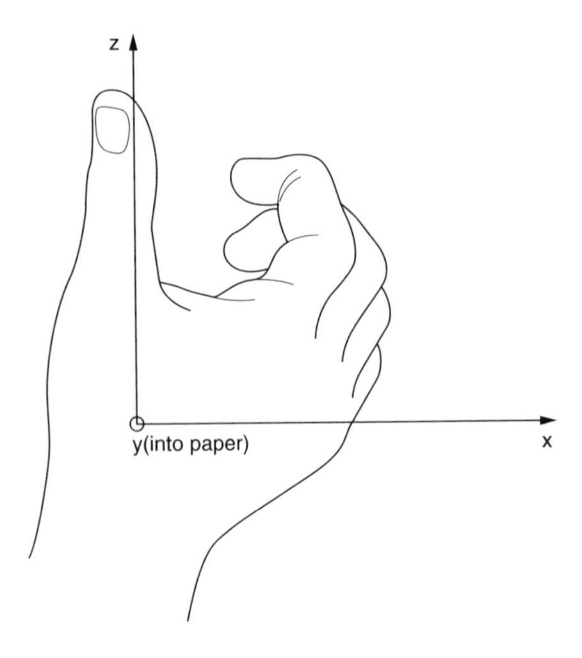

Figure 5 Conventions for a system of axes.

So far we have looked at the moment in 3D created by one force, **F**. When calculating the moment about a point in the low back, for instance, the weights of several body segments have to be taken into account. One method is to add up the moment that each of these forces creates about the low back. This is summarised by the equation:

$$\mathbf{M} = \mathbf{r} \wedge \mathbf{F} + \mathbf{r}_1 \wedge m_1\mathbf{g} + \mathbf{r}_2 \wedge \mathrm{m}_2\mathbf{g} + \cdots$$

where **F** is an external force acting on the body, such as a weight at the hands; **r** is the vector running from the low back to the point of application of **F**; m_1, m_2 are the masses of body segments (kg); $\mathbf{r}_1$, $\mathbf{r}_2$ are the vectors running from the low back to the centres of mass of body segments; and **g** is the acceleration due to gravity (9.81 m/s^2, downwards).

A second method to calculate **M**, which is useful when moments at several joints of the body are already known, is to calculate the moment round the wrist, then use this result and add the effect of the forearm weight to work up to the elbow, and so on to the shoulder, until the low back is reached. This second method, although it may seem less immediate, is more economical if the moments at the wrist, elbow and so on were required anyway. The following equation is used (symbols are shown on Figure 6).

$$\mathbf{M} = \mathbf{r}_{\mathrm{cm}} \wedge m\mathbf{g} + \mathbf{M}_{\mathrm{adj}} + \mathbf{r}_{\mathrm{adj}} \wedge \mathbf{R}_{\mathrm{adj}}$$

where **M** is the moment at the selected joint; $\mathbf{M}_{\mathrm{adj}}$ is the moment at the adjacent joint; $\mathbf{r}_{\mathrm{adj}}$ is the vector running from the selected joint to the adjacent one; m is the mass of the segment between these two joints; $\mathbf{r}_{\mathrm{cm}}$ is the vector running from the selected joint to the centre of mass of the segment; and $\mathbf{R}_{\mathrm{adj}}$ is the resultant force calculated at the adjacent joint.

For example:

$$\mathbf{R}_{\mathrm{adj}} \text{ at the wrist is } \mathbf{F} + m_{\mathrm{hand}}\mathbf{g}$$
$$\mathbf{R}_{\mathrm{adj}} \text{ at the elbow is } \mathbf{R}_{\mathrm{wrist}} + m_{\mathrm{forearm}}\mathbf{g}$$

and so on, if **F** is a force on the hand.

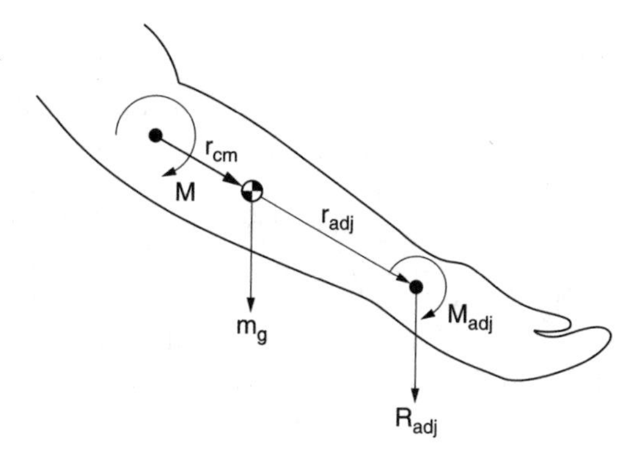

Figure 6 The moment M_{adj} at the wrist can be used to calculate the moment M at the elbow.

Moments in dynamic tasks

Previous sections focused on calculations for static tasks, but body segments undergoing accelerations experience inertial forces that must be added to the moments at the joints. The main difficulty in performing an analysis of a dynamic task is recording instantaneous accelerations throughout the task. Early work by Ayoub and El Bassoussi (1978) included a prediction of accelerations in a dynamic computer model: it is approximated as a function of the angle of each limb at the end of the lift and the duration of the lift. Motion tracking systems now offer accurate measures of the position, velocity, and acceleration of targets at frequencies up to about 500 Hz.

The most common simplification is to ignore accelerations, and treat the problem as a static one. This will lead to errors for many tasks. McGill and Norman (1985) carried out static and dynamic evaluations of L4–L5 intervertebral joint loading. Results with the dynamic analyses were on average 19% higher than with the static approximation, and some up to 52% higher. Garg *et al.* (1982) found that dynamic evaluations were two to three times higher than static ones. These differences are probably due to differences in lifting speed, method, and weight lifted. McGill and Norman (1985) proposed a quasi-dynamic model, in which the only dynamic component was the acceleration of the object lifted. Tsuang *et al.* (1992) have shown that differences between static, quasi-dynamic and dynamic analyses increase mainly with the speed of the lift. Danz and Ayoub (1992) found that peak vertical hand forces at the initiation of a fast lift were 3.0–3.5 times the magnitude of the load being lifted.

The moment at a particular joint varies throughout the motion due to changes both in lever arms and in accelerations. At any instant, the moment depends on the value of the linear acceleration of the centre of mass of each segment, on the direction of this acceleration, and also on the angular acceleration of each segment. The resistance to rotation that an object has depends on its mass and shape and is described by its moment of inertia.

The general equation for the moment **M** at a joint at a particular instant is:

$$\mathbf{M} = \mathbf{r}_{cm} \wedge m\mathbf{g} + \mathbf{M}_{adj} + \mathbf{r}_{adj} \wedge \mathbf{R}_{adj} + \mathbf{r}_{cm} \wedge m\mathbf{a} + \mathbf{I} \wedge \ddot{\boldsymbol{\theta}}.$$

The first three terms have already been described (see Figure 6) in the previous section on static moments. Symbols used in the additional terms are: **a** is the linear acceleration of the centre of mass of the segment ($m\,s^{-2}$); $\ddot{\boldsymbol{\theta}}$ is the angular acceleration of the segment about its centre of mass (degrees s^{-2}); and **I** is the moment of inertia of the segment about its centre of mass ($kg\,m^2$).

A note on units for moments

In the previous example, the lever arm was expressed in metres (m), the force in Newtons (N) and so the moment was obtained in Newton-metres (Nm). These are SI units and therefore recommended; however the following units also can be found in the literature.

The kg-force or kilopound is the force exerted by a mass of 1 kg due to gravity; it is equal to 9.81 N. Thus if a 10 kg object is held at the hand, it exerts a force of 98.1 N. Its weight is 98.1 N, while its mass is 10 kg. One can also come across the pound (lb). This is equal to 0.4536 kg, and multiplying by 9.81 we obtains 4.450 N as the force exerted by 1 lb. Sometimes moments are expressed in inch-pounds, or foot-pounds. With 1 inch = 0.025 m and 1 foot = 0.30 m, we obtain the equivalence: 1 inch-pound = 0.1112 Nm and 1 foot-pound = 1.3349 Nm.

Prediction of strength and task feasibility using moments

Many experimental studies have been carried out to measure the maximum voluntary static strength (overall body strength and the strength of individual joints) of men and women. The prediction of the percentage of the population capable of performing a task can be useful for pre-placement strength testing, return to work assessments, and especially by job and product designers when simulating high exertion tasks during the early part of the design process to avoid costly redesigns (see Chaffin, 1999).

One approach used for task assessment is to evaluate the moments required at each joint by a task and compare them with joint strength data from the population. At least 95% of the population should be accommodated by a given task design. Chaffin (1988) gives examples of this method. One of these applies to pulling carts for moving stock, and it is shown how the percentage of women capable of this drops as the force required increases. In another example, it is shown that most of the working population is capable of performing a particular lifting task (however evaluation of loads on the spine shows the task actually is hazardous). A note of caution is warranted. Since these strength data were collected during static postures, they overestimate strength capacities during dynamic activities. Though valuable in design, any interpretation of results should consider the nature of the task.

Reliability of maximum joint strength data

A problem with using joint strength data is the reliability of the data on a particular joint's maximum strength. A wide range of results can be found in the literature, and this may be due to different testing methods as well as to human variability. For instance, elbow strength results are sometimes given as the force people are able to exert using their hand. This method is not satisfactory as the wrist may be the weak link in this task. To avoid this problem there have been experiments in which the participants exert a force on a device attached proximal to the wrist. These results are incomplete if the distance between this point and the elbow is not quoted. The most useful data are those where elbow strength is expressed directly as a moment and the method for its determination is noted. Low-back strength is particularly difficult to define, as there is no obvious point from which to define moments when testing for strength.

Dependence of joint strength with joint angles

The next limitation concerns joint angles. The force a muscle can exert depends on its length, so moment capabilities depend on joint angles. For example, elbow strength depends not only

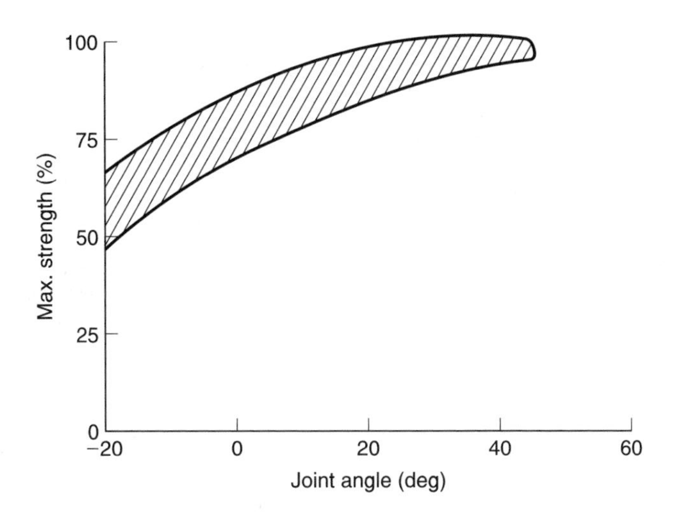

Figure 7 Back extensor strength as a function of trunk angle. Results from 4 studies, normalized by denoting the top value of each curve as 100%. (From Svensson, 1987).

on the elbow angle but also on the shoulder angle, as muscles span across both joints. Yet many results on elbow moments do not report the shoulder angle.

For a compilation from the literature of moment-angle curves of major joints, the reader is referred to Svensson (1987). Figure 7 has been adapted from one of these results and shows the dependence of trunk extensor moment and trunk angle. The hatched area includes curves from four different studies.

A word of caution is in order when referring to published results concerning joint angles: the field of biomechanics does not appear to have any angle conventions and the position of the zero angle varies across studies. One useful standard may be that set by the British Orthopaedic Association (1966) in their booklet describing terminologies used in joint motion. Their method is the 'Zero Starting Position': to accept the 'anatomical position' of a limb as zero degrees. For instance (see Figure 8) the elbow angle is zero for the extended straight arm, and its range of movement is from about 150′ flexion to 10′ hyperextension.

Interpretation of results

Readers can refer to prediction equations in Chaffin *et al.* (1999, p. 263) for maximum moments as a function of joint angles. However, as a wide range of values can be found from other sources, a rough compilation of these ranges is presented in Table 1. It is only intended to give the reader an order of magnitude with which to compare evaluated moments, and further work is needed in this area. The ranges given in Table 1 include all those found in some of the literature, and they all refer to the so-called 'fit and healthy' volunteer. The angle notation in Table 1 follows the British Orthopaedic Association convention described earlier. Results refer to moments along a single axis, for instance pure flexion or pure abduction, so caution must be exercised in using them for a task in which moments in several directions are combined.

An additional word of caution is warranted as antagonistic muscles (i.e., muscles that have opposite effects) often contract simultaneously to the agonist(s). For example at some elbow angles, voluntary elbow flexion recruits not only the biceps (flexor), but also to some extent the triceps (extensor). This co-contraction is often believed to have a joint stabilising role, as

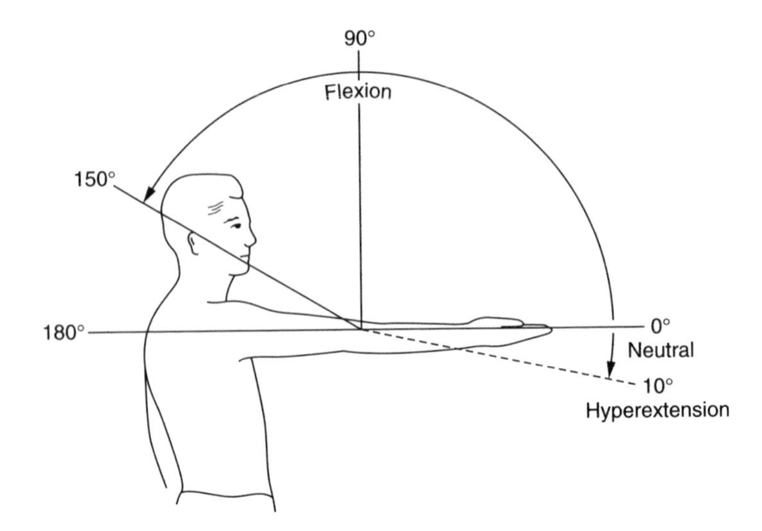

Figure 8 The elbow — flexion and hyperextension (from British Orthopaedic Association. 1996).

one effect is increased joint stiffness. Therefore, it is possible that prediction models overestimate a person's strength if co-contraction is beyond the level during strength testing.

At this stage there are very few data on maximum moments exerted dynamically, probably because of the problem in recording moments that cover a wide enough maximum dynamic range of angles, velocities and accelerations. In conclusion, comparing the moments required by a task with data on maximum capabilities provides some useful guidance in terms of orders of magnitude, but the method should be used with caution for the reasons noted.

Forces on the low back

A study by Chaffin (1988) was mentioned earlier, in which a particular lifting task was analysed. From the moments at the joints it was estimated that most of the working population would be capable of performing the task; however evaluation of loads on the spine indicated that the task could put the back at risk. This section discusses criteria that are available for assessing such risk. A simple 2D model to calculate the compression force on the spine was given earlier in the chapter (Figure 3). This section will describe other models, including 3D ones for asymmetrical postures.

Guidelines from low-back forces

The most commonly used guideline for task assessment is the value of the compression force between vertebrae. Experiments on cadaveric spines have shown that fractures appear above certain levels of compression. The level is lowest for older people; female spines are, as a general rule, weaker than male spines (Hutton and Adams, 1982). The National Institute for Occupational Safety and Health (NIOSH, 1981) concluded that tasks causing a compression on the lumbo-sacral joint greater than 6400 N are above the 'maximum permissible limit': they are unacceptable and engineering controls are required. On the other hand compressions under 3400 N can be tolerated by most young, healthy workers (over 75% of women and over 99% of men). The revised NIOSH equation uses a single 3400 N criterion (Waters *et al.*, 1993). It must be noted that these guidelines relate to lifting in the sagittal plane, and the spine may be much more vulnerable under axial rotation or hyper-flexion (Adams and Hutton, 1981). The majority of compression tests have sought the ultimate compression strength, but work by

Table 1 Maximum Voluntary Joint Strengths (Nm) from Some of the Literature. The ranges Presented Include the Ranges from These Studies

Joint strength	Joint angle (degrees)	Range of moments (Nm) of subjects from several studies		Variation with joint angle
		Men	Women	
Elbow flexor	90	50–120	15–85	Peak at about 90°
Elbow extensor	90	25–100	15–60	Peak between 50° and 100°
Shoulder flexor	90	60–100	25–65	Weaker at flexed angles
Shoulder extensor	90	40–150	10–60	Decreases rapidly at angles less than 30°
Shoulder adductor	60	104	47	As angle decreases, strength increases then levels at 30° to −30°
Trunk flexor	0	145–515	85–320	Patterns differ among authors
Trunk extensor	0	143	78	Increases with trunk flexion
Trunk lateral flexor	0	150–290	80–170	Decreases with joint flexion
Hip extensor	0	110–505	60–130	Increases with joint flexion
Hip abductor	0	65–230	40–170	Increases as angle decreases
Knee flexor	90	50–130	35–115	In general, decreases with knee flexion but some disagreement with this, depending on hip angle.
Knee extensor	90	100–260	70–150	Minima at full flexion and extension
Ankle plantarflexor	0	75–230	35–130	Increases with dorsiflexion
Ankle dorsiflexor	0	35–70	25–45	Decreases from maximum plantar flexion to maximum dorsiflexion

Brinckmann *et al.* (1987) provides data relating to repetitive tasks and the strength of intervertebral joints under cyclic loading. For instance they have shown that for a cyclic load of about half the ultimate compression strength, the probability of a fatigue fracture after 100 cycles is nearly 50%. The value of the compression force may not be the most relevant parameter related to back injury, and guidelines should be used with their limitations in mind, especially at extremes of trunk motion and in dynamic tasks.

Jäger and Luttmann (1999) argue that the 3400 N criterion does not have sufficient epidemiological or biomechanical justification. Jäger and Luttmann (1997) recommended compression limits between 1.8 kN and 6.0 kN, depending upon the age and gender of the worker. There is not widespread consensus regarding the use of a specific criterion, and questions have been raised regarding the applicability of cadaveric data to set *in vivo* limits (Dempsey, 1998). Factors such as the temperature of the specimens, thawing effects, specimen fixation, and the testing environment (Adams, 1995) effect the strength of spinal segments.

Applications

The value of lumbar spine compression is the most frequently used criterion in the evaluation of tasks that may put the back at risk. It has been used extensively in the analysis of lifting tasks, for example, to compare lifting techniques (Chaffin, 1999), evaluate patient handling techniques (Gagnon *et al.*, 1986) or to determine maximum acceptable weights (Hutton and Adams, 1982; Jäger and Luttmann, 1986). More examples of the use of low-back modelling in the analysis of industrial tasks can be found in Norman and McGill (1999).

Due to the inherent difficulty collecting postural data in the workplace, one approach is to use simulation to generate 2D motion data from four frames of a standard videotape of a lift (Chang *et al.*, 2003). The postures in the four frames are matched to a mannequin in a computer programme to generate the initial input data. The use of four frames was suggested by Hsiang *et al.*, (1998) as an optimum number (fewer frames degraded accuracy, whereas more did not significantly enhance it). A motion pattern prediction algorithm (Hsiang *et al.*, 1999) is then used to simulate reasonably accurate motion data for the entire lift, and temporal information is retrieved from the videotape (based on frame rate of input video). The advantage of this approach is that the analysis approximates closely the results provided by a dynamic analysis conducted in the laboratory, the main limitation being that this approach is currently restricted to 2D analyses.

Ergonomists often assess peak loads in individual tasks. Peak loads may be appropriate for some analyses, but cumulative loads may be more relevant, for instance with respect to low-back pain (Norman *et al.*, 1998). Assessment of cumulative loads requires the the continuous recording of information on all tasks in the job, including those with no MMH. Few investigators have assessed cumulative loads (Jäger *et al.*, 2000; Kumar, 1990; Norman *et al.*, 1998), likely because the data collection and reduction costs are not trivial.

Models of the low back

There were several early studies to evaluate forces on the spine through *in vivo* measurement (e.g., Nachemson and Morris, 1964; Nachemson and Elfström, 1970). Following this, intradiscal pressure measurements have been used mostly in conjunction with electromyography (EMG) and intra-abdominal pressure measurements to verify the validity of low-back models (Schultz *et al.*, 1982). Such direct measurement is rarely practical or allowed by ethics committees, leading to modelling efforts discussed below.

In the simple example of Figure 3, one set of back muscles, situated posterior to the spine, was used to resist a trunk flexion moment. This created a compression force on the spine. This is only one model of the low back: many others, varying in complexity, have been developed. The rest of this section describes how to evaluate muscle and compression forces with 2D and 3D models. It has been written for readers who wish to carry out calculations themselves.

A simple 2D model

We will now complete the simple 2D model used at the beginning of the chapter (Figure 3). It allowed back muscles (representing the erector spinae group) to resist a trunk flexion moment. If trunk extension is to be resisted, these muscles can relax and abdominal muscles (rectus abdominis) take over. Figure 9 summarises the results of this model. The lumbar spine compression is shown as a function of the low-back moment. The vertical force on the hands also compresses the spine, as shown from the parallel lines for 0 N, 1000 N upwards and 1000 N downwards.

The equations describing this 2D model, and from which Figure 9 was obtained, are now explained. A flexion moment is provided by the rectus abdominis (R), and an extension moment is provided by the erector spinae (E) (see Figure 10). E acts at a distance y_E from the centre of the spine, and the lever arm for R is y_R.

If M_x is the flexion or extension moment which these muscles must provide:

$$\text{if } M_x > 0 \quad \text{(flexion required)}$$

$$M_x = (-y_R) \times (-R)$$

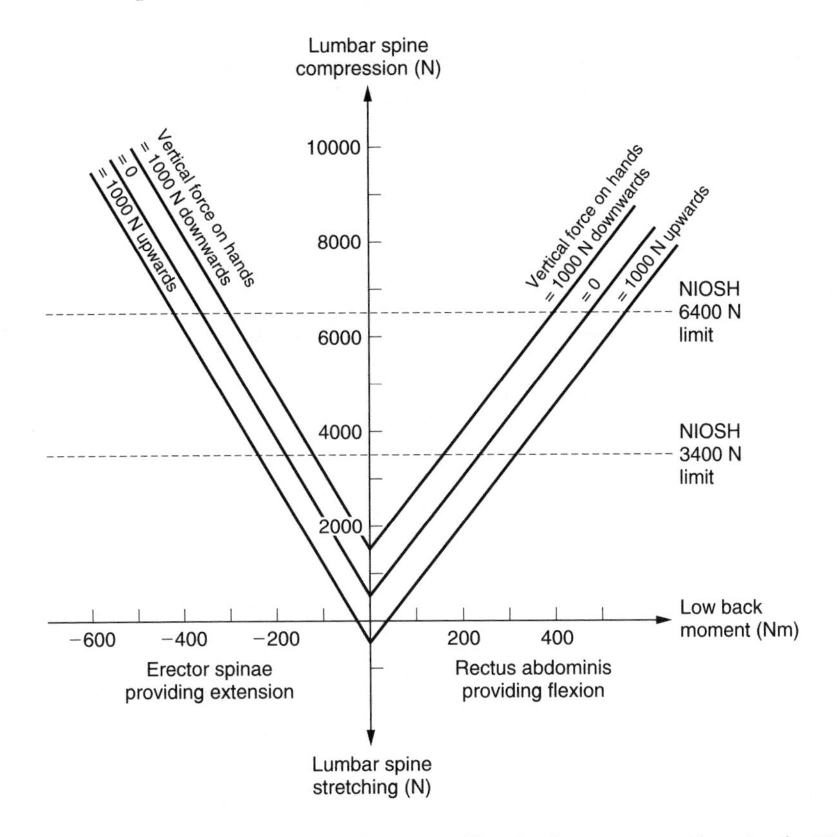

Figure 9 Lumbar spine compression as a function of low-back moment, with a simple 2D model.

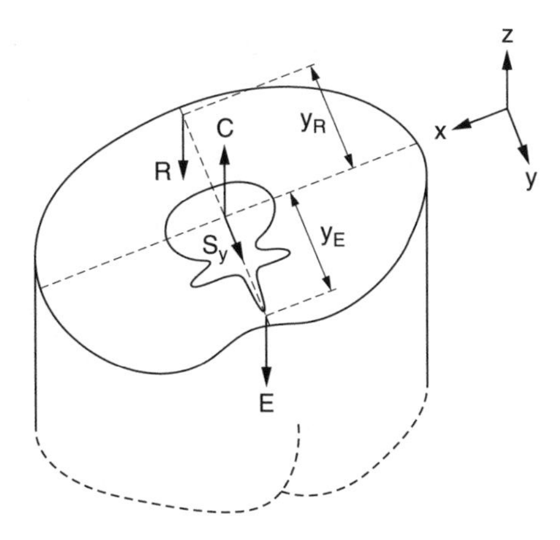

Figure 10 A simple 2D-model represented on a section of the low-back (R=rectus abdominis, E=erector spinae, C=compression, y_R, y_E=lever arms).

(both the y and z axes are in an opposite direction to y_R and R, hence the minus signs), so

$$M_x = y_R R$$

else if $M_x < 0$ (extension required)

$$M_x = -y_E E$$

Finally, E, R and C must add up to a vertical force F_z that counteracts the weight of the body and any external downwards force acting at the hands:

$$F_z = -(\text{sum of body weights and vertical hand force})$$

$$F_z = C - R - E$$

so the value of C, the compression, is obtained.

In the same way, any horisontal force at the hands will be compensated by a horisontal shear force S_y at the intervertebral joint.

For flexion/extension moments, this crude model gives estimates that are very similar to those given by a more detailed model. Returning to the graphical summary in Figure 9 for flexion/extension, the slopes on the graph are $1/y_R$ for positive moments and $1/y_E$ for negative moments, with $y_R = 8.0$ cm and $y_E = 5.8$ cm. (These numbers are average values: see the later section on 'Low-back geometry'.)

3D models

The previous model did not have any muscles accounting for lateral flexion or axial rotation of the trunk. A model described by Chaffin *et al.* (1999, p. 239) will be briefly summarised. The model consists of six muscles (Figure 11): the rectus abdominis (R) and erector spinae (E), provide flexion and extension, respectively; the vertical components of the left and right obliques (VL and VR) provide lateral flexion to the left and to the right. Finally, the horisontal components of the left and right obliques (HL and HR), provide respectively positive (anticlockwise) and negative axial moments round the z axis. Other low-back forces in the model are the compression force C on the inter-vertebral joint (if $C < 0$, the force is on the contrary an extension force), and the lateral and antero-posterior shear forces, S_x and S_y on the intervertebral joint. One more force not mentioned so far is the force P due to intra-abdominal pressure: it is believed that the rise in pressure in the abdominal cavity, that occurs during heavy manual handling, supports the trunk and effectively produces an extensor moment. This moment is equivalent to a force P acting on the centre of the diaphragm. This topic will be discussed in more detail later.

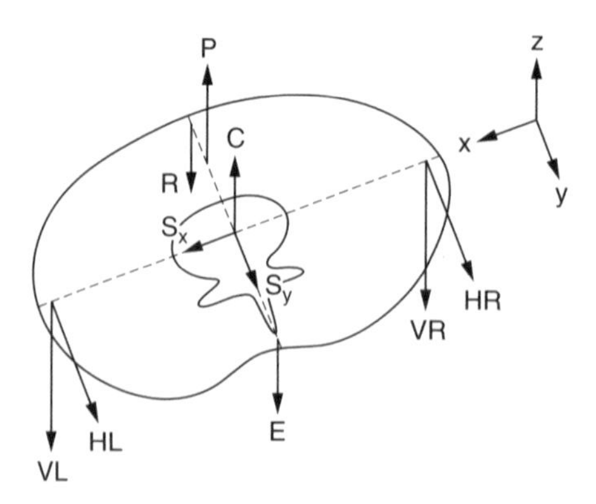

Figure 11 A schematic diagram of a simple 3D model, adapted from Chaffin and Andersson, 1984. (R, E, C = rectus abdominis, erector spinae and compression; P = force due to intra-abdominal pressure; VL, HL are the vertical and horisontal components of the obliques on the left side of the body; VR, HR on the right side).

Table 2 Equations for the Simple 3D Model (adapted from Chaffin and Andersson, 1984)

IF	M_x	$\geq$	0	THEN E	=	0	(flexion required)
	M_x	$\leq$	0	THEN R	=	0	(extension required)
IF	M_y	$\geq$	0	THEN VR	=	0	(flexion to left required)
	M_y	$\leq$	0	THEN VL	=	0	(flexion to right required)
IF	M_z	$\geq$	0	THEN HR	=	0	(anti-clockwise rotation required)
	M_z	$\leq$	0	THEN HL	=	0	(clockwise rotation required)
F_x		=	S_x				
F_y		=	$S_y + HL + HR$				
F_z		=	$C + P - R - E - VL - VR$				
M_x		=	$-y_R P + y_R R - y_E E$				
M_y		=	$x_p (VL - VR)$				
M_z		=	$-x_p (HL - HR)$				

The equations for this model are presented in Table 2. Antero-posterior lever arms are denoted by y, lateral lever arms by x, so y_E is the distance of the erector spinae force E behind the spine, for instance, x_O is the lateral lever arm of the obliques. M_x, M_y, M_z are the low-back moments to be provided by the muscles. F_x, F_y, F_z are the forces provided by the low back to counteract body weight and hand-forces.

One limitation of this model is that the vertical and horisontal components of the obliques are made to act independently, whereas oblique muscles in reality always pull simultaneously in the vertical and horisontal directions. As this model allows an oblique to provide purely a horisontal force, the compression on the spine may be underestimated for tasks involving axial rotation. We will discuss later a 10-muscle model by Schultz and Andersson (1981), which models the obliques in a more realistic way, with internal obliques acting posteriorly downwards and external obliques acting anteriorly downwards.

This 10-muscle model, and others involving more muscles require a computer to carry out a particular mathematical procedure (linear optimisation); such calculations are not trivial. Accordingly a 'micromodel' was proposed (Tracy, 1988), to model the obliques more realistically and produce results that are closer to those of models requiring linear optimisation.

The rule for oblique action is as follows: Internal and external obliques pull respectively posteriorly and anteriorly downwards, as in Figure 12. Suppose a clockwise axial rotation moment must be provided ($M_z < 0$): this can be done by the external on the left (XL) and by the internal on the right (IR). If lateral flexion to the left is also required ($M_y > 0$), XL and IL will be in action. With this model, XL acts strongly to provide both M_z and M_y, and either IR or IL act, depending on which moment is the largest. So if axial rotation is more important than lateral flexion, XL and IR are active.

The equations for this model are given in Table 3. Its predictions come close to those of the 10-muscle model which is to follow, but a computer is not needed. Its main limitations are that the erector spinae and the rectus abdominis (E and R) are placed in the mid-sagittal spine, whereas in reality they are groups of muscles situated to the left and the right. They contribute to lateral flexion moments, whereas the simple models seen here only allow obliques to do this.

3D models requiring optimisation

If the erector spinae and the rectus abdominis are placed as separate forces on either side of the sagittal plane, and if any other trunk muscles are also represented, special techniques are required to estimate forces exerted by the different muscles. Lateral flexion to the right, for

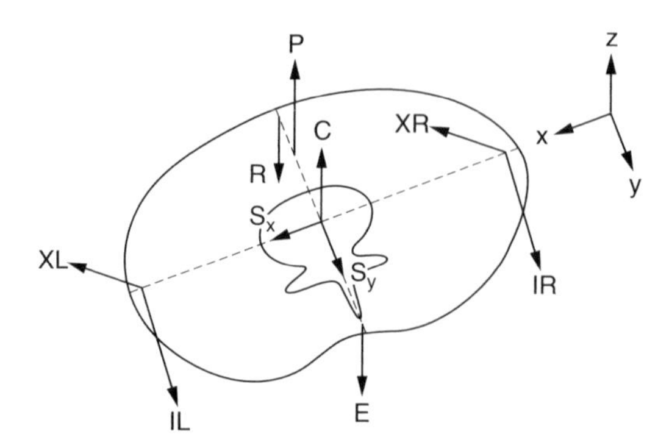

Figure 12 Schematic diagram of the micro-model—a simple 3D model with more realistic representation of the obliques. (Symbols described in Table 3).

instance, can now be provided by the right obliques, the right erector spinae, or the right rectus abdominis. Mathematically, there are more variables (forces) than equations. The indeterminacy must be solved by making assumptions, and giving rules as in Tables 2 or Tables 3 is not possible or practical when many muscles are involved.

A technique is then to use linear programming: this optimises one variable while making sure a number of equations are satisfied. Schultz and Andersson (1981) and Schultz *et al.* (1983) have established models with 10 to 22 muscles, based on the assumption that the spinal compression C is to be at a minimum. A linear programming routine will ensure that all muscle forces provide the required moments and do not exceed a maximum capability of 100 N/cm^2 and that at the same time the forces have been distributed so that C is as low as possible. Although the problem may seem complicated, linear programming routines are in principle straightforward to use. For asymmetrical tasks, models such as the ones developed by Schultz *et al.* (1983) are far superior to simple calculator-based models, as shown by Schultz *et al.* in validation experiments.

The condition to minimise the compression force will have the effect that a muscle with a larger lever arm will provide a force in preference to one with a smaller lever arm. Muscles close to the spine act only if other muscles have reached their maximum capability. Bean *et al.* (1988) have proposed a more sophisticated assumption: first use linear programming to minimise not compression but muscle intensities (force per cross-sectional area): the largest of all the muscle intensities, $\mathbf{I}^*$ must be as small as possible. This ensures that no muscle is giving its maximum while other muscles which could also contribute are inactive. The next step is to solve the problem all over again, minimising C, but with the condition that no muscle intensity exceeds $\mathbf{I}^*$. This is a double linear optimisation approach.

Hughes and Chaffin (1995) compared the Bean *et al.* (1988) approach to that of minimising the sum of the cubed muscle stresses using electromyography (EMG) data collected from eight participants while resisting extension and torsion moments. The authors did not find one approach to be superior for all muscles, as the sum of cubed stress model was best predictor of rectus abdominis activity, whereas the Bean *et al.* approach predicted left erector spinae, right and left external oblique, and left latisimus dorsi activity better.

In a related study, Nussbaum *et al.* (1995) compared the ability of an artificial neural network (ANN) to predict muscle activity as compared to the approaches discussed in the previous paragraph. The ANN results were more highly correlated with the EMG activity, indicating that this approach provides an alternative to optimisation, or actual EMG input.

Table 3 Equations for the Micro-model Shown in Figure 12

Connections	
R	rectus abdominis
E	erector spinae
XL, XR	left and right external obliques, acting in the (y, z) plane, downwards and towards the ventral part of the trunk, at 45° to the transverse plane
IL, IR	left and right internal obliques, acting in the (y, z) plane, downwards and towards the dorsal part of the trunk, at 45° to the transverse plane
P	force due to intra-abdominal pressure
C, S_x, S_y	compression and shear forces on the intervertebral joint
Axes (x, y, z)	centered on the spine, (x, y) in the transverse plane with x to the left of the body, y directed posteriorly, z directed upwards
F_x, F_y, F_z	resultant reaction forces at the level of the section
M_x, M_y, M_z	resultant reaction moments at the level of the section
ABS (x)	absolute (positive) value of x
SUM =	$\frac{M_y + M_x}{2(x_o \cos 45)}$
DIFF =	$\frac{M_y - M_z}{2(x_o \cos 45)}$
Equations	
$F_z = C + P - E - R - (IL + IR) \cos 45 - (XL + XR) \cos 45$	
$F_y = (IL + IR) \sin 45 - (XL + XR) \sin 45 + S_y$	
$F_x = S_z$	
$M_x = -y_E E + y_R R - y_P P$	
$M_y = x_o(IL - IR) \cos 45 + x_o(XR - XL) \sin 45$	
$M_z = x_o(IL - IR) \sin 45 + x_o (XR - XL) \sin 45$	
If $M_x \geq 0$ then $E = 0$	
If $M_x < 0$ then $R = 0$	
If $M_y > 0$ and $M_z > 0$	
and ABS $(M_y) \geq$ BS (M_z) then	$IR = 0$ $XR = 0$ $IL =$ SUM $XL =$ DIFF
and ABS $(M_y) <$ ABS (M_z) then	$IR = 0$ $XL = 0$ $IL = -$SUM $XR = -$DIFF
If $M_y \geq 0$ and $M_z \leq 0$	
and ABS $(M_y) \geq$ ABS (M_z) then	$IR = 0$ $XR = 0$ $IL =$ SUM $XL =$ DIFF
and ABS $(M_y) \leq$ ABS (M_z) then	$IL = 0$ $XR = 0$ $IR = -$SUM XL $= -$DIFF

(*Continued*)

Table 3 Continued

If $M_y < 0$ and $M_z < 0$	
and ABS (M_y) $\geq$ ABS (M_z) then	$IL = 0$
	$XL = 0$
	$IR = -$ SUM
	$XR = -$ DIFF
and ABS (M_y) < ABS (M_z) then	$IL = 0$
	$XR = 0$
	$IR = -$ SUM
	$XL =$ DIFF
If $M_y < 0$ and $M_z \geq 0$	
and ABS (M_y) $\geq$ ABS (M_z) then	$IL = 0$
	$XL = 0$
	$IR = -$ SUM
	$XR = -$ DIFF
and ABS (M_y) < ABS (M_z) then	$IR = 0$
	$XL = 0$
	$IL = -$ SUM
	$XR = -$ DIFF

3D models requiring 'biological input'

Some biomechanical models require the input of EMG signals, or 'biological input'. This approach bypasses the optimisation and ANN predictions discussed above, and will only be mentioned briefly. EMG-assisted modells are often very complex and require exceptional expertise not only due to the biomechanical modelling, but also to properly acquire, treat, and interpret EMG data from people performing real tasks. As such, the models are rarely used by non-researchers. An overview of the model developed at the University of Waterloo, and additional references, is provided by McGill (1999). Details of the model developed at Ohio State, and additional references, can be found in Marras and Granata (1997).

There are tradeoffs when choosing between EMG data and optimisation approaches. Although optimisation requires less data collection and equipment, EMG may be able to provide a more realistic view of activation patterns, especially given the high between-subject differences in trunk muscle co-activation that have been observed (e.g., Perez and Nussbaum, 2003). EMG and optimisation approaches can also be combined into an EMG-assisted optimisation approach (Cholewicki *et al.*, 1995).

Inputs to biomechanical calculations

Posture input

Biomechanics is sometimes used as a predictive tool, on a posture that has not been observed but has been estimated as a likely posture for a task. However, the posture may not be realistic and it may be worth ensuring that the body is in balance by checking that the resultant of all external forces lies in the area between the two feet. If the posture is asymmetric, each leg can take a different proportion of the resultant force at the feet, and this needs to be measured with a force plate, which brings us back to the laboratory. Therefore, biomechanics must be employed as a predictive tool with caution.

Table 4 Segment Masses and Locations of Centre of Gravity, from Pheasant (1986)

Segment	Mass (percentage body mass)	Location of centre of gravity
1. Head and neck	8.4	57% of distance from C7 to vertex
1a. Head	6.2	20 mm above tragion
2. Head and neck and trunk	58.4	40% of distance from hip to vertex
2a. Trunk	50.0	46% of distance from hip to C7
2b. Trunk above lumbo-sacral joint	36.6	63% of distance from hip to C7
2c. Trunk below lumbo-sacral joint	13.4	Approximately at the hip joint
3. Upper arm	2.8	48% of distance from shoulder to elbow joints
4. Forearm	1.7	41% of distance from elbow to wrist joints
5. Hand	0.6	40% of hand length from wrist joint (at centre of an object gripped)
6. Thigh	10.0	41% of distance from hip to knee joints
7. Lower leg	4.3	44% of distance from knee to ankle joints
8. Foot	1.4	47% foot length forward from the heel (half height of ankle joint above the ground) — mid-way between ankle and ball of foot at the head of metatarsal **[I]**

	Total body mass* (kg)							
	Men				Women			
Percentiles	5th	50th	95th	S.D.	5th	50th	95th	S.D.
British (19–65 years)	55.3	74.5	93.7	11.7	44.1	62.5	80.9	11.2

* Masses (in kg) to be multiplied by 9.81 to obtain weights (or forces in N) for the calculation of moments.

Body segment weights

Table 4 summarises masses and the locations of the centre of gravity of body segments, compiled by Pheasant (1986). Other anthropometric data used in modelling, such as link lengths, can be found in the same reference. Segment mass data originate from very small, poorly representative samples, so may be a source of error in biochemical calculations.

Low-back geometry

There have been recent improvements in the data available for low-back models. Both computed axial tomography (CAT) and magnetic resonance imaging (MRI) have been used to measure muscle lever arms and cross-sectional areas, whereas previously values came from a limited sample of cadaveric data. Some of the results needed for the models described in this chapter can be found in Table 5. These are lever arms for 96 females (Chaffin *et al.*, 1990) and at the L3–L4 level for 26 males (Tracy *et al.*, 1989). Data for more muscles and at other lumbar sections can be found in these same studies. Lever arms for both sexes have also been measured by Nemeth and Ohlsen (1986) at the lumbo-sacral joint, and by Kumar (1988) at L3 and L4 (as well as T7 and T12). Daggfeldt and Thorstensson (2003) reported MRI results for lever arms between T12/L1 and L5/S1 for four participants in two different postures.

Table 5 Lever Arms of Some Muscles at L3–L4 Level, from a CT Study of 96 Women (Chaffin *et al.*, 1990) and in MRT Study of 26 Males (Tracy *et al.*, 1989). Standard Deviations in Parentheses)

	Erector spinae	Rectus abdominis	Oblique
Females			
Antero-posterior lever arm (mm)	52 (4)	70 (19)	20 (10)
Lateral lever arm (mm)	34 (4)	43 (11)	113 (16)
Males			
Antero-posterior lever arm (mm)	58 (5)	80 (18)	17 (12)
Lateral lever arm (mm)	38 (3)	34 (10)	122 (11)

Role of intra-abdominal pressure

The most widespread theory of the role of intra-abdominal pressure (IAP) in low-back force production is that the pressure supports the trunk, and its action on the diaphragm and pelvic floor is equivalent to a force for trunk extension. According to this model, the force produced by IAP is calculated by multiplying the pressure by the area of the diaphragm. The extensor moment created by this force is the product of the force with the lever arm of the centroid of the area on which IAP acts. Using this model, IAP reduces lumbar compression by 4–30% according to Schultz *et al.* (1982), 2–8% according to Leskinen and Troup (1984). Recently, Daggfeldt and Thorstensson (2003) reported that IAP produced 9–13% of the extension torque during static exertions.

The reduction in spinal compression due to IAP may be underestimated, because calculations of moments and lumbar loads, ignoring IAP, sometimes still result in excessive compression values although no structural failure is observed (Jones, 1983; Chaffin *et al.* 1999). On the other hand, both Krag *et al.* (1985) and Nachemson *et al.* (1986) have argued from experimental evidence that IAP does not reduce lumbar compression; EMG readings showed that trunk extensor muscle action was not reduced when the abdominal cavity was voluntarily pressurised.

A number of theories for the role of IAP have been put forward, and the reader is referred to Aspden (1987) for a review of these various theories. More recently, McGill (1999) indicated that IAP may not have a direct role in reducing spinal compression, but rather may act to increase trunk stiffness to prevent tissue strain. Until the controversies on the role of IAP are resolved, one approach is to represent IAP as an extensor force, as in Figure 11. Shown in Table 6 are some values found in the literature for various tasks, but many authors choose to ignore IAP and set the force to zero. It has been possible to measure IAP with a swallowed radio-pill for some time (Davis and Stubbs, 1977), but some experience is required to use this technique. Chaffin *et al.* (1999, p. 227) have published a prediction equation for IAP, using the hip moment and angle, for lifting in the sagittal plane.

Angle of discs

So far we have referred to forces on 'the lumbar spine' without specifying a particular vertebra or disc. The low-back models discussed are too crude to differentiate between different vertebrae, and the main difference between calculations at L3 or at L5–S1 is the weight of the trunk above it. There is one other difference, though, and that is the angle of the inter-vertebral discs. The force referred to as the compression force earlier on is actually partly compression and partly shear if the intervertebral joint is not perpendicular to the line of action of the erector spinae or rectus abdominis. Unfortunately there is very little information on disc angles for various postures, and as there are large variations in the degree of lordosis in the

Table 6 Intra-abdominal Pressure for Various Tasks (1 kPa = 7.6 mmHg)

Task	IAP (kPA)	Force (N) developed by IAP over 299 cm^2
Schultz *et al.* (1982)		
Relaxed standing	1.0	30
Uprights, arms in, holding 8 kg in both hands	1.5	45
Flexed 30°, arms out	4.2	125
Flexed 30°, arms out, holding 8 kg in both hands	4.4	130
Davis and Stubbs (1977)		
Breathing	1	30
90 mmHg 'safe limit'	12	560
Grieve and Pheasant (1982)		
Competitive weight lifting	40	1196
Nachemson *et al.* (1986)		
Valsalva manoeuvre	4	120

Estimate for diaphragm area: 299 cm (Leskinen and Troup, 1984).

Estimate for IAP lever arm: 48 mm (Schultz *et al.*, 1982).

population, predictions on disc angles are associated with a large uncertainty. One approach is to infer the shape of the spine from the shape of the surface of the back (Stokes and Moreland, 1987; Tracy *et al.*, 1989) but this work is still somewhat inconclusive. Until more information is available, one solution is to make the approximation that L3 remains perpendicular to the line of action of the erector spinae and rectus abdominis, whatever the posture (Schultz *et al.*, 1983, for instance). For other approaches, and examples of how these angles have been derived for modelling see Chaffin *et al.* (1999, p. 225). In any case, the uncertainty will mean that some of the compression force evaluated may in fact be a shear force, and vice versa.

Uses and limitations of biomechanics

This chapter has surveyed the application of biomechanics in the ergonomics field. Calculating moments at joints provides an estimate of the demands of the task. Moments calculated in 3D will also highlight possible twisting efforts which could be eliminated. Biomechanical calculations allow applied forces or posture to be varied, so that problems and solutions can be identified.

Repetitive work and fatigue

Biomechanics cannot on its own answer questions of the type: 'What force can be applied safely and without fatigue x times a minute for y hours, given that n rest pauses of m minutes are provided?' Biomechanics is better suited at providing estimates of instantaneous forces. Physiology can be helpful when evaluating the fatigue potential of tasks (localised and whole-body). Even less can be said at the moment about dynamic work, presumably because of the large number of variables in the problem. In general, whether the work is static, intermittent or dynamic, biomechanics cannot on its own give reliable answers, except in extreme cases where the task can be shown to be so strenuous it can only be performed occasionally.

Safe limits

Human variability is the main problem when determining acceptable limits, especially where back pain is concerned. Limits on lumbar spine compression, such as those used by the NIOSH equation (Waters *et al.*, 1993), are based on results of ultimate compression strength tests performed on cadaveric spines. However, as discussed in this chapter, these do not usually produce the effects observed in real back injuries. Additionally, less than 6% of workers compensation claims costs attributed to materials handling and less than 1% of the total number of claims affect the discs in the low back (Dempsey and Hashemi, 1999).

There may also be some confusion on how often a task can be repeated if it creates a spine compression of the order of magnitude of the ultimate compression of strength of cadaveric specimens. These tests imply that one single exertion would damage the spine, yet the NIOSH guidelines allow this type of task to be done quite frequently and these tasks are indeed regularly performed in industry. This highlights one of the problems with a focus solely on spinal compression. As mentioned earlier, studies have shown that considering integrated forces across a job is important (Jäger *et al.*, 2000; Kumar, 1990; Norman *et al.*, 1998), and may present a more prudent approach than considering instantaneous peak loadings.

Sources of inaccuracy

Results of low-back forces vary across models, so it is useful to bear in mind the assumptions and simplifications a model employs. Biomechanical models should not be used unless the user is well aware of the simplifying assumptions and how they impact interpretation of the results.

Some inputs to low-back calculations are subject to uncertainty. Data acquisition can be an initial source of error, as no posture recording system is completely accurate. The problems of predicting or measuring muscle activity to deal with indeterminacy were discussed earlier, as some error will be present regardless of the approach used. At this stage, some estimates of the weights and centers of mass of body segments are often required, and these are necessarily inexact estimates based primarily on studies of cadavers.

The role of IAP remains unclear, and the lack of direct measurement will lead to estimation errors. However, the effect of this potential inaccuracy may be attenuated since comparative calculations are prone to the same error.

There are individual variations in spinal shape between people, and there is further uncertainty on disc angle changes with trunk motion. The largest compression estimates will be obtained with discs that are perpendicular to the line of action of the muscles. There are quite large interperson variations in some muscle lever arms; for instance in Table 5, the lever arm of the erector spinae has a standard deviation of nearly 10% of the mean value. Accordingly the uncertainty about the force provided by the erector spinae is also represented by a standard deviation of about 10% of the force calculated with the mean lever arm.

All these sources of uncertainty are not usually critical to the interpretation when results are used to compare tasks (unless the magnitude of the error differs due to the condition being investigated), but are useful to bear in mind when results are used as absolute numbers, and perhaps evaluated against guidelines. There is a mathematical method of evaluating the effect of all the uncertainties on the final result (Barford, 1985), but another way is to experiment with different values of the input parameters to ascertain the sensitivity of the model.

Conclusions

Biomechanics is a useful tool to evaluate manual handling tasks, highlight problems, and compare possible alternate task designs. Models used can be more or less sophisticated; some require computers while a lot can be achieved just with a calculator. Often times, practical constraints in the workplace limit the sophistication of biomechanical analyses. Although more sophisticated methods can have higher validity, practical considerations can limit their applicability to some circumstances.

Results of any analysis should be interpreted with a basic knowledge of the simplifications and uncertainties that were involved in the calculations. Even if one uses a software package available commercially, an understanding of the limitations of the model and assumptions required is necessary for a competent application. Other methods usefully drawn in to complement biomechanics include physiological measurements, EMG, injury statistics, discomfort charts and questionnaires.

References

Adams, M.A. (1995). Mechanical testing of the spine: an appraisal of methodology, results, and conclusions. *Spine*, **20**, 2151–2156.

Adams, M.A. and Hutton, W.C. (1981). The effect of posture on the strength of the lumbar spine. *Engineering in Medicine*, **10**, 199–202.

Aspden, R.M. (1987). Intra-abdominal pressure and its role in spinal mechanics. *Clinical Biomechanics*, **2**, 168–174.

Ayoub, M.M. and El Bassoussi, M.M. (1978). Dynamic biomechanical model for sagittal plane lifting activities. In: C.G. Drury (ed.), *Safety in Manual Materials Handling*. (Cincinnati, OH: US Department of Health, Education and Welfare), pp. 88–95.

Barford, N.C. (1985). *Experimental Measurements: Precision, Error and Truth*, 2nd edition (New York: John Wiley).

Bean, J.C., Chaffin, D.-B. and Schultz, A.B. (1988). Biomechanical model calculation of muscle contraction forces: a double linear programming method. *Journal of Biomechanics*, **21**, 59–66.

Brinckmann, P. (1986). Injury of the annulus fibrosus and disc protrusions. An *in vitro* investigation on human lumbar discs. *Spine*, **11**, 149–153.

Brinckmann, P., Johannleweling, N., Hilweg, D. and Biggemann, M. (1987). Fatigue fracture of human lumbar vertebrae. *Clinical Biomechanics*, **2**, 94–96.

British Orthopaedic Association (1966). *Joint motion. Method of measuring and recording*. Published by the American Academy of Orthopedic Surgeons, reprinted by the British Orthopaedic Association, 1966.

Chaffin, D.B. (1988). A biomechanical strength model for use in industry. *Applied Industrial Hygiene*, **3**, 79–86.

Chaffin, D.B. (1999). Static biomechanical modeling in manual lifting. In: W. Karwowski and W.S. Marras (eds) *The Occupational Ergonomics Handbook*. (Boca Raton: CRC Press).

Chaffin, D.B. and Andersson, G.J. and Bernard, B.J. (1999). *Occupational Biomechanics*, 3rd edn. (New York: Wiley-Interscience).

Chaffin, D.B., Redfern, M.S., Erig, M. and Goldstein, S.A. (1990). Lumbar muscle size and locations from CT scans of 96 women of age 40 to 63 years. *Clinical Biomechanics*, **5**, 9–16.

Chang, C.-C., Hsiang, S., Dempsey, P.G., and McGorry, R.W. (2003). A computerized video-based biomechanical analysis tool for lifting tasks: model development and software design. *International Journal of Industrial Ergonomics*, **32**, 239–250.

Cholewicki, J., McGill, S.M., and Norman, R.W. (1995). Comparison of muscle forces and joint load from an optimization and EMG assisted lumbar spine model: Towards development of a hybrid approach. *Journal of Biomechanics*, **28**, 321–331.

Daggfeldt, K., and Thorstensson, A. (2003). The mechanics of torque production about the lumbar spine. *Journal of Biomechanics*, **36**, 815–825.

Danz, M.E., and Ayoub, M.M. (1992). The effects of speed, frequency, and load on measured hand forces for a floor to knuckle lifting task. *Ergonomics*, **35**, 833–843.

Davis, P.R. and Stubbs, D.A. (1977). Safe levels of manual forces for young males. *Applied Ergonomics*, **8**, 141–150; **8**, 219–228; **9**, 33–37.

Dempsey, P.G. (1998). A critical review of biomechanical, epidemiological, physiological and psychophysical criteria for designing manual materials handling tasks. *Ergonomics*, **41**, 73–88.

Dempsey, P.G., and Hashemi, L. (1999). Analysis of workers' compensation claims associated with manual materials handling. *Ergonomics*, **42**, 183–195.

Drury, C.G., Roberts, D.P., Hansgen, R. and Bayman, J.P. (1983). Evaluation of a palletising aid. *Applied Ergonomics*, **14**, 242–246.

Gagnon, M., Sicard, C. and Sirois, J.P. (1986). Evaluation of forces on the lumbo-sacral joint and assessment of work and energy transfers in nursing aides lifting patients. *Ergonomics*, **29**, 407–421.

Garg, A., Chaffin, D.B. and Freivalds, A. (1982). Biomechanical stresses from manual load lifting: a static vs dynamic evaluation. *IIE Transactions*, **14**, 272–281.

Grieve, D.W. (1987). Demands on the back during minimal exertion. *Clinical Biomechanics*, **2**, 34–42.

Grieve, D.W. and Pheasant, S.T. (1982). Biomechanics. In: W.T. Singleton (ed.) *The Body at Work—Biological Ergonomics*. (Cambridge: Cambridge University Press), pp. 71–161.

Hsiang, S.M., Brogmus, G.E., Martin, S.E., and Bezverkhny, I.B. (1998). Video based lifting technique coding system. *Ergonomics*, **41**, 239–256.

Hsiang, S.M., Chang, C.C., and McGorry, R.W. (1999). Development of a set of equations describing joint trajectories during para-sagittal lifting. *Journal of Biomechanics*, **32**, 871–876.

Hughes, R.E. and Chaffin, D.B. (1995). The effect of strict muscle stress limits on abdominal muscle force predictions for combined torsion and extension loadings. *Journal of Biomechanics*, **28**, 527–533.

Hutton, W.C. and Adams, M.A. (1982). Can the lumbar spine be crushed in heavy lifting? *Spine*, **7**, 586–590.

Jäger, M., Jordan, C., Luttmann, A., and Laurig, W. (2000). Evaluation and assessment of lumbar load during total shifts for occupational materials handling jobs within the Dortmund Lumbar Load Study—DOLLY. *Int. J. Ind. Ergon.*, **25**, 553–571.

Jäger, M. and Luttmann, A. (1986). Biomechanical model calculations of spinal stress for different working postures in various workload situations. In: E.N. Corlett, J.R. Wilson and I. Manenica (eds) *The Ergonomics of Working Postures*. (London: Taylor & Francis), pp. 144–154.

Jäger, M. and Luttmann, A. (1997). Assessment of low-back load during manual materials handling. In: P. Seppälä, T. Luopajärvi, C.H. Nygård, and M. Mattila (eds) *Proceedings of the 13th Triennial Congress of the International Ergonomics Association*, Volume 4, (Helsinki: Finnish Institute of Occupational Health), pp. 171–173.

Jäger, M. and Luttmann, A. (1999). Critical survey on the biomechanical criterion in the NIOSH method for the design and evaluation of manual lifting tasks. *Int. J. Ind. Ergon.*, **23**, 331–337.

Jones, D.F. (1983). Back injury research: have we overlooked something? *Journal of Safety Research*, **14**, 53–64.

Krag, M.H., Gilbertson, L. and Pope, M.H. (1985). Intra-abdominal and intra-thoracic pressure effects upon load bearing of the spine. *31st Annual Meeting Orthopedic Research Society*, Las Vegas, Nevada.

Kumar, S. (1988). Moment arms of spinal musculature determined from CT scans. *Clinical Biomechanics*, **3**, 137–144.

Kumar, S. 1990. Cumulative load as a risk factor for back pain. *Spine*, **15**, 1311–1316.

Leskinen, T.P.J. and Troup, J.D.G. (1984). The effect of intra-abdominal pressure on lumbosacral compression when lifting. *Computer-aided Biomedical Imaging and Graphics Physiological Measurement and Control: Proceedings* (Aberdeen: PMCS), p. 4.

MacDonald, E.B. (1984). Back pain, the risk factors, and its prediction in work people. Occupational aspects of low back disorders. *Society of Occupational Medicine, Symposium Proceedings*, pp. 1–17.

Marras, W.S. and Granata, K.P. (1997). The development of an EMG-assisted model to assess spine loading during whole-body free-dynamic lifting. *Journal of Electromyographic Kinesiology*, **7**, 259–268.

McGill, S.M. (1999). Dynamic low back models: theory and relevance in assisting the ergonomist to reduce the risk of low back injury. In: W. Karwowski and W.S. Marras (eds) *The Occupational Ergonomics Handbook*. (Boca Raton: CRC Press).

McGill, S.M. and Norman, R.W. (1985). Dynamically and statically determined low-back moments during lifting. *Journal of Biomechanics*, **18**, 877–885.

McGorry, R.W., Chang, C.C., Teare, P.R. and Dempsey, P.G. (2002). The flexible handheld ergonomics evaluation tool. *Ergonomics in Design*, **10**, 5–11.

Nachemson, A. and Elfström, G. (1970). Intravital dynamic pressure measurements in lumbar discs. *Scandinavian Journal of Rehabilitation Medicine* (Suppl. 1), 1–40.

Nachemson, A. and Morris, J. (1964). *In vivo* measurements of intradiscal pressure. *Journal of Bone and Joint Surgery*, **46A**, 1077–1092.

Nachemson, A.L., Andersson, G.B.J. and Schultz, A.B. (1986). Valsalva maneuver biomechanics: effects on lumbar trunk loads of elevated intra-abdominal pressures. *Spine*, **11**, 476–479.

Nemeth, G. and Ohlsen, H. (1986). Moment arm lengths of trunk muscles to the lumbosacral Joint obtained *in vivo* with computed tomography. *Spine*, **11**, 158–160.

NIOSH (National Institute for Occupational Safety and Health) (1981). *A Work Practices Guide for Manual Lifting*. Cincinnati: DHHS (NIOSH) publication no. 81–122.

Norman, R., Wells, R., Neumann, P., Frank, J., Shannon, H., and Kerr, M., OUBPS Group (1998). A comparison of peak vs cumulative physical work exposure risk factors for the reporting of low back pain in the automotive industry. *Clinical Biomechanics*, **13**, 561–573.

Norman, R.W. and McGill, S.M. (1999). Selection of 2-D and 3-D biomechanical spine models: issues for consideration by the ergonomist. In: W. Karwowski and W.S. Marras (ed.) *The Occupational Ergonomics Handbook*. (Boca Raton: CRC Press).

Perez, M.A. and Nussbaum, M.A. (2003). Principle components analysis as an evaluation and classification tool for lower torso sEMG data. *Journal of Biomechanics*, **36**, 1225–1229.

Pheasant, S. (1986). *Bodyspace. Anthropometry, Ergonomics and Design*. (London: Taylor & Francis).

Schibye, B., Hansen, A.F., Hye-Knudsen, C.T., Essendrop, M., Bocher, M., and Skotte, J. (2003). Biomechanical analysis of the effect of changing patient-handling technique. *Applied Ergonomics*, **34**, 115–123.

Schultz, A.B. and Andersson, G.B.J. (1981). Analysis of loads on the lumbar spine. *Spine*, **6**, 76–82.

Schultz, A., Andersson, G., Örtengren, R., Haderspeck, K. and Nachemson, A. (1982). Loads on the lumbar spine. *Journal of Bone and Joint Surgery*, **64A**, 713–720.

Schultz, A., Haderspeck, K., Warwick, D. and Portillo, D. (1983). The use of lumbar trunk muscles in isometric performance of mechanically complex standing tasks. *Journal of Orthopaedic Research*, **1**, 77–91.

Stokes, I.A.F. and Moreland, M.S. (1987). Measurement of the shape of the surface of the back in patients with scoliosis. *Journal of Bone and Joint Surgery*, **69A**, 203–211.

Svensson, O.K. (1987). On quantification of muscular load during standing work. A biomechanical study. Dissertation from the Kinesiology Research Group, Department of Anatomy, Karolinska Institute, Stockholm, Sweden.

Tracy, M.F. (1988). Strength and posture guidelines: a biomechanical approach. Ph.D. Thesis, University of Nottingham.

Tracy, M.F., Gibson, M.J., Szypryt, E.P., Rutherford, A. and Corlett, E.N. (1989). The geometry of the lumbar spine determined by magnetic resonance imaging. *Spine*, **14**, 186–193.

Troup, J.D.G., Foreman, T.K., Baxter, C.E. and Brown, D. (1987). The perception of back pain and the role of psychophysical tests of lifting capacity. *Spine*, **12**, 645–657.

Tsuang, Y.H., Schipplein, O.D., Trafimow, J.H. and Andersson, G.B.J. (1992). Influence of body segment dynamics on loads at the lumbar spine during lifting. *Ergonomics*, **35**, 437–444.

Waters, T.R., Putz-Anderson, V., Garg, A., and Fine, L.J. (1993). Revised NIOSH equation for the design and evaluation of manual lifting tasks. *Ergonomics*, **36**, 749–776.

Wells, N. (1985). *Back Pain* (Office of Health Economics).

Chapter 18

The definition and measurement of mental workload

Ted Megaw

Introduction

In the industrially developed countries of the world, there has been an increasing tendency since the middle of the 20th century for many tasks to be dominated by mental rather than physical task components. Most of us have encountered tasks of this kind, tasks such as driving a car or interacting with computers and computer-based technology. In addition, many jobs in the so-called 'high risk' occupations are characterised by their mental rather than physical demands. Typically, these include the jobs of pilots, train drivers, medical personnel and process control operators. Intuitively, it is reasonable to assume that it is essential in such jobs that the mental demands of the job are within the mental capabilities of those performing the jobs to ensure acceptable system performance. It is this obvious assertion that has lead over the last 40 years to the increasing effort to develop reliable, valid and practical measures of mental workload (MWL).

There has been much discussion surrounding the question of providing an acceptable definition of MWL in order to structure the measurement process. As with other areas of ergonomics, for example, visual fatigue, this has proven difficult. This can be appreciated if one considers the wide array of definitions of MWL offered by the various contributors to the first significant collection of papers devoted to the subject, Moray (1979). And things have not got that much better! In a more recent collection of papers, Linton *et al.* (1989) state 'The simple fact of the matter is that nobody seems to know what workload is. Numerous definitions have been proposed, and many of them seem complete and intuitively "right". Nevertheless, current definitions of workload all fail to stand the test of widespread acceptance or quantitative validation.' [p. 22]. The culmination of this uncertainty is reflected in the contents of the international standard (ISO 10075, 1991) entitled Ergonomic Principles Related to Mental Work-Load, in particular, Part 1: General Terms and Definitions. This standards document offers no definition of MWL, but rather defines terms such as mental stress and mental strain in the context of the simple engineering or stimulus-based approach to stress where stress(ors) is defined in terms of the influences impinging upon a person and strain as the effects of those stressors on the individual (Cox, 1978)—but see Chapter 19 of this book.

Rather than pursue a succinct definition of MWL, it is more profitable to provide a simple framework to help understand the implications for the actual measurement process of MWL. The most important feature of the framework is its dynamic nature. As shown in Figure 1, there are three main components to the framework and, as will be described in the main part of this chapter, measurements can be obtained to a greater or less effect in respect of them. However,

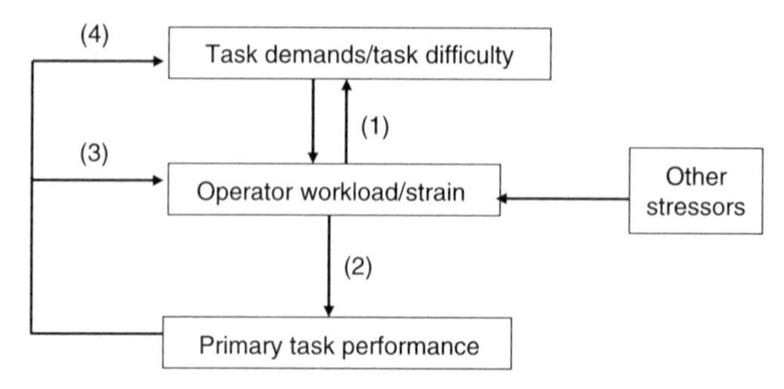

Figure 1 A framework for mental workload definition and evaluation.

the essence of MWL and its measurement is reflected in the relationships between the three components.

The task demands reflect the characteristics of the tasks undertaken by a person. Trying to quantify these objectively is impossible except in the case of very simple tasks. The operator workload is conceived in terms of the operator performing the task and, if the engineering approach to stress research is adopted, is equivalent to measures of operator strain or effort. A majority of MWL measurement is concerned with measures of this kind. Primary task performance is self-explanatory, and most frequently is described in terms of speed and errors. However, the complexity of the measurement problem is only appreciated when one realises that there are few, if any, linear or even monotonic relationships between measures of these three MWL components. That is to say, contrary to what might be expected, as the task demands increase there is not necessarily an increase in operator workload or decrease in task performance. An indication of why such simple relationships do not exist emerges when one considers the various relationships as numbered in Figure 1.

1. Operator workload is not simply a function of task demands but is influenced by how the task is perceived by the operator, sometimes referred to as cognitive appraisal. Most importantly it is influenced by factors such as levels of operator skill, and amount of practice and training. These will determine the strategies that the operator adopts, so that the operator workload is effectively reduced if efficient strategies are adopted. In addition, motivation and arousal levels and fatigue can influence operator workload, in some cases by causing the adoption of less effective strategies. These factors indicate that measures of operator workload will not necessarily reflect objective measures of task demands. For example, just because a learner driver records a high level of workload does not imply that the driving task is imposing too many demands and should, therefore, be modified, but rather that there is need of further practice and training.
2. Although it is often assumed that there is a close relationship between operator workload and task performance, with high operator workload being associated with relatively poorer performance, dissociations between workload and performance frequently occur.
3. Operators monitor their own performance, often unconsciously, as well as being provided with numerous sources of feedback. This may change the way they perceive the task, alter their performance strategies and have motivational consequences, and hence modify their workload.
4. Performance outcomes can modify the tasks themselves so that task demands are altered. For example, in the case of a process control operator diagnosing a complex fault, it is possible that by following some inappropriate diagnostic strategy, the subsequent task demands are increased.

These last two interactions highlight the fact that there are likely to be temporal fluctuations in MWL levels while performing essentially the same job. However, these fluctuations are not easy to assess.

Figure 1 also indicates that workload is influenced by factors outside the task itself (external stressors). These influences are often discussed with reference to an inverted-U relationship relating task performance to levels of physiological arousal, whereby performance is relatively poor under conditions of low (under arousal) or high (over arousal) levels. Physiological arousal levels are a function of both task demands and the impinging external stressors. By applying the contentious Yerkes–Dodson law (Yerkes and Dodson, 1908) to the inverted-U relationship, a number of predictions can be made about the influences of external stressors on primary task performance. Vigilance related tasks, characteristic of many monitoring tasks, might be enhanced by the addition of external stressors such as music-while-you-work or white noise (Smith, 1989). Such tasks pose low workload demands and hence yield low levels of intrinsic arousal. The role of the external stressors would, therefore, be to increase arousal levels and thus improve performance. Taking an example at the other extreme, although performance in respect of a fairly high demanding task might be satisfactory under certain training conditions, one might find that such performance levels do not transfer to real operational conditions. This is because the increased arousal resulting from stressors such as personal fear and a range of environmental factors can lead to conditions of over arousal. Thus external stressors may cause both performance enhancement and degradation, depending on the task demands. Parallel effects can be postulated when comparing the effects of external stressors on extroverted against introverted personality types and on trained against untrained operators.

Underlying psychological processes

Much of our current understanding of MWL has been influenced by what were originally described as models of information processing. According to these early models (Welford, 1968) humans are characterised by possessing a limited amount of information processing capacity such that performance deteriorates if task demands exceed this capacity. In the early models this concept of limited capacity remained fairly ill-defined, though attempts to measure the capacity through the application of information theory were prevalent (e.g., as reflected in the work of Hick (1952), and Fitts (1954)). More recently, a number of key refinements have been made by authors such as Kahneman (1973) and Wickens (Wickens and Hollands, 1999) which have important implications for the definition and measurement of MWL. Most significantly, the term 'limited processing capacity' has been replaced by the term 'attentional resources'. These resources have to be shared between a number of basic psychological processes such as perception, working memory, and response execution, depending on the task demands. This is illustrated in the general model proposed by Wickens which is shown in Figure 2. MWL can then be conceived in terms of the relationship between the resources supplied to these various processes and the task demands in the manner depicted in Figure 3 (adapted from Wickens and Hollands, 1999). On the left of the figure is a region where task performance is satisfactory because the resources available are in excess of the task demands, and there is spare capacity or spare attentional resources. This region reflects conditions of relatively low task demands where workload is inversely related to the amount of spare resource capacity. Moving to the right of the figure, one enters a region where insufficient resources are available to meet the task demands because the limit in available resources has been reached. This is a region where task demands are relatively high and where workload is inversely related to primary task performance.

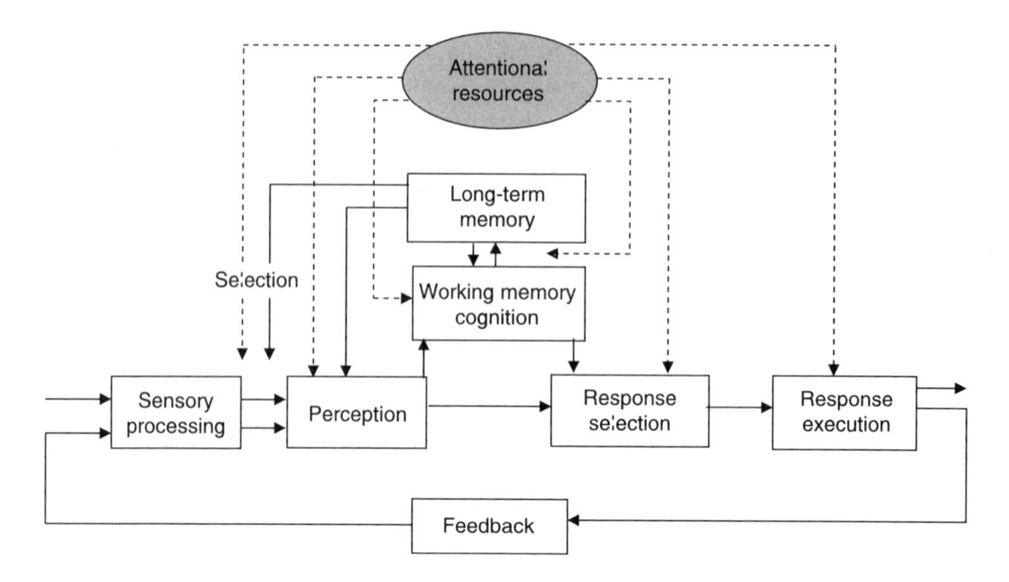

Figure 2 A general model of human information processing adapted from Wickens and Hollands (1999).

A number of other important refinements, some more contentious than others, need to be addressed in the light of Figure 3 in order to get a fuller understanding of the relationship between performance and resources invested.

1. *Controlled versus automatic processes.* Arising out of the research by Schneider and Shiffrin (1977) and Treisman and Gelade (1980) on visual search, it was found that for tasks where elements were consistently mapped together in relation to stimulus characteristics and required responses, performance was not degraded by increasing task difficulty. The consistent mapping is usually achieved after an individual has received plenty of experience of the task characteristics. However, with inconsistent mapping, even after practice, performance decreases with increased task difficulty. In the former case, it can be concluded that the task does not require the investment of attentional resources, while in the latter it does. Another example of automatic processes is the execution of so-called motor programmes. Many motor activities, although extremely complex, appear not to require the investment of attentional resources. A good example of this is walking. After an extensive learning phase, walking is controlled in an open-loop rather than a closed loop manner whereby a command is issued to the brain instructing a walking motor programme to be initiated. Of course, there are some peripheral closed-loop activities to overcome, for example, unevenness in the walking surface, but these are carried out unconsciously. However, if the unevenness becomes too great, for example when walking across a rocky surface, then conscious closed-loop control, requiring attentional resources, occurs.
2. *Data-limited versus resource-limited performance.* Norman and Bobrow (1975), in a somewhat hypothetical paper, discuss the relationships that may exist between attentional resources invested and performance. They postulated that some aspects of performance are not determined by the amount of resources invested because the performance is limited by the quality of the data (data-limited performance), while other aspects of performance are improved by increasing the amount of resources invested (resource-limited performance). Examples of data-limited aspects of task

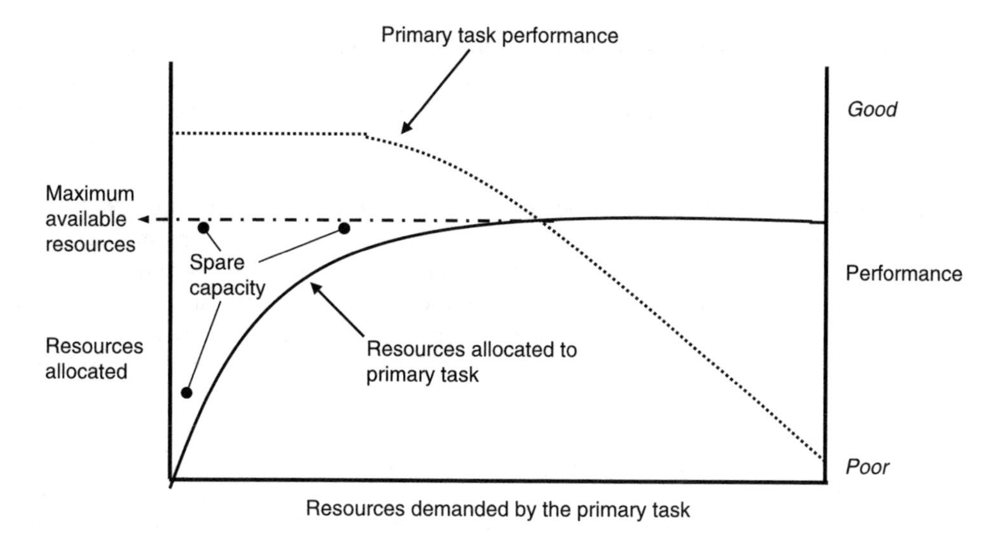

Figure 3 The relationship between the resources allocated to the primary task and the resources demanded by the primary task (—), and the relationship between primary task performance and the resources demanded by the primary task (. . .)

performance are visual detection and memory retrieval. In the former case, performance is limited by the discriminability or conspicuity of the target and, in the latter case, by the quality of the memory representation. Naturally, many tasks will include both data- and resource-limited components.

3. *Single or multiple resources.* This and the next section relate to the fact that many tasks we undertake are made up of a number of subtasks, all of which may impose different demands. At the same time, it is clear that some of these subtasks can be executed in a parallel fashion while others can only be executed serially and, therefore, require task switching. Whether or not these tasks can be carried out in parallel or serially to a large extent reflects physical limitations. For example when driving a car it is impossible to monitor the outside conditions while at the same time looking down at the radio display: hence the interest in developing head-up displays (HUDs). On the effector side, it is impossible to hold a mobile phone while manually changing gear and leaving at least one hand on the steering wheel. When there are no such physical limitations, it is clear that tasks can be carried out in parallel. For example, one can hold a conversation with a passenger or over a mobile phone while driving. However, although driving performance may appear to be unaffected by the parallel task, there may be hidden deficiencies in the driving performance resulting in lower situational awareness resulting in drivers being less able to anticipate potentially dangerous events. The above observations provide the context for the current debate on the use of mobile phones while driving, including hands-free sets.
The last example implies that it is not only physical limitations which determine when these subtasks can be carried out in parallel or not. This is seen in Wickens' (2002) multiple-resource model. According to this, rather than attentional resources being generally available to support any of the various underlying psychological processes, resources are reserved so that only a certain percentage of the total resources can be allocated to any particular psychological process. The dimensions on which the resources are divided remain debatable. There is evidence, often intuitive, that there are separate resources for the two main input modalities: auditory and visual, and for the two main output modalities: vocal and manual. Two other closely related

dimensions originally proposed by Wickens reflect the coding of perceptual information and the coding of information within working memory. For example, visual information can be coded spatially when observing a road map but verbally when reading text. At the same time when manipulating or reasoning about perceptual information which for example has been initially coded spatially, the information may subsequently be coded verbally. Thus when one attempts to plan a route from observing a map, one may end up with a series of memorised verbal instructions of the kind 'follow the main road and take the third turning on the left'. According to the multiple resource model, the extent to which tasks or task elements can be executed in parallel will depend on the extent they compete for the same specific attentional resources. This approach is discussed further in the section on secondary task techniques of MWL assessment.

4. *Task switching*. It is clear from the previous section that under several circumstances, depending on the specific underlying workload demands, certain tasks or subtasks cannot be efficiently carried out in a parallel manner. In such cases, it is necessary, therefore, for operators to switch between tasks or subtasks. This process can be thought of as task management. A critical aspect of the introduction of automation is to ensure operators adopt optimum task management strategies. A typical task where switching is important is the monitoring of multiple displays in police and process control rooms, hospital operating theatres and aircraft and other vehicle workplaces. How operators should allocate the time they spend observing each display was first discussed by Senders (1964). While task switching is an essential component of efficient performance, it should be noted that switching does impose its own demands. These have been investigated under the guise of 'switch costs' (Wylie and Allport, 2000; Rubenstein *et al.* 2001) reflected by an increase in reaction times in tasks following a task switch. Further problems may arise from the extra demands imposed by task switching, resulting in operators being reluctant to switch and spend too much time on low priority tasks at the expense of higher priority tasks, a process referred to as 'cognitive lock-up'. The three-dimensional model of MWL developed by the TNO Human Factors Group (Neerincx, 2003) specifically includes task switching as one of the MWL dimensions, the other two being the percentage of available time needed to carry out the task and the level of processing demands.

From the above discussion, it is not difficult to realise why it has been difficult to arrive at an acceptable definition of MWL. To a certain extent the problem is one of the proverbial chicken and egg. Are we trying to use our knowledge, albeit incomplete, of underlying models of psychological processes to derive definitions and measures of MWL, or are we using MWL measures to investigate models of psychological processes? In reality, a bit of both, I think.

Classification of mental workload techniques and measures

The classification shown in Table 1 is based on the framework provided by Hill *et al.* (1987). Although probably not exhaustive, the list does include what are currently considered the most important measures of MWL. The main division of the techniques is into analytic and empirical measures, the latter requiring operators to perform the task(s) under investigation unlike the former where no such participation is required. The analytic techniques are frequently used to provide predictive estimates of MWL, and are often incorporated into the overall system design process.

Table 1 A Classification of Mental Workload Measurement Techniques

Analytical techniques
- Comparability analysis
- Mathematical models
- Expert opinion
- Task analytic methods
- Simulation models

Empirical techniques
- Primary task performance
  - time related
  - error related
  - indices (combined measures)
  - strategy related
- Operator opinion/subjective techniques
  - rating scales: single dimensional or multi-dimensional, absolute or relative
  - interviews
  - questionnaires
- Secondary (concurrent or dual) task techniques
  - loading task
  - subsidiary task
  - adaptive task
  - embedded task
- Physiological or psychophysiological techniques
  - cardiac activity
  - heart rate, heart rate variability, blood pressure
  - brain activity
    - electroencephalographic activity (EEG)
    - cortical evoked response/event-related brain potentials (ERPs)
    - magnetoencephalographic activity (MEG)
    - functional magnetic resonance imaging (fMRI)
    - positron emission tomography (PET)
    - brain metabolism
  - electrodermal activity/galvanic skin response (GSR)
  - respiratory function
  - eye function
    - visual occlusion
    - eye movements
    - eyeblinks
    - pupil response
  - hormonal analysis
  - muscle activity (EMG)
  - blood glucose levels

Criteria for MWL techniques and measures

A number of attempts have been made to establish the appropriate criteria for MWL measurement techniques. Typical of these are those specified by Wickens and Hollands (1999):

- *Sensitivity.* The technique should provide measures that are sensitive to changes in task demands or in required attentional resources.
- *Diagnosticity.* The technique should provide measures that allow the cause of variation in MWL to be identified. In this context, the application of multiple-resource

theory could lead to improved system design by ensuring that specific resources are not over-demanded.

- *Selectivity*. The technique should provide measures that allow variations in MWL to be distinguished from possible confounding factors including physical workload and emotional stress. Also included here are the confounding effects resulting from environmental factors such as noise and illuminance.
- *Obtrusiveness/Intrusiveness*. The measurement technique should not interfere or disrupt primary task performance whose workload is being assessed when there are safety risks, for example in the case of performing process control tasks or driving tasks under real operational settings.
- *Bandwidth*. The technique should provide measures that allow fairly rapid variations in MWL to be tracked over time, without risk of the measures saturating.
- *Reliability*. The technique should provide measures that can be replicated.

Additional criteria are sometimes cited (O'Donnell and Eggemeier, 1986):

- *Implementation requirements*. These relate to factors such as the instrumentation required (including costs) and the amount of participants' training.
- *Participants' acceptance*. This relates to the extent to which participants will follow the instructions they are given and co-operate with the measurement technique as required.

A broad indication of the extent to which the various measurement techniques comply with the above criteria is provided by O'Donnell and Eggemeier (1986). Additionally, the following review of MWL techniques makes frequent reference to selection criteria issues.

Review of some of the more popular MWL measurement techniques

Clearly, it is not possible to cover all the techniques in this chapter. I have, therefore, decided to concentrate on the more popular techniques, mainly at the expense of providing detailed coverage of the analytic techniques.

Analytic techniques

Comparability analysis

This technique is used to prospectively evaluate performance and workload associated with systems that are under development by making comparisons with the performance and workload associated with some reference system. Naturally, it is important that the systems under development have as many similar characteristics as the reference system as possible. It is difficult to find many references to the use of this technique, but it is often used informally within the design process. However, the use of the technique or tool is encouraged within the MANPRINT framework with regard to supporting the Program Initiation Phase (Malone *et al.* 1988).

Mathematical models

Information theoretic, manual control and queuing theory models are the most commonly used mathematical models of MWL. Information theoretic models are based on the application of information theory (Shannon and Weaver, 1959) to early models of limited processing capacity as described by Welford (1968). The major problem with this approach has been the failure to be able to quantify in information terms for anything more than the most simple of tasks, typified by choice reaction times (Hick, 1952) and simple manual aiming movements (Fitts, 1954). Manual control models are based on classical models of feedback control within engineering systems. They have been reasonably successfully applied to a variety of tracking tasks, both compensatory and pursuit. These tasks form some of the essential components of vehicle driving, aircraft piloting and videogames. Thus, they allow assessments to be made of system controllability within such tasks. However, as already implied, the models are restricted to continuous tasks where the number of input and output variables is limited. Queuing theory models were developed in the context of Operations Research and have received some success in the context of modelling visual scanning behaviour (Senders and Posner, 1976).

Expert opinion

This technique involves taking advantage of experts who have knowledge both of MWL and the work context in which the MWL is to be assessed. A typical example is the use of Pro-SWAT. SWAT stands for Subjective Workload Assessment Technique and, as will be seen in the section on subjective methods, comes in several guises. In the case of Pro-SWAT, the technique parallels the use of the walkthrough technique employed in usability evaluation. An example of the technique is reflected in the study of Kuperman (1985). The context of the study was the conceptual development of a number of alternative fighter aircraft crewstation designs. Several expert pilots were asked to predict workload using the three rating scales typically associated with SWAT (see Table 3) in relation to the alternative designs and to a number of typical mission scenarios segments.

Task analytic methods

These methods, probably the most popular of all the analytic methods, initially involve carrying out a task analysis and then generating a time-line analysis on which the workload estimation can be made. The initial task analysis can be made by using a task specification (task synthesis) or by collecting data from operators performing the task. A common practice is to base the initial task analysis and workload estimation on operators performing the task, and then to propose certain changes to the system design if unacceptably high MWL is identified, so that the resulting changes in estimated workload can be predicted. What is important to realise is that the validity of the workload estimation relies on the quality of the task analysis as well as on the workload estimation procedures. Three task analytic models have dominated the literature. They are TLAP (Time-Line Analysis and Prediction — Parks and Boucek, 1989), VACP (Visual, Auditory, Cognitive, Psychomotor — Aldrich *et al.* 1989) and W/INDEX (Workload Index — North and Riley, 1989). In essence, these three models differ in certain critical underlying assumptions:

- *Channels used.* In completing the time-line analysis, TLAP makes use of five channels: vision, audition (both hearing and speech), hands, feet and cognition. VACP uses the channels implied by the acronym, while W/INDEX uses channels derived from multiple-resource theory, typically visual perception, auditory perception, spatial cognition, verbal cognition, manual response and voice response.
- *Red-line threshold.* Both TLAP and VACP models have red-line threshold values above which performance starts to degrade, whereas W/INDEX assumes that performance decrement is a function of task similarity.
- *Demand level.* TLAP assumes demands for any channel are either zero or totally taken up. VACP and W/INDEX permit graded demands.
- *Interaction of workload components/channels.* Both TLAP and VACP assume that the channels represent independent workload components. However, according to the W/INDEX model, resources overlap with each other so that interference can occur across channels as specified by the values found in a so-called conflict matrix.

Over recent years there have been numerous attempts to validate the various models and to change the various underlying assumptions, particularly in relation to the choice of channels and to the values in the conflict matrix. Examples of studies into validation are provided by Wickens *et al.* (1989) and Sarno and Wickens (1995). The results are not that convincing. One of the more recent task analytic models is PUMA (Performance and Usability Modelling in ATM) developed by the National Air Traffic Services for estimating workload associated with air traffic control (ATC) activities (Kilner *et al.* 1998). PUMA relies heavily on the use of video both for the initial task analysis and for representing workload variations over time. It also incorporates a version of the W/INDEX model.

Simulation models

These models attempt to predict human performance by modelling underlying human psychological processes. To a large extent the validity of such models is a function of how well the models simulate psychological processes. Earlier discussion in this chapter has revealed that, although there is plenty of uncertainty in this area, a number of broad principles have emerged and these form the basis for the simulation models. Probably the most popular simulation modelling tool for ergonomists is MicroSaint (Laughery *et al.* 2000). It is a general-purpose tool, rather than just a simulation of human behaviour, and can be used to model any process that can be represented by a network of tasks or subtasks. More specifically, MicroSaint models the flow of tasks based on specifying factors such as the times to complete tasks, the preconditions under which tasks can be executed, the permitted pathways through the task networks and so on. As with traditional task analysis, the level of detail in the analysis can be decided by the researcher. Typical applications of MicroSaint to MWL assessment can be found in See and Vidulich (1998) and Liao and Moray (1993).

Empirical techniques

Primary task performance

When using empirical techniques, it is only natural that measures of primary task performance will be obtained. However, there are a number of obvious limitations with relying solely on these measures. First, while poor human performance can be indicative of task demands being too high, acceptable performance does not necessarily reflect task demands. This is because acceptable performance can be achieved over a range of task demands resulting in varying amounts of spare mental capacity or attentional resources, as illustrated in Figure 3. The amount of available spare capacity is often a critical factor as it determines the available resources to perform other tasks concurrently, particularly those encountered under emergency conditions, that may not be encountered during the MWL estimation procedures. Second, if performance measures are obtained over comparatively short periods of time, or periods of time known by the operators, operators can allocate extra resources or engage in task management strategies to cope with periods of exceptionally high task demands. However, because such resources cannot be maintained over extended periods of time, performance will at some point decline. In this respect, MWL is analogous to physical workload where one can only sustain short bouts of very high physical workload before becoming fatigued. Third, and contrary to the previous assertion, there are certain tasks where the allocation of increased resources or effort will not lead to improvements in performance. As mentioned earlier, performance in these tasks is referred to as data-limited (Norman and Bobrow, 1975) and are typical of memory retrieval and visual search tasks. Finally, there are problems with the selection and interpretation of performance measures. In some cases performance measures may be very crude (aircraft lands safely or not safely). In most cases, a range of performance measures is obtained, commonly reflecting performance time and errors. The question is, therefore, how to combine these measures to provide a meaningful performance index. Often indices derived to reflect the speed-error trade-off are based on arbitrary assumptions.

Secondary task techniques

The development of secondary task techniques, also referred to as dual task or concurrent task techniques, reflects an attempt to quantify MWL by estimating the attentional resources or mental effort invested in primary task performance. It was realised that this could best be done by requiring operators to perform a second task, the secondary task, while at the same time performing the primary task, the task under consideration. The technique aims to quantify, indirectly, spare processing capacity or spare attentional resources (see Figure 3). In the original applications of this technique, a comparatively simple view of processing capacity was held, namely that the processing capacity was limited and that it could be allocated to any task processing component (e.g., perception, response selection, motor control and so on). Two basic forms of the secondary task technique were initially developed. These are the loading task paradigm and the subsidiary task paradigm.

Loading task. With this paradigm the instructions given to participants emphasise the need to give priority to the secondary task so that performance attains the same level as under a control condition where only the secondary task is given. The workload associated with the primary task is then interpreted in terms of the relative impairment to primary task performance under conditions of concurrent performance compared to when only the primary task is performed. The concurrent secondary task is seen as 'loading' the processing capacity and thus leaving less spare capacity available to perform the primary task. Hence, the greater the processing capacity demands associated with the primary task the greater will be the decrement in primary task performance under dual task conditions.

A typical example of the application of this paradigm is illustrated by the results of a study by Dougherty *et al.* (1964). Under conditions of normal primary flight, performance using a traditional or a newly developed pictorial JANAIR display was indistinguishable. However, adding a secondary task, in this case a simple digit read out task, to which participants had to give priority, was found to impair flight performance much more with the traditional display than with the pictorial display, and this became more evident the greater the difficulty of the loading task. Hence the addition of the loading task was able to demonstrate that operators had more spare capacity when using the pictorial display, and therefore, that the pictorial display was less demanding than the conventional one. The important implication of this result is that the use of the pictorial display 'freed up' processing capacity which could be invaluable under emergency or other high demand situations.

However, the shortcomings of the technique are very obvious. The technique contravenes the criteria of obtrusivness because the very aim of the technique is to cause a performance decrement to primary task performance. For this reason, the technique has to be limited to simulator or simulation conditions to avoid the safety consequences of serious errors. Even under these conditions, participants may be reluctant to sacrifice primary task performance by giving priority to a seemingly irrelevant secondary task (hence the importance of obtaining control data from having participants perform both the primary and secondary tasks alone). Second, there is the question of interpreting measures of primary task performance. This has been mentioned in a previous section. However, additional problems of interpretation occur if the conditions being compared do not employ the same primary task performance measures. Schouten *et al.* (1962), when comparing the workload demands of a number of basic psychomotor tasks, tried to overcome this problem by expressing the decrement in primary task performance under dual task conditions as a percentage of the performance under single task conditions. However, is it then correct to assume that, say, a 10% increase in the number of errors in the case of one primary task is equivalent to a 10% increase in the time to perform a different type of primary task?

Despite these criticisms, the loading task is a potentially useful technique to assess MWL in the early stages of system design when a realistic prototype is available.

Subsidiary task. The origin of this paradigm is attributed to Brown and Poulton (1961). With this technique, participants are instructed to give priority to the primary task during concurrent performance. MWL is then assessed in terms of the impairment to secondary task performance compared to when the secondary task is performed alone. Thus, it is assumed that the more demanding the primary task, the less spare capacity there is available to allocate to the secondary task and, therefore, the greater the related decrement in secondary task performance. The advantages over the loading task paradigm are that this technique allows participants to perform primary tasks under real operational conditions and that MWL can be quantified using the same performance measures that are characteristic of the chosen secondary task. Brown and Poulton (1961) compared the MWL demands of driving a car through a quiet residential area with driving through a busy shopping area. An auditory task was chosen as the secondary task that required the drivers to listen to sequences of 8-digit numbers presented every 4 seconds. Each sequence of numbers differed from the preceding sequence in respect of one of the 8 digits, and it was the task of the driver to identify which number had changed. Not only was it found that the number of errors on this secondary task increased when the task was performed while driving but also that the number of errors was significantly greater when driving through the shopping area compared with the residential area, indicating the relative increased workload.

The major difficulty encountered with the application of both these paradigms is that in most cases interference is found between secondary and primary performance which makes quantification of MWL difficult. For example, in the study of Brown and Poulton (1961), under dual-task conditions it was found that slight changes occurred in primary driving performance measures compared with when the driving was performed alone, reflected in reduced driving speeds and changes in control usage. These interference effects are not surprising, and the degree to which they occur is a feature of the extent the primary and secondary task performance involve the same underlying psychological processes. Thus, when comparing the MWL demands of two different primary tasks, one can get different results depending on the particular secondary task selected. It is this feature which gives the secondary task methodology two key characteristics, namely high diagnosticity and high obtrusiveness. Typical secondary tasks include single reaction time, choice reaction time, manual tracking, monitoring, memory, mental arithmetic, speech shadowing, and time estimation.

In extreme cases, where the primary and secondary tasks do not share any of the same psychological processes, no interference often occurs, so that neither primary nor secondary task performance is degraded under dual-task conditions. Allport *et al.* (1972) found that skilled pianists could attend to and repeat back continuous speech (speech shadowing) while at the same time sight reading piano music without any degradation in performance to either task compared with the tasks executed alone. Results of this kind lead to the loss of favour of the concept of a single generalised resource of information capacity to be replaced by the development of models based on multiple processing channels and multiple resources (Wickens and Hollands, 1999). These models are also reflected in the development of the task analytic techniques of MWL assessment.

Before leaving this category of techniques, two variations should be mentioned that have been developed to overcome some of the problems. These are the adaptive task and the embedded task techniques.

Adaptive task. This technique is a variant of the subsidiary technique whereby interference to primary task performance is controlled by varying the difficulty of the secondary task. The cross-adaptive technique developed by Kelly and Wargo (1967) is an example of this. The primary task was a continuous tracking task while the secondary task was a discrete monitoring task. Primary task performance was continuously monitored and, if performance fell below a certain level, the secondary task was turned off until performance on the primary task returned to a reference level. A number of tracking task parameters were varied and the demands placed by these were then assessed in terms of the extent to which the secondary task had to be turned off in order to maintain the same levels of primary task performance. Apart from the technical difficulties of introducing the cross-adaptive technique, the technique is limited to the assessment of primary tasks that are characterised as having a continuous performance component.

Embedded task. This is another technique to minimise interference to primary task performance and can be used in operational settings as well as in simulations or laboratory environments. An embedded task is a calibrated task which is already a component of the operator's primary task but which is assumed to take only a secondary role and, therefore, is unlikely to interfere with more important primary task components. Shingledecker *et al.* (1980) suggested using radio communications activities as the embedded task for fighter pilots. Communications activities chosen for workload estimation required a number of verbal responses and manual radio switching activities to be executed by the pilots. The total time to complete the communications tasks was used as the secondary task performance measure and this was found by Shingledecker and Crabtree (1982) to be sensitive to the loadings imposed upon the primary tracking task performed in a low-fidelity flight simulator. Vidulich and Bortolussi (1988a, b) describe a more recent application of this technique. Interestingly, the results from their study demonstrated that results based on secondary task performance conflicted with those from using a subjective evaluation technique based on the AHP (Analytic Hierarchy Process) technique (see later section on subjective measures). While the performance data demonstrated benefits from the introduction of speech input devices into a helicopter, the subjective evaluation indicated additional workload imposed by its introduction, possibly from the need to monitor verbal feedback as well as to organise speech inputs.

In general, because of their high obtrusiveness, secondary task techniques are not particularly well suited to the assessment of MWL in realistic operational settings. On the other hand, reflected in their high diagnosticity and sensitivity, the techniques can offer useful insights within laboratory and simulation settings into spare processing capacity and underlying workload components. The use of the embedded secondary task should probably receive some more attention.

Psychophysiological techniques

The original assumption underlying the use of psychophysiological measures of MWL is very simple: as workload is increased, so there is a corresponding increase in the operator's level of arousal (often referred to as an intervening variable) reflected in the activity of the autonomic nervous system. The level of arousal can be recorded by a number of psychophysiological techniques. While no general pattern of changes in the various psychophysiological variables to known changes in workload has emerged, it has been shown that certain measures demonstrate relative specificity to different workload components. Because psychophysiological measures are not generally obtrusive, in that participants are not required to execute any extra overt behaviour, and are capable of measuring fluctuations in MWL over time, the techniques offer some distinct advantages over secondary task measures when used as a diagnostic tool. Against this, psychophysiological techniques demonstrate poor selectivity as measures are easily confounded by a variety of external factors including physical workload, emotional stress and environmental factors such as illuminance and noise, previously discussed in terms of the inverted-U relationship between arousal and performance. This means that the techniques are not naturally suited to being used in real operational settings where these variables are hard if not impossible to control for. Despite the enormous reductions in the cost of recording and data analysis hardware and software, the costs are still comparatively high, especially in terms of human support and expertise. Also, problems associated with low signal-to-noise ratios are frequently encountered. In addition, procedures for measuring and interpreting various psychophysiological measures have not been universally agreed. This is typical of the measurement and interpretation of heart rate variability (sinus arrhythmia). Excellent reviews of psycholophysiological techniques are given by Kramer (1991) and Wilson and Eggemeier (1991), and many of the general issues associated with psychophysiological recording are discussed in Chapter 21 by Kumashiro.

Cardiac activity

Over the years, cardiac activity has provided the most popular psychophysiological measures. To a large extent, this has been because cardiac measures are relatively easy and cheap to obtain, particularly measures of mean heart rate and heart rate variability. Studies showing the possible sensitivity of mean heart rate to changes in workload include those by Wilson and Fullenkamp (1991) and Bonner and Wilson (2002). In both cases substantial percentage changes in heart rate above resting levels were associated with different operational segments of a typical flight scenario. In passing, it is worth mentioning that heart rate estimates of MWL did not always correlate with those obtained from subjective evaluations. Additionally, results are not always convincing, to the extent that some studies have failed to demonstrate reliable influences on mean heart rate (e.g., Casali and Wierwille, 1983, 1984). The lack of consistency using mean heart rate can be attributed to at least two major factors. First, the data can be seriously confounded by the effects of accompanying physical workload components, both dynamic and static. This is often likely to be the case as there is evidence that the increase in mean heart rate attributable to purely MWL components is very small (Zwaga, 1973). Second, and perhaps more significantly, there is a lack of understanding as to why or how mean heart rate should increase with increasing mental demands. One explanation for these inconsistent results is in relation to the so-called intake-rejection hypothesis proposed by Lacey and Lacey (1978). According to this, heart rate increases with the intake of information (visual detection, scanning, listening, etc.) but decreases with the rejection of information (memory retrieval, problem solving, etc.), thus offering some potential diagnosticity to heart rate measures.

Because of the lack of consistency with the results based on mean heart rate, more recent interest has been devoted to measures of heart rate variability (sinus arrhythmia) particularly as they are more reflective of the underlying control of heart rate by the autonomic system (both sympathetic and parasympathetic). There are many measures of heart rate variability, ranging from simple standard deviation and some rather *ad hoc* measures (Luczak and Laurig, 1973; Kalsbeek, 1971), to measures derived from time-series analysis (Mulder and Mulder, 1981), particularly from the application of standard spectral analysis programs. In the case of the latter, the major frequency band of interest has been the low frequency components around 0.1 Hz, taken to be indicative of underlying blood pressure control mechanisms. This band has been found to show a decrease in power with an increase in the amount of effort allocated to a task (e.g. Mulder and Mulder, 1981; Vicente *et al.*, 1987). There is also some evidence that the spectral analysis of heart rate may yield some diagnosticity (Sirevaag *et al.*, 1988). However, problems of reliability of spectral analysis measures remain. For a more recent discussion on the use of sinus arrhythmia as a measure of pilot's mental workload, readers are referred to Backs (1995).

Event-related brain potentials (ERPs)

The ERP reflects a transient series of oscillations in brain activity that can be recorded via surface electrodes in response to discrete external events. Data analysis requires the relationship between the ERP and the eliciting stimulus to be distinguished from background electroencephalographic (EEG) activity. As with EEG, the ERP has several components but is analysed in the time domain rather than the frequency domain characteristic of EEG analysis. Because the ERP is a response to a specific stimulus, it is often necessary to introduce this stimulus as a separate secondary task, although it is possible to analyse the ERPs related to discrete stimulus events embedded within the primary task.

The most commonly analysed component of the ERP is the P300 to either a visual or auditory stimulus. Isreal *et al.* (1980) recorded the P300 component in a secondary visual discrimination task, while participants performed a simulated ATC task (the primary task). Results showed that as the number of elements to be monitored in the ATC task increased, there was a decrease in the amplitude of the P300 component elicited by the secondary task. That the amplitude of the P300 component exhibits some diagnosticity is confirmed by the results of Ragot (1984) who demonstrated that it is insensitive to motor processing load but sensitive to perceptual and central processing loads. Other components of the ERP have been found to show both selectivity and diagnosticity.

While the use of the ERP is promising, there are a number of problems with the technique. First, there is the problem of interference with primary task performance resulting from the secondary task that cannot easily be overcome by using an irrelevant (or unattended) stimulus probe technique. Second, the signal-to-noise ratio is low, requiring the recording of several replications and sophisticated signal processing methods. Third, the technique is mainly restricted to use in controlled laboratory conditions.

Eyeblink activity

Potentially useful indications of the extent of demands on the visual mode of information processing are measures of endogenous eyeblinks. These are blinks that are not made reflexly to specific stimuli. The most frequently used measures are blink frequency and blink duration. Blink frequency can be recorded by electro-oculography (EOG) and by means of video recording. However, using normal scan rates video is insufficient for measuring blink duration.

Blink rate has been found to decrease under conditions of high workload both in driving tasks (Lecret and Pottier, 1971) and in flight tasks (Wilson *et al.* 1987; Wilson and Fullenkamp, 1991). However, the relationship is not always as expected so that in some cases blink rate is higher under relatively high workload conditions, possibly because there is a tendency to blink after taking in visual information. Another study failed to reveal any effects on blink rate (Casali and Wierwille, 1983). In their study, auditory rather than visual load was varied.

Blink duration has also been found to decrease with increased visual load, for example, in a study using a B-52 flight simulator (Stern and Skelly, 1984). These results can be interpreted, perhaps naively, by arguing that as the visual load increases there is a corresponding need to optimise eye fixation time to input the relevant information. Similar results have been reported by Sirevaag *et al.* (1988).

Certainly, there is considerable evidence to suggest that eye blink measures are sensitive to visual processing components of the task. The inconsistent results relating to blink rate may indicate a diagnostic potential for the measure with the acquisition of visual information being associated with lower blink rates, and visual cognitive processes with higher rates.

Pupil diameter

Over the centuries, market traders have assessed a person's interest in an object they are trying to sell by observing his or her pupil diameter. With increased interest and arousal, pupil size increases. The use of pupil size as a measure of MWL goes back to Kahneman (1973) who demonstrated that it was closely associated with the number of items held in short-term or working memory. Since then, a number of studies have shown pupil size to be sensitive to a number of workload components including perceptual (Beatty, 1982a, b), cognitive (Casali and Wierwille, 1983) and response related (Richer and Beatty, 1985) components. Thus, while pupil size appears to meet the sensitivity criterion, it demonstrates little diagnosticity. It is claimed, however, that it can distinguish data-limited processing from resource-limited processing (Beatty, 1982a, b).

The main problems with pupil size measurements are that they are confounded by ambient illuminance and by the combination of illuminances of various light sources within the field of view. Additionally, pupil size is influenced by eye accommodation and vergence processes associated with depth perception. For these reasons, the use of pupil diameter as a measure of workload in real operational settings is likely to remain limited.

While I have not covered all the available psychophysiological measures, the above should be sufficient to highlight some of their relative advantages and disadvantages. Two final points should be made in support of their continuing investigation. First, the measures show good correlations with subjective measures (Hart and Hauser, 1987; Wilson and Fullenkamp, 1991), despite the results from one study mentioned before (Bonner and Wilson, 2002) and therefore have some face validity. Second, the measures appear to offer diagnosticity. For example, within the aviation domain, although take-off and landings are both associated with comparable increases in pilot heart rate, eyeblink rate is lower during landing than during take-off, suggesting a greater need to take in visual information during landing (Wilson and Eggemeier, 1991). A more recent example of the use of multiple measures is provided by Veltman (2002).

Subjective/operator opinion techniques

Finally, we come to the ergonomist's favourite class of techniques! Subjective measures of mental workload (MWL) keep growing in popularity. Not only are they easy and cheap to administer, but they are also characterised by high face validity — when somebody tells you that they feel they have had to put a lot of effort into performing a task, one cannot help but assume that this experience reflects the execution of processes underlying task performance. In this context the participants are judging the interaction between themselves and the task demands. On the other hand, true validity remains relatively elusive. The general use of subjective techniques in ergonomics, including rating scales, is covered in Sinclair's chapter, while Eggemeier and Wilson (1991) present an excellent review of the development of subjective measures of MWL.

Three rating scales have dominated the literature. The first of these is the Cooper–Harper scale (Cooper and Harper, 1969) which employs a single 10-point scale, each point having a verbal descriptor, while the other two are multi-dimensional scales, NASA-TLX (Hart and Staveland, 1988) and SWAT (Reid and Nygren, 1988). A number of modifications have been made to the Cooper–Harper scale, the two most frequently used versions being the Modified Cooper–Harper Scale (Wierwille and Casali, 1983) and the Bedford Scale (Roscoe, 1987). Numerous other scales have been developed, incorporating various degrees of methodological rigour, and some of these will be described in addition to those already mentioned. The different subjective techniques can be distinguished by a number of features: (1) single or multiple dimension scales, (2) absolute or relative judgements, and (3) retrospective or instantaneous data collection. In general, Gopher and Braune (1984) have provided good evidence of the potential high consistency of subjective estimates of workload.

Single dimensional scales

Cooper–Harper Scale (Cooper–Harper Aircraft Handling Characteristics Scale), Cooper and Harper, 1969. This scale is still widely used within the aircraft industry, for whom it was designed. It is especially tailored towards performance and controllability in relation to aircraft characteristics. The 10-point scale is accompanied by verbal descriptors for each of the points. For example, a pilot rating of 6 is equivalent to the verbal descriptor 'adequate performance requires extensive pilot compensation.' This is taken to indicate that the aircraft has very objectionable but tolerable deficiencies and that these warrant improvement. An obvious feature of the scale is that no references are made in the verbal descriptors to workload.

Modified Cooper–Harper Scale, (Wierwille and Casali, 1983). In this scale, the descriptors have been re-worded in terms of the mental effort required to perform tasks and, therefore, can be applied to tasks other than controlling an aircraft. For example, the verbal descriptor for point 6 of the scale now reads 'maximum operator mental effort is required to attain adequate system performance' indicating that MWL is high and should be reduced.

The Bedford Rating Scale, (Roscoe, 1987). The Bedford Scale was also developed for application within the flight environment. Again, the 10 original workload level descriptors from the Cooper–Harper Scale have been modified. Interestingly, a majority of the descriptors make reference to spare capacity rather than or as well as effort invested. Referring to point 6 of the scale, the verbal descriptor reads 'little spare capacity: level of effort allows little attention to additional tasks' indicating that workload needs to be reduced. Although descriptors of this kind match up with some definitions of MWL, it is quite difficult to see how participants can estimate their own spare capacity, especially as spare capacity seems likely to be a function of the specific processing resources involved. Nevertheless, numerous studies using the scale have confirmed that it can be used to differentiate between different levels of loading in simulated flight environments (e.g. Tsang and Johnson, 1989).

Two other reasonably well validated absolute rating scales have been developed at the University of Stockholm. They include the Stockholm 9-Point Scale for Workload Assessment (Bratfisch *et al.* 1972) and the Stockholm 11-Point Scale for Workload Assessment (Dornic, 1980).

Multi-dimensional scales

Because most contemporary definitions of MWL refer to its multi-dimensional characteristic, it is not surprising that attempts have been made to capture this aspect by developing multi-dimensional assessment procedures. As well as the possibility of developing more valid assessments of workload, such procedures have the additional potential of providing a diagnostic tool.

NASA-TLX (Task Load Index), (Hart and Staveland, 1988). The NASA-TLX is probably the most widely used of all MWL scales, not only because of its multi-dimensional feature but also because it is comparatively easy to administer. The technique requires participants to complete ratings in relation to 6 different scales, each scale represented by a 10 cm line divided into 20 intervals. The ratings on each scale are converted to values of 0 to 100. The scales were chosen by Hart and Staveland (1988) as a result of an extensive programme of laboratory research. The definitions of the scales are given in Table 2. A notable feature of the scales is that a physical workload dimension is included. Following the ratings, which are

Table 2 The NASA-TLX Rating Scale Definitions

Dimension	Endpoints	Descriptors
MENTAL DEMAND	Low/High	How much mental and perceptual activity was required (e.g., thinking, deciding, calculating, remembering, looking, searching, etc.)? Was the task easy or demanding, simple or complex, exacting or forgiving?
PHYSICAL DEMAND	Low/High	How much physical activity was required (e.g., pushing, pulling, turning, controlling, activating, etc.)? Was the task easy or demanding, slow or brisk, slack or strenuous?
TEMPORAL DEMAND	Low/High	How much time pressure did you feel due to the rate or pace at which the tasks or task elements occurred? Was the pace slow and leisurely or rapid and frantic?
PERFORMANCE	Good/Poor	How successful do you think you were in accomplishing the goals of the task set by the experimenter (or yourself)? How satisfied were you with your performance in accomplishing these goals?
EFFORT	Low/High	How hard did you have to work (mentally and physically) to accomplish your level of performance?
FRUSTRATION	Low/High	How did you feel during the task? Insecure, discouraged, irritated, stressed and annoyed versus secure, gratified, content, relaxed and complacent.

made retrospectively, participants are required to weight the relevance of the 6 dimensions to the task under investigation by completing a paired comparison procedure. All pairs of the dimensions are compared, 15 in all, so that a particular dimension could end up with a weighting of a value between 0 and 5. The overall workload index is computed by combining the ratings and the weights: each of the 6 ratings is multiplied by the corresponding weight to give adjusted ratings, these are then summed and, finally, this sum is divided by 15 to give a value between 0 and 100. The need to include the weighting procedure, which naturally prolongs the data collection time, has been queried. Results from Byers *et al.* (1988) and Dickinson *et al.* (1993) suggest the weighting protocol is unnecessary as very high correlations were found between weighted and unweighted scores, although, as Dickinson *et al.* (1993) also point out, the correlations were not perfect, suggesting the weighting protocol may make a small but significant contribution.

SWAT (Subjective Workload Assessment Technique), (Reid and Nygren, 1988). Compared with the NASA-TLX, the SWAT technique is comparatively time consuming, often requiring an hour or so to implement fully. A near unique feature of SWAT is that it is based on a psychological model of how judgements of mental workload are formed by participants. The technique requires participants to rate on three workload dimensions, each rating scale having 3 points. These are described in Table 3. However, before this is done, participants have to develop an underlying scale of MWL. This is achieved by asking participants to sequence 27 cards in relation to the workload level they represent. Each card is given a unique combination of the three levels of each of the three dimensions shown in Table 3, with 1–1–1 naturally being allocated the lowest position in the sequence (lowest workload) and 3–3–3 the highest position (highest workload). While performing this scaling procedure to order the remaining 25 cards, participants have to imagine a task or work situation that they feel corresponds to the various combinations of levels of the three

Table 3 The SWAT Rating Scale Definitions

1. **Time Load**
   1. Often have spare time. Interruptions or overlap among activities occur infrequently or not at all.
   2. Occasionally have spare time. Interruptions or overlap among activities occur frequently.
   3. Almost never have spare time. Interruptions or overlap among activities are very frequent, or occur all the time.

2. **Mental Effort Load**
   1. Very little conscious mental effort or concentration required. Activity is almost automatic, requiring little or no attention.
   2. Moderate conscious mental effort or concentration required. Complexity of activity is moderately high due to uncertainty, unpredictability, or unfamiliarity. Considerable attention required.
   3. Extensive mental effort and concentration are necessary. Very complex activity requiring total attention.

3. **Psychological Stress Load**
   1. Little confusion, risk, frustration, or anxiety exists and can be easily accommodated.
   2. Moderate stress due to confusion, risk, frustration, or anxiety noticeably adds to workload. Significant compensation is required to maintain adequate performance.
   3. High or very intense stress due to confusion, frustration, or anxiety. High to extreme determination and self-control required.

dimensions. The sequence of cards is then subjected to a conjoint measurement technique developed by Krantz and Tversky (1971) which produces an interval scale of workload ranging from 0 to 100. This means that each of the 27 combinations of levels of the three workload dimensions now has a value of between 0 and 100 associated with it. Once this has been done, all that remains is for the participant to rate the task under examination in relation to the three dimensions so that the corresponding workload value can be read off from the previously developed underlying workload scale.

The need to go through the conjoint measurement procedure for each participant results from the finding that participants do not give equal weighting to the three dimensions during this procedure, some participants weighting time load as being the more important dimension while others rate effort or stress as being the more important (Reid and Nygren, 1988). Hence, the weighting procedure, like that used by NASA-TLX, can be seen as a way of reducing individual differences. However, Biers (1995) has claimed that simply using the composite scores from the three SWAT scales and omitting the weighting procedure (i.e., so that workload is effectively rated on a 7-point scale with unitary values going from 3 to 9) does not lead to a decrease in sensitivity. Biers' study involved fighter pilots and manipulated cockpit configuration and mission task. In its favour, the use of the conjoint measurement procedure may enable the SWAT technique to be used in a predictive manner, as in the case of Pro-SWAT.

Relative judgements

All the scaling techniques described so far in this section have involved absolute judgements which means there is considerable uncertainty concerning the characteristics of the underlying scales. In most cases it would seem safest to conclude that these scales are ordinal, although protagonists of SWAT have argued that they reflect interval scales. The use of relative judgement techniques is one way to be more confident that the underlying scales are indeed interval ones. This can be achieved by using paired comparison and magnitude estimation techniques. An example of the use of paired comparisons is the Analytic Hierarchy Process (AHP) developed by Saaty (1980) and implemented for workload assessment by Vidulich and Tsang (1987). As Vidulich (1989) points out, AHP does not '. . . use raters as workload meters'. On the other hand AHP is used to elicit '. . . the conscious decision-making and experiential knowledge to extract expert judgements about workload'. The AHP forms the basis for the subjective technique developed by Vidulich and his colleagues (Vidulich 1989; Vidulich *et al.* 1991) entitled SWORD (Subjective WORkload Dominance). The SWORD technique involves three stages: (1) collecting the paired comparison data, (2) constructing a judgement matrix, and (3) calculating the SWORD workload ratings. The paired comparisons technique involves raters comparing all pairs of tasks or design alternatives on a 17-point scale where point 9 represents equal rating of the two alternatives. From the subsequent judgement matrix, workload ratings are calculated by obtaining the geometric mean for each row of the matrix and then normalising the means. A consistency value is also obtained which assesses the extent to which the underlying workload scale represents an interval scale. Vidulich and his colleagues were particularly interested in using SWORD as a projective or predictive technique. One of the examples used by Vidulich *et al.* (1991) was the evaluation of a variety of proposed formats for aircraft HUDs. Results showed that the ratings of predictive MWL by pilots correlated very well with retrospective ratings and that both types of rating correlated well with performance under simulation conditions. The authors argue that the SWORD technique may be a viable alternative to task analytic techniques, but with the added advantage of being considerably less labour-intensive.

Instantaneous judgements

All the scaling techniques mentioned so far have involved retrospective or prospective judgements. While this feature does not appear to have invalidated their use, many investigators have wanted to be able to see how workload varies during the course of performing a task without resorting to psychophysiological techniques. For this to be achieved, the ratings need to be completed at frequent intervals and as quickly as possible while the primary task is being performed. In a sense, the rating procedure can now be seen as a concurrent secondary task.

POSWAT is an example of this technique where pilots have to make judgements every minute using a kneeboard with 10 keys, key 1 corresponding to little or no workload and key 10 to excessive workload where the pilot can only just maintain flying the aircraft. The pilot is prompted to make his or her judgements by a tone presented over the pilot's headset and a red light appearing on the kneeboard. Unfortunately, Mallery (1987) does not present the results on a minute-to-minute basis, but rather he averages ratings for each of the flight scenario segments. These average scores were found to be sensitive to the different demands associated with the segments. ISA (Instantaneous Self Assessment) is a further example of this technique and was developed by the National Air Traffic Services (UK) and Eurocontrol to provide continuous ratings of MWL during ATC tasks. The scale is a 5-point one and uses a small keypad with the five keys corresponding to 'excessive', 'high', 'comfortable', 'relaxed' and 'under-utilised'. Participants are cued to make a rating by a visual signal presented on the radar console.

A major concern with these techniques, particularly if they are to be used under real operational conditions, is that the execution of the ratings may act as an extra concurrent task, not only influencing primary task performance but also leading to an overall increase in workload ratings. To see if primary task performance is influenced by concurrent workload judgements, Tattersall and Foord (1996) investigated the use of ISA in a laboratory setting. Participants made judgements every 2 minutes while performing a simple pursuit tracking task with three levels of task difficulty. Half the participants used a keypad to make their judgements while the other half spoke their responses. In addition to using the ISA technique, retrospective subjective ratings were taken using a modified SWAT procedure where the weighting protocol was excluded. Essentially the results from the ISA technique correlated very closely with the ratings on the three SWAT scales. However, most importantly, the results showed that the tracking error was increased at the time the judgements were made. Contrary to the predictions of multiple resource theory, verbal responses caused the same decrement in the primary tracking performance as the manual responses, suggesting the added demand created by the rating procedure was not simply one of competing output modes, in which case one should have observed more interference when using the keyboard to input the workload ratings.

To avoid the possible interference between the rating process and primary task performance, Jensen (1995) describes a technique he called C-SAW (Continuous Subjective Assessment of Workload). Rather than have pilots continuously assess workload while they were performing the primary flight tasks under normal operational conditions, the primary task was filmed and the pilots made their judgements while watching the replay. Jensen used a 10-point scale, each point with a verbal descriptor equivalent to those used in the Bedford Scale, and a small keypad to input the ratings. Preliminary results showed that pilots could make workload judgements at a rate of one every three seconds when prompted to do so. The results also demonstrated that the ratings were what would have been expected from the accompanying timeline analysis. To a very large extent, this technique resembles Pro-SWAT described earlier as an example of a task analytic technique.

Comparison between rating scales

Numerous studies have compared the use of the different scales, mainly in relation to sensitivity. At this stage one can do little more than echo the remarks of Eggemeier and Wilson (1991) when they conclude that there is insufficient evidence to favour any one particular scale. However, there are the obvious benefits of employing multiple dimensional ratings in that they have some diagnostic value as well as aiding predictive workload evaluation. There are also the claims made that, if the weighting procedures are employed, both SWAT and NASA-TLX demonstrate less individual variations in ratings than are found with the single dimensional scales. When it comes to participants' acceptance, there is some evidence from the results of Byers *et al.* (1988) that when the weighting procedures are omitted from both NASA-TLX and SWAT, NASA-TLX is more acceptable than both SWAT and the Modified Cooper–Harper Scale. In an early comparison study by Wewerinke (1974), reported by Moray (1982), three popular single dimension scales were compared: a simple non-verbal 10 cm line requiring an estimate of 'effort you spend performing the task' from a value of 0 to 10, the original Cooper–Harper Scale, and the modified Cooper–Harper Scale. The correlations between the ratings from the three scales were nearly perfect, suggesting the verbal descriptors used in the Cooper–Harper Scales were adding little or nothing to the assessment process.

Validity of subjective measures

It has generally been agreed that the validity of subjective measures should be discussed mainly in terms of sensitivity to variations in task demands and but also in relation to primary task performance. There are limitations with these criteria. First, sensitivity is usually assessed in relation to the investigators 'expert' evaluation of task difficulty or task loading. However, definitions of MWL frequently express workload not in relation to task loading per se but to loading of the individual, often in relation to some underlying model of information processing. Similarly, it has been argued that primary task performance may not necessarily reflect workload to the extent that workload and subjective ratings may become dissociated. That is to say, it does not follow that as levels of subjective workload increase, primary task performance will become correspondingly impaired. A number of these dissociations have been described by Yeh and Wickens (1988). In principle, these dissociations are derived from the reasonable assumption that subjective ratings are reflecting the amount of attentional resources or effort a person is investing in performing a particular task:

1. In cases of task underload, where available resources far outweigh the resources required by the task demands, individuals may invest unnecessarily greater resources, thinking this will improve their performance. An example of this kind of argument is supported by the results when using the NASA-TLX to assess the MWL associated with vigilance tasks. These tasks are traditionally classified as undemanding, yet the results of Warm *et al.* (1996) confirm that surprisingly high workload ratings are reported especially in relation to the mental effort and frustration scales. The authors suggest that extra resources above those required by the task demands are invested to overcome the tedium accompanying such tasks. An alternative interpretation of this phenomenon is that performance is, in effect, data- rather than resource-limited (Norman and Bobrow, 1975). In the case of overload, where resources are fully invested, any increase in task demands will be reflected in decreased performance but no increase in subjective assessment because resources are fully invested. In other words, the subjective measures have saturated.
2. Because greater investment of resources should yield better performance for resource-limited tasks there is an inherent dissociation whereby increased resource investment leads to superior performance. This can be achieved by increasing an individual's motivation, for example, by providing performance feedback.
3. Dissociation will occur when an easy dual task configuration is compared to a difficult single task. The example cited by Yeh and Wickens (1988) is one of introducing automation to aid primary task performance. Although improvement in performance is likely to be found, subjective rating may be higher for the aided situation because of the need to invest increased resources as a result of the time sharing aspect of the aided task, most probably from an increase in working memory load; that is to say an increase in task management demands. This is confirmed by Aretz *et al.* (1996). Using the NASA-TLX, they demonstrated that out of a number of potential task demands, the number of concurrent tasks had the largest impact on pilots' subjective workload ratings.
4. According to multiple resource theory (Wickens and Hollands, 1999), performance in multiple task environments is likely to be worse when the tasks compete for similar resources. This rivalry for resources will not be reflected in the subjective measures because the overall amount of resources invested may be the same as when there is not such high competition for resources.

A strong argument, therefore, can be put forward for placing limited reliance on the exclusive use of subjective ratings of MWL in multi-task conditions and on using such ratings in a predictive manner.

Conclusions

Readers can be forgiven for concluding that MWL evaluation presents a potential minefield and that it is necessary to proceed with great caution. What I have tried to do in this chapter is provide a very general framework (Figure 1) for the evaluation process in the absence of a universally agreed definition of MWL and in a climate of continuing research into underlying psychological processes. I have overviewed a majority of the measurement techniques in relation to the framework, and have indicated their relative advantages and disadvantages according to a number of well established criteria. What can be concluded with confidence is that in order to gain insight into MWL issues there is a need to use a range of measurement techniques rather than 'putting all your eggs into one basket'.

References

Aldrich, T.B., Szabo, S.M. and Bierbaum, C.R. (1989). The development and application of models to predict operator workload during system design. In: G.R. McMillan, D. Beevis, E. Salas, M.H. Strub, R. Sutton and L. van Breda (eds) *Applications of Human Performance Models to System Design* (New York: Plenum), pp. 65–80.

Allport, D.A., Antonis, B. and Reynolds, P. (1972). On the division of attention: A disproof of the single channel hypothesis. *Quarterly Journal of Experimental Psychology*, **24**, 225–235.

Aretz, A.A., Johannsen, C. and Ober, K. (1996). An empirical validation of subjective workload ratings. In *Proceedings of the Human Factors and Ergonomics Society 40th Annual Meeting* (Santa Monica, California: The Human Factors and Ergonomics Society), pp. 91–95.

Backs, R.W. (1995). Going beyond heart rate: Autonomic space and cardiovascular assessment of mental workload. *International Journal of Aviation Psychology*, **5**, 25–48.

Beatty, J. (1982a). Phasic not tonic pupillary responses vary with auditory vigilance performance. *Psychophysiology*, **19**, 167–172.

Beatty, J. (1982b). Task-evoked pupillary responses, processing loads, and the structure of processing resources. *Psychological Bulletin*, **91**, 276–292.

Biers, D.W. (1995). SWAT: Do we need conjoint measurement. In *Proceedings of the Human Factors and Ergonomics Society 39th Annual Meeting* (Santa Monica, California: The Human Factors and Ergonomics Society), pp. 1233–1237.

Bonner, M.A. and Wilson, G.F. (2002). Heart rate measures of flight test and evaluation. *International Journal of Aviation Psychology*, **12**, 63–77.

Bratfisch, O., Borg, G. and Dornic, S. (1972). Perceived item difficulty in three tests of intellectual performance capacity. Report No. 29, Institute of Applied Psychology, University of Sweden.

Brown, I.D. and Poulton, E.C. (1961). Measuring the spare 'mental capacity' of car drivers by a subsidiary task. *Ergonomics*, **4**, 35–40.

Byers, J.C., Bittner, A.C., Hill, S.G., Zaclad, A.L. and Christ, R.E. (1988). Workload assessment of a remotely piloted vehicle (RPV) system. In *Proceedings of the Human Factors Society 32nd Annual Meeting* (Santa Monica, California: The Human Factors Society), pp. 1145–1149.

Casali, J.G. and Wierwille, W.W. (1983). A comparison of rating scale, secondary task, physiological, and primary task workload estimation techniques in a simulated flight emphasising communications load. *Human Factors*, **25**, 623–641.

Casali, J.G. and Wierwille, W.W. (1984). On the measurement of pilot perceptual workload: A comparison of assessment techniques addressing sensitivity and intrusion issues. *Ergonomics*, **27**, 1030–1050.

Cooper, G.E and Harper, R.P. (1969). The Use of Pilot Rating in the Evaluation of Aircraft Handling Qualities. Report No. NASA TN-D-5153, Moffett Field, California: Ames Research Centre.

Cox, T. (1978). *Stress* (London: MacMillan Press).

Dickinson, J., Byblow, W.D. and Ryan, L.A. (1993). Order effects and the weighting process in workload assessment. *Applied Ergonomics*, **24**, 357–361.

Dornic, S. (1980). Spare capacity and perceived effort in information processing. Report No. 567, Institute of Applied Psychology, University of Sweden.

Dougherty, I., Emery, J.H. and Curtin, J.G. (1964). Comparisons of perceptual work load in flying standard instrumentation and the contact analog vertical display (JANAIR-D228–421–019), Bell Helicopter Co., DTIC No. AD610617.

Eggemeier, F.T. and Wilson, G.F. (1991). Performance-based and subjective assessment of workload in multi-task environments. In: D.L. Damos (ed.) *Multiple-Task Performance* (London: Taylor & Francis), pp. 217–278.

Fitts, P.M. (1954). The information capacity of the human motor system in controlling the amplitude of movement. *Journal of Experimental Psychology*, **47**, 381–391.

Gopher, D. and Braune, R. (1984). On the psychophysics of workload: Why bother with subjective measures? *Human Factors*, **26**, 519–532.

Hart, S.G. and Hauser, J.R. (1987). Inflight application of three pilot workload measurement techniques. *Aviation, Space, and Environmental Medicine*, **53**, 54–61.

Hart, S.G. and Staveland, L.E. (1988). Development of the NASA-TLX (Task Load Index): Results of empirical and theoretical research. In: P.A. Hancock and N. Meshkati (eds) *Human Mental Workload* (Amsterdam: North-Holland), pp. 139–183.

Hick, W.E. (1952). On the rate of gain of information. *Quarterly Journal of Experimental Psychology*, **4**, 11–26.

Hill, S.G. *et al.* (1987). Analytic techniques for the assessment of operator workload. In *Proceedings of the 31st Annual Meeting of the Human Factors Society* (Santa Monica, California: The Human Factors Society), pp. 368–372.

International Organization for Standardization (1991). ISO 10075, *Ergonomic Principles Related to Mental Work-Load, Part 1: General Terms and Definitions* (Geneva: ISO).

Isreal, J.B., Chesney, G.L., Wickens, C.D. and Donchin, E. (1980). P300 and tracking difficulty: Evidence for multiple resources in dual task performance. *Psychophysiology*, **17**, 259–273.

Jensen, S.E. (1995). Developing a flight workload profile using Continuous Subjective Assessment of Workload (C-SAW). In: R. Fuller, N. Johnston and N. McDonald (eds) *Human Factors in Aviation Operations* (Aldershot, Hampshire: Ashgate Publishing), pp. 307–312.

Kahneman, D. (1973). *Attention and Effort* (Englewood Cliffs, New Jersey: Prentice-Hall).

Kalsbeek, J.W.H. (1971). Sinus arrhythmia and the dual task method in measuring mental load. In: W.T. Singleton, J.G. Fox and D. Whitfield (eds) *Measurement of Man at Work: An Appraisal of Physiological and Psychological Criteria in Man–Machine Systems* (London: Taylor & Francis), pp. 101–113.

Kelly, C.R. and Wargo, M.J. (1967). Cross-adaptive operator loading tasks. *Human Factors*, **9**, 395–404.

Kilner, A., Hook, M., Fearnside, P. and Nicholson, P. (1998). Developing a predictive model of controller workload in air traffic management. In: M.A. Hansen (ed.), *Contemporary Ergonomics 1998* (London: Taylor & Francis), pp. 409–413.

Kramer, A.F. (1991). Physiological metrics of mental workload: A review of recent progress. In: D.L. Damos (ed.) *Multiple-Task Performance* (London: Taylor & Francis), pp. 279–328.

Krantz, D.H. and Tversky, A. (1971). Conjoint measurement analysis of composition rules in psychology. *Psychological Review*, **78**, 151–169.

Kuperman, G.G. (1985). Pro-SWAT applied to advanced helicopter crewstation concepts. In *Proceedings of the 29th Annual Meeting of the Human Factors Society* (Santa Monica, California: The Human Factors Society), pp. 398–402.

Lacey, J.I. and Lacey, B.C. (1978). Two-way communication between the heart and the brain: Significance of time within the cardiac cycle. In: E. Meyer and J. Brady (eds) *Research in the Psychobiology of Human Behavior* (Baltimore: Johns Hopkins University Press), pp. 99–113.

Laughery, R., Archer, S., Plott, B. and Dahn, D. (2000). Task network modelling and the Micro Saint family of tools. In *Proceedings of the 44th Annual Meeting of the Human Factors and Ergonomics Society* (Santa Monica, California: The Human Factors and Ergonomics Society), pp. 721–724.

Lecret, F. and Pottier, M. (1971). La vigilance, facteur de securite dans la conduite automobile. *Le Travail Humaien*, **34**, 51–68.

Liao, J. and Moray, N. (1993). A simulation study of human performance deterioration and mental workload. *Le Travail Humain*, **56**, 321–344.

Linton, P.M. *et al.* (1989). Operator workload for military system acquisition. In: G.R. McMillan, D. Beevis, E. Salas, M.H. Strub, R. Sutton and L. van Breda (eds) *Applications of Human Performance Models to System Design* (New York: Plenum Press), pp. 21–45.

Luczak, H. and Laurig, W. (1973). An analysis of heart rate variability. *Ergonomics*, **16**, 85–97.

Mallery, C.J. (1987). The effect of experience on subjective ratings for aircraft and simulator workload during IFR flight. In *Proceedings of the Human Factors Society 31st Annual Meeting* (Santa Monica, California: The Human Factors Society), pp. 838–841.

Malone, T.B., Perse, R.M., Heasley, C.C. and Kirkpatrick, M. (1988). MANPRINT in the programme initiation phase of systems acquisition. In *Proceedings of the 32nd Annual Meeting of the Human Factors Society* (Santa Monica, California: The Human Factors Society), pp. 1108–1112.

Moray, N. (ed.) (1979). *Mental Workload: Its Theory and Measurement* (New York: Plenum Press).

Moray, N. (1982). Subjective mental workload. *Human Factors*, **24**, 25–40.

Mulder, G. and Mulder, L.J.M. (1981). Information processing and cardiovascular control. *Psychophysiology*, **18**, 392–405.

Neerincx, M.A. (2003). Cognitive task analysis: Allocating tasks and designing support. In: E. Hollnagel (ed.) *Handbook of Cognitive Task Design* (Mahwah, New Jersey: Lawrence Erlbaum Associates), pp. 283–305.

Norman, D. and Bobrow, D. (1975). On data-limited and resource-limited processing. *Journal of Cognitive Psychology*, **7**, 44–60.

North, P.A. and Riley, V.A. (1989). W/INDEX: A predictive model of operator workload. In: G.R. McMillan, D. Beevis, E. Salas, M.H. Strub, R. Sutton and L. van Breda (eds) *Applications of Human Performance Models to System Design* (New York: Plenum), pp. 81–89.

O'Donnell, C.R.D. and Eggemeier, F.T. (1986). Workload assessment methodology. In: K.R. Boff, L. Kaufman and J.P. Thomas (eds) *Handbook of Perception and Human Performance, Volume II* (New York: Wiley and Sons), pp. 42/1–42/49.

Parks, D. and Boucek, G. (1989). Workload prediction, diagnosis, and continuing challenges. In: G.R. McMillan, D. Beevis, E. Salas, M.H. Strub, R. Sutton and L. van Breda (eds) *Applications of Human Performance Models to System Design* (New York: Plenum), pp. 47–64.

Ragot, R. (1984). Perceptual and motor space representation: An event related potential study. *Psychophysiology*, **21**, 159–170.

Reid, G.B. and Nygren, T.E. (1988). The subjective workload assessment technique: A scaling procedure for measuring mental workload. In: P.A. Hancock and N. Meshkati (eds) *Human Mental Workload* (Amsterdam: North-Holland), pp. 185–218.

Richer, F. and Beatty, J. (1985). Pupillary dilations in movement preparation and execution. *Psychophysiology*, **22**, 204–207.

Roscoe, A.H. (1987). In-flight assessment of workload using pilot ratings and heart rate. In: A.H. Roscoe (ed.) *The Practical Assessment of Pilot Workload.* AGARD-AG-282, pp. 78–82.

Rubenstein, J.S., Meyer, D.E. and Evans, J.E. (2001). Executive control of cognitive processes in task switching. *Journal of Experimental Psychology: Human Perception and Performance*, **27**, 763–797.

Saaty, T.L. (1980). *The Analytic Hierarchy Process: Planning, Priority Setting, Resource Allocation* (New York: McGraw-Hill).

Sarno, K.J. and Wickens, C.D. (1995). Role of multiple resources in predicting time-sharing efficiency: Evaluation of three workload models in a multiple-task setting. *International Journal of Aviation Psychology*, **5**, 107–130.

Schneider, W. and Shiffrin, R.M. (1977). Controlled and automatic human information processing I: Detection, search, and attention. *Psychological Review*, **84**, 1–66.

Schouten, J.F., Kalsbeek, J.W.H. and Leopold, F.F. (1962). On the evaluation of perceptual and mental load. *Ergonomics*, **5**, 251–260.

See, J. and Vidulich, M.A. (1998). Computer modelling of operator mental workload and situational awareness in simulated air-to-ground combat: An assessment of predictive validity. *International Journal of Aviation Psychology*, **8**, 351–375.

Senders, J. (1964). The human operator as a monitor and controller of multidegree systems. *IEEE Transactions on Human Factors in Electronics*, **HFE-5**, 2–6.

Senders, J.W. and Posner, M. (1976). A queuing model of monitoring and supervisory control. In: T.B. Sheridan and G. Johannsen (eds) *Monitoring Behavior and Supervisory Control* (New York: Plenum Press), pp. 245–259.

Shannon, C.E. and Weaver, W. (1959). *The Mathematical Theory of Communication* (Urbana, Illinois: The University of Illinois Press).

Shingledecker, C.A. and Crabtree, M.S. (1982). Subsidiary radio communications tasks for workload assessment: II. Task sensitivity evaluation. Report No. AFAMRL-TR-82–57, Wright Patterson Airforce Base, Ohio.

Shingledecker, C.A., Crabtree, M.S., Simons, J.C., Courtright, J.F. and O'Donnell, R.D. (1980). Subsidiary radio communications tasks for workload assessment in R & D simulations: I. Task development and workload scaling. Report No. AFAMRL-TR-80–126, Wright Patterson Airforce Base, Ohio.

Sirevaag, E., Kramer, A., de Jong, R. and Mecklinger, A. (1988). A psychophysiological analysis of multi-task processing demands. *Psychophysiology*, **25**, 482.

Smith, A. (1989). A review of the effects of noise on human performance. *Scandinavian Journal of Psychology*, **30**, 185–206.

Stern, J.A. and Skelly, J.J. (1984). The eye blink and workload considerations. In *Proceedings of the 28th Annual Meeting of the Human Factors Society* (Santa Monica, California: The Human Factors Society), pp. 942–944.

Tattersall, A.J. and Foord, P.S. (1996). An experimental evaluation of instantaneous self-assessment as a measure of workload. *Ergonomics*, **39**, 740–748.

Treisman, A.M. and Gelade, G. (1980). A feature integration theory of attention. *Cognitive Psychology*, **12**, 97–136.

Tsang, P.S. and Johnson, W.W. (1989). Cognitive demands in automation. *Aviation, Space, and Environmental Medicine*, **60**, 130–135.

Veltman, J.A. (2002). A comparative study of psychophysiological reactions during simulator and real flight. *International Journal of Aviation Psychology*, **12**, 33–48.

Vicente, K., Thorton, D. and Moray, N. (1987). Spectral analysis of sinus arrhythmia: A measure of mental effort. *Human Factors*, **29**, 171–182.

Vidulich, M.A. (1989). The use of judgement matrices in subjective workload assessment: The Subjective WORkload Dominance (SWORD) technique. In *Proceedings of the Human Factors Society 33rd Annual Meeting* (Santa Monica, California: The Human Factors Society), pp. 406–1410.

Vidulich, M.A. and Bortolussi, M.R. (1988a). A dissociation of objective and subjective workload measures in assessing the impact of speech controls in advanced helicopters. In *Proceedings of the 32nd Annual Meeting of the Human Factors Society* (Santa Monica, California: The Human Factors Society), pp. 1471–1475.

Vidulich, M.A. and Bortolussi, M.R. (1988b). Speech recognition in advanced rotocraft: Using speech control to reduce manual control overload. In *Proceedings of the American Helicopter Society National Specialists' Meeting — Automation Applications in Rotocraft* (Atlanta, Georgia: Southeast Region of the American Helicopter Society).

Vidulich, M.A. and Tsang, P.S. (1987). Absolute magnitude estimation and relative judgement approaches to subjective workload assessment. In *Proceedings of the Human Factors Society 31st Annual Meeting* (Santa Monica, California: The Human Factors Society), pp. 1057–1061.

Vidulich, M.A., Ward, G.F. and Schueren, J. (1991). Using the subjective workload dominance (SWORD) technique for projective workload assessment. *Human Factors*, **33**, 677–691.

Warm, J.S., Dember, W.N. and Hancock, P.A. (1996). Vigilance and workload in automated systems. In: R. Parasuraman and M. Mouloua (eds) *Automation and Human Performance: Theory and Applications* (Mahwah, New Jersey: Lawrence Erlbaum Associates), pp. 183–200.

Welford, A.T. (1968). *Fundamentals of Skill* (London: Methuen).

Wewerinke, P.H. (1974). Human operator workload for various control conditions. In *Proceedings of the 10th Annual Conference on Manual Control* (Ohio: Wright-Patterson Air Force Base, Ohio), pp. 167–192.

Wickens, C.D. (2002). Multiple resources and performance prediction. *Theoretical Issues in Ergonomics Science*, **3**, 159–177.

Wickens, C.D. and Hollands, J.G. (1999). *Engineering Psychology and Human Performance*, 3rd Edition (Upper Saddle River, New Jersey: Prentice Hall).

Wickens, C.D., Larish, I. and Contorer, A. (1989). Predictive performance models and multiple tasks performance. In *Proceedings of the Human Factors Society 33rd Annual Meeting* (Santa Monica, California: The Human Factors Society), pp. 96–100.

Wierwille, W.W. and Casali, J.G. (1983). A validated rating scale for global mental workload measurement applications. In *Proceedings of the Human Factors Society 27th Annual Meeting* (Santa Monica, California: The Human Factors Society), pp. 129–132.

Wilson, G.F. and Eggemeier, F.T. (1991). Psychophysiological assessment of workload in multi-task environments. In: D.L. Damos (ed.) *Multiple-Task Performance* (London: Taylor & Francis), pp. 329–360.

Wilson, G.F. and Fullenkamp, P. (1991). A comparison of pilot and WSO workload during training missions using psychophyiological data. In *Proceedings of the Western European Association of Aviation, Vol. II, Stress and Error in Aviation*, pp. 27–34.

Wilson, G.F., Purvis, B., Skelly, J., Fullenkamp, P. and Davis, I. (1987). Physiological data used to measure pilot workload in actual flight and simulator conditions. In *Proceedings of the 31st Annual Meeting of the Human Factors Society* (Santa Monica, California: The Human Factors Society), pp. 779–783.

Wylie, G. and Allport, A. (2000). Task switching and the measurement of 'switch costs'. *Psychological Research*, **63**, 212–233.

Yeh, Y.Y. and Wickens, C.D. (1988). Dissociation of performance and subjective measures of workload. *Human Factors*, **30**, 111–120.

Yerkes, R.M. and Dodson, J.D. (1908). The relationship of strength of stimulus to rapidity of habit formation. *Journal of Comparative Neurological Psychology*, **18**, 459–482.

Zwaga, H.J.G. (1973). Psychophysiological reactions to mental tasks. *Ergonomics*, **16**, 61–67.

Chapter 19

The nature and measurement of work-related stress: theory and practice

Tom Cox and Amanda Griffiths

Introduction

In a civilised society people should have the right to work in order to earn an adequate living to support themselves and their families. It is the duty of their society to provide appropriate opportunities for them to do so, to ensure that such work is without detriment to health or risk to safety, and to support those, who through no fault of their own, cannot work. The fact that these things are not always possible presents a challenge to society. Work-related stress is part of that challenge. Surveys of working people in Europe, North America and elsewhere have identified work-related stress as a major cause of ill-health, sickness absence, early retirement, and as a threat to safety.

In England and Wales, data from the Labour Force Surveys of 1990 and 1995 showed that stress and stress-related illness were second only to musculo-skeletal disorders as the major cause of ill-health for working people (Hodgson *et al.*, 1993; Jones *et al.*, 1998). At that time, it was estimated that this resulted in about 6.5 million working days lost to industry and commerce each year. In terms of annual costs, estimated within the 1995–96 economic framework, the financial burden to British society, its organisations and people, was at least £3.7–3.8 billion. Data from the subsequent survey carried out in 2001/2002 revealed that the estimated prevalence of stress and stress-related conditions had increased twofold from 1990 levels (Health and Safety Executive, 2003). Although musculo-skeletal disorders were more prevalent than stress and stress-related conditions in England and Wales, the latter were responsible for more lost working days (13.4 million as compared to 12.3 million). The average number of days lost per case for stress, depression or anxiety (just over 29 days per case) was significantly higher than for any other work-related illnesses.

Such figures are typical across many countries. It is not surprising therefore that there is a legal imperative in many countries for employers to address the challenge of work-related stress. In Britain and the European Union, this is now relatively long established as part of the general provision in law for health and safety.

In 1989, the European Commission published its Framework Directive on the Introduction of Measures to Encourage Improvements in the Safety and Health of Workers at Work (EC, 1989). All Member States of the European Union were required then to transpose these measures into their respective national legislation. The Directive required employers to avoid risks to workers' safety and health, to evaluate those risks that could not be avoided, to combat such risks at source (Article 6: 2) and to 'consult workers and/or their representatives and allow them to take part in discussions on all questions relating to safety and health at work' (Article 11: 1). Employers were charged with developing a 'coherent overall prevention policy which covers technology, organization of work, working conditions, social relationships and the influence of factors related to the working environment' (Article 6: 2). They had to 'make an assessment of the risks to safety and health at work' and to 'decide on the protective measures to be taken' (Article 9: 1). It was stated that risk assessment should involve 'a systematic examination of all aspects of the work undertaken to consider what could cause injury or harm, whether the hazards could be eliminated, and if not what preventive or protective measures are, or should be, in place to control the risks' (European Commission, 1996). In other words, employers in the European Union have a responsibility to take reasonable steps to protect their employees from all aspects of work or the working environment that could be detrimental to their safety and health. These include those related to the experience of stress.

In addition, in England and Wales in the 1990s, precedents were established in common law for employees allegedly damaged by employers' failure to fulfil their duty of care, in relation to the design and management of work, to sue for reparation. In a landmark case in 1994, John Walker, a social services manager, obtained a judgment against his former employers for failing to protect him from a health-endangering workload (Industrial Relations Law Reports, 1995). The judgement made it clear that there was no reason why psychological damage should be excluded from the scope of an employer's duty of care while issues of work design and management have always been within the scope of the law. Previously, 'damage' was largely understood in terms of physical harm. Since the Walker case, there have been other cases in Britain where employees have been awarded financial settlements for claims that can be said to be stress-related.

This chapter provides an introduction to nature and measurement of work-related stress. Both are central to its study and management. The issues involved here are not straightforward given both the nature of the phenomenon (Cox, 1993) and the complexity inherent in undertaking scientifically 'acceptable' research in the context of real working life (Griffiths, 1999). Partly as a result, measurement issues have given rise to much debate between academics (Cox *et al.*, 2003).

Nature and definition of stress

There is now broad agreement on the nature of 'work-related stress' and on the psychological processes involved. This consensus is reflected in European legislation and related guidance. For example, in 1995 the British Health & Safety Executive (Health & Safety Executive, 1995) described work stress in its guidance for employers as:

> . . . the reaction people may have to excessive pressures or other types of demand placed upon them.

Two years later, the European Commission's Working Group on Stress (European Commission, 1997) concluded that:

> Work-related stress is the emotional and psycho-physiological reaction to aversive and noxious aspects of work, work environments and work organisations. It is a state characterised by high levels of arousal and distress and often by feelings of 'not coping.'

Both statements were consistent with a more detailed and theoretically based position set out by the authors in Cox (1993, p. 17):

> Stress arises when the person perceives that he or she cannot adequately cope with the demands being made on them or with threats to their well being, when coping is of importance to them, and when they become anxious or depressed as a result. The experience of stress is therefore defined by, first, the person's realisation that they are having difficulty coping with demands and threats to their well-being, and, second, that coping is important and that the difficulty in coping worries or depresses them.
>
> This approach allows a clear distinction between, say, the effects of lack of ability on performance and those of stress. If a person does not have the necessary knowledge or level of skill to complete a task then their performance will be poor. They may not realise this or if they do it might not be felt to be of importance or give rise to concern. These are not 'stress' scenarios. However, if the person (a) does realise that they are failing to cope with the demands of a task, and (b) experiences concern about that failure because it is important, then this is a 'stress' scenario.

Contemporary definitions of work stress have their roots in the different approaches to its study that developed during the latter half of the 20th century. These are discussed below. The current consensus represents a move away from early engineering and physiological approaches, towards a more psychological approach that involves an understanding of the importance of people's interactions with their environment, and of their perceptions and their emotional reactions to work.

Theoretical approaches and models

In 1978, Cox described three different theoretical approaches to the study of work stress: (i) the engineering or stimulus-based approach, (ii) the physiological or response-based approach, and (iii) a psychological approach. This particular architecture of theory has been taken up and used by many subsequent authors including Cooper *et al.* (2001).

The engineering approach conceptualises work stress as a stimulus, an aversive or noxious characteristic of the work environment and as a cause of strain. In contrast, the physiological approach defines stress in terms of the common physiological effects of a wide range of aversive or noxious stimuli and as a response to a threatening environment. Both approaches were conceptually rooted in the stimulus–response (S–R) psychology of the immediate pre- and post-war years (1940s to early 1970s). The psychological approach is more contemporary and conceptualises work stress in terms of the dynamic interaction between individuals and their work environment. When studied, stress is either inferred from the existence of problematic person–environment interactions or measured in terms of the cognitive processes and emotional reactions that underpin those interactions. This is the approach to measurement that is expounded here. Cox (1978, 1993) suggested that the main psychological approaches and models could be divided into (i) the interactional theories which describe the architecture of person–environment interactions (French *et al.*, 1982) person-environment-fit model, Seigrist's effort–reward imbalance model (1996), and Karasek's job-demands/job control model (Karasek *et al.*, 1981; Karasek and Theorell, 1990), and (ii) the transactional theories, that describe the processes involved in such interactions, and include appraisal theories of stress and coping (Lazarus, 1966; Cox, 1978; Lazarus and Folkman, 1984).

Engineering approach

The engineering approach treated stress as a stimulus characteristic of the person's environment and as an independent variable. It was usually conceived in terms of excessive load placed on the individual, or some other aversive or noxious element of that environment (Cox, 1978; Cox and Mackay, 1981; Fletcher, 1988). Occupational stress was thus treated as a property of the work environment, and usually as an objectively measurable aspect of that environment. In 1947, Symonds wrote, in relation to psychological disorders in RAF flying personnel, that '... stress is that which happens to the man, not that which happens in him; it is a set of causes not a set of symptoms'. Thirty years later, Spielberger (1976) argued, in the same vein, that the term stress should refer to the objective characteristics of situations. According to this approach, stress was said to produce a strain reaction which, although often reversible, could prove to be irreversible and damaging. The concept of a stress 'threshold' grew out of this approach and became part of the description of individual differences in 'resistance to stress' or vulnerability.

Physiological approach

The physiological approach to the definition and study of stress received its initial impetus from the work of Selye (1950, 1956). He defined stress as 'a state manifested by a specific syndrome which consists of all the non-specific changes within the biologic system' that occur when it is challenged by aversive or noxious stimuli. Stress was treated as a generalised and non-specific physiological response and as a dependent variable. Selye argued that the physiological response was tri-phasic in nature involving an initial alarm stage (sympathetic-adrenal medullary activation) followed by a stage of resistance (adrenal cortical activation) giving way, under some circumstances, to a final stage of exhaustion (terminal reactivation of the sympathetic-adrenal medullary system and death). Repeated, intense or prolonged elicitation of this physiological response, it was suggested, increased the wear and tear on the body, and contributed to what Selye (1950, 1956) described as the 'diseases of adaptation'. This apparently paradoxical term arises from the contrast between the immediate and short-term advantages bestowed by physiological responses to stress (energy mobilisation for an active behavioural response) and their long-term disadvantages (increased risk of certain 'stress-related' diseases).

Criticisms of engineering and physiological approaches

Three types of criticism of these early approaches have been offered: (i) they no longer account for enough of the available empirical data, (ii) they are dated and conceptually limited, and (iii) they represent an over-medicalisation of the subject. The latter is particular to the physiological approach. These criticisms are outlined in more detail below.

The first criticism is that the engineering and physiological models no longer adequately account for the existing data. Consider, for example, the effects of noise on performance and comfort. The engineering model would seem to suggest that exposure to loud noise (as 'stress') would cause, in a direct and simple manner, impairment of performance, discomfort and possibly ill-health. However, the effects of noise on task performance, for example, are not a simple function of its loudness nor of its frequency but are subject to its nature (type of noise) and to both individual differences and situational effects (Cox, 1978; Melamed *et al.*, 1993). Further, noise levels that might normally be disruptive may help maintain task performance when subjects are tired or fatigued (Broadbent, 1971), while even higher levels of music may be freely chosen in social or leisure situations. The simple equating of demands (such as loud noise) with stress has been associated with the erroneous belief that a certain amount of 'stress' is associated with maximal performance (Welford, 1973) and possibly good health.

Scott and Howard (1970) wrote 'certain stimuli, by virtue of their unique meaning to particular individuals, may prove problems only to them; other stimuli, by virtue of their commonly shared meaning, are likely to prove problems to a larger number of persons'. This statement implies the mediation of strong cognitive as well as situational (context) factors in the overall stress process. Such a point also has been made by Douglas (1992) with respect to the perception of risk. Such perceptions and related behaviour, she argued, are not completely explained by the natural science of objective risk but are strongly determined by group and cultural biases.

The physiological model is equally open to criticism. Both the non-specificity and the time course of the physiological response to aversive and noxious stimuli have been shown to be different from those described by Selye (1950, 1956) and required by this model. This was obvious by the early 1970s. Mason (1968, 1971), for example, showed that some noxious physical stimuli do not produce the stress response in its entirety. In particular, he has cited the effects of heat. About the same time, Lacey (1967) argued that the low correlations observed among different physiological components of the stress response are not consistent with the notion of an identifiable response syndrome. There is also a difficulty in distinguishing between those physiological changes that represent stress and those that do not, particularly as the former may be dissociated in time from the stressor. By the 1980s, there was much research suggesting that if the stress response syndrome did exist, it was not non-specific. Subtle but important differences in the overall pattern of response have been demonstrated (e.g., Dimsdale and Moss, 1980; Cox and Cox, 1985; Cox *et al.*, 1985).

The second criticism of the engineering and physiological models of stress is that they are dated and conceptually limited, being set within the relatively simple stimulus-response paradigm of the immediate pre- and post-war years. They largely ignore individual differences of a psychological nature, together with the perceptual, cognitive and emotional processes that might underpin them (Sutherland and Cooper, 1990). They treat the person as a passive vehicle for translating the stimulus characteristics of the environment into psychological and physiological responses. In large part, they ignore the interaction between the individual and their environment that is an essential part of systems-based approaches to biology, behaviour and psychology. They ignore the psychological, social and organisational contexts to work stress.

The third criticism is particularly focused on the physiological model. The ideas promoted by Selye (1950, 1956) and his advocates dictated a 'medicalisation' of the subject area. This was associated with an attendant narrowing of focus within the definition and practice of stress management. This approach has encouraged strategies that concentrate on individuals and their responses to 'stress', independent of the context within which the problem occurs. Partly as a result of this, we have witnessed the development of band-aid 'solutions' (relaxation or aromatherapy, for example) to stress problems in the workplace. Such views also tended to encourage the attribution of responsibility for any weakness or problems to the individual. Attempts to re-work the physiological approach have not resulted in a more coherent or clearer account of stress research or stress management. The development of notions of 'good' stress (eustress) and 'bad' stress (distress) have not resolved any of the existing problems, but have simply added to the semantic confusion. They have further encouraged the misguided view that 'some stress is good for you'. This erroneous belief in optimal levels of stress has been used on occasions to justify or excuse poor work design and management practice.

Psychological approaches

The third approach to the question of definition is a psychological one, and offers what is essentially a cognitive–emotional model of stress. In doing so, it attempts to overcome some of the criticisms that the other two types of model have attracted.

Variants of this psychological approach dominate contemporary stress theory, and among them two distinct types can be identified: the interactional and the transactional (Cox, 1978, 1993). The former focuses on the architecture or structure of individuals' interactions with their work environment, while the latter is more concerned with the psychological processes underpinning those interactions. Transactional models are primarily concerned with cognitive appraisal, emotion and coping. In a sense they represent a development of the interactional models, and are largely consistent with them. This chapter adopts a transactional approach based on Cox (1978, 1985, 1993). It proposes the unifying concept of a stress process, beginning with (i) its antecedent factors (in relation to work) and (ii) the cognitive perceptual processes involved and the emotional experience of stress; it then considers (iii) the health correlates of that experience. This approach provides a framework for measurement.

Stress as a psychological phenomenon

Stress has been defined as a cognitive state (Cox, 1985a, 1993) which is part of a wider process reflecting the person's perception of and adaptation to the demands of the (work) environment. Arguably, its defining aspect for the individual is essentially emotional. This approach emphasises the person's cognitive appraisal (Lazarus, 1966) of the situation, and their emotional reaction to it. It treats the whole process of perceiving and reacting to stressful situations within a problem solving context (Cox, 1987). Stress is not treated as a dimension of the physical or psychosocial environment: it cannot be defined solely in terms of workload (see Chapter 18) or the occurrence of events determined by consensus to be stressful. Equally, it cannot be adequately defined solely in terms of responses that are sometimes correlates of stress, such as physiological mobilisation or performance dysfunction. Framed in this way, the study of work stress is largely about normal people failing to cope with the problems that face them in their working lives and the experience and consequences of that failure.

A central feature of this approach to work stress is the process of cognitive appraisal (Lazarus, 1966), which Holroyd and Lazarus (1982) defined in terms of being 'the evaluative process that imbues a situational encounter with meaning'. Later refinements of this theory (Folkman *et al.*, 1980) suggest that there are primary and secondary components to the appraisal process. Primary appraisal involves a continual monitoring of the person's transactions with his or her environment, focusing on the question, 'Do I have a problem?' Secondary appraisal is contingent upon the recognition that a problem exists and involves a more detailed analysis and the generation of possible coping strategies (see below). It has been suggested by Cox (1978, 1987, 1993) that the overall process of appraisal involves a continual monitoring of at least four aspects of individuals' transactions with their environment, and a continual evaluation of the balance between them (Cox, 1987). Cognitive appraisal appears to take account of individuals' perceptions of:

1. The demands on them
2. Their ability to cope with those demands
3. The constraints under which they have to cope, and
4. The support they receive from others in coping.

Demands

Demands are requests for action or adjustment, whether cognitive, emotional, behavioural or physiological in nature. They require some degree of decision making and the exercise of skill. They may be imposed by the external environment, say as a function of work or the work-home interface, or may be internal, reflecting the person's needs: material, social or psychological. There may be several important dimensions describing external (job) demands, for example, 'pleasantness/unpleasantness' and 'ease/difficulty' (see Cox, 1985b). Demands usually have a time base, the effects of which may be amplified by an acute sense of time urgency (type A behaviour: Zyzanski and Jenkins, 1970).

Ability to Cope

The absolute level of demand would not appear to be the important factor in determining the experience of stress. More important is any discrepancy which exists between the perceived level of demand and individuals' perceptions of their ability to meet that demand. The size of this discrepancy appears to be an important determining factor in the stress process. However, the relationship between the discrepancy and the intensity of the stress experience may be curvilinear rather than linear (Cox, 1978). Within reasonable limits stress can arise through either overload (demand > abilities) or through underload (demand < abilities), or through some combination of the two. A person's ability to cope with such an imbalance may be constrained or supported in different ways and to different extents.

Constraints

Constraints operate as restrictions or limitations on free action or thought, often reflecting a loss or lack of discretion or control over actions. These may be imposed externally. For example, constraints may be imposed by the requirements of specific jobs or by the rules of the organisation. They may also be role related or reflect the beliefs and values of the individual.

There is some discussion in the literature as to whether the effects of job demands and constraints (job discretion) are additive or subject to a true interaction. Karasek (e.g., Karasek, 1979; Karasek *et al.*, 1981) has argued that a true interaction exists, while others, for example Warr (1990), have found evidence only for a simple additive effect. Despite this, the notion that high job demands and constraints create a disproportionately risky condition for worker health remains popular particularly among policy makers.

Support

Support can be made available in different ways, most essentially through 'social interaction': through advice and information, through practical assistance or by providing understanding and declaring empathy. It is possible that women need and are more sensitive to social support than men (Cox *et al.*, 1983, 1984).

Environmental antecedents of work stress

The concepts of demand and ability, constraint and support are high level ones and, in the reality of work, are combined and represented in a wide variety of different ways. The way they combine largely reflects the way work is designed and managed in its social and organisational contexts and the work characteristics that these processes determine (see Figure 1). Much of the current discussion of work stress is therefore focused on what are essentially failures of work design and management.

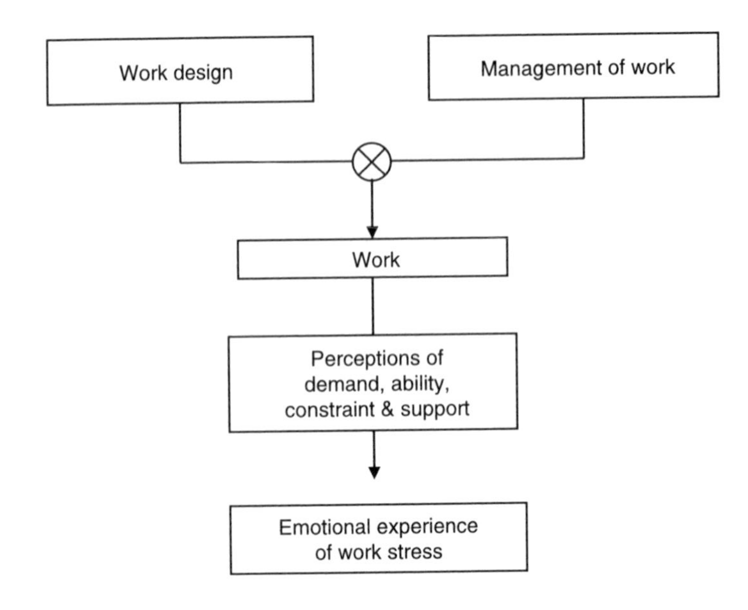

Figure 1 Antecedents of work stress.

Those work characteristics that have the potential for causing harm to the individual are 'hazards' and those that do so and are associated with the experience of stress are 'stress-related hazards' or 'stressors'. Sometimes, these stressors have been termed 'psychosocial and organisational' hazards. They are now well represented in the academic literature and in guidance published for employers by governments, unions, employers' bodies, insurers and the media. There has been much research into such sources of stress at work, and many taxonomies have been published (see, e.g., Cooper and Marshall, 1976). There appears to be broad agreement across these taxonomies. An example is provided in Table 1.

Individual differences and coping

A key component of psychological theories of stress is the concept of 'coping'. It concerns people's attempts to make a potentially stressful transaction less stressful (Lazarus and Folkman, 1984) and reduce the emotional experience of stress (Ferguson and Cox, 1997). These attempts are context-specific and are based on initial perceptions of the situation (primary appraisal) together with an evaluation of the possible ways of dealing with it (secondary appraisal). It is clear that people vary in their coping styles and preferences (Semner, 2003) as well as in their flexibility. Optimists, for example, have been shown to be more likely to use coping strategies that are appropriate for the situation than are pessimists (Carver and Scheier, 1999).

There are many different ways in which a person's coping resources might be conceptualised; however, it could be useful to think of them in terms of energy, knowledge, attitudes, skills and behavioural style. Here the idea of skill has to be extended beyond traditional conceptualisations in terms of psychomotor and technical skills, to include social and cognitive skills. The attitudes and behavioural style as well as personal knowledge and skills can be developed both through formal education and training and more informally through untutored experience. Furthermore, several elements of this 'package' of resources are subject to change, influenced by factors such as time of day, fatigue, and state of health. In addition to any consideration of its cognitive and perceptual elements, the state of stress is

Table 1 Psychosocial and Organisational Hazards (Adapted from Cox, 1993)

Content of work	
Job content	Lack of variety or short work cycles, fragmented or meaningless work, under use of skills, high uncertainty, continuous exposure to people through work
Workload and work pace	Work overload or under load, machine pacing, high levels of time pressure, continually subject to deadlines
Work schedule	Shift working, night shifts, inflexible work schedules, unpredictable hours, long or unsociable hours
Control	Low participation in decision making, lack of control over workload, pacing, shift working, etc.
Environment and equipment	Inadequate equipment availability, suitability or maintenance; poor environmental conditions such as lack of space, poor lighting, excessive noise
Context to work	
Organisational culture and function	Poor communication, low levels of support for problem solving and personal development, lack of definition of, or agreement on, organisational objectives
Interpersonal relationships at work	Social or physical isolation, poor relationships with superiors, interpersonal conflict, lack of social support
Role in organisation	Lack of participation; role ambiguity, role conflict, and responsibility for people
Career development	Career stagnation and uncertainty, under promotion or over promotion, poor pay, job insecurity, low social value to work
Home-work interface	Conflicting demands of work and home, low support at home, dual career problems

often defined by the person's experience of negative emotion, unpleasantness and general discomfort.

The effects of stress

Many of the effects of stress are short-lived and cause no lasting harm, but if intense or sustained can be more damaging. The experience of stress is usually accompanied by changes in the way people feel, think, behave, and in their physiological function. Anxiety and depression are common symptoms, as are irritability, impaired memory and attention span. Interpersonal relations may be damaged. Evidence suggests that stress is associated with cardiovascular disease, gastro-intestinal disorders, musculo-skeletal disorders and impaired immune function (Cox *et al.*, 2000). In addition, work-related stress is thought to account for a significant proportion of the variance in many maladaptive health behaviours such as smoking, substance abuse and sleep patterns (Shirom, 2003). Ultimately, work stress may affect the healthiness and productivity of the organisation (Cox and Thomson, 2000) through a variety of pathways: increased absence and intention to leave, poor morale and commitment, lack of trust; poor productivity and poor quality of work; and poor safety performance. It can therefore be costly to organisations beyond the cost to the individual.

Measuring work-related stress: a framework

This chapter takes the stress process and its key elements, as described above, as its framework for measurement. It incorporates aspects of risk assessment (see also Chapters 32 and 34).

Elsewhere, the authors have described and discussed a full risk assessment methodology for work-related stress (Cox *et al.*, 2002, 2003). This framework is applicable to group study and risk assessment purposes in the workplace rather than to the clinical assessment of individuals. It is a nomothetic method rather than an idiopathic one. It is not possible to offer a comprehensive review of the plethora of measures which might be used in such an approach. The discussion here is focused on the Nottingham methodology as presented in its Work Environment Survey, and two self-report instruments — the Stress Arousal Checklist (SACL) and the General Well-being Questionnaire (GWBQ).

Triangulation

Essentially the approach to measurement set out here is built on the inter-relatedness of data from three different sources: an assessment of the work environment (antecedent factors: stressors), the measurement of the emotional experience of stress, and the assessment of health and wellbeing. Relating data from three different sources has been referred to as triangulation (see below) and, for best practice, such data should be collected using different methods. However, in reality, both in research and practice, this is the exception rather than the rule and much depends on self report.

Ideally, the principle of triangulation should be applied both within and between domains. Across domains, data collection could take the form of continuous or repeat monitoring and thus be capable of mapping and cross referencing changes in all domains. This should help overcome the problem of missing data and help resolve inconsistencies in the data given that these are not extreme. Within domains, several different measures should be taken and preferably across different measurement modalities to avoid problems of common method variance. This may be most relevant and easiest to achieve in relation to the measurement of changes in the third domain: behaviour, physiology and health status. There is no available evidence to suggest that the various measures from the different domains can be statistically combined into a single and defensible 'stress index'.

Measurement of the stress process

What has to be measured is the stress process: 'antecedents in the work environment → the experience of stress → its psycho-physiological and health correlates'. This might be simplified conceptually to the basic health and safety equation of: 'work hazards–stress–harm'. In this context, the overall measurement process can be said to resemble a risk assessment. This approach underlines both the complexity of measurement and the inadequacy of asking for or using single one-off measures of stress (however defined).

Somewhat similar approaches can be found in the literature. For example, Bailey and Bhagat (1987) recommended a multi-method approach to the measurement of stress consistent with the concept of triangulation. They have argued in favour of balancing the evidence from self-report of work and experience, physiological and unobtrusive measures. Their unobtrusive measures relate to what Folger and Belew (1985) and Webb *et al.* (1966) called non-reactive measures, and include: physical traces (such as poor house-keeping), archival data (such as that on absenteeism), private records (such as diaries), and non-intrusive observation and recordings.

Assessing the work environment

The Work Environment Survey is a methodology for collecting data on the possible antecedents of the experience of stress in failures of work design and management. It uses situational reasoning and focuses on building a consensus model of the problems inherent in the work environment. It operates first through a rolling series of interviews, with the work group of concern, to build a model of work and any failures of work design or management or problems in its social or organisational contexts.

From this model, a quantitative assessment instrument is developed and tailored to the situation of the work group being assessed. This is essentially a device for assessing the risk to health, broadly defined, of these different problems and the role of the experience of stress in mediating or moderating that relationship. In this way, the three key elements of the stress process (see above) are tied together in overall measurement process. The instrument is validated through discussions with the work group of concern. The instrument is then applied to the whole assessment group.

Measuring the experience of stress

Within the measurement framework described above, the experience of stress is defined as a psychological state which is part of and reflects a wider process of interaction between individuals and their (work) environment. That experience is essentially emotional in nature. Within the present framework, the measurement of stress should be based primarily on self-report measures which focus on the associated emotional experience (Cox, 1978, 1985a, 1993).

Stress arousal checklist (SACL)

The measurement of mood may offer one direct method of tapping the individual's experience of stress, and there was a surge of interest in this issue in the late 1970s and in the 1980s as witnessed by a series of articles in the *British Journal of Psychology* (Cox and Mackay, 1985; Cruickshank, 1984; King *et al.*, 1983) and elsewhere (Burrows *et al.*, 1977; Ray and Fitzgibbon, 1981; Russell, 1979, 1980; Watts *et al.*, 1983). Most of these studies have employed the 'Stress Arousal Checklist' (SACL) developed by Cox and Mackay and originally published in the *British Journal of Social and Clinical Psychology* (Mackay *et al.*, 1978).

The SACL is an adjective checklist and was developed using factor analytical techniques (Cox and Mackay, 1985; Gotts and Cox, 1990; Mackay *et al.*, 1978) for the measurement of self-reported mood. It presents the respondent with 30 relatively common mood describing adjectives, and asks to what extent they describe their current feelings. The model of mood which underpins the checklist is two dimensional. One dimension appears to relate to feelings of unpleasantness/pleasantness or hedonic tone (stress) and the other to wakefulness/drowsiness or vigour (arousal). Such a model is well represented in the relevant psychological and psychophysiological literature (see, e.g., Mackay, 1980; Russell, 1980). The split half reliability coefficients for the two scales which tap into these dimensions have always proved acceptable: for example, arousal 0.82 and stress 0.80 (Watts *et al.*, 1983; Gotts and Cox, 1990). Both were conceived of and developed as state measures, and are thus seen as transient in nature. The statistical device of test–retest coefficients is therefore not appropriate as a test of reliability.

Consistent with the transactional model (see above), it was suggested by Mackay *et al.* (1978) that the stress dimension may reflect the perceived favourability of the external environment, and thus have a strong cognitive component in its determination. Arousal, it was suggested, might relate to ongoing autonomic and somatic activity, and be essentially psychophysiological in nature. It became obvious that stress may partly reflect how appropriate the level of arousal is for a given situation, and the effort of compensating for inappropriate levels (Cox *et al.*, 1982).

Together the two dimensions can be used to describe a four quadrant model of mood within which characteristic emotions and related states may be identified: high arousal and high stress (anxiety), high arousal and low stress (pleasant excitement), low arousal and high stress (boredom), and finally, low arousal and low stress (relaxed drowsiness). There are now many reported studies using the SACL, and reporting data from its two scales (Burrows *et al.*, 1977; Cox *et al.*, 1982; King *et al.*, 1983; Ray and Fitzgibbon, 1981; Watts *et al.*, 1983). There have also been a number of studies which have used modified versions of the checklist (Cruickshank, 1982, 1984), although locally inspired changes in the instrument cannot always be defended (Cox and Mackay, 1985).

A third scale has been suggested based on the use of a '?' category on the response scale associated with the different mood adjectives. This category signifies, in part, uncertainty about whether the adjective given currently describes the respondent's mood. A score based on the frequency of '?' responses might reflect an inability to report feelings, and this may be symptomatic of a disordered psycho-physiological state. Such a scale has an acceptable split half reliability coefficient: 0.89 (Cox and Mackay, 1985).

A compilation of the available British and Australian data has allowed the publication of mean levels for different groups, broken down by country of origin, age, sex and occupation (Gotts and Cox, 1990). Some of these 'normative' data are presented in Table 2.

Measuring health at work

Fortunately, it appears that the experience of work-related stress is more likely to be associated with changes in the level of general malaise than in the incidence of death, or of disease, disability or injury (see Cox *et al.*, 2000b). For most studies, general malaise has to be the focus of measurement. General malaise has also been referred to as sub-optimal health.

A theory of sub-optimal health

Health has been defined as a changeable state along a continuum from complete healthiness to death (World Health Organisation, 1946; Rogers, 1960). This broad definition implies that health cannot be equated with mere absence of obvious disease, injury or disability, and is not restricted to the sound physical condition of the body; it also has psychological and social aspects. In this context, well-being relates to individuals' experience of their health.

An important watershed in the health continuum is represented by the point where disease, disability or injury become obvious to the person, and are usually represented by objectively verifiable and clinically significant signs and symptoms. Some of these will be

Table 2 Some Normative Data for the SACL (derived from Gotts and Cox, 1990)

	Dichotomised scores						*Q*		
	Stress			Arousal					
Sample	*x*	SD	*n*	*x*	SD	*n*	*x*	SD	*n*
Mixed population	6·0	4·6	1027	6·4	3·2	1040	4·2	4·2	1079
Males: mixed sample	6·0	4·7	296	6·6	3·2	297	4·9	4·6	266
Females: mixed sample	6·0	4·6	731	6·3	3·3	743	3·9	4·1	584
Students	6·3	4·9	515	5·7	3·6	518	4·7	4·1	535
Ages 16 to 30	6·2	4·6	466	6·0	3·2	469	5·0	4·3	379
Ages 31 to 45	5·9	4·9	344	7·2	3·3	353	3·5	4·2	334
Age more than 45	5·1	4·2	122	6·4	3·3	123	3·7	4·1	1132

More complete normative data have been published as part of a manual for the SACL (Gotts and Cox, 1990). Further information can be obtained from the authors.

diagnostic of a particular condition; others will more generally reflect the impact of that condition. Some will inevitably represent the effects of stress experienced in relation to being ill.

The zone between complete physical, psychological, and social healthiness on the one hand, and obvious disease, disability or injury, on the other, has been termed sub-optimal health (Rogers, 1960). It has been suggested that sub-optimal health may be represented as an experiential pool of signs and symptoms of general malaise, each, on its own, of no particular clinical significance and certainly not diagnostic of any particular condition. Such signs and symptoms may or may not be precursors to disease, injury or disability depending on the operation of a wide range of health risk and salutogenic factors. At any time, different groups of signs and symptoms within the experiential pool will imperfectly predict particular ill-health outcomes. As a condition develops, the predictive group will refine itself, attract new signs and symptoms, and the prediction itself may strengthen. As a condition weakens, or the person recovers, then the reverse process will occur. It is 'sub-optimum health', as a concept and an experience, that equates most closely to the popular understanding of 'wellbeing' and the two terms are used more-or-less interchangeably here.

A person considered to be normally healthy, by themselves or others, or more particularly judged not to be ill, will still experience something by way of such signs and symptoms of general malaise, and the possible pool of such experiences will be formally present at the group level. At this level, structural modelling will reveal a pattern and clusters of signs and symptoms reflective of an underlying normal experience model. However, it has been suggested that the normal experience of well-being may both reflect the experience of stress as one mediator of the effects of life and working conditions, and also in turn affect other responses to stress, such as self-reported mood (see Mackay *et al.*, 1978; Cox and Mackay, 1985).

The General Wellbeing Questionnaire was developed within the framework of this theory of suboptimal health.

Development of the general well-being questionnaire (GWBQ)

It was in the mid 1980s that the Nottingham group began to build a measurement tool for well-being based on the self-report of signs and symptoms of general malaise (Cox and Brockley, 1984; Cox *et al.*, 1983, 1984). There were several different questionnaire instruments available at the time that by the nature of their scales and internal structure offered some description of that area of health (Crown and Crisp, 1966; Derogatis *et al.*, 1974; Goldberg, 1972; Gurin *et al.*, 1960). However, none of these were judged to be exactly what was required for use with a more-or-less healthy working population in Britain and for the purpose of assessing work-related stress.

Initially a compilation of non-specific symptoms of general malaise was produced from existing health questionnaires (see above) and from diagnostic texts. These symptoms included reportable aspects of cognitive, emotional, behavioural and physiological function, none of which were clinically significant in themselves. From this compilation, a prototype checklist was designed with each symptom being associated with a five point frequency scale ('never' through to 'always') which referred to a six month response window. In a series of classical factor analytical studies, on British subjects, variously reported (Cox *et al.*, 1983, 1984), two clusters of symptoms or factors were identified (see Table 3). These factors were derived as orthogonal. The first factor (GWF1) was defined by symptoms relating to tiredness, emotional lability, and cognitive confusion; it was colloquially termed 'worn out'. The more cognitive items would appear to imply difficulties in decision making (in the specific context of feeling 'worn out'): (a) Has your thinking got mixed up when you have had to do things quickly? (b) Has it been hard for you to make up your mind? and (c) Have you been forgetful?

Table 3 Items Defining the GWBQ Scales (International Version)

GWF1	
	Have your feelings been hurt easily?
	Have you got tired easily?
	Have you become annoyed and irritated easily?
	Have your thinking got mixed up when you have had to do things quickly?
	Have you done things on impulse?
	Have things tended to get on your nerves and wear you out?
	Has it been hard for you to make up your mind?
	Have you got bored easily?
	Have you been forgetful?
	Have you had to clear your throat?
	Has your face got flushed?
	Have you had difficulty in falling or staying asleep?
GWF2	
	Have you worn yourself out worrying about your health?
	Have you been tense and jittery?
	Have you been troubled by stammering?
	Have you had pains in the heart or chest?
	Have you unfamiliar people or places made you afraid?
	Have you been scared when alone?
	Have you been bothered by thumping of the heart?
	Have people considered you to be a nervous person?
	When you have been upset or excited has your skin broken out in a rash?
	Have you shaken or trembled?
	Have you experienced loss of sexual interest or pleasure?
	Have you had numbness or tingling in your arms or legs?

These may have implications for personal problem solving and coping (see Cox, 1987). The second factor (GWF2) was defined by symptoms relating to worry and fear, tension and physical signs of anxiety; it was colloquially termed 'up tight and tense'. This model of sub-optimum health appeared to have some face validity in that it was acceptable to a conference audience of British general practitioners and medical and psychological researchers (see Cox *et al.*, 1983).

It is therefore suggested by the authors that sub-optimum health, the 'grey area' between complete healthiness and obvious illness, is made up of two states, one related to being 'worn out' or exhausted, and the other related to being 'up tight and tense'. The former has an interesting cognitive component, possibly related to decision making and coping, while the latter is partly defined by physical symptoms of anxiety and tension. It has been shown that people vary in the extent to which they report these feelings, both between individuals and across time, and it has been suggested that this variation may not only (a) reflect the experience of stress, but also (b) affect other responses to stress, such as self-reported mood (see Mackay *et al.*, 1978; Cox and Mackay, 1985). There is evidence that in workplace studies the worn-out or exhaustion scale shows greater utility in workplace studies than does the tense and uptight scale; it demonstrates a more consistent relationship with other, non-health measures of interest within the workplace (Cox *et al.*, 2000).

This research culminated in the publication in English of the General Well-Being Questionnaire (Cox and Gotts, 1987).

Table 4 Some Normative Data for the GWBQ (Derived from Unpublished Data of Cox and Gotts) for Mixed Populations

INTERNATIONAL VERSION (1987)						
	'Worn out'(12 Items)			'Up tight' (12 Items)		
Sample	*X*	S.D.	*n*	*X*	S.D.	*n*
All	16·7	8·3	2300	10·7	7·4	2312
Males	15·9	7·8	1031	8·2	6·5	1042
Females	17·4	8·6	1262	12·8	7·4	1262
British sample by age (years)						
16–20	16·5	8·7	141	11·5	7·9	141
21–25	16·9	9·2	147	11·3	7·6	147
26–30	15·6	8·4	236	10·2	7·5	236
31–35	17·2	8·6	239	9·0	6·5	239
36–40	16·1	8·1	201	9·2	7·5	201
41–45	15·5	8·6	199	10·4	7·7	199
46–50	16·0	8·3	175	9·7	7·7	175
51–55	14·5	8·0	174	9·1	7·4	174
56–60	13·7	8·0	127	7·7	6·6	127
>60	13·5	6·4	26	4·8	5·8	26

Further information can be obtained from the authors.

In the late 1980s new data were collected through a series of linked studies in Britain and Australia. These data were re-analysed, and the model and its associated scales were amended to increase their robustness in relation to this international sample, and also to a diversity of homogeneous samples (see Table 4).

A number of symptoms (items) were deleted from the original scales, but no new symptoms were added. The two new international scales were each defined by twelve symptoms but retained their essential nature: worn out and tense and uptight. The deleted symptoms were among the weaker ones in terms of scale definition and item loadings. The early questionnaire was revised, new norms were computed and an international version was published and has been in use since then (e.g., Cox and Griffiths, 1995, Cox *et al.*, 2000).

Recent developments

The more recent development of the GWBQ has focused on two issues: the question of cultural-linguistic difference and the further development of the scales.

The possibility of cultural-linguistic differences in the experience and report of well-being has been explored in working populations in Taiwan (Ruey-Fa, 1993) and in Singapore (Ho, 1996). For example, bilingual Taiwanese school teachers (English and Mandarin) completed English and Mandarin versions of the GWBQ. The data clearly showed the emergence of identical two factor models from these data sets with teachers' scores on the two versions being very highly correlated. The Taiwanese (Mandarin) model was indistinguishable from the UK English language version. Despite this extreme test, the cultural-linguistic interchangeability of the GWBQ should not be taken for granted, and is a matter of empirical test between its English origins and other cultural-linguistic situations.

Recent research by the authors has suggested that feelings of being worn out or exhausted are commonly associated with exposure to poor work design or problems with the management of work and often moderate the relationship between such failures and their health effects. Qualitative data collected during these studies has allowed the authors to begin both extending and refining the measurement of being 'worn out' or exhausted with the possibility of developing new sub-scales within the existing measurement framework. This development should allow a greater sensitivity of measurement and, hopefully, a greater depth of understanding of the effects of work stress on well-being.

Organisational healthiness

In addition to measurements of individual health and stress, there are various measures of 'organisational' healthiness (Cox and Thomson, 2000) that may be used in studies of the correlates of stress at work. Their choice will depend on the case in question. In some organisations, for example, absence from work may be a key measure. In others, where there may be a tendency to work through illness, or to work at home, it will be meaningless.

Concluding comments

This chapter has attempted to describe the nature of the problem of work-related stress in a way that points up its relevance to both individual workers and to their organisations. It is an important contemporary issue in occupational health and safety. The current consensus defines stress as a psychological state with an important emotional component. Furthermore, there are now theories which can be used to relate the experience and effects of work stress to exposure to work hazards and to the harmful effects on health that such exposure might cause. These theories can be used to place the measurement of stress within a health and safety framework and link it to risk assessment. This approach facilitates both the measurement and the management of work-related stress.

The inadequacy of single one-off measures is widely recognised in the literature but despite this they continue to be used. This diversity may account for much of the disagreement within stress research on measurement. Part of the solution to this problem lies with agreeing the theoretical framework within which measurement is made, but part lies with the development of a more adequate technology of measurement based in 'good practice' in a number of areas including occupational health psychology, psychometrics, knowledge elicitation and knowledge modelling. A forced standardisation of measurement is not being argued for here and should be resisted for its effects on scientific progress. What is being argued for throughout is better measurement processes, conforming to recognised good practice in relevant areas, and applied within a declared theoretical context.

The approach offered here is based on the stress process and is consistent with risk assessment. It considers: the antecedents of stress in failures of work design and management, the experience of stress, and its effects of health or well-being. These three key elements in the stress process can be brought together in a particular way. First, the relationship between exposure to the failures of work design and management, on the one hand, and possible effects on health or well-being, on the other must be explored. Then the question of whether this relationship is mediated or moderated by the experience of stress is examined. Logically this establishes a description of the stress process that is secure using the principle of triangulation. This may well be sufficient for the Courts of Law as well as those of applied science.

References

Bailey, J.M. and Bhagat, R.S. (1987). Meaning and measurement of stressors in the work environment. In: S.V. Kasl and C.L. Cooper (eds) *Stress and Health: Issues in Research Methodology*, (Chichester: Wiley & Sons).

Broadbent, D.E. (1971). *Decision and Stress.* (New York: Academic Press).

Burrows, G.C., Cox, T. and Simpson, G.C. (1977). The measurement of stress in a sales training situation. *Journal of Occupational Psychology*, **50**, 4–51.

Carver, C.S. and Scheier, M.F. (1999). Optimism. In: C.R. Synder (ed.) *Coping: The Psychology of What Works.* (New York: Oxford University Press), pp. 182–204.

Cooper, C.L., and Marshall, J. (1976). Occupational sources of stress: a review of the literature relating to coronary heart disease and mental ill health. *Journal of Occupational Psychology*, **49**, 11–28.

Cooper, C.L, Dewe, P. and O'Driscoll, M. (2001). *Organisational Stress: A Review and Critique of Theory, Research and Applications.* (Thousand Oaks, CA: Sage Publications).

Cox, S., Cox, T., Thirlaway, M. and Mackay, C.J. (1985). Effects of simulated repetitive work on urinary catecholamine excretion. *Ergonomics*, **25**, 1129–1141.

Cox, T. (1978). *Stress.* (London: Macmillan).

Cox, T. (1985a). The nature and measurement of stress. *Ergonomics*, **28**, 1155–1163.

Cox, T. (1985b). Repetitive work: occupational stress and health. In: C.L. Cooper and M. Smith (eds) *Job Stress and Blue Collar Work.* (Chichester: Wiley & Sons).

Cox, T. (1987). Stress, coping and problem solving. *Work and Stress*, **1**, 5–14.

Cox, T. (1993). *Stress Research and Stress Management: Putting Theory to Work.* (Sudbury: HSE Books).

Cox, T. and Brockley, T. (1984). The experience and effects of stress in teachers. *British Educational Research Journal*, **10**, 83–87.

Cox, T. and Cox, S. (1985). The role of the adrenals in the psychophysiology of stress. In: E. Karas (ed.) *Current Issues in Clinical Psychology.* (London: Plenum Press).

Cox, T., and Gotts, G. (1987). *The General Well-Being Questionnaire Manual.* (University of Nottingham: Institute of Work, Health and Organisations).

Cox, T., and Griffiths, A. (1995). The nature and measurement of work stress: theory and practice. In: J. Wilson and N. Corlett (eds), *The Evaluation of Human Work: A Practical Ergonomics Methodology* (London: Taylor & Francis), pp. 783–803.

Cox, T., Griffiths, A. and Rial-González, E. (2000). *Research on Work-Related Stress.* (Luxembourg: Office for Official Publications of the European Communities).

Cox, T., Griffiths, A., and Randall, R. (2003). A risk management approach to the prevention of work stress. In: M.J. Schabracq, J.A.M. Winnnubst & C.L. Cooper (eds) *Handbook of Work and Health Psychology* (Second Edition) (Chichester: Wiley & Sons), pp. 191–206.

Cox, T. and Mackay, C.J. (1981). A transactional approach to occupational stress. In: N. Corlett and J. Richardson (eds) *Stress, Work Design and Productivity.* (Chichester: Wiley & Sons).

Cox, T. and Mackay, C.J. (1985). The measurement of self-reported stress and arousal. *British Journal of Psychology*, **76**, 183–186.

Cox, T., Randall, R. and Griffiths, A. (2002). Interventions to control stress at work in hospital staff. (Sudbury: HSE Books).

Cox, T., Thirlaway, M. and Cox, S. (1982). Repetitive work, well-being and arousal. In: H. Ursin and R. Murison (eds), *Biological and Psychological Basis of Psychosomatic Disease. Advances in the Biosciences*, **42**, 115–135 (Oxford: Pergamon Press).

Cox, T., Thirlaway, M., Gotts, G., and Cox, S. (1983). The nature and assessment of general well-being. *Journal of Psychosomatic Research*, **27**, 353–359.

Cox, T., Thirlaway, M. and Cox, S. (1984). Occupational well-being: Sex differences at work. *Ergonomics*, **27**, 499–510.

Cox, T. and Thomson, L. (2000). Organizational healthiness: Work-related stress and employee health. In: P. Dewe, M. Leiter and T. Cox (eds) *Coping, Health and Organizations.* (London and New York: Taylor & Francis), pp. 173–190.

Crown, S. and Crisp, A.H. (1966). A short clinical diagnostic self-rating scale for psychoneurotic patients. The Middlesex Hospital Questionnaire (MHQ). *British Journal of Psychiatry*, **112**, 917–923.

Cruickshank, P.J. (1982). Patient stress and the computer in the waiting room. *Social Science and Medicine*, **16**, 1371–1376.

Cruickshank, P.J. (1984). A stress and arousal mood scale for low vocabulary subjects. *British Journal of Psychology*, **75**, 89–94.

Derogatis, L.R., Lipman, R.S., Rickels, K., Uhlenhuth, E.H. and Convi, L. (1974). The Hopkins Symptom Checklist (HSCL). In: P. Pichot (ed.), *Modern Problems in Pharmacopsychiatry*, Volume 7. (Basel: Karger).

Douglas, M. (1992). *Risk and Blame*. (London: Routledge).

Dimsdale, J.E. and Moss, J. (1980). Short-term catecholamine response to psychological stress. *Psychosomatic Medicine*, **42**, 493–497.

European Commission (1989). Council Framework Directive on the Introduction of Measures to Encourage Improvements in the Safety and Health of Workers at Work. 89/391/EEC. *Official Journal of the European Communities*, **32**, No L183, 1–8.

European Commission (1996). *Guidance on Risk Assessment at Work*. (Brussels: European Commission).

European Commission (1997) The Advisory Committee for Safety, Hygiene and Health Protection at Work: Report on Work-related Stress. CE-V/4-97-015-EN-C. Luxembourg: European Commission.

Ferguson, E. and Cox, T. (1997) The functional dimensions of coping scale: theory, reliability and validity. *British Journal of Health Psychology*, **2**, 109–129.

Fletcher, B.C. (1988). The epidemiology of occupational stress. In: C.L. Cooper and R. Payne (eds) *Causes, Coping and Consequences of Stress at Work*, (Chichester: Wiley & Sons).

Folger, R. and Belew, J. (1985). Non-reactive measurement: a focus for research on absenteeism and occupational stress. In: L.L. Cummings and B.M. Straw (eds) *Organizational Behaviour*, (Greenwich, CT: JAI Press).

French, J.R.P., Caplan, R.D., and van Harrison, R. (1982). *The Mechanisms of Job Stress and Strain* (New York: Wiley & Sons).

Griffiths, A. (1999). Organizational interventions: facing the limits of the natural science paradigm. *Scandinavian Journal of Work, Environment and Health*, **25**, 589–596.

Goldberg, D.P. (1972). *The Detection of Psychiatric Illness by Questionnaire*. Maudsely Monograph No. 21 (London: Oxford University Press).

Gotts, G. and Cox, T. (1990). *Stress and Arousal Checklist: A Manual for Its Administration, Scoring and Interpretation* (Melbourne, Australia: Swinburne Press).

Gurin, G., Veroff, J. and Feld, S. (1960). *Americans' View of Their Mental Health* (New York: Edinburgh).

Health and Safety Executive (1995). *Stress at Work — An Employer's Guide*. (Sudbury: HSE Books).

Health and Safety Executive (2003) http:/www.hse.gov.uk/statistics

Ho, J. (1996). School organizational health and teacher stress in Singapore. Unpublished doctoral dissertation, University of Nottingham, United Kingdom.

Hodgson, J.T., Jones, J.R., Elliott, R.C. and Osman, J. (1993). *Self-reported Work-related Illness*. (Sudbury: HSE Books).

Holroyd, K.A. and Lazarus, R.S. (1982). Stress, coping and somatic adaptation. In: L. Goldberger and S. Breznitz (eds) *Handbook of Stress*. (New York: Free Press).

Industrial Relations Law Reports (1995). Walker v Northumberland County Council, 35.

Jones, J.R., Hodgson, J.T., Clegg, T.A. and Elliot, R.C. (1998). *Self-reported Work-related Illness in* 1995. (Sudbury: HSE Books).

Karasek, R.A. (1979). Job demands, job decision latitude and mental strain: implications for job redesign. *Administrative Science Quarterly*, **24**, 285–308.

Karasek, R. and Theorell, T. (1990). *Healthy Work: Stress, Productivity and the Reconstruction of Working Life*. (New York: Basic Books).

Karasek, R.A., Baker, D., Marxer, F., Ahlbom, A., and Theorell, T. (1981). Job decision latitude, job demands and cardiovascular disease. *American Journal of Public Health*, **71**, 694–705.

King, M.G., Burrows, G.D. and Stanley, G.V. (1983). Measurement of stress and arousal: validation of the stress arousal checklist. *British Journal of Psychology*, **74**, 473–479.

Lacey, J.I. (1967). Somatic response patterning and stress: some revisions of activation theory. In: M.H. Appley and R. Trumbull (eds) *Psychological stress* (New York: Appleton-Century-Crofts).

Lazarus, R. (1966). *Psychological Stress and the Coping Process*. (New York: McGraw-Hill).

Lazarus, R. and Folkman, S. (1984). *Stress, Appraisal and Coping*. (New York: Springer Publishing).

Mackay, C.J. (1980). The measurement of mood and psychophysiological activity using self-report techniques. In: I. Martin and P. Venables (eds) *Techniques in Psychophysiology.* (Chichester: Wiley & Sons).

Mackay, C.J., Cox, T. Burrows, G. and Lazzerini, T. (1978). An inventory for the measurement of self-reported stress and arousal. *British Journal of Social and Clinical Psychology,* **17**, 283–284.

Mason, J.W. (1968). A review of psychoendocrine research on the pituitary-adrenal cortical system. *Psychosomatic Medicine,* **30**, 576–607.

Mason, J.W. (1971). A re-evaluation of the concept of non-specificity in stress theory. *Journal of Psychiatric Research,* **8**, 323–333.

Melamed, S., Harari, G. and Green, M. (1993). Type A behaviour, tension, and ambulatory cardiovascular reactivity in workers exposed to noise stress. *Psychosomatic Medicine,* **55**, 185–192.

Ray, C. and Fitzgibbon, G. (1981). Stress, arousal and coping with surgery. *Psychological Medicine,* **11**, 741–746.

Ruey-Fa, L. (1994). Organisational healthiness, stress, and well-being. Unpublished doctoral dissertation, University of Nottingham, United Kingdom.

Rogers, E.H. (1960). *The Ecology of Health.* (New York: Macmillan).

Russell, J.A. (1979). Affective space is bipolar. *Journal of Personality and Social Psychology,* **37**, 345–346.

Russell, J.A. (1980). A circumplex model of affect. *Journal of Personality and Social Psychology,* **39**, 1161–1178.

Scott, R. and Howard, A. (1970). Models of stress. In: S. Levine and N. Scotch (eds) *Social Stress.* (Chicago: Aldine).

Selye, H. (1950). *Stress.* (Montreal: Acta Incorporated).

Selye, H. (1956). *Stress of Life.* (New York: McGraw-Hill).

Siegrist, J. (1996). Adverse health effects of high effort-low reward conditions at work. *Journal of Occupational Health Psychology,* **1**, 27–43.

Semner, N. (2003). Individual differences, work, stress and health. In: In M.J. Schabracq, J.A.M. Winnnubst and C.L. Cooper (eds) *Handbook of Work and Health Psychology* (Second Edition), (Chichester: Wiley & Sons), pp. 83–120.

Shirom, A. (2003). The effects of work stress on health. In: M.J. Schabracq, J.A.M. Winnnubst and C.L. Cooper (eds) *Handbook of Work and Health Psychology* (Second Edition), (Chichester: Wiley & Sons), pp. 63–82.

Spielberger, C.D. (1976). The nature and measurement of anxiety. In: C.D. Spielberger and R. Diaz-Guerrero (eds) *Cross-Cultural Anxiety.* (Washington, D.C.: Hemisphere).

Sutherland, V.J. and Cooper, C.L. (1990). Understanding stress: psychological perspective for health professionals. *Psychology and Health, Series: 5* (London: Chapman & Hall).

Symonds, C.P. (1947). Use and abuse of the term flying stress. In: *Air Ministry, Psychological Disorders in Flying Personnel of the Royal Air Force, Investigated during the War,* 1939–1945 (London: HMSO).

Watts, C., Cox, T. and Robson, J. (1983). Morningness-eveningness and diurnal variations in self-reported mood. *Journal of Psychology,* **113**, 251–256.

Warr, P.B. (1990). Decision latitude, job demands and employee well-being. *Work and Stress,* **4**, 285–294.

Welford, A.T. (1973). Stress and performance. *Ergonomics,* **16**, 567–580.

World Health Organization (1946). *Constitution of the World Health Organisation (3).* (Geneva: WHO).

Zyzanski, S.J. and Jenkins, C.D. (1970). Basic dimensions within coronaryprone behaviour pattern. *Journal of Chronic Diseases,* **22**, 781–795.

Chapter 20

Vision and visual work

P.A. Howarth and M.A. Bullimore

Introduction

This chapter is concerned firstly with the assessment of visual performance, and secondly with issues of visual discomfort and fatigue. The vast majority of our knowledge about the world is obtained through vision, and in the work context the increasing use of technology-based systems has highlighted the role that vision plays in providing us with information. The ubiquitous use of visual displays, whether in the form of a computer screen, a television, or a simple digital clock, illustrates the primary role played by the eye at the person–machine interface. Against this background, we also need to consider the issue of visual comfort, because no matter how good a display is in terms of visual performance, if it causes discomfort or fatigue it will be unsatisfactory.

Visual performance

Let us begin by considering what is meant by visual performance and why we might want to assess it. By visual performance we mean how well people can take in visual information. This can involve a range of levels of complexity from simply detecting a light to integrating complex qualitative and quantitative information from a display or visual scene. Visual performance is a function of:

(a) the abilities of the observer — the inherent capacities and limitations of the human visual system, individual idiosyncracies and special characteristics like levels of arousal and fatigue;
(b) the characteristics of the observed objects — 'displays' — how bright, how much contrast, how big and for how long viewed;
(c) the characteristics of the visual environment in which viewing takes place.

We will deal here principally with the first two aspects: characteristics of the observer and of the viewed objects. For each we will consider the important aspects which affect visual performance and describe ways of assessing it. Environmental factors affecting visual performance are dealt with in Chapter 24. We recommend strongly that these two chapters be read together, as it is rarely possible to deal with a practical issue of visual performance without some concern also for environmental characteristics.

Except for a few special circumstances, such as camouflage, the ultimate purpose of assessment will most often be to optimise a visual task. In an ergonomics framework we are concerned with issues such as:

Design

How should this display be designed so that, say, bleary-eyed nightshift nurses will be able to read, from the other side of the bed, how much of the drug has been infused into a patient's arm?

Trouble shooting

Why do quality controllers/inspectors continually miss flaws in the seal of this milk powder packaging? How can we improve their performances?

Evaluation

It may be important to know about the visual function of specific people in consideration of the needs of a particular population or to check an experimental sample, e.g., what proportion of a group show colour vision deficiencies and of what type?
These scenarios might lead an ergonomist to ask questions about visual performance, such as:

How accurately can people absorb this kind of visual information? How much information do they miss?
Would performance improve if we changed the way information was displayed?
What kinds of error are common?
Does this performance deteriorate over time or improve with practice or experience?

Visual discomfort

Turning now to visual discomfort and fatigue, we first need to clarify our terminology. Vision itself cannot be 'fatigued' and when people claim to have 'visual' fatigue it is not their vision which is tired, but rather it is the person themselves. There may well be specific muscular reasons underlying these complaints — for example, a person who needs reading spectacles will complain of 'fatigue' when extensive near work is performed without them, but it is generally the symptoms of discomfort which give rise to the complaint and not the reduced ability of the eye to focus near objects. Often environmental changes can bring about not only improvements in visual performance, but also improvements in comfort. In circumstances where light levels are too low, an increase in illumination can not only improve visual performance but, as we shall see, can also reduce the size of the pupil of the eye, and could thereby reduce the strain on the visual system.

In this chapter we will cover basic information about the human visual system and the important characteristics of objects and displays which affect visual performance and visual comfort. This will provide the reader with information about what might be important in particular contexts, and help in deciding what to measure and how. We aim to clarify what is feasible for the general ergonomist to attempt in the way of measurement; some kinds of visual performance assessment can be carried out fairly readily using easily obtainable equipment but other kinds are complex and are more appropriately performed by vision experts. The chapter will also refer to other sources which give more detail about particular issues raised here.

The human visual system

In this section we will give an overview of the human visual system and discuss how its physiology can influence visual performance and comfort. The basic structure of the human eye is shown in Figure 1.

Light enters the eye through the transparent cornea, where the majority of the refraction, or focusing of the light, occurs. It then passes through the pupil, the circular aperture in the iris which regulates the amount of light reaching the retina. The pupil normally looks black because little or no light comes back from inside the eye (however in a flash photograph it can look red because the inside of the eye is lit up by the light). The light is then refracted further by the lens, to form an inverted image on the retina. (For a detailed treatise of retinal image formation see Charman, 1983, or Atchison and Smith, 2000.) Constriction of a muscle within the eye, the ciliary muscle, modifies the shape of the lens, and hence its power, so that objects at various distances from the eye can each be brought into focus on the retina, a process termed *accommodation*. Children and young adults possess large amounts of accommodation, and hence have no difficulty in focusing on an object as close as 10 cm although sustained viewing at this distance may cause fatigue. An observer's ability to accommodate will, however, decrease with age (see Figure 2). This is termed *presbyopia* and most people over the age of 40–45 years will require spectacles for near vision.

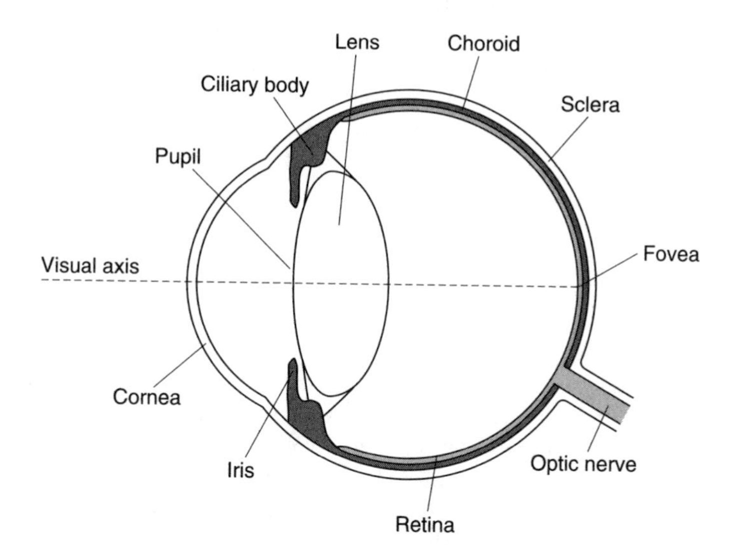

Figure 1 Horizontal cross-section through the human eye.

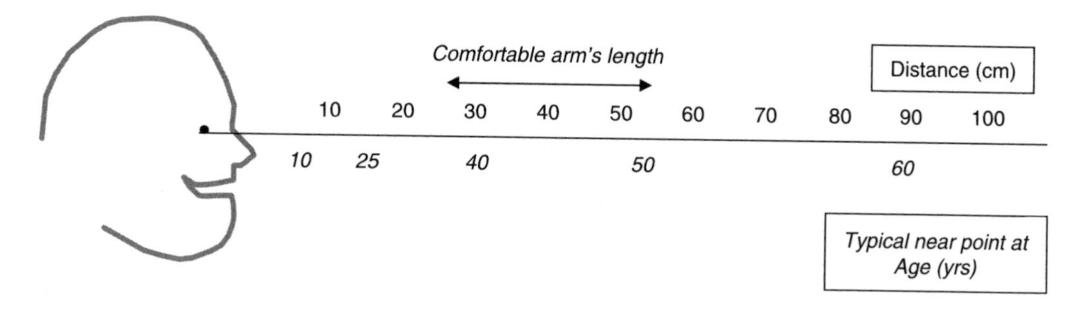

Figure 2 The change in Near Point with age when no near-vision spectacles are worn.

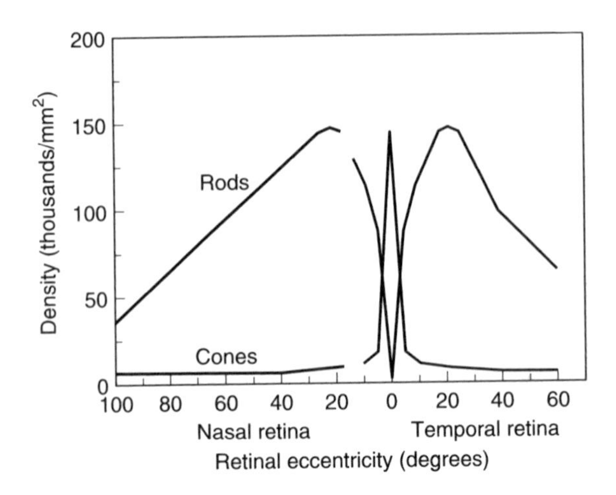

Figure 3 Density of rods and cones across the human retina in the horizontal meridian. The gaps in the functions are due to the optic nerve.

Sensitivity to light

Light falling on the retina stimulates light sensitive cells called *photoreceptors*. These convert the light energy into electrical signals which are transmitted to the visual cortex, in the rear portion of the brain. The complex processing which occurs in the retina, visual pathways and the cortex is discussed elsewhere (e.g., De Valois and De Valois, 1988; Zeki, 1993, Sekuler and Blake, 2002, Bruce *et al.*, 2003). The photoreceptors are divided into two kinds, *rods* and *cones*, which have different characteristics and properties. The distribution of rods and cones across the retina is shown in Figure 3. The cones, which are responsible for vision at higher light levels, discrimination of fine detail and the perception of colour, are most abundant in the central or foveal region. When we 'look at' something we turn the eye so that the image of what we are interested in falls on the fovea — an eccentricity of 'zero' in Figure 3. The rods, which are responsible for vision at low levels of illumination, are found in greater numbers in the peripheral retina. The implications of these relative distributions will be considered later. The visual system is unable to detect light at levels below $10^{-6}\,cd\,m^{-2}$ and in the *scotopic* range, between 10^{-6} and $10^{-3}\,cd\,m^{-2}$, only rods are functioning. At light levels above $3\,cd\,m^{-2}$, the *photopic* range, cones play the major role in vision. The area between 10^{-3} and $3\,cd\,m^{-2}$ is called the *mesopic* range wherein both rods and cones are operating. The mesopic range corresponds roughly to dusk, at which time colours are hard to make out. There is good evidence to suggest that rods are actually still active at the lower end of the photopic range (e.g., the steady-state spectral sensitivity of the pupil mimics that of rods, not cones (see Berman, 1987)) although at these light levels it is clear that cones are the dominant photoreceptors in terms of visual perception.

In the dynamic visual environment the eye has to adapt to changing light levels and does so in three different ways. First, the pupil can change size; however the maximum area change is less than 100 times (10^2 fold) — far too small to account for the eye's immense range of sensitivity (over 10^9 fold). Second, small rapid changes in neural sensitivity take place in the retina. These occur in milliseconds and compensate for small changes in light levels, e.g., walking in and out of shade. The third mechanism involves slow changes in the photopigments in the rods and cones and is seen, for example, in the slow adaptation after entering a cinema or a photographic darkroom. The process of dark adaptation may be observed by measuring the eye's increasing ability to detect a dim light over a period of time in darkness. The dark adaptation curve is a bi-phasic function (see Figure 4) where the first

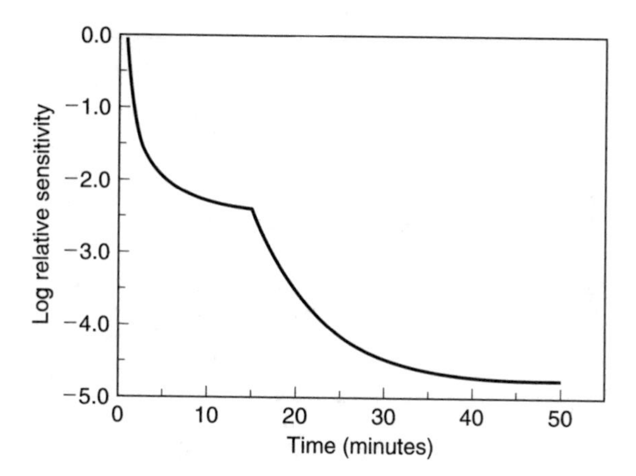

Figure 4 The dark adaptation curve. The parameters of the function (e.g., the time at which the rod-cone break occurs, and the time to reach total dark adaptation) depend crucially on the pre-adaptation conditions.

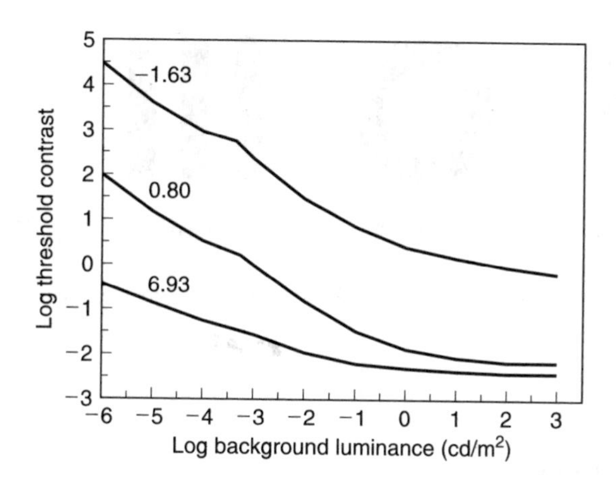

Figure 5 Threshold contrast as a function of luminance for three target sizes (after Blackwell, 1946). Target sizes are displayed in log mrad^2.

portion represents changes in the sensitivity of cones, which can take up to about 10 min, and the second portion shows changes in visual sensitivity mediated by rods.

The processes above relate to the eye's *absolute* sensitivity. It may be more relevant, in a practical context, to consider the luminance of the target *relative* to the background. Extensive experimental studies (e.g., Blackwell, 1946) have examined the eye's ability to detect small circular targets against a uniform background, with the detection threshold described in terms of *contrast,* defined as $\Delta L/L$ — where L is the background luminance and ΔL is the difference between the target luminance and the background luminance (although alternative definitions of contrast may be employed in other circumstances). Threshold contrast was found to be dependent on the adaptation level of the retina, thresholds being lowest at photopic luminances (see Figure 5).

Furthermore, contrast detection thresholds decrease with increasing stimulus size (see Figure 5) and with increasing presentation time (see Figure 6).

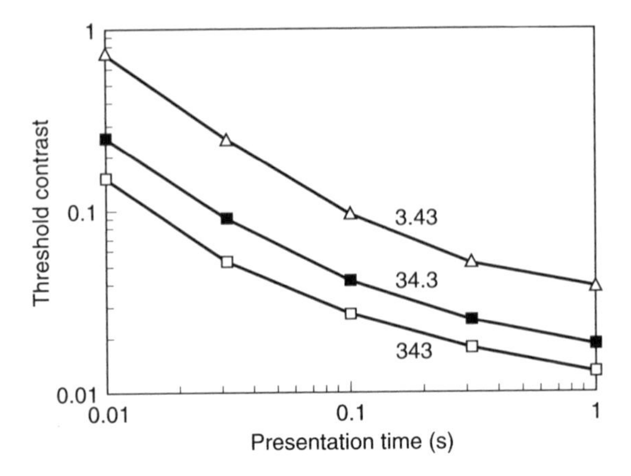

Figure 6 Threshold contrast as a function of presentation time for a 4 min arc target and for three background luminances (after Blackwell, 1959).

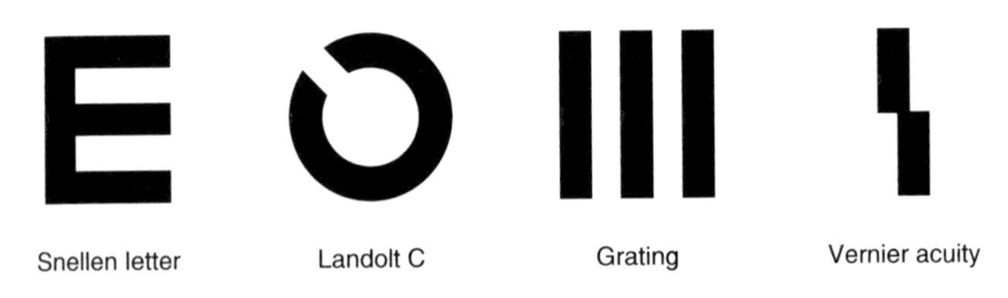

Figure 7 Various targets which may be used in the measurement of visual acuity.

Spatial aspects of vision

We are usually concerned not just with detecting an object, but also with discriminating detail. This attribute of the visual system is normally referred to as *visual resolution* or visual acuity (VA). Visual acuity is usually defined as the minimum angular separation between two lines which is necessary to perceive two lines rather than one. A variety of targets can be employed in its measurement, such as letters, Landolt Cs or gratings (see Figure 7).

Visual acuity values may be expressed in terms of minutes of arc (min arc) or as a Snellen fraction, e.g., 6/6 (or 20/20 in the USA). The fraction is more commonly used by clinicians where the numerator refers to the test distance in metres (or feet) and the denominator signifies the distance at which the limbs of the letter would subtend 1 min arc. Under optimal conditions the range of normal visual acuity is 6/4 to 6/6 (0.67 to 1.00 min arc). Not surprisingly, a reduction in luminance or contrast will result in a decrease in visual acuity (see Figure 8).

A more complete picture of the visual system's spatial capabilities may be determined by testing people's ability to detect light and dark bars (luminance sine wave 'gratings'). The grating is defined in terms of its contrast and of its spatial frequency — the number of cycles (a light and dark bar) per degree. Threshold contrast is determined as a function of spatial frequency, to yield the *contrast sensitivity function* (CSF) — a graph of (the reciprocal of) threshold contrast as a function of the spatial frequency. (This name is somewhat misleading as the function describes sensitivity to size [spatial frequency] and not sensitivity to contrast). Although such a function may at first sight appear of little practical interest, any object or scene can be represented as a series of sine waves of different contrast and spatial frequency and hence its visibility can be predicted from known contrast sensitivity values (see Sekuler and Blake, 2002, or Bruce *et al.*, 2003).

More important than this in a practical sense, however, is the way in which this function can explain visual problems which people experience even though they have good visual acuity. Such cases have become more common with the advent of refractive surgery, and complaints have been heard of post-operative vision being 'sharp' but still 'not right'. These could be explained by a change in the contrast sensitivity function, as shown in figure 9 and figure 10.

A further important aspect of the eye's spatial sense is its extraordinary ability to detect the misalignment of two lines. Berry (1948), Westheimer (1979a) and others have shown that

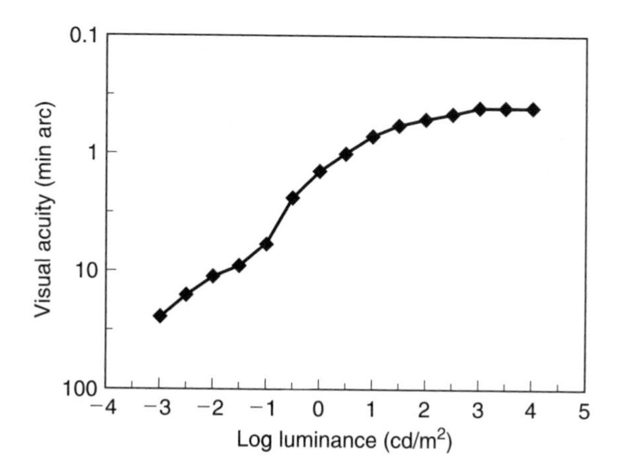

Figure 8 Visual acuity (min arc) as a function of luminance.

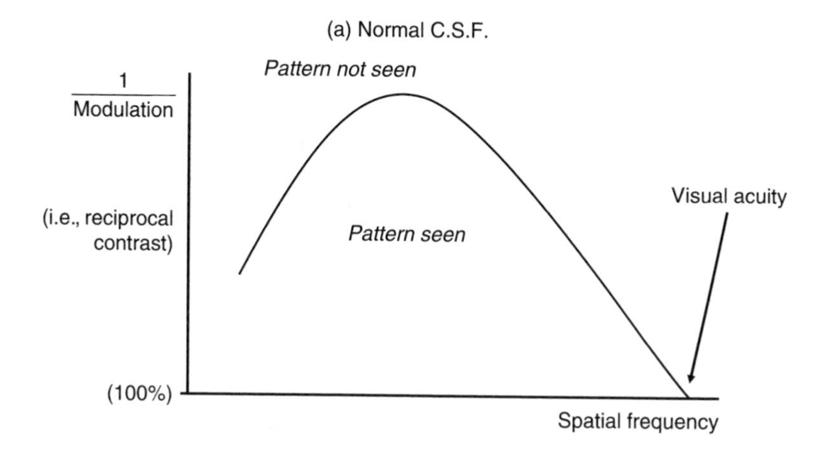

Figure 9 (a) The normal contrast sensitivity function. (b) Reduced visual acuity (e.g., for someone who is short-sighted). (c) Normal visual acuity, but reduced mid-range sensitivity; under these circumstances the whole world would look a little 'misty'. (d) A possible change in the contrast sensitivity function following refractive surgery.

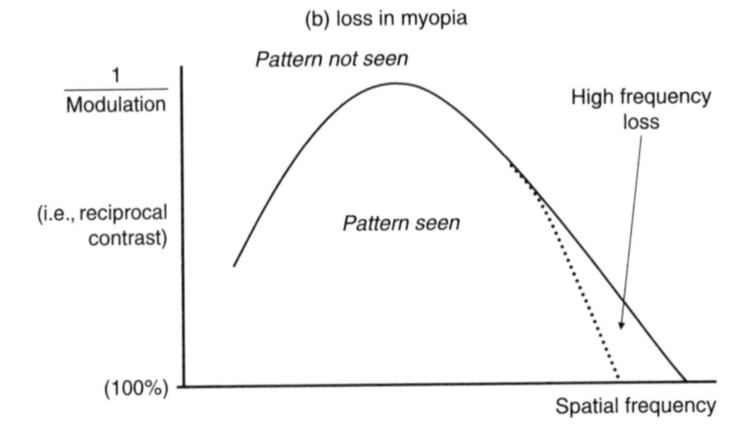

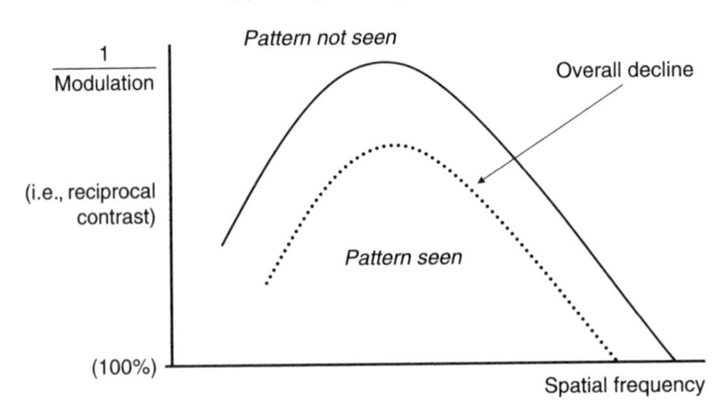

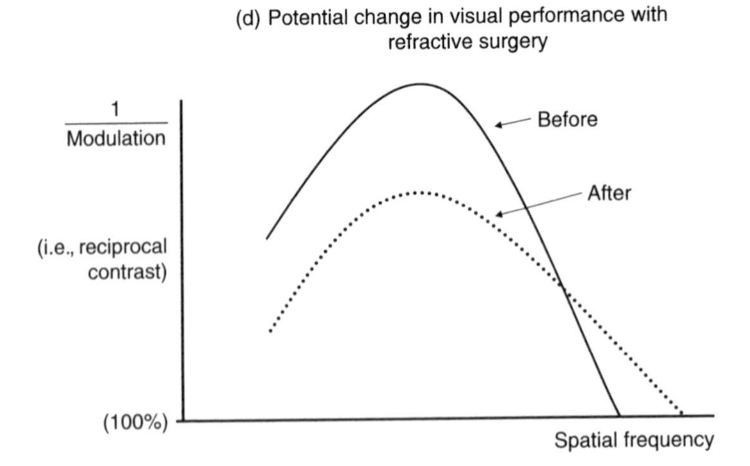

Figure 9 Continued.

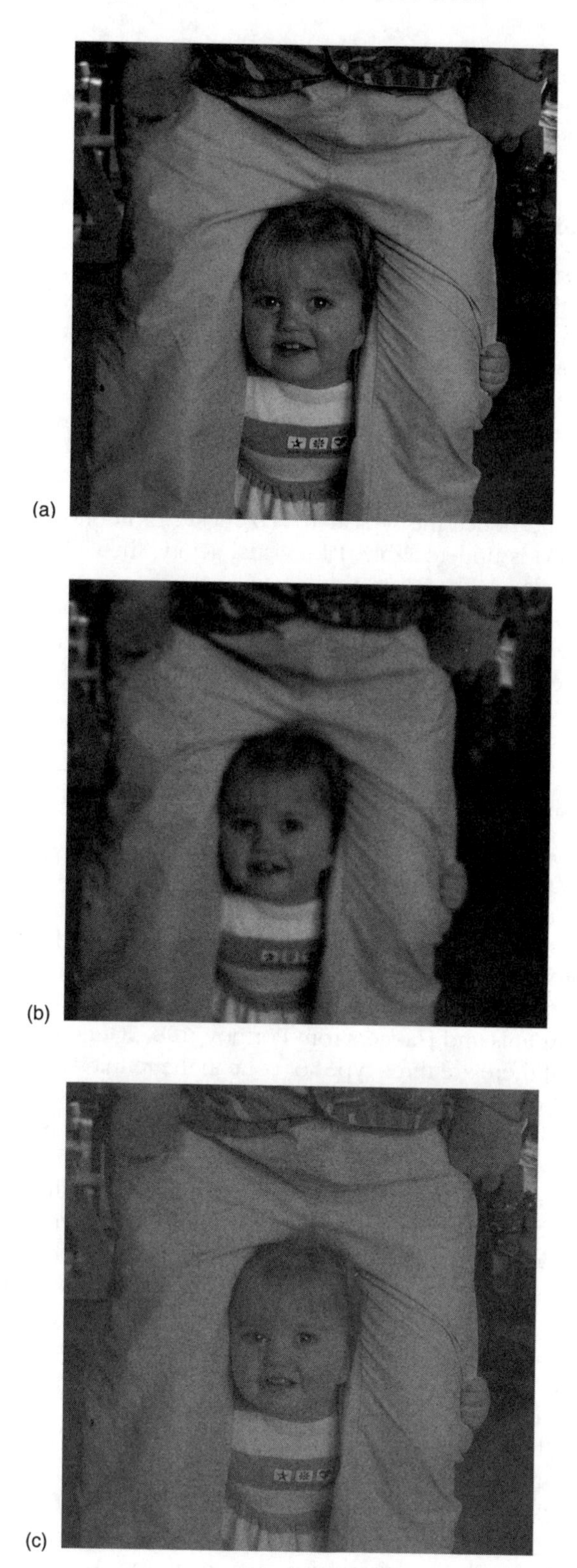

Figure 10 Effect of blur and contrast deduction. (a) Normal image. (b) Blurred image (as seen by an uncorrected mild myope). (c) Reduced contrast image (as would be seen in the presence of scatter).

observers can identify misalignment to an accuracy of 5 seconds of arc. This threshold is referred to as *vernier acuity* and is relevant to the reading of micrometers, slide rules and other tasks where the precise judgement of alignment is required.

Temporal aspects of vision

The human visual system is good at detecting a target that is changing with respect to time. Observers can detect that an object is moving for velocities as low as 7 min arc per sec with no frame of reference (Boyce, 1965) or 1 min arc per sec with a reference frame (Salaman, 1929). Further work has shown that target velocities of up to 5 deg per sec have little influence on visual acuity or vernier acuity (Westheimer and McKee, 1975).

The visual system is also very sensitive to detection of flicker and two thresholds are important. The *critical fusion frequency* (CFF) is the maximum temporal frequency (in Hz) at which flicker can be detected. Under photopic conditions the CFF is around 60 Hz and hence the typical 100 Hz flicker of fluorescent lights in the UK and Europe (120 Hz in USA) is undetectable. Like visual acuity, the CFF declines with luminance and at low light levels, such as in the cinema, flicker of a given frequency is less detectable. The second threshold which may be of interest is the minimum modulation (or change) in luminance required for the detection of flicker. This is termed *temporal contrast sensitivity* and has been shown to a function of both temporal frequency and luminance (de Lange, 1958).

Colour vision

An important feature of our visual system is the ability to discriminate colour. Colour can be a powerful tool in the design of visual displays and it may be defined in terms of hue and saturation. Hue describes the perceived colour, e.g., red or blue, and saturation describes how pale or how dark the colour is.

The nature of human colour vision is discussed more fully in specialised articles and texts (e.g., Hurvich, 1981; Adams and Haegerstrom-Portnoy, 1987, Gegenfurtner and Sharpe, 1999) but suffice it to say that there are three types of cone in the retina with peak sensitivity to short (S-cones), medium (M-cones) and long wavelengths (L-cones). These receptors are sometimes called blue, green and red cones respectively, based on the peak sensitivity to colours but these terms are misleading because the cones themselves are not coloured, nor do they actually signal these colours. Our ability to discriminate between different colours arises from the fact that a given wavelength of light will stimulate each cone type to a different extent, in the same way that colour televisions produce a range of colours by varying the luminance ratio of the blue, green and red pixels. Optimally, we are able to discriminate between colours as close as 2 nm in wavelength (see Figure 11) although our ability to discriminate between desaturated (pale) colours is poorer. Colour vision is, however, defective in some individuals. This will be discussed later.

The optical power of the eye is dependent upon the wavelength of light and objects of different wavelengths are focused at different points within the eye. This phenomenon is called 'chromatic aberration', and while its practical consequences are generally not severe, focusing difficulties can occur when wavelengths from extremes of the spectrum are viewed together. To generalise, this means that if a red and a blue object are adjacent, one may seem to be blurred compared with the other. Also, they may appear to be at different depths, a phenomenon known as chromeostereopsis, because of the eyes' chromatic aberration (Thibos *et al.*, 1990).

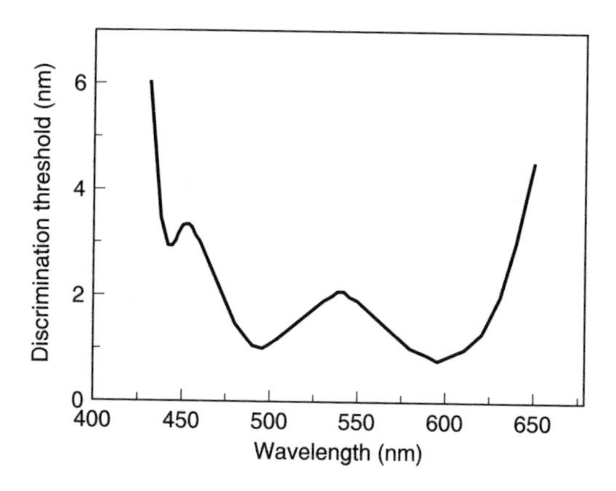

Figure 11 The variation in wavelength discrimination with test wavelength.

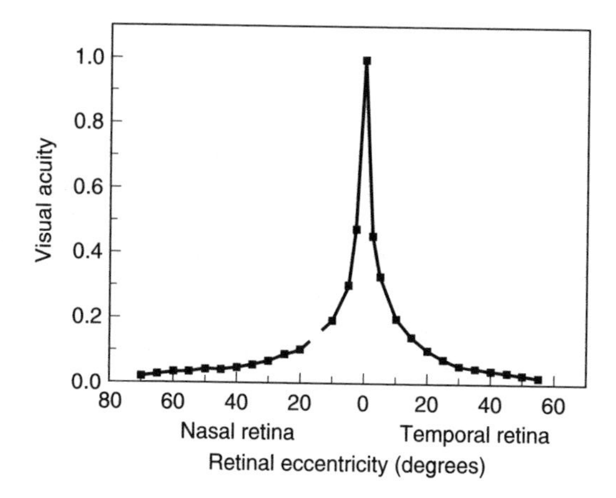

Figure 12 The variation in visual acuity (normalised) with retinal eccentricity (after Wertheim, 1891).

The visual field and visual search

The majority of the aspects of visual performance discussed so far have concerned optimal or foveal viewing. Each eye has, however, a wide field of vision extending 100 degrees temporally, 50 degrees nasally, 60 degrees superiorly and 90 degrees inferiorly from the visual axis. Our visual capabilities vary across the visual field; for example, visual acuity is best at the fovea (see Figure 12, and compare this with the cone distribution shown in Figure 3).

Our ability to detect a static object is in part a function of its position within the visual field. The probability of detection, within a single fixation pause, may be plotted against eccentricity to yield the characteristic *visual detection lobe* (see Figure 13). The visual lobe varies with exposure time and target size, hence the detectibility of a peripheral target can be improved by increasing its size (see Figure 13). The visual lobe is an important concept in visual search and inspection tasks since most detection takes place away from the visual axis.

Unlike static visual performance, detection of a dynamic target can actually be better in the periphery. CFF and temporal contrast sensitivity do not decline rapidly with eccentricity. On the contrary, CFF is actually higher in parts of the peripheral visual field than the central field. This can be demonstrated easily with a conventional TV or VDU. If you look directly at the screen you can probably not perceive the flicker whereas if you shift your gaze to the left or right of the screen it may appear to be shimmering.

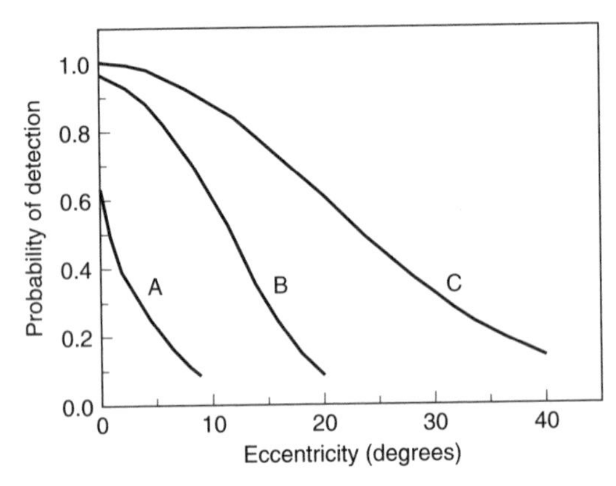

Figure 13 The visual detection lobe, the probability of detection of a target within a single fixation pause as a function of eccentricity, for three targets (C is the largest and A is the smallest).

Stereopsis and eye movements

One of the most important attributes of the human visual system is that we have two eyes which can move together. Possessing two eyes enables us to perceive the world in three dimensions — an ability termed stereopsis — with which we can make extremely accurate judgements of the relative distance of objects from ourselves. Stereopsis is usually described in seconds of arc, and thresholds of 10 sec arc or less are possible: that is, an object 1 m away can be detected as being closer than another which is 1.00075 metres away.

The muscles which move the eyes are controlled by visual feedback so that with both eyes open they remain pointed towards the object of interest. Covering one eye will break the feedback loop — there is no visual information as to where that eye is directed — and the eye may take up a different position. This change in eye position is termed *heterophoria* and while it is quite normal for someone to have a small degree of heterophoria, for some individuals this can lead to discomfort and symptoms such as headaches.

The eyes can make rapid movements — saccades — to enable the object of interest to be imaged on the fovea. This is particularly important, for example, in the context of visual search. Alternatively the eyes can track a moving object — a pursuit movement — in order to keep the image on the fovea. Furthermore, the eyes can move rapidly in order to compensate for voluntary and involuntary movements of the head. For all of these 'version' movements the eyes move left or right together. The eyes can also make 'vergence' movements, which alter the angle between the visual axes, in order to look at objects at different distances. The eyes converge in order to view a near object and diverge to view a more distant object. An important characteristic of the vergence system is that it fatigues relatively easily and hence sustained convergence or frequent changes in vergence may produce discomfort. Most people will be able to converge closer than 10 cm: hold a pencil in front of your nose and bring it towards you. When it appears double, you've passed your *near point of convergence.*

Inter-subject variations in visual performance

We have now considered the major characteristics of the normal human visual system. We must also consider factors which will decrease the visual capabilities of the observer. Disease and poor health, for example, may influence visual performance, as will the intake of tobacco, prescribed (and nonprescribed) drugs and age (see, e.g., Adams *et al.*, 1978; Gilmartin, 1987, Weale, 1992).

In many individuals the optical components of the eye do not form a clear image on the retina due to a *refractive error*. There are three types of refractive error, the most commonly considered being *myopia* or near-sightedness. Myopia affects 20–25% of the working population and, because the cornea and lens are too powerful or the eye is too long, the image of a distant object is brought to focus in front of the retina. Because of this, distant objects appear blurred. Myopes can, however, see near objects clearly and hence uncorrected myopia may not decrease visual performance in the near environment. Myopia is corrected with concave spectacle or contact lenses. In 10–15% of people the eye is too short, or the optical components are not powerful enough, and the image will be focused behind the retina, a condition termed *hyperopia* (hypermetropia) or far-sightedness. Unlike myopia, the visual effects of hyperopia are often not obvious. This is because many hyperopes can exert their accommodation in order to bring distant objects and, if the hyperopia is not too severe, near objects into focus. Convex spectacle or contact lenses will be required for clear and comfortable vision in the older hyperope and the younger hyperope performing sustained visual tasks. The third class of refractive error, which affects the majority of the population, is *astigmatism.* Like myopia, this produces a decrease in visual performance which cannot be compensated for by accommodation, but unlike either myopia or hyperopia it is equally detrimental to distance and near vision. In the astigmatic eye, lines of different orientations are focused at different positions relative to the retina. For example, an astigmat may see the horisontal poles of a scaffold clearly while the vertical poles appear blurred. Virtually everyone has *some* astigmatism, and it is only when the amount is large that visual performance is affected. As for myopia and hyperopia, astigmatism may be corrected with spectacles or contact lenses.

The visual performance of an observer may change considerably with age (Figure 2). Not only does the over 45-year-old have to come to terms with their decreased ability to accommodate (presbyopia) but they will also have a reduced pupil size, and changes in the crystalline lens, which will result in less light reaching the retina. Hence people over 50 may require higher levels of illumination in order to perform as well as their younger colleagues and take longer to adapt to changes in illumination. Furthermore, the changes in the crystalline lens may increase their susceptibility to disability glare (see Chapter 24) as there will be increased scatter within the eye. Finally, the correction of presbyopia with bifocals or multifocals may cause focusing problems if people are looking through the wrong part of the spectacle lens (see, e.g., Howarth, 2005).

Earlier we discussed the essential characteristics of normal colour vision. It should be acknowledged, however, that some 8% of males and 0.5% of females have 'defective' colour vision that differs from the accepted norm to a greater or lesser extent. The relative frequencies and characteristics of the various types of defect are given in Table 1. All congenital colour deficiencies are due to anomalies in the retina, the most common type being anomalous trichromacy where one of the three cone types in the retina is abnormal. A more severe defect is dichromacy where one of the cone types is absent. The most dramatic defect occurs in the rod monochromat who has no colour discrimination, a scotopic spectral sensitivity function and reduced visual acuity, but these people make up a minute proportion of the population. Mild differences from the norm generally have little, if any, effect on day-to-day life, and anomalous trichromats are often unaware of their abnormality.

Characteristics of tasks and viewed objects which affect visual performance

In the visual working environment we are generally concerned with more than the detection of simple spots of light, distinguishing single characters, or discriminating two colours. We are concerned with the acquisition of visual information from various sources. These are usually complex rather than simple stimuli and are most often well above threshold levels for visual detection.

Table 1 Prevalence and Properties of Colour Vision Defectives in the Male Population. Although the Prevalence of Each Type is Much Less in the Female Population, the *Relative* Proportions are Similar

Type of defect		Spectral colour discrimination	Prevalence (%)
Anomalous Trichromacy			
	Protanomalous	Reduced for green, yellow,	1.0
	Deuteranomalous	orange and red	5.0
	Tritanomalous	Reduced for blue-green cyan, and blue	0.001 (?)
Dichromacy			
	Protanope	Absent for green, yellow,	1.0
	Deuteranope	orange and red	1.0
	Tritanope	Absent for blue-green cyan, and blue	0.001 (?)
Rod Monochromacy		Little or no discrimination	0.003 (?)

This section describes briefly characteristics of visual tasks and viewed objects which affect how well they can convey information to people through the visual system. Here 'viewed objects' means all those things from which people receive visual information. These may be 'displays' in the traditional sense, they may be the focus of an inspection task, or they may be any other kind of material such as printed documents or vehicles on the road.

Types of visual task

What constitutes good visual performance depends on the requirements of the task. In thinking about assessment of visual performance it is important to appreciate the nature of the tasks being carried out — this may affect the type of assessment which is appropriate. Three examples will illustrate task differences:

1. *Detection.* Some visual tasks require simply that an observer detects the presence or absence of something or finds out where something is. Examples are detecting that a warning light has come on, checking a manufactured unit for breaks in a seal, or finding the cursor on a computer screen. Here good visual performance merely requires only that the observer see the object against its background — no other discrimination is needed.
2. *Recognition.* Most often a visual task will require that an observer detect *and* recognise what something is — this demands a higher level of discernment because there has to be discrimination between stimuli. This is the case, for example, in obtaining information from graphic displays and text or carrying out complex inspection tasks.*

**Visual inspection.* Visual inspection, usually associated with monitoring product quality, represents a specific kind of visual task notable for its sustained and invariable nature. The same general principles affecting visual performance and assessment of other tasks also apply to visual inspection. Visual inspection is, however, an area of industrial ergonomics which, because of the direct impact of its performance on profitability, has received special attention over the last 20 years. While we make reference to inspection in a general way, for more detailed discussion we refer the reader to writings dedicated to the subject (Smith and Lucaccini, 1977; Drury, 1973; Drury and Addison, 1973; Megaw, 1979).

Here good performance involves being correct in the judgement of what it is that has been detected. In counting out change, for example, it is important to distinguish between different coins.

3. *Interpretation.* Most tasks also require observers to interpret what they have seen in terms of what it means for their subsequent actions. Examples include establishing which of a row of warning lights has come on, the significance of a blemish on a photograph, what a dial is indicating and what a text message written in 'text-speak' actually means.

Here we are not considering this cognitive level of extracting meaning from visual information, but rather the sensory capacity to obtain information, such that cognitive factors can begin to play. We should not, however, lose sight of the fact that both sensory and cognitive factors are important in the design and evaluation of visual material; no matter how lucid and interesting the prose, it will be without value if it is written in tiny grey characters on a grey page; conversely, no matter how legible the message it is useless if it makes no sense. Beware! — *cognitive* problems can be mistaken for problems of *visual* performance.

Types of visual display

Visual information comes to us in the form of light either reflected by objects or emitted by them. Nowadays more and more visual information, specifically from displays in the working environment, comes to us via sophisticated emissive technologies in the form of cathode-ray tubes (CRTs), light emitting diodes (LEDs), backlit liquid crystal displays (LCDs), plasma displays, etc. (Travis *et al.*, 1992, National Research Council Committee on Vision, 1983; ANSI, 1988). Before new technologies were widely available, visual information was displayed principally by the traditional reflective technologies of inks and paper, printed labels for electro-mechanical dials and some simple emissive displays like warning lights. There is a rather large body of knowledge associated with these traditional media in terms of guidelines for good design for performance (e.g., McCormick and Sanders, 1987; Helander, 1987; Boff and Lincoln, 1988).

It is more difficult to produce robust standards and guidelines appropriate for this new class of display. This makes assessment of visual performance more important since it is often not possible to refer with confidence to guidelines about the physical characteristics of images on the displays; the chances are that there are none researched sufficiently for the particular quality of display with which you are concerned. This fact is reflected in the current emphasis of the International Standards Organisation's (ISO) efforts to produce ergonomics standards for visual displays which attempt to define standard test procedures for visual performance rather than physical characteristics (e.g., the ISO Standard 9241/3). Nevertheless, although existing guidelines may be inappropriate for state-of-the-art displays there will always be work environments where people are exposed to examples of outdated technology.

Basic principles of appropriate design for visual performance

It is the intention in this section to explain underlying principles which govern the suitability of visual information for the human visual system, rather than to present exhaustive guidelines for design or evaluation. The section is divided into two. In the first part, simple physical aspects of the display are considered, while the second part considers aspects which involve some cognitive component.

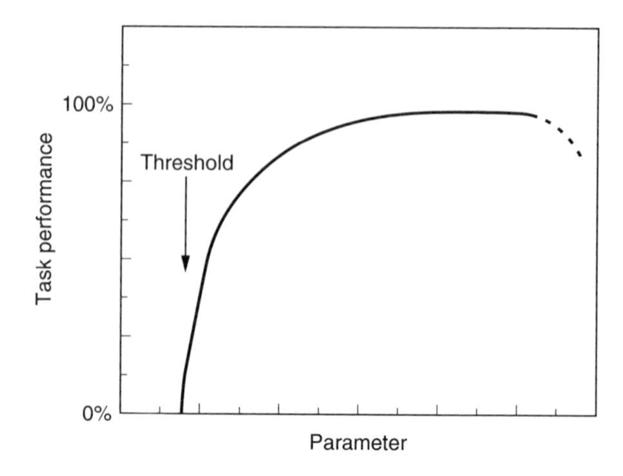

Figure 14 The influence of parameter on visual performance. The 'parameter' may be size, contrast, luminance, or time. The broken portion of the line signifies a potential decrease in performance.

Physical aspects

As a general rule, visual performance is better the brighter the ambient lighting, the greater the contrast between object and background, the larger the object and the longer the viewing time. The influence of these physical parameters on visual performance can be seen in a general model presented in Figure 14. If the parameter value is too low (e.g., the size is too small) then the task will be below threshold. As the parameter increases threshold is reached and subsequent increases will improve performance until the optimal level is reached. In some cases though, if the parameter continues to increase performance will eventually decline. We shall now consider the relevant parameters.

Illumination

Within the normal range of illumination levels which we encounter naturally, or produce artificially, visual performance is improved by increasing illumination. This is because the eye is relatively more sensitive to change at higher illumination levels. Chapter 24 discusses the importance of maintaining a relatively constant illumination level within the visual field so that visual performance is not affected by adaptation to either of the extreme levels. There are some rare occasions when illumination levels can become too high, for example, in a visual environment combining bright sunshine and wide expanses of snow, visual performance deteriorates.

Contrast

While it is generally true that the greater the luminance contrast the better the task performance, it is important to qualify this statement. For a moment consider, as an example, driving at night. Performance in the task of detecting oncoming vehicles is enhanced by their displaying bright headlights, but the contrast between these and the rest of the scene commonly causes discomfort and disability and hence a reduction in performance of the visual driving task as a whole. This example demonstrates the need to think about the whole task context rather than simply parts of it. It is also important to think of contrast both in terms of bright objects against dark backgrounds and of dark objects against light backgrounds. For some tasks there are advantages in illuminating the background to improve the observer's ability to see a stimulus — for example, in checking for flawed items using

Figure 15 The text is more conspicuous because of the pattern of lines, an effect which is not predictable from simple parameters of luminance, size or contrast.

backlighting or shadows. Conversely, for tasks involving written character recognition it seems that *in general* dark text and symbols on a light background ('positive contrast') is preferable to light text on a dark background ('negative contrast') (Gould *et al.*, 1987a).

Size

Generally the larger the object (the greater the angular subtense at the eye) the more easily it will be seen and discriminated from other objects. For a resolution task this generalisation holds for sizes greater than 1 min arc (the normal resolution threshold) up to the point where optimum performance is reached — the value for which depends upon other factors discussed here. As demonstrated by the model in Figure 14, above a certain point increasing the size will not improve performance and might in fact degrade it. Imagine standing directly in front of a large advertising hoarding and trying to read it! Applied to alphanumeric characters, size recommendations are that character heights should be large enough to subtend an angle of around 20 min arc at the observer's eye, i.e., about 4 min height at a viewing distance of 600 mm.

Exposure time

Visual performance is better the longer an observer gets to look at, or look for, something. This has implications both for the design of tasks — it is important to ensure that presentations are for an adequate length of time — and for the measurement of performance. Indeed, 'time to detect' can be used to assess the visibility of a stimulus.

Aspects which include a cognitive component

All the above physical parameters apply to the basic sensory processes of the human visual system. The parameters considered now all have, in addition, some cognitive aspects to them, and require a more sophisticated appreciation of their interactions. Consider Figure 15 where the physical aspects of the stimulus do not easily reveal the way the visual system responds to the pattern elements in this illustration.

Dynamic aspects

Most often when we consider criteria for the design of visual material we are concerned with stable and static images — for most tasks this is an optimum condition. Images might, however, be unstable due to vibration of the observer or the viewed object or characteristics of the display itself. Furthermore, dynamic displays are becoming more widespread. Care must be taken to compensate for the effects of these movements by ensuring that speed of movement is controlled and that illumination, contrast, image size and viewing time are increased above that which permits adequate performance with static and stable images. The conspicuity of an object, i.e., its capacity to attract our visual attention, is in part a function of its dynamic characteristics — movement or intermittency. A moving or flashing stimulus is more conspicuous than a static one.

Change and comparison of visual stimuli

Relative judgements are easier to make than absolute judgements. For example, you can easily detect that a vehicle brake light has come on if you notice the increase in intensity; however, if you miss the brightness change then because they are the same colour it is difficult to know whether you are looking at brake lights or rear lights. In this example there is no external brightness reference to compare the lights with, and the task is extremely difficult. A reference item for comparison will improve performance in many circumstances, such as inspection tasks and monitoring tasks, and the use of reference lines and markers are of significant value in assisting visual search and judgements.

Pattern recognition and coding

We are adept at pattern recognition and tend to group visual information in terms of similarities of its physical appearance: colour, brightness, shape, size and orientation. These factors can be used to enhance visual performance by helping the viewer organise visual information. Designing a bank of dials so that the pointers all line up in the same direction (particularly either horisontally or vertically) when status is normal makes it easier to detect when one of them is registering an abnormal condition. We are better able to see and distinguish objects if they have unbroken lines and boundaries and whole regular shapes. These factors are exploited in the design of camouflage, where a major principle is to break up boundaries and outlines with colour or shading. As another example, in the design of alphanumeric characters the implication, and empirically supported wisdom, is that it is important to use clear, non-slanting and simple fonts without serifs, such as are recommended for improved performance by dyslexics.

Redundancy

Visual performance can sometimes be enhanced by providing observers with redundant information. The detection and discrimination of warning lights for different functions, for example, can be improved by making them a different colour *and* a different shape. Similarly, by being colour- and size-coded, British paper money scores over its US counterpart for visual discriminability. It is sometimes appropriate to use other senses as a redundant cue to aid visual performance, for example auditory cues will improve the detection of visual warnings.

Use of colour

Colour has an important role as a coding device in separating and grouping elements in a design. When colour is used for coding, or for grouping information, it is important not to use too many colours, although authorities differ in the maximum number advisable (e.g., Grether and Baker (1972) recommend using no more than 10 colours but preferably 3!). Of course, many more colours and shades can be used to render form and depth in visual displays. The saturation level of colour can be important, and desaturated (pale) colours should be avoided. In using colour for coding the colour discrimination abilities of the user population must be considered. An example of failure to do so is the use of self-administered glucose tests for diabetics, where early tests involved colour matching, even though diabetes is known to cause colour vision defects.

Colour is also important in providing contrast and it can be used to make objects conspicuous. The success of a colour used as a highlight depends upon the visual context in which it is used; all else being equal, orange has better contrast with green, for example, than it does with red. The concept of generic 'high visibility' colours can be misleading. Those colours which we generally refer to as 'high visibility', such as bright and fluorescent yellows, oranges and yellow-greens are effective in many environments because on average they contrast well with their backgrounds. In other specific background circumstances these colours can also be 'low visibility'. For example, red flags by the roadside will be highly conspicuous as few natural scenes are bright red whereas the referees assistants at soccer matches no longer use red flags because they merge in with red garments in the crowd and with red seats in some stadiums.

Display location, size and area

The best location for display of visual material is roughly perpendicular to the observer's line of sight, unobstructed and preferably requiring a minimum of eye movement. The most frequently accessed information is, therefore, best placed centrally. Standardising location of specific types of information is useful in reducing search time. The appropriate size to make a display depends on characteristics of the task, mainly the amount of information which must be displayed and its relative importance. If the display area is too small and too dense this will increase search time and decrease legibility. Clutter and complexity in layout reduce performance, and issues such as these have received extended attention in the user-interface design literature.

Conclusion on task and object characteristics and visual performance

This section has reviewed basic general principles of the design of visual displays and tasks which affect people's visual performance. It is worth bearing in mind, in the following section about methods, that a widely used and economical method of assessing visual performance is to compare the particular circumstances of interest with standards and guidelines of good practice. Some sources for these guidelines have been referenced here. Applying these principles, with discretion, can often save a great deal of time and energy by preventing assessments of performance which effectively repeat other people's work.

Characteristics of tasks, viewed objects, and individuals that affect reports of visual discomfort

Visual fatigue

One of the problems in considering visual fatigue is that the term itself is used in a variety of contexts. Some authors use visual fatigue to describe subjective complaints of discomforts while others apply the term to changes in visual function. This led the National Research Council's Committee on Vision (1983) to conclude: 'The terms *visual fatigue* and *eyestrain* are frequently used in ill-defined and differing ways. These terms do not correspond to known physiological or clinical conditions. We suggest instead that researchers and others use terms that specifically describe the phenomena discussed, such as *ocular discomfort, changes in visual performance* and *changes in oculomotor functions*'.

A model of visual discomfort is displayed in Figure 16. The model is centred around the idea that the overall aim of the visual system is to produce single, clear vision. The presence of any refractive errors along with the binocular coordination of the two eyes, will influence how successful the visual system is in achieving this aim. Errors of refraction (e.g., hypermetropia) and errors of binocular co-ordination (e.g., heterophoria) can lead to symptoms of visual discomfort, which are thought to be the result of the visual systems efforts to compensate for such errors, in an attempt to produce single, clear vision. The model argues for the presence of a pain mechanism somewhere within the CNS, which respond to the demands being made upon the visual system. For example, in the uncorrected hypermetropic individual (who has to accommodate at distance in order to see clearly, with near tasks requiring further accommodative effort) conducting near work, clear, single, vision is only being achieved with the use of sustained muscular effort. This, in turn, may lead to symptoms of visual

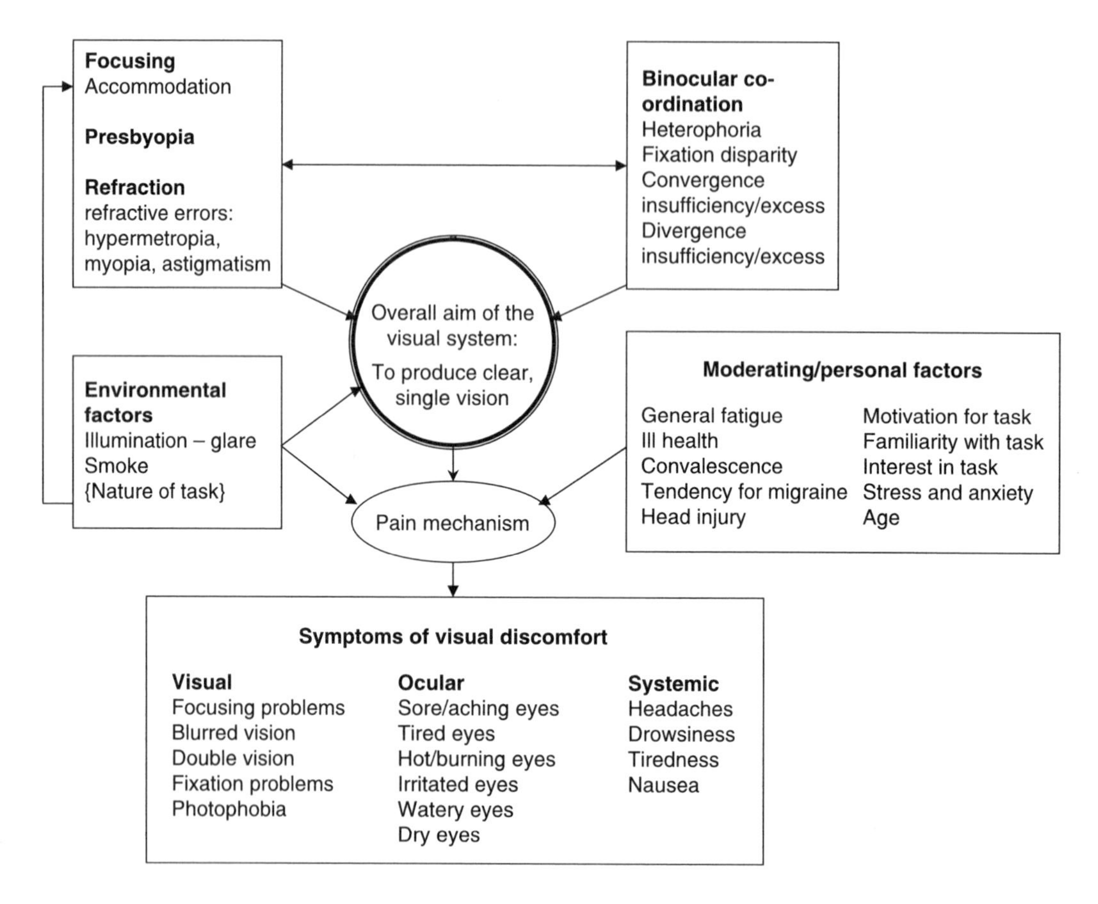

Figure 16 Model of visual discomfort. The arrows indicate interactions (e.g., presbyopia could blur vision, and it could also affect heterophoria which could in turn interfere with single vision).

discomfort. An important part of the model however, is the influence of moderating factors (e.g., general well-being), which can have a direct influence on the pain mechanism. If the uncorrected hypermetropic individual, for example, is reading an extremely interesting story book, or is playing an enjoyable computer game on a VDU, symptoms of visual discomfort may go unnoticed. However, if the same individual is reading uninteresting material, or is conducting a monotonous task on a VDU, the symptoms of visual discomfort may develop relatively quickly.

At this moment it is not clear to what extent reports of visual discomfort are moderated by more general, non-ocular, feelings of fatigue and discomfort. This idea is supported by earlier views of Duke-Elder and Abrams (1970) who have reported that general fatigue and ill health increase the likelihood of reports of visual discomfort. According to Duke-Elder and Abrams, anxiety and emotional strain also increase the probability of reports of visual discomfort. Heaton (1968) has also stated that visual discomfort is very common in convalescence after serious illness and after a head injury. Other moderating factors include motivation for the task and familiarity with the task. North (2001), for example, has reported that trainee VDU operators are more likely to report symptoms of visual discomfort, than fully trained workers, due to the high level of concentration required to learn a new task. More recently, Clemes and Howarth (2002, 2003) noted that reports of variation in visual discomfort over the menstrual cycle were correlated with reports of visually-induced nausea when subjects played a VR game whilst wearing a head-mounted display, but there was no change in visual discomfort over the menstrual cycle in two other VDU experiments which did not induce general discomfort and nausea.

Although it is tempting to think of visual discomfort simply as arising as a consequence of viewing a visual stimulus, the above findings indicate the importance of considering the influence, or presence, of any moderating factors before drawing conclusions.

Near vision

Without a doubt, the most common complaint that the optometrist encounters is that of difficulty with near vision. These generally come about because of presbyopia — the functional consequence of the reduction in the eyes amplitude of accommodation (i.e., how much it can increase its power in order to focus on near objects) which happens when the lens hardens with age (Figure 2). Apart from the loss of visual clarity, the main consequence is an increase in discomfort. This comes about principally because the ciliary muscle (which changes the eye's power by altering the shape of its lens, Figure 1) has to use a greater proportion of its total muscular power to bring about a given change in the harder lens. When it uses a large proportion of the total muscular power available to it for an extended period of time, fatigue sets in accompanied by discomfort.

An increase in light level will help. This will decrease the pupil size which will, as photographers would be able to tell you, increase the depth of field of the eye. By doing this, a lesser burden placed on the ciliary muscle because a smaller change in the eye's power is needed to bring the object into the 'clear enough' (rather than 'precise') focus that we require.

Methods for assessment of visual performance

Assessment of visual performance may be important in a number of different circumstances. Just what it is appropriate to measure, and how, will depend on these circumstances. Assessments are generally necessary either to troubleshoot unsatisfactory conditions or as a tool in design and research.

When something is wrong there is often objective and quantifiable evidence of poor performance. It may be evident from people making errors (e.g., confusing alphanumeric characters or failing to detect flaws) or performing more slowly than anticipated (e.g., taking longer to do specific tasks or spending more time idle). Unsatisfactory conditions can also become evident as the result of subjective complaints from people about fatigue, discomfort (e.g., glare, 'eyestrain'), or general matters (e.g., ill-health). In all cases the assessor's job is to try to identify the causes of poor performance and usually to devise and evaluate ways of improving it. This might involve assessment of:

1. *Individual's visual functional abilities.* For example, are specific individuals displaying decrements in particular aspects of vision related to their work? Is their visual system deficient in any way?
2. *Characteristics of the viewed objects and visual task.* For example, are the contrast and luminance values too close to threshold levels for optimum performance? Does spatially reorienting the task improve the situation?
3. *Visual environmental factors.* For example, is the spectral output of overhead lighting adversely affecting colour discrimination?
4. *Non-visual factors.* For example, is the general health of employees good? How satisfactory are social and organisational factors within the workplace? Are other

environmental factors, heat and humidity or vibration for example, having a detrimental influence on visual performance?

In design and research it is sometimes important to assess what level of visual performance can be expected from a specific group of people or from a specific design of task or display. In these circumstances the most appropriate assessment will involve:

(a) *measuring abilities of people* doing tasks typical of those they might be required to do,
(b) *reference to guidelines* for comparison of the characteristics of a task or display,
(c) *evaluation tests* of the display/task with a sample of people typical of the likely user population.

In sampling from populations, assessment of visual function is sometimes necessary to describe or screen a group of people and decide whether they represent the abilities of a specific population. This is important in selecting people to take part in empirical assessments of visual stimuli. Whenever assessment is necessary the common elements are the person, the task and the environment. The assessment of the visual environments and of non-visual factors is dealt with in other chapters of this book. The other two elements, namely assessment of personal visual function and assessment of tasks and displays, are discussed separately here, although some overlap will be evident.

Assessment of task/display

Performance based measures

Performance-based measures involve the observation and measurement of people's performance on visual tasks, either in their natural environment (e.g., factory, driving cab or office) or in laboratory settings where selected attributes of the task can be simulated and examined under more controlled circumstances. These measures can, and have been, employed extensively in the evaluation of lighting conditions, display quality and the effects of prolonged visual performance. However, there are a number of problems in designing and interpreting these measures, and a variety of approaches have been taken to account for these problems.

First, performance usually involves both speed and errors and people often make complex trade-offs between them. For tests that allow subjects to establish their own criteria for time and errors, a slight change in error rate could be reflected in a relatively large change in speed. This can make the use of performance-based measures extremely difficult unless an underlying model of the trade-off is available. The problem can sometimes be overcome in the design of the tests themselves. For example, people can be allowed to take as long as they want, and the performance measure will then be accuracy alone; this approach is seen in the reading of the optometrist's letter chart. As an alternative, the performance measure could be the time taken to achieve a certain level of accuracy: an example of where the criterion is 100% accuracy is how long it takes someone to locate a particular town on a map. Similarly, other accuracy levels could be fixed by rejecting trials where the subject's error rate is greater or less than a predetermined level, and then using speed alone as the performance measure.

A second problem with performance-based measures is that, in short-term studies, visual performance may differ in unpredictable ways from when the task is performed on a prolonged or permanent basis. This is of particular concern in long-term inspection tasks when vigilance and tiredness may be involved. While this is a general problem in ergonomics, particular difficulties can come about in visual tasks because of the long-term demands on accommodation and convergence.

Finally, there is a complex relationship between the visual demands imposed by the task, the amount of effort and attention allocated and the resultant performance levels. Again, it may be possible for the design of tests to control subjects' arousal and attention to some extent, for example, by rewarding good and penalising poor performance or by employing secondary tasks.

Analytical and empirical approaches

Two types of performance method can be identified, the 'analytical' approach and the 'empirical' approach (Hopkinson and Collins, 1970; Boyce, 2003). In the analytical approach the performance of simple contrived tasks is observed so that a quantitative model may be developed to relate visual performance to visual conditions. For example, Blackwell (1946) describes a method which involves detecting a spot of light against a darker or lighter background. In this way the relationship between contrast, luminance and visual performance can be modelled, and this model can subsequently be applied to more complex tasks such as the legibility of characters on a given background. This approach has a great deal of merit when comparing tasks or displays which differ only with respect to one or two variables.

In the empirical approach, the speed and accuracy with which a task is performed is measured under real or simulation conditions. Weston (1945), like Blackwell (1946), investigated the relationship between task contrast and visual performance. He used a large number of Landolt Cs (C's oriented in various directions with the subject's task being to identify the location of the gap; see Figure 7) to test people's speed and accuracy under a variety of contrast and light levels. Various other types of tasks have been employed such as reading text (Carmichael, 1948; Kruk and Muter, 1984; Nordqvist *et al.*, 1986; Gould *et al.*, 1987b), simulated inspection (Brozek *et al.*, 1950; Murch, 1983), and search (Bodmann, 1962; Neisser, 1964). Modifications of these tasks have been used to examine the effects of contrast, luminance and size (Khek and Krivolilavy, 1966; Boyce, 1974; Stone *et al.*, 1980).

In many real life situations the empirical approach has great practical value because it allows for comparison between a number of options where multiple variables are involved and where there are not resources to develop a complex model to help predict performance. For example, suppose that a choice must be made between three different liquid crystal displays for use on a chemical analysis machine. If these displays vary in a single dimension, say the sizes of character that they can support, then a good decision can be made confidently on the basis of an analytical approach. Knowing the range of distances from which chemists will need to read results it is possible to select the most appropriate display sizes. However, if the choice had to be made between three different displays, one of which was liquid crystal, one was a vacuum fluorescent display and the other an LED display, there are many more variables differentiating them: e.g., display colour, luminance, contrast, character form, size and effective viewing angles. There is no ready model to help make the decision about which would be best. Who knows what the appropriate weightings are for each variable? This choice can be made empirically in a user test by comparing the legibility of characters on each of the displays in ambient lighting conditions, and from angles and distances to the display, which cover the ranges expected in the machine's use. The advantage of a performance based experiment like this is that it provides useful and sound predictive information for the specific application. The disadvantage of the approach is that it adds little to the body of theoretical knowledge about visual performance, since it has compared the performance of discrete complex objects but revealed nothing quantitative about the interactions between the many variables which were involved.

The empirical approach is also particularly useful when visual performance is being affected by higher level factors beyond the simple physical attributes of the task (such as size and contrast) or the physiological attributes of the visual system. These factors range from the legibility of characters to the organisation of visual information in certain ways to capitalise on our visual pattern recognition abilities. For example, an analytical approach could help in improving inspection performance in the detection of stitching irregularities in the seams of jeans, by increasing ambient illumination levels and changing the lights' spectral characteristics to enhance colour contrast between stitching and cloth and/or allowing inspectors longer to look at each pair of jeans. There might be vast scope for improvement in the visual performance of railway timetable-enquiry clerks, even if they are using full-colour high-resolution visual displays. For example, it might be helpful to organise the listings graphically and to introduce different grouping and colour coding conventions on to the screens. Here an analytical approach would not be a suitable way to assess different ways of organising the visual layout. An empirical approach, simulating the clerks' search tasks and measuring visual performance with each of several layout options, would be a more appropriate way to assess potential improvements.

Task evaluation

How can we assess the effects of changing the physical attributes of a task? Suppose we know from the analytical approach that an increase in illumination level might be expected to improve the performance of someone reading documents. How can we assess first, whether there is any need for improvement, and second whether the strategy we have adopted has been successful?

In assessing visual tasks and performance the approach taken by the Commission Internationale de l'Eclairage (CIE, 1972, 1981) was to use the parameter of contrast to define a measure they termed 'visibility level'. By determining what contrast reduction is necessary to reduce the task to threshold you effectively determine how far above threshold the task was in the first place. The higher above threshold, the more 'visible' the task. The approach has been successfully applied to a variety of lighting situations and to paper-based tasks (Boyce, 2003). To use this approach you need to have some means of reducing contrast without affecting overall luminance, and this is the function of a 'visibility meter'. In assessing a visual task, a vision or a lighting specialist would probably either use this approach or would measure the physical attributes of the task and then apply the values obtained to an existing visual performance model.

But what if you haven't got a visibility meter or a sophisticated photometer? As discussed earlier in this chapter there is a general relationship between task performance, and each of the physical attributes of size, contrast, illumination and viewing time. The relationship between performance and any of these parameters is shown graphically in Figure 14, and knowledge of this function provides us with a simple, yet elegant, means of assessing whether a visual task is adequate. If we knew for a given parameter how far above threshold the performance is optimal, then we could devise a strategy to reduce the parameter by that amount. If the visual task was originally above the optimum level, then this reduction would still leave the task above threshold. On the other hand, if the task was sub-optimal, then this reduction would leave the task below threshold!

Size is an appropriate candidate for this approach, and as a good rule of thumb, if the task is reduced in size by a factor of three and can still be performed, then the initial conditions were acceptable for adequate visual performance. The integrative aspect of this simple approach can be seen by considering that by reducing *any* parameter such as contrast, task luminance or illumination, the whole curve relating performance to size (size being the 'parameter' in Figure 14) will be altered. This alteration 'parameter' could then take the task below threshold, and it could not be seen. This size reduction can be achieved easily by increasing the distance from the eye to the task by a factor of three (Bailey, 1987).

A major advantage of this simple strategy is that the match between the task and the individual performing it can be assessed by using the person themself as the observer. Alternatively, the task alone can be assessed by using an observer with good eyesight. A word of caution is in order! It is important that size is the only parameter that is being varied and that the measurement procedure itself should not affect the task. Since, in this example, the task is moved further away, the focusing demands on the observer are less. Supposing the person who normally performs the task wears spectacles designed to focus at the task distance and not at further distances. The task itself could be quite acceptable but when the viewing distance is increased detail could become unclear due to focusing rather than image size reasons. This could lead to an incorrect conclusion that at the normal working distance the task was inadequate.

Subjective reports

Another class of methods involves the use of subjective measures based on questionnaires, interviews or informal discussion. Typically, this approach has been used both to assess visual performance and to investigate complaints of visual discomfort. The advantage of these methods for environments and tasks outside the laboratory (where testing procedures can be strictly controlled) is that the effect of complex variables can often be rapidly assessed. These are relatively easy and economical methods but are not without their drawbacks. Subjective reports should not always be taken at face value — people are often mistaken in their assessment of their own visual system and its performance.

Subjective reports are also likely to be biased by popular beliefs and topical misconceptions. A complaint about glare on screens, for example, could be prompted by a belief that VDUs would be better if provided with a special anti-glare screen rather than because there is a real performance problem. Sheedy *et al.*, (2003) have recently highlighted the issue of disability glare on VDUs by investigating the effect of filters on visual performance, and have shown that this belief may not be factually-based. The investigator needs to develop methods to avoid being misled by the subjects' analysis; a brief investigation with placebo treatments or use of subjective reports from people other than those who were party to the original analysis would be useful techniques to adopt as controls.

Subjective reports are of different kinds; they can be more or less structured and are often most reliable when they are most structured and specific. For example the choice between specific options — which of these fonts is more legible — is more likely to yield useful results than asking an open-ended question. On the other hand, open-ended questions can often reveal unforeseen problems which might affect performance, for example, with equipment cleaning and maintenance practices or seasonal variations in light levels.

A good example of the use (and misuse) of subjective reports is the literature concerning reports of ocular discomfort and visual display units. The increasing use of VDUs in the early 1970s brought with it a plethora of studies reporting a high incidence of complaints of visual discomfort. However, reviews of these early studies are invariably critical: Helander *et al.* (1984), for example, stated that 21 of the 28 studies they surveyed had serious design faults. These flaws included a lack of control groups and biased samples (there is anecdotal evidence that, in at least one early study, subjects were encouraged to over-report difficulties by one of the participants because this would bring problems to the attention of the management). Subsequently, Howarth and Istance (1986) suggested that in many of these studies of visual discomfort the use of one-off questionnaires was inappropriate. If groups are well-matched and appropriate measures used (such as *change* in discomfort over the day, rather than simply discomfort at the end of the day), then no difference is found between VDU users and non-users (Howarth and Istance, 1985). This finding does not negate results of studies which show significant differences between VDU users and non-users (e.g., Knave *et al.*, 1985), but rather indicates that these reports are demonstrating problems other than the use of VDUs *per se*.

Assessment of personal visual function

In certain circumstances we may wish to assess a person's visual capabilities. This could be because we suspect that an individual's poor visual performance has a physiological basis. Alternatively, we may wish to evaluate a task or display and need to ensure that the group of observers to be used are 'normal'. In the same way, assignment of subjects to experimental groups may be on the basis of their visual capabilities. Many people will be able to tell you something about visual problems they have; however, they may be totally unaware of visual disabilities such as minor colour vision defects. Finally, a change in visual function could itself be the metric of interest. This section is divided into two parts. In the first, basic tests of visual function are described which a competent ergonomist should be able to perform. In the second, visual functions needing elaborate (and expensive) equipment not generally available to the non-specialist are reviewed.

Basic tests of visual function

Test charts are available that allow visual acuity measurements to be made easily. Distance visual acuity charts contain rows of letters which decrease in size down the chart. These letter sizes are labelled by the distance at which they would subtend 5 mins arc (and the limbs, 1 mins arc) at the eye. Hence, an 18 m letter on the chart would be three times as large as a 6 m letter. Most charts are designed for use at 6 m and this distance should be adhered to wherever possible in order to avoid confusion. To use the standard Snellen chart, the observer is instructed to read as far down it as possible and the lowest line in which most of the letters are read may be taken as the threshold (the visual acuity values are marked clearly on most charts). Visual acuity may also be measured for near vision with appropriate charts, which usually employ lower case Times Roman print. An observer with normal visual acuity should have no difficulty in reading 5 point (N5) print at 40 cm. This raises an important point, which is that 6/6 distance visual acuity does not itself guarantee good intermediate or near vision, particularly in observers over 40 years of age. Hence visual acuity should always be assessed at a distance relevant to the task or display that the observer is or will be using. Also, as well as measuring both eyes together visual acuity should be measured for each eye separately since an imbalance may be contributing to any reported symptoms. Careful attention should also be paid to the luminance of the test chart (see Figure 8): there is a variety of international standards for chart luminance and as a guideline we recommend a value of between 80 and 300 cd m^{-2}.

The standard Snellen chart described earlier has been used for many years with little change. Over the last 15 or so years a number of new charts have been developed. These range from charts consisting of luminance sine waves at various orientations to letters embedded in random-dot noise. An interesting recent development has been the introduction of low-contrast test charts (see, e.g., Bailey and Bullimore, 1991; Reeves *et al.*, 1993; Regan and Neima, 1983). This appears to be very much the future for visual performance assessment, as shown by the work of Haegerstrom-Portnoy *et al.* (2000) investigating visual performance and age.

The type of high contrast test-chart we recommend was introduced by Bailey and Lovie (1976) and consists of rows of five black letters on a white background. The size of the letters on each row is related logarithmically to the rows above and below, and with this chart we record the *logarithm of the minimum angle of resolution* (logMAR). The work of Westheimer (1979b) and Hallden (1972) suggests that a logMAR scale is a perceptually equal-interval scale. Being logarithmic, the scale does not have a 'true' zero, although 6/6 is recorded as a logMAR of zero, and so the scale can be taken to be at an interval level of measurement, but not at a ratio level. The chart has five letters on each line, and each letter correctly read increases the person's score by 0.02 log units. The person reads as much of the chart as they can, and their vision is then scored according to the number of lines and letters they correctly identified. Despite the scientific advantages of the logMAR charts, the standard Snellen chart is more likely to be encountered. This latter chart is quite adequate for most purposes, however keep in mind that we *cannot* assume that measurements using this chart are at a measurement level higher than ordinal (or possibly ordered metric). The practice of averaging vision scores is, therefore, incorrect.

The ergonomist should be aware of the contribution of accommodation and vergence problems to visual discomfort. A subject's near point of accommodation (NPA) can be measured with a near vision chart (or even a newspaper!). The print is moved slowly towards the observer until they report it beginning to blur — this point is the near point of accommodation. As mentioned earlier (Figure 2) the person's accommodative ability declines with age, and so while a NPA of < 10 cm might be normal for a teenager, a 35-year-old might not be able to focus much closer than 20 cm from their eyes. It is desirable that an individual should have a near point of accommodation significantly closer than their required viewing distances for a display — as noted above it is not desirable for the person to use a large amount of their available accommodative power for an extended period of time. Reading glasses and bifocals alter a subject's near point, and it will be more appropriate to take this measurement with the subject wearing their spectacles. The near point of convergence can be measured in a similar fashion using no more than a pencil. This is held vertically and moved towards the observer until it first appears 'double' — this represents the near point of convergence. Most people, irrespective of their age, should have near points of convergence no further than 8–10 cm and any value much beyond this range may give rise to symptoms. Do not confuse the near points of accommodation and convergence: in the former you are looking for *blurring* of the target, while in the latter you are looking for the target to appear *double*.

Colour vision may also be assessed relatively simply by the ergonomist. The simplest and most common type of test uses pseudo-isochromatic plates. These are book tests of numbers, letters or symbols in which the background camouflages the task for the colour defective. The Ishihara Plates are an example of this type of test. These tests are fairly efficient at detecting colour defectives and will often differentiate between protan- and deutan-type defects. It is important, if the correct standard illuminant (Illuminant C) for which the tests were designed is not available, that daylight is used to illuminate such tests. If daylight is not available either, then cool fluorescent tubes can be used. Other light sources, e.g., incandescent lights, will unacceptably alter the apparent colour of the plates, possibly producing incorrect results.

Stereopsis is the final visual function that one can realistically assess without a large amount of equipment. Inexpensive tests are readily available, such as the Titmus Fly Test and the TNO Test. In these tests, the two eyes are dissociated with either crossed polarising filters or red and green filters, and a composite picture or pattern (e.g., of random dots) is placed in front of the person. Because of the filters employed the two eyes will see different images, in the same way that the two eyes see slightly different views of a 3D object. If the person has stereopsis, a 3D pattern will be seen coming out from, or going into, the page.

Instead of using the above techniques, the ergonomist may use a 'vision screener' in order to evaluate the vision of an observer. These instruments are based on the principle of the

Wheatstone stereoscope, and eyes are tested either singly or together. In most instruments the following aspects of vision are assessed:

1. Distance and near visual acuity.
2. Colour vision.
3. Heterophonia.
4. Stereopsis.

Several vision screeners are commercially available including the Keystone Telebinocular and the Titmus Vision Screener. Although instrument norms are available, the quantitative results obtained from these machines should be treated with caution since their false alarm rate is generally high. As a screening instrument they are, however, generally excellent and will usually detect people who should be referred for expert evaluation. Individuals should normally be referred to an optometrist for a visual examination, who, on request, will provide a written report. Although there may be a charge for this service, it may be the most economical way to solve problems.

Specialised tests

A variety of visual functions have been assessed in the evaluation of visual workload. We shall examine briefly some of the techniques described in the literature although the practising ergonomist may not have the resources to perform most of them. It is important that when measuring these functions we understand the relevance of any recorded changes. Indeed, there is clearly a need to distinguish 'fatigue', as described in the literature, from an adaptation process.

Malmstrom *et al.* (1981) employed an objective optometer to measure the accommodative response to a far and near sinusoidally moving target. They found that the accommodative response diminished significantly over a $6\frac{1}{2}$ min period and propose that this is due to fatigue of the accommodation system. There is an abundance of literature demonstrating that a period of sustained near vision can induce changes in the accommodation and vergence systems (e.g., Fisher *et al.*, 1987; Gilmartin and Bullimore, 1987; Jaschinski-Kruza and Schubert-Alshuth, 1992; Ostberg, 1980; Owens and Wolf-Kelly, 1987; Pigion and Miller, 1985). Ostberg (1980) demonstrated that 2 hr of close work induced a proximal shift in both the resting focus and the farpoint of accommodation, although Murch (1983) could not replicate these findings. There is, however, no evidence that such changes represent fatigue rather than simply the adaptability of the human visual system. Fisher *et al.* (1987) found that although symptomatic and asymptomatic individuals showed accommodative adaptation of similar magnitudes, there were significant differences in the baseline measures and the temporal characteristics of the adaptation.

Haider *et al.* (1980) demonstrated that distance visual acuity decreased from 0.93 to 1.22 min arc following 3 h of near work whereas Dainoff *et al.* (1981) found no change in distance visual acuity for 23 subjects who undertook near work. Jaschinski-Kruza (1984) demonstrated that contrast sensitivity for high spatial frequency gratings presented at 5 m was significantly reduced after 3 h of near work and that the results of these studies were due to optical effects. It should be noted that these changes are for distance visual acuity and may not imply any change in visual function for near work nor explain any associated discomfort. Conversely, Lunn and Banks (1986) showed that after reading text presented on a VDU, contrast sensitivity was reduced for a limited range of spatial frequencies. In this instance the reduction was neural rather than optical in origin, hence it can be seen that a change in contrast sensitivity does not itself tell us anything about causal factors.

It has been suggested that visual fatigue can result in a change in eye movement behaviour. Megaw (1986) and Megaw and Sen (1984) were able to demonstrate effects of continuous VDU viewing on some eye movement parameters, although the effects also showed significant intersubject variations. Wilkins (1986) has suggested that the presence of 50 Hz flicker causes the eye to overshoot its target and therefore increases the frequency of corrective saccades, but he was unable, however, to explain any relationship between these changes and reports of visual discomfort. Leermakers and Boschman (1984) have shown that fixation times and the length of primary saccades are determined by the contrast of the text and that these effects are correlated with subjective reports of comfort. It is unlikely, therefore, that changes in saccadic behaviour reflect anything other than the difficulty in extracting information from the display.

A variety of other methods have been used in an attempt to evaluate visual performance. These include measuring changes in critical fusion frequency, pupil size and blink rate. The plethora of discrepant tests suggested for assessing visual workload indicates that no single visual function adequately reflects visual work.

In summary, therefore, there are a number of tests that an ergonomist can perform in a work setting, both objective and subjective, to assess a person's visual performance and to evaluate their comfort, or discomfort, whilst doing so. Some measures are outwith the scope of the ergonomist, relying as they do on specialised equipment, but the fundamental principles underlying these measures should be readily understood so that the ergonomist can interpret any values produced for them by the specialist.

References

Adams, A.J., Brown, B., Flom, M.C., Jampolsky, A. and Jones, R. (1978). Influence of socially used drugs on vision and vision performance. *AGARD Conference Proceedings*, No. 218, C5. 1–11.

Adams, A.J. and Haegerstrom-Portnoy, G. (1987). Colour deficiency. In: J.F. Amos (ed.) *Diagnosis and Management in Vision Care.* (Boston: Butterworths), pp. 671–713.

ANSI (1988). *American National Standard for Human Factors of Visual Display Terminal Workstations.* ANSI/HI'S 100-1988 (Santa Monica, CA: Human Factors Society).

Atchison, D.A. and Smith, G. (2000). *Optics of the Human Eye.* (Oxford: Butterworth Heinemann).

Bailey, I.L. (1987). Mobility and visual performance under dim illumination. In: *Night Vision: Current Research and Future Directions*, National Research Council Committee on Vision (Washington, D.C.: National Academy Press), pp. 220–230.

Bailey, I.L. and Bullimore, M.A. (1991). A new list for the evaluation of disability glare. *Optometry and Visual Science*, **68**, 911–917.

Bailey, I.L. and Lovie, J.E. (1976). New design principles for visual acuity letter charts. *American Journal of Optometry and Physiological Optics*, **53**, 740–745.

Berman, S.M. (1987). Pupillary size differences under incandescent and high pressure sodium lamps. *Journal of the Illuminating Engineering Society*, Winter 1987; **16**, 3–20.

Berry, R.N. (1948). Quantitative relations between vernier, real depth, and stereoscopic depth acuity. *Journal of Experimental Physiology*, **38**, 708–715.

Blackwell, H.R. (1946). Contrast thresholds of the human eye. *Journal of the Optical Society of America*, **36**, 624–643.

Blackwell, H.R. (1959). Specification of interior illumination levels. *Illumination Engineering*, **54**, 317–353.

Bodmann, H.W. (1962). Illumination levels and visual performance. *International Lighting Review*, **13**, 41–47.

Boff, K.R. and Lincoln, J.E. (1988). *Engineering Data Compendium: Human Perception and Performance*, Volumes I, II and III (New York: John Wiley).

Boyce, P.R. (1965). The visual perception of movement in the absence of a frame of reference. *Optica Acta*, **12**, 47–52.

Boyce, P.R. (1974). Illumination, difficulty, complexity and visual performance. *Lighting Research and Technology*, **6**, 222–226.

Boyce, P.R. (2003). *Human Factors in Lighting* (London: Taylor & Francis).

Bruce, V., Green, P.R. and Georgeson, M.A. (2003). *Visual Perception: Physiology, Psychology and Ecology*, 4th edition. (Hove: Psychology Press).

BS ISO 8995: 2002 Lighting of indoor workplaces (Standard prepared jointly by CIE-TC 3-21 and ISO/TC 159/SC 5).

Brozek, J., Simonson, E. and Keys, A. (1950). Changes in performance and in ocular functions resulting from strenuous visual inspection. *American Journal of Psychology*, **63**, 51–66.

Campbell, F.W. and Durden, K. (1983). The visual display terminal issue: a consideration of its physiological, psychological and clinical background. *Ophthalmic and Physiological Optics*, **3**, 175–192.

Carmichael, L. (1948). Reading and visual fatigue. *Proceedings of the American Philosophical Society*, **92**, 41–42.

Charman, W.N. (1983). The retinal image in the human eye. In: N. Osborne and G. Chader (eds.) *Progresses in Retinal Research*, Volume 2 (Oxford: Pergamon), pp. 1–50.

Clemes, S.A. and Howarth, P.A. (2002). *A field study investigating the relationship between visual discomfort and the menstrual cycle in VDU workers*. In: P.T. McCabe (ed.) *Contemporary Ergonomics* (London: Taylor & Francis), pp. 189–194.

Clemes, S.A. and Howarth, P.A. (2003), Habituation to virtual simulation sickness when volunteers are tested at weekly intervals. In: D. De Ward, K.A. Brookhuis, S.M. Sommer, and W.B. Verwey (eds) *Human Factors in the Age of Virtual Reality* (Maastricht, The Netherlands: Shaker Publishing).

Commission Internationale de L'Eclairage (CIE) (1972). *A Unified Framework of Methods for Evaluating Visual Performance Aspects of Lighting*. Publication CIE 19 (TC 3.1) (Paris: International Commission on Illumination).

Commission Internationale de L'Eclairage (CIE) (1981). *An Analytic Model for Describing the Influence of Lighting Parameters upon Visual Performance*. Publication CIE 19.21 and 19.22 (Paris: International Commission on Illumination).

Dain, S.J., McCarthy, A.K., and Chan-Ling, T. (1988). Symptoms in VDU operators. *American Journal of Optometry and Physiological Optics*, **651**, 162–167.

Dainoff, M.J., Happ, A. and Crane, P. (1981). Visual fatigue and occupational stress in VDU operators. *Human Factors*, **23**, 421–428.

de Lange, H. (1958). Research into the dynamic nature of the human foveacortex systems with intermittent and modulated light: 1. Attenuation characteristics with white and coloured light. *Journal of the Optical Society of America*, **48**, 777–784.

DeValois, R.L. and DeValois, K.K. (1988). *Spatial Vision*. (New York: Oxford University Press).

Drury, C.G. (1973). The effect of speed working on industrial inspection accuracy. *Applied Ergonomics*, **4**, 2–7.

Drury, C.G. and Addison, J.L. (1973). An industrial study of the effects of feedback and fault density in inspection performance. *Ergonomics*, **16**, 159–169.

Fisher, S.K., Ciuffreda, K.J., Levine, S. and Wolf-Kelly, K.S. (1987). Tonic adaptation in symptomatic and asymptomatic subjects. *American Journal of Optometry and Physiological Optics*, **64**, 333–343.

Gegenfurtner K.R. and Sharpe L.T. (eds) (1999) *Color Vision: From Genes to Perception* (New York: Cambridge University Press).

Gilmartin, B. (1987). The Marton Lecture: ocular manifestations of systemic medication. *Ophthalmic and Physiological Optics*, **7**, 449–459.

Gilmartin, B. and Bullimore, M.A. (1987). Sustained near-vision augments inhibitory sympathetic innervation of the ciliary muscle. *Clinical Vision Sciences*, **1**, 197–208.

Gould, J.D., Alfaro, L., Finn, R., Haupt, B. and Minuto, A. (1987a). Reading from CRT displays can be as fast as reading from paper. *Human Factors*, **29**, 497–517.

Gould, J.D., Alfaro, L., Barnes, V., Finn, R., Grisclikowsky, N. and Minuto, A. (1987b). Reading is slower from CRT displays than from paper: attempts to isolate a single-variable explanation. *Human Factors*, **29**, 269–299.

Grether, W.T. and Baker, C.A. (1972). Visual presentation of information. In: H.P. Van Cott and R.G. Kincade (ed.) *Ergonomic Aspects of Visual Display Terminals* (Washington, D.C.: American Institutes for Research), pp. 41–121.

Haegerstrom-Portnoy, G., Schneck, M.E., Lott, L.A. and Brabyn, J.A. (2000). The relation between visual acuity ond other spatial vision measures. *Optometry and Visual Science*, **77**, 653–662.

Haider, M., Kundi, M., and Welsenbock, M. (1980). Worker strain related to VDUs with differently coloured characters. In: E. Grandjean and E. Vigliam (ed.) *Ergonomic Aspects if Visual Display Terminals* (London: Taylor & Francis), pp. 53–64.

Hallden, U. (1972). Notes on the statistical treatment of the visual resolution. *Acta Ophthalmologica*, **50**, 47–57.

Helander, M.G. (1987). Design of visual displays. In: G. Salvendy (ed.) *The Handbook of Human Factors* (New York: John Wiley), pp. 507–549.

Helander, M.G., Billingsley, P.A. and Schurick, J.M. (1984). An evaluation of human factors research on visual display terminals in the workplace. In: F.A. Muckler (ed.) *Human Factors Review: 1984.* (Santa Monica, CA: The Human Factors Society), pp. 55–129.

Hopkinson, R.G. and Collins, J.13. (1970). *The Ergonomics of Lighting* (London: MacDonald).

Howarth, P.A. (2005) Role of Vision in falls. In: Haslam R.A. and Stubbs D.A. 2005 (eds) *Understanding and preventing falls* (London: Taylor & Francis).

Howarth, P.A. and Istance, H.O. (1985). The association between visual discomfort and the use of visual display units. *Behaviour and Information Technology*, **4**, 131–149.

Howarth, P.A. and Istance, H.O. (1986). The validity of subjective reports of visual discomfort. *Human Factors*, **28**, 347–351.

Hurvich, L.M. (1981). *Color Vision* (Sunderland, MA: Sinauer Associates).

Jaschinski-Kruza, W. (1984). Transient myopia after visual work. *Ergonomics*, **27**, 1181–1189.

Jaschinski-Kruza, W. and Schubert-Alshuth, E. (1992). Variability of fixation disparity and accommodation when viewing a CRT visual display unit. *Ophthalmic and Physiological Optics*, **12**, 411–419.

Khek, J. and Krivohlavy, K. (1966). Variation of incidence of error with visual task difficulty. *Light and Lighting*, **59**, 143–145.

Knave, B.G. (1983). The visual display unit. *Ergonomic Principles in Office Automation.* (Stockholm: Ericsson Information Systems), pp. 11–41.

Knave, B.G., Wiborn, R.I., Voss, M., Hedstrom, L.D. and Berqvist, U.O. (1985). Work with video display terminals among office employees: 1. Subjective symptoms and discomfort. *Scandinavian Journal of Environmental Health*, **11**, 457–466.

Kruk, R.S. and Muter, P. (1984). Reading of continuous text on video screens. *Human Factors*, **26**, 339–345.

Leermakers, M.A.M. and Boschman, M.C. (1984). Eye movements, performance and visual comfort using VDTs. *IPO Annual Progress Report*, **19**, 70–75.

Lunn, R. and Banks, W.P. (1986). Visual fatigue and spatial frequency adaptation to video display of text. *Human Factors*, **28**, 457–464.

Malmstrom, F.V., Randle, R.J., Murphy, M.R., Reed, L.E. and Weber, R.J. (1981). Visual fatigue: the need for an integrated model. *Bulletin of the Psychonomic Society*, **17**, 183–186.

McCormick, E.J. and Sanders, M.S. (1987). *Human Factors in Engineering and Design*, 6th edition (New York: McGraw-Hill).

Megaw, E.D. (1979). Factors affecting inspection accuracy. *Applied Ergonomics*, **10**, 27–32.

Megaw, E.D. (1986). VDUs and visual fatigue. In: D.J. Oborne (ed.) *Contemporary Ergonomics 1986*, (London: Taylor & Francis), pp. 254–258.

Megaw, E.D. and Sen, T. (1984). Changes in saccadic eye movement parameters following prolonged VDU viewing. In: E. Grandjean (ed.) *Ergonomics and Health in Modern Offices* (London: Taylor & Francis), pp. 352–357.

Murch, G.M. (1983). Visual fatigue and operator performance with DVST and raster displays. *Proceedings of the Society for Information Display*, **14**, 53–61.

Muter, P., Latremouille, S.A., Treurniet, W.C. and Beam, P. (1982). Extended reading of continuous text on television screens. *Human Factors*, **24**, 501–508.

National Research Council Committee on Vision (1983). *Video Displays, Work and Vision* (Washington, D.C.: National Academy Press).

Neisser, U. (1964). Visual search. *Scientific American*, **210**, 94–100.

Nordqvist, T., Ohlsson, K. and Nilsson, L. (1986). Fatigue and reading text on videotext. *Human Factors*, **28**, 353–363.

North R.V. (2001). *Work and the Eye*, 2nd edition (Oxford: Butterworth-Heinemann).

Oborne D.J. (1982). *Ergonomics at Work* (New York: John Wiley).

Ostberg, O. (1980). Accommodation and visual fatigue in display work. In: E. Grandjean and E. Vigham (eds) *Ergonomic Aspects of Visual Display Terminals*. (London: Taylor & Francis), pp. 41–52.

Owens, D.A. and Wolf-Kelly, K. (1987). Near work, visual fatigue and variations in oculomotor tonus. *Investigative Ophthalmology and Visual Science*, **28**, 743–749.

Pigion, R.G. and Miller, R.J. (1985). Fatigue of accommodation: changes in accommodation after visual work. *American Journal of Optometry and Physiological Optics*, **62**, 853–863.

Reeves, B.C., Wood, J.M. and Hill, A.R. (1993). The reliability of high and low contrast letter charts. *Ophthalmic and Physiological Optics*, **13,** 17–26.

Regan, P. and Neima, D. (1983). Low-contrast letter charts as a test of visual function. *Ophthalmology*, **90**, 1192–1200.

Salaman, M. (1929). *Some Experiments on Peripheral Vision*. MRC Special Report 136 (London: HMSO).

Sekular, R. and Blake, R. (2002) *Perception*, 4th edition (New York: McGraw-Hill).

Sheedy, J.E., Subbaram, M.V., and Hayes J.R. (2003). Filters on computer displays — effects on legibility, performance and comfort. *Behaviour and Information Technology*, 22 (6), 427–433.

Smith, R.L. and Lucaccini, L.F. (1977). Vigilance research: its application to industrial problems. In: S.C. Brown and J.N.T. Martin (eds) *Human Aspects of Man-Machine Systems* (New York: Open University Press).

Snyder, H.L. (1988). Image quality. In: M. Helander (ed.) *Handbook of Human–Computer Interaction*, (Amsterdam: North Holland), pp. 437–474.

Stone, P.T., Clarke, A.M. and Slater, A.I. (1980). The effect of task contrast on visual performance and visual fatigue at a constant luminance. *Lighting Research and Technology*, **12**, 144–159.

Thibos, L.N., Bradley, A., Still, D.L., Zhang, X. and Howarth, P.A. (1990). Theory and measurement of ocular chromatic aberration. *Vision Research*, **30**, 33–49.

Travis, D.S., Stewart, T.F.M. and Mackay, C. (1992). Evaluating image quality. *Displays*, **13**, 139–146.

Weale R.A. (1992). *The Senescence of Human Vision* (Oxford: Oxford University Press).

Werthelm, T. (1891). Peripheral visual acuity (Translated by I.L. Dunsky, 1980). *American Journal of Optometry and Physiological Optics*, **57**, 915–924.

Westheimer, G. (1979a). The spatial sense of the eye. *Investigative Ophthalmology and Visual Science*, **20**, 893–912.

Westheimer, G. (1979b). Scaling of visual acuity measurements. *Archives of Ophthalmology*, **97**, 327–330.

Westheimer, G. and McKee, S.P. (1975). Visual acuity in the presence of retinal image motion. *Journal of the Optical Society of America*, **65**, 847–850.

Weston, H.C. (1945). *The Relation between Illumination and Visual Performance*, Report No. 87 (London: Industrial Health Research Board).

Wilkins, A. (1986). Why are some things unpleasant to look at? In: D.J. Oborne (ed.) *Contemporary Ergonomics 1986*. (London: Taylor & Francis), pp. 259–263.

Zeki, S. (1993). *A Vision of the Brain*. (Oxford: Blackwell Scientific Publications Ltd).

Chapter 21

Practical measurement of psychophysiological functions for determining workloads

Masaharu Kumashiro

Importance of the relationship between workload and work burden

The stress and fatigue induced by work are psychosomatic responses appearing as a result of factors associated with accomplishing one's duties and an imbalance between work conditions and the work environment and the adaptive capacity of the people who work there. When the psychosomatic response induced by work exceeds a person's range of tolerance, a special effort is required for that person to continue accomplishing his or her daily work. Typical related phenomena include, for example, the need for rests. The state where one has lapsed into this condition is referred to as work burden, manifested by symptoms such as stress and fatigue. Consequently, workload expresses the degree of qualitative and quantitative load induced by work and is closely related to stress and industrial fatigue (see, for example, Chapter 18 for consideration of workload and Chapter 19 for a comparison of a number of views and definitions of stress).

The objective of workload-related research is to seek optimisation of the work task. Specifically, this will contribute to the design of work conditions and work environments that impose minimal stress and industrial fatigue due to work.

A work system designed from an ergonomics perspective achieves a skilful balance between workload and work burden, resulting in reduced stress and fatigue in workers. A decrease in work burden not only prevents negative health conditions arising from that work, but can also prevent the occurrence of work accidents and incidents, as well as promote positive health conditions. Therefore, the relationship between workload and work burden must be understood before ergonomic design can take place. Also, this must be followed by an examination of ways of measuring and evaluating work burden — the root cause of stress and fatigue.

However, a review of a great many workload-related studies conducted as part of past ergonomic research leads to a certain level of apprehension. Past research has pursued the relationship between external tasks and internal responses and much of this research has sought to minimise external tasks. Such an approach would best be described as the intervention of risky ergonomics. When ergonomists intervene in workload research, it must always be from a position of consistent consideration of creating a balance between external tasks and internal responses. Optimisation of work tasks does not mean minimising workload, in fact it means providing flexibility for creating a balance between task and load. Viewing work among certain coherent units as a total, we find the relationship between task and load

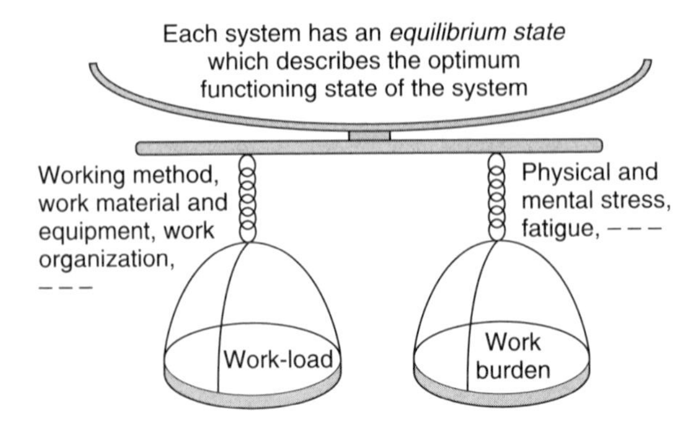

Figure 1 The holistic and multi-disciplinary approach to the work related problems (total workloads) is an ergonomics principle which is based on this system thinking method.

to be balanced, that is, ±0. When we view changes over time, we find the relationship sometimes favors the plus side while at other times it favors the minus side; that is to say, the relationship requires flexibility (Figure 1). The premise for this concept is that one must investigate work burden. In general, we may separate the types of practical technology for studying work burden into the following:

- *Subjective aspect.* Symptoms of self-conscious fatigue and other psychosomatically related complaint symptoms and physical conditions.
- *Objective aspect.* Operational behavior such as subsidiary behavior, performance and human error.
- *Psychophysiological function aspect.* Maladjustment of psychophysiological functions such as the lowering of cerebration.
- *Others.* Transitions in fatigue phenomena in the workplace, such as work history, labour conditions, adaptation symptoms, patient's history and sequela, behavioral and cognitive proclivity to contract disease, physical strength, physical constitution, age and sex.

Types of industrial fatigue due to evolving workplace

When selecting work burden measurement tools, we must focus on the relationship between workload and work mode. Work modes change along with production equipment and production technology advances, and may bring about many changes in the fatigue phenomena of workers subjected to such changes. As might be expected, work mode plays a large part in the selection of measurement tools. When we survey production systems at manufacturing sites we find a transition from work in a human–tool system and work in a human–machine system to work in a human–computer system. Today, there is a continuing shift towards human–machine–information–environment systems, and changes in fatigue phenomena will result.

Work and fatigue phenomena under the human–tool system

In the human–tool system, indeed in the classic human–machine system, the production system itself is manual, relying on physical operations performed by workers. In terms of the sources of fatigue, the work could generally be classified as full-body, physical work. Such work requires considerable consumption of calories by workers, because of a high relative metabolic rate. The fatigue phenomenon is thus physical fatigue primarily related to muscular fatigue.

The mechanism for the onset of this type of physical fatigue is relatively simple and its evaluation is not overly difficult. In other words, such work causes the immediate appearance of fatigue conditions. Its occurrence is easily sensed in the form of worker exhaustion or hunger and the intake of high calorie foods. A short rest or brief sleep in response to physiological demands serves as an effective means of relieving the fatigue.

This is a somewhat bold explanation, but the acute fatigue resulting from rigorous physical labour can be readily comprehended when considered as a result of the balance between the amount of physical energy consumption and the prescribed calories. Putting it another way, the fatigue conditions brought about by muscle-related work in the human–tool system have a very simple make-up compared with those which are encountered today.

This type of work is categorised as muscular work. Classical physiological techniques, such as relative metabolic rate (RMR), oxygen consumption, heart rate (HR) and electromyograms are generally used as effective work strength indices. Hence fatigue is pursued from the perspective of human body physiology.

Work and fatigue phenomena under the human–machine system

Machines or automatic machines have replaced tools as the primary means of production and a semi-automatic method of production has come into use. Along with this, people's work has become more sensori-motor in nature. Fatigue phenomena occurring in this work system have shifted to localised muscular fatigue and psychological fatigue.

The work of human–machine systems has led to paced work methods that inevitably compel the worker to cope with increasingly controlled and high-speed work. Under such paced work methods, worker freedom has been reduced, causing tension under higher speed conditions. Moreover, the work itself has also been configured in a fixed operational structure, where the same work is performed repetitively, thus increasing the worker's work load further. In other words, the mechanical repetition of motions causes workers to lose zest in their consciousness periphery. This increases a sense of monotony and weariness, eventually leading to abnormalities in the central nervous system, and, as a result, gives rise to psychological nervous fatigue.

From these developments it has become necessary to understand fully the fluctuations in psychophysiological functions, thereby giving greater consideration to the psychological aspects of work. At the same time, the observation of subsidiary behavior has become an effective evaluation tool.

Work and fatigue phenomena in the human–computer system

Here the focus is increasingly placed on automation of work through computer control, thus increasing work involving information input and processing systems requiring human recognition–judgement–control. Decision making and information processing capabilities involving such characteristics as accuracy, speed and immediate actions are required more than ever before. Having arrived at this stage, the relationship between workload and work burden changes from possessing a mental and physical aspect to having a cognitive aspect, creating a need for cerebral-physiological studies. In particular, there is a need to focus on fluctuations in the levels of cerebral cortex activity, involving techniques to ascertain physiological phenomena using multiple channel recorders and there is a need to find an optimum approach to determining psychophysiological functions. Visual display terminal work in particular, in process control and other operational systems, is a classic example of this.

A number of studies have been undertaken since the beginning of the 1980s in relation to the impact of such human–computer system work on the human body. Many of these studies have pointed to the occurrence of visual fatigue (see Chapter 20) and an increase in the static muscular load (see Chapter 16). This sort of work has the potential to affect visual function, ultimately leading to damage to the visual information processing ability that plays a role in the vision–cerebrum–peripheral function system. This mode of fatigue (which is also increased in human–machine systems work), thus induces increased central nervous system stress and increases the nervous sensation type of fatigue.

The occurrence of a stress phenomenon that triggers these sorts of computer work-related fatigue and mental fatigue problems was referred to by Brod (1984) as 'technostress', and it now has considerable impact on today's industrial society.

Work at present and in the near future; human–machine–information–environment system

It is considered likely that there will be a significant increase in work based on computer processing in which a worker looks at several pieces of visual data appearing on a large multi-screen display while processing the information on a personal computer. It is also likely that there will be an increase in the introduction of virtual reality (VR) devices. Thus, it is necessary to take into consideration the impact on health which working in a computer-generated virtual environment will have. The author conducted two separate experiments to study impacts on operators; the results of these studies are referred to below as a basis for the following discussion (Kumashiro, 2000).

One form of virtual reality environment is typically produced by a system in which a goggle-type display worn on the head is used to project three-dimensional images, and a kind of data input device called a data glove and a spatial position detection device enable an individual to perform activities in a virtual space created using computer graphics. Attempts have already been made to employ such VR systems in a variety of areas, but there have been few studies on the physiological effects of such systems. In the first study, a VR system employed in the specification design stage of a manufacturing line was used in simulations of assembly tasks for 36 minutes by ten workers, and the physiological effects of this experience were studied. It was found that despite the short working time of 36 minutes there were effects on visual functions, including a decrease in the frequency of blinking and a shortening of dark focus. Measurement of heart rate variability (HRV) indicated a preponderance of sympathetic nerve activity during the VR work, and three of the ten participants complained of motion sickness due to the virtual environment. Various other physiological effects were also indicated (Figure 2).

The second study used 11 participants in 'virtual city' searches extending up to a maximum of 60 minutes. A single session was 10 minutes, and consisted of two minutes in a walk-through state and 8 minutes in a flying state. This was repeated six times. As a result, two of the 11 participants abandoned the programme 10 minutes after starting, one 15 minutes after starting, two 30 minutes after starting and one 33 minutes after starting, making a total of six who failed to complete the programme. In all cases, the reason for failure to complete the programme was severe motion sickness caused by the virtual environment.

Observations of EEG fluctuations during the burden of an imposed task revealed that, compared with a reference value recorded with the subject at rest, the quantity of α waves emitted during performance of the task was decreased, and moreover, a characteristic peak within the α wave band disappeared. Both the values measured by a gravicoder and the scale

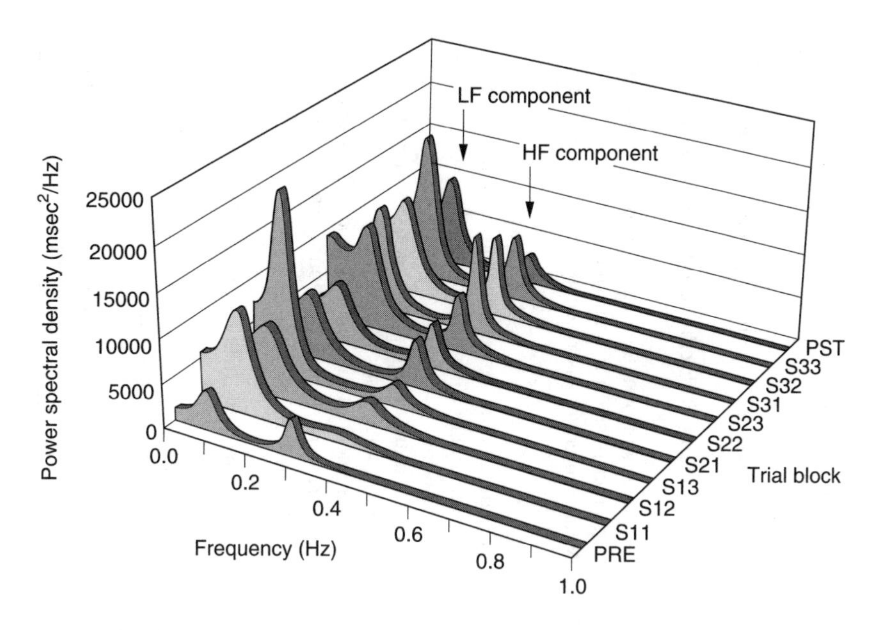

Figure 2 Example of HRV spectra, at rest before virtual reality activity (PRE); during VR (S11-S33); and at rest after activity (PST), four minutes each. Compared with the at-rest state prior to work, the working state saw an increase in LF component, taken to be an index of sympathetic nervous activity.

of head swaying increased with the passage of time. No significant changes were observed in blood pressure variability (BPV), while changes were observed in HRV.

On the other hand, an increase in the blink rate was observed with the passage of time. Thus, virtual environments, which are likely to be introduced with increasing frequency into industrial activities and also into ordinary daily situations, may well have new impacts on health, which cannot be anticipated within the parameters of our knowledge of the health impacts of video display terminal work.

In the light of such changes in industrial technology, work systems and jobs, with consequent changes in fatigue phenomena, ergonomics assessment needs techniques from sensory physiology, psychophysics, and anatomic nerve function measurement. This is in addition to conventional physiology techniques, subjective assessment, psychophysiology, and performance measures. We also need to consider personality and behavioral type effects.

Methods for investigating psychophysiological functions

Outline of investigative methods

Various methods are used to investigate psychophysiological functions in order to evaluate the degree of fatigue at the workplace. Table 1 lists several of these. Sensibility function tests are techniques to produce qualitative and quantitative responses by loading a certain fixed performance requirement onto the participants under investigation. One of the methods classified under muscle function investigations in Table 1 is the knee jerk percussion threshold reflex test, but this is not categorised as a sensory performance test because it measures an involuntary muscle response.

Physiological function techniques can only be used after having previously investigated the conditions that are likely to affect the results obtained from the test technique adopted. In other words, through the test results it must be possible to attach

Table 1 List of the Main Methods for Examining Psychophysiological Functions

1. Sensory function test	Vision function	Accommodation test (far point and near point distance, contraction and relaxation time)	Measurement of recognised percussion value
	Auditory function	Minimum audible pure tone percussion value test	Measurement of recognised percussion value
	Skin sensibility	Two point touch discrimination percussion value method	Measurement of discriminated percussion value
	Parallel functions	The body's centre of gravity test	Measurement of automatic oscillation phenomenon in erect position maintenance condition
2. Psychological function test (including information processing function test)		CFF test	Measurement of discriminated percussion value
		Reaction time test (simple and choice types)	Measurement of information processing capability
		Blocking test	Measurement of performance under uncontrolled conditions
		TAF test	Measurement of overall functions as one of the performance tests
		Dual task method	Measurement of information processing capability
3. Muscle function test	Muscular strength	Gripping power test Back muscle strength test Leg strength test	Measurement of body motion functions using a dynamometer
	Coordinated motion	Finger tapping ability test Two-handed co-ordination test	Measurement of motion control function
	Knee jerk	Test for hyper- or hypoactive	Measurement of patellar tendon reflex (PTR)
4. Respiratory function test		Lung capacity test Maximum air ventilation test	Measurement of body motion function
5. Autonomic nerve testing			Tests such as cold pressure rise test are available, but inappropriate for industrial sites.

a psychophysiological meaning to the changes in the psychosomatic function of the participant observed.

Muscle function tests are different from continuous observation methods such as recording the heart rate, brain waves, and electromyograms. The work must be interrupted to set a time for the tests, which should take place at least before and after the work and, as far as possible, at intervals while the work is being performed.

For respiratory function tests, the condition of the test equipment must be calibrated constantly using a fixed standard.

With autonomic nerve tests, the participant(s) must be proficient with respect to the test equipment and methods.

As a rule, when conducting psychophysiological function tests at the workplace, a great effort is made to promptly observe and measure any physiologically and psychologically abnormal phenomena of the participants. Usually more attention is directed to the results obtained by the various test equipment used rather than to how the interactions and combinations of the various psychophysiological functions in response to the work are deduced. For this reason, the true character of the workload and of the industrial fatigue arising as a result of it are not investigated in depth, and background factors that induce and amplify the task and its cause and effect relationship are neglected. Instead, credence is given only to the phenomena as depicted by the test equipment and, as a result, tasks and fatigue at workplaces are not accurately pinpointed. Thus, a deceptive image may be conveyed to those responsible for job design and health management. Consequently, in order also to make an accurate evaluation of the effects of the task on people who are working, the true character of fatigue and the background factors must be considered from the perspective of a worker group and its working conditions, rather than just considering individual fluctuations. Moreover, the correlation between fatigue and conditions must be studied in detail at the time of analysis.

Electrophysiological measurement

Electrophysiological measurement is a method of measuring on the body surface area extremely small levels of activity potential (approximately 100 mV) generated from cells during various human physiological phenomena. It should be noted that electric potential measured on body surface area ranges from μ(micro)volt through to millivolt units.

Methods of measurement well known in the field of ergonomics like EEG, EMG, and ECG are used for this method of measurement. Table 2 shows standard criteria for electrical potential and frequencies of these three kinds of electrophysiological signals.

It is most common for machines used for electrophysiological measurements to use an AC power supply. Because bioelectric phenomena are captured by attaching electrodes to the body, safety management is of the highest priority so that electric shock and other kinds of accidents do not occur.

The next precaution is the prevention of artifact interference. As has been explained, electrophysiological measurement is the measurement of microscopic bioelectric signals. Thus, one must ensure that any other type of signal besides those one is measuring (artifacts) does not interfere.

The most problematic of all artifacts is AC noise, also called ham. AC noise affects the contact impedance of the electrode. In order to lower, or even stabilise, contact impedance it is important to take great care when the electrodes are attached. For example, one basic rule of bioelectric measurement is that oil on the skin is removed with an alcohol disinfectant.

More care is required when dealing with small electrical signals such as brain waves and induced brain waves compared to relatively larger muscular electrograms and electrocardiograms.

Table 2 Properties of Electrophysiological Signals

	ECG	EEG	EMG
Standard sensitivity	1 mV/10 mm	50 μV/5 mm or 50 μV/7 mm	50 μV/10 mV 50 μV
Calibrated voltage	1 mV	50 μV	
Frequency range	0.05 Hz (time constant 3.2 s)	0.5 Hz (time constant 0.3 s) 60 Hz	5 Hz–2000 (time constant 0.03 s)
Standard feed speed	25 mm/s	30 mm/s	1–5 cm/s (may vary depending on measurement conditions)

A Comparison of Healthy Subjects with Patients after Myocardial Infarction. *Circulation*, **91**, 1918–1922.

Heart rate variability

From heart rate (HR) to heart rate variability (HRV)

Measurement of heart rate (HR) fluctuation has long been used as an index for comprehending workloads, and observation of HR is a traditional means for measuring the functions of living bodies. HR sharply and faithfully expresses the reaction of the functions of the body to stimulation and has thus remained until today as an effective index that is difficult to discard. When used as a means of evaluating workload, however, unless we can express quantitatively the type and amount of load which is being imposed on workers, it is difficult to evaluate the reaction of the living body as indicated by the HR.

As demonstrated by a great deal of past data, HR is sensitive to both physical and psychological loads and shows the same reaction to both. In order to eliminate such ambiguity, it is usually used in conjunction with observation of work behavior. Figure 3 shows a superimposition of HR fluctuation and activity records taken for a single day's work of one type at a certain worksite, thereby allowing us to draw a correspondence between work behavior and HR.

Although these sorts of data are convenient for observing the general workload situation, they are not adequate for determining work burden quantitatively. In the latter half of the 1960s, a new approach recognised the importance of HR but took into consideration also its inadequacies for this type of research. This approach focused attention on the R–R (interbeat) interval even during HR observation. For example, a great deal of research was reported on converting the amount of cardiac output into an index as a means of evaluating the psychological stress reaction that paced work systems had on the human body. Hokanson *et al.* (1971) pointed to a rise in the blood pressure value for conveyor-paced work as compared with self-paced work systems. In the same manner, Salvendy and Knight (1970) also indicated an increase in the work task load under conveyor-paced work systems using fluctuations in blood pressure values and arrhythmia as indices.

Manenica (1977) reported that even if the average prescribed time value per unit obtained from a (pacing) free work system were established as the conveyor feed speed, the average value and the fluctuation of the R–R interval of HR during conveyor work still increases in comparison with that for self-paced work systems. It has been recognised that the phenomenon of increase in the cardiac response under paced work systems is present regardless of differences in the characteristics of the work itself (Sharit *et al.*, 1982).

Subsequently, from about 1980, serious studies began on the practical use of HRV (heart rate variability) which took the analyses of this R–R interval one step further.

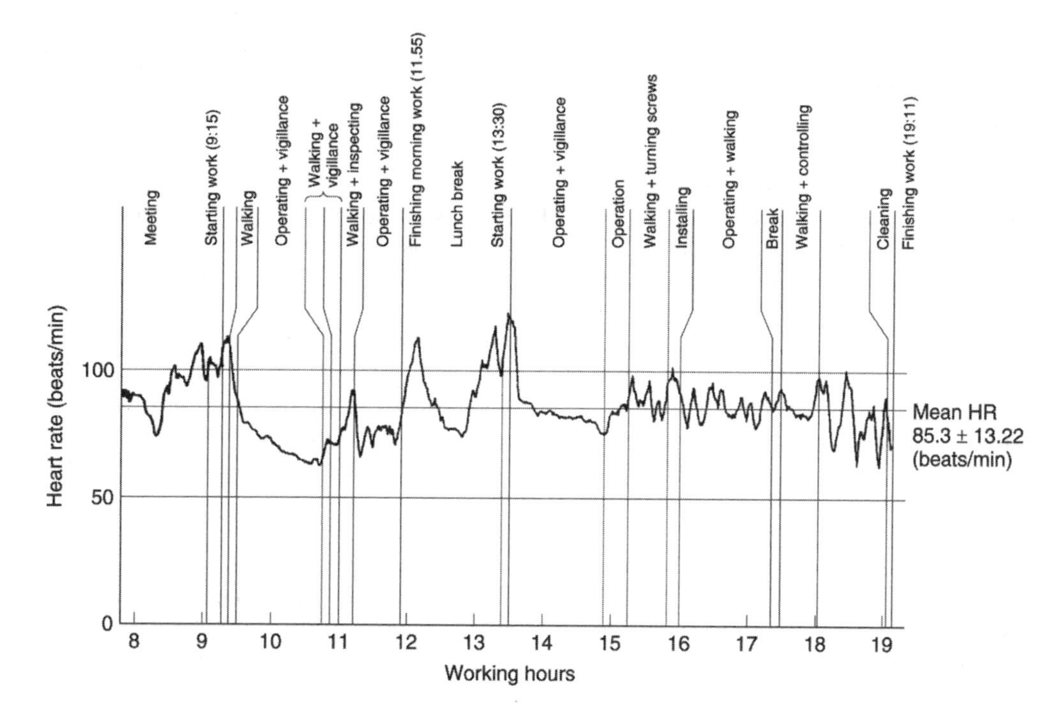

Figure 3 Fluctuation of heart rate with working hours in a day of a control valve operator at a plant (Kumashiro, unpublished, 1988 company report).

Heart rate variability (HRV) measurement

Generally, since Kalsbeek (1971) had pointed out that respiratory sinus arrhythmia was variable depending on quantity of workload, studies have been conducted on the use of HRV as a quantitative index for mental workload.

Among variations in heart rate, variation generated by 'fluctuation' brought about by automatic nerves in a sinus node is called heart rate variability (Hayano, 1996). Whereas in healthy people there is a 2–5% fluctuation in heart rate (R–R interbeat) while at rest, this kind of fluctuation stops completely as a result of automatic nerve shutdown caused by drugs or at times of heart transplant (Pomeranz *et al.*, 1985; Sands *et al.*, 1989).

The high frequency component of HRV is brought about by respiratory changes in heart nerve activity in response to respiratory sinus arrhythmia. Heart nerve activity — the main mechanism which causes respiratory fluctuation, is as follows (Hayano *et al.*, 1996). The activity of vagus nerve efferent neurons is greatly suppressed during inhalation and is stimulated during exhalation as a result of direct interference from the respiratory centre. Vagus nerve stimulated input from a pressure receptor or top centre is shut down during inhalation as a result of input from stretch receptors in the lungs (inspiratory gating). It is thought that the cause of the low frequency component of HRV is fluctuation in sympathetic nerve activity and cardio-parasympathetic nerve activity caused by pressure receptor reflex of Mayer waves in blood pressure variation (Madwed *et al.*, 1991). Mayer waves are said to be fluctuations in arterial pressure occurring in approximately 10 second cycles that can be seen physiologically and which are caused by a delay in the blood pressure control system brought about by pressure receptor reflex. (Madwed *et al.*, 1991).

The causes of the very low frequency and ultra low frequency components of HRV are not yet understood. It is thought that various types of reflex activity in the circulatory–respiratory centre and activity of the hypercentre are reflected in these components.

Historically, basic research on HRV has been conducted since the 1960s. Practical use was attempted during the 1980s but until now there has been little clear applied research. Today, however, interest in HRV has been enthusiastically revived as an effective means for the evaluation of mental workload.

Heart rate variability is normally measured by variations in the R–R intervals in an electrocardiogram. Because by definition HRV does not include fluctuations in heartbeat brought about by types of arrhythmia such as extrasystole and ectopic rhythm, QRS waves from non-sinus rhythm must be excluded from any analysis. This is particularly the case when analysis of HRV is made from a halter electrocardiogram, and because in general accurate automatic recognition of supraventricular extrasystole is difficult, at this present point in time it is common for a threshold value (it is normal for the extrasystole to be more than 20–30%) to be set for the variation ratio of the R–R interval so that it can be distinguished from sinus rhythm. Accordingly, the R–R interval from continuous sinus rhythm QRS waves are used for the analysis of HRV, and this kind of R–R interval is referred to as the N–N interval in HRV analyses.

Two main spectral components have been reported in HRV. These are the RSA component that is seen at high frequencies of about 2.5 Hz and the MWSA component that is seen at low frequencies of about 0.1 Hz. The RSA component corresponds to respiratory sinus arrhythmia and serves as an index of parasympathetic nerve activity. The MWSA component refers to Mayer waves that appear in HRV through the aid of the reflex mechanism of a pressure receptor, and is modified by the parasympathetic nerve activity, but is primarily thought to be an index showing sympathetic nerve activity.

The high frequency component of HRV is not only transmitted by vagus nerves, but this amplitude reflects the level of cardio-vagus nerve activity (Sands *et al.*, 1989). Cardio-vagus nerve activity is measured by the amount of compaction of the R–R interval when a sufficient volume of atropin is administered when there is cardio-sympathetic nerve shutdown caused by a beta-blocker. From these results the power of the high frequency component, average compaction and CCV quantitatively reflect cardio-vagus nerve activity. It has been learned, however, that this is not reflected by a normalised unit.

The relationship between the low frequency component and autonomic nerve activity is somewhat complex. Whereas in roughly half of young people aged 35 and under there is an increase in the power of the low frequency component when they stand erect or at a head-up tilt, this increase can be suppressed by beta-blockers (Lipsitz *et al.*, 1990). In contrast, no change is observed in persons above the age of 35 (Lipsitz *et al.*, 1990; Shannon *et al.*, 1987). Furthermore, there is an increase in the power of the low frequency component as well as the high frequency component during night-time sleep (Vanoli *et al.*, 1995), with a decrease in both at times of dynamic movement (Hayano *et al.*, 1994). Accordingly, whereas the power of the low frequency component is affected by sympathetic nerves, it is thought that it mainly reflects cardio-vagus nerve activity. Moreover, the normalised unit of the low frequency component and low frequency/high frequency are used as indices for sympathetic nerve activity and also for a balance between sympathetic nerves and parasympathetic nerves (Pagani *et al.*, 1986).

Most studies have relied on the Fourier transform algorithm. However, an autoregressive (AR) approach is more useful in calculating HRV, and an example of the calculation of HRV by AR power spectral analysis is provided in this chapter. AR algorithms can furnish the number, amplitude, and centre frequency of the oscillatory components automatically, without requiring *a priori* decisions. Because short segments of data are more likely to be stationary, the AR algorithms, which are capable of operating efficiently even on shorter series of events, appear to provide an additional advantage (Malliani *et al.*, 1991). Data obtained through ECG (CM5 induction) measurement are A/D converted at a sampling frequency of 100 Hz and stored on a floppy disk or similar device. The data are saved for the purpose of a detailed

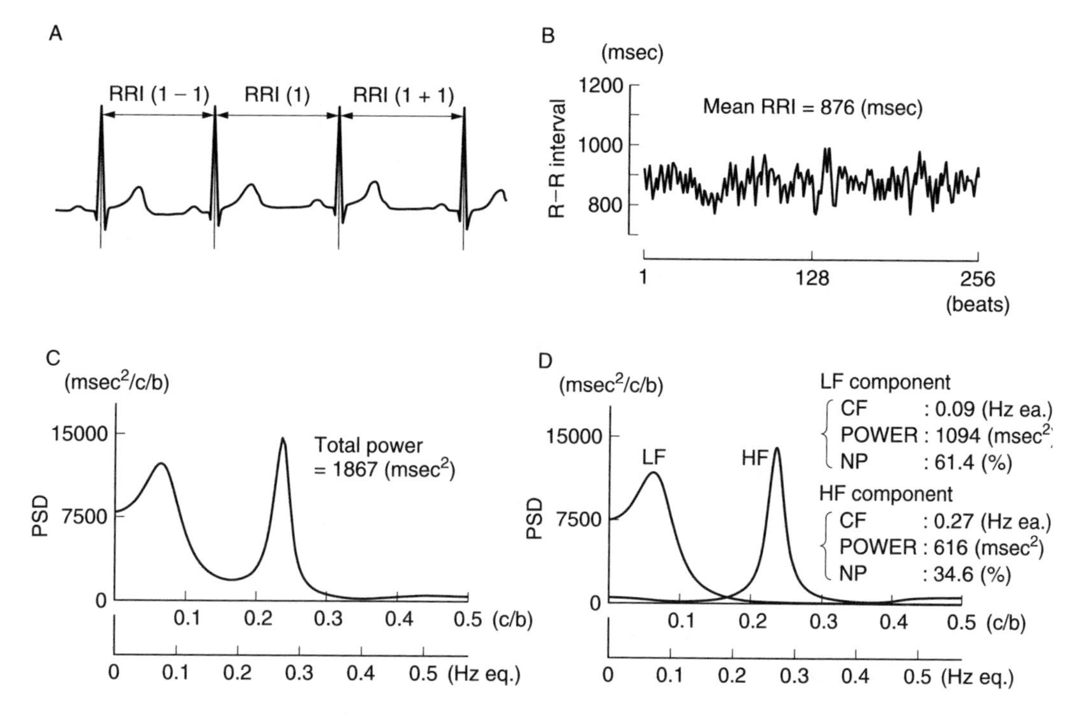

Figure 4 Schematic outline of autoregressive spectral analysis of R–R interval (RRI) variability. A: from surface electrocardiogram a time series of RRIs was calculated as function of beat number. B: RRIs trendgram. C: autoregressive power spectrum from B. D: individual spectral components. Note: PSD, power spectral density; c/b, cycles/beat; Hz eq, Hertz equivalent; HF, high frequency; LF, low frequency; CF, centre frequency, NP, normalized power. (Inoue *et al.*, 1990).

analysis. Thereafter, the first and second steps are to produce a time series and trend graph for the R–R interval over 256 heartbeats, not including artifacts, using data such as premature ventricular contractions (Figure 4a, b). The third step is to determine the AR power spectrum of the trendgraph (Figure 4c). The fourth step is to compute the middle frequency of each spectral component (Figure 4d). The power of each spectral component is standardised by dividing it by the total power minus the DC component (if there is one). The optimum AR model was determined by minimising the value of final prediction error (Akaike, 1970). Stationarity was confirmed by the pole diagram analysis (Baselli *et al.*, 1987; Pagani *et al.*, 1986). These procedures permit the spectral components to be seen.

Figure 5 shows an example of each spectral component of normal healthy males at rest, lying face up. Two main spectral components are recognised, the LF component (middle frequency: 0.09 Hz eq, 0.08 cycles/beat, power: 1,094 ms^2, standardised power: 61.4%) and the HF component (middle frequency: 0.27 Hz eq, 0.24 cycles/beat, power: 616 ms^2, standardised power: 34.6%). These components can be studied by normalising them with such expressions as %LF, %HF and LF/HF.

As demonstrated in the example it is possible to deduce the LF and HF components of HRV. Thus, research using this approach to evaluate mental workload has gradually been promoted in the domain of ergonomics. Considering the number of years that have passed since HRV began to draw attention, however, the number of research reports is quite small. Up to the present time, there have been very few reports of studies where HRV has been applied to field studies, while conversely, the number of laboratory reports is strikingly high. The reason for this is attributed to the difficulty of quantifying such factors as

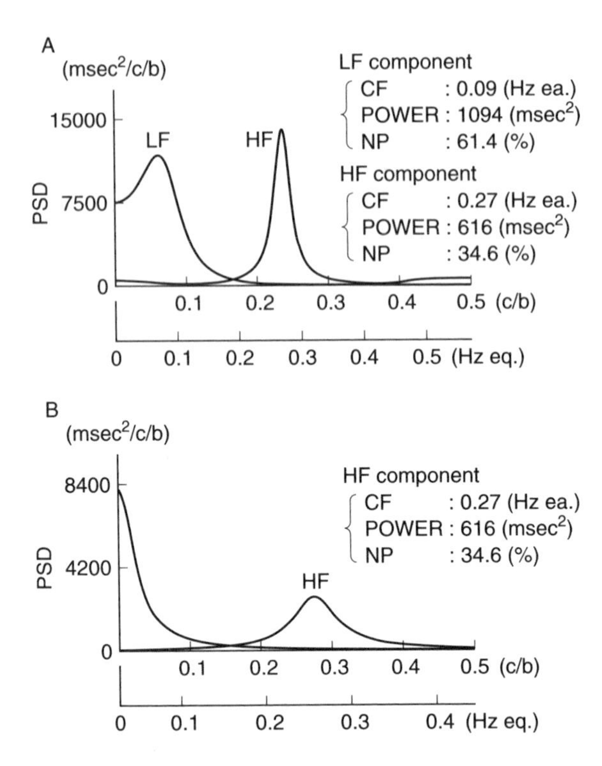

Figure 5 Individual spectral components of subjects at rest in supine position. In healthy male (A) there were 2 major spectral components, i.e., LF component and HF component. On the contrary, in neurological complete quadriplegic male (B), HF component was observed, whereas LF component was observed. See caption of Figure 4 for definitions of abbreviations. (Inoue *et al.*, 1990).

respiratory control or the complex body motions of workers when targeting actual work. Nevertheless, if one quantitatively observes motion and behavior during work and painstakingly removes the respiratory influence during EGG analysis, this problem can be solved to some extent. HRV is an index of workload that can be useful, with this sort of effort. There are many reports from laboratory research where mental arithmetic has been used in tasks to study the HRV response (for example, Cerutti *et al.*, 1988; Miyake *et al.*, 1990; Sloan *et al.*, 1991; Zwiener *et al.*, 1982). In addition to this, there are reports on memory research (Aasman *et al.*, 1987), tracking (Vicente *et al.*, 1987; Backs *et al.*, 1991), target detection (Pagani *et al.*, 1991) and VDT data entry work (Itoh, 1988).

Application for measuring changes in work burden caused by the adoption of a head-up tilt by young people and elderly people (Akatsu *et al.*, 1999), and a psychophysiological evaluation of test drivers have been conducted (Richter *et al.*, 1998). Also, 24-hour measurement using an electrocardiogram and studies using a SDNN index have been conducted for evaluating work-related stressors (Van Amelsvoort *et al.*, 2000). The results in all these cases appear to suggest that HRV can be an effective index for evaluating mental workload.

Cases such as the following are available as examples of the use of HRV in industrial field studies or related domains: highway driving (Egelund, 1982), punch card operation work (Kamphuis and Frowein, 1985) and piloting the work of aircraft pilots using flight simulators (Itoh *et al.*, 1990). These various reports suggest that low cycle components of 0.1 Hz serve as useful indices of mental workload.

In general, it is felt that the mental workload can be investigated from fluctuations in sympathetic activities (Pomeranz *et al.*, 1985; Pagani *et al.*, 1986) where this low cycle region primarily is significant, by focusing on the 0.1 Hz components referred to as the Mayer waves.

On the other hand, due to the fact that the HF component corresponds to the respiratory sinus arrhythmia and reflects only parasympathetic activities (Akselrod *et al.*, 1985), it is thought that the degree of imbalance in the autonomic nervous system could also be made an index using LF and a comparison with LF — the LF/HR ratio.

The range of HRV applications is broad and it is felt that HRV will come into increasing use in the future for the evaluation of occupational workload.

Electroencephalograph (EEG)

Sources of electrical signals emitted as brain waves are the centres of life activity such as brain stem reticular formation of the thalamus and hypothalamus in the centre of the cerebrum, and cerebrum association areas, as well as sensory centres such as visual and auditory centres. Accordingly, measurement of the cerebrum is useful as a tool for obtaining a wealth of information related to the activities of the cerebrum.

The discovery of brain waves

Changes in electric potential of several tens of microvolts containing frequency components from 0.5–100 Hz can be attained by placing an electrode on a person's scalp and amplifying using a bioamp. Since Galvani (1791) observed that electrical current is generated when muscles constrict it had been thought that similar electrical changes also occurred when the brain is active. The first recording of electrical activity from the brain of an animal was carried out by Caton in England in 1875. He recorded electrical activity thought to be direct current from the cerebral cortex of domestic rabbits and monkeys, and came to the conclusion that there was a relationship between this activity and brain function. Although this was followed by observations of electrical activity of the cerebral cortex by many researchers, it was Berger (1929–38) who was the first person to record and accurately record electrical activity of the human brain.

He began his research into electrical activity of the human brain in 1924 and it was he who gave the name alpha waves to regular waves of 10 Hz with an amplitude of about 50 μV observed largely on the occipital and vertex when a normal person is at rest with eyes closed. He also recognised that these alpha waves disappear when a person opens their eyes and gazes at an object, and that these waves are replaced by waves of between 18–20 Hz with amplitude of 20–30 μV. He called these waves beta waves. He published 14 papers between this time and 1938, in which he recorded his observations of a wide range of phenomena, such as brain waves, epilepsy, brain tumors and brain waves of other psychoneurotic illnesses very much like those studied in clinical brain wave studies today (1929–1938).

Principles of methods of brain wave measurement

A summary of the principles of methods for recording brain waves is given below. Electrical activity in the brain is conducted to an electrode block through electrodes attached to the scalp and along their lead wires (guide wires), from where it is conducted to an electroencephalograph. An electroencephalograph consists of two parts — an amplifier and a recording mechanism. Minuscule electrical variations generated from the brain in microvolt units are amplified between one million and two million times and then recorded by the recording mechanism.

Features of an electroencephalograph

Because brain waves can have a fairly wide range of frequencies, from less than 1 Hz through to more than 100 Hz, an electroencephalograph must be able to faithfully record phenomena in a fixed frequency range. The greater the time constant (the time required for the amplitude of applied calibrated voltage to decrease to 37%; the smaller the time constant the greater the omission of slow phenomena) of low frequency waves the more possible it is to record slow phenomena. It is usual for a time constant of 0.3 to be used for recording brain waves, and when necessary a time constant of 0.1 is used.

Electrodes

There are two broad categories of electrodes used for measuring brain waves on the scalp — disc electrodes and needle electrodes. Various kinds of metals are used to make disc (plate) electrodes and needle electrodes. However, when choosing materials for electrodes one must be mindful of the problem of artifacts caused by polarisation in an electrode. Silver, metal, sun platinum and pewter are used in disc electrodes and stainless steel for needle electrodes. However, because in theory electrodes made from silver chloride are considered to have the least occurrence of polarisation, silver chloride is generally used in the manufacture of plate electrodes used for measuring brain waves that are available on the market.

Brain wave noise (artifacts)

Because when measuring brain waves input into an electroencephalograph which becomes amplified consists of minute electrical variations measuring in the tens of microvolts, and its amplified frequency band includes alternating current with a frequency of 50 or 60 Hz, it is important to take measures to counter various forms of noise (artifacts).

Internal noise is generated by the amplifier itself, and it is common for it to be generated from the amplifier's first circuit. It can be generated from a transistor or resistance and is essentially unavoidable, or it can be caused by faulty parts or faulty contact of wires, in which case it can be remedied. One means of determining whether noise is internal noise or not is to make a short circuit between the input terminals (between the arbitrarily numbered terminals of the electrode block in the electroencephalograph) of the amplifier, and if it still persists then it is generally safe to regard the noise as internal noise.

External noise is noise which enters the amplifier from outside and creates more of a problem than internal noise in terms of measuring brain waves. There are many kinds of external noise, including leak current noise, noise caused by electrostatic induction, noise caused by electromagnetic induction, noise caused by modulated high frequency, microphonic noise, noise caused by faulty earthing, and faulty earthing of electrodes.

Methods for leading electrodes

A minimum of two electrodes is required for recording brain waves, and the brain waves are recorded as the potential difference between the two electrodes. There are two main methods of leading brain waves; one is a standard method which uses standard electrodes, while the other is a bipolar method which uses probe electrodes instead of standard electrodes.

On one part of the body there is a site where potential is measured at zero, and if it is possible to place one of the electrodes on this part and determine potential variation as the difference between zero and that shown by the second electrode, an absolute value can be recorded. However, because in practice it is not always possible to use this kind of site, a site which lies at a distance from the brain is used. But because if an electrode is placed on the trunk or limbs it is not possible to read the brain waves due to interference from the ECG (which is a thousand times bigger than the brain waves) it is usual for the ear lobes, tip of the nose and mastoids to be used for this purpose. An electrode which is placed at a site where the potential approximates zero is called a reference electrode.

In contrast, an electrode which is placed on the scalp for the purpose of recording brain waves themselves is called an exploring electrode. In cases where in a particular montage the reference electrode is the same for all leads, it is called a common reference electrode.

Methods of positioning electrodes

In the past various individual methods have been used insofar as the number and placement of electrodes on the scalp is concerned. However, since a proposal for a standard method was put forward at an international symposium on brain wave studies in 1958, the system which was proposed has been adopted all around the world.

The ten–twenty electrode system which was adopted as a result of this symposium (Jasper, 1958) involves covering both hemispheres of the cerebrum with a total of 19 electrodes spaced more or less the same distance apart, with an electrode on each ear lobe making a total of 21 electrodes altogether.

Classification of brain waves

Broadly speaking there are two types of brain waves — spontaneous brain waves (EEG, electroencephalogram) and evoked potential (EP).

Spontaneous brain waves are not, in a strict sense, sine waves even though they are phenomena close to sine waves, so factors such as cycles (frequency of occurrence, frequency), amplification and phases need to be specified, as is the case with sine waves, in order to record these waves (Figure 6). Also, because brain waves are not strictly sine waves, in addition to recording frequency the duration of waves and waveforms must also be recorded. Spontaneous brain waves are divided into four categories according to their frequency, with the fastest being beta (β) waves (14–30 Hz), followed by alpha (α) waves (8–13 Hz), theta (θ) waves (4–7 Hz), and delta (δ) waves (0.5–3 Hz). Beta waves are continuous waves and are said to represent a heightened level of consciousness, as well as excited states. Alpha waves, on the other hand, are most commonly used as an indicator for ascertaining wakefulness.

Research to date has shown that an increase in intracranial nervous activity causes potential variation to shift from delta waves to theta waves to alpha waves, whereas beta waves occur when alpha waves lose periodicity (alpha-blocking) as a result of psychological burden and the like. Recently, considerable attention has been drawn to frontal midline (Fm) theta waves (7–8 Hz waves observed around the centre of the forehead). These Fm theta waves are a useful index of mental workload, and more specifically they are brain waves which react to the degree of difficulty of work tasks and levels of stress.

Evoked potential is transient potential fluctuation which is recorded at various sites between the time when various sensory forms of stimulation are input to receptors and the time when they arrive at the cerebral cortex. Among the different types of evoked potential there is cerebral evoked potential such as visual, audible, and somatosensory evoked potential, brain stem evoked potential, and spinal evoked potential. It has recently been discovered that electric potential does not consist only of an electrical response which occurs

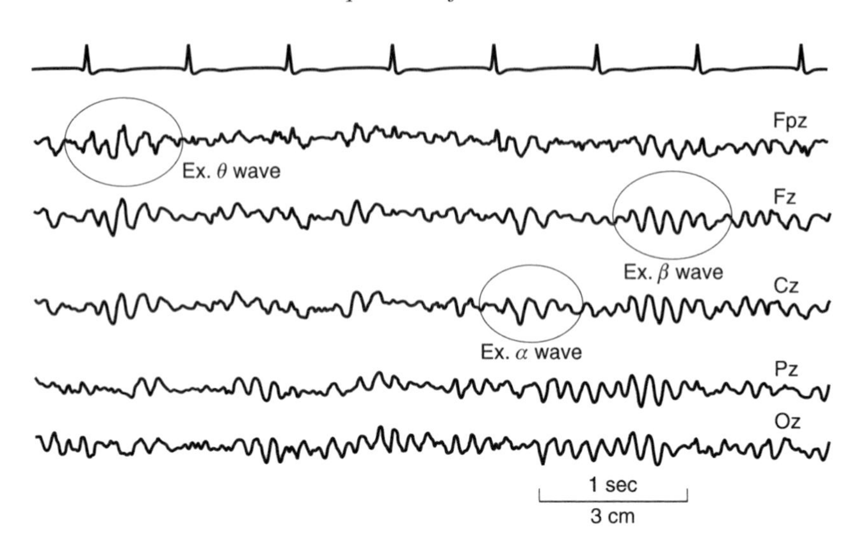

Figure 6 Typical EEG recording; the subject was a 24-year-old male with eyes closed. The Fpz, the area measured in the diagram, is mid-point between Fp1 and Fp2. In the same way, Oz is the mid-point between O_1 and O_2. Other electrodes are in accordance with the ten–twenty electrode system.

passively in response to sensory stimulation, but that there is also a component (p. 300, contingent negative variation, movement-related potential) that varies in accordance with psychological activity such as attention, recognition, problem-solving and autokinesia (Sutton *et al.*, 1965; Picton *et al.*, 1974). These are referred to as event-related potential (ERP).

Methods of brainwave analysis

The following elements will be discussed in terms of the analysis of brain waves. Where spontaneous brain waves are concerned, frequency and amplitude will be considered, and in the case of evoked brain waves we will consider latency between the presentation of stimuli and potential variation and the amplitude and polarity of potential variation.

There are two methods for measuring the frequency of spontaneous brain waves. These are the waveform recognition method in which the waveforms drawn on the chart are visually measured according to a dedicated scale, and a method which requires a power spectrum for analysing frequencies from measured data using a computer.

In waveform recognition, the cycle of each waveform is recorded on a dedicated scale and is converted to a frequency (0.5 intervals). The converted frequencies are organised according to rate of occurrence, and a frequency distribution is obtained at the time of measurement. For example, when determining sleep, 30 seconds to 1 minute is one block and the rate of occurrence of each frequency is obtained for each block.

In recent times the use of computers for conducting an analysis of brain wave frequency has become the norm. The fast Fourier transformation (FFT) method is generally used. Frequency analysis entails breaking down measured data into sine waves for each set frequency and expressing their intensity in each frequency band. This is called a power spectrum, with frequency represented horisontally and power value represented vertically. The power value is the square root of the amplitude. Recent increases in the speed of computer processing have made it possible to undertake colour mapping of frequency distribution over the entire scalp as part of a processing method which is called 'topography', and which is based on the power value obtained from FFT processing. This method of topographic processing makes it possible to simultaneously analyse information related to brain waves such as the sites of occurrence, frequency, power and time.

Utilisation of brain waves in the field of ergonomics

Brain waves are used in the field of ergonomics in order to obtain various kinds of information. Spontaneous brain waves and evoked potential have been used recently in ergonomics research, such as in research on an index for psychological fatigue (Lorist *et al.*, 2000), a work burden index (Lamberts *et al.*, 2000; Yagi *et al.*, 1998; Kiroy *et al.*, 1996; Humphrey *et al.*, 1994), a sleep quality index (Takahashi *et al.*, 1998; Bonnet *et al.*, 1994), a vigilance index (Kuller *et al.*, 1998), a recognition evaluation index (Hohnstein *et al.*, 1998), and a precautionary attention index (Yamada, 1998).

Electromyogram (EMG) method: especially surface electromyography

The method of using an electromyogram is a technique with which ergonomists are very familiar. EMG may be defined as data of activity potential which arises when muscles contract voluntarily, involuntarily or reflexively, which is recorded using an electromyograph.

Humans have a skeleton which is made up of 206 bones and approximately 500 muscles which rest on the skeleton. Muscle contraction occurs when signals from the motion centre of the brain stimulate the motion cells of the spinal cord, thus exciting motion nerves which then become electrical signals which contract muscle fibers. Before using an EMG, ergonomists must first understand the relationship between the task being tested and its relationship with the body's muscles. This requires basic anatomical and physiological knowledge. The next requirement is a full grasp of the workload of the task being subjected to EMG measurement. In other words, ergonomists must observe work conditions, work postures, and work movement of the worker. When analysing the waveforms of an EMG all this is needed in order to fully understand the properties of the EMG signals which have been obtained (Figure 7). It is this knowledge which makes it possible to distinguish useful signal data from noise and artifacts. Accordingly, this procedure requires careful calibration, use of equipment, data processing, analysis, and the formulation of a test plan which does not contradict the hypotheses inherent in the relationship between EMG and muscle function.

There is an extremely wide range of electrodes which may be used to record the activity potential of muscles. Although it is possible to use special small electrodes, laser-etching electrodes, as well as needle electrodes used for clinical diagnosis, none of these is practical when using EMGs for ergonomic purposes. The most widely used and most practical electrodes, and which are easy to use, are surface electrodes and minute wire electrodes (Nelson *et al.*, 1983). In occupational ergonomics it is most common for a worker in a workplace to be the subject of such testing. It is in this practical environment that surface electrodes are recommended as a fail-safe, non-invasive and practical type of electrode. When using surface electrodes, two plate electrodes measuring between 8–10 mm in diameter are attached to the surface of the skin directly above the targeted muscle. Two electrodes are attached 2–3 cm apart, and parallel to the muscle fibers, around the point where nerves located at what is called a motor point enter the muscle. Just as when undertaking other electrophysiological measures, care must be taken to ensure that the earth wire is firmly in place and that the electrode is tightly fastened to the skin. Needle electrodes are mainly used in the area of clinical medicine. Needle electrodes are used when muscle potential is conducted from an area which is smaller than a surface electrode. To be more specific, they are used when testing is confined to just one muscle or a deep muscle. Needle electrodes are

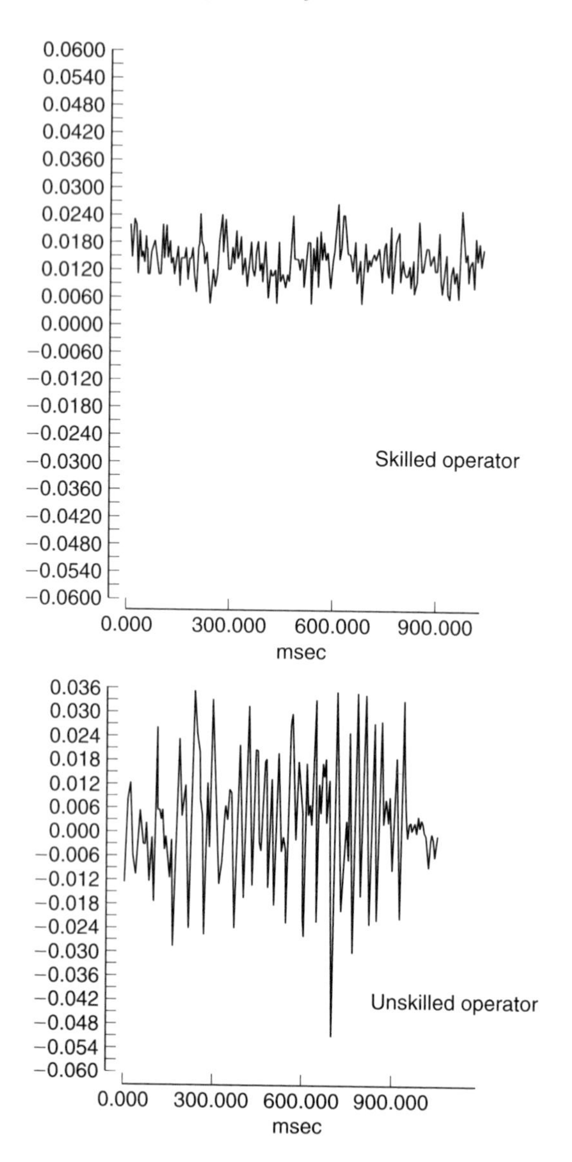

Figure 7 Recording of a typical EMG. These are recordings of the upper right biceps of overhead crane operators made during the operation of the equipment. The image at top shows the measurements for a 59-year-old skilled male worker. The image at bottom shows the measurements for a 27-year-old skilled male worker.

unsuitable for making separate observations from a motor unit, and what is more they have the disadvantage of intensifying pain felt by the participants. In general, a surface EMG is used by integrating the recorded waveforms. This is referred to as integrated EMG (IEMG). Some of the most commonly used methods when using an EMG to undertake an assessment of musculoskeletal load involve the percentage of maximum voluntary contraction (MVC), root mean square, and amplitude. The former is a method of making an index showing muscle power at the time when a task is being done where muscle power is represented as a percentage of the participant's maximum voluntary contractions. One drawback of this system is that depending on the psychological state of the participant at the time when MVCs are recorded to create a standard from which to work, there may be a variance in the degree of

MVC. The latter method has been reported as having been used in an assessment of musculoskeletal load when testing muscle tension and fatigue (Roman-Liu *et al.*, 2001).

The use of surface EMG in the field of rehabilitation ergonomics

Because, as previously mentioned, compared to needle EMG surface EMG is able to record the state of activity of many motor units from a wider range of muscles, it is indispensable in the area of rehabilitation ergonomics. Surface EMG is used in the field of rehabilitation when undertaking motion analysis, gait analysis, and thoracic limb motor analysis. However, when using surface electrodes for this kind of research one must check the effect of muscle potential crosstalk from nearby muscles and prevent motion artifact caused by the skin, electrodes and lead wires. Surface EMG is useful in the area of rehabilitation medicine for evaluating the movement of patients with abnormal muscular tension. In addition, it may also be used as an effective means for determining states in which articular movement in posture control is controlled by central nervous system strategies. That is to say, it can be used for deciding which muscle groups to select for executing movement targeted by the central nervous system, as well as for finding out how the central nervous system anticipates changes in posture, etc. It has many uses in a clinical setting, and is used, for example, for easing muscle tension and in exercises for strengthening muscle power (Brucker and Bulaeva, 1996), the effects of muscle exercises (Andersson *et al.*, 1998), and for assessing muscle fatigue in chronic low back patients (Kankaanpaa, *et al.*, 1998).

Use of surface EMG in the field of ergonomics

When using EMG for assessing muscle fatigue, it is normal to take a surface EMG. Surface EMG is an important tool for workload in ergonomics (Hagg *et al.*, 2000). Recently, long-term recording has been undertaken, including one instance where four hours of vehicle driving was recorded using this method (Duchene and Lamotte, 2000). When measuring using a surface EMG, the sebum cutaneum in the middle of the belly of the muscle being targeted must be removed and two electrodes attached about 1 cm apart from each other along the direction of motion of the muscle. A surface EMG involves very small potential ranging from about 500 μV to about 10 mV. It is for this reason that removal of the sebum cutaneum and attaching of the electrodes must be carried out with great care. If due care is not taken with these procedures, there will inevitably be problems with artefacts.

There are generally two methods for evaluating an EMG. One method involves measuring the intensity of muscle contraction from EMG amplitude and the size of muscle potential, from which one can evaluate the extent of muscle burden from the intensity of this muscle contraction. The other method involves estimating muscle fatigue from changes in the frequency components of the EMG. That is to say, when muscles are fatigued the EMG power spectrum moves to a low frequency band.

EMGs have been used widely for many years in research on muscle fatigue. Muscle fatigue may be defined as when there is a lowering in the power of muscles. When evaluation of muscle fatigue is carried out using an EMG it is common for researchers to reduce the power density in the high frequency band of EMG signals and to increase it for the low frequency band when muscle contraction induces fatigue (Kaiser *et al.*, 1962; Kogi *et al.*, 1962; Kadefors *et al.*, 1968; Kwatny *et al.*, 1970). The median frequency and central frequency of the power density spectrum are values used for recording frequency shifts related to fatigue. Lindstrom and colleagues (1970) have shown that the central frequency of the power spectrum is proportionate to transmission speed. Furthermore, Lindstrom and Petersen (1981) have shown that when the central frequency decreases at times of isometric and isotonic contraction, it more or less corresponds to the index curve determined by its time constant.

One of the most traditional forms of research using an EMG is research studying the impacts which methods of carrying loads, the weight of loads, and the length of time loads are carried have on the activities of shoulder, lower back and leg muscles. For instance, Bobet and Norman (1982) have reported that it is only the reciprocal effect of the combination of the three factors of method of carrying, weight and duration that have an effect on muscle activity. Even today, in the field of occupational health and ergonomics much research is been carried out in relation to manual material handling. Marras and Davis (1998) and Granata *et al.* (1999) cite spinal load as one indicator of workload.

Also, the activity level of an EMG is often used for assessing the impact of work posture and workplace layout. One of the most researched areas at the workplace is that conducted on sitting posture (see Chapter 16). Spinal load has been used as one workload indicator with regard to sitting posture, Van Dieen *et al.* (2001) have carried out research on evaluating three different kinds of chairs with regard to the relationship between sitting posture and lower back pain. Incidentally, this is representative of the type of research conducted in the field of ergonomics in occupation health. Those concerned with the design of chairs will find themselves involved in traditional ergonomic research. There is research by Soderberg and colleagues (1986) from some time ago. They assessed the impact which postures on a flat chair and on a chair inclined forwards had on the activities of these muscles. Data input work was used for their study. The research plan took into account worker posture, the length of continuous sitting (time), and the reciprocal effects of posture and time. As part of this research the MVC of back muscles was recorded while the participants lay on their stomachs, and this was used to standardise the data. The standardised data was analysed using ANOVA. The results clearly showed that when sitting on chairs with different designs there were differences in the levels of activity of the erector muscles of the spine.

There has also been research undertaken from a ergonomic perspective on the long-debated issue of setting up work regions for production engineering or industrial engineering. Such research has been undertaken by Dean and colleagues (1999) in which test conditions were established with three different arm's lengths and applied to setting up work areas for sit-down work. Soderberg and colleagues (1986) investigated activity at three sites of spinal erector muscles when sitting in a chair with the seat inclined. Activity level on a EMG is often used in evaluating tool design. One can cite the examples of product evaluation of ballpoint pens (Udo *et al.*, 2000), and the usability of screwdrivers (Freund *et al.*, 2000). In addition, EMGs have also been used in establishing ergonomic guidelines (Villanueva *et al.*, 1998) for flat panel displays for VDT work. The examples that have been provided above are but a few examples of the use of the EMG. Of the three means of measurement discussed in this chapter (HRV, EEG, EMG), the EMG is such a familiar method that it doesn't bear comparison with the other two, resulting in the publication of thousands of reports over the years. Furthermore, using the internet and other means it is easy to find examples of the use and analysis of EMG corresponding to readers' research objectives.

References

Aasman, J., Mulder, G. and Mulder, L.J.M. (1987). Operator effort and the measurement of heart-rate variability. *Human Factors*, **29**, 161–170.

Akaike, H. (1970). Statistical predictor identification. *Annals of the Institute of Statistics and Mathematics*, **22**, 203–217.

Akatsu, J., Kumashiro, M., Miyake, S., Komine, N., Takahashi, Y., Hashimoto, M., Togami, H. and Inoue, K. (1999). Differences in heart rate variability between young and elderly normal men during graded head up tilt. *Industrial Health*, **37**, 68–75.

Akselrod, S., Gordon, D., Madwed, J.B., Snidman, N.C., Shannon, D.C. and Cohen, R.J. (1985). Hemodynamics regulation: investigation by spectral analysis. *American Journal of Physiology*, **249** (Heart and Circulation Physiology 18), H867–H875.

Andersson, E.A., Ma, Z. and Thorstensson, A. (1998). Relative EMG levels in training exercises for abdominal and hip flexor muscles. *Scandinavian Journal of Rehabilitation Medicine*, **30**, 175–183.

Backs, R.W., Ryan, A.M. and Wilson, G.F. (1991). Cardiorespiratory measures of workload during continuous manual performance. In *Proceedings of the Human Factors Society 35th Annual Meeting*, pp. 1495–1499.

Baselli, G.S., Cerutti, S. and Civardi, F. (1987). Heart rate variability signals processing; a quantitative approach as an aid to diagnosis in cardiovascular pathologies. *International Journal of Biomedical Computing*, **20**, 51–70.

Berger, H. (1929–1938). Über das Elektrenkephalogramm des Menschen (I–XIV Mitteilungen). Archives of Psychiatric.

Bobet, J. and Norman, R.W. (1982). Use of the average electromyogram in design evaluation investigation of a whole-body task. *Ergonomics*, **25**, 1155–1163.

Bonnet, M.H. and Arand, D.L. (1994). The use of prophylactic naps and caffeine to maintain performance during a continuous operation. *Ergonomics*, **37**, 1009–1020.

Brod, C. (1984). *Technostress* (Reading, MA; Addison-Wesley Publishing). Original: *Technostress*, Addison-Wesley Publishing, Massachusetts.

Brucker, B.S. and Bulaeva, N.V. (1996). Biofeedback effect on electromyography responses in patients with spinal cord injury. *Archives of Physical Medicine and Rehabilitation*, **77**, 133–137.

Caton, R. (1875). The electric currents of the brain. *British Medical Journal*, **4**, 278.

Cerutti, S., Fortis, G., Liberati, D., Baselli, G., Civardi, S. and Pagani, M. (1988). Power spectrum analysis of heart rate variability during a mental arithmetic task. *Journal of Ambulatory Monitoring*, **1**, 241–250.

Dean, C., Shepherd, R. and Adams, R. (1999). Sitting balance I; trunk-arm coordination and the contribution of the lower limbs during self-paced reaching in sitting. *Gait Posture*, **10**, 135–146.

Duchene, J. and Lamotte, T. (2001). Surface electromyography analysis in long-term recordings; application to head rest comfort in cars. *Ergonomics*, **44**, 313–327.

Egelund, N. (1982). Spectral analysis of heart rate variability as an indicator of driver fatigue. *Ergonomics*, **25**, 663–672.

Freund, J., Takala, E.P. and Toivonen, R. (2000). Effects of two ergonomic aids on the usability of an in-line screwdriver. *Applied Ergonomics*, **31**, 371–376.

Galvani, L. (1791). De viribus electricitatis in motumusculari commentarius. Memoris of the Institute of Sciences, Bologna, Italy.

Granata, K.P., Marras, W.S. and Davis, K.G. (1999). Variation in spinal load and trunk dynamics during repeated lifting exertions. *Clinical Biomechanics*, **14**, 367–375.

Hagg, G.M., Luttmann, A. and Jager, M. (2000). Methodologies for evaluating electromyographic field data in ergonomics. *Journal of Electromyography and Kinesiology*, **10**, 301–312.

Hayano, J., Yasuma, F., Okada, A., *et al.* (1996). Respiratory sinus arrhythmia; Phenomenon improving pulmonary gas exchange and circulatory efficiency. *Circulation*, **94**, 842–847.

Hayano, J., Taylor, J.A., Mukai, S., *et al.* (1994). Assessment of frequency shifts in RR interval variability and respiration with complex demodulation. *Journal of Applied Physiology*, **77**, 2879–2888.

Hohnsbein, J., Falkenstein, M. and Hoormann, J. (1998). Performance differences in reaction tasks are reflected in event-related brain potentials (ERPs). *Ergonomics*, **41**, 622–633.

Hokanson, J.E., Degood, D.E., Forrest, M.S. and Brittain, T.M. (1971). Availability of avoidance behaviours in modulating vascular-stress responses. *Journal of Personality and Social Psychology*, **10**, 60–68.

Humphrey, D.G. and Kramer, A.F. (1994). Toward a psychophysiological assessment of dynamic changes in mental workload. *Human Factors*, **36**, 3–26.

Itoh, Y. (1988). The relation between the changes of biophysiological reactions and subjective mental workload in the typing tasks under time pressures. *The Japanese Journal of Ergonomics*, **24**, 253–260.

Itoh, Y., Hayashi, Y., Tsukui, I. and Saito, S. (1990). The ergonomic evaluation of eye movement and mental workload in aircraft pilots. *Ergonomics*, **33**, 719–733.

Jasper, H. (1958). Ten–twenty electrode system of the International Federation. *Electroencephalography and Clinical Neurophysiology*, **10**, 371–375.

Kadefors, R., Kaiser, E., *et al.* (1968). Dynamic spectrum analysis of myopotentials with special reference to muscle fatigue. *Electromyography*, **8**, 39–74.

Kaiser, E. and Petersen, I. (1962). Frequency analysis of action potentials during tetanic contraction. *Electroencephalography and Clinical Neurophysiology*, **14**, 955.

Kalsbeek, J.W.H. (1971). *Measurement of Man at Work* (New York: Van Nostrand Reinhold).

Kamphuis, A. and Frowein, H.W. (1985). Assessment of mental effort by means of heart rate spectral analysis. In: J.F. Orlebeke, G. Mulder and L.J.P. van Doornen (eds) *Psycho-Physiology of Cardiovascular Control: Models, Methods, and Data.* (New York: Plenum), pp. 841–853.

Kankaanpaa, M., Taimela, S., Laaksonen, D., *et al.* (1998). Back and hip extensor fatigability in chronic low back pain patients and controls. *Archives of Physical Medicine and Rehabilitation*, **79**, 412–417.

Kiroy, V.N., Warsawskaya, L.V. and Voynov, V.B. (1996). EEG after prolonged mental activity. *International Journal of Neuroscience*, **85**, 31–43.

Kogi, K. and Hakamada, T. (1962). Showing of surface electromyogram and muscle strength in muscle fatigue. *Report of the Institute for Science of Labor*, **60**, 27–41.

Kuller, R. and Laike, T. (1998). The impact of flicker from fluorescent lighting on wellbeing, performance and physiological arousal. *Ergonomics*, **41**, 433–47.

Kumashiro, M. (2000). New technology in the workplace and their impact on health. *26th International Congress on Occupational Health*, Keynote addresses, 67–76.

Kwatny, E. and Thomas, D.H., *et al.* (1970). An application of signal processing techniques to the study of myoelectric signals. *IEEE Transactions on Biomedical Engineering*, **17**, 303–312.

Lamberts, J., van Den Broek, P.L., Bender, L., van Egmond, J., Dirksen, R. and Coenen, A.M. (2000). Correlation dimension of the human electroencephalogram corresponds with cognitive load. *Neuropsychobiology*, **41**, 149–153.

Lindstrom, L., Magnusson, R., *et al.* (1970). Muscular fatigue and action potential conduction velocity changes studied with frequency analysis of EMG signals. *Electromyography*, **10**, 341–356.

Lindstroln, L. and Petersen, I. (1981). Power spectra of myoelectric signals; Motor unit activity and muscle fatigue. In: E. Stalberg and R.R. Young, (eds) *Clinical Neurophysiology*, (London: Butterworths Publishers), pp. 66–87.

Lipsitz, L.A., Mietus, J., Moody, G.B., *et al.* (1990). Spectral characteristics of heart rate variability before and during postural tilt; Relations to aging and risk of syncope. *Circulation*, **81**, 1803–1810.

Lorist, M.M., Klein, M., Nieuwenhuis, S., De Jong, R., Mulder, G. and Meijman, T.F. (2000). Mental fatigue and task control; planning and preparation. *Psychophysiology*, **37**, 614–625.

Madwed, J.B., Albrecht, P., Mark, R.G., *et al.* (1991). Low-frequency oscillation in arterial pressure and heart rate; a simple computer model. *American Journal of Physiology*, **256**, H1573–H1579.

Malliani, A., Pagani, M., Lombardi, F. and Cerutti, S. (1991). Cardiovascular neural regulation explored in the frequency domain. *Circulation*, **84**, 482–492.

Manenica, I. (1977). Comparison of some physiological indices during paced and unpaced work. *International Journal of Production Research*, **15**, 261–275.

Marras, W.S. and Davis, K.G. (1998). Spine loading during asymmetric lifting using one versus two hands. *Ergonomics*, **41**, 817–834.

Miyake, S., Inoue, K., Kamada, T. and Kumashiro, M. (1990). Cardiovascular response in long term mental arithmetic. *The Japanese Journal of Ergonomics*, **26**, 142–143.

Nelson, R.M. and Soderberg, G.L. (1983). Laser etched bifilar fine wire electrode for skeletal muscle motor unit recording. *Electroencephalography and Clinical Neurophysiology*, **55**, 238–239.

Pagani, M., Lombardi, F., Guzzetti, S., *et al.* (1986). Power spectral analysis of heart rate and arterial pressure variabilities as a marker of sympatho-vagal interaction in man and conscious dog. *Circulation Research*, **59**, 178–193.

Pagani, M., Mazzuero, G., Ferrari, A., Liberati, D., Cerutti, S., Vaitl, D., Tavazzi, L. and Malliani, A. (1991). Sympathovagal interaction during mental stress: a study using spectral analysis of heart rate variability in healthy control subjects and patients with a prior myocardial infarction. *Circulation*, **83** [suppl II], II-43–II-51.

Picton, T.W. and Hillyard, S.A. (1974). Human auditory evoked potentials. II; Effects of attention. *Electroencephalography Clinical Neurophysiology*, **36**, 191–199.

Pomeranz, B. *et al.* (1985). Assessment of autonomic function in humans by heart rate spectral analysis. *American Journal of Physiology*, **248**, H151–H153.

Richter, P., Wagner, T., Heger, R. and Weise, G. (1998). Psychophysiological analysis of mental load during driving on rural roads; a quasi-experimental field study. *Ergonomics*, **41**, 593–609.

Roman-Liu D., Tokarski, T. and Kaminska, J. (2001). Assessment of the musculoskeletal load of the trapezius and deltoid muscles during hand activity. *International Journal of Occupational Safety and Ergonomics*, **7**, 179–793.

Salvendy, G. and Knight, J.L. (1979). Physiological basis of machine-paced and self-paced work. *Proceedings of the Human Factors Society 23rd Annual Meeting*, pp. 158–162.

Sands, K.E.F., Appel, M.L., Lilly, L.S., *et al.* (1989). Power spectrum analysis of heart rate variability in human cardiac transplant recipients. *Circulation*, **79**, 76–82.

Shannon, D.C., Carley, D.W. and Benson, H. (1987). Aging of modulation of heart rate. *American Journal of Physiology. Heart and Circulatory*, **253**, H874–H877.

Sharit, J., Salvendy, G. and Deisenroth, M.P. (1982). External and internal attentional environments: I. The utilization of cardiac deceleratory and acceleratory response data for evaluating differences in mental workload between machine-paced and self-paced work. *Ergonomics*, **25**, 107–120.

Sloan, R.P., Korten, J.B. and Myers, M.M. (1991). Components of heart rate reactivity during mental arithmetic with and without speaking. *Physiology and Behavior*, **50**, 1039–1045.

Soderberg, G.L., Blanco, M.K., *et al.* (1986). An EMG analysis of posterior trunk musculature during flat and anteriorly inclined sitting. *Human Factors*, **28**, 483–491.

Sutton, S., Baren, M., Zubin, J. and John E.R. (1965). Evoked-potential correlates of stimulus uncertainty. *Science*, **150**, 1187–1188.

Takahashi, M. and Arito, H. (1998). Sleep inertia and autonomic effects on post-nap P300 event-related potential. *Industrial Health*, **36**, 347–353.

Udo, H., Otani, T., Udo, A. and Yoshinaga, F. (2000). An electromyographic study of two different types of ballpoint pens; investigation of a one hour writing operation. *Industrial Health*, **38**, 47–56.

Van, Amelsvoort L.G., Schouten, E.G., Maan, A.C., Swenne, C.A. and Kok, F.J. (2000). Occupational determinants of heart rate variability. *International Archives of Occupational and Environmental Health*, **73**, 255–262.

van Dieen, J.H., de Looze, M.P. and Hermans, V. (2001). Effect of dynamic office chairs on trunk kinematics, trunk extensor EMG and spinal shrinkage. *Ergonomics*, **44**, 739–750.

Vanoli, E., Adamson, P.B., Ba-Lin, *et al.* (1995). Heart rate variability during specific sleep stages. *Circulation*, **91**, 1918–1922.

Vicente, K.J., Thornton, D.C. and Moray, N. (1987). Spectral analysis of sinus arrhythmia: a measure of mental effort. *Human Factors*, **29**, 171–182.

Villanueva, M.B., Jonai, H. and Saito, S. (1998). Ergonomic aspects of portable personal computers with flat panel displays (PC-FPDs); evaluation of posture, muscle activities, discomfort and performance. *Industrial Health*, **36**, 282–289.

Yagi, A., Imanishi, S., Konishi, H., Akashi, Y. and Kanaya, S. (1998). Brain potentials associated with eye fixations during visual tasks under different lighting systems. *Ergonomics*, **41**, 670–677.

Yamada, F. (1998). Frontal midline theta rhythm and eyeblinking activity during a VDT task and a video game; useful tools for psychophysiology in ergonomics. *Ergonomics*, **41**, 678–688.

Zwiener, U., Bauer, R. and Scholle, H.Ch. (1982). The influence of mental arithmetic on autospectra, coherence, and phase spectra of autonomic rhythms in man. *Automedica*, **4**, 113–121.

Part IV

Assessment and design of the physical workplace

The physical environment comprises what must be the most widely explored of any group of ergonomics variables. These variables are components of most situations we investigate and affect people and their performance in a multitude of ways.

In spite of the long history of environmental assessments, the methods and techniques are still developing. This is, in part, because of the increasing impact of environments and information and communications technology in general, but it is also because our understanding of environment the psychophysiological and physiological effects of environmental stressors is improving.

Moreover, the array of methods used reflects these various effects. Different aspects of the physical environment have been said to have effects — positive and negative — on all the facets of individual well-being described earlier in Chapter 1. All environments at the extreme will affect workers' health; what we need are data to tell us at what levels such effects as, for instance, noise-induced hearing loss, or heat stress, or musculo-skeletal diseases might be found. Here, of course, we need to know not just the physical stimulus measurements but also exposure times, and also what alleviating effect different job, task or equipment designs, or personnel selection, might have.

Ideally, of course, we are concerned with keeping harmful environmental variables at inside the levels where health effects are found. Therefore we would wish to reduce causes of annoyance, discomfort or dissatisfaction, such as perceived flicker from fluorescent lights, discomfort glare on a VDU, irritating air draughts, noise annoyance, cramped workspaces etc, and obviously eliminate all serious health and safety problems.

The third type of possible effects of the physical environment is upon performance. This can occur because of the first two, health problems or discomfort and dissatisfaction, but may also be produced more directly. Examples are disability glare on a VDU, speech interference noise levels, cold conditions leading to loss of dexterity, or reduced vigilance through distractions from poor seating. Possible performance decrement has been proposed in terms of output and errors, for mental and physical tasks, and with respect to both direct task-related measures and also systemic ones such as absenteeism or labour turnover. It must be said that, in general, results on performance effects of the environment are more equivocal than those on health or discomfort/dissatisfaction.

The final proposed outcome of working environments, good or poor, is on the attitudes that such conditions might engender in the workforce. This might be seen in such opinions (spoken or unspoken) as: 'If management think so little of us to give us such conditions to work in then why should we co-operate with what they want'. Consequences will be resistance to change, lack of innovation or dynamism, a tendency to cure rather than prevention, and generally an unwillingness to give of skills and time. Working environments perceived as good on the other hand may evoke opposite reactions.

This section opens with four chapters reviewing all the effects described above in the context of the visual, climatic and auditory environments; consequently methods of assessment are described which will allow us to evaluate working conditions and to initiate the correct remedial action where necessary. Parsons (Chapter 23), Howarth (Chapter 24), and Haslegrave (Chapter 25) explain, for the unwary, the multiplicity of physical characteristics and units of measurement of environmental stimuli, the different human responses of relevance, and the consequently diverse methods and techniques of measurement. They also describe evaluation criteria such that assessments can be made, and provide examples of practical working environment assessment. Before this, in Chapter 22, Parsons describes the type of general environmental survey that might be conducted prior to in-depth assessment of any particular aspects of the environment

The last three chapters in this section have close links, all having anthropometry as a major component. Pheasant and Steenbekkers (Chapter 26) outline the subject, the sources of data and the methods for applying them to real problems in workplace layout and design. Superficially the procedures are simple, but in fact there are many poor applications of anthropometry, which arise to a great extent through insufficient recognition of the reasons for the choice of the various dimensions or percentiles used. Pheasant and Steenbekkers give clear guidance in this respect.

The next chapter (27) by Corlett deals with a particular workspace design application, industrial seating, which has important anthropometric components. Many people believe they are competent to design seats, and the wide variety of seats available does, in part, support their view. The comments of seat users, particularly where the seat is a necessary part of a workplace, give clear evidence that a lot of this confidence is misplaced! The seat evaluation chapter brings into study the utility and acceptability of the seat, as well as its dimensions. In fact these two factors are relevant for any workplace, and variations on the methods in this last chapter could be used more generally. One of the reasons for incorporating this chapter within the group is to point out these broader aspects, and to round off what might otherwise have been seen as primarily technical matters with a reminder that the ergonomist is interested more broadly than in just the relationships between physical dimensions.

Computer databases and graphics, largely within the family of computer aided design (CAD) systems, have made major strides in recent years and permit the simulation of workplaces at much less cost than physical mock-ups (Gallwey and O'Sullivan, Chapter 28). As is emphasised in the chapter, this does not obviate the need for physically testing the design, but it does bring a good design more quickly to realisation and allows the rapid exploration of a greater number of different designs before investing in mock-ups or prototypes. As computer technology advances we can expect even more sophistication in these methods, including the calculation of physiological and psychological parameters in addition to the physical ones.

Chapter 22

The environmental ergonomics survey

Ken Parsons

Introduction

This chapter provides practical advice on how to carry out a survey to determine the effects of the environment on people and make recommendations for improvement. A frequent request to ergonomics practitioners is to conduct an environmental survey to determine why people are complaining. Why is the environment uncomfortable? Are productivity and health affected? What can we do about it? Both general principles and practical examples are provided such that readers will be able to design and conduct an environmental survey of their own and have an understanding of the general issues. A more profound discussion of principles and practice in the areas of thermal environment lighting, and acoustics is provided in Chapters 23, 24 and 25. The environmental survey can be regarded as the first and often sufficient response to a practical question.

Environmental ergonomics

Ergonomics can be defined as the application of knowledge of human characteristics to the design of systems. People in systems operate within an environment and environmental ergonomics is concerned with how they interact with the environment from the perspective of ergonomics. Although there have been many studies of human responses to the environment (light, noise, heat, cold, etc.) over hundreds of years and much is known, it is only with the development of ergonomics as a discipline that the unique features of environmental ergonomics are beginning to emerge. In principle, environmental ergonomics will encompass the social, psychological, cultural and organisational environments of systems, however to date it has been viewed as concerned with the individual components of the physical environment. Typically, ergonomists have considered the environment in a mechanistic way, in terms of the lighting or noise survey for example, rather than an integral part of ergonomics investigation. That is, for example, if cold distracts the worker what are the consequences for the overall system?

Global activity

The establishment of the study of human responses to the physical environment has paradoxically inhibited the development of environmental ergonomics as it has produced associated institutions that provide inertia to the acceptance of an ergonomics approach. Examples include learned societies and conferences on specific aspects of the environment such as noise, lighting or vibration. The International Society for

Environmental Ergonomics first met in Bristol in 1984 and has since then held successful biannual conferences around the globe. The original intention was to provide a forum for environmental ergonomists. However, it very soon became specifically concerned with human responses to heat and cold. In fact it could be regarded as the forum for that subject. This provided a clear demonstration that there are few researchers and institutions that consider human responses to environments as a whole, rather than in terms of its component parts. The International Standards Organisation (ISO), and more recently European Standards Organisation (CEN), have made significant contributions in the area of environmental ergonomics. However, the existence of established standards committees in noise, vibration, lighting and others has hampered progress, as these bodies often take a product- or manufacturer-orientated perspective which is not human centred and not conducive to an integrated ergonomics approach. This position is not static however and it has become increasingly recognised that people experience total environments and that ergonomics methods are essential for effective practical application. Much knowledge exists and new approaches will allow that knowledge to contribute to environmental ergonomics as a major and essential contribution to ergonomics investigation.

Practical effects of environments on people

There is a continuous and dynamic interaction between people and their surroundings that produces physiological and psychological strain on the person. This can lead to discomfort, annoyance, subtle and direct effects on performance and productivity, affects on health and safety, and death. Examples would include discomfort in offices due to glare, noisy equipment, draughts, or smells. In the cold people experience frostbite and die from hypothermia. In the heat they collapse or die from heat stroke. People exposed to vibrating tools have damage to their hands. Performance can be dramatically affected by loss of manual dexterity in the cold, noise interfering with speech communication or work time lost because the environment is unacceptable or distracting. Accidents can occur due to glare on displays, missed signals in a warm environment or disorientation due to exposure to extreme environments.

Environment and people: principles

Most of the energy that makes up our environment originally comes as electromagnetic radiation from the Sun. Around 1373 $W m^{-2}$ (the solar constant) enters the outer limits of the Earth's atmosphere and this arrives on the Earth in modified form where it is transformed from place to place and from one form to another (heat, mechanical, light, chemical, electrical). The wide diversity of environments to which people are exposed are therefore defined by that energy which varies in level, characteristic and form. It is the human condition to interact and survive in those environments and part of that has been the creation of 'local' optimum environments; for example, buildings.

The human body is not a passive system that responds to an environmental input in a way that is monotonically related to the level of the physical stimulus. Any response depends upon a great number of factors. If viewed in engineering terms the 'transducers' of the body (sensors — eyes, ears, etc.) have their own specification in terms of responses to different types of physical stimuli (e.g., the eyes have spectral sensitivity characteristics). In addition, the body does not behave as a passive system; for example, the body responds to a change in environmental temperature by reacting in a way consistent with maintaining internal body temperature (e.g., by sweating to lose heat by evaporation). The body therefore senses the environment with a 'transducer' system that has its own characteristics and it reacts in a dynamic way to environmental stimuli.

The above engineering model is simplistic. There are many other factors involved. For example, the way in which a stimulus is perceived and hence any response to it will depend upon that person's past experience, their emotional state at the time and other factors. It is with consideration of these physical, physiological and psychological factors that the environmental ergonomist must provide a practical solution to the problems of how a human occupant will respond to an environment.

An additional factor that must be considered is that of individual differences. These can be conveniently divided into inter-individual differences that are differences between people (e.g., males and females, tall and short people) and intra-individual differences that are differences that occur in the same person over time (e.g., emotional state, menstrual cycle changes in females). There are ways in which design can be made for specific individuals. However, it is usual in practice to design for a population of users. It is often adequate therefore to describe individual differences in terms of statistical parameters of the population (e.g., mean and standard deviation of responses), see Chapter 26.

Environmental ergonomics methods

There are four principal methods of assessing human response to environments. These are: subjective methods, where those representative of the user population actually report on the response to the environment; objective measures, where the occupant's response is directly measured (e.g., body temperature, hearing ability, performance at a task); behavioural methods; where the behaviour of a person or group is observed and related to responses to the environment (e.g., change posture, move away, switch on lights); and modelling methods. Modelling methods include those where predictions of human response are made from models that are based on experience of human response in previously investigated environments (empirical models) or rational models of human response to environments that attempt to simulate the underlying system and hence can be used to relate cause and effect.

Subjective methods would include the use of simple rating scales, of thermal comfort for example, and also more detailed responses and questionnaires; they could also include discourse analysis and focus groups. They have the advantage of being relatively easy to carry out and are particularly suited to assessing psychological responses such as comfort and annoyance. They can also usefully be used when the contributing factors to a response are not known. They have the disadvantage of being difficult to design having a number of potential methodological biases. In addition subjective methods are often not appropriate for assessing such things as affects on health; for example a person cannot always detect when they are under a great deal of physiological strain. Also an environmental stress can interfere with a person's capacity to make a reliable subjective assessment. A further disadvantage is that subjective methods often require the use of a representative sample of the user population being exposed to the environment of interest. This is cumbersome if used in initial design.

Objective methods have the advantage of providing direct measures of human response. This could include measures of body temperature, transmitted acceleration to the head from vibration inputs, etc., as well as direct measures of performance at a task. The main disadvantages are that a representative sample of the user population is required to be exposed to the environment of interest (not useful for design), the measuring instruments can interfere with what they are used to measure and objective measures cannot easily predict states such as comfort.

Behavioural methods are probably underused in environmental ergonomics. They can have the unique advantage of not interfering with what they are attempting to measure. They include changes in posture, changing clothing, adjusting the environment, moving away, working faster or slower, and so on. A model is needed to interpret the reason for any behaviour. Observer training is required. These methods are particularly suited for studying some people with disabilities, children, or other special populations, or contexts where other methods would be inappropriate. A difficulty is determining cause and effect. Did the person change posture because they were too hot or was the chair uncomfortable or the line of sight obscured? So called 'adaptive opportunity' provides an indication of the possibility for people to adapt (move away, adjust clothing, etc.). An otherwise unacceptable environment may become acceptable if it is designed with adaptive opportunity.

Models of human response to environments have the advantage of being consistent in their response, are easy to use, give a quick response and can be used in both design and evaluation. The main disadvantages are that the models provide only approximate responses when designing for individuals and inevitably there will be factors in any real environment, which the models will not consider.

The method that is most appropriate will depend upon the specific investigation under consideration. A simple environmental survey may take a checklist approach, where the ergonomist completes a list of questions about the environment. This may be complemented by interviews or stand-alone questionnaires completed by people who occupy the environment and the use of simple measuring instruments to provide a first indication of the nature of the environment (e.g., temperature, humidity, noise levels, light levels).

If the environment is new or a specific question is being asked (e.g., comparison of two rooms; assessing different lighting systems, etc.) then a more formal approach using human participants may be required. The sections below present practical advice on how to conduct trials with human participants and a practical assessment of a room where workers have complained about their environment.

Human participant trials

The design of any test or trial using human participants will depend upon the specific aims of that test or trial. However, there are general principles and these are outlined below. A typical trial involves exposing people to environments of interest in a controlled way, measuring environmental conditions and recording the responses of people.

Specify the aim

An optimum trial design will achieve its aim with efficient use of resources. To achieve this it is important to be clear about the specific aim or aims. For example, if the aim is to compare three types of environments then a repeated measures design, where all participants are exposed to all environments (in a balanced order) in identical conditions, may provide the best comparison. Contrast this with the evaluation of an index where a wide range of environmental conditions may be optimum. If both aims need to be met then both must be met in the design. It is necessary therefore to be specific about the aims of the trial.

Which participants and how many?

A valid method of evaluating environments would be to use a panel of experts. This technique is used in wine tasting, for example where acknowledged experts give opinions concerning the quality of wines. This technique depends upon identifying unbiased acknowledged experts. This is usually not possible in the area of environmental assessment and the trial designs should specifically avoid bias. It is usual to identify a random sample of participants as representatives of the population of interest. This is a question of statistical sampling and relevant factors such as age, gender, experience, and anthropometry could be identified and influence selection. The number selected will depend upon the aim and experimental design. A calculation can be made based upon the power of a statistical test; that is, the probability of accepting the alternative hypothesis (for example, environment A is more comfortable than environment B) given that it is true. This is a rather academic approach and requires assumptions to be made about the strength of effect you expect, which is rather circular as this is what you are trying to find out. The allocation of participants to treatments will be of practical importance. If there are three rooms and three types of glazing being compared (that is nine conditions) then nine participants would allow a 9×9 Latin square design. That is where each participant is exposed to each condition in a different, balanced, order. A repeated measures design is where all participants are exposed to all conditions.

Although not statistically rigorous, other pointers are useful. It is generally considered that, for normally distributed responses, increasing the number of participants provides a diminishing return in terms of a sample representing a population. Numbers of greater than eight are often considered as an acceptable sample size. It is also useful to consider approximate probability. For example, if two vehicles were compared by four people then the probability of all four preferring vehicle A to vehicle B due to chance (when there is actually no difference in comfort between the vehicles) is 1/2 to the power of $4 = 1/16 = 6.25\%$. So a sample of four would not be sufficient to make a decision at a 5% level even in the case of an extreme result. An example of practical significance is whether the experimenter would be satisfied that if all their participants preferred A to B then this is considered sufficient evidence that A is more comfortable than B. It is useful therefore to estimate how many participants it would take for practical significance to be established. It may be that statistical significance may be established with the use of large groups but the effect may be small and not of practical significance. A more rigorous statistical approach can be taken in any particular experiment, however the ‘rules of thumb’ above can be useful. The use of experimental trials in a formal assessment of a specific hypothesis is particularly useful in environmental design (which heating systems to select etc).

Case study: environmental survey of an office where people refused to work

The following is a practical example of an environmental survey which is an actual case. It describes what was done and what was found and includes actual materials used.

The problem

The ergonomist received a request for assistance from managers of an office complex where in a particular large open plan office workers were refusing to work and some were refusing to enter the room. The request arrived via the internet with attached photographs of the buildings and office. The workers complained of feeling unwell with unusual odours, tastes, facial tensions, headaches and lethargy. Medical opinion had recommended tests for carbon monoxide build up which had proven negative. Because of the seriousness of the case the manager had been directed to seek expert assistance. Do they have a sick building?

The environmental ergonomics survey

It was agreed that the ergonomist would make a 'first shot' environmental ergonomics survey and make recommendations for a solution and for further action. The ergonomist conducted a one day assessment that included arrival and briefing by the managers, observation and measurements in the office, checklist completion by the ergonomist, completion of stand-alone questionnaires by the staff, interview of individual workers and debriefing of managers. A report and recommendations were provided soon after the assessment.

The 'expert' checklist

The expert checklist used by the ergonomist is shown in Figure 1. The checklist is designed to force the expert to systematically address important questions and to provide direct views concerning what is significant (e.g., what is the best aspect?)

PHYSICAL ENVIRONMENT AND HUMAN PERFORMANCE CHECKLIST

Complete this checklist in the context of the organisational culture and mission, the job requirements of the staff and how the physical environment may affect their performance and productivity.

General impression (one sentence and one word descriptors)

Good points (include what is the best aspect)

Bad points (include what is the worst aspect)

Air quality (immediate impression on entering - stuffy, smelly, dusty?)

Thermal environment (hot, cold, humid, draughty, hot/cold surfaces, sweaty)

Lighting and visual environment (easy to see, lighting levels, clean windows and lights, glare, general appearance)

Noise and vibration (detect vibration, footfall, background noise level, interruptions, interference with task, annoying, noise sources)

Furniture (appearance and condition, fit for purpose, fit to persons size, telephone, chair)

Computer equipment (correctly positioned and adjusted, glare on screen, reflections, orientation)

Overall layout (storage space, organisation of work, filing system, coats and accessories)

Adaptive opportunity (clothing adjustment, move around, open window, control over conditions, level of activity, take breaks)

Distraction (sources of distraction from task that cease work or interfere with performance)

Overall conclusion : Environment optimum for performance? Yes or No.

Recommendations:

Figure 1 Expert checklist used by the ergonomist as part of the environmental ergonomics survey.

It can be seen that the office appeared well equipped with a good layout and modern workstations. The fabric of the building was poor and there was a fusty smell. The best aspect was the modern attractive equipment with good layout and that the room was quiet. The worst aspect was that there was poor building fabric and the smell. Air quality seemed poor with some dampness. The thermal environment was acceptable although a supplementary heater was noted. Daylight was present but when dull outside the lights provided poor colour rendering and the strip light was not working. It was a quiet office with good appearance and possibility to open windows. Although caution must be taken concerning 'expert' subjective opinion, it was evident that air quality was a problem.

Questionnaire

A simple questionnaire (Figure 2) was given to each of the workers at their workplaces (although some workers refused to enter the room and one of the (female) workers was immediately ill on entering and removed themselves, providing a modification to the context in which the questionnaire was completed. The workforce answered questions on how they generally felt, and how they felt now.

Overall the workers were generally satisfied with their thermal, lighting and accoustic environments. They found the room very smelly, they were not satisfied with the air quality and identified 'musty, damp' smells and chemical smells. Some workers reported feeling dizzy, headaches, tiredness with irritation to the eyes and throat. Symptoms persisted when away from work and led to absenteeism.

One of the workers had been greatly troubled by the problem and had kept a diary of her experiences while working in the room. This included periods of illness attributed to working in the office.

Interviews with staff

Individual interviews with staff confirmed the subjective reports. Some staff had been moved to the new office from what had previously been a better social environment. The problem had been going for a number of months and unions and medical personnel had been involved. Some staff felt that they had been identified as 'complaining about nothing'. The interview detected a genuine interest in solving the problem and a genuine level of illness and frustration.

Environmental measurements

The air temperature and humidity were measured at each workplace with a whirling hygrometer. The air movement was observed by blowing (children's) bubbles at the workplaces. Horizontal illuminance was recorded with a light meter and noise levels with a sound level meter. The air temperature was around 21°C with 50% rh. There was very low air movement apart from when the windows were open. The light levels were around 300 lux on the workspaces (500 lux near windows) and noise levels were around 55 dB(A). None of these values would be expected to cause severe discomfort, although the lack of air movement was noted.

Wrap up meeting

As the problem appeared to be one of air quality, an inspection of the building and building services was made with the Health and Safety and Building Services Managers. A final discussion was then held with the managers. They were not clear if they had an environmental problem or an organisational problem. In any event the situation had persisted sufficiently for a managerial solution to be essential. Calling in experts will have helped the situation and the assessment was sufficiently clear to conclude that there was an air quality issue.

Report

A full environmental survey was not necessary as the problem was focussed around air quality. Lighting maintenance and improved colour rendering would provide some improvement. Three areas of advice were given for moving towards a solution. These are presented below in the actual letter sent to the managers from the ergonomist.

What do YOU think of your WORKPLACE environment?

Please answer the following questions concerning YOUR COMFORT and SATISFACTION with your environment.

Thermal environment

1. Please indicate on the scale how **YOU** feel now

Hot
Warm
Slightly warm
Neutral
Slightly cool
Cool
Cold

2. Please indicate how **YOU** would like to be **NOW**

Warmer   No change   Cooler

3. Are you generally satisfied with your thermal environment at work?

Yes   No

4. Please give any additional information or comments which you think are relevant to the assessment of your **THERMAL** environment at work (e.g. draughts, dryness, clothing, suggested improvements etc)

Lighting and visual environment

5. Please indicate on the following scale how **YOU** find your **VISUAL** environment **NOW**

Very uncomfortable
Uncomfortable
Slightly uncomfortable
Not uncomfortable

6. Please indicate any sources of glare **YOU** can see in your **VISUAL** environment **NOW**

7. Are you generally satisfied with your **LIGHTING AND VISUAL** environment at work?

Yes   No

8. Please give any additional information or comments which you think are relevant to the assessment of your **VISUAL** environment at work (e.g. glare, visual scene and view, general visual impression, flicker, colour)

Figure 2 Questionnaire completed by the occupants of a room at their workplace.

Noise

9. Please indicate on the following scale how **YOU** find the **NOISE** in your environment **NOW**

Very annoying
Annoying
Slightly annoying
Not annoying

10. Please indicate any particular sources of **NOISE** that **YOU** can hear in your environment **NOW**.

11. Are you generally satisfied with the **NOISE** level in your environment at work?

Yes No

12. Please give any additional information or comment which **YOU** think are relevant to the assessment of the **NOISE** in your environment at work (e.g. machines, talking, outside noise etc).

Air quality

13. Please indicate on the following scale how **YOU** find the **AIR QUALITY** in your environment **NOW**

Very smelly
Smelly
Slightly smelly
Not smelly

14. Please indicate any particular sources of pollution that contribute to the **AIR QUALITY** in your environment **NOW**.

15. Are you generally satisfied with the **AIR QUALITY** in your environment at work?

Yes No

16. Please give any additional information or comment which **YOU** think are relevant to the assessment of the **AIR QUALITY** in your environment at work (e.g. smells from smoke, chemicals, machines etc).

General

17. Do you suffer from persistent symptoms at work such as dry lips, eyes and throat, runny nose etc.

Yes No

18. Are you generally satisfied with your environment at work?

Yes No

19. Please give any additional information or comments which YOU think are relevent to the assessment of your environment at work including sugestions for improvement.

Figure 2 *Continued.*

To: Health and Safety Manager

John,

Environmental ergonomics survey of large open plan office

Very pleased to meet with you yesterday. I present below a brief report of my findings. As you know I was in the building for only one day and made only a preliminary assessment. However I felt that the problem was identified with reasonable confidence that air quality was the issue and not other environmental components. Although there are always organisational issues associated with this sort of problem, I did not get the impression that they were severe in this case — for example, in some organisations it is clear that the workers are not on board with the aims of the organisation and are not interested in solving the problems. In your case my impression was that there was a genuine interest and concern and it is worth keeping the workers on board.

As briefly discussed in our wrap-up meeting with Ted, my view is that there are environmental ergonomics, occupational hygiene, building services and management issues. I would suggest that all of these need to be addressed.

1. Environmental assessment — you could sample the air in the room and identify exactly what is in it. You could then compare the results with the Threshold Limit Values (TLVs) and other limits. The environmental air can include chemicals, biological contaminants and particulates. I would recommend taking advice from a certified laboratory/expert consultant in this area. If you can find an experienced consultant he or she may be able to put their finger straight on the problem.
2. Building services — you should try and identify the source(s) of the problem. The source did not appear to be in the room although there was damp at the windows. The carpet is often the culprit but it seemed new. Does it get steam cleaned and are chemicals used? May be useful to have a word with the cleaners. Subjectively that whole area of the building did seem to exhibit odour, however, it was very much concentrated in the room. It is interesting that the room was the last on a lower level in the corridor. There seemed to be sources of chemicals in the building, from the toilets for example and an analysis of air flows within the building would be of interest. You could take expert advice on this and also make your own survey.
3. Management issues — The implementation of the above investigations will be useful in showing that the problem is being taken seriously. The workers involved clearly have a problem and a management solution is inevitable. We discussed this at our meeting and there are a range of possibilities.

Who to contact

I will make some investigations into who to contact. However, it would be useful for you to contact the British Occupational Hygiene Society (BOHS) and the Chartered Institute of Building Services Engineers (CIBSE) for advice and a list of certified consultants. If you do contact a consultant be very clear about what you want otherwise you may get a specialist analysis of part of the problem. You want a whole solution. You also do not really want someone to say you have a problem/sick building, etc. We have moved on from there.

I hope that the above is of some use.

The above example describes an actual case where air quality was a problem. The checklist and questionnaire approach would also have identified other problems in the environment had they occurred. Simple on-the-spot solutions may then have been possible. If solutions were not obvious then for thermal environments, noise, vibration and lighting, a more detailed investigation could be recommended based upon the guidance provided in Chapters 23, 24 and 25.

Chapter 23

Ergonomics assessment of thermal environments

Ken Parsons

Introduction

Thermal environments can be divided conveniently into hot, neutral (or moderate) and cold conditions. Applied ergonomics methods for assessing thermal environments include objective methods, subjective methods and methods using mathematical (usually computer) models. Objective methods include measuring the physiological response of people to the environment. Responses in terms of sweat rate, internal body temperature, skin temperatures and heart rate are useful measures of body strain. Performance measures at simulated or actual tasks can also be useful. Subjective measures are particularly helpful when assessing psychological factors such as thermal comfort and satisfaction. They can also be useful in quantifying the effects of moderate cold or moderate heat stress. Mathematical models have become popular in recent years because, although often complex, they can be easily used in practical applications, employing computers. Some of the more sophisticated rational (or causal) models involve an analysis of the heat exchange between people and their environment and also include dynamic models of the human thermoregulatory system. Empirical models can provide useful mathematical equations which 'fit' data obtained from exposing human subjects to thermal conditions (see also Chapter 9).

The aim of this chapter is to present the principles behind practical methods for assessing human response to hot, moderate and cold environments and to present a practical approach to assessing thermal environments with respect to human occupancy.

The principles

The following brief discussion provides the underlying principles behind assessing human response to thermal environments. For a fuller discussion and references the reader is referred to a standard text such as Parsons (2003).

Relevant measures

It is now generally accepted that there are six important factors that affect how people respond to thermal environments. These are air temperature, air velocity, radiant temperature, humidity and the clothing worn by, and the activity of, the human occupants of the environment. In any practical assessment, instruments and methods for quantifying these factors may be used.

Thermoregulation

People are homeotherms, that is they react to thermal environmental stimuli in a manner which attempts to preserve their internal body ('core') temperature within an optimal range (around 37°C). If the body becomes too hot, vasodilation (blood vessel expansion) allows blood to flow to the skin surface (body 'shell') providing greater heat loss. If vasodilation is an insufficient measure for maintenance of the internal body temperature then sweating occurs resulting in increased heat loss by evaporation. If the body becomes too cold then vasoconstriction reduces blood flow to the skin surface and hence reduces heat loss to the environment. Shivering will increase metabolic heat production and can help maintain internal body temperature.

The physiological reaction of the body to thermal stress can have practical consequences. A rise or fall in internal body temperature can lead to confusion, collapse and even death. Vasoconstriction can lead to a reduction in skin temperature and complaints of cold discomfort and a drop in manual performance. Sweating can cause 'stickiness' and warmth discomfort and 'mild heat' can provide a drop in arousal.

Thermal indices

A useful tool for describing, designing and assessing thermal environments is the thermal index. The principle is that factors that influence human response to thermal environments are integrated to provide a single index value. The aim is that the single value varies as human response varies and can be used to predict the effects of the environment. A thermal comfort index for example would provide a single number which is related to the thermal comfort of the occupants of an environment. It may be that two different thermal environments (i.e., with different combinations of various factors such as air temperature, air velocity, humidity and activity of the occupants) have the same thermal comfort index value. Although they are different environments, for an ideal index identical index values would produce identical thermal comfort responses of the occupants. Hence environments can be designed and compared using the comfort index.

A useful idea is that of the standard environment. Here the thermal index is the temperature of a standard environment that would provide the 'equivalent effect' on a subject as would the actual environment. Methods of determining equivalent effect have been developed. One of the first indices using this approach was the effective temperature (ET) index (Houghten and Yagloglou, 1923). The ET index was in effect the temperature of a standard environment (air temperature equal to radiant temperature, still air, 100% relative humidity for the activity and clothing of interest) which would provide the same sensation of warmth or cold felt by the human body as would the actual environment under consideration.

Heat balance

The principle of heat balance has been used widely in methods for assessing human responses to hot, neutral and cold environments. If a body is to remain at a constant temperature then the heat inputs to the body are balanced by the heat outputs. Heat transfer can take place by conduction (K), convection (C), radiation (R) and evaporation (E). In the case of the human body an additional heat input to the system is the metabolic heat production (M) due to the burning of food in oxygen by the body. Using the above, the following body heat equation can be proposed:

$$M \pm C \pm R \pm K - E = S \qquad (1)$$

If the net heat storage (S) is zero then the body can be said to be in heat balance and hence internal body temperature can be maintained. The analysis requires the values represented in equation (1) to be calculated from a knowledge of the physical environment, clothing and activity.

Rational indices

Rational thermal indices use heat transfer equations (and sometimes mathematical representations of the human thermoregulatory system) to 'predict' human response to thermal environments. In hot environments the heat balance equation (1) can be rearranged to provide the required evaporation rate (E_{req}) for heat balance ($S=0$) to be achieved, e.g.,

$$E_{req} = (M - W) + C + R \quad (2)$$

(K can often be ignored and W is the amount of metabolic energy that produces physical work). Because sweating is the body's major method of control against heat stress, E_{req} provides a good heat stress index. A useful index related to this is to determine how wet the skin is; this is termed skin wettedness (w) where:

$$w = \frac{E}{E_{max}} = \frac{\text{actual evaporation rate}}{\text{maximum evaporation rate possible in that environment}} \quad (3)$$

In cold environments the clothing insulation required (IREQ) for heat balance can be a useful cold stress index based upon heat transfer equations.

Heat balance is not a sufficient condition for thermal comfort. In warm environments sweating (or skin wettedness) must be within limits for thermal comfort and in cold environments skin temperature must be within limits for thermal comfort. Rational predictions of the body's physiological state can be used with empirical equations which relate skin temperature, sweat rate and skin wettedness to comfort.

Empirical indices

Empirical thermal indices are based upon data collected from human subjects who have been exposed to a range of environmental conditions. In hot environments curves can be 'fitted' to sweat rates measured on individuals exposed to a range of hot conditions. There has been little research of this kind for cold conditions, however a wind chill index was developed based upon the cooling of cylinders of water in outdoor conditions. Wind chill provides the 'trade-off' between air temperature and air velocity. Comfort indices have also been developed entirely empirically from subjective assessments over a range of environmental conditions.

Direct indices

Direct indices are measurements taken on a simple instrument which responds to environmental components similar to those to which humans respond. For example a wet, black globe with a thermometer placed at its centre will respond to air temperature, radiant temperature, air velocity and humidity. The temperature of the globe will therefore provide a simple thermal index which, with experience of use, can provide a method of assessment of hot environments. Other instruments of this type include the temperature of a heated ellipse and the integrated value of unaspirated wet bulb temperature, air temperature and black globe temperature (Wet Bulb Globe Temperature — WBGT).

Measuring instruments

Air temperature is traditionally measured using a mercury in glass thermometer, although more recently thermocouples and thermistors have been used. An advantage of electronic instrumentation is that values can be continuously recorded and fed into computers for later analysis. The dry bulb of a whirling hygrometer gives a value of air temperature. If there is a large radiant heat component in the environment then it will be necessary to pass air over the sensor (e.g., by rapid whirling) or shield the air temperature transducer (e.g., using a wide mouthed vacuum flask). Air humidity can be found from the wet and dry bulb of a whirling hygrometer. Other methods include capacitance devices and hair hygrometers.

Radiant temperature is usually quantified in the first analysis by measuring black globe (usually 150 mm diameter) temperature. Correcting the globe temperature for air temperature and air velocity allows a calculation of mean radiant temperature. If more detailed analysis is required then instruments for measuring plane radiant temperatures in different directions should be used. Correction factors may be necessary to allow for the shape of the human body; however use of a globe thermometer provides a satisfactory initial measurement method.

Air velocity should be measured down to about $0.1\,m\,s^{-1}$ in indoor environments. Generally cup or vein anemometers (i.e., masses rotated by moving air) will not measure such low air speeds. Suitable instruments are hot wire anemometers, where the cooling power of moving air over a hot wire, corrected for air temperature, provides air velocity, or Kata thermometers, where air movement cools a thermometer and cooling time is related to air velocity. Kata thermometers are however cumbersome and time consuming to use in practical applications.

In more recent years integrating systems have been developed which detect all four environmental parameters and integrate measurements into thermal indices which predict, for example, thermal comfort. The instruments are usually simple to use and provide practical solutions for the non-expert. They are often relatively expensive however. Another development has been the use of transducers connected to digital storage devices, to allow recording of environmental conditions over long periods of time. The devices usually allow easy interfacing with digital computers where sophisticated analysis can be performed.

Subjective methods

Thermal comfort is usually defined as 'that condition of mind which expresses satisfaction with the thermal environment'. The reference to 'mind' emphasises that this is a psychological phenomenon and hence the importance of subjective assessment methods. Subjective methods range from simple thermal sensation votes to more complex techniques where semantic or cognitive models of human perception of thermal environments can be determined. In a simple practical assessment of thermal environments two types of scale are generally used. One type is concerned with thermal, sensation and the other is concerned with acceptability (i.e., a value judgement). Specific questions regarding general satisfaction, draughts, dryness and open-ended questions asking for other comments provide a useful brief subjective form, especially for measuring the acceptability and comfort of an environment. There are biases and errors which can occur in taking subjective measures, but it should not be forgotten that the best judges of their thermal comfort are the human occupants themselves (see Chapter 4 on indirect observation and Chapter 22 on the environmental survey).

Behavioural and adaptive methods

In environmental assessment the ergonomist should be aware that people are not simply passive receptors of the environment, but they will respond to avoid discomfort and thermal strain. Any environmental assessment therefore should consider the 'adaptive opportunity' for people to alter their exposure to an environment. Can people move away from the environment, adjust clothing, adjust work rate, open windows, adjust heating controls, turn on fans and so on? Designing work to include adaptive opportunity can be part of an ergonomist's recommendations. Examples where adaptive opportunity may not exist will be in hot environments where encapsulated protective clothing must be worn and the worker cannot easily leave the work (e.g., boiler stripping); in cold environments where activity level cannot be increased and in moderate environments where there is a strict dress code (e.g., uniform); in an emergency room where workers cannot leave the task or if windows cannot be opened or closed. Of particular concern will be populations who have limited adaptive opportunity due to their condition. These may include people with physical disabilities, old people, children and babies.

Human performance

Despite a number of studies having been carried out on the effects of thermal environments on human performance there is no specific information of practical value. Human performance can be considered in terms of physical (e.g., manual dexterity) and psychological (e.g., behavioural, cognitive) effects. If heat or cold stress is sufficiently severe that internal temperatures pass beyond limits at which major physiological effects occur (e.g., causing collapse or hallucinations) then clearly performance will be impaired. Within such limits, effects are influenced by factors such as motivation, level of proficiency at the task and individual differences.

The state of practical knowledge is such that it is not yet possible to predict reliably effects on manual or cognitive performance in hot environments or effects on cognitive performance in cold environments. Major effects do occur, however, on manual dexterity, cutaneous sensitivity and strength in cold environments. Hand skin temperature can be used to predict effects and it is generally agreed that to maintain manual dexterity the hands should be kept warm. If workers are sufficiently distracted to take time off the task then this will clearly affect productivity. The distraction effects of thermal stress can also degrade performance, particularly in cold environments.

Clothing

Clothing can be worn for protection against environmental hazards and for aesthetic reasons as well as for thermal insulation. In thermal terms a microclimate is produced between the human body and the clothing surface. The applied ergonomist should ensure that the microclimate allows the body to achieve desirable physiological and psychophysical objectives. Physiological objectives may include the maintenance of heat balance for the body, and the preservation of skin temperatures and sweating at levels which allow for comfort. An interaction between thermal aspects and material type should also be considered (e.g., the effect of 'scratchy' materials being exacerbated by sweating).

The dry thermal insulation of clothing is greatly affected by how much air is trapped within clothing layers as well as within clothing. There are two common units which quantify dry clothing insulation; these are the CLO and the TOG. The CLO value is a clothing insulation value which is intended to be a 'relative unit', compared to a normal everyday costume necessary for thermal comfort in an indoor environment. For example, a 'typical' business suit (including underclothes, shirt, etc.) is often quoted as having a thermal insulation of 1.0 CLO. A nude person has zero CLO. In terms of thermal insulation 1 CLO is said to have a value of $0.155\,m^2\,°C\,W^{-1}$. The TOG value is a unit of thermal resistance and is a property of the material. It can be measured on a heated flat plate, for example, in terms of heat transfer. It does not necessarily relate to the thermal insulation provided to a clothed person. For comparison, 1 TOG is equal to $0.1\,m^2\,°C\,W^{-1}$. When clothing becomes wet, due to sweating or external conditions, then the clothing insulation is altered (usually greatly reduced). There are methods which estimate the thermal properties of wet clothing; however, these crude and thermal insulation values for wet clothing are not well documented. Additional factors which can greatly affect the thermal insulation of clothing include pumping effects due to body movement, ventilation and wind penetration.

In application it is not usually necessary to have a detailed description of clothing insulation properties. The important point is whether the clothing achieves its objectives. The objectives for clothing will be determined in an overall ergonomics systems analysis, involving a description of the objectives of the organisation, task analysis, allocation of functions, job design and so on. This will include the objectives in terms of thermal insulation which will also involve the design of the clothing to include pockets to keep hands warm and devices to keep workers cool in hot environments (e.g., ice jackets, etc.).

User tests and trials will provide important information about whether clothing meets its objectives. Objective measures such as sweat loss, skin temperatures and internal body temperatures or subjective measures of thermal sensation, comfort or stickiness can be of great value. Performance measures at actual or simulated tasks can also be used to evaluate whether clothing has achieved its thermal and other objectives.

Safe surface temperatures

The applied ergonomist is also interested in the effect there will be on the body from physical contact between the human skin and surfaces in a workplace; for example, what sensation is caused by bare feet on a 'cold' floor or whether brief contact with a domestic product (e.g., a cooker, oven door, knobs or a kettle) will result in pain or burns. There are many factors involved in determining human response. These include the type and duration of contact, the material and condition of the surface and the condition of the human skin. However, a simple model based upon the heat transfer between two semi-infinite slabs of material in perfect thermal contact can provide a practical method of assessment.

The principal point is that there is a contact temperature which exists between the skin and the material surface which is dependent upon the physical properties of the skin and the material and which will influence the effect on the person. For example, if one touched a metal slab at 100°C then the contact temperature would be of the order of 98°C and the metal would be felt to be extremely hot. If, however, one touched a cork slab at 100°C then the contact temperature would be around 46°C and the cork would feel much less hot than the metal. Contact temperature can therefore be used to predict effects on the body, and can be calculated from the following equation:

$$T_{con} = (b_1 T_1 + b_2 T_2)/(b_1 + b_2) \qquad (4)$$

where T_1 and T_2 are the initial surface temperatures (°C); T_{con} is the contact temperature (°C); and b_1 and b_2 are thermal penetration coefficients calculated from the following equation:

$$b = (Kpc)^{1/2} \mathrm{JS}^{-1/2}\,\mathrm{m}^{-2}\,{}^{\circ}\mathrm{C}^{-1}$$

where K is the thermal conductivity; p is density; and c is specific heat. To calculate T_{con} the values of b for human skin and for different materials are required. These are provided in Table 1.

McIntyre (1980) argues that if one assumes a skin temperature of 34°C and a simple reaction time for an individual of 0.25 s then, from the work of Bull (1963), whose data suggest a partial burn at a skin temperature of about 85°C for 0.25 s contact, one can estimate temperatures which would produce partial burns. Some of these temperatures are provided in Table 1 (from McIntyre, 1980).

A practical approach to providing maximum surface temperatures for heated domestic equipment is provided in British Standard 4086 (BSI, 1966). The Standard considers three categories of material type and three types of contact duration. A summary of the limits is provided in Table 2.

Because of the complexity of the problem and ethical considerations regarding experimentation involving pain and burns on human subjects, knowledge is incomplete in this area. In addition, specifications for safety of manufacturing products often involves other costs and benefits to be considered (see Chapter 37 in this book). The limits provided in BS 4086 (BSI, 1966) have recently been reviewed and an additional document, BS PD 6504 (BSI, 1983), has been produced which provides background medical information and data in terms of discomfort, pain and burns. There is, however, little information concerning the inter- and intrasubject variation in response, effects of skin condition and effects on different populations (e.g., the aged, children) and other variables important for practical application. Further work in this area has led to European and International Standards

Table 1 Thermal Penetration Coefficients (b)

Material	b ($\mathrm{JS}^{-1/2}\,\mathrm{M}^{-2}\,{}^{\circ}\mathrm{C}^{-1}$)	Momentary contact surface temperature for burn threshold (°C)
Human skin	1000	
Foam	30	–
Cork	140	450
Wood	500	187
Brick	1000	136
Glass	1400	121
Metals	>10000	90

Table 2 Maximum Surface Temperatures (°C) for Heated Domestic Equipment (BSI, 1966)

	Handles (kettles, pans, etc.)	Knobs (not gripped)	Momentary contact
Metals	55	60	105
Vitreous enamelled steel and similar surfaces	65	70	120
Plastics, rubber or wood	75	85	125

(EN 563, ISO 13732). For a full discussion of human skin reaction to contact with hot, moderate and cold surfaces see Parsons (2003).

International standards

There have been many national and international standards concerned with thermal comfort, heat stress and cold stress. Thermal comfort standards can define conditions for thermal comfort and indicate the likely degree of discomfort of occupants of thermal environments. Standards for heat stress and cold stress attempt to specify conditions that will preserve health and often comfort and performance. Standards can also provide guidance on environmental design and control, they standardise methods to allow comparison and they contribute to assessment and evaluation.

Influential institutions throughout the world that produce standards or guidelines include the American Society of Heating, Refrigerating and Air Conditioning Engineers (ASHRAE) in the USA, the Chartered Institute of Building Services Engineers (CIBSE) in the UK, National standards bodies, the ISO and also European Standardization under CEN, the World Health Organisation (WHO), the American Conference of Governmental Industrial Hygienists (ACGIH), the International Labour Organisation (ILO) and more. Many standards have defined limits or at least methods in terms of rational or empirical thermal indices or physiological condition (e.g., body 'core' temperature). There are also standards that provide techniques and methods (e.g., physiological and subjective assessment methods).

ISO standards

For the assessment of hot environments a simple method based on the WBGT (wet bulb globe temperature) index is provided in ISO 7243. If the WBGT reference value is exceeded a more detailed analysis can be made (ISO 7933) involving calculation, from the heat balance equation, of sweating required in a hot environment. If the responses of individuals, or of specific groups, are required (for example in extremely hot environments) then physiological strain should be measured. Methods of measuring mean skin temperature, heart rate, internal body ('core') temperature and mass loss are all described in ISO 9886 (Figure 1).

ISO 7730 provides an analytical method for assessing moderate environments and is based on the Predicted Mean Vote and Predicted Percentage of Dissatisfied (PMV/PPD)

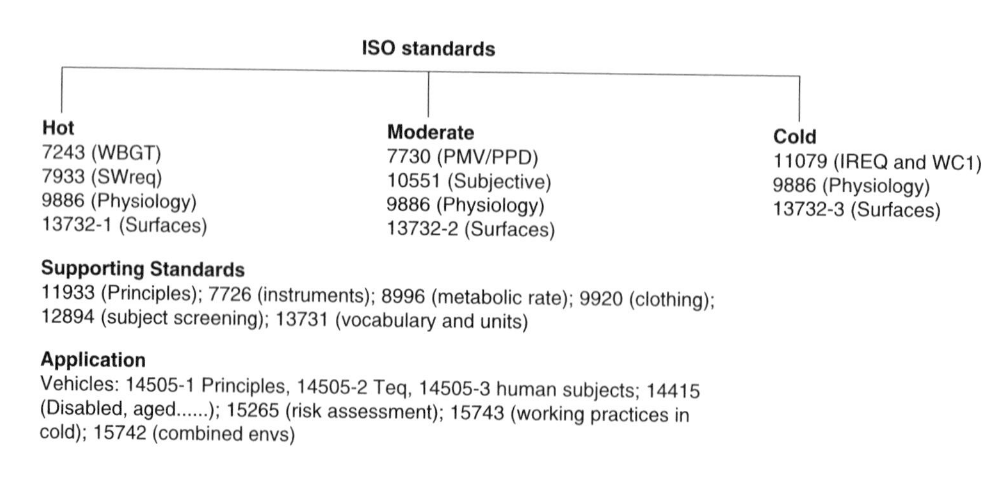

Figure 1 ISO Standards for assessing thermal environments.

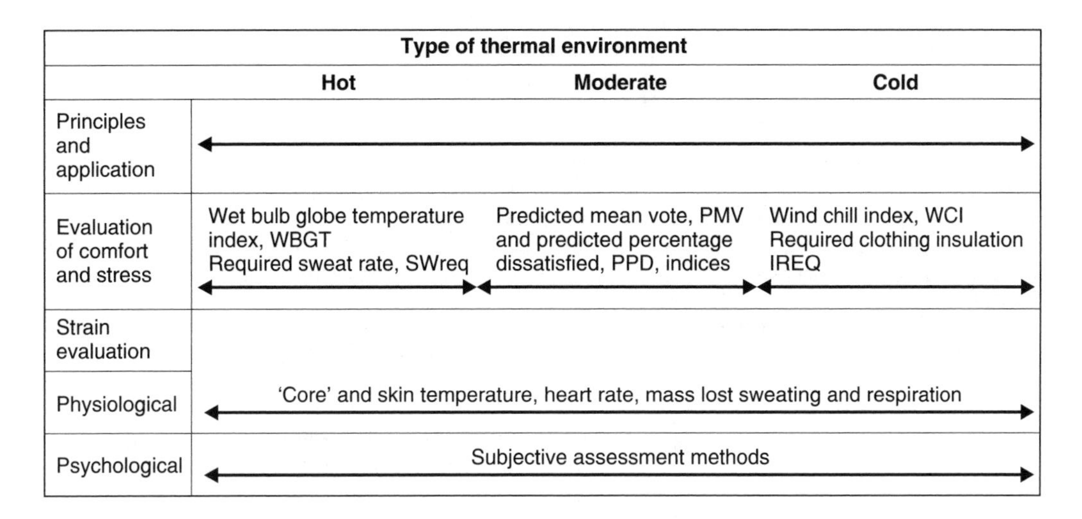

	Type of thermal environment		
	Hot	**Moderate**	**Cold**
Principles and application	←	→	→
Evaluation of comfort and stress	Wet bulb globe temperature index, WBGT Required sweat rate, SWreq	Predicted mean vote, PMV and predicted percentage dissatisfied, PPD, indices	Wind chill index, WCI Required clothing insulation IREQ
Strain evaluation			
Physiological	'Core' and skin temperature, heart rate, mass lost sweating and respiration		
Psychological	Subjective assessment methods		

Figure 2 The assessment of thermal environments using the series of ISO standards.

index and on criteria for local thermal discomfort, in particular draughts. If the responses of individuals or specific groups are required, then subjective measures should be used (ISO 10551).

ISO TR11079 (Technical report) provides an analytical method for assessing cold environments involving calculation of the clothing insulation required (IREQ) from a heat balance equation. This can be used as a thermal index or as a guide to selecting clothing.

Supporting standards include an introductory standard (ISO 11399) and standards for estimating the thermal properties of clothing (ISO 9920), metabolic heat production (ISO 8996) and a standard for definitions symbols and units (ISO 13731). Other standards consider instruments and measurement methods (ISO 7726) and medical supervision of individuals exposed to hot or cold environments (ISO 12894). ISO work on contact with solid surfaces is divided into hot, moderate and cold surfaces. Ergonomics data to establish temperature limit values for hot touchable surfaces are provided in European standard EN 563 (1994). This has been extended to the three part standard ISO 13732 (Part 1 — hot surfaces, Part 2 — moderate surfaces and Part 3 — cold surfaces). Newly developed standards include a three-part standard for the assessment of vehicle thermal environments ISO 14405, a standard for people with special requirements (e.g., disabled, aged, ISO TS 14515) and a standard for risk assessment (ISO 15265). Working practices for cold environments are presented in ISO 15743 and there in a proposed new standard that will consider the integration and interaction of total environments (ISO 15742).

ISO 11399 (1993) presents a description of principles and methods of application of the series of ISO standards and should be consulted for an initial overview. The ISO working system showing how the collection of standards can be used in practice is presented in Figure 2.

Thermal models

Practical models which simulate how people respond to hot, moderate or cold environments can be used to assess and design thermal environments. They can also be integrated into larger computer-based expert and knowledge-based systems for use by the ergonomics practitioner.

Table 3 Aims, Tide and Status of ISO Standards Concerned with the Ergonomics of the Thermal Environment

Aims of the standard		Title of the document	Status
General presentation of the set of standards in terms of principles and application		Ergonomics of the thermal environment: principles and application of International Standards	ISO DIS 11399
Standardisation of quantities symbols and units used in the standards		Ergonomics of the thermal environment: definitions symbols and units	New work item
Comfort and thermal stress evaluation			
Thermal stress evaluation in hot environments	Analytical method	Hoț environments — Analytical determination and interpretaion of thermal stress using calculation of required sweat	ISO 7933
	Diagnostic method	Hot environments — Estimation of the heat stress on working person, based on the WBGT-index (wet bulb globe temperature)	ISO 7243
Comfort evaluation		Moderate thermal environments — Determination of the PMV and PPD indices and specification of the conditions for thermal comfort	ISO 7730
Thermal stress evaluation in cold environments		Evaluation of cold environments — Determination of required clothing insulation, IREQ	ISO TR 11079 Technical report
Data collection standards	Metabolic rate	Ergonomics — Determination of metabolic heat production	ISO 8996
	Requirements for measuring instruments	Thermal environments — Instruments and methods for measuring physical quantities	ISO 7726
	Clothing insulation	Estimation of the thermal insulation and evaporative resistance of a clothing ensemble	ISO 9886
Evaluation of thermal strain using physiological measures		Evaluation of thermal strain by physiological measurements	ISO 9886
Subjective assessment of thermal comfort		Assessment of the influence of the thermal environment using subjective judgement scales	ISO DIS 10551
Selection of an appropriate system of medical supervision for different types of thermal exposure		Ergonomics of the thermal environment — Medical supervision of individuals exposed to hot or cold environments	ISO CD 12894
Contact with hot, moderate and cold surfaces		Documents in draft form	New work items
Comfort of the disabled		Documents in draft form	New work item
Design of work for cold		Documents in draft form	New work item
Long-term assessment of environmental quality		Documents in draft form	New work item
Vehicle environments		Documents in preparation	New work item

Examples of models which simulate the human response to thermal environments are provided by Haslam and Parsons (1987) and Parsons (2003). Models of the human thermoregulatory system controlling a passive body (e.g., made up of cylinders and a sphere with thermal properties similar to those of the human body) can be used to predict changes in temperature within and over different parts of a clothed body. These models can then

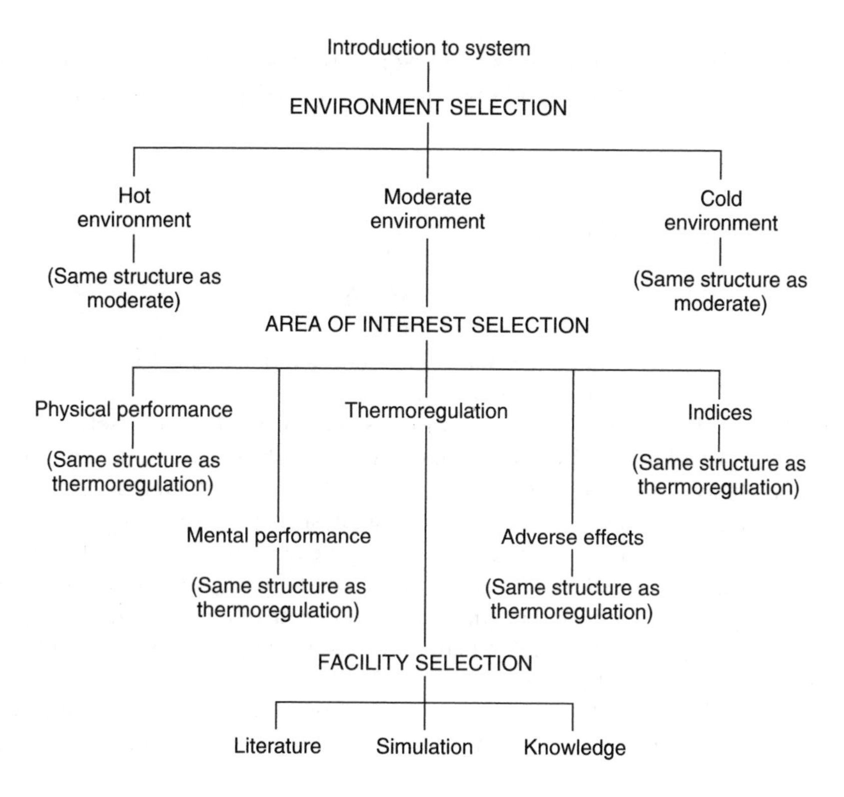

Figure 3 System tree diagram.

simulate how persons could respond in terms of heat stress or cold stress in outdoor environments or in terms of thermal comfort indoors. Investigations of the nature of expertise used in assessing the human response to thermal environments, coupled with the requirements of ergonomics practitioners and simulations using such models as are described here, can provide the input to expert systems. An example of the structure of such a system as described by Smith and Parsons (1987) is provided in Figure 3. Such systems have become valuable tools in integrating the principles and knowledge involved in the assessment of human response to thermal environments for practical application.

The practice

The principles mentioned above can provide a practical methodology which can be used to assess thermal environments with respect to effects on their human occupants. Almost by definition practical assessments will have factors specific to particular applications and one universal method is therefore difficult to provide.

Practical assessment of hot environments

The assessment of hot environments is particularly important as danger to health can occur rapidly. For environments where experience has been gained in monitoring workers, working practices can be developed and conditions monitored using simple thermal indices (e.g., wet bulb globe temperature (WBGT) index).

The WBGT index is used in ISO 7243 (ISO, 2000) as a simple method for assessing hot environments. For conditions inside buildings and outside buildings without solar load:

$$\mathrm{WBGT} = 0.7t_{nw} + 0.3t_g \tag{5}$$

and outside buildings with solar load:

$$\mathrm{WBGT} = 0.7t_{nw} + 0.2tg + 0.1t_a \tag{6}$$

where t_{nw} is the natural wet bulb temperature (i.e., not 'whirled'); t_g is 150 mm diameter black globe temperature; and t_a is the air temperature.

Acclimatisation programmes, before workers begin work, are useful. It is particularly important that workers do not become unacceptably dehydrated (e.g., greater than 4% of body weight lost in sweat) or have an unacceptably elevated internal body temperature (e.g., greater than 38.0–38.5°C). A more detailed analysis can be provided by using rational assessments of the environment (ISO 7933). Allowable exposure times based on such factors as predicted elevated internal body temperature or dehydration can be provided. It is important to remember however that there are individual differences in workers and that knowledge of heat transfer for the human body is incomplete. Experience is therefore required in the use of rational models.

If individuals are exposed to extremely hot environments then individual physiological measures of heart rate, internal body temperature and sweat loss should be taken and each worker observed closely.

Practical assessment of cold environments

Similar general guidelines apply to cold environments as were described for hot environments. A simple index such as the wind chill index can be used when experience has been gained with its use, where:

$$\mathrm{WCI} = (10\sqrt{v} + 10.45 - v)(33 - t_a) \tag{7}$$

where v is the air velocity ($\mathrm{m\,s^{-1}}$) and t_a is the air temperature (°C). The effects associated with different values of the wind chill index are:

WCI	Effect
200	Pleasant
400	Cool
1000	Cold
1200	Bitterly cold
1400	Exposed flesh freezes
2500	Intolerable

When the WCI value is calculated it is often useful to calculate the t_a value which would provide the same wind chill in 'calm' air ($v = 1.8\,\mathrm{ms^{-1}}$).

Clothing is important and a compromise must be reached between thermal insulation and clothing design to reduce effects on worker performance and safety. Of particular interest is the temperature of the body's extremities (hands and feet). Prolonged exposure may lead to thermal injury but severe discomfort distraction and loss of manual dexterity are the most commonly occurring effects. Although there is some debate, 'back of hand' temperatures above 20–25°C should maintain some comfort and performance. Hand temperatures of less than 10–15°C are usually unsatisfactory, although they should not produce injury. Performance effects will depend upon duration of exposure. Cold can produce severe discomfort and this has behavioural and distractive effects. Distraction may reduce manual and cognitive workload capacity.

Rational indices can be used to predict the required clothing insulation for heat balance and thermal comfort and also allowable exposure times based upon a drop in body 'core' temperature. Physiological measures of heart rate, mean skin temperature and body core temperature should be used if assessing individuals in cold environments. Any drop in body core temperature is unsatisfactory, but 36°C is a lower working limit. A core temperature of below 35°C is defined as hypothermia. Medical screening of subjects should take place before exposure to either hot or cold environments.

Practical assessment of moderate environments

A common request to the environmental ergonomist is to assess an indoor climate such as found in an office. The practical method used in an actual case is outlined here, although some adjustment to the results has been made to illustrate points.

The workers in a large office were complaining that their thermal environment was unacceptable. The ergonomist was asked to assess the environment, quantify the problem and make recommendations for improvement if necessary. One day was allowed for the assessment and a total of four days for the whole project, including both analysis and final report, a not unusual time restriction.

Worker relations

Complaints about working environments can be stimulated by other work related problems and it is important for the ergonomist to gain an impression of the physical, social and organisational environment in general. In addition it is useful to have the co-operation and understanding of management and workers. The worker representative was therefore contacted and the ergonomist introduced. It was explained that the ergonomist was attempting to improve the thermal environment conditions. The physical and subjective measures which were to be taken were also demonstrated to the workers' representative who then passed on the information to the office occupants.

Where, when and what to measure

The question of where and when to measure is a question of statistical sampling. The thermal environmental conditions will vary throughout a space and also with time (during the day, night and seasonal variations). The more measuring points in the room and the more measuring times, in general, the more accurately the environment can be quantified. This then is a question of resources. Only one day was allowed for measurement so a plan of the office was obtained and individual workplaces identified. Measurements should be taken at the positions of the workers. Ankle, chest and head heights were chosen as measuring points at each workplace. Ten workplaces were chosen as the sample, 'evenly spread' throughout the office. Measurements were taken over a 3 h period under what had been established as 'typical' conditions throughout the morning, a time when complaints had been received. The ventilation systems were identified and set to normal working. Outside weather conditions were noted.

A 150 mm diameter globe thermometer was placed at each workplace (only two were available so they had to be moved around) for at least 20 min before readings were taken. Using a hot wire anemometer, air velocity and air temperature were measured at ankle, chest and head height of the worker. A whirling hygrometer was used at chest height to measure wet and dry bulb temperatures (dry bulb was used as a cross check for air temperature with the air temperature sensor on the hot wire anemometer). The workers' clothing and activity were noted, and movements throughout the room were also noted.

Subjective assessment forms (see Appendix) were handed to each worker and collected centrally (i.e., the working position was noted but a degree of anonymity was maintained). The subjective forms allowed some information to be collected regarding time variations (i.e., outside the survey time) and general satisfaction.

Analysis

Physical measures

Analysis of physical measurements takes place in two parts. The first part is to obtain for each measurement point air temperature, mean radiant temperature, relative humidity and air velocity from the instrument measures and also to determine metabolic heat production and clothing insulation values. The second part is to predict the degree of discomfort. The subjective measures are analysed separately and complement the physical measures.

The air temperature and air velocity were measured directly using the hotwire anemometer. The mean radiant temperature t_r is obtained from globe temperature t_g corrected for air temperature (t_a) and air velocity (v). If the mean radiant temperature is within a few degrees of room temperature then McIntyre (1980) suggests that:

$$t_r = t_g + 2.44\sqrt{v}(t_g - t_a) \qquad (8)$$

where temperatures are in °C and air velocity in m s^{-1}. Relative humidity is calculated from the dry bulb (air temperature) and aspirated (whirled) wet bulb of the whirling hygrometer. Table 4 provides typical values. Table 5 provides metabolic heat production values for typical activities and Table 6 provides clothing insulation values for typical clothing. Useful information is provided by presenting all physical data in a table, or on the office plan in the final report.

Table 4 Relative Humidity (%) from Dry Bulb and Aspirated Wet Bulb Temperature

	Aspirated wet bulb temperature (°C)									
Dry bulb temperature °C	12	14	16	18	20	22	24	26	28	30
12	100									
14	79	100								
16	62	81	100							
18	49	64	82	100						
20	37	51	66	83	100					
22	28	40	54	68	83	100				
24	20	31	43	56	69	84	100			
26	14	24	34	45	58	71	85	100		
28	9	18	27	37	48	59	72	85	100	
30	5	12	21	30	39	50	61	73	86	100

Table 5 Estimates of Typical Metabolic Heat Production Values

Activity	Metabolic heat production ($W\,m^{-2}$)
Seated, at rest	58
Standing, relaxed	70
Standing, light arm work	100
VDU operation	70
Driving	70–100

Table 6 Estimates of Typical Clothing Insulation Values ($1\,CLO = 0.155\,m^2\,°C\,W^{-1}$)

Type of clothing	Clothing insulation (CLO)
None	0
Light summer clothing (briefs, shorts, short sleeved shirt, light socks, light shoes	0.3
Light work clothing (light underwear, cotton long sleeved workshirt, light long trousers, socks, shoes)	0.65
Light business suit (including underclothing etc.)	1.0
Heavy business suit (including underclothing etc.)	1.5

Prediction of whole-body thermal discomfort

Despite some theoretical limitations, one of the most useful thermal comfort indices is the predicted mean vote (PMV) of Fanger (1970) which is used in ISO 7730. Air temperature, mean radiant temperature, air velocity, humidity, clothing and activity values can be integrated to predict the mean thermal sensation vote of a large group of people on a seven point thermal sensation scale (as used on the subjective assessment form in the Appendix). The values range from PMV = 3 (hot) through PMV = 0 (neutral) to PMV = −3 (cold). PMV = 0 provides comfort conditions. From the PMV value a predicted percentage of dissatisfied (PPD) value can be calculated. This is related to the percentage of people likely to complain about the thermal conditions. Values of PMV and PPD are presented in Tables 7 and 8 for typical environmental conditions.

The thermal discomfort results for all ten workplaces will not be presented here; however the PMV and PPD values for each workplace were calculated and labelled on a copy of the plan of the office, for the final report. This showed the predicted whole-body thermal sensation (comfort) pattern over the office and those areas of likely complaint.

The following is an example of calculations for one workplace; the physical measurements were: $t_a = 18°C$; $t_r = 18°C$; $v = 0.15\,m\,s^{-1}$; relative humidity = 50%; clothing insulation = 0.65 CLO; metabolic rate = $70\,W\,m^{-2}$.

Using Tables 7 and 8:

$$PMV = -1.7 \quad \text{and} \quad PPD = 62\%.$$

Table 7 PMV Values for Air Temperature, Clothing and Activity (assume: mean radiant temperature = air temperature, air velocity = 0.15 m s^{-1} and relative humidity = 50%)

		Air temperature (°C)						
Clothing (CLO)	Activity (W m^{-2})	16	18	20	22	24	26	28
0.65	58	–	−2.7	−2.0	−1.3	−0.6	0.0	0.8
1.0	58	−2.1	−1.6	−1.1	−0.5	0.0	0.6	1.2
1.5	58	−1.1	−0.7	−0.3	0.2	0.6	1.1	1.5
0.65	70	−2.2	−1.7	−1.2	−0.6	0.0	0.5	1.0
1.0	70	−1.3	−0.9	−0.5	0.0	0.4	0.9	1.3
1.5	70	−0.5	−0.2	0.2	0.5	0.9	1.2	1.6
0.65	100	−0.9	−0.5	−0.1	0.3	0.6	1.0	1.4
1.0	100	−0.3	0.0	0.3	0.6	1.0	1.3	1.6
1.5	100	0.3	0.5	0.7	1.0	1.3	1.5	1.8

Table 8 Interpretation of PMV Values in Terms of Thermal Sensation and Predicted Percentage of Dissatisfied (PPD)

Sensation	Cold	Cool	Slightly cool	Neutral	Slightly warm	Warm	Hot
PMV	−3	−2	−1	0	1	2	3
PPD (%)	–	75	25	5	25	75	–

Therefore the prediction is that, on average, a person will be between slightly cool and cool at this position. Also it can be seen that for all other conditions remaining the same an increase in air temperature from 18 to 24°C (or an increase in clothing insulation to around 1.6 CLO) will provide a PMV value of 0 required for comfort. This could be a recommendation, or a recommendation could be made in terms of increased activity level or a combination etc.

Local thermal discomfort

As well as overall or whole-body thermal sensation thermal conditions can produce effects on local areas of the body. For example, cold air moving around the workers ankles may cause a draught. The most common forms of local discomfort are caused by cooling due to air movement, heat losses due to asymmetric radiation (e.g., a radiant draught caused by workers sitting next to cold walls or windows) and thermal gradients. There is some debate about conditions which produce discomfort, but cool air movements (especially, if fluctuating) should be avoided above 0.15 m s^{-1}, and particularly for exposed skin areas and if the subject is already cool. Radiant asymmetry should not exceed 10°C (less in the case of heated ceilings) and vertical temperature gradients should not be greater than 3°C. General observation of the workplaces, air velocity measures at the three heights (ankle, chest and head), and mean radiant temperatures will provide an indication of possible local thermal discomfort.

Dryness is probably related to air velocity, humidity and air and radiant temperatures, and is usually due to the evaporation of fluids from the eyes, nose and mouth which can lead to various problems, for example, with contact lenses. Local discomfort and other factors such as dryness and overall satisfaction should also be examined using subjective methods.

Subjective responses

Analysis of subjective responses involves determining the average of, and variation in, response. The responses of how workers felt at the time of measurement can be compared with predicted responses. In general subjects in the office example used, gave a wider range on the scale than the predicted measures. The subjective measures were also presented on a plan of the office in the final report. On average, workers were between slightly cool and cool with some subjects cold and some neutral. Draughts were reported in some areas. Most workers wished to be warmer. Responses regarding general sensation at work were similar to responses made about the conditions when they were measured. Most people were generally dissatisfied with the thermal environment.

Concluding remarks and recommendations

The above measurement and analysis allowed recommendations to be made in a final report which were related to the original objectives. An average increase in air temperature was recommended with some specific recommendations about draughts for particular workstations. It was also noted that the high level of dissatisfaction indicated may be due to general work or workplace dissatisfaction and not simply related to thermal conditions.

References

BSI (1966). *Recommendations for Maximum Surface Temperatures of Heated Domestic Equipment.* (London: British Standards Institution).

BSI (1983). *Medical Information on Human Reaction to Skin Contact with Hot Surfaces.* (London: British Standards Institution).

Bull, J.P. (1963). Burns. *Postgraduate Medical Journal*, **39**, 717–723.

Fanger, P.O. (1970). *Thermal Comfort.* (Copenhagen: Danish Technical Press).

Haslam, R.A. and Parsons, K.C. (1987). A comparison of models for predicting human response to hot and cold environments. *Ergonomics*, **30**, 1599–1614.

Houghten, F.C. and Yagloglou, C.P. (1923). Determining equal comfort lines. *Journal of American Society of Heating and Ventilation Engineering*, **29**, 165–176.

ISO 7726: 1985. *Thermal Environments—Instruments and Methods for Measuring Physical Quantities.* (Geneva: International Standards Organisation).

ISO 7243: 2000. *Hot Environments — Estimation of the Heat Stress on Working Man, Based on the WBGT-index (Wet Bulb Globe Temperature).* (Geneva: International Standards Organisation).

ISO 7933: 1989. *Hot Environments—Analytical Determination and Interpretation of Thermal Stress using Calculation of Required Sweat Rate.* (Geneva: International Standards Organisation).

ISO 8996: 1990. *Ergonomics — Determination of Metabolic Heat Production.* (Geneva: International Standards Organisation).

ISO 9886: 1992. *Evaluation of Thermal Strain by Physiological Measurements.* (Geneva: International Standards Organisation).

ISO 7730: 1993. *Moderate Thermal Environments — Determination of the PMV and PPD Indices and Specification of the Conditions for Thermal Comfort.* (Geneva: International Standards Organisation).

ISO 9920: 1993. *Estimation of the Thermal Insulation and Evaporative Resistance of a Clothing Ensemble.* (Geneva: International Standards Organisation).

ISO 12894: 2001. *Ergonomics of the Thermal Environment Medical Supervision of Individuals Exposed to Hot or Cold Environments.* (Geneva: International Standards Organisation).

ISO 10551: 1995. *Assessment of the Influence of the Thermal Environment Using Subjective Judgement Scales.* (Geneva: International Standards Organisation).

ISO 11399: 1995. *Ergonomics of the Thermal Environment: Principles and Application of International Standards.* (Geneva: International Standards Organisation).

ISO TR 11079: (Technical Report): 1993. *Evaluation of Cold Environments Determination of Required Clothing Insulation, IREQ.* (Geneva: International Standards Organisation).

ISO 11092: 1993 *Textiles—Physiological Effects—Measurement of Thermal and Water Vapour Resistance Under Steady-State Conditions (Sweating Guarded-Hotplate Test)*, Geneva: International Standards Organization.

ISO TS 13732-2: 2001. *Method for the Assessment of Human Responses to Contact with Surfaces* (ISO DTR 13732) Part 2: *Human Contact with Surfaces at Moderate Temperature.* (London: BS).

ISO CD 13732-3: 2001. *Ergonomics of the Thermal Environment—Touching of Cold Surfaces*, Berlin: DIN.

ISO TS 14415: 2003. *Ergonomics of the Thermal Environment: The Application of international standards for people with special requirements.* (Geneva: International Standards Organization).

ISO CD 15265: 2002. *Strategy for Risk Assessment and Management and Working Practice in Cold Environments.* (London: BS).

ISO 7933: 2004. *Ergonomics of the Thermal Environment—Analytical Determination and Interpretation of Heat Stress Using Calculation of the Predicted Heat Strain.* ISO, Geneva.

McIntyre, D.A. (1980). *Indoor Climate.* (London: Applied Science Publishers).

Parsons, K.C. (1993). *Human Thermal Environments.* (London: Taylor & Francis).

Parsons, K.C. (2003). *Human Thermal Environments*, 2nd edition. (London: Taylor & Francis) ISBN 0415237920.

Smith, T.A. and Parsons, K.C. (1987) The design, development and evaluation of a climatic ergonomics knowledge based system. In: E.D. Megaw (ed.), *Contemporary Ergonomics 1987.* (London: Taylor & Francis), pp. 257–262.

Appendix

The following subjective form was used in a moderate office environment where workers had been complaining about general working conditions. Various details about workers' characteristics and location were collected separately. The form was handed to workers for completion at their workplace. Question 1 determines the workers' sensation vote on the ISO scale. Note that this can be compared directly with the measured PMV. Question 2 provides an evaluation Judgement. For example, question 1 determines subject's sensation (e.g., warm). Question 2 compares this sensation with how the subject would like to be. Questions 3 and 4 provide information about how workers generally find their thermal environment. This is useful where it is not practical to survey the environment for long durations. Questions 5 and 6 are catchall questions about workers' satisfaction and any other comments. Answers to these questions will provide information about whether more detailed investigation is required. Answers will also indicate factors which are obvious to the workers but not obvious to the investigator.

Please answer the following questions concerned with YOUR THERMAL COMFORT.

1. Indicate on the scale below how you feel Now.
   Hot
   Warm
   Slightly warm
   Neutral
   Slightly cool
   Cool
   Cold
2. Please indicate how you would like to be NOW
   Warmer No change Cooler

3. Please indicate how you GENERALLY feel at work:
   Hot
   Warm
   Slightly warm
   Neutral
   Slightly cool
   Cool
   Cold
4. Please indicate how you would GENERALLY like to be at work:
   Warmer No change Cooler
5. Are you generally satisfied with your thermal environment at work?
   Yes No
6. Please give any additional information or comments which you think are relevant to the assessment of your thermal environment at work (e.g., draughts, dryness, suggested improvements, etc.).

Chapter 24

Assessment of the visual environment

Peter A. Howarth

Introduction

The scope of this chapter

Vision provides us with more information than all of our other senses combined. In the context of work, an increasing number of jobs involve the use of visually-displayed information and, as a consequence, the environmental conditions necessary to optimise the eyes' performance are of paramount importance.

While the visual sense is in some ways exquisitely fine, in other ways it is dreadfully misleading. Although we can detect an offset in a line of as small as 5″ of arc (the width of a pencil viewed at a distance of 300 m), when indoors under artificial lights we can think that two pieces of cloth are the same colour — to discover on going outside that they are quite different. In the first of these examples, visual performance is dependent upon the quantity of light, whereas in the second it is the quality (the spectral characteristics) of the light which limits performance.

There are many different aspects of the visual environment to consider when trying to produce the conditions necessary for good visual performance, not all of them immediately obvious. Consider an example: visual displays incorporating liquid crystal alphanumeric characters are to be installed in a self-service petroleum pump. What visual considerations are there? Some spring to mind immediately, such as character readability and the positioning of the display so that it can be seen by drivers of different heights, sitting or standing. But the relevant elements of the visual environment are not simply those related directly to the visual task in isolation from its surroundings. Other questions need to be asked: Is the display going to be used both during the day and night? If so, how is it going to be lit, and will supplementary lighting be needed? What is known of the spectral characteristics of the light that will fall on the display? Will it be lit by monochromatic sodium road lighting, and what account needs to be taken of this in the use of colour in the design? If supplementary lighting is going to be used, how should it be positioned to avoid producing veiling reflections in the display, and will discomfort be produced from glare? Is the display going to be covered with either glass or plastic, and if so what reflections will be produced in this cover by artificial lights, by car headlights, or by the sun as it crosses the sky? Clearly the issues involved in assessing the visual environment are not restricted to the ambient light levels!

This chapter discusses conditions of the visual environment that may be of concern to the ergonomist, and indicates why and how they may need to be assessed. Of necessity, the chapter is general in nature: the aim is to provide the reader with sufficient information to enable him or her to approach a wide range of problems with an adequate understanding of the key issues. With the background information provided here, texts dealing with specific issues — such as the CIBSE Code for Lighting (CIBSE, 2002) — should be more readily understood. Finally, further information on the issues considered here, along with full references, can be found in standard texts such as Boyce (2003).

The nature of vision

It is often said that vision runs on light. While it is true that the eye's photoreceptors respond to light, they do not signal absolute light level — adaptation removes this information at the earliest stage of the visual process. Because of its adaptation systems the human eye has a tremendous total operating range, and the illuminance at midday on a Florida beach may be 10^9 times higher than on a dark and cloudy night on the Scottish moors. However, although the eye can operate in either place, at any one time it is restricted to a small part of this range. It is this adaptation that makes us so poor at absolute visual judgements; on the other hand we are usually extremely good at relative judgements. Rather than responding to light level per se, the visual system responds to variation (temporal, spectral or spatial) in light. It is more appropriate to say that vision runs on contrast — a luminance, or a chromatic, change over either space or time.

Contrast describes a variation in light, and this chapter begins with a review of the nature of light and how it affects the visual system. This is followed by a description of the characteristics of natural and artificial light sources, and a discussion of which aspects of the visual environment deserve attention. Finally, those parameters which it is realistic and useful to assess in a practical situation are considered.

How light affects the visual system

Light is a small portion of the electromagnetic radiation spectrum. This spectrum includes radio waves, microwaves, ultraviolet radiation, infra-red radiation, and x-rays. What is special about light is that we can see it. While this seems an obvious statement, there is an important fact to be gleaned from it—namely, 'light' is defined by the human visual system, and not by the light source. If you cannot see it, it's not light. The sun, for example, radiates a wide range of the electromagnetic spectrum, some of which passes through the atmosphere and some of which is filtered out by it. However, the limits of what we call light are those of the human eye [approximately 380–760 nm. (10^{-9} m)] not those of the sun. The spectral sensitivity of the eye, termed the $V\lambda$ function, is shown as a solid line in Figure 1. A photometer measuring *light* will, in effect, measure the total amount of radiation present over the spectrum as weighted by this function. This is done by passing the radiation through filters, so that the combination of the filter and the radiation sensor has the same response characteristics as the human eye.

The effectiveness of electromagnetic radiation in producing the sensation of vision depends upon how sensitive the eye is to the wavelengths present.

The *spectral characteristics* of the light (i.e., the relative amount of radiation at each wavelength) give it its characteristic appearance (i.e., its colour), and the weighted sum of the radiation present at each wavelength determines the *amount of light* present.

A potentially troublesome point to note here is that the spectral sensitivity of the eye is not totally independent of the light level. The eye has two kinds of photoreceptor, which have different operating ranges. One kind are called cones, of which there are three types (to give us colour vision), the other kind are called rods. At normal light levels the cones are the active photoreceptors, and the rods are essentially saturated (their photopigment is highly bleached). However at lower light levels, e.g., around dusk, there is not enough light for cones to operate, and rods become the active photoreceptors. The spectral sensitivity of

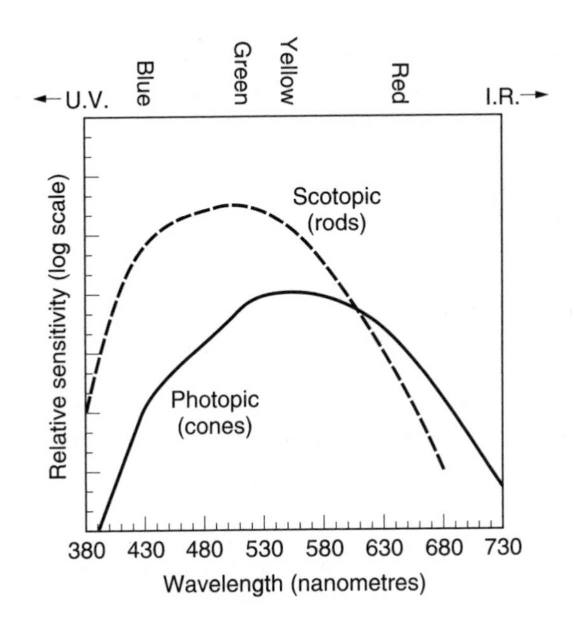

Figure 1 The spectral sensitivity of the eye for daytime light levels (solid line), known as the Vλ function, and the sensitivity for nighttime light levels (dashed line), known as the V$'\lambda$ function. The colour names describe the appearance of each portion of the spectrum (cf., a rainbow).

the rod system differs from that of the normal cone system, and the change in sensitivity of the eye as the light level changes from photopic (cone) to scotopic (rod) levels is termed the 'Purkinje shift'. Purkinje, a Czech physician, noticed in 1825 that while red and blue paint on signposts looked the same brightness during daylight, at night the blue looked much brighter than the red. There are two aspects of the Purkinje shift to note from Figure 1: going from cone vision (solid curve) to rod vision (dashed curve) (1) the peak of the function moves from 555 nm. to 507 nm. and (2) the overall sensitivity of the eye increases.

From the functions shown in Figure 1, we can see that the absolute sensitivity of rods and cones is very similar at long wavelengths. This shows why long-wavelength light is used to 'preserve' dark-adaptation, e.g., on ship's bridges at night: at these long wavelengths, when the light level is high enough for cones to operate there is much less bleaching of rods than there would be at shorter wavelengths.

Other sources of variation in spectral sensitivity exist, and these include differences between stimulus conditions and between people. To standardise light measurement, a number of 'standard observers' with defined spectral sensitivities for particular stimulus field sizes, field positions, and adaptation levels have been specified by the CIE (Commission Internationale de l'Eclairage). The 'standard observer' is a useful concept for defining units and standards and although individuals differ from it, these differences are generally too small to be of practical significance. Unless otherwise indicated, the calibration of a light meter will use the normal photopic function shown in Figure 1.

The three cone types present in the normal eye each have a different spectral sensitivity (the Vλ graph of Figure 1 is a composite function for the whole eye). A given wavelength will stimulate each cone type by a different amount, and it is the relative stimulation of each that gives rise to the percept of a particular colour. The individual colours of a rainbow are seen because each wavelength produces a different ratio of cones stimulation. However, the same ratio (and hence the same colour appearance) can be produced by different combinations of wavelengths. A mixture of long wavelength (red) light and short-middle wavelength (green) light can provide the same relative stimulation as a middle wavelength (yellow) light, and in appearance the two will be indistinguishable. (This is a *metameric* match: two objects (in this case the light sources) appear to be identical in colour even though their spectral composition differs.) However, the colour rendering properties of these two light sources will be very different: under the mixture a rose will look red and a leaf will look green, whereas under the single light both would look yellow.

The *temporal characteristics* of the light source may be of interest when artificial lights are used, either because flicker is noticeable in the environment or because of interactions between the light source and the equipment being used. The visual system integrates light over a finite period of time, and a light flickering faster than the integration time will be perceived as steady rather than flickering. This integration time varies with light level, and so modern films, for example, appear to be continuous rather than flickering, even though their frequency is below the maximum that humans can detect under optimal conditions.

The *spatial characteristics* of tasks and lighting are the final aspect of the visual environment considered here. Luminance and chromatic variation can occur across a room, across a workplace, and at different heights from the ceiling. Directional lighting can facilitate tasks such as inspection. Also, lighting may be deliberately arranged within a room so that features stand out; this can be for safety (highlighting fire extinguishers and exits) as well as for aesthetic reasons.

Spatial variation is not always advantageous, however. At any given moment the eye has a limited operating range, and a large range of light levels within the environment is to be avoided because visual performance will be degraded at the extremes. This effect can be seen in large rooms when sunlight shines through a window, providing lots of light for the areas near the window but leaving the areas away from the window relatively gloomy. Here, without supplementary lighting the details in the shadows may be below threshold, and cannot be seen.

Light sources

It is appropriate here to consider light sources as being either natural or artificial.

Natural light

The sun is the main source of natural light. Direct sunlight, light which has been scattered by the atmosphere to become skylight, and light which has been bounced off the moon all originate from the sun. Three aspects of light from the sun are considered here.

First, the wavelength spectrum of light from the sun is broad, and contains no large peaks. The spectrum of an overcast northern sky is illustrated in Figure 2(a). At different times of the day, and at different places in the sky, the relative amount of energy at each wavelength changes. The most noticeable example of this occurs late in the day, when the emission spectrum of the western sky is predominantly long-wavelength. For colour rendering purposes, daytime skylight has been accepted as 'normal' light, with a clear northern sky providing the spectral standard.

Second, the sun is giving off energy in a manner which can be considered to be continuous. There is essentially no rapid temporal variation in sunlight, with slow natural variations being caused by clouds, by the earth's rotation, and by eclipses.

Third, the sun as a source is very small in angular terms, while the sky as a source is very large, unless walls and ceilings block it out. Although direct sunlight is a lot more intense than skylight, because of this size difference the major component of natural lighting in buildings comes from the sky and not from the sun, even on a cloudless day. This is particularly true in those parts of the world where the sun is reputed never to shine!

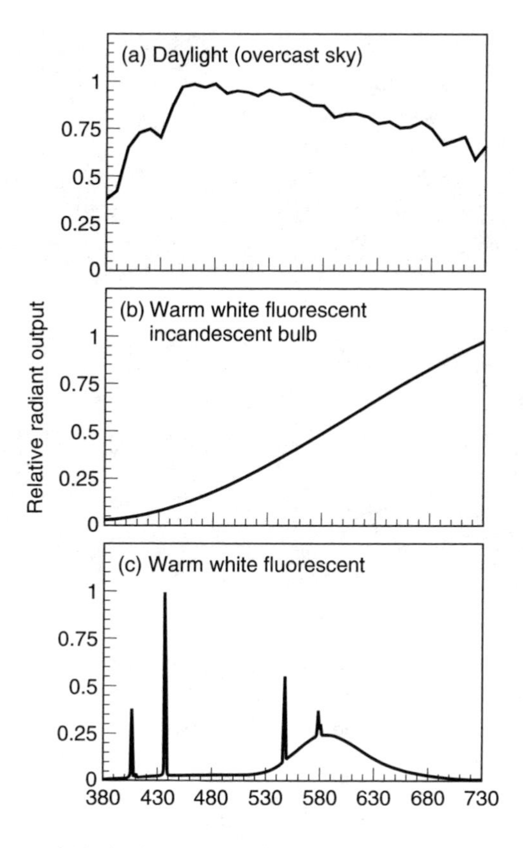

Figure 2 Emission spectra of (a) daylight, and two artificial sources: (b) an incandescent bulb, and (c) a fluorescent tube. Each graph has been normalised so that its peak equals 1.

Incandescent light

When an object is heated, it gives off electromagnetic radiation. If you could take a metal ball and paint it perfectly matt black, then at normal room temperature it would neither reflect nor emit any light. These properties are what is meant by the term 'black body'. If you started heating the ball, it would eventually change colour and start to glow, i.e., it would emit radiation within the visible spectrum. An electric element of a cooker is like this — turn the cooker on and the black element starts glowing red (and also gives off lots of long-wavelength radiation as heat). The metal ball would vary in colour as its temperature changed, and at different temperatures it would have a characteristic (broad band) emission spectrum and, consequently, a characteristic colour. Because of this known variation in colour with temperature, light sources are often specified in terms of their 'colour temperature', which is the temperature at which a black body would have the same colour appearance as the source.

In the same way that a cooker element radiates light as well as heat, a thin piece of wire heated by passing electricity through it will give off light. Again, the emission spectrum of the light will depend upon the temperature to which the filament has been heated. An incandescent light is simply a thin piece of tungsten wire, surrounded by inert gas within a glass container, heated to make it glow.

Figure 2(b) shows the spectral emission of a typical, commercial light bulb, which is yellow-white in appearance. The hotter the wire, the more it glows and the relatively greater amount of light it emits — particularly at short wavelengths (with a consequent rise in its colour temperature). However, the hotter the wire, the more evaporation there is from its surface and the shorter its life expectancy. So there is a trade-off between lamp temperature

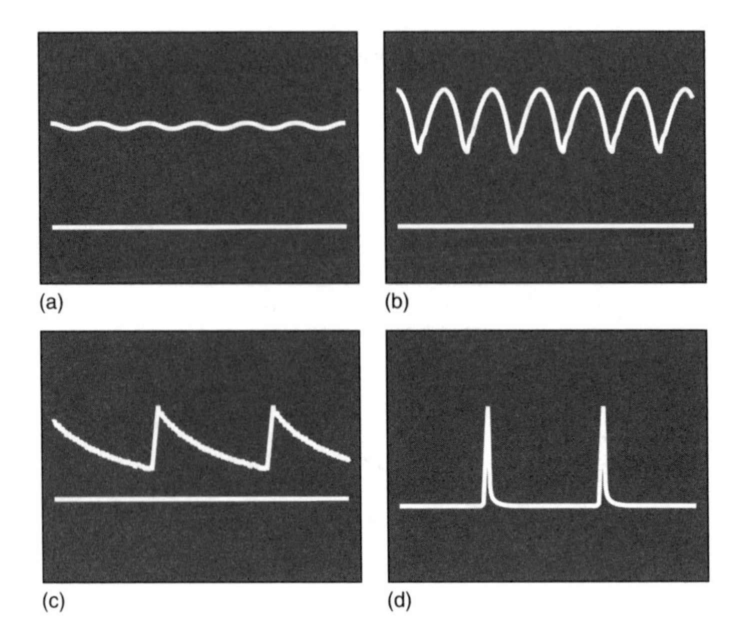

Figure 3 Temporal variation of light output from four different sources: (a) incadescent bulb, (b) fluorescent tube, (c) green phosphor (P31) VDU, (d) VDU. In each case the height above the baseline indicates intensity. Each trace shows variation over 1/20th second: the mains frequency was 60 Hz and the VDUs were driven at 50 Hz.

(and hence light output) and lamp life. A recent improvement in this trade-off has come with the use of halogen gases within the lamp. These combine with the evaporating tungsten during the lamp's operation, and the evaporated tungsten does not become deposited on (and blacken) the inner wall of the bulb. Also, after evaporating into the halogen gas the tungsten is deposited back onto the filament. This increases the lamp's life expectancy, and it can then be run at a higher temperature with both a higher luminous efficacy (i.e., it gives off more light per unit of electricity) and also a spectrum more closely approaching that of skylight.

What about the temporal characteristics of incandescent lamps? An incandescent bulb run from the mains supply (50 Hz in the UK, Australia and Europe, 60 Hz in the USA) will show a temporal variation in its light output. However, Figure 3(a) shows that this variation is small — there is low luminance modulation. Although the filament will alter its temperature with the variation in the electricity, once the light is switched on the filament remains at a high temperature and never has a chance to cool down. Because the electrical variation produces only a small temperature change, there is little (<5%) variation in the light output over each cycle and so flicker from incandescent lights is not a problem.

Discharge lamps

The common fluorescent tube is an example of a discharge lamp. These work on a very different principle from that used in an incandescent lamp. Electrons emitted from a cathode at one end of the tube pass through a pressurised gas to an anode at the other end of the tube. Some are captured by atoms of the gas, thereby raising the atoms' energy level. However, like many high-energy objects these atoms are unstable. They subsequently release electromagnetic energy, and if this energy is within the visible spectrum, it is called light. Unlike the light from incandescent lamps, the emission from discharge lamps is usually at discrete wavelengths. The use of different gases, and different gas pressures, produces a variety of emission spectra.

The two gases most commonly used in commercial discharge lamps are mercury, which is found in the ubiquitous fluorescent tube, and sodium, found in road lights. Because discharge lamps generally emit at only a few discrete wavelengths, this emission by itself does not give satisfactory colour appearance. To improve the colour-rendering properties of fluorescent tubes, the inside of the tube is coated with a phosphor. This coating absorbs energy emitted by the lamp at discrete wavelengths (including the invisible ultraviolet) and re-emits light across a broad band of wavelengths within the visible spectrum. In Figure 2(c) the peaks are caused by the gas emission while the lower, continuous, background is produced by the phosphor. The choice of phosphor will determine the appearance of the emission spectrum, and lamp manufacturers are capable of producing green, blue, and even red fluorescent tubes. Normally tubes are produced with a warm appearance (yellowish-white) or with a cool appearance (blue-white), the use of which depends on the application and taste of the user. For example, to improve the product appearance, store meat counters are often lit by a 'warmer', redder light than would be acceptable for normal interior applications.

Low pressure sodium lights have long been used for roadway lighting. These produce 'sodium yellow' light — almost monochromatic light of a wavelength just below 600 nm. These lamps are now being superseded by high pressure sodium lights which, although they have a broader spectral output, are still spectrally centred around the sodium lines, and appear yellow. These high pressure lamps have a higher luminous efficacy than normal incandescent bulbs, and can provide over 100 $\mathrm{lm\,W^{-1}}$ as opposed to 17 $\mathrm{lm\,W^{-1}}$. For economic reasons they are being used increasingly in applications where accurate colour rendition is not necessary. However. because of the limited spectral output of these lamps they may not always be subjectively satisfactory when used as a sole light source. This situation may be remedied by providing both yellow-white high pressure sodium and blue-white cool fluorescent lamps, which together provide a broader overall spectrum. The combination can be subjectively very pleasant, mimicking as it does the combination of yellow sunlight and bluer skylight.

Discharge lamps may also differ from incandescent lamps in their temporal characteristics. As seen in Figure 3, the modulation of a discharge lamp is much higher (generally about 50%) than that of an incandescent bulb. With alternating current of 50 Hz the electrical signal driving the discharge lamp is, in effect, rectified when light is produced, and the light level rises and falls 100 times per second. This frequency is well above the maximum detectable visually, which for large fields is usually no higher than 60 Hz and, therefore, fluorescent light flicker should not be a problem. Experience tells us otherwise: before the introduction of Visual Display Units (VDUs) into the workplace, fluorescent tubes were the principal topic of visual environmental complaints amongst office workers. However, one reason for these complaints is that sometimes other frequencies are also present in the emission. A mains frequency flicker can occur at the ends of the tubes; this can be remedied with appropriate shielding. Also, after thousands of hours of use, sub-harmonic flicker may be seen over the whole of the tube. This is remedied by tube replacement.

One solution of the problem of fluorescent tube flicker is to stagger electronically the phase of a bank of tubes. When some are fully on others are not, and this strategy reduces the overall modulation. However, this solution is being superseded, and recent developments in ballast (the circuitry driving the tube) technology have enabled tubes to be driven at much higher frequencies (e.g., 20 kHz) than that of the mains, effectively eliminating all flicker. As well as reducing operating costs (by increasing the luminous efficacy of the lights), these high-frequency lamps have been reported to reduce complaints of headaches in offices lit by fluorescent lamps (Wilkins *et al.* 1988).

The other main difference between incandescent lamps and fluorescent tubes is their spatial extent. Although not all discharge lamps are large (some are the size of a normal light bulb) long tubes are commonly found in domestic, industrial and commercial applications. The amount of light given off by any point on the tube is far less than that given off by a point on a naked incandescent bulb providing the same illuminance. For this reason, we might expect fluorescent tubes to give rise to fewer complaints about glare (see later in the chapter). However, because it appears less bright a fluorescent tube will often be left exposed while an incandescent bulb will usually be incorporated into a glare-reducing fitting, and so the comparison is not a fair one.

Self-luminous tasks

Traditionally, the visual environment has been considered as consisting of two elements: the objects being viewed (such as paper copy, the road, instruments and dials) and their lighting (such as daylight, room lights or task-specific lights). However, the increasing use of self-luminous sources, such as VDUs, TVs, FPDs, and LEDs, has led to a reconsideration of this viewpoint because the visual characteristics of the object are far less dependent upon the illuminating light than previously. As an illustration, VDUs are discussed briefly below.

VDUs essentially use the same technology as televisions and other cathode ray devices, where an electron beam scans rows of phosphor dots. The beam is turned on and off, and the phosphor glows or does not glow. Again, the spectral, temporal, and spatial characteristics of the display should be considered.

Spectral

The main monochrome phosphors in present use are amber (Phosphor P134), white (Phosphor P4), and green (Phosphor P31 or P138). If luminance-matched, none has an intrinsic spectral advantage over the others. (The claim that because green is closer to the peak of the human spectral sensitivity curve it is preferable from a user's viewpoint is incorrect if the screens are equiluminant. The advantage of having a phosphor which peaks where the eye is most sensitive is that less radiation is needed to produce a given luminance).

After the phosphor has been excited by the electron beam it emits radiation within the visible spectrum. The VDU phosphor, like that of fluorescent lights, does not decay instantaneously. Rather, it glows for a while, with ever decreasing intensity, after the beam has passed over it. The time course of this persistence varies between phosphors, so the temporal characteristics of all screens are not identical. For example, green (p31) phosphor has a slower decay than white (P4). Whether or not this is preferable depends upon the application and upon personal choice. Figures 3(c) and (d) show the change in luminance over time of a single character on a green and on a white VDU, respectively. In each case, the discrete event of the phosphor excitation and subsequent decay is apparent. The greater persistence of the green phosphor [the long tail of Figure 3(c)] results in a lower average luminance variation over time, or modulation, when compared with the white phosphor. For an application where characters remain for a while in the same place on the screen this reduced modulation may be preferable because screen flicker will be less apparent. However, for an application where characters are being moved frequently the longer-persistence phosphor is often considered annoying, because of the 'ghost image' left briefly behind.

The *type* of ambient illumination, incandescent or discharge, has no spectral effect on the performance of screen-based tasks. (However, there may be spatial differences between them because the larger the area of a lamp (e.g., a fluorescent tube) the more likely it is to be reflected in a screen.)

Temporal

Apparent flicker (see later) is the major identifiable complaint arising from the temporal variation of VDUs. Often, you can see flicker from a VDU or a TV when looking to the side of the screen, but the flicker disappears when you look directly at it. This is because the eye is generally more sensitive to flicker in its periphery than in the centre of its visual field. Also, generally the larger the stimulus the more sensitive you are to flicker, and so it is more apparent on a screen with dark characters and a bright background than on a screen with lit characters and a dark background. (Sensitivity to flicker is context-dependent, and while frequencies as high as 90 Hz can be detected under optimum conditions, 60 Hz is generally taken as the maximum detectable under normal lighting conditions.) The flicker may be less obvious if the screen luminance is reduced, however this may be at the cost of decreasing the visibility of the screen characters, with a consequent decrease in visual performance. The long-term solution to this flicker problem is for the screen refresh rate to be raised.

As well as inherent flicker, older screens often show two problems which are sometimes considered as spatial rather than temporal. These are 'jitter', and 'swim', and each is well-described by its name. The first is a fast oscillatory movement of a character around a central point, the second is a slow drift of the characters, as if the screen was being viewed through moving water. Both are caused by circuit instabilities and inadequate filtering. FPDs do not suffer from this problem.

Spatial

The main issue with VDUs, like other self-luminous tasks, is that there may be conflicting requirements of the ambient lighting. Generally, the higher the ambient light level, the lower the contrast on a screen between a task and its background: the luminance of each will be raised by the same numerical amount, and so the ratio between them will be lowered. However if an operator is reading paper documents, these will require a higher illuminance than is generally recommended for VDU work. If a problem exists, and a reasonable compromise cannot be achieved, the solution could be to provide specific task lighting on the documents. (This solution will also meet the aim of having a higher illumination level on the immediate tasks, with a lowering of illumination level on the surrounding areas.)

Evaluation

Whilst it is possible to evaluate VDU problems objectively, a subjective assessment of the seventy of the problem is generally a more realistic approach. There is no need here for intricate measurements, objective or subjective, as the only real issue to be faced is whether the screen is subjectively satisfactory for the user. This information may be determined from the operator by interview or questionnaire, and a satisfactory investigation of the issue could include both these user responses and an 'expert evaluation' of the severity of the problem. Of course, the ultimate solution to any substantial problem is to replace the screens.

Parameters of interest and their assessment

There are three main reasons why measurement of the visual environment is usually undertaken.

1. Health and safety. The lighting of an area should be adequate to ensure that people can live safely, and it should not in itself be a health hazard. Measurement of the visual environment can provide information as to whether or not these criteria are met.
2. Visual performance. There is a large body of knowledge telling us how visual performance is affected by variables in the visual environment, such as illuminance and contrast. Assessment of these variables can provide information about the expected performance in the location considered.
3. Aesthetic reasons. A pleasant environment is conducive to well-being, and will usually result in less stress and better task performance.

Laws, standards, and guidelines exist to provide information about the requirements for the first two reasons. The third is often a matter of personal preference, although many guidelines and recommendations for aesthetically pleasing lighting seem to have good general support.

The following section discusses the various parameters which might be relevant, and provides background information about each; the taking of measurements is discussed further in the final section.

Amount of light

In the recent past, measurement and specification of light has been complicated by the use of different units in different countries, and by different groups of people within the same country. The text below uses currently accepted SI units. Because alternative units will still occasionally be encountered, for example in instructions for older instruments, and in older texts, descriptions of some previous units, and their conversion factors, have been included here.

Because of the adaptability of the human visual system we are very poor at making absolute judgements (but very much better at making *relative* judgements) and the eye itself is not a particularly trustworthy meter. Hence, light levels are usually measured objectively (although their subjective appearance should not be ignored), and Figure 4 illustrates the different concepts involved. There are two distinct aspects to consider. The first is the question of how much light is present to 'lighten things up' — how much light is falling on this book, or how much light goes into your eye. This is termed *illuminance*. The second is how much light is being given off by something, and this is termed *luminance* (or luminous intensity if the source is small). The difference between illuminance and luminance can be illustrated by closing this book (after reading the next sentence!). The amount of light falling on the book is constant—the illuminance doesn't change—but overall the page reflects more light than the book cover, and so the page has a higher average luminance.

Illuminance

If we took a photocell which faithfully recorded every photon which landed on it and placed it on a desk, then the photocell output would depend upon the number of photons falling on it, in other words the illuminance on the desk surface. Illuminance measures tell you how much light is falling within the photocell catchment area, and nothing else. This amount of light could have been produced by a single bulb, a bank of fluorescent tubes, or by daylight—the source is irrelevant here, the measurement simply tells you how much light there is.

Illuminance varies with the position of the source. Imagine you had a very small light bulb hanging from a wire above a desk. If the bulb was radiating equally in all directions (and no light was reflected from the walls or ceiling) the number of photons falling on the photocell would depend upon three factors:

1. The area of the photocell's collecting surface.
2. The distance from the source to the photocell.
3. The angle between the desk surface and the light source.

Consider these in turn.

1. The larger the photocell area, the greater the number of photons falling on it; make the catchment area twice as big and twice the number of photons are captured. However, the number of photons captured *per unit area* would be constant. Unless there is

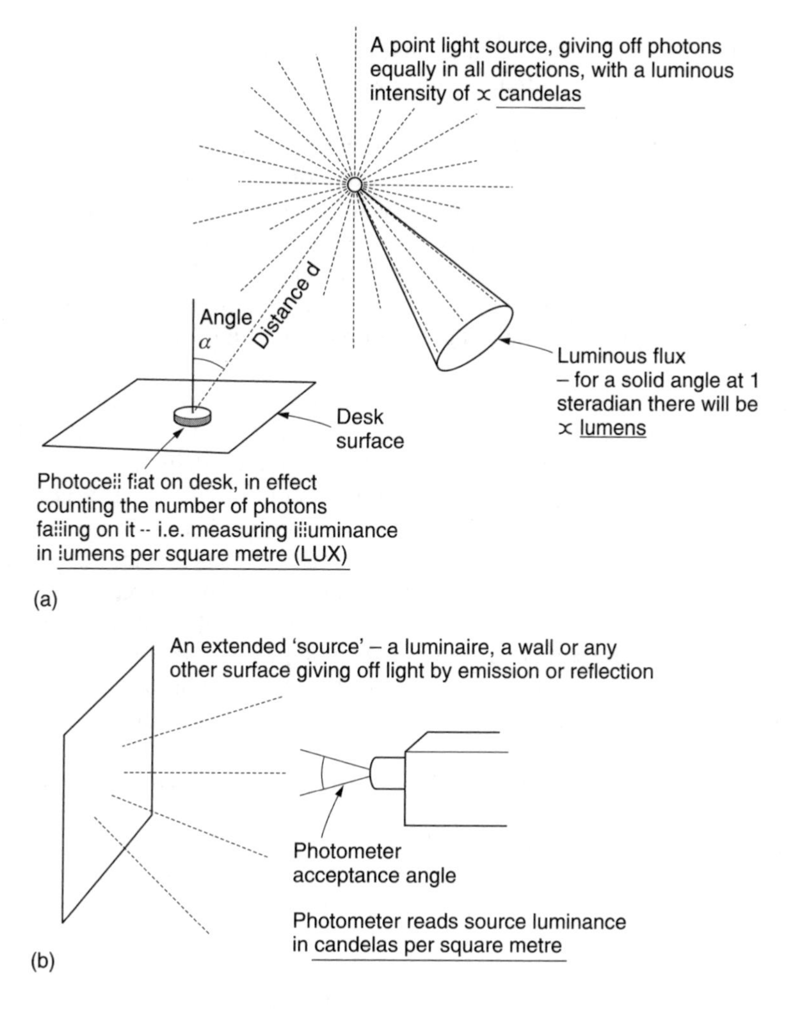

Figure 4 The measurement of light: (a) illuminance, (b) luminance.

variation across the desk surface, the size of the collecting area of the photocell is irrelevant. We can think of an illuminance measurement as being a measure of how many photons per unit area are falling on the desk, and the unit used for measurement ('lumens per square metre' which for convenience is termed 'lux') reflects this fact.

2. The further away the photocell is from the source the fewer photons will fall on it, and we can easily determine the quantitative relationship between the source distance and the illuminance value. Imagine that the photocell had an area of $1\,\text{cm}^2$ and was directly below the tiny light bulb at a distance of 1 m. Because the source radiates equally in all directions the amount of light falling on the photocell is proportional to the 'solid angle' made by the photocell at the source. The solid angle is the area of the photocell divided by the square of the distance from it to the source, and is measured in steradians (sr). So here the $1\,\text{cm}^2$ photocell at 100 cm distance would subtend $1/100^2 = 1 \times 10^{-4}$ sr.

   The point to note here is that the solid angle varies inversely with the *square* of the distance, and from the previous deduction this means that the number of photons falling on the photocell varies inversely in proportion to the *square* of the distance from it to the source. Because the illumination on a surface is expressed in terms of the amount of light falling on it per unit area, the inverse square law follows: the illuminance on the surface varies with the square of the distance between the surface and the source.

To illustrate that *illuminance* depends on the distance from the source to the object being illuminated, imagine a torchlight with a diverging beam shone directly at a wall from a couple of inches away. If you walked away from the wall two things would happen: the circle of light on the wall would increase, and the amount of light falling at any point on the wall would decrease. Double the distance from the torch to the wall and you increase by four the area of wall illuminated, decreasing by four the illuminance (i.e., the amount of light per unit area of the wall) falling on the lit portion of the wall.

3. The above examples have assumed that the photocell was positioned perpendicularly to the source. What if the photocell were angled—which is what would happen if it was placed on an inclined document holder rather than flat on the desk—with the light source directly above it? Here the amount of light per unit area falling on the photocell would be less than if it were perpendicular to the source, and the illuminance varies with the angle that the photocell makes with the perpendicular. If we start with the photocell directly perpendicular to the source the illuminance will be at a maximum value. If we now rotate the photocell so that its face is no longer perpendicular to the source, the illuminance on the face will be reduced by the cosine of the angle of rotation. For example, when the photocell is rotated 60° the illuminance will be reduced by a factor of two, as $\cos 60° = 0.5$.

Illuminance units. The amount of light falling on a given surface area, the luminous flux per unit area, is the illuminance at a point on the surface. The SI unit is the 'lux' which is one lumen per square metre ($\mathrm{lm\,m^{-2}}$). That is to say, when luminous flux (see below) of 1 lumen is spread over a surface area of $1\,\mathrm{m}^2$, the illuminance is 1 lux. An alternative name, not in common use, for the lux is the 'metre candle', while the 'phot' is the illuminance when 1 lumen is spread over $1\,\mathrm{cm}^2$.

The *foot-candle* is a unit still frequently encountered, especially on old illuminance meters; it is an illuminance of $1\,\mathrm{lm\,ft^{-2}}$, and is equal to 10.76 lux.

The *troland* is a unit used in vision research, when an eye with a $1\,\mathrm{mm}^2$ entrance pupil views a surface with a luminance of $1\,\mathrm{cd\,m^{-2}}$ (see below) then the retinal illuminance is 1 troland.

For illustration, Table 1 provides some representative indoor illuminance values recommended by the CIBSE.

Luminous intensity (point sources)

In the previous section we considered the light source to be very small, and to give off light equally in all directions. If this source was so small that it effectively had no area, it would be called a 'point source'. In fact, this is a theoretical concept as all real sources have a finite area to them. However, the idea of radiation being emitted from a single point equally in all directions is a useful one in explaining the concepts of luminous intensity and luminous flux. The more intense a source is, the more light it gives off in all directions: luminous intensity is an attribute of the light source. Now consider a cone with a point source at its centre: the more intense the source the more photons there would be within this cone; in other words, the greater 'luminous flux' there would be within the cone, see Figure 4(a). So luminous flux describes the light itself, and not the source.

An ideal point source gives off radiation equally in all directions. Although real sources can sometimes be assumed to have negligible area (consider the angular subtense of a

Table 1 Examples of Recommended Illuminance Values (from the CIBSE Code, 2002)

Condition	Recommended value (lux)
Rarely visited locations, with limited perception of detail required, e.g., railway platforms	50
Continuously occupied areas, with limited perception of detail required e.g., waiting rooms	200
General offices Airport ticket counter	500
Drawing boards Bench and machine work (fine detail)	750
Cloth inspection Assembly work (fine detail, e.g., watchmaking)	1500
Inspection of extremely fine detail (e.g., small instruments)	2000

motorcycle headlight viewed from a mile away) the property of directionality does not generally hold. Interior and exterior light fittings are often designed with reflectors or shades to produce a certain pattern of light, and the 'hot spots' of a headlight are a good example of this.

Given that the luminous flux varies with direction, it follows that the luminous intensity of a real (as opposed to a theoretical) source also varies with direction.

Luminous intensity units. Luminous intensity is a measure which describes the amount of light emitted by a point source. The SI unit is the *candela* (cd). This basic unit of light is actually defined in terms of the emission of a black body at the freezing point of platinum, 2040°K.

Luminous flux units. When light is emitted from a point source, the 'density of photons' within a cone centred at the source is termed the luminous flux. The number of photons within the cone will be a function of both the source intensity and the cone solid angle, and the SI unit of luminous flux is the *lumen* (lm). This is defined as the luminous flux emitted through a solid angle of 1 sr from a point source of intensity 1 cd. A point source with a luminance of 1 cd will emit overall a total of 12.57 (i.e., 4π) lm.

Luminance (extended sources)

In practice, one generally needs to consider extended sources rather than point sources. Usually a fluorescent tube, a light fitting, or some other source such as the sky subtends an appreciable angle at the eye, and cannot be considered to be infinitesimally small. Hence, luminance is specified in candelas per square metre of the source rather than in candelas. To take the measurement, a light meter with a small acceptance angle is used which will collect light within, e.g., 1°, 0.5°, or 1′, depending on the instrument. It makes no difference how far away the source is, as long as it is larger than the meter's acceptance angle. Consider a luminance meter pointed at a wall, collecting and recording all of the light within its acceptance angle. If we move the meter closer to the wall more photons will be collected by the photocell from any given point. But because the acceptance angle of the meter is fixed, the portion of the wall from which photons are being collected by the photocell is also reduced. Because the increased number of photons collected from each point is balanced by the decreased area of the wall within the meter's acceptance angle the luminance reading remains constant. So luminance, unlike illuminance, is independent of the measurement distance.

Luminance units. Luminance describes how much light is emitted from a surface, and the SI unit is the candela per square metre ($cd\,m^{-2}$). (This unit in the past has been called a 'nit' but, to the chagrin of schoolchildren everywhere, this name is infrequently used). The other unit that may be encountered is the stilb, which is $1\,cd\,cm^{-2}$.

Reflectance

Illuminance and luminance are closely linked. The amount of light falling on a point on the wall is its illuminance, and the amount of reflected light coming back from the wall is its luminance. If all of the light that fell on the wall was reflected, then the values of the illuminance and the luminance would be the same, using appropriate units. If some of the light was absorbed, then the values would differ. The reflectance of the wall may be found by comparing the illuminance and the luminance values, as explained below.

An important point to consider before going further is the directionality of reflections. A perfect matt white surface will reflect light equally in all directions, irrespective of the direction of the ambient light, and is called a perfectly diffuse reflecting surface (Figure 5(a)). On the other hand, a highly faithful reflector, like a mirror, will only reflect light at the same angle to the normal that the ambient light makes (Figure 5(b)). Most surfaces lie somewhere between these two extremes, with a higher luminous flux occurring at the angle of reflection than elsewhere (Figure 5(c)). (Try moving this book and see if you can detect specular reflection on the paper from any light sources.) Finally, if a surface reflects light back along its own path it is termed a retro-reflector.

The greater the proportion of unwanted specular reflection from a surface, the more likely a person is to experience annoyance, discomfort and degraded visual performance. Where specular reflection is likely to be a problem, the positioning of light fittings must take this into account: areas of relatively high luminance should be avoided at places where they might be

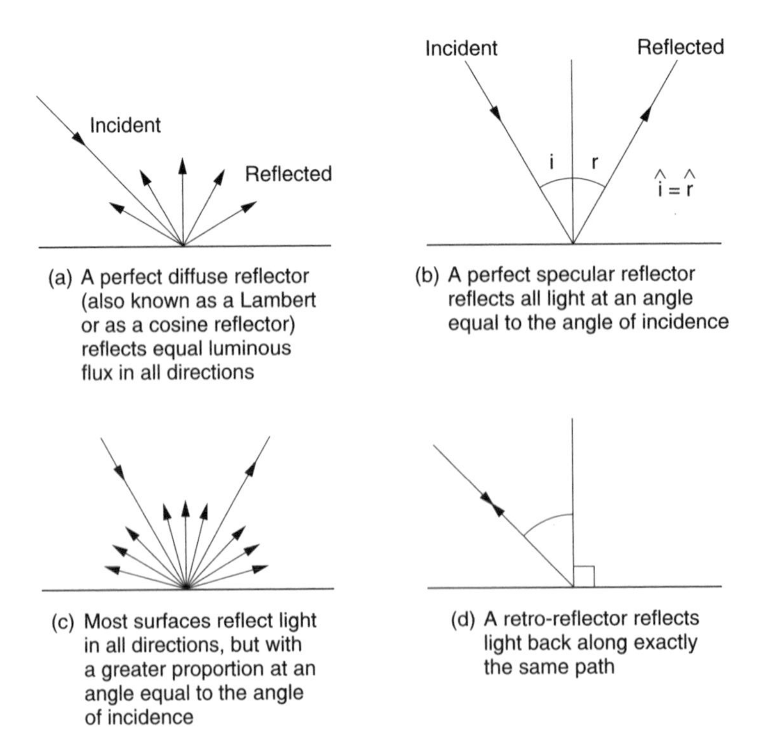

Figure 5 The four reflection: (a) diffuse, (b) specular, (c) mixed, (d) retro-reflection.

reflected into the eye. Where new equipment is used in a previously-designed room, such as new VDUs in an old office, the aim should be to allow for adjustability in the equipment so that specular reflections from the screen can be reduced.

Reflectance units. Irrespective of its angle of incidence, all of the light falling on a perfect diffuse reflecting surface is emitted equally in all directions. No light is lost, and it is intuitively appealing to have units which are defined in such a way that the values for luminance and illuminance are identical here. This is the rationale behind the old (now deprecated) luminance units *apostilb* and *foot-lambert.* When the illuminance falling on a perfect diffuse reflector is 1 lux, the luminance of the surface will be 1 apostilb (asb). It follows that if the surface does not reflect all of the light, then the luminance in apostilbs will be equal to the reflectance of the surface, multiplied by the illuminance in lux. The same rationale applies for imperial units, and with an illuminance of one foot-candle (i.e., $1\,\mathrm{lm\,ft^{-2}}$), the luminance of a perfect diffuse reflector will be one foot-lambert. (It follows that $10.76\,\mathrm{asb} = 1$ foot-lambert.)

The accepted SI luminance unit is the candela per square metre, and for a perfectly diffuse reflecting surface there is a simple relationship between it and the apostilb:

$$1\,\mathrm{cd\,m^{-2}} = \pi\,\mathrm{asb}.$$

So if the illuminance on a wall of reflectance 0.8 is 220 lux, then the luminance of the wall will be:

$$(220 \times 0.8) = 176\,\mathrm{asb} \tag{1}$$

or

$$\frac{220 \times 0.8}{3.14} = 56\,\mathrm{cd\,m^{-2}}. \tag{2}$$

The first calculation is more straightforward, so why is the use of the apostilb deprecated? The reason is that in practice most surfaces are *not* perfectly diffuse, but rather have some specular component. Because of this the wall luminance will vary, and the measured value will depend upon the observer's position.

Instead of using the reflectance of the wall to determine the wall luminance, a quantity called the 'luminance factor' is used. The luminance factor is the ratio of the luminance of a reflecting surface, viewed in a given direction, to that of a perfect white diffusing surface identically illuminated. If the reflecting surface is itself a perfect diffuser, then the value of the luminance factor is the same as the reflectance, is independent of the viewing position, and cannot be greater than one. On the other hand, if the surface does have specular reflections, then the luminance factor will vary with viewing position and at the angle of reflection it could be greater than one. Hence:

$$\text{Luminance }(\mathrm{cd\,m^{-2}}) = \frac{\text{illuminance (lux)} \times \text{luminance factor}}{\pi}$$

The reflectance value of a surface is normally found for 'light', i.e., the whole of the visible spectrum. It is also possible (although not normally needed in practice) to determine the reflectance values for different wavelengths. If this were done, then it would be seen that different coloured walls had different reflectance spectra. For example, a wall which looks red might reflect a large proportion of long-wavelength light, but little medium- or short-wavelength light, while one which looks green might reflect a large proportion of medium-wavelength light but little long- or short-wavelength light.

Flicker

Flicker, noticeable rapid fluctuations in light level, can be a serious problem in artificial environments. Unfortunately, objective measurement of flicker is not simple because it requires rapid-response equipment, normally available only to a lighting specialist.

Subjective assessment of flicker is, however, much more feasible and both the area of noticeable flicker and the degree of noticeable flicker can be adequately assessed by descriptive means. Also, because the periphery of the eye is more sensitive than the central area to flicker, subjective assessment may actually be a more relevant method. Hence, when dealing with flicker the precise circumstances under which it is seen, such as the luminance and the position of the source in the visual field, should be noted.

In considering the subjective assessment of flicker, we should take a lesson from noise assessment (see Chapter 25 of this book). In the same way that the psychological effect of noise can be independent of the sound level (think about the dripping of a tap) flicker can have an annoyance or a distractive effect out of all proportion to its physical magnitude. A subjective assessment of the flicker should not only consider the physical aspects of the stimulus, such as the perceived flicker strength, but should also evaluate the psychological effect that the flicker is having on the person. The positive side to flicker is that because it is very attention-getting, its use is a very good visual method of conveying warning information.

Two further aspects need to be considered. First, some people are especially sensitive to flicker. Epileptics are an extreme example, and for them flicker (particularly at frequencies around 10 Hz) can provoke seizures. Second, flicker which is not visually detectable may still affect parts of the visual system. The human retina responds to flicker at high frequencies (over 100 Hz) even though the light appears steady, and no flicker is seen. It remains to be determined whether other parts of the human visual system are affected by high flicker frequencies, and whether performance or comfort are affected.

Colour

Again, a full spectral assessment of the environment requires specialised equipment. However, this assessment is very rarely needed.

Different light sources have different colour rendering properties, and the colour of an object is determined both by the spectral composition of the light source, and by the spectral reflectance properties of the object and its surround. In this way two objects, like a shirt and tie, can appear identical in colour under one set of lights (a metameric match) but can be noticeably different from each other under another set of lights. The eye is easily fooled under these circumstances. However, while changes in colour appearance like this do occur, the adaptability of the visual system is such that most objects will retain their correct appearance under a wide variety of light sources. This 'colour constancy' can be seen outdoors when a cloud passes over the sun—although the spectral composition of the light falling on the grass and the trees changes drastically, the colours of the objects do not seem to change.

When considering the chromatic aspects of the environment, both the colour rendering properties of the light sources and the pleasantness (or otherwise) of the lighting and the environmental colours must be included. The first of these is likely to have been considered by the lighting designer in an environment where it is important, e.g., where colour discrimination or colour matching are included in the tasks performed in that location. This occurs more often than casual thought would suggest, for example, text and instructions are often colour coded. There is an important point to stress here: colour coding is valueless if the ambient light does not reveal the colour differences. Consider sodium light, which is monochromatic. In this light the three cone types are always stimulated in the same ratio. All objects appear to be the same colour, yellow, and depending on how much of the light they reflect, some are darker than others. Hence, the colour coding of road signs has to be revealed at night by means of supplementary lighting.

Good colour rendering is not always necessary, and other criteria may take precedence. In some situations (such as industrial exteriors and warehouse interiors) the lights may have a simple safety function. Good colour discrimination is not required, for example, if the lighting only has to reveal the presence or absence of objects. Here the cost of running the lights may be a more important criterion than their pleasantness.

In an environment where people have to spend a large proportion of their working time the colour rendering properties of the lights, or combination of the lights, plays an increasingly important role. Lamps with poor colour rendering are generally considered to provide a less pleasant environment than those with good rendering. Of particular importance here is the appearance of skin tones: the better the appearance of skin, the more subjectively preferable is the light. Although 'warm' lamps are generally considered preferable to those with a 'cool' appearance, lamps with a narrow spectral band (such as high pressure sodium) generally provide an unacceptable environment for long-term working. Also of interest here, and a factor where little research has been undertaken, is how people performing different occupations prefer different spectral combinations. Draughtspeople and designers, for example, may tend to use cool, blue-white light, while office work is more commonly performed under warmer, yellower light.

Daylight coming from a northern sky is broad-band (see Figure 2) and is the reference illuminant used for colour-vision testing. Different lamps are compared with this standard as far as their colour rendering properties are concerned, and the CIE have devised a 100 point scale, the 'colour rendering index' (CRI) for lamps. As a generality, the higher the value of the CRI the better the lamp performs (e.g., incandescent lamps may have a CRI of 99, an artificial daylight fluorescent lamp has a CRI of 93, while a white fluorescent lamp has a CRI of 56). However, these are overall values for the lamps, and a lamp with a high score does not necessarily perform well over all of the spectrum, although it should give good rendering of most colours.

Colours will often have specific culturally-based meaning associated with them (e.g., red = stop, green = go).* Here assessment of the use of colour involves measuring the performance associated with the colour rather than the colour itself. Also, visual performance can sometimes be enhanced by using colour to provide a contrast between a task and its surround. Unfortunately, while the concept of colour contrast has some intuitive meaning, (most people would consider the colour contrast between red and blue to be greater than that between yellow and green), as yet there is no widely accepted objective metric by which it can be evaluated. At present, this assessment has to be made on a subjective basis.

*Nature has developed colour coding, as seen in the colouring of wasps and bees: yellow and black hoops warn of a stinging insect (or an insect impersonating one). While 'yellow' is a human sensation (the insect itself is not yellow/black, but rather we sense it that way) we can infer from the colour coding that whatever might otherwise prey on these insects have some form of colour vision. Similarly, the use of colour in camouflage depends upon the colour vision of the observer: two objects may be metamerically matched to a person with normal colour vision, yet to someone with a colour vision defect (or a normal person looking through a coloured filter) they will look different. Because of this, people with colour vision defects have been used in the past to 'spot' camouflaged objects.

Glare

A number of quite distinct lighting-related visual problems, such as discomfort and reduced visual performance, have been grouped together under the heading of 'glare'. These problems have in common the fact that they are all associated with light levels that are relatively high when compared with the ambient light levels. Although different forms of glare may occur simultaneously, they are essentially independent because they do not have the same underlying physiological mechanism. It is not surprising, then, that it is possible to have discomfort without disability, and vice versa, even though both will be often found occurring together.

Discomfort glare

When a portion of the visual field has a much higher luminance than its surround, a feeling of discomfort may occur around the eyes and brow. While this effect has been studied empirically for many years, the physiological mechanisms involved are still unknown. The facial and the iris muscles have been suggested as the site of the discomfort, but possibly neither are involved. When a glare source is turned on, people tend to wrinkle their brow and partially close their eyelids; concurrently the iris sphincter muscle contracts and the pupil constricts. Whether these are simply responses to the increased light and the associated discomfort, or whether they cause the discomfort is not known.

Although the mechanism of discomfort glare is unknown, the conditions under which discomfort occurs have been well established for a number of years. The early work of Hopkinson, Petherbridge, and Guth (Hopkinson and Collins, 1970) revealed that:

Discomfort *in*creases with:

an increase in the luminance of the glare source,
an increase in the angular size of the glare source at the eye.

Discomfort *de*creases with:

an increase in the luminance of the background,
an increase in the angular position of the glare source relative to the line of sight.

By definition discomfort is subjective, and discomfort glare is not easily quantified. A given physical configuration of lights will not only give rise to different reported amounts of discomfort from different people, but also to different reported amounts of discomfort from the same person on different occasions. Subjective assessment in this situation is not particularly reliable. On the other hand, the *physical* parameters of different lighting configurations are, in theory, easy to determine. If it is known how the above four factors (glare source luminance, size, position, and background luminance) interact, it should be possible to measure aspects of the environment and determine on an objective scale how good or how bad the environment is. This is the rationale behind the various glare indices established in different countries—they say little about how an individual will respond, but they do allow an objective evaluation to be made of the lighting configuration. Calculation of the Glare Index involves determining the luminance, size and position of the glare source(s). Full details for the calculations involved may be found in CIBSE, SLL and IESNA publications (see the information list at the end of this chapter).

Despite the apparent validity of these studies in carefully-controlled experimental conditions, the evaluation of the glare indices in practice is complicated. While the luminance and size of the glare source(s) are easily measured, there are real difficulties in determining the value of both the background luminance and the glare source position. First, the luminance of different walls and different parts of the ceiling will vary, and the problem arises of what value constitutes the 'true' background luminance. In practice, an 'average' value has to be estimated. Second, there is not one unique glare index for one room position, but rather one glare index for each position in the room and each position of the eyes.

As someone looks around a room, the glare index will vary. One of the parameters mentioned above is the angle between the line of sight and the glare source. So while a head and eye position may be specified for measurement purposes, e.g., the person may be assumed to be looking down at their desk, or perhaps to be looking straight ahead, each position is arbitrarily chosen and is not one which the person necessarily adopts or has problems with.

Not surprisingly, there is a poor correlation between subjective reports of discomfort and the glare index. The advantage of the index is that it does provide an objective description of the environment. The disadvantage of the index is that this figure does not in itself describe well the subjective discomfort of an individual subjected to that glare.

An alternative way of evaluating glare is to use an 'expert observer' approach, where an individual experienced in glare assessment can evaluate the degree of discomfort they feel in a certain situation. The simplest way of initially determining whether there is a problem is to shield the suspected glare source(s) and see if there is an improvement in comfort. If so, then you can safely assume that there was some discomfort present in the first place. This can then be evaluated more precisely. A number of descriptive scales have been produced, perhaps the most useful of which is that proposed by Hopkinson (Hopkinson and Collins, 1970). Four criterion points were used, each of which describes a transitory position of subjective feeling:

1. Just perceptible.
2. Just acceptable.
3. Just uncomfortable.
4. Just intolerable.

If necessary, the scale can be extended by including end points (imperceptible, and intolerable) and by allowing descriptions between the criterion points (e.g., not quite acceptable, but not yet uncomfortable). This will produce a nine point scale. An alternative method is simply to rate the amount of discomfort on a rating scale, with defined anchor points having specific, known, rating values, e.g., a rating scale where imperceptible = 0, just uncomfortable = 5, and intolerable = 10. (See Chapter 3 for general discussion of rating scales.)

There are a number of ways to reduce discomfort glare. Consider in turn the four factors mentioned previously. Reducing the glare source luminance will reduce the discomfort. This solution may not always be feasible, because the source could be providing illumination for a different part of the room. Increasing the angle between the viewer's line of sight and the glare source, by moving either the source or the person, will decrease the discomfort. Another, less obvious, solution is to *increase* the background luminance. It seems counterintuitive that increasing the amount of light present reduces the discomfort, but this manoeuvre reduces the contrast between the source and the background. A practical way of doing this is by painting the walls and ceilings so that they reflect more light, thereby increasing the ambient light level. Finally, the solid angle of the glare source can be reduced by shielding the source. Changing the luminaire fitting to one with a narrower cut-off angle will decrease the visible extent of the source.

These solutions are not necessarily independent. For example, by reducing the size of the glare source the illuminance it provides for some other purpose may be decreased, necessitating an increase in the source luminance. In the same way, a bare light bulb or a luminaire with a small fitting or shade may be improved by increasing the size of the fitting. Although the glare source solid angle is increased, the source luminance is lowered, with a consequent reduction in discomfort. These are all solutions which are imposed on the person by outside manipulation of the light sources. Where feasible, it is preferable to allow individuals control over their immediate visual environment, with directional local lighting, and in this way enable them to position the light sources in such a way that they are comfortable.

As a final point, the long-term effects of small amounts of discomfort glare are unknown. It is reasonable to suppose that someone working all day under conditions where glare is just perceptible will have a build-up of discomfort over the day. The need for improvement here is greater than in the situation where someone is briefly exposed to 'just uncomfortable' glare once or twice per day. Similarly, someone's idea of how tolerable glare is will depend upon what they are doing, and how interested and involved they are in it. A person watching television may choose to have a low ambient light level but a bright screen — environmental conditions which in other circumstances would be considered intolerable. The point to remember here is that the glare assessment cannot be considered in isolation from the context in which it is made.

Disability glare

While the physiological mechanisms involved in discomfort are not understood, the ways in which an extraneous light source can affect visual performance are quite clear. All involve contrast reduction. The degradation of visual performance occurs in one of two ways: a glare source can act directly by reducing the contrast between an object and its background, or indirectly by affecting the eye.

'Direct' disability glare can occur because of a discrete reflection, such as the specular reflection of a light source from the surface of a VDU screen. Here the luminance of both the object (the characters being viewed) and the background (the surrounding screen) are raised by the addition of the extra light but the contrast is reduced. It can also occur because of a diffuse reflecting veil over the whole of the task, as is seen, for example, when a car windscreen mists up. The whole of the scene looks grey and washed out, and both luminance contrast and colour contrast are diminished. These two examples have in common that the contrast between the object and the background is decreased, with a consequent reduction in object visibility. Hence to reduce the disability one should raise the contrast between the task and the background.

'Indirect' disability glare affects the eye and not the visual task. It is seen, for example, when a car approaches at night with its headlights on full beam and your eyes get dazzled. The disability in this situation has two sources:

(a) there is scatter within the eye reducing the retinal image contrast, and
(b) the adaptation level of the eye is raised as the car approaches. After the car has passed it takes a little while for the eye to re-adapt to the ambient light level.

The fact that scatter within the eye decreases visual performance (see Chapter 20) is particularly pertinent when one considers age-related changes to the eye. The media within the eye generally become more opaque with age, and in particular the lens tends to go 'cloudy'. Although these changes will increase the effect of direct disability glare to a certain extent, the real problems are seen in the presence of indirect disability glare. This is because for the older person opacities within the media will scatter the light to a much greater extent than is found for a younger person. Hence an older person who is able to perform perfectly well under normal lighting conditions may be severely disabled in the presence of indirect disability glare and the reason for this disability may not be visually apparent to a younger person.

Both sources of disability are present in other situations, such as when an object is seen against or near a bright window. They have been termed 'indirect' here to emphasise that they act by affecting the eye rather than affecting the scene. Indirect disability glare is usually found in circumstances where discomfort will also be present because in both cases the eye itself is affected by the glare. Like discomfort, the disability is often reduced by raising the light level. On a city street a car headlight, even when it is not dipped, causes less disability than the same light on an unlit road because the extra ambient light raises the eyes' adaptation level. In the city, the headlight is less bright in comparison with its surroundings, and we can take this reasoning further by considering that the same headlight would produce virtually no disability during daylight.

There is a problem encountered less often where performance is reduced, which can still be considered as glare, and this is where there is an area of low light as compared with the ambient light level to which the eye is adapted. The problem faced here by the visual system is exactly the same as that faced by a photographer encountering deep shadows. If the exposure is set manually for the ambient light the details in the shadows are lost, whereas if the camera is set to record these details everything else in the photograph will be overexposed. The eye's adaptation system acts like an automatic camera, but it isn't possible to vary it in the way you can with a manual override camera. The automatic setting the eye will adopt is appropriate for the ambient light, and not for the shadows, and so details in the shadows are lost to the visual system.

Contrast rendering, and directionality of light

The effectiveness of a lighting configuration is not dependent only upon whether any noticeable disability is present. Despite this, quantification of lighting effectiveness is an area in which laboratory-based research has been applied in the past largely to lighting design rather than to environmental assessment.

The relative positions of the light source, the visual task, and the observer determine how effectively the task contrast is rendered, and recently a measure of lighting effectiveness, the Contrast Rendering Factor (CRF) has been devised. The CRF has been used mainly in regard to paper-based tasks, which is where it is at its most useful. If the task lighting in an office is suspected to be deficient, then measurement of the CRF would be an appropriate way to investigate the problem. Ideally, the CRF is measured by comparing the contrast of the object under the ambient lighting with its contrast under reference lighting (completely diffuse, unpolarised illumination). It is important to realise that the CRF is specific to a particular target, a particular location, and a particular observer position and is not a measure which describes the lighting alone. So the CRF of writing on matt paper will be different from that of writing on glossy paper under otherwise identical conditions. The CIE Publication 19/2 (and Boyce, 2003) provides details of assessing the CRF if access to reference lighting is available.

If reference lighting is not available, do not despair, as two approximate methods have been described by CIBSE. For simplicity, these methods assume 'standard tasks', and, interestingly, because the task is no longer a variable, they do both provide a quantitative description of the lighting configuration. In the first method, the illuminance at different planes at the workplace is measured, and the value of the CRF is found from a calibration graph. In the second method, a CRF gauge is used at the workplace, and the CRF value is read directly.

Generally, the higher the CRF, the more acceptable the visual performance. This might lead one to suppose that the reference lighting conditions, uniform diffuse illuminance, could be considered as 'ideal'. However, this is not the case as CRF is not the only criterion by which the directionality of lighting is judged. Uniform, shadow-free lighting gives an extremely bland appearance to an interior, with the solidity of objects being less readily apparent because of the lack of relief, as shown in Figure 6.

Figure 6 Object viewed under (a) flat and (b) directional light. Taken from (Howarth 2005) with permission.

The directionality and modelling of the lighting at any point in a room may be quantified by combining two measures. One is the scalar illuminance, the other is the illumination vector. These may be measured by suspending a cube (at any orientation) at the position of interest and measuring the illuminance value of each face. Incidentally, in the absence of a cube, this measurement may still be taken: improvise by imagining that you have a cube, and then hold the illuminance meter against each imaginary face. Averaging the values of the six cube faces gives the scalar illuminance. The illumination vector is then calculated by constructing a vector diagram of the differences within each of the three pairs of opposing faces, as illustrated in Figure 7.

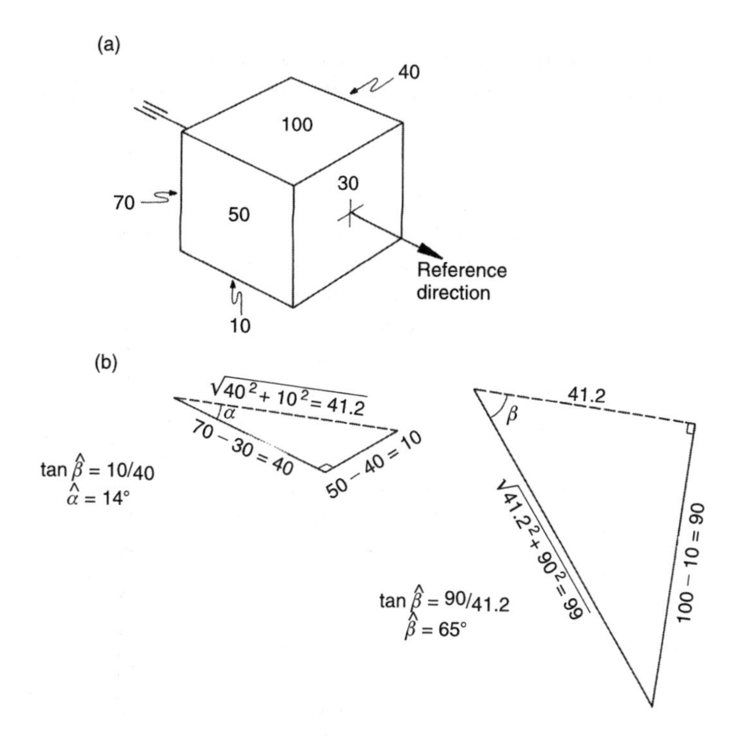

Figure 7 Calculation of the scalar and vector illuminances. (a) The illuminance values are measured on each face of a cube suspended in space, and the scalar illuminance is their arithmetic average (50 lux). (b) A vector diagram can be constructed by first determining the difference between each pair of faces (90, 40, and 10 lux). Then the angle of the illumination vector in the horizontal meridian relative to the reference direction shown in (a) is determined as shown in the left-hand portion of (b); tan $\alpha = 10/40$ so $\alpha = 14°$. The magnitude of this vector is the square root of $40^2 + 10^2$ (the two measurements taken in the horizontal plane) which equals 41.2. The angle in the vertical meridian, relative to the direction just determined, can now be found, as shown in the right-hand portion of (b). Tan $\beta = 90/41.2$ and so $\beta = 65°$. The magnitude of the illuminance vector is the square root of the sum of the squares of these two sides: the square root of $(41.2^2 + 90^2)$ which equals 99 lux. *Answer*: In (a) the illumination vector has a magnitude of 99 lux, in a direction 14′ to the left of the reference direction, and 65′ below this direction. The vector : scalar ratio = 99/50.

Combining the illumination vector value with the scalar value gives the vector/scalar ratio. A value of 0.5 is considered essentially shadow-free, while a value of 3.0 provides very strong modelling, with the details within the shadows often being invisible. Pleasant modelling of human facial features, which is generally an extremely important criterion in relation to the overall acceptability of the visual environment, occurs between values of 1.2 and 1.8.

The directionality of lighting can also be used to reveal relief, and as a means of directing attention. For example, the inspection of textured items, such as cloth, may be enhanced by providing directional lighting which will cause shadows to be cast. The uniformity of these shadows can be easily assessed-defects showing up as an imprecision in their pattern. The elegant role that the lighting plays here is to facilitate performance by changing the nature of the visual task.

A further consideration in lighting uniformity is the illuminance distribution over a workplace. Variation in light level across a workplace will both differentially direct the attention to different areas, and will provide a more pleasing visual environment. For example, the whole surface of a desk may have an apparently-satisfactory illuminance level according to published guidelines, but if there is no spatial variation across the desk it appears uninteresting. This situation is greatly improved by providing slightly higher illumination on the area just in front of the person, where they put their immediate work. When the task is illuminated to a higher level than the surround, the appearance is much more interesting and pleasant.

Daylight

The need for windows in buildings, providing natural light, has come into question recently because of their cost in terms of heat-loss and energy conservation. The scientific evidence for a physiological need for windows is, at best, unproven, however the psychological evidence is clear. A small, windowless room can easily be considered cell-like and restricting, while the presence of a window provides visual access to the outside world. Larger rooms are considered less restrictive, and in a well-controlled environment the absence of windows becomes less important. However, in these windowless environments the information, such as the time of day, and the variety provided by the changing outside light is still absent.

As well as the illuminance variation over the day, the provision of daylight can also provide spatial and spectral variation within a room, again decreasing the monotony of the environment. The illumination variation may be quantified as a change in the 'daylight factor' across a room. The daylight factor is the ratio of the illuminance from the skylight measured on a horisontal surface within the room to the illuminance from the skylight (not direct sunlight) measured on a horisontal plane which has an unobstructed access to the hemisphere of the sky. At different positions within the room the daylight factor will vary, and if required this variation may be assessed over a room. For interiors where the daylight factor variation is not large, such as rooms with skylights, or rooms that are not too deep, and when the average daylight factor is 5% or greater, an interior will appear generally to be well day-lit. Also, if the illuminance from the sky is not known, the relative daylight factor at different positions within the room will give information about the spread of daylight within the environment.

The pleasantness of variation in spectral content and luminance level is not restricted to natural lighting. On the contrary, in some parts of the world the design of mood lighting provides a considerable source of revenue for interior designers and fixture designers alike.

Overview of measuring the visual environment

Standards and guidelines

Lighting guidelines and standards come in two forms, either statutory, where certain legal requirements have to be met, or recommendatory, where the recommendations provided are given as examples of 'good current practice'. In neither case should the values given be assumed to be 'ideal'. For example, in the case of a legal *minimum* requirement for a certain illuminance at a workplace, the minimum required is not necessarily the *optimum* value. Illuminance values which were considered good practice 20 years ago are now, with the advent of more efficient lamps, often considered inadequate.

Most countries have their own standards and guidelines, and specific values will not be presented here. Rather, the reference and further information sections contain details of documents which contain the necessary information.

What measures are needed?

An assessment of the visual environment is going to be needed in one of two situations. The first is a design situation where the concepts discussed previously will have to be considered in a somewhat abstract manner. The example given in the introduction, that of designing a petroleum pump display, is a case in point. The second situation is a more immediately practical case, where an environment already exists and has to be evaluated in some way.

Generally the assessment will be needed because problems are known to be present, and the appropriateness of the measures to be taken depends upon what these problems are.

The preceding sections have discussed the variety of factors of concern in the visual environment. Outside of the laboratory, many of these lend themselves to subjective assessment and flicker is a good example. Objective physical quantification of the frequency and modulation of flicker is complicated, however it is usually unnecessary. The appropriate question may be 'Is there unpleasant flicker present?, a question that has to be answered on subjective rather than on objective grounds.

In an assessment of an existing environment, the people who live and work in that environment are an important resource. As well as providing information about what environmental problems exist, their severity, and the appropriateness of the measures being taken to alleviate them, people familiar with the environment can also provide information which reveals problems which are not immediately obvious. Problems which may occur at different times of the day, or only under certain circumstances (such as when an infrequently used piece of equipment is operating) fall into this category. Once the problem has been established, then the relevant measures can be taken.

An example

Suppose you are approached by a client who feels that the average illuminance within an office is too low at the end of the day, and that some of the desks are inadequately lit. You are asked to make an assessment of the environment, and to suggest improvements. Off you go to the office armed with a luminance meter, an illuminance meter, and a tape measure. What do you then do?

Consider this problem in the following four stages:

Average illuminance within the room

The first issue to be considered in determining the average illuminance within the room is 'how many measurement points are needed, and where should the measurements be taken?' The answers to these questions depend upon the size and dimensions of the room, and the accuracy required. The following method will give accuracy to within 10%.

The first step is to determine the dimensions of the room, including the ceiling height, and to sketch a scale plan of a cross-section of the room. The positions and dimensions of the luminaires should be marked on this sketch. Next, the 'Room Index' (an overall size-measure of the room parameters) is found from the ceiling height and the length of one wall. For a square room, it is numerically half the ratio of the wall length to the ceiling height. If the room is almost square, then take the measurement for the largest wall in determining the wall : ceiling ratio. The minimum number of measurement points can then be found from Table 2.

If the room cannot be considered as square, an extra calculation is involved. First measure the smaller wall length, and the ceiling height. From these figures you can determine the minimum number of points needed within the square defined by the dimension of the smaller

Table 2 Minimum Number of Measurement Points Needed for Different Sized Rooms

Ratio wall length : ceiling height	Room index	Minimum number of points
<2	<1	4
2–4	1–2	9
4–6	2–3	16
>6	>3	25

wall. This then tells you how far apart the measurements points need to be. By applying this value to the room as a whole the required number of measurement points may be calculated. Take the example of a room which is 20 m × 30 m with a 4 m ceiling. If the room were 20 m × 20 m then the wall : ceiling ratio would be 5, and from the above table a minimum of 16 measurement points would be needed. So in the 20 m × 30 m room, which is 1.5 times as large, 24 measurement points would be needed to maintain the distance between measurement points.

Greater accuracy than 10% can be achieved by increasing the number of measurement points, and when time allows the more points taken the better. Doubling the number of points halves the error, and accuracy to within 5% can then be claimed.

Once you have determined the number of measurement points, the next step is to divide the plan of the room into a grid consisting of a number of squares each equal in area. If necessary, the number of measurement points should be increased to ensure that the grid is symmetrical. The idea is to take illuminance measurements at the centre of each of these squares. However, before doing this check from the plan that the measurements are not going to be biased by having a disproportionate number of measurements taken directly underneath luminaires. The illuminance measurements should be taken in the horisontal meridian, at around desk height, without shadows being cast on the meter and the values should be recorded directly onto the sketch plan of the office. In this way, you will also have a visual indication of the illuminance variation over the room to use in conjunction with the arithmetic average for the room as a whole.

The desk illumination

Again, it is a good idea to sketch a plan of what is being measured, in this case each desk. While doing this, note down what measurements need to be determined, and under what conditions they need to be taken. One position on the desk may be identified as being the location where the primary tasks are performed, with other areas being either where objects used less are placed, or simply storage space. If objects on the desk are inclined, in the way that a document holder may be, then the horisontal plane may not be the appropriate place to take the measurement and this should be noted on the sketch. Measurements should be taken with any occupants in place, so that any shadows usually cast by the person are present.

As well as the illuminance falling on the desk, the luminance of the various points of interest can be determined, as measured from the occupant's usual position. Knowing the illuminance on an object, and its luminance, the 'luminance factor' may be calculated (see earlier).

What other measurements are needed?

Depending on the circumstances, further measures could be appropriate. In the present example, the luminances and reflectances of the walls, floor and ceiling should be determined. Wall reflectance is usually important to the lighting of small rooms, but less so for large rooms where the area immediately adjacent to the wall is the only part of the room affected by light reflected from it. On the other hand, the larger the room the greater the importance of the ceiling reflectance. While a small room can have a satisfactory appearance with a darker ceiling, a large room needs to have a large proportion of the light incident on the ceiling reflected back into the room.

Optionally, a subjective 'expert observer' assessment of discomfort glare could be taken from each workplace; the relative daylight factor could be measured at each grid point; the contrast rendering factor could be established; measurements could be repeated to determine the variation over the day, and so on. While these measures might all be outside the original brief, where time and enthusiasm permits, they enable a much more complete description of the workplace to be made and provide a sounder basis on which to make recommendations about appropriate changes.

Assessment of the environment

Now that the measurements have been taken, the values obtained can be compared with values known to represent good practice, as detailed in the CIBS Code. Here the recommendation for standard service illuminance (the mean illuminance over the lifetime of the lamps) for a general office is 500 lux, with a minimum of 200 lux for any working surface. The Code also gives recommendations which are applicable here for illuminance ratios:

(a) The ratio of the minimum illuminance to the average illuminance over the task area should not be less than 0.8. While a greater degree of non-uniformity is acceptable between task areas and areas of the room where no tasks are performed, the illuminance on areas adjacent to the tasks should still not be less than 0.3 of the task illuminance.
(b) In an interior with general lighting, the ratio of the average illuminance on the ceiling to the average illuminance on the horisontal working plane should be within the range 0.3 to 0.9.
(c) In an interior with general lighting, the ratio of the average illuminance of any wall to the average illuminance on the horisontal working plane should be within the range of 0.5 to 0.8.
(d) In an interior with localised or local lighting, the ratio of the illuminance on the task area to the illuminance around the task should be not more than 3 : 1.

By comparing the measured values with the above recommendations, the extent of any problem present in the office may be evaluated, and remedial action may be suggested. Here the extra measurements suggested may be helpful in differentiating between the available solutions. For example, if the wall and ceiling reflectances are acceptable, then the solution may lie in modification of the luminaires, whereas if these reflectances are low the situation could be improved by painting the walls and ceiling.

This example is, of necessity, short and addresses a specific and relatively standard issue. Assessment of other visual environments will involve different circumstances and problems. For example, the criteria by which the design of visual aspects of a car interior is judged will be rather different from those applicable in an office. With this in mind, the next section contains a checklist to indicate the range of problems which might be present in any visual environment, and the questions that may need to be asked.

Checklist

What are the problems?
What measures are needed?
What guidelines are available?

Consider each of the following if appropriate:

Light levels

Luminance and illuminance:

Where is the light needed?
What variation is there?

across the room?
across the workplace?

Is supplementary light needed anywhere?

at any particular time of the day?
at any particular time of the year?
for particular purposes related to the immediate task?
for purposes unrelated to the tasks, e.g., safety lighting.

Surfaces

What are the reflectances of the various surfaces?

the walls,
the ceiling,
the floor.

What are the reflectance ratios between them?

Glare

Discomfort glare:

Is it a problem?

Subjective or objective assessment needed?
Can it be relieved by modifying the environment?
Disability glare:

Is it affecting performance?
Can it be relieved by moving or shielding the lights?

Temporal aspects

Is flicker apparent:

in the visual task itself, e.g., a VDU?
in the environment, e.g., from fluorescent tubes?

Chromatic considerations

Is the colour rendering of concern?

Is the colour rendering of the lights acceptable?
Is the colour appearance of the lights acceptable?

Is colour discrimination a factor for concern?

How good is the colour rendition of the present lights?
Is supplementary light with better colour rendition needed for the task, to enhance colour discrimination? Is colour coding present, and if so are the lights adequate for the colours to be discriminated?

Spatial considerations

Directionality:

Is there a specific need for directional lighting?
What variation is there over the room?
What variation is there over the day?

How is the daylight supplemented by artificial light, and how does this vary over the day?

Are shadows a problem under working conditions?

Highlighting:

Are there any features that need special consideration to increase their conspicuity?

Are there any special-purpose needs, such as safety lighting with a backup power supply?

Users

Are there particular user groups with specific requirements?

Non-visual considerations of the visual environment

How can the environment be maintained under the desired conditions:

What maintenance is needed?

What non-visual effects occur, e.g., heat from luminaires?

As well as the above sources, information can be obtained from the bodies which produce standards in individual countries (e.g., British Standards, American ANSI standards, German DIN standards). Also, International Standards have been established in many areas by the International Standards Organisation (ISO).

References and information sources (annotated)

Boyce, P.R. (2003). *Human Factors in Lighting* (London: Taylor & Francis). This is the primary book for anyone interested in the area.

BS ISO 8995: 2002 Lighting of indoor workplaces (Standard prepared jointly by CIE-TC 3-21 and ISO/TC 159/SC 5.

CIBSE Code for Lighting (2002). This book replaces the old CIBS Code for Interior Lighting. See http://www.cibse.org/

CIBSE publications: These cover a wide variety of issues, e.g., glare, lighting for VDUs, lighting guides for various applications such as engineering, libraries, hospitals. The section of CIBSE dealing with these issues is now known as the Society for Light and Lighting. Information can be obtained from: Chartered Institution of Building Services Engineers, 222 Balham High Road, London SW12 9BS or from http://www.cibse.org/

CIE (Commission Internationale de l'Eclairage) publications: These also cover a wide range of issues, such as colour rendering, light measurement, and visual performance. Information on their availability can be obtained from lighting organizations of most countries (e.g., CIBSE, IES) or from http://www.cie.co.at

Cronly-Dillon, J., Rosen, E.S. and Marshall, J. (1985). *Hazards of Light: Myths and Realities, Eye and Skin* (Oxford: Pergamon Press).

Egan, M.D. (2001). *Concepts in Architectural Lighting* 2nd edition (New York: McGraw-Hill). A comprehensive book covering vision as well as lighting in an easily understood volume.

Galer, I. (Ed.) (1987). *Applied Ergonomics Handbook*: Chapter 9, The Environment-Vision and Lighting (London: Butterworths). Although short and a little outdated, this is still a useful reference chapter dealing with the principles of good lighting, and practical design considerations.

Hopkinson, R.G. and Collins, J.B. (1970). *The Ergonomics of Lighting* (London: Macdonald). Also outdated in parts, but still covers the basic principles well.

Howarth P.A. (2005). Role of vision in falls. In Chapter 5, Haslam R.A. and Stubbs D.A. (eds) *Understanding and Preventing Falls* (London: Taylor & Francis).

IESNA Lighting Handbook. 9th edition (New York: Illuminating Engineering Society of North America). This handbook is described by IESNA as 'an indispensable reference for industry professionals ...known as the "Bible of Lighting."' It currently costs $425 to non-members of IESNA.

IESNA Lighting Ready Reference which is a 'compendium of the most often used materials from the *IESNA Lighting Handbook,* 9th edition' and is less expensive (currently $75) The Society can be contacted at: Illuminating Engineering Society of North America, 345 East 7th Street, New York, NY 10017 or at http://www.iesna.org

Pritchard, D. C. (1999). *Lighting,* 6th edition (London and New York: Longman). This inexpensive paperback is fairly up to date, and is highly recommended for the more technical aspects of lamps and lighting.

Society for Light and Lighting: see CIBSE.

Smith N.A. (2000). *Lighting for Health and Safety* (Oxford: Butterworth Heinemann). An excellent book, written at a level appropriate for students and practitioners. Primarily deals with the issues from a lighting viewpoint rather than from a human factors viewpoint.

Wilkins, A.J., Nimmo-Smith, I., Slater, I.A. and Bedocs, L. (1988). *Fluorescent lighting, headaches and eye-strain.* Presented at the National Lighting Conference, Cambridge, March 1988.

Wyszecki, G. and Stiles, W.S. (1982). *Colour Science: Concepts and Methods, Quantitative Data and Formulae,* 2nd edition (New York: John Wiley). A standard reference work on colour: comprehensive, but not easy reading.

Chapter 25

Auditory environment and noise assessment

Christine M. Haslegrave

Introduction

It is rare to experience silence in the modern world and our auditory environment is very complex. Sound can bring pleasure and information but unnecessary sound or too much sound is annoying, distracting and possibly harmful. Sound in our environment can be generated by transport of all forms, by equipment and machinery, and by people themselves. In the community, this is mainly due to traffic, to neighbours, and to televisions, radios and other domestic equipment. In an office, sound comes from people talking, telephones, computers and printers. In manufacturing industry, it comes mostly from motors, fans, pumps and compressors, moving machinery, contact between tool and workpiece, and resonant plates or housings. Traffic and machinery can also set up vibrations in the structure of the building. All these have to be considered when investigating the sources of unwanted or harmful noise and in finding ways of reducing the noise levels (through suppression or shielding) or of providing hearing protection.

It is not always possible to separate the positive and negative responses to sound. It is quite possible for a sound to be wanted by one hearer and unwanted by many others, and conflicts may arise between differing interests in both living and working environments. Ergonomists often need to evaluate auditory environments which may include sounds which are unwanted although they are not so loud as to be harmful, or sounds which are essential or enjoyable to one person while distracting or annoying to another. For this reason, we tend to distinguish between sound and noise. Kryter (1985) has defined noise in terms of its effects on people as 'audible acoustic energy that adversely affects the physiological or psychological well-being'. As he says, this is consistent with the usual definition of noise as 'unwanted sound'.

Noise levels therefore may be measured to assess whether there is a problem in terms of health, safety, communication, working performance, comfort or annoyance. Equipment may be needed to measure sound levels of individual sources and of the environment, exposure over a period of time, hearing ability and parameters relevant to risk of hearing damage. Correspondingly, an equally wide range of measures have been defined for such evaluations. This chapter describes the methods used for measuring sound (or noise) and some of the more common measures and criteria which have been proposed for assessing noise. The aim is to introduce various techniques used by ergonomists, but without attempting to cover specialist applications or the concerns of acoustic engineers who deal with noise control, architectural acoustics or design of communication equipment. Similarly, there is no intention of covering either the physics of sound or physiology of hearing, other than to introduce the necessary concepts and terms.

For a more detailed study of the ergonomic aspects of the auditory environment, recommended texts dealing with hearing, perception of sound and the effects of noise on work and health include Jones and Chapman (1984), Kryter (1985, 1994), Loeb (1986), and Sanders and McCormick (1992). Guidance on acoustic treatments and techniques for reducing noise levels can be found in a series of case studies published by the UK Health and Safety Executive (1995) as well as in technical noise control handbooks.

Measurement of sound

Sound is a variation of pressure in the air (or in any other elastic medium) which the human sense of hearing detects. It is transmitted in the form of pressure waves as a series of compressions and rarefractions travelling outwards from the source of the sound. The speed of the wave depends on the medium, but in air at 20°C sound waves travel at a velocity of 344 ms^{-1}. The amplitude of the sound wave is the fluctuation above or below the ambient air pressure. Our sensation of loudness however does not come directly from the air pressure but from the intensity of the sound wave, which is the rate at which sound energy is transmitted in the wave. This is defined in terms of the energy passing through a unit area in unit time (or sound power per unit area). Air pressure can be measured directly and it is usual to measure sound levels in terms of sound pressure, and then to calculate sound intensity from these measurements, as described later.

Auditory stimuli at the ear result from the combined signals received from all sources in the environment. The waveforms (or variation in amplitude with time) of typical sound signals are shown in Figure 1. A pure tone (which might be obtained from a tuning fork or a vibrating string) oscillates at a single frequency and can be represented as a sine wave, but most sounds are made up of complex signals containing many frequencies. Complex signals may be analysed into their component sine waves (each with a characteristic frequency, amplitude and phase) to understand the content of the signal. This process is known as frequency or Fourier analysis, giving a frequency spectrum that indicates the principal frequencies contained within the signal and their relative intensities (energies). When a sound is made up of frequencies covering most of the audible sound spectrum, it is known as white or broad-band noise. An impulsive signal (such as a door slamming or a hammer blow) is a single pressure pulse with a very fast rise time (around 25 ms or less) to the initial peak amplitude, followed by small pressure oscillations decaying over about 1 s.

The audible range of sound levels is enormous, from a quiet whisper to a warning siren (which is close to the pain threshold), representing a range of power in the sound signal of over one to one billion as can be seen in Table 1. Similarly, the ear is sensitive to a large range of frequencies — the audible range is approximately 20 Hz–20 kHz, with greatest sensitivity between 2–5 kHz. Most speech is between 300 and 700 Hz, with all vowel sounds below 1 kHz, but sibilant consonants may be higher than 5 kHz. Low frequency sounds below 20 Hz, usually termed infrasound, only seem to cause adverse effects (resonant vibration in chest, throat and nasal cavities, and pain in the ear) at very high intensity (Kryter, 1994) and this frequency range is not usually considered when assessing the auditory environment.

Definition of units

Several units are used for sound measurement, and it is helpful to give some brief definitions to show the relationships between the various basic measures. A fuller description can be found in texts such as Kohler (1984).

Sound is described by intensity and by frequency. (The corresponding sensations in human hearing are termed loudness and pitch.) Intensity or power of the oscillations in the air

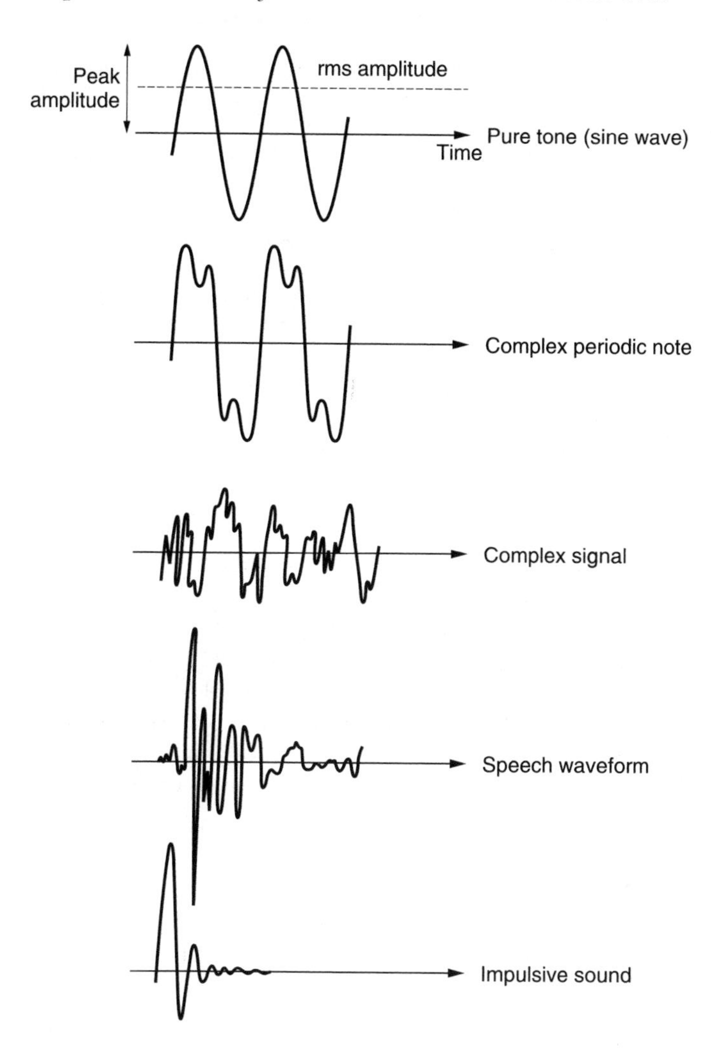

Figure 1 Waveforms of various sound signals.

is measured in watts per metre^2 ($W\,m^{-2}$), while pressure is measured in newtons per metre^2 ($N\,m^{-2}$); frequency is measured in hertz (Hz).

The total sound power of a sound source (such as a machine) can be measured and is expressed in watts (W).

In a sound wave emitted from a source, the power and pressure are related according to the following equation:

$$\text{Power/unit area} = \text{energy flow/unit area} = \frac{\text{pressure}^2}{\text{density of air} \times \text{speed of sound}}$$

The sound power level is therefore proportional to the square of the sound pressure.

From the sound signals in Figure 1, it may be seen that the sound amplitude fluctuates rapidly over time, so the root mean square value (rms) of the pressure is the measure normally used. This is derived from the time average of the squared values of the instantaneous pressures over the duration of the measurement.

These are the units used for physical measurements of sound, but the auditory characteristics of the human ear are such that the corresponding sensations of loudness and

Table 1 Range of Intensity and Pressure of Audible Sound

Intensity $W m^{-2}$	Pressure $N m^{-2}$	Decibel level	
10^{-12}	0.00002	0	Hearing threshold
3×10^{-6}	0.04	65	Conversation
10^{-4}	0.2	80	Town traffic
10^{-2}	2	100	Workshop
3	36	125	Jet at take-off (60 m away)
100	200	140	Pain threshold

pitch are not linearly related to the intensity and frequency of the sound. Measures of these sensations are therefore given separate units (which are defined later): loudness is normally measured in phons, and pitch is measured in mels.

Comparison of sound levels

Sound levels are usually measured relative to other sound levels, or to a base reference level. For this, a unit called the decibel (dB) has been defined as the ratio of their intensities or powers. (The original unit was the bel, which represented a 10-fold increase in intensity, but this was found to be too large in practical applications.) The decibel is a logarithmic unit and thus corresponds quite closely to the non-linear response of the human ear.

The base reference sound intensity is chosen as the power of a standard vibration in the air which is just on the threshold of hearing, or one billionth of a watt per square metre ($10^{-12} W m^{-2}$). Thus, a sound level of NdB is related to its intensity of $I W m^{-2}$ by

$$N \text{ dB} = 10 \log_{10} \frac{I}{I_0}$$

where I_0 is the reference level intensity at the threshold of hearing.

Given the relationship between sound intensity and pressure,

$$N \text{ dB} = 10 \log_{10} \frac{I}{I_0} = 10 \log_{10} \frac{p^2}{p_0^2} = 20 \log_{10} \frac{p}{p_0}$$

where p is the sound pressure level and p_0 is the reference level amplitude at the threshold of hearing ($2 \times 10^{-5} N m^{-2}$).

The values of intensity, pressure and decibels for some typical sounds within the audible range are shown in Table 1. The relationships between the three scales are shown in Table 2. A sound which is 10 times louder than another has an intensity level which is greater by 10 dB. Since the decibel scale is logarithmic, a sound 100 times louder is said to be 20 dB louder. The smallest change in loudness that the human ear can discriminate is 1 dB, but in practice the minimum difference needed to recognise a sound above the background noise level is 3 dB, which represents a doubling in intensity or loudness. A doubling of the sound pressure level is equivalent to an increase of 6 dB in sound level.

Since the various units are related in this way (with loudness or sound power proportional to the square of sound pressure), a useful guide to differences in sound levels is given by:

$$+3 \text{ dB} = 1.4 \times \text{sound pressure level} = 2 \times \text{sound power level}$$

$$+6 \text{ dB} = 2 \times \text{sound pressure level} = 4 \times \text{sound power level}$$

$$+20 \text{ dB} = 10 \times \text{sound pressure level} = 100 \times \text{sound power level}$$

Table 2 Decibel Scale

Ratio of intensities (powers) of two sounds	Ratio of pressures of two sounds	Difference in decibels between the two sound levels
1	1	0
10	3.16	10
100	10	20
200		23
400		26
600		28
800		29
1000	31.6	30
10000	100	40
100000	316	50
1000000	1000	60
1	1	0
1/10	1/3	−10
1/100	1/10	−20

Table 3 Addition of Two Sound Levels in Decibels

Difference (dB) between sounds	Add to the louder sound (dB)
0	3
1	2.5
2	2
4	1.5
6	1
8	0.5
10	0

Effects of several sources of sound

What then happens when two sounds are heard at the same time? How are their effects added? Sound levels are not directly additive when they are measured in logarithmic dB units. If two sounds are received at the ear simultaneously, the resultant is the sum of the energies $(I_1 + I_2)$ in the two sounds and so the combined sound power level becomes:

$$\text{combined sound power level (W m}^{-2}\text{)} = I_1 + I_2.$$

Thus, the decibel level is:

$$\text{combined sound level (dB)} = 10 \log \frac{(I_1 + I_2)}{I_0}.$$

The combined loudness of sounds from two or more sources can therefore be calculated by using this formula, but a simpler rule of thumb (accurate to 0.5 dB) is given in Table 3. If the two sounds are of equal intensity (difference 0 dB), their combined intensity is 3 dB higher than the intensity of either. If the difference between the sounds is 5 dB, the combined sound is approximately 1 dB higher than the louder of the two. If the difference between the sounds is 10 dB, the combined sound is approximately equal to the louder sound and there is no noticeable difference in the intensity of the resultant sound. For example:

two sounds of 75 dB combine to give a sound level of 78 dB,
two sounds of 75 dB and 80 dB combine to give a sound level of 81 dB,
two sounds of 75 dB and 85 dB combine to give a sound level of 85 dB.

As a practical consequence of this, the sound of a machine or piece of equipment can be measured even in the presence of background noise, provided that the background noise is at least 10 dB below the machine noise level. If the background noise is greater than this, a correction factor for the measured machine noise level can still be calculated as shown above. It is also useful to note that the ratio of two sound levels is calculated by subtracting the level in decibels: this gives a signal-to-noise ratio as the difference in decibel levels between the desired signal and the unwanted noise.

Measures of noise

It is obvious from this brief review that measurements of sound or noise levels will be influenced by the temporal, intensity and spatial characteristics of the signals. The temporal characteristics can include both the frequency spectrum and fluctuations in the overall sound level with time. A variety of measures have therefore been developed for specific purposes and the most important are discussed below.

Subjective measures

Since the response characteristics of the ear are non-linear, both frequency and intensity affect our perception of loudness of a sound. For example, we do not perceive a sound level at a frequency of 50 Hz arriving at the ear to be as loud as if the signal frequency had been at 1000 Hz. Curves of equal subjective loudness can be plotted, giving the sound levels which provide constant perceived loudness at various frequencies and indicating the combinations of intensity and frequency which appear equally loud to a human listener. This shows that the 50 Hz tone must have a loudness of about 85 dB to give the same subjective loudness as a 1000 Hz tone at 50 dB.

Subjective loudness is measured in a unit called a phon. This unit is equivalent to the decibel at 1000 Hz and is also logarithmic. While the phon measures the subjective equality of sounds, a unit called the sone was defined to measure the relative loudness of sounds. One sone is defined as the loudness of a 1000 Hz tone of 40 dB (40 phons). A sound of 2 sones is one judged to be twice as loud. The phon and sone scales are related by the following formula:

$$\text{phons} = 40 + 10 \log_2 \text{sones}$$

Every increase of 10 phons then doubles the loudness in sones, so that, for example, 50 phons is equivalent to 2 sones.

There is also a subjective measure of pitch, which corresponds to frequency for a pure sinusoidal tone and to the fundamental frequency for a complex waveform. Perceived pitch is given a unit called a mel — defined by a pure tone of frequency 1000 Hz at a sound pressure level of 60 dB, which is said to have a pitch of 1000 mels. Any two tones separated by a given number of mels appear equally far apart in pitch, regardless of their frequency.

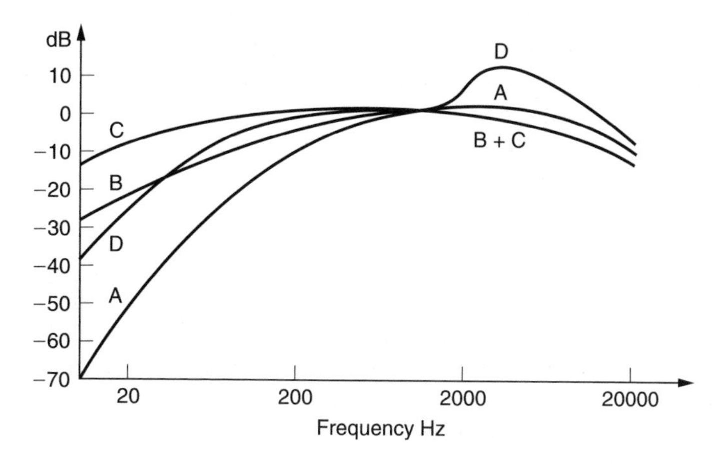

Figure 2 Frequency weighting characteristics.

Weighted measures

In view of the non-linearity of hearing response, most instruments are designed to measure sound levels on a decibel scale which is weighted to match the characteristic of the ear. Several scales have been developed for different purposes and their response characteristics have been standardised internationally. The A-, B- and C-weighted scales were developed to match the responses for sounds of low, moderate and high intensity. The most commonly used is the dBA scale, which gives the best correlation with subjective tests of perceived loudness, and also with ratings of noise annoyance. The A-scale was in fact designed to match the 40 phon equal loudness contour.

This means that the actual measurement of the sound pressure level is converted to a weighted dBA sound level, in accordance with the response characteristic shown in Figure 2 which takes account of the middle range of frequencies to which the human ear is most sensitive. The other scales shown in Figure 2 are less commonly used. The B-scale was designed to match the equal loudness contour at 70 dB, and the C-scale for a flat rating across the frequency scale. There are also D-weighted scales which were designed as equal noisiness scales for assessing aircraft noise. It is normally only for frequency analysis that unweighted sound levels are used.

Empirical measures

In most everyday situations noise contains sounds from various sources, of differing frequencies and intensities, and also extending over different periods of time. In order to describe the noise environment, various statistical distribution measures are used. These are based on noise levels measured on a dBA scale. The most important are defined as follows.

Equivalent level of sustained noise (L_{eq}). This is the average level of sound energy over a given period of time, which integrates all the fluctuating noises to represent them as an average steady level. It therefore takes account of short but high peaks.

Median noise level (L_{50}). This is the noise level which is exceeded for 50% of the time.

Background noise level (L_{90}). This is the 10th percentile level — the level that is exceeded for 90% of the time.

Peak noise level (L_1 or L_{10}). This is the 99th or 90th percentile level — the level that is exceeded for 1% or 10% of the time.

All these indices can be used to give a measure of the total environment in, say, an office or a factory. The last three give a feel for the range of the noise — the 'average' level, the low background level and the highest level heard over a period of time. They are measured by taking a continuous recording of the dBA level over a known time period and then performing a statistical analysis of the record.

Two other measures are also commonly used.

Day-night equivalent level (L_{dn}). This is a 24 hour L_{eq} used for assessing community noise exposure, where the value of L_{eq} is measured over 24 hours but the night-time readings between 22.00 and 6.00 hours are weighted with the addition of 10 dB.

Sound exposure level (SEL or L_{AE}). This is useful for comparing unrelated noise events in terms of their total acoustic energies: the energy is integrated over the duration of each event and expressed as the equivalent level over 1s. It is also often used to describe the noise energy of a single event such as a passing car.

Frequency analysis

Frequency analysis is used when detailed information is needed about a complex sound signal. It can help to isolate possible sources of noise in machinery or in the wider environment. It is also used to evaluate the relative contributions of different frequency components when assessing the risk of damage to hearing. The most common analysis is to plot a frequency or power spectrum, which displays the sound energy across the range of frequencies. The frequency range can be split into frequency bands (usually one octave or 1/3 octave wide) which are analysed separately.

An octave is an interval which represents a doubling in frequency: the upper frequency bound is twice the lower frequency bound and the centre frequency is taken as the geometric mean of the two bounds. The octave bandwidths used for industrial sound analysis are normally similar to those forming the musical scale, but the two sets of octave bands are not identical. The centre frequencies of the two sets are:

Musical octaves
32 64 128 256 512 1024 2048 4096 Hz

Industrial sound octaves
37.5 75 150 300 600 1200 2400 4800 9600 19200 Hz

When more accurate frequency analysis is required, $\frac{1}{3}$ octave bands can be used by logarithmically dividing each octave band into three. Narrow band analysis (using a spectrum analyser) gives still more detail. The bandwidth is usually defined as a fixed percentage of the frequency to be analysed. For example, a 6% frequency band at centre frequency 500 Hz would have a bandwidth of 30 Hz covering the range between 485 Hz and 515 Hz.

Noise measurement procedures

Noise measurements can be taken in order to survey the levels in a working area (producing a noise map, as illustrated in Figure 3) and identify any hazards to hearing for people in that area, or to assess the noise level produced by a particular source.

General guidance on procedures is given below but standardised test procedures have also been developed to ensure accurate and comparable measurements for specific applications. ISO 3740 (ISO, 2000) is a guide to noise problems, covering the general procedures used in the

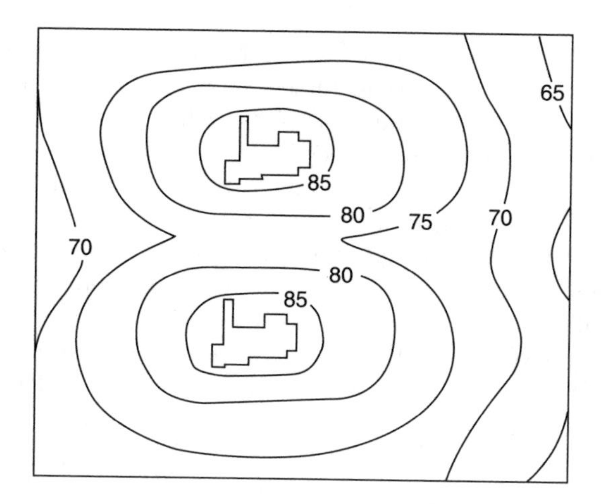

Figure 3 Noise map in machine shop.

measurement of noise and the evaluation of its effects on human beings. This document lists ISO standards which are applicable to specific problems. In addition to international standards, national and industrial standards may apply.

Instruments

Sound level meters are the most commonly used instruments for measuring the auditory environment. Some incorporate filter sets giving octave band or $\frac{1}{3}$ octave band frequency analysis. They record the sound level in each filter band to give the frequency content of the sound. In specialist applications, wave analysers may be used for more detailed narrow band frequency analysis. Integrating sound level meters are capable of measuring over longer periods of time and of providing statistical distribution analysis of the temporal characteristics of the sound levels.

Dosemeters are small integrating sound level meters which can be carried in a pocket or attached close to the wearer's ear and are used to measure the total personal exposure to noise during a period such as a working day. These give the total A-weighted sound energy received during the measurement period and may be used to calculate the L_{eq} value. The dose is often displayed as the proportion of the maximum permitted 8 hour dose and there may also be an indication of whether a standard peak noise level has been exceeded.

Probe microphones are used for measuring sound in the ear, for instance for evaluating the effect of hearing protectors. An alternative method of measuring the sound in earphones is to use an acoustic coupler (artificial ear), which is a standard cavity having an acoustic impedance similar to that of the ear.

Field measurements

Noise of individual machines or products may be measured in the laboratory or in an anechoic chamber if precise measurements are required. For most purposes, however, this is not necessary and it is sufficient to use a convenient period in the workplace itself. Environmental measurements obviously have to be carried out in the field and under normal conditions.

There are obviously problems in measuring individual noise sources in a workplace or in a town environment, due to the presence of other noise sources. It may be possible to choose a quiet period at night or at the weekend, when very little other machinery is working, but it is unusual to have no background sources of noise. Even at night there is likely to be traffic noise, and in the built environment there will be heating, air conditioning, refrigeration or other plant running. However, reference to Table 3 shows that this will usually have a small or negligible effect on the noise measurement. If the difference between noise source and background is less than 3 dB (in any frequency band), or if the noise level is below that of the background, the noise level cannot be measured reliably. If the difference is between 3 dB and 10 dB, the effect of the noise source can usually be calculated to an acceptable degree of accuracy by comparison with the background level measured with the noise source switched off. This is more difficult when background noises are intermittent or fluctuating, and in any field survey care has to be taken to monitor this during the measurement period.

Outdoors several factors may affect the measurements. Wind is the most common problem, since this causes air turbulence around the microphone and low frequency noise. It is usual to shield the microphone with open-cell foam when taking measurements outside, although this is not fully effective at higher wind speeds. Wind, temperature and humidity can also affect the actual sound levels, as they change attenuation over distance and can create shadow zones. A fuller discussion of these atmospheric effects, and the attenuation produced by surroundings and barriers or walls, can be found in Beranek (1971).

Laboratory measurements

Although background noise is likely to be less of a problem in the laboratory, measurements there are liable to be affected by the structure of the enclosed space (by walls or other objects in the acoustic field). Sound may be reflected off hard surfaces or alternatively may be absorbed by fabric surfaces or acoustic tiles. When sound is reflected, the distribution of sound levels in the room will depend on the phase relationships between the incident and reflected sound waves. This may increase noise levels or interfere with speech and other signals. Sound can be reflected several times from the walls of a room and cause reverberation. (Reverberation time is the time taken for the resultant sound to decay.)

In enclosed spaces, there are three distinct regions around a noise source:

Near field where there may be interference between the sound wave being emitted and reflections from the room, so that the sound level varies slightly with the location of the meter. The near field is considered to extend over a distance approximately equal to the wavelength of the lowest frequency emitted or to twice the greatest dimension of the source machine (whichever is the greater).

Far field, which approximates to free field conditions. Here the noise intensity decreases with distance from the source, following an inverse square law. This region can be identified by checking whether the sound level measurements obey this law.

Reverberant field close to reflecting surfaces, where the reflected noise is diffuse and noise levels are higher than in the free field and depend on the geometry of the room and the absorption properties of the surfaces. Measurements of the noise levels from the source will not be accurate in this region.

Two specialist types of noise laboratories are found: the anechoic chamber and the reverberation chamber. In an anechoic chamber (such as the one shown in Figure 3), all surfaces are covered with highly sound absorbent material so that there are no reflections or echoes. This simulates the free-field conditions experienced outdoors. A reverberant room is one in which the walls completely reflect sound energy and where no walls or surfaces are parallel. This gives a diffuse sound field in which sound energy is uniformly distributed. This is most suitable for measuring total power output of a noise source. Both the anechoic chamber and the reverberation chamber can be used to determine the sound power of a source, but an anechoic chamber can also be used to determine the reflectivity of surfaces.

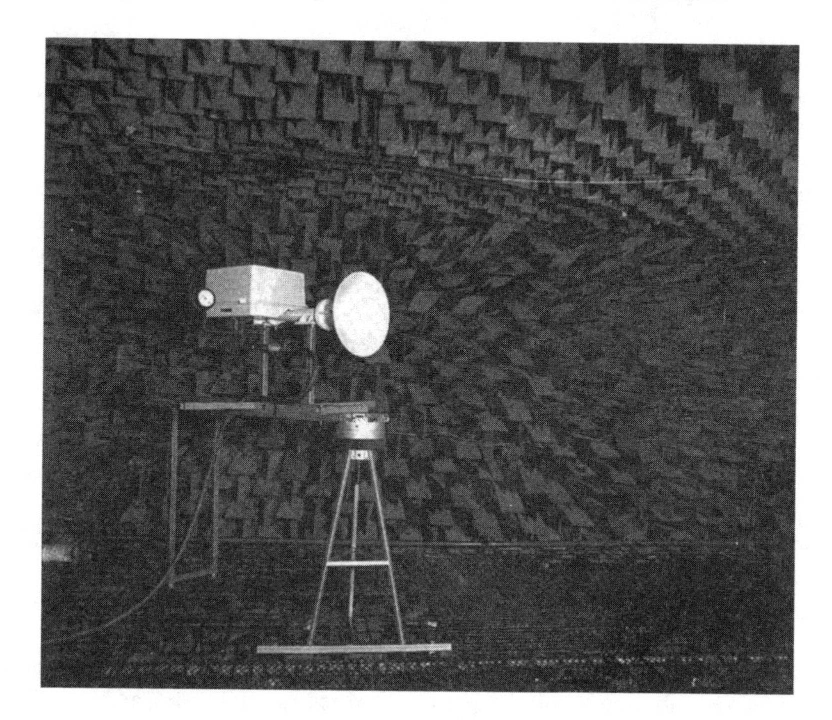

Figure 3 Anechoic chamber measurement of the noise level of a ship's whistle (Reproduced with permission from the Motor Industry Research Association, Nuneaton, UK).

Most measurements are in fact made in a semi-reverberant room, where there is a mixture of direct and reflected sound. To ensure accuracy, measurements of noise levels from a particular source should be made in the far field region.

Location of microphone

The person taking noise measurements and the microphone of the sound level meter itself can both interfere with the acoustic field being measured by blocking or reflecting sound waves, and this can cause significant errors (as large as 6 dB) in the measurements. If possible, the sound level meter should be left mounted on a stand. If it has to be held in the hand, it should be held at arm's length.

In general, whether outdoors or in an anechoic chamber, a directional (free-field or frontal incidence) microphone is used and should be pointed towards the noise source. The alternative random-incidence microphones are designed to record sound from all directions and are normally used in reverberant or diffuse sound fields. However, it should be noted that some test standards prescribe procedures different from these. If a random-incidence microphone is used in free-field conditions, its readings are most accurate when it is orientated at an angle of 70°–80° to the source.

The microphone is normally located in a position representative of a hearer's ear but without any person present. If the hearer would be standing, their ear height is usually assumed to be 1.5 m. In a seated workplace, the seated ear height of an average person can be used for positioning the microphone. However, when assessing noise levels in a workplace, the measurements should be made at all positions in which people may be exposed to the noise, taking account of all tasks, operating conditions and working practices.

In the working environment, it is sometimes necessary for the machine operator to be present while the noise measurements are made. The microphone must then be positioned as close to the operator's head as possible, while avoiding reflections from the head or body or absorption by clothing. The microphone should be at least 50 mm from the side of the operator's head (and preferably further away). Where miniature instruments are attached to the hearer's collar or helmet, or placed inside hearing protectors, it is necessary to take account of such factors and to make corrections to the measured values (which would normally be specified in the instrument manuals).

Identifying noise sources

Various methods can be used to identify or isolate specific sources of noise in a complex environment. Where possible, the different components of the machine or environmental noise should be switched on or off separately and investigated in turn. If necessary, highly directional probe microphones can be used to determine the direction of sound propagation and thus identify the noise source, but simpler methods can also be used. For example, potential sources can be isolated by surrounding them with a suitable absorbent material, such as lead sheet, which can easily be shaped around components. Alternatively, they can be shielded by barriers or temporary walls of attenuating materials.

In isolating noise sources it may be necessary to consider whether reflection or reverberation are contributing to the noise level. In some cases housing and panels may act as a sounding box, either through vibration or by reflecting noise from other sources. If so, it may help in the investigation to fill the air spaces and damp the reverberations temporarily. Weights may be added to increase damping while testing the effects of different components.

Measurement of hearing ability and risk of hearing damage

Measuring hearing ability

A person's hearing ability is measured in terms of the minimum threshold of perception, using the normal psychophysical techniques such as the 'method of limits'. The measurements are usually made with an instrument called an audiometer and presented in the form of an audiogram. An audiometer generates pure tones over the range of audible frequencies and, when measuring hearing, scans through the range of intensities at each frequency. The tones are presented through headphones and the signal can be directed separately or simultaneously to the two ears. If sounds are presented through speakers instead of headphones, there may be some differences in the measured threshold (perhaps up to 6 dB lower) due to differences in the acoustic field and the binaural threshold is likely to be lower than the monaural threshold (Loeb, 1986).

In order to avoid any influence from other sounds, no words should be spoken and the person whose hearing is being tested should be asked to indicate the presence or absence of a noise by a silent signal such as the movement of a finger. The measurements should obviously take place in a very quiet room. An initial period should be allowed for the person to adapt to the low noise level and recover from any temporary threshold shift due to previous exposure to noise. The US regulations on occupational noise exposure (OSHA, 1983) specify that the first baseline audiograms taken in a hearing conservation programme should be taken after at least 14 hours without exposure to workplace noise, although subsequent annual audiograms are permitted at any time during the working day.

'Normal' thresholds have been established (given for example in ISO, 1990a) and loss or impairment of hearing is usually defined as a minimum change of 10 dB or 15 dB in the threshold. Hearing level is taken as the amount by which the average threshold is raised in an individual, so that a positive hearing level represents hearing ability worse than the defined 'normal' level at any frequency. In an audiogram, the reference level of 0 dB represents the normal threshold of hearing (that of people who have no hearing disability or age deterioration) at each frequency over the auditory range.

Measuring masking thresholds

Although absolute hearing threshold is measured in a quiet environment, it is often important to know the hearing threshold for a signal or for speech in a noisy environment. This is termed the masking threshold, which defines the threshold of detection (not intelligibility) against the background level of noise. The effect is measured by determining the absolute threshold of the sound when presented alone, then measuring it in the presence of the masking sound. The amount of masking is then defined as the difference in decibel level by which the threshold of audibility is raised above the absolute threshold.

The probability that a signal will be detected increases with the level of the signal above the background. So the masked threshold is sometimes defined as the level at which there is a given probability (say 75%) of correct detection of the signal. This is not an absolute measure since it depends on factors such as the level of expectancy of the hearer and the rise time, duration and temporal shape of the signal (Sorkin, 1987).

Assessing the risk of hearing damage

Although a single impulsive sound can cause damage to the ear-drum, this is rare and hearing loss is usually caused by long-term exposure to noise. The exposure does not have to be continuous to cause damage, since the effects of intermittent exposure are cumulative. Evidence of the risk of hearing damage was presented in ISO Standard 1999 (ISO, 1990a), which clearly showed that repeated exposure to noise of over 90 dB for 8 hours a day will produce significant deafness and that the potential for damage increases with the frequency of the noise source. It is now generally recognised that a risk of hearing damage exists at lower levels of noise and most countries have established regulations for a maximum acceptable level for workplaces based on a time-weighted average sound level over an 8 hour working day (noise dose). Usually three action levels are defined, at which employers must introduce progressively stricter administrative and engineering controls to protect their employees. For example in the European Union, the first two action levels are at 85 dBA and 90 dBA, which will be reduced by 2006 to 80 dBA and 85 dBA together with addition of an upper limit value of 87 dBA (Commission of the European Communities, 1986; Commission of the European Union, 2003).

Some standards also set an upper limit for peak sound pressure level (SPL). In the European Union an upper limit of 200 Pa or 140 dBC (relative to a 20 μPa baseline) is currently set for impulsive noise exposure (Commission of the European Communities, 1986), and by 2006 this will be tightened by adding two lower action levels of 112 Pa and 140 Pa (Commission of the European Union, 2003). In America exposure to any steady noise level is not permitted above 115 dBA (OSHA, 1983). However, there is not a consensus on standards for exposure to impulsive sound or to infrasonic or ultrasonic frequencies. The risks from the complex effects of impulsive noises are reviewed in both Loeb (1986) and Kryter (1985), who quote damage risk contours giving an indication of acceptable maximum peak SPL values for impulsive noises of different durations and repetition frequencies.

The main indicator used to assess hearing damage for an individual is their hearing threshold. If this is measured before and after exposure to loud noise, the threshold shift may be determined by the difference between the two thresholds:

Temporary threshold shift (TTS) is a measure of the short-term and reversible change experienced after exposure to loud noise, which may persist for minutes or hours depending on the exposure. Since recovery starts as soon as the noise ceases, it is necessary to specify the time at which the TTS is determined. This is normally measured 2 minutes after exposure.

Noise-induced permanent threshold shift (NIPTS) is a measure of the long-term effect of exposure to noise, which is not reversible.

The mechanisms of TTS and NIPTS are not necessarily identical but there seems to be a correlation between them, and Kryter (1985) has suggested that the TTS of a group of workers after 8-hour exposure might be used as a criterion for the risk of long-term damage. More usually, hearing loss is assessed by the drop below the population norm. In industrial hearing conservation programmes, the operators at risk are assessed at the start of their employment and changes in hearing are monitored by annual tests. Kryter (1994) provides an extensive review of the evidence for the relative influences on hearing thresholds of presbycusis, sociocusis and nosocusis (physiological ageing, exposure through activities of everyday living, and damage through disease or trauma, respectively).

Since industrial workers are exposed to noise which may vary considerably over a working period, it is important when assessing the risk of hearing damage to calculate the noise levels over the whole period of exposure (usually over an 8 hour working day). The noise dose is calculated from the measured L_{eq} value (equivalent level of sustained noise). In measuring this, it should be remembered that operating conditions, especially communication signals, may contribute to the total noise dose. When this occurs in occupations where the signals are presented through headphones (as for aircraft pilots), the noise levels can be recorded at the ear, if necessary using a miniature or probe microphone, and the noise dose can be obtained by later analysis of the recording (Glen, 1976).

According to Davies and Jones (1982), most standards apply the 'equal-energy principle', which suggests that exposure to higher intensities can be permitted for short periods providing that the total energy within an 8 hour period does not exceed the normally permitted dose (as in ISO 1999 (ISO, 1990a)). This allows for quiet periods during the day, but does not take into account any recovery which may occur during these periods. For example, the equal energy principle assumes that a continuous 4 hour exposure followed by 4 hours in a quiet environment has the same effects as four 1 hour exposures to the same level of noise separated by quiet periods of 1 hour. On this assumption, halving the duration of exposure permits the sound level to be increased by 3 dB. The OSHA standard 1910.95 (OSHA, 1983) uses a less conservative 5 dB halving rule. However to date, the extent of recovery during quiet periods is not fully understood, so that guidelines cannot yet be given for exposure/rest cycles.

Measurement of effects of noise on comfort and performance

International and industry-wide standards have been drawn up for assessing comfort and the effects of noise on performance in different auditory environments. The effects of noise on performance are task specific and must often be assessed by experimentation, although general guidelines can be found in most ergonomics textbooks. Butler *et al.* (1999) provide a review of current knowledge on these effects. Techniques for assessing two important aspects of communication (speech interference and design of auditory signals) are discussed later.

People vary enormously in their responses to noise, being influenced by factors such as intrusion into privacy or whether the sound is intermittent or unexpected, and probabilistic or statistical measures are needed to make any assessment of annoyance and performance effects of noise. Some of the reasons for individual differences are discussed by Jones and Davies

(1984). People can adapt well to a continuous background of sound, and experiments therefore have to be designed carefully to include an adequate degree of realism in the experimental conditions. A corollary to this is that measurements of the level of annoyance may be influenced by factors other than the noise levels which are the subject of the investigation: noise annoyance may just be a symptom of poor morale or stress.

Annoyance

Many different measures are used to assess noisiness, annoyance and intrusiveness, some specific to environmental, community or transport noise. It is important to realise that loudness or noisiness cannot be equated with annoyance and investigators have found large differences in assessments of noise sources judged on these criteria (Kryter, 1985). Various aspects of perceptions of noisiness have been studied: judgements of noisiness or loudness of the neighbourhood, dissatisfaction with present noise level, frequency with which annoyance is felt, degree of annoyance, and interference with activities (Jones and Davies, 1984). These should be carefully distinguished in any survey which is conducted, since measures taken simply to reduce noise levels may not be appropriate to reduce the level of annoyance. Noise annoyance is obviously a multi-factorial problem. For example, at least thirteen primary and derived measures have been used to assess community noise (Sanders and McCormick, 1992). A few of the evaluation methods which can be used are discussed below.

The annoyance level of noise is frequently measured by means of surveys and questionnaires, but it must be remembered that the wording of questions or of instructions given to the respondents can have a considerable influence on their responses and ratings. Advice on the design of appropriate rating scales may be found in Loeb (1986) (and in Chapter 4 of this book in general terms).

Kryter (1985) developed scales of annoyance or perceived noisiness (in units called noys) which have been used for measures of aircraft and traffic noise. These were based on experimentally derived equal annoyance curves (pN dB) analogous to the equal loudness curves (phons). However, annoyance has additional psychological, social and economic dimensions so that Kryter noted that the threshold of perceived annoyance can vary with previous exposure, location indoors or outdoors, time of day, and the impulsive or startle characteristics of the noise.

Many indices have been used to quantify disturbance caused by fluctuating noise (for example, noise rating (NR), noise and number index (NNI), traffic noise index (TNI), noise pollution index (NPI)), and these are discussed in Loeb (1986) and Kryter (1985). The values of these indices are often calculated by statistical analyses of noise over long duration measurement periods.

Schultz (1978) produced a dosage response curve as a predictive tool. This relates noise exposure (in terms of day-night average sound level L_{dn} in dBA) to the level of community annoyance, based on large scale surveys from various countries (mostly related to traffic noise). Others suggest that short duration noises can be particularly annoying and have proposed measures using peak levels of noise events, as for aircraft overflights (Fidell, 1984). Other special measures have been developed to evaluate aircraft noise exposure (ISO, 1978) and a discussion of some of these can be found in Ollerhead (1973).

Preferred noise criteria (PNC) curves were introduced as design criteria for background noise in offices, rooms or halls (Beranek, 1971). They were based on the speech interference level (SIL), which is a measure of speech intelligibility and is described in the next section. The PNC curves are a set of arbitrary sound spectra for steady noises, serving as a reference for rating noise environments. They are also useful in deciding where in the frequency spectrum the greatest benefits could be obtained by noise reduction treatments. PNC curves are based on the need for acceptable speech communication and specify that the loudness in phons should not exceed SIL by more than 22 units. The spectrum of the measured background noise is plotted over the noise criteria curves and the noise is rated by the number of the curve which equals or just exceeds the noise spectrum at any point. This rating is then compared to a table of recommended ratings to evaluate the suitability of the noise level for the particular room environment.

It is not only loud noises that can be annoying. The effects of intrusiveness of low-level (or infrequent) noise exposure have been investigated by Fidell and others (Fidell *et al.* 1979; Fidell and Teffeteller, 1981). They found that the L_{eq} measure is insensitive to these types of noises and developed methods of predicting the level of intrusion or annoyance from measures of signal detectability.

Speech intelligibility

Noise can mask speech, but the degree of masking depends on various factors since there is a large measure of redundancy in speech, especially when the hearer is familiar with the subject matter. Thus, the assessment of speech intelligibility requires more than measuring the signal-to-noise ratio. Speech intelligibility can be measured most directly by listening tests, presenting different types of standardised material, such as nonsense syllables, phonetically balanced word lists which contain all speech sounds (or phonemes), and sentences. The A-weighted decibel scale is normally used to measure sound level in these applications.

Several methods have been developed to predict speech intelligibility from physical measures of the speech signal, noise environment and task parameters (for example speaker-hearer distance). Three of the methods are given here and discussions of their relative merits can be found in Kryter (1985) and other textbooks.

1. *The Articulation Index* (ANSI, 1969) predicts the likelihood of difficulties with speech communication, and is calculated from the differences between the SPLs of the speech and the masking noise in various frequency bands. These are weighted according to their importance in the intelligibility of speech and then summed to give the articulation index. Originally 20 bands were used between 250 Hz and 7000 Hz, but it is now calculated from octave or $\frac{1}{3}$ octave analysis.

   The relationship between the articulation index and intelligibility has been determined experimentally for different types of spoken material, and is expressed in a series of graphs in the ANSI Standard which indicate the percentage of syllables, words or sentences that will be correctly understood. Corrections are given in the standard to take account of factors such as reverberation or noise interruption. The following values of articulation index give an indication of the difficulty of speech communication:

   <0.4 difficulties likely
   0.4–0.7 some difficulties may occur
   >0.7 good speech communication possible

2. *Speech Interference Level* (SIL) is a simpler method (ISO, 2003) which does not require direct measurement of the speech level. The SIL is calculated from the mean of noise SPL in octave bands centred at 500, 1000, 2000 and 4000 Hz. The resulting value is compared with tables relating it to speech intelligibility for different voice efforts (normal, raised, shouting) and for various distances between speaker and listener. Preferred speech interference level is a similar measure giving the maximum distance over which speech communication is possible, assuming that the speaker's voice

effort is appropriate (ISO, 1974). ISO standard ISO 9921 (ISO, 2003) gives a procedure for assessing speech interference level when the hearer is wearing hearing protectors.

3. *Direct Measurements of SPL* (in dBA) have also been used as an index of speech interference, predicting SIL. Loeb (1986) suggested a relationship of

$$\text{SIL} = \text{SPL} - (9 \text{ or } 10\,\text{dB})$$

Webster (1979) produced a chart relating dBA, voice effort and speaker-hearer distance to quality of communication, so that either SPL or SIL could be used in assessing this.

Design of warning and information signals for safety and audibility

The effects of masking on warning and alarm signals are discussed in Webster (1984), giving a method for predicting the level at which a pure tone signal will be audible. However, additional criteria are needed to ensure that the signal also attracts attention and is recognisable. The perception of a masked signal varies with frequency, signal-to-noise ratio and absolute level of background noise.

Four main factors need to be taken into account in designing auditory warnings and signals: audibility, noise dose, startle and discriminability (Coleman *et al.*, 1984). Reports from Patterson and Milroy (1979) and Coleman *et al.* (1984) describe procedures which have been used to design the appropriate sound level for auditory warnings in aircraft and in the mining industry, respectively.

The first stage of this process is the measurement or prediction of the masked threshold as described by Coleman *et al.* (1984, 1998). The general method employed to identify the 'design window' for a particular work environment (as shown in Figure 4) is to analyse a recording of the background noise and perform a detailed narrow-band spectral analysis. This can then be used to calculate the masked threshold, incorporating a 97.5th percentile

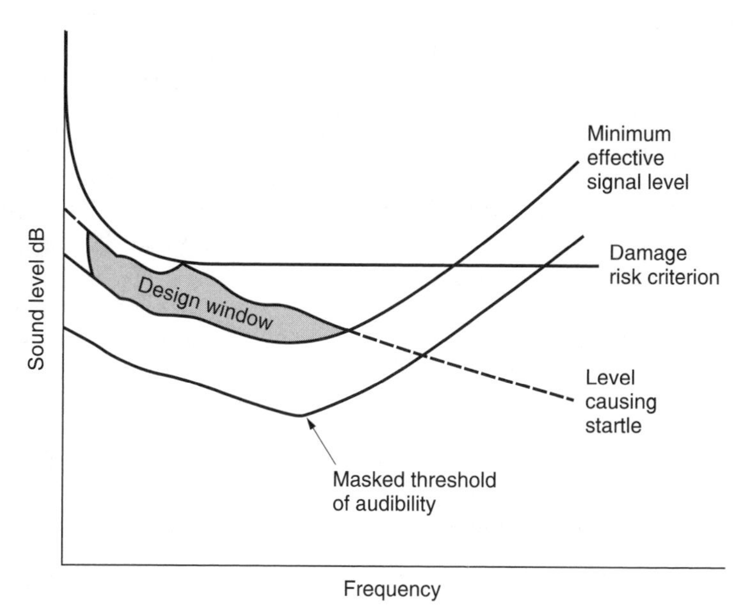

Figure 4 'Design window' method for specifying signals to be used in noisy environments (after Coleman *et al.* 1984).

absolute threshold criterion. This 'design window' defines the upper and lower boundaries of frequency and intensity within which the signal components must lie in order both to be audible and to attract attention without causing startle or risk of hearing damage.

Figure 4 shows how the various factors are considered in specifying the 'design window' for a warning signal. This is constructed from knowledge of the hearing abilities of the workforce, the nature of the masked threshold due to the background noise and the levels above threshold which are necessary for the signal to be audible without causing startle. It would also be possible to take into account the attenuation effects when hearing protectors are used. The threshold could also be set for different population criteria: the maximum level to avoid startle should be determined for people with normal hearing, while the minimum effective level should be set for the people in the population who have the poorest hearing ability.

Assessment of personal hearing protectors

When hearing protection is necessary in a noisy environment, criteria used in the selection of appropriate equipment must include the effect on communication (both for speech and for auditory signals) as well as the level of protection provided. New forms of both active and passive hearing protection devices are being developed (Casali and Berger, 1996) and these may require new assessment criteria (including assessment of usability in relation to adjustable tuning and attenuation). Behavioural issues also need to be considered and Hughson *et al.* (2002) have suggested ways in which workers exposed to noise can be encouraged to wear hearing protection.

Evaluating attenuation of noise levels

Noise attenuation curves are normally measured by the manufacturers and are supplied with hearing protection equipment. The attenuation may be measured by methods such as those in ANSI standard S3.19-1974 (ANSI, 1974). The US Environmental Protection Agency (1979) requires manufacturers to test hearing protectors to this standard and to label them with a noise reduction rating (NRR). The NRR is calculated from $\frac{1}{3}$ octave band analysis of measured attenuation. This can then be used to estimate the noise exposure (in dBA) for the wearer: either as the workplace sound level in dbC less NRR, or as the sound level in dBA plus 7 dB less NRR (OSHA, 1983).

Sutton and Robinson (1981) reviewed various other procedures which can be used to estimate the level of protection given to a wearer (usually defined as the reduction in dBA level at the ear) from a knowledge of the frequency response of the protector and the frequency spectrum of background noise. The most accurate method of calculating the protection provided is given in ISO 1999 (ISO, 1990b), where attenuation (minus a variance correction) is subtracted from the workplace sound levels measured with octave band analysis. It should be noted, however, that these methods indicate the protection provided under optimum conditions and do not allow for poor fitting of the protector or noise leakage due to hair or other factors. Studies have indicated that the methods may well overestimate the actual protection when measured in workplace conditions (Berger, 1983; Lempert, 1984).

A subjective method of measuring attenuation (by threshold shift) can also be used, as outlined in ISO 4869-1 (ISO, 1990b). A minimum of 16 people should be tested (preferably more), using normal audiometric techniques. When using the hearing protection, they should have a hearing threshold level in either ear which is not worse than 15 dB at frequencies below 2 kHz and not worse than 25 dB at frequencies above 2 kHz. The test signals specified are pink noise filtered through $\frac{1}{3}$ octave bands from 63–8000 Hz. (Pink noise contains the full range of audible frequencies but is weighted towards the lower frequencies so that the sound pressure spectral density is inversely proportional to frequency.) This method can be used to compare or rank protection equipment and to evaluate design features which may affect performance.

The Health and Safety Executive (1998) guidance on noise regulations explains how to calculate the 'assumed protection' from the attenuation values provided by equipment manufacturers or recorded in tests on a group of people. The 'assumed protection' is taken as the mean attenuation at each frequency minus the standard deviation over the tests (probably of the order of 5 dB), in order to allow for the variation in protection between wearers and to ensure that the majority of wearers will have at least the predicted noise reduction.

Simulated ('dummy') headforms and miniature microphones can also be used to measure attenuation provided by hearing protectors, although the results may differ from subjective measurements because of the effects of bone conduction leakage. A semi-objective technique, in which a miniature microphone is attached to the wearer's ear, can be used to investigate the effects of hair, spectacles or helmets worn at the same time as hearing protectors.

Evaluating audibility of communications

In most work situations, the effects on speech communication need to be considered when assessing the effectiveness of hearing protectors. Evaluation measures include the articulation index, the speech interference level and intelligibility of messages (assessed by experimental subjective tests). Although the criteria defined for the articulation index usually assume that the hearers have normal hearing ability, the articulation index can be assessed against reduced hearing thresholds and a measure of 'articulation index incorporating hearing ability (AIIHA)' was suggested by Coleman *et al.* (1984). They also proposed a technique for selecting protectors which would take into account the interaction between the attenuation characteristics and hearing ability, the range of noise and the speech conditions.

Concluding remarks

It is apparent from the wide variety of techniques introduced in this chapter that one of the most important considerations in any noise assessment is the choice of appropriate criteria for the evaluation. It is rarely sufficient simply to measure the sound pressure level or acoustic power at a single location, since the effects of noise on hearers are considerably more complex. Both perception and individual responses or preferences are usually influenced by the social or occupational context in which the sound is heard as well as by the auditory characteristics of the human ear.

When assessing a noisy environment, ergonomists may need to consider the effects on auditory health, safety, performance, transmission of information/communication, annoyance and pleasure for different groups of hearers. These are evaluated in different ways and a range of criteria may be needed to assess the environment. It is hoped that this chapter gives helpful guidance on the most appropriate methods to choose for assessments of community, transport and industrial environments.

References

ANSI (1969). *American National Standard for Calculation of the Articulation Index.* ANSI S3.5-1969. (New York: American National Standards Institute).

ANSI (1974). *American National Standard Method for the Measurement of Real-ear Protection of Hearing Protectors and Physical Attenuation of Ear-muffs.* ANSI S3.19-1974. (New York: American National Standards Institute).

Beranek, L.L. (1971). *Noise and Vibration Control.* (New York: McGraw-Hill).

Berger, E. (1983). Using NNR to estimate the real world performance of hearing protectors. *Sound and Vibration*, **18**, 26–39.

Butler, M.P., Graveling, R.A., Pilkington, A. and Boyle, A.L. (1999). *Non-Auditory Effects of Noise at Work: A Critical Review of the Literature Post 1988.* (Sudbury: HSE Books).

Casali, J.G. and Berger, E.H. (1996). Technology advancements in hearing protection circa 1995: active noise reduction, frequency/amplitude-sensitivity, and uniform attenuation. *American Industrial Hygiene Association Journal*, **57**, 175–185.

Coleman, G.J. (1998). The signal design window revisited. *International Journal of Industrial Ergonomics*, **22**, 13–318.

Coleman, G.J., Graves, R.J., Collier, S.G., Golding, D., Nicholl, A.G.McK., Simpson, G.C., Sweetland, K.F. and Talbot, C.F. (1984). *Communications in Noisy Environments.* Technical Memorandum TM/84/1 (EUR P.74). (Edinburgh: Institute of Occupational Medicine).

Commission of the European Communities (1986). *Council Directive of 12 May 1986 on the Protection of Workers from the Risks Related to Exposure to Noise at Work.* Council Directive 86/188/EEC. Official Journal of the European Communities, No. L137/29, 24 May 1986.

Commission of the European Union (2003). *Council Directive of 6 February 2003 on the Minimum Health and Safety Requirements Regarding Exposure of Workers to the Risks Arising from Physical Agents (Noise).* Council Directive 2003/10/EC. Official Journal of the European Union, No. L42/38, 15 February 2003.

Davies, D.R and Jones, D.M. (1982). Hearing and noise. In: *The Body at Work*, edited by W.T. Singleton (Cambridge: Cambridge University Press), pp. 365–413.

Environmental Protection Agency (EPA) (1979). Noise labeling requirements for hearing protectors. *Federal Register*, **42**, 56139–56147.

Fidell, S. (1984). Community response to noise. In: *Noise and Society*, edited by D.M. Jones and A.J. Chapman (Chichester: John Wiley), pp. 247–277.

Fidell, S. and Teffeteller, S.R (1981). Scaling the annoyance of intrusive sounds. *Journal of Sound and Vibration*, **78**, 291–298.

Fidell, S., Teffeteller, S.R., Horonjeff, R.D. and Green, D.M. (1979). Predicting annoyance from detectability of low-level sounds. *Journal of the Acoustical Society of America*, **66**, 1427–1434.

Glen, M.C. (1976). The contribution of communications signals to noise exposure. *Applied Ergonomics*, **7**, 197–200.

Health and Safety Executive (1995). *Sound Solutions — Techniques to Reduce Noise at Work.* (Sudbury: HSE Books).

Health and Safety Executive (1998). *Reducing Noise at Work. Guidance on the Noise at Work Regulations 1989.* (Sudbury: HSE Books).

Hughson, G.W., Mulholland, R.E. and Cowie, H.A. (2002). *Behavioural Studies of People's Attitudes to Wearing Hearing Protection and How These Might Be Changed.* Health and Safety Executive Research Report No. 028. (Sudbury: HSE Books).

ISO (1974). *Acoustics — Assessment of Noise with Respect to its Effect on the Intelligibility of Speech.* ISO Technical Report TR 3352 (Geneva: International Standards Organisation).

ISO (1978). *Acoustics — Procedure for Describing Aircraft Noise Heard on the Ground.* ISO 3891 (Geneva: International Standards Organisation).

ISO (1990a). *Acoustics — Determination of Occupational Noise Exposure and Estimation of Noise-induced Hearing Impairment.* ISO 1999 (Geneva: International Standards Organisation).

ISO (1990b). *Acoustics — Hearing Protectors — Part 1: Subjective Method for the Measurement of Sound Attenuation.* ISO 4869-1 (Geneva: International Standards Organisation).

ISO (2000). *Acoustics — Determination of Sound Power Levels of Noise Sources — Guidelines for the Use of Basic Standards.* ISO 3740 (Geneva: International Standards Organisation).

ISO (2003). *Ergonomics — Assessment of Speech Communication.* ISO 9921 (Geneva: International Standards Organisation).

Jones, D.M. and Chapman, A.J. (1984). *Noise and Society.* John Wiley & Sons: Chichester.

Jones, D.M. and Davies, D.R (1984). Individual and group differences in the response to noise. In *Noise and Society*, edited by D.M. Jones and A.J. Chapman (Chichester: John Wiley), pp. 125–153.

Kohler, H.K. (1984). The description and measurement of sound. In *Noise and Society*, edited by D.M. Jones and A.J. Chapman (Chichester: John Wiley), pp. 35–76.

Kryter, K.D. (1985). *The Effects of Noise on Man*, 2nd Edition. (London: Academic Press).
Kryter, K.D. (1994). *The Handbook of Hearing and the Effects of Noise.* (San Diego: Academic Press).
Lempert, B.L. (1984). Compendium of hearing protection devices. *Sound and Vibration*, **18**, 26–39.
Loeb, M. (1986). *Noise and Human Efficiency.* (Chichester: John Wiley).
Ollerhead, J.B. (1973). Noise: how can the nuisance be controlled? *Applied Ergonomics*, **4**, 130–138.
OSHA (Occupational Safety and Health Administration) (1983). Occupational noise exposure; hearing conservation amendment; final rule. *Federal Register*, **48**, 9738–9783.
Patterson, R.D. and Milroy, R. (1979). *Existing and Recommended Levels for Auditory Warnings on Civil Aircraft.* Civil Aviation Authority Contract Report (Contract No. 7D/S/0142). (Cambridge: Medical Research Council, Applied Psychology Unit).
Sanders, M.S. and McCormick, E.J. (1992). *Human Factors in Engineering and Design*, 7th Edition. (New York: McGraw-Hill).
Schultz, J. (1978). Synthesis of social surveys on noise annoyance. *Journal of the Acoustical Society of America*, **64**, 377–405.
Sorkin, R.D. (1987). Design of auditory and tactile displays. In *Handbook of Human Factors*, edited by G. Salvendy (New York: John Wiley), pp. 549–576.
Sutton, G.J. and Robinson, D.W. (1981). An appraisal of methods for estimating effectiveness of hearing protectors. *Journal of Sound and Vibration*, **77**, 79–91.
Webster, J.C. (1979). Effects of noise on speech. In *Handbook of Noise Control*, 2nd edition, edited by C.M. Harris (New York: McGraw-Hill).
Webster, J.C. (1984). Noise and communication. In *Noise and Society*, edited by D.M. Jones and A.J. Chapman (Chichester: John Wiley), pp. 185–220.

Chapter 26

Anthropometry and the design of workspaces

Stephen T. Pheasant and L.P.A. (Bea) Steenbekkers

Introduction

Anthropometry deals with body measurements — particularly those of size, shape and body composition. Biomechanics is the application of mechanical principles to the study of the structure and function of the human body. As applied to ergonomics, the two are closely linked since the science of biomechanics commonly provides the criteria for the application of anthropometric data to the problems of design.

This chapter is concerned with anthropometry and the methods for using anthropometric data. More information about this subject can be found in Chaffin and Anderson (1992), Clark and Corlett (1984), Pheasant (1986) and Steenbekkers and van Beijsterveldt (1998).

Anthropometric data

The variability of most bodily dimensions can be described, to a tolerable degree of accuracy, by a mathematical function known as the *normal distribution*. This name does not imply that it describes the distribution of a dimension of 'normal people' — whatever that might mean anyway — but rather as meaning 'the distribution which you will find most useful in practical affairs'. To deal with this slight semantic difficulty, it is sometimes referred to as the Gaussian distribution, after Johann Gauss (1777–1855), the German mathematician and physicist who first explored its mathematical properties in detail. The fact that bodily characteristics such as stature (i.e., standing height) are also normally distributed is an empirical observation due to the English anthropologist and geneticist Sir Francis Galton (1822–1911).

The distribution of stature of British men is shown in Figure 1. Frequency is plotted vertically and stature is plotted horisontally. Beneath the horisontal axis is a second scale showing percentiles of stature. In any particular characteristic, n% of the population concerned is smaller than the nth percentile, i.e., the nth percentile is the value which is exceeded in $(100 - n)$% of cases. Note that the percentiles are close together in the center of the distribution and widely spread in the tails.

The curve is symmetrical about the mid-point. The 50th percentile value is also the most common value and the frequency declines systematically as you enter the tails of the distribution.

Figure 2 shows exactly the same data, plotted in a slightly different way. This is the cumulative form of the normal distribution, sometimes known (because of its shape) as the *normal ogive*. Plotted horisontally is stature; plotted vertically are percentiles of stature. This is a particularly useful way of presenting the data for our present purposes because it allows us

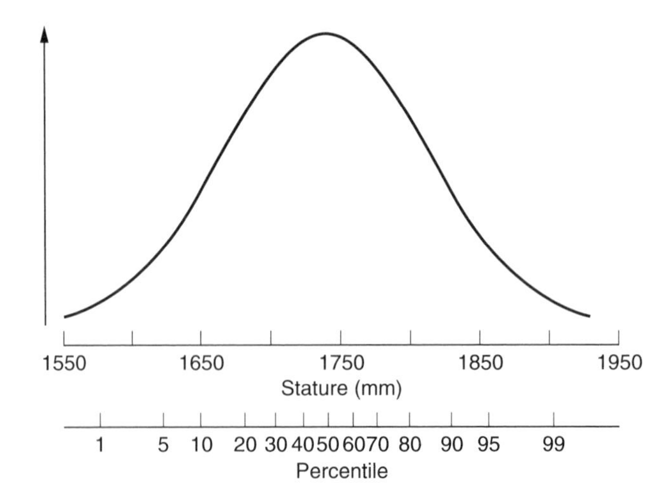

Figure 1 The normal distribution of the stature of British men. After Pheasant (1986).

to read off directly the percentage of people who will be *accommodated* (i.e., satisfied) with respect to a particular anthropometric criterion. To take a slightly trivial example, Figure 2 tells the percentage of men who would be able to pass under a doorway of a given height, without running the risk of banging their heads. The normal distribution is completely described by two parameters, the mean and the standard deviation. The mean is the same thing as the familiar arithmetical average and for normally distributed variables it is equal to the 50th percentile. The standard deviation is a measure of dispersion; it describes the extent to which an individual might be expected to differ from the mean. So we might say, for example, that the mean stature of a carefully selected sample of high-jumpers was greater than the mean stature of the population at large, but that their standard deviation was less.

If the mean (m) and the standard deviation (s) of a normally distributed variable are known, then any percentile (X_p) which we might happen to require may be calculated from the following equation:

$$X_p = m + sz$$

where z (the standard normal deviate) is a factor for the percentile concerned. Values of z for some commonly used percentiles (p) are given in Table 1, which is the cumulative version of the normal distribution plotted in tabular form. British men have a mean stature of 1740 mm with a standard deviation of 70 mm, whilst the values for British women are $m = 1610$, $s = 62$; for American men $m = 1755$, $s = 71$; and American women $m = 1625$, $s = 64$. That large differences between populations exist becomes clear when looking at the data of e.g., Dutch adults. The mean stature of Dutch males is 1848, $s = 80$ and of Dutch adult women $m = 1686$ and $s = 66$. More complete versions of Table 1 are given in Pheasant (1986, 1990), as is a further discussion of the statistical basis of anthropometrics.

User populations

The anthropometric characteristics of any given human population will depend upon a number of factors. The most important ones, from the point of view of ergonomics, are sex, age, ethnicity and occupation — usually in that order.

When applying anthropometric data to any particular design problem, the first step will generally be to define the *target population* of users for whom the product (workstation,

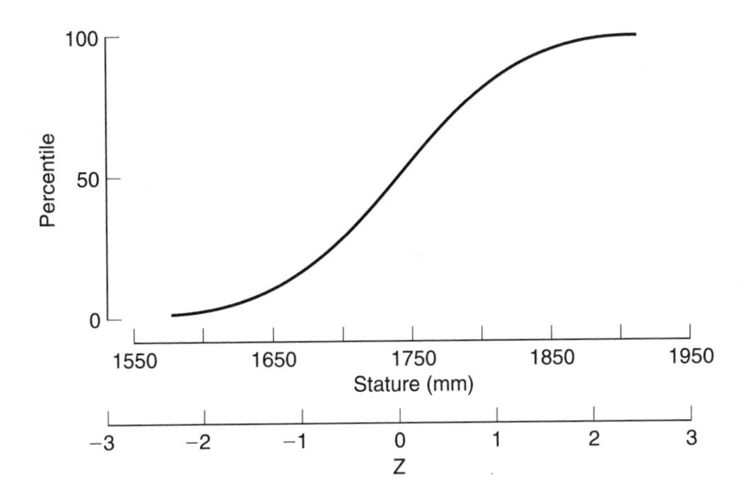

Figure 2 The normal distribution of the stature of British men, plotted in cumulative form. After Pheasant (1986).

Table 1 Selected Values of z and p

P	z	p	z
1	{−}2.33	99	2.33
2.5	{−}1.96	97.5	1.96
5	{−}1.64	95	1.64
10	{−}1.28	90	1.28
15	{−}1.04	85	1.04
20	{−}0.84	80	0.84
25	{−}0.67	75	0.67
30	{−}0.52	70	0.52
40	{−}0.25	60	0.25
50	0.00	50	0.00

environment) is intended, and to locate a source of anthropometric data for the population concerned (or for one which resembles it as closely as possible in relevant respects). The consequences of using inappropriate data will be that fewer people will be satisfied than intended — sometimes quite drastically so — or even worse, the performance of the eventual users will be impaired.

In most adult populations a difference of about 7% between the average heights of men and women (and a somewhat larger difference in their standard deviations) will be found. On average, men will also be larger in most other respects, although the magnitude of the difference will vary from dimension to dimension. The most important exception to this rule is hip breadth. In addition to the more obvious differences in shape between men and women, it is worth noting that men have proportionally greater limb lengths. That is, if we were to compare a man and a woman of *equal stature,* we should expect the man to have longer arms and legs, bigger hands and feet, and so on. There will also be proportional differences in dimensions which have a substantial soft tissue component. This is partly because men have (on average) greater muscle bulk whereas women have greater bodily fat; and it is partly due to sex differences in fat distribution.

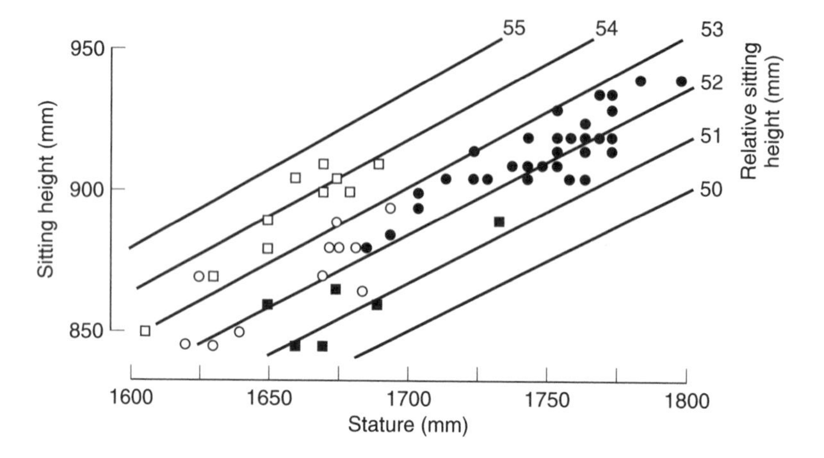

Figure 3 Ethnic differences in the relationship between average sitting height and average stature in samples of adult men. (● = European, ○ = Indo-Mediterranean, □ = Far Eastern, ■ = African). After Pheasant 1986).

People also change in shape as they get older. In our society at least, they tend to put on weight; and after the age of about 55, muscle bulk begins to decrease and the spine begins to shorten due to changes in the properties of the intervertebral discs and the foot bow. Their body stopped growing a long time ago, with exception of the nose and the ear. The cartilage of nose and ear continue to grow during all our live.

Anthropometric differences due to the ageing process itself are confounded with differences due to long-term historical processes known as secular trends. In Europe and North America, people have been getting steadily taller. In the previous century there has been an upward trend in adult stature at about 10 mm/decade. In Europe and North America there is some evidence that the trend has now come to a halt, but in Japan it is still under way.

The ethnic groups of the world differ in both size and shape. Figure 3 shows sitting height (i.e., the distance from the seat to the crown of the head) plotted against stature. The oblique lines on the chart show sitting height divided by stature. Each of the major ethnic divisions of the world include both tall and short ethnic groups, but they tend to have characteristic body proportions. Black Africans have relatively long legs for their stature; far eastern peoples have relatively short legs (particularly so the Japanese). Europeans and Indo-Mediterraneans (who together make up the so-called Caucasoid division of mankind) are somewhere between the two extremes.

Within populations body proportions also differ remarkably. In the Delft Gerontechnology study 26 body dimensions were measured on 750 persons in the ages between 20 and 30 and over 55 years (Steenbekkers and van Beijsterveldt, 1998). From this data set, the persons (all women) were selected with a stature between (arbitrarily) 166.1 and 166.8 cm ($n=36$). In Figure 4 shoulder height, fist and popliteal height are presented for these persons with (approximately) the same stature. It shows that where the stature varies with 0.8 cm (0.42% of the total dimension), shoulder height varies with 6.9 cm (i.e., 5% of the total dimension), fist height with 9.8 cm (13%) and popliteal height with 6.9 cm (15%).

For these participants the breadth dimensions vary even more. For example shoulder breadth varies between 39.6 and 48.5 cm (20%) and hip breadth sitting between 35.0 and 47.2 cm (29.6%).

This implies that when a person has a 'mean stature', he/she will most probably not have a mean shoulder height, hip breadth or whatever dimension. This requires the possibility of individual adaptation of workplaces.

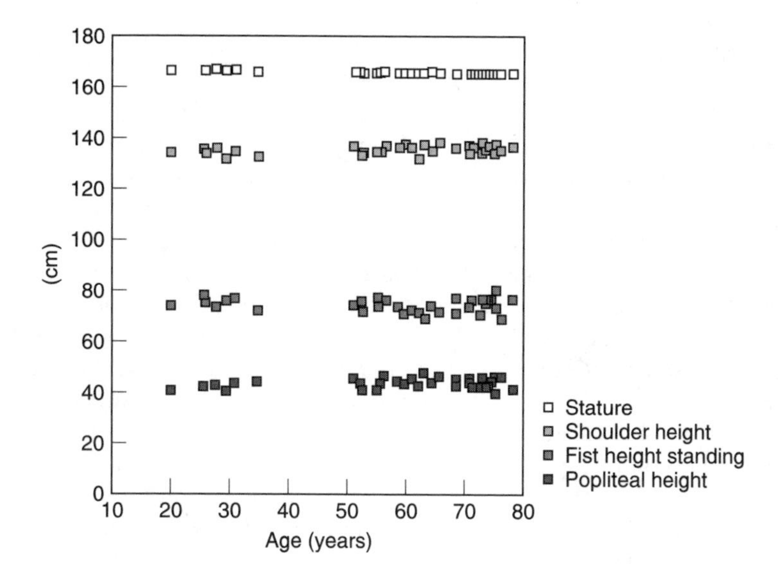

Figure 4 Shoulder, fist and popliteal height of women with the same stature.

Different sources of data exist. Relevant to a user is whether the data are original measurements or estimations based on other sources. Relatively few data sources are based on original measurements of large samples, more often they are estimations based on original data in other populations or estimations on the basis of rather small samples. It is relevant to know the date and the population of these original data before being able to assess the value or relevance of the data for the given design problem. Lately many sources have been put together in overviews of data of children, adults and elderly persons, from all over the world: Child data (Norris and Wilson, 1995), Adult data (Peebles and Norris, 1998), and Older adult data (Smith *et al.*, 2000).

Measurement of body dimensions

Until recently mostly traditional methods (subject in a standard posture with the head in the Frankfurt plane) were used to assess body dimensions of subjects. The equipment used was the anthropometer, balance and different kinds of calipers. For some years a new technique has been used: 3-D surface anthropometry. In this method laser scanning is used to assess and compute 3-dimensional dimensions of the body.

The 3-D Surface Anthropometry extends the measurement of the body to detailed, high-resolution measurement of the surface of the body. The latest automated surface anthropometry has many advantages over the traditional technologies. It is faster, less expensive, provides detail about the surface shape as well as 3-D locations of measures relative to each other, enables easy transfer to Computer Aided Design (CAD) or Manufacturing (CAM) tools, and results in a scan that is independent of the measurer, making it easier to standardise. (www.sae.org/technicalcommittees/caesumm.htm, d.d. 21–10–2002).

Computerised analysis and data storage makes it possible to enlarge the usability of data. This method can for example be used to make clothing that exactly fits the body. (Vannier and Robinette, 1995).

Criteria

Before pursuing some of the methods for using anthropometric data we must look at how we decide whether what we want to do will be adequate for the purpose. We need criteria both to guide us in our applications and to provide a framework against which we can test our decisions.

In the context of ergonomics, a criterion is a standard of judgement which defines the extent to which a particular product (workstation, environment) is appropriately matched to its human users. For a criterion to have any practical (or scientific) value, it must be possible to specify the operation which an investigator would have to perform, in order to determine whether the criterion had indeed been satisfied in any particular case. A criterion which may be defined in this way is known as an *operational criterion*. There is not much point in saying that a product should be 'easily usable' unless we can define usability in terms of how it could be measured (see also Chapters 11 and 13). Some product standards work in this way; BS 6652, *Packages Resistant to Opening by Children*, defines a set of procedures for conducting an experiment to determine how easily people are able to open a particular container. An experiment of this kind is called a *user trial*. An ergonomics criterion of this kind indicates whether an existing product is to be regarded as satisfactory but it does not reveal much about how the product should be designed. This has both advantages and disadvantages.

Design criteria are hierarchical. At the highest level there are very general concepts: comfort, efficiency, safety, usability and so on. In themselves, these are not easy to define in operational terms — at least not in ways which are directly useful to the designer. To get round this problem, we need to break down these high-level criteria into subordinate criteria at successively lower levels in the hierarchy. Consider the criteria which might define an 'ergonomically-designed chair'. One of these would obviously be comfort. Subordinate to this would be more specific design principles, such as the provision of adequate postural support and the avoidance of pressure hot-spots on supporting surfaces. These in turn could be broken down into component parts dealing, for example, with the angle of the backrest, the height of the seat, and so on. At this lower level of the hierarchy, it should be relatively easy to provide operational design criteria. For example, the criterion, 'the user should be able to sit with the feet on the floor without experiencing undue pressure on the underside of the thighs', leads directly to 'the height of the seat should not be greater than the lower leg length of a short user' and in turn leads directly to the recommendation that the seat height should not be greater than 400 mm.

The criteria which define a successful outcome to the design process fall into three main groups: comfort, performance, and health and safety.

Case studies show that those ergonomics measures which increase comfort and well-being are also likely to improve productivity (and vice versa). For example, Ong (1984) studied a group of data entry operators at an airline computer centre in Singapore, before and after ergonomic improvements to the design of their workstations. He found both an increase in productivity and a dramatic reduction in the symptoms of visual and muscular fatigue and discomfort. Performance as measured by keystrokes per hour increased by 25% — but at the same time, the error rate dropped from an overall 15% (1 character in 66 incorrect) to 0.1% (1 character in 1000 incorrect). Dainoff and Dainoff (1986) describe experiments on data-entry operators in which comparisons were made between a workstation which was designed according to commonly accepted ergonomics guidelines and one which deliberately broke most of the rules. (The workstations differed with respect to seat design, keyboard height, screen location, task lighting, glare, etc.) Performance was 25% better at the ergonomically-designed workstation. When the differences in lighting were eliminated there was still a performance difference of 17.5%. The participants also expressed a preference for the ergonomically-designed workstation and experienced less back and shoulder pain. Aside from any humanitarian considerations involved, performance differences of this magnitude amply justify the costs of the ergonomics intervention.

In many problems of workstation design, the immediate objective will be to achieve appropriate muscular efforts for the performance of a given task. We may reasonably assume that this will have both short- and long-term benefits. In the short-term it will reduce fatigue — and by so doing, improve both performance and subjective comfort. In the long-term, it will reduce the incidence of conditions such as back pain, neck pain and repetitive strain injuries. The sickness absence which results from these common musculoskeletal disorders is economically costly, both for the organisation concerned and for society as a whole.

Physiological and biomechanical principles

To achieve the desirable outcomes described earlier we need some rational procedures and principles. To help us to decide the broad arrangements of equipments, displays and controls, and layouts in general, there are four principles, stated by McCormick and Sanders (1987, p. 343), which might be seen as a normalisation of common sense; because they are so evidently ignored in many situations they should not be underrated.

1. *Importance principle.* Those components which are most essential to safe and efficient operation should be in the most accessible positions.
2. *Frequency of use principle.* Those components which are used most frequently should be in the most accessible positions.
3. *Function principle.* Components with closely related functions should be located close to each other.
4. *Sequence of use principle.* Components which are often used in sequence should be located close to each other and their layout should relate logically to the sequence of operation.

Note that the term 'accessible' relates not only to physical accessibility (such as ease of reach) but also to visibility and to other more abstract characteristics.

Further sets of principles are needed to deal with the direct relationships between the person and the workplace. These will arise from consideration of the physiological, biomechanic and sensory (information transfer) relationships required if the people concerned are to maintain the comfort, performance and health aspects which were stated earlier to be the basis of our design requirements. Reference to ergonomics texts, appropriate to the aspects being designed, will give information on the requirements to be met if good performance is to be possible.

Within industry it has been common to rely on concepts such as the principles of easy movement, still listed in many methods texts, for decisions on workspace arrangements. These principles are limited in how they accord with the needs of human physiology and psychology, and Corlett (1978) proposed a set which was more in line with current knowledge (Table 2). These provide more specific guidance for the choice of anthropometric dimensions to meet our basic principles.

Application of anthropometric data and criteria

Design limits

The slope of the normal ogive is steepest at the mean and it decreases steadily in the tails of the distribution. This has an extremely important consequence for ergonomics in that it is increasingly difficult to accommodate extreme individuals. Let us consider a specific example: the desirable range of adjustment for the height of a working chair. For the purposes of argument we shall ignore things like the table with which the chair will be used and the task which will be performed (both of which are, in practice, very important) and we shall consider the chair taken in isolation. A seat which is too high causes undue pressure on the undersides

Table 2 New Principles for Workplace Layout

1.	The worker should be able to maintain an upright and forward facing posture during work.
2.	Where vision is a requirement of the task, the necessary work points must be adequately visible with the head and trunk upright or with just the head inclined slightly forward.
3.	All work activities should permit the worker to adopt several different, but equally healthy and safe, postures without reducing capacity to do the work.
4.	Work should be arranged so that it may be done, at the worker's choice, in either a seated or standing position. When seated, the worker should be able to use the backrest of the chair at will, without necessitating a change of movements.
5.	The weight of the body, when standing, should be carried equally on both feet, and foot pedals designed accordingly.
6.	Work activities should be performed with the joints at about the mid-point of their range of movement. This applies particularly to the head, trunk and upper limbs.
7.	Where muscular force has to be exerted it should be by the largest appropriate muscle groups available and in a direction co-linear with the limbs concerned.
8.	Work should not be performed consistently at or above the level of the heart; even the occasional performance where force is exerted above heart level should be avoided. Where light hand work must be performed above heart level, rests for the upper arms are a requirement.
9.	Where a force has to be exerted repeatedly, it should be possible to exert it with either of the arms, or either of the legs, without adjustment to the equipment.
10.	Rest pauses should allow for all loads experienced at work, including environmental and information loads, and the length of the work period between successive rest periods.

of the thighs; one which is too low makes standing up and sitting down needlessly difficult and encourages a slumped position of the spine. Most ergonomics books would recommend therefore that the height of a chair should be a little below the popliteal height of its user. (Popliteal height is the vertical distance from the floor to the crease at the back of the knee.) It so happens that the popliteal height of British adults (ignoring sex differences) has a mean of 455 mm and a standard deviation of 30 mm. If we assume the optimal height of a seat to be 40 mm less than popliteal height, then the distribution of optimal heights will have a mean of $455 - 40 = 415$ mm, with a standard deviation of 30 mm. By calculating percentiles of this distribution, we can calculate the percentage of the target population who would be matched by any given range of height adjustment. From Table 1 we find that the 25th and 75th percentiles are 0.67 standard deviations below and above the mean, respectively. Hence to satisfy the 50% of users who are between these limits, we would need to make our chair adjustable by $0.67 \times 30 = 20$ mm, on either side of the mean value, i.e., from 395 to 435 mm. Repeating this calculation for other percentiles, the following results are obtained:

Percentage satisfied	50	60	70	80	90	95	98
Millimetres adjustment	40	50	62	77	98	118	140

We are in a situation of diminishing returns in which each additional unit of adjustment yields less benefit in terms of the percentage of satisfied users.

In practical terms we have to set limits on our attempts to satisfy the largest possible number of users. Where should we set these *design limits*? By convention we usually choose to accommodate (i.e., satisfy) the range of users who are between the 5th and 95th percentile of whatever characteristic we are dealing with. In some problems (like the one we have just considered) this will accommodate 90% of people; in others (as we shall see shortly) it will accommodate 95%. This is an arbitrary choice based on expediency. Beyond these limits the situation of diminishing returns begins to tell against us rather dramatically. In some cases however (such as when dealing with health and safety issues) we will need to set wider limits. The rule of thumb for making such decisions is to consider the worst possible consequences of a mismatch.

Fitting trials

A fitting trial is an experimental investigation of the relationships between the dimensions of a product (workstation, environment) and the dimensions of its users. In general, the experimental subjects will be asked to try out an adjustable mock-up of the product concerned. Critical dimensions of the product will be adjusted through a range of values, and the subjects will be asked to express their preferences with respect to comfort, ease of use, and so on. A fitting trial is therefore a special kind of psychophysical experiment. It is important that the subjects in the experiment should be a representative sample of users of the product — both with respect to their body dimensions and with respect to their general fitness and anything else which might be relevant. Suppose we wished to conduct an experiment to determine the narrowest gap between two obstructions, of a given height, that subjects could pass through without experiencing undue inconvenience. We could do the experiment simply by shifting furniture around, so as to create gaps of different widths. We could start with a gap which was so narrow that no one could get through; and then systematically widen it (perhaps in 100 mm increments) until we had a gap which everyone could get through. If we plotted a graph of gap width (on the horisontal axis) against the percentage of people who could get through (on the vertical axis), we should expect to get a relationship which looked something like Figure 2, i.e., the cumulative form of the normal distribution. We could make it look exactly like Figure 2 by calculating the mean and standard deviation of the minimum gap widths through which each individual subject could pass, and then use these parameters to calculate percentiles, using the z values in Table 1. (In essence we are smoothing our data by fitting a normal distribution.) The experiment could be made more sophisticated by observing whether the subjects passed through crab-wise or head on, or by asking them whether it was easy or difficult, and so on. In fact, many different criteria could be used, and if the data were plotted for each of these criteria in turn, a series of normal distributions, spaced out along the horisontal axis, would be expected.

To conduct a fitting trial to determine the optimal height for a control (like a door handle or a light switch) or for a working surface would be slightly more complicated. A gap between obstacles can be too narrow for convenience, but it cannot reasonably be too wide; whereas a door handle or a work bench can be too low, too high or just right. We could start with a position which was too low for everyone; then as we raised the level of the object, we would expect the percentage of subjects who judged it to be too low to decrease steadily. As it did so, the percentage of people saying it was just right would climb to a peak value, before beginning to fall again — as an increasing number of people began to say that it was too high. The characteristic form which the results of such experiments are generally found to take, is shown in Figure 5.

The latter experiment has a considerably greater element of subjective judgement than the former (which is much more of a go/no-go situation, in which you can either get through the gap or else you cannot). This has a number of important consequences. Characteristically, it is found that the range of optima reported in ascending trials (i.e., from low to high) is placed higher than the range reported in descending trials (i.e., from high to low). This is true of psychophysical experiments in general.

If we conduct our fitting trial with each subject running through the range of possible positions of the door handle, from low to high and back to low again, and plot the results, a graph similar to that of Figure 6 would result. Subjects have experienced all the

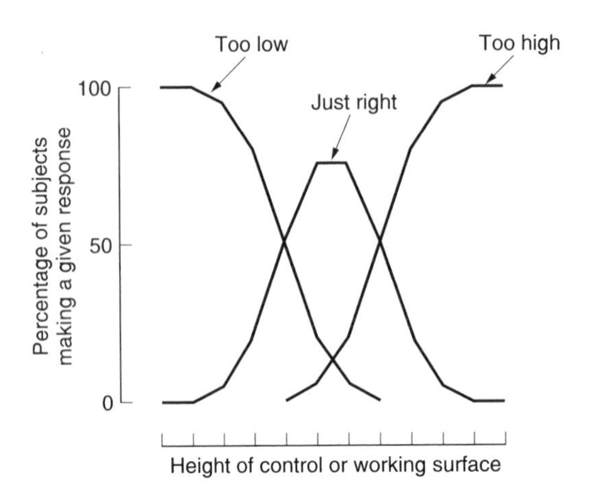

Figure 5 Characteristic form taken by the results of a fitting trial.

positions from too low to too high, and a section in between which they say is satisfactory. The test is conducted, for each person, with the height changing up and down, to incorporate the effects of ascending and descending trials as described above. We use subjects who are at the extremes of the body dimensions which are relevant to the dimension of the object which we are testing; in the example of Figure 6 these are the 5th and 95th percentiles. We may include some others if we wish, and 50th percentiles were also used here.

If a line can be drawn through all the 'satisfactory' dimensions, then this indicates that there is one level of this dimension which suits everyone. If this cannot be done, then a line through the bottom of the highest 'satisfactory line', and one through the upper end of the lowest, demonstrate the minimum range of adjustment needed to suit the population concerned. The method due to Jones (1963), finally needs all the separately selected dimensions putting together and checking with the sample of subjects, since dimensions may interact with each other, which would only be revealed by their final grouping and checking.

It is also worth noting that the form of results shown in Figure 5 also turns up in quite different areas of ergonomics, for example, in studies of thermal comfort (Fanger, 1973; Grandjean, 1988; and see Chapter 23). In fact it is likely to be relevant to studies of subjective preference in general. The lines on the graph could represent all sorts of things: too soft and too hard; too cool and too warm; too light and too dark; too young and too old, etc.

The anthropometric method of limits

The process by which we establish final design recommendations with respect to anthropometric and biomechanical criteria, is known as the *method of limits* (e.g., Woodworth and Schlosberg, 1954). The term is borrowed from psychophysics. The anthropometric method of limits is essentially a model or analogue of the fitting trial, in which the anthropometric data stand as substitutes for the experimental subjects. You could say that, in applying this method, we are attempting to predict, using pencil and paper methods, what the result of a fitting trial would be if we were to perform one. (Note that the fitting trial is a special case of the psychophysical method of limits.)

In applying this method to any particular design problem, we are attempting to establish those boundary conditions, which make an object too big, too small, and so on, with respect to certain design criteria. In the simplest cases we may do this by inspection. This requires us to identify the *limiting user*— that is a hypothetical individual, who by virtue of his or her extreme bodily characteristics, is particularly difficult to accommodate with respect to the

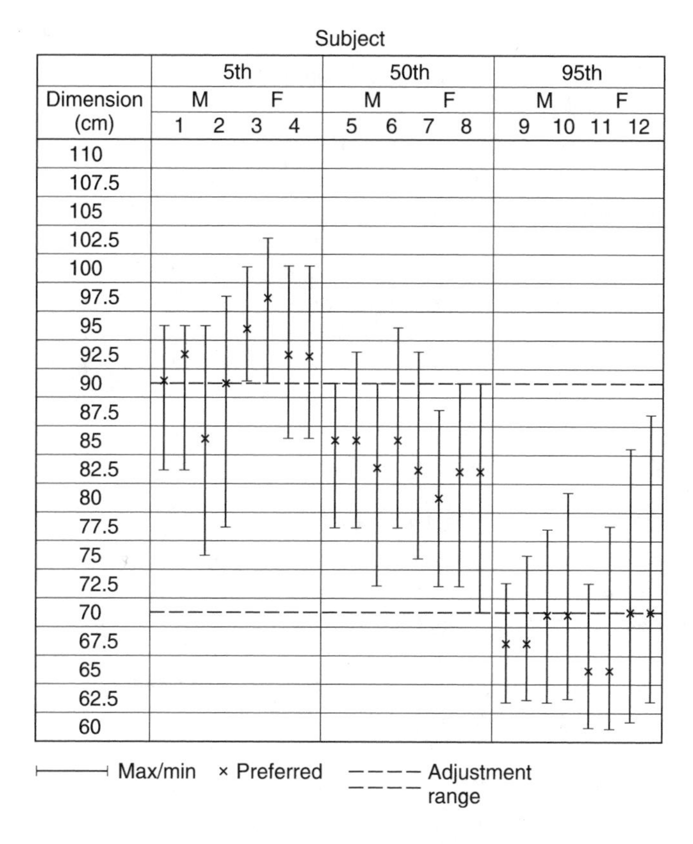

Figure 6 Graphical display of the results of a fitting trial, from which the extent of adjustment to the dimension may be deduced. 12 subjects were used, 6 male and 6 female, in the 3 percentile groups shown.

criteria concerned. If the limiting user is accommodated, it necessarily follows that the majority of the population, who are less demanding in their requirements, will be accommodated as well.

Anthropometric criteria for workspace design

Anthropometric criteria fall into three principal categories: clearance, reach, and posture.

Clearance

Clearance problems include those which relate to head room, knee room, elbow room and so on, as well as those concerned with access through passageways, around and between equipment and into equipment for maintenance purposes. These are amongst the most important issues in workspace design since mismatches in these respects may be particularly hazardous. Unresolved clearance problems may also have knock-on effects in terms of unsatisfactory working postures, which the user is forced to adopt. For example, if the distance between the surface of the seat and the underside of a table does not provide adequate clearance for the thighs and knees of the users, then they may have to adapt to the situation by pushing the chair backwards (and leaning forward excessively to perform the task) or by perching on the front of the seat (and hence losing the support of the backrest). This is quite a common problem at service counters and cash tills.

Clearance should be adequate for the largest user. For practical purposes it will usually be expedient to set the design limits at the 95th percentile hence by definition accommodating 95% of the target population.

As an example, suppose we wish to find the minimum clearance required between the arms of a chair. We look up a suitable table of anthropometric data. Searching through this table we find the dimension 'hip breadth', which for our target population (British adults) has 95th percentile values of 405 mm for men and 435 mm for women. We adopt the larger figure. This would be an exact fit for the hips of the limiting user, but it would not allow for her clothing (since anthropometric data are usually quoted for unclad people) nor would it allow her any leeway. Making a commonsense correction for these we arrive at a round figure of 500 mm. (We could, if we wished, formalise this last stage, by specifying exactly what clothing ensemble she is likely to be wearing, and how much leeway we wish to give, but in reality (and for this problem) this degree of precision is rarely realistic.)

The process we have gone through could be summarised as follows:

Criterion: seat breadth > hip breadth.
Limiting user: 95th percentile woman = 435 mm hip breadth.
Corrections: clothing and leeway = 65 mm.
Design recommendation: seat breadth > 500 mm.

This application of the method of limits is in many respects equivalent to the first fitting trial we discussed in the previous section; and if we plotted a graph of the seat breadth (horisontally) against the percentage of users accommodated (vertically) we should again get a normal ogive, as shown in Figure 2.

Reach

Reach problems include those which are concerned with the location of controls in the workspace as well as things like the height of a seat (where it is necessary for the feet to reach the floor) and the height of visual obstructions. Seat depth also falls into this category since it is necessary for the user to 'reach' the backrest without undue pressure on the backs of the knees.

The procedure for dealing with reach problems is the same as the one used for clearance problems, except in this case the limiting user will be a small member of the target population, usually a person who is 5th percentile in the relevant characteristic. Note that both the clearance and the reach criteria impose limits in one direction only. The clearance criterion indicates when an object is too small, but not when it is too large — and conversely for reach. Both clearance and reach therefore are *one-tailed constraints*. There may of course be other constraints acting in the opposite direction. In the case of clearance and access problems, these might include economy of space, or the reduction of distances travelled. In laying out working areas, we often find that clearance problems (e.g., elbow room) interact with reach problems (e.g., the accessibility of controls). The interaction of two opposing sets of constraints creates a situation which is equivalent to the one which is described in Figure 5.

Note also that there is an important class of design problems (concerned with the safeguarding of machinery) in which the conventional criteria of clearance and reach are reversed. In these cases we actively wish to prevent access to hazardous areas, or to place the hazards out of reach. So the limiting user might, for example, be a person with long slender limbs who could reach the furthest distance through the smallest aperture.

Posture

Posture problems are inherently more complicated. For example, a working surface which is too high is just as undesirable as one which is too low. This limits the design in two directions to give a *two-tailed constraint* of the kind shown in Figure 5. In this situation there are two options open to us. We either have to provide an adjustable workstation, so that each user may set it to his or her own optimum dimensions (as discussed previously under 'Design limits'); or else we have to settle on a single overall compromise value which will maximise the number of users who are accommodated and minimise the inconvenience suffered by the remainder. Supposing our problem concerns the height of a workbench to be used by a standing person for performing a certain manipulative task. (It is assumed that men and women will use the bench.) A suitable height is between 50 and 100 mm below elbow height. It follows from the shape of the normal distribution that the single overall compromise value which will accommodate the greatest number of people is 75 mm below the average (or 50th percentile) elbow height. Looking up an appropriate table of data, we find the 50th percentile elbow height is 1090 mm for men and 1005 mm for women. This gives an overall average (i.e., for both sexes) of 1048 mm, to which we should add 25 mm for shoes, giving 1073 mm. Subtracting the 75 mm and rounding up, gives us a final recommendation of exactly 1000 mm. We may express this formally as follows:

Criterion: (elbow height — 100 mm) < working surface height < (elbow height — 50 mm).
Optimal compromise: 50th percentile elbow height — 75 mm = 973 mm.
Corrections: 25 mm for shoes.
Design recommendation: working surface height = 998 mm, rounded 1000 mm.

At this point it will be pertinent to consider how many users will be mismatched with respect to the criteria and how seriously inconvenienced they will be. This will enable us to decide whether a single compromise height will indeed be satisfactory, or whether an adjustable workstation will be necessary.

For a further discussion of the method of limits and further examples of its application see Pheasant (1986, 1990). For a compilation of design standards and guidelines concerned with anthropometric and ergonomic issues, see Pheasant (1987) and Steenbekkers and van Beijsterveldt (1998).

Conclusions

In this chapter some of the basic uses of anthropometric data have been described, together with an introduction to sources. When choosing anthropometric data it is necessary to read the small print. Who have been measured? What ages? Both sexes? How many in each sample? How old are the data? Listings of anthropometric data and their diagrams look very impressive, but this is no substitute for reliability and statistical validity.

When preparing a design for a particular situation, e.g., in a factory or office, and you are not sure whether or not tabulated data are suitable, it is valuable to measure some major dimensions for a sample of potential users (e.g., their stature). By comparing the mean and standard deviations of relevant dimensions for the situation you are concerned with, to the tabulated dimensions in published data, simple statistical tests will confirm whether or not they are from the same populations. If so, then you can use the tabulated values with confidence.

It must be emphasised that ergonomics is not synonymous with anthropometrics; it is an unfortunate belief in some design offices that this is the case, and that an ergonomic situation is achieved by choosing dimensions in conjunction with anthropometric tables. Anthropometrics are a necessary, but by no means sufficient, contribution to ergonomic design and evaluation. If the reader keeps in mind that we are dealing with a whole person, with needs for comfort, performance, interest and all those other things which make people so fascinating, then we are unlikely to reduce them to a simple matter of linear dimensions, no matter how vital the statistics.

References

Chaffin, D.B. and Anderson, G.B.J. (1992). *Occupational Biomechanics*, 2nd edition (New York: John Wiley & Sons).

Clark, T.S. and Corlett, E.N. (1984). *The Ergonomics of Workspaces and Machines: A Design Manual* (London: Taylor & Francis).

Corlett, E.N. (1978). The human body at work: new principles for designing workspaces and methods. *Management Services*, May, 20–52.

Dainoff, M.J. and Dainoff, M.H. (1986). *People and Productivity — A Manager's Guide to Ergonomics in the Modern Office* (Toronto: Holt, Rinehardt and Winston).

Fanger, P.O. (1973). *Thermal Comfort* (New York: McGraw-Hill).

Grandjean, E. (1988). *Fitting the Task to the Man — An Ergonomic Approach*, 4th edition (London: Taylor & Francis).

Jones, J.C. (1963). Fitting trials. Architects Journal Information Library, February, p. 321.

McCormick, E.S. and Sanders, M.S. (1987). *Human Factors Engineering and Design*, 5th edition (New York: McGraw-Hill).

Norris, B. and Wilson, J. (1995). *Childdata, The Handbook of Child Measurements and Capabilities — Data for Design Safety* (Nottingham: Department of Trade and Industry).

Ong, C.N. (1984). VDT workplace design and physical fatigue: a case study in Singapore. In *Ergonomics and Health in Modern Offices*, edited by E. Grandjean. (London: Taylor & Francis), pp. 484–494.

Peebles, L. and Norris, B. (1998). *Adultdata, The Handbook of Adult Anthropometric and Strength Measurements — Data for design safety*, (Nottingham: Department of Trade and Industry).

Pheasant, S.T. (1986). *Bodyspace — Anthropometry, Ergonomics and Design* (London: Taylor & Francis).

Pheasant, S.T. (1987). *Ergonomics — Standards and Guidelines for Designers* (London: British Standards Institution).

Pheasant, S.T. (1990). *Anthropometrics — An Introduction*, 2nd edition (London: British Standards Institution).

Smith, S., Norris, B. and Peebles, L. (2000). *Older Adultdata, The Handbook of Measurements and Capabilities of the Older Adult-Data for Design Safety*, (Nottingham: Department of Trade and Industry.

Steenbekkers, L.P.A. and van Beijsterveldt, C.E.M. (eds) (1998). *Design Relevant Characteristics of Ageing Users — Backgrounds and Guidelines for Product Innovation*. (Delft: Delft University Press).

Vannier, M.W. and Robinette, K.M. (1995). *Proceedings '95 Biomedical Visualization*, October 1995, pp. 1–8, IEEE Computer Society Press.

Woodworth, R.S. and Schlosberg, H. (1954). *Experimental Psychology*, 3rd edition (London: Methuen and Co).

www.sae.org/technicalcommittees/caesumm.htm, d.d. 21–10–2002.

Chapter 27

The evaluation of industrial seating

E.N. Corlett

Introduction

The work seat is as much a tool as a keyboard or any other piece of equipment in the workplace. Its design and functioning will be influenced by the tasks to be done, the environment and, of course the anthropometry of its users. It is obvious that there will not be one seat appropriate for all jobs, although well designed adjustments can extend the application range of a seat. Nevertheless, a seat is not 'ergonomic' of itself but only as a result of how well it is designed to allow its users to achieve their work objectives.

Seat requirements

Some of the important requirements for the adequate functioning of a work seat are given in Table 1. They are based on the seat as a full body support rather than as a temporary perch, for whilst the latter has utility, most of the body weight is on the feet, which is against its use for the longer term.

If more than one third of the body's weight is carried via the feet the sitters will complain of leg discomfort (Corlett and Gregg, 1994, Eklund *et al.*, 1982); their leg muscles will be actively supporting considerable body weight which would otherwise be transmitted through the seat and backrest, and this will be a major source of discomfort. The requirement to allow changes of posture is a good one and also a practical necessity in many jobs. A high seat at a supermarket checkout can allow the users to reach further to the sides if they can move their legs to gain support and this muscular activity is beneficial in itself.

In many cases a work seat has to resist other forces than just the body weight, and these must be transmitted through the sitter to the seat and hence to the ground. The backrest is one channel for this, particularly for pushing forces, for if no backrest is present the back muscles have to be in tension to increase the rigidity of the spine to transmit these forces via the seat and legs. Even without these horisontal forces the weight of the trunk and head is partly channelled through the backrest. (Corlett and Eklund, 1984).

The evaluation of seating

A work seat is for working from, thus the function must influence the form. The behaviour and subjective judgements of the users must be part of the evaluation and to use dimensions alone is not adequate. Table 2 shows methods used by various workers to evaluate some of the aspects of chair design listed in the previous table. The choice of methods will depend on the investigation's purpose and, particularly, on whether the work is done in the field.

Table 1 Functional Factors in Sitting

The task
Seeing
Reaching
Exerting forces
The sitter
Support weight
Resist accelerations
Under-thigh clearance
Trunk-thigh angle
Leg loading
Spinal loading
Neck/arms loading
Abdominal discomforts
Stability
Postural changes
Long-term use
Acceptability
Comfort
The seat
Seat height
Seat shape
Backrest shape
Stability
Lumbar support
Ajustment range
Ingress/egress

In what follows several methods which have been found useful are presented, based on the seat model of Table 3, which links functions, effects and the required measures.

A more comprehensive overview of seating evaluation would extend the methods section of Table 3 to include such research techniques as X-rays, optical recording systems and a wider range of physiological areas of study. Such fundamental research is complimentary to that discussed here, seeking to understand why discomforts arise, discs are damaged or particular postures adopted. It gives the basis for the practical evaluations described in this chapter, where the acceptability, security and effectiveness of a seat are the primary focus.

Dimensional evaluation

As noted earlier, body dimensions are not sufficient on their own for designing a seat, but they are, of course, essential and the applications of anthropometric data are described in earlier chapters. The length and height of a seat to give adequate clearances have been discussed by Akerblom (1954), Floyd and Roberts (1958) and Murrell (1965). Briefly, for a chair in use at a table, the seat length must be short enough to clear the calf of a 5th percentile female user, and the height allow light contact under the thighs when the sitters' feet are flat on the floor. Obviously height adjustments will be necessary if a 95th percentile male and a 5th percentile female are intended to use the same chair.

For decisions on adjustment, Jones (1963) devised a simple and effective fitting trials procedure; see the chapter on anthropometry and workspaces. When running such trials we might keep Branton's (1969) comment in mind, that people older than 30–35 years are likely

Table 2 Some of the Methods Used in Assessing the Functional Qualities of Industrial Seating

	Methods										
Functional factors	Dimensional measurements	Fitting trials	Force or pressure measurements	Bio-mechanics calculations	Observations or timing of behaviours	Subjective judgements overall	Subjective judgements body parts	Scaled Checklists	Cross-modality	Reach/ force/ stability	Stature changes
Seeing	√	√				√					√
Reaching	√	√		√		√	√				√
Seat	√		√	√	√	√	√	√		√	√
Backrest	√		√	√	√	√	√	√		√	√
Adjustments	√	√						√		√	
Ingress/egress	√	√		√	√	√		√			
Stability		√		√		√				√	
Support weight			√	√		√	√		√	√	
Under-thigh clearance	√		√			√	√	√			√
Trunk-thigh angle	√			√	√						
Leg load			√	√		√	√	√			√
Spinal load				√		√	√				√
Neck/arm load					√	√	√	√			√
Posture changes			√		√	√					
Long-term use					√	√	√	√			√
Acceptability				√		√		√			
Comfort			√		√	√	√	√	√		
Lumbar support	√	√	√				√	√		√	√

Table 3 Seating Model for Assessment of Industrial Seats (adapted from Eklund, 1986)

		Responses and effects	
Functional factors		Initial	Subsequent
The task The sitter The seat	(detailed items as in Table 1)	Postures Loads Pressures Influences on blood flow Discomfort Preferences	Discomfort Pain Disease Reduction in performance
Measures			
Workplace dimensions Work weights Work forces Work reaches Work time patterns Anthropometry Strength		Biomechemical load EMG Statute change Rating Ranking Dilations of body parts Linear measurements Posture	Rating Ranking Clinical examination Epidemiological studies Performance

to be more sensitive to discomfort than younger people. The evidence in fitting trials is in favour of experimentation since the variability in both subjects and tasks will require evidence for how easy it is to see, to reach, to exert forces effectively and to maintain a stable position on the seat.

From the fitting trials procedure direct evidence for the need for adjustments and the necessary extent of these is obtained. LeCarpentier (1969) found variations by subjects on preferred dimensions on different days during his studies of comfort but this is a subjective and individual matter, discussed later in this chapter. Where dimensions have direct influence on activities the same range of variability does not appear to exist, allowing decisions on dimensional features to be chosen with confidence.

Ingress and egress

These factors are especially important in two cases in particular, seating for the elderly and sit-stand seats. The former is not part of this chapter but studies by Shipley *et al.* (1969) showed that accessibility of chairs for the elderly was best assessed by the times taken to get into and out of them, coupled with observations of the subjects' behaviour during these manoeuvres. However, for people without disabilities time might be only one measure, perhaps not sufficient for separating a range of seats.

For sit-stand sitting the advantages of a saddle are reduced when its use is observed. They are awkward for women in skirts and posture variations are limited. The times of use for what is often a hard seat surface which supports only part of the buttocks, as well as observations of how users get on and off them, are probably the best methods of assessment. Seat and user stability, discussed later, are also important, as are the problems of getting on and off low and high seats, again matters for observation, where a pilot trial to identify important factors and prepare a pro-forma record sheet can be valuable.

Table 4 Coding of Sitting Postures (Branton and Grayson, 1967)

Head	
Free of support	1.
Against headrest	2.
Against side wing	3.
Supported by hands	4.
Trunk	
Free from backrest	1.
Against backrest	2.
Lounging/slumped back	3.
Arms (one or both)	
Free from armrest	1.
Upon armrest	2.
Legs	
Free, both feet on floor	1.
Crossed at knee	2.
Crossed at ankle	3.
Stretched forward	4.

Observational methods

It is an obvious point that the investigator should look at what the users do, the other side of the coin to gathering their experiences of using the chairs. Kember (1976) utilised an adaptation of Benesh Notation, a choreographer's tool, to record in detail the postures and movements of a chair user. The method is capable of recording on a time base and, providing the three months intensive training suggested by Kember is not prohibitive, will record sitting activities in great detail. For most studies this investment in training will be unacceptable, for there is still need for detailed analysis after recording.

In general, workers have used simpler methods, of which that by Branton and Grayson (1967) is typical. It may be modified to include actions and other factors relevant to the problem under investigation with little difficulty. For their easy chair comfort study they selected positions of the head, trunk, arms and legs, with separate numbers in each body part for the important postures, as shown in Table 4. If, for example, the head was supported by the hands, or the legs were stretched forward, the relevant number on the recording sheet was ringed. The recording system is easy to learn, to use and to analyse, by computer if desired.

Whilst the whole range of observational techniques can be called upon, the time for analysis must be kept in mind in relation to the value of the results and to understanding the problem. Activity sampling methods, which are often simple to use, (Branton, 1969, Grandjean, 1980) can condemn the user to many hours of tedious analysis. If body markers can be used, electronic methods are very practical and in the laboratory are well used, see Chapter 16 on posture recording. Whilst video recording is becoming more acceptable it is sometimes not desired by subjects and, as yet, analysis can also be a long process.

The length of observation periods should bear a relationship to the period of use. Where work involves frequently moving from the chair then short periods of observation may be appropriate but if a 'desk-bound' job is under study assessments at intervals during the whole day are more suitable. Workers have reported significant differences between morning and afternoon sitting behaviours (Branton and Grayson, 1967, Shipley *et al.*, 1969) whilst there are differences between men and women in the sitting postures adopted as well as in the length of sitting periods, noted by Shipley *et al.* (1969) but almost certainly applicable to situations other than the elderly.

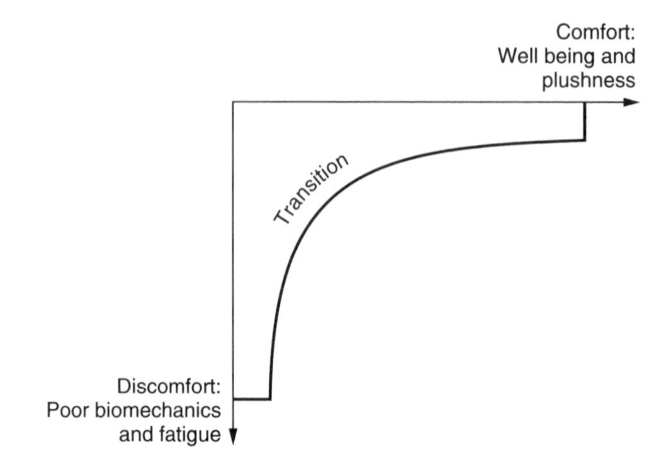

Figure 1 A conceptual model of sitting comfort and discomfort.

Subjective methods

When using these methods we are utilising the subjects' experience of the seat to identify things about the seat; our purpose is to seek information to lead to improvements. One area where these methods have been used extensively is in studies of comfort and discomfort, key factors in the acceptability of a chair.

Work in the investigation of comfort by Zhang and Helander (1992), Zhang *et al.* (1996), Helander and Zhang (1997) demonstrates that comfort is closely related to the softness and appearance of the seat whilst discomfort relates to the length of use and is fatigue related. This is an extreme simplification of their findings but, for our purposes in seat assessment we will use their model, Figure 1, which shows that whilst the absence of discomfort does not necessarily get a chair assessed as comfortable, and vice versa, an uncomfortable chair can never be considered comfortable. Thus both comfort and discomfort should be measured, for which these workers showed an instrument developed from Shackel *et al.*'s (1969) General Comfort Rating Scale and their Chair Evaluation Checklist, (see later). This checklist (Figure 2) (Helander and Zhang, 1997) produces a profile of the subjects' experiences. They found the Comfort section provided only small changes over the day except for the items 'I feel relaxed', 'I feel refreshed' and the overall item 'I feel comfortable'. For the discomfort side there were significant time period effects, making it necessary to run the Checklist at intervals throughout the day. The two scales can be presented together, their trials found no effect from their proximity.

Hence it would appear that these scales, for comfort and discomfort, are relatively independent and can be used to assess each dimension using the combined instrument of Figure 2. It should be noted however that Helander and Zhang (1997) report that the items 'feeling relaxed' and 'feeling refreshed' did show some time effects. Whilst their association of this to fatigue is probably correct the comment by Shipley (1980) should be noted. This was that discomfort varies with arousal and attention; periods when attention 'turns inward towards the condition of the self' can cause what has been ignored to receive attention. Hence, for example, the work conditions can be an influence on the perception of comfort, should the external environment come to be viewed adversely.

The Helander and Zhang checklist of Figure 2 gives a profile of the users' feelings; any combination of these checklists over a number of subjects may well give a broad variance for the data. A similar design of checklist has been used to assess the characteristics of a chair (Figure 3). This Chair Feature Checklist (CFCL) will give a profile of the chair features,

Discomfort factors are rated below

	Not at all			Moderately			Extremely		
I have sore muscles	1	2	3	4	5	6	7	8	9
I have heavy legs	1	2	3	4	5	6	7	8	9
I feel uneven pressure from seat pan or seat back	1	2	3	4	5	6	7	8	9
I feel stiff	1	2	3	4	5	6	7	8	9
I feel restless	1	2	3	4	5	6	7	8	9
I feel tired	1	2	3	4	5	6	7	8	9
I feel uncomfortable	1	2	3	4	5	6	7	8	9

Comfort factors are rated below

	Not at all			Moderately			Extremely		
I feel relaxed	1	2	3	4	5	6	7	8	9
I feel refreshed	1	2	3	4	5	6	7	8	9
The chair feels soft	1	2	3	4	5	6	7	8	9
The chair is spacious	1	2	3	4	5	6	7	8	9
The chair looks nice	1	2	3	4	5	6	7	8	9
I like the chair	1	2	3	4	5	6	7	8	9
I feel comfortable	1	2	3	4	5	6	7	8	9

Figure 2 The final Chair Evauation Checklist. *Instructions to subjects*: Please rate the discomfort and comfort of this chair. Discomfort factors are listed to the left and comfort factors to the right. Mark an 'X' on each line at the point that best describes your feelings or impressions. Note: 1 = Not at all; 9 = Extremely.

showing which cause difficulties for users, and will provide results which can be used for directly modifying chair design, showing which features are most disliked or difficult to use. What is more, for a given range of anthropometric dimensions the variance of the results is likely to be acceptably narrow. Note, incidentally, that this kind of scaled feature checklist can be used for many other investigative purposes than for chairs.

The General Comfort Rating Scale of Shackel *et al.* (1969) (Figure 4) has been shown to give useful results. It was initially used by Shackel *et al.* over ten chairs and distinguished between them appropriate to the uses to which they were being put. Tests run by Drury and Coury (1982) used this same scale at half-hour intervals to investigate a series of seats for office workers. They found it effective and, additionally, useful to compare their results with those of Shackel *et al.*, giving them some idea of the relative merits of their chairs to those previously tested.

In using subjective methods the form of the questions is obviously important. Habsburg and Mittendorf (1980) note the tendency of some subjects to try to make a general judgement rather than one for themselves alone. They recommend that the phrase, 'is it for me/not for me' should be part of the test instructions. Another point about such tests is that different groups will have different standards of judgement, there is no standard for comfort. Aircrew tested by Wachsler and Learner (1960) reported feelings of comfort when their buttocks and backs were comfortable, and were willing to overlook other discomforts, whilst key punch operators studied by Langdon (1965) described their chairs as comfortable even when some were as high as 480 mm.

A widely used method of assessing discomfort is the 'body map' proposed by Corlett and Bishop (1976), to get a scaled measure, over time, of body part discomfort (BPD) (see also Chapter 16). By associating the results from this measure with the postures adopted to achieve

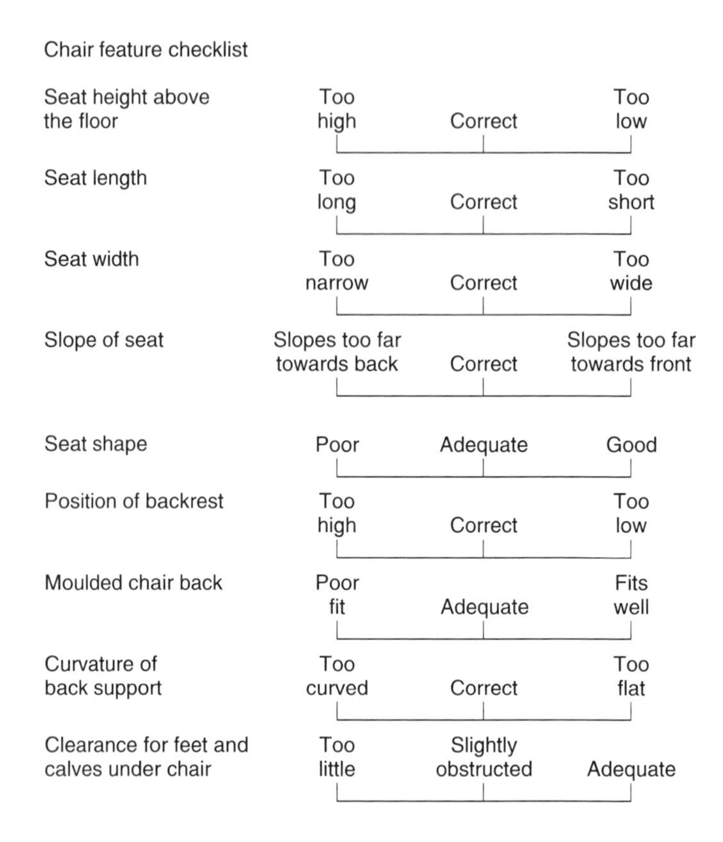

Figure 3 Chair feature checklist of Shackel *et al.* (1969), modified by Drury and Coury (1982).

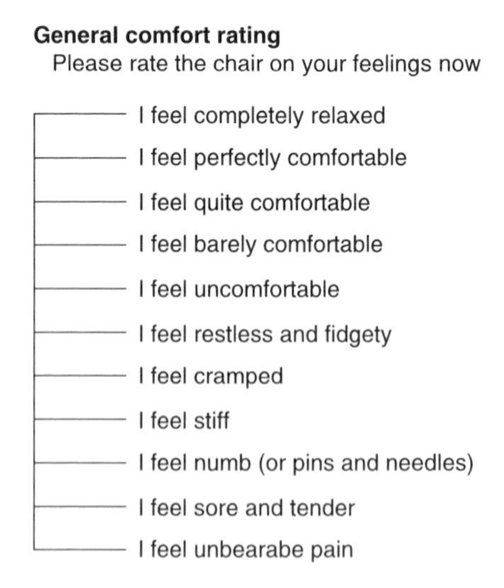

Figure 4 General comfort rating scale of Shackel *et al.* (1969).

the work output the sources of the reported discomforts can often be identified readily and can be corrected. This procedure, together with several of those mentioned earlier can be used before and after the changes to the seating or workplace, demonstrating the improvements (or not!) which have been achieved for the users.

How long to test?

The UK Furniture Industry Research Association will test chairs, often until destruction, but that is not an option for ergonomists! A practical procedure is to give the users a brief period to get used to the chair, adjusting it to their own dimensions, then run trials of two or more hours, gathering information at regular intervals. Both Shackel *et al.* (1969) and Drury and Coury (1982) give details of such procedures. LeCarpentier (1969) found no changes in preferred easy chair settings after 20 min even though his subjects used them for 4 hours. On the other hand, Jones (1969) found that some levels of discomfort did not appear until after three hours or more, though Branton (1966) identified a pattern of sitting behaviour in less time than this.

Cross modality matching (CMM)

Cross modality matching (Stevens, 1975; Pepermans and Corlett, 1983) is the procedure by which an experimenter asks subjects to judge the level of one stimulus by comparing it with another in another modality. Thus, an attempt by Branton (1969) to use a hand dynamometer to indicate the level of bodily tension experienced by a sitter. These trials were unsuccessful but he considered that the procedure had some value and should be explored. This was done by Gregg (2000) in an extensive series of studies of methods for chair evaluation. In one study (Gregg, personal communication) he used a sphygmomanometer cuff as the comparison stimulus for the pressure at different points on a seat surface, the subjects inflating the cuff to a point where they considered it to represent the perceived pressure on the seat. In another (Gregg, 2000) he used the cuff as a stimulus for subjects to identify levels of pressure which would match the five levels of the BPD diagram. This was not to calibrate the BPD scale but to see if certain levels of CMM when the pressure cuff was used could be identified with discomfort levels used in the BPD system. He demonstrated that a subject could identify a level of discomfort related to a pressure, suggesting that the procedure could be useful for linking values of the cuff as a CMM stimulus with a discomfort scale. He also noted that the memory for a pressure was very transient.

Magnitude estimation, (Stevens, 1975), could be seen as a special case of CMM. In this technique subjects are asked to state a number which they see as representative of the presented stimulus, there is no constraint on the range of numbers used. Gregg (2000) demonstrates that subjects can be consistent in the way they use their chosen scale and, if the geometric mean of the estimates of a group of subjects is used the values across all subjects can be combined and form a valid metric.

To demonstrate this he used the pressure cuff, with pressures presented in a random order and with two trials with a time interval between them, and asked subjects to state a number representing each pressure. Using the geometric mean he found that the results for the two trials were not significantly different and that the results for all nine subjects could be combined.

The procedure is very practical for fieldwork. Gregg (2000) used it in a trial of two different sets of seats in a sewing factory. He tested twenty workplaces and asked for an evaluation of 16 chair features using magnitude estimation, where a higher number represented an improvement. He found that the operators could do the required estimation with ease, and rapidly, wthout undue interference with their work.

Posture change and stability

Attempts have been made to use the number of posture changes in a given time as a measure of seat discomfort. Rieck (1969), when looking at automobile seating, noted that the measurements of small movements in the seats were not related to perceptions of discomfort. Of course changes in posture are frequently necessary in the conduct of a job, which increases the difficulty of interpreting the metric. Branton (1966), when looking at different seat shapes, found gross changes of posture, larger than the 'fidgets' which are the target of this method.

In a trial by Gregg (2000) using small pressure transducers, he ran a one and a half hour trial with one subject, recording a trace from two transducers over the whole period. With a computer programme to analyse the traces he found that an increase in fidgets over the time period was present and concluded that the use of even just two transducers, (one under the right ischial tuberosity and the other under the thigh), would be a practical procedure for fieldwork where the work activity required very little movement.

Stability has usually been considered in terms of the chair tipping over, or its resistance to movement during work. Branton (1969) suggested that 'if the seat does not allow postures which are both stable and relaxed, the need for stability seems to dominate that for relaxation', requiring muscle activity to secure stabilisation. It may appear self-evident that people will not relax on a seat if, by doing so, they would fall out of it, but some proposals for forward-sloping seats have certainly increased the efforts needed to maintain stability at the expense of relaxation. This is not to say that, in some cases, the trade off has not been acceptable to some users; it is just necessary to note that muscular effort is a concomitant of such designs.

Changes in stature

Since one of the factors in the seating model proposed earlier is a reduction in spinal loading it is necessary to devise a procedure to measure these loads. Eklund *et al.* (1986) examined several methods for calculating the effectiveness of a seat in these terms. Biomechanical analysis, in conjunction with a force platform (Eklund *et al.*, 1983), or an instrumented chair (Eklund, 1986; Eklund *et al.*, 1987; Gregg, 2000) could assess the loads on the spine, but provided no direct evidence of the validity of the values, or their actual effects on the sitter.

Eklund and Corlett (1986) reported a study to compare BPD, biomechanical analysis and stature change measures for industrial seats. The study was conducted in a laboratory and involved a force production task, a sideways viewing task (as in some fork-lift truck driving) and a sit-stand seat. They noted that all three methods were effective but contributed different areas of information. The theoretical evaluation of the load, using biomechanics, was quick and inexpensive. The stature change method needed at least 30 min of exposure per subject and may need several subjects but was better for major load situations. It had the advantage of giving a numerical measure of the actual effect of the load on the spine. Discomfort assessment was inexpensive, sensitive and suitable for field work but it, too, required long exposure of the subjects. Trials with the precision stadiometer had demonstrated its practicability in the industrial setting.

Evaluation procedures

The procedures adopted depend on the purposes of the evaluation. Several different test procedures are given here, each for a particular purpose; the seekers for a procedure to adopt may adapt what follows to suit their needs.

1. Shackel *et al.* (1969) wished to evaluate a range of chairs for their utility in various situations. Their procedure had three approaches:

   (a) ranking of the chairs for preference whilst sitting on them but without being able to see or touch them;

(b) long term sitting whilst working, with regular completion of a GCR and a BPD ranking;
(c) completion of a CFCL at the end of the session.

This procedure showed where chairs were inadequate, either in dimensions or in relation to tasks, and which chairs were preferred.

2. Drury and Coury (1982) required a test for a single chair to see if it was suitable for certain jobs and to provide information to the maker as to any necessary alterations. The procedure they adopted was:

   (a) comparison with anthropometric data, standards and principles;
   (b) subjects' opportunity to adjust the chair until their feelings of comfort were maximised, typically taking about 5 minutes;
   (c) a sitting and working period of about 2.5 hours, with GCR and BPD scales given every half hour.
   (d) CFCL given at the end of the sitting session.

3. Yu *et al.* (1988) used a fractional factorial experiment to evaluate seven variables in seat design for a sewing task. These were seat height, seat angle, whether or not the seat rocked, whether or not it swivelled, backrest distance, backrest height and backrest angle. Their dependent variables were general discomfort, BPD, stature change and electromyography (EMG).

   Two females spanning the 90% range of female stature were subjects. In a summary table of results they showed that BPD and stature change were the most informative, with the results supporting each other. The other two methods did not give any additional information for the selection criteria for the chair. The tests demonstrated the economy and effectiveness of this procedure.
4. Wotzka *et al.* (1969) were concerned with recommending a seat for four large auditoria at a university. First they did an activity sampling study, with a questionnaire, about the seat and desk characteristics. From this they deduced a range for the variables of importance and built five seats to test their effectiveness. A number of subjects attended one session of about 20–30 minutes. First, the seat was adjusted for each subject 'until the most comfortable posture had been achieved', when each subject assessed six body parts for 'uncomfortable, medium or comfortable' on a questionnaire. The final test was in an auditorium, using the same seats and the same questionnaire.

   Here, the questionnaire was linked with observed behaviours and body part judgements in a long study requiring a large number of subjects, a procedure which increases the confidence in the results but at an increase in time and cost. Other workers have used similar procedures. Hunting and Grandjean (1976) used a large number of subjects, activity sampling and bi-polar questionnaires on discomfort of particular body parts, in the comparison of seats, and deduced preferences and changes from the results.
5. In the development and evaluation of a new design of seat Gregg (2000) employed some procedures which have not been common in seat investigations. The seat, Figure 5, had a curved surface designed to allow the sitter to be higher than is conventional, with an open angle at the hip and both feet firmly on the ground; (see Corlett and Gregg, 1994; Corlett, 1999).

   Two approaches were taken. The first dealt with methods to assess various specific aspects of the seat itself whilst the second concerned the use of the seat in

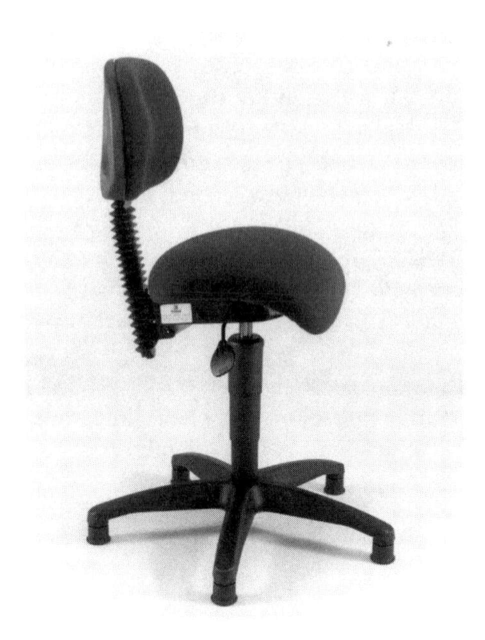

Figure 5 Photograph of Nottingham Seat.

workplaces; each used some methods specific to the approach concerned.

To assess the effects of features of the seat, spinal loading with the curved seat was compared with that from a conventional, 90 degree sitting, version, using a precision stadiometer (Eklund and Corlett, 1986). Distribution of the body weight between seat surface and floor was examined, via an instrumented chair and a force plate in the floor, for various seat shapes produced by vacuum consolidation, to devise the appropriate curvature (Gregg, 2000; Corlett and Gregg, 1994). A modified BPD, together with CMM trials using a sphygmomanometer cuff, were used to investigate the pressure distributions across the seat surface for various seat heights and angles. Modifications of the CFCL and GCQ, illustrated earlier in this chapter, as well as a Magnitude Estimation protocol, were developed for use in the field trials.

In the major field trial the procedure was such that the experimental seat could be compared to that in current use in the plant. For seat discomfort the BPD method was used, with an additional insert identifying the buttock area, together with a five point scale to describe the intensity of the discomfort at each point. This was done at intervals over the working day.

At the end of the shift the seat features were investigated using a version of the CFCL, the tests covering both the new and the conventional seat. Also at the end of the work period, Magnitude Estimation was used to evaluate sixteen factors, comparing the two seats. There was duplication between the two ranges of tests, but this was deliberate to determine the effectiveness of the two methods. In addition, perceived pressure and perceived stability were recorded subjectively on linear scales at intervals through the days of the trials. A group of twenty workers were involved, ten with the new seat, and after one week the seats were switched so that those with the conventional seat then used the new seat.

Not only did Magnitude Estimation fully support the other methods, it presents its information in a ratio format. This showed that, apart from two items, perceived weight on the seat and under thigh pressure, which were equal for both seats, the new seat was ahead in all other fourteen measures. The procedure (see Gregg, 2000; Stevens, 1975; Kvalseth and Shackel, 1983). was simple to use, subjects did not seem to have any difficulties in alloting numbers to their feelings and the results were also simple to analyse. As a procedure to obtain reliable data in a statistically manipulable form it is to be recommended for its ease, speed and simplicity.

Conclusions

As with much ergonomics testing there are two sides to the work. There is always a need to understand the characteristics of the details of the situation, the impact of seat angle on spinal shape or pressure distribution across the seat surface. But there is also a need to remember that the seat is only part of the problem, which is the effects of the activities of doing the work. Thus any field trial has to consider if a task analysis is initially necessary, what the worker is trying to do, the time, effort, movements, etc. which are involved and what can be done to alter other things than the seat. So an initial survey should reveal the structure for the study.

The previous section has suggested procedures for various situations when a seat, or seats, have to be tested and for a minimum number of measures which will provide the data from which to make decisions. After these decisions a follow-up study is necessary, but not always conceded by industrial partners. Yet without such a study the value of what has been done cannot be understood.

In spite of many years of study it appears that there will still be great need for chair research, both in the field and in the laboratory. A tour of any supermarket, office, or factory, or a discussion with a physiotherapist or orthopaedic surgeon will tell us how far there is to go. But the increasing recognition of the serious nature of back problems, and their commonality has put the health and safety situation in the spotlight. The major weakness in the computerised office is the worker's back and this is what will keep seating high on the ergonomist's series of problems.

References

Akerblom, B. (1954). Chairs and sitting. In *Symposium on Human Factors in Equipment Design*, edited by W.F. Floyd and A.T. Welford. (London: HK Lewis).

Branton, P. (1966). The comfort of easy chairs. Interim report, Furniture Industry Research Association, UK.

Branton, P. (1969). Behaviour, body mechanics and discomfort. *Ergonomics* **12**, 316-327.

Branton, P and Grayson, G. (1967). An evaluation of train seats by observation of sitting behaviour. *Ergonomics* **10**, 35–51.

Corlett, E.N. (1999). Are you sitting comfortably? *International Journal of Industrial Ergonomics* **24**, 7–12.

Corlett, E.N. and BIishop, R.P., (1976). A technique for assessing postural discomfort. *Ergonomics* **19**, 175–182.

Corlett, E.N. and Eklund, J.A.E. (1984). How does a backrest work? *Applied Ergonomics* **15**, 111–114.

Corlett, E.N. and Gregg, H. (1994). Seating and access to work. Chapter 25 in: *Hard Facts about Soft Machines*, edited by R. Lueder and K. Noro (London: Taylor & Francis).

Corlett, E.N., Wilson, J. and Manenica, I. (eds) (1986). *The Ergonomics of Working Postures*, Section 5, Seats and sitting (London: Taylor & Francis).

Drury, C.G. and Coury, B.G. (1982). A methodology for chair evaluations. *Applied Ergonomics* **13**, 195–202.

Eklund, J.A.E. (1986). Industrial seatng and spinal loading. PhD thesis, University of Nottingham.

Eklund, J.A.E. and Corlett, E.N. (1986). Experimental and biomechanical analysis of seating. Chapter 28 in: *The Ergonomics of Working Postures*, edited by E.N. Corlett, J. Wilson and I. Manenica (London: Taylor & Francis).

Eklund, J.A.E. and Corlett, E.N. and Johnson, F. (1983). A method for measuring the load on the back of a seated person. *Ergonomics* **26**, 1063–1076.

Eklund, J.A.E., Houghton, C.S. and Corlett, E.N. (1982). Industrial seating, a report on some pilot studies. Internal report, University of Nottingham.

Eklund, J.A.E., Ortengren, R. and Corlett, E.N. (1987). A biomechanical model for the evaluation of spinal load in seated work tasks. In *Biomechanics XB*, edited by B. Jonsson (Champaign, IL: Human Kinetics Publishers).

Floyd, W.F. and Roberts, D.F. (1958). Anatomical and physiological principles in chair and table design. *Ergonomics* **2**, 1–16.

Grandjean, E. (1980). *Fitting the Task to the Man* (London: Taylor & Francis).

Gregg, H. (2000). A new work seat for commerce and industry, design and evaluation strategies. PhD. University of Nottingham.

Habsburg, S. and Mittendorf, I. (1980). Calibrating comfort: systematic studies of human responses to seating. In: *Human Factors in Transport Research*, edited by D.J. Oborne and T.A. Levis. (New York: Academic Press).

Helander, M.G. and Zhang, L. (1997). Field studies of comfort and discomfort in sitting. *Ergonomics* **40**, 895–915.

Hunting, W. and Grandjean, E. (1976). Sitzverhalten und subjektives Wohlbefinden auf Schwenkbaren und fixierten Formisitzen. Zeitschrift fur *Arbeitswissenschaften* **30**, 161–164. (Quoted in report by E. Grandjean and W. Hunting from Federal Institute of Technology, Zurich, of 20.10.1976.)

Jones, J.C. (1963). Fitting trials. *Architects Journal* **137**, 321–325.

Jones, J.C. (1969). Methods and results of seating research. *Ergonomics* **12**, 171–181.

Kember, P. (1976). The Benesh movement notation used to study sitting behaviour. *Applied Ergonomics* **7**, 133–136.

Kvalseth, T.O. and Shackel, B. (1983). The implementation of ergonomics. Chapter 16 in *Ergonomics of Workplace Design*, edited by T.O. Kvalseth. (London: Butterworths).

Langdon, F.J. (1965). The design of card punches and the seating of operators. *Ergonomics* **8**, 61–65.

LeCarpentier, E.F. (1969). Easy chair dimensions for comfort — a subjective approach. *Ergonomics* **12**, 328–337.

Lueder, R.K. (1983). Seat comfort: a review of the construct in the office environment. *Human Factors* **25**, 701–711.

Murrell, K.F.H. (1965). *Ergonomics* (London: Chapman & Hall).

Pepermans, R.G. and Corlett, E.N. (1983). Cross-modality matching as a subjective assessment technique. *Applied Ergonomics* **14**, 169–176.

Rieck, A. (1969). Uber die Messung des Sitzkomforts von Autositzen. *Ergonomics* **12**, 206–211.

Shackel, B., Chidsey, K.D. and Shipley, P. (1969). The assessment of chair comfort. *Ergonomics* **12**, 269–306.

Shipley, P. (1980). Chair comfort for the elderly and infirm. *Nursing*, Supplement 20. Sleep and Comfort.

Shipley, P., Haywood, J., Furness, W. and Rose, J. (1969). Testing easy chairs for the elderly. Report to the Research Institute for Consumer Affairs, London.

Stevens, S.S. (1975). Chapter 4 in *Psychophysics* (New York: John Wiley & Sons).

Wachsler, R.A. and Learner, D.B. (1960). An analysis of some factors influencing seat comfort. *Ergonomics* **3**, 315–320.

Wotzka, G., Grandjean, E., Burandt, U., Kretschmar, H. and Leonhard, T. (1969). Investigations for the development of an auditorium seat. *Ergonomics* **12**, 182–197.

Yu, C-Y., Keyserling, W.M. and Chaffin, D.B. (1988). Development of a work seat for industrial sewing operations: results of a laboratory study. *Ergonomics* **31**, 1765–1786.

Zhang, L. and Helander, M.G. (1992). Identifying factors of comfort and discomfort: A multi–dimensional approach. *Advances in Industrial Ergonomics and Safety IV*, edited by S. Kumar, pp. 395–402. (London: Taylor & Francis).

Zhang, L., Helander, M.G. and Drury, C.D. (1996). Identifying factors of comfort and discomfort in sitting. *Human Factors* **38**, 377–389.

Chapter 28

Computer aided ergonomics

Tim J. Gallwey and Leonard W. O'Sullivan

With the advent of the personal computer and improvements in its power and memory there has been a growing interest in developing computer-based techniques in ergonomics. In the early stages various operating systems were employed and different programming languages were used. But in recent years the technology has matured to the point where Windows NT has become the operating system of choice, and C++ is very largely the software language preferred. These two convergences mean that software developers now have a stable environment in which to work and there are immense opportunities for a rapid growth in market offerings.

Introduction

Ergonomists have used computers for a long time to carry out the tedious calculations of statistics such as Analysis of Variance (ANOVA) on data collected in the laboratory or the field. But the computer offers much more. In addressing design issues large bodies of data have to be consulted (e.g., anthropometric values) and the computer is very useful for storing such data and providing sophisticated methods of speedy and easy access. However design involves much more than looking up data. In addition complex and time consuming calculations are required over and above those of statistics, say the solution of sets of equations. But when design involves ergonomics it is very different from the traditional, rather simplistic image of the engineer, who solves a number of equations in a closed form model and then produces an answer, e.g., take the first derivative and set it to zero.

Ergonomics poses particular problems because of the large numbers of different variables to be assessed, their often imprecise nature, and the variety of conditions that have to be satisfied, with complex inter-relationships between them. Optimal solutions in such situations are not a realistic quest but rather ergonomists have to work towards a 'good' or better solution. Therefore the process is usually an iterative one, and the quality of the solution depends on the time and money available. Very often professional judgement has to be exercised to devise one or two solutions that look promising, and then some sort of evaluation has to be carried out in order to choose. Fitting trials are an obvious simple example of this type of approach. In most cases in the past very few alternatives could be evaluated, and unconventional ones were probably excluded because the resources could not be risked on work that might be too much of a long shot. Similarly, the time constraints have meant that the evaluation process had to be somewhat superficial.

By using the computer it is now possible to build much more complex models, to build a lot more of them, with a greater variety of approaches, and to evaluate them much more extensively. For example, anthropometric databases are now an integral part of human-models, which usually involve biomechanics equations and the use of inverse kinematics;

computers are the only convenient means for harnessing the power of such models. But, in isolation, these human-models are limited, as they also need computer representation of the workplace or product. Then the user can try different configurations of the workplace components, or their layout, or the design of the product, or some combination(s) of these. Huge numbers of workplace components are available and details of these can be stored in a directly accessible database, in the appropriate electronic form, and from these the designer can highlight potential problems, e.g., due to the postures required. But the performance of the human in the system, or the performance of the system itself, must be predicted. The knowledge of how to estimate these is available in a variety of sources and software can be used to find them, and get them in a form that is convenient for the user, in a widely used computer format. Furthermore on-line data needs to be captured in the laboratory and in the field.

However it is not only ergonomists who are involved in the design of systems involving people. Very often there are planning industrial, design or manufacturing engineers, or health and safety personnel, who need access to ergonomics information. It can be made available to them, and they can be given advice on its relevance, limitations, and places of application. Clear explanations and definitions can be provided and these more peripheral users can receive some indication of the situations where expert guidance or participation is needed, from ergonomists. The net result is that computer aided ergonomics widens the range of information available to designers, makes it more accessible, provides a means to do more evaluations, more extensively, and with greater variety.

Development

The first major book on the subject was by Karwowski *et al.* (1990), which was a collection of 39 papers. The development was stimulated further by a series of international conferences on Computer Aided Ergonomics and Safety (CAES). The first was held in Tampere, Finland on May 18–20, 1992 (Mattila and Karwowski, 1992), and was followed by others in Barcelona (Mondelo *et al.*, 1999) and Hawaii (Karwowski *et al.*, 2001). These provide a valuable source of developments over the time with some demonstrations of software, but usually no code as such, although sources are specified at times. In many cases the software described was not available commercially then, and is probably not available yet, but rather it was under development in some academic or research establishment. Hopefully the growth of the GNU General Public License movement (Free Software Foundation, 1991) will result in much more of this material becoming available and will stimulate contributions from other developers to improve and enhance what is offered. Making these available on the LINUX operating system will stimulate further developments and hence make improvements due to ergonomics much more widely available, especially in industrially developing countries.

There are three main markets for this type of software (Bonney, 1999) and hence three main types of software available. The first and most important market consists of the normal, mainly large company, commercial user. The product has to be very professionally developed and packaged and extensive support services have to be provided, so it is usually rather expensive, e.g., €20,000 to €200,000. The second market consists of small and medium sized enterprises (SMEs), often manufacturing based, that need similar software but, for cost and training reasons, need it to be considerably simpler with a lower level of service. The third market consists of research groups, often working in universities, where the software needs to be very open so that new developments and algorithms can be added easily. Ideally this material should be provided free through the GNU General Public License system (e.g., on the Internet) or for a nominal fee to cover copying and documentation costs or the cost of a book (e.g., on LINUX) with a CD-ROM. This last group could be a source of material from which further developments and adaptations can be made to meet specialist user needs.

This chapter presents what the authors consider to be representative examples of all three types of software rather than attempting an exhaustive approach that is not possible within the limits of a single chapter. Mattila (1996) looked at the possible benefits to be gained and concluded that this is the systematic and effective way to integrate ergonomics expertise into design. Recent overviews have been given by Feyen *et al.* (2000), Landau (2000), and Chaffin (2001).

Anthropometry

All the CAD systems for human-modelling contain one or more databases of anthropometric data but for Small and Medium-sized Enterprises (SMEs) something simpler and cheaper is more appropriate.

Peoplesize

Datasets come from surveys completed in 1991 to 1993 in the US and 1996 in the UK. It also includes data from the 1990s for French, German, Italian and Netherlands populations and has a composite European population along with 1990s data for Japanese people and 1980s data for Chinese. These data are divided by age groups and include UK infants from birth, and UK and US children from 2 years upwards. The number of dimensions available depends on the population, which includes 280 dimensions for German, UK, and US adults.

It is available in a Professional version, which automatically sets percentiles to fit a required percentage of people and can build composite populations, and an Easy version which is approximately half the price. Both are in the 'cheap' end of the market. Selected data can be exported to a file or clipboard and the user can choose word-processing, simple text or spreadsheet format with matching illustrations for report purposes. It is marketed by Open Ergonomics. Megaw (1996) reviewed an early version and found it convenient to use but that the database was limited and that it needed more features to take advantage of the possibilities offered by the computer. The current version has some of these.

Discrete event simulation

General implementations

Probably the first people to address this topic from an ergonomics perspective were Siegel and Wolf (1969) and their book is a very valuable starting point. The basic idea of this technique is that a number of probabilistic events interact or interfere with each other, and they are modelled by means of a precedence diagram or a network. Traditional simple examples arise in queueing situations where customers arrive at various time intervals and the time to serve them varies with their needs; Poisson distributions are often used. In more complex situations a probability distribution may be constructed from empirical data collected in the field or, if that is not available, one is often assumed using estimated values for the distribution parameters. Random samples are then taken at intervals of time as defined in the model, and the combined effects of the values obtained are then determined, e.g., success or failure of the system. If a very large number of samples is taken, and the results from these runs of the model are combined, a good estimate can be obtained of how the real situation will behave, if the model is realistic enough. Trying to study such processes by mathematical equations quickly becomes quite impossible, and discrete event simulation is the only sensible way to tackle it. It is also sometimes called Monte Carlo simulation.

The big problem is getting the data. Human operator performance times also vary according to probability distributions but the question is which ones and what values. One way to estimate is from MTM but those values lack variability. Knott and Sury (1987) reported coefficients of variation from 0.22 to 0.57 and skewness from 3.9 to 0.3. Gallwey and O'Muirthile (1989) used empirical values from an assembly line which they modelled using MicroSaint (see below). They obtained a good fit with the Weibull distribution, but the Beta values varied widely between operators. More recently Hoffman and Lim (1997) applied Fitts Law and Hick's Law to derive suitable functions but more work is needed before really good models can be built.

Obvious examples arise in human reliability (see Chapter 32). There is the probability that the person perceives the situation correctly, makes the correct decision of what to do about it, and executes the action correctly. While the human is going through these steps the system controlled is itself subject to a range of possible changes, and the time in which these happen limits the time available for the human to respond. Also, if the human has made a mistake in information processing, there is a probability distribution for the likelihood that of realising the mistake. But most of the probability data available are in the form of point estimates. Similarly there is a time element. In many cases each task has to be performed within a given time if adverse effects are to be avoided and, if a mistake is made, there is usually a limited time to correct that mistake before adverse effects occur or control is lost. Which time distributions to use and which parameter values to put into the model have to be decided. The state of knowledge of human performance makes it rather difficult to implement models such as THERP, HEART or SHERPA but Crawford and Gallwey (1999) implemented THERP by means of a binary event tree approach. Where data were absent, they suggested using a triangular distribution with estimates of optimistic, expected or pessimistic values but they were unable to test their approach on real problems. A recent example of this type of approach is given by Zulch *et al.* (2003).

Micro saint

Although not oriented exclusively towards ergonomics this package has significant potential for applications in the area with a number of features specifically tailored for it. It is designed to perform Discrete Event Simulations of tasks represented in a network, and was developed from an earlier package called SAINT, which ran on mainframe computers. Originally developed for PC use in analysing crew activities in military helicopter operations, it has been extended to handle a greater variety of applications.

The major option of interest to ergonomists is the Integrated Performance Modelling Environment (IPME) for analysing human system performance. Amongst its special features are: an environmental model for workplace climate, operator traits and states, performance shaping functions to modify dynamically operator task times and probability of failure, estimation of operator workload, and a measurement suite to define multiple runs with different initial values. At the heart of the package is the Human Operator Simulation Engine. It includes such features as Monte Carlo simulation, work surface definition which includes a large number of specific features, micro models of human behaviour, modelling of the effects of task failure, and a variety of specialist advanced modelling features. IPME runs on Silicon Graphics machines or on LINUX on any hardware. A smaller specialist tool is WinCrew which is used for studying high cognitive workloads. A description by the developers can be found in Laughery *et al.* (2001) and Vidulich (1998) gives an example of an application of Micro Saint.

General ergonomics analyses

EDS (Ergonomics Database System)

The idea of this package is to bring together tools and data for a variety of ergonomics evaluations, and some of its features are described in Schmidtke and Jastrzebska-Fraczek (2000). It uses ACCESS and requires 30 Mb of space on the hard disk. Due to the graphics features it also requires a Pentium PC with good graphics and sound cards with at least 16 Mb of working memory. It incorporates anthropometric data for 108 dimensions and 46 cases of body forces exerted.

It has a graphical user interface which starts with a main menu divided into a basic module, and single modules. The basic module is divided into Technical Components (transport, lifting, conveying, etc.), Environmental Factors, and Work Tasks and has the option of a data search for these fields. The single modules are divided into types of workplaces (VDU and office work, process control, monitoring, production work, assembly work, building work) with a data search facility, and a set of evaluation modules for: body forces and measures, stress/strain analysis (including the NIOSH equation), ergonomic examination and evaluation, checklist for task analysis, and ergonomics terms and definitions. These are supplemented by a literature search facility. It is available from Knowledge Soft Design and more details are available on www.e-ksd.com.

ERGO 2000

Very much at the cheap end of the market, this set of programmes is included with Konz and Johnson (2000). It has been criticised by a reviewer as appearing to have been written by students, with poor presentation and not offering a great deal. Nevertheless you do get what you pay for and it does offer a variety of options. At the start it has pull-down menus for the following: forms, anthropometry, environment, lift/move, stat/math, time, and units. Some material is just for information, some is in the form of checklists, some gives formulae, and some of it provides calculations. Each breaks down into quite a number of parts and these parts refer the user to the appropriate part of the text. It would appear to be attractive to quite a number of people, particularly among students.

ERGOEX

The ERGOnomics EXpert system was developed by Gilad and Karni (1999) with the intention of ameliorating the conditions for workers at a workplace. The user starts by choosing the design situation to be addressed (e.g., workplace, equipment, etc.) and then enters anthropometric data about the worker. The user then defines the task (such as 'standing–light assembly') and then workplace features such as chair, or footrest. The software then recommends dimensions for the equipment followed by a drawing with front and side views. An intermediate menu is then selected for advisory responses, which take the form of 'No footrest' or 'Add armrests'. A drawing can then be produced for the recommended layout. A further feature provides for lighting information to be entered, again resulting in a series of recommendations. It is intended for both expert and non-expert practitioners.

Observer

One of the really awkward and time-consuming tasks in ergonomics has been the analysis of data collected on video tape. It is a commonly used medium for posture studies, group work, and workplace evaluation but fortunately it can now be speeded up and made a lot easier by the use of the Observer package (Noldus *et al.*, 2000), which is marketed by Noldus

IT. Originally developed for use with analogue tape the latest version (Observer Video-Pro) can also handle digital tape which is becoming more common. A number of behaviours or activities are selected and assigned a key code and these keys are pressed to record the occurrence of an event during an observation session. These behaviours can be divided into classes to allow the calculation of statistical data for events occurring simultaneously. In the next phase the VCR is run at recording speed and the software time code can then be used to collect such information as the time between occurrences of the behaviours, duration of the behaviours, frequency, and so on. Noldus *et al.* (2000) present an example of its application to a study of work-related musculo-skeletal disorders (WMSDs) in an automobile assembly plant, and another example of a similar application is given by Carey and Gallwey (1998).

A typical setup for use with analogue or digital tape includes a professional video cassette recorder with built-in RS-232 interface to allow the VCR to be controlled by the computer, a time code generator and reader, and a video overlay board to display the video image in a window on the computer screen. The time code reader is either a plug-in board or an external unit. The latter allows a notebook computer to be used as a video analysis station. It runs on Windows NT or 95 and higher. The PC should have at least a 133 MHz Pentium processor, 32 Mb of RAM and 25 Mb of free hard disk space. For use with digital tape only, the PC needs only a video capture board (e.g., an MPEG encoder). Software support is provided for all common digital video formats.

Powerlab

Available from www.adinstruments.com is a wide range of software and hardware for the life sciences, and hence for the physiology end of ergonomics. Software is available to use with transducers for force, temperature, pressure and so on, or with front-ends for amplifiers, bridges, and special purpose applications such as spirometry, GSR, and blood pressure. The software controls data collection and is used also for subsequent analysis. There are both PC and Mackintosh versions.

Human-Modelling CAD systems

An overview of six such packages can be seen in Das and Sengupta (1995); three of these are presented below, i.e., JACK, MANNEQUIN and SAMMIE. The paper contains a very useful table comparing the packages on twelve criteria and makes general recommendations for potential users.

ANTHROPOS

The original version of this package was called ANYBODY and was developed by Lippmann (1986). It was largely a computer manikin with an interface to the CADKEY design package which tended to limit its adoption to a somewhat narrow user base. Subsequently it was adapted to be usable with AutoCAD as well, so that it became much more widely acceptable. It was updated at intervals and further versions of it were developed, such as *Virtual* ANTHROPOS, and the ability to work with 3D Studio was added. These developments were reviewed by Lippmann (2000). It has now been purchased by Techmath and is incorporated into the RAMSIS package.

HUMAN

Sengupta and Das (1997) developed this package using the features available in AutoCAD. The manikin can be shifted from one location to another, can be rotated, and the angle between two body segments can be altered using the Posture command. The workstation is evaluated using the Reach and Vision commands. Within Reach the user selects the left or right arm and moves it to a point; the software makes the changes provided the angles are within the ranges of motion of the joints. Vision produces vision cones from the eye points with the line of sight determined from the inclination of the ear–eye plane; alternatively it can use AutoCAD's Camera command. The Analyse command gives quantitative information on the positions of various body segments. Sengupta and Das (1997) provide examples of applications to a supermarket checkout, and an office workstation.

JACK

The original development was done at the University of Pennsylvania. Body dimensions are taken from the 1988 Anthropometric Survey of U.S. Army Personnel. The human-model has 69 segments, 68 joints, a 17-segment spine, 16-segment hands, coupled shoulder-clavicle joints, and 135 degrees of freedom. Joint limits are derived from NASA data, and human types can be chosen from large, medium and small frames as defined by SAE and other data sources. Alternatively, the human shape and size can be set by using human figure scaling, e.g., specify the extreme dimension(s) for a segment(s) such as sitting height. Body segments can be manipulated and JACK has inverse kinematics in the software to move the linked segments and joints as a human body would be expected to move. Included in this is the ability to define parameters such as positioning and bending of the torso, pelvic orientation, or arm positions. The human-model can be assigned tasks, and simulations can then be run to examine a variety of scenarios such as the effects of changes in spatial relationships, timing or clearances. These can be used with a variety of Virtual Reality tools.

The human-models (male and female) can be placed in CAD images of a large variety of typical work workplaces and equipment to simulate the activities of the workers. Results obtainable include what the human-model can see, what he/she can reach, whether the designed work situation can accommodate a variety of sizes of human. JACK computes the forces on joints and segments for a given posture and can provide animations and still-frame images for making presentations. An example of the latter is shown in Figure 1. There is a task analysis toolkit which includes tools such as NIOSH lifting analysis, low back analysis, predetermined times, and recovery and fatigue requirement estimation. There is a motion capture toolkit to facilitate virtual reality devices.

JACK can import 3-D graphics data based on VRML, IGES, stereolithography (STL) and inventor (IV) formats and has options to use utilities from OpenGL Optimiser and Decimate. It has a set of basic hand tools, texture mapping and lighting capabilities, and a 'camera' that can swing horisontally and vertically and zoom. More information is available from JACK (2002). JACK was used by Sundin *et al.* (2000a) to evaluate the design of a space station in virtual form. They examined some 4000 requirements and in only nine of these did they find JACK unsuitable; they reported that it was 'most suitable' in this application. In another case they used it in the design of workstations in a bus assembly plant (Sundin *et al.*, 2000b).

MANNEQUIN PRO

Biomechanics Corporation of America was the original company marketing this tool that was then known simply as Mannequin. It is now available from Nexgen Ergonomics (see http://www.nexgenergo.com). It is more of a tool to be used in conjunction with other graphics software, such as AutoCAD or 3D Studio, and therefore uses common file formats. It contains anthropometric data from eleven populations, including the 1988 Natick Report on the US Army (U.S. Army) and NASA-STD-3000. The human-models are claimed to be ergonomically

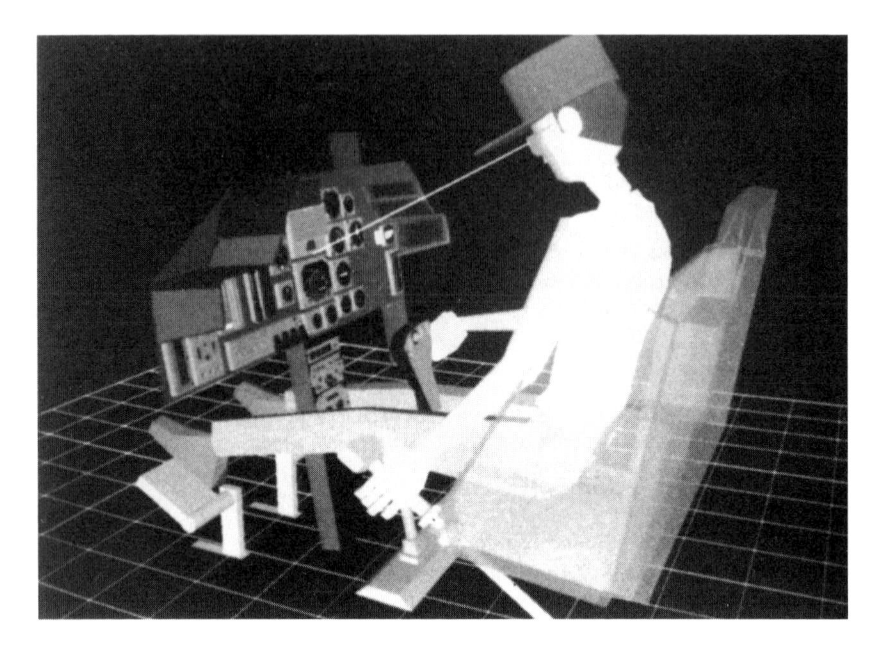

Figure 1 JACK. Colour shaded image of a cockpit model.

correct for a number of ethnic groups, percentiles and body types. Body parts can be articulated within human ranges of motion and there is a library containing pre-defined body and hand positions. There are three somatotypes for adult male, adult female, and child, and manikins for 2.5, 5, 50, 95, and 97.5 percentiles (see Figure 2). Alternatively the manikins can

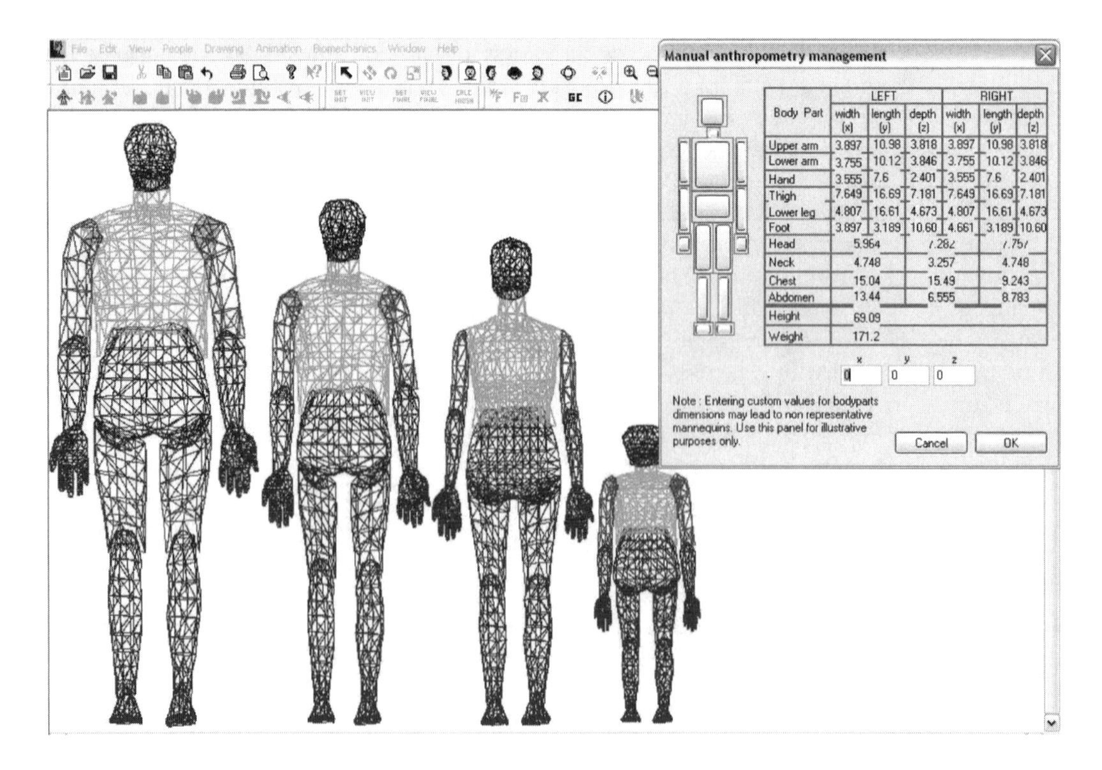

Figure 2 Mannequin models [Anthropometric.bmp].

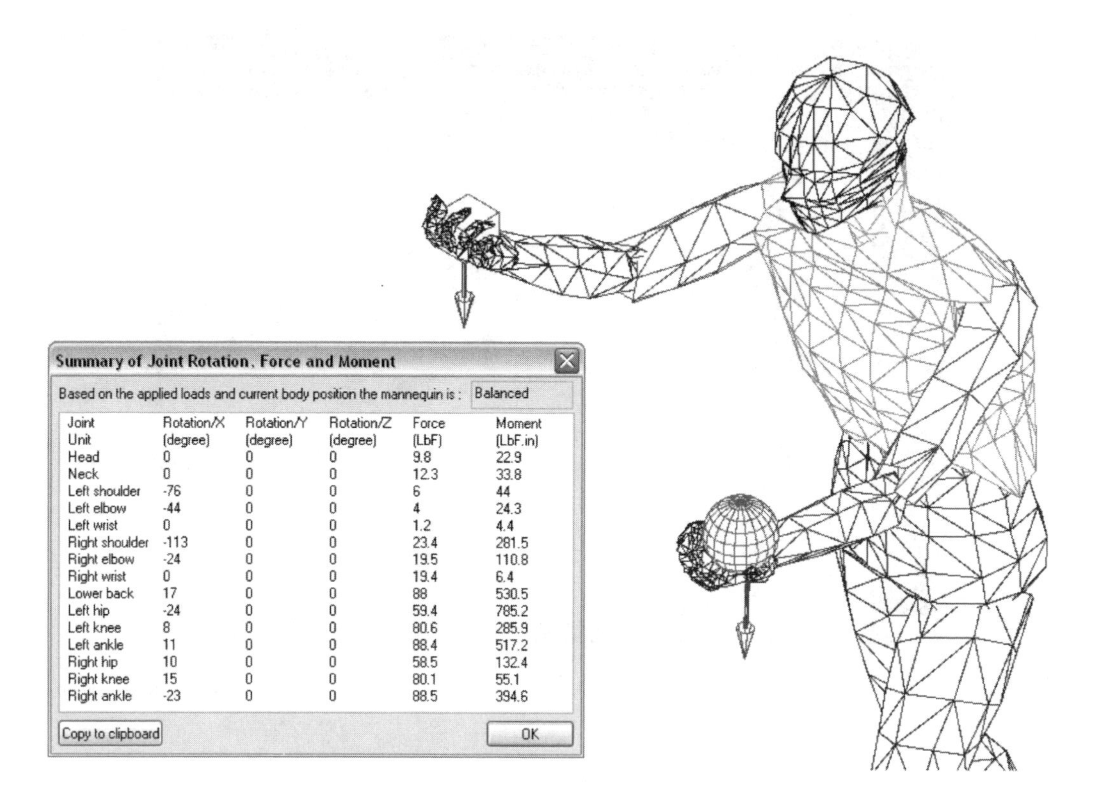

Joint	Rotation/X	Rotation/Y	Rotation/Z	Force	Moment
Unit	(degree)	(degree)	(degree)	(LbF)	(LbF.in)
Head	0	0	0	9.8	22.9
Neck	0	0	0	12.3	33.8
Left shoulder	-76	0	0	6	44
Left elbow	-44	0	0	4	24.3
Left wrist	0	0	0	1.2	4.4
Right shoulder	-113	0	0	23.4	281.5
Right elbow	-24	0	0	19.5	110.8
Right wrist	0	0	0	19.4	6.4
Lower back	17	0	0	88	530.5
Left hip	-24	0	0	59.4	785.2
Left knee	8	0	0	80.6	285.9
Left ankle	11	0	0	88.4	517.2
Right hip	10	0	0	58.5	132.4
Right knee	15	0	0	80.1	55.1
Right ankle	-23	0	0	88.5	394.6

Figure 3 Joint force and balance calculations for lifting [Forces.bmp].

be customised and there are five levels of detail. It can calculate body balance and simulate lifting, pushing and pulling and has some animation features (see Figure 3).

Outputs obtainable include field of vision cones, reach envelopes for hands and feet, reach to selected points, and what will be 'seen' from selected eye points for different positions of the 'camera'. It includes the NIOSH 1991 lifting equation (Waters *et al.*, 1993) and calculates the reaction joint force and torque due to the applied load and posture. On being exported to 3-D Studio enhanced rendering and animation are possible. An example of its output is shown in Figure 4.

It requires an IBM compatible PC (486 with co-processor, or Pentium) and any of the following operating systems: Microsoft Windows 95, Windows 98, Windows Me, Windows NT and Windows 2000 Professional. CAD drawings can be imported via DXF format (e.g., drawings of the workplace and equipment with which the manikin will interact as in Figure 5), and export is available in a number of formats. A review of an earlier version is given in Stewart (1992). He found it quite easy to develop 3D figures in a variety of poses and to customise the figure or pose, and particularly liked its ability to animate and record body motions. He stated that the data came from Humanscale and Bodyspace, but serious doubts were cast by Kroemer (1992) on these sources; he asked questions about the algorithms used for the equations of motion and the biomechanical torque calculations. Whether or not these have been improved subsequently is not clear.

PCMAN

The idea here is to provide a relatively cheap tool for obtaining dynamic posture data without the use of complex equipment. It requires only two video cameras (preferably digital but not

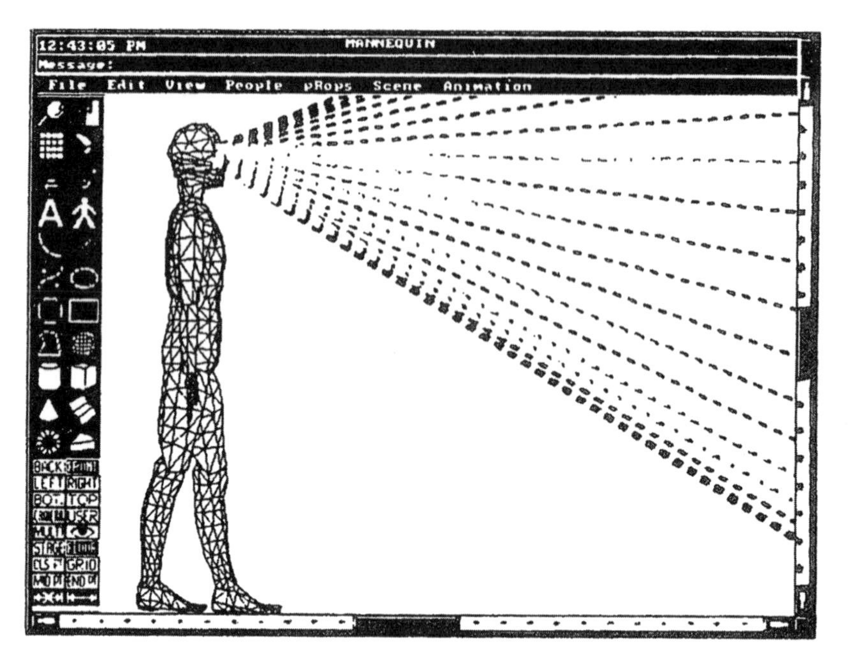

Figure 4 MANNEQUIN. A screen image showing field of view analysis and the system's iconic interface (sourced from HUMANCAD, 1991).

Figure 5 CAD drawing and Mannequin model of assembly workstation [Factory.bmp].

necessarily) which can be set up without expensive calibration fixtures and the like; the process here merely requires a wire frame-like box to be situated in the visual field. From this the software carries out the calibration process. It was developed some years ago at the Technical University of Munich and has been uprated at intervals.

Preferably it should be used in conjunction with RAMSIS. In that case data on the body observed on the video is transferred to RAMSIS, which then uses its database information to fit an almost exact manikin to it. Much more detailed analysis and output data then become available. Seitz and Bubb (2000) give a description of the main features of this package and some additional information is available in Seitz and Bubb (2001). Independent evaluations do not appear to be available in the literature but a demonstration of an application to modern ballet is very impressive.

RAMSIS

The developers of this package claim to have 70% of the world market within the automobile industry. It was developed at the Technical University of Munich specifically for the design of automobiles and, not surprisingly, is particularly suited to those applications. It is available in a stand-alone version or integrated with CATIA, which is used widely for automotive tooling design and manufacture. Figures 6 and 7 demonstrate the use of RAMSIS during automobile design.

The model can be of wire frame form or with surfaces and has a shoe model. It has ninety anthropometric types for each database for adults with a number of different databases, plus data for a wide range of children. There is quite a range of animation functions with task-driven posture and motion simulation using probability calculations. Other features include the analysis of health and comfort, vision with integrated eye motion, seat belt issues, and reach limits. Figure 8 shows a sample vision envelope. The CATIA version can be linked to a number

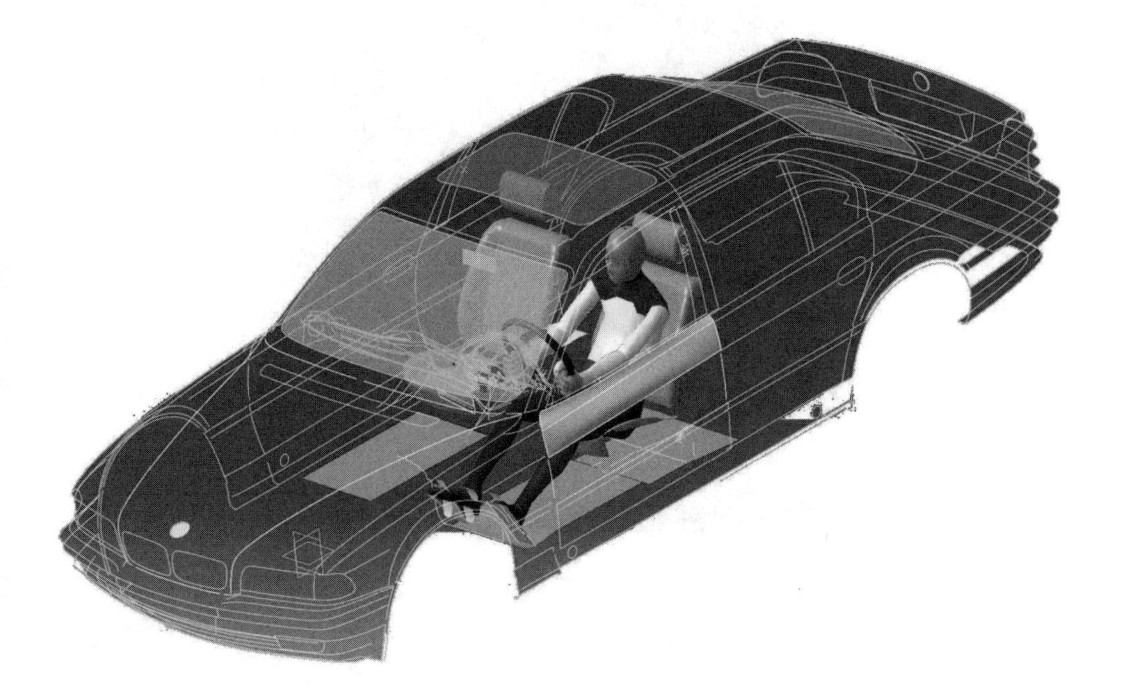

Figure 6 RAMSIS model use in automobile design [Carlv2.bmp].

Figure 7 RAMSIS in a CAD model of an automobile [Car2.bmp].

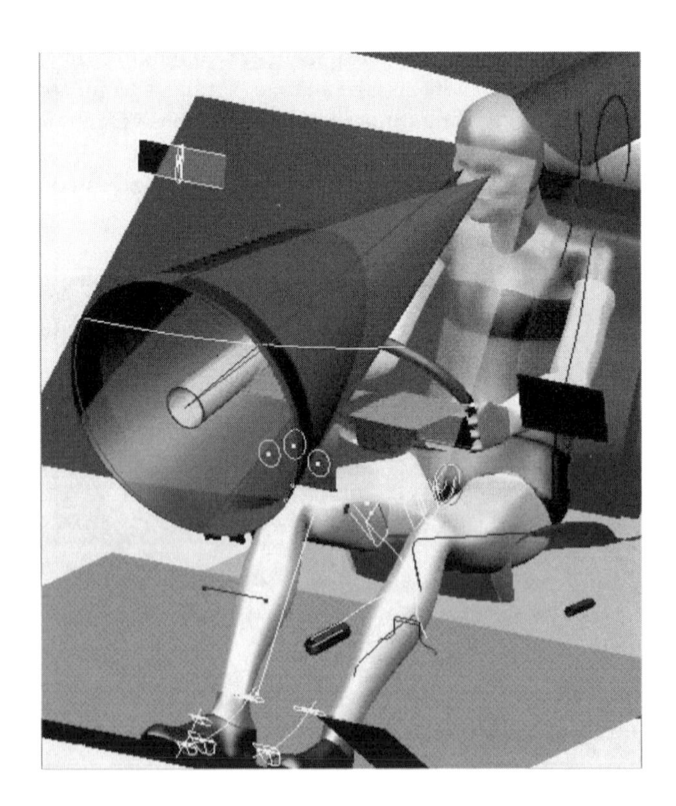

Figure 8 Vision envelope controls [vision2.bmp].

of SAE tools. The BodyBuilder option provides additional features for anthropometric investigations, automatic generation of test samples according to design requirements, and statistical analysis of conventional and 'multi-dimensional' percentiles, etc.

The stand-alone and BodyBuilder options are available for use on Hewlett Packard, Silicon Graphics, and SUN Ultrasparc UNIX machines and for Windows NT. The CATIA integrated version is available only for Hewlett Packard, IBM and Silicon Graphics systems. An application of the package in automobile design is given by Geuss (2000). Attempts to find other examples in the literature drew a blank. The package is marketed by Tecmath and more information is available on their web page www.techmath.com. At the time of writing its suitability for workplace evaluations is somewhat questionable but later versions are planned to change this.

SAMMIE

One of the earliest packages to provide a computerised man-model is SAMMIE (System for Aiding Man–Machine Interaction Evaluation). It was developed at University of Nottingham and then Loughborough University of Technology, and has been described by Case *et al.* (1990) and Porter *et al.* (1999). For some time it was only available for UNIX operating systems (e.g., SUN or Silicon Graphics) but is now also available for Windows NT and Windows 2000 from SAMMIE CAD (see http://www.sammiecad.com). It can also be adapted for some DEC, HP, and IBM machines and is X-windows based so can be run across a network.

The system includes pre- and post-processors for DXF and IGES to enable import of CAD data from other systems and to enable SAMMIE model data to be exported. The human models have 23 body segments and 21 constrained joints and are claimed to be capable of the full range of normal human movement. The joint movement ranges can be constrained to reflect acceptable or preferred comfort ranges inside the normal joint range, or to reflect the effects of restrictive clothing or physical disability.

It is claimed to be able to evaluate the fit of human-models in a workspace, reach areas and volumes, the view 'seen' by human-models, vehicle mirror design, and postural comfort with an output of joint angles. The software enables a large variety of workplace models and equipment to be created for evaluation purposes. It has an interactive graphical user interface, surfaces can be rendered or shaded, and it has a variety of plot/display output forms and CAD data exchange facilities. An example of its application can be seen in Figure 9. Feeney *et al.* (2000) used it for a contract from the British Department of the Environment to make recommendations on the accessibility of buildings for disabled people.

SAFEWORK

The manikin used here has 148 degrees of freedom with 100 links and 104 variables. It is claimed to have fully articulated models for the hand, hip, shoulder and spine with limits on joint mobility. The anthropometric data come from a number of surveys including those from the US Army in Natick (see References). Features include vision analysis, postural and comfort angle analysis, reach plus lifting, lowering and carrying analysis, and provision for animation, collision detection, virtual reality and clothing. Major CAD interfaces are available and the system runs on Hewlett Packard, IBM and Silicon Graphics workstations. It includes the NIOSH equation, Snook and Ciriello equations, and RULA. The package is marketed by DELMIA and some very limited information is available from their website www.safework.com.

Figure 9 SAMMIE. An adjustable computer workstation model.

Human motion evaluation

It is debatable whether computer packages for motion evaluation should be a separate topic or should be included with posture analysis. They are presented separately on the basis that they should be regarded as dynamic posture analysis tools whereas the others can be seen as tools for static situations. They are all at the expensive end of the price range.

CODA 3

Three in-line rotating mirror scanners are used to detect up to twelve prismatic reflective markers on this portable system. The differently coloured markers enable individual identification of landmarks in a wide range of indoor and outdoor lighting conditions. The two outermost scanners are mounted at a distance of one meter apart to provide triangulation, and, due to the rigid frame, calibration is almost eliminated. It was originally developed for sports applications but is suitable for industrial measurements and for static and dynamic ergonomics analysis of designs. Output options include graphics displays of stick figures with or without force vectors, and co-ordinates vs time, and there are estimates of joint centres and the calculation of forces and moments in joints.

PEAK

The original aim of these products was to develop analysis tools to help athletes prepare for the Los Angeles Olympic Games. The product of interest here is the Peak Motus® motion measurement system operating on the Windows system. It uses video and hardware to capture the co-ordinates of moving points, either from the video itself or from passive markers (infra red or visible light) using multiple cameras. There are three levels of sophistication within the system with the possibility of subsequent upgrading. The basic system cannot exceed collection rates of 30 Hz with NTSC format or 25 Hz with PAL so is only suitable for lower speeds of movement. An optional add-on is an automatic acquisition module, which is used to track reflective markers automatically. The 3D optical capture system is a faster, marker-only system, that uses from two to twelve cameras to collect 3D co-ordinates, and there are various add-ons to provide specialist data collection options. The usual operating rate is around 50 Hz but with special cameras this can go up to 1000 Hz.

The equipment can be linked with other analogue devices such as force plates and EMG units. Analysis options include linear and angular distances, velocities and accelerations, gait analysis, and export to animation packages. It requires at least a Pentium 3 with a 133 MHz processor, at least 40 Gb hard drive, 256 Mb RAM and a 16*10*40 CD re-writer. Data can be exported to EXCEL for analysis or a special Peak module can be used. The web page is: http://www.peakperform.com.

SIMI

Most of the typical uses mentioned in the company literature refer to applications in sports science so some adaptation could be necessary for using it to assess industrial or office workplaces. However they do say that they are developing a system which will use a pattern matching algorithm to detect points without body markers, which will be much more useful for industrial users. But the package does offer a range of sophisticated tools.

There are a number of options, one of which is SIMI Motion. It uses non-invasive 3D data acquisition with video technology, and they claim that digitisation of movement data is simple with manual or automatic tracking. It can integrate data from such devices as force plates and EMG equipment and can import data from third party devices. It can use up to four cameras with speeds from 100 Hz to 240 Hz and the camera data can be synchronised with outputs from force plates, and/or from EMG. Analysis software allows the calculation of the coordinates of points, as well as their velocities and accelerations, and calculates angles and distances between points. Further analysis features include the solution of Fast Fourier Transforms and interpolation, and derivation using splines. EMG analysis includes median and mean frequency, and Root Mean Square with optional filters. Data can be captured from goniometers and accelerometers. Various versions of Windows are suitable but it needs a Pentium III PC with 128 Mb of RAM and a hard disk with at least 15 Gb free. Cameras can be either digital or video with a suitable adaptor board. More information is available on the www.simi.com website. A database search failed to find any papers that mention this equipment in the title.

VICON

There are several versions of this package available with various levels of sophistication. They generally use special CCD cameras, with visible or invisible strobes, which provide 1.3 M pixels, and have speeds from 50 to 250 frames per second. In addition special miniature cameras are available for use in restricted environments with a visible light strobe. The system uses retro-reflective markers and the users claim that the numbers of these that can be used are unlimited. Many of their applications have been in sports and biomechanics but some have involved very detailed data collection of data on parts of the hand. The data can be synchronised with that from other sources such as force plates and EMG systems with up to 12 cameras in one option or up to 24 in another. Calibration is dynamic and done automatically while it calculates the co-ordinates of the camera position.

The computer runs on Windows NT or 2000 and can display data as markers, stick figures, or fully rendered skeletons in 3D workspace. Cubic spline interpolation is available along with hands-off reconstruction of marker trajectories. Output information can include positions of points, distances, and linear and angular velocity and acceleration. The web site www.vicon.com contains further information.

From company literature it is clear that the emphasis is on such areas as gait analysis, sports science, biomechanics, rehabilitation medicine, and biomedical engineering. There appears to be a plentiful number of papers on these topics where this equipment has been used, but an example related to ergonomics is Chadwick and Nicol (2000). A very useful looking paper giving an overview, especially of upper limb movement analysis, is that by Rau *et al.* (2000). It emphasises that most work to date has been on the lower limb, which involves much less complex issues, and so upper limb work requires these rather more complex technologies.

Lighting and noise

Lighting

Software for lighting calculations was developed as Radiance in the late eighties on a UNIX platform with support from the US Department of Energy and the Swiss federal government. It is part of a larger suite of programmes called ADELINE (Advanced Daylighting and Electric Lighting Integrated New Environment) that is part of Task 12 of the Solar Heating and Cooling Program of the International Energy Agency. Participants in its development came from seven European countries and the US, co-ordinated by Erhorn and Stoffel (1996).

The ADELINE developers also produced Desktop Radiance (1999). It is a modified and enhanced version for Windows and AutoCAD platforms running on a PC. Input files include scene geometry (from AutoCAD), materials specification, luminaire data, and sky conditions for daylight calculations. Calculated values include luminance and colour, and simulation results can be displayed as colour images, numerical values, or iso-lux plots. According to Ward (1994) 'the simulation blends deterministic and stochastic ray-tracing techniques to achieve the best balance between speed and accuracy in its local and global illumination methods'. Ward also reported that significant commercial users favoured Radiance over other similar products.

Noise

Early work on computer modelling of noise distribution in a work environment was carried out by Shield (1980) who developed NOISESHIELD as a FORTRAN programme. Subsequently she co-operated with Jones (Jones and Shield, 1988) and later with Dance (Dance and Shield 1997, Dance and Shield 1999), and they extended their software to the point where it gives estimates to within 2 dB to 3 dB of the correct value for a rectangular box shaped volume.

Combined

Hayes and Gallwey (2001) developed a package which combines AutoCAD and Desktop Radiance with routines for noise provided by Dance. It is written in C++ and runs on Windows NT and uses the ACCESS database as an easy means of inputting and updating the lighting and acoustic properties of materials in the workspace. EasySurf was added to provide a means of producing iso-noise lines. One of the shortcomings is that at the time of development Desktop Radiance was only available in the Beta-3 version so there was a shortage of user documentation, and programme crashes were a problem.

Desktop Radiance contains its own routines for contour lines. There is a Radiance menu-item in AutoCAD and sub-menu-items activate different functions in Radiance. But initial functions have to be performed to specify the surface material, types of window glazing, and standard furnishings, and luminaires have to be selected from a luminaires library. Analysis of a scene is initiated by establishing the 'camera' position and view (e.g., at the point in space of the operator's eye). The simulation process then produces a rendered image of the scene with shadows and other lighting effects. Quantitative analysis can give an image of the scene with iso-lux lines, or else false-colours can be generated to code the lighting levels.

Physical effort

3DSSPP

The Three Dimensional Static Strength Prediction Program (3DSSPP) was developed by Chaffin and his associates and implements the static strength model described in Chaffin *et al.* (1999). It is PC based and provides a biomechanical analysis whereby the user can estimate the loads on specific joints due to the force exerted by the hand(s), especially in lifting. It assumes that the effects of acceleration and momentum are negligible so it is limited to 'slow' movements.

Input data can be in metric or imperial units and the posture information can be entered by a joint angle method or by inverse kinematics. The first requires fifteen body joint angles and the second requires the location of the hands to be specified in three dimensions. Load in each hand has to be specified with the force vector angles. The output can give a number of different views, which can represent the human figure in stick, skeleton, or flesh form, and the 'camera' can view from any distance and angle. Output options include such things as posture, joint locations, joint moments, spinal forces and moments, strength capabilities and sagittal plane lowback analysis. The percent of people capable of the force is displayed with comment as to whether the balance required is acceptable, unacceptable or critical, and the required coefficient of friction at the feet is given.

The package has been widely used in ergonomics, often as a check on the safety of proposed experimental or investigation work. It has also been incorporated into a number of other commercially available packages.

EEPP

The Energy Expenditure Prediction Program (EEPP) is another PC based offering from the University of Michigan. Input data in terms of load height and mass are required in imperial units and energy outputs are in old metric units, i.e., kilocalories. The types of tasks addressed are: lifting and lowering in stoop or squat postures or with one hand; holding, walking and carrying with the load in a number of possible positions relative to the body; horisontal arm work forward in the sagittal plane, in one of four postures; lateral arm work; pushing and pulling at two heights; general hand and arm work for various situations; and an approximation for throwing of loads. Some of these use data from the literature, and the developers derived others.

NIOSH lifting equation

Quite a number of the more extensive packages include this as one of part of the suite, especially those designed for biomechanics work. One programme available for free download can be found on this web site: www.purswell.com/niosh.exe. It runs on Windows with a convenient interface.

RNGP

The Revised NIOSH Guide Program was developed by Garg (1993) of the University of Wisconsin-Madison. The programme determines the Recommended Weight Limit (RWL) for the NIOSH equation and the Lifting Index (LI). It has other features which identify the percent of males and females capable of coping with the lifting or lowering task for two-handed work, and a number of different analyses. It works in both metric and imperial.

Posture analysis

Most of the human-modelling packages claim to provide posture analysis but they are usually more in the nature of outputs of the resulting posture in terms of a figure. Ergonomists need more than this. The data from the biomechanical manipulations need to be analysed to determine whether or not the resulting postures are acceptable or harmful. Such a feature must look at the joint angles relative to the corresponding person's range of motion (ROM), the forces or loads involved, the repetitiveness or frequency of the movements, the time available for recovery, and any elements of static load.

ERGONOM

Swat and Krzychowicz (1996) present a set of three programmes to evaluate loads that result from workplace postures where standing and walking positions dominate. ERGONOM 1 describes postural zone boundaries for comfortable manual operations, ERGONOM 2 is for the evaluation of a machine at an early design stage, and ERGONOM 3 provides a working posture evaluation scheme. There are four postural zones. Zone I is the comfort zone where the operation does not require awkward postures. Zone II is the area above and below the comfort zone with arms above the heart level or trunk bent at more than 20 degrees; it has one force moment. Zone III demands maximal arm raising and standing on tiptoe; it has two force moments. Zone IV involves extremely high postural stress; three force moments occur. The evaluation produces a numerical value which classifies the posture as comfortable, significantly uncomfortable, very uncomfortable, or extremely uncomfortable.

OWAS

It was developed to investigate fairly quickly a large number of working postures in a Finnish steelworks, Ovako Oy, in co-operation with the Finnish Institute of Occupational Health (FIOH, 1992). The letters stand for Ovako Work Analysis System. The Tampere University of Technology in Finland developed computerised versions called OWASCO and OWASAN. They are designed for live use on a laptop PC, or for analysis of a video of a person performing the tasks of their job. For each element of the job numeric codes are entered to classify the features of the tasks, and then the PC uses them to calculate the action required (if any) for the job assessed (see more detail in Chapter 16). The video approach is much easier as it allows more time to choose the appropriate codes. A search of journal paper titles did not reveal any papers on either of these two programmes, but students have found them very acceptable in the field.

PATsI

As part of a larger exercise Carey developed add-on software to use with ERGOMAN and the ERGOPlan packages. It was called Posture Analysis Techniques for Industry and was described in Carey and Gallwey (1999). It is written in BUILDER C++ for Windows NT. The programme takes joint angle data generated by ERGOMAN and converts the values to one of four zones defined by Drury (1987). These zones correspond to percentages of ROM for the joints of interest, assuming that the person involved conforms to a set of averages published by NASA. There is also a feature to capture the same kind of data from video tape of a person doing the job dressed in black although this feature has been somewhat unreliable. (ERGOMAN is no longer on the market and ERGOPlan has been modified and revised to become the new package Process Engineer.)

PEO

The Swedish National Institute for the Working Life developed this tool. It aims to collect real-time data on a specific category of subjects and to compare the physical loads between two or more groups, or to do before and after studies. Along with actual posture data it categorises loads into four groups. It can give an overview of loads during a full working week by obtaining time-limited data followed by a weighting procedure. Fransson-Hall *et al.* (1995) describe an early version of it.

RULA

McAtamney and Corlett (1993) developed Rapid Upper Limb Assessment (RULA)—see Chapter 16. It is similar to OWAS but provides a more detailed breakdown and more analysis. It has been widely used in a variety of work situations, and a number of more general packages include this as one of the suite of programmes available. An on-line version is available from Osmond on www.ergonomics.co.uk and from this site it is also possible to download a coloured guide sheet to assist with scoring.

Thermal environment

For a good overview of CAD models in this area readers should consult Parsons (1993a). Also see Parsons (1995) and Chapter 23 in this book for comments on computer models used to evaluate clothing issues.

CLOMAN

This is really intended for research workers and is available from the developers on a diskette for a copying fee (Lotens and Havenith, 1991). It is written in FORTRAN and runs on a PC. As the name implies it is concerned with the effects of clothing insulation and how to model them.

Haslam and Parsons

These researchers obtained a contract from the UK Army (APRE) to evaluate four different models for predicting human responses to hot and cold environments. For a variety of reasons they made some adaptations to the originals, but basically they evaluated the Pierce lab. 2-node model, the Stolwijk and Hardy 25-node model, Givoni and Goldman's model for rectal temperature response, and the ISO/DIS 7933 model. For the evaluation process they wrote code in FORTRAN and compared the predicted values with those obtained from experiments that had been published in the literature. The results and source code are given in Haslam and Parsons (1989). They found that the 25-node model gave the most accurate predictions, and the 2-node model was often accurate but at times was poor for exercise conditions. The rectal response model usually over-estimated deep body temperature except for very hot or heavy exercise conditions, while the ISO/DIS model gave allowable exposure times that were often reasonable but would not have protected subjects for some exercise conditions.

It was possible to obtain a copy of the documents from APRE. A number of researchers/developers have used this code to devise their own versions, e.g., Gallwey, Maher *et al.* (1999). Users can choose to do an evaluation using any one of the four models. Their TEESI programme is written in C++ and runs on Windows NT, but needs some enhancements to facilitate users.

IH calculator

A variety of calculations are available by using this programme which is aimed at Industrial Hygienists, hence the name. It is written for DOS and is very cheap but not very convenient to use, as results from one calculation have to be saved manually if needed for subsequent calculations. In addition to some basic aspects of the thermal environment, it has routines to address ventilation calculations. It is available from Industrial Hygiene Services and is related to a very useful book on ventilation (ACGIH, 1995).

Parsons' programmes

Appendix 2 of Parsons (1993b) contains source code in TURBO BASIC for three useful programmes to calculate and interpret the following:

1. Heat stress index according to ISO 7833 (1989)
2. PMV/PPD thermal comfort index according to Fanger (1970) and ISO 7730 (1984)
3. IREQmin and WCI

Some explanation of the calculation process is given in Parsons (1992) and see also Chapter 23.

Overview

From the above it will be seen that there is a wide range of computer aids and tools available in ergonomics, and that they cover a wide range of cost and of degrees of development. At the present stage quite a lot of computer expertise is needed to use many of them. However human-models, for example, have received a lot of attention so that many of them are now quite usable. However many purists have doubts about the validity of their results, so even these have to be treated with some caution. Other areas of ergonomics such as human reliability are still at a relatively early stage of development and are dependent on researchers obtaining much more data on human functioning before adequate computer models can be constructed. As always the human is much more complex than many people realise, and the number of variables is huge, so there is still a formidable task ahead and a lot more research and development is needed. But useful progress has been made and there is now a much better realisation of the possibilities available to designers of products and work systems.

References

ACGIH, American Conference of Governmental Industrial Hygienists Inc. Cincinnati, Ohio.

ACGIH (1995). *Industrial Ventilation* (22nd ed.) American Conference of Governmental Industrial Hygienists Inc. Cincinnati, Ohio.

APRE, Army Personnel Research Establishment, Hampshire, U.K.

Bonney, M.C. (1999). Possible markets and software. *Developers Meeting of IDEA Project*, University of Nottingham, U.K.

Carey, E.J. and Gallwey, T.J. (1998). The use of video techniques to analyse postural stress, In S. Kumar (ed.) *Advances in Occupational Ergonomics and Safety 2. Proceedings of the XIIIth Annual International Occupational Ergonomics and Safety Conference* (Amsterdam: IOS Press), pp. 198–201.

Carey, E.J. and Gallwey, T.J. (1999). PATsI: computerised Postural Analysis Techniques for Industry. In P. Mondelo, M. Mattila, and W. Karwowski (eds), *Proceedings of the International Conference on Computer-Aided Ergonomics and Safety, CAES '99.* Universitat Politecnica de Catalunya, Barcelona.

Case, K., Porter, J.M. and Bonney, M.C. (1990). SAMMIE: a man and workplace modelling system. In W. Karwowski, A. Genaidy and S.S. Asfour (eds), *Computer-Aided Ergonomics* (London: Taylor & Francis), pp. 31–56.

Chadwick, E.K.J. and Nicol, A.C. (2000). Elbow and wrist joint contact forces during occupational pick and place activities. *Journal of Biomechanics*, **33**, 591–600.

Chaffin, D.B. (2001). *Digital Human Modelling for Vehicle and Workplace Design* (Warrendale, PA: Society of Automotive Engineers, Inc.).

Chaffin, D.B., Andersson, G.B.H. and Martin, B.J. (1999). *Occupational Biomechanics* (3rd ed.) (New York: Wiley).

Crawford, J.W. and Gallwey, T.J. (1999). BETSSy: a Binary Event Tree Simulation System for safety modelling. In P. Mondelo, M. Mattila, and W. Karwowski (eds), *Proceedings of Computer Aided Ergonomics and Safety Conference*, Universitat Politecnica de Catalunya, Barcelona.

Dance, S.M. and Shield, B.M. (1997). The complete image-source method for the prediction of sound distribution in non-diffuse spaces, *Journal of Sound and Vibration*, **201**, 473–489.

Dance, S. and Shield, B.M. (1999). Modelling of sound fields in enclosed spaces with absorbent room surfaces, Part 1: performance spaces. *Applied Acoustic*, **58**, 1–18.

Das, B. and Sengupta, A.K. (1995). Computer-aided human modelling programs for workstation design. *Ergonomics*, **38**, 1958–1972.

Desktop Radiance (1999). Beta-3 Version of *Lighting Estimation Software*, Lawrence Berkeley National Laboratory and MarinSoft Inc., Berkeley, CA.

Drury, C.G. (1987). A biomechanical evaluation of the repetitive motion injury potential of industrial jobs. *Seminars in Occupational Medicine*, **2**, 41–49.

Erhorn, H. and Stoffel, J. (eds) (1996). *ADELINE 2.0 User's Manual.* (Stuttgart, Germany: Fraunhofer-Institut fur Bauphysik).

Fanger, P.O. (1970). *Thermal Comfort.* (Copenhagen: Danish Technical Press).

Feeney, R., Summerskill, S., Porter, M. and Freer, M. (2000). Designing for disabled people using a 3D human modelling CAD system. In K. Landau (ed.) *Ergonomic Software Tools in Product and Workplace Design* (Stuttgart: Ergon Verlag), pp. 195–203.

Feyen, R., Liu, Y., Chaffin, D., Jimmerson, G. and Joseph, B. (2000). Computer-aided ergonomics: a case study of incorporating ergonomics analysis into workplace design. *Applied Ergonomics*, **31**, 291–300.

FIOH (1992). OWAS: a method for the evaluation of postural load during work. Finnish Institute of Occupational Health, Helsinki, Finland.

Fransson-Hall, C., Gloria, R., Kilbom, A., Winkel, J., Karlqvist, L. Wiktorin, C. and the Stockholm MUSIC 1 Study Group (1995). A portable ergonomic observation method (PEO) for computerised on-line recording of postures and manual handling. *Applied Ergonomics*, **26**, 93–100.

Free Software Foundation (1991). *Copyright Document*, Boston MA.

Gallwey, T.J., Maher, D.A. and Crawford, J.W. (1999). Thermal Environment Estimation System for Industry (TEESI). In P. Mondelo, M. Mattila, and W. Karwowski (eds) *Proceedings of the International Conference on Computer-Aided Ergonomics and Safety CAES '99*, Universitat Politecnica de Catalunya, Barcelona.

Gallwey, T.J. and O'Muirthile, M. (1991). Work-time distributions for paced assembly lines, In M. Pridham, and C. O'Brien (eds) *Production Research: Approaching the 21st Century* (London: Taylor & Francis).

Garg, A. (1993). *Revised NIOSH Guide Program for Manual Lifting,* (Brown Deer, WI: Arun Garg, self published).

Geuss, H. (2000). Optimizing the product-design process by computer aided ergonomics. In K. Landau (ed.) *Ergonomic Software Tools in Product and Workplace Design* (Stuttgart: Ergon Verlag), pp. 132–140.

Gilad, I. and Karni, R. (1999). Architecture of an expert system for ergonomics analysis and design. *International Journal of Industrial Ergonomics,* **23**, 205–221.

Haslam, R.A. and Parsons, K.C. (1989). Models of human response to hot and cold environments, Volumes 1 and 2. Human Modelling Group, Department of Human Sciences, University of Technology, Loughborough, U.K.

Hayes, E.J. and Gallwey, T.J. (2001). Estimating contours of noise and lighting levels in a work environment. *CAES 2001, Proceedings of the Computer Aided Ergonomics and Safety Conference,* Maui, Hawaii.

Hoffmann, E.R. and Lim, J.T.A. (1997). Concurrent manual-decision tasks. *Ergonomics,* **40**, 293–318.

Industrial Hygiene Services Inc., 11760 Westline Industrial Drive, St. Louis MO 63146, USA and on http://ourworld.compuserve.com/homepages/ihsi.

Jack (2002). http://www.plmsolutions-eds.com/products/efactory/jack/

Jones, C.J. and Shield, B.M. (1988). A computer model for the prediction of noise levels in factories. *Proceedings of the Institute of Acoustics,* 10, Part 5.

Karwowski, W., Genaidy, A. and Asfour, S.S. (eds) (1990). *Computer Aided Ergonomics: A Researcher's Guide* (London: Taylor & Francis).

Karwowski, W., Mondelo, P., Das, B., and Mattila, M. (2001). *CAES 2001, Proceedings of the Computer Aided Ergonomics and Safety Conference,* Maui, Hawaii.

Knott, K. andf Sury, R.J. (1987). A study of work-time distributions on unpaced tasks. *IIE Transactions,* March, 50–55.

Konz, S. and Johnson, S. (2000). *Work Design: Industrial Ergonomics* (5th ed.) (Scottsdale: Holcomb Hathaway).

Kroemer, K.H.E. (1992). Letter to the Editor. *Bulletin of the Human Factors Society,* Santa Monica, CA, p. 13.

Landau, K. (ed.) (2000). *Ergonomic Software Tools in Product and Workplace Design* (Stuttgart: Ergon Verlag).

Laughery, R., Archer, S., Plott, B. and Dahn, D. (2001). Task network modelling and the Micro Saint family of tools. *Proceedings of the IEA 2000/HFES 2000 Congress,* San Diego, 1-721–1-724.

Lippmann, R. (1986). Arbeitsplatzgestaltung mit Hilfe von CAD. (Workstation design with help from CAD.) *REFA Nachrichten,* **3**, 13–16.

Lippmann, R. (2000). ANTHROPOS quo vadis? ANTHROPOS human modelling past and future. In K. Landau (ed.) *Ergonomic Software Tools in Product and Workplace Design* (Stuttgart: Ergon Verlag), pp. 156–172.

Lotens, W.A. and Havenith, G. (1991). Calculation of clothing insulation and vapour resistance. *Ergonomics,* **34**, 233–254.

Mattila, M. (1996). Editorial: Computer-aided ergonomics and safety—a challenge for integrated ergonomics. *International Journal of Industrial Ergonomics,* **17**, 309–314.

Mattila, M. and Karwowski, W. (eds) (1992). *Computer Applications in Ergonomics, Occupational Safety and Health, CAES '92,* (Amsterdam: North-Holland).

McAtamney, L. and Corlett, E.N. (1993). RULA: a survey method for the investigation of work-related upper limb disorders. *Applied Ergonomics,* **24**, 91–99.

Megaw, E.D. (1996). PeopleSize. *Applied Ergonomics,* **27**, 140–140.

Mondelo, P., Mattila, M. and Karwowski, W. (eds), 1999. *Proceedings of the International Conference on Computer-Aided Ergonomics and Safety, CAES '99.* Universitat Politecnica de Catalunya, Barcelona.

Nexgen Ergonomics Inc., 6600 Trans Canada Highway, Suite 750, Pointe Claire, Quebec, Canada,H9R 4S2. More details are available on: http://www.nexgenergo.com.

Noldus Information Technology b.v., Costerweg 5, P.O. Box 268, 6700 AG Wageningen, The Netherlands.

Noldus, L.P.J.J., Trienes, R.J.H., Hendriksen, A.H.M., Jansen, H. and Jansen, R.G. (2000). The Observer Video-Pro: new software for the collection, management and presentation of time-structured data from video tapes and digital media files. *Behavior Research Methods, Instruments & Computers*, **32**, 197–206.

Open Ergonomics Ltd., Loughborough Technology Centre, Epinal Way, Loughborough, Leicestershire LE11 3GE, U.K.

Parsons, K.C. (1992). The thermal audit. In E.J. Lovesey (ed.) *Contemporary Ergonomics* (London: Taylor & Francis), pp. 85–90.

Parsons, K.C. (1993a). *Human Thermal Environments*. (London: Taylor & Francis), pp. 287–319.

Parsons, K.C. (1993b). *Human Thermal Environments*. (London: Taylor & Francis), pp. 327–336.

Parsons, K.C. (1995). Computer models as tools for evaluating clothing risks and controls. *Annals of Occupational Hygiene*, **39**, 827–839.

Porter, J.M., Freer M.T. and Case, K. (1999). Computer aided ergonomics. *Engineering Designer*, **25**, 4–9.

Rau, G., Disselhorst-Klug, C. and Schmidt, R. (2000). Movement biomechanics goes upwards; from the leg to the arm. *Journal of Biomechanics*, **33**, 1206–1216.

SAMMIE CAD Limited, Arwen House, 7 The Pingle, Quorn, Leicestershire, LE12 8FQ, UK. Also see http://www.sammiecad.com for more details and a wide selection of publications.

Schmidtke, H. and Jastrzebska-Fraczek, I. (2000). The ergonomic database system (EDS) — an example of computer-aided production of ergonomic data for the design of technical systems. In K. Landau (ed.) *Ergonomic Software Tools in Product and Workplace Design* (Stuttgart: Ergon Verlag), pp. 214–229.

Seitz, T. and Bubb, H. (2000). Anthropometry and measurement of posture and motion. In K. Landau (ed.) *Ergonomic Software Tools in Product and Workplace Design* (Stuttgart: Ergon Verlag), pp. 28–36.

Seitz, T. and Bubb, H. (2001). An approach for a low-cost alternative for full-body posture and movement measurement and analysis. In W. Karwowski, P. Mondelo, B. Das and M. Mattila, *CAES 2001, Proceedings of the Computer Aided Ergonomics and Safety Conference*, Maui, Hawaii.

Sengupta, A.K. and Das, B. (1997). Human: an Autocad-based three dimensional anthropometric human model for workstation design. *International Journal of Industrial Ergonomics*, **19**, 345–352.

Shield, B.M. (1980). A computer model for the prediction of factory noise. *Journal of Applied Acoustics*, **13**, 471–486.

Siegel, A.I. and Wolf, J.J. (1969). *Man-Machine Simulation Models* (New York: Wiley).

Stewart, J.R. (1992). Ergonomic analysis with Mannequin. *Bulletin of the Human Factors Society*, Santa Monica, CA, pp. 9–10.

Sundin, A., Ortengren, R. and Sjoberg, H. (2000a). Proactive human factors engineering analysis in space station design using the computer manikin Jack. *Proceedings of SAE Conference on Digital Human Modelling DHMC 2000*, Dearborn, Michigan.

Sundin, A., Christmansson, M. and Ortengren, R. (2000b). Use of a computer manikin in participatory design of assembly workstations. In K. Landau (ed.) *Ergonomic Software Tools in Product and Workplace Design* (Stuttgart: Ergon Verlag), pp. 204–213.

Swat, K. and Krzychowicz, G. (1996). ERGONOM: Computer-aided working posture analysis for workplace designers. *International Journal of Industrial Ergonomics*, **18**, 15–26.

U.S. Army, Natick Research and Development Laboratories, Natick, MA.

Vidulich, M.A. (1998). Computer modelling of operator mental workload and situational awareness in simulated air-to-ground combat: An assessment of predictive validity. *International Journal of Aviation Psychology*, **8**, 351–375.

Ward, G.J. (1994). The Radiance lighting simulation and rendering system, Computer Graphics, *Proceedings of the SIGGRAPH Conference*, July, 459–472.

Waters, T.R., Putz-Anderson, V., Garg, A. and Fine, L.J. (1993). Revised NIOSH equation for the design and evaluation of manual lifting tasks. *Ergonomics*, **36**, 749–776.

Zulch, G., Kruger, J., Schindeler, H. and Rottinger, S. (2003). Simulation-aided planning of quality-orientated personnel structures in production systems. *Applied Ergonomics*, **34**, 293–301.

Part V

Analysis design and evaluation of work systems

One of the most obvious changes in ergonomics practice during the past two decades or so has been the expansion in its areas of application. Most in the field would now accept that the human-machine system embraces more than the interface controls and displays, the workplace and environment. We must be interested in the wider context of work, the psychosocial environment and the impact of organizational factors such as production technology, organization structure and finance. In some quarters the need to emphasise such wider concerns has led to establishing a 'macroergonomics' movement, especially in North America. In Europe there is the tendency more to assume that of course ergonomics includeds such organisational emphasis, and more to talk of a systems ergonomics approach as reflecting the wider, interacting and systemic nature of the field of study and practice. This part of the book then, and Part VII following, takes a fairly broad systems view of work systems and of ergonomics/human factors.

In the first two chapters we reflect the tendency for work (and other) systems to be increasingly collaborative and distributed, and for human factors to take a systems perspective in order to understand all relevant influences and their interations. For many people working in ergonomics/human factors, they are interested in large-scale, complex and networked systems of people, computers and administrative procedures and structures, their operations distributed over time and space. Many of the modern theories and approaches of (or adopted by) human factors — such as distributed cognition, joint cognitive systems, naturalistic decision making, situated action and cooperative work — are focussed on the structure, operation and understanding of such systems. However, probably the first such coherent approach, albeit much modified in the years since the 1960s, is that of socio-technical systems (STS). It is this topic that opens up the this section in Chapter 29 from Waterson, covering not just the philosophy and approach of STS, but also techniques for their analysis.

The major implementation from STS theory has been the development and running of work teams, originally as (semi)autonomous work groups in such companies as Volvo and Saab, then as self directed or self managed work or project teams, and now talked of as team working. From being in the province of limited field experiments in manufacturing industry, via considerable interest in the military, team working is now often the organisational form of choice, and teams can be formed, permanently or temporarily, for mission, project, process purposes, and to be co-located or distributed (virtual). The issues surrounding the use and design of teams, and particularly concerning their evaluation, are addressed in Chapter 30 by Salas, Priest and Burke. A number of tools are described, based on general methods as described earlier in this book, including rating scales, modelling and protocol analysis, but adapted for use with teams. Specially developed performance measurement methods are also described, including event-based scales and automated performance measurement. All the difficulties of measurement for individuals will be faced with the addition of those specific to understanding groups of people and their combined performance.

The next three chapters, by Haslegrave and Corlett (31), Kirwan (32) and van der Schaaf and Wright (33) concentrate upon assessments which can be made in order to evaluate the 'quality' or success of a work system — performance measurement — and also to identify areas for system improvement through redesign. They are most of all three different approaches to risk assessment. Recent changes in legislation and in the underlying attitudes within industrial society have given impetus to the search for ergonomics risk assessment methodologies, especially in the context of manual handling and exposure to risk of musculoskeletal disorders. Haslegrave and Corlett review the background, need, basis and contemporary approaches to such risk assessment.

Along with recognition of the critical need for improved human–computer interaction in determining the success of computer systems, arguably the other major boost to the perceived importance of human factors worldwide was the Three Mile Island nuclear power plant incident in 1979, and subsequent concern for nuclear power plant safety generally. This concern has continued through the occurrence of a number of well publicised disasters or near disasters involving complex systems in power plants, chemical processes, and transport systems, exemplified most recently at least in the UK by the rail industry. Fundamental to many of these incidents and to safety in such systems is the potential for human error and for organisational failure. Therefore there is great interest in developing methods for the analysis and measurement, and subsequent enhancement, of human reliability. Kirwan in Chapter 32 discusses this as a component of a ten part generic methodology, and provides very much the perspective of an active practitioner.

There is a strong link between the consideration of human reliability and the use of accident and incident reporting and analysis techniques. In Chapter 33, van der Schaaf and Wright show how theories of human error may well be very productive in terms of ergonomics measures for safety improvements. They describe how we can learn from different types of incident reporting — confidential, anonymous etc. — and also describe frameworks that can be used to develop, implement and manage such systems. Many of the points they make about reporting and analysis reflect cautions in the context of event observation and use of archives generally (see Part 1 of this book).

Finally, in Chapter 34 Hollnagel examines the causes and consequences of organisational failure. The learning that takes place in the light of understanding failure is vitally important, and is examined at individual, workgroup and organisation levels, and beyond, which provides a methodological approach all of its own. The relationship between accidents and near misses is explored, and in this way Chapter 34 synthesises nicely with Chapters 32 and 33.

Chapter 29

Sociotechnical design of work systems

Patrick Waterson

Introduction

This chapter is concerned with the sociotechnical systems (STS) approach to the design of work systems. The term 'sociotechnical systems' was coined by Emery and Trist (1960) in order to explain and understand the complex interrelationship between humans, machines and the environmental aspects of work systems. In its simplest form a core feature of the STS approach is the distinction between social and technical subsystems in organisations, and the proposal that there should be joint optimisation and parallel design of the two (Parker and Wall, 1998).

Proponents of the STS approach argue that effective performance (defined in terms of production output, quality, employee well-being, job satisfaction, etc.), is a function of the degree to which design of the social and technical systems are considered to be complementary, as well as the relative effort placed upon the parallel design of both systems. If more effort is placed on the design of one system (e.g., the technical) as compared to the other, then sub-optimal performance of the overall system is predicted to be the net outcome. The STS approach is often seen as an attempt to avoid technology-led approaches to work design (Blackler and Brown, 1986) and at the same time, place emphasis upon the value and benefits of human-centred design. Biases towards technological, as compared to human concerns, are frequently viewed as a partial explanation for the under performance of technology in the workplace (Landauer, 1997), as well similar problems that apply to the use of new manufacturing practices (e.g., Total Quality Management, Business Process Re-engineering) within industry as a whole (Waterson *et al.*, 1999; Mumford and Hendricks, 1996).

Sociotechnical studies of the workplace have a long history stretching back to the earliest days of research in human factors and ergonomics in the immediate period following World War II. In addition, STS has influenced many other areas including large parts of organisational psychology, systems theory and industrial sociology (van Eijnatten, 1993). Within the field of human factors and ergonomics the list of applications of the STS approach is extensive covering all of the aspects associated with work systems including the design of technology, jobs and arrangements for teamworking. Likewise, the range of work domains in which STS has been applied is also long. These include established areas such as human-computer interaction and occupational safety (e.g., Eason, 1988; Barling *et al.*, 2002), through to more recent forms of work involving knowledge management and electronic trading (e.g., Damodaran and Olphert, 2000; Coakes *et al.*, 2002).

In this chapter we will first look at the background history of the STS approach alongside its main principles, constructs and components in the second section, before moving on to examine some STS methods and tools for the analysis and design of work systems in the third section. The penultimate section describes an integrated example of the application of STS methods and tools to a particular type of work and its associated design requirements (i.e., task allocation within naval command and control systems). A final section briefly describes some of the outstanding challenges for STS in the future and provides suggestions for further reading.

Sociotechnical systems theory

Background history

STS theory can trace its origins back to the pioneering work of a group of researchers at the Tavistock Institute of Human Relations in London during the early 1950s. One researcher (Ken Bamforth), an ex-miner and a researcher at the Tavistock was given the opportunity to visit the mine in South Yorkshire where he had previously worked. During the visit Bamforth noticed a new form of work organisation that was operating in the coal mine and that had come about because of the difficulties of using normal equipment for extracting coal. In what has become something of a classic paper, Trist and Bamforth (1951) describe how the workers had organised themselves into small, relatively autonomous work groups consisting of eight miners, who were responsible, as a group, for a full cycle in the coal extraction process. What was particularly interesting about this form of work organisation was that it was more efficient as compared to the prevailing method of working prescribed by management of the mine and which tended to emphasise a 'one best way', or Tayloristic (i.e., strict demarcation of roles and responsibilities with few opportunities for employee involvement), method of working.

Throughout the 1950s and 60s a number of other STS studies were conducted worldwide and in numerous contexts and work settings (e.g., the Indian textile industry — Rice, 1958; clothing factories in Norway — King, 1964). Most of these studies continued the line of Bamforth and the other Tavistock research workers in examining the value of work redesign as initiated and implemented by workers and pointed to the benefits of their participation in the design of work systems. In addition, the studies also demonstrated the value of forms of work organisation that were a mixture of formal and informal design and where work groups were given a degree of ownership and authority over the way in which work was carried out. These so-called semi-autonomous, or autonomous work groups (AWGs) were viewed amongst followers of the STS approach as being one of the most optimal ways of maintaining high levels of productivity and output, and at the same time ensuring that the psychological needs of the individual worker were met (e.g., by providing a satisfying job and more generally a better quality of working life). In more recent years, AWGs and other similar forms of teamworking arrangements (e.g., high performance work teams — Buchanan, 1987; flexible work groups — Kelly, 1982) have been the subject of a number of other evaluations. For example, studies have shown that this form of work organisation can bring about greater employee satisfaction (Trist *et al.*, 1977), as well as reductions in operational costs due to employee innovation (Walton, 1977).

Alongside, recommendations concerning work groups and employee participation, the STS approach also makes use of a number of other principles, constructs and approaches to understanding work design.

STS constructs and components

From the point of view of the STS approach work design is seen as being influenced by many factors that depend upon the context in which it takes place (i.e., design is situation specific). There is therefore a need for a high degree of flexibility and choice in terms of what parts of the work system (e.g., individual jobs, managerial responsibilities, aspects of the technology being used) should be redesigned and precisely how this should be done.

The STS approach also encompasses a set of values that apply to work design of all types (rich or lean technological environments, manufacturing or service industrial sectors, etc.), and involve all of its various components (i.e., technology, people, environment). Mumford and Axtell (2002) for example, describe how a great deal of emphasis was put early on within the work of the Tavistock group, on the potential of STS interventions to improve the quality of working life of the individual worker. A central concern of the STS approach is that individuals involved with the work system should be treated with dignity and respect, and their aspirations and needs catered for. A similar concern applies within more recent Tavistock work (e.g., Heller, 1997), which emphasises the need to take into account the views of people when environmental change is being introduced.

One other important aspect of the STS approach that should be mentioned is the stress it place upon the value of participation and involvement in proposed changes to an existing work system, or the design of a new one. Proponents of the STS approach argue that the best way to achieve successful change is by involving all those likely to be affected by change in the development process. Participation in the change process is beneficial since it increases levels of ownership of the change process amongst employees, as well as reducing the likelihood that the proposed changes may be rejected by end users or other stakeholder groups. Amongst STS practitioners there is also an awareness that participation has itself to be managed carefully and in many cases supported by tools and methods specifically designed to facilitate involvement and collaborative design (Wilson and Haines, 1997; Greenbaum and Kyng, 1991 and Chapter 36 in this book).

Whilst there have been few critics of the underlying philosophy and motives of STS proponents and researchers, there have been other criticisms aimed at the apparent lack of theoretical coherence within the approach (e.g., Kelly, 1978 — a theme we shall return to later on in the chapter). These criticisms have to some extent resulted in a number of attempts to systematically describe some core principles for STS.

Principles of sociotechnical design

In a series of papers published in 1976 and 1987 the late Albert Cherns laid out what are generally recognised as some of the important principles of the STS approach (Table 1).

The principles take the form of a set of guidelines that can be used to help to design a safe and healthy work system and made up of jobs that are rewarding and satisfying for employees. The work system is typically made up of one or more persons, some form of technology in computerised (e.g., desktop computers) or non-computerised form (e.g., a factory production line), as well environmental characteristics and influences surrounding the overall work system. For example, principle 2 (minimal critical specification) is an attempt to deal with the problems of over designing the work system so that, for example, individual roles and/or tasks to be performed by the technology (i.e., automated) are overly prescribed by the designer. Instead, Cherns suggests that work systems should be designed so that they are flexible, 'minimally specified' and where possible, designed partly by the individuals themselves within their work teams (principle 3). Other principles aim to improve the well-being and job satisfaction of individuals (e.g., principles 4, 7 and 8), as well as aspects of the management of team boundaries and communication requirements (principles 5 and 6).

More recently, there have been other attempts to provide principles and guidelines for sociotechnical design. Clegg (2000) for example, has provided a set of principles that aim to update the earlier list from Cherns, as well as apply more directly to the domain of systems and technological design. Clegg has expanded the original nine principles into three

Table 1 Examples of Cherns' (1976, 1987) Principles of Sociotechnical Design

Principle	Summary
Compatibility	The process of design must be compatible with its objectives.
Minimal critical specification	No more detail in design than is needed, but design must express the essential requirements.
Sociotechnical criterion	Control is local and should be given to the immediate work team – the aim is therefore to make supervision minimal
Multi-function	Individuals and groups need a range of tasks to provide satisfying jobs and for redundancy and flexibility.
Boundary location	Boundaries are political boundaries and should be allocated managerial resources.
Information flow	Information should flow initially to the prime user or group and should avoid intermediaries where possible.
Support congruence	Systems should be established within a framework of social support for desired behaviour.
Design and human values	Emphasis in design is placed upon the quality of working life
Incompletion	Design is iterative and continuous.

different overlapping groups making a total of 19 principles in all. The first group of principles described by Clegg fall into the category of what he terms 'meta-principles', that is, principles which cover the philosophy or worldview underpinning design activity. The second and third groups cover the content and process of design respectively. Table 2 outlines some examples of the three types of principles.

In some respects the term 'principles' is misleading since it might be taken to imply that the contents of Tables 1 and 2 should be applied in a mechanistic, or prescriptive manner. However, as Clegg (2000, p. 474) points out STS principles are not designed as blueprints that should be strictly adhered to, but rather as 'ideas for debate, providing rhetorical devices through which detailed design discussions can be opened up and elaborated'. The principles are also designed to be used alongside methods and tools that support different types of sociotechnical activity, for example, forming an overview of a particular design or carrying out the evaluation of design options. Some examples of these tools are described in the next section of the chapter.

Tools and methods for STS design and analysis

Overview

As we mentioned earlier a characteristic of the STS approach is its flexibility. One consequence of this is that there are no hard and fast rules or guidelines regarding what types of methods should be used in a study, or where and when these should be applied. The types of methods to be used are largely up to the researcher and may be determined by factors such as: the specific nature of the problem to be addressed; the degree of access granted the researcher (e.g., access to company personnel, documentation and other data); the timescale of the study; the views and requirements of study participant regarding the expected outcome from the study; and, the type of feedback that is to be given to the participants involved in the study. This latter point is especially important since the STS approach often involves what is known as 'action research' (Foster, 1972).

Action research typically involves the STS researcher adopting the role of a participant in the change process (as compared to a disinterested 'scientific' researcher). This role also involves the commitment that the researcher will seek practical solutions to the problem

Table 2 Examples of Clegg's (2000) Principles of Sociotechnical Systems Design

Principle	Summary
Meta-principles	
Design is systemic	All aspects of system design are inter-connected. Leaving out one part (e.g., human aspects) will inevitably lead to sub-optimal performance of the whole system.
Values and mindsets are central to design	Attitudes and values on the part of designers, managers, users etc., will shape the outcomes of design. It may be helpful at the outset of design to state clearly any possible biases and leanings toward possible design options.
Design is socially shaped	Design is subject to social movements and trends, these may sometimes manifest themselves as fads and fashions.
Content principles	
Systems should be simple in design and make problems visible	Design simplicity includes ease of system use, understanding and learnability.
Design entails multiple task allocations between and among humans and machines	Task allocations should be systematically analysed and evaluated. In the case of deciding to allocate tasks to a team or AWG then decision-criteria should be used (e.g., Medsker and Campion, 1997).
Problems should be controlled at source	Problems should be tackled by those closest to the problem since this has motivational and logistical advantages (e.g., people like to have control over the problems they face).
Process principles	
Evaluation is an essential part of design	Evaluation, although rarely undertaken, has several advantages, the chief one being that an organisation can learn from its mistakes and successes.
Design involves multidisciplinary education	There is a need for a diverse range of expertise and skills within design in order to bring about innovation, as well as viewing design from several perspectives.
Resources and support are required for design	Successful design comes at a price and in order to follow some of the principles there will be a need for organisations to invest resources into the process.

being addressed, as well as examining more theoretical issues. For example, a researcher involved in the study of the introduction of a new IT system within a large company might try to help the company find ways of involving users in the change process (e.g., by helping to run user workshops), whilst at the same time seeking to address research questions such as how the technology changes or re-structures patterns of working in the light of theories within the scientific literature.

As well as involving a practical objective, the STS approach also typically employs more than one method of study when carrying out a research project. Wall *et al.* (1986) for example, carried out semi-structured interviews during the course of their study of the introduction of AWGs within a company manufacturing sweets, whilst at the same time asking shopfloor workers to complete a questionnaire survey during the early and later phases of the change process (a so-called 'quasi-experimental field study' — see Cook *et al.*, 1990 for further details of this research method). The important point to note is that STS studies often employ a variety of methods and these are not exclusively limited to either quantitative or qualitative approaches. Furthermore, application of the STS approach may also involve many of the methods described in other chapter in this volume (e.g., task analysis — Chapter 6; knowledge elicitation techniques — Chapter 8).

The next sections review methods which are slightly less well known and do not easily fit within the quantitative/qualitative distinction between research methods, nevertheless, they are often used within STS studies. A rough distinction can be made between methods that support the analysis and planning of design based upon STS (e.g., soft systems methodology and the use of scenarios), and those that are intended to provide help in evaluating the outcomes of design and making it possible to suggest areas of the work system that can be redesigned (e.g., checklists and decision-criteria).

Methods used to analyse and plan STS design of work systems

Soft systems methodology

Soft systems methodology (SSM) is a technique that was designed in order to facilitate the analysis of complex work systems (e.g., departments within a company, whole organisations involving many stakeholders) and to make it easier to carry out the task of planning and implementing change within these systems (Checkland and Scholes, 1999; Clegg and Walsh, 1998). SSM has been successfully applied within a variety of contexts involving sociotechnical issues, these include the management of IT-based changed within healthcare and civil service departments (see for example Chapters 4 and 5 in Checkland and Scholes, 1999), the elicitation of user requirements within new IT systems (Atkinson, 2000), as well as studies of the cognitive representations of system developers (Waterson *et al.*, 1999).

Typically SSM is used within a set of workshops where representatives from the various groups involved with change (e.g., managers, end users, system specialists) are gathered together and discuss different views and characterisations of the new system. These workshops often take place over a period of time and are usually facilitated by one or more analysts who have gathered data regarding the problem or change under consideration. The data is then presented in a format that makes it simple for workshop participants to visualise and discuss (e.g., flowcharts, pictures or cartoons). This format is typically called the 'rich picture' within the SSM framework. The process of analysing, evaluating and implementing these representations typically involves some combination of these stages.

Tables 3 and 4 describe an example of the application of SSM that was part of a case study that took place within an engineering company that were implementing a set of CADCAM (computer-aided design, computer-aided manufacturing) tools (based upon the work of Symon and Clegg, 1991a,b).

SSM can prove to be a very useful method for gathering a wide variety of opinions regarding a new system. The method has several strengths, including the fact that it helps to include a wider set of perspectives (e.g., by involving stakeholders in the analysis and implementation of change) than is perhaps normally the case, in the overall process of managing change. Second, SSM provides a systematic means of organising complex data and representing it in a simple and concise manner. Finally, the methodology encourages a participative and systemic approach to design (i.e., it is congruent with some of the sociotechnical principles described earlier in Tables 1 and 2).

Needless to say, SSM also has some disadvantages (Clegg and Walsh, 1998, pp. 231–232), including the fact that in many cases the language and terminology used by SSM (e.g., rich picture, root definition) can be complex to use and describe. Furthermore, SSM also offers little guidance in the details of how to implement some of the outputs from its use, perhaps for this reason it is best used in combination with some of the other methods described in this section, rather than in isolation.

Table 3 Stages of Applying SSM to the Analysis of the Implementation of the CADCAM Tools within the Engineering Organisation (Symon and Clegg, 1991a, b, c)

Stage	Description	Summary of activities and background
1.	Problem Situation	• Aerospace engineering company employing 2000 people. • Complex organisational structure, three major functions: design, manufacture and corporate engineering. • Design and manufacture functions operate autonomously, in the past there has been little success in integrating the two functions. • Senior managers decide to invest in CADCAM in order to: reduce costs and improve quality; improve integration between design and manufacturing. • Project team is set up to manage implementation, researchers are asked to investigate the implementation process and make recommendations for improvements from an organisational (as compared to technical) point-of-view.
2.	The 'rich picture'	• Research carried out over period of 18 months. • Variety of data collected including interviews n=35), periods of observation and attendance at meetings and questionnaire survey (n = 36). • Main concern of the staff at headquarters was getting the technology implemented at a minimum cost. • Little attention was paid to the wider organisational impact of CADCAM or the opportunities the implementation provided for reorganising work in the two functions (i.e., design and manufacturing).
3.	Relevant systems and 'root' definitions	• The researchers decided to describe both the existing relevant system that they believed was guiding the implementation, as well as an alternative system and its root definition. • The researchers argued that the project was seen and managed as a technology implementation project (as compared to an opportunity for organisational change).
4.	Conceptual model	• In order to meet the demands of the alternative root definition (i.e., the implementation as an organisational change), the researchers developed a conceptual model. • The model consisted of seven elements in a flowchart, these were: Theorise-> Resource-> Plan-> Participate-> Educate-> Support-> Evaluate • The new model was used as the basis for a set of recommendations to the company.
5.	Comparison and agenda	• Comparison between the new model and the real events going on in the company resulted in the identification of some major gaps. • These gaps led the researchers to generate an agenda for change. • Some of the key items in the agenda included — investing more resources in the change programme, encouraging more participation by end users and placing more emphasis on wider education, awareness and communication for end users and others.
6.	Debate	• An open report describing the finding, root definitions and agendas for change was presented to all interested parties within the companies. • End users expressed satisfaction with the outcomes and acknowledged the technology-led aspect of the implementation. • Senior managers however, were less convinced and disagreed with many of the conclusions from the report.
7.	Implementation	• Some of the recommendations from the report were implemented. • In particular, a CADCAM user group was set up in order to address issues related to end user participation, education and training for example.

Table 4 Elements in the Root Definition of the CADCAM Implementation

	Existing system	Alternative system
C — Customers (or Clients) — i.e., the people who receive what the system does	Corporate engineers and designers in the different product groups	Same
A — Actors — i.e., the people who carry out what the system does	Corporate engineering (senior managers)	Everyone involved with the change (e.g., managers, end users) — i.e., a participative approach to the implementation
T — Transformation — i.e., what the system changes, from one state to another	Replacing a manual system for design and manufacture with a CADCAM system	Designing a new organisational system, integrating design and manufacturing and assisted by the CADCAM system
W — Weltanschauung — i.e., the underlying worldview	Emphasis is upon technical issues and problems, the view being that organisational issues can be handled later	Joint (i.e., sociotechnical) consideration of technical, strategic and organisational issues
O — Owners — i.e., those who have power over the system, for example, to cause it to cease to exist	Corporate engineering (senior managers)	Everyone involved with the change (e.g., managers, end users) — i.e., a participative approach to the implementation
E — Environmental i.e., constraints — constraints that the system has to take as given	A minimum of scarce company resources (i.e., time and money)	Process will require appropriate resources to operate (in some cases more money will have to be invested in addressing the organisational issues)

Scenario-based methods and tools

One of the most common techniques used by STS practitioners, as well as others drawn from a variety of backgrounds (e.g., human factors, software engineering, human–computer interaction), are scenarios. Carroll (2000) defines a scenario as an 'informal narrative description' and the term is widely used to convey some kind of story-like representation that can serve as the basis for discussion and debate amongst groups such as system end users and other stakeholders. One of the main advantages of scenarios is that they are a natural way for people to explain what they are doing or how something can be achieved (Preece *et al.*, 2002). Within the context of generating the requirements for a new system or a new set of work-based roles for example, scenarios offer a simple way of describing the existing way in which the work system is operating, as well as planning the design of a new system that is sensitive to the needs of people working within it.

Nadin *et al.* (2001) used scenarios to help redesign a set of jobs and tasks carried out by shopfloor workers in a company producing photographic film and paper. At the time of the study the company was planning to introduce a new computer-based scheduling system, and the likelihood was that the new system would change the work of two main types of jobs ('solutions' and 'melting' personnel) within its chemical processing department (e.g., changing the tasks they performed and the way in which they interacted). One part of the study involved generating a set of scenarios that described the existing way in which work was carried out by the two job types, as well as alternative involving different levels of skill distribution and interaction between the two types of personnel. The scenarios were generated over a series during a series of three workshops involving shopfloor workers and managers and were structured around three main headings (the scope and structure of the scenario and the roles involved) — see Table 5.

The scenarios proved to be popular amongst the participants at the workshops largely because they helped to structure discussion centred on the change process, as well as having the added benefit that all of those likely to be effected by the change were involved in generating the scenarios. In many cases the use of scenarios can prove worthwhile since they help participants to consider issues that would have been otherwise ignored or overlooked. This is especially important given the stress placed upon joint consideration of both social and technical aspects of change management, as well as other systemic issues (see principle 1 in Table 2), within the STS approach.

Table 5 A Summary of the Scenarios Generated Using the Tool in the Workshops (Nadin *et al.*, 2001)

	SCENARIO 1	SCENARIO 2	SCENARIO 3
Scope	Existing structure	Partial multi-skilling	Complete multi-skilling
Structure	Maintain present boundaries: Solutions personnel work upstairs	Maintain boundaries but Melters share some tasks with Solutions personnel	No boundaries. Abandon distinction between Solutions and Melting personnel.
			Work as one crew and rotate tasks
	Melting personnel work downstairs		
Roles	Specialised and separated	Overlap of tasks. Melters to become partly multi-skilled.	Single role for all crew members combining all tasks
	Solutions to carry out existing tasks		

Tracer studies

The analysis of a complex work system often requires the researcher to gather data about the flow of information between different people and the technologies they may be using. This type of analysis may be helpful in identifying communication 'bottlenecks' between individuals or larger groups within an organisation. Similarly, the analysis may highlight areas where a technology frequently breaks down or causes problems (e.g., where faulty machinery results in poor quality of manufactured parts on a production line). One way in which this information can be systematically gathered is by using what is known as a 'tracer study'. Hornby and Symon (1994, p. 167) define the tracer study as 'a method of identifying and describing organisational processes (such as decision-making and communication) across time and stakeholder group by the use of 'tag(s)' (such as documents or meetings)'. In this case 'tags' can be taken to mean anything that records the participation of individuals during the overall tracing process (e.g., documents such as meeting minutes, print outs from machinery, timesheets). The use of tracer studies typically involves five main stages (Table 6).

Tracer studies have been used successfully in a number of case studies (see, e.g., Hornby and Symon, 1994, pp. 174–185). One of the advantages of the method is that it can be used

Table 6 Stages involved in using a tracer study (based upon Hornby and Symon, 1994, pp. 171–1974).

Stage	Description	Summary of activities

1.	Identification of 'tags'	• Listing all possible tags associated with the process to be traced.
		• In many cases there will be more tags available than the researcher can practically or usefully use. In this case some clear selection criteria may be useful, these include:
		  • Theoretical framework: some tags may be more appropriate when examining specific theories (e.g., decision-making) or hypotheses.
		  • Level of analysis: is the researcher examining the whole process (e.g., decisions taken throughout the organisation) or more specific processes (e.g., decisions taken in one department)?
		  • Informal processes: tags can be developed to examine less formal processes such as meetings that occur spontaneously (e.g., consultations between peer groups).

2.	Formulating sampling criteria	• Formulating specific criteria for sampling specific cases of the process under investigation (e.g., tackling a research question within the context of one particular group in the organisation).
3.	Examination of specific tag(s)	• Placing the tag within the context it is used by posing questions such as: what is the history of the tag? What is its espoused function? Where does it originate? How much time is spent on the tag in one place?
4.	'First pass' identification of relevant informants	• Matching the use of the tags to individuals using them. • Selecting a sample of tag users and interviewing them • Using other methods (e.g., observation) to further investigate use of the tag.
5.	Ending the tracer study	• Deciding when to complete the study, usually after the researcher has identified the start and end positions of the tag and carried out sufficient data collection such that no further useful information is forthcoming ('saturation point' — Glaser and Strauss, 1967).

to examine in detail factors that have contributed to the design of a work system over time from the point of view of both the process and the people involved. Tracer studies, in common with other STS methods are frequently used in tandem with other methods both quantitative and qualitative (for an example in the domain of system development see Axtell *et al.*, 1997).

Methods used to evaluate STS (re-)design of work systems

Checklists and decision criteria

At the heart of the STS approach is a concern that the design of work systems should go beyond the planning and generation of design options, but should also critically involve their systematic evaluation and subsequent refinement (cf., the principles in Tables 1 and 2). A number of STS evaluation methods exist and some of these are aimed at covering the larger work system (e.g., the organisation as a whole or a department within the organisation — see Hendrick, 1997 for examples) or specific work sub-systems (e.g., interfaces between people and technology — Mumford, 1995). Furthermore some evaluation methods focus on specific areas or domains of design activity in more detail (e.g., usability evaluation — Nielson, 1997; requirements engineering — Eason *et al.*, 1996). Two of the most common evaluation methods that can be used to evaluate sociotechnical aspects of the whole work system, as well as its parts, are the use of checklists and decision criteria.

Checklists can be used by researchers, as well as others outside of the human factors and ergonomics profession (e.g., managers), to assess the degree to which an existing or planned work design compares to established measures, or indicators, of good practice. Campion and Medsker (1992; Medsker and Campion, 1997) for example, describe a checklist that can be used to degree to which teams have been successfully implemented within an organisation (Table 7).

Checklists can also be used to benchmark the progress of changes to a work system as they occur over time. For example, they can sometimes be turned into other material, such as survey questionnaires, and used as an important means of gaining feedback from shopfloor workers and other personnel involved in the change process. It should also be noted that these types of methods are often more successful when they are accompanied by further guidance covering strategies and other advice that can be used to implement changes to the

Table 7 Example Questions in the Form of a Checklist for the Design of Teams (adapted from Campion and Medsker, 1992)

1. Are worker's tasks highly interdependent, or could they be made to be so? Would this interdependence enhance efficiency or quality?
2. Do the tasks require a variety of knowledge, skills, abilities such that combining individuals with different backgrounds would make a difference in performance?
3. Is cross-training desired? Would breadth of skills and work force flexibility be essential to the organisation?
4. Could increased, motivation, and effort to perform make a difference in effectiveness?
5. Can social support help workers deal with job stresses?
6. Could increased communication and information exchange improve performance rather than interfere?
7. Could increased cooperation aid performance?
8. Are individual evaluation and rewards difficult or impossible to make or do workers mistrust them?
9. Could common measures of performance be developed and used?
10. Is it technically possible to group tasks in a meaningful, efficient way?

work system (Clegg, 2000; Parker and Wall, 1998). A final set of methods that can be used to evaluate options for work design are the various forms of decision criteria that exist for STS evaluation.

Methods that use decision criteria typically involve a group of stakeholders involved with the design process collectively evaluating design options using a pre-defined categorised list of issues that are known to be of relevance within the design domain. The advantage of the pre-defined list is that aspects of the design of the system that either be overlooked or ignored can be systematically addressed and evaluated. Table 8 outlines a categorised list of decision criteria that were developed for use within the domain of task allocation, but can also be used for wider STS work design (Waterson *et al.*, 2002).

A typical situation in which the decision criteria in Table 8 are used is where there are a number of design options for a particular set of work tasks. Stakeholders work through the criteria and consider the design options in the light of questions that are raised as a result of consulting the criteria. For example, stakeholders might examine the category 'Cultural/environmental issues' and pose the question — 'do any of the design options have cultural/environmental concerns?' In order to provide stakeholders with some help the category is then broken down into criteria that should be taken into account and posed as sub-questions (e.g., 'do any of the design options raise data security issues?'). Further examples of the use of decision criteria are described in more detail in the penultimate section.

Table 8 Examples of Decision Criteria (Waterson, Older Gray and Clegg, 2002)

Category	Criteria
Goal issues	Goals of the system; goals of the organisation; other goal issues.
Organisational issues	Existing/required systems and procedures; existing/required organisational norms and practices; organisational requirements; organisation structures; accountability; other organisation issues.
Cultural/ environmental issues	Cultural issues; legal requirements; health and safety; data security; national/ international political considerations; other cultural/environmental issues.
Resource issues	Technology; people; money; other investments (e.g., time); knowledge, skills and education; other resource issues.
People issues	Level, type and number; knowledge, skills and education; trust; training; social acceptability; other people issues.
Task issues	Speed of response; accuracy; frequency; physical demands; cognitive demands; emotional demands; reliability; efficiency; flexibility; operational criticality (likelihood and implications); safety criticality; uncertainty of occurrence; uncertainty of the type of problem; uncertainty of the correct response; variability in performance; task interdependencies; redundancy; other task issues.
Job design/work organisation issues	Control (e.g., autonomy, responsibility, variance handling, decision making); accountability; variety; skill use and development; workload; clarity of goals and requirements; feedback on performance; communication and social contact; health and safety (mental and physical); flexibility; information exchange and knowledge sharing; situation awareness; understanding of the system; mutual adjustment/team collaboration; role ambiguity and overload; motivation; satisfaction; stress; performance; other job design/work organisation issues.
Technology issues	Feasibility of automation; cost of automation; maintainability; reliability; level of performance; trust in technology; level, type and amount of technology; other technology issues.

Other methods

Aside from the methods we have described there exist a number of others that can be used for STS analysis, planning and evaluation. Many of these have much in common with soft systems methodology, scenarios and the use of decision criteria. Some of the most important include the group of methods that can be subsumed under the title of 'stakeholder analysis' (Burgoyne, 1994), of which SSM is a close relative. In addition, in the last few years methods that encourage and support greater user involvement in the design process (e.g., 'rapid prototyping' — Mayhew, 1999; Scandinavian participatory design — Bodker *et al.*, 1991) have become popular. Likewise, a technique that is similar in its analytical aims to that of tracer studies, known as 'variance analysis' (Davis and Wacker, 1982), is often suggested as an effective means of designing work systems that utilises the STS approach.

In the next section we will examine an integrated example of the use of some of themethods described above within the application domain of the design and evaluation of naval command and control systems. This example is based upon the work of Waterson *et al.* (2002).

An integrated STS method for the design of work systems

Background

The method was developed in order to help human factors personnel, as well as other stakeholders involved in work system design, to make decisions regarding how work should be allocated between and amongst humans and machines. Traditionally, this type of decision-making has been referred to as 'task allocation' and has tended to refer mainly to allocations between human and machines (Sharit, 1997; Older *et al.*, 1997). However, the allocation of work between humans, for example in the form of the design of jobs or forms or work organisation, is also relevant to the design of complex human and computer systems. The method therefore aims to cover both the human–machine and human–human aspects of the allocation of work within complex systems.

Description of the method and a worked example of its use

The method is designed along the lines of a flowchart (Figure 1) and users are encouraged to work through the stages in the chart (Stages A–G) using a set of smaller scale methods and tools. Some of these apply to individual stages within the overall method (e.g., Stage A — which uses scenarios and is similar in operation to SSM and stakeholder analysis), whilst others are intended for generic use (e.g., decision criteria — Stages B, C and D).

Users of the methods begin by adopting a top-down description of the system under construction (Stage A). Once the overall view of the system has been established, it is possible to identify the characteristics of the system necessary to meet this view. This part of the process involves the use of existing techniques for requirements analysis and specification (labelled 'Specification of Requirements' in Figure 1). It should now be possible to produce a detailed task description of all the tasks to be performed by the system and the relationships between them in order to achieve its objectives (labelled 'Specification and Analysis of Tasks' in Figure 1). Once this has been accomplished users of the method are in a position to begin allocating tasks.

In all cases where allocations take place, reference is made to the decision criteria described earlier (Table 8). For example, in the case of mandatory allocations (Stage B in Figure 1), users may consult the decision criteria and find that one of the issues in the list of criteria suggest a mandatory allocation. Examples of decision criteria that indicate a mandatory allocation include feasibility of automation (i.e., if it is not technically feasible to allocate a task

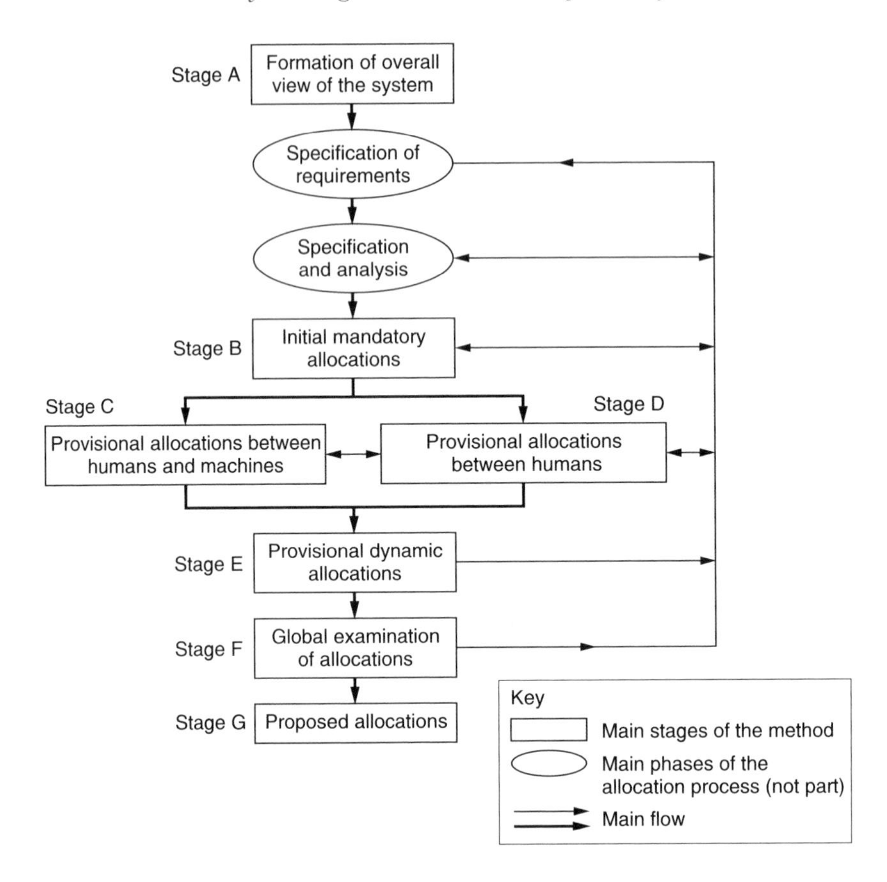

Figure 1 Flowchart outlining the stages in the integrated method.

to a machine, then it must be performed manually) and legal requirements (i.e., if legislation exists stating that the task much be performed by a human or a machine, then the task must be allocated accordingly).

Stages C and D involve similar forms of decision making regarding allocations between humans and machines and allocations between humans. Stage E applies to dynamic allocations where responsibility for carrying out the task may be shared by humans and machine (e.g., in the case of the autopilot in an aircraft where part of its functioning may be automatic and part under manual control). The remaining stages in the methods (Stages F and G) involve procedures for further checking and recording the results of the allocation process.

The following sections of the chapter describe Stages A, C and D in more depth, further details of the other stages in the methods are available in Waterson *et al.* (2002). An example of the use of the method is also described for each of Stages A, C and D. The example is drawn from a workshop that was run with a group of participants made up of three human factors experts, two end users and four system designers, all of whom were active in the area of naval command and control systems. The context in which the method was used involved the redesign of the management and operation of the bridge sub-system within a naval warship, in particular the sub-system dealing with navigation and collision avoidance.

Stage A: Formation of overall view of the system

The manner in which stage A works is similar to the types of scenario-based methods describe earlier on in the chapter. Users of the method are first asked to develop a number of alternative choices for how the system could work, describing them using a set of structured headings. These include: a description of the scope and boundary of the choice; its underlying vision and reasons for the vision; the organisational structure and roles involved in the choice; benefits and costs associated with the choice; an indication of the implications and overall rationale behind each choice; and finally, a preference rating for each choice. If there is an existing system, the structured headings are used first to describe the way in which the system currently works. The users then identify the potential advantages and disadvantages of each choice, using the decision criteria (Table 8) to ensure that all relevant issues are examined for each option.

Following consideration of the decision criteria, users are required to indicate the order of preference between each option, and the rationale underlying their decision. Table 9 describes the results of allocation decision making during Stage A of the method.

Participants at the workshop first described and summarised the existing ways of working (Choice 1); they then brainstormed around different organisational choices that were possible, or likely to be possible in the future, in terms of the human and technological characteristics of the bridge sub-system. Choice 1 provides an outline description of the existing sub-system, and is based on the traditional distribution of work amongst humans and machines. Choice 2 is an alternative based on technology as an aid to the human. Choice 3 incorporates a greater degree of technology, and Choice 4 is a description of the sub-system with full automation (with the human acting as supervisor). In this case, Choice 2 was the preferred option on the basis that it is consistent with existing role structures, the human remains in control, and the criterion of a more balanced workload is achieved.

Stage C: Between humans and machines

In this stage allocations are made between humans and machines. For each task or function, all the decision criteria are identified that might indicate a preference for how it may be allocated. The criteria are then listed and in the case of a criterion indicating that one of the options is not suitable, the rationale is recorded and only the remaining options are evaluated. Evaluation involves comparing the qualitative notes that are made under the headings (i.e., rationale, other details, issues arising) for each task and then deciding upon an allocation preference. Finally, users are required to summarise the main decision underlying the order of preferences for allocations under the heading 'Overall Rationale'. An example of the output from Stage C of the method is summarised in Table 10.

Participants at the workshop initially worked through the task analysis of the navigation and collision avoidance sub-system and in particular, the operational task of 'taking visual bearings'. Participants then worked through the decision criteria (Table 9) and selected categories of criteria and individual criteria that they viewed as particularly relevant to the task. For example, in the case of the category 'Task' they selected the criteria 'Speed of Response' as being of particular importance. In terms of this use of the decision criteria, as in all of the other examples shown in Figure 6, all of the allocation options were possible. For example, it was possible to imagine a situation where human, machine and dynamic allocations could be applied to the task of taking visual bearings.

Having worked through the decision criteria and evaluated the allocation options and considered the underlying rationale, participants then assigned a preference rating to each of the options. In this example participants allocated the task to the machine for three main reasons: greater accuracy; reduced workload for the human; and a quicker speed of response.

Stage D: Between humans

The fourth stage in the method, Stage D, involves allocating work amongst the humans working in the system. In this case users first identify the different human roles within the

Table 9 Example Outcome from Stage A of the Integrated Method

	Choice 1	Choice 2	Choice 3	Choice 4
Scope/boundary	Navigation & Collision avoidance	Same	Same	Same
Vision	Traditional	Use currently developed technology to support humans	Semi-auto decision-support	Unmanned bridge
Reason for vision	Career structure	Slightly reduced manpower	Reduced manpower; human confirms decision	Reduced manpower; human monitors
Level of automation	Minimal	Mild	High	Full
Organisation structure	Hierarchical	Similar	Multi-tasking; very simple structure	Multi-task monitoring
Roles	Specialised roles; clearly defined	Still specialist roles	Local validation and monitoring	Remote validation and monitoring
Benefits	Proven	Reduced workload; improved operational performance; less errors	Less people at risk; technology spurt, less humans.	Less people at risk; technology spurt, less humans.
Costs	High indirect costs	Some increased training; high indirect costs	High indirect costs	High indirect costs (procurement, monitoring, maintenance)
Implications	Inefficiencies; higher manpower	Reduced workload	Lose situation awareness; high technical risk	Legal implications; rely on indirect vision (sensors); high technical risk
Preference	3	1	2	4
Rationale	Need to reduce manpower; high training	More balanced workload	Reduced manpower; loss of situational awareness	Does not meet legal requirements

Table 10 Example Outcome from Stage C of the Integrated Method

Task no.	Task Name	Category of criteria	Criteria	Options H	Options M	Options H-M	Rationale	Other details	Issues arising
5.1.2.1.1	Take 3 Visual bearings	Task	Accuracy	√	√	√			
		Job design/work organisation	Workload	√	√	√			
		Technology	Cost of automation	√	√	√			
		Task	Speed of response	√	√	√			
		Job design/work organisation	Skill use	√	√	√			

Preference H	Preference M	Preference H-M	Overall Rationale
3			Lower cost of automation
	1		Greater accuracy; lower workload for human; high speed of response
		2	Optimum workload

KEY: H = human only; M = machine only; H-M = dynamic.

system, beginning with the way in which the work is currently allocated if there is an existing system. Then they develop a range of alternative role designs (i.e., some alternative allocations), which are evaluated using those criteria that indicate a preference (in particular drawing on the job design/work organisation criteria). As before, the users indicate their preference between the options, and present the rationale behind their choice, based on the results of their evaluation.

Table 11 illustrates use of this tool on the same 'taking visual bearings' example used during Stage C of the method. The tasks considered here are those associated with 'manoeuvring', including 'navigating' and 'managing ship avoidance'. As in previous stages participants worked through the decision criteria and then evaluated the allocation options on the basis of specific criteria before assigning an allocation preference.

The individual tasks and their existing role allocations are listed in Choice 1 — these tasks are currently undertaken by four different people acting in different roles with no supporting automation. Choice 2 allocates the tasks across three roles, also with no automation. In Choice 3, some of the tasks are shared between two human roles, while the remainder are allocated to machine. In this case Choice 1 (the existing way of working) was judged to be problematic due to heavy reliance upon manpower (reductions in manpower being one of the main drivers for redesign of the bridge sub-system). Choice 3 by contrast was judged to be risky since it relied too much upon machine support and this could be dangerous in the event of malfunction or breakdown of the technology. Similarly, Choice 3 was viewed as being less socially acceptable since it would reduce the control and skill variety of human operators on the bridge. Overall, Choice 2 was judged to be the most preferred option since it allowed dynamic allocation between the two operator roles and could potential increase level of job satisfaction and responsibility. Choice 2 was also preferred since it allowed for a degree of redundancy between the two operators (i.e., if one operator made an error the other operator could potentially identify the error and correct it).

Summary

A major advantage of the method is that it draws upon a much wider set of sociotechnical concerns as compared to previous attempts that either focus on task allocation or job design in isolation. By contrast, the method facilitates consideration of these issues in parallel and places emphasis upon a systemic approach to the design of the work system as a whole (cf., the principles in Table 2). The method also brings together some of the various methods and tools we have described earlier (e.g., scenarios and decision criteria). Finally, individual stages within the methods can be adapted for use by designers and stakeholders, for example, stage A might be used to 'brainstorm' alternative design options (e.g., in combination with elements from SSM).

Future challenges for STS and further reading

Future challenges

There are at least three major challenges facing researchers within the area of STS in the future. First, and perhaps most importantly, there is a need to re-examine some of the theoretical underpinnings of the STS approach as a whole. Much of the theoretical work in this area is now thirty or so years old and many of the core constructs and components (e.g., optimisation of social and technical systems, the concepts of redundancy and optimisation) need to be reappraised in the light of new theories and concepts. Much of the STS approach concerns itself with technological work and it is not clear how STS theory compares with more recent

Table 11 Example Outcome from Stage D of the Integrated Method

Roles	Choice 1	Choice 2	Choice 3
Role 1	Visual sighting; identify other vessel	Visual bearing	Question position; action
Role 2	Open ocean manoeuvre; report to command; ship avoidance; detect, identify, determine course, speed of other vessel; visual bearing; radar; calculate CPA; right of way; action	Question position	Question position; action (shared role)
Role 3	Visual bearings; radar ranges; radio aids; GPS; determine position; question position	Visual sighting; identify other vessel; open ocean manoeuvre; report to command; ship avoidance; detect, identify, determine course, speed of other vessel; radar' calculate CPA; right of way; action; radar ranges; radio aids; GPS; determine position; detect, determine course of other vessel; examine radar; calculate CPA (combined role)	
Role 4	Detect, determine course of other vessel; examine radar, calculate CPA		
Machine			Visual sighting; identify other vessel; open ocean manoeuvre; report to command; ship avoidance; detect, identify, determine course, speed of other vessel; visual bearing; radar; calculate CPA; right of way; visual bearings; radar ranges; radio aids; GPS; determine position
Preference	2	1	3
Rationale	Manpower intensive	Greater job satisfaction; appropriate support; more responsibility; acceptable degree of change	Greater risk; less socially acceptable

attempts to understand the overlap between social, organisational and technological issues in these types of work domains (e.g., actor/network theory — Latour, 1992, 1990; distributed cognition — Hutchins, 1995; structuration theory — Orlikowski, 1992, 2000; for a recent discussion of these issues see the paper by Kaghan and Bowker, 2001).

The second major challenge for the STS approach relates to its practical application. In common with just about every type of method, technique or tool within human factors and ergonomics, there is a need for more evaluation studies of the application of the STS approach. Typical questions that need to be provided with answers include: what are the exact benefits of the approach relative to others, what are the drawbacks, and can a company demonstrate a return on investment by using the approach? Some preliminary evaluations exist (e.g., Waterson *et al.*, 2002), alongside more detailed, although now rather dated, large-scale meta-analyses of STS studies and interventions (e.g., Pasmore *et al.*, 1982), however, much more of these of these need to be undertaken in the future.

Finally, there is a need to understand the application of the STS approach from a theoretical and practical standpoint as it is applies within new work contexts (e.g., distributed working across different time zones and locations — see Olson and Olson, 2000; call centres and the service sector of the economy more generally). Such work needs to go beyond the statement of 'principles' (e.g., Clegg, 2000) or the description of 'barriers' to such forms of working (e.g., Damodaran and Olphert, 2000), if it is to make a difference to the success or failure of these initiatives. STS still represents a useful and powerful approach to the understanding and redesign of work systems, however, there remains much to do within the area of future research and development.

Further reading

A good source of up-to-date research using the STS approach is available in the form of papers regularly published in the journal *Human Relations*. For more details of qualitative methods (including some of those described in this chapter) see the extensive treatment given in the volume edited by Denzin and Lincoln (2000), as well as Cassell and Symon (2003). Further details of the application of quasi-experimentation to work system design are available in a special edition of the journal *Personnel Psychology* (Autumn, 2002, Volume 55, Issue 3). Alternative ways of applying scenarios to work design are described by Axtell *et al.* (2001) and Weidenhaupt *et al.* (1998), as well as in the volume edited by Carroll (1995) and in a special edition of the journal *Interacting with Computers* (2000, Volume 13). Descriptions of other types of methods for task allocation are given in a special edition published in 2000 of the journal *International Journal of Human Computer Systems* (Volume 52, No. 2 — see especially the papers by Grote *et al.* and Dearden, Harrison and Wright).

References

Atkinson, C.J. (2000). Sociotechnical and soft approaches to information requirements elicitation in the post methodology era. *Requirements Engineering*, **5**, 67–73.

Axtell, C.M., Waterson, P.E. and Clegg, C.W. (1997). Problems integrating user participation into software development. *International Journal of Human–Computer Systems*, **47**, 323–345.

Axtell, C.M., Pepper, K., Clegg, C.W., Wall, T.D. and Gardner, P. (2001). Designing and evaluating new ways of working: The application of some socio-technical tools. *Human Factors and Ergonomics in Manufacturing*, **11**, 1–18.

Barling, J., Kelloway, E.K. and Zacharatos, A. (2002). Occupational Safety. In P.B. Warr (Ed.), *Psychology at Work* (5th Edition) (Harmondsworth: Penguin Books).

Blackler, F. and Brown, C. (1986). Alternative models to guide the design and introduction of new technologies into work organisations. *Journal of Occupational Psychology*, **41**, 211–21.

Bodker, S., Greenbaum, J. and Kyng, M. (1991). Setting the stage for design as action. In J. Greenbaum and M. Kyng (Eds.), *Design at Work: Cooperative Design of Computer Systems* (Hillsdale, NJ: LEA Press).

Burgoyne, J.G. (1994). Stakeholder analysis. In C. Cassell and G. Symon (Eds.), *Qualitative Methods and Analysis in Organizational Research* (London: Sage Publications).

Campion, M.A. and Medsker, G.J. (1992). Job design. In G. Salvendy (Ed.), *Handbook of Industrial Engineering* (New York: Wiley).

Carroll, J.M. (Ed.) (1995). *Scenario-Based Design: Envisioning Work and Technology in System Development.* (New York: Wiley).

Carroll, J.M. (2000). Introduction to the special issue on 'Scenario-Based Systems Development'. *Interacting with Computers*, **13**, 41–42.

Cassell, C. and Symon, G.J. (Eds) (2003). *Qualitative Methods in Organizational Research: An Essential Guide* (London: Sage Publications).

Checkland, P. and Scholes, J. (1999). *Soft Systems Methodology in Action: A 30 Year Retrospective.* (Chichester: John Wiley & Sons).

Cherns, A. (1976). The principles of sociotechnical design. *Human Relations*, **29**, 783–792.

Cherns, A. (1987). Principles of sociotechnical design revisited. *Human Relations*, **40**, 153–162.

Clegg, C.W. (2000). Sociotechnical principles for system design. *Applied Ergonomics*, **31**, 463–477.

Clegg, C.W. and Walsh, S. (1998). Soft systems analysis. In G. Symon and C. Cassell (Eds.), *Qualitative Methods and Analysis in Organizational Research.* (London: Sage Publications).

Coakes, E., Willis, D. and Clarke, S. (Eds.) (2002). *Knowledge Management in the Sociotechnical World.* (London: Springer-Verlag).

Cook, T.D., Campbell, D.T. and Peracchio. L. (1990). Quasi-experimentation. In M.D. Dunnette and L.M. Hough (Eds.), *Handbook of Industrial and Organizational Psychology.* (Palo Alto, CA.: Consulting Psychologists).

Damodaran, L. and Olphert, W. (2000). Barriers and facilitators to the use of knowledge management systems. *Behaviour and Information Technology*, **16**, 405–413.

Davis, L.E. and Wacker, G.L. (1982). Job design. In G. Salvendy (Ed.), *Handbook of Human Factors.* (New York: Wiley).

Dearden, A., Harrison, M. and Wright, P. (2000). Allocation of function: Scenarios, context and the economics of effort. *International Journal of Human Computer Systems*, **52**, 289–318.

Denzin, N.K. and Lincoln, Y.S. (Eds.) (2000). *Handbook of Qualitative Research* (2nd Edition). (London: Sage Publications).

Eason, K.D., Harker, S.D.P. and Olphert, C.W. (1996). Representing socio-technical systems options in the development of new forms of work organization. *European Journal of Work and Organizational Psychology*, **5**, 399–420.

Emery, F.E. and Trist, E.L. (1960). Sociotechnical systems. In C.W. Churchman and M. Verhulst (Eds.), *Management Science* (Volume 2) (Oxford: Pergamon Press).

Foster, M. (1972). An introduction to the theory and practice of action research in work organizations. *Human Relations*, **25**, 529–556.

Glaser, B. and Strauss, A.L. (1967). *The Discovery of Grounded Theory: Strategies for Qualitative Research.* (Chicago: Aldine).

Greenbaum, J. and Kyng, M. (Eds.) (1991). *Design at Work: Cooperative Design of Computer Systems.* (Hillsdale, NJ.: LEA Press).

Grote, G., Ryser, C., Wäfler, T., Windischer, A. and Weik, S. (2000). KOMPASS: A method for complementary function allocation in automated work systems. *International Journal of Human Computer Systems*, **52**, 253–265.

Heller, F. (1997). Sociotechnology and the environment. *Human Relations*, **50**, 605–624.

Hendrick, H.W. (1997). Organizational design and macroergonomics. In G. Salvendy (Ed.), *Handbook of Human Factors* (2nd Edition) (New York: John Wiley & Sons).

Hornby, P. and Symon, G. (1994). Tracer studies: A methodology for examining organisational processes, In C. Cassell and G. Symon (Eds.) *Qualitative Methods in Organizational Research: A Practical Guide.* (London: Sage Publications).

Hutchins, E. (1995). *Cognition in the Wild.* (Cambridge, MA: MIT Press).

Kaghan, W.N. and Bowker, G.C. (2001). Out of the machine age?: Complexity, sociotechnical systems and actor network theory. *Journal of Engineering and Technology Management*, **18**, 253–269.

Kelly, J. (1978). A reappraisal of sociotechnical systems theory. *Human Relations*, **31**, 1069–1099.

Kelly, J. (1982). *Scientific Management, Job Re-Design and Work Performance.* (London: Academic Press).

King, S.D.M. (1964). *Training within the Organization.* (London: Tavistock Institute).

Landauer, T.K. (1995). *The Trouble with Computers.* (Cambridge, MA: MIT Press).

Latour, B. (1992). Where are the missing masses? The sociology of a few mundane artefacts. In W.E. Bijker and J. Law (Eds.), *Shaping Technology/Building Society: Studies in Sociotechnical Change.* (Cambridge, MA: MIT Press).

Latour, B. (1999). *Pandora's Hope: Essays on the Reality of Science Studies.* (Cambridge, MA: Harvard University Press).

Mayhew, D.J. (1999). *The Usability Engineering Lifecycle.* (San Francisco: Morgan Kaufmann).

Medsker, G.J. and Campion, M.A. (1997). Job and team design. In G. Salvendy (Ed.), *Handbook of Human Factors* (2nd Edition), (New York: John Wiley & Sons).

Mumford, E. (1995). *Effective Systems Design and Requirements Analysis: The ETHICS Method.* (London: Macmillan).

Mumford, E. and Hendricks, R. (1996). Business process re-engineering RIP. *People Management*, May 1996.

Mumford, E. and Axtell, C.M. (2002). Tools and methods to support the design and implementation of new work systems. In D. Holman, T.D. Wall, C.W. Clegg, P. Sparrow and A. Howard (Eds.), *The New Workplace: A Guide to the Human Impact of Modern Working Practices.* (London: John Wiley & Sons).

Nadin, S.J., Waterson, P.E. and Parker, S.K. (2001), Participation in job redesign: An evaluation of the use of a sociotechnical tool and its impact. *Human Factors and Ergonomics in Manufacturing*, **11**, 53–69.

Nielson, J.K. (1997). Usability testing. In G. Salvendy (Ed.), *Handbook of Human Factors* (2nd Edition), (New York: John Wiley & Sons).

Older, M.T., Waterson, P.E. and Clegg, C.W. (1997). A critical assessment of task allocation methods and their applicability. *Ergonomics*, **40**, 151–171.

Olson, G.M., and Olson, J.S. (2000). Distance matters. *Human–Computer Interaction*, **15**, 139–179.

Orlikowski, W. (1992). The duality of technology: Rethinking the concept of technology in organizations. *Organization Science*, **3**, 398–427.

Orlikowski, W. (2000). Using technology and constituting structures: A practice lens for studying technology in organizations. *Organization Science*, **11**, 404–428.

Parker, S.K. and Wall, T.D. (1998). *Job and Work Design: Organizing Work to Promote Well-Being and Effectiveness.* (London: Sage Publications).

Pasmore, W.A., Francis, C., Haldeman, J. and Shani, A. (1982). Socio-technical systems: A North American reflection in the empirical studies of the seventies. *Human Relations*, **35**, 1179–1204.

Preece, J., Rogers, Y. and Sharp, H. (2002). *Interaction Design: Beyond Human–Computer Interaction.* (London: John Wiley & Sons).

Rice, A.K. (1958). *Productivity and Social Organization.* (London: Tavistock Institute).

Sharit, J. (1997). Allocation of functions. In G. Salvendy (Ed.), *Handbook of Human Factors* (2nd Edition), (New York: John Wiley & Sons).

Symon, G.J. and Clegg, C.W. (1991a). Technology-led change: A study of the implementation of CADCAM. *Journal of Occupational Psychology*, **64**, 272–290.

Symon, G.J. and Clegg, C.W. (1991b). Implementation of a CADCAM system: The management of change at EML. In K. Legge, C.W. Clegg and N.J. Kemp (Eds.), *Case Studies in Information Technology, People and Organisations.* (Oxford: NCCC Blackwell).

Trist, E.L. and Bamforth, K. (1951). Some social and psychological consequences of the long-wall method of coal-getting. *Human Relations*, **4**, 3–38.

Trist, E.L., Susman, G. and Brown, G.W. (1977). An experiment in autonomous group working in an American underground coal mine. *Human Relations*, **30**, 201.236.

van Eijnatten, F.M. (1993). *The Paradigm that Changed the Workplace.* (Stockholm: Swedish Centre for Working Life (Arbetslivcentrum)).

Walton, R.E. (1977). Work innovations at Topeka: After six years. *Journal of Applied Behavioural Science*, **13**, 422–433.

Wall, T.D., Kemp, N.J., Jackson, P.R. and Clegg, C.W. (1986). An outcome evaluation of autonomous work groups: A long–term field experiment. *Academy of Management Journal*, **29**, 280–304.

Waterson, P.E., Clegg, C.W., and Axtell, C.M. (1999). 'Getting boxes on desks' — An exploratory study of the use of soft systems as a technique for eliciting cognitive representations. *Institute of Work Psychology Memorandum*, Mushroom Lane, University of Sheffield, Sheffield, United Kingdom.

Waterson, P.E., Clegg, C.W., Bolden, R.I., Pepper, K., Warr, P.B. and Wall, T.D. (1999). The use and effectiveness of modern manufacturing practices: a survey of UK industry. *International Journal of Production Research*, **37**, 2271–2292.

Waterson, P.E., Older Gray, M.T. and Clegg, C.W. (2002). A sociotechnical method for designing work systems. *Human Factors*, **44**, 376–391.

Weidenhaupt, K., Pohl, K., Jarke, M. and Hammer, P. (1998). Scenarios in system development: Current practice. *IEEE Software*, **15**, 34–45.

Wilson, J.R. and Haines, H.M. (1997), Participatory ergonomics. In G. Salvendy (Ed.), *Handbook of Human Factors* (2nd Edition). (New York: John Wiley & Sons).

Chapter 30

Teamwork and team performance measurement

Eduardo Salas, Heather A. Priest and C. Shawn Burke

Introduction

Seventy-three percent of all business organisations within the United States have employees that are members of teams (Carey, 1995). During the early 1980s organisations began to use teams to handle environmental complexity and competition from global markets. Initial assumptions were that the implementation of teams should automatically lead to better performance and productivity. Managers and workers assume that with a collected group of people, there are more skills and resources available to solve complex problems. However, as the use of teams increased it became evident that they are not automatically effective and often their potential is not realised. As a consequence, research into understanding the factors contributing to team effectiveness and their training has risen over the past twenty years.

Leading the forefront in understanding these issues has been research conducted within the aviation community. More specifically, a great deal of research in the past two decades has been conducted by human factors researchers observing aviation teams (Guzzo and Dickson, 1996; Salas *et al.*, 2001). In the mid 1970s, the focus of aviation research shifted from a purely technical examination of flight safety and operations to one that also began to focus on team dynamics in the cockpit (Orlady and Orlady, 1999). The evaluation of accidents and safety records led to the exploration of other possible causes of incidents rather than the technical errors of individual crewmembers. This line of work has contributed a great deal of knowledge to increasing our understanding of teams and more specifically, their training requirements (see crew resource management training (CRM), Wiener *et al.*, 1993).

In addition to the aviation community, the military has also been at the forefront of investigating issues related to team, team training, and the requisite measurement requirements (Salas *et al.*, 1995). The erroneous downing of the *USS Vincennes* by a U.S. Aegis Cruiser during the Persian Gulf conflict led to an international incident and further highlighted the importance of understanding team performance within complex environments. This incident led to a programme of research aimed at understanding team decision making under stress (see Cannon-Bowers and Salas, 1998). This programme has produced a wealth of information about teams, teamwork, and team training.

Although the aviation and military communities have heavily invested in research on team effectiveness, team training, and team performance organisational researchers have also greatly contributed to this knowledge base (e.g., Belbin, 2000; West *et al.*, 1998; West, 2003; Katzenbach and Smith, 1993; Katzenbach, 1998). As we have learned much over the past two decades, and the use of teams is only expected to increase, this chapter will briefly review a representative sample of what is known about teams (more specifically teamwork) and team performance measurement. First, we define teams and teamwork. Second, we discuss the measurement requirements and assumptions that are essential to the systematic development of effective team performance measurement system(s). Third, we discuss a representative sample of methods and tools that can be used to measure teams. Finally, we conclude with a few remarks about the next frontier in teams and team performance measurement. For a more detailed review the reader is referred to Salas and Cannon-Bowers (2001), Guzzo and Dickson (1996), Illgen *et al.*, (1993), and Swezey and Salas (1992).

What is teamwork?

All teams are not created equal. Teams, groups, units, and collectives vary in organisation, degree of task interdependance, objectives, missions, life-span, and communication requirements. For our purpose, teams can be defined as two or more individuals who must perform distinct, complementary, or interdependent tasks in pursuit of a common, specified goal (Brannick *et al.*, 1997). Teams must communicate, as well as share information and resources in order to meet their goal(s). The coordination demands within teams are such that members must perform their duties in a timely and integrated fashion. Further, the ability of each individual member to adapt and adjust through reliance on other team members determines the level of a team's coordination, and thus teamwork. This definition highlights the importance of communication, synchronisation, adaptation, and integration of information and action in order for the team to be successful.

Evidence from laboratory studies, as well as investigations of teams performing in complex contexts (i.e., in the wild), leads to conclusion that teamwork is a multidimensional and dynamic phenomenon which is sometimes deceiving and difficult to observe and capture (see Salas and Cannon-Bowers, 2000). At its core, teamwork comprises a set of interrelated knowledge, skills/behaviours, and attitudes (i.e., KSAs) that taken together form the competencies necessary for effective team performance (Salas *et al.*, 1992). KSAs are, in other words, what team members 'think, feel, and do' in team settings (Salas and Cannon-Bowers, 2000). Within the context of teams, knowledge can be defined as concepts, facts, and ideas that individual members have that may affect the way teams perform (Cannon-Bowers and Salas, 1997). These areas of knowledge include, but are not limited to, team members: (a) holding shared task models (i.e., common models of situations and strategies of coping with task demands), (b) understanding the importance of teamwork skills, and (c) possessing knowledge of their teammates' characteristics, and their mutual familiarity (Cannon-Bowers and Salas, 1998). There has been much evidence that indicates the importance of cognition as a necessary component of teamwork (Guzzo and Dickinson, 1996; Orasanu, 1990; Webber *et al.*, 2000; Weiner *et al.*, 1993).

Skills or 'what members do' can be defined as behavioural and cognitive sequences, including procedures, which must be present to complete a task (McIntyre and Salas, 1995). Behavioral components of teamwork have typically received the most attention within the team literature. A few of the more prominent ones will be briefly described (i.e., adaptability, leadership, communication, mutual performance monitoring, back-up behaviour, see Cannon-Bowers and Salas, 1997; Oser *et al.*, 1989). Adaptability or flexibility involves the dynamic reallocation of functions between team members. Teams must use compensatory behaviour and reallocation of resources based on information taken from the environment (Prince and Salas, 1993) in order to adapt to the dynamic interdependencies present within teams, as well as within their environments. A second behavioural dimension, team leadership, has just recently begun to be investigated, but has been argued to be a key determinant of team effectiveness (Stewart and Manz, 1995). Team leadership requires the leader to provide structure, support, and direction to other team members. Leaders provide information about team members and tasks (Fussell *et al.*, 1998) and serve to create and maintain shared affect, behaviour, and cognition (Kozlowski *et al.*, 1996). Communication has also been argued to be an important skill within teams. Specifically, closed-loop communication is an important part of team interaction, coordination, feedback, and orientation. Closed-loop communication refers to the exchange of information between two or more team members, typically for the purpose of clarifying or acknowledging the receipt of information (Orasanu, 1990). Finally, the dimensions of mutual performance monitoring and the resultant back-up behaviour will be briefly discussed. Mutual performance monitoring occurs when team members observe each other's activities and performance, typically involving feedback and backup behaviour (Cannon-Bowers *et al.*, 1995). Backup behaviour occurs when team members have an understanding of other team member's tasks and are able to provide and ask for assistance (Brannick *et al.*, 1997). This is observed when someone is unable to perform a task and another team member fills in and completes the task. It can also involve a team member correcting another's mistake.

Finally, attitudes can be defined as the way members 'feel'. Attitudes that have been argued as important within teams include, (a) collective orientation (i.e., the extent to which team members feel that the team approach is superior to the individual approach), (b) collective efficacy (i.e., the extent to which team members believe that the team can perform effectively as a whole), (c) mutual trust, and (d) team cohesion (Salas and Cannon-Bowers, 2000).

Now that the types of KSAs that impact teamwork have been identified, the next question is, 'how do these KSAs interact to facilitate teamwork?' The answer to this question is illustrated in Figure 1. Team members can hold KSAs, both at the individual and team level. For example, team members can have self-efficacy (i.e., their feeling about how well they will do individually) as well as shared team efficacy (i.e., how well the team can do as a whole). These individual and team KSAs are what make teamwork possible. We contend that communication, team leadership, mutual performance monitoring, back-up behaviour

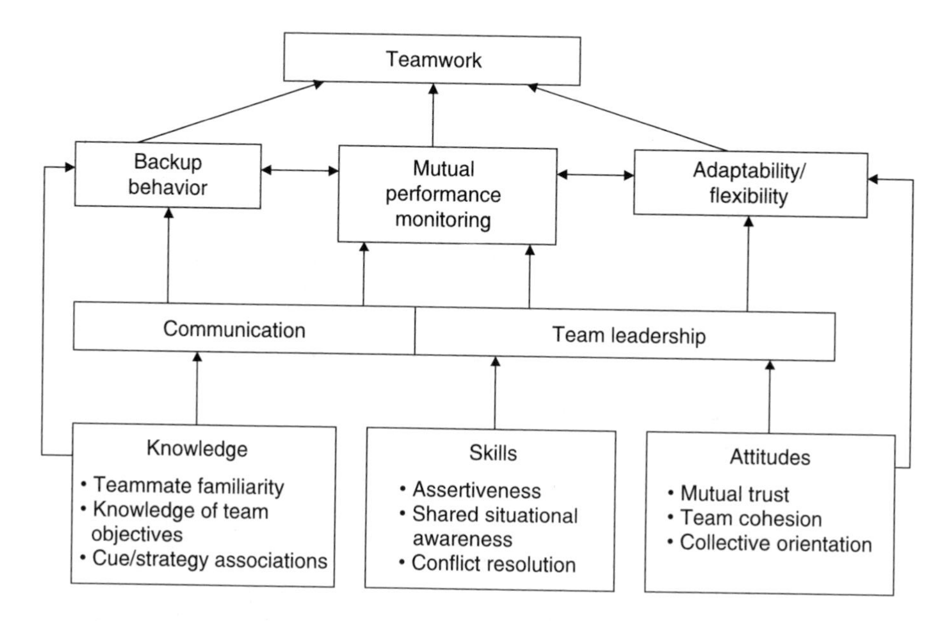

Figure 1 Illustration of how KSAs contribute to teamwork.

and adaptability/flexibility form the core foundation for teamwork. While these KSAs appear to be the most influential when determining whether a team performs successfully, other KSAs listed in Figure 1 have an impact on teamwork.

Adding to the complex nature of teamwork, there is also a high level of interdependence between the KSAs. For example, shared situational awareness contributes to a higher level of adaptability while also being a necessary component of mutual performance monitoring, which leads to team members exhibiting backup behaviour. All of this leads to more effective teamwork. The relationship between KSAs and teamwork are not linear, but can more accurately be described as a cumulative, curvilinear process.

What are the requirements for team performance measurement?

Meister (1985) argued that measurement is paramount for understanding human behaviour. This argument can be extrapolated to all levels of human behaviour, individual as well as team. Therefore, as the development and use of systematic team performance measures is vital to effective teamwork and team performance, we feel that some basic assumptions should be covered regarding the development of such systems (see Table 1). These assumptions form the foundation for the design, development, and testing of team performance measurement systems. Understanding of these assumptions leads to the recognition that there are certain requirements that underlie effective team measurement systems.

There are three basic requirements for designing team performance measurement systems (Cannon-Bowers and Salas, 1997). First of all, a team measure must be able to assess multiple levels of measurement. More specifically, as team performance is a compilation of the performance of individual team members, measurement systems must capture both individual and team level data in order to accurately evaluate a team. Second, team performance measurement systems must capture process as well as outcomes. Team processes are those moment-to-moment actions, behaviours, and events that contribute to the team outcomes. Both types of information are needed in that, while outcome measures reflect 'bottom line' performance (i.e., quality, quantity), they do not provide the diagnostic information needed for feedback and developmental purposes (Cannon-Bowers and Salas, 1997). Process measures need to be collected to gather information pertaining to the 'how' or steps by which team members arrived at the final outcome. Third, team

Table 1 ***Representative*** Sample of Basic Team Measurement Assumptions (based on Meister, 1985; Brannick *et al.*, 1997)

	Basic team measurement assumptions
1.	The construction of team measures must be based on team-based theoretical models.
2.	Team performance and processes are reflected through the actions and observable outcomes of individual team members.
3.	The goal of measurement is to allow observations of behavior and changes in behavior of team members, which represent underlying processes.
4.	Certain team processes and outcomes are more readily available for measurement than others.
5.	Processes not easily measured must be matched to a corresponding observable behavior or quantified through feedback from team members.
6.	Individual performance of team members can be considered within the context of the team and used, in combination with other factors, to assess team performance.

performance measurement systems must describe, evaluate, and diagnose performance. They must describe what the team members do, create standards of acceptable team performance, and be diagnostic enough to help with remedies—this will facilitate feedback and allow corrections to be made in order to improve overall team performance (Brannick *et al.*, 1997).

What challenges face those responsible for the development of team performance measurement systems?

The requirements described already drive several challenges that one faces when developing a team performance measurement system. These challenges need to be confronted to ensure that developed measurement systems meet the intended needs of the user. We will briefly discuss six such challenges.

The first challenge facing those who develop measurement systems is to define the purpose of measurement. Why? Team performance measurement systems can be used to accomplish many objectives, such as: (a) team member selection, (b) certification, (c) facilitation of problem diagnosis, (c) feedback for training, (d) team training evaluation, and (e) research purposes. In addition, measurement systems can range from paper-and-pencil surveys, which are inexpensive once developed, to large-scale computer simulations that can be extremely expensive. This is significant because the purpose of the measurement drives the definition of the kind of data one needs to collect from the team. And the kind of data required drives what resources are needed to measure team performance (Meister, 1985). For example, if the purpose of measurement is to get crews' reactions on a training programme, a simple paper-and-pencil survey may suffice. However, if the purpose is to ensure the crew has the needed competencies, a more elaborative, labour-intensive measurement scheme may be needed. Therefore, purpose of measurement must be carefully defined and considered.

A second challenge is the choice of the appropriate stimulus or scenario(s) to use when measuring team performance. Stimuli can range from an actual task the team is assigned to perform to a written survey the team members are asked to complete. The choice of stimuli can have an impact on the team's behaviours and can determine what attributes will be elicited while conducting team measurement procedures. The stimuli in team measurement are typically embedded in the task to be performed and, therefore the situation in which teams will be assessed must also be considered. Likewise, the attributes and behaviours that you want the stimuli or scenario to elicit are important considerations in measuring team performance. Attributes and behaviours are intricately linked, in that attributes consist of behavioural skills. In order to measure attributes, one must measure the observable behaviours. Therefore, the attribute of interest or the attribute one chooses to study determines what behaviour must be observed.

A third challenge for those developing team performance measurement systems is how to ease the collection of observational data. Dickinson and McIntyre (1997) pointed out that it takes a team to measure a team accurately. In other words, a single person can rarely if ever collect all the information regarding teamwork in a given situation. It takes multiple subject matter experts (SMEs) to observe everything going on when teams interact, as there are multiple instances of teamwork in any one scenario. It takes a team of observers to pick up on all the instances of teamwork. The number of SMEs needed depends on several factors, the including number of team members, complexity of the task, physical environment of the task, and amount of communication and interaction required to complete the task. In fact, Dickinson and McIntyre recommended no more than two team members per rater in complex team settings, because they are required to interact frequently and team members are close in proximity. Finally, in collecting observational data one needs to be careful not to overload those conducting the ratings/observations. For example, it has been argued that raters can only accurately distinguish between four to five dimensions when rating team performance (Smith-Jenstch *et al.*, 1998). If the dimensions to be observed in teamwork are too numerous, redundancy starts to appear as the lines between dimensions begin to blur.

A fourth challenge related to team measurement is how to quantify the response(s) of teams. There are several methods that can be applied. When team behaviours have to be observed, behaviours can be scored through direct observation or rated from videotape, but some system of quantification must be developed and assigned to behaviours of interest. The instruments used to measure these behaviours can range from global to specific. Global instruments are generally subjective and could consist of questions such as: did the team members communicate clearly? Specific instruments are generally objective and can consist of questions such as: did the pilot contact the tower when instructed to do so?

A fifth consideration is the timing of measurement. Teams mature and evolve over time and this has implications in terms of performance measurement. Similarly, Brannick and Prince (1997) have noted that how long members have worked together appears to influence the performance of team members. Additionally, due to the dynamic nature of teams it is inherently difficult for measurement systems to slice out portions of events in order to target the specific processes of interest. The changing, dynamic nature of teams requires that a researcher be very conscious of what transpires within the team and between team members at every moment on the continuum of a task in order to obtain an accurate picture of what they want to observe. Inappropriate timing in team measures may yield results that are not representative of team processes. Furthermore, multiple measurements should be taken throughout performance to gather a truly representative picture.

Finally, while each of these challenges must be appropriately addressed, one challenge universally effects all team measurement. What do we measure? What variables must be observed or measured in order to accurately assess team performance? First of all, as noted, measurements cannot simply measure outcomes or products of a team. Outcomes should be evaluated. What processes and outcomes to focus on is a function of the resolution of the above challenges, resources, the underlying hypotheses and research questions, and the measurement methods available. With these challenges in mind and based on what has been learned about teams in the last two decades, many measurement methods have been developed and implemented over the years. The following section offers representative examples and insight into methods of measurement based on the literature.

What team measurement tools are available?

There are numerous team measurement tools available to assess a team's knowledge, skills, and attitudes. Some of these, which will be discussed here, are rating scales, behavioural checklists, event-based measures, cognitive measures, attitudinal/affective measures, and automated performance measures. Countless examples of these tools applied to team performance can be found in recent literature (Jones and Harrison, 1996; Lynn and Reilly, 2000). Furthermore, while several of these tools can be used to measure a combination of skills, cognition, and/or affect depending on the exact measurement content, they can generally be divided into categories based on their primary use: assessing skills, cognition, or affect.

Assessment of behaviour

Rating scales

Rating scales use successive intervals to rate a team's performance on certain competencies or teamwork skills (see also Chapter 4). The first step to using a rating scale to measure teamwork is to compile a description of the behaviours of interest that represent components of teamwork and the keywords that describe those behaviours. The descriptions are the foundation for a set of rules that the scales are based on. One way to do this is to apply the critical incident technique. Critical incident techniques refer to a method often used in the early stages of an evaluation, where events and incidents that are memorable are evaluated to extract the negative and positive things which happened in a specific situation (Chell, 1998). Simple observations of teamwork or expert opinions can also be used to create the descriptions.

Once behaviours have been identified and defined there are many ways in which the scale can be presented. A few variations that can be used include: (a) either continuous or discontinuous lines can be used to graphically represent the continuum of behaviours, (b) descriptive words can be used to represent different positions along the scale, and (c) verbal descriptions, lines, and numbers can also be used (Meister, 1985). The actual design of the scale is up to the discretion of the evaluator as long as it fits the desired task and provides a representation of different degrees of behaviour for the raters. Once the scale has been created, raters must be trained in the use of the scale as the information they collect represents the actual data the researcher or practitioner has to work with. Their training is of paramount importance or the results obtained will be meaningless and organisational resources will have been wasted. In addition to an expert rating team performance, rating scales can also be used as self-reports to gather a team member's perception of their own or other team members' behaviours. An example of a portion of a finished rating scale is shown in Figure 2.

Behaviorally anchored rating scales (BARS)

The BARS is a widely used rating scale that was first introduced by Smith and Kendall (1963). This scale is similar to the rating scale, but BARS use behavioural anchors along with the Likert scales often used in normal rating scales. The anchors provided for raters are concrete and serve as a guide to compare to the observed behaviour. The scale provides examples of each level of a given team component and each level is often defined or classified (e.g., Fair, Good, and Excellent). This scale is beneficial because it helps control for any individual idiosyncrasies of the raters (Salas *et al.*, 2002). An example is provided in Figure 3.

Event-based measurement scales

Event-based measurement consists of taking *a priori* defined cues and events and embedding these into scenarios. These embedded events then serve as triggers for participants to exhibit competencies of interest. Within these types of measurement scale, as with all good

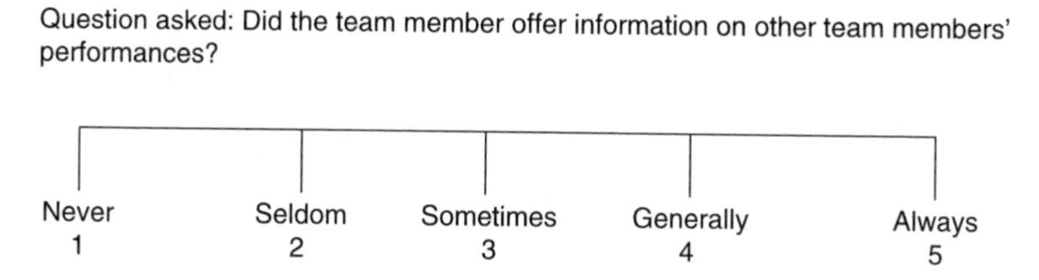

Figure 2 Example graphic rating scale measuring feedback.

FEEDBACK
DEFINITION: Feedback refers to the supply of information pertaining to a team member's performance. Feedback entails a free exchange of useful information between team members and involves not only giving information, but also accepting and seeking out information from other team members when necessary.
KEY WORDS: seeks, accepts, provides, performance, information
DECISION RULES: Feedback is present when team members monitor each other's performance and recognize when assistance is needed. Furthermore, feedback exists through a team member's ability to respond to another team member's request for information, which includes having the knowledge to do so, and to in turn ask for performance information when needed. Feedback also requires the acceptance of positive and negative information provided by team members.

Question asked: Did the team member offer information on other team members' performances?

Figure 3 Example graphic rating scale measuring feedback.

measurement, the competencies that are of interest to practitioners drive the development of the performance measurement tool. Performance measurement systems of this type allow practitioners to tie measurement directly back to targeted competencies. This type of scale construction also increases the ease of assessment in that raters know *a priori* where the cues are embedded within the scenario, giving them an indication of when targeted competencies should be exhibited. In addition, in order to maintain task consistency, scripts are developed that detail the introduced events and communications that should occur between outside agents or personnel who participate in the scenarios. These scripts ensure that scenarios are consistent across teams and training will be standardised while also providing a highly diagnostic assessment tool. The above properties lead to event-based measurement scales typically containing exceptional psychometric properties.

Figure 4 illustrates an example of a event-based measurement scale, targeted acceptable responses to generated events or tasks (TARGETS—Fowlkes *et al*., 1994). As is illustrated, events or cues embedded within the simulation or training exercise serve to prompt targeted behaviours. Raters have a list of cues and expected behaviours that are theoretically determined *a priori*. Raters must then determine and record whether or not participants exhibited the appropriate behaviour at the appropriate time. Event-based measurement techniques such as TARGETs have been used quite successfully, especially in military teams who have regimented events and behaviours that are easy to predict.

Automated performance measurement

Automated performance measurement can be anything, from as simplistic as an automatic counter to as complex as a high tech computer simulation of jet flight. Automated performance measures can offer several advantages over non-automated measures, but, at the present time because they cannot replace non-automated measures, they should be used in conjunction with such measures. Automated performance measures can be less obtrusive than human observers and perform more uniformly across different teams and tasks. Automated performance measurement also is devoid of the many biases and idiosyncrasies

Flight segment	Event	Target	Hit
Approaching checkpoint	Navigation	Report distance to checkpoint.	
		Report outbound heading from checkpoint.	
At checkpoint	COMMUNICATION from base: SIGMET (SIGnificant METeorlogical information) over checkpoint.	Note implications of SIGMET (e.g., cannot land at checkpoint).	
		Obtain additional information (e.g., from flight service).	
		Make new plan – choose alternates from: -Return to base (best choice) -Pleasanton -Kelly AFB -Randolph AFB	

Figure 4 Events and associated TARGETS for a short-hop segment in which helicopter crew is on a training maneuver (Fowlkes *et al.*, 1994).

that human raters display. However, automated data collection only detects overt responses (Meister, 1985). Other methods would have to be used to measure cognitive and attitudinal components of teamwork. In addition, automated performance measurements can be extremely expensive and require maintenance that can tax organisational resources.

Assessment of team cognition

Concept mapping

Concept mapping has been described as graphical representations of the knowledge structure and content of an individual within a particular domain (Swan, 1995). Like all good knowledge elicitation methods (see Chapter 8), concept mapping must go beyond the measurement of knowledge accuracy and measure the content, structure, and richness of individuals and teams (Cooke *et al.*, 2000). Concept mapping helps researchers to create representations of domain-specific concepts that typically represent mental models. While initially used with individuals, it has been increasingly adapted for use in team performance measurement (i.e., to measure such cognitive constructs as shared mental models, shared situational awareness, and other types of shared team knowledge). Concept mapping is considered a knowledge elicitation method, rather than a method to assess the actual performance outcomes of a team (Cooke *et al.*, 1997). This method creates a map of cognitive concepts that are linked together based on their association with the other concepts in the map. There are several different types of map due to varying types of associations the links are based on. Links between maps can be based on causality or continuity, among other things. Cognitive maps can be developed by requesting the data from team members through interviews or surveys or by post hoc examination of data (Mohammed *et al.*, 2000). In addition, there are several variations of this technique. For example, it can be used: (a) the researcher or practitioner provides the concepts which participants fit into a pre-defined structure, (b) the participants come up with the concepts and place them into pre-defined structures, or (c) participants develop both the concepts and the linkages among them. This measure is useful in dealing with knowledge because it shows the relation of one concept to another through the links and can tell researchers a great deal about the shared strategic thinking of teams. Concept maps show the underlying knowledge that can lead to team success.

Pathfinder

Pathfinder is a measurement technique used to elicit similarity judgments and produce the applicable scaling to represent the structure of the concepts (Mohammed *et al.*, 2000). Pathfinder is similar to concept mapping, in that it allows one to measure cognitive frameworks or structural representations of cognition. However, the method by which this is accomplished is slightly different than that used for concept maps. Specifically, relevant concepts are identified by the researcher for inclusion in the rating process, usually through a team task analysis. Then participants are instructed to make pairwise ratings among all combinations of concepts. Goldsmith *et al.* (1991) recommend between 15 to 30 concepts be used and argue that the more concepts that are available, the more productive the resulting structure used in Pathfinder. However, as participants are required to make pairwise ratings, practitioners should realise that as the number of concepts included increases, Pathfinder can become very time intensive. After ratings are made, Pathfinder uses an algorithm that takes psychological proximity data and represents this in the form of a structural map containing constructs and the links between them. Pairs of concepts with a high degree of similarity (as indicated by participant ratings) are shown as direct links, more distal relations are shown as indirect links. If there is no relationship between two concepts no link is presented. In addition, to the production of a structural representation of knowledge, Pathfinder also produces a numerical index of coherence (i.e., indicating the consistency of a set of similarity ratings). For more information on this technique see Goldsmith and Kraiger (1996) and Cooke and Schvaneveldt (1988).

Probed protocol analysis

Probed protocol analysis (see Chapter 7) is a way to map the knowledge and thought processes of team members. In measuring teams, participants or team members would be likely to be provided with the necessary steps in performing some task that exhibits the team component being studied. Participants would then be asked why they would perform a certain step after another step, why they performed a task, and what other team members should be doing while they perform a task. This measure can give researchers and managers a better understanding of individual and team comprehension for not only when they and their teammates perform tasks, but also why they are doing what they are doing.

Attitudinal/affective measures

Behavior and knowledge are not the only targets of team measurement. Attitudinal or affective components of teams are also of interest in measurement. Attitude can be defined as 'an internal state that influences an individual's choices or decisions to act in a certain way under particular circumstances' (Cannon-Bowers *et al.*, 1995). Team attitudinal components include, but are not limited to, satisfaction, collective efficacy, cohesion, trust, and group potency. A more in depth discussion of these can be found in Burke *et al.*, (1993).

Attitude measures are almost exclusively self-report and are typically done using a Likert scale (see Chapter 4) or some similar paper and pencil survey. Therefore, rating scales are also commonly used to measure affect. To develop a measure of attitude, the affect of interest (e.g., satisfaction) must be defined. Statements or questions are then developed based on the definition of the construct. Subject matter experts (SMEs) can help ensure that the measure is relevant to the attitudinal component of interest. Attitude measures are extremely common because they are easy to administer and are inexpensive. It is important to keep in mind that since attitudes are reported by participants and are, therefore, colored by their own biases and experiences, the results should be examined accordingly. However, these

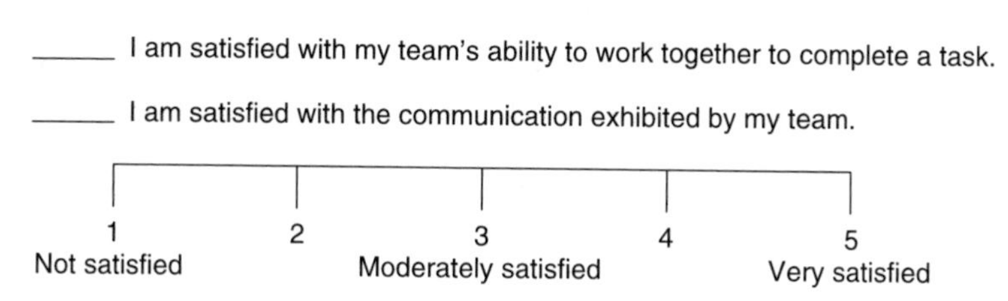

Figure 5 Attitudinal measure of member satisfaction.

measures can still be useful and can contribute significantly to the study of team processes. Two example statements and a scale to measure collective efficacy are shown in Figure 5.

Summary

We have briefly reviewed a representative sample of the types of tools that can be applied to team performance measurement in order to illustrate the variety of tools available, as well as the fact tools may differ in their suitability, dependent on the task and attributes that one is interested in measuring. Numerous examples of applying tools and methods of measurement to team components using a theoretical framework can be found in the literature (Cannon-Bowers and Salas, 1997; Johnston *et al.*, 1999; Orasanu, 1990; Paris *et al.*, 1999; Salas *et al.*, 1997).

What is the future of team measurement?

Distributed virtual teams

With the growing technology available to academics, industry, government, and the military in recent years, new teams have begun to develop. These new types of teams lead to interesting new directions for team measurement. While new technology has led to additional resources for measurement of teams, it has also led to a new set of questions for those who seek to measure teamwork. One of the more prominent questions regards the challenges posed by distributed virtual teams.

Distributed teams refer to teams who meet all the requirements of a typical team, but are geographically distributed. With e-mail and real time communication aids (e.g., instant messaging, teleconferencing), the capabilities of teams to function with team members dispersed nationally and even internationally will continue to grow. Virtual teams are distributed teams that use the technology available to accomplish work that once was accomplished by extensive travel and phone conversations in distributed teams (i.e., virtual teams use state of the art technological advances—Geber, 1995). As with teams in general, virtual and distributed teams may vary in terms of their life span. While some teams are long term and will be kept intact indefinitely, others are brought together for a short time and are often cross-functional. This trend towards virtual teams also results in reducing the normal amount of face-to-face daily interactions that occur within the office. Telephones, fax machines, e-mail, teleconferencing, and videoconferencing have replaced the daily interaction of teams in the office, in halls, and around the water cooler. While distributed or virtual teams can be immensely beneficial, this new breed of teams have also led to new problems that must be tackled by human factors and ergonomics researchers and practitioners (Gundry, 2001). Some of these challenges deal with promoting team effectiveness in general, while others are related specifically to team measurement.

In terms of measurement, distributed teams pose the question of 'how do you measure teams who have little, if any, face-to-face interaction?' Can they be measured like any other team, despite the fact that they may have never met? Outcome measures can obviously still be applied. However, measuring processes becomes more difficult. Individual KSAs can be evaluated, but other processes like collective orientation, mutual trust, team cohesion, backup behaviour, and teammate familiarity may be difficult to measure or may not even be present. These KSAs have long been established as foundations of successful teamwork for co-located teams (Annett *et al.*, 2000; Cannon-Bowers *et al.*, 1995; Miles, 2000). It is a challenge to future team measurement to develop new ways of measuring teams who may redefine teamwork by using new technological resources which change team dimensions now taken for granted.

Human performance modelling and dynamic assessment

As teams are continuously confronted with more and more technological advances, environments in which they interact become more dynamic and ambiguous. Furthermore, with more technology, fewer people are being used in certain areas (i.e., the military). To counteract a possible decline in team effectiveness, a relatively new method of performance assessment and prediction is being explored—human performance modelling. While neither modelling nor models of information processing are new (Zachary *et al.*, 1998), human performance modelling seeks to combine the two to develop a system for predicting human behaviour based on a model of human mental activities, particularly when humans are interacting with a system. Human performance models seek to 'explain' human performance by depicting major processes or characteristics of a system or performance scenario and the relationship between these processes or categories. Effective models seek to predict how the processes or systems of interest will result in a certain behaviour. Models can then be used as a tool to generate testable hypotheses that, ideally, can be replicated by real world human performance (Sanders and McCormick, 1993). Human performance models can be used to predict how teams will perform in certain situations and how their different characteristics and processes will interact when presented with certain cues.

Zachary *et al.* (1992) reported a developing human performance model called COGNET or cognition as a network of tasks, which models the performance of watchstanders in the Combat Information Center. These researchers recommend using human performance modelling, in conjunction with emerging technology, to allow teams to evolve with the times. Successful use of human performance modelling can assist researchers in identifying behaviours that lead to the desired performance.

Human performance modelling, used in combination with simulations can be useful in training and measurement of team performance in a number of areas. Areas where modelling is currently being applied and will continue to be applied in the future include mission space, mission tasks, strategy, tactics, intelligent systems emulating human decision making processes, and optimal resource utilisation. Based on these models, simulations can be designed to enable measurement of team performance that most closely approximates real world settings and scenarios (Pharmer and Campbell, 2000). Furthermore, human performance modelling and simulation can be used to generate and evaluate team designs and evaluate system designs. Dynamic assessment is another important issue in team performance measurement. It requires researchers to track, assess, and interpret team performance moment-to-moment, as it changes (Cannon-Bowers and Salas, 1997). It is difficult to capture online team performance, but it must be explored. Current research is evaluating just how real these simulations are to team members, through evaluation of presence and immersion in simulated environments (Pandzic *et al.* 2001; Witmer and Singer, 1998).

Conclusion

With the growing number of teams in all facets of business, industry, and the military, the need for team performance measures is evident. Human factors and ergonomics researchers and industrial and organisational psychologists have discovered a great deal about the process of measuring teams effectively. However, more must be done to improve the translation of methods and evaluations to those who actually need them, mainly the practitioners that would benefit from applying them. Explaining more thoroughly how these measures were developed and what they are designed to do is one step towards closing the gap between researchers and practitioners.

Another important point to keep in mind is that the methods covered in this chapter, along with virtually any other method, must be applied across the life span of a team in order to be effective. As mentioned early in the chapter, teams constantly evolve and change and must be re-evaluated. Also, it is important to assess both individuals and teams as a whole to contribute the most to the measurement of a team. Most of all, measures must be applied with diligence to ensure all the considerations and challenges, including purpose of the measurement and outcome or process measures, are addressed.

This chapter is by no means an exhaustive exploration of team performance and measurement. We simply sought to raise awareness within the field of human factors and ergonomics of the complexity of assessing team performance and some of the methods and tools that exist for measuring team performance. By addressing some of the basic questions regarding the construction of and issues surrounding team measures, more can be done to ensure effective team performance measurement.

References

Achille, L.B., Schulze, K.G. and Schmidt-Neilson, A. (1995). An analysis of communication and the use of military teams in navy training. *Military Psychology*, **7**, 95–107.

Alliger, G.M., Tannenbaum, S.I., Bennett, W. Jr., Traver, H. and Shotland, A. (1997). A meta-analysis of the relations among training criteria. *Personnel Psychology*, **50**, 341–361.

Annett, J., Cunningham, D. and Mathias-Jones, P. (2000). A method for measuring team skills. *Ergonomics*, **43**, 1076–1094.

Baker, D.P., Salas, E. and Cannon-Bowers, J. (1998). Team task analysis: Lost but hopefully not forgotten. *The Society for Industrial and Organizational Psychology*, **35**, 79–83.

Bal, J. and Gundry, J. (1999). Virtual teaming in the automative supply chain. *Team Performance Management: An International Journal*, **5**, 174–193.

Belbin, R. M. (2000). *Beyond the Team*, (Oxford, Butterworth-Heinemann).

Brannick, M.T. and Prince, C. (1997). Overview of team performance measurement. In M.T. Brannick, E. Salas and C. Prince (Eds.), *Team Performance Assessment and Measurement: Theory, Methods, and Applications* (pp. 3–16). (Mahwah, NJ: Lawrence Erlbaum Associates).

Brannick, M.T., Salas, E. and Prince, C. (1997). *Team Performance Assessment and Measurement: Theory, Methods, and Applications*. (Mahwah, NJ: Lawrence Erlbaum Associates).

Burke, C.S., Volpe, C., Cannon-Bowers, J.A. and Salas, E. (1993). *So what is teamwork Anyway? A synthesis of the team process literature*. Paper presented at the *39th Annual Meeting of the Southeastern Psychological Association*, Atlanta, GA.

Campbell, G.E., Cannon-Bowere, J.A. and Villalonga, J. (n.d.) Achieving training effectiveness and system affordability through the application of human performance modeling. Retrieved January 18, 2002 from http://www.manningaffordability.com/S&tweb/PUBS/ TrngEfft.html.

Cannon-Bowers, J.A. and Salas, E. (1997). A framework for developing team performance measures in training. In M.T. Brannick, E. Salas and C. Prince (Eds.), *Team Performance and Measurement: Theory, Methods, and Applications*. (pp. 45–62). (Mahwah, NJ: Lawrence Erlbaum Associates).

Cannon-Bowers, J.A. and Salas, E. (1998). *Decision Making Under Stress*. Washington, D.C.: American Psychological Association.

Cannon-Bowers, J.A., Tannenbaum, S.I., Salas, E. and Volpe, C.E. (1995). Defining competencies and establishing team training requirements. In R.A. Guzzo and E. Salas (Eds.), *Team Effectiveness and Decision Making in Organizations*. (pp. 333–380). (San Francisco, CA: Jossey-Bass).

Carey, R. (1995). It takes teamwork. *Sales and Marketing Management*, **147**, 13–14.

Chell, E. (1998). Critical incident technique. In G. Symon and C. Cassell (Eds.), *Qualitative Methods and Analysis in Organizational Research: A Practical Guide*. (Thousand Oaks CA: Sage Publishing).

Cooke, N.J. and Schvaneveldt, R.W. (1988). Effects of computer programming experience on network representations of abstract programming requirements. *International Journal of Man–Machine Studies*, **29**, 407–427.

Cooke, N.J., Salas, E., Cannon-Bowers, J.A. and Stout, R.J. (2000). Measuring team knowledge. *Human Factors*, **42**, 151–173.

Cooke, N.J., Stout, R. and Salas, E. (1997). Broadening the measurement of situational awareness through cognitive engineering methods. Paper presented at the *41st Annual Meeting of the Human Factors and Ergonomics Society*.

Dickinson, T.L. and McIntyre, R.M. (1997). A conceptual framework for team measurement. In M.T. Brannick, E. Salas and C. Prince (Eds.), *Team Performance and Measurement: Theory, Methods, and Applications*. (pp. 19–43). (Mahwah, NJ: Lawrence Erlbaum Associates).

Driskell, J.E. and Salas, E. (1992). Collective behavior and team performance. *Human Factors*, **34**, 277–288.

Dube, L. and Pare, G. (2001). Global virtual teams. *Communications of the ACM*, **44**, 71–73.

Flanagan, J.C. (1954). The critical incident technique. *Psychological Bulletin*, **51**, 327–358.

Fowlkes, J.E., Lane, N.E., Salas, E., Franz, T. and Oser, R. (1994). Improving the Measurement of team performance: The TARGETs methodology. *Military Psychology*, **6**, 47–61.

Fussell, S.R., Kraut, R.E., Lerch, F.J., Scherlis, W.L., McNally, M. and Cadiz, J.J. (1998). *Coordination, Overload and Team Performance: Effects of Team Communication Strategies*. Paper presented at the meeting of the Computer-Supported Cooperative Work Conference, Cambridge, MA, San Diego.

Gawron, V.J. (2000). *Human Performance Measures Handbook*. (Mahwah, NJ: Lawrence Erlbaum Associates).

Geber, B. (1995). Virtual teams. *Training*, **32**, 36–40.

Gibson, M. (1983). Formative evaluation: The effect of group development. *Management Education and Development*, **14**, 153–167.

Goldsmith, T.E., Johnson, P.J. and Acton, W.H. (1991). Assessing structural knowledge. *Journal of Educational Psychology*, **83**, 88–97.

Goldsmith, T.E. and Kraiger, K. (1996). Applications of structural assessment to training evaluation. In J.K. Ford and Associates (Eds.), *Improving training effectiveness in Organizations* (pp. 73–96). (Hillsdale, NJ: Lawrence Erlbaum Associates).

Goldstein, I.L. (1993). *Training in Organizations* (3rd ed.) (Belmont, CA: Wadsworth).

Gundry, J. (2001). The human factor: Psychological implications of m-work. *Knowledge Ability Limited*. Retrieved on January 18, 2002 from www.knowab.co.uk/wbwmwork.

Guzzo, R.A. and Dickson, M.W. (1996). Teams in organizations: Recent research on performance and effectiveness. *Annual Review of Psychology*, *47*, 307–338.

Horner, M. (1997). Leadership theory: Past, present, and future. *Team Performance Management*, **3**, 270–287.

Illgen, D.R., Major, D.A., Hollenbeck, J.R. and Sego, D.J. (1993). Team research in the 1990s. In M.M. Chemers and R. Ayman (Eds.), *Leadership theory and research: Perspectives and directions* (pp. 245–270). New York: Academic Press, Inc.

Ingram, H., Teare, R., Scheuing, E. and Armistead, C. (1997). A systems model of effective teamwork. *The TQM Magazine*, **9**, 118–127.

Johnston, J.H., Paris, C. and Smith, C.A. (1999). Toward assessing the impact of TADMUS decision support system and training on team decision making. Retrieved January 4, 2002 from the Department of Defense Command and Control Research Program Web site: http://www.dodccr-p.org/1999CCRTS/pdf_files/track_6/107johns.pdf.

Jones, M.C. and Harrison, A.W. (1996). IS project team performance: An empirical assessment. *Information and Management,* **31**, 57–65.

Katzenbach, J.R. and Smith, D.K. (1993). *The Wisdom of Teams: Creating the High–Performance Organization.* (New York: Harper Business).

Katzenbach, J.R. (1998). Making teams work at the top. *Leader to Leader,* **7**, 32–38.

Kimball, L. and Eunice, A. (1999). The virtual team: Strategies to optimize performance. *Health Forum Journal,* **42**, 58–63.

Kozlowski, S.W.J., Gully, S.M., Salas, E. and Cannon-Bowers, J.A. (1996). Team leadership and development: Theory, principles, and guidelines for training team leaders and teams. *Advances in Interdisciplinary Studies of Work Teams,* **3**, 253–291.

Lynn, G.S. and Reilly, R.R. (2000). Measuring team performance. *Research Technology Management,* **43**, 48–57.

McIntyre, R.M. and Salas, E. (1995). Measuring and managing for team performance: Emerging principles from complex environments. In R.A. Guzzo and E. Salas (Eds.), *Team Effectiveness and Decision Making in Organizations* (pp. 9–45). (San Francisco: Jossey Bass).

Meister, D. (1985). *Behavioral Analysis and Measurement Methods.* (New York: John Wiley & Sons).

Meister, D. (2001). Basic premises and principles of human factors measurement. *Theoretical Issues in Ergonomics Science,* **2**, 1–22.

Miles, J.A. (2000). Relationships of collective orientation and cohesion to team outcomes. *Psychological Reports,* **86**, 435–444.

Mohammed, S., Klimoski, R. and Rentsch, J.R. (2000). The measurement of team mental models: We have no shared schema. *Organizational Research Methods,* **3**, 123–165.

Orasanu, J. (1990). *Shared mental models and crew performance.* (Technical Report #46) Laboratory of Cognitive Science, Princeton University, NJ.

Orlady, H.W. and Orlady, L.M. (1999). *Human Factors in Multi-crew Flight Operations.* (Brookfield, VT: Ashgate Publishing Company).

Oser, R., McCallum, G.A., Salas, E. and Morgan, B.B. (1989). *Toward a Definition of Teamwork: An Analysis of Critical Team Behavior* (Tech Rpt 89-004). US Naval Training Systems Center, Human Factors Div, Orlando, FL.

Pandzic, I., Babski, C., Capin, T., Lee, W., Magnenat-Thalmann, N., Musse, S.R., Moccozet, L., Seo, H. and Thalmann, D. (2001). Simulating virtual humans in networked virtual environments. *Presence,* **10**, 632–646.

Paris, C.R., Salas, E. and Cannon-Bowers, J.A. (1999). Human performance in multi-operator systems. In P.A. Hancock (Ed.), *Human Factors and Ergonomics* (pp. 329–386). (San Diego, CA: Academic Press).

Pharmer, J.A. and Campbell, G.E. (2000, December). Verification and validation of three human performance models within the context of a field experiment. Retrieved May 15, 2002, from the Office of Naval Research International Field Office Web site: http://www.ehis.navy.mil/tp/humanscience/masakowski/dera/Nov00/HSI/Presentations/human%20perform%20m odels%20-%20pharmer%20&%20campbell.ppt.

Prince, C. and Salas, E. (1993). Training and research for teamwork in the military aircrew. In E. L. Wiener, B.G. Kanki and R.L. Helmreich (Eds.), *Cockpit Resource Management* (pp. 337–366). (New York: Academic Press).

Salas, E., Bowers, C.A. and Cannon-Bowers, J.A. (1995). Military team research: 10 years of progress. *Military Psychology,* **7**, 55–75.

Salas, E., Bowers, C.A. and Edens, E. (2001). An overview of resource management in organizations: Why now? In E. Salas, C.A. Bowers and E. Edens (Eds.), *Improving Teamwork in Organizations: Applications of Resource Management Training.* (Mahwah, NJ: Lawrence Erlbaum Associates).

Salas, E., Burke, C.S. and Cannon-Bowers, J.A. (1997). Methods, tools, and strategies for team training. In M.A. Quinones and A. Eheenstein (Eds.), *Training for a Rapidly Changing Workplace: Applications of Psychological Research.* (Washington, D.C.: APA Press).

Salas, E., Burke, C.S. and Cannon-Bowers, J.A. (2000). Teamwork: Emerging principles. *International Journal of Management Review,* **2**, 339–356.

Salas, E., Burke, C.S., Fowlkes, J.E. and Priest, H.A. (2003). *On Measuring Teamwork Skills.* In J.C. Thomas (Ed.), *Comprehensive Handbook of Psychological Assessment, Volume 4: Organizational Assessment* (pp. 1105–1149). (New York: John Wiley & Sons).

Salas, E. and Cannon-Bowers, J.A. (2000). The anatomy of team training. In S. Tobias and J.D. Fletcher (Eds.), *Training and Retraining: A Handbook for Business, Industry, Government, and the Military* (pp. 312–335). (New York: Macmillan Reference USA).

Salas, E. and Cannon-Bowers, J.A. (2001). The science of training: A decade of progress. *Annual Review of Psychology, 52,* 471–499.

Salas, E., Dickinson, T.L., Converse, S.A. and Tannenbaum, S.I. (1992). Toward an understanding of team performance and training. In R.W. Swezey and E. Salas (Eds.), *Teams: Their Training and Performance* (pp. 3–29). (Stamford, CT: Ablex Publishing Corp).

Salas, E., Rhodenizer, L. and Bowers, C.A. (2000). The design and delivery of crew resource management training: Exploiting available resources. *Human Factors,* **42**, 490–511.

Sanders, M.S. and McCormick, E.J. (1993). *Human Factors in Engineering and Design* (7th ed.). (New York: McGraw-Hill Inc.).

Smith, P.C. and Kendall, L.M. (1963). Retranslations or expectations: An approach to the construction of unambiguous anchors for rating scales. *Journal of Applied Scales,* **47**, 149–155.

Smith-Jenstch, K.A., Johnston, J.H. and Payne, S.C. (1998). Measuring team-related expertise in complex environments. In J.A. Cannon-Bowers and E. Salas (Eds.), *Decision Making Under Stress: Implications for Individual and Team Training* (pp. 39–60). (Washington, D.C.: American Psychological Association).

Stewart, G.L. and Manz, C.C. (1995). Leadership for self-managing work teams: A typology and integrative model. *Human Relations,* **48**, 747–770.

Swan, J.A. (1995). Exploring knowledge and cognitions in decisions about technological innovations — Mapping managerial cognitions. *Human Relations,* **48**, 1241–1270.

Swezey, R.W. and Salas, E. (1992). Guidelines for use in team-training development. In R.W. Swezey and E. Salas (Eds.), *Teams: Their Training and Performance* (pp. 219–245). (Stamford, CT: Ablex Publishing Corp.).

Tziner, A. and Haccoun, R.R. (1991). Personal and situational characteristics of transfer of training improvement strategies. *Journal of Occupational and Organizational Psychology,* **64**, 167–178.

Vidmar, N. and Hackman, J.R. (1971). Interlaboratory generalizability of small group research: An experimental study. *The Journal of Social Psychology,* **83**, 129–139.

Webber, S.S., Chien, G., Payne, S.C., Marsh, S.M. and Zaccaro, S.J. (2000). Enhancing team mental model measurement with performance appraisal practices. *Organizational Research Methods,* **3**, 307–322.

West, M.A., (2003). Innovation implementation in work teams. In P.B. Paulus (Ed.), *Group Creativity: Innovation Through Collaboration* (pp. 245–276). (London: Oxford University Press).

West, M.A., Borrill, C.S. and Unsworth, K.L. (1998). Team effectiveness in organizations. In C.L. Cooper and I.T. Robertson (Eds.), *International Review of Industrial Organizational Psychology.* (Chichester: John Wiley).

Wiener, L., Kanki, B.G. and Helmreich, R.L. (Eds.) (1993). *Cockpit Resource Management.* (San Diego, CA: Academic Press).

Witmer, B.G. and Singer, M.J. (1998). Measuring presence in virtual environments: A presence questionnaire. *Presence,* **7**, 225–240.

Zachary, W.W., Ryder, J.M. and Hicinbothom, J.H. (1998). Cognitive task analysis and modeling of decision making in complex environments. In J.A. Cannon-Bowers and E. Salas (Eds.), *Making Decisions Under Stress: Implications for Individual and Team Training* (pp. 315–344). (Washington, D.C.: American Psychological Association).

Zachary, W., Ryder, J.M., Ross, L. and Weiland, M. (1992). Intelligent human–computer Interaction in real time, multi-tasking process control and monitoring systems. In M. Helander and M. Nagamachi (Eds.), *Human Factors in Design for Manufacturability* (pp. 377–402). (New York: Taylor & Francis).

Chapter 31

Work conditions and the risk of injury

Christine M. Haslegrave and E. Nigel Corlett

Introduction

With the current awareness of the contributions of work conditions to industrial injuries, it is seen as increasingly important to identify such contributions and work to reduce their impact. Legislation in this direction puts the onus on management to identify and modify hazardous situations, and to be seen to have taken the necessary steps to create the safest practical working conditions. In this context, risk assessment is a term well known to many managements.

In some industries, e.g., nuclear engineering, risk assessment is a relatively sophisticated activity and pursued at design, manufacture and operation levels. Some of this sophistication can be recognised in other chapters of this book, dealing with human reliability assessment and with incident reporting (Chapters 32 and 33), which have relevance for this chapter and which will be referred to.

If we are going to discuss risk and its assessment it is as well to be clear what we mean by 'risk'. It is a term which is used widely in ordinary speech and with various meanings. For our purposes we will use as a definition that a risk is the overlap between a hazard and a vulnerability. People are vulnerable to deafness if working in a noisy environment but if the working conditions are very quiet, then there is no risk. Equally, a workplace requiring frequent heavy lifting represents a hazard to which people are vulnerable, some more than others, hence a risk exists and action is necessary.

To evaluate work conditions so that the risks can be assessed will thus require us to identify the hazardous conditions and the related vulnerable nature of those exposed to them. Where the vulnerabilities are known and their characteristics understood in relation to a hazard it is often possible to survey widely across a plant to find the extent of the hazard and thus the probable costs of actions to mitigate it. Many problems in handling, in manipulation and in repetitive tasks lend themselves to this treatment. However, where the contribution of a hazard to illness or accident is less well documented, more detailed examination of the situation may be needed and caution should be exercised before the results are transferred to other situations.

From the above it will be evident that risk assessment is not yet, if it ever will be, at a stage where simple techniques will provide a number which can confirm the level of a risk. Expert judgements by a variety of experts, ergonomists, engineers, medical and safety personnel are used to decide the acceptable levels above which the various hazards require attention. This and other chapters describe several of these procedures, but the implication of this situation is that the participative involvement of people with several areas of knowledge, including the operators themselves, is desirable if further errors are not to be introduced or the situation made worse (Corlett, 1991; Noro and Imada, 1991; Wilson, 1991, 1994).

Recognising the presence of a hazard

In field work it is rarely possible to do an experiment, so we are confined to examining the existing situation and the past history. Historical data can come from accident and injury records, company medical and personnel records (where personal confidentiality MUST be observed), national health and safety data, and publications in relevant journals. The use of archival data is discussed in Chapter 3 and incident reports in Chapter 33, which presents models of accident occurrence and of human error causation contribute to the understanding of sources of potential mismatch between situations and the people involved in them, a major source of hazards. Such mismatches, indications of an inadequate workplace, can cover the total working environment and not just the physical fit between worker and tools. The inability to recover from an operational error, an inhibition against reporting a fault or being unable to pursue recommended practices due to time or other pressures are all examples where environmental pressures, be they social, organisational or physical, contribute to the inadequacy of a workplace.

Assessment of risk

A good understanding of ergonomics provides the basis for a sound analysis of the current situation. As mentioned above, what is being sought is where the mismatches lie between the tasks, environment, etc. and the people doing the jobs. Having identified these mismatches, then some estimate of their severity is required.

A rapid overview of a particular office or factory can be gained by the use of Åberg's Loads and Causes Survey (Åberg, 1981). One version of the recording form is shown in Figure 1. In the use of this form a group of technical and operational people with experience of the work decide on the appropriateness and interpretation of the various causes, listed along the top of the form, and the numerical grading which they will use for judging the severity of the contribution of each cause to the intensity of each load factor. Three grades of severity are usually enough for the judgements required, and these are entered into the body of the table for each job under investigation. In the left-hand column, headed 'magnitude', a judgement of each load factor is entered, using the same scale of levels as for the contribution of the causes. After discussion amongst the analysing group, to agree any variations in their judgements, the sums given at the foot of the form give values which suggest the priorities for change, with the numbers in the left hand column against each load factor giving an indication of which are the most severe in their effect.

For some circumstances, the direct and long term observation of operators is appropriate. Branton (1970) watched four operators of capstan lathes over a period of sixteen weeks to identify potential causes of accidents. He created simple forms to assist in recording the movements of operators, the purposes of their movements and their end points. From the data he was able to show how injuries might occur, to get an estimate of frequency of exposure and to make proposals for hazard and injury reduction. Branton's recognition of hazard came not just from his observations, which would not have required much ergonomics, but also from his understanding of the information processing — or motor control — difficulties in locating components and in locating and operating certain controls. Such recognition permitted him to make more valid proposals for improvement, due to his deeper understanding of the causes of the errors he saw.

This would be true, also, for the use of techniques such as OWAS or RULA, described in Chapter 16. The purpose of such techniques is to identify postures, combined with forces and repetitions, which may indicate a risk of low back or upper limb injury. However, the posture or muscle loading as recorded by these two techniques is but one factor in the

Load factor		Cause factor								
Magnitude	Type	Technology/ process	Room, buildings	Material flow	Lay out	Machines	Tools, aids	Work organization	Work method	Workplace design
	Noise									
	Dust, smoke									
	Climate									
	Body									
	Space									
	Circulatory load									
	Information load									
	Constraint monotony									
	Contact possibilities									
	Co-operation possibilities									
	Disturbance of others									
	Others disturbing									
	Sum									

Figure 1 Åberg's Loads and Causes Survey form.

context of hazard. A lack of co-worker or management support, as for example in the case of a nurse lifting a patient or when pressure for output encourages the reduction of rest breaks, could be key factors in tipping the situation from the undesirable but feasible to the dangerous. This would not be recognised just from the study of record charts, but could be identified by a skilled and informed observer.

In any study, observation is often the best start and gives a much better indication of where mismatches lie than is possible just from hypothesising about the problem. Seeing what people really do and talking to them about what the work feels like, and what they are trying to accomplish, leads to a clearer knowledge of the links between people and their work. The work, indeed, may only be a minor factor in risk; the organisation, workplace structure or other factors may be of greater importance.

Checklists and questionnaires

If a good understanding has been developed from the preliminary surveys, questionnaires can be useful for the more detailed investigation (Chapter 4 gives essential advice on questionnaire design and use). We must repeat the caution, however, that to pluck an instrument from a publication and apply it without considering its original source and purpose may be unwise, and thought should be given to whether it is appropriate for the proposed new purpose. Clearly, questionnaires and their cousin the checklist are valuable in the context of hazard investigation, although their use on their own is usually insufficient.

Environmental factors

(a) Are levels of noise sufficient to cause mental stress or interfere with communications or safety?
(b) Does music impose rhythmic patterns that are inappropriate for the task?
(c) Are lighting levels causing operators to adopt awkward postures to avoid shadows or to see properly?
(d) Are flickering lights causing stress to operators?
(e) Is the air temperature avoidably low at any time of the year?
(f) Is protective clothing issued because of the environment, constraining posture, or do gloves affect grip?
(g) Is poor environment a source of discontent among operators?
(h) Are there chemicals in the air that might be affecting the operators' coordination or muscular system?

Figure 2 Extract from checklist for the identification and reduction of work-related upper limb disorders (HSE, 1990). Crown copyright, reproduced with the permission of the Controller of HMSO.

Checklists which require only a 'yes' or a 'no' response are widely used as memory aids, being valuable to check that all aspects of a situation have been covered. In the context of hazard analysis they are also valuable to check whether all practical or known measures have been taken.

Figure 2 shows an extract from a broadly based set of questions in a checklist published by the UK Health and Safety Executive, in a guide to preventing work-related upper limb disorders (HSE, 1990). The extract deals with environmental factors and the questions can be answered by 'yes' or 'no.' They draw attention to a range of environmental factors deemed to be relevant, but do little more than this. It is clear, however, that most questions could be broken down to give the components of the factors concerned, which could then be given a response scale e.g., 'very often' to 'not at all' for question (d). The resulting answers would reveal, with a little more effort, the approximate extent of the mismatches in the environment.

Rather than ask questions about what is wrong in a situation, many checklists present what a good situation should be, leaving the user to identify the departure from this ideal in the case under investigation. Sometimes the wording is imprecise, indicating goodwill but little guidance. It is of little use to say that loads should be minimised or trunk deflection kept low. The set of guidelines given as a checklist on manual handling, shown as Figure 7, has its share of these goodwill statements but also gives specific statements against which situations can be tested, such as:

> Handling loads at heights below the knees or above the shoulder should be eliminated wherever possible.

or

> Use grasps which are two-handed, with the load or force evenly distributed between both hands.

These examples give specific requirements, so that a hazard assessment could document where these were breached, and present the reasons. Moreover, a study of operator behaviours 'before' and 'after' intervention could then demonstrate a reduction (or otherwise!) in the hazard.

The scaled form of checklist is exemplified by the Chair Feature Check List (CFCL) illustrated in Chapter 27. Each feature of the chair relevant to its use has been provided with a scale running from one extreme of inadequacy to the other, e.g., 'too high' to 'too low.' Each question can be answered by introspection whilst seated on the chair, and a profile of the chair's qualities drawn from the results. What needs to be done to improve the chair can be identified with only a modest knowledge of ergonomics, probably enabling the sitter to make improvements even if they are not the best that could be achieved.

The application of such a form of checklist can be developed for other areas. Figure 3 shows one designed to explore the match between an office job and an individual's needs. From a knowledge of the jobs surveyed, and the mismatch on the cognitive dimension, sources of error could be surmised which could be compared with known errors. The risk of error is taken to be relative to the severity of the mismatch, even though the exact relationship is not known. Nevertheless, the analysis of such a checklist, in the light of concerns regarding the severity of the effects of various types of errors, would present a starting point for further investigations.

A technique based on the checklist approach but further developed to prioritise modifications and redesigns, based on their criticality and cost, has been proposed by

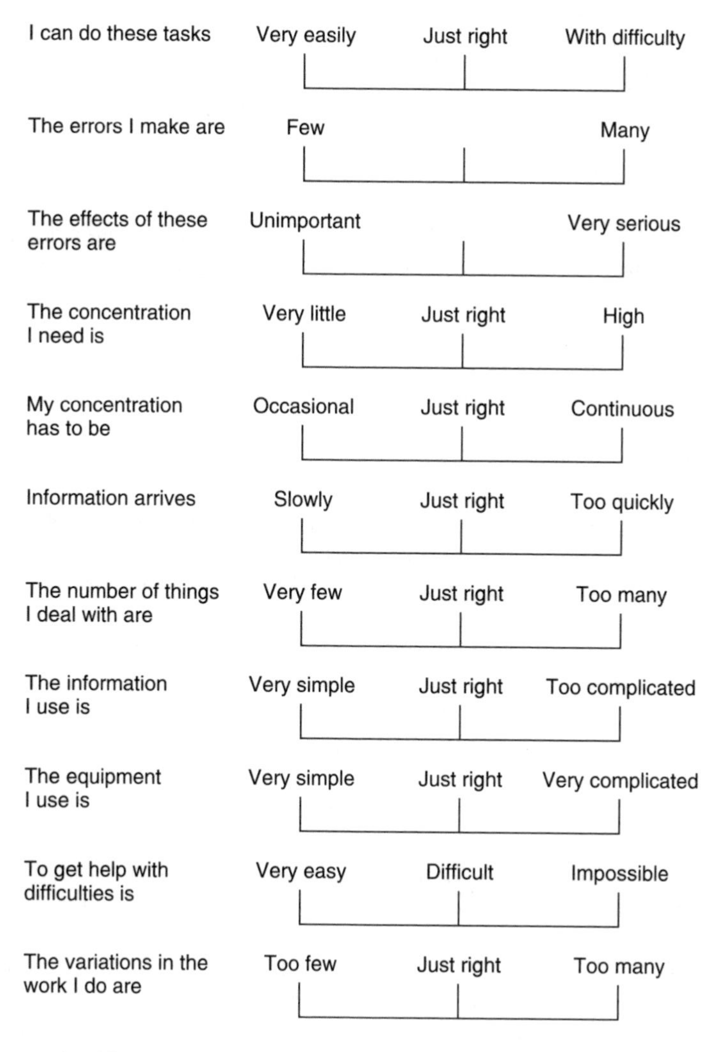

Figure 3 Feature checklist to investigate the match between an office job and the individual's needs.

Yeow and Sen (Yeow, 2002; Sen and Yeow, 2003). Whilst undertaking studies in the electronics industries in Malaysia, Sen and Yeow (2003) recognised the contributions of poor ergonomics to many of the losses arising in manufacture. They required a procedure which would assess the reasons for the component and assembly failures which arose due to the inability of workers to pursue the correct procedures for manufacture. Further, they wished to classify the procedural failures so that they could deal with the most important first, the importance including the costs to company and customers.

Their approach was to develop what they called Ergonomic Weighted Scores (EWSs) to evaluate the extent to which incorrect operations were the causes of the faulty work, and the reasons why operators were unable to perform correctly. They proposed the method outlined in Figure 4. The procedure was to create a Critical Instruction (CI) list, using a group of factory people familiar with the design and manufacture of the products, together with an ergonomist. Critical Instructions were those key instructions for the manufacture of the product which an examination of past failures demonstrated had been responsible for the faulty work. They used five criteria for selecting the CIs.

(i) The CI must be directly related to the production defect.
(ii) CIs must follow and support the requirements of the Malaysian ISO 9002:1994 Quality Standard System (SIRIM, 1994).
(iii) CIs must follow Occupational Health and Safety requirements.
(iv) CIs must follow the defined requirements and procedures of the manufacturing process.
(v) CIs must support the use of correct materials, tools and instruments.

Having identified where a CI related to a particular production defect, it was weighted by the product of the following four factors, to give a total weighting:

Factor A is the weighting on the difficulty of detecting the defect
Factor B is the weighting of the difficulty of repairing the defect
Factor C is the weighting on the cost of repairing the defect
Factor D is the percentage which the particular defect contributed to the total of the production defects.

Factors A, B and C were assessed on a scale of Low (1), Moderate, (2) or High (3). Factor D was assessed from three months records of the defect in question compared to all defects.

The total weighting was calculated from the product of the four factors since criticality of a defect is dependent on all four factors.

To operate the procedure, after an introductory session, the team set out the CIs which they considered were relevant to each defect. The CIs were assessed against the five criteria (i)–(v) above, and the CIs which did not fulfil them were omitted, decided on the basis of a consensus amongst the team. The CIs were then weighted using the above factors A–D, again solving difficulties by consensus. The result was that each defect had its own group of CIs, weighted according to their contributions to the defect and its costs. The weightings of the CIs associated with each defect were added, giving a total score for each defect which prioritised the importance of the defect for solution against its contribution to company losses.

However, the defect in question might not arise from non-observance of all the identified CIs, so shop floor personnel were trained to study the suspect operation to identify which CIs were not being observed. These were the ones which then went towards the summation of the Total Score for the operation concerned. From these results the key factors which were responsible for the fault were identified and attacked, highest value first, using ergonomics to make the operation possible, together with any technical changes

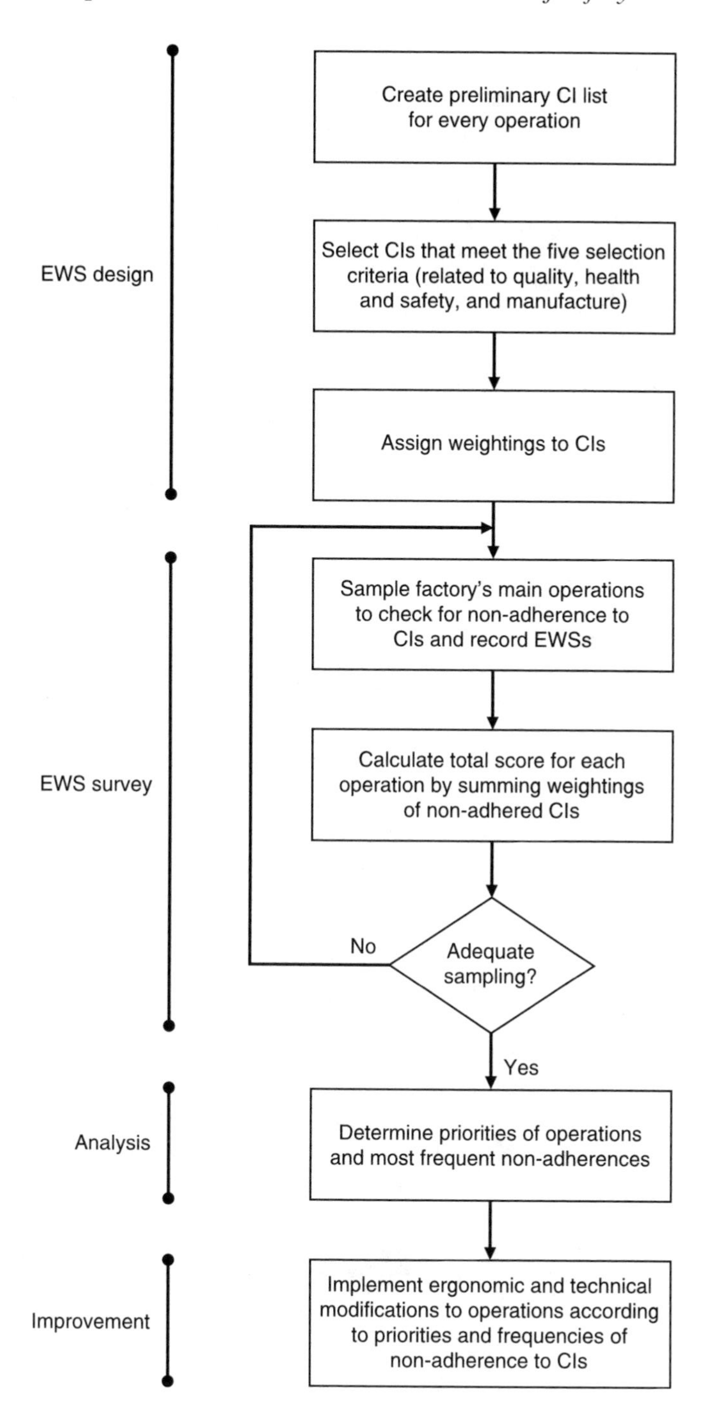

Figure 4 The EWS (Ergonomic Weighted Scores) method of prioritising manufacturing operations for ergonomic improvements, based on surveying adherence to Critical Instructions (CIs) (adapted from Yeow (2002)).

which seemed appropriate. The aims were both to reduce production defects, customer complaints and rejection costs, and to improve the health and safety of the workforce.

Several of these studies in the electronics industry achieved major savings. In printed circuit board assembly, the company was reported to have saved $571,000 in the first year whilst the customer saved $139,000 (Yeow, 2002, pp. I-246, I-247). Another study, on component dropout from printed circuit boards, made a potential saving in a year of over $2 million (Yeow, 2002, p. I-218).

Psychosocial aspects

As mentioned earlier, psychosocial factors can influence performance and work demands, changing the level of a hazard and thus affecting the risk of industrial injuries. Examples of psychological work demands, which can be associated with physiological indicators of strain, are the levels of monotony, boredom, job satisfaction, autonomy (which can include factors frustrating job performance), and social support. The section, in Chapter 32, on performance shaping factors (PSFs) is also relevant here.

The general approach to assessing risk must be to understand first, by investigation, the holistic nature of the work. It is therefore consonant with our approach to investigating work conditions and risks to look at psychosocial work aspects from the perspective of a mismatch between what exists and what people expect. It is also necessary to note that what exists may well be 'what is perceived to exist', in the sense that people's perceptions of a situation are what is relevant if we are speaking of a mismatch. This point has been well illustrated by Kjellstrand (1974) who showed that, as workplace constraints such as pacing increased, perceptions of environmental stress increased, although the actual environmental situation was substantially the same.

The psychosocial factors which are usually of interest to the ergonomist are work stress (a term needing definition before it is measured — see Chapter 19), fatigue and attention, work attitudes and social support or cohesion. Apart from identifying the presence and strength of these factors for research purposes, there is interest in their contributions to general stress levels and, of course, to the potential mismatches mentioned above.

Questionnaires may be suitable for investigating some of these aspects. Karasek's (1998) Job Content Questionnaire and the NIOSH Generic Job Stress Questionnaire (Hurrell and McLaney 1988) provide measures of psychological demands and stresses in a job, intellectual discretion and authority over decisions, together with indicators of mental strain symptoms (exhaustion, depression, job satisfaction and life satisfaction). Where work pressures exist, job stress is suspected or alienation is evident, which are all factors liable to lead to personal risk, such questionnaires can give indications of mismatch.

In general, questionnaires will be suitable where mismatches are believed to exist and where features relevant to the work can be answered by reference to things as they are at the time of the enquiry, or in the recent past. These aspects are then separated into their component features, the features making part of the 'model' or concept of the work and its hazards, which are under investigation. It is important that the investigator has such a model, for single issue studies are unlikely to be very useful, since most psychosocial responses are multicausal. Investigations which ignore this will be sub-optimal.

Moreover, the simple use of psychosocial data collected by questionnaires can be unwise. Answers can only be interpreted where the questionnaire is based on an appropriate model of the effects of the psychosocial factors, and where the model is interpreted with some understanding of the culture of the site of the study, and of its organisational and social structures. Many investigators stress the desirability of complementing questionnaires and other instruments with interviews, so that a clearer knowledge of the meaning of the responses is then available.

The reader will be aware that there are difficulties in compressing methods for investigating psychosocial factors into part of a single book chapter! There is no substitute for a deeper study in this area. Nevertheless, there are some well developed instruments, with normative values, which can be introduced. One widely used technique is the Stress Arousal Check List (SACL) of Cox and McKay, (1985) described in Chapter 19 of this book. The underlying model and robustness of the technique are discussed there and in the quoted references.

Fatigue is both a physical and a psychological state, and earlier chapters have discussed the former in terms of muscular and cardiovascular symptoms. Several fatigue inventories exist to explore people's feelings of fatigue and one which is well established is that of the Industrial Fatigue Research Committee of the Japan Association of Industrial Health (1970). It consists of thirty one items covering feelings of physical states such as stiffness, drowsiness or difficulties of concentration. This inventory is often used by Japanese researchers within a battery of tests and measures. In a study of checkout operators in supermarkets, Kishida (1991) looked at job structures, using observation and a questionnaire survey, and responses on the Industrial Fatigue Inventory. He analysed the results to identify correlations between fatigue responses and work times and activities, highlighting aspects of the work which gave rise to the peak responses.

Job attitude questionnaires have a long history, as have studies of the effects of social support in the workplace and outside it. Even more than for the previous two areas, the use of the available instruments and more especially their interpretation are strongly linked to the models on which they are based. It is deemed inappropriate, therefore, to expand on these, beyond making the cautionary point that, when studying the literature, the reader should keep in mind that there is a strong cultural dimension which may make the transfer of methods gleaned from the literature unsuitable for direct transfer to a local problem.

Of course, this caution applies across the whole of the research field; make sure that you agree with the model's applicability to the problem before you accept the conclusions.

Quantifying the hazards

The purpose of hazard investigation is to define, and, as far as possible, quantify the hazards, and then to present what should be done to deal with the problems. Clear presentation of the analysis is essential, but the analyses themselves must reveal the picture. So, having used workplace surveys, questionnaires, checklists or interviews, the investigator then needs to reduce the data and highlight the key factors of the hazard.

Moreover, where hazards have been identified, some indication of their importance is needed and decisions have to be taken on how to eliminate or control them. If the hazard can be eliminated with no significant effect on the rest of the system, then that is the decision to be taken. If this is not the case, then a risk assessment is needed, so that undue effort is not put into highly unlikely events with minimal human risk or commercial impact.

We present here an outline of a procedure which can be used for hazard assessment and for deciding on subsequent actions. Its purpose is to provide a framework around which a company can build its own assessment procedure to meet its particular requirements, an example being shown in Figure 5.

The approach is to assemble a group of relevant people within the company and ask them to assess each hazard which has been identified. Their database will be the information collected using some of the techniques already described, together with a listed sequence of the work activities, for instance from a method study, which should be augmented with a list of all the associated or nearby equipment, such as the presence of fork-lift truck movements or high pressure lines. The group studies these for potential hazardous scenarios

Hazard severity	**Probability of occurrence**		
	3	2	1
3	Stop job and eliminate	Eliminate as high priority	Eliminate as priority
2	Eliminate as high priority	Eliminate; guard	Eliminate; guard
1	Eliminate; guard; protect against	Eliminate; guard; protect against	Eliminate; guard; protect against

Note: Levels of hazard, probability of occurrence and action should be decided beforehand in relation to experience with the plant.

Hazard severity: Could be defined in terms of consequences.
Thus: 3 = Death or serious injury; high cost; major damage
2 = Minor injury; moderate cost; moderate damage
1 = No injury; minor cost

Probability of occurrence:
3 = Frequent
2 = Infrequent, e.g. once or twice a year
1 = Rare, unlikely but possible

Actions, in the body of the matrix:
The sequence should be: (i) eliminate, (ii) guard, (iii) protect against, e.g. special clothing. (iv) train for special work behaviours. In all cases knowledge of the hazard and warning notices should be provided. Actions may, of course, be combined as needed. They are not to be seen as independent.

Figure 5 Initial assessment of levels of hazard — an action matrix.

or conditions, possibly aided by a checklist compiled by a person expert on the potential for malfunction of various pieces of the equipment. The possibility of increases in the hazards arising from night shift working or from situations such as poor access for maintenance or flooding should be taken into account.

When looking at these data, the task demands for physical or mental effort can have important influences, and required activities should always be assessed with the worst case situation in mind. As the group pursues the assessment, a note is made of each identified hazard, and possible ways of overcoming it. However, all known hazards are not equal, and some priority has to be determined to see that an effective programme is established. This is done by 'triage', the process of prioritising the risks so that the important ones are dealt with first. To gain some common metric for this, the probability of an occurrence and the severity of its effects on personnel as well as on the company are combined, often by multiplication, to give an opportunity to put the situation in some sort of order.

The consequences can then be presented as a sequenced list, in order of increasing severity, perhaps starting with 'no delay or cost' and extending through 'product delay', 'damage', etc. to injuries of varying severity. A suggested classification for this process is given in Figure 5, which includes some proposed responses to the combinations of severity and probability of occurrence.

It is recommended that this process is conducted by a group of people having various areas of background knowledge and familiar with the situation, the requirements of the activities and the personnel involved. Their discussions and agreements when setting up and conducting the process and deciding on the action strategies should be recorded so that, as experience is gained, they can be improved. The agreed procedure should also, preferably, be pilot tested and reviewed before any full scale implementation.

Reference was briefly made earlier to Performance Shaping Factors. These are modifiers of performance and will affect the importance of any objectively identified hazard. For example, oil on a floor is a hazard, but it is an increased hazard when workers are under time pressure and tempted to run. Hence, having assessed as objectively as possible the inadequacies and mismatches of a situation, these should be considered in conjunction with a table of Performance Shaping Factors. For example, one feature which gives managers and safety officers much concern is risk-taking behaviour. Risk-taking can be generated by a company's social or managerial climate, familiarity with the situation, a personality trait or other reasons. The influence of such behaviour on a particular hazard can be to increase its severity, and a company's hazard analysis report should take these factors into account.

So far, in this discussion, we have dealt with hazards and risks, with indications of where vulnerabilities lie. National health and safety bodies publish material on the human vulnerability to chemicals, to noise and to several other environmental stressors, whilst reference to many psychological vulnerabilities will be found in the chapters on stress, work organisation and mental work load assessment, with their lists of references. The next section will deal particularly with methods for investigating the hazards in many musculoskeletal activities, where the hazard, the vulnerability and the consequent risk are, in some cases, dealt with in the procedures.

Techniques for assessing the effects of work activities on musculoskeletal problems

Many industries recognise the hazards associated with load handling and with repetitive, short cycle, activities. This concern tends to focus on problems associated with the lower back and with injuries to the upper limbs. At the same time it is increasingly recognised that there is a contribution to these injuries from the psychosocial environment. Hence the conditions of this environment are now taken into account and the material earlier in this chapter should be kept in mind when exploring the hazards in these areas.

People vary in their susceptibility to injury from muscular activities and, since there is no way of identifying accurately a level of susceptibility, the proposals for action are intentionally conservative. At the same time, the absence of serious recorded injuries should not be taken as evidence that no risk exists. Damage in the musculoskeletal areas is cumulative as well as occurring sometimes after a single event, so the results of finding, when using the following techniques, that a risk exists should be taken as a signal that something should be done; the criteria associated with the methods do, in many cases, provide a graded scale of actions.

The adoption or maintenance of a posture in order to exert a force alerts us to the fact that both static and dynamic work is done, and information in earlier chapters, particularly Chapters 15, 16 and 17 should be reviewed. Where the methods deal with a particular part of the body, e.g., a particular limb, an understanding of the observations cannot rest just on some numerical assessment of the postures but must be interpreted in the light of the contributions of other parts of the body to the task and not just those immediately involved.

Assessment techniques

There is a wide range of useful techniques, many of which are listed in Figure 6 and it will be seen that they can be divided into those which are measured with instruments and those for which the data arises from observation or subjective self reports. In most practical cases it is desirable to survey the situation to assess the scale of the problem, and several of the techniques listed are suitable for this.

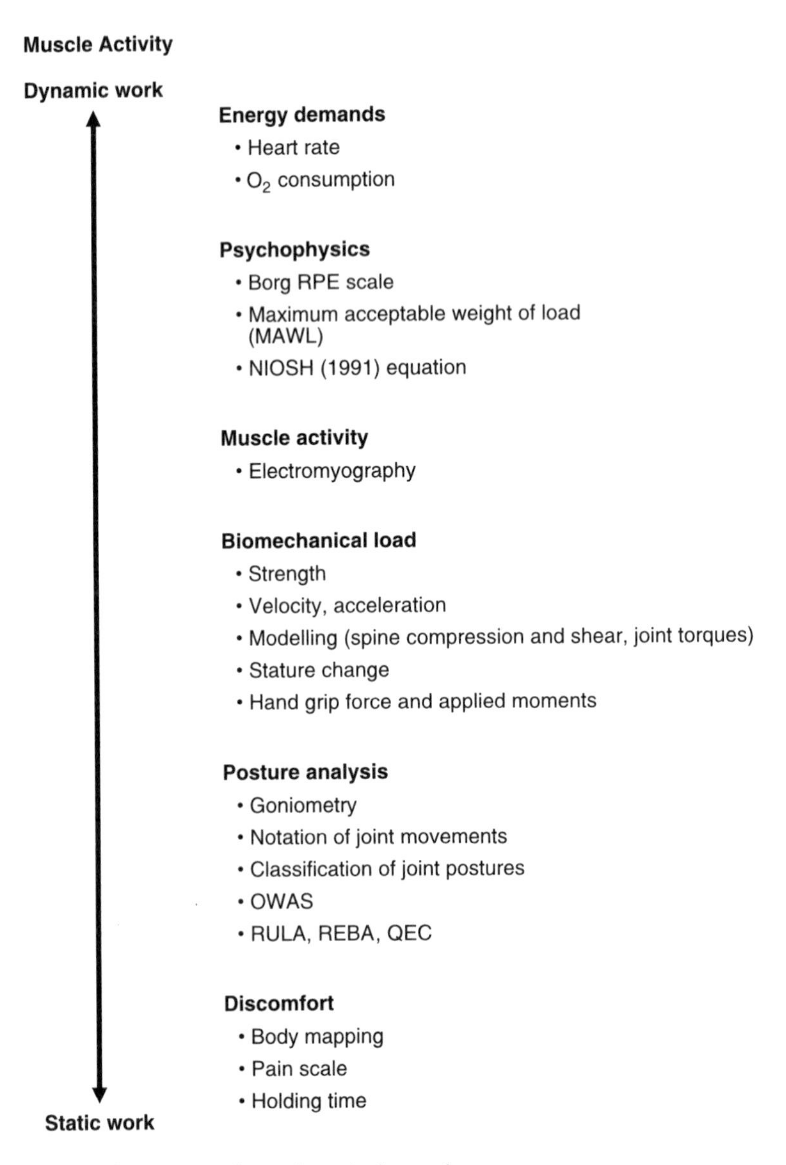

Figure 6 Methods of assessing physical work demands.

Risk assessment requires tools which can be used in the field rather than those designed for the laboratory, and methods for surveying large populations or groups of workstations. The factors contributing to the development of musculoskeletal disorders are now well recognised, but dose-response relationships are on less sure ground and it has to be acknowledged that our current understanding of the risks borne by individual workers is still limited. It may not be necessary to provide measures of absolute risk but it is important to have comparative measures which assist in establishing priorities for making changes and redesigning work.

Discomfort, pain and fatigue can be precursors of injury, indicating where the body is experiencing strain from which it cannot recover in the course of the working day. Hence the use of a discomfort survey employing a body map may highlight points of overload, from which an examination of the work activities should reveal those which are introducing the discomforts or indicate areas for further detailed study.

Where upper limb problems are suspected, the Nordic Questionnaire (Kuorinka *et al.* 1987), as modified by Dickinson *et al.* (1992) for UK use gives major data for reducing the problem to a size which focuses on looking at the problem areas. This is also possible with the procedure of Kemmlert and Kilbom (1986) which is presented in the same chapter. This complements the personal reports collected in the Nordic Questionnaire by linking such reports to factors in the workplace which may be causing the muscular stress.

For manual handling problems the assessment procedure of Birnbaum *et al.* (1993) is straightforward and can identify major sources of hazard. It is conducted under the headings of (a) the task, (b) the worker, (c) the workplace, (d) job design and (e) the organisation. Later in this chapter an assessment method (given as an appendix) is discussed, which has been adapted from one by the UK Health and Safety Executive (HSE, 1998). From such surveys, where the causes of the problems are clear, an action list can be compiled and decisions on solutions and timing of implementation can be drawn up.

Investigating work-related upper limb musculoskeletal disorders

These disorders arise in many forms and the symptoms are often non-specific. They are variously described as a work-related upper limb disorder (WRULD), cumulative trauma disorder (CTD), work-related musculoskeletal disorder (WMSD), repetitive strain injury (RSI), occupational overuse syndrome (OOS) or other term. Whilst it is recognised that there are many contributors to such a disorder, there are three main physical factors implicated in its occurrence. These are the force exerted in relation to the strength of the muscles involved, the posture of the body segments involved and the repetitive nature of the actions. This last factor is most important in relation to the body's ability to recover from exertions, so the fourth factor in any investigation is the time between the groups of repeated calls on the body tissues and structures concerned, both within the task cycle and between periods of work activity.

A survey method designed to investigate workplaces where upper limb disorders are suspected is RULA, Rapid Upper Limb Assessment (McAtamney and Corlett, 1993). This is described in detail in Chapter 16. It provides a method of assessing posture, whether seated or standing, and pays particular attention to the neck, trunk and upper limbs. It also assesses the contribution of the muscular effort, whether arising from exerting external force, from maintaining a posture or from muscle loading in the task, such as for holding tools. RULA is intended to be used by non-ergonomists but requires some training or previous experience in posture observation. It can be used as a tool for surveying large numbers of workplaces in a company to identify tasks where a risk of upper limb disorders might exist and it can prioritise the changes and redesigns which may be required. It is also valuable for the investigation of individuals and their workplaces to identify where modifications are needed. There are two other survey methods for investigating musculoskeletal disorders. One is REBA (Rapid Entire Body Assessment, Hignett and McAtamney, 2000), which deals with major body postures such as those used by nurses when handling patients, and the other is QEC (Quick Exposure Check, Li and Buckle, 1998) for the rapid survey of the conditions in a plant. There is also a checklist for identifying factors in the workplace associated with upper limb disorders designed by Keyserling *et al.* (1993).

Recording motions and forces of hand, wrist and arm

Video recordings are valuable for investigating hand and wrist movements and angles, providing the line of view of the cameras is perpendicular to the motion. From video records the times of cycles and actions can also be identified. Whilst goniometers give very precise angular information they are not usually suitable for field use.

There are several notations for recording postures, some of which have been mentioned in Chapter 16. The notation of Armstrong *et al.* (1982) is useful for hand-arm recordings. It allows the arm posture and the force exertion to be recorded concurrently and, if needed, plotted on a time base which helps to give an understanding of the duration of task elements and rest pauses.

Forces exerted on equipment in pulling and pushing can be recorded using low-priced transducers but grip and squeezing forces are difficult since the transducers may alter the shape of the control surface itself. These forces have been measured under laboratory conditions and McGorry (2001) has demonstrated the possibility of instrumenting hand tools to record grip forces during their use. In field work it is often enough to find whether a change has increased or reduced the forces involved, rather than a numerical value for the force, and this can sometimes be found from the use of Borg's RPE technique, which is described in Chapter 15. An interesting application of this method was in a comparison of force exertions using a butcher's knife at different working heights. Subjects could exert very similar forces, on a single occasion, at various heights but when asked to judge the effort on Borg's scale the differences between the various directions and heights of force exertion were very clear.

Repetition in tasks

There is little guidance on what represents repetition in tasks. Silverstein *et al.* (1986) suggest that a 'high' repetition rate might be taken as one in which the cycle time is 30 seconds or less, or where repeating sub-cycles occupy more than 50% of the fundamental cycle. As Kilbom (1994) has noted, the shoulder may be less tolerant of repetitive movements than the elbow and wrist, which in turn are less tolerant than the fingers.

The degree of repetition within manual work has been assessed by amplitude probability distributions of EMG activity levels (Jonsson, 1988), where an increase in the lower frequencies is identified as an increase in muscle fatigue, and by observer ratings on a repetition scale characterising both speed of hand movement and amount of recovery time in the last cycle (Latko *et al.*, 1997).

From the physiological point of view repetitiveness has several components. Moore and Wells (1992) propose the following measures: the amount of tissue movement, the number of repetitions, the cycle time and an estimate of 'sameness.' For them, 'sameness' of muscle involvement is when a muscle is repeatedly used even though the action observed may appear to be different. Kivi (1984) points out that sub-cycles in a long cycle job may be repeated, providing a lack of variation for the musculature despite the fact that the job may appear variable.

To assess the 'sameness' of actions or of muscle involvement occurring in a work cycle Moore and Wells (1992) utilise the autocorrelation function on records which give continuous posture and time histories. This is calculated by taking one cycle of movement and cross-correlating it, step by step using constant time intervals, against the whole of the work record for that factor. From this they obtain the duration, or the frequency, of each movement and a measure of its 'sameness' from the value of the autocorrelation function.

Assessing manual handling tasks

To some extent manual handling tasks are a special case of musculoskeletal injuries; there is no clear dividing line between work causing cumulative damage from long repeated exposures and high force exertions which may risk overloads on body structures. In the case of back pain, a major concern for those involved in manual work, this may be manifest after a single exertion or arise as a result of long exposure to muscular effort and poor posture, and it is often difficult to determine which.

Since many jobs combine both situations, any assessment procedure must examine whether poor posture and/or excessive load are present. Reference particularly to Chapters 16 and 17 will alert the reader to what is involved. In general, biomechanical loading and muscle strength are factors to investigate, together with posture, where infrequent lifts of heavy objects are concerned. Where objects of a moderate weight are concerned but short periods of frequent lifting are present, psychophysical methods of determining acceptable load are useful. If such work is done over several hours, and fatigue arises, physiological methods can come into play.

As with most assessment methods, a preliminary study should find where the major loadings are occurring in the working day. Thus tasks which present the greatest risk are studied. However, the whole work day should be surveyed since, again, repetitions are a source of fatigue and injury even when the loads appear light. Also, in many manual handling jobs, loads vary greatly and the additional stresses which may occur at peak periods of activity should not be overlooked.

Identifying hazardous postures

The principal risk factors to be considered are:

- working height
- reach distance
- bending of the trunk, in any direction
- twisted, awkward or constrained postures

All of these are aggravated by any force exertion.

The OWAS procedure, described in Chapter 16, is useful in this context. It was designed for analysing heavy work in a steelworks but does draw attention to posture in a simple way, with or without force exertion.

Many checklists have been devised for posture and load evaluation, and these should cover the task, the worker, the workplace, the design of the job and the organisational framework within which the work is done. A summary of the points to consider when investigating workplaces is given in Figure 7, whilst Figure 8 gives guidance on routes to possible solutions to the identified risks.

An assessment checklist, slightly modified from that presented by the UK Health and Safety Executive (HSE, 1998), is given in the appendix. The modification deals with the organisational component, which can have serious impact on the hazard. The assessment identifies the areas of hazard and the consequent problems, which are marked off on the checklist and then combined into a list of concerns under each checklist heading, from which a list is created for specifying future actions. The HSE (1998) publication gives extensive guidance for the use of the procedure.

Any action list compiled from an assessment should be specific about what should be done, based on the severity of the hazard. It should be clear about which actions should be done immediately, which ought to be done in the near future and which, because they are of lesser severity, can be incorporated in long term changes. As with all action lists, delivery dates should be specified and follow-up procedures defined, with named persons taking responsibility.

Estimating biomechanical loads

Biomechanical analysis is discussed in Chapter 17, with methods for recording postures for analysis in Chapter 16. Whilst two-dimensional calculations can be useful for symmetrical

1. Avoid having to lift or transfer wherever possible.
2. The number of trunk flexion movements required to perform a task and the range of rotation and side bending should be kept to a minimum.
3. Any requirements for a load to be supported or force applied over a substantial part of the work cycle should be minimized.
4. Handling loads at heights below the knees or above the shoulder should be eliminated wherever possible.
5. Minimize congestion and confined work spaces to ensure that an adequate base can be used to perform the task.
6. Use grasps which are two-handed, with the load or force evenly distributed between both hands/arms. Adopt grasps with maximal body contact area and where the upper arms and elbows are by the side, elbow joints are in mid-range and wrists are straight.
7. When performing two-person lifts or transfers, wherever possible select a colleague of similar height. Ensure that your colleague is trained in the techniques you will use.
8. Avoid jolting or sudden movements, especially with a high load or force, or with trunk flexion or rotation.
9. Wherever possible, use the trunk and arm muscles to stabilize the load and posture, and use the leg muscles (especially the thigh muscles) to provide the force or movement required.
10. Do not attempt to lift or transfer a weight which *may* be beyond your capacity, or is likely to be unpredictable, without suitable mechanical or human assistance.
    This can arise in difficult workplace conditions (e.g. slippery floor) or, in health care, when handling an uncooperative human or animal patient. If in doubt, put the object/person in a safe position and fetch assistance.

Figure 7 Ten guidelines on manual handling.

Is the task necessary?
Can the workstation be relocated to reduce carrying?
Can the work height/reach distance be improved?
Can the load be reduced, for example by using smaller units or counterbalancing the weight?
Would a handling aid help?
Can the task be mechanised?
Should job rotation be introduced to allow periods for the musculoskeletal system to rest and recover?

Figure 8 Checklist suggesting potential solutions.

postures, particularly for comparison of various proposed solutions, for asymmetric postures three-dimensional analysis is needed. Few techniques available at present take full account of the dynamic nature of the work, and assessments of the inertial forces are particularly difficult. For general work, these analysis methods are useful for the comparative analysis of alternative layouts, for evaluating the effects of loads of moderate weight and for the assessment of the designs of manual handling aids. Because biomechanical analysis is time consuming its full power is probably best used for the major problems and areas of high risk.

Whilst muscular overuse injuries can arise at any skeletal joint or muscle, the low back is the focus of attention for most studies. It is also the only joint area where numerical maximum load criteria have been set, which have been used for load handling limits in many countries. The evidence for these criteria was reviewed in detail by Waters *et al.* (1993) during revision of the NIOSH guidelines, when they stated that a compressive load on the lumbar spine of 3.4 kN represented a risk of low back injury. They also noted that shear and torsional forces were

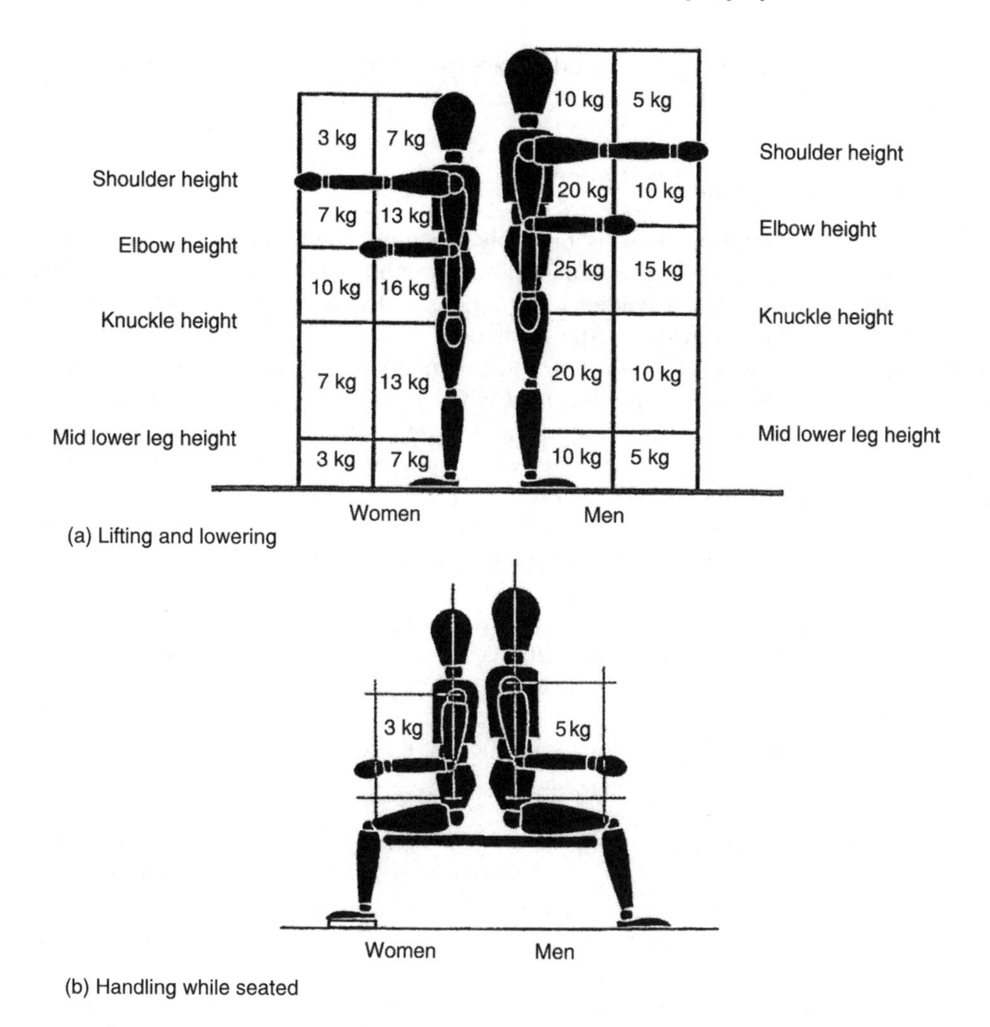

Figure 9 Guideline figures for identifying when manual lifting and lowering operations may not need a detailed assessment (after HSE (1998)). (a) lifting and lowering. (b) handling while seated.

important but there are no widely accepted methods, as yet, to incorporate these forces so that their contributions can be specified.

Interpretations of the NIOSH guidelines, supplemented by subsequent experience and studies, are used for indications of maximum acceptable loads, such as those stated by HSE (1998). The diagrams shown in Figure 9 are useful as indicators of when a hazard probably exists; they should not be taken as safe limits since particular circumstances may well modify these values, hence the factors given in the checklist in the appendix should be kept in mind when employing these diagrams.

Figure 9(a) gives guidelines for the maximum acceptable weight under ideal conditions for lifting and lowering. If the hands enter more than one box then the lower of the values in the boxes should be used. The values are for tasks performed at no more than 30 operations per hour; as a rough guide operations at up to twice a minute should have the values reduced by 30% and for greater frequencies, a reduction of 50% should be made. For twelve times a minute or more, then an 80% reduction is called for. If the person does not control the pace of work, if few rest pauses or changes of activity are possible, or if the load is supported for any length of time, then a full assessment is called for. These guidelines can be used where carrying is done provided that the load is carried no further than ten metres, has good handholds and is carried close to the body. For seated operations, if work is required outside the box shown in Figure 9(b), then a full assessment should be done.

The NIOSH (1991) equation

Whilst the NIOSH (1991) equation is probably the most comprehensive attempt so far to integrate the physiological, biomechanical and psychophysical aspects of load handling, combined with epidemiological evidence from musculoskeletal injury rates of low back pain, the long term collection of evidence still continues.

The NIOSH criterion is set in terms of a 'Recommended Weight Limit', which is expressed in a formula taking into account: the height at which the lift commences, the vertical travel in the lift, the reach distance and the frequency of lift. As these factors are likely to interact to increase the risk, the formula takes this into account; the equation is shown in Figure 10 and is discussed in detail in Waters *et al.* (1993, 1994).

As is evident from the form of the equation, it takes a baseline limit of 23 kg and modifies it according to the conditions under which the load is moved. It is a formula for calculating a weight limit for the movement of loads and should not be used outside this situation. Furthermore, the 23 kg basic load is that estimated from United States epidemiological data and may not apply to some other populations. The stated assumption of the results of the calculation is that 90% of healthy US workers should be able to operate with this load over the stated work period.

The limitations recognised for the application of this formula, according to Waters *et al.* (1993), are as follows:

1. It applies to lifting and lowering tasks but not to carrying or other manual handling activities, nor to those which require significant energy expenditure (since the physiological criteria were based only on the need to restrict expenditure to avoid fatigue); nor does it apply to the lifting of people, to shovelling, or to supporting handling aids such as barrows.
2. It only applies to standing tasks and not to tasks involving sitting or kneeling.
3. It does not take account of sudden or unpredicted conditions such as shifts in load distribution, an unexpectedly heavy or light load, or foot slip.
4. It is not designed to assess lifting single-handed, constrained workplaces, poor thermal environments or unusual loads such as contaminated objects.

Although the criteria are stated to be conservatively set, this does not eliminate the use of knowledge and judgement in relation to the individuals undertaking the tasks. In particular the effects of the organisational and environmental conditions on the workers as well as their state of health, age and similar factors should be borne in mind. There are also practical limitations to its use, since it is difficult to apply when a variety of load weights are being handled and when tasks vary within a job.

Lifting indices for identifying hazardous jobs

The ratio of the actual load to be lifted and the NIOSH (1991) equation's Recommended Weight Limit, calculated as shown in Figure 10, was developed as a Lifting Index, by Waters *et al.* (1993), to identify tasks which pose a significant risk of low back injury. With the Index greater than 1 it is likely that some fraction of the workforce will be at risk of such injury, whilst an Index of 3 or greater indicates a serious level of risk even for fit and experienced workers. However, there is not, as yet, sufficient knowledge to determine a quantitative relationship

Recommended Weight Limit (RWL) in kg =
$$23\text{ kg} \times HM \times VM \times DM \times AM \times FM \times CM$$
where

Horizontal multiplier HM is $\frac{25}{H}$
with H the horizontal distance (in cm) of hands from mid-point between the ankles. Measure at the origin and the destination of the lift.

Vertical multiplier VM is $1 - (0.003\,|V - 75|)$
with V the vertical distance (in cm) of the hands from the floor. Measure at the origin and destination of the lift.

Distance multiplier DM is $0.82 + \frac{4.5}{D}$
with D the vertical travel distance (in cm) between the origin and the destination of the lift.

Asymmetric multiplier AM is $1 - (0.0032\,A°)$
with A the angular displacement (in degrees) of the load from the sagittal plane. Measure at the origin and destination of the lift.

Frequency multiplier FM is input from Table A, based on the average frequency rate of lifting measured in lifts/min and duration(≤ 1 hour, ≤ 2 hours, ≤ 8 hours assuming appropriate recovery allowances).

Coupling multiplier CM is input from Table B, on the basis of the degree of coupling between hand and load.

Table A Frequency multiplier

Frequency lifts/min	≤ 1 hour		≤ 2 hours		≤ 8 hours	
	V < 75 cm	V ≥ 75 cm	V < 75 cm	V ≥ 75 cm	V < 75 cm	V ≥ 75 cm
0.2	1.00	1.00	0.95	0.95	0.85	0.85
0.5	0.97	0.97	0.92	0.92	0.81	0.81
1	0.94	0.94	0.88	0.88	0.75	0.75
2	0.91	0.91	0.84	0.84	0.65	0.65
3	0.88	0.88	0.79	0.79	0.55	0.55
4	0.84	0.84	0.72	0.72	0.45	0.45
5	0.80	0.80	0.60	0.60	0.35	0.35
6	0.75	0.75	0.50	0.50	0.27	0.27
7	0.70	0.70	0.42	0.42	0.22	0.22
8	0.60	0.60	0.35	0.35	0.18	0.18
9	0.52	0.52	0.30	0.30	0.00	0.15
10	0.45	0.45	0.26	0.26	0.00	0.13
11	0.41	0.41	0.00	0.23	0.00	0.00
12	0.37	0.37	0.00	0.21	0.00	0.00
13	0.00	0.34	0.00	0.00	0.00	0.00
14	0.00	0.31	0.00	0.00	0.00	0.00
15	0.00	0.28	0.00	0.00	0.00	0.00
>15	0.00	0.00	0.00	0.00	0.00	0.00

Table B Coupling multiplier (CM)

Couplings	Coupling multipliers V < 75 cm	Coupling multipliers V ≥ 75 cm
Good	1.00	1.00
Fair	0.95	1.00
Poor	0.90	0.90

Figure 10 NIOSH (1991) equation.

from which an estimate could be made of the level of risk of low-back disorders for a given Lifting Index.

Two other indices, similar in concept, are the Job Severity Index (Ayoub and Mital, 1989) and the Lifting Strength Rating (Chaffin, 1974). As with the former index, these aim to assess the job demand against the capacity of the workers under the job conditions. The Job

Severity Index uses the job demands, identified through task analysis, and the worker's capacity, measured either by psychophysical tests or on the basis of published data. The Lifting Strength Rating is calculated from the weights handled in the job and the predicted strength of a strong person in the postures observed for handling the weights.

Whilst both of these indices can be used to assess the lifting demands of a task in which the weight of load or the workplace factors vary, they do not quantify the degree of risk involved. They can be used, however, to compare the relative severity of different jobs for the purpose of evaluating or modifying the conditions.

Psychophysical measurements of maximum acceptable weight of load

Snook (1978) developed the psychophysical technique of asking operators themselves to judge what they consider to be a maximum acceptable load for a given work period, (not to be equated with a safe load). From a methodological point of view, the technique is based on the psychophysical Method of Adjustments, but with the test standard being the subject's own conception of maximum load. The subjects are asked to adjust the weight of the load, (or any other task variable such as frequency of lift), according to their perception of the strain and fatigue involved, the final workload being taken as the Maximum Acceptable Weight of Load (MAWL) under the working conditions.

The method has the advantage that the operator's judgement takes into account the whole job, integrating biomechanical and physiological factors. The disadvantage is that the subjects have to extrapolate their judgements from the short period of the trial to the whole of a regular working day. Some validation studies have been conducted (Legg and Myles, 1981, Ljungberg *et al.*, 1982, Mital, 1983, Karwowski and Yates, 1986), but the results are somewhat contradictory. Many subjects overestimate their capacity but some underestimate it. The tests should be made with experienced workers and an initial training period is important if results are to be reliable. The tests should be repeated and the results of at least two trials averaged to ensure the repeated results lie within 15% of each other.

Snook and Ciriello (1991) have collected a large database of psychophysical judgements for lifting tasks, as well as for a few other forceful activities, under a variety of workplace arrangements. This is useful for assessing load weights on the criterion of acceptability rather than of safety.

Conclusions

On the basis, stated at the beginning of this chapter, that a risk is the overlap between a hazard and a vulnerability, we have endeavoured to show how some work hazards may be assessed and how the possible vulnerabilities of those exposed to such hazards can expose them to risk. Methods for revealing hazardous conditions, and for quantifying them in some cases, have been presented and human limits beyond which people may become vulnerable have also been shown. This last area is one where we are still weak; the transfer of research knowledge into measures useable in practice without heavy qualification is growing, but slowly.

The methods of ergonomics, combined with experienced judgement, medically recorded evidence and worker insight still offer the best procedure for recognising the major points of mismatch between work, its environment and worker capacity, and in dealing with their consequences. We can recognise the factors associated with work-related injuries and have appropriate methods to deal with many of them. The setting of criteria is developing all the time, particularly under the pressure of litigation and Health and Safety legislation.

Where such pressures lead to a maintained improvement in working conditions, the record shows that workplace injuries fall and efficiency improves.

References

Åberg, U. (1981). Techniques in redesigning routine work. In *Stress, Work Design and Productivity*, edited by E.N. Corlett and J. Richardson. (Chichester: John Wiley & Sons), pp. 157–163.

Armstrong, T.J., Foulke, J.A., Joseph, B.S. and Goldstein, S.A. (1982). Investigation of cumulative trauma disorders in a poultry processing plant. *American Industrial Hygiene Association Journal*, **43**, 103–116.

Ayoub, M.M. and Mital, A. (1989). *Manual Materials Handling*. (London, Taylor & Francis), pp. 198–209.

Bigos, S.J., Battie, M.C., Spengler, D.M., Fisher, L.D., Fordyce, W.E., Hansson, T.H., Nachemson, A.L. and Wortley, M.D. (1991). A prospective study of work perceptions and psychosocial factors affecting the report of back injury. *Spine*, **16**, 1–6.

Birnbaum, R., Cockcroft, A. and Richardson, B. with Corlett, N. (1993). *Safer Handling of Loads at Work—A Practical Guide*. The Insititute for Occupational Ergonomics, University of Nottingham, UK.

Branton, P. (1970). A field study of repetitive manual work in relation to accidents at the workplace. *International Journal of Production Research*, **8**, 93–107.

Chaffin, D.B. (1974). Human strength capacity and low back pain. *Journal of Occupational Medicine*, **16**, 248–254.

Corlett, E.N. (1991). Ergonomics fieldwork: an action programme and some methods. In: *Towards Human Work*, edited by M. Kumashiro and E.D. Megaw. (London, Taylor & Francis), pp. 179–185.

Cox, T. and Mackay, C.J. (1985). The measurement of self-reported stress and arousal. *British Journal of Psychology*, **76**, 183–186.

Dickinson, C.K., Campion, K., Foster, A.F., Newman, S.J., O'Rourke, A.M.T. and Thomas, P.G. (1992). Questionnaire development: an examination of the Nordic Musculoskeletal Questionnaire. *Applied Ergonomics*, **23**, 197–201.

H. S. E. (1990). *Work Related Upper Limb Disorders — A Guide to Prevention* (London: HMSO).

H. S. E. (1998). *Manual Handling, Guidance on Regulations — Manual Handling Operations Regulations 1992. HSE Publication Guidance on Regulations L23*. (Sudbury: HSE Books).

Hignett, S. and McAtamney, L. (2000). Rapid entire body assessmant, (REBA). *Applied Ergonomics*, **31**, 201–205.

Hurrell, J.J. and McLaney, M.A. (1988). Exposure to job stress: a new psychometric instrument. *Scandinavian Journal of Work, Environment and Health*, **14**, 27–28.

Industrial Fatigue Research Committee of the Japan Associaton of Industrial Health (1970). The inventory for subjective symptoms of fatigue (revised 1970). *Digest of Science of Labour*, **25**, 12–33.

Jonsson, B. (1988). The static component in muscle work. *European Journal of Applied Physiology*, **57**, 305–310.

Karasek, R., Brisson, C., Kawakami, N., Houtman, I., Bongers, P. and Amick, B. (1998). The job content questionnaire (JCQ): an instrument for internationally comparative assessment of psychosocial job characteristics. *Journal of Occupational Health Psychology*, **3**, 322–355.

Karwowski, W. and Yates, J.W. (1986). Reliability of the psychophysical approach to manual lifting of liquids by females. *Ergonomics*, **29**, 237–248.

Kemmlert, K. and Kilbom, Å. (1986). *Method of Identification of Musculo-skeletal Stress Factors which may have Injurious Effects*. National Board of Occupational Safety and Health, Solna, Sweden.

Keyserling, W.M., Stetson, D.S., Silverstein, B.A. and Brewer, M.L. (1993). A checklist for evaluating ergonomic risk factors associated with upper extremity cumulative trauma disorders. *Ergonomics*, **36**, 807–831.

Kilbom, Å. (1994). Repetitive work of the upper extremity: Part 1. Guidelines for the practitioner. *International Journal of Industrial Ergonomics*, **14**, 1–2, 51–57.

Kishida, K. (1991). Workload of workers in supermarkets. In: *Towards Human Work*, edited by M. Kumashiro and E.D. Megaw (London: Taylor & Francis), pp. 269–279.

Kivi, P. (1984). Rheumatic disorders of the upper limbs associated with repetitive tasks in Finland in 1975–1979. *Scandinavian Journal of Rheumatology*, **13**, 101–107.

Kjellstrand, L. (1974). Quality of life at the workplace. In: *The Quality of Life at the Workplace, Proceedings of the Regional Trade Union Seminar*, OECD Paris, pp. 33–42.

Latko, W.A., Armstrong, T.J., Foulke, J.A., Herrin, G.D., Rabourn, R.A. and Ulin, S.S. (1997). Development and evaluation of an observational method for assessing repetition in hand tasks. *American Industrial Hygiene Association Journal*, **58**, 278–285.

Legg, S.J. and Myles, W.S. (1981). Maximum acceptable repetitive lifting workloads for an 8 hour day using psychophysical and subjective rating methods. *Ergonomics*, **24**, 907–916.

Li, G. and Buckle, P. (1998). A practical method for the assessment of work related musculoskeletal risks — Quick Exposure Check (QEC). In: *Proceedings of the Human Factors and Ergonomics Society 42nd Annual Meeting*, Chicago, Volume 2. (Santa Monica), Human Factors and Ergonomics Society: pp. 1351–1355.

Ljungberg, A.S., Gamberale, F. and Kilbom, Å. (1982). Horizontal lifting — physiological and psychological responses. *Ergonomics*, **25**, 741–757.

McAtamney, L. and Corlett, E.N. (1993). RULA: a survey method for the investigation of work-related upper limb disorders. *Applied Ergonomics*, **24**, 91–99.

McGorry, R.W. (2001). A system for the measurement of grip forces and applied moments during hand tool use. *Applied Ergonomics*, **32**, 271–279.

Mital, A. (1983). The psychophysical approach in manual lifting — a verification study. *Human Factors*, **25**, 485–491.

Moore, A.E. and Wells, R. (1992). Towards a definition of repetitiveness in manual tasks. In: *Computer Applications in Ergonomics, Occupational Safety and Health*, edited by M. Mattila and W. Karwowski. (Amsterdam: North-Holland), pp. 401–408.

Noro, K. and Imada, A. (Eds) (1991). *Participative Methods*. (Chichester, Wiley).

Sen, R.N. and Yeow, P.H.P. (2003) Ergonomic weighted scores to evaluate critical instructions for improvements in a printed circuit assembly factory. *Human Factors in Ergonomics and Manufacturing*, **13**, 1–17.

SIRIM (1994) MS ISO 9002:1994 *Quality Systems — Model for Quality Assurance in Production, Installation and Servicing*. Standards and Industrial Research Institute Malaysia (SIRIM), Malaysia.

Silverstein, B.A., Fine, L.J. and Armstrong, T.J. (1986). Hand, wrist and cumulative trauma disorders in industry. *British Journal of Industrial Medicine*, **43**, 779–784.

Snook, S.H. (1978). The design of manual handling tasks. *Ergonomics*, **21**, 963–985.

Snook, S.H. and Ciriello, V.M. (1991). The design of manual handling tasks: revised tables of maximum acceptable weights and forces. *Ergonomics*, **34**, 1197–1213.

Waters, T.R., Putz-Anderson, V., Garg, A. and Fine, L.J. (1993). Revised NIOSH equation for the design an evaluation of manual lifting tasks. *Ergonomics*, **36**, 749–776.

Waters, T.R., Putz-Anderson, V. and Garg, A. (1994). *Applications Manual for the Revised NIOSH Lifting Equation*. DHHS (NIOSH), Cincinnati, Ohio.

Wilson, J.R. (1991). Design decision groups: a participative process for developing workplaces. In: *Participative Methods*, edited by K. Noro and A. Imada. (Chichester: Wiley), 81–96.

Wilson, J.R. (1994). Devolving ergonomics: the key to ergonomics management programmes. *Ergonomics*, **37**, 579–594.

Yeow, P.H.P. (2002). Ergonomic improvements in the printed circuit assembly factories of the multimedia industries in Malaysia. PhD thesis. Multimedia University, Malaysia.

Appendix

An assessment procedure for the manual handling of loads, adapted from HSE (1998)

When these questions are considered they should be answered, where possible, using a scale of 'low', 'medium', or 'high'. The problems arising from the work should be noted against each item, the notes giving information for possible remedial action. The actions decided should then be agreed and implemented as discussed in this chapter.

The task(s)

Do they involve:

holding loads away from the trunk
twisting
stooping
reaching upwards
large vertical movements
long carrying distances
strenuous pulling or pushing
unpredictable movement of load
repetitive handling
insufficient rest or recovery
a work rate imposed by the process
special equipment which reduces the ease of access to the work

The loads

Are they:

heavy
bulky and/or unwieldy
difficult to grasp
unstable and/or unpredictable
intrinsically harmful, e.g., sharp or hot

The working environment

Are there:

constraints on posture
restrictions on foot space
poor floors
variations in levels
hot/cold/humid conditions
strong air movements
poor lighting conditions

Individual capability

Does the job:

require unusual capability/body size
hazard those with a health problem
hazard those who are pregnant
call for special information
require special equipment

The organisation

Does it provide:

appropriate training
appropriate cooperation between workers
effective worker/management cooperation and support
adequate time for the tasks or is overtime required/are there rush periods
opportunity to discuss/arrange job changes

Chapter 32

Human reliability assessment

Barry Kirwan

Introduction

Accidents such as at Three Mile Island, Chernobyl and Bhopal demonstrated unequivocally the importance of considering human error in high risk systems (e.g., see USNRC, 1980; USSR State Committee, 1986; Bellamy, 1986). Such accidents led to the development and implementation of the approach of Human Reliability Assessment (HRA). However, accidents attributed to 'human error' continue to occur, in industries ranging from nuclear power, chemical and petrochemical, to transport industries, and medical sectors, and 'human error' even occasionally contributes to financial disasters in the stock markets. This suggests that the approach needs further implementation and dissemination to other areas of human work, as well as further enhancements to the methods used themselves for assessing and reducing human error. This chapter seeks to provide an overview of the area of HRA: what it does, how it does it, and how it can help to make work situations safer and less prone to human error.

HRA is a hybrid area, arising out of the disciplines of engineering and reliability on the one hand, and psychology and ergonomics on the other. The former disciplines require human error probabilities to fit neatly into the logical mathematical framework of *probabilistic safety analysis* (PSA). The latter urge more detailed and theoretically valid modelling of the complexity of the human operator and human performance (Wagenaar, 1986), and the need to consider human performance in its proper context (Dougherty, 1993; Hollnagel, 1993). PSA is the quantitative statement defining the expected frequencies of accidents, and hence it determines whether or not a plant's risk compares favourably or otherwise against pre-defined risk criteria (Green, 1983; Cox and Tait, 1991; Kirwan, 1994). Thus HRA must be incorporated into PSA if risk is to be properly estimated. However, as discussed later, HRA can be carried out independently of PSA, in order to learn where the human error vulnerabilities are, to prioritise them, and to reduce them. Furthermore, although HRA is used traditionally to consider and manage risks, it can be equally applied to improve productivity and/or quality. It is basically a methodology for systematically examining (and improving) human performance from an error perspective.

A generic approach or framework for HRA is presented (see Kirwan, 1994). Although this framework has 10 steps within it (as will be described), there is a core of three goals, namely:

1. *Human error identification*. What can go wrong?
2. *Human error quantification*. How often will a human error occur?
3. *Human error reduction*. How can human error be prevented from occurring impact on the system reduced?

Most research in the first major decade of HRA (the 1980s) focused on (2), the development of human reliability quantification techniques. From the 1990s however, there has been a gradual shift towards item (1), which is logically at least (if not more) important, since unless all significant errors are identified a HRA will underestimate the impact of human error. Whilst a good deal of research has been carried out in the field of human error identification and particularly error classification, and many techniques have appeared over the recent years, in practice there remain only a few that have proven to be of regular use to practitioners. These will be the ones focused on in this chapter, although others will be referred to briefly for interest.

Human error reduction (item 3 above) has recently become more important since there is an obvious need for this capability once HRA is being applied in earnest, as some unacceptably probable human errors are bound to be identified, which must in some way be reduced in expected frequency. This is where the role of the ergonomist (or human factors practitioner) in HRA is clearest and most useful, since the ergonomist can usually specify a number of ways to improve the reliability of operators' performance. The ergonomist can of course do this without HRA, but HRA, in the context of PSA, enables the determination and prioritisation of the critical human involvements from a systems viewpoint. These are the sometimes complex and subtle human error and hardware failure combinations that other ergonomics approaches or engineering reliability analysis methods alone would not normally find.

This chapter concentrates mainly on human error identification and quantification, as these have been most heavily researched in HRA, and other chapters in this volume deal with methods such as task analysis and error reduction (or performance improvement). The chapter does not consider some of the more complex mathematical issues underlying the integration of HRA into PSA (see, for example, Apostolakis *et al.*, 1987; or Park, 1987). These would require a chapter in themselves, and are not required for an appreciation of how HRA works and is applied. Furthermore, as quickly becomes apparent to the practitioner, human reliability analysis is far from being a precise science (and nor is probabilistic safety assessment itself; Nicks, 1981*)*, but is a useful means of identifying and prioritising plant safety vulnerabilities to human error, and thereby reducing the frequency of accidents. For more general references on HRA approaches, see Swain and Guttmann (1983), Dhillon (1986), Dougherty and Fragola (1988), Humphreys (1988), Swain, 1989, Kirwan (1994), Embrey *et al.* (1994), and Gertman and Blackman (1994). For more general references on human error see Rasmussen *et al.* (1987), Reason (1990; 1998), Senders and Moray, (1991), Hollnagel (1993), and Woods *et al.* (1994).

The following section details the generic HRA methodology. Following this, there is a brief discussion of future areas of investigation and development needs within the field of HRA.

Human reliability assessment: a generic methodology

Once it is decided that a human error or human reliability problem requires analysis, a means of systematically solving this problem is required. The 'problem' may range from possible errors during a nuclear power plant emergency, to a desire for improved surgical performance (patient survival and recovery rate) in a heart operation, or avoidance of large-consequence errors in stock market operations. Whatever the objective of the assessment, a systematic methodology of HRA will help ensure that the problem itself is dealt with reliably, minimising biases or errors distorting the analysis.

Each of the 10 steps of the methodology defined in Figure 1, from initial problem definition to final documentation of the results, is briefly explained and further references are given on available guidelines and current research in these areas.

1. Problem definition. To define precisely the problem and its setting in terms of the system goals and the overall forms of human-caused deviations from those goals.

2. Task analysis. To define explicitly the data, equipment, behaviour, plans, and interfaces used by the operators to achieve system objectives, and to identify factors affecting human performance within these tasks.

3. Human error analysis. To identify all significant human errors affecting performance of the system, and ways in which human errors can be recovered.

4. Representation. To model the human errors and recovery paths in a logical manner such that their impact on the system can be quantitatively determined. This usually necessitates integrating human errors with hardware failures in a fault or event tree.

5. Screening. To define the level of detail and effort with which the quantification will be conducted, by defining all significant human errors and interactions, and ruling out insignificant errors which can be effectively ignored by the study.

6. Quantification. To quantify human error probabilities and human error recovery probabilities, in order to define the likelihood of success in achieving the system goals.

7. Impact assessment. To determine the significance of human reliability with respect to the achievement of the system goals, to decide whether improvements in human reliability are required, and if so what are the primary errors and factors negatively affecting system reliability.

8. Error reduction. To identify error reduction mechanisms, means of supporting error recovery likelihood, and ways of improving human performance in achieving system goals, so that an acceptable level of system performance can be achieved.

9. Quality assurance. To ensure that the enhanced system satisfactorily meets system performance criteria, and will continue to do so in the future.

10. Documentation. To detail all information necessary to allow the assessment to be understandable, audible, and reproducible.

Figure 1 The relationship between various steps.

Problem definition

There are two ways of defining the problem, dependent upon whether it is being considered as a problem in its own right, or as an integral part of a larger risk assessment. When a human reliability 'problem' has been identified (e.g., a desire to improve safety, or to assess risk or productivity), and defined in its system context (e.g., to assess the risks of offshore platform evacuation by lifeboat in severe weather), discussions should occur with system design and plant engineers, and if possible with operational and managerial personnel. Discussions around the problem will help define more precisely the scope of the project, and the range of conditions and scenarios that must be considered in order to resolve the problem satisfactorily, and/or fulfil the goals of a general safety assessment of a plant.

It is useful at this stage to define the system goals, at various levels, for which operator actions are required. This will in turn define higher level goals towards which the operators are aiming (e.g., maintain reactor core cooling; achieve highest output). It is also important to gain some understanding of these high level goals, and to determine in particular where and how the goal of safety fits in. Ideally there will be clear criteria via which the operators know when to 'drop' their production goals in favour of safety goals. If such criteria do not exist, then production goals may compromise safety goals and it is possible that the roots of a major human reliability problem are already inherent in the system. Therefore, with existing systems it is well worth investigating this 'safety culture' aspect of the plant, as it can influence human reliability predictions dramatically, and is an important aspect of the problem definition. If there are no clear safety criteria, it must not be assumed that operators will necessarily make the right decision in the heat of the moment.

If the HRA is being carried out as part of an overall risk assessment, the human reliability analyst will probably be given a set of scenarios that have been chosen for the risk analysis, and asked to consider human contributions to risk within these scenarios. This set of scenarios is sometimes known as a '[system] fault schedule'. In this case there are five types of human-system interaction which the analyst should consider with respect to an incident scenario (Spurgin *et al.*, 1987):

1. maintenance/testing errors affecting safety system availability (latent errors),
2. operator errors initiating the incident,
3. recovery actions by which operators can terminate the incident,
4. errors of commission (e.g., misdiagnosis) by which operators can prolong or even aggravate the incident, and
5. emergency repair actions by which operators can restore initially unavailable equipment and systems.

Consideration of these types of interaction, and discussions with the system risk analysts at the problem definition stage will enhance the smooth integration of the human reliability analysis into the system risk analysis. It may also identify new important scenarios that the system analysts had not initially considered. This is important because the problem may otherwise be defined in too limited a scope. As an example, Kirwan's (1987) assessment of an emergency offshore depressurisation system originally was only intended to look at operational failures during emergency scenarios and not maintenance aspects. However, when investigated further, a highly significant maintenance error was identified which had potentially dramatic effects on the whole platform. This error was probably of more significance than the. entire set of operational failures put together, and so was brought into the scope of the risk assessment.

At the end of the problem definition stage the problem to be addressed should be explicitly defined in its system context. A list of scenarios to be addressed and, within each scenario, a list of overall tasks required to achieve system and safety goals should also have been identified. This sets the scene for the task analysis phase. Figure 2 shows a brief example of the results of the problem definition phase.

Task analysis

The object of task analysis is to provide a complete and comprehensive description of the tasks that have to be performed by the operator(s) to achieve the system goals. There are many forms of task analysis (Drury, 1983; Kirwan and Ainsworth, 1992). Sequential task analysis looks at operator actions as they occur in chronological order. Hierarchical task analysis considers tasks in terms of the hierarchy of goals the operator is trying to achieve (Shepherd, 1986, 2001). Tabular task analysis (Pew *et al.*, 1987; Kirwan and Ainsworth, 1992) concentrates on performance as a function of the information available, and operators' knowledge, expectations and beliefs about the situation (see also Kirwan, 1994). Other forms of task analysis are discussed within an overview of its approach and techniques in Chapter 6.

If a detailed HRA is being carried out, then a task analysis is essential since it provides a detailed description of the operators' tasks from which it will be possible to identify errors in the next phase. The basic methods of deriving information for the task analysis are: observation; structured and unstructured interviews with operators, maintenance personnel, supervisors, managers and system designers; analysis of procedures; incident analyses;

Problem:	To effect emergency shutdown (ESD) of a chemical plant during a loss of power scenario.
Problem Setting:	A computer-controlled, operator-supervised plant suffers a sudden loss of main power. The VDU display system will also fail and so the operator, backed up by the supervisor, must initiate ESD manually using hardwired controls in the Central Control Room. However, due to valve failures on plant, these actions are only partially successful, and so the operator must send out another operator onto plant to determine which ESD valves have not closed. The CCR operator, via engineering drawings, can then determine which manual valves must be closed on plant. The outside operator must then go to close these valves, completing this action successfully within 2 hours from the onset of the scenario.
System Goals:	The overall system goals are safe shutdown of all feeds to the plant within 2 hours of loss of power. In this scenario there are no production goals once the event occurs since safety is clearly under threat (there is an effective prioritised visual and auditory alarm system available to several operators and supervisors). Prior to the event, the operator is concerned with achieving steady feed throughout via monitoring the top two levels of a VDU display hierarchy, and notifying the supervisor of any alarms higher than level 2. The outside operator (on plant) will have various duties associated with maintenance tasks.
Overall Human Error Considerations:	No operator initiating events were identified, and maintenance errors were not relevant except that identifying and moving the local manual valves could prove difficult. Recovery actions involve identifying the appropriate valves to close and closing them. Errors of failing to realise that ESD has not been 100% effective, and of mis-identifying the valves, appear most important from a risk perspective. Loss of power is so evident that misdiagnosis or failure to diagnose this event is not considered to be credible. A possible error of commission could be that a key emergency valve was locked into the wrong position (regular site inspections should prevent this).

Figure 2 Example of problem definition.

structured walkthroughs of procedures (where the operator talks the analyst through the procedure); and examination of system documentation such as engineering/process flow diagrams. In practice it is important not to rely on procedures/operating instructions as the sole source for defining the task, since actual operating practices often differ somewhat from the formal written documents.

For a proceduralised task in which the operator (or supervisor, or maintenance person) is using familiar skills or written/remembered rules, a hierarchical task analysis is probably most appropriate. An example is shown in Figure 3, part of a task analysis for the task of starting up a plant, and considering the sub-task 'warm up furnace'. Sequential information is represented in this form of hierarchical task analysis, via the numbers on the boxes at each level which determine the order in which the tasks should occur. Plan 4 in the diagram shows a more complex plan the operator must use while monitoring temperature and pressure (adopted from Shepherd, 1986).

Figure 4 illustrates another form of task analysis using a tabular format showing the type of information that can be recorded by the analyst during the task analysis. This type of task analysis may be more useful when the operator is in either a highly dynamic situation and/or one in which problem solving or diagnostic behaviour is required. The format focuses on the events as they occur in time, since the order in which events occur partly determines behaviour. However, the analyst must also be aware of the hierarchy of goals the operators are trying to achieve, which may shift during an emergency situation. In detailed analyses it may be necessary to utilise both hierarchical and tabular task analysis formats. Wherever possible the task analysis should be verified, if only by asking operations personnel to review it. Having defined the operators' tasks in detail, it is then appropriate to determine what can

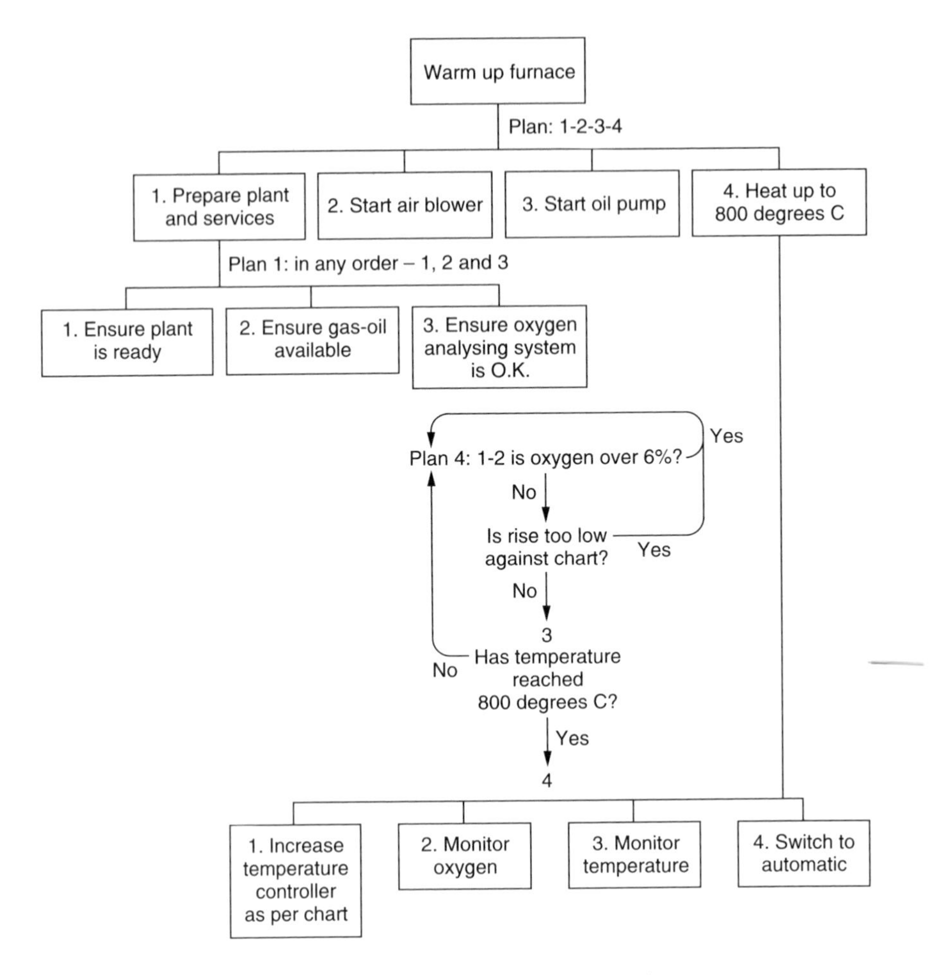

Figure 3 Example of hierarchical task analysis after Shepherd (1986).

go wrong, in terms of what human errors can occur. Examples of task analyses carried out as part of real HRAs are given in Kirwan and Ainsworth (1992).

Human error analysis

Human error analysis is arguably the most critical part of a human reliability analysis, since if a significant error is omitted at this stage then it will not appear subsequently in the analysis and hence the results may seriously underestimate the effects of human error on the system.

The most basic, but still useful approach is to consider the following possible 'external error modes' (Swain and Guttman, 1983) at each step in the procedure defined in the task analysis:

Error of omission:	act omitted (not carried out)
Error of commission:	act carried out inadequately
	act carried out in wrong sequence
	act carried out too early/too late
	error of quality (too little/too much/wrong direction)
Extraneous error:	wrong (unrequired) act performed

T	Opr	System Status	Info Available	Operator Expectation	Procedure (written, memorised)	Decision/ Communications Act	Equipment/ location	Feedback	Secondary duties; Distractions; Penalties	Comments
0:30 mins	SS 00 CRO	ESD; Cell 23 full of gas; mixture beyond explosion point at present; platform at muster status.	Gas cloud; Loud roaring noise.	Looking for leak in pipe union, flange, seal etc. Check near the gas detectors which were alarming.	1) Locate source and isolate if possible. 2) Maintain personal safety (use breathing apparatus) 3) Prevent ignition of gas.	Ops search for leak in compressor module using sound and visual cues, as well as the gas detectors. Communicate to CCR.	Cell 23 gas detectors.	Noise and smell of gas tactile cue if flesh exposed to gas jet path; if cold enough may see white gas plume from leak.	Maintain personal safety and avoid causing ignition. Extra delays if depressurise.	Search will be more difficult in breathing apparatus. Deluge would make search safer, though less likely to succeed.
0:40	OFM CCRO SS	As at 0:30	Panel indications and from operators now outside of cell 23.	Gas leak now confirmed. Ignition possible.	Minimise chances of ignition.	CCR operator reviews vessel pressures on VDU system: gives OIM status report and confirms significant gas leak occurrence. CCR opts to consider whether to depressurise or not. Main concern is with the liquid in the vessels which can only be removed if gas pressure remains in the compressors. However, no remote controls exist and it is considered too hazardous to try the operation. Decide therefore to muster personnel at other end of the platform until deluge cools down compressors and pressure drops.			Many enquiries may block communica-tion channels, and must be responded to by CCR operators.	Consider merits of removing compressor liquid – How long would the vessels withstand fire if ignited? Results of hazard analyses should be immediately transmitted to offshore personnel from onshore emergency centre to update their knowledge and enhance decision-making.
0:42	CCRO	As at 0:30	Deluge available (from light on console).	Ignition still likely.		Tells outside operators to clear cell 23. Muster points broadcast on public address system. Deluge activated.				

Figure 4 Example of tabular task analysis.

This approach is rudimentary but nevertheless can identify a high proportion of the potential human errors that can occur, as long as the assessor has a good knowledge of the task and a good task description of the operator system interactions.

Another method for human error analysis is embedded within the Systematic Human Error Reduction and Prediction Approach (SHERPA: see Embrey, 1986). This human error analysis method consists of a computerised question-answer routine that identifies likely errors for each step in the task analysis. The error modes identified are based on the 'skill rule and knowledge' model (Rasmussen *et al.*, 1981), and Generic Error Modelling System (GEMS: Reason, 1990). Table 1 shows the psychological error mechanisms underlying the SHERPA system. An example of the tabular output from such an analysis is shown in Figure 5 (Kirwan and Rea, 1986). This 'human error analysis table' has similarities to certain reliability engineering approaches to identifying the failure modes of hardware components. One particularly useful aspect of this approach is the determination of whether errors can be recovered immediately, at a later stage in the task, or not at all, information useful if error reduction is required later in the analysis. This particular tabular approach also attempts to link error reduction measures to the causes of the human error, on the grounds that treating

Table 1 Psychological Error Mechanisms in SHERPA

1. *Failure to consider special circumstances.* A task is similar to other tasks but special circumstances prevail which are ignored, and the task is carried out inappropriately
2. *Short cut invoked.* A wrong intention is formed based on familiar cues which activate a short cut or inappropriate rule
3. *Stereotype takeover.* Owing to a strong habit, actions are diverted along some familiar but unintended pathway
4. *Need for information not prompted.* Failure of external or internal cues to prompt need to search for information
5. *Misinterpretation.* Response is based on wrong apprehension of information such as misreading of text or an instrument, or misunderstanding of a verbal message
6. *Assumption.* Response is inappropriately based on information supplied by the operator (by recall, guesses, etc.) which does not correspond with information available from outside
7. *Forget isolated act.* Operator forgets to perform an isolated item, act or function, i.e., an act or function which is not cued by the functional context, or which does not have an immediate effect upon the task sequence. Alternatively it may be an item which is not an integrated part of a memorised structure
8. *Mistake among alternatives.* A wrong intention causes the wrong object to be selected and acted on, or the object presents alternative modes of operation and the wrong one is chosen
9. *Place losing error.* The current position in the action sequence is misidentified as being later than the actual position
10. *Other slip of memory* (as can be identified by the analyst)
11. *Motor variability.* Lack of manual precision, too big/small force applied, inappropriate timing (including deviations from 'good craftsmanship')
12. *Topographic or spatial orientation inadequate.* In spite of the operator's correct intention and correct recall of identification marks, tagging, etc., he unwittingly performs a task/act in the wrong place or on the wrong object. This occurs because of following an immediate sense of locality where this is not applicable or not updated, perhaps due to surviving imprints of old habits, etc.

the 'root causes' of the errors will probably be the most effective way to reduce error frequency.

A recent addition to the Human Error Identification field has been the TRACEr technique (Technique for the Retrospective Analysis of Cognitive Errors: Shorrock and Kirwan, 1998, 2002). This technique attempted to extend approaches such as SHERPA, and draw from other techniques as well, to be used in the Air Traffic Control (ATC) sector of industry.

A significant aspect was that the approach was intended to be 'bi-directional', i.e., useful for analysing events that have happened, as well as for predictive analyses. Such a technique would then stay 'current', as the users would be aware of the real events and errors in the industry at the same time as predicting errors for future designs, etc. This allows a form of organisational learning to take place inside a company or industrial application area.

TRACEr uses a number of taxonomies for classifying the error and recording the contextual factors that contributed to the error (and its recovery if applicable). As well as having a set of external error modes, TRACEr also has internal error modes, psychological error modes, and performance shaping factors, as well as classifications of equipment, and the air traffic control tasks being carried out, etc. The overall structure of TRACEr is shown in Figure 6. As with the original SHERPA tool, TRACEr utilises a flowchart structure to enhance inter-user reliability, and an example is given in Figure 7.

TRACEr has been developed after extensive analysis of errors in air traffic control, and has had a number of applications, both retrospective and predictive, as well as tests of internal and external validity (e.g., see Shorrock and Kirwan, 1998; Evans *et al.*, 1999; Shorrock and Kirwan, 2002). TRACEr represents a significant 'upgrade' of some of the 'older' techniques in

Task 51: Terminate supply and isolate tanker. (Sequence of remotely-operated valve operations)

Task step	Error type	Recovery step	Psychological mechanism	Causes, consequences and comments	Recommendations		
					Procedures	Training	Equipment
51.1	Action too late	No recovery	Place losing error	Overfill of tanker resulting in dangerous circumstance	Operator estimates time/records amount loaded	Explain consequences of overfilling	Fit alarm-timing/ volume/tanker level
51.2.1	Action omitted	5.2.4	Slip of memory	Feedback when attempting to close closed valve. Otherwise alarm when liquid vented to vent line			Mimic of valve configuration
51.2.2	Action too early	5.2.2	Place losing error	Alarm when liquid drains to vent lines	Specify time for actions	Operator to count to determine time	
	Action omitted	5.2.2	Slip of memory	As above and possible over pressure of tanker (see step 51.2.3)			Mimic of valve configuration
51.2.3	Action too early	No recovery	Place losing error	If valve closed before tanker supply valve overpressure of tanker will occur		Stress importance of sequence and explain consequences	Interlock on tanker vent valve
	Action omitted	5.2.6	Slip of memory	Automatic closure on loss of instrument air			Mimic of valve configuration
51.2.4	Action omitted	5.2.2	Slip of memory	Audio feedback when vent line opened		Explain meaning of audio feedback	Mimic of valve configuration
51.3	Action omitted	No recovery	Slip of memory	Latent error	Add check on final valve positions before proceeding to next step		Mimic of valve configuration

Figure 5 Example extract of human error analysis (Kirwan and Rea, 1986).

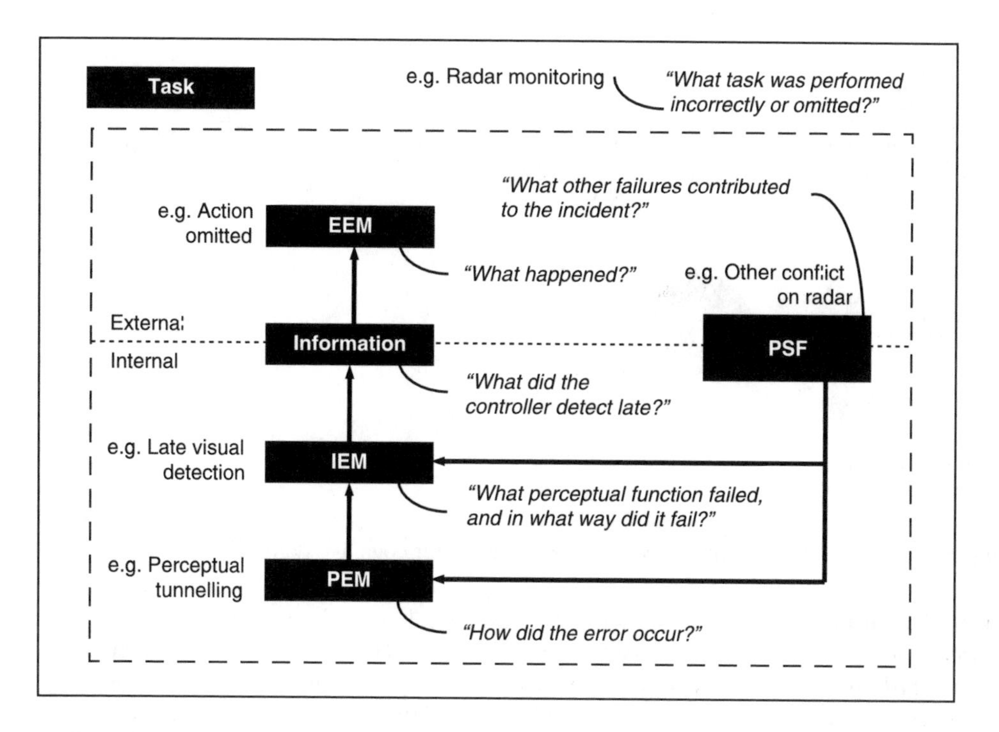

Figure 6 Relationship between the TRACEr classification systems.

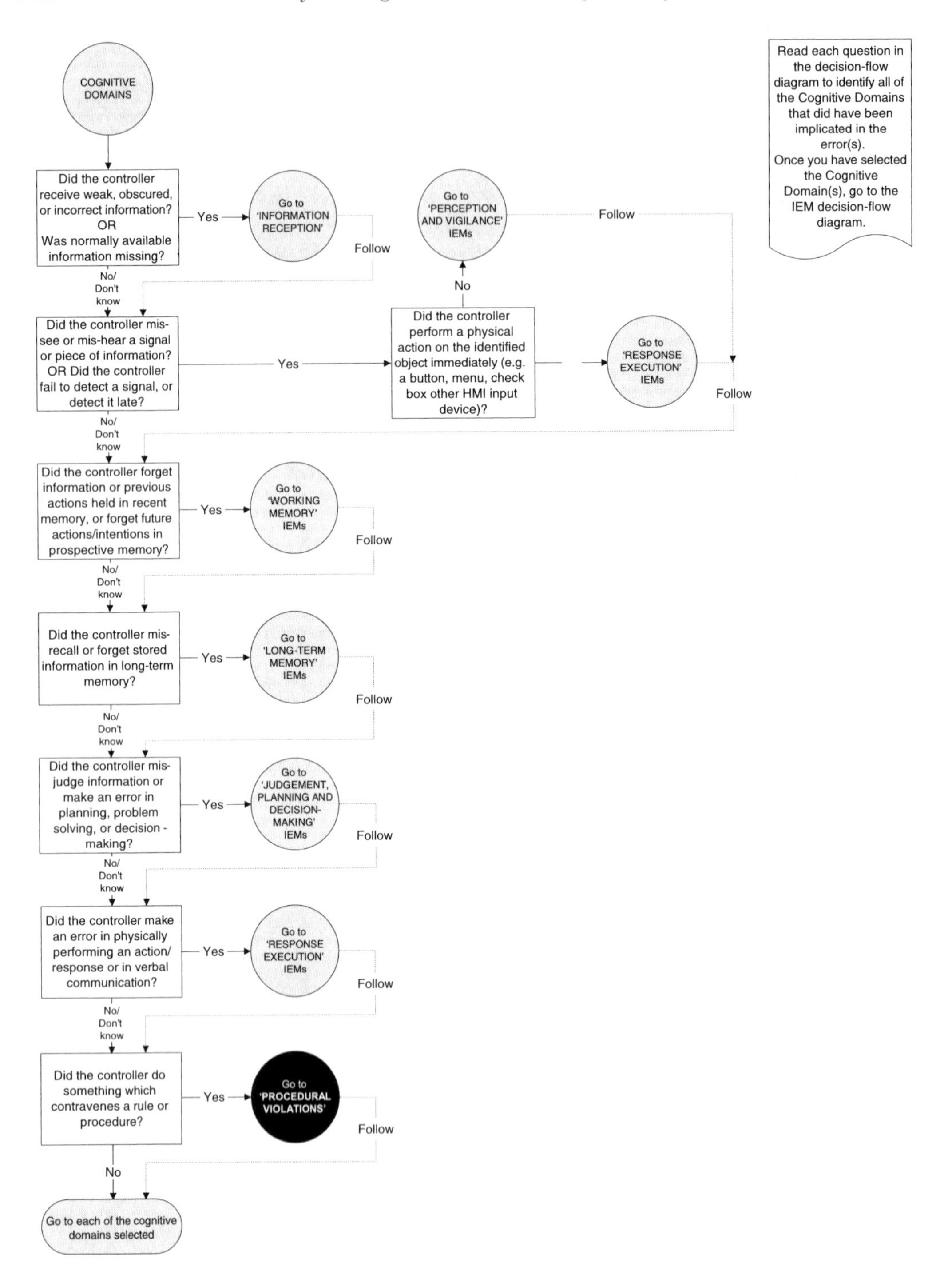

Figure 7 Extract from TRACEr decision-flow diagram for cognitive domains.

error identification. It is also influencing error and incident analysis more generally in air traffic management, since it was the basis for the European HERA project (Human Error in ATM (Isaac *et al.*, 2002) — a retrospective human error analysis technique). HERA is being applied to analyse and learn from errors in several European Air traffic organisations, and has recently been merged with a US system to generate a common US-European approach called JANUS

(Isaac and Pounds, 2001). Such developments are useful for HRA since when the HRA 'community' uses common methods, at least in the same industry, then more can be learned to improve risk and safety. Methods such as TRACEr, being used in UK ATC, have allowed for the first time a direct and explicit comparison between problems found with existing systems, and considerations of how future systems can resolve such problems without creating new ones. This lends a 'coherence' to the area of human error analysis and management, which should benefit safety, and help 'build' safety into the design of future systems. In a more general sense, such an approach, where implemented, will enable practical organisational learning to take place.

Methods such as SHERPA and TRACEr are 'single-assessor' methods, i.e., only one assessor is required to apply the technique. Another approach when trying to identify errors that have not yet happened, is to harness the expertise of several experienced operational people. This follows the adage 'several heads are better than one'. Another useful technique for error identification, therefore, borrowed from the field of Reliability Engineering, and adapted for use in HRA, is the HAZOP technique (Hazard and Operability Study) technique (see Kletz, 1984; Kirwan and Ainsworth, *op. cit*). The Hazard and Operability (HAZOP) study approach was widely used in the process industries and recently been extended to address other types of system (e.g., programmable electronic systems, safety management systems etc.). Variants on the HAZOP approach are described in detail in Kennedy and Kirwan (1998). At a basic level, HAZOP is an 'inquiry' method, essentially asking 'what-if?' questions. It requires the following:

- An experienced chair-person to lead the HAZOP exercise.
- A representation of the system to be 'HAZOP-ed' (e.g., a specification drawing or description or a task analysis) and supporting information (e.g., pictures).
- A hybrid team of HAZOP personnel (e.g., designers, controllers, a HF practitioner) who collectively have knowledge of the intended system and how controllers work.
- A set of guidewords to structure the inquiry process (see Table 2).
- A recording format.
- A secretary (usually one of the active participants) to record the process and actions arising.

The purpose of HAZOP is to identify deviations away from the intended functioning of the system. Therefore, for instance, consider the task of an air traffic controller using a new computer-supported display system. The controller, instead of having to contact the pilot via radio-telephone, can select a new flight level for the aircraft and send it digitally to the aircraft cockpit using data-link technology currently being developed. In such a case, the controller

Table 2 HAZOP Guide Words and Definitions

Guide word	Definition
No	No part of the intention was achieved (omission of an action/part of a task step)
More	Too much effect resulted from the action compared to the intention
Less	Too little effect resulted from the action compared to the intention
As Well As	The intention was successfully achieved but other effects also resulted
Part Of	Only part of the intention was achieved by the action
Reverse	The action resulted in the opposite effect to the intention
Other Than	The original intention was completely substituted by another action or intention
Early	The action was conducted earlier than was appropriate
Late	The action was conducted later than was appropriate

may select a flight level from a menu provided on the screen. If the guide word 'no' was applied to the selection of a 'menu' in this air traffic controller task, a deviation such as 'no [aircraft] heading entered into system' would be identified. In turn, for each deviation, the HAZOP study group would go on to identify the consequences of the error on the system, indications that the error occurred, system defences, and ways in which such an error would be recovered or reduced. In this particular case, the guideword 'other' might be more safety critical, i.e., if the controller entered the wrong flight level into the system — in such a case, the aircraft could be placed at the same level as another converging aircraft. An example of HAZOP output is given in Table 3. This shows that it is necessary to consider the consequences, and also early possibilities for error reduction.

HAZOP applied to human error identification or analysis (often called Human HAZOP to distinguish it from the more general HAZOP used in systems reliability) has been found to be useful and insightful in a number of different industries (e.g., nuclear power, offshore petrochemical, and air traffic management). It can also consider some more difficult error forms such as rule violations (Mason, 1997). There are however, some disadvantages. The major limitation of HAZOP is summarised tartly by the phrase, '*garbage in, garbage out*'. If the members of the group are not appropriately qualified or experienced, or if the session is not run effectively, or if the group does not 'mix well', the results may be unsound.

Another limitation, or constraint, is that HAZOP can be resource-costly in terms of human resources — it takes some time to get the group working effectively together, and can take a long time to analyse a system comprehensively. HAZOP should be targeted at safety critical systems, or systems or interfaces that require high usability or reliability. All that can be said of this limitation or constraint, is that other industries have found the investment to be worthwhile. Use of HAZOP and a technique such as SHERPA or TRACEr represents an effective error identification approach, since while there will be some overlap, the two together will be comprehensive.

Techniques such as SHERPA and Human HAZOP, and even new ones such as TRACEr, represent proven approaches that work in practice. They are useful for identifying simple errors such as 'slips or lapses' (Reason, 1990), and for more complex rule-following errors. Two areas where there is still need for further development in Human Error Identification are those of diagnosis and errors of commission. Two techniques found to be of use in identifying misdiagnoses are Fault Symptom Matrix Analysis and Confusion Matrix Analysis (see examples in Kirwan, 1994). These techniques consider the signals that a human operator is faced with, and determine the degree of 'confusability' of the signals given the possible diagnoses and the anticipated frequency of the events to which they correspond. This general approach has existed for some time (see Potash *et al.*, 1982), and is still used today (e.g., see Kirwan *et al.*, 2001). Nevertheless, there is still the need for better understanding of this complex area, with some researchers approaching the problem instead from the domain of cognitive simulations that can replicate faulty diagnostic performance (see Kirwan and Hollnagel, 1998). Therefore, despite Three Mile Island having occurred some twenty odd years ago, this is still an area where further development is required.

The area of errors of commission refers to doing something that is not required. This can include a rule violation or a misdiagnosis, but it also can include simple slips that leave a system disabled, for example, after maintenance. This area, like misdiagnosis, is a difficult one, but for different reasons. With misdiagnosis in nuclear power plants, for example, the assessment problem is usually made difficult because of the sheer complexity of the scenarios being considered. With more general errors of commission (e.g., leaving a safety system in test mode after maintenance), the problem is one of diversity — there are so many possible errors that it is difficult to be comprehensive in identifying them. Also, by definition such errors are unintended, and so it is not clear why they should happen.

Table 3 Output from ATC-based HCI HAZOP Study (from Kennedy *et al.*, 2000).

Function	Guide Word	Cause	Consequence	Indication	System Defences	Human Recovery	Recommendations
Highlight Object (aircr-aft label) on display	No	Another item preventing access to target aircraft	Difficulty in hooking target aircraft	No highlighting of target	None	Drag blocking object out of way; Strategic management of screen items	Design objects to 'roll around' each other; use Height filtering; Flip system to move between object on top and the one beneath; Highlight background
	Other	Clustering results in different aircraft being highlighted instead of target	Instruction may be given to wrong aircraft on the system	As Above	Highlighting is colour coded to indicate direction of travel; Call sign is displayed on all menus	As Above	As Above
Track Pointer	No	Mouse cursor moves off menu	Menu automatically closes when cursor moves away	Loss of menu on screen — controller needs to look at screen and not keyboard	Put gate around object to prevent loss of menus	If noticed, the cursor can be repositioned over the function and reclicked for mouse to reappear	As Above
	Other	Mouse cursor knocked-off intended function during keyboard entry	Input made into wrong function due to controller using keyboard whilst not looking at the screen	Different functions have different consequences (e.g., heading and speed); Aircraft not behaving as expected; Conflict alert may indicate error		Mouse use results in head-up selection of menus on screen - thus increasing likelihood of identifying an error	Training to encourage mouse rather than key input; Provide head-down display of input function to prevent head-down errors; Lock object so it does not slip out of focus

Nevertheless they do, and have been a significant concern in the nuclear industry for some years, leading to research programmes to address the issue (e.g., see Cooper *et al.*, 1996).

The need to address errors of commission is similar to the need to consider misuse of products in product design, only in the former it is usually a much larger and more complex system or 'product' under consideration. It is likely that an approach in this area could benefit from considering what the system 'affords' the user to do, and then determining what is likely. Such an approach has been taken by Baber and Stanton (the TAFEI technique — 1996), for example, and seems a reasonable approach for products and perhaps small systems. Whether such an exhaustive analysis is tractable for a large system such as a nuclear power plant, however, remains an open question.

In summary on human error identification approaches, they have moved on in terms of development and application considerably since the first version of this chapter in 1990. Nevertheless, despite a plethora of methods being developed (see Kirwan, 1998(a, b), still some of the more basic approaches remain those that are seen as most useful and most used in practical applications. There has, however, been an improved focus on 'context' as driving performance (e.g., Hollnagel, 1998), so that errors are increasingly identified as a function of contextual factors and considerations. This is probably also a reflection of a slow paradigmatic shift in Human Factors more generally towards naturalistic decision-making. Such a shift is a positive one, as it increases the chances of being able to determine error reduction mechanisms later, and it may also help in the quantification of the likelihood of the human error.

The next step following identification of these error forms is to represent them in some logical format so that their effects on the system goals can be evaluated.

Representation

In order to determine the cumulative risk of all the various failure 'paths', whether involving human performance or not, and in order to establish which of these failure paths contribute most towards system risk, all failures and recoveries must be integrated into a logical framework which can be quantified. This process is called representation, and can occur via several approaches. The two most common in PSA are fault trees and event trees, and these are briefly described below. An alternative approach is via mathematical simulation methods. Such methods are more complex and less frequently used, and so are not described here (however, see Cacciabue, 1998).

A fault tree is a logical structure which defines what events (human errors, hardware/software faults, environmental events) must occur in order for an undesirable event (e.g., an accident) to occur (Henley and Kumamoto, 1981; Cox and Tait, 1991; Kirwan, 1994). The undesirable event or outcome, usually placed at the top of the 'tree' and hence called the *top event*, may for example be '*failure to launch a lifeboat successfully at first attempt*,' or 'failure to achieve re-circulation of primary coolant', etc. The tree is constructed primarily by using two types of 'gate' by which events at one level can proceed to the next level up until finally they reach the top event. The first type of gate is an 'OR' gate, and the event above this gate occurs if *any* one of the events joined below it by this gate occur. An event above an 'AND' gate only occurs if *all* the events joined below it by this gate occur. An example is shown in Figure 8.

A fault tree can be used to represent a simple or complex pattern of system failure paths, and may comprise human errors alone, or a mixture of human, hardware, and/or environmental events, depending upon the scenario. Once structured, the events (including human error probabilities) must be quantified to determine the overall top event frequency (or probability), and the relative contributions of each error to this undesirable event.

Another type of 'tree' is the event tree, and in HRA parlance this is sometimes called an operator action tree (OAT: see Figure 9). The OAT proceeds from an initiating event, usually

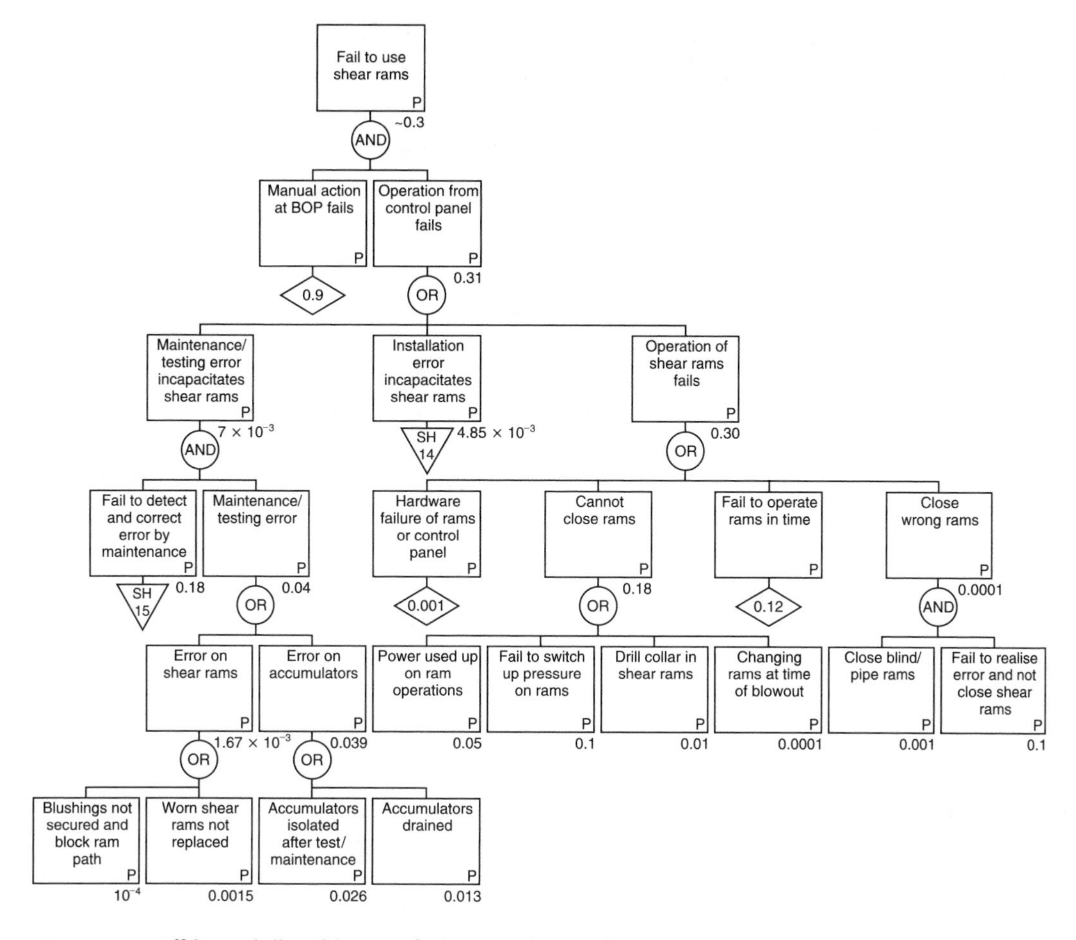

Figure 8 Offshore drilling blowout fault tree sub-tree: fail to use sher rams to prevent blowout.

placed at the left-hand side of the tree (e.g., loss of power causes ESD (emergency shutdown) demand). The tree is then developed to consider a set of sequential events each of which may or may not occur, causing the tree to branch in a binary fashion at each event 'node'. The events and branching continue until an 'end state' is reached for each path, which is either success in terms of achieving system safety, or else failure in terms of lost production, plant or equipment damage, injury, or fatality. OATs are especially useful when considering dynamic situations, and in general are more readily understandable than fault trees when human performance is dependent upon previous actions/events in the scenario sequence (Hall *et al.*, 1982).

The above are formal methods used for representing moderately complex patterns and sequences of failures. Whilst the examples given here are simple, in practice such trees can become quite complex, with many 'nodes' and events, and in the case of event trees, a large number of possible final outcomes. It is something of an art to develop such trees so that the human errors are adequately and accurately represented, without letting the trees become too complex and unwieldy. Nowadays, however, a number of mature software packages are available for fault and event tree modelling to assist the user.

If in a particular assessment the number of errors is small, and their effect on the system goals is very simple, such representation may be unnecessary. Furthermore, some analyses may stop at this point if their objective was merely qualitative in nature, i.e., simply to

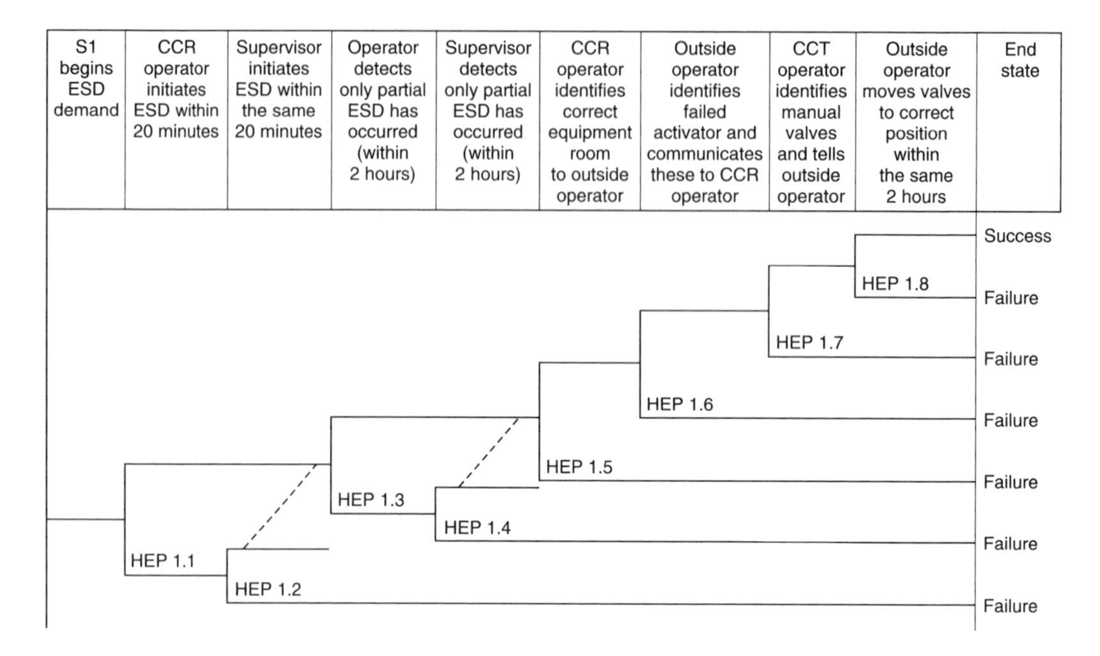

Figure 9 Operator action tree for ESD failure scenario (Kirwan, 1988; 1994).

identify human error modes without quantifying their probability or consequential effect on the system. In all other cases however, quantification of the human error probabilities will be necessary. Due to resource limitations screening may be applied to limit the amount of quantification required. If screening is not utilised, then quantification is the next step.

Screening

A screening analysis identifies where the major effort in the quantification analysis should be applied. There may for example be particular tasks which are theoretically related to the system goals being investigated, but which in fact make little contribution to risk if they fail (due to diverse reliable backup systems which adequately compensate for human error, or to the trivial nature of the tasks themselves). It is efficient to expend little effort on such tasks and instead focus on those in which human reliability is critical. The identification of those errors that can be effectively ignored by the rest of the study is the purpose of a screening analysis.

The *systematic human action reliability procedure* (SHARP) methodology defines three methods of screening logically structured human errors (see Spurgin *et al.*, 1987). The first method 'screens out' those human errors which can only affect the system goals if they occur in conjunction with an extremely unlikely hardware failure or environmental event. The second method involves allocating each human error a probability of 1.0, and examining the effects of the various errors on the top event frequency. Those that have a negligible effect even with a probability of unity may not be considered further. The third method assigns broad probabilities to the human errors based on a simple categorisation (such as the one shown in Table 4). This method works in the same way as the previous method but is a 'finer-grained' analytic method.

With most screening methods (particularly the third method above) there is a danger of ruling out of the study important errors and interactions, balanced against a need to reduce to a manageable level the complexity of the analysis and the required resources. As a general

Table 4 Generic Human Error Probabilities

Category	Failure probability
Simple, frequently performed task, minimal stress	10^{-3}
More complex task, less time variable, some care necessary	10^{-2}
Complex, unfamiliar task, with little feedback and some distractions	10^{-1}
Highly complex task, considerable stress, little performance time	3×10^{-1}
Extreme stress, rarely performed task	10^{0}

rule when applying any screening technique at any level in the study — if in doubt, leave the human error in the fault/event tree. The next step following screening, or following error analysis if screening was not applied, is quantification of the human errors in the fault or event trees.

Quantification

Human reliability quantification techniques quantify the human error probability (HEP), which is the metric of human reliability assessment. The HEP is defined as

$$\text{HEP} = \frac{\text{(Number of errors occurred)}}{\text{Number of opportunities for error to occur}}$$

Thus, if when buying a cup of coffee from a vending machine, on average one time in a hundred tea is accidentally purchased, the HEP is taken as 0.01. It is somewhat educational to try and identify HEPs in everyday life with a value of less than once in a thousand opportunities, or even as low as once in ten thousand.

In an ideal world there would be many studies and experiments in which HEPs were recorded. In reality there are few such recorded data (however, see Taylor-Adams and Kirwan, 1995; Gibson *et al.*, 1999; Basra and Kirwan, 1998). The ideal source of human error 'data' would be from industrial studies of performance and accidents, but at least three reasons can be deduced for the lack of such data:

(a) difficulties in estimating the number of opportunities for error in realistically complex tasks (the so-called denominator problem);
(b) confidentiality and unwillingness to publish data on poor performance;
(c) lack of awareness of why it would be useful to collect data in the first place (and hence lack of financial incentive for such data collection).

Nevertheless, some data are shown in Table 5 as an example of the type of HEPs that have been collected. Such HEPs at least can give the assessor a 'feel' for relative likelihoods of certain tasks.

Table 5 Example Human Error Data

Category	Failure probability
Simple, frequently performed task, minimal stress	10^{-3}
More complex task, less time variable, some care necessary	10^{-2}
Complex, unfamiliar task, with little feedback and some distractions	10^{-1}
Highly complex task, considerable stress, little performance time	3×10^{-1}
Extreme stress, rarely performed task	10^{0}

Overall, however, there remains a 'data shortage'. This is one reason for not being able to quantify HEPs directly, e.g., by reference to an exhaustive database of HEPs. However, even if there was a sound datum, e.g., for a chemical plant operator failing to respond to an alarm in scenario X, how can this HEP be generalised to scenario Y? In other words, what defines the 'generalisability' of data from one situation to another? It would depend on the similarity of the *context* in scenarios X and Y — but what aspects of the context would have to be similar, and exactly how similar would they have to be? Such difficult and as yet unresolved issues as these dogged the early years of HRA (the 60s and 70s), when a number of database approaches were tried and failed to succeed (see Topmiller *et al.*, 1984).

This data shortage and data extrapolation problem led to the development of non-data-dependent approaches, namely to the use of expert opinion. This is by no means necessarily a bad thing. Expert opinion has been used successfully in other areas (e.g., Murphy and Winkler, 1974; or Ludke *et al.*, 1977), and in any case is used occasionally in reliability engineering and probabilistic safety assessments (Nicks, 1981) where similar problems often exist.

The human error quantification techniques described below all contain an element of expert judgement, even though some in particular give the appearance (not necessarily intended) of being based on 'hard' well-founded empirical data.

There are a number of reviews of HRA quantification techniques (Humphreys, 1988; Swain,1990; Kirwan, 1994). Human reliability quantification techniques were qualitatively assessed. Seven of the main HRA techniques used are as follows:

Absolute Probability Judgement (APJ) (Seaver and Stillwell, 1983) *Paired Comparisons* (PC) (Hunns, 1982)
Technique for Human Error Rate Prediction (THERP) (Swain and Guttmann, 1983)
Human Error Assessment and Reduction Technique (HEART) (Williams, 1986, 1988, 1992)
Success Likelihood Index Method (SLIM) (Embrey *et al.*, 1984)
Human Cognitive Reliability Model (HCR) (Spurgin *et al.*, 1987)
Justification of Human Error Data Information (JHEDI: Kirwan, 1990, 1994, 1997a)

Three of the techniques (APJ, PC, and SLIM) use a group of expert judges and formal expert elicitation procedures to evaluate HEPs. APJ and PC largely leave the judgement task to the experts, with some help from the analyst or 'facilitator' who may point out inconsistent judgements or biases in the judgement-making process. SLIM also uses expert judges, but the judges are asked to consider the factors that affect performance, and from the assessment of these factors and modelling of their influence on performance, they then determine the human error probability. The experts are assisted by the analyst in creating a quantitative causal model of the influence of these factors on the HEP. Typical performance shaping factors (PSF) utilised are stress, quality of interface design, quality of procedures, and degree of training.

THERP and HEART, in contrast, either include a database of HEPs or specify procedures for generating numerical HEPs directly. These techniques require only one analyst, rather than a group of experts. The data on which they rely are a mixture of field experience and judgement in the case of THERP, and a mixture of judgement together with data from ergonomics and psychological performance literature in the case of HEART.

The HCR model (also called the time reliability correlation (TRC) approach) attempts to quantify diagnostic/cognitive errors as a function of time elapsed since the onset of the incident, and assumes that the likelihood of successful diagnosis and, consequently action, increases as the time available increases. This approach is a mixture of judgement and simulator data. However, this approach fell out of favour when simulator exercises trying to validate the technique found contrary evidence (Dolby, 1990; Kantowitz and Fujita, 1990). Essentially, although a HCR or TRC approach is very appealing (since it is easy to gain the input data of time to respond, and easy to fit the results into a PSA), human performance cannot be predicted accurately based on this single parameter.

JHEDI (see Kirwan, 1997a) is also a data-based technique, and belongs to a particular UK company (British Nuclear Fuels), and is used exclusively in the nuclear fuel reprocessing industry. It was based on data collected in that industry over a thirty-year period, and is only currently used within that context. It is mentioned here since it requires less judgement on the part of the analyst than HEART or THERP. Instead, the judgement is built into the system via PSF analysis of the data-set within the system. The assessor is therefore asked factual questions about the quality of procedures for example, rather than asking to give a rating of them, or to interpret how their degree of adequacy will impact the HEP. JHEDI is therefore more structured (and computerised).

Within the scope of this chapter it is not possible to review all these techniques. Instead, therefore, three are reviewed, namely SLIM, HEART and THERP. The first two are probably of most interest to the ergonomist, and the latter exemplifies the reliability engineering oriented approach, and is also the technique that has been most widely used to date. For a review of the others, see Humphreys (1988), Swain (1989), Kirwan (1994) or the indicated source references. The reader is referred to Swain (1989) in particular for an in-depth review of fourteen HRA approaches.

Success likelihood index method (SLIM)

SLIM can best be explained by means of an example human reliability assessment, in this case an operator de-coupling a filling hose from a chemical road tanker. The operator may forget to close a valve upstream of the filling hose, which could lead to undesirable consequences, particularly for the operator. The human error of interest is 'failure to close V0204 prior to decoupling filling hose'. In this case the decoupling operation is simple and discrete, and hence failure occurs catastrophically rather than in a staged fashion.

PSF identification

The 'expert panel' would typically comprise, for example, two operators with 10 years experience, one human factors analyst, and a reliability analyst familiar with the system who also has some operational experience.

The panel is initially asked to identify a set of *performance shaping factors* (PSFs), which are any factors relating to the individual(s), environment, or task, which affect performance positively or negatively. The expert panel could be asked to nominate the most important or significant PSFs for the scenario under investigation. In this example it is assumed the panel identify the following major PSFs as affecting human performance in this situation: training, procedures, feedback, perceived risk, and time pressure.

PSF rating

The panel are then asked to consider other human errors possible in this scenario (e.g., mis-setting or ignoring an alarm), and for each one, to decide to what extent each PSF is optimal or sub-optimal for that task in the situation being assessed, on a scale of 1 to 9, in this case with 9 as optimal. For the three human errors under analysis, the ratings obtained are as follows:

			Performance Shaping Factors		
Errors	Training	Procedures	Feedback	Perceived risk	Time pressure
V0204 Open	6	5	2	9	6
Alarm mis-set	5	3	2	7	4
Alarm ignored	4	5	7	7	2

PSF weighting

If each factor was equally important, one might simply add each row of ratings and conclude that the error with the lowest rating sum (alarm mis-set) was the most likely error. However, this expert panel, as with most panels, does not feel the PSFs are all equal. In this particular case (and with this particular panel of experts), the panel feels that perceived risk and feedback are most important, and are in fact twice as important as training and procedures, which are in turn one and a half times as important as time pressures. (As it is a routine operation, time is not perceived by the panel to be particularly important). Weightings for the PSFs can be obtained directly from these considered opinions, as follows, normalized to sum to unity:

Perceived risk	0.30
Feedback	0.30
Training	0.15
Procedures	0.15
Time Pressure	0.10
Sum	1.00

SLIM, and the decision analysis technique it is based upon, called *simple multi-attribute rating technique* (SMART: Edwards, 1977) propose simply that preference can be derived as a function of the sum of the weightings multiplied by their ratings for each item (human error). SLIM does this and calls the resultant preference index a *success likelihood index* (SLI). This is illustrated using a table of weightings (W) × ratings (R): (SLI = Sum $W \times R$) (see Table 6).

In this case, the lowest SLI is 4.3, suggesting that 'alarm mis-set' is still the most likely error. However, due to the weightings used, the likelihood ordering of the other two errors have now been reversed (close inspection of the figures reveals that this is because feedback is held

Table 6 SLI Calculation

Weighting	PSF	Weighting × Rating	V0204	Alarm mis-set	Alarm ignored
0.30	Feedback	(0.3 × 2)	0.6	0.6	2.1
0.30	Perceived risk	etc.	2.7	2.1	2.1
0.15	Training		0.9	0.75	0.6
0.15	Procedures		0.75	0.45	0.75
0.10	Time		0.60	0.40	0.2
	SLI (Total)		5.55	4.30	5.75

to be important, and there is ample feedback for 'alarm ignored' but not for 'V0204 open'). Clearly at this point, a designer would realise that increased feedback about the position of V0204 to the operator might be desirable.

However, the SLIs are not yet probabilities. Rather, they are indications of the relative likelihoods of the different errors. Thus the SLIs show the ordering of likelihood of the different errors, but do not yet define the absolute probability values. In order to transform the SLIs into HEPs, it is necessary to 'calibrate' the SLI values. (Note: the paired comparisons technique also requires this calibration using the same basic formula.) Two earlier studies by Pontecorvo (1965) and Hunns (1982) have derived such a calibration relationship, both suggesting a logarithmic relationship of the form:

$$\mathrm{Log}_{10}\,(\mathrm{HEP}) = a\,\mathrm{SLI} + b.$$

If two tasks for which the HEPs are known are included in the task/error set which are being quantified, then the parameters of the equation can be derived via simultaneous equations, and the other (unknown) HEPs can be quantified. If in the above example, two more tasks (*A* and *B*) were assessed which had HEPs of 0.5 and 10^{-4} respectively, and were given SLIs of 4.00 and 6.00, respectively, then the equation derived would be:

$$\mathrm{Log}\,(\mathrm{HEP}) = -1.85\,\mathrm{SLI} + 7.1.$$

The HEPs would then be: V0204 = 0.0007; alarm mis-set = 0.14; alarm ignored = 0.0003.

This is the body of the rationale underlying SLIM, but in practice SLIM is more complex and is computerised to facilitate its ease of use and to prevent bias, often found in the elicitation of expert opinions. The computerised version, known as SLIM-MAUD (SLIM using Multi Attribute Utility Decomposition: Embrey *et al.*, 1984), due to the mathematics in the software which is present partly to avoid such bias, will produce slightly different values (HEPs) than the hand calculated method used above. In particular the simple summary of weightings and ratings is refined in several ways according to the more detailed mathematical requirements of multi-attribute utility theory. However, the above is the general rationale of SLIM, and enables the reader to understand more easily how SLIM works.

Human error assessment and reduction technique (HEART)

This technique is of particular interest to ergonomists as it was based on the human performance literature. It has been designed by its author (Jerry Williams) as a relatively quick method for HRA, to be simple to use and easily understood. Its fundamental premise is that in reliability and risk equations one is interested in ergonomics factors which have a large effect on performance, e.g., causing a decrement in performance by a factor of three or more. Thus, whilst there are many well-studied ergonomics factors and consequent guidelines (e.g., lighting recommendations), many of these factors actually have (in reliability terms) a negligible effect on operator performance. HEART therefore concentrates on those factors that have a significant effect.

This point is important because in part it underlies something of a communications gap between engineers and ergonomists. Engineers and designers designing a plant cannot spend unlimited funds on the optimisation of ergonomics aspects, and often ask how important (in quantitative terms) an ergonomics recommendation is. Frequently the ergonomist is unable to answer this question, which to the engineer is fundamental. HEART in particular, and some of

Table 7 Generic Classifications (HEART, after Williams, 1986)

Generic task	Proposed nominal human unreliability (5th–95th percentile bounds
(A) Totally unfamiliar, performed at speed with no real idea of likely consequences	0.55 (0.35–0.97)
(B) Shift or restore system to a new or original state on a single attempt without supervision or procedures	0.26 (0.14–0.42)
(C) Complex task requiring high level of comprehension and skill	0.16 (0.12–0.28)
(D) Fairly simple task performed rapidly or given scant attention	0.09 (0.06–0.13)
(E) Routine, highly-practiced, rapid task involving relatively low level of skill	0.02 (0.007–0.045)
(F) Restore or shift a system to original or new state following procedures, with some checking	0.003 (0.0008–0.007)
(G) Completely familiar, well-designed, highly practiced, routine task occurring several times per hour, performed to highest possible standards by highly-motivated, highly-trained and experienced person, totally aware of implications of failure, with time to correct potential error, but without the benefit of significant job aids	0.0004 (0.00008–0.009)
(H) Respond correctly to system command even when there is an augmented or automated supervisory system providing accurate interpretation of system stage	0.00002 (0.000006–0.0009)
(M) Miscellaneous task for which no description can be found	0.03 (0.008–0.11)

the other techniques (e.g., SLIM, JHEDI), allow the human reliability analyst to answer this question quantitatively.

The first part of the HEART assessment process is to refine the task in terms of its generic proposed nominal human unreliability, as shown in Table 7. Thus the task is first assigned a nominal human error probability by classifying it according to whether it is a complex task, a routine task, and so on. The next stage is to identify error-producing conditions (EPCs) which are evident in the scenario and would negatively influence human performance. A table of the major EPCs in HEART is shown in Table 8.

Example

As a hypothetical example of how HEART is used to quantify a human error probability for a task, taken from Williams (1988), we assume that a safety, reliability, or operations engineer wishes to assess the nominal likelihood of an operative's failing to isolate a plant bypass route following strict procedures. The scenario necessitates a fairly inexperienced operator applying an opposite technique to that which he normally uses to carry out isolations and involves a piece of plant, the inherent major hazards of which he is only dimly aware. It is assumed that the man could be in the seventh hour of his shift, that there is talk of the plant's imminent closure, that his work may be checked and that the local management of the company is desperately trying to keep the plant operational despite the real need for maintenance because of its fear that partial shutdown could quickly lead to total permanent shutdown.

Using a simplified HEART, the safety reliability and operational engineer's assessment could look something like this:

Type of Task=F			*Nominal Human Unreliability*=0.003
EPC Description	*Total HEART Affect*	*Engineer's Assessed Proportion of Affect (from 0 to 1)*	*Assessed Affect*
Inexperience	× 3	0.4	$(3-1) \times 0.4+1=1.8$
Opposite technique	× 6	1.0	$(6-1) \times 1.0+1=6.0$
Risk misperception	× 4	0.8	$(4-1) \times 0.8+1=3.4$
Conflict of objectives	× 2.5	0.8	$(2.5-1) \times 0.8+1=2.2$
Low morale	× 1.2	0.6	$(1.2-1) \times 0.6+1=1.12$
Assessed nominal likelihood of failure $0.003 \times 1.8 \times 6.0 \times 3.4 \times 2.2 \times 1.12=0.27$			

Time-on-shift effects would be ignored as there is no indication of monotony.

Similar calculations may be performed if desired for the predicted 5th and 95th percentile bounds, which in this case would be 0.07–0.58. As a total probability of failure can never exceed 1.00, if the multiplication of factors takes the value above 1.00 the probability of failure has to be assumed to be 1.00 and no more.

The relative contribution made by each of the error producing conditions to the amount of unreliability modification is as follows:

	% contribution made to unreliability modification
Technique unlearning	41
Misperception of risk	24
Conflict of objectives	15
Inexperience	12
Low morale	8

Thus an HEP of 0.27 (just over one in four) is calculated, which is a very high predicted error probability, and unlikely to be acceptable. In this case technique unlearning is the major contributory factor to this poor performance, and so clearly either some form of retraining, or else redesign to make the isolation procedures consistent across plant, must be considered. However HEART goes further than other techniques in error reduction, as for each EPC it gives corresponding suggested error reduction approaches, e.g., for the task above, the remedial measures in Table 9 would be proposed.

Thus HEART offers a quick and simple human reliability calculation method which also gives the user (engineer or ergonomist) suggestions on error reduction. HEART has recently been reviewed qualitatively by its author and is under further development (Williams, 1992).

The conditions 18 to 26 in Table 8 are presented simply because they are frequently mentioned in the human factors literature as being of some importance in human reliability assessment. To a human factors engineer, who is sometimes concerned about performance differences of as little as 3%, all these factors are important, but to engineers who are usually concerned with differences of more than 300%, they are not very significant. The factors are identified so that engineers can decide whether or not to take account of them after initial screening.

Table 8 HEARTS EPCs (Williams, 1986)

Error producing conditions		Maximum predicted nominal amount by which unreliability might change going from 'good' conditions to 'bad'
1.	Unfamiliarity with a situation which is potentially important but which only occurs infrequently or which is novel	× 17
2.	A shortage of time available for error detection and correction	× 11
3.	A low signal-to-noise ratio	× 10
4.	A means of suppressing or overriding information or features which is too easily accessible	× 9
5.	No means of conveying spatial and functional information to operators in a form which they can readily assimilate	× 8
6.	A mismatch between an operator's model of the world and that imagined by a designer	
7.	No obvious means of reversing an unintended action	× 8
8.	A channel capacity overload, particularly one caused by simultaneous presentation of non-redundant information	× 8
9.	A need to unlearn a technique and apply one which requires the application of an opposing philosophy	× 6
10.	The need to transfer specific knowledge from task to task without loss	
11.	Ambiguity in the required performance standards	× 6
12.	A mismatch between perceived and real risk	
13.	Poor, ambiguous or ill-matched system feedback	× 5.5
14.	No clear direct and timely confirmation of an intended action from the portion of the system over which control is to be exerted	× 5
15.	Operator inexperience (e.g., a newly-qualified tradesman, but not an 'expert')	× 4
16.	An impoverished quality of information conveyed by procedures and person/person interaction	× 4
17.	Little or no independent checking or testing of output	× 4
18.	A conflict between immediate and long-term objectives	
19.	No diversity of information input for veracity checks	
20.	A mismatch between the educational achievement level of an individual and the requirements of the task	× 3
21.	An incentive to use other more dangerous procedures	× 3
22.	Little opportunity to exercise mind and body outside the immediate confines of a job	
23.	Unreliable instrumentation (enough that it is noticed)	× 3
24.	A need for absolute judgements which are beyond the capabilities or experience of an operator	× 2.5
25.	Unclear allocation of function and responsibility	
26.	No obvious way to keep track of progress during an activity	× 2.5
		× 2
		× 2
		× 1.8
		× 1.6
		× 1.6
		× 1.6
		× 1.4

Table 9 A Subset of HEART Remedial Measures (Williams, 1986)

1. Technique unlearning (× 6)	The greatest possible care should be exercised when new techniques are being considered to achieve the same outcome — they should not involve adoption oÈ opposing philosophies
2. Misperception of risk (× 4)	It must not be assumed that a user's perception of risk is the same as the actual level — if necessary a check should be made to ascertain where any mismatch might exist and what its extent is
3. Objectives conflict (× 2.5)	Objectives should be tested by management for mutual compatibility, and where potential conflicts are identified these should either be resolved to make them harmonious or made prominent so that a comprehensive management control programme can be created to reconcile such conflicts as they arise, in a rational fashion
4. Inexperience (× 3)	Personnel criteria should contain specified experience parameters thought relevant to the task-chances must not be taken for the sake of expediency
5. Low morale (× 1.2)	Apart from the more obvious ways of attempting to secure high morale, by way of financial reward for example, other methods involving participation, trust and mutual respect, often hold out at least as much promise — building up morale is a painstaking process, which involves a little luck and great sensitivity — employees must be given reason to believe

Technique for human error rate prediction (THERP)

THERP is in itself a total methodology for assessing human reliability. The quantification part of THERP comprises the following:

(a) a database of human errors which can be influenced by the assessor to reflect the impact of PSFs on the scenario;
(b) a dependency model which calculates the degree of dependence between two operator actions (e.g., if an operator fails to detect an alarm, then failure to carry out appropriate corrective actions reliably cannot be treated independent from this failure);
(c) an event tree modelling approach to combine HEPs for steps in a task into an overall task HEP;
(d) the assessment of error recovery paths.

The basic THERP approach is shown in Figure 10, from Bell (1984).

Probably because of its database and similarities to reliability engineering approaches, THERP has been used more than any other technique in industry applications. This small section can only cover the rudiments of the technique, and for more information the reader is referred to Swain and Guttmann (1983), Bell (1984) and Kirwan *et al.* (1988).

An example of the type of basic event data given in THERP is shown in Table 10.

A human reliability analysis event tree (HRAET, as shown in Figure 11) can be used to represent the operator's performance. Alternatively, as shown in Figure 12, an operator action event tree can be used (Whittingham, 1988). In each case the event tree represents the sequence of events and considers possible failures at each branch in the tree (omission, commission, and so on). These errors are quantified and error recovery paths are then added to the tree where appropriate. The quantification takes place by usage of a set of human error data tables. These tables are derived from a mixture of industrial data (e.g., from the US defence sector) and judgement of the main author of the technique (Swain).

Phase 1: Familiarisation

Plant visit

Review Information From System Analysts

Phase 2: Qualitative Assessment

Talk-Through or Walk-Through

Task Analysis

Develop HRA Event Trees

Phase 3: Quantitative Assessment

Assign Nominal HEP's

Estimate the Relative Effects of Performance Shaping Factors

Assess Dependence

Determine Success and Failure Probabilities

Determine the Effects of Recovery Factors

Phase 4: Incorporation

Perform a Sensitivity Analysis, if Warranted

Supply Information to Systems Analysts

Figure 10 Outline of a THERP procedure for HRA (adapted from Bell, 1984).

Figure 13 shows the PSFs which can be used in a THERP study, although often in a THERP study only one or at most a few of these are utilised quantitatively (e.g., stress). For each datum there is an allowable range for the HEP (e.g., within a factor of three or ten), and the assessor uses the consideration of PSF to decide what value to use within that range. Additionally, for diagnosis errors, there is a time reliability curve, with associated large uncertainty bounds.

It is important to model recovery, particularly when investigating highly proceduralised sequences, since often an operator will be prompted by a later step in the procedures to recover from an earlier error in a previous step. For example, if an operator omits a step to turn the power on, and then a second step is attended to which involves checking certain power-supplied instruments, then the operator will rapidly recover the first error. If recoveries in such highly proceduralised situations (for which THERP is typically used) are not identified, then human error may be overestimated. THERP is in fact the only technique that emphasises error recovery in this way. Error recovery paths are shown on the operator action tree in Figure 12 as dashed lines, and in this particular tree these are largely recoveries of one person's error by another person.

THERP also models dependency between human errors. If, for example, an operator is in a high stress situation trying to carry out a procedure quickly, or is de-motivated or fatigued, etc., then a whole section of procedures and recovery steps within these procedures may be carried out wrongly. This is an example of a 'human dependent failure'. An important example of dependency modelling concerns one operator checking another operator. It is unlikely that the action by operator 1 and the check by operator 2 will be completely independent, since the first operator may assume that the second will detect any faults, and the second may assume the first did the job properly. THERP is one of the few techniques which actually quantitatively models this dependency between actions/errors. It uses a simple five-level model of dependency from zero dependence (i.e., total independence) through low, medium and high dependence levels to complete dependence. The effects of these levels of dependence are mathematically calculated in the HRAET/OAT. Since THERP's model of

Table 10 Task Analysis — Initiation of Flow via Stand-by Train (Webley and Ackroyd in Kirwan *et al.*, 1988)

Task identifier*	Task of interest	Error identifier	Human errors	Human error probabilities (range of median values given in USSR Scute Committee, 1986)	
A	Identity loss of flow via duty train	A_t	Fail to identify loss of flow	10^{-4}–0.25	Alarm response model
B	Start correct procedure	B_t	Fail to start procedure	10^{-3}–10^{-2}	Procedure used
				10^{-2}–5×10^{-2}	Procedure not used
C	Roving operator opens correct value	C_t	Error of omission — verbal order	10^{-3}	
			Error of commission — select incorrect value	10^{-3}–10^{-2}	
D	1st operator start stand-by pump via remote control	D_t	Error of omission — written procedures available	10^{-3}–10^{-2}	Procedure used
				10^{-2}–5×10^{-2}	Procedure not used
		D_2	Error of commission — select incorrect value	10^{-3}–10^{-2}	
E	Supervisor checks 1st operator	E_t	Error of omission — written procedures available	10^{-3}–10^{-2}	Procedure used
				10^{-2}–5×10^{-2}	Procedure not used
		E_2	Error of commussion — select incorrect value	10^{-3}–10^{-2}	
G	Shift Manager checks activation	G_1	Fail to initiate checking function	10^{-3}–10^{-2}	
			Fail to identify errors	10^{-2}–5×10^{-2}	

* Task F no errors identified for this step.

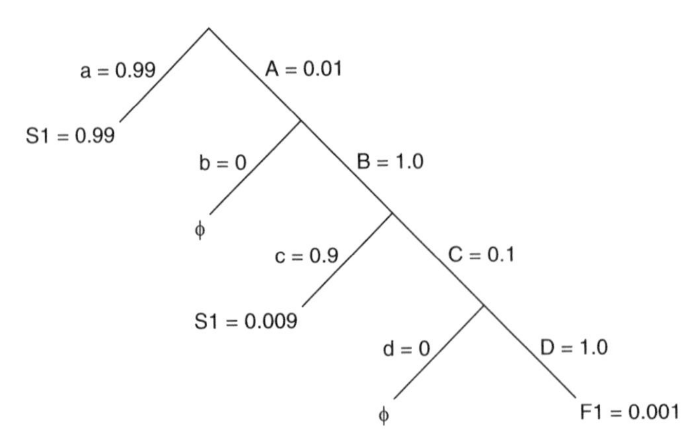

Figure 11 HRA event tree of hypothetical calibration task (adapted from Swain and Guttmann, 1983).

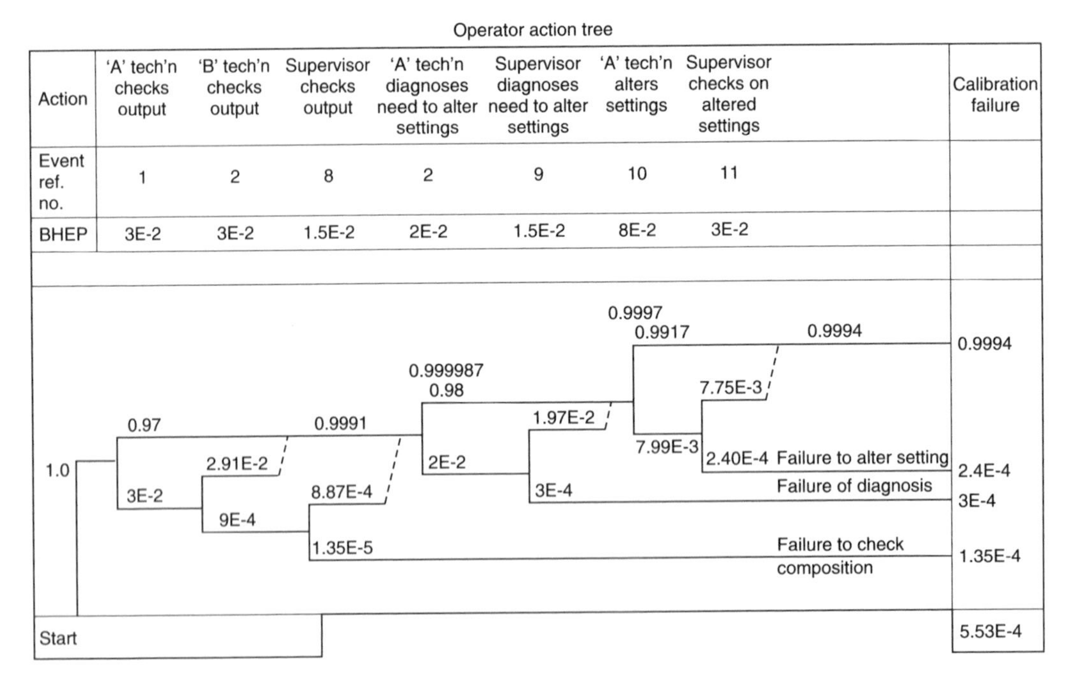

Figure 12 Supervisory check on altered settings (Whittingham, 1988).

dependence is one of the few available, other techniques sometimes 'borrow' this part of THERP when carrying out risk assessments (e.g., see Kirwan *et al.*, 2001).

As a more general point, human dependent failures can be extremely important in any human reliability assessment, and care should be taken to identify these where possible (e.g., if an operator might become incapacitated, or a misdiagnosis might occur, or if production pressures might totally overrule safety considerations). For a deeper treatment of the dependence issue, see Kirwan (1994).

External PSFs		Stressor PSFs	Internal PSFs
Situational characteristics:	Task and equipment characteristics:	Psychological stressors:	Organismic factors:
Those PSFs general to one or more jobs in a work situation	Those PSFs specific to tasks in a job	PSFs which directly affect mental stress	Characteristics of people resulting from internal and external influences
Architectural features Quality of environment: temperature, humidity, air quality, and radiation Lighting Noise and vibration Degree of general cleanliness Work hours/work breaks Shift rotation Availability/adequacy of special equipment, tools, and supplies Manning parameters Organisational structure (e.g. authority, responsibility, communication channels) Actions by supervisors, co-workers, union representatives, and regulatory personnel Rewards, recognition, benefits	Perceptual requirements Motor requirements (speed, strength, precision) Control-display relationships Anticipatory requirements Interpretation Decision making Complexity (information load) Narrowness of task Frequency and repetitiveness Task criticality Long and short term memory Calculational requirements Feedback (knowledge of results) Dynamic vs. step-by-step activities Team structure and communication Man-machine interface factors: Design of prime equipment, test equipment, manufacturing equipment, job aids, tools, fixtures	Suddenness of onset Duration of stress Task speed Task load High-jeopardy risk Threats (of failure, loss of job) Monotonous, degrading or meaningless work Long, uneventful vigilance periods Conflicts of motives about job performance Reinforcement absent or negative Sensory deprivation Distractions (noise, glare, movement, flicker, colour) Inconsistent cueing	Previous training/ experience State of current practice or skill Personality and intelligence variables Motivation and attitudes Emotional state Stress (mental or bodily tension) Knowledge of required performance standards Sex differences Physical condition Attitudes based on influence of family and other outside persons or agencies Group identifications
Job and task instructions:		Physiological stressors:	
Single most important tool for most tasks		PSFs which directly affect physical stress	
Procedure required (written or not written) Written or oral communications Cautions and warnings Work methods Plant policies (shop practices)		Duration of stress Fatigue Pain or discomfort Hunger or thirst Temperature extremes Radiation G-force extremes Atmospheric pressure extremes Oxygen insufficiency Vibration Movement constriction Lack of physical exercise Disruption of circadian rhythm	

Figure 13 Comprehensive list of PSFs for THERP analysis (adapted from Swain and Guttmann, 1983).

THERP has a derivative called ASEP (Accident Sequence Evaluation Procedure: Swain, 1987) which is a quicker and more conservative (i.e., pessimistic) version of THERP. ASEP can be used as a screening approach, and then THERP can be used to quantify the significant errors (i.e., the ones that matter in the risk assessment). Such an approach can be more efficient in terms of assessment resource usage.

Advantages and disadvantages of SLIM, HEART, and THERP

SLIM-MAUD is a highly structured approach to the use of expert opinion, and often has high credibility with the experts taking part. The computerised version enables the user to investigate how improvements, in particular PSF (e.g., interface design) can affect the HEP, allowing the cost effectiveness of error reduction strategies to be investigated (see Impact assessment). However, SLIM relies on 'experts' who may be difficult to find and verify as experts, and on calibration data (at least two known HEPs). SLIM is an exhaustive technique and hence can use up a relatively large amount of personnel resources and time. Perhaps the largest problem for SLIM is finding adequate calibration data — a review of the validation of SLIM found mixed evidence for the approach's empirical validity (Kirwan, 1997), mainly due to problems of calibration.

HEART is one of the quickest techniques available, and does not require experts, the EPCs instead being based on extensive analysis of the human performance literature. It also offers means of determining error reduction strategies and investigating the cost effectiveness of such strategies. However it does not consider possible interactions between the various EPCs. Perhaps the largest problem for HEART is that it requires significant input from the analyst, yet there is little currently in the way of training for the technique. Many of the EPCs are also difficult to assess for a plant at the design stage. Nevertheless, HEART has performed well in several empirical validation exercises, and is a relatively popular approach. (see Kirwan, 1997; Kennedy *et al.*, 2001).

THERP is similar in appearance to conventional reliability assessment approaches, yet it can bring PSFs into the equation, it explicitly models human errors and error recovery, and considers dependency between errors. THERP has also received widespread application in a number of different countries, primarily in the nuclear power domain. However, THERP also requires a good deal of resources, albeit in terms of a single well-trained assessor rather than a panel of experts, and often two different assessors may carry out THERP in rather different ways (see Kirwan, 1988). THERP has performed reasonably well in a number of validation exercises, and despite an apparent reduction in popularity, remains a major HRA technique.

Selection of an appropriate quantification technique

Comprehensive reviews by Humphreys (1988) and Swain (1989) have attempted to aid the process of selecting which quantification techniques to use in a particular application. The first review qualitatively analysed the eight techniques mentioned earlier, drawing conclusions from the published literature. It assessed the techniques against a set of criteria, namely:

Accuracy: numerical accuracy (in comparison with known HEPs); consistency between experts and assessors.

Validity: the technique's theoretical basis in human performance theory; its empirical validity; its 'face' validity as perceived by assessors, experts, etc.; and its comparative validity (comparing results of one technique with results of another for the same scenario).

Usefulness: qualitative usefulness in determining error reduction mechan; sensitivity analysis capability, allowing assessment of effects on HEPs of error reduction mechanisms.

Effective use of resources: equipment and personnel requirements; data requirements, e.g., SLIM and PC require calibration data (at least two known and contextually similar HEPs); training requirements of assessors and/or experts.

Acceptability: to regulatory bodies; to the scientific community; to assessors; auditability of the quantitative assessment (e.g., by the regulatory authorities).

Maturity: the current maturity of the technique.

Eight techniques are assessed against these criteria in Table 11, and a second selection guideline table based on perceived user requirements is presented in Table 12.

Table 11 Summary of Evaluation of Techniques (adapted from Kirwan *et al.*, 1988)

	APJ	PC	TESEO	THERP	HEART	IDA	SLIM	HCR
Accuracy	Moderate	Moderate	(Low)†	Moderate	Moderate	(Low)	Moderate	(Low)
Validity	Moderate/high	Moderate	Low	Moderate	Moderate	Moderate	Moderate	Low
Usefulness	Moderate/high	Low/Moderate	Moderate/High	Moderate	High	Moderate/high	High	Low/Moderate
Effective use of resources*	Moderate	Low/Moderate	High	Low/Moderate	High	Low/moderate	Low/Moderate	Moderate
Acceptability	Moderate	Moderate/high	Low	High	Moderate/High	(Moderate)	Moderate/high	Low/Moderate
Maturity	High	Moderate	Low	High	Moderate	Low/Moderate	Moderate/high	Low

*A rating of high on this criterion means the technique is favourable with respect to effective use of resources (i.e., resource requirements are low), and vice versa.

†Rating in parentheses were based on the subjective opinions of the authors and contributions to the document, since insufficnet empirical evidence was available to justify a rating from applications alone.

Note: This table was constructed prior to the experimental study assessing accuracy (Kirwan, 1988).

Table 12 Selection matrix (from Kirwan *et al.*, 1988)

	APJ	PC	TESEO	THERP	HEART	IDA	SLIM	HCR
Is the technique applicable for:								
Simple and proceduralised tasks?	Y	Y	Y	Y	Y	Y	Y	N
Knowledge-based, abnormal tasks?	Y	Y	Y	N	Y	Y	Y	Y
Misdiagnosis which makes a situation worse?	Y	Y	N	N	N	Y	Y	N
Are qualitative recommendations possible?	Y	N	Y	Y	Y	Y	Y	Y
Is sensitivity analysis possible?	N	N	Y	Y	Y	Y	Y	Y
Does the technique have:								
Requirements for calibration data?	N	Y	N	N	N	N	Y	N
Requirements for experts (judges)?	Y	Y	N	N	N	Y	Y	N

With these two tables it is possible for the user to decide which technique or set of techniques to utilise.

A detailed analysis of twenty-two validation exercises carried out in various countries over a 20 year period, is reported in Kirwan (1997b, c). With respect to the techniques discussed, APJ, THERP (and ASEP), JHEDI and HEART all appear to have a reasonable degree of validity. SLIM and PC have fared less well in such studies, though this appears to be primarily due to problems with calibration or the calibration data themselves (e.g., see Kirwan, 1994). Overall, however, there is sufficient empirical evidence of accuracy to justify acceptance of conclusions from HRA studies carried out by competent practitioners.

Impact assessment

Once human error probabilities have been quantified, the system risk, or reliability, can be calculated and compared to an acceptable level to see if improvement is necessary. If so, it will first be necessary to determine whether human error was a major contributor to inadequate system performance. This involves carrying out an analysis of the importance of each 'event' in the fault tree, to see which events (quantitatively) most affect the predicted top event (accident) frequency. Most fault tree computer analysis packages will automatically determine the most important events.

If human errors do not significantly affect the predicted frequency of the event, then reduction of risk will be best effected by hardware/software improvements to the system. Otherwise, and particularly if human error 'dominates' the undesired event frequency, error reduction mechanisms should be investigated. Identification of the major human errors will enable the most effective error reduction mechanisms to be specified in the next phase. The positive effects of these mechanisms can be calculated and factored back into the quantitative analysis until system performance reaches an acceptable level. Similar calculations and 'sensitivity analysis' can also be carried out for event-tree-based assessments.

If human error cannot be reduced to an acceptable level, even with additional hardware recommendations, then significant redesign of the system and/or its operation will be required. Usually however, an effective combination of human and hardware modifications can be found to achieve an acceptable level of risk.

Error reduction

Error reduction mechanisms will be unnecessary if human reliability is adequate, or not the most effective means of achieving system performance, or if not within the scope of the assessment. If on the other hand error reduction is required then either particular human errors identified in the analysis may be reduced in impact or frequency, or else a more general error reduction strategy may be developed to improve overall task performance.

In the case of specific identified critical errors, there are several ways of reducing their impact on the system (see also Kirwan, 1990, 1994):

Prevention by hardware or software changes: Use of interlock devices to prevent error; automate the task, etc.

Increase system tolerance: make the system hardware and software more flexible or self-correcting to allow a greater variability in operator inputs that will achieve the intended goal.

Enhance error recovery: enhance detection and correction of errors by means of increased feedback, checking procedures, supervision and automatic monitoring of performance.

Error reduction at source: reduction of errors by improved procedures, training, and interface or equipment design.

The first two measures require collaboration between the ergonomist (or human reliability analyst) and system design personnel, and may well prove expensive. Improved error recovery probabilities are often the simplest to implement, but may not reduce error probabilities by a sufficient amount, and may not be feasible with all critical errors. Error reduction 'at source' therefore may well be the primary means of improving human reliability. This often will require consultation with a human factors specialist, although some quantification techniques (e.g., HEART) do prescribe specific error reduction mechanisms for identified errors. Error reduction measures can also be identified by considering the error causes and the PSFs identified in the human error analysis phase (e.g., via TRACEr or a comparable technique). Such measures should be effective as they are aimed at the root causes of the human error.

A second additional analysis of the results will be possible only with quantification methods which use a structured performance shaping factor (PSF) approach (e.g., SLIM, HEART, THERP). With these approaches it is possible to determine the contributions of individual PSFs to human error goals (see the earlier HEART worked example). For example, the most significant PSF in a particular scenario may be 'quality of procedures' and, therefore, error reduction measures aimed at improving the quality of procedures will be most effective at reducing error likelihood. The potential reduction however, can actually be calculated since with these methods error probability is a direct function of the PSFs. Thus it may be that improving the quality of procedures for a particular error makes the error ten times less likely in a particular assessment, which may render the error probability 'acceptable'. If however, the error probability is still not low enough, then clearly other additional PSFs should be investigated. Furthermore, if for example quality of procedures is the most important PSF for a number of human errors, this then suggests that a single global error reduction strategy generally to enhance performance can be specified. This type of investigation of the results will enable the cost effectiveness of potential error reduction strategies to be assessed.

If an overall improvement in performance is required, then the principal PSFs identified in the task analysis, human error analysis, and quantification phases should be addressed. It may be, for example, that training was continually cited as a prominent factor affecting performance in a particular process plant emergency scenario, and in fact was rated the most important factor when quantification took place. Clearly, focusing on training in this particular scenario should have an overall positive effect on performance, and may be the best single way of reducing the impact of human error on the system goals. Some case studies of PSAs, including the derivation of error reduction strategies, are given in Humphreys (1988), and also a nuclear power plant HRA is extensively documented in Kirwan *et al.* (1995).

Where possible, the positive effects of the error reduction strategy adopted should be factored back into the quantitative analysis and it should be checked that the HEPs and overall system risk calculated become acceptable. This requires not only the precise operational definition of each aspect of the error reduction strategy, but also a method of ensuring that the strategy is properly implemented and maintained throughout the remaining life of the plant. This is part of the quality assurance phase.

Quality assurance

If the assessment has generated specific error reduction mechanisms the implementation and effectiveness of these should be ensured. For example, if as a result of an analysis equipment design recommendations are accepted, the successful implementation of the design changes should be monitored, and if possible the effects on performance verified at a later stage. In the UK, for example, measures derived from a risk assessment and implemented in a nuclear power plant, will be assessed initially after implementation, and then some time later (e.g., 2 years) to ensure that they are still working and having the desired effect on performance. This is necessary, as new measures may have a 'novelty value' and so improve performance but only temporarily — after the novelty wears off, performance (or error rate) may slip back to an undesired level. It is therefore necessary to check that the improvement measures are having 'sustainable impact'.

A continual performance monitoring system represents a powerful quality assurance system. A risk assessment generally predicts reliability for many years ahead, and much can change in this time: a gradual degradation in performance standards; an increasing maintenance loading; the loss of personnel involved in the design; impromptu changes during commissioning and start-up of the plant; and increasing 'retrofit' changes to the plant. Due to such factors the plant risk level may slowly and almost imperceptibly rise above the initial predicted and acceptable level to one that is unacceptable. Long-term performance monitoring (i.e., recording human errors, near-misses, incidents, and accidents) will detect this. A performance monitoring system can identify the point in time at which the assumptions and results of human reliability analysis are no longer applicable, and signify the need for a further analysis to justify the acceptability of the risk of the plant. This type of quality assurance system can avoid the gradual erosion of the safety barriers (the Bhopal syndrome — see Kirwan, 1994).

A performance monitoring system also has three other major advantages. First, operational error concerning errors can be used to detect trends towards new error modes or system failure paths, which may not have been predicted. The system then becomes an early warning system, and can detect problems before an incident or accident happens. Such an approach was used recently in Air Traffic Control (ATC) to detect a developing problem with hand-over practices (between one controller and another), and to correct them before a serious incident occurred (NATS, 2001). Second, such a system can help fine-tune the human error analysis system, since errors will be captured and recorded, and so error classification can learn from this process, making future periodic risk assessments more closely allied to reality. Third, an operational monitoring system can be used to collect quantitative data over a period of years, so that HEP information can be gained, either to be used directly, or to calibrate future risk assessments. Setting up such operational monitoring and recording systems represents an investment in terms of resources and gaining commitment from management and workers alike to use and trust the system, but the medium and long term payoffs in terms of improved performance, and resilience against accidents, will be worthwhile.

Documentation

It is important for the auditability and justifiability of the results of the study that it is fully documented. This will enable personnel unconnected with the assessment to understand how it was carried out, including all assumptions and judgements made during the study, and allow the study to be independently examined, updated, or even reproduced if necessary. It will also provide a potentially useful database from which to monitor future progress of the system's performance in comparison to the predictions made in the analysis. It is also entirely possible that at some stage in the future an accident investigation may take recourse to reviewing the predictions of an earlier safety assessment, to assess why the PSA failed to predict the occurrence of the accident that has now occurred. This perhaps would at least allow HRA to learn from its own mistakes, should they occur.

Future directions in HRA

Some areas are noted below in which human unreliability can affect system risk, but which are not currently adequately addressed by HRA techniques. These are therefore predicted to be the most likely future directions for human reliability developments.

New application areas for HRA

The HRA field has mainly concerned itself with the high risk, high technology industry sector, including nuclear power plants and chemical plants. There exist however, a large number of other lower technology sectors, e.g., mining, which often incur a high risk via a large number of 'small' accidents (say one or two fatalities), rather than (high risk technology) industries where the high risk is caused by a very small probability of an accident with many and serious consequences. This is clearly an area where applied human reliability should be able to help reduce risk. Some workers have already begun to use HRA approaches in such areas (Collier and Graves, 1986).

Some other areas of application have also arisen for the application of HRA approaches, such as medical risk (e.g., human errors during surgery), and even stock market applications, as well as more diverse applications in the transport sector, particularly in aviation. Not all of these applications use the whole HRA 'framework' from task analysis to quantification, however. In such areas, the approach is one of determining where the significant vulnerabilities lie in the system, as a function of human performance, and hence the focus is on identification of errors and then addressing them to make the system more robust. This is still HRA, but does not necessarily involve quantification, nor representation in fault or event trees. In this respect, it needs to be remembered that quantification is a means of prioritising human error issues, and is not an end in itself, so there can still be great value in HRA or the application of human error analysis approaches, even if quantification does not take place. Indeed, it is likely that the greatest growth area for HRA in the future will be in low technology or non-complex-system risk areas, as well as the transport sector.

Management, organisational and sociotechnical contributions to risk

Three of the most salient accidents emphasising the impact of human error in the past two decades remain the Bhopal (1984), Chernobyl (1986), and Challenger Space Shuttle (1986) disasters. Each of these contained a significant human error contribution, without which the disasters would not have occurred. However, it is arguable that current HRA techniques would have had difficulty in predicting them, because the fundamental types of error leading to the accidents were neither procedural nor diagnostic in the conventional sense (Watson and Oakes, 1988). Rather, all three involved a management and organisational error component. The Bhopal plant was possibly subject to economic pressures, resulting in management decisions which may have been more in favour of production than safety (Bellamy, 1986). The Challenger Space Shuttle disaster also appeared to contain a significant element of management-type decision error, also influenced by economic considerations (Rogers *et al.*, 1986). Lastly, the Chernobyl disaster (USSR State Committee, 1986) contained such a bizarre series of events affected by management that, had an analysis predicted this scenario prior to the accident, it possibly would have been treated with derision. Nevertheless, even such accidents as Chernobyl can be rationalised after the event, and indeed similar sequences can and have occurred (Reason, 1988).

Errors may also occur when dealing with small groups, at a 'lower' management and organisational level and caused by individual, organisational, and sociotechnical factors (e.g., social pressures, personality conflicts; Bellamy, 1983). These factors and the errors they cause are rarely assessed in HRAs, yet clearly can influence risks; this may happen via risk-taking: decisions to ignore procedures; failure to communicate information due to personality conflicts, etc. Techniques could therefore be developed in the future which identify, quantify and specify how to reduce these types of error. There is a current focus on teams and team performance in several industries, and an understanding of how team errors occur would be useful, especially for distributed team performance in critical incidents.

It is clear that HRA must eventually consider not only the operators in the control room, but also the management running the plant, since the latter can have a profound effect on safety. Few approaches exist, though one approach recently has focused on adapting the HAZOP approach for assessment of 'safety culture failures' (Kennedy and Kirwan, 1998).

In broad terms HRA is becoming an integral part of risk assessment, and its usage is steadily increasing. In the future it would seem likely that this will continue and the 'science' of HRA will become an accepted tool for use in many areas, not only those dominated by the need to assess high risk systems.

Error Recovery

It would be remiss not to point out that error recovery remains an under-researched area in HRA, despite its importance, and despite the fact that humans are very good at recovering from most of the errors they commit. Of all the techniques of HRA, THERP is the one that has always emphasised the importance of error recovery modelling, although the approach to such modelling has always remained 'implicit' rather than a clear methodological approach. There has been recent attention directed to the development of basic considerations of error recovery in HRA or incident analysis approaches (e.g., see Bove, 2001; Shorrock and Kirwan, 2002). Such work is trying to explicate the different reasons for recovery, whether self-recovery, or recovery induced by the system or another co-worker. Clearly, recovery is partly to do with one's own monitoring of performance, and partly (possibly mainly) a function of context and events in that context. This is an area deserving fundamental research, because it possibly offers a significant advantage in system reliability terms, and also because it keeps the human operator where (s)he belongs, as the most flexible and critical safety item in a complex system.

Context and HRA

The HRA approach is grounded in Information Processing Theory, with models such as Wickens (1982) under-pinning a number of approaches (e.g., see Isaac *et al.*, 2002). More recently, there has been an emphasis towards the treatment of human performance as a function of context, for example in the CREAM technique (Hollnagel, 1998). Such an approach aims to 'overhaul' the perhaps simplistic approach to performance shaping factors used by many HRA techniques, and determine more precisely how behaviour occurs as a function of context. It is too early to decide if such an approach is arguably better or worse than the existing approaches, but both 'paradigms' are limited by our understanding of human performance and human behaviour, and what drives us.

In particular, since HRA aims to be explicit, sometimes to the point of quantification of HEPs, both 'philosophies' are extremely limited by a lack of data that could tease out the subtle relationships and inter-dependencies between PSF or contextual factors and human performance in specific tasks and/or contexts. The end result, whatever basic philosophy underpins a HRA approach, and whether using THERP or CREAM, is a need to simplify human performance concepts, and to shore up data needs with expert judgement. All that can be said of current approaches as discussed in this chapter, is that they have served a useful function for some time, and continue to do so. Whilst their theoretical under-pinnings may at times be unfashionable and not robust academically, their empirical validity has been demonstrated on numerous occasions.

Therefore, whilst new contextual approaches or others will occur as Human Factors and Psychology continually evolve, they should not replace existing methods until they are shown to be at least as good as what they seek to replace. In the end, HRA is very much a tool in a market place. When something more useful or more efficient or more effective comes along, then it will be replaced. Until that time, many HRA practitioners and tool developers grounded in the Information Processing theoretic domain, are trying to treat context more formally through the human error identification phase of HRA, rather than the quantification phase (e.g., see Shorrock and Kirwan, 2002; Isaac *et al.*, 2002).

Ecological validity and HRA

A number of researchers in the human error field have recently been questioning the approach of HRA from an 'ecological' perspective (e.g., Hollnagel, 1998; Paries, 2001; Amalberti, 2001). The essence of this argument is that errors are too often seen in HRA as being discrete events, failures in performance, as opposed to successful performance — and that this is simply a wrong view of real performance. Instead, human performance includes making mistakes — that is how we act in life, and this is part of our adaptability. If we commit no errors at all, then in fact we lose a valuable feedback mechanism for our own behavioural control, and significant errors may then arise.

This author has no argument with this philosophy, and awaits the development of approaches based on it. In the meantime, there are some important implications for HRA and system safety. First, there should not be an attempt to completely eradicate all human errors from a system, as this is simply not natural, and in any case is not really achievable in complex systems. Second, the emphasis for research should be on how humans control their own performance and keep it within acceptable boundaries (the so-called 'performance envelope'), whether this is a single human operator, or a team, or an organisation. This aspect reinforces a need to better understand error recovery processes. Third, risk assessments themselves tend to focus on isolated and independent failures that occur randomly or stochastically, and then coincidentally occur to lead to system failure. Many accident investigators have a problem with this largely stochastic analysis of events. They feel, rather, that certain events were 'waiting to happen', or else seem to feel that 'Murphy's Law' (if it can go wrong, it will) 'took over' and led to a disastrous chain of events. It may be that our 'performance control mechanisms', including those at a team level, can break down, leading to events such as Chernobyl or Challenger, in ways we still do not yet fully understand. Such larger-scale and more integrative failure mechanisms could be more damaging to systems than 'discrete' human errors, and may be under-addressed in risk assessments. This is therefore an area deserving further research.

Looking both ways — the JANUS concept, and learning

HRA deals with human error. So does accident and incident investigation. The two are undeniably conceptually linked, yet the two areas of prediction and retrospective analysis have remained distant cousins for a long time in most industries, and the two fields often appear to virtually ignore each other. This seems to be due to different organisational origins of the two approaches, rather than different conceptual bases. In practice, those carrying out risk assessment are often in a different part of a company than those analysing incidents, and the personnel often have different backgrounds — reliability engineers and human factors people, versus often operationally seasoned people involved in incident investigation.

Recently, there has been more discussion occurring between these 'separated twins'. This has led to a number of approaches trying to adopt the logically obvious approach of linking prediction to retrospective analysis. As noted earlier, a HRA practitioner in a company or industry that is linked to an incident analysis function, can remain aware of what factors are really causing incidents and human errors, and so remains 'current' and 'relevant'. Some companies have no doubt done this for many years, but recently there have been attempts to create more explicit links between prediction and retrospective analysis, by making the analytical elements (the error and factor taxonomies) the same. This is most clearly occurring in the air traffic control domain (see Isaac and Pounds, 2001; Shorrock and Kirwan, 2002; Isaac *et al.*, 2002) via the TRACEr, HERA and JANUS approaches. The latter approach, named after the Roman god that could look both at the past and the future (the etymological origin of the month January), has merged two incident analysis systems from US and European ATC (Isaac and Pounds, 2001). The aim is to be able to compare incident causes from both sides of the Atlantic, and learn more about human error problems in ATC. HERA and TRACEr both are predictive tools, as well as retrospective approaches.

The advantage of a JANUS approach is that a company can learn more usefully about how to manage error in its own context. Events and errors that have occurred in the past, can be prevented in future system designs, if there is a degree of homogeneity or isomorphism in the error approaches in a company. If this happens, then rather than having two independent approaches of retrospective analysis and prediction, the company has a unified and stronger approach of *error management*.

Furthermore, from an academic perspective, if there are more approaches with this 'double view', then our understanding of human error and its causes will increase, perhaps one day to the point when we are not so data-limited, and can develop much better approaches, contextual or otherwise.

Summary

This chapter has overviewed the field of human reliability assessment largely as a tool for reducing the risk of large-scale accidents, although it has other uses (e.g., in improving productivity). A broad generic methodology has been outlined, detailing the various steps in HRA and its interactions with PSA, and the more prominent methodologies and issues in practical HRA have been outlined. From the discussions in this chapter it is apparent that HRA is a relatively mature approach, and has been expanding into other domains than those with which it was traditionally associated.

There are a number of interesting areas of research at the moment, such as more contextual approaches to HRA, and the need for a better understanding of error recovery, and of performance monitoring and control. Such areas could enrich the HRA approach, or derive alternative and better paradigms and techniques — time will tell.

However, it is likely that the focus on human error from a structured and systematic viewpoint, and the ability to merge such analysis with a modelling of how the rest of the system components and sub-systems can fail, will mean that HRA is used for some time to come. HRA, whether quantitative or qualitative, remains a tool that the ergonomist should have available in his or her toolkit, to help solve applied problems in the evaluation of human work situations.

References

Amalberti, R. (2001). The paradoxes of almost totally safe transportation systems. *Safety Science,* **37**, 109–126.

Apostolakis, G.B., Beer, V.M. and Mosleh, A. (1987). A critique of recent models for human error rate assessment. Paper presented at the *International POST-SMIRT 9th Seminar on Accident Sequence Modelling,* Munich, Germany. Reprinted in *Reliability Engineering and System Safety* (1988), **22**, 201–217.

Baber, C. and Stanton, N.A. (1996). Human error identification techniques applied to public technology: predictions compared with observed use. *Applied Ergonomics,* **27**, 119–131.

Basra, G. and Kirwan, B. (1998). Collection of offshore human error probability data. *Reliability Engineering and System Safety,* **61**, 77–93.

Bell, B.J. (1984). Human reliability analysis for probabilistic risk assessment. *Proceedings of the 1984 International Conference on Occupational Ergonomics,* pp. 35–40.

Bellamy, L.J. (1983). Neglected individual, social and organisational factors in human reliability assessment. *Proceedings of the 4th National Reliability Conference,* Reliability 85, NEC, Birmingham.

Bellamy, L.J. (1986). The safety management factor: an analysis of human aspects of the Bhopal disaster. Paper presented at the *1986 Safety and Reliability Symposium,* Southport.

Bellamy, L.J., Kirwan, B. and Cox, R.A. (1986). Incorporating human reliability into probabilistic risk assessment. Paper presented at *5th International Symposium in Loss Prevention and Safety Promotion in the Process Industries,* Société de Chimie Industriale.

Bello, G.C. and Columbari, V. (1980). The human factors in risk analysis of process plants: the control room operator model, TESOC. *Reliability Engineering,* **1**, 3–14.

Bove, T. and Andersen, H.B. (2001). Types of error recovery in air traffic management. *3rd Conference on Engineering Psychology and Cognitive Ergonomics,* Edinburgh, edited by D. Harris, Aldershot, U.K.: Ashgate.

Cacciabue, P.C. (1998). Modelling and Simulation of Human Behaviour in System Control. (New York: Springer).

Collier, S.G. and Graves, R.J. (1986). Improving human reliability: practical ergonomics for design engineers. In *Proceedings of the 9th Advances in Reliability Technology Symposium,* University of Bradford.

Comer, M.K., Seaver, D.A., Stillwell, W.G. and Gaddy, C.D. (1984). *Generating Human Reliability Estimates Using Expert Judgement.* USNRC Report Nureg/CR-3688 (Washington, D.C.: USNRC).

Cooper, S.E., Ramey-Smith, A., Wreathall, J., Parry, G.W., Bley, D.C., Luckas, W.J., Taylor, J.H., and Barriere, M.T. (1996). A technique for human error analysis (ATHEANA). Brookhaven National Laboratory, Upton NY, US NRC Nureg/CR-6350, Washington, D.C. 20555, May.

Cox, S.T. and Tait, N.R.S. (1991). *Reliability, Safety and Risk Management* (London: Butterworth-Heinemann).

Dougherty, E. (1993). Human reliability analysis and context. *Reliability Engineering and System Safety*, **41**, 25–47.

Dolby, A.J. (1990). A comparison of operator response times predicted by the HCR model with those obtained from simulators. *International Journal of Quality and Reliability Management*, **7**, pp. 19–26.

Drury, C.G. (1983). Task analysis methods in industry. *Applied Ergonomics*, **14**, 19–28.

Edwards, W. (1977). How to use multi-attribute utility measurement for social decision-making. *IEEE Transactions on Systems, Man, and Cybernetics*, SMC-7-5.

Embrey, D.E. (1986). SHERPA: a systematic human error reduction and prediction approach. Paper presented at the *International Topical Meeting on Advances in Human Factors in Nuclear Power Systems*, Knoxville, Tennessee.

Embrey, D.E. (1987). Error Analysis in Proceduralised Tasks. Material presented in the *Human Reliability Analysis Course*, SRD, UKAEA, Culcheth, Cheshire.

Embrey, D.E. and Kirwan, B. (1983). A comparative evaluation study of three subjective human reliability qualification techniques. In *Proceedings of the Ergonomics Society's Conference 1983*, edited by K. Coombes (London: Taylor & Francis), pp. 137–141.

Embrey, D.E., Humphreys, P., Rosa, E.A., Kirwan, B. and Rea, K. (1984). *SLIM-MAUD: An Approach to Assessing Human Error Probabilities Using Structured Expert Judgement*. USNRC Report Nureg/CR-3518 (Washington, D.C.: USNRC).

Embrey, D.E., Kontogiannis, T. and Green, M. (1994). *Preventing Human Error in Process Safety*. Centre for Chemical Process Safety (CCPS), American Institute of Chemical Engineers (New York: CCPS).

Evans, A., Slamen, A.M, and Shorrock, S.T. (1999). *Use of Human Factors Guidelines and Human Error Identification in the Design Lifecycle of NATS' Future Systems*. Paper presented at the FAA-Eurocontrol Interchange Meeting, Toulouse, France, April.

Gertman, D.I. and Blackman, H. (1994). *Human Reliability and Safety Analysis Data Handbook*. Chichester: John Wiley & Sons.

Gibson, H., Basra, G. and Kirwan, B. (1999). Development of the CORE-DATA database. *Safety and Reliability Journal*, Safety and Reliability Society, Manchester, **19**, 6–20. ISBN 0961-7353.

Green, A.C. (1983). *Safety Systems Reliability* (Chichester: John Wiley).

Hall, R.E., Fragola, J.R. and Wreathall, J. (1982). Post-event human decision errors: operator action tree/time reliability correlation. NUREG/CR-2010, USNRC, Washington, D.C. 20555.

Henley, E.J. and Kumamoto, H. (1981). *Reliability Engineering and Risk Assessment* (New Jersey: Prentice Hall).

Hollnagel, E. (1993). *Human Reliability Analysis: Context and Control*. (London: Academic Press).

Hollnagel, E. (1998). *Cognitive Reliability and Error Analysis Method: CREAM*. (Oxford, UK, Elsevier).

Humphreys, P. (1988: Ed.). *The Human Reliability Assessor's Guide*. Report RTS 88/95Q. NCSR, UKAEA, Culcheth, Cheshire.

Hunns, D.M. (1982). The method of paired comparisons. In *High Risk Safety Technology*, edited by A.E. Green (Chichester: John Wiley).

Isaac, A., Kirwan, B. and Shorrock, S. (2002). Human Error in European Air Traffic Management: the HERA Project. *Reliability Engineering and System Safety*, **75**, 257–272.

Isaac, A. and Pounds, J. (2001). Development of an FAA-Eurocontrol approach to the analysis of human error in ATM. Paper presented at the *FAA Eurocontrol R&D Seminar*, Santa Fe, December 3–7.

Kantowitz, B. and Fujita, Y. (1990). Cognitive theory, identifiability, and human reliability analysis. *Reliability Engineering and System Safety*, **29**, 317–328.

Kennedy, R. and Kirwan, B. (1998). Development of a HAZOP-based method for identifying safety culture failure mechanisms. *Safety Science*, **30**, 249–274.

Kennedy, R., Kirwan, B., Summersgill, R. and Rea, K. (2000). Validation of HRA techniques — déjà vu five years on? In *Foresight and Precaution, Proccedings of SREL 2000*, Edinburgh, May 15–17, Eds. Cottam, M.P., Pape, R.P., Harvey, D.W. and Tait, J. Balkema, Rotterdam. ISBN 90-5809-140-6.

Kirwan, B. (1986). *Techniques to Aid Human Error Identification in Human Reliability Assessment, Human Reliability Course Manual* (London: IBC Technical Services Limited).

Kirwan, B. (1987). Human reliability analysis of offshore emergency blowdown system. *Applied Ergonomics*, **18**, 23–34.

Kirwan, B. (1988). A comparative evaluation of five human reliability assessment techniques. In *Human Factors and Decision Making*, edited by B.A. Sayers (Oxford: Elsevier), pp. 87–104.

Kirwan, B. (1990). A resources flexible approach to human reliability assessment for PRA, *Safety and Reliability Symposium*, Altrincham, September (London: Elsevier Applied Sciences), pp. 114–135.

Kirwan, B. (1992a). Human error identification in human reliability assessment. Part 1: overview of approaches. *Applied Ergonomics*, **23**, **5**, 299–318.

Kirwan, B. (1992b). Human error identification in HRA. Part 2: Detailed comparison of techniques. *Applied Ergonomics*, **23**, **6**, 371–381.

Kirwan, B. (1994). *A Guide to Practical Human Reliability Assessment* (London: Taylor & Francis).

Kirwan, B. (1994). *A Guide to Practical Human Reliability Assessment*. (London: Taylor & Francis).

Kirwan, B. (1997a). The Development of a nuclear chemical plant human reliability management approach. *Reliability Engineering and System Safety*, **56**, 107–133.

Kirwan, B. (1997b). Validation of Human Reliability Assessment Techniques: Part 1 — Validation Issues. *Safety Science*, **27**, 25–41.

Kirwan, B. (1997c). Validation of human reliability assessment techniques: Part 2 — Validation Results. *Safety Science*, **27**, 43–75.

Kirwan, B. (1998). Human error identification techniques for risk assessment of high risk systems — Part 1: Review and evaluation of techniques. *Applied Ergonomics*, **29**, 157–177.

Kirwan, B. (1998). Human error identification techniques for risk assessment of high risk systems — Part 2: Towards a framework approach. *Applied Ergonomics*, **29**, 299–318.

Kirwan, B. and Ainsworth, L.A. (Eds) (1992). *A Guide to Task Analysis* (London: Taylor & Francis).

Kirwan, B. and James, N.J. (1989). The development of a human reliability assessment system for the management of human error in complex systems, in *Reliability 89*, Brighton Metropole, June, pp 5/A/2–15/A/2.10.

Kirwan, B. and Rea, K. (1986). Assessing the human contribution to risk in hazardous materials handling operations. Paper presented at the *First International Conference in Risk Assessment of Chemicals and Nuclear Materials*, Surrey University.

Kirwan, B., Martin, B.R., Rycraft, H. and Smith, A. (1990). Human error data collection and data generation. *International Journal of Quality and Reliability Management*, **7**, 34–66.

Kirwan, B., Scannali, S. and Robinson, L. (1996). Applied HRA: a case study. *Applied Ergonomics*, **27**, 289–302.

Kirwan, B. and Hollnagel, E. (1998). *The Requirements of Cognitive Simulations for Human Reliability and Probabilistic Safety Assessment.* CSEPC 1, Kyoto.

Kirwan, B. and Kennedy, R. (2001). Assessing Safety and Usability of Air Traffic Management (ATM) Systems Using a HAZOP Approach. In *Human Error and Safety System Development (HESSD 2001)*, Linkoping, Sweden, June 11–12.

Kirwan, B., Kennedy, R. and Hamblen, D. (2001). Human Reliability Assessment in Probabilistic Safety Assessment — Guidelines on Best Practice for Existing Gas-Cooled Reactors. *IBC Conference on Probabilistic Safety Assessment*, London, 26–27 November.

Kletz, T. (1984). *HAZOP and HAZAN – Notes on the Identification and Assessment of Hazards* (Rugby: Institute of Chemical Engineers).

Ludke, R.L., Strauss, F.F. and Gustafson, D.H. (1977). Comparison of five methods for estimating subjective probability distributions. *Organisational Behaviour and Human Performance*, **19**, 162–179.

Mason, S. (1997). Procedural violations — causes, costs and cures. In *Human Factors I Safety Critical Systems*, F. Redmill and J. Rajan (Eds) (Oxford: Butterworth-Heinemann).

Murphy, A.H. and Winkler, R.L. (1974). Credible interval temperature forecasting: some experimental results. *Monthly Weather Review*, **102**, 784–794.

Nicks, R. (1981). Probabilistic approach to problems in reactor safety risk assessment. Paper presented at *Convegno Internazionale sul Fondamenti della Probabilita e della Statistica*, Luino, Italy.

Paries, J. (2001). Personal communication.

Park, K.S. (1987). *Human Reliability* (Oxford: Elsevier).

Pew, R.W., Miller, D.C. and Feehrer, C.G. (1987). *Evaluation of Proposed Control Room Improvements Through Analysis of Critical Operator Decisions* (Palo Alto, CA: Electric Power Research Institute).

Pontecorvo, A.B. (1965). A method of predicting human reliability. *Annals of Reliability and Maintenance*, **4**, 337–342.

Potash, L., Stewart, M., Dietz, P.E., Lewis, C.M. and Dougherty, G. (1981). Experience in integrating the operator contribution in the PRA of actions operating plants. In *Proceedings of the ANS/ENS Topical Meeting on PRA*, New York, American Nuclear Society.

Rasmussen, J., Pederson, O.M., Camino, A., Griffon, M., Mancini, C. and Gagnolet, P. (1981). *Classification System for Reporting Events Involving Human Malfunctions.* Report Riso-M-2240, DK-4000 (Roskilde, Denmark: Riso National Laboratories).

Rasmussen, J., Duncan, K.D. and Leplat, J. (Eds.) (1987). *New Technology and Human Error.* (Chichester: John Wiley & Sons).

Reason, J. (1988b). Human fallibility. *Consortium of Local Authorities Proof of Evidence in the UK Hinkley 'C' Power Station Enquiry*, COLA 22.

Reason, J.T. (1990). *Human Error* (Cambridge: Cambridge University Press.).

Reason, J.T. and Embrey, D.G. (1986). *Human Factors Principles Relevant to the Modelling of Human Errors in Abnormal Conditions of Nuclear and Major Hazardous Installations.* Report for the European Atomic Energy Community (Lancs: Human Reliability Associates Ltd).

Reason, J.T. (1990). *Human Error.* (Cambridge, UK: Cambridge University Press).

Reason, J. (1998). *Managing the Risks of Organisational Accidents.* (Aldershot, England: Ashgate).

Rogers, W.P. *et al.* (1986). *Report of the Presidential Commission on the Space Shuttle Challenger Accident*, June 6.

Seaver, D.A. and Stillwell, W.G. (1983). *Procedures for Using Expert Judgement to Estimate Human Error Probabilities in Nuclear Power Plant Operations.* Nureg/CR-2743. (Washington, D.C.: USNRC).

Senders, J.W. and Moray, N.P. (1991). *Human Error: Cause, Prediction and Reduction.* (New Jersey: Lawrence Erlbaum Associates).

Shepherd, A. (1986). Issues in the training of process operators. *International Journal of Industrial Ergonomics*, **1**, 49–64.

Shepherd, A. (2001). *Hierarchical Task Analysis.* (London: Taylor & Francis).

Shorrock, S.T. and Kirwan, B. (1998). TRACEr — a technique for the retrospective analysis of cognitive errors in Air Traffic Management. In D. Harris (Ed) *Engineering Psychology and Cognitive Ergonomics*, Vol. 3, pp. 163–171, (Aldershot, UK: Ashgate).

Shorrock, S.T. and Kirwan, B. (2002). Development and application of a human error identification technique for ATM. *Applied Ergonomics*, *33*, 319–336.

Spurgin, A.J., Lydell, B.O., Hannaman, G.W. and Lukic, Y. (1987). Human reliability assessment-a systematic approach. In *Reliability 87*, NEC, Birmingham.

Swain, A.D. (1987). Accident sequence evaluation program human reliability analysis procedure. NUREG/CR-4722. Washington, D.C.-20555: USNRC.

Swain, A.D. (1989). Comparative evaluation of methods for human reliability analysis, GRS-71, Gesellschaft fur Reaktor Sicherheit, GRSmbH, Koln, Germany.

Swain, A.D. and Guttmann, H.E. (1983). *A Handbook of Human Reliability Analysis with Emphasis on Nuclear Power Plant Applications.* Nureg/CR1278 (Washington, D.C.: USNRC).

Taylor-Adams, S.T. and Kirwan, B. (1995). Human Reliability Data Requirements. *International Journal of Quality and Reliability Management*, **12**, 24–46.

Topmiller, D.A., Eskel, J.S. and Kozinsky, E.J. (1984). Human reliability databank for nuclear power plant applications. NUREG/CR-2744, USNRC, Washington, D.C. 20555.

USNRC (1980). *Three Mile Island: A Report to the Commissioners and to the Public* (The Rogovin Report), USNRC Report Nureg/CR-1250-V. (Washington, D.C.: USNRC).

USNRC (1984). *Seabrook Station Probabilistic Safety Analysis. Section 10: Human Actions Analysis.* (Washington, D.C.: USNRC).

USNRC (1985). *Loss of Main and Auxiliary Feedwater at the Davis-Beese Plant on June 9, 1985.* Nureg-1154. (Springfield, VA: National Technical Information Service).

USSR State Committee (1986). *The Utilisation of Atomic Energy: The Accident at the Chernobyl Nuclear Power Plant and Its Consequences.* Information compiled for the IAEA Experts' Meeting, 19–25 August, Vienna (Part 1).

Wagenaar, W. (1986). *The Cause of Impossible Accidents.* Speech delivered on the occasion of the 6th Dujiker Lecture, 18 March, Netherlands Institute of Psychologists, University of Amsterdam.

Watson, 1. and Oakes, F. (1988). *Human Reliability Factors in Technology Management.* SRD Report, UKAEA, Culcheth, Cheshire.

Webley, K. and Ackroyd, P. (1988). Process control study. In Humphreys, P. (Ed.) The Human Reliability Assessor's Guide. Report RTS 88/95Q. NCSR, UKAEA, Culcheth, Cheshire.

Whittingham, R.B. (1988). The design of operating procedures to meet targets for probabilistic risk criteria using HRA methodology. Paper presented at *IEEE Conference on Human Factors in Nuclear Power*, Monterey, CA, pp. 303–310.

Wickens, C. (1992). Engineering psychology and human performance. New York: Harper-Collins.

Williams, J.C. (1986). HEART — a proposed method for assessing and reducing human error. In *Proceedings of the 9th Advances in Reliability Technology Symposium*, University of Bradford.

Williams, J.C. (1988). A data-based method for assessing and reducing human error to improve operational performance. Paper presented at the *IEEE Conference on Human Factors in Nuclear Power*, Monterey, CA, pp. 436–450.

Williams, J.C. (1992). Towards an improved evaluation analysis tool for users of HEART. In *Proceedings of the International Conference on Hazard identification and Risk Analysis, Human Factors and Human Reliability in Process Safety* (Centre for Chemical Process Safety, AIChemE), January 15–17, Orlando, Florida.

Woods, D.D. *et al.* (1994). Behind human error: cognitive systems, computers and hindsight. CSERIAC State of the art report. Ohio: Wright Patterson Air Force Base. October.

Chapter 33

Systems for near miss reporting and analysis

T.W. van der Schaaf and L.B. Wright

Introduction

This chapter focuses on how to design, implement, maintain and operate a system for near miss reporting and analysis in the domains of industry and transportation. Factors such as the type of event to be reported, the goals of the reporting system, design specification and analysis are described and discussed. The experiences of the authors are described with reference to two case studies of near miss reporting systems, and what has been learned from these experiences.

The first step in designing and implementing a reporting system is to decide on the type of event that the system will receive. The range from accidents to near misses can all be reported via a reporting system. As the definitions of accidents and near misses are central to the design of reporting systems they are explicitly defined here. There are many definitions for these classes of event, however unless they are clearly defined it is possible that confusion may arise between professionals (making comparisons across industry difficult) and those responsible for reporting the accidents and near misses (leading to inappropriate or under reporting). Heinrich (1931) defines an accident as 'an unforseen, improper or non-planned occurrence — an accident need not necessarily lead to an injury'; Brown (1979) similarly uses the following definition of accident 'the unplanned outcome of inappropriate behaviour'. A near miss is, according to Van der Schaaf (1991) 'any situation in which an ongoing sequence of events was prevented from developing further and hence preventing the occurrence of potentially serious (safety related) consequences'.

The relationship between accidents and near misses may be illustrated by the following simple model of accident causation (see Figure 1).

This chapter concentrates on near misses as the focus of the reporting system for a number of reasons:

1. They occur in greater numbers than actual accidents, thus providing a larger data set (or the same size database in much less time) on which to base prevention measures
2. There is no need to wait until an actual accident has occurred to understand the causal process and begin prevention measures (near misses as precursors to later accidents)
3. The successful recovery actions which prevented the near miss from escalating to a fully-fledged accident can be used to help encourage recovery behaviour (as an additional safety strategy besides just preventing initial errors/failures)
4. As near misses are non-consequential there are less problems and barriers for free, complete, unbiased reporting.

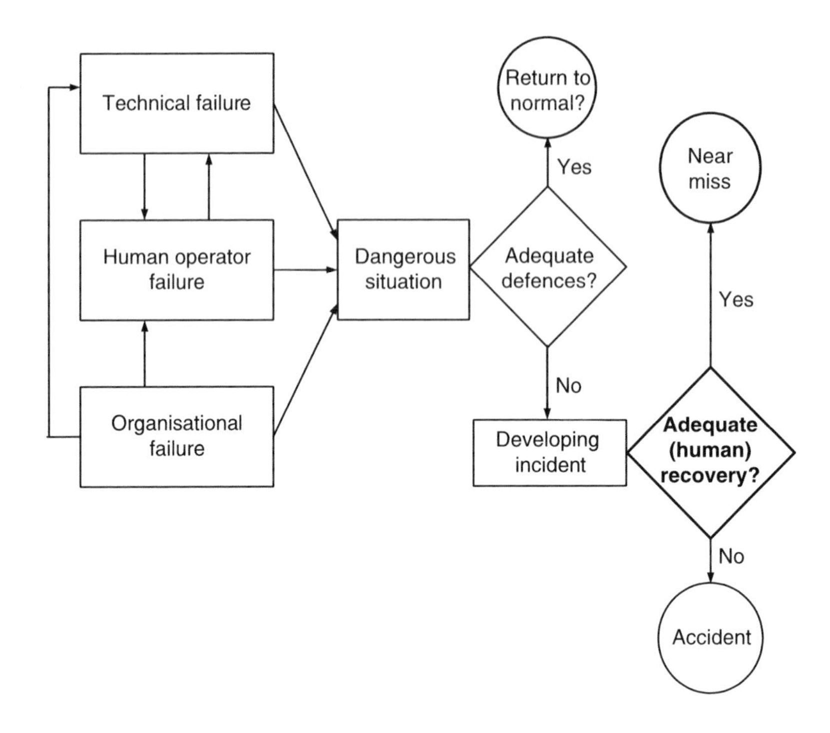

Figure 1 The PRISMA incident causation model (van der Schaaf, 1992).

Goals of near miss reporting systems

The goal or goals of a reporting system should be determined from the outset. The goal(s) of the system then drives the way the system is designed and organised in terms of data input (the types of event that should be reported), data processing (the method selected to analyse the data, and how the data are selected for analysis) and output (implemented changes, monitoring and feedback to staff).

Van der Schaaf (1991) outlines three purposes for the collection of near misses:

Modelling and learning about new failure modes

Within this objective, van der Schaaf suggests that the specific goal is to 'identify likely factors or system elements' qualitatively in order to identify precursors of future accidents. From this qualitative insight there are two ways to reduce the likelihood of future incidents and accidents. First by eliminating error-inducing factors, and secondly by introducing or strengthening recovery-promoting factors. Recovery factors are those factors which contribute to a full or partial recovery of a situation after an error or failure has occurred, and so prevent or reduce the negative consequences of the error or failure. In order for successful recovery actions to be strengthened, they must be identifiable and part of the modelling process (see Van der Schaaf, 1988; van der Schaaf and Kanse, 2000 for a detailed explanation). Further, another type of qualitative insight may be the identification of new or unusual combinations of factors which resulted in an incident or accident. As new types of events are always possible (often what we fail to predict does happen), this means that the set of possible accident scenarios is increased. However, a completely pre-coded system is unlikely to be able to include such events. This is not unrealistic — no taxonomic system can possibly list all possible failure modes. This type of analysis then supplements the pre-coded system.

Monitoring the frequency of known failures

This objective requires causal pattern recognition across a number of minor accidents or of more major accidents into pre-coded categories. Using such a system it is possible to determine which factors or combinations of factors occur most frequently. This would then form the basis for a rational decision making process about where to concentrate resources in an attempt to improve safety in the most efficient way. If such a system were also linked to actions undertaken, then the effectiveness of interventions could also be monitored i.e., has the intervention led to a decrease in the incidence of the particular factors it was aimed at?

Maintain alertness and motivate staff

After long periods where there are no visible signs of inherent risks (e.g., people injured; fires; major damages) near misses can be used to maintain alertness to risks. Descriptions of near misses can be used for training and discussion purposes, and for safety promotion campaigns. Near misses can also be used as reminders of the usefulness or necessity of existing barriers and defences; and of the importance of *ad hoc* unplanned recovery.

Designing reporting systems

Prior to designing the near miss reporting system, decisions need to made regarding the organisational structure of the system. Near miss reporting systems may be administered in-house or by an external (trusted) third party (an honest broker). There are advantages and disadvantages of both. In-house systems have the advantages of being an integral part of company systems, operated by staff who understand the company culture and ways of working. Data from the near miss system can easily be compared with data from other management systems within the company, and ownership of the data rests with the company. If the administrators of the system are trusted by staff, then the system can flourish. However, if the company has a punitive approach to near misses then an in-house system may not be trusted or used by staff. Further, the near miss system should not be an ad-hoc addition to current management systems — specific staff should be assigned to deal with near misses or the workload associated with an in-house system may result in under-staffing or under-utilisation of the data. Third party systems have the advantage of being perceived as external to the company — this is especially useful where the company is discipline focussed. Guarantees of confidentiality and no-blame are more likely to be believed when an external broker is involved. However, in the case of externally run systems, the data are normally not owned by the individual company and this can make comparisons with other safety management systems within the company difficult. The choice between in-house and third party systems is dependent ultimately upon the company culture, willingness to report and discipline systems. There is no one best way to administer near miss reporting systems, and the choice should be of the one which is most beneficial to the company.

Van der Schaaf's (1992) Table 1 provides an overview of the modules necessary for a near miss framework. This framework outlines the various purposes and goals of a reporting system, and details the various modules necessary to fulfil these goals.

This table represents the inputs to the system (modules 1–3), the way these inputs are processed (modules 4–6) and the output and evaluation which follows the preventive measures.

Table 1 An overview of Different Versions of the Basic Framework According to Different Purposes (from van der Schaaf, 1992 p. 37)

Module	*Modelling* purpose	*Monitoring* purpose	*Alertness* purpose
1. *Detection* recognition and reporting	Everything	Known problems only	Recognising and reporting
2. *Selection* of events for analysis according to purpose(s)	New reports only	*Not relevant*	Convincing
3. *Description* of all relevant hardware, human and organisational factors	Detailed	*Not relevant or very superficial*	Detailed examples of new and old hazards
4. *Classification* according to socio-technical model	Flexible: looking for new root causes	Routine: standard set of root causes	Not relevant
5. *Computation* statistical analysis to uncover certain (patterns of) factors	*Not relevant.* Only single events considered	Periodic analysis of updated large database	Not relevant
6. *Interpretation and Implementation* translation of statistical results into corrective and preventive measures	Finding (new) ways of improving prevention and recovery	*Not relevant/Already prescribed by module 4*	Near misses as precursors; focus on recovery mechanisms
7. *Evaluation* measuring the effectiveness of proposed measures after their implementation	Not relevant	Comparing predicted and actual effects of implemented measures	Not relevant

Detection

In terms of input, the designers of the reporting system need to decide on the type of events that the system is being designed to collect. There are a variety of choices, including for example, major accidents, minor deviations and near misses, events that were successfully recovered, particular types of event such as SPADs (signals passed at danger — on railways), LOCA (loss of coolant accident — in nuclear power plants) or unsafe practices and situations. A near miss reporting system concentrates on near misses (including recovery actions), and often also on unsafe situations and practices as well. This chapter concentrates on near miss reporting systems although the principles of the design and implementation apply equally to reporting systems which aim to collect information on actual accidents and their consequences. The decision on which type of event to collect via the reporting system is dependant upon the goals of the system and the types of events which are already collected and analysed by the company or department.

The three purposes — of modelling, monitoring and alertness — help the system designer in deciding what types of report should be collected. When the purpose is modelling, then the reporting system is interested in all reports of near misses. If the goal of the reporting system is to monitor the occurrence of known failures or problems, then the goal of the reporting system, and the nature of the particular failure that is being monitored, should be clearly explained to staff. In this case, the reporting system is only interested in obtaining and analysing reports on this particular issue, and all other types of report are ignored. The third purpose, alertness, aims to maintain staff awareness of hazards and near misses and so is dependant upon decisions regarding modelling or monitoring: staff alertness to all near misses or to particular types of near miss may be raised and maintained via the reporting system.

In practice one may often find all three purposes combined into one integrated near miss system: a near miss system may collect all near misses and focus on specific situations or events for part of its operation. However, in the initial stages of designing and implementing the reporting system it is wise to have clarity on the purpose and type of reports in order to motivate and encourage staff to report.

Giving a wide definition of near misses encourages staff to make reports and makes it easier for staff to decide to make a report. Further, to encourage staff to make reports they must be aware of the system, how it operates, how to make a report and what benefits can be gained from reported near misses.

Decisions also need to be made on the route via which reports can be made. This decision may be constrained by the type of system that is implemented (e.g., users of a fully anonymous system may be unlikely to use a telephone route out of office hours as this might require the reporter to leave their telephone number on an answering machine). There are a number of possible routes that are suitable for making reports, via a form, electronically, via telephone, or in person.

When a report is made on paper, the form may contain a number of questions for the reporter to answer (such as date, time, equipment being used, place of event, etc.), or it may contain a fairly open question such as 'describe the event you wish to report and how/why you think it happened'. If the reporting scheme is not anonymous then there should be space for the name and contact details of the reporter. Thought should be given to the method of eliciting reports — to make it as easy as possible for reporters reply paid envelopes could be provided along with a supply of forms. Forms should be available in places that are easy for potential reporters to access and not on request from senior managers.

Similarly, if the report is to be made electronically, decisions on the design of the questions to be answered must be made. There are internet-based near miss reporting systems where reports are made electronically and are anonymous — however it is likely that an individual could be traced via the internet, and further assurances may be required in such cases. Also, unless reporters have access to user-friendly computer technology, an electronic reporting system is likely to have limited success.

There are a number of advantages and disadvantages to telephone reporting. The advantages relate to the ease and rapidity of making a report. Further, the person making the report is directly in contact with someone from the reporting system and thus there is a greater sense of involvement and personalisation, and possibly more information provided by the reporter. The disadvantages are that if the reporter calls out of hours or when the analysts are unavailable, then they may not leave a message or a contact number. Telephone reporting is likely to increase the number of staff required to man the reporting system. However, should telephone reporting be an attractive option, then it is made more attractive to potential reporters if reporting a near miss involves no cost to the reporter (e.g., Freephone/Tollfree number).

Reporting in person (to a trusted colleague), has similar advantages and disadvantages as telephone reporting. It is immediate and both quicker and easier than filling in a form. However, depending on the size of the potential reporting pool it can be time consuming and require a large number of trained staff.

Selection

Once a reporting system receives reports, decisions are required on which reports to analyse in depth. The selection of reports for analysis becomes an increasingly important task with an increase in report numbers. Reporting rates have been known to increase dramatically once a 'safe' channel for reporting is offered, often 4–10 times the previous rate (Kaplan *et al.*, 1998).When a reporting system is first launched it is likely that initially all reports can be analysed (in terms of manpower), however, this situation is likely to change rapidly as reporting rates increase. Hence, the analysts and the managers need to clearly identify the incident reports that will be analysed in depth, those that will be analysed in a minimal way and those which are identified simply as repeat incidents. Freitag and Hale (1997) call this a compromise between 'generating enough events to learn from and avoiding swamping the analysis system with too much work which will cost more than its added value'. The rationale for deciding on the appropriate level of analysis (from full analysis to minimal analysis) should be clear and easy to follow by all analysts and may be determined by seriousness of potential consequence, by type of event (e.g., all incidences of maintenance errors) or by new events only.

In the case of modelling, the system is only interested in selecting new near misses for the description and analysis phase. The modelling goal may be short term (or continued until no new near misses are detected), and formalised into the goal of monitoring at a later date. For monitoring, known near misses (e.g., those identified from the modelling phase, or predetermined event types) are routinely classified and compiled in a database to detect any statistically significant trends in terms of improvements in error prevention or recovery promotion. For the goal of alertness, near misses are selected that are examples of old well known recurring problems, and new 'impossible' combinations of factors. These are then used to provide convincing illustrations of the fact that an absence of accidents should not be equated with perfect hazard control. Possible serious effects of seemingly minor deviations must be stressed as should the factors and measures which have acted as effective recoveries and barriers and defences.

Description

We have already stated that the three purposes of a near miss reporting system may be combined. This has implications for the description and analysis of reports. Should the purpose be confined solely to modelling, with no intention to implement monitoring at a later date, then analysis may be performed on an individual basis without taking into account the need to be able to compare across multiple near misses. If however, the ultimate aim is to monitor event causes across multiple near misses, then a structured and standard approach is required. In either case, enough detailed description of the reported near miss is required to enable the analysis to take place. In the case of maintaining alertness, enough detail is required to allow others to learn from the near misses.

The investigation of any report selected for further processing must lead to a detailed, complete neutral description of the sequences and events leading to the near miss situation. For example, an analysis based on Fault Tree techniques (Hyos and Zimolong, 1988) will help the investigator to describe all relevant system elements (technical failures, management decisions, operator errors, operator recoveries, etc.) in a tree-like structure. This description will show all causal elements in their logical order and chronological sequence.

Classification

The classification of root causes of near misses may be done on an individual basis without the use of a standardised taxonomy only if modelling alone is the goal of the system. Assigning causal factors on an individual basis (in a very rich 'narrative') rather than using a taxonomy does not allow the comparison of multiple events and is concerned with determining the range of unique, non-recurring causes present in events. The examination and analysis of single events does not allow for such a comparison and learning, prevention and improvements can therefore only take place from individual events. Such a reactive policy may fail to reduce the most frequently occurring causes and tackle only those causes that have resulted in high profile accidents.

If however, the goal is to enable statistical analysis of factors via the build up of a data base of causes, then a taxonomy of causes will allow reliable patterns of results to emerge and structural factors in the organisation to be identified over time. Using such a system it is possible to determine which factors or combinations of factors occur most frequently. This would then allow preventive measures to be based on the factors that occur most frequently, rather than on one (usually) major incident.

There is a wide range of taxonomies available (e.g., SRK, Rasmussen, 1982; GEMS, Reason 1990; PRISMA, Van der Schaaf, 1991; Feggetter's checklist, 1982; CIRAS, Wright *et al.*, 2000). The choice of taxonomy will to some extent reflect the appropriateness of the coding structure and availability. However, in order to aid decision making, Wright (2002) has compiled a list detailing the minimum requirements of a taxonomy. These are shown below:

1. Based on a suitable underlying theory of human behaviour.
2. Should include the full range of possible causes, i.e., technical components, human behaviour, and organisational and managerial causes.
3. Be revealing i.e., distinguish between events and underlying causes.
4. Should be comprehensive in a qualitative sense (i.e., should be able to handle accidents, damages and near misses; cover both failures and recoveries).
5. Be quantitative, enabling results to be computed and accumulated across many accidents.
6. Be consequential. Suggest ways of influencing and rectifying the root causes identified.
7. Be reliable i.e., that various independent analysts should reach the same conclusions.

A number of additional criteria are suggested by Shorrock (2002). These are aspects which are viewed as beneficial and optional to ensure that the taxonomy will aid in the process of assigning cause and helping to improve the safety of a system and may be used in determining whether to institute a system (e.g., cost factors, training). The implementation of a method will, of course, depend to some extent upon cost. However, turning this into a minimal requirement may detract from the efficacy and applicability of the analysis process selected, as currently all minimal requirements are viewed as equally important. The validity aspect (called predictive accuracy by Shorrock (*op cit*)) is as yet an undetermined measure for taxonomies which retrospectively assign causes to incidents and accidents. As such it is not included in this list of minimal requirements. However, as robust methods develop to check the validity of analysis methods then this item could be added. It is viewed here as an attractive addition but not as a requirement. It is suggested that taxonomies should at least meet the minimal requirements (and preferably some of the optional ones) before being used in a 'real life' setting. Optional criteria for taxonomic approaches are therefore.

Validity
Auditability
Resource efficiency (training)
Resource efficiency (usage)
Usability
Life-cycle applicability
Flexibility

Once a taxonomy has been selected those performing the analysis and classification of root causes need to have sufficient training in the use of the taxonomy.

Computation

The root causes as identified by analysis and classification must be entered into a database for further statistical analysis. This means that a near miss reporting system is not meant to generate ad hoc reactions by management after each and every serious near miss, but rather that factors contributing to multiple events are analysed periodically (e.g., each quarter) to identify structural factors in the organisation and work practices. Computation can only be performed where a set of standard root causes (taxonomy) is used to classify root causes.

Interpretation and implementation

Having identified the structural factors (root causes) the reporting system should aid the interpretation of them, i.e., suggest argumented, theoretical/logical ways of influencing the factors, to eliminate or diminish error factors and to introduce recovery opportunities in the organisation or system. Implementation of prevention strategies needs to be co-ordinated and publicised to staff. The implementation of a prevention strategy directly based on the database they helped to build also signals to staff that their reports are taken seriously and do not simply disappear into a 'black hole' (see below).

Evaluation

Once the reporting system has been designed and implemented, review and evaluation of the system is necessary to ensure that the system fulfils the goal(s) that were the impetus for initiating the system. However, evaluating the effectiveness of a reporting system and the resulting improvement measures (in the case of the Monitoring goal) is difficult and complicated for the following reasons:

- Multiple interventions usually take place in parallel; ideally to test the effectiveness of a particular corrective measure it would be implemented in isolation and the effect could be measured; but as corrective measures also take place at the same time as many other company initiatives and rule changes it is very difficult to assess exactly how successful any one corrective measure is.

- The time lag for changes to occur and take effect is unknown — hence multiple interventions may have taken place and it is difficult to be sure which intervention is responsible for the changes measured.
- Organisations and their culture and rules change over time.

Nevertheless, evaluation is necessary and can be achieved by monitoring the types of incidents and causes that occur over time and whether such causes increase or decrease. Evaluation can also take place at the level of the reporting system including assessment of:

- The number of reports.
- Other's errors or one's own errors.
- Level of potential risk.
- Checks for reporting biases.
- Measure of reporting culture as a precursor to performance.
- Whether reports are dealt with in a timely fashion.

Table 2 highlights the decisions that must to be made for each of Van der Schaaf's seven modules.

However, simply following the design specification and implementing the seven modules does not guarantee the success of a near miss reporting system. After implementing the design steps there is still effort required to ensure the success of the system. Any reporting system hinges on the willingness of staff to make (unbiased) reports (see Van der Schaaf and Kanse, 2004, for an overview of barriers to reporting). Thus after implementation the reporting rate needs to be monitored and steps taken to ensure staff trust the system. The success of a reporting system also depends upon reporting being perceived as meaningful by the reporters, and so management have to be willing to make system improvements and implement prevention measures based on the analysis of the data received. Such action will also stimulate reporting.

Implementation preconditions

Although the above tables provide a summary of all the steps and decisions required for the design of a successful reporting system, there are even more critical issues that need to be addressed before implementation of a near miss reporting system. We have termed these 'implementation preconditions' as the items should be determined prior to the launch of the system. These include management support; no-blame near miss reporting; training and supporting safety staff and defining feedback routes to all staff. These implementation preconditions are discussed below.

Management support

In order for any reporting scheme to be successful, to generate and act on reports, it is necessary that it is given full unconditional support at the highest level of the company (Lucas 1991). Ives (1991) provides evidence of management 'killing off' a reporting system through the unnecessary re-organisation of a successfully operating system. It is therefore important that management understand the purposes of the reporting system and foster its use rather than create barriers.

Table 2 The Near Miss Modules and the Decisions at Each Stage (Wright, 2002)

Van der Schaaf's seven modules	Decisions necessary at each module
Module 1: Detection (recognition and reporting)	1. What type of reports are to be collected: e.g., near misses, unsafe practices, deviations, recovered events. 2. Publicity and training to ensure staff are made aware of the type of events to report. 3. Designing of the reporting route.
Module 2: Selection of events for deeper analysis	1. As it is unlikely that all reported incidents can be analysed to the same level, the system needs to devise rules that clearly identify the events that should be analysed deeper, and those which have a cursory analysis. 2. Based on potential rather than actual consequences.
Module 3: Detailed description and deeper study	1. Select a method which enables detailed description and deeper analysis (e.g., Fault Tree analysis). 2. This relies on detailed investigation and interviewing and therefore adequate training for analysts.
Module 4: Classification of root causes	1. Select taxonomy for the classification of root causes. The taxonomy should be based on a number of prerequisites (see Wright, 2002, Shorrock, 2002 for an overview). 2. Analysts require training in the methodology 3. Inter-rater reliability performed regularly. 4. Results in a set of classifications and not simply the 'main cause'.
Module 5: Computation to recognise patterns and/or priorities	1. Design data base to enable statistical analysis. Can database handle text and numbers or numbers alone? 2. Perform analysis to determine reliable patterns which enable structural factors to be identified.
Module 6: Interpretation and implementation of recommendations	1. Statistical results used to generate recommendations that can be implemented. 2. Design group that meets to discuss the feasibility of implementing changes. 3. Feedback channel to provide route for organisational learning (e.g., publications, input from system to 'toolbox talks' or briefing and training sessions).
Module 7: Evaluation	Decide on ways in which the above recommendations to improve the system will be evaluated, and time frame (bearing in mind that it may take some time to build up a sufficiently large database).

Management attitude to near miss reporting schemes

Lucas (1991) cites three factors that are directly under management control which are vital for the success of near miss reporting (not necessarily only for a third party scheme, but also for in-house near miss reporting initiatives). These are anonymity, forgiveness and feedback.

Anonymity

Whether a near miss reporting scheme should be confidential or anonymous depends on the goals of the system, the depth of information required, the safety culture of the organisation and the way in which information is reported. If the reporting system allows anonymous reports to be submitted, then information gained from the first contact is the sum and total of all information that can be gained-an anonymous system prevents re-contact with the reporter. Confidential systems may result in the reporter becoming anonymous once the report has been fully processed. This is the case for example in both ASRS (the US Aviation Safety Reporting System) (http://asrs.arc.nasa.gov/) and CIRAS (see later). The confidentiality in effect allows the reporter to remain anonymous as far as the company disciplinary systems are concerned.

Both confidential and non-confidential reporting schemes share the same analytical objectives, but confidentiality provides the reassurance that people need if they are to report many of the events or circumstances of most interest to risk management. Reporting initially often takes a great deal of courage, to overcome the fear of ridicule, embarrassment and retribution (Van der Schaaf and Kanse, 2004). This makes maintaining the confidence and trust of reporters essential. Harrison (1991) gives the following advantages of confidential reporting schemes:

1. Improves the completeness of accident and incident reports.
2. Helps overcome the barriers associated with near miss reporting.
3. Increases the information to build up a human error database.
4. Increases suggestions for improvements.

There has been much discussion in the literature about the issue of confidentiality versus anonymity in reporting systems. Like confidential schemes, anonymous reporting schemes aim to overcome the barriers associated with self-reporting of errors and accidents. However, the disadvantages of anonymous reporting outweigh the benefits gained. This is because anonymous reports cannot be followed-up to obtain further details, nor can the source be verified. Therefore such a system is more likely to attract 'nonsense' reports or personally motivated reports designed to settle scores. An anonymous system is reliant upon the initial report for all details — any missing information remains missing. This can have a deleterious effect on analysis and therefore upon subsequent attempts to address the causes of incidents. Conversely, a system that provides confidentiality can verify the source and perform interviews with reporters to elicit further details. There are, however, limitations to the confidential approach which also apply to anonymous reporting: reports cannot be verified by witnesses or by other staff who may have been involved in the incident, and in the case of third party systems technical evidence cannot be requested from the company. Such limitations however, are balanced by the willingness to report and the type of information provided by reporters.

The majority of third party reporting systems, e.g., CHIRP (the UK Confidential Human factors Incident Reporting Programme, http://www.chirp.co.uk), ASRS and MARS (Marine Accident Reporting System:http://www.nautinst.org/marineac.htm) all provide confidentiality rather than anonymity for the reasons stated above.

Forgiveness

In terms of forgiveness, this means that management should not take punitive action against reporters no matter the circumstances of the report. When the reporting system is confidential, this is taken as a given. Much attention has been given to the punitive aspects of reporting near misses and accidents via a company or third party reporting system. Confidential systems are introduced in order to circumvent the disciplinary process in an attempt to gain information about what was happening 'at the coal face' rather than to find out who was breaking the rules or not performing adequately. Such systems are often no-blame, in so far as individual reporters are guaranteed not to be disciplined if they report something that they have been involved in only via the confidential system. However, if they are involved in an incident which is known to management and for which disciplinary action would be the normal outcome and also report it to the confidential system, they are not guaranteed protection from discipline. In other words, whilst confidential reports are treated confidentially and there is no blame or discipline attached to such reports, should the incident come to light via other means, then the individual will be treated as they would normally, whether a confidential report is filed or not. In fact, as management is never provided with the names of individuals making reports in a truly confidential system, there is no opportunity to cross reference confidential reports with reports or incidents that come to light through normal channels or investigations.

When ASRS was first introduced it provided immunity from prosecution for those submitting reports. Of course this had the effect of generating reports on the same incident by a number of individuals present at the time of the incident — all thereby ensuring that they were immune from the disciplinary process. Immunity from discipline can be counter-productive to the aims of a confidential reporting system as it distorts the number and quality of reports received (duplicate reports are counted separately). This results in statistical evidence that is based not on the number of near misses actually occurring but on the number of individuals who witnessed or were involved in the same incident. Thus immunity from discipline should not be an integral part of any near miss reporting system.

Much discussion has taken place as to whether a reporting scheme should be no-blame, provide immunity or simply exist within an enlightened culture. Berman (1996) suggests that a no-blame culture is both difficult to achieve and potentially self-defeating, and proposes instead a culture of enlightened response, although he fails to clearly identify how this could be achieved. Reason (2001) also discusses the need for an enlightened or 'just' culture specifically in relation to the railway industry. He suggests that a no-blame culture is neither feasible nor desirable. He does however, acknowledge the difficulty of reaching the state of a just culture: which depends upon understanding motivations as well as actions. In such a situation, managers have to decide which acts deserve punishment and which do not, and where to draw the line. This can in itself create what appears to be an unjust system. Furthermore, unless staff are able to discuss their motivations and actions openly, how can managers and those in a position to make judgements on the underlying motivations and required punishments, perform their function without prejudice or bias. Marx (2001) attempts to address these issues and provide guidance on establishing policies for a just culture.

The desirability of a just culture is not disputed here, but the feasibility of moving directly from a blame culture to a just culture (for example in the railway industry) seems overly optimistic. Until it is clearly demonstrated that a just culture exists, and until employees trust their managers with information without the fear or recrimination, near miss reporting will remain an under-utilised resource. A number of authors (Adams and Hartwell, 1977, Webb *et al.*, 1989) have linked under-reporting of incidents and suppression of information to the apportionment of blame and disciplinary action i.e., to a blame culture. In the UK railway industry a combination of the blame culture and staff perceptions of the utility of making reports has resulted in under-reporting. Clarke (1998) found that the pattern of intended under-reporting indicated a 'risk management' cultural approach (Lucas, 1991) in the industry, which serves to emphasise specific types of incident (e.g., signals passed at danger — SPADs) which should be reported to the detriment of a broader range of potential problems. Clarke (*op. cit.*) also found that only 3% of train drivers would report rule breaking by a colleague. Confidential reporting is one way of re-gaining the trust of staff and establishing a reporting culture. A 'just' culture is hopefully the next step.

Feedback

The third factor which Lucas (*op. cit.*) identifies as vital for the success of near miss reporting is feedback. Reporting schemes may be readily accepted and embraced by staff — reports may be forthcoming in the start up phase, some of them perhaps generated to 'test' the system. As a minimum, a confirmation after receiving each report should be sent to the reporter ; furthermore, it should be made clear if and when the reporter could be asked to answer follow-up questions for a more complete picture of the incident: in that case the reporter might even be added to the analysis team. However, if staff are to continue to use the system and keep making reports then they have to see concrete results and receive feedback on what actions their collective reports have generated. This may be done through individual feedback or via a publication which all potential reporters receive. This ensures that reports are not seen as entering a 'black hole' and disappearing without trace. Multiple feedback channels (such as a publication that corresponds to 'toolbox talks' and is consistent with individual feedback) provide the best way of ensuring that staff are aware that reports are being taken seriously and acted upon, thus helping to ensure the success of the system.

Dissemination and explanation of the system

In order for staff to use a near miss reporting system with confidence they have to be aware of the aims, objectives and goals of the system. There are a variety of ways in which staff can be introduced to the near miss reporting system. These include:

- The distribution of letters from management outlining and supporting the new system
- Presentations and briefings by administrators of the system or by supervisors/trainers.
- Videos which show how the system works.
- A demonstration/pilot with a small sample of incidents using critical incident interviews can show how reports are processed and learned from.

The greater the personalisation of the information and the greater the opportunity to ask questions and understand the system the more likely staff are to trust and use the system. A combination of methods is more likely to have the most success in disseminating the system. Whatever method(s) is used to promote the system the following points should be included:

- The rationale for introducing the system; this should include the relationship of near miss causes and accident causes, i.e., that near misses are precursors to actual incidents (for an empirical study on this see Wright, 2002; Wright and Van der Schaaf, 2004).
- The importance of understanding recovery actions and the usefulness of recovery actions for colleagues. The system can then be used as a channel for 'hero' stories (Kanse, 2004).
- How the system works.
- How reports can be made.
- How reports are processed.
- What happens to the information (who has access to the reports, how they are analysed, who receives the information contained in reports).
- Feedback.
- No blame policy.

Incentives for reporting

A near miss reporting scheme may wish to increase the number of reports it receives by providing incentives for staff to submit reports. Such promotional campaigns should not be undertaken lightly, and the dangers of doing so should be taken into account. The rewarding of reports may lead to biases in the data which would not otherwise be present; trivial reports may be made, or fallacious reports may be generated in order to receive rewards. Also after stopping the rewards, reporting rates may fall — thus confounding the reasons for the fall in reports. An example may help to highlight the dangers of rewarding employees for making reports concerning safety. Makin and Sutherland (1991) report

> ...the international health organisations undertook a concerted campaign to eradicate smallpox...the 'front line troops' for the campaign were health visitors. Each of these had a geographical area for which they were responsible. In order to motivate the health visitors a bonus scheme was introduced. Arguing that the final goal was eradication of smallpox, a scheme was devised whereby each visitor was rewarded according to the absence of smallpox in their area. However, although the visitors consistently earned good bonuses, smallpox remained endemic. When considered from the visitors' perspective the reasons for this apparently paradoxical situation becomes clear. If you are rewarded for lack of cases, the incentive is to turn a blind eye. When in doubt don't report. The system is obviously open to abuse. Management finally realised the potential for abuse, and the reward system was turned on its head. Instead of being rewarded for the absence of cases, visitors were now rewarded for finding cases. The results were dramatic, undiscovered cases now came to the attention of the authorities and could be treated.

This case highlights the dangers of providing rewards for making reports, and therefore such reward systems should not be part of an event reporting system.

Barriers to (voluntary) reporting

Several researchers have provided guidelines for implementing (voluntary) reporting schemes from the perspective of organisational design: Lucas (1991) has already been mentioned earlier, and also Reason (1997, p. 197) lists a number of intuitively convincing factors for 'engineering a reporting culture'. However, much less is known about the *individual reporter's perspective*: under which circumstances are people prepared to share their experiences of a work-related incident with their colleagues, their boss, etc., by reporting it to a formal scheme? A recent literature survey by Van der Schaaf and Kanse (2004) showed that barriers to reporting such incidents voluntarily may be grouped as follows:

Fear: both of disciplinary action by management, and of the reactions of one's colleagues in the form of embarrassment;
Risk acceptance: such incidents are seen as 'part of the job' and cannot be prevented therefore;
Uselessness of reporting: management is perceived as having an attitude of taking no notice of such reports anyway, and will thus not take preventive action;
Practical reasons: if reporting is seen as too time-consuming, or too difficult.

From this list it is clear that it will be the *combination* of the scheme's design, it's implementation, and the local 'safety culture' which will determine it's perceived value (and therefore the willingness to provide input) by the workforce.

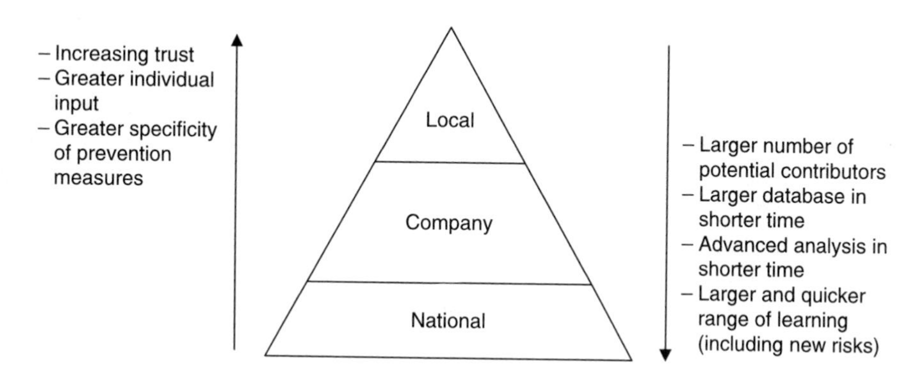

Figure 2 Advantages and disadvantages of local, national and company systems.

Multiple domains and levels of reporting-based organisational learning

Near miss reporting schemes may operate at a departmental level, at a company level, at a regional level or even at a national level. They may involve one company or many companies operating in a similar line of business. The level of reporting (departmental, company, national etc.) should be decided prior to the implementation and design of the system, if possible, as this reduces future problems of incompatibility (between local systems moving to a company or national system), due to collection, investigation, and analysis differences which may occur at different levels. Another important dimension is that of use in a single domain (e.g., just passenger railway transport), or over a number of different domains (e.g., passenger transport by rail, airline, ferry, bus, etc.). Again, early on it should be decided whether such different domains will have or keep their own specific tailor-made reporting and analysis schemes, producing databases that cannot be compared, or whether at a (inter-)national level perhaps an integrated view will be required in order to make balanced policy decisions on, in this case, our future transportation priorities. Obviously, the key question here is about *methodological standardisation or not:* of data collection methods, incident analysis, data base structure etc. In a recent report on Patient Safety Data Standards for US healthcare a hierarchical design of near-miss reporting systems is presented combining the two earlier mentioned dimensions: it may start at a single hospital department for a single medical specialty/domain, via a medium level (a single hospital-wide system or a domain-specific national system), to a nation-wide system for multiple domains (Aspden *et al.*, 2004, p. 171). The advantages of such a gradual build-up of standardisation would be: (a) direct comparability of databases between departments/hospitals (benchmarking); and (b) the possibility to aggregate these databases to provide an overview at a higher level and based on much more data.

Figure 2 outlines the advantages and disadvantages associated with the different levels of reporting.

The triangle model represents the number of potential contributors to a near miss reporting system, with a national system having the largest number of potential contributors and a local system having the least. Local systems are likely to have a greater trust and therefore greater input to the system per head than more distant and less trusted national systems. However, as a national system has a greater number of potential contributors it is likely to build a large database in a shorter period of time than a local reporting system, thus learning from events can take place quicker with a larger range of learning being possible. Conversely local systems are likely to learn very specific lessons and so implement very specific measures.

Operational systems

In this section two practical examples of operational near miss reporting systems are discussed, one in transportation in the UK and the other at an industrial plant in the Netherlands. The former is a good example of a system 'run' by an outside, neutral party, while the latter functions entirely within the same company which provides the reports.

Confidential incident reporting and analysis system (CIRAS)

The confidential incident reporting and analysis system (CIRAS), for rail, was established as a local system for one railway company in 1996, with a potential reporting pool of 700 drivers. In the next five years it expanded to become a national system, with ten times as many potential reporters. Here we describe the local system only, in the context of the seven design modules and implementation preconditions.

The CIRAS system was originally designed to fulfill the three purposes of 'modelling', 'monitoring' and 'alertness' as described above. Accordingly, the aim of the system was to collect reports from individuals of near misses, incidents and error promoting conditions, which would not normally be reported through official channels, and to use this information to enhance existing safety management systems. The system was introduced following a background study by Heybroek (1995) who suggested that the blame culture in the UK industry prevented disclosure of information and therefore prevented learning. Heybroek (op cit) suggested a third party system would overcome the barriers to reporting. CIRAS is not intended to replace existing reporting procedures, and is a complementary system that operates in parallel with existing reporting channels. CIRAS is confidential and 'blame free', and therefore staff can report not only technical failures but also human errors, without fear of recrimination and discipline. For a full description of the system see Davies *et al.* (2000).

Table 3 describes the process of the local CIRAS system by using the seven design modules from Table 1.

Detection

The local CIRAS system was developed to collect voluntary reports of near misses and unsafe acts and practices. The system does not accept reports of alcohol or drug related incidents or of accidents as they happen (real time incidents). There are company channels for these types of reports and CIRAS encourages reporters to use these channels. In order to ensure that staff are aware of the existence of the reporting system and can successfully detect the type of incidents that may be reported, briefing and training were required.

A training pack was devised, accompanied by leaflets and a short video that described the system and how a report would be processed. The training especially highlighted the confidentiality of the system. The team administering the system were unable to personally brief all staff, and so a cascade briefing system was utilised, which took a period of three months. The cascade briefs were then followed up by CIRAS visits to staff depots to ensure the briefing had been successful, that potential reporters understood how the system worked and could ask questions of those administering the system.

Selection

All near misses and unsafe situations or practices reported to CIRAS are selected for description and analysis.

Table 3 The Local CIRAS System in Terms of van der Schaaf's Seven Modules

Module	CIRAS local
1. Detection: recognition and reporting	All near misses and unsafe acts/practices. Not real time reports, drug/alcohol reports. Briefing 700 staff. Devised training pack, leaflets and video. Cascade briefing system backed up by individual depot visits to ensure staff understood how to use the system.
2. Selection: of events for analysis according to purpose(s)	Everything.
3. Description: of all relevant hardware, human and organisational factors	Initial report followed by detailed interview including 'follow up' form.
4. Classification: according to socio-technical model	Causal factors assigned according to the CIRAS human factors taxonomy which includes Technical, Human and Organisational causes.
5. Computation: statistical analysis to uncover certain (patterns of) factors	Statistical analysis performed on a three monthly basis.
6. Interpretation and Implementation: translation of statistical results into corrective and preventive measures	Recommendations made — individual companies decided whether to implement preventive measures. Records kept of implemented measures.
7. Evaluation: measuring the effectiveness of proposed measures after their implementation	Record of all changes kept: not evaluated for effectiveness.

Description

In order to obtain as complete a description as possible of the near miss situation, the CIRAS system supplements the initial report with a telephone or personal interview. At this interview the 'what', 'how', 'where', 'when' and 'why' of the near miss are described. Further information is obtained via a standard 'follow up' form. The interview is central to maintaining confidentiality.

Classification

The classification system used drew heavily on the work of Rasmussen (1982) and Reason (1990) incorporating the Skill-Rule-Knowledge distinctions and organisational and human causal factors. Further, as perception of signals is fundamental to the role of the driver, perception was also included in the model. Shappell and Wiegmann (2001) also recognise the need for a perceptual code and have included it in their taxonomy HFACS (human factors analysis and classification system). A supervisory layer was added to the human and organisational causes as the railway industry is extremely hierarchical. Finally a specific set of technical codes was developed in order to locate the area where technical problems occurred (on sets, signals etc) rather than to determine the nature of the technical fault. Drawing on the work of van der Schaaf (1991) recovery actions were also included in the model. A full description of the CIRAS classification system can be found in Wright *et al.* (2000). The original taxonomy met the majority of minimal requirements as proposed by Wright (2002).

Computation

Although information regarding incidents is provided to the individual companies on an ongoing basis, statistical analysis of amalgamated data is performed periodically, on a three monthly basis, and accompanied by a formal report.

Interpretation and implementation

The CIRAS system has no particular method for translating statistical results into corrective measures. However, recommendations based on statistical results and individual events are discussed with the individual company and other relevant parties and the decision to implement these decisions rested with the individual company. Records were maintained of any changes implemented and also of the reasons for not implementing changes.

Evaluation

The evaluation of corrective measures is an area where much improvement can be achieved. The local CIRAS system has never evaluated the effectiveness of implemented corrective measures.

Training in causal taxonomy

A number of analysts were trained in using the causal coding taxonomy; this included training in the taxonomy, use of the taxonomy and inter-rater reliability trials. This training was completed in parallel with maintaining the system and analysing reports. New analysts began coding real reports after gaining an adequate level of inter-rater reliability (70%) with colleagues.

Conclusions

The local CIRAS system provided more information on near misses, unsafe practices and situations than were reported through the normal reporting channels and provided a unique learning opportunity for the company involved. Subsequently, the system expanded to include all safety-related staff on the UK railway network in 2001. This led to major changes in the way in which CIRAS was organised. CIRAS provides important information that can be used as part of the overall safety management system and brings to light information that companies may not receive via other reporting systems.

The Lyondell Chemicals near-miss system

Compared to CIRAS, this section describes a somewhat smaller system, operated by and within a single chemical process industry plant in the Netherlands, part of the US multinational Lyondell. It was started in 1994 as a result of the need to have an incident reporting system as prescribed by the company's 'Manufacturing Excellence' standards. From the very beginning, the Safety Management Group at Eindhoven University of Technology (which was very familiar with industrial safety in the chemical industry) were involved in the design, implementation, and evolution of the 'NM' system.

Table 4 summarises the operational aspects of the NM system in terms of the seven-module design framework (van der Schaaf, 1991).

Detection

The NM system specifically focuses on near-misses, although it can handle actual damages and injuries as well. The entire population of potential reporters consists of 250 Lyondell employees and 80 workers continuously employed at the plant from outside companies (contractors), and together they produce some 700 reports per year.

Table 4 The Lyondell 'Near Miss' System in Terms of van der Schaaf's Seven Modules

Module	Lyondell
1. Detection: recognition and reporting	700 reports per year (30% by contractors!) 50% on safety, 20% on reliability, 15% on health, 10% on environment, 5% on quality-related issues Very short and simple initial reporting form to maximise detectability; anonymity as an option: hardly ever chosen (15/4000 reports)
2. Selection: of events for analysis according to purpose(s)	Informal risk-based decision by Daily Coordination Meeting Team: for deep(PRISMA) analysis (6.5% of reports), or for simple identification of 'main direct causes' only Occasionally special, short, focussed projects: e.g., in the 4 weeks of the 4-yearly 'Turn-around' inspection the extra 400 contractors are urged to report every possible deviation (generating in '97 an extra 334 reports)
3. Description: of all relevant hardware, human and organisational factors	For deep analysis: building a PRISMA causal tree summarises all relevant factors in (chrono-)logical visual overview Each Near-Miss tree has a failure side and a recovery side, thus generating two causal databases in parallel
4. Classification: according to socio-technical model	According to Eindhoven Classification Model of System Failure, using <20 (sub-)categories to cover all technical, organisational and human 'root' causes Model regularly expanded to incorporate new insights (e.g., into management and recovery factors)
5. Computation: statistical analysis to uncover certain (patterns of) factors	Simple, descriptive statistics at the database level: histogram of relative frequencies of all types of root causes to determine dominant factor(s) Also looking for clusters of related causes within reports: e.g., ergonomic design problems associated with skill-based errors, or: correlations between poor procedures for infrequent tasks, a culture of ignoring them, and errors in execution of such tasks
6. Interpretation and implementation: translation of statistical results into corrective and preventive measures	Most countermeasures driven by database patterns, but still 30% as a direct reaction to a single report, usually one with obvious, far-reaching consequences (e.g., discovering a design fault in a widely used piece of equipment, or an inconsistency in a common procedure)
7. Evaluation: measuring the effectiveness of proposed measures after their implementation	By measuring changes in relative frequency of targeted type of root cause (e.g., the 'check and double check' rule-based error was reduced from 11% to 4% in 3 years) In terms of safety indicators (e.g., the Recordable Case Rate of medical treatments was halved to 0.4 per 200.000 hours worked) In terms of availablity of equipment: increased to a very high 99.5% In terms of environmental indicators (e.g., number of leakages went down from 60 in 1994 to 20 in 2000) The bottom line: operational and maintenance cost per unit of invested capital did *not* increase during this period

In the course of time, these employees have expanded the safety orientation of their reports to the system to include almost all types of process deviations: reliability of equipment, environmental issues, health issues, and even quality or logistic problems.

Selection

Every incoming report is estimated for its severity: potential consequences of the near-miss as well as the expected frequency are the basis for this decision by the multidisciplinary Daily Coordination Meeting team. When selected for a deep analysis (according to the *Modelling purpose*), a report is subjected to the PRISMA method (van der Schaaf, 1992), which means that multiple root causes are identified, classified, and interpreted in modules 3–6 (see below).The majority of reports however are used for the *Monitoring purpose* ('it' has happened again) and contribute to understanding trends for known weak points of the plant: only 1 or 2 direct causes are identified and classified in these cases.

Description

When selected for PRISMA analysis, the incident is analysed step-by-step to identify two types of causal factors: failure factors (why things went wrong initially) and recovery factors (why and how the resulting deviations were detected, understood and corrected in time, to prevent actual consequences).

Classification

Each identified failure or recovery factor is then labelled according to the Eindhoven Classification Model, which covers all types of technical/design, organisational/management, and human behaviour errors. Here, the same models by Rasmussen (1982) and Reason (1990) are used as for the CIRAS taxonomy.

Computation

Every classified root cause is added to a database of many analysed incidents, and each quarter a PRISMA-profile is made: a histogram of the relative frequencies of all types of root cause categories. This allows management to locate the probable dominant root cause(s) as target for specific countermeasures. Also, incidents may be checked for recurring clusters of meaningfully related types of causal categories: these will be evaluated for a joint treatment by preventive measures.

Interpretation and implementation

PRISMA guides the direction of countermeasures for a specific target root cause by providing a matrix (again based on the Rasmussen and Reason models) which links every causal category to the most effective type of countermeasure. This prevents management from relying on choosing a popular panacea (like re-training, warning to 'pay more attention and think twice next time', more procedures to circumvent a fundamental design problem by forcing workers to do illogical things, etc.).

Evaluation

After implementation of the preventive countermeasure suggested by the matrix, the effectiveness of this intervention may be estimated by monitoring the changes in relative frequency of the targeted root cause: a failure factor should gradually, over the following months, decrease, while a recovery factor should increase in frequency. Also at the level of performance output parameters (usually the actual types of consequences which were prevented in the reported near misses) improvement may be measured, although this usually is made difficult by the rarity of such actual consequences.

Implementation issues

Over a period of 6 months, the following preparations were carried out before the actual green signal was given to start reporting. An intense communication and training campaign was held to make clear to all potential reporters the why, what and how of the new system. Two failed attempts in earlier years to launch NM reporting were publicly analysed, and the differences of the new approach were stressed: one of the earlier attempts had not provided for fast and relevant feedback to the reporters, and the other had lacked a meaningful causal analysis method. Management committed itself to a strict 'no-blame' policy for genuine near-miss reports: the urge to punish workers for their errors was considered less important than the opportunity to establish a safe communication channel from the workfloor upwards, and thus learn about the actual carrying out of tasks and of their minimal requirements.

A pilot study was carried by researchers from EUT using highly confidential interviews with a cross-section of the workforce, about a recent near miss (of their own choice) they had been involved in. The resulting small database, without any identifiers as to informants, was used to demonstrate the complete processing of reports (all seven modules of the framework) later in the new system. A coordinator was appointed who would actually 'run' the system. This person (an ex-operator who now worked in the safety department) was trusted by both workfloor and supervisory staff, and was thoroughly trained in the PRISMA technique and its theoretical backgrounds.

An elaborate system for providing feedback to reporters was built: every report's reception was to be acknowledged by the coordinator within 24 hours by thanking the reporter, who was also to be updated on every decision based on his report. A weekly overview (without identifiers) was posted everywhere in the plant; every month the most important near-misses were also shared with the neighbouring chemical plants; every quarter a PRISMA-profile was calculated, and posted everywhere, along with the database analysis results and management reactions. Awards were given for the 'Golden Near Miss of the Year by Lyondell, and for the 'best near miss of the month' by the contracting companies. Finally, the yearly evaluation by management of the system of NM reporting was shared with all.

Conclusions

With a few ups and downs, the Lyondell NM system has not only survived for over a decade now, but has been expanding, incorporating new insights (in organisational failure and recovery factors) and new groups of reporters(it is very unusual just how much the contractors cooperate in this reporting scheme). Although measurement of (safety) culture is still highly debatable and difficult, the open atmosphere created over the course of years is also widely seen as a valuable result of the scheme.

Some of the remaining weak points are: the fact that only one person does the bulk of the analysis precludes the usual inter-rater reliability checks, but there are (infrequent) independent checks by EUT researchers in the course of their projects in the plant; first-line supervisors (middle management) still have a hard time in consistently acting in the spirit of the no-blame policy, as they sometimes feel sandwiched between the workfloor and the management team; a continuous repetition of the need for bias-free reporting is necessary, as the percentage of reports on one's own behaviour is still very low, and even successful recoveries are not always reported.

General conclusions

This chapter has outlined the variety of situations and purposes under which near-miss reporting systems may function. It provides a frame work for designing, implementing, maintaining and operating such useful but vulnerable safety management approaches. Their relative success in industry and transportation is now also being recognised in other sectors, particularly in the healthcare domain. The two brief case studies presented here show that basically similar mechanisms and requirements can be adopted with respect to near-miss reporting, but they also highlight the important cultural differences which must be taken into account.

References

Adams, N.L. and Hartwell, N.M. (1977). Accident reporting systems: a basic problem area in industrial society. *Journal of Occupational Psychology*, **50**, 285–298.

Aspden, P., Corrigan, J.M., Wolcott, J., Erickson, S.M. (eds.) (2004). *Patient Safety: Achieving a New Standard for Care.* (Washington, DC: The National Academies Press).

Berman, J. (1996). Confidential event reporting (HF/GNSR/5009). Final report. Document reference: 112/C/03/il 1996 — A report submitted under the 1995/1996 IMC GNSR Programme. *Commercial-in-confidence.*

Brown, I.D. (1976). Psychological aspects of accident causation: theories, methodology and proposals for future research. Unpublished report prepared for the Medical Research Council, Environmental Medicine Committee's Working Party on Specific Aspects of Accident Research.

Clarke, S. (1998). Safety culture on the UK railway network. *Work and Stress*, **12**, 6–16.

Davies, J.B., Wright, L., Courtney, E. and Reid, H. (2000). Confidential incident reporting on the UK railways: The CIRAS system. *Cognition, Technology and Work*, **2**, pp. 117–125.

Feggetter, A.J. (1982). A method for investigating human factors aspects of aircraft accidents and incidents. *Ergonomics*, **25**, 1065–1075.

Freitag, M. and Hale, A. (1997). Structure of event analysis In Hale, A., Wilpert, B. and Freitag, M. (eds.) *After the Event: from Accident to Organisational Learning.* (Oxford: Pergamon).

Harrison, P.I. (1991). Harnessing operational experiences and learning lessons: the value of confidential incident reporting schemes. In: *Proceedings of the 11th Annual Safety and Reliability Society Symposium*, (September 18–19, Sutton Coldfield, England, pp. 42–56.

Heinrich, H.W. (1931). *Industrial Accident Prevention.* (New York: McGraw-Hill).

Heybroek, R. (1995). *Improving Safety Training: Human Factors Discrepancies Report.* Vosper Thornycroft (UK) Limited, MSC Division. Commercial-in-confidence.

Hyos, C.G. and Zimolong, B. (1988). *Occupational Safety and Accident Prevention.* (Amsterdam: Elsevier).

Ives, G. (1991). 'Near miss' reporting pitfalls for nuclear plants. In Van der Schaaf, T.W., Lucas, D.A. and Hale, A.R. (eds.). *Near Miss Reporting as a Safety Tool.* (Oxford: Butterworth-Heinemann).

Kanse, L. (2004). *Recovery Uncovered: how people in the chemical process industry recover from failures.* PhD thesis, Eindhoven University of Technology, Eindhoven.

Kaplan, H.S., Battles, J.B., van der Schaaf, T.W., Shea C.E. and Mercer, S.Q. (1998). Identification and classification of the causes of events in transfusion medicine. *Transfusion*, **38**, 1071–1081.

Lucas, D.A. (1991). Organisational aspects of near miss reporting. In Van der Schaaf, T.W., Lucas, D.A. and Hale, A.R. (eds.). *Near Miss Reporting as a Safety Tool.* (Oxford: Butterworth-Heinemann).

Makin, P. and Sutherland, V. (1991). A fatal inversion? *Occupational Safety and Health*, Nov. 1991, 40–42.

Marx, D. (2001). Patient safety and the 'just culture': A primer for health case executives. (New York: Columbia University).

Rasmussen, J. (1982). Human errors, a taxonomy for describing human malfunction in industrial installations. *Journal of Occupational Accidents.* **4**, 311–333.

Reason, J.T. (1990). *Human Error.* (Cambridge: Cambridge University Press).

Reason, J.T. (1997). *Managing the Risk of Organisational Accidents.* (Hampshire: Ashgate Publishing).

Reason, J.T. (2001). *Human Error Reduction Operation.* (for Railway Safety Limited) The Vision Consultancy, England.

Van der Schaaf, T.W. (1988). Critical incidents and human recovery: some examples of research techniques. In Goossens L.H.J. (ed.) *Human Recovery: Proceedings of the COAST A1 Working Group on Risk Analysis and Human Error,* Delft University of Technology, 13 October 1987.

Van der Schaaf, T.W. (1991). A framework for designing near miss management systems. In Van der Schaaf, T.W., Lucas, D.A. and Hale, A.R. (eds.). *Near Miss Reporting as a Safety Tool.* (Oxford: Butterworth-Heinemann).

Van der Schaaf, T.W. (1992). *Near miss reporting in the Chemical Process Industry.* Eindhoven: Technische Universiteit Eindhoven. PhD Thesis.

Van der Schaaf, T.W. and Kanse, L. (2000) Errors and error recovery. In Elzer, P.F. Kluwe, R.H. and Boussoffara, B. (eds.), *Human Error and System Design and Management.* (Lecture notes in control and information sciences, 253, pp. 27–38). (London: Springer-Verlag).

Van der Schaaf, T.W. and Kanse, L. (2004). Biases in incident reporting databases: an empirical study in the chemical process industry. *Safety Science,* **42**, pp. 57–67.

Shappell, S.A. and Wiegmann, D.A. (2001). Unravelling the mystery of general aviation controlled flight into terrain accidents using HFACS. *Presented at the 11th International Symposium on Aviation Psychology.* Columbus, OH: The Ohio State University.

Shorrock, S. (2002). Error classification for safety management — Finding the right approach. Presented at *Investigation and Reporting of Incidents and Accidents,* University of Glasgow, Glasgow, 17th–20th July 2002.

Webb, G.R., Redmand, S., Wilkinson, C. and Sanson-Fisher, R.W. (1989). Filtering effects in reporting work injuries. *Accident Analysis and Prevention,* **21**, 115–123.

Wright, L.B. (2002). The analysis of UK railway accidents and incidents: a comparison of their causal patterns. PhD Thesis, University of Strathclyde, Glasgow.

Wright, L.B., Davies, J.B., Courtney, E. and Reid, H. (2000). CIRAS: Collecting and analysing human factors data from the UK rail industry. *Proceedings of ESREL 2000, SARS and SRA-Europe Annual Conference, Foresight and Precaution.*

Wright, L.B. and van der Schaaf, T.W. (2004). Accident versus near-miss causation: a critical review of the literature, an empirical test in the UK railway domain, and their implications for other sectors. *Journal of Hazardous Materials,* 111, pp. 105–110.

Chapter 34

Learning from failures: a joint cognitive systems perspective

Erik Hollnagel

Introduction

It is an indisputable fact of everyday life that failures do take place in the sense that every now and then something happens that was neither intended nor desired. The term failure in itself means the omission of occurrence or performance of something, or in other words that a person did not achieve what s/he had hoped for or intended. (The word comes from Latin *fallere,* which means to deceive or disappoint.) The term is commonly applied to both humans and artefacts since technology obviously can also fail — both in the hardware and in the software — and we usually say that something fails if it does not function as it should. It is often assumed that failures that can be attributed to technology in some sense are simpler to deal with than failures that are attributed to human functions, although this is a serious mistake (Leveson, 1995; Perrow, 1984). To avoid getting into this debate, the focus of this chapter will be on failures at work that can be attributed to humans, and with some suggestions for how it is possible to learn from such failures. The chapter is thus about failures of actions rather than failures of technology.

Having said that, it is necessary immediately to point out that the distinction between failures of actions and failures of technology are but two sides of the same coin. It can be argued that all failures in principle can be attributed to human actions at some level. Technology — and especially software — does not fail by itself, but only fails because some action has been deficient at some stage. When action failures are considered in ergonomics, the thought usually goes to operations or work at the sharp end, i.e., the here and now of work where people (operators, practitioners) cope directly with the demands of the situation. But according to the contemporary view of accidents, that which happens at the sharp end is not just the result of the current situation but also affected by what has happened in the past — by what is removed in time and space. The blunt end is often associated with regulators, administrators, economic policymakers, and technology suppliers who affect and control the resources and conditions that confront the practitioners at the sharp end, sometimes leading to conflicting goals and demands. But the blunt end may equally well comprise failures that happened previously but were not completely mitigated or recovered at the time, i.e., failures that come from the sharp end in the past, so to speak. While such failures often are due to people occupying other roles, ranging from maintenance to management, they may also be due to the practitioners themselves. The distinction between actions at the sharp and actions at the blunt end is thus not just one of abstraction or organisational 'distance', but also one of time.

The annoying thing about a failure is not just that it happens, but also the consequences it may have. Depending on the type of work and the specific conditions of work, the consequences may range from a mere nuisance to a disaster. If a failure prevents a person from achieving something that was intended or wanted, it may be annoying and perhaps even disturbing. But if a failure violates a safety function or a protecting system, such as the barriers or defences that are parts of most industrial processes, then the consequences can be far worse, ranging from incidents over accidents to disasters. For instance, if you forget the shopping list when going to the supermarket, you may miss an important ingredient when it is time to cook dinner. But if a train driver passes a red signal without stopping, an accident may easily occur.

In order to be able to learn from failures three things are needed, as a minimum. First, it must be realised that a failure has occurred (failure detection). Second, the failure must be characterised and the likely causes must be identified. Finally, some form of response or countermeasure must be thought of and carried out. The response is of importance in itself as the remedial or corrective action needed to keep the work going, but that is not the concern here. Moreover, the response is also important for learning, since the outcome confirms or disproves whether the situation was correctly understood, i.e., whether the fault was interpreted correctly.

Each of the three steps has been the subject of volumes of work, in relation to topics such as early detection and alarm systems, diagnosis and accident/incident analysis, and procedures and response management. Rather than try to summarise that body of knowledge, this chapter will focus on seeing the three together, to provide an overview and assess them in relation to how they affect the ability to learn from failures.

If anything can fail . . .

The legendary example of a failure is the event that gave rise to the formulation of the well-known 'Murphy's Law'. Captain Edward Murphy, Jr., of Edwards Air Force Base, was one of the engineers on the rocket-sled experiments that the United States Air Force carried out in 1949 to test human tolerance to high-deceleration stress. In one experiment, which involved a set of 16 accelerometers mounted to different parts of the body, the subject had just exceeded the existing record of thirty-one times the force of gravity on the rocket sled. Unfortunately, nobody could say by how much since the gauges had not worked. There were two ways each sensor could be glued to its mount and on inspection — after the experiment, of course — Murphy found that a technician had installed each of them backward. From this experience Captain Murphy drew the lesson that 'if there's more than one way to do a job and one of those ways will end in disaster, then somebody will do it that way'. The test subject, who was Murphy's boss, Major John Paul Stapp, quoted Murphy's pronouncement at a news conference a few days later (Tenner, 1996, p. 22–23). Over time this became 'If anything can go wrong, it will,' which made it into the Webster's dictionary in 1958.

Edward Murphy was, of course, not the first to notice that people can do things wrongly. In 1711 Alexander Pope, in 'An Essay on Criticism', noted that 'To err is human, to forgive Divine', rephrasing the Roman saying 'Errare humane est' (which, by the way, continues 'sed in errore perseverare turpe est' — meaning that it is disgraceful to persist in the error.) But in order to avoid the disgrace of persisting in the error, it is necessary to know that something has been done wrongly and use that knowledge to find a better way. In other words, it is necessary to be able to learn from failures.

Although failures are inevitable and unavoidable, they should not be shunned at any cost. Failures can also be useful, since they provide an opportunity to learn. If everything always went according to plan and if nothing ever failed, then it would be impossible to learn anything. (One might counter, of course, that if everything did go as planned, then there would be no need to learn, i.e., there would be perfect knowledge from the start. However, that condition belongs in a philosophical Utopia.) This has been expressed as the 'Fundamental regulator paradox'.

> The task of a regulator is to eliminate variation, but this variation is the ultimate source of information about the quality of its work. Therefore, the better the job a regulator does the less information it gets about how to improve.
>
> (Weinberg and Weinberg, 1979, p. 250).

(Note the delightful double meaning of 'a regulator'; it could be a technical regulating mechanism consisting of hardware or software, or it could be an organisation or authority.) In plain terms the fundamental regulator paradox means that if something rarely or never fails, then it is impossible to know how well it works. We may, for instance, in a literal or metaphorical sense, be on the right track but also be precariously close to the limits. Yet if there is no indication of how close, it is impossible to improve performance.

From a commonsense point of view we may fail to notice that something is wrong because we cannot see it or because we cannot distinguish or differentiate it sufficiently well. The lack of a failure indication can either mean that everything works correctly — because we keep to the middle of the road, so to speak, that the measurements or signals are too weak, or that the criteria are too imprecise, as shown by the simple logic of Figure 1.

Accepting the fact that failures do occur, we should try to make the most of them, which means that we should try to learn from them. This has been most clearly expressed in the engineering sciences, probably because the relation between design decisions and functioning is more obvious than in behavioural sciences, i.e., in situations of human work. A good example of that is the following quote from Henri Petroski:

> It is imperative in the design process to have a full and complete understanding of how failure is being obviated in order to achieve success. Without fully appreciating how close to failing a new design is, its own designer may not fully understand how and why a design works. A new design may prove to be successful because it has a sufficiently large factor of safety (which, of course, has often rightly been called a 'factor of ignorance'), but a design's true factor of safety can never be known if the ultimate failure mode is unknown. Thus the design that succeeds (i.e., does not fail) can actually provide less reliable information about how or how not to extrapolate from that design than one that fails. It is this observation that has long motivated reflective designers to study failures even more assiduously than successes.
>
> (Petroski, 1994, p. 31)

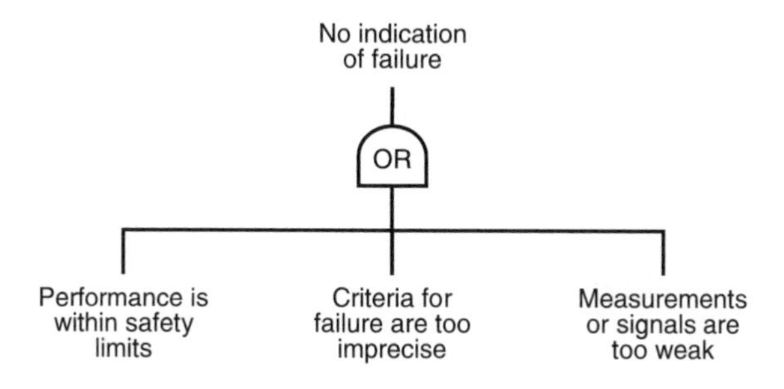

Figure 1 Interpretations of 'no failure'.

A similar sentiment is expressed by Tenner:

> It is both a sad and a happy fact of engineering history that disasters have been powerful instruments of change. Designers learn from failure. Industrial society did not invent grand works of engineering, and it was not the first to know design failure. What it did do was develop powerful techniques for learning from the experience of past disasters. It is extremely rare today for an apartment house in North America, Europe, or Japan to fall down. Ancient Rome had large apartment buildings, too, but while its public baths, bridges, and aqueducts have lasted for two thousand years, its big residential blocks collapsed with appalling regularity. Not one is left in modern Rome, even as a ruin.
>
> (Tenner, 1996, pp. 23–24)

Both authors emphasise that it is paramount to have a 'full and clear understanding' of what went wrong, i.e., of what the possible causes of the failure were. In the engineering world a 'full and clear understanding' is considered achieved when it can be determined that some component or structure has failed. In ergonomics and human factors engineering, and in behavioural sciences in general — i.e., in fields where the subject of study is human work or involves humans in some irreducible way — the 'full and clear understanding' is more difficult to achieve, both because it is much harder to understand human performance and social systems and because there is less agreement about what constitutes a correct and complete explanation.

The event that led to the formulation of Murphy's Law is a good example of that. If the cause was determined to be the failure of the hapless technician to install the accelerometer gauges correctly, or rather that the technician had installed the gauges incorrectly, then the remedy might be to censure that person in some way, perhaps fire him, demote him, or send him to a refresher course, and that would be the end of it. (One might, of course, also try to understand why the technician had failed to install the gauges correctly, hence consider the working conditions and potential precipitating factors.) If on the other hand it were realised that the gauges could be installed in two different ways although only one of them was correct, then the lesson would be one of design. The solution would therefore be to redesign the gauges, so that they could only be installed correctly. Perhaps the analysis might even try to understand why the design team had failed to consider this obvious failure mode, although such a course of action would be less likely.

Indeed, the case of the unfortunate Major Stapp illustrates that it is essential really to understand what went wrong in order to learn from the failure, but also that this is no simple matter. (There is no information about what actually happened in the specific case, although it would be interesting to know.) It is the purpose of this chapter to discuss how this may be done in situations of work, i.e., within the wide field of ergonomics. Lest the reader should be disappointed, let it be made clear from the start that there are no easy methods for doing so. It is nevertheless useful to go through a number of issues that should be kept in mind, and which may help to improve learning from failures.

Purposes and levels of learning

The purpose of learning from failures is not just to avoid that the same failure will happen again in the future — i.e., to be safe from risk and dangers and to avoid accidents — but also to learn how to do things better. There is a significant organisational concern for the former as seen by the discussion about organisational safety and safety culture, which has blossomed since the late 1980s. There is also considerable concern for the latter particularly if it can be related to tangible short-term goals and increased efficiency. This view was reflected at least as early as 1931 when Heinrich formulated the ten axioms of industrial safety (Heinrich *et al.*, 1980), of which the last pointed out that 'the safe establishment is efficient productively and the unsafe establishment is inefficient' (*Ibid.*, p. 21). The enthusiasm is somewhat reduced if it refers to long-term goals such as a learning organisation (or possibly even a high-reliability organisation), innovation and creativity. Yet rather than engage in a discourse about the potential conflicts between safety and efficiency goals, and how safety and efficiency may depend upon each other, this chapter shall maintain the focus on the practical and theoretical issues of learning from failure.

In principle, the possibility of repeating a failure can be removed simply by eliminating the corresponding task or activity. This is, however, not a realistic option except in very exceptional cases, and even then only for as long as the failure 'lives' in individual or institutional memory. The only practicable alternative is therefore to learn how to do things better, which means to understand what went wrong and make the appropriate changes. From a pragmatic perspective such learning can be described as taking place on several levels, of which we here shall distinguish between the following three: the individual level, the workgroup or team level, and the organisational level. The levels represent different ways of looking at the work situation and the entities involved. Yet rather than view each level as an additional layer, as in an onion model or a Russian doll, or a classic model of ergonomics — see Chapter 1, the levels represent different types of joint systems.

As shown by Figure 2, the individual level comprises the worker and the process (or task), and the context or environment is represented by the workgroup and the workplace. The workgroup level comprises the individual work system (process and worker) and the workplace, within the context or setting of the company. The organisational level comprises the workgroup and the company (the organisational processes), within the context set by the regulator. The point is that the joint system at each level (individual, workgroup, organisational, etc.) is seen as a system in its own right and with certain functional characteristics, for instance in terms of dynamics and delays as well as interactions and couplings (to use the terms from Perrow, 1984). The description at each level the joint system

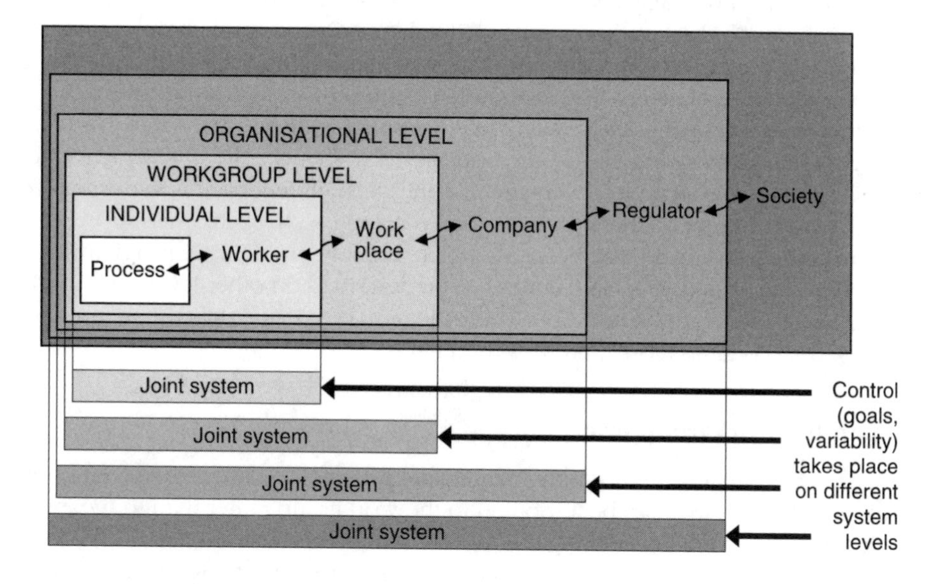

Figure 2 Levels of joint systems at work.

is relative to a context or environment and therefore depends on how the system's boundaries are defined. The level of description must always be chosen so that it is appropriate for the purpose. In other words, there is no *a priori* sense in which one level of description is better than any other.

Learning on the individual level

On the individual level the focus is on the person who does the work, and who therefore usually also directly experiences the effects of the failure. In a commonly used terminology this is referred to as the person at the sharp-end, i.e., the time and place where things do happen, where failures are made and noticed, and where the harmful effects often are released. (There may sometimes be exceptions to last point, for instance when the consequences remain as latent conditions in the system or the environment). The individual level has always been of concern to ergonomics and human factors, not least in relation to the various disciplines that deal with workplace design, interface design, automation, etc. The individual level or the sharp end is by definition the time and place where the failures unfold, although this usually is the consequence of a number of preceding events. It is common to think of this as a chain of actions, such as the description in the First Axiom of Industrial Safety, which was formulated in the first edition of Heinrich's book on *Industrial Accident Prevention* published more than 70 years ago:

> The occurrence of an injury invariably results from a completed sequence of factors — the last one of these being the accident itself. The accident in turn is invariably caused or permitted directly by the unsafe act of a person and/or a mechanical or physical hazard.
>
> (Heinrich, Petersen and Roos, 1980, p. 21).

The second axiom goes on to point out that the majority of accidents are caused by the unsafe acts of persons. This view is still commonly held in most fields of human endeavour, although it has been seriously criticised since the beginning of the 1980s.

Learning on the individual level is fundamental because the individual has a natural motivation to improve his or her ways of working. Quite apart from issues such as responsibility and culpability, no sane person would want to subject themselves to the possible consequences of failures. The individual level is also important because this is where any kind of learning eventually must take place. Learning is direct or enactive, which means that the person learns by doing something and then experiences the consequences of their actions. This has the considerable advantage that the delay in feedback is very small or even negligible. The consequences are experienced on the spot, so to speak, which particularly in the case of near misses or incidents means that there are few problems in imagining or seeing what could have happened. Even if failures can be described on other levels, the workgroup and the organisation are also made up of individuals. Thus, if individuals do not learn, the workgroup and the organisation cannot learn either.

Learning on the workgroup level

The workgroup or team level refers to the behaviour of the group rather than of the individual, and therefore to failures that can be attributed to the workgroup rather than to the individual. In all work situations worth considering, people are part of a group or a team. Even the single astronaut in the space capsule, or the engineer in a train are part of a team, in the sense that they would be unable to do their work or accomplish their function unless they had the support of the team and unless they collaborated with it. Very often an organisation prescribes a set of roles that are meant to support the individual at work, even though the workgroup at times may not be physically present. The learning of the workgroup therefore facilitates individual learning, and the social climate and norms may often influence the performance of the individual significantly.

Learning on the workgroup level can be direct or enactive, but also indirect or vicarious. It is direct when it refers to failures of primary functions on the workgroup level, such as communication or coordination. It is, of course, the individuals who experience the effects of that but the workgroup may function as a closely-knit unit or a collective entity — a joint system — and thereby benefit from enactive learning in the same way that an individual does. In addition to direct learning, a workgroup also offers the possibility of indirect or vicarious learning, i.e., learning by observing others or learning in the sense that consequences provide information rather than directly strengthening or weakening behaviour (cf. the description of second-loop learning below). Vicarious learning was praised by the Chinese philosopher Confucius (551–479 BC), who recommended that one should 'incorporate others' good points into oneself so that one improves; self-reflect upon others' bad points so that one does not commit the same mistake'. The advantages of vicarious learning were more recently espoused by the theories of social learning (Bandura, 1971), sometimes also called model-based learning, which describe how a person can learn simply by watching others and noting the consequences of behaviour. Indeed, it has been suggested that vicarious or observational learning is more effective than direct experience. By watching what others do, a person can witness the correct behaviour sufficient to achieve the desired results without having to go through indeterminate attempts of trial-and-error. There is little wasted effort, which makes learning more efficient. Vicarious learning is clearly an advantage for a workgroup, and can in many ways be seen as the basis for formal training and instruction programmes.

Learning on the level of the workgroup can be as fast as on the level of the individual, but may also include types of learning where there is some delay, for instance when an experience has to be processed by the workgroup leading to, e.g. a change in formal or informal rules and norms of work. The delay is due firstly to the fact that the basis requires a number of cases rather than a single event, and secondly because the deliberations needed take time. Because of this, learning on the workgroup level may end up being less effective as single-loop learning.

Learning on the organisational level

The organisational level refers to the organisation as an entity, as a joint system by itself above the workgroup, and therefore to something that is characteristic or important across individuals and workgroups. Just as individuals can be replaced in a workgroup by someone else who carries out the same function, so can the workgroup be replaced by one that is functionally isomorphic — i.e., one which functions in the same way — without affecting the way in which the organisation functions. Learning on the organisational level refers to what is 'remembered' by the organisation rather than by individuals, although it must be in a form that is amenable to or accessible by each person.

Although the interest in organisational learning goes back at least to 1970s (Argyris and Schön, 1978), the issue has become more important in the last decade or so since organisational shortcomings have emerged with increasing frequency as a central factor in accident explanations. This has made it more pressing to understand how organisations learn and change when faced with accidents, or even to address how society learns and how to regulate safety within given contexts (e.g., types of process industries, transport networks, healthcare).

Table 1 Prevention and Recovery Information System for Monitoring and Analysis (PRISMA)

Detection	Usually on the basis of voluntary reporting by employees.
Selection	Filtering of those reports with the highest informative value.
Description	Details of the selected events, by means of qualitative fault tree techniques.
Classification	Description of each of the many basic causes, according to the a system failure model.
Computation	Statistical analysis of large database of incidents.
Interpretation and Implementation	Translation of the statistical results, to come to theoretically supported suggestions for management actions.
Evaluation	Analyse the effectiveness of such implemented actions by means of an explicit feedback loop.

When an accident occurs, the response usually follows a common form: there is first a look for someone to blame, then there are attempts to try to understand why it happened, followed by attempts to learn and take precautions for the future. In the sharp-end, blunt-end framework, organisations are removed from the sharp-end, at least when it comes to the primary work process. (In a very real sense, the functions and services of the organisation also constitute a primary work process, which defines an organisational sharp end. That corresponds to shifting the perspective and focus of the description, as outlined by the joint systems approach, cf. Figure 2. Although it is quite legitimate to do so, this chapter will keep the focus on the individual and the discussions will be related to that.)

Whereas learning on the individual level is direct and therefore usually fast and efficient, learning on the workgroup and organisational levels taken together is more complex. This is partly because learning is vicarious rather than enactive, hence has to be mediated, and partly because a number of people and/or organisational functions may be required for learning to take place. The relative complexity of learning on the organisational level is evident from the seven-step process in Table 1, proposed by Dye and van der Schaaf (2002).

In this proposal the progression from detection to response comprises a number of steps and requires a number of functions, in particular a more formal analysis using some type of system failure model. Learning is thus based on a generalised experience rather than a single case. This proposal puts emphasis on the detailed analysis of the available data, aiming at finding effective responses. Other suggestions (e.g., Phimister *et al.*, 2001) emphasise the dissemination of information, both of events and results, to the organisation at large.

Learning from failures on the organisational level is typically indirect or vicarious, since those who are responsible usually are several steps removed from the sharp-end. This means that their experiences are second-hand at best, and also that there may be considerable delays in the process. Another characteristic is that the result of the learning is more indirect, i.e., expressed in terms of new procedures, guidelines or training programmes, and in improvements to the design of the technological work environment, rather than in the direct modification of actions or adjustment of norms. Organisational learning is also typically double-loop, as described below. Finally, the effectiveness of learning on the organisational level, and to a lesser degree on the workgroup level, depends on the socio-cultural context (e.g., Owen, 2001).

Learning beyond the organisational level

Since a description in terms of joint systems is recursive, it is possible to go on to mention additional levels, for instance the level of the regulator — which is another kind of organisation — or even the level of society. (The five levels illustrated by Figure 2 are obviously for the purpose of illustration only, and should not be taken as normative.) On the other hand, it does not make much sense to go 'below' the level of the individual, at least not for ergonomics. (It is assumed that a discussion in terms of neuroergonomics (e.g., Parasumaran, 1998) is irrelevant to a discussion of learning from failures.)

When a regulator learns from failures, it is usually the failures of others and only exceptionally of the regulator themselves. This, indeed, is almost part of the definition of a regulator, since this is someone who makes regulations (rules or directions) for others to follow. The problems that regulators face in carrying out their work have been discussed extensively by Reason (1997) and will not be repeated here. Society may also learn, but this process is even slower and usually unregulated. A concrete example of learning on the level of society is the regulation of occupational health and safety passed by national parliaments. An example on a multi-national level is the Seveso directive, which refers to the lessons learned from the accident at the chemical plant at Seveso in 1976. It also illustrates the delay in learning on this level. The European Council issued the first Seveso directive in 1982; the second Seveso directive replaced it in 1996, twenty years after the accident took place.

Single-loop and double-loop learning

Learning from failures involves the detection that a failure has occurred, the identification of the likely causes, and the formulation of appropriate countermeasures in one way or the other. (Note that even if the failure is correctly detected — which it normally is — an incorrect identification or analysis may lead to wrong countermeasures, hence to an inefficient response.) We tend to think of this learning as something that happens directly and in a single loop, in the sense that the corrections are devised to be applied under the same conditions. In other words, the countermeasures are developed to work without changing the overall conditions. It is, however, also possible to change the performance conditions themselves, i.e., improve the conditions for work. Since this is a change that may affect the more direct countermeasures, it is called double-loop learning (see Figure 3).

> When the error detected and corrected permits the organisation to carry on its present policies or achieve its presents objectives, then that error-and-correction process is *single-loop* learning. Single-loop learning is like a thermostat that learns when it is too hot or too cold and turns the heat on or off. The thermostat can perform this task because it can receive information (the temperature of the room) and take corrective action. Double-loop learning occurs when error is detected and corrected in ways that involve the modification of an organisation's underlying norms, policies and objectives.
>
> (Argyris and Schön, 1978, pp. 2–3)

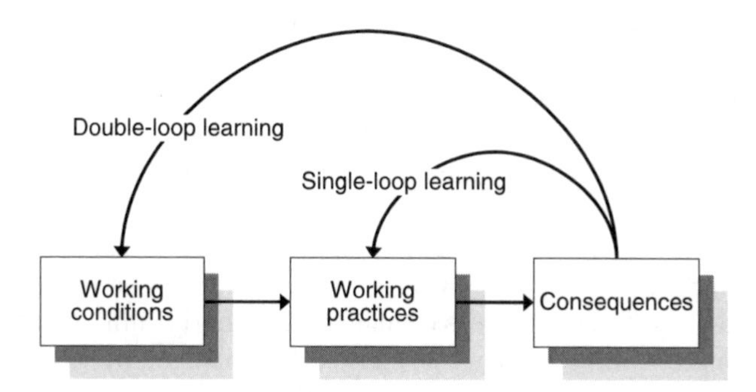

Figure 3 Single and double-loop learning.

Single-loop learning is characteristic of situations where goals, values, frameworks and strategies are taken for granted. Any reflection is directed toward making the strategy more effective. Single-loop learning is thus fast, since it can focus on the concrete situation without the need of more elaborate analyses. Double-loop learning, in contrast, means that the role of the framing and learning systems, which underlie actual goals and strategies, is questioned. In the first case reflection is very concrete, in the second it is more fundamental and involves higher-level considerations and longer time spans. The feedback is therefore also both delayed in time and more generic in content.

Years of studies in social science show that the greater the need is and the more that is at stake, the more likely are individuals — and organisations — to rely on single-loop learning. This is not really surprising since efficiency under most conditions is valued higher than thoroughness, but it does mean that less is learned from the failure than possibly could be. What is achieved is an immediate, and possibly short-lived, correction, rather than a more thorough assessment and possible revision of the conditions for work. The difference between single-loop and double-loop learning is thus very similar to the principle of the Efficiency-Thoroughness Trade-Off (ETTO; cf. Hollnagel, 2002), which can be found in so many examples of human endeavour.

Summary

On the individual level, learning from failures is a question of building up one's own experience and adopting safer work practices. The motivation to do so is usually high, and few resources are required except perhaps time. Learning on the individual level can be facilitated if it is possible for people to mitigate minor incidents or near misses and if the feedback from actions taken in general is clear and unambiguous. It is also useful if opportunities are given to discuss personal experiences with others, both to realise that the failure may not be unique and to learn from similar experiences. In this sense there can be some vicarious learning even at the individual level. On the team level (the workgroup or work system) learning is a question of establishing safe practices. This involves a compilation of individual experience as well as cooperation in extracting the lessons to be learned. Learning can take place through improvisation, where unsuccessful improvisations are weeded out while successful ones are integrated into the repertoire of methods. On the organisational level, safe practices are usually codified in procedures and rules, as well as in the sets of barrier systems and barrier functions that are put in place. Learning is always vicarious, and may usually take a long time. The main characteristics of learning at different levels are summarised in Table 2.

Realising that a failure has occurred

The detection that a failure has occurred is the logical precursor to doing anything at all. Without an indication that something has gone wrong, without the feedback signal that a failure has occurred, a system will continue with its actions as they were started. And if, indeed, these actions bring the system to the desired goal state, there is no practical need to learn anything, except perhaps as a positive reinforcement that the approach taken was the right one.

Technological developments have provided many ways in which the detection of failures can be aided or performed automatically. The multitude of alarm systems and alarm philosophies that exist in different process domains are a good example of that, although they are never designed with learning in mind. In many cases, particularly where the time constant of the process is small, the automated detection may be followed by a partly automated response or correction. The danger of automatic correction is, of course, that the practitioner

Table 2 Characteristics of Learning on Different Levels

Level of learning	Threshold to learn	Delay in learning	Type of learning	Resources needed
Individual	Low (personal experiences)	Negligible (on the spot)	Enactive (direct)	Trifling (mostly time to reflect)
Workgroup	Intermediate (aggregated experiences)	Short (days, weeks)	Mixed enactive and vicarious	Small
Organisation	High (statistics, formal analyses)	Long (months, years)	Vicarious (indirect)	Considerable (time, manpower, technology)

never learns anything, cf. the fundamental regulator paradox described above. In studies of automation in human-machine systems this is often referred to as the phenomenon of de-skilling (Kontogiannis, 1997).

In considering learning from failure, the focus is on detections that are made and realised by the practitioner, even though they may well be aided by technology, such as an alert or a warning. Logically, the detection that something has gone wrong can happen at the moment when the failure occurs; immediately after the failure has occurred but before there is a recognisable change in the system state; and after there has been a recognisable change in the system state. People may, however, respond to a failure in a number of ways. Apart from the important issue of how recovery from failures can take place (e.g., Kanse and van der Schaaf, 2000), people may also take more or less notice of the failure. There is a threshold for learning, which depends on how large a risk the failure appears to carry. If the risk is seen as small, people may simply tolerate the failure as an event of little significance. If the risk is above a given threshold, people will need to take notice of the failure and try to learn. Learning from failures can be seen as analogous to the regulation of action described by the Theory of Risk Homeostasis (Wilde, 1982), but applied in the long-term rather than in the short-term.

It is important to note that detection and learning depend on each other. One consequence of learning from failures is that the detection of failures becomes more precise. The simplest example of that is the case of learning a complex skill where the trainer or instructor can point out failures and deviations that the trainee may be unable to detect. This means that the learning-detection loop becomes self-reinforcing, at least until it reaches a point where the gain from improved detection is marginal, in the sense that nothing important is learned. Detection is furthermore never a passive process or an exclusively reactive affair, where the indication that something has gone wrong mechanically prompts the person to respond and hopefully learn. Detection is rather a proactive process, a constant monitoring of performance to catch failures as they occur. The detection threshold is dynamic, i.e., it depends on the current conditions, the needs, and the resources that the system has. This also affects the identification, which may be more or less shallow.

Further conditions to be able to learn from failures are that the nature or cause of the failure is correctly determined and that this happens sufficiently fast, i.e., that the feedback is not delayed. It is, however, not a condition that the response to the failure is correct. On the contrary, it is not unreasonable to propose that more is learned from failures for which the response is incorrect than for failures for which the response is correct. The reason for that is simple: if the response is correct, then the whole affair can be forgotten as a deviation that was correctly handled. It is dealt with and leaves no trace, hence there is no urgency or motivation for actually learning anything. On the other hand, if the response is inadequate or insufficient, then the failure conditions will continue to exist or perhaps even deteriorate. This means that it becomes necessary to pay more attention and effort to it, and specifically to find out — hence learn — why the response was inadequate. One possibility is that the failure was not detected early enough or well enough, which brings us to the issue of failure detection.

Opportunities for learning: from near-misses to accidents

Failures come in many shapes, and are often characterised according to the severity of their outcomes. It is common to distinguish between incidents and accidents, as a minimum. An accident can be defined, for instance, as an unforeseen event or occurrence, which results in serious property damage or injury, possibly even loss of life. An incident can in a similar fashion be defined as an unforeseen event or occurrence, which results in only minor injury or property damage. An incident is generally also understood as an event which might have progressed to become an accident, but which did not do so for one reason or another. The difference between incidents and accidents is thus in the magnitude and severity of the outcome and the criterion for classifying an event as one or the other may vary among domains such as aviation, train systems, health care, nuclear power generation, etc. Common to both incidents and accidents is that they are clearly noticeable, and that it therefore normally is futile to hide or suppress them — with notable exceptions, of course.

In addition to incidents and accidents there are also events, which under the right — or rather, wrong — conditions might have become incidents and accidents. These events are usually called near misses (or near miss events), which are defined as any event or occurrence that could have resulted in some kind of injury or property damage, but which did not. (A related term used by the UK Health and Safety Executive is 'dangerous occurrences' defined as something which does not result in a reportable injury, but which clearly could have done.) Further down the scale is the category of unsafe acts, i.e., actions that in some way broke a principle for safe behaviour, for instance as a minor violation. Unsafe acts are included among the failure types because they constitute a necessary link between actions and failures in most theories of human performance. Table 3 shows examples of different types of failures in different domains.

There are two essential and important differences between near misses and incidents/accidents, though all are failures in some way. The first has to do with how conspicuous they are: an incident or accident is very obvious and is objective or intersubjective in the sense that two or more people may note it. The consequences are usually also quite evident. Near misses, on the other hand, are not nearly as obvious, neither in how they appear nor in their consequences. It may, indeed, often only be the person to whom the near miss happens that notice them.

The second difference has to do with the criterion or definition. Although definitions of incidents and accidents may differ, and in particular that the classification of what is an

Table 3 Examples of Failure Types

Type of failure	At work	In the traffic	At home
Accident	Being injured or killed	Being killed or seriously injured	Fire or water leakage
Incident	Being hit but not injured	Being hit by a vehicle	A blown fuse; breaking a window
Near miss	Something falling down close to	Almost being hit by a vehicle	Forgetting to lock the door
Unsafe act	Not wearing a hard hat	Walking against a red light	Standing on a chair instead of a ladder

incident and what is an accident may vary among domains, there is never any doubt that an incident/accident has occurred. In the case of near misses, the definition is less precise and much is left to personal judgment. One proposed definition is, for instance, that a near miss is something that others can learn from or an opportunity to improve practice. That clearly calls for a subjective evaluation or judgement.

It is common knowledge that there are many more near misses than accidents, and each of us experiences near misses regularly. For this reason the study of near misses has become very important as opportunities of learning (Dye and van der Schaaf, 2002; Jones *et al.*, 1999; van der Schaaf *et al.*, 1991). The importance of using near misses as opportunities for learning is based on the assumption that near misses are failures that do not have serious consequences, but which by their nature (and explanation) are similar to the events that do turn into failures. Were near misses of a completely different aetiology than accidents, then there would be no reason to take an interest in them. (A consequence of this is that incidents/accidents and correct actions all come from the same type of performance, hence that the notion of 'error-as-cause' is superfluous.)

The common point in all of this is that failures come in many forms and degrees of severity, and that there may be much to learn from failures that are less severe than incidents and accidents. This is often formulated as a pyramid, or sometimes as an iceberg model, as shown in Figure 4. The analogy is that accidents only are the top of the iceberg, i.e., that there are many things which are not observable and which in some sense are below the water level. The pyramid depicts the relationship of accidents to other types of unsafe situations. One type is incidents with consequences that could have resulted in injury or damage that meets accident report thresholds, but that failed to do so. Another type is near misses, or almost incidents, which raised the danger of an accident or incident, but which due to fortunate circumstances failed to develop further. A final type is unsafe acts, i.e., actions where the performance variability almost reached the threshold of the near miss classification.

The numbers in Figure 4 are based on the analysis of 1,753,498 reported accidents representing 21 different industrial groups (Bird, 1974; from Heinrich *et al.*, 1980). Assuming that the 1 : 10 : 30 : 600 ratios are correct, it is clearly insufficient to direct the total effort at the relatively few events terminating in serious or disabling injury when there are in addition 640 property damage or no-loss incidents occurring that provide a much larger basis for more effective control of total accident losses. For one thing, one need not wait so long between opportunities to learn. Second, the learning would be less costly because the experience is direct and the consequences smaller. These ratios have achieved an almost mythical status, often quoted as 'studies show that approximately 300 near misses occur for every accident'. The numbers may nevertheless be very different in reality. Although it is nearly impossible to

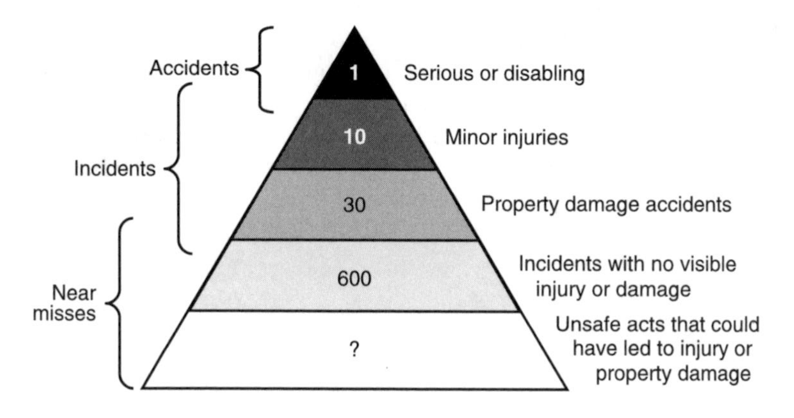

Figure 4 The symbolic iceberg model.

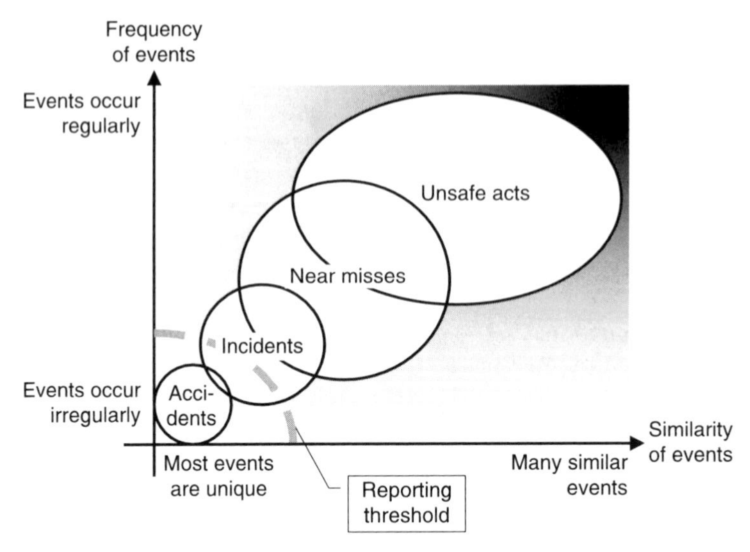

Figure 5 Failure types and the reporting threshold.

get reliable numbers for the near misses, recent statistics on work injuries in the UK show that in 1999/2000 there were 220 fatal injuries and 28,652 non-fatal injuries, which gives a ratio of 1 : 130 for the first two categories alone (rather than 1 : 10). If we, just for the sake of illustration, scale up the ratios derived from Figure 4 by a factor 13, the result is that there should be 8,320 lesser events — and that number may not even include the true near misses.

These calculations should not be taken seriously as far as the actual numbers are concerned, but certainly should be in terms of their meaning. One reason to be careful is that the number of cases in each category depends on the definitions used by the event reporting or data collection system. This is an area where there is a notorious lack of agreement and consistency (Brazier, 1994), due in no small measure to the difficulty of accounting for situations where humans are involved. A second reason is that there is a reporting threshold, which partly is tied to how unique the manifestations of the events are, cf. Figure 5. This means that while we can be fairly certain about the number of accidents, failures with lesser consequences are reported with less consistency and reliability. Bourne and van Ours (2002), for instance, found that workplace accident rates, with the exception of fatal accidents, were inversely related to both the level of unemployment and the change in unemployment. Near misses are only rarely reported, and unsafe acts are as a rule not reported at all.

Near miss reporting is problematic as a safety tool since there are a number of practical problems in establishing a good reporting system (van der Schaaf, 1995 and Chapter 33). This means that it may be difficult to use near misses to learn from failures at the organisational level. The problems of recognising and using near misses are far smaller on the workgroup level and practically negligible on the individual level. The organisation may still need to encourage people to pay attention to near misses, and perhaps even provide them with some help in doing so. As long as this is done for the purpose of personal learning rather than reporting, there should be few practical or practical problems. Training people to learn from near misses may conceivably also serve as a good introduction for a reporting system at a later time.

Conclusions

This chapter has presented some basic issues that must be kept in mind when the possibility of learning from failures is considered. Apart from the fact that the ability to learn from failures

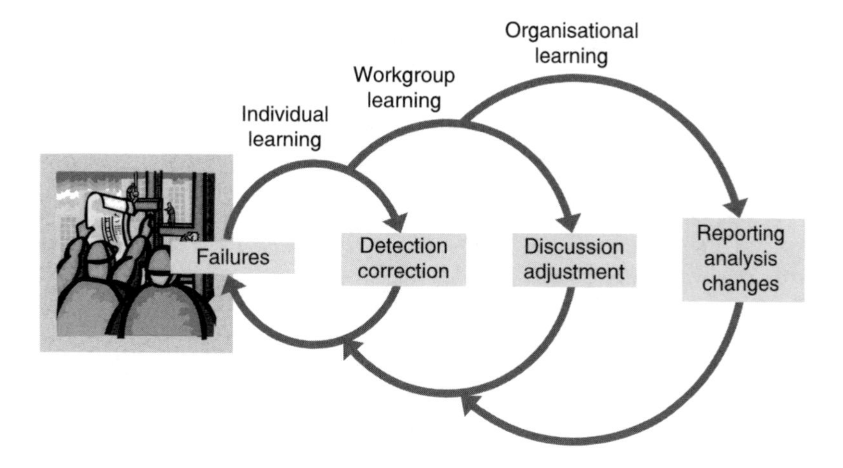

Figure 6 Coupling between levels of learning.

provides the very basis for human development in general, the need to learn from failures is growing as fast, if not faster, than the complexity of the systems we use. Learning from failures can be considered on three different levels: the individual level, the workgroup level, and the organisational level. These levels differ in a number of important ways, as listed in Table 2. The couplings between the three levels are shown in Figure 6. On the individual level people learn directly from their failures, very often as an afterthought to the detection-correction activities that ensue. Learning usually takes place at the same timeframe and at the same place as the failure, so delays are minimal. Even though learning is on a personal level, several things can be done to facilitate the process as mentioned above. On the workgroup level, learning reflects the experiences of the individuals as well as of the workgroup as a whole. More time and resources are obviously needed, but learning is usually closely coupled to the actual work and therefore very concrete. Learning may, however, also slowly become more formalised and subject to established principles and procedures. The main difference on the workgroup level is whether learning is enactive or vicarious, where the latter requires more time and resources. Finally, learning on the organisational level can be a laborious and slow process, which in many cases takes months and years rather than days and weeks. It is based on generalised experience and will for that reason alone take much longer. The outcome is usually directed at the organisation's norms and policies, rather than the work activities directly, i.e., it is double-loop rather than single-loop learning.

Ergonomics and human factors have recently paid much attention to the problems of learning on the organisational level and to safety management. While this is a welcome development, it should not make us forget that learning from failures must occur on several levels at the same time in order to be effective and that each level has different needs of support. Learning from failures is as much a question of individual development as it is one of event reporting and risk management, and the current preoccupation with the latter should not make us neglect the former.

References

Argyris, C. and Schön, D. (1978). *Organizational Learning: a Theory of Action Perspective.* (Reading, Mass: Addison-Wesley).

Bandura, A. (1971). *Social Learning Theory.* (New York: General Learning Press).

Bird, F. (1974). *Management Guide to Loss Control.* (Atlanta, GA: Institute Press).

Boone, J. and van Ours, J. (2002). *Cyclical Fluctuations in Workplace Accidents.* CEPR Discussion Paper no. 3655. (London: Centre for Economic Policy Research). http://www.cepr.org/pubs/dps/DP3655.asp.

Brazier, A.J. (1994). A summary of incident reporting in the process industry. *Journal of Loss Prevention in the Process Industries*, **7**, 243–248.

Dye, J. and van der Schaaf, T.W. (2002). PRISMA as a quality tool for promoting customer satisfaction in the telecommunications industry. *Reliability Engineering and System Safety*, **75**, 303–311.

Hale, A., Wilpert, B. and Freitag, M. (Eds.) (1997). *After the Event: From Accident to Organisational Learning.* (Mahwah, NJ: Elsevier).

Heinrich, H.W., Petersen, D. and Roos, N. (1980). *Industrial Accident Prevention: A Safety Management Approach* (Fifth Edition). (New York: McGraw-Hill Book Company).

Hollnagel, E. (2002). Understanding accidents — from root causes to performance variability. In J.J. Persensky, B. Hallbert and H. Blackman (Eds.), *New Century, New Trends. Proceedings of the 2002 IEEE 7th Conference on Human Factors and Power Plants*, September 15–19, Scottsdale, AZ, USA (New York: IEEE).

Jones, S., Kirchsteiger, C. and Bjerke, W. (1999). The importance of near miss reporting to further improve safety performance. *Journal of Loss Prevention in the Process Industries*, **12**, 59–67.

Kanse, L. and van der Schaaf, T.W. (2000). Recovery from failures — Understanding the positive role of human operators during incidents. In: D. De Waard, C. Weikert, J. Hoonhout, and J. Ramaekers (Eds.), *Proceedings Human Factors and Ergonomics Society Europe Chapter Annual Meeting 2000*, Maastricht, Netherlands, November 1–3, 2000, pp. 367–379.

Kontogiannis, T. (1997). A framework for the analysis of cognitive reliability in complex systems: A recovery centred approach. *Reliability Engineering and System Safety*, **58**, 233–248.

Leveson, N.G. (1995). *Safeware — system safety and computers.* (Reading, MA: Addison-Wesley).

Owen, C.A. (2001). The role of organisational context in mediating workplace learning and performance. *Computers in Human Behavior*, **17**, 597–614.

Parasuraman, R. (1998). *Neuroergonomics: The Study of Brain and Behavior at Work* (http://www.cua.edu/students/org/csl/neuroerg.htm).

Perrow, C. (1984). *Normal accidents: Living with High-risk Technologies.* (New York: Basic Books).

Petroski, H. (1994). *Design Paradigms: Case Histories of Error and Judgment in Engineering.* (Cambridge, MA.: Harvard University Press).

Phimister, J.R., Oktem, U., Kleindorfer, P.R. and Kuhnreuter, H. (2001). *Near-miss Management Systems in the Chemical Process Industry.* http://opim.wharton.upenn.edu/risk/downloads/01-03-JP.pdf

Tenner, E. (1996). *When Things Bite Back: Technology and the Revenge of Unintended Consequences.* (New York: Vintage Books).

van der Schaaf, T.W. (1995). Near miss reporting in the chemical process industry: An overview. *Microelectronics Reliability*, **35**, 1233–1243.

van der Schaaf, T.W., Lucas, D. and Hale, A. (Eds.) (1991). *Near Miss Reporting as a Safety Tool.* (Oxford: Butterworth-Heinemann).

Wilde, G.J.S. (1982). The theory of risk homeostasis: Implications for safety and health. *Risk Analysis*, **2**, 209–225.

Part VI

Implementation and organisation analysis

How often, in our professional or personal lives, have we had what we believed to be an excellent idea, or proposal turned down, and how often has that happened not because of lack of intrinsic merits or due to flaws in the plan but because of the way we introduced it to our colleagues or friends? Likewise the success or failure of many systems, or at least the degree of their successful utilisation, will be determined by how they are implemented. Eason, in Chapter 35, looks at this issue generally and with relevance for all the other content of this book. Even before we can carry out such evaluation we need to have initiated change, for which we need to understand how best to manage any ergonomics intervention in industry. How do we get agreement for our investigation or development, how best do we manage the initiative, how can we communicate with all concerned and, again, how can we best aid successful implementation of our suggested changes?

The type of implementation strategy adopted, as identified by Eason, will determine the degree of participation that is possible. Concepts such as the user 'champion', user representation, and full user involvement will be more relevant and feasible to follow in a strategy of incremental implementation than in a 'Big Bang' scenario. Whilst there are some downsides — for instance the time and other resources required, participant coping problems and so on — Eason generally argues for user-orientation in systems implementation, which includes development and introduction. In doing so he provides apt commentary for the concerns of the whole of this book. Concern for and involvement of the people who work in offices, factories, transport and services etc., must underpin ergonomics methods and techniques.

With this in mind, many ergonomists a development and promotion actively use (and research in) participatory methods and processes. Although this philosophy and approach can have difficulties and unwanted side effects, generally it is *the* appropriate methodology for ergonomics *practice*; Wilson, Haines and Morris in Chapter 36, review types, processes and methods of participation.

In Chapter 1 it was noted that ergonomics can have aims which relate to organisational as well as individual well-being, although the strong linkage between these was stressed. Whilst many of us would, in an ideal world, put individuals' interests as our major priority, we must work in the real world. In view of this, Oxenburgh and Simpson argue (in Chapter 37) an economic case generally must be made before an organisation will fund, or even play host to, an ergonomics investigation. This consideration can be extended; a powerful motivation for ergonomics input in consumer product design for instance is the economic threat of strict product liability provisions. Thus, we do require some means of showing the economic returns on our efforts, and Oxenburgh and Simpson present some relevant techniques.

If we are to take a true systems approach in ergonomics/human factors, and if we are to have any meaningful impact in our work, then we must understand much about the organisation which is the site of our investigations. We must be able to assess, for instance, how the way in which an organisation is structured, and its management philosophy and its willingness to change, may influence behaviour or may affect any developments we initiate. Shipley in the second edition of this book provided a view on such issues which was food for thought for ergonomics practitioners, presenting a powerful argument about *how* ergonomists should tackle investigations into people's work, whatever the particular study focus. In this edition, Siemieniuch and Sinclair take a more structured and analytical viewpoint on organisations, describing how process analysis examines sets of output-producing activities, and addressing how work is organised. Techniques such as scenario and link analysis, and influence diagrams, are described. The discussion of soft systems methodology and of expertise pick up on earlier discussions, for instance in Chapters 8, 10 and 29. Finally, in a short end Chapter 39, Annett overviews where ergonomics and human factors science is now and where it is going, especially as regards its methodological base.

Chapter 35

Ergonomic interventions in the implementation of new technical systems

Ken Eason

Introduction

One of the most frequent sources of change in working life is the implementation of new technology. This can have many effects on the work system; it may threaten existing jobs, may be a major cause of disruption to working patterns or may offer people much better and more effective tools with which to work. Whatever the effects the introduction of new technology is a time when the work system is at its most malleable; it is in a state of change and other kinds of change which would not be countenanced in times of stability, may become possible. There are needs and opportunities to make ergonomic interventions at times of technical systems change and the purpose of this chapter is to explore these opportunities in different contexts of technical change.

A systems context for ergonomic interventions

The perspective adopted in this chapter is that a work system — whatever product it makes or service it provides — can be treated as an 'open socio-technical system' (Katz and Kahn, 1966). It is an open system in that it engages in transactions with other systems, i.e., receives inputs, raw materials etc from suppliers and delivers its products to its market. It is a socio-technical system (Emery and Trist, 1960) in that the resources that do the work are people organised into a social system consisting of work roles and technology organised into more or less integrated technical systems to support the work process (see also Chapter 29 for more on socio-technical systems). Three concepts from socio-technical systems theory will be helpful when considering the impact of changing the technical sub-system on the total work system:

- *Co-optimisation.* The technical and social systems are tightly coupled and changes in one have implications for the other. For the overall work system to function effectively the social and technical system must be 'co-optimised' so they work in harmony together.
- *Adaptive systems and quasi-stationary equilibrium.* Work systems are open to changes in their environment and must adjust, for example, to changes in their market, to new regulations, to new technology opportunities etc. The work system must therefore be capable of adaptation. Another way of expressing this is to say the work system is in 'quasi-stationary equilibrium'; a stable way of doing today's business but containing adaptive elements which will enable it to reform to meet new challenges.

- *The principle of minimal critical specification.* This principle established by Cherns (1976) is a prerequisite for an adaptive system. It says that the socio-technical system that comprises the work system should not be fully specified. This is a rather counter-intuitive concept in systems design. It means that in any system that contains human beings as system components there should be freedom or autonomy left in the system so that the workforce can adopt new work practices, use technology in different ways etc., as occasion demands. The workforce is therefore a major mechanism for recognising the need for change and finding adaptive ways of responding. If the work system is so tightly specified that the workforce is not empowered to make changes, the system may be unable to respond adaptively as circumstances change.

With this systems perspective in place we can examine what kinds of ergonomics interventions may be appropriate when a change in the technical system of a work system is planned.

There are many purposes an ergonomist might have when making a contribution to a work system. In this chapter we will consider four aims which are summarised in Figure 1. The aims are expressed as immediate and enduring. The immediate aims are to contribute to the new work system so that it enhances the performance, satisfaction and health of the work force. The enduring aims are that in the longer term the resultant work system is in a better position to adapt and develop as circumstances change.

With specific reference to the separate aims, the distinction between immediate and enduring mean the following:

1. *Socio-technical systems integration.* A new technical system has many implications for social sub-systems which have to be addressed if an integrated system is to result. In the short term this often means considering, for example, issues of human-machine interaction, to ensure the new technical system has the utility and usability necessary for the workforce to achieve their tasks (see Chapters 11 and 13). In the longer term the technical system and social system may need ongoing adjustments in, for example, the deployment of the technology, job structures, recruitment, training, reward systems etc. to sustain integration as the enterprise adjusts to new business demands.
2. *Safety and health.* A new technical system can raise new questions about the safety and health of the workforce both directly, e.g., in the operation of machinery, and indirectly e.g., in the changing nature of work patterns, the creation of new physical work environments, etc. In the short term an ergonomic intervention may consist of a risk assessment of the new technical system and the design of suitable work environments, work practices etc. to protect safety and health (see Chapter 31). In the longer term there will be continuing changes to the work system and there will be a need to establish a safety and health culture which can maintain an ongoing review of safety practice and modify it to sustain best practice.
3. *Learning and adaptation.* A new technical system may require the workforce to learn new skills and there may also be new work practices to master. The development and use of relevant training techniques when the system is implemented will therefore be a necessary intervention. However, in a world of continuous change, individual learning needs to be continuous. The current preoccupation of the business world with 'learning organisations' (Senge, 1990) and 'adaptive systems' (Haeckel, 1998) is testament to the enduring nature of these demands on work systems. Meeting not just the immediate need for new skills but the enduring need

Aim	Immediate	Enduring
Socio-technical systems integration	To create effective forms of human-machine interaction	To sustain an effective socio-technical system
Safety and Health	To design work settings that are safe to work in	To create a safe and healthy work culture
Learning and Adaptation	To develop competence for current work goals	To sustain the work system as an adaptive, learning organisation
Motivation and Participation	To engage the workforce in developing the current work system	To establish jobs and forms of empowerment which promote enduring forms of motivation

Figure 1 Aims of ergonomic interventions.

to keep refreshing the competence of the workforce is a major challenge for an ergonomic intervention.

4. *Motivation and participation.* It is widely acknowledged that technical change is much more successful when employees are motivated to accept it. Two related routes are advocated. First, if the workforce can participate in the change process with real power to influence events, they can then develop a sense of ownership of the change (Damodaran, 1996 — and see Chapter 36 of this book). Second, by designing work with new technology so that it meets requirements for satisfying, worthwhile jobs, people will have something tangible to gain from the change process. However, people continue to develop in their work and careers and the process of change is continuous. The enduring need is not therefore a one-stage participation to produce better jobs but the establishment of ongoing participative mechanisms to support the long-term empowerment of the workforce in the evolution of the work system of which they are a part.

One feature that is common across all four kinds of intervention is the difference between meeting immediate needs and creating a work system that is adaptive in the longer term. In the former case, the intervention may be to ensure that the new socio-technical system which is implemented is effective, safe and healthy for the workforce. In the latter case, the aim is to improve the capability of the work system by enhancing its adaptive mechanisms so that it can cope with future demands for change. In a survey of social science and ergonomic interventions made in organisations in the UK and Germany, Klein and Eason (1991) found this distinction to be a concern of many practitioners of the human and social sciences. Many of them were given the opportunity to make contributions to the current system within limited timescales and terms of reference. They were worried that their work would be transient, swept away when the next wave of change came along. Many of the practitioners were working to establish practices in the organisation which would sustain good working practice in the longer term. Klein and Eason call this 'institutionalising' the practice so that it becomes part of normal procedure within the organisation. A good example is the 'devolution of ergonomics' approach (Wilson, 1994). An ergonomist might, for example, be invited to design the workstation and physical environment etc associated with a new technical system. If the ergonomist uses the opportunity of the new technical system implementation to train some of the workforce in the rudiments of auditing workstations etc. these members of staff may be able to monitor future circumstances and trigger actions when necessary to sustain good ergonomic practice.

Although the technological sub-systems of work systems are increasingly sophisticated it remains the case that the learning and adaptive part of the work system is the staff who occupy roles in the social sub-system. In helping staff in organisations to develop their capacities to cope with change Bridger (1990) refers to the need to manage 'the double task'. The first task is to get the current job done. Task two is to review progress in task one and modify it in accordance with new objectives, new learning etc. Unless the staff of the work system have the capability to undertake task two on a regular basis and facilitators in the work system to encourage these reviews and enable actions to be taken as a result, the learning and adaptive capability of the system may be inadequate to cope with the pace of change. In this chapter therefore we are concerned with ergonomic interventions which help staff in work systems with both task one (undertaking their work when the new technical system is implemented) and with task two (taking ownership of reviewing and changing the work system so that it matures and adapts over time).

Scoping the organisational context

Every opportunity to make an ergonomics intervention is different, with its own constraints and challenges. The purpose of this chapter is to examine the way in which the context of technical system change shapes the way an ergonomic intervention can be undertaken. The initial stage of an intervention therefore requires some form of scoping to establish the context that exists in order to determine what can be done and how it needs to be done. Eason, Harker and Olphert (1996) have set out the following process for scoping a new organisational opportunity. This is part of the ORDIT methodology which is a process for helping an organisation make effective socio-technical systems changes. In the ORDIT methodology, scoping involves four activities.

1. *Analysis of the current system.* This is an outline of the current socio-technical system which is likely to be affected by planned changes. This analysis helps to establish the boundaries to the system and to get a perspective on history, work culture, current work practices, etc.
2. *The planned change.* This is an assessment of the drivers for change (the objectives being sought), the planned form of the change (e.g., it might be a new technical system) and the state of development of the plan.
3. *Key stakeholders and principle requirements.* An analysis of the stakeholders within the work system who are likely to be affected by any changes e.g., the management who commissioned the change, the workforce who will use the future technical system, the technical staff who are implementing it, etc. For each key group of stakeholders it is useful to identify key requirements in relation to the change. The managers may, for example, expect improvements in system performance. The workforce may have more defensive requirements, for example, to keep their jobs, to sustain current work practices, etc.
4. *The contract for the intervention: explicit and tacit.* The practitioners will have been commissioned to play some role in the technical system implementation and scoping is a phase when this can be explored, made more explicit and to some extent negotiated. The request may, for example, be to contribute on safety and health but it may be desirable to extend this to consider aspects of systems integration, for example how safety relates to incentives or to training.

For the purposes of this chapter it is assumed that scoping has shown that a major technical system is to be implemented and has revealed the strategy by which it is to be

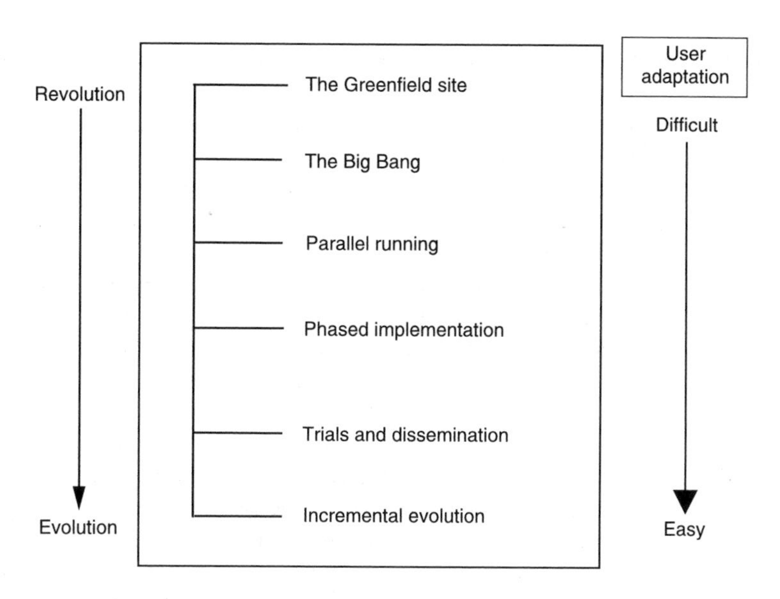

Figure 2 Implementation strategies.

implemented. It is likely that the strategy will be one of the six discussed below (or some combination of them).

Technical system implementation strategies

The six major strategies used for technical system implementation listed in Figure 2 are derived from Eason's (1988) examination of strategies for the implementation of computer systems. The strategies are arranged on a dimension from the most revolutionary to the most evolutionary. At one extreme we have the major start up of a complete new system overnight and, at the other, the steady introduction of technical facilities over an extended period of time.

The aim in this section is to identify the main principles of each strategy and the circumstances in which it might be used. We will then consider the implications for the ergonomic intervention in relation to the four aims for ergonomic interventions.

The 'greenfield site' implementation

This is when a new technical system is part of a complete new start up, i.e., a new factory is built, a new shop is opened. The work process and the technical system may be based on existing systems elsewhere but designers have a chance to introduce new processes and technologies. Once the new system is in place there is often a lengthy commissioning period when the whole process is tested before the factory produces products, the shop is open for business etc.

From one perspective this is an excellent opportunity for an ergonomic intervention. Many features of the work system are being newly created and it may be possible to pursue both aims 1 and 2. There could be opportunities to work with the designers of the new process on many issues; on workstation layout, the physical environment etc. to ensure good human-machine interaction and safe and comfortable working conditions. It may also be possible to apply many of the principles of job design (Emery, 1980; Aberg, 1981; Parker and Wall, 1998) to create work with meaning, variety, feedback etc. which will be satisfying and motivating for the workforce. Klein and Eason (1991) give an example of the development of a new chocolate factory in which the social system was based on autonomous work groups and planned on the basis of job design principles.

From another perspective, however, the Greenfield site is a problem for ergonomic interventions because the workforce may not be appointed in time to play a part in the planning process. Typically, the work force will be newly recruited probably in time for training before the new system goes live. Decisions may need to be taken about the technical system well before the workforce arrive and they may have consequences for the jobs, the work process, the work stations etc that the workforce will encounter. Designers may, of course, have the best of intentions with respect to the health and welfare of the workforce and may have access to good ergonomic advice. However, there may be little opportunity to engage the workforce in the planning process and begin the process of building a learning organisation in which staff will 'own' the work system and seek continuous improvements in its performance. For these purposes it is better to recruit staff early and/or leave many of the decisions about working practices and organisational procedures as late as possible so that the workforce can play their part in the planning process. The literature on the planning of change emphasises the importance of providing the workforce with these opportunities to influence the working practices they will have to operate. As will be seen this is common when the workforce already exists. When it does not exist the desire to make decisions for future staff has to be resisted if the long term aims of the intervention are to be realised.

The big bang

One of the most difficult kinds of implementations to undertake effectively is when an existing work system is discontinued in its entirety and replaced immediately with a new system. This sudden and dramatic change is necessary when an organisation needs to sustain the continuity of its business but, because of system interdependencies, has to change a large 'critical mass' of activities to bring in innovations. One of the best known examples is the overnight switch of the London Stock Market to electronic trading in 1986. Stock market trading involves complex interactions by many trading organisations and it was necessary for everyone to make the switch at the same time. Obviously this can be a high risk strategy because all the parts of the new system have to function well when the system is switched on if the normal level of business is to be sustained. This is a difficult objective to achieve and there is often an 'initial dip' in performance before the gains expected of the new system are achieved. Sometimes the dip can be deep and long. When the UK Passport Office switched to a new passport issuing system in 1999 (National Audit Office, 1999) performance dropped to 27% of the previous issue rate and did not recover for 18 months.

Some of the problems in 'big bang' implementations may be technical because it may be the first time the complete system has been tested under a full load. The London Ambulance Service introduced a new computer based ambulance despatch system in 1992 (Page *et al.*, 1993; Wastell and Cooper, 1996) and the technical system could not cope with the calls made to it and it had to be discontinued. However, new systems introduced in this way also place a major burden on the social system of the organisation. The new technical system may require new jobs, new administrative procedures etc. as well as the development of new skills. If a major change of this kind is undertaken very quickly without full preparations, the classical symptoms for resistance to change exist. In addition to realistic concerns about their jobs, the workforce will probably know little of what is intended and there may be many unfounded rumours. The techniques for avoiding resistance to change are well documented, e.g., Buchanan and Huczenski (1997). They focus on the participation of the workforce in the planning of new work practices, full consultation so that they have positive attitudes towards the change and prior training to ensure they have the skills and competencies to cope in the new situation. If there is detailed participation before the change there should be many opportunities to ensure good job design, examine health and safety issues, etc. When it is well managed there are advantages to a big bang implementation; it becomes the big event and, if everybody is committed to its success and shares in the achievement, it can be a shared experience of great significance for the organisation. When the conditions are right the collective energy and excitement of such an event can carry the new system through many teething troubles and create lasting bonds between members of the workforce.

The one danger of a successful instant changeover of this kind is that it will be seen as heralding a period of stability when there is no further need for developmental activities. And yet there will be a continuing need for systems adaptation and for further learning and staff development. One way of facilitating these longer-term goals is to keep some of the participative structures created for the planning of the change in place after the change. Staff can, for example, take responsibility for summative evaluations of various attributes of the new system, from productivity to safety and health.

Parallel running

A popular way of minimising the risks to business productivity when introducing a new technical system, especially if it is a computer based system, is to run the new system alongside the old system. When it is 'mission critical' that no loss of quantity or quality in output should occur during the system change, this can be a secure way of proceeding. In socio-technical systems terms what happens in this form of implementation is that the social system operates with two technical systems at the same time. The old one is used to deliver products and services whilst the work is repeated with the new system in order to detect and remove any technical problems and build the competence and confidence of staff in using the new system.

As a process of change this approach can have positive effects on management, the workforce and designers. It gives everybody the time to test the new system in a field setting without having to rely on it. It is not, however, without its drawbacks. Operating both systems, for example, involves extra work and it is rarely the case that extra human resources are available. If the new system keeps failing, there may be a reaction against the extra work and a desire to retain tried and trusted methods. On the other hand, if the new system works well and saves a lot of time and energy, it can be frustrating to continue using the old system. In the freight forwarding case study described in Klein and Eason (1991), a new system provided a way of capturing data once and then using those data for a wide range of clerical functions. The old system required the users to repeat the data capture for every clerical function. After a short period of parallel running the staff pleaded to be allowed to use the new system for real tasks.

Within the parallel running approach there is an opportunity to work with the workforce and the designers on the new technical system, interface issues and the workstations, etc. What is more problematic is to work on any job or organisational changes that might be appropriate when the new system is fully in place. The continued operation of the old system requires the existing organisational structures and procedures to remain in place. This can reinforce the usual organisational inertia to keep things as they are. It may mean that the opportunity to change the entire socio-technical system for the benefit of the staff and the business is missed. A number of authors have advocated a participative approach to considering work organisation and job design questions when technical change is planned, e.g., Mumford (2000) and Eason and Sell (1981). One difficulty with these approaches is that the workforce often does not have sufficient understanding of the new technical system to be able to examine its consequences for alternative forms of work organisation. In parallel running the workforce develop a detailed knowledge of the new technical system and are well able to participate in job design discussions. Planning organisational change to take place at the point when the technical system has been fully commissioned can be an effective way of ending a period of parallel running.

Phased implementation

Making changes to massive work systems can be made more manageable by phasing in the changes over a period of time. There are two ways in which large-scale changes can be sub-divided to facilitate phased introduction. First, the functionality of the technical system can be introduced in phases so that, for example, basic facilities can be introduced first and more advanced ones at a later date. Secondly, it may be possible to introduce the system in different parts of the organisation at different times. It is this latter approach, often called 'roll out', which will be examined in this section. Phased implementation is an attractive option for organisations that have many semi-autonomous sections, e.g., branches of a bank, building society or government agency, a chain of shops, etc. If a similar technical system is to be implemented across the organisation it can be done as a series of 'little bangs' taking each branch or section at a time. The disturbance of the change is localised by this strategy and has little impact on the overall productivity of the organisation. Another advantage is that an implementation team can be assembled with all the necessary expertise and they can give each branch their full attention for a short period to introduce this new system with maximum efficiency.

This approach has many advantages if the aim is to make ergonomic interventions. If ergonomic competence is built into the multi-disciplinary implementation team there will be opportunities to influence the new forms of human-machine interaction, workstations, the physical environment, the training, etc. Since these activities are repeated in each branch there will be an opportunity to influence the standards for the activities across the whole organisation. As with the other strategies there are dangers. A 'roll out' of a technical system of this kind is most often a centrally planned process in which the new system is largely imposed on each branch. There is often an imperative to get the system implemented as quickly as possible and therefore little opportunity for the workforce at each branch to do more than learn the new work practices. And yet for each workforce this 'little bang' is just as much upheaval, uncertainty and rejection of current practices as it is for the workforce in the 'big bang' implementation. This can lead to resistance as the roll out proceeds. As Tapscott (1982) observes in his 'second law of office systems' the 'difficulty of implementation increases with every new implementation'. This is often counter-intuitive to the implementation team who, having solved the technical problems and worked out efficient procedures for the changeover, may expect each new implementation to be faster and smoother. For the workforce, however, the experience is of the swift imposition of procedures established elsewhere with little consultation or attention to their local conditions or practices.

One way to avoid these outcomes is to apply the sociotechnical 'principle of minimum critical specification' at the central planning stage of development, i.e., pre-define only the features of the system that have to be standardised to permit inter-branch co-operation and control. This can leave substantial discretion for local workforces to vary the way the system is implemented to meet local conditions. Individual branches or units typically differ from one another in significant ways, e.g., in customer or caseload, in geographical area, in premises, in staff profiles, etc. There may be quite substantial and legitimate varieties in work practice to accommodate these differences. If the new technical system allows some flexibility it is possible for local staff to work out the specific ways in which they will adopt the new system, e.g., personalising some of the interfaces, establishing local job design, matching workstation layouts to the premises, etc. However, these issues cannot be fully addressed in the limited time period (in many cases, one week) when the implementation team are in the branch or department to implement the new system. A local working party needs to be convened ahead of implementation to plan how the implementation can best proceed for their branch. An ergonomist can play a key role in helping the workforce exploit system flexibility so that the system meets their needs and they can develop a sense of ownership of it. Eason (1995), for example, describes the implementation of an office system in a number of departments of a legal practice. In each department the clerical staff planned specific forms of training, chose their own furniture and established their own workstation layout. These were important areas of participation and influence for the staff and contributed to the rapid adoption of the new system. Once established such local working groups can be sustained to review the way systems are working and plan future developments.

Trials and dissemination

A common way to initiate a major organisational change is to make a trial or pilot implementation. This differs from a prototype test or an off-line simulation by being a live running of an intended new system. In some respect this is like the first implementation of a phased implementation programme except that some aspects of the system are still under test and there may be further development before full dissemination. If the concept of a trial has real significance it should mean that, following a review, a decision could be taken to abort dissemination or undertake major redevelopment before the change programme is implemented. Trials or pilot implementations are often undertaken in major organisations with, for example, many branches or who have critical performance standards to sustain and have to reduce as many risks as possible before major commitment.

The fact that this is explicitly a learning strategy makes it ideal as a vehicle for ergonomic interventions. As in the other examples, the fact that it is a live implementation means it is an opportunity to test whether forms of human-machine interaction are fit for purpose, usable by the workforce, etc. and that the physical workstations conform to health and safety requirements. The staff themselves can, if possible, play a part in the detailed matching of the technical system to their particular circumstances. However, the formal fact of a trial raises other possibilities. A technical system embedded in a work process and interacting with the existing culture of the social system will have many implications; for jobs, for work practice, for rewards, power and influence, etc. Many of these issues are unsuspected by those creating the new technical system but the trial can make them evident. An evaluation of the wider organisational consequences of the technical system can often turn what was expected to be a proving trial for a technical system into a socio-technical systems reappraisal. In the freight forwarding system cited earlier (Klein and Eason, 1991) a trial implementation of the new system was made involving three neighbouring branches that were to use the system as a common database to consolidate loads before export. The trial demonstrated that whilst the database was extremely useful to the clerks, its use for load consolidation raised many questions for management — 'whose load was it', 'who got what share of the revenue', etc. After the trial the organisation went ahead with a branch based database system but discontinued the idea of load consolidation because it raised too many organisational issues.

There are three common issues which affect the usefulness of a trial implementation from the perspective of the aims of the ergonomic interventions:

- *Ensuring the learning.* At its best the trial is a strong evidence base for 'action learning '(Revans, 1982), i.e., it constitutes experimental actions from which members of the organisation can learn and improve performance. However, it is often the case that lip service only is paid to the trial aspect of the implementation. If it is treated as a technical system proving trial and there are few problems, there is a tendency to proceed quickly to full dissemination without reviewing the implications for the workforce or organisational practice. Often the dates for the implementation 'roll out' have already been pencilled in. If the trial includes two features this outcome can be avoided. First, an agreed programme of evaluation is needed in which data will be gathered systematically about performance, usability, organisational implications etc. Second, the body assessing the trial, preferably including management, design team and representatives of the workforce, needs terms of reference to receive and comment on the evaluation before dissemination can continue.
- *Recognising the 'Hawthorne effect'.* It is often the case that a trial implementation goes very well, lulling the organisation into a false sense of security and there is a rude awakening when dissemination proves much more difficult. Two factors can cause this to occur. First, the organisation may have chosen a part of the organisation for the trial that fits the technical system well and has an amenable workforce. A more useful test would have been to do the opposite but it would be a very brave project leader who would take this risk. The second issue is that a trial is tailormade for the 'Hawthorne Effect' (Rothlisberger and Dickson, 1939). This is the well-known situation where a change proves very effective in terms of productivity not because of the change itself but because the workforce find themselves 'in the spotlight' and respond to the extra attention they are receiving. Being chosen for a trial can be exciting or nerve wracking; either way it is not an ordinary occasion and senior managers and specialists may be very much in evidence. Work staff may also be asked for their opinions and changes may be made on the basis of their views. All of these factors may encourage strong motivation and extra effort. None of these factors may be present in later implementations when it has all been done before and no one is interested in the views of the staff. This is not an argument for avoiding the 'Hawthorne Effect'. Rather it is a statement that it will occur, that its effects need to be recognised and, if possible, reproduced in later stages of the dissemination.
- *Preparing a dissemination strategy.* The trial needs to be used to plan the dissemination procedures. There are some obvious ways of doing this, for example, using it to establish who needs to be trained in what ways. However, the trial can also be used to start the process of involving the rest of the workforce. In one major Government department, for example, user representatives from around the country are invited to the trial site to talk to their colleagues about the experience of the new system. They are then asked to start planning for their own local implementation. If there is some freedom to adopt the system for local needs and practices there will be a real opportunity for local design and ownership which will both improve the integration of the socio-technical system and avoid the dangers of resistance to change.

Incremental implementation

The final implementation strategy is one that is becoming increasingly common as generic forms of technology become widely available as consumer products. Whereas the kinds of technical systems involved in 'Greenfield Sites'; or 'Big Bang' implementations are often custom built, or at least put together in unique ways, it is now quite easy for organisations to purchase standard pieces of technology for rapid installation. This is very common in the information and communication technology industry where 'shrink wrapped' products — new software packages, PCs, telephone networks — can be acquired and implemented without detailed planning and a lengthy development process. The effect is that, over a period of time, an organisation can acquire a sophisticated set of technical facilities through a series of incremental steps without having engaged in an organisation wide process of planned technical change.

For the workforce there are many advantages to this form of technical capability acquisition. First, it is likely to be a distinctly user-centric process. It may well be that it is the individual or a work team that is choosing the technology to meet a particular need. Even if this is not the case, it is quite likely that the workforce has a lot of discretion as to whether they do or do not make use of the facilities on offer. In incremental implementation the new facility is often a specific upgrade or an addition to technology already available and may not be essential to the conduct of the job. Another gain for the workforce is that there is often no major learning to be done and no major changes in the work process.

Ethnographic studies in recent years have demonstrated the unique cultural microclimate that exists in work systems and the particular work practices the workforce has evolved, for example, Hutchins (1995) in studies of team work in the docking of ships, and Brown and Duguid's (2000) study of photocopier maintenance staff. Selecting their own new technology may enable staff to sustain useful work practices whereas a central development, unaware of the significance of local practice, may implement technology that makes such activities difficult to sustain.

If this is truly a user controlled, consumer-oriented approach to technical change it may be that there is no need for an ergonomic intervention. As long as the human factors issues of the generic product are attended to by the supplier perhaps there is no need for attention at the point of implementation? Unfortunately, the evidence suggests otherwise. The piecemeal and localised development of technical capability can cause many problems. The incremental addition of equipment to a workstation, for example, can turn what might have been a well-planned work base into something disorganised and possibly dangerous. The workforce or its advisors may not choose equipment in a planful way and may find it not fit for function or usable. They may also underestimate the amount of learning required. Bajer (1998) in a study of software upgrade implementations, for example, found that most staff assumed the new system would work in the same way as the old one and, only after many errors, did they discover they had to develop new skills. The cumulative effect of these factors and the discretionary nature of the users can mean that very little of the new functionality gets used.

Another set of problems arising from this piecemeal approach is how to sustain a cohesive socio-technical system. Technical support staff often find, for example, that it is difficult to support the wide variety of technology being acquired and it is difficult to sustain interoperability between different kinds of equipment. Similarly, the organisation may become less cohesive as each section develops its own practices. These factors suggest there is a real need for ergonomic interventions in the incremental approach to technical change. However, because it is fragmented over time and across the organisation there is often no clear structure within which an ergonomist can operate. One strategy that we have used in a number of projects with this character (e.g., Eason *et al.*, 2000) is an action research approach. In this approach localised and incremental developments are encouraged but there is an overarching steering body on which user communities are represented. A programme of regular evaluations are conducted to identify the positive and negative outcomes of changes being made and the steering committee has the responsibility of reviewing the impact on the integrity of the entire work system. They then undertake work to ensure the necessary degree of cohesion where possible without infringing the autonomy of local units. This strategy provides the opportunity for ergonomists to work with the steering committee and to contribute to the ongoing evaluations. They are then able to assist with implementation issues and, because the staff are already engaged in both design and review activities, use the steering committee as a forum for establishing a continuous learning process supporting longer term system objectives.

Conclusions

There are many human and organisational issues associated with technical change and there are a wide variety of contributions that ergonomics can make. How that contribution can best be made depends upon the organisational context and, in particular, the strategy by which the technical system is being implemented. To be successful the ergonomist has to 'start where the organisation is' and work with the leverage and opportunities that it provides. Three overriding themes emerge from this analysis which influence how this is to be done.

- *Engaging with the workforce.* It is fundamental to ergonomics that practitioners work with the people who will use the equipment, occupy the workstations, etc. The different implementation strategies provide different opportunities for this to take place. The more user centred strategies give good opportunities not only to work with the workforce but to trial systems so that changes can be made before implementation. In other strategies this engagement may be more problematic either because the workforce does not exist as in the Greenfield site or because there may be a very limited time period for the engagement, e.g., in the Big Bang or Phased Implementation. It becomes important in these circumstances either to create the opportunity for this engagement in the early stages of the development or to defer detailed design until local decisions can be taken.
- *The participation of the workforce.* Aim 4 in Figure 1 was to create the motivation in the workforce to avoid resistance to change and get a commitment to the new system. The aim is to avoid the 'them imposing on us' view and develop a 'this is what we want' attitude. This can only be achieved by giving the workforce the opportunity to influence what is being implemented. This is very much the case with Incremental Evolution which tends to be user led. Trials and Dissemination and Parallel Running also give some opportunities for the workforce to be influential. In other strategies most of the workforce could be marginalised and resistance to change may result. It may be necessary to look for leverage in these strategies to provide opportunities for the workforce to have influence over the issues that matter to them.
- *Working for long term institutionalisation.* It is often the case that the ergonomics intervention is short term and that the expectation is that it will contribute directly to the design of the system to be implemented. However, if the long term aim of creating an adaptive, learning system is to be pursued part of the work must be to create the learning and reviewing mechanisms that can foster adaptation and transfer useful knowledge to staff. The Trials and Dissemination approach has a review mechanism built into it but most strategies do not. Again the requirement will be to look for opportunities to create and perpetuate mechanisms by which staff can engage with developments and to use them as learning opportunities wherever possible.

The context of technical change is very influential in determining what kind of ergonomics intervention is necessary. However, this is not to say that the ergonomist has to be a passive recipient of the constraints inherent in any particular situation. If the short and long term aims are to be pursued it is likely that the ergonomist will have to be proactive and imaginative to create the conditions necessary for a successful intervention.

References

Aberg, U. (1981). Techniques in redesigning routine work. In Corlett, E.N. and Richardson, J. (eds) *Stress, Work Design and Productivity* (Chichester: John Wiley), pp. 157–164.

Bajer, J. (1998). *An investigation of the human costs of software upgrades in organisations.* Doctoral Thesis, Loughborough University, Loughborough, Leics. UK.

Bridger, H. (1990). Courses and working conferences as transitional learning institutions. In E. Trist and H. Murray (eds) *The Social Engagement of Social Science*, Volume 1 (Philadelphia: University of Pennsylvania Press), pp. 221–245.

Brown, J. and Duguid, P. (2000). *The Social Life of Information* (Boston, MA: Harvard Business School Press).

Buchanan, and Hucznski, A. (1997). *Organizational Behaviour*, Third Edition (London: Prentice Hall).

Cherns, A.B. (1976). The principles of socio-technical design. *Human Relations*, **28**, 783–792.

Damodaran, L. (1996). User involvement in the systems design process — a practical guide for users. *Behaviour and Information Technology*, **15**, 363–377.

Eason, K.D. (1988). *Information Technology and Organisational Change* (London: Taylor & Francis).

Eason, K.D. (1995). User-centred design: for users or by users? *Ergonomics*, **38**, 1667–1673.

Eason, K.D. and Sell, R.J. (1981). Cases studies in job design for information processing tasks. In Corlett, E.N. and Richardson, J. (eds) *Stress, Work Design and Productivity* (Chichester: John Wiley), pp. 195–208.

Eason, K.D., Harker, S.D. and Olphert, C.S. (1996). Representing socio-technical systems options in the development of new forms of work organization. *European Journal of Work and Organizational Psychology*, **5**, 399–420.

Eason, K.D., Yu, L. and Harker, S.D.P. (2000). The use and usefulness of functions in electronic journals: The experience of the SuperJournal Project. *Program*, **34**, 1–28.

Emery, F.E. (1980). Designing socio-technical systems for 'Greenfield' sites. *Journal of Occupational Behaviour*, **1**, 19–27.

Emery, F.E. and Trist, E.L. (1960). Socio-technical systems. In C.W. Churchman and M. Verhulst (eds) *Management Science Models and Techniques*, Vol. 2 (London: Pergamon).

Haeckel, S.H. (1999). *Adaptive Enterprise* (Boston: Harvard Business School Press).

Hutchins, E. (1995). *Cognition in the Wild* (Cambridge Mass: MIT Press).

Katz, D. and Kahn, R.L. (1966). *The Social Psychology of Organizations* (London: Wiley).

Klein, L. and Eason, K.D. (1991). *Putting Social Science to Work* (Cambridge: Cambridge University Press).

Mumford, E. (2000). Technology and freedom: a socio-technical approach. In E. Coates, D. Willis and R. Lloyd-Jones (eds) *The New Socio-Tech: Graffiti on the Long Wall* (London: Springer-Verlag), pp. 29–38.

National Audit Office (1999). *The Passport Delays of Summer 1999.* Report HC Session 1988–99, HC 812 (London: The Stationery Office).

Page, D., Williams, P. and Boyd, D. (1993). *Report of the Inquiry into the London Ambulance Service.* Report commissioned by South Thames Regional Health Authority, 40 Eastbourne Terrace, London, W2 3QR.

Parker, S.K. and Wall, T.D. (1998). *Job and Work Design* (London: Sage).

Revans, R.W. (1982). *The Origins and Growth of Action Learning* (Bromley, UK: Chartwell Bratt).

Rothlisberger, F.J. and Dickson, W.J. (1939). *Management and the Worker* (Cambridge, MA: Harvard University Press).

Senge, P. (1990). *The Fifth Discipline* (New York: Doubleday).

Tapscott, D. (1982). *Office Automation: A User Driven Method* (New York: Plenum).

Wastell, D.G. and Cooper, C.L. (1996). Stress and technological innovation: A comparative study of design practices and implementation strategies. *European Journal of Work and Organizational Psychology*, **5**, 377–397.

Wilson, J. (1994). Devolving ergonomics: the key to ergonomics management programmes. *Ergonomics*, **37**, 579–594.

Chapter 36

Participatory ergonomics

John Wilson, Helen Haines and Wendy Morris

Introduction

Throughout this book the issue of participation is raised, both implicitly and explicitly. In a sense it is impossible to carry out 'evaluation of human work' without at least some participation from those job holders actually doing the work. However, when we are talking about participation within systems and product design (Chapters 10 and 11) investigations and interventions studies (Chapter 31) or systems implementation (Chapter 35) we need something more than user surveys or trials or the direct observation of people at work. This chapter concerns the area that has come to be known as participatory ergonomics (PE). Since the 1980s (although some suggest earlier) there has been a growing interest in the means by which 'non-experts' can become involved in applying ergonomics within their own organisations. This interest is evidenced in part by the increasing numbers of journal articles, reports and conference papers on the topic. This chapter draws upon these and upon several sources written by the present authors (Wilson and Haines, 1997; Haines and Wilson, 1998; Haines, 2003; Morris *et al.*, 2004).

If we look in more detail what is meant by PE, several authors have advanced their own definitions. Imada (1991), describing PE as one perspective in macroergonomics, states that 'participatory ergonomics requires that end-users (the beneficiaries of ergonomics) be vitally involved in developing and implementing the technology.' (p. 30). A slightly more elaborate definition is provided by Nagamachi (1995) for whom PE is 'the worker's active involvement in complementary ergonomics knowledge and procedures in their workplace... supported by their supervisors and managers in order to improve their working conditions and product quality' (p. 371). Whilst other definitions can be found in the literature, for the purposes of this chapter Wilson's (1995a) statement that PE concerns 'the involvement of people in planning and controlling a significant amount of their own work activities, with sufficient knowledge and power to influence both processes and outcomes in order to achieve desirable goals' is taken as the reference. (p. 1071). In understanding this definition it is important to emphasise that it is not suggested that participants always have a 'blank cheque'; there will almost always be limits on what they can do and on the suggestions they would like to implement. However, in order to do justice to the participatory process and to manage the expectations of participants, certain points will need to be clarified. Limits on what is achievable — in terms of what is a 'significant amount' of the activities, what are the 'desirable goals' and thus what 'knowledge and power' are needed for participants to be sufficiently influential — should be made clear at the outset.

In reviewing the PE literature, it could be argued that the growth of PE has largely been achieved through 'doing'. That is to say a high proportion of the literature concerns reports of PE endeavours in which different researchers describe the execution of one participatory project or another. Whilst this growth may well have served to establish the term within the ergonomics community, it has also led to the term being used to cover a fairly broad range of different approaches and ideas Looking beyond ergonomics, a similar debate can be found within some of the other domains using participatory approaches (see, e.g., Reuter, 1987). Indeed, within the management literature, participation has been referred to as a 'fuzzy concept' (Cotton, 1993).

For Garrigou (2002) participation is 'both a rich and complex process linked with philosophical ideas such as democracy in the workplace' (p. 3). However, it is not the intention in this chapter to explore further the philosophical or theoretical basis for participation at work. Nor will the vast amount of work on employee participation in the arena of industrial democracy, co-operatives or even in a wider political sense be touched on other than indirectly. The concentration is upon participative programmes and techniques which are of immediate relevance to mainstream ergonomics — that is in the analysis, design, implementation and evaluation required to improve an individual's tasks, equipment, interfaces, workplaces, jobs and work organisation. For a discussion of some of the wider issues in participation the reader is referred to Cotton (1993), Glew *et al.* (1995), or Mumford (1991).

In this chapter the development and growth of participatory ergonomics as a discipline is considered along with the potential advantages and disadvantages associated with taking a participatory approach. The focus then moves on to the methods used within PE, before a framework is presented along with a discussion of the requirements for a successful participatory ergonomics intervention. Finally, in an appendix to the chapter, a case study undertaken by one of the authors is included in order to illustrate the application and methods of PE in more detail.

Participation and ergonomics

There can be little doubt that PE has enjoyed a steady growth over the course of the last two decades. If we look for the reasons for this growth, it is possible to identify a number of important influences. These include: human centred work systems with emphasis upon an involved, responsible workforce; the recognition that motivation and performance are complex issues; increased Trade Union acceptance of participation; more decentralised and 'flatter' IT-based organisations, greater emphasis on the attributes of quality, flexibility and customer service; and the continued introduction of new technical systems and methods of organising them.

Overall, it can be argued that a much more participative culture has developed within many western countries. Reactions against the autocratic and non-participative ethos of scientific management (or 'Taylorism') are evident in both work organisation theory and practice. Today's high performing companies are said to have incorporated employee empowerment and teamwork into their working practices as well as 'people involvement' programmes.

Developments in European ergonomics and health and safety legislation are also germane to PE. The European Framework Directive (89/391/EEC) on 'the introduction of measures to encourage improvements in the safety and health of workers at work' contained specific provisions about consultation and participation of workers. Legislation now requires employers to consult employees, who are not already represented by trade union safety representatives, about matters that affect their health and safety, with especial impact in countries where such consultation was not the norm (e.g., the UK) compared to those where it was (e.g., Scandanavia). Furthermore, the need for employee involvement and consultation is also emphasised in guidance and standards. For example, in the UK, publications include 'Successful health and safety management, HSE, 1993; 'Upper limb disorders in the workplace, HSE, 2002 and British Standard BS 8800:1996 'Occupational Health and Safety Management Systems.

It is increasingly recognised that there is a need for organisations to be more self-reliant in ergonomics. Most organisations simply do not have the resources to have a consultant ergonomist examine every one of their work processes (even if there were enough ergonomists to go around). Moreover, such an 'outside oriented' approach does not fit with the concept of a health and safety culture of which internal ownership of ergonomics can form an important part (HSE, 1996).

Looking beyond these wider influences, for ergonomists, participation — as a philosophy and a process — has always been of interest due, in part, to the strong ergonomics emphasis on application. People-centred approaches are endemic, from user trials in consumer product testing to user-centred design in human-computer interaction (Chapter 13) to human-centred technology and sociotechnical programmes in industry (Chapter 29). Most ergonomists also take a broad view of work redesign and of improving the work environment, arguing that the physical aspects of work neither can nor should be divorced from a consideration of psychosocial and organisational factors. Even when a work redesign programme establishes physical factors as being of greatest concern, to ignore other aspects of work may lead to only partial acceptance and success of any changes made, due to the central role of workforce attitudes (Wilson and Grey, 1990).

If we look at how participatory ergonomics has been applied, it becomes apparent that participatory initiatives can operate at a number of different levels. In the literature we can find examples ranging from one-off interventions tackling workplace or equipment design (see, e.g., van der Molen *et al.*, 1997; Maciel, 1998; Imada and Stawowy, 1996; Moore, 1994) through series of multiple interventions, such as those associated with quality circles or continuous improvement schemes, (see, e.g., Haims and Carayon, 1998; De Looze *et al.*, 2000; Bohr *et al.*, 1997) to full participatory ergonomics management programmes (see, e.g., Lanoie and Tavenas, 1996; Loisel *et al.*, 2001; Liker *et al.*, 1991; Mansfield and Armstrong, 1997) or even participatory initiatives concerned with organisational systems influencing an entire industry sector (Moore and Garg, 1998; Kardborn, 1998; De Jong and Vink, 2002). This diversity was illustrated by a recent collection of case studies assembled as part of a European research project (Morris *et al.*, 2004). Some of these cases concerned very small-scale participatory projects, such as a work team of four employees developing, with a manufacturer, a motorised truck to transport equipment around a hospital site. However, at the other end of the scale, there were examples of industry-wide interventions such as a paper industry initiative, which was established following a review of accidents and injuries within the industry. This initiative used a participatory approach to improve both the safety culture and aspects of machinery design within the entire paper manufacturing industry.

In order that workforce participation in the application of ergonomics methods be most successful, there must be a spreading of relevant skills and confidence throughout the company. There are growing numbers of efforts to train the workforce in ergonomics. This training might be aimed at any or all of shop-floor workers, supervisors, industrial, production or design engineers, health and safety personnel, but the most complete programmes certainly include shopfloor staff in order to help them assess work and workplaces and propose solutions for ergonomics problems. As with ergonomics generally, the limited resources available to do this must be recognised, and as a consequence we might establish 'cascaded' training and awareness programmes, which are to some extent further extensions of a 'Train the Trainers' approach (see Nyran, 1991; Silverstein *et al.*, 1991 for examples). In essence, to train the trainers meets the need to spread ergonomics more widely, including to non-specialists. Ergonomists train a core team in ergonomics and in ergonomics education techniques, the training being largely practically-based and emphasising 'learning by doing'. Initial training is passed on to others in the workforce by this core team.

An important very early series of studies by a group of French ergonomists followed the line of providing tools, training, knowledge and confidence for workers to analyse their own work problems and to develop feasible alternatives. The focus of these studies varied from chemical plant control room design including display specifications (Boel *et al.*, 1985), to posture and health analyses (Montreuil and Laville, 1986), to introduction of new shift work systems (Teiger and Laville, 1987). Their approach was to set up study groups comprising workers, ergonomists and technical staff; the worker representatives were equipped with the tools and skills needed to analyse and make recommendations about particular problems. In some of their work, a task force was formed to oversee the whole project, keeping all levels in the company in touch with current developments. The researchers succeeded in achieving recognition amongst the workforce of the value of an ergonomics perspective and problem-solving process; through their involvement in the administration of questionnaires and direct observation, workforce groups managed to engage the interest of many colleagues in the issues and in finding solutions. Overall the researchers felt that the role of ergonomics had been widely accepted in these companies, not only in order to provide solutions but also to help the workers themselves iterate slowly towards solutions (see also Daniellou *et al.*, 1990).

Arguments for and against participation

As the field of PE has developed, the potential gains associated with using this approach have been reported and discussed. These gains may be direct or indirect (to some extent systemic). It has long been suggested that end user participation in the design of equipment or workplaces will lead to the development of better design solutions, as the practical experience and expertise of end users is bought into the process (Imada 1991; St Vincent *et al.*, 1997; De Looze *et al.*, 2000). However it is important to clarify what is meant by 'better' design as, depending upon the product, the context and the participants, this may mean quite different things. For example, in the case of a piece of industrial machinery, better design can mean one that is more efficient, produces items of a higher quality, is safer to use, is more easy to use or may be used by a wider range of the working population; better design for a consumer product may be entirely to do with usability.

As well as improved solutions the second main advantage for participative processes is said to be the greater acceptability of these solutions for the stakeholders (Van der Molen *et al.*, 1997; De Jong and Vink, 2002). The reasoning is that if people (or their peers) have been involved in generating a solution for change then they are likely to be more committed to making the change work, to be less resistant to change and to be more satisfied as a result (e.g., Imada and Robertson, 1987; Wilson, 1995b). If so, and given also that we expect a better designed solution, more fit for purpose, then the process of implementation of change should be more effective and of higher quality. Involvement in a development or implementation process can mean faster and deeper learning of a new system and hence decreased training costs and improved performance (e.g., see Wilson and Grey Taylor, 1995; Kuorinka and Patry, 1995; Zink, 1996). The whole participation process can be a learning experience for 'designers' and 'users' alike (e.g., see Sanoff, 1985; St Vincent *et al.*, 1997). It can also lead to improvements in processes and employee well-being (e.g., see Maciel, 1998).

Looking more widely, benefits from the use of participatory approaches may include improved industrial relationships (Lanoie and Tavenas, 1996), opportunities for personal development amongst those involved and a spread of interest and expertise in ergonomics

(Noro, 1991). Participative processes can help embed an ergonomics perspective within the company if people are allowed to make a genuine contribution and influence affairs (St Vincent *et al.*, 1997; Wilson, 1991a, b).

The relationship between participative processes and the development and gains from working in groups or teams is of great interest. Health and safety and workplace ergonomics investigations can be carried out through work groups. Self-directed team working anyway is often an appropriate mechanism to implement principles of enriched job design, and social support is seen as an effective reducer of strain resulting from stress at work. If work is to be organised on the basis of self-directed teams then it makes good sense if the groups that develop the ideas for such work organisation and consequent physical workplace requirements will then subsequently become the work teams themselves. In this way the process of change enhances the content. This also meets the sociotechnical principle of comparability (between process and outcome).

Whilst much of the literature on participatory ergonomics is positive, it would be a mistake not to consider some of the difficulties that an ergonomist may encounter when trying to use such an approach within an organisation. It is perhaps understandable that most of the reports in the literature present a positive outcome from the use of participatory methods. After all, organisations that may have encountered difficulties from using such an approach may well be resistant to publicise what may be seen as failure. And, even where an initiative has been successful, in the authors' experience, participants can often identify aspects of the process or the methods employed that had a negative impact either on themselves or their organisation.

There are a number of reasons why a participatory approach may fail or encounter difficulties. First of all, for whatever reason, some people will just not want to participate (Neumann, 1989). It is also possible that management might see participation as a threat to their right to manage rather than as an aid (Mumford, 1991) that some people may perceive it as an attempt to weaken trade unionism (Forrester, 1986). Moreover participation, as philosophy or process, is neither an easy option, nor, perhaps always the most appropriate one to choose (Mossink, 1990). A half-hearted or cynical application will reap a harvest of problems, now or in the future. Use of participation within manufacturing industry has certainly often been viewed with suspicion by managers, trade unions and work forces (Garrigou *et al.*, 1995; Garrigou, 2002). Side-effects or systemic consequences may also give problems, for instance increasing the autonomy of assembly work groups may affect the work of the purchasing or maintenance departments; enlarged job content for word processing may reduce requirements for personal assistants, clerks or even information specialists; improvements in one groups's skill levels or learning may make workers in other offices or departments envious or dissatisfied. Related to this last point, the very process of participation may have unexpected systemic effects, in that other groups may wish to be similarly involved but this may not always be feasible or desirable.

Another concern about the use of participation is the quality of outcomes from a process which is based upon group decision making. Some may argue that such a process will only choose the safe option or one that may be viewed as the lowest common denominator. It is therefore important to have established appropriate structures and methods for the participatory group process. Additionally the development of new workplaces or systems using a participatory approach may take more time and personnel resources and therefore be more costly in the short term. The counter argument to this is, however, that the cost advantages from the introduction of a more efficient and effective solution or design will prove beneficial in the long run and will outweigh extra development costs. Finally, a key issue of work systems change, participative or not, concerns the issue of when the change agent or facilitator should leave the scene. If they leave too late, then true participation may be stifled; if they leave too early, a vacuum may be created in which nobody feels any ownership for the process or change (Wilson, 1995b).

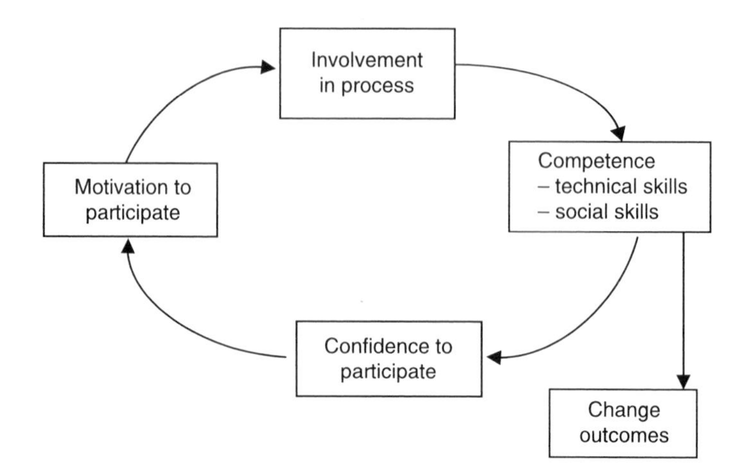

Figure 1 The participation cycle (Haines and Wilson, 1998).

From the preceding discussions it can be seen that participation should not be seen as an easy option. There will be dysfunctional consequences for any participative initiative that is in reality a sham or poorly planned and managed. However, participative initiatives that are well designed and implemented have been shown to lead to a number of benefits both for the organisation and for the participants themselves. The main possible gains for participation are summed up in the participation cycle portrayed in Figure 1.

Methods used in participatory ergonomics

At the heart of any well designed project is the identification and use of appropriate methods. According to Imada and Nagamachi (1995) the role of participative tools and techniques is 'to channel participants' unique knowledge and skills to solve problems and make improvements' (p. 309). In the literature, a whole range of different tools and methods has been reported as being useful within participatory ergonomics initiatives. However, if we examine this work more closely, it becomes apparent that only some of these tools and methods have been developed or adapted specifically with participatory ergonomics in mind. Many of the others have been borrowed, either from more traditional ergonomics initiatives or from other relevant fields of practice, and then applied within a participative exercise.

One way of thinking about these methods is as a toolkit for participatory ergonomics. Like any toolkit, it contains instruments which are useful for one particular job in the participatory process as well as others that are much more flexible. With this in mind, Tables 1 and 2 show some of the tools and methods reported in the literature, grouped according to the job for which they might be used within a PE intervention (even though, in reality, of course, there will be a certain degree of overlap between the different categories). Table 1 provides examples of methods developed or adapted with participatory ergonomics or design specifically in mind, whereas Table 2 summarises some of the methods and techniques which, although developed elsewhere, have been applied within participatory ergonomics initiatives.

The participatory tools and methods may differ significantly in the resources they require: not just in terms of time and money but also in the expertise that will be required. We can find a number of cases where the analysis of problems is largely tackled by 'experts', rather than by workers or end users (e.g., Kourinka and Patry, 1995; Vink *et al.*, 1995). These experts

Table 1 Tools and Methods Developed for Participatory Ergonomics (table adapted and extended from Haines and Wilson, 1998)

Method or technique	Primary purpose	Reference (for examples of where the method or technique has been described or used)
'Statistical analysis measurements' (SAM)	Problem analysis	Liker *et al.*, 1991
JDLC chart	Problem analysis	Nagamachi, 1995
Posture analysis tool	Problem analysis	Nagamachi, and Tanaka, 1995; Nagamachi, 1991
'Five ergonomic viewpoints'	Problem analysis	Imada, 1991
Autoconfrontation	Problem analysis	Kuorinka, 1997
PSIDAR	Problem analysis	Hanse and Forsman, 2001
Activity analysis	Problem analysis and situation prediction	Garrigou *et al.*, 1995
Round robin questionnaire	Creativity stimulation and idea generation	O'Brien, 1981; Wilson, 1991a
Word map	Creativity stimulation and idea generation	O'Brien, 1981; Wilson, 1991a
AKADAM	Solution generation	McNeese *et al.*, 1995
Silent drawing/assessment	Creativity stimulation and solution generation	O'Brien, 1981; Wilson, 1991a; Kuorinka *et al.*, 1994
Scenario driven discussions	idea generation and solution development	Snow *et al.*, 1996
Dynamic space manipulation	idea generation and solution development	Snow *et al.*, 1996
Design Decision Group	Idea generation and concept evaluation	Wilson, 1991a; Lehtela and Kukkonen, 1991; Imada and Stawowy, 1996
Ergonomics design review process (series of analytical tools)	Problem analysis; solution generation	Aikin *et al.*, 1994
The Hourglass model	Problem analysis, idea generation	Akselsson *et al.*, 1999
Problem Solving Group	Idea generation and concept evaluation	Wilson, 1995b
Intervention ideas	Concept evaluation	Moir and Bucholz, 1996

may well use some of the more complex and technical analytical tools that are available (for some analytical methods — such as some computer-based modelling and simulation tools — the sophistication lies not so much in their use as in the interpretation of the results produced). However, as Liker *et al.* (1989) point out, the ability of organisations to remain self-sufficient goes down as the sophistication of ergonomics analysis tools increases. And as one of the aims of participatory ergonomics may be to free institutions from reliance on the outside expert, the use of highly complex tools may prove self-defeating.

Design decision groups

Some of those working within the field of PE have combined a number of different participatory tools and techniques in order to develop specific approaches to undertaking

Table 2 Tools and Methods Developed Elsewhere but Applied within Participatory Ergonomics Initiatives (table adapted and extended from Haines and Wilson, 1998)

Method or technique	Primary purpose	Reference (for examples of where the method or technique has been described or used
Team formation, building	Preparation and support	Caccamise, 1995
Team training	Preparation and support	Gjessing *et al.*, 1994; Loisel *et al.*, 2001; Garmer *et al.*, 1995; Mansfield and Armstrong, 1997
"Train-the-trainers"	Preparation and support	Corlett, 1991a; Silverstein *et al.*, 1991
Breakthrough Thinking®	Preparation and support	Haims and Carayon, 1998
Task analysis, functional task decomposition	Problem analysis	McNeese *et al.*, 1995; Halpern and Dawson, 1997
Work study techniques (e.g. MTMM)	Problem analysis	Kuorinka and Patry, 1995; Noro, 1991
Trace method	Problem analysis	Kourinka *et al.*, 1994
Pareto analysis	Problem analysis	Imada, 1991; Lewis *et al.*, 1988
Cause-and-effect diagram	Problem analysis	Imada, 1991; Lewis *et al.*, 1988
Biomechanical analysis	Problem analysis and solution evaluation	Liker *et al.*, 1991
Requirement analysis	Idea generation	Zink, 1996
Link analysis	Problem analysis	Imada, 1991
Videos/photographs	Problem analysis and solution generation	Lewis *et al.*, 1988; Liker *et al.*, 1991; Algera *et al.*, 1990; Halpern and Dawson, 1997
Computer aided diagrams or drawings	Problem analysis	Berling *et al.*, 1998; Bengtsson *et al.*, 1997
Brainstorming techniques	Problem analysis and solution generation	Lewis *et al.*, 1988; Moore, 1994; Liker *et al.*, 1991; Gjessing *et al.*, 1994
Discussion groups	Problem analysis and solution generation	Buckle and Ray, 1991; Gontijo and Odenrick, 1999; Zink, 1996; Maciel, 1998; Cohen, 1997
Focus groups	Idea generation and concept development	Caplan, 1990
Equipment testing	Concept evaluation	Van der Schaaf and Kragt, 1992
Prioritised voting or other prioritisation techniques	Problem analysis and concept evaluation	Moore, 1994; Lewis *et al.*, 1988; Gjessing *et al.*, 1994; Van der Schaaf and Kragt, 1992; Westlander *et al.*, 1995
Scale modelling	Solution generation	Van der Schaaf and Kragt, 1992; Algera *et al.*, 1990
Virtual reality	Concept evaluation	Snow *et al.*, 1996; Akselsson *et al.*, 1999

(continued)

Table 2 Continued

Method or technique	Primary purpose	Reference (for examples of where the method or technique has been described or used
Layout modelling and mock-ups	Concept evaluation	Wilson, 1991a; St Vincent *et al.*, 1998; Kuorinka *et al.*, 1994; Algera *et al.*, 1990; Nagamachi, 1991
Role playing techniques and simulation games	Idea generation and concept evaluation	Ruohomaki, 1995;
Sampling plans	Problem analysis	St Vincent *et al.*, 1998
Work site observation	Problem analysis	Jones, 1997; Vink *et al.*, 1995; Loisel *et al.*, 2001
Interviews, questionnaires	Problem analysis, idea generation	St Vincent *et al.*, 1998; Van der Schaaf and Kragt, 1992; Algera *et al.*, 1990; Vink *et al.*, 1995
Checklists	Problem analysis and concept evaluation	Rawling, 1991; Keyserling and Hankins, 1994; Moore and Garg, 1997; St Vincent *et al.*, 1998; Gjessing *et al.*, 1994; Vink *et al.*, 1995; Laitinen *et al.*, 1997

participatory interventions (e.g., Vink *et al.*, 1995; 1997, Garrigou, 2002). An example of what might be described as some kind of broader participatory method can be found in the work of one of the authors and his colleagues. Design decision groups (or DDGs) combine a number of the techniques mentioned earlier in this section — word maps, round-robin questionnaires, layout modelling and mock-ups — into a systematic approach for workplace layout and design (Wilson, 1991a). This concept is derived from the work of Dennis O'Brien (of the UK Home Office) into methods for running creative meetings and for involving users in systems design (e.g., O'Brien, 1981). O'Brien adapted theories and techniques from market research and the literature on creativity and innovation, in particular using non-directive and group discussions, developing 'thinking tools' and encouraging participants to talk about, write down or draw, attitudes experiences, visions or opportunities (see also Caplan, 1990). O'Brien described a number of techniques or tools within what he called Shared Experience Events (SEEs); amongst these are: structuring of contributions from sub-groups whilst others listen in silence, only later criticising; use of drawing as far as possible, which enhances concept development, communication of ideas, and also helps in structuring the process; and very careful pre-planning and preparation. His ideas were used in a study of fire brigade control room design (Langford, 1982).

Wilson and his colleagues went on to utilise design decision groups in a number of applications, including the design of retail checkouts and library issue desks (Wilson, 1991a). The DDG process is summarised in Figure 2 and further description and illustration of the stages appear in the appendix to this chapter.

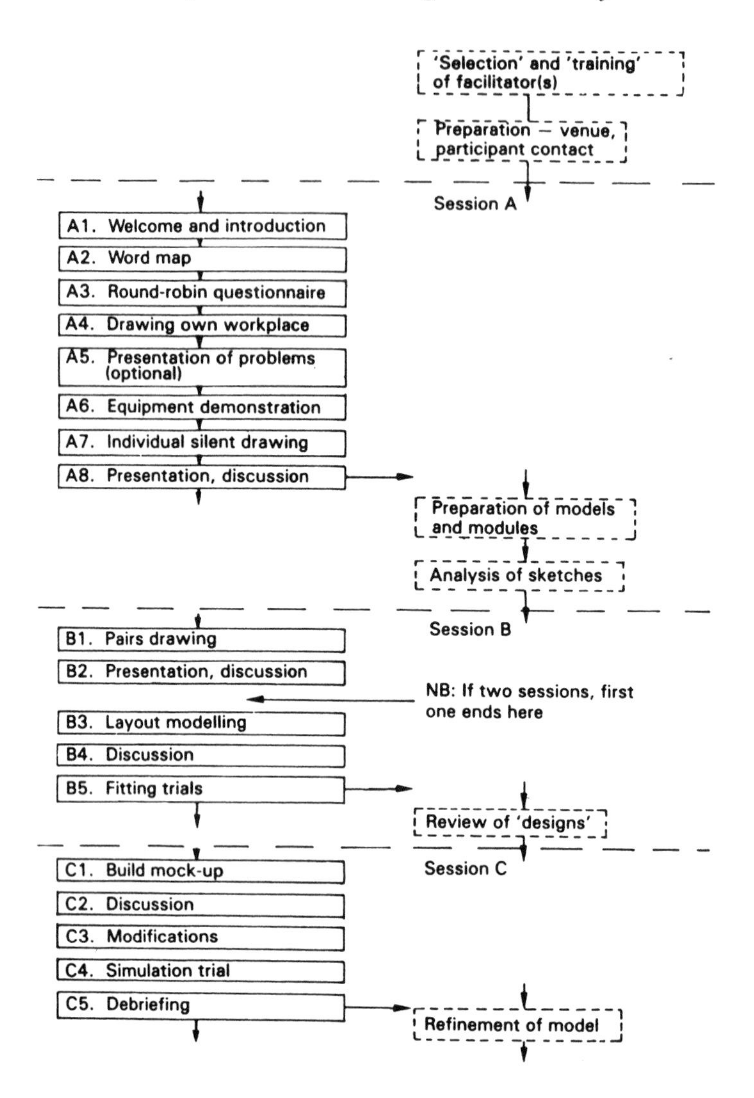

Figure 2 Summary of the design decision group process.

An extension of the DDG process

One criticism often made of DDG and similar techniques is that they are usually only employed with white collar or technical staff. Exceptions can be found in an adapted DDG as reported by Wilson (Wilson, 1995b). Other examples are also in Lehtela and Kukkonen (1991) and Aikin *et al.*, (1994). The workers concerned were crane drivers operating from a control room in an incinerator plant. Early observation identified that they had severe problems related to viewing the task, the design of joystick controls and a deadman's handle, seating and postural inadequacies, and general environmental impacts. A form of Design Decision Group technique was adapted for use with the drivers; critically this involved them in sourcing and costing their preferred design solutions within a budget set by management. The process is illustrated in Table 3. It will be seen that the final stage is that of continual improvement — the embedding, in this small company, of participation as *the* way to go about ergonomics problem assessment and solution.

Use of a participative process here probably did not lead to identification of any different problems and causes than would have been the case with a traditional ergonomics approach.

Table 3 Stages of the Procedure (from Wilson, 1995)

DOC	
Stage One:	Familiarisation: of investigator with the work and workplace and of the participants with ergonomics and the aims of the PSG
Stage Two:	Field visits: by participants to 'similar' sites
Stage Three:	Design Decision Group (A): elements of session A of the DDG process
Stage Four:	Design Decision Group (B): elements of session C (+B) of the DDG process
Stage Five:	Lighting simulation: using screens, coverings etc. in the control room, to reduce reflection problems, etc.
Stage Six:	Sourcing and costing of solutions: crane drivers finding products or adaptation, which best meet their preferred solutions, whilst staying within budget
Stage Seven:	Continual improvement: after the facilitators have left

However, the perspective on these problems and the priorities given them was different, and this guided subsequent improvements in directions which best met the needs and interests of the workers. The solutions which were eventually agreed were, at best, satisfying; that is, it could not be argued that they were the best that could have been developed but they did entail a satisfactory answer to the problems. Since there is always the question of what is 'best' in ergonomics anyway, and since the process of continual change has led to further improvements, the outcome was considered to be successful. The change agents withdrew from the process as soon as possible, with the consequence that the proposed redesigns were only partially implemented and therefore from their viewpoint there are still some ergonomics difficulties. However, this seems to be outweighed by finding not only that the changes made have been completely accepted, widely welcomed and have given rise to a dramatic fall in complaints and postural problems, but that the group has continued to discuss and implement further workplace and work improvements, demonstrating that an ergonomics perspective has been embedded for that group of workers to at least a limited extent.

Participatory ergonomics framework

The fact that PE can operate at a number of different levels and encompasses a range of different approaches can make the study of such initiatives quite a challenge. In part to address this issue, a framework for the guidance and review of participatory ergonomics projects has been developed (Haines *et al.*, 2002). The Participatory Ergonomics Framework (the PEF) describes a number of different dimensions across which participatory initiatives may vary. Each dimension has two or more associated categories that further define this feature of the initiative (see Table 4).

When using the framework there are two important points to consider. The first is to identify the number of different participatory groups that may exist within an organisation. For example, a participatory initiative may involve cell based quality circles, departmental health and safety committees and structured organisational ergonomics teams. In such cases it is suggested that these each of these groups should be evaluated separately. Second, within each of these groups it is important to clarify whose perspective is being considered when looking at the different categories in each of the dimensions. This is because the evaluation may well differ for each stakeholder. For example, whilst operators may

Table 4 The Participatory Ergonomics Framework (PEF) — Haines *et al.* 2002

Dimension	Categories
Permanence	Ongoing — Temporary
Involvement	Full direct participation — Direct representative participation — Delegated participation
Level of influence	Group of organisations — Entire organisation — Department — Work group/team
Decision making	Group delegation — Group consultation — Individual consultation
Mix of participants	Operators — Line management — Senior Management — Internal specialist/technical staff — Union — External advisor — Supplier/purchaser — Cross industry organisation
Requirement	Compulsory — Voluntary
Focus	Physical design/specification of equipment/workplaces/work tasks — Design of job teams or work organisation — Formulation of policies or strategies
Brief	Problems identification — Solution development — Implementation of change — Set-up/structure process — Monitor/oversee process
Role of ergonomics specialist	Initiates and guides process — Acts as expert — Trains participants — Available for consultation — Not involved

volunteer to take part in the participatory group if there is only one line manager then they may be required to participate.

In addition to allowing different participatory interventions to be compared or evaluated, the PEF may also have a role in guiding the planning of a participatory project (Haines *et al.*, 2002). It is suggested that the various dimensions and their categories can be used to structure discussions during the planning phase of such a project. As a result, the process to be used, the boundaries of the group, definitions of participants' roles etc. may be agreed at the outset. Such considerations should ensure that as much attention is given to the process of participation as to the content of the intervention.

Requirements for Participatory Initiatives

One remaining question for those considering the use of PE might be 'what do we need for a successful participatory initiative?' This issue has been considered by a number of authors and, more recently, was the focus for discussion at a workshop held in Brussels (Morris *et al.*, 2002) as part of a European project that considered participation of end users in design. During the workshop, the experts attending were asked to consider what might be the requirements for successful participatory ergonomics interventions. The following factors were identified as particularly important:

- Commitment to the project at all levels of the organisation.
- A champion to support and/or facilitate the process.
- A sense of urgency — a reason to take a participative approach.
- Clear definition of actors and their roles — who will be involved?
- Structures to support the process — how will the participation be managed?
- Appropriate levels of knowledge for all participants.
- Previous good experience of participation.
- Trade Union involvement.

- Involvement of end users in all stages of equipment design.
- Preferably involvement of manufacturers from the beginning of the process.
- Keeping the project simple — well defined and well targeted.
- Keeping the client's needs in focus.

Some of these factors had previously been reported by Haines and Wilson (1998). In addition they identified a number of other points, including the need for training in general ergonomics principles or in teamwork and interaction skills (see also Hasle *et al.*, 1997; St Vincent *et al.*, 1997; Ingelgard and Norgren, 2001). The role of the ergonomist in a participatory ergonomics initiative was also highlighted as requiring careful thought and definition. Their input may vary from one of 'expert' to group facilitator, either between different participatory projects or within a single project (Haims and Carayon, 1998; Wilson, 1995b). Finally the issue of sufficient resources for the project is a vital one. Senior management agreement for the financial, personnel and time resources required is essential for the both the setting up and running of the participatory project.

In summary, some of the requirements for a successful PE initiative are shown in Table 5 (Wilson and Haines, 1997).

Conclusions

Efforts to enable the widespread and successful implementation of ergonomics at work have multiplied in recent years; strenuous efforts are being made to take ergonomics out of the pages of text books and make it more applicable in practical use. We cannot just rely upon 'ergonomics experts' — there are not enough of these, it will not be cost effective for most smaller organisations and is, anyway, not the right way to approach the challenge. This challenge is to:

- Motivate people to become involved in ergonomics within their own organisation as part of a process that has meaning and influence.
- Give them tools and techniques and related training to make ergonomics interventions, ensuring that training comes at the right time for all relevant groups.
- Ensure they have enough understanding to know when and how to use these tools and techniques, *and when they cannot and must call in specialist assistance.*

As Corlett (1991b) argues:

> (We must) give ergonomics away...transfer our knowledge and methods to others who are closer to the places where changes have to be made, so that they do much of the ergonomics for themselves... Of course, there will be occasions when ergonomics is miss-used, or badly used, but with experience it will improve and the knowledge itself is a good protection from abuse. Until ergonomics is widely practised by other than professional ergonomists, it is likely to remain something to be added on at the end. What is more, until it is more widely recognised what ergonomics really is, we will not get it introduced into design specifications, where human performance and requirements become a normal part of the design criteria. Until this becomes the accepted practice, the implementation of ergonomics will always be an uphill fight (p. 418).

Table 5 Requirements for Participatory Ergonomics (Source: Adapted from Wilson, 1995b and Wilson and Haines, 1997)

Climate for participation	Avoid times of uncertainty, hostility, crisis Better in open, non-hierarchical, noncentralised organisations. Support if group work and enriched jobs in place already Devolve Ergonomics: Foundation/Spread/Embed
Support and resources	Support throughout the organisation and unions Key decision-maker commitment Expect resistance — rational and irrational Allocate sufficient finance and personnel No unreasonable time constraints Set initial flexible goals and criteria
Facilitator	Is not necessarily project "owner" Sensitive yet robust and decisive Unbiased yet knowledgeable Flexible and consistent Respected by participants and organisation Knows when to leave the process
Set-up	Timing and starting point are critical. Process has to have initiation and evolution Interdependence of motivation, ability and confidence of participants Participation to be voluntary where possible Potential benefits should be promoted during set-up
Processes and methods	Participants can be self-selecting; otherwise use volunteers, elections, or selection Size of group: 6–10 Need facilitator, chairperson, resources assistant Methods to be well costed, cost-effective and flexible Process to be iterative, relaxed, and allow compromises Methods to promote creativity, learning, confidence, and decision making Include solution costing methods Need to have direction without being overtly directed
Outcomes	Testable solutions or plans Continual improvements Spread of expertise in the organisation Process to become internal and ordinary, not external and extraordinary

There is a strong argument for a participative approach to ergonomics: both participation at an individual level — whereby workers are involved in workplace and work assessment and in providing solutions to any health and safety problems; and also participation at an organisational level — whereby engineering, production, medical and other staff develop their own programmes for ergonomics awareness and action. There is a danger that participation may be seen as an easy option when, in reality, it requires careful planning and implementation. It is important to identify why a participatory ergonomics approach should be used — some guidance as to how to plan and develop such initiatives is now available. The wide range of case studies in the literature provide examples of projects within different industries and opportunities for lessons to be learnt from others' experiences. Of course there are limits to PE. There are certain problems and solutions that will require professional (often outside) ergonomics expertise. At least a part of the training we give should be aimed to allow companies to understand not just what they can do internally but also what they should not attempt. In this respect, an ergonomics programme should be no different to many company activities involving a mixture of in-house and outside expertise. In fact, at a more general level, there is little in an ergonomics and participative approach which is not a part of what should be good practice within management and organisational culture.

References

Aikin, C., Rollings, M. and Wilson, J.R. (1994). Providing a foundation for ergonomics: Systematic ergonomics in engineering design (SEED). In *Proceedings of the Twelfth Congress of the International Ergonomics Association*, **5**, 216–218.

Algera, J.A., Reitsma, W.D., Scholtens, S., Vrins, A.A.C. and Wijnen, C.J.D. (1990). Ingredients of ergonomic intervention: How to get ergonomics applied. *Ergonomics*, **33**, 557–578.

Akselsson, R., Hornyánszky Dalholm, E., Johansson, C., Kihlberg, S., Mathiasson, S., Odenrick, P., Wikström, T. and Winkel, J. (1999). A Centre for Research on People, Technology and Change at Work (Change@Work). In *Proceedings of the Conference on TQM and Human Factors*, CMTO, Linkopings Universitet, pp. 439–448.

Bengtsson, P., Johansson, C. and Akselsson, K.R. (1997). Planning working environment and production by using paper drawings and computer animation. *Ergonomics*, **40**, 334–347.

Berling, C., Blome, M., Johansson, C., Odenrick, P. and Rassner, F. (1998). Methods for introducing improvements at work. *Human Factors in Organisational Design and Management — VI*, P. Vink, E. Koningsveld and S. Dhondt (eds.) (Elsevier Science B.V.) pp. 555–565.

Boel, M., Daniellou, F., Desmores, E. and Teiger, C. (1985). Real work analysis and workers' involvement. In *Ergonomics International 85, Proceedings of the 9th Congress of the International Ergonomics Association*, Bournmouth, pp. 235–237.

Bohr, P.C., Evanoff, B.A. and Wolf, L.D. (1997). Implementing participatory ergonomics teams among health care workers. *American Journal of Industrial Medicine*, **32**, 190–196.

BSI 8800 (1996). *Guide to Occupational Health and Safety Management Systems.* (British Standards Institution).

Buckle, P.W. and Ray, S. (1991). User design and office workers — an evaluation of approaches. In: *Contemporary Ergonomics, Proceedings of the Ergonomics Society's 1991 Annual Conference*, Southampton, England.

Caccamise, D.J. (1995). Implementation of a team approach to nuclear criticality safety: The use of participatory methods in macroergonomics. *International Journal of Industrial Ergonomics.* **15**, 397–409.

Caplan, S. (1990). Using focus group methodology for ergonomic design. *Ergonomics*, **33**, 527–533.

Cohen, R. (1997). Ergonomics program development: prevention in the workplace. *American Industrial Hygiene Association Journal*, **58**, 145–149.

Corlett, E.N. (1991a). Ergonomics fieldwork: an action programme and some methods. In *Towards Human Work: Solutions to Problems in Occupational Health and Safety*, M. Kumashiro and E.D. Megaw (eds.). (London: Taylor & Francis).

Corlett, E.N. (1991b). Some future directions for ergonomics. In: M. Kumashiro and E.D. Megaw (eds) *Towards Human Work: Solutions to Problems in Occupational Health and Safety.* (London: Taylor & Francis).

Cotton, J.L. (1993). *Employee Involvement. Methods for Improving Performance and Work Attitudes.* (Newbury Park: Sage).

Daniellou, F., Kerguelen, A., Garrigou, A. and Laville, A. (1990). Taking future activity into account at the design stage: participative design in the printing industry. In *Work Design in Practice*, C.M. Haslegrave, J.R. Wilson and E.N. Corlett (ed.) (London: Taylor & Francis).

De Jong, A.M. and Vink, P. (2002). Participatory ergonomics applied in installation work. *Applied Ergonomics*, **33**, 439–448.

De Looze, M., van Rhijn, G., Tuinzaad, B. and van Deursen, J. (2000). A participatory and integrative approach to increase productivity and comfort in assembly. *Ergonomics for the New Millennium. Proceedings of the XIVth Triennial Congress of the International Ergonomics Association and 44th Meeting of the Human Factors and Ergonomics Society*, San Diego, California, USA July 29–August 4, 2000. Human Factors and Ergonomics Society, Santa Monica, Vol. 5, pp. 142–145.

Forrester, K. (1986). Involving workers: participatory ergonomics and the trade unions. In *Contemporary Ergonomics*, D.J. Oborne (ed). (London: Taylor & Francis).

Garmer, K., Dahlman, S. and Sperling, L. (1995). Ergonomic development work: Co-education as a support for user participation at a car assembly plant. A case study, *Applied Ergonomics*, **26**, 417–423.

Garrigou, A. (2002) Participatory ergonomics: a risky activity between commitments and reality. French National Report. TUTB-SALTSA project. www.etuc.org/tutb

Garrigou, A., Daniellou, F. Carballeda, G. and Ruaud, S. (1995). Activity analysis in participatory design and analysis of participatory design activity. *International Journal of Industrial Ergonomics, Special Issue: 'Participatory Ergonomics'*, **15**, 311–329.

Gjessing C.C., Schoenborn, T.F. and Cohen, A. (1994). Participatory Ergonomics Interventions in Meatpacking Plants, *DHHS (NIOSH)* Publication No. 94–124.

Glew, D.J., O'Leary-Kelly, A.M., Griffin, R.W. and Van Fleet, D.D. (1995). Participation in organizations: a preview of the issues and proposed framework for future analysis, *Journal of Management*, **21**, 395–421.

Gontijo, L. and Odenrick, P. (1999). Supporting continuous participatory processes by using ergonomics. In *Proceedings of the Conference on TQM and Human Factors*, CMTO, Linkopings Universitet, pp. 248–252.

Haims, M.C. and Carayon, P. (1998). Theory and practice for the implementation of 'in-house' continuous improvement participatory ergonomics programs. *Applied Ergonomics*, **29**, 461–472.

Haines H.M. (2003). *Understanding Participatory Ergonomics — Developing Theory and Practice.* PhD Thesis unpublished, University of Nottingham

Haines, H.M. and Wilson, J.R. (1998). *Development of a Framework for Participatory Ergonomics.* (HSE Books).

Haines, H.M., Wilson, J.R., Vink, P. and Koningsveld, E. (2002). Validating a framework for participatory ergonomics (the PEF). *Ergonomics*, **45**, 309–327.

Halpern, C. and Dawson, K. (1997). Design and implementation of a participatory ergonomics program for machine sewing tasks. *International Journal of Industrial Ergonomics*, **20**, 429–440.

Hasle, P., Limborg, H.J., Hvenegaard, H. and Bruvik-Hansen, A. (1997). A critical evaluation of bottom up strategies in participatory ergonomics. *From Experience to Innovation – IEA'97. Proceedings of the 13th Triennial Congress of the International Ergonomics Association*, Tampere Finland, June 29–July 4 1997, Seppala P, Luopajarvi T, Nygaard CH, Mattila M (Eds) Finnish Institute for Occupational Health, Helsinki, Volume 1, pp. 364–366.

Hanse, J. and Forsman, M. (2001). Identification and analysis of unsatisfactory psychosocial work situations: a participatory approach employing video-computer interaction, *Applied Ergonomics*, **32**, 23–29.

Health and Safety Executive (1993). *Successful Health And Safety Management* HS(G)65 (London: Health and Safety Executive Books).

Health and Safety Executive (1996). *A Guide to the Health and Safety (Consultation with Employees) Regulations* (London: Health and Safety Executive Books).

Health and Safety Executive (2002). *Upper Limb Disorders in the Workplace* HS(G)60(rev) (London: Health and Safety Executive Books).

Imada, A.S. (1991). The rationale and tools of participatory ergonomics. In *Participatory Ergonomics*, Noro, K. and Imada, A.S. (eds.) pp. 30–51. (London: Taylor & Francis).

Imada, A.S. and Nagamachi, M. (1995). Introduction to a special Issue on participatory ergonomics, *International Journal of Industrial Ergonomics*, **15**, 309–310.

Imada, A.S. and Robertson, M.M. (1987). Cultural perspectives in participatory ergonomics. *Proceedings of the Human Factors Society 31st Annual Meeting*, pp. 1018–1022.

Imada, A.S. and Stawowy, G. (1996). The effects of a participatory ergonomics redesign of food service stands on speed of service in a professional baseball stadium. *Human Factors in Organizational Design and Management*, Brown, Jr., V.O. and Hendrick, H.W. (eds.) (Elsevier Science B.V.) pp. 203–208.

Ingelgard, A. and Norrgren, F. (2001). Effects of change strategy and top-management involvement on quality of working life and economic results. *International Journal of Industrial Ergonomics*, **27**, 93–105.

Jones, R.J. (1997). Corporate Ergonomics Program of a Large Poultry Processor. *AIHA Journal*, **58**, 132–137.

Kardborn, A. (1998) Inter-organisational participation and user focus in a large-scale product development programme: The Swedish hand tool project. *International Journal of Industrial Ergonomics*, **21**, 369–381.

Keyserling, W.M. and Hankins, S.E. (1994). Effectiveness of Plant-Based Committees in Recognizing and Controlling Ergonomic Risk Factors Associated with Musculoskeletal Problems in the Automotive Industry, *Rehabilitation*, **3**, 346–348.

Kuorinka, I., Cote, M-M, Baril, R., Geoffrion, R., Giguere, D., Dalzall, M.-A. and Larue, C. (1994). Participation in workplace design with reference to low back pain: a case for improvement of the police patrol car. *Ergonomics*, **37**, 1131–1136.

Kuorinka, I. and Patry, L. (1995). Participation as a means of promoting occupational health. *International Journal of Industrial Ergonomics*, **15**, 365–370.

Kuorinka, I. (1997). Tools and means of implementing participatory ergonomics, *International Journal of Industrial Ergonomics*, **19**, 267–270.

Laitinen, H., Saari, J. and Kuusela, J. (1997). Initiating an innovative change process for improved working conditions and ergonomics with participation and performance feedback: A case study in an engineering workshop. *International Journal of Industrial Ergonomics*, **19**, 299–305.

Langford, J.B. (1982). West Sussex Fire Brigade: control room design. Scientific Research and Development Branch, Home Office.

Lehtela, J. and Kukkonen, R. (1991). Participation in the purchase of a telephone exchange — a case study. Designing for Everyone. *Proceedings of the Eleventh Congress of the International Ergonomics Association*, Y. Queinnec and F. Daniellou (eds) (London: Taylor & Francis).

Lanoie, P. and Tavenas, S. (1996) Costs and benefits of preventing workplace accidents: the case of participatory ergonomics, *Safety Science*, **24**, 181–196.

Lewis, H.B., Imada, A.S. and Robertson, M.M. (1988). Xerox leadership through quality: merging human factors and safety through employee participation. *Proceedings of Human Factors Society 32nd Annual Meeting.*

Liker, J. K., Nagamachi, M. and Lifshitz, Y. R. (1989). A comparative analysis of participatory ergonomics programs in U.S. and Japan manufacturing plants. *International Journal of Industrial Ergonomics*, **3**, 185–189.

Liker, J.K., Joseph, B.S. and Ulin, S.S. (1991). Participatory ergonomics in two US automotive plants. In *Participatory Ergonomics*, Noro, K. and Imada, A.S. (eds.), pp. 97–139. (London: Tayor and & Francis).

Loisel, P., Gosselin, L., Durand, P., Lemaire, J., Poitras, S. and Abenhaim, L. (2001). Implementation of a participatory ergonomics program in the rehabilitation of workers suffering from subacute back pain. *Applied Ergonomics*, **32**, 53–60.

McNeese, M.D., Zaff, B.S., Citera, M., Brown, C.E. and Whitaker, R. (1995). AKADAM: Eliciting user knowledge to support participatory ergonomics. *International Journal of Industrial Ergonomics, Special Issue: 'Participatory Ergonomics'*, **15**, 345–365.

Maciel, R. (1998). Participatory ergonomics and organizational change. *International Journal of Industrial Ergonomics*, **22**, 319–325.

Mansfield, J.A. and Armstrong, T.J. (1997). Library of Congress Workplace Ergonomics Program. *American Industrial Hygiene Association Journal*, **58**, 138–144.

Moir, S. and Buchholz, B. (1996). Emerging Participatory Approaches to Ergonomic Interventions in the Construction industry. *American Journal of Industrial Medicine*, **29**, 425–430.

Montreuil, S. and Laville, A. (1986). Cooperation between ergonomists and workers in the study of posture in order to modify work conditions. In *The Ergonomics of Working Postures*, edited by E.N. Corlett, J.R. Wilson, I. Manenica (London: Taylor & Francis), pp. 293–304.

Moore, J.S. (1994). Flywheel truing — a case study of an ergonomic intervention. *American Industrial Hygeine Association Journal*, **55**, pp 236–244.

Moore, J.S. and Garg, A. (1997). Participatory ergonomics in a red meat packing plant, part 1: Evidence of long-term effectiveness. *American Industrial Hygeine Association Journal*, **58**, 127–121.

Moore, J.S. and Garg, A. (1998). The effectiveness of participatory ergonomics in the red meat packing industry. Evaluation of a corporation. *International Journal of Industrial Ergonomics*, **21**, 47–58.

Morris, W., Wilson, J.R. and Koukoulaki, T. (2002). *Participatory Design of Working Equipment and Standardisation Process in Europe — Workshop Report.* Unpublished. SALTSA — BTS, Brussels.

Morris, W., Wilson, J.R. and Koukoulaki, T. (2004). Developing a participatory approach to the design of work equipment: Assimilating lessons from worker's experience. (Brussels:TUTB).

Mossink, J.C.M. (1990). Case History: design of a packing workstation. *Ergonomics*, **33**, 399–406.

Mumford, E. (1991). Participation in systems design - what can it offer? In *Human Factors for Informatics Usability*, B. Shackel & S.J. Richardson (eds.) (Cambridge University Press), pp. 267–290.

Nagamachi, M. (1991). Application of participatory ergonomics through quality-circle activities. In *Participatory Ergonomics*, Noro, K. and Imada, A.S. (eds.), pp. 139–165. (London: Taylor & Francis).

Nagamachi, M. (1995). Requisites and practices of participatory ergonomics. *International Journal of Industrial Ergonomics, Special Issue: 'Participatory Ergonomics*, **15**, 371–379.

Nagamachi, M. and Tanaka, T. (1995). Participatory Ergonomics for Reengineering in a Chemical Fiber Company. *Proceedings of the Human Factors and Ergonomics Society 39th Annual Meeting*, pp. 766–770.

Neumann, J. (1989). Why people don't participate when given the chance. *Industrial Participation*, No. 601 (Spring), pp. 6–8.

Noro, K. (1991). Concepts, methods and people. In *Participatory Ergonomics*, Noro, K. and Imada, A.S. (eds.), pp. 3–30. (London: Taylor & Francis).

Nyran, P.I. (1991). Cost effectiveness of core-group training. In: *Advances in Industrial Ergonomics and Safety III*, edited by W. Karwowski and J.W. Yates (London: Taylor & Francis).

O'Brien, D.D. (1981). Designing systems for new users. *Design Studies*, **2**, 139–150.

Rawling, R.G. (1991). Participative approaches to the design of physical office environments. In *Participatory Ergonomics*, K. Noro and A. Imada (eds) (London: Taylor & Francis), pp. 53–72.

Reuter, W. (1987). Procedures for participation in planning, developing and operating information systems. In *System Design for Human Development and Productivity: Participation and Beyond*, P. Doherty *et al.* (eds.), pp. 271–276. (Amsterdam: North-Holland).

Ruohomaki, V. (1995). A simulation game for the development of administrative work processes. In *The Simulation and Gaming Yearbook*, Saunders, D. (ed.) (London: Kogan Page), pp. 264–270.

Sanoff, H. (1985). The application of participatory methods in design and evaluation. *Design Studies*, **6**, 178–180.

Silverstein, B.A., Richards, S.E., Alcser, K. and Schurman, S. (1991). Evaluation of in-plant ergonomics training. *International Journal of Industrial Ergonomics*, **8**, 179–193.

Snow, M.P., Kies, J.K., Neale, D.C. and Williges R.C. (1996). Participatory design, *Ergonomics in Design*. **4**, 18–24.

St-Vincent, M., Fernandez, J., Kuorinka, I., Chicoine, D. and Beaugrand, S. (1997). Assimilation and use of ergonomic knowledge by non-ergonomists to improve jobs in two electrical product assembly plants. *Human Factors and Ergonomics in Manufacturing*, **7**, 337–350.

St-Vincent, M., Chicolne, D. and Beaugrand, S. (1998). Validation of a participatory ergonomics process in two plants in the electrical sector. *International Journal of Industrial Ergonomics*, **21**, 11–21.

Teiger, C. and Laville, A., 1987. How ergonomic methods can be used by non-ergonomists. *Paper presented at 2nd International Occupational Ergonomics Symposium. Applied Methods in Ergonomics.* Zadar, Yugoslavia.

Van der Molen, H.F., Vink, P. and Urlings, I.J.M. (1997). A participatory ergonomic approach to the redesign of scaffolders' work. *From Experience to Innovation — IEA'97. Proceedings of the 13th Triennial Congress of the International Ergonomics Association*, Tampere Finland, June 29–July 4 1997, Seppala P, Luopajarvi T, Nygaard CH, Mattila M (Eds) Finnish Institute for Occupational Health, Helsinki, Volume 1, pp. 450–452.

Van der Schaaf, T.W. and Kragt, H. (1992). Redesigning a control room from an ergonomics point of view: a case study of user participation in a chemical plant In *Enhancing Industrial Performance*, H. Kragt (ed.), (London: Taylor & Francis), pp. 165–178.

Vink, P., Peeters, M., Grundemann, R.W.M., Smulders, P.G.W., Kompier, M.A.J. and Dul, J. (1995). A participatory approach to reduce mental and physical workload. *International Journal Industrial Ergonomics*, **15**, 389–396.

Vink, P., Urlings, I.J.M. and van der Molen, H.F. (1997) A participatory approach to redesign work of scaffolders. *Safety Science*, **26**, 75–85.

Westlander, G., Viitasara, E., Johansson, A. and Shahnavaz, H. (1995). Evaluation of an ergonomics intervention programme in VDT workplaces. *Applied Ergonomics*, **26**, 83–92.

Wilson, J.R. (1991a). Design Decision Groups — A participative process for developing workplaces. In *Participatory Ergonomics*, K. Noro and A. Imada (eds). (London: Taylor & Francis).

Wilson, J.R. (1991b). A framework and a foundation for ergonomics? *Journal of Occupational Psychology*, **64**, 67–80.

Wilson, J.R. (1995a). Ergonomics and participation. In Wilson, J.R. and Corlett, E.N. (eds) *Evaluation of Human Work: A Practical Ergonomics Methodology*. 2nd and Revised Edition. (London: Taylor & Francis).

Wilson, J.R. (1995b). Solution ownership in participative work design: The case of a crane control room. *International Journal of Industrial Ergonomics*, **15**, 329–344.

Wilson, J.R. and Grey, S.M. (1990). But, what are the issues in work design? In C. Haslegrave, J.R. Wilson, E.N. Corlett (Eds) *Work Design in Practice* (London: Taylor & Francis), pp. 7–15.

Wilson, J.R. and Grey Taylor, S.M. (1995). Simultaneous engineering for self directed work teams implementation: A case study in the electronics industry. *International Journal of Industrial Ergonomics*, **16**.

Wilson, J.R. and Haines, H.M. (1997). Participatory ergonomics, In *Handbook of Human Factors and Ergonomics* (2nd edition) Salvendy, G. (ed.) (Chichester: Wiley), pp. 490–513.

Zink, K.J. (1996). Continuous improvement through employee participation. Some experiences from a long-term study in Germany. *Human Factors in Organizational Design and Management*, Brown, Jr., V.O. and Hendrick, H.W. (eds.) (Elsevier Science), pp. 155–160.

Appendix

Extract from: Participatory Ergonomics, Chapter 5, Design Decision Groups — A Participative Process for Developing Workplaces by John R. Wilson; edited by K. Noro and A. Imada (London: Taylor & Francis) 81–96.

Design decision groups procedure

Scene setting

In order to enable (not force) people to participate in meetings, exercises, investigations, etc., certain general principles can be observed. They should feel that there is some purpose to participating, that outcomes will be of benefit to themselves and/or to others. They should feel that they have enough information and knowledge to make a useful input, and should feel confident enough to ask for or to obtain such knowledge where they feel it is lacking. One important but tricky requirement is that the group must not be overtly directed or feel that this is the case; however, the group must have direction and be purposeful; disruptive or domineering members must be subtly reined in and not allowed to dominate meetings.

The above requirements for the DDG process can be achieved in several ways. We felt that holding meetings at an independent site (our laboratory) and emphasising our independence and the confidentiality of the details of the process were important. Meetings were facilitated rather than run, generally with one investigator acting as introducer, prompter, interpreter and general resource, with a second investigator on hand as technical resource, including making and serving refreshments, distributing or showing graphical and visual aids, supplying materials and so on. The second investigator was also responsible for recording the proceedings, using video and photography (as unobtrusively as possible), and for note taking. All sessions were allowed to run in a relaxed manner, with everybody being gently encouraged to contribute, and with a very flexible agenda. At the same time, the investigators were well aware of certain tasks or exercises which had to be completed, and certain information which had to be derived, within particular time-scales. We have found that, although certain guidelines for doing this can be followed and can help, it is not surprising that it also takes a certain sort of person to facilitate the groups. Not everyone has the ability to communicate and subtly direct, without dictating or dominating.

All potential participants were contacted at their place of work and informed that they would take part in confidential discussions about the design of their workplace, how much they would be paid, and the dates and times when we would need them. Library staff largely came for evening sessions; retail staff came on their afternoons off. All participants were collected and returned by taxi. Payment, which compared very favourably with their normal wages, was made in one lump sum at the end of all sessions.

Session A

Introduction

This includes a welcome, introductions, an explanation of who the design meeting organisers are and a general overview of what is expected of the design sessions. Refreshments are served and every effort is made to establish a comfortable, relaxed environment.

Word map

'Word maps' or 'visual maps' are constructed by every participant, volunteering words, which they feel are connected with the particular design topic for a 5–10 min period. This technique defines the total problem space and helps produce the dynamic agenda. In our case we simply asked participants to make an inventory of objects and equipment required at their workplace. This acts as an 'ice-breaking' device and also participation is such an easy, non-stressful exercise it breaks down any inhibitions that quieter members of the gorup might have about expressing their ideas in front of their colleagues. Participation was encouraged by going 'round the table' several times (see Figure 3).

Round-robin questionnaire

Participants are presented with a series of very simple open-ended questions such as: 'A good work counter is . . .?' or 'Problems at my workplace are . . .?'. Eacn question is printed on a separate sheet, and these are passed around all participants with each attempting to complete the sentence with a different ending to that chosen by the others (Figure 4). This method is preferred to the alternative of each participant filling out an individual questionnaire for two reasons. Firstly, it makes the exercise a little more productive, in that each participants has to think of a different ending from those given by the others, thus providing more information by reducing duplication. Secondly, it can act as a stimulus. If, for example, participants had

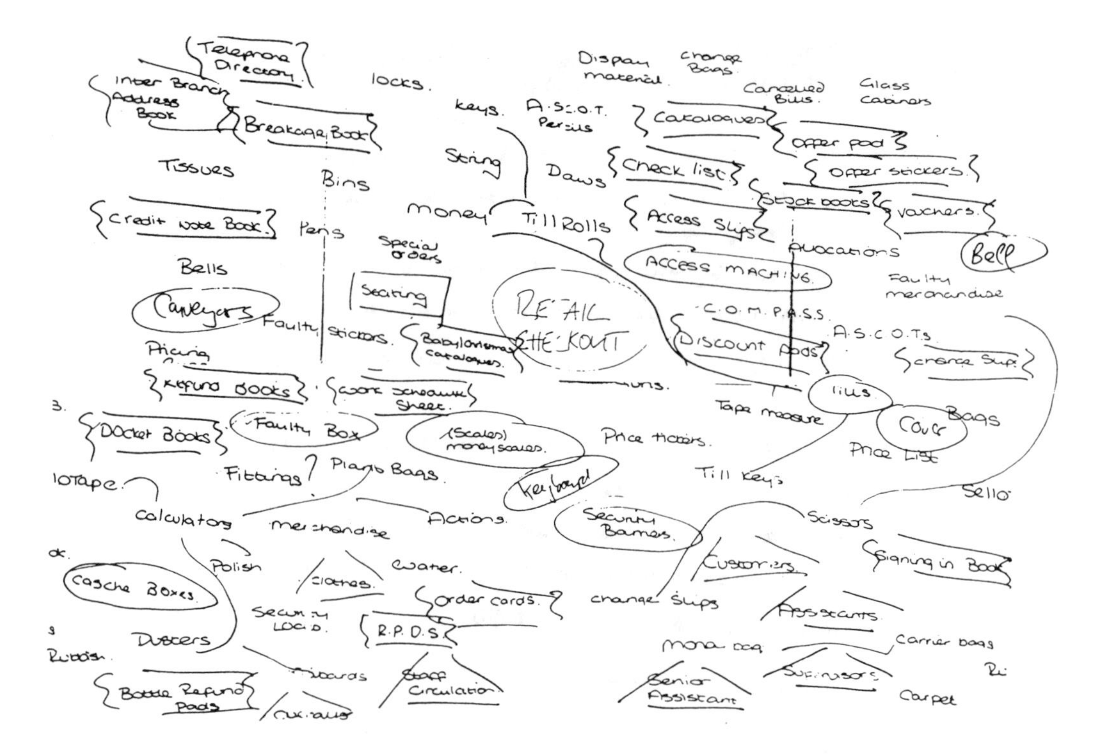

Figure 3 Part of a simple word map.

missed the point of the exercise, or were unable to think of any ideas for a certain sentence, then seeing what had been written by others may trigger a line of thought that could otherwise have lain dormant.

Drawing of own workplace

As a further ice-breaking exercise, as preparation for the more important drawing exercises to come, and to provide investigators with extra information, participants were asked to draw their existing workplace, indicating its good and bad points.

Slide presentation of ergonomics problems

Slides are selected from associated fieldwork, covering a range of topics: different types of layout, different types of equipment, etc., and, most importantly, a series showing how these pieces of equipment have been poorly installed in certain environments. The slides emphasise such features as poor seat height, lack of leg room, poor work-surface height, lack of work space and bad positioning of equipment and displays.

This stage is intended to provide a logical progression from the round-robin questionnaire. At that stage, the participants are asked to think about what makes good and bad designs, but only have their own experience to draw upon. The slide show is intended to make participants aware of systems and layouts that they may not have encountered before, and also to emphasise problems that they may not have considered whilst being influenced so heavily by their current system. Slides of a variety of different workplace and systems can also emphasise the point that we are required to design something entirely new, and that suggested designs should be unimpeded by a narrow range of experience. We are conscious that such a

Seeing Rubbish and not being able to find anything
big piles of Faulty goods stack up in one corner

When I'm working, the things I dislike in a checkout are: having
to stand and look down on customers,
not having enough room to write and
bag items.

not having the right equipment at hand to
deal with customers paying by ~~[illegible]~~ credit card

Having to stand, waiting for someone to answer
the bell.

To show a customer where something is. (point).
Having papers etc hanging around.
merchandise hanging around under your feet.
Having catalogues on the top near the tills.
Having no spare till rolls.

People eg [illegible]

when you can't find price list or a price in book
when someone wants a item reaching over or looking at [illegible]

Not Having enough Time To Talk To Customers
when there are no customers to serve but you can't come off till another person comes on
when you cant find anything when there are so many draws.

Figure 4 Example of a round-robin questionnaire.

procedure must not direct our participation in any particular direction, either of criticism of existing systems or of ideas for new ones. It may not be appropriate in all circumstances.

Equipment demonstration

When systems or equipment are to be installed which may be new to any or all participants it is advisable to demonstrate these and their operation. Actual examples should be used where possible; if not illustrations could be used.

Individual silent drawing

Each participant is asked to produce sketches on A4-sized paper of both plan and perspective views of their ideas for a good new workplace, incorporating any new systems. These should detail proposed layouts for people, work counters, modules and equipment (Figure 5). At this stage, only very general ideas need to be produced, and participants are discouraged from becoming too involved in detail (e.g., precise measurements of work surfaces, or exact location of documents). To call this stage a 'silent drawing exercise' is perhaps overstating the point, but it is beneficial to discourage communication in the hope that a variety of different ideas and designs will be produced for use in the later discussion stage. Each individual is

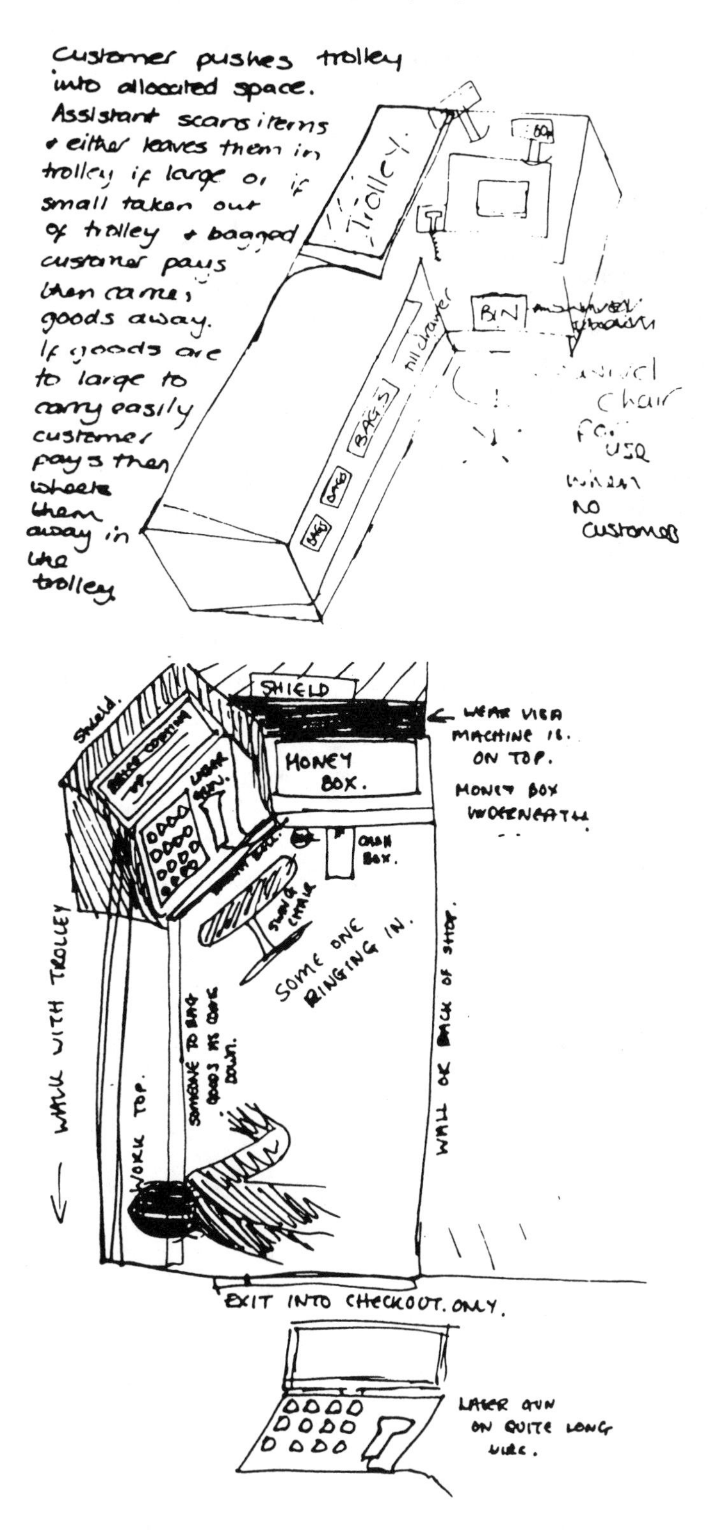

Figure 5 Three examples of individually drawn proposed work spaces, showing individual differences in design ideas and how these can be illustrated and communicated. Considerable colour coding is used in the originals.

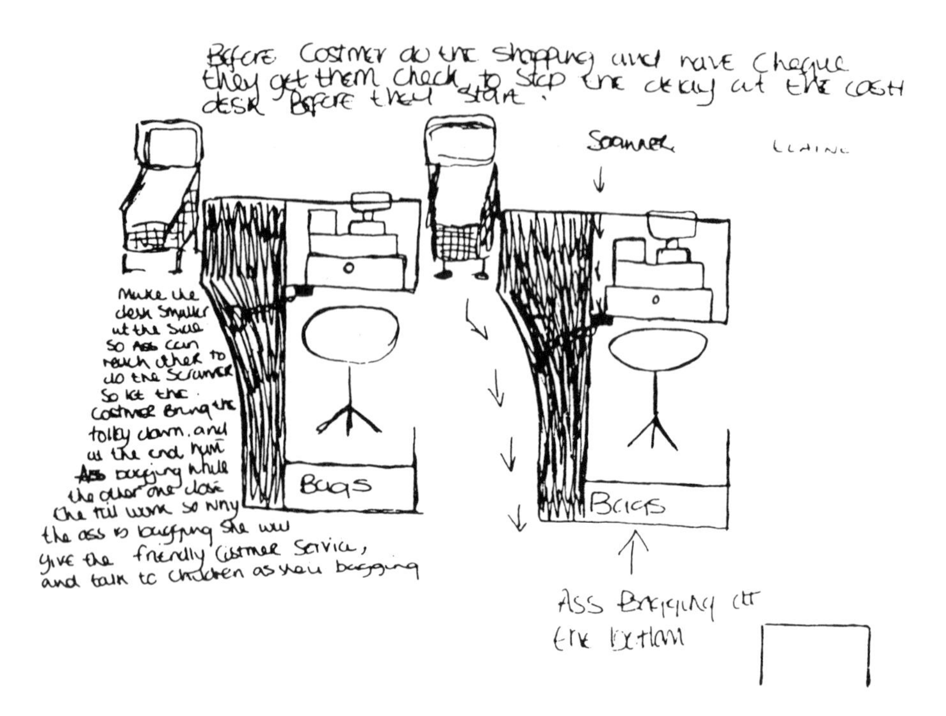

Figure 5 Continued

encouraged to try out several rough ideas before producing a final sketch suitable for presentation. Different coloured pens are available for colour coding various items of equipment or furniture.

Individual presentation and discussion

Each participant is asked to present their final sketch to the rest of the group and to explain the reasoning behind the design. The designs are discussed and gently (the role of the facilitator) criticised.

After each of the sketches has been discussed many problems should be highlighted, and certain designs may be seen as 'superior'. The role of the facilitator is then to encourage a general discussion, concentrating on one or two particular designs and bringing up any points that may not have been previously considered, probably by referring back to issues arising from the word-map and round-robin stages.

Session B

Drawing in pairs

At the beginning of the session the facilitator briefly runs through some of the points arising from the sketches made in session A. The group is then divided into pairs or threes, according to the similarity of their ideas, and asked to produce a larger sketch, on A3 paper, of the plan for the workplace. This time more emphasis is placed on detail. Positioning of equipment is to be specified more precisely, as are seating and furniture design (Figure 6).

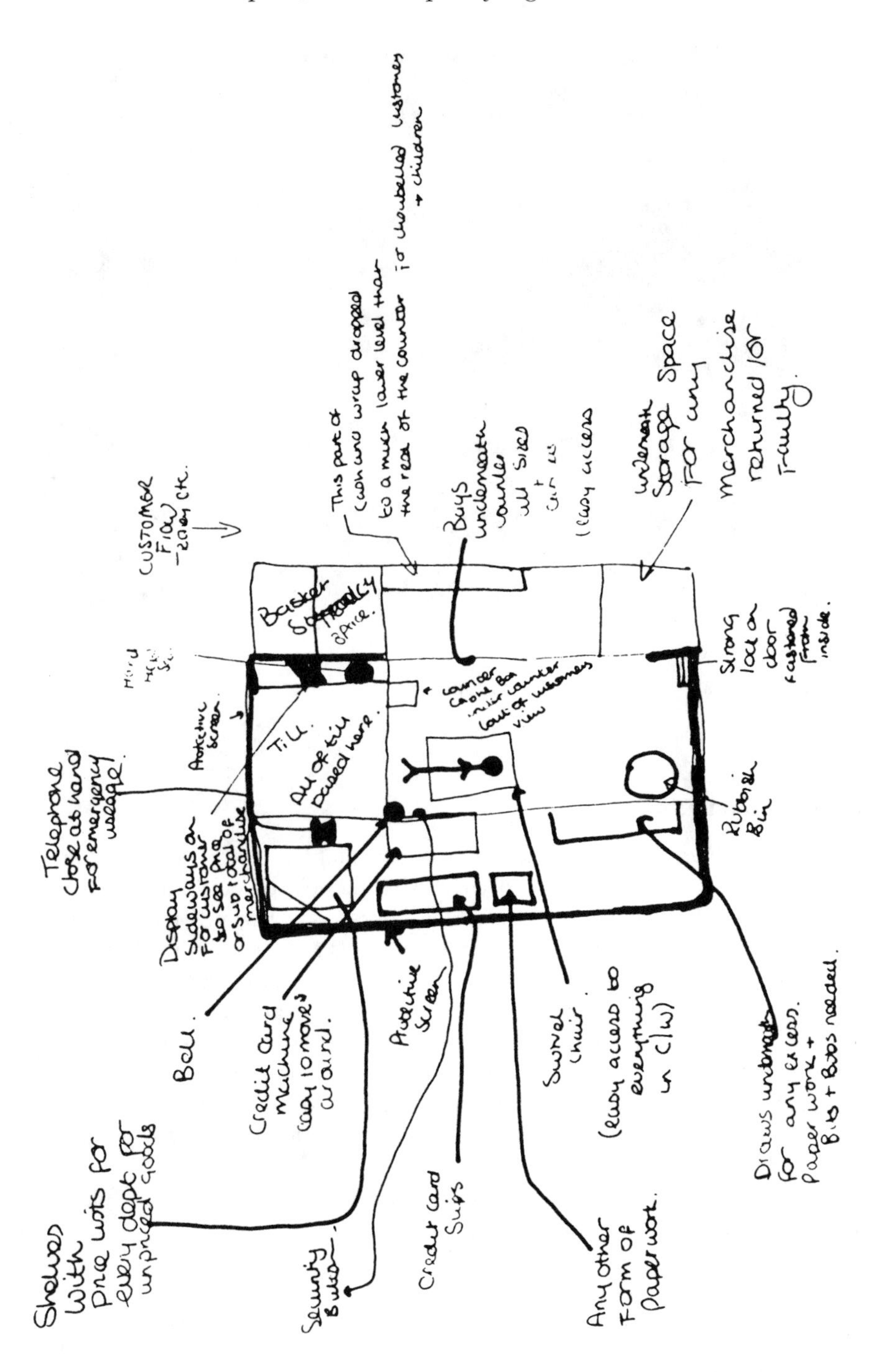

Figure 6 Example of a workplace developed by a pair of participants and used as the basis for a model-building exercise.

Presentation and discussion

Each drawing is presented to the rest of the group by its authors. Discussion at this point is geared toward choosing the 'best' drawing, but no firm decisions need to be reached as each group with get a further chance to demonstrate their ideas using models of the system.

Figure 7 Initial planning for layout: two groups.

Simple layout modelling

Participants are shown empty spaces of approximately the size available at the future workplace site. Very strong corrugated cardboard models of equipment are provided, in addition to a supply of mock-up counters, tables, desks, consoles and other modules. A variety of actual seats, tables and benches is also provided, as is a stock of string, flat card, modelling knives and heavy tape. Although the use of card models may seem, at first glance, to be amateurish, there are several compelling advantages. Models and mock-ups are more portable and participants are not frightened of breakage so building model workstations becomes a fast rapidly changing exercise. Extra pieces of equipment or new pieces can be found or created very quickly. Work-station, work-counter and module sizes and positioning can be changed rapidly and easily. We have not had any participants who felt that they were taking part in a worthless or 'cheap' game.

Participants, still in subgroups of two or three, build a three-dimensional full-size model of their idea of the optimum workplace (Figures 7 and 8).

Discussion

Discussion involving all participants is firmly geared toward eliminating any inferior design(s) and forming a consensus of opinion as to the optimal layout. This is often an amalgam of the various ideas represented by the different mock-ups. Limited 'walkthroughs', testing aspects of the workplace, may be undertaken at this stage.

Fitting trials

On reaching this stage, a general workplace layout and the positioning of all the equipment should largely have been agreed. The next crucial factor to be tackled is that of work-surface heights and depths. These are determined through fitting trials, with the aid of adjustable metal

Figure 8 Discussion of the juxtaposition of equipment.

stand holding the cardboard models, using all group members as subjects. In addition, however, the facilitator must ensure, at the time or later, that anthropometric characteristics of other potential users are taken into account.

Session C

Building a mock-up

On returning for the third session, participants build a complete mock-up (or two if there is still internal disagreement) of the preferred workstation, integrating ideas on the placement and juxtaposition of equipment with work-surface height and depth data (Figure 9).

Discussion

The discussion entails detailed consideration of the operational requirements at each particular workstations; if two designs have been modelled, each sub-group must defend its ideas against those of the other group.

Modifications

As a result of the discussion, any modifications to the general structure of the workstations are made. If more than one design is still being considered, the best alternative is generally selected at this stage.

Simulation trial and assessment

The trials are aimed at simulating as many as possible of the scenarios, operations and tasks which are or might be undertaken in actual use. Participants take turns in playing the roles of staff and customers (Figure 10), including children (Figure 11). Final adjustments to the workplace model are made at this point.

Figure 9 Building a mock-up workplace.

Figure 10 A simulation trial.

Debriefing

Debriefing takes place after each session but this is the most important time of all. Participants are encouraged to discuss the value of the exercise and the perceived quality of their designs.

Refinements are made at the end of each stage, with the investigators accounting for constraints, equipment or tasks which may have been omitted (deliberately or accidentally) in the DDG sessions, or which may have come to attention later.

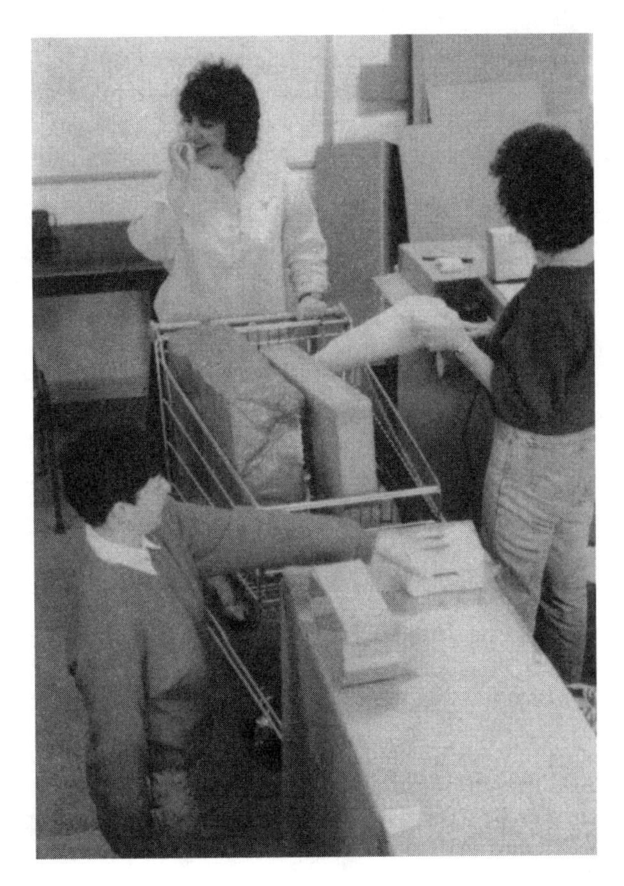

Figure 11 A trial showing (initially sceptical!) participant simulating a child.

Design decision group outcomes

The outcomes of a DDG study, or other similar work, are varied. At the very least we have managed to find out, from the people who should know most about it, the problems or benefits associated with existing systems, the potential problems with the new system, and a number of concepts to be fed into the development process. At best, one or more feasible design alternatives for the whole workstation is provided; this is an invaluable outcome, regardless of whether these designs are 'new' to the investigators or are combinations of ideas already developed. The process and setting also allow flexible and iterative testing of design concepts, whether derived beforehand or during the process.

It has been noticed in all out work with DDGs that the creative, yet relaxed, atmosphere and the resultant concentration of minds seems to encourage the expression of opinion or facts about other aspects of work redesign. During the libraries study especially we learned much of interest about training and support, feedback and other communication, autonomy, role clarity and other job-content and organisation issues. This allowed us to make a more informed analysis of current situations, and to make better suggested changes for improvement. Although not used by us in such a context, it is suggested that the spin-off benefits will be considerable when such an exercise is conducted at a local level with people whose own technology, workplace, and jobs are to be changed. Awareness of what should be done to improve jobs and workplaces and of the criteria which should be applied, and the confidence to promote their own ideas can all help to create a more committed workforce, as well as healthier, safer, more effective and more satisfying jobs. The use of DDGs can thus give rise to direct gains through its outcomes in terms of design ideas, and can also give rise to systemic gains through the very application of the process.

Chapter 37

Economic analysis in ergonomics

Maurice Oxenburgh and Geoff Simpson

Introduction

It has been strongly argued that there is a need to emphasise economic arguments in the promotion and justification for health and safety changes in the workplace. As Alexander (1985) stated: 'A manager may allow ergonomics to be tried, but without bottom-line improvements (or other equally convincing measures of effectiveness) the use of ergonomics will soon diminish and then disappear'. A fundamental problem within this context is that health and safety research and implementation is still seen in many organisations as largely altruistic with no significant, or even tangible, returns on investment. This is remarkable when you consider the total cost of (un)safety to industry as given in, for example, a paper by Rimmington (1993).

> The studies we [the Health and Safety Executive] have done suggest that the non-injury costs of relevant incidents in work activity cost somewhere between £2 billion and £6 billion per annum, which when added to the injury costs adds up to between 5% and 10% of all UK companies gross trading profits or somewhere between 1% and 2% of gross domestic product.

Despite such evidence it often remains true that, at local levels, management does not fully appreciate the economic implications of (un)safety and while ergonomics continues to be considered as primarily a health and safety discipline it will, inevitably, face similar problems of irrelevance.

Unfortunately, ergonomics, like its collaborators within health and safety, has tended to shy away from economic argument. In part this is due to genuine and justifiable reservations on the assumptions necessary to cost issues in health and safety. However, it is also in part due to a lack of realisation of the extent to which equally debatable assumptions are made in the apparently more legitimate accountancy field and, in part, through a lack of awareness of the techniques and data which can be used to build an economic justification.

Ergonomics may have an additional problem in contrast to the 'traditional' health and safety disciplines, as Simpson (1993) has suggested:

> ... the very existence of (or the apparent need for) ergonomics is an implicit "threat" to other disciplines. For example if you need to invoke a "new" discipline to overcome the shortcomings of designers, managers, existing safety specialists, occupational physicians, etc. to improve safety/health/performance at work, then it can be inferred, implicitly at least, that such disciplines have failed.

If there is even a hint of truth in this, it becomes even more important that ergonomics encompasses all the arguments which can substantiate its contribution.

So, what arguments do we use? There are the arguments used by government agencies and departments that concern legislation. An early example of using the state or nation-wide costing was made by Oxenburgh and Guldberg (1993) in the late 1980s. The study was to answer the question whether or not the introduction of appropriate legislation would be cost effective in reducing back injuries due to manual handling. An interesting outcome of this study was that much, if not most, measures taken by industry to reduce back injuries were for economic reasons and not for health and safety ones. A recent Government example of the use of economics to implement ergonomics is the Ergonomics Rule of the Washington State, USA (2002). The stated purpose of the proposed ergonomics rule (WAC 296-62-051) is to reduce employee exposure to workplace hazards that can cause or aggravate work-related musculoskeletal disorders (WMSDs). In workplaces where these hazards exist, employers must reduce them. Doing so will prevent WMSDs such as tendinitis, carpal tunnel syndrome and low back disorders. The relevant government department made a thorough study of the economics of the proposed legislation in order to back up their health and safety argument.

The problem with these types of studies is that they do not relate to individual workplaces where reduction of injuries must take place. Even with a system of workplace inspectors it is only the larger workplaces that will be visited whereas the vast majority of workers worldwide work in small workplaces. The probability that a small workplace will be visited by an inspector is so remote that the employer can ignore safely legislation. It is of little use to tell an employer that, on the average, the costs to comply with the legislation are outweighed by the benefits by a factor of four; the retort one is likely to get will be ‘so what?’ or even ‘we are different’.

One does not even get much help from the economists as the prevailing mode in rational economics is that if a system is more productive it will be implemented without the necessity for legislation. That is, an employer will institute the means that will give the greatest profit; if improving safety does not increase profits then there is no rational reason to do such.

In order to use economic arguments we must look at industry (whether production or service) from the point of view of the employer. He or she is there to make a profit — but is that the only motivation?

Frick (1997) has compared the organisational aspects of quality control with health and safety. Until recently quality was delegated to the operational level and it was not until management realised that it was a management responsibility that quality could improve. Occupational health and safety is too often regarded as a low level activity delegated to an operational level of supervision. Frick goes on to discuss the other aspects of a company culture and, indeed, the interests of individual managers within an enterprise.

Unless occupational health and safety in general and ergonomics in particular can be linked to the various aspects of management including the company culture, organisation and individual interests, it will remain peripheral and ineffective.

The purpose of this chapter, therefore, is to identify methods that can be seen to be in the interests of individual managers and in the interests of the enterprise as a whole. We will see how to build an economic case for an intervention and some of the procedures and calculations which can be used. By way of conclusion, studies which have used economic analysis of ergonomics interventions are quoted to show that not only is the objective advocated desirable, but it is economically achievable.

The basic requirements for an economic analysis of ergonomics

Traditionally industrial ergonomics has been justified on the basis of health and safety with occasional, though usually vague and somewhat embarrassed, references to production improvements. For some areas of ergonomics, e.g., product design, human computer interaction, this is, perhaps, less true where commercial advantage has accepted some tenets of ergonomics.

The first step necessary in developing an economic base for ergonomics is the resurrection of production improvements as a legitimate objective. Campbell (1993) has pointed out 'safety does not pay, it is production that pays, but safe production pays better'.

The second step is the acceptance that health and safety issues can be legitimately considered as loss prevention topics. Most enterprises have, in recent years, undergone some form of rationalisation in pursuit of a 'leaner and fitter' operation. This has almost inevitably involved staff reductions and often the breakdown of skill boundaries to promote an increasingly multi-skilled workforce. While the 'leaner and fitter' organisation is perhaps more productive and profitable under normal circumstances, it is also less resistant to disruptions in its reduced workforce. If disruptions occur as a result of, for example, 'organisationally self-inflicted' sickness such as repetitive strain injury, then clearly it is in the organisation's financial interest to remove that drain on resources. In a similar fashion, the increasing use of casual and temporary workers may lead to a loss of knowledge and an increase in unsafe practices accompanied by an increase in injury rates (Quinlan *et al.*, 2001). In a nutshell, organisational aspects that have an effect on the health of employees and the health of the enterprise are legitimate concerns for occupational health and safety.

Having defined some of the costing approaches relevant to ergonomics, the third requirement is to identify the kinds of data which can be used in costing calculations.

When discussing the various parameters for costing interventions Corlett (1988) noted:

> In the nature of things any evaluation is only partial. This can be from inadequacies and insufficiencies in models, in variables and in methods. What is unnecessary is that they should be deliberately insufficient in their perspective. Such a state can arise when models are imposed on situations, not adapted to the cases under investigation. To leave cost-benefit analysis to accountants, to economists, even to ergonomists, is to build in a cause for failure. The utility of professionals, particularly in combination, is undoubted but the understanding of what has to be measured can come only with the aid of those involved. This applies particularly to the involvement of lower management and relevant segments of the work-force, including representatives.

Thus participatory ergonomics, in order to derive relevant data, must be seen to be an essential ingredient in determining the economics of an intervention.

Finally, the fourth stage is familiarisation with models and procedures appropriate to the analysis of ergonomics interventions in economic terms. It is unfortunate that many of those engaged in the various aspects of occupational health and safety are not trained in economics and often shy away from such concerns. Sometimes it is due to a lack of knowledge but also it may be due to a misplaced view of the role of health and safety — it is above mere money matters.

The language of management

The reality of the situation is that ergonomists and managers are often talking different *languages*. Whether or not ergonomists feel comfortable with the idea, enterprises are profit-making centres and every job has been created as a necessary part of a larger profit-making machine. An ergonomist, like any other person in the industry, will therefore be expected to help the organisation achieve its goals. Management's remit is centred around profit-making, product quality and meeting production time-scales and they are likely to see the benefits of ergonomics change as, at best, peripheral to their objectives. The ergonomist then gets frustrated by management's lack of immediate enthusiasm for their ideas and wanders off wondering why management cannot understand them. The simple fact is that if ergonomists want to communicate effectively with people in industry, then it is up to them to try and learn industry's *language*. This does not mean that the ergonomist's role has to concentrate exclusively on production or service-related issues but getting recommendations implemented which improve the health and safety of the workforce will be much easier if the ergonomist talks the industry's language.

For example, if ergonomists conduct a study of a group of workers who are standing at a bench all day assembling small components, it is highly probable that they will conclude that the workstations should be redesigned so that the workforce can be seated. If they then approach the works manager and argue that money should be spent on modifying benches and buying seats to improve working postures, the chances are very slim that the changes will be authorised. The manager may say that there is no apparent problem, after all people have been working like that for as long as can be remembered with no complaints, that people are not paid to be comfortable, and that anyway the budget is not available. However, investigation may show that the work study department had previously assessed the job and, when working out the appropriate 'relaxation allowance', had added 3% of time to cover the employees standing all day. By pointing out that the manager was paying these workers an extra 3% to stand up, the manager may then authorise the ergonomist to change the workstations to allow operators to work seated. The end result was exactly what the ergonomist wanted but the route was through productivity and economics — talking the same language.

Making investment return predictions for ergonomics

The production of predictive costing figures is essentially part of the marketing exercise of presenting the proposal itself. Professionals in the marketing field emphasise the importance of presenting a product in the most favourable light without actually telling lies. The latter point is crucial, for if you exceed what is credible the proposal is likely to be lost and it is possible that all future proposals will be viewed sceptically. Inevitably given a shortage of some data, assumptions have to be made. It is therefore important to specify the assumptions made and always to be conservative in terms of predicted achievements. It is better to offer little and deliver more, than to offer a great deal and risk disappointing people by delivering less than they expected. Anyone who has ever attempted a predictive cost-benefit analysis is fully aware of the dangers. For example, you predict a saving of £10,000, but in reality provide only £8000; you can almost guarantee the reaction which will not be (as you had hoped), 'Thanks, that £8000 is very useful', but rather 'What happened to the other £2000'!

High levels of accuracy are not crucial. In fact it is probably advantageous if the calculations are seen to be only as reasonable estimates, as any subsequent debate will almost certainly improve the calculations. For example, if the manager does not agree with your approximation of, say, overheads on salary cost and proposes a different figure, when this is incorporated into the costing there can be little further argument since it is the manager's own figure. Some examples of economic cost arguments for specific interventions are developed below.

Life-cost accounting

A very useful step forward in using costing to implement occupational health and safety was made by Dahlén (Dahlén and Wernersson, 1995). He compared the concept of life-cost accounting (in engineering practice; the life of a machine) with employment of workers (long-term employment). He used activity-based costing to derive the cost drivers for injury and thus the allocation of those costs. Although he was looking mainly at the costs of absenteeism, labour turnover and training, he did introduce some productivity factors including re-working, increased scrap and overtime due to the use of unskilled labour.

A more traditional way to use life-cost accounting is by calculating the contribution of poor ergonomics to the total cost of a machine or group of machines (e.g., the accumulated costs throughout the life cycle of the machine). This framework can also be useful in 'loss prevention' arguments. The case study examines the economics of a coal winning machine (a shearer).

Case study 1 — coal mining

An estimate of the total life-cost to the organisation can be derived covering all that type of machine in the industry. This is approximated by the equation:

$$\text{Total life cost} = n(x + Lt)$$

where, L = the life expectancy of the machine, n = the number of machines in use, x = the capital cost per unit, t = the operating cost of a machine per annum.

For shearers, the figures were: $L = 8$ years; $n = 540$; $x = £250,000$, $t = £500,000$. Using this equation, the total life cost of shearers in UK mining was £2.3 billion.

Predetermined time and motion systems (e.g., MTM 1; Maynard *et al.*, 1948) are very powerful tools for the industrial ergonomist. Where these cannot be easily applied, even simple conservative estimates of improvements on a breakdown of the various operating costs may be all that are required to argue for change.

Ideally the performance implications of particular ergonomics limitations should be known, in this way the overall job implications can be built up from the component task limitations. Unfortunately, given the complexity of industrial jobs and the specific context in which they are carried out, this ideal approach is only possible retrospectively using current data. For predicting the influences of ergonomic design deficiencies on performance a number of avenues can be used:

- an assessment of the machine or sample of machines,
- if relevant knowledge is not available and pilot studies are not possible then task synthesis techniques may be used (e.g., Annett *et al.*, 1971; Maynard *et al.*, 1948),
- other aspects of the ergonomics literature can be used either by careful analogy with the machine under consideration or by the use of task-oriented human error/reliability data (e.g., Swain and Guttmann, 1983 and see Chapter 32).

This information can then be related to the operational cycle of the equipment involved and conservative estimates derived of the performance improvements likely to arise. Table 1 shows that ergonomics improvements in shearer design are likely to save 6.4 min each shift. These machines are available for production for 330 min each shift and therefore the current ergonomic limitations lose 2% of the total potential operating time. In operating cost terms this is a loss of £10,000 per machine per year or, over the life of the machine, about one-third of the capital cost of the machine.

Table 1 Potential Savings from Improved Ergonomics of Coal Winning Machines

Shift breakdown	Average duration (min)	Estimated percentage saving through application of ergonomics of coal winning machines	Potential saved time (min)
Travelling time	89	0	0
Preparation and meals	26	1% through better design, reducing preparation	0.26
Machine running	107	2% resulting from improved performance	2.14
Ancillary time	29	1% through better design	0.29
Operational time	63	0	0
Lost time			
Electrical	12	5% through diagnostics	0.6
Mechanical	31	10% through access and handling improvements	3.1
Mining/geological	77	0	0
Total min/shift	434		6.39

It should be noted that these estimates include a potential saving through improving the maintainability features of shearers. Maintenance aspects of machinery are often overlooked; but as maintenance costs are typically 30% of the costs of ownership, substantial savings can usually be found simply through recommending improved access to those components requiring frequent attention during routine maintenance operations.

Long term analysis

The plant studied by Aaras over a 15 year period was primarily concerned with the assembly and wiring of telephone switching panels and employed a predominantly female workforce (Spilling *et al.*, 1986). The workstations involved considerable muscular loading and awkward postures and this was reflected in high sickness absence records for the plant.

During an extensive ergonomics redesign particular emphasis was placed on the need to give each operator greater flexibility allowing, for example, for both seated and standing operation. Several other changes were also made, including improved seating, improved tools and major changes to both lighting and ventilation. A number of the factors including sickness absence and labour turnover were monitored throughout the period.

Prior to the improvements, musculoskeletal sickness absence was running at 5.3% of the production time available; in the seven years after the improvements it had dropped to 3.1%, a difference which was significant. The analysis of labour turnover was even more marked. Prior to the intervention employee turnover was running at about 30%, whereas in the period after the change it had been reduced to an average of slightly over 7.5%. Obviously the labour turnover could have been influenced by many factors other than the ergonomics improvements but in interviews with the staff the improved working conditions were the most frequently mentioned issue. These reductions in labour turnover also created additional financial benefits beyond the immediate production improvements; for example, there were attendant savings in terms of both training and recruitment costs.

Amortising the costs and benefits over the life of the experiment, the savings were over six times the cost of the intervention.

Workplace analysis

Although the concept of life-cost accounting is a well accepted engineering practice and is suitable when purchasing capital equipment, most workplaces are not engineering enterprises and many ergonomics interventions do not require the purchase of capital equipment. It seems unlikely that the long-term experiment of Aaras will, or even can, be repeated and other economic models are required, particularly those that can be used in individual workplaces.

Liukkonen has been instrumental in helping Swedish industry to examine the costs of occupational health and safety since the 1980s and we believe that she was the first to codify these costs in a form that could be used to improve working conditions at the workplace and work station level. Most of her work was published in Swedish but she has summarised some of her ideas in English (Kupi *et al.*, 1993).

Although we suspect that cost-benefit analysis models for occupational health and safety have been developed by some large enterprises for their internal use they have not been published or otherwise made generally available. It seems that there are only two commercially available computer programmes based on the original work of Liukkonen; The Potential (Miljödata, 1999) and the Productivity Assessment Tool (Oxenburgh, 2004a).

Although generic cost-benefit analysis tools are in short supply in the literature it does not mean that some form of costing is not used by occupational health and safety practitioners. Morrissey (2002) surveyed, by questionnaire, occupational health and safety practitioners in Australia. These practitioners came mostly from larger companies and he found that about 75% of respondents had used some form of costing. However, on closer analysis the costing used was mostly based on the direct wage costs with the indirect costs (lowered productivity, overtime, product damage and so on) often ignored. It is frequently the indirect costs, rather than the direct costs, that are major cost factors and it is from savings in these indirect costs that benefits can flow from an intervention. Morrissey's survey points to the unfortunate but well-known fact that cost-benefit analysis is not generally used by occupational health and safety practitioners.

To illustrate the major features when using economics to assist in the implementation of ergonomics, or any other occupational health and safety intervention, we will use the Productivity Assessment Tool. This is the model with which we are most familiar but it may be assumed that all cost-benefit analysis programmes developed in the future will have many of these same features.

The following case study illustrates how to use cost-benefit analysis as an aid to implementing an ergonomics intervention and is derived from Oxenburgh *et al.* (2004b). One salient point is that this intervention was not initiated because of concern over injury but due to concern with other cost factors. A good ergonomics workplace intervention need not only be in response to injury.

Case study 2 — commercial cleaning

The workplace

This case study is a typical example of management increasing productivity via ergonomics improvements. The enterprise is a university where there was considerable pressure to outsource cleaning, it being thought that this would be more cost effective than the in-house cleaning staff.

The building selected to illustrate this case is an office/classroom/laboratory building of about 8,600 square meters (93,000 square feet). The basic methodology employed by the maintenance manager was to measure the area of each space (offices, corridors, toilets, etc.) along with the type of floor coverings and fittings and determine the standard and frequency for cleaning each area. There are standard tables within the industry that give time bases for cleaning offices, toilets and so on with an allowance for rest breaks.

Work organisation

Over a period of some years the work loads of the cleaners had become skewed; on some days the workload was light and on other days heavy. To even out the workload would make better and more effective use of the cleaners' time and, as work overload is a well known risk factor for musculoskeletal injury, to even out the workload would reduce the injury risk.

This is, of course, a relatively low-cost organisational improvement and is accommodated within the normal management cost structure. However, it still takes time to determine new methods, which is time away from other management duties; the cost of management time is included in the case study costing.

Individual cleaning tasks

As in all industries there are fads and fashions; in the cleaning industry small back-pack vacuum cleaners have become popular. This type of vacuum cleaner is very useful for small areas, particularly those with poor access such as staircases, under desks and tiered seating in lecture theatres, but their use had been extended to all areas even those with large expanses of carpet.

In the larger carpeted areas of the building the back-pack vacuum cleaners were replaced by floor model self-propelled vacuum cleaners which were more efficient. Simply by going from a 9″ (23 cm) width head of the back-pack cleaner to 27″ (69 cm) width head of the floor model provided an increased cleaning efficiency and a saving in time of not less than three fold. Removing the back-pack from the backs of the cleaners removed a potential source of injury.

Other widely used equipment were floor polishers. In this building the corridor and laboratory floors were all vinyl tiles requiring frequent polishing. The cheap and popular rotary polisher was slow to use and was replaced by a straight line polisher which, in the model chosen, is three times faster per unit area than the rotary polisher.

Working and exposure time

Previously it took 27 person-hours each day to clean this particular building and, after the work re-organisation and equipment improvements, the time was reduced to 19 person-hours each day; effectively a reduction from four to three cleaners per day.

Although there were clear productivity and cost gains due to the reduction in time required to clean the building, was there an increase in injury risk? After all, three people were now doing the work of four.

Musculoskeletal stress and injury has several causes including load, daily duration of exposure to the load, posture of the individual and so on. Reduction of the time spent at any particular task implies a reduction in exposure and thus a reduction in injury risk. The time the cleaners now spend with a vacuum cleaner on their backs is considerably reduced to only a few difficult-to-clean areas, and similarly the change to the straight line polisher reduced the time spent using the machine by two-thirds. With the organisational changes eliminating overload periods by evening out the daily tasks, the risk of musculoskeletal injury had been reduced considerably.

Cost benefit analysis

Table 2 is a summary of the analysis made by the maintenance manager of the situation before he made the changes (the 'initial case') and the situation after the changes ('improved equipment').

Table 2 Cost Benefit Analysis for the Improved Equipment and Work System

		Initial case	Improved equipment
Number of full-time cleaners		4	3
Paid time per cleaner (38 hour week)	hour/year	1,976	1,976
Wage paid directly to each cleaner	$/hour	14.13	14.13
Employment cost (wages plus overheads of supervision and administration)	$/hour	17.81	17.81
Total employment cost for the cleaners	$/year	140,800	105,600
Intervention costs (cost of equipment and management time)	$/year	—	10,800
Savings (assumed to occur in one year)	$/year	—	35,200
Pay-back period	months	—	< 4

The cost for the intervention was $10,800 which included management time and new equipment. The total employment cost for the cleaners, including wages and some overheads, was reduced from approximately $140,000 to $105,000 per year, with a saving of $35,000 per year. The pay back period was less than four months.

This case study illustrates that, by the application of ergonomics along with good management, it is possible to increase productivity effectively and at the same time reduce the risk of a musculoskeletal injury.

Data needed for a cost-benefit study

After the work area to be examined has been defined wages and other direct costs are relatively easy to come by. Complications may arise due to employee turnover during the year and the use of part-time staff but an estimate may be made by using the average costs during a one-year period. One can also take the proportionate costs of supervision, whether direct or indirect — direct being managers intimately concerned with the day-to-day running of the selected work area and indirect being managers and support staff that may be concerned with the larger enterprise (e.g., human resources, maintenance and computer support staff). Other data, such as overtime, outsourcing and training and especially that related to employee turnover, are usually readily available. If significant, injury costs can also be added.

The costs that are more difficult to assess, yet are often more important from the cost benefit point of view, relate to productivity. Productivity may relate to output, quality, energy costs, machinery maintenance, materials wastage and other factors that may be affected by the proposed intervention. Clearly each case is different and the investigator must be aware of those factors that may be affected by the proposed intervention and ensure that they are assessed.

Although problems of industrial relations can occur for a variety of reasons, many of which may be outside the scope of ergonomics, poor and/or unsafe working conditions may be a contributor. An apparent cause of an industrial relations difficulty is, in fact, often a symptom of a wider problem. For example, complaints from VDU operators about eyestrain through poor lighting, although often the case, may also arise from other sources, e.g., unhappiness at organisational changes. Solving industrial relations problems is difficult enough without being side-tracked into negotiating on the wrong issues.

However, the costs due to poor ergonomics when there is no injury are not so easy to ascertain. Our contention is that such losses may be more extensive and costly than

Table 3 List of Factors that may Relate to an Ergonomics Intervention

Readily available costs	Costs needing investigation
Wages	Deriving the optimum productivity in output
Taxes and insurance payable on wages (social costs)	Re-work and error costs
Paid time not worked (vacation and statutory holidays)	Warranty costs
Paid time not worked (sickness absences)	Losses due to insufficient skills or training
Injury and medical costs	Energy and other utility losses
Overtime	Losses due to unsuitable hand tools and equipment
On-going training	Losses due to poor design of work station layout or design
Direct supervision	Equipment down time due to maintenance
Indirect supervision	Unsuitable work interface (computer software)
Employee turnover costs (administration)	Losses due to employee tiredness (physical and/or mental)
Training for new employees	Mater Materials waste costs
Out-sourcing	
Head office overheads	

even injury losses. Oxenburgh (1991, 2004b) has examples of such losses due to poor ergonomics and one of these examples, commercial cleaning, is reproduced here.

Table 3 illustrates some of the types of data needed. Clearly, the more data that are entered then a more accurate picture will emerge of the actual costs of employment but the essential data are those which will change due to the intervention.

The first column in the table lists the basic costs of employment and the second column lists data mostly relevant to production. Deming (1982) has pointed out that quality of output is a critical factor and wrote:

> A plant was plagued by a huge amount of defective product. "How many people have you on this line for rework of defects made in previous operations?" I asked the manager. He went to the blackboard and put down three people here, four there, and so on — in total, 21 percent of the work force on the line.
>
> Defects are not free. Somebody makes them, and gets paid for making them. On the supposition that it costs as much to correct a defect as to make it in the first place, then 42 percent of his [the managers'] payroll and burden was being spent to make defective items and to repair them.

Quite a lot of the productivity data, if not in the enterprise's records, can be obtained by questioning the supervisors and other employees. These data may not be accurate but it forms the basis of the costing for the intervention. Providing that the management accept the 'loss of productivity' data as relevant, even if not accurate, this is sufficient for adding to the cost side.

Of course, a major difficulty with production is what to measure? In the example quoted from Deming the output was physical items; for productivity one can just count the items produced in a given time and compare that with what the output should be given optimum conditions. But what about service industries? These include the hospitality industry (restaurants, hotels), office work, call centres and so on. Ingenuity may be needed; for example, call centres often use the number of calls made or the number of sales made in a given time as measures of productivity but used alone these figures may be misleading; to these data must be added the factors of absence, turnover and training to give a better measure of the costs of employment and productivity. With such measures the present employment costs and the expected improvements due to the intervention may be costed and compared — the cost benefit approach.

Unfortunately, there is no generic method for measuring productivity. Each work station needs it own measure and this must not be confused with the accounting practice of 'cost centres'. These latter may not relate to the intervention nor to the parameters that need to be measured. The important point is that the intervention must be able to be measured, or at least estimated, and is acceptable to management.

The benefit of this work station or work area approach is that it is specific to the enterprise and does not depend upon nebulous nation-wide figures that may not relate to an actual enterprise.

Conclusions

If ergonomics is as important to industry as ergonomists believe, then the extent to which it is not routinely used must suggest some failing in the way it is presented to management. It would seem reasonable to presume that, as most managers already have plenty of problems, they are unlikely to be enthusiastic when someone turns up telling them they have more, especially if they are of the previously unheard-of ergonomics variety! Given that they already know they have safety problems, health problems, productivity problems and cash flow problems amongst others, it can hardly be surprising if they 'turn a deaf ear' to someone telling them they also have ergonomics problems. Moreover, a manager's job is to solve problems not to go looking for them.

The acceptance of these simple points immediately identifies a new approach to the promotion of ergonomics. First, there are no such things as ergonomics problems; there are however a wide range of problems including safety problems, occupational health problems, productivity problems and labour turnover problems in which ergonomics knowledge may be of assistance. Second, ergonomics must be seen increasingly as the provider of solutions rather than as the identifier of problems. Once it is accepted that the role of ergonomics is to aid in the solution of accepted industrial problem areas, then the development of economic arguments to support and justify the expenditure on ergonomics becomes much easier. The presentation of ergonomics as relevant to accepted problem areas supported with an economic bias immediately gives the ergonomists the advantage of talking the same language as the manager they are trying to convince. The increasing use of economic benefit in support of ergonomics change will in no way undermine the objective of improving health and safety. In many circumstances, it may in fact be the only way to ensure that improvements are implemented.

This chapter has shown the kind of data which can be used to develop a cost-benefit approach to ergonomics and some of the procedures available to utilise such data. All that remains is for the use of economics in ergonomics to become the rule rather than the rare exception. This will hopefully lead to more publications on the economic analysis of ergonomics benefits, which will in turn enable more ergonomists to present their proposals and results in a form of immediate interest to the industries who fund them.

End note

Recently, *Applied Ergonomics* (vol 34, 5, September 2003) had a Special Issue devoted to cost effectiveness.

Beevis, whose 1970 article on costs and benefits is reproduced in this Special Issue, concludes in his new article that:

> ... comparatively few studies provide detailed information on the costs and benefits, or improvements in effectiveness from ergonomics applications. Cost-benefit studies are difficult to conduct for any discipline and it remains very difficult to assign costs and benefits to ergonomics applications.

We would agree that, in general, detailed studies are difficult to perform, and most of the articles in the Special Issue supported the difficulty in accumulating sufficiently detailed data. But one needs to ask how much data is sufficient and how much detail is really needed?

There seems to be a general impression that business looks far ahead and plans work to the last detail. This may be correct in the planning stages of very large scale operations but most work that people do is day-to-day and the vision of management is usually this year's profitability. In the article by Seeley and Marklin a case was made for improving and replacing hand tools in certain manual tasks. They expressed surprise that the pay-back period was only four months yet they still noted that they had difficulty estimating costs and benefits. Did they determine the level of data needed to show their case prior to starting the intervention?

Hendrick's article briefly described several interventions that were cost effective and usually showed this with limited data.

We should get away from the idea that to make a financial case for an ergonomics intervention in industry is an exact science: data will always be limited, and often inexact, but that is the way industry mostly is. We need to work within the limits of the 'clients'.

Unfortunately, none of the articles in the Special Issue proposed a generic method that could be used by ergonomists and others wishing to implement ergonomics solutions. Each author took their case study and showed its financial effectiveness, but the methodology was unique to that system.

Perhaps the most important general conclusion from the Special Issue is that ergonomists need to use a 'business model' when they are contemplating an intervention; discussion of health and safety alone is not sufficient. Ergonomists, just like everyone else, need to illustrate the financial role that they can play in an enterprise. Once this is accepted, the best techniques to suite our arguments will emerge. On the other hand if we do not accept the imperative, we will find ourselves doing little more than complain that 'its too difficult' or that 'it shouldn't apply to us'.

References

Alexander, D.C. (1985). Making ergonomics pay: adopting a business approach. *Industrial Engineering*, July, 32–39.

Annett, J., Duncan, K.D., Stammers, R.B. and Gray, M.J. (1971). *Task Analysis*. Training Information Paper No. 6 (London: HMSO).

Campbell, B. (1993). Towards safe production in the Ontario mining industry. In *Proceedings of MineSafe International 1993*. (Perth, Western Australia: WA Chamber of Mines).

Corlett, E.N. (1988). Cost-benefit analysis of ergonomic and work design changes. In Obourne (ed.) *International Review of Ergonomics* (London: Taylor & Francis), vol. 2, pp. 85–104.

Dahlén, P.G. and Wernersson, S. (1995). Human factors in the economic control of industry. *International Journal of Industrial Ergonomics*, **15**, 215–221.

Deming, W.E. (1982) *Out of the Crisis* (Cambridge: Cambridge University Press), p. 11.

Ergonomics Rule of the Washington State, USA (2002). http://acess.wa.gov and search Ergonomics Rule.

Frick, K. (1997). How organised conservatism prevent managers from seeing the profits of improved ergonomics. IEA, Tampere.

Kupi, E., Liukkonen, P. and Mattila, M. (1993). Staff use of time and company productivity. *Nordisk Ergonomi*, **4**, 9–11.

Maynard, H.B., Stegmarten, G.J. and Schwarb, J.L. (1948). *Methods — Time Measurement* (New York: McGraw-Hill).

Miljödata (1999). *The Potential.* www.miljodata.se.

Morrissey, M. (2002). Assessment of the extent that OHS professionals use financial analysis to evaluate OHS programs. Masters thesis, School of Safety Science (Sydney: University of New South Wales).

Oxenburgh, M. (1991). *Increasing Productivity and Profit Through Health and Safety* (North Ryde, NSW: CCH Australia Ltd).

Oxenburgh, M. and Guldberg, H. (1993) The economic and health effects on introducing a safe manual handling code of practice. *International Journal of Industrial Ergonomics*, **12**, 241–253.

Oxenburgh, M. and Matchbox Software (2004a). *The Productivity Assessment Tool.* www.product Ability.co.uk

Oxenburgh, M., Marlow, P. and Oxenburgh, A. (2004b). *Increasing Productivity and Profit Through Health and Safety*: The Financial Returns from a Safe Working Environment, Second Edition (CRC Press: Boca Raton, Florida and Taylor & Francis London. ISBN 0-415-24331-9).

Quinlan, M., Mayhew, C. and Bohle, P. (2001). The global expansion of precarious employment, work disorganisation and occupational health: a review of the recent literature. *International Journal of Health Services*, **35**, 335–414.

Rimmington, J. (1993). Does health and safety at work pay? *Safety Management*, September, 59–63.

Simpson, G.C. (1993). Applying ergonomics in industry: some lessons from the mining industry. Keynote Address to the Ergonomics Society Annual Conference. In *Contemporary Ergonomics 1993*, edited by E.J. Lovesey (London: Taylor & Francis).

Spilling, S., Eitrheim, J. and Aaras, A. (1986). Cost benefit analysis of work environment investment at STK telephone plant at Kongsvinger. In *The Ergonomics of Working Postures*, edited by E.N. Corlett, J. Wilson and I. Manenica (London: Taylor & Francis), pp. 380–397.

Swain, A.D. and Guttmann, H.E. (1983). *Handbook of Human Reliability Analysis with Emphasis on Nuclear Power Plant Applications.* Report NUREG/CR-1278 (Washington, DC: US Nuclear Regulatory Commission).

Chapter 38

The analysis of organisational processes

C.E. Siemieniuch and M.A. Sinclair

Introduction

This chapter is structured as follows. First there is a detailed description and definition of what a process is and why there is advantage in taking a process viewpoint when attempting to introduce change into an organisation. Then a range of methods for exploring, representing and analysing both existing and evolving processes are presented. Characteristically, all these methods involve two phases: an initial data-capture phase, followed by a formal analysis and representation phase. We discuss sources of detailed information about current processes, the kinds of data that can be elicited and a range of methods for doing so, and then focus more on methods for representing process activities (e.g., Rich Pictures) followed by a range of basic process analysis techniques (e.g., Hierarchical Task Analysis, Process Charts, Petri Nets, etc).

After this we take a step back and consider the whole process of process analysis and the requirement for the techniques described in the two previous sections to fit into a coherent approach. This section provides a brief description of Checkland's Soft Systems Methodologies and short summaries of other potentially useful system development methodologies. The chapter finishes with some solid advice about the truths and practicalities of process analysis, giving the reader some indication of the gaps between theory and reality and a pointer towards some of the hurdles and pitfalls to be avoided.

This chapter considers some of the approaches and techniques identified throughout the book but with the particular focus of organisation and analysis.

Background to the need for process analysis

Why do we need to carry out process analysis?

We need to do this because, after identifying the performance goals of the overall process(es), the set of activities or sub-processes that make up the whole form the foundation for much of the analysis and research that we do as ergonomists, social psychologists, business consultants, and industrial engineers. Once you have identified the process, you can allocate functions, analyse health and safety issues, do job design, identify and develop training and selection procedures, consider communications and the concomitant information technology and telecommunications infrastructural requirements, and worry about team and organisational cultures. Other areas of impact include location and design of workstations, environmental issues, hardware design, user input/output mechanisms and so on.

There is one other strong reason for us to use the process as the basis of our work in organisations, and that is because thinking about organisations as a structured collection of processes has become the received wisdom for all the professional groups involved in organisational design, whether it be factories, hospitals, or transport systems that are the subject of study. Since any innovations or changes to the structure and operations of the organisation will require a team effort from different professionals to bring them about, it is sensible to adopt a common viewpoint or framework from the outset. However, you should be aware that other viewpoints are possible as well; one could view the organisation as a psychosocial driver of human well-being and the quality of life, or as a knowledge engine, or in several other viewpoints. The point here is that a process viewpoint is very useful, but not uniquely valid.

What is a process?

Unfortunately, there are several definitions, depending on whether you are, for example, a lawyer, a computer scientist, or a business manager. We have adopted the business management definition for this chapter, because many of us work in businesses and will try to improve business performance.

A well-recognised, generic definition comes from Davenport (1993):

> A process is 'a structured, measured set of activities designed to produce a specified output for a particular customer or market. It implies a strong emphasis on how work is done within an organisation'.

This definition implies the following characteristics:

- A process begins and ends with customers (they may be internal or external).
- The process is made up of ordered and partially-ordered activities intended to be performed repetitively and either in sequence and/or in parallel (otherwise, it's a one-off project).
- The activities create value for the customers; most often, this is either a wanted product or a needed service.
- The activities are carried out by combinations of technology (machines, software) and people.
- A process involves several organisational units, either within a company or across several companies.
- A process is instantiated by the allocation of goals, resources, responsibilities, and authority. This includes the authority to improve the process.
- Resources typically comprise space, money, machinery, software, communications, people and knowledge.

These characteristics are those of a semi-autonomous entity; only the identification of goals is missing from the list above. Note also that although the organisation (for example a hospital, automotive assembly factory, charity) can usefully be viewed as a collection of processes, the characteristic that a process runs across organisational units implies that a lot of small, localised processes are not recognised as such; they are below the conceptual horison. This does not mean that they are unimportant; but that some sub-processes are best left undefined and in the hands of the localised team, group or community. Those that must be defined (e.g., for safety reasons) can be labelled as procedures. We discuss this later.

There are many classes of processes; for this chapter we define three:

- Operational processes, that deliver products to external customers (e.g., mortgages, automobiles, beans, bills), and bring in the revenue to support the company. Typically, these are very repetitive processes, where possible formalised, standardised, automated and documented
- Support processes, that deliver products to internal customers (e.g., personnel, budgets and expenditures, catering). These provide the replenishment services to the operational services
- Design and development processes, that deliver both new products and new capabilities to make them (e.g., design of a new model car, with designed process, procedures, and people requirements). Where the first two processes are knowledge-utilising processes, this class of process is knowledge-generating.

There are likely to be formal and informal versions of all the above classes of process being used simultaneously within an organisation. Individuals and groups will evolve ways to circumvent difficulties, take short cuts, do things 'better, faster and cheaper' in the firm belief that the formalised version was not meant for either this particular context or for 'people of our experience'. This formal/informal split has relevance to issues raised later and must always be considered when trying to represent and evolve a current 'as is' process.

The next point to make is that processes are generic, until they are instantiated; thus, you can have a Business Process for 'order fulfilment' (in most organisations, this is the key revenue-gaining process), but not have anything other than documents to see. When a manufacturer (such as Ford Motor Corporation) creates a facility to deliver vans and another to deliver automobiles to customers, both of these will be examples of order fulfilment, but there will be differences between them. Each instantiation of the process is then known as a 'business case'.

Realising a 'Business Case'

Some of the basic points are given below:

- A business process is instantiated as a 'business case'. For example, the 'Order fulfilment BP' is instantiated differently for Ford cars and Ford vans.
- A business process must have goals. Usually, these are:

    — Generation of 'profit' (for non-profit organisations, this may be goodwill, or something similar)
    — Identification and maintenance of the customer base (to make sure that the process answers their needs, and that it continues to do so)
    — Acquisition of new customers
    — Delivery of value: on time, in full, to cost and to expected quality (note that these may lead to customer satisfaction; in this new century it will be customer delight that is the goal; giving the customer more than was expected for the price – preferably something novel)
    — Efficient and effective operations for delivery
    — Satisfaction and enthusiasm of its stakeholders (not just the customers, but everyone who has a 'stake' in the process – operators, maintainers, managers, and so on).

- A business process must be interfaced, resourced, and managed within an environment (in other words, business processes are 'open' systems; we consider this in the next section).
- A business process must be capable of self-improvement ('autopoiesis').

Clearly, then, process analysis is a significant activity to be carried out early in any definition of a business process, and with a considerable demand for wisdom, skill and effort. The effort must come from within you; the wisdom and skill will have to be learnt. Remember that both of the latter include learning from mistakes. You will make many of these; the important thing is to be quick to find and recover them. You will do this best by being open about your work, and listening to what people say about it.

The environment of a business process, and the effects of environmental complexity on processes

As stated above, business processes and their instantiations as business cases are open systems, and this applies to the sub-processes within them. Since they are affected by and are responsive to their environments, it is necessary to consider some characteristics of the environment and its effect on processes.

The most important characteristic of the environment is that it is non-stationary. It is continually changing, as customer demands change, as other suppliers move in and out of the market, and as the company itself is re-organised with changes to strategy, policies, skill bases and capability distribution. This constitutes a 'complex' environment, and it is now necessary to consider some aspects of complexity because of its effects on processes.

Complexity has become of great interest in business circles for its strategic implications, though it has been discussed for many years by many other authors, usually under different titles (e.g., Ashby, 1962; Becker, 1992). Since then, work by researchers associated with the Santa Fé Institute (e.g., Kauffman, 1989; Allen, 1998) has inspired many authors to analyse organisations from this perspective (e.g., Waldrop, 1992; Marion, 1999; Rycroft and Kash, 1999; Sawhney and Prandelli, 2001). A set of characteristics that are believed to cause complex behaviour in enterprise systems have emerged from the studies above. Some of these are:

- Many processes, with lots of agents of different kinds (e.g., software, individuals, teams).
- Some degree of behavioural autonomy (i.e., the processes and agents can elect to change their behaviour).
- Multiple steady states for agents (e.g., the individual in the separated roles of machine expert, team leader, and holiday-maker; for a process, it could be in any state from stripped down for maintenance, or it could be producing different products at different times).
- Lots of interactions between agents and processes in an environment (e.g., reaching design decisions with engineers at different sites).
- Effects of co-evolving processes and people within them (e.g., two interacting agents change their behaviours as a result of training and changing external pressures, and as a result become either more co-operative or, sometimes, more challenging to each other).

If several of these are present, then as the theory predicts and practical experience supports, the overall behaviour of the business process is likely to be unpredictable. This unpredictability may be beneficial, in that it opens a window of opportunity for the process, or it may be harmful. This has far-reaching consequences; the first and most important conclusion is that it is no longer sensible to believe that the extended enterprise, the organisations that comprise it, and the processes which the extended enterprise executes, will operate 'normally', and that the role of management is to restore the systems to normality. Instead, 'normality' will require unremitting effort for its maintenance (Gregg, 1996; Rycroft and Kash, 1999). This requires a clear understanding of its processes. In turn, this provides the basis for near-continuous measurement, planning and control of processes and of the organisational structures that resource and support them in order to maintain organisational homeostasis (seen as organisational predictability) and a semblance of coherence for periods of time. Furthermore, homeostasis and coherence are likely to be sustainable only for short periods, followed by rapid change; an example is in the automotive industry, where new models entail drastic changes to the assembly processes every five years or so.

A second conclusion is that it is unlikely that centralised control and adaptation will be a successful strategy, given co-evolution and parallelism among the agents and the environment, and that the adoption of a distributed, self-organising systems philosophy is probably the only strategy likely to be successful. However, the main problem that such an approach offers is that of maintaining organisational coherence; as one senior manager in the automotive industry (Cullen, 1995) expressed it, reporting recent experience: 'When you blur management boundaries, you had better be very clear about the process, otherwise you will soon be in deep trouble'.

The nature of the change in the competitive environment has some other effects, too. As Martin (1998) has pointed out, mass customisation is the accepted way forward for products. What has prevented mass customisation from happening earlier was a lack of knowledge about the customer, and an inability swiftly to transfer and utilise information about the customer — as Stewart (1997) elegantly expressed it, we have now moved from a time of information deserts, to a time of information floods. This increase in information has enabled more complex processes (because of the greater ease of technology diffusion), delivering higher quality goods more swiftly to customers (because of more widespread competition). Furthermore, the use of the internet for business-to-business and for business-to-customer transactions ('e-business') is leading to the end of the bundled product. Products are bundled for two reasons; lack of knowledge about the customer (necessitating a blunderbuss, 'one size fits all' approach), and the need to cross-subsidise products. e-Business can expose these cross-subsidies, by concentrating cheaply on provision of one of the products (because information costs are heavily reduced), and by customising the service, at minimal cost. This kind of attack on ones' processes can only be fought off by having a deep understanding of these processes, and making the necessary changes before one's competitors can mount their attack.

Processes as essential components of organisational learning and good governance

In the discussion above, we concentrated on operational aspects of processes. Here, we discuss some control aspects. Firstly, when you document a process as a result of process analysis, you capture knowledge about the process and how it works, the skills and knowledge that people need to operate the process, health and safety issues, and, if the analysis has been carried far enough, how to retrieve errors in the process. This constitutes core organisational learning, that can be used to train new people, and which can be cannibalised for new processes. Given that all processes can be improved—more efficient, less costly, quicker—the current process can be used as a baseline, against which the enhancements can be measured, in an incremental exercise in enhancing 'process maturity'. And, if the process is 'world class', it can be used as a benchmark in the design of new

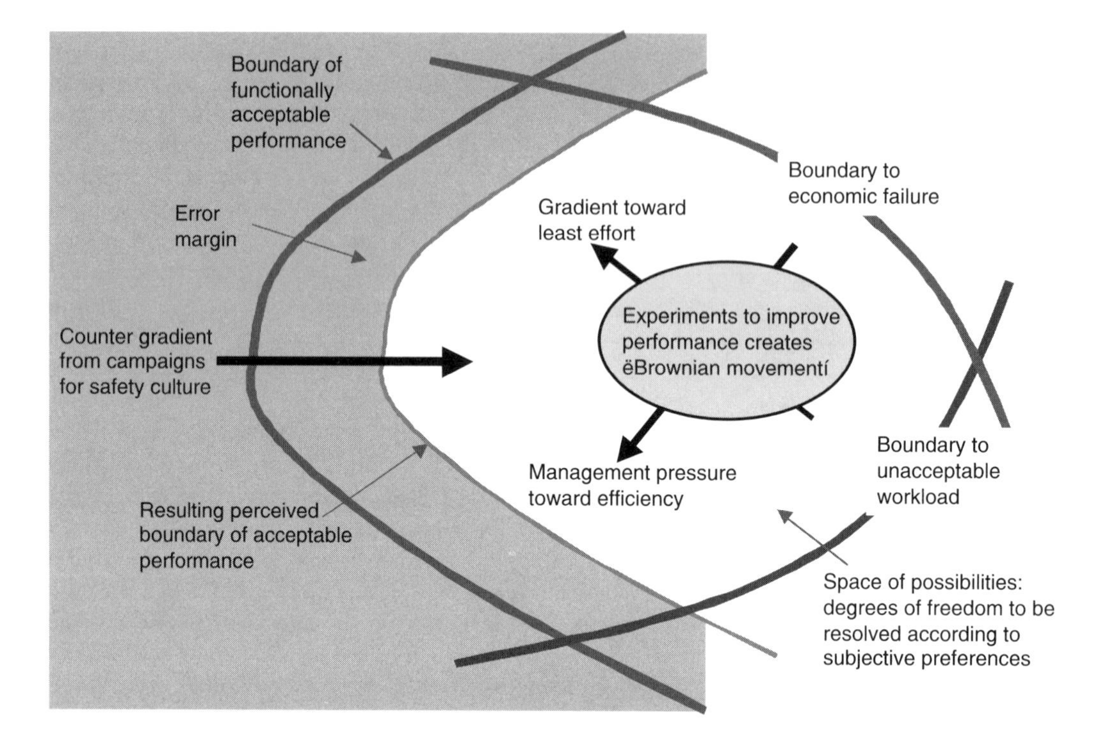

Figure 1 The 'drift to disaster' in systems and processes. In the centre is the (designed) envelope of safe performance. Clockwise from the top, the boundaries are economic performance (which can move, depending on competition), unacceptable workload (moves, for example due to 'down-sizing'), and disasters. The latter has an 'error margin', where near-misses and other events happen. Typically, systems/processes drift to the left over a period of years, due to management pressures and to the development of 'short cuts' to ease the workload. The counter-measure, according to Rasmussen, is safety campaigns and the development of a 'safety culture'. From Rasmussen (1997, 2000).

processes. We are discussing here knowledge management which is too big an issue for this chapter; further discussions will be found in Prahalad and Hamel (1990); Senge (1992); Stewart (1997); Morton (1998); Siemieniuch and Sinclair (2000, 2002).

In turn, knowledge of the processes and the management of this knowledge is a key requirement for avoiding the 'drift to disaster', as exemplified by Rasmussen (1997, 2000), in Figure 1.

However, the development and maintenance of a safety culture depends critically on 'good governance', as outlined in Turnbull (1999). Other, international, sources about governance will be found at:

http://www.oecd.org/topic/0,2686,en_2649_37439_1_1_1_1_37439,00.html
http://www.worldbank.org/html/fpd/privatesector/cg/
http://www.icgn.org/

Three of the key questions (there are many more) for good governance are:

- Have the processes been engineered as 'best in class'?
- Are the processes operated as 'best in class'?
- Does the company have processes for process auditing; capability acquisition; change management?

It will be seen immediately that satisfactory answers to these will depend on process knowledge and management of that knowledge, as outlined, for example, in Siemieniuch and Sinclair (1999, 2002). And that knowledge must be gathered and systematised by people like you, having the proficiency to do so.

Exploring and capturing process data

Questions that process analysis tries to address

The questions that one usually tries to answer are all related to the issues discussed above, but are usually phrased much more simply:

- *What's going on?* This is partly an exploration question; it is often surprising to outsiders how little an organisation knows about what really happens within it as a process is executed, though there may be many documents stating what ought to happen. It is also partly a question about process measurement, in order to control the process, and to establish the necessities for operators to have optimal 'situation awareness' (Endsley and Kiris, 1995; Endsley and Kaber, 1997; Endsley, 1998; Sandom, 2001).
- *What can be formalised?* This question is to do with knowledge capture for training, process re-design, and for enhancing process and product maturity (see, e.g., March, 1986; Humphrey, 1989; Curtis, Hefley *et al.*, 2002).
- *Who does what?* Typically, this question is concerned with organisational structures and roles, the allocation of functions or activities to these roles (whether human or machine) and, especially, the interfaces between them. This is a basic step in Human Reliability Assessment, concerned with identifying responsibilities and authority—see Chapter 32.
- *Where are the problems?* Process problems may be due to variability in the products with which the process must deal and which are necessitated by mass customisation, stoppages due to quality problems, process sensitivities due to a lack of process maturity, bottlenecks arising from product enhancements that weren't originally planned, or any of a host of other causes. The intention here is usually to improve performance, as is indicated in the Japanese 'just in time' philosophy.
- *Who's in charge of what?* This is a precursor to human reliability assessment and to process improvement, usually intended to be achieved by reorganisation of the people involved in the process, and usually featuring more automation and a reduction in the numbers of people employed. Issues of autonomy, the distribution of responsibility and authority, workload, team size and structure, selection and training based on the skills needed, are addressed here.
- *Who is talking to whom, and about what?* The purpose is usually to identify requirements for communications infrastructures and applications, but also to identify weaknesses in process designs (especially when things are not running smoothly), and to understand some of the sociotechnical aspects of process control.

Methods for capturing data about processes

The methods discussed below are all general methods, often adapted from their original purposes. Only brief descriptions are discussed; the reader is invited to use the references for further information. Good compendia of information are Kirwan and Ainsworth (1992); Hunt (1996); and Salvendy (1997), in addition to the contents of this book.

Characteristically, all the methods involve two phases; an initial data-capture phase (discussed in this section), followed by a formal analysis and representation phase (discussed in the next section). For this reason, the discussion that follows has been structured to outline what data are captured, sources for these data, some general issues in collection and analysis, and then brief discussions of the various representation methods that utilise these data.

The basic information to be captured is shown in Table 1.

Table 1 Basic Information to be Captured in Process Analysis. These Classes will Usually be Sufficient for One Person/One Workstation/One Process Analyses. More Complicated Scenarios (e.g., a Process Executed by a Team) will Require More Classes

Type of information	Definition	Examples
Process tasks	The activities that must be undertaken to achieve the goal. Usually these can be arranged both hierarchically, and sequentially.	Exit the room: Open door; Walk through doorway; Close door
Controls	That which initiates, or otherwise affects, the performance of the activity	'Start'; 'Stop', 'Pause while . . .', 'Repeat until . . .'
Inputs	What the task transforms	A piece of paper on which an invoice is to be printed; a block of metal which is to be machined into a component; data to be converted into a graph; electricity.
Outputs	The result of the activity	The invoice; the component; scrap metal; heat.
Resources	What is needed to enable the task to be performed. There is an implication that resources are unchanged by incorporation in the task; otherwise, they are inputs	Knowledge; people; tools; schedules of work; plans; IT infrastructures; work surfaces; control interfaces
Location	Geographical position where the activity takes place	Shopfloor; office; 'left hand does this here, and right hand does that there';
Communications	Signals between activities or people that co-ordinate the execution of the process.	Messages; acknowledgements; queries; timing information.
Constraints	Exigencies of the process and its environment that affect the conduct of the process	'Wheelchair access is required'; 'Power cuts are frequent'; 'Whole-body protective clothing must be worn'.
Strategy, which dictates the goals for the process	Senior managers, formal Vision Statements, engineering resource plans, etc.	Importance of this information is proportional to the time period for which your final analysis is expected to be relevant.
Operational policies	Senior managers, policy documents, notices to employees, etc.	Policies often conflict, and may be uneven in their application

(*Continued*)

Table 1 Continued

Type of information	Definition	Examples
Standards	Internal: design standards; process engineering standards; product engineering standards External: Supplier proprietary standards, industry standards, national standards, international standards	Note that there may be several conflicting sets, if the process serves customers in different nations.
Process constraints	Engineering design, process engineering, health & safety, facilities management, process operation, distribution	Some of these will be documented (e.g., 'do not run this machine faster than...'); others will be experiential (e.g., 'do those before lunch, if you want them posted today')
Standard process instructions	Documents, line management, designers, supervisors, operators	Statements of what needs to be done in order to deliver a product. May not have sufficient detail, and may not reflect current practice.
Standard operating instructions	Documents, line management, designers, supervisors, operators	Statements about how to operate particular machines, applications, tools, etc. May be out of date, and may not be correct.
Process characteristics and behaviour	Operators, supervisors, managers	Tends to be held in human heads, not in documents.
Process failures	Maintenance records, Health & Safety records	Records may not be in a useful format for your purposes, nor record the detail you need
Process quality	Process quality records, notes of Quality Circle meetings, 'Voice of the customer' projects	The numbers are important; even more so, the interpretations of these
Process limitations	People, case study references, anecdotes, design documents	Many undocumented and with individual perspectives

General sources of this information are given below. It will be seen that a very wide range of sources can provide useful information; none of them will be sufficient by itself. It is always good practice to utilise many sources, and to accept that all of them will contain errors of one sort or another and that some may reflect the 'informal' or 'personalised' version rather than the formal documented one. It is also necessary to understand that nothing in processes stays the same over longer periods of time; the organisation's people come and go, society values and expectations change, competitors keep competing, and systems change incrementally. Consequently, whatever data are collected, their provenance must be assessed before usage.

Typical sources of data are shown in Table 2.

Actually collecting the data that you need will obviously require a number of different methods. The usual methods are shown in Table 3; however, unorthodox methods may be very useful too. An example comes from Webb *et al.* (1966); in a museum, they wanted to know which exhibits appealed to children of different age groups. In those days, all the exhibits were kept in glass cabinets. So, at the end of the day they went round and counted the number of nose prints on the glass, and measured how high they were. This provided information of sufficient accuracy to enable them to redesign the layout.

Table 2 Sources of Information for Process Analysis. Note the Number of Different Sources, and hence the Need for Using Many Data Capture Methods

Organisational structure	Line management documents, human resource management documents	Formal statements regarding responsibilities and reporting structures, but may not match what actually happens
Organisational culture and ways of working: 'the way we actually do things here'	People	Almost certainly undocumented, and the biggest determinant of actual process operation — linked to quality, failures and limitations
Benchmarks	Can be any combination of the above	A crucial source of information. Other non-competitors in other sectors may have similar processes, and may be best-in-class, with whom you can share knowledge. Competitors may share information as well; for example, in the pharmaceutical industry, competitors share information about the design of pilot plants

Analysis and representation of process data

Process representation techniques

Having used a variety of methods to capture the data, there remains the problem of extracting meaning from them. An important step here is cross-checking your data, as indicated in Table 3 above. Frequently, this will involve returning to the people and checking the implications of the data, and getting more in-depth explanations.

Assembling the many kinds of data is difficult. Clearly, it could all be held in relational databases, and applications used to extract meaning. However, one is often left with the problem of the provenance of the conclusions that are presented. Unless the process is significantly complicated, it may be best to use a manual method, using a large wall and felt-tipped pens (covering the wall with paper first is usually a good idea!). The particular advantage of the manual method is that the wall serves as a memory surface to extend your working memory, and is a surface common to all viewers. One can stand back and see the whole picture; equally, one can stand close and look at detail. While computer methods have other advantages, the restrictions of screen size mean that one has only a small porthole through which to study a large field of data.

One particular technique that is very useful here is the 'Rich Picture', discussed next. This is often used in conjunction with the techniques discussed below.

Rich pictures

As the name suggests, these are pictures rich in detail, and can be conjectured as attempts to marshal the data collected above. As such, they are 'memory surfaces', enabling both you and the stakeholders to comprehend and discuss the problem and its issues, and to wonder about solutions. Rich pictures originated in Checkland's Soft Systems Methodology (SSM) (Checkland, 1981, 1995; Checkland and Scholes, 1990), which is a comprehensive, generic approach to understanding processes and their environments, and designing solutions to the perceived problems (usually software solutions to IT issues). The methodology is discussed later.

Table 3 Classes of Methods for Data Capture.

Method	Comments
Literature searching	Useful for old records, paper records, library information, and benchmarking. Records may not be in a suitable structure, and may omit important information for your purposes. What you find in the professional and technical literature is likely to have been refereed, and is likely to be trustworthy.
Web-based searches	Search should involve both the intranet and internet. Access to a good search engine is vital. What you find may not be trustworthy or directly relevant to the context of interest.
Interviews with stakeholders	Main source of undocumented information. It is always wise to interview as many people as possible, to discover alternative viewpoints, identify errors, and to corroborate what you have been told or discovered by other methods.
Surveys	Only useful for benchmarking, for process snapshots, and for establishing baseline performance indices.
Observation	Many different classes of observational techniques; all of them need practice. Offer a bottom-up view of the process. What you see is physical; provides little guidance on cognitive aspects of the process.
Traceability studies	Essentially involves following a product, document, data or decision through the process of interest. Allows you to check that what is supposed to happen does happen, but may be subject to one-off events.
Simulations or storyboards	May be the only way to analyse new processes. Techniques can range from virtual reality, 'wizard of Oz' simulations, mock-ups using cardboard boxes, paper based models etc.

In Checkland's SSM, he introduced the acronym, CATWOE embracing customers/clients, actors, transformations, world views, owners and environment (see later).

Rich Pictures should attempt to include all of these. Usually, one starts with a large circle, representing the boundary of the system and the problem. One then identifies the people involved, their relationships and their activities; then the assumptions and environmental effects, by placing bubbles or boxes on the diagram, joined by labelled arrows.

We present two examples of Rich Pictures in Figures 2 and 3, and both biased towards manufacturing industry (see also Chapter 10). It will be seen immediately how convenient such diagrams are, both to arrange one's own thinking, and to convey the complexities of real-life processes to others. What is not immediately obvious is that different people will draw different diagrams, and it is in the reconciliation of these different views of the world that much effort will be expended, and on which the quality of the final solution will depend.

Having assembled one's data and extracted meaning, it is now possible to use formal methods to represent the data. A particular advantage of using these methods is that they are rule-based; these rules allow internal checks on the coherence of the process that you describe. A wide range of methods exist and we consider only a few of these in this chapter.

Basic process analysis techniques

These are all predicated on the notion of hierarchical task analysis (Chapter 6), but with different emphases. They may be undertaken either when a process is being designed, or when some aspect of an existing process is being documented. Where they differ is in the

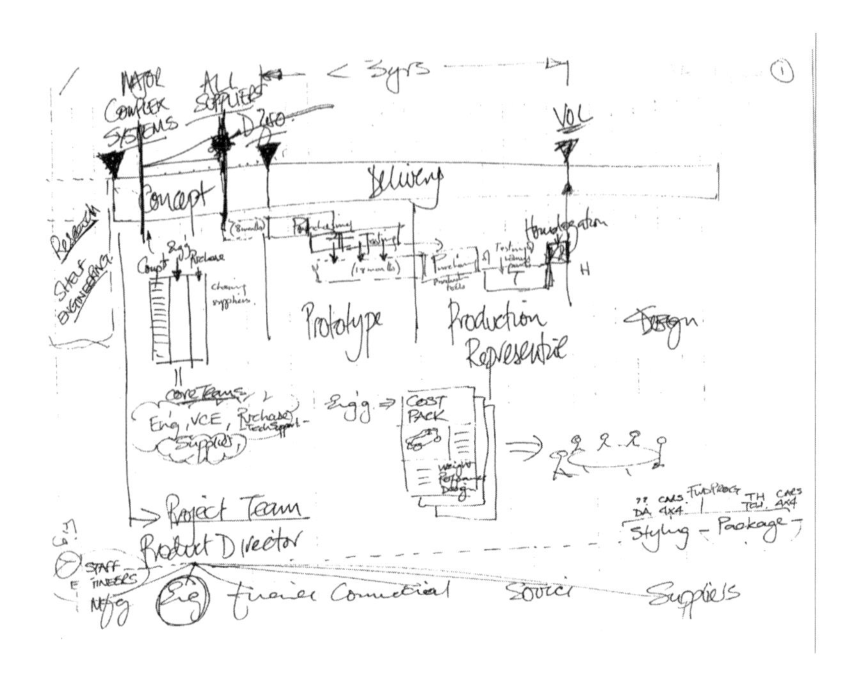

Figure 2 Rich picture of automotive design and development. This was created by the Project Chief Engineer of a project which produced a very successful vehicle in three years from concept. It is a high-level diagram, showing the core process, and how this process is put into effect.

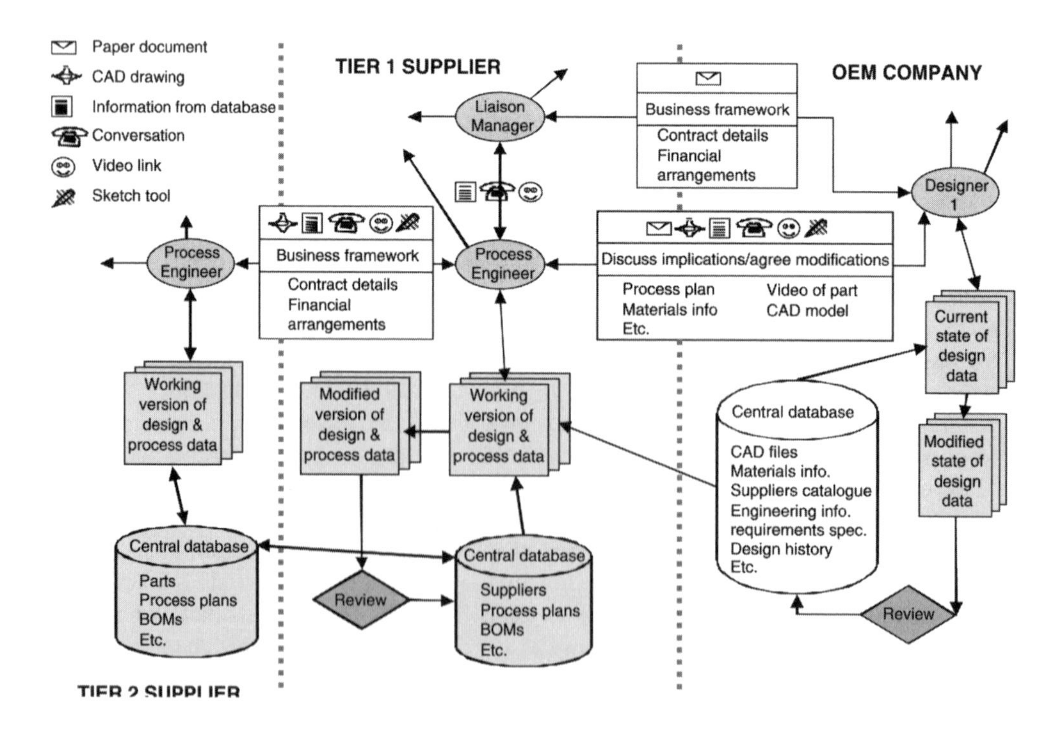

Figure 3 A more formal version of a Rich Picture, based partly on Figure 2. This is part of a much larger diagram, showing how a design issue in the OEM company might be solved in a supply chain distributed across Europe. The focus was on the nature of the communications between the people involved, and the support they would need.

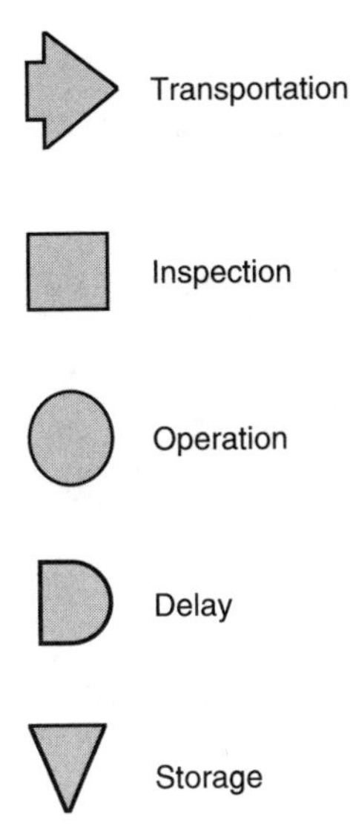

Figure 4 The five elements from which processes can be constructed. Definitions of these elements, and a full set of rules for their combination, will be found in ASME (1972); and in ISO 5807 1985 Information processing — documentation symbols and conventions for data, programme and system flowcharts, programme network charts and system resource charts. Other sources are Kirwan and Ainsworth (1992) and Salvendy (1997).

presentation for the user of the basic information that has been captured; what they capture is very similar from technique to technique. The other common characteristic is that time is an important variable in the presentation; usually running from left to right, or top to bottom.

Only a few of the common techniques are discussed below, out of a huge range, for reasons of space. We discuss process charts as being the simplest technique, followed by three widely-used similar techniques (HTA, $IDEF_0$ and Petri Nets). This is to demonstrate how different representations emphasise different aspects of the data collected. Then we discuss Link Analysis, Influence Diagrams, and Scenario Analyses as additional techniques for shedding light on processes.

Process charts

A good example of these is the basic process chart, with five types of task elements, as shown in Figure 4. These can be assembled as shown in Figure 5 to provide a time-sequenced process (time from top to bottom, represented by the black line)—in this case, a high-level chart of your future.

These charts have two particular advantages:

- They are easily understood by all stakeholders.
- They show clearly where effort could be placed to improve efficiency ('Why do we have so many delays/transportations/storages?').

The main disadvantages of the method are:

- It does not differentiate between human and machine activities (though this can be overcome by splitting the page into columns and moving the symbols into the appropriate columns).
- It does not link in resources for the activities, nor information flows, nor controls over them. This is a fairly major drawback.

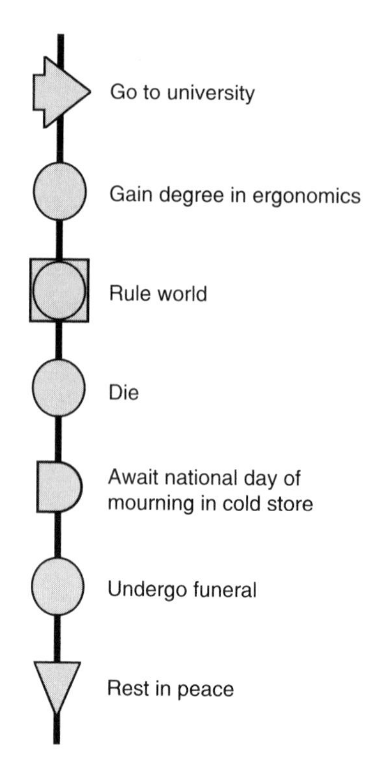

Figure 5 Example of flow process chart. This is a high level example; a similar chart could be created for making a cup of coffee. Different charts can be merged, for example by a Transportation and an Operation ('Assemble X to Y'); they can also be split, for example by an Inspection.

Hierarchical task analysis (HTA)

This technique unfortunately has the same name as a whole class of techniques that involve the decomposition of activities. Good descriptions of it will be found in Kirwan and Ainsworth (1992), Shepherd (1986) and in Chapter 6. It is widely used by ergonomists, but is not common among engineers. An example is given in Figure 6.

Advantages of the method are:

- The method is very clear in its structure, distinguishing between goals, plans and activities, and having a strong, logical layout. This provides a good check on the data that has been collected, enabling the analyst to see quickly where there are omissions or insufficiencies.
- It is also possible to move on to training needs analyses and other analyses from the representation.

However, there are disadvantages:

- Once one adds resources, inputs and outputs to the representation, it does become cluttered.
- The representations usually require some explanation before the stakeholders will feel comfortable with them; they are not immediately intuitive as are other representations.
- In particular, engineers (who do more to change our world than most professions) are not familiar with this methodology, and usually prefer more common, engineering representations.

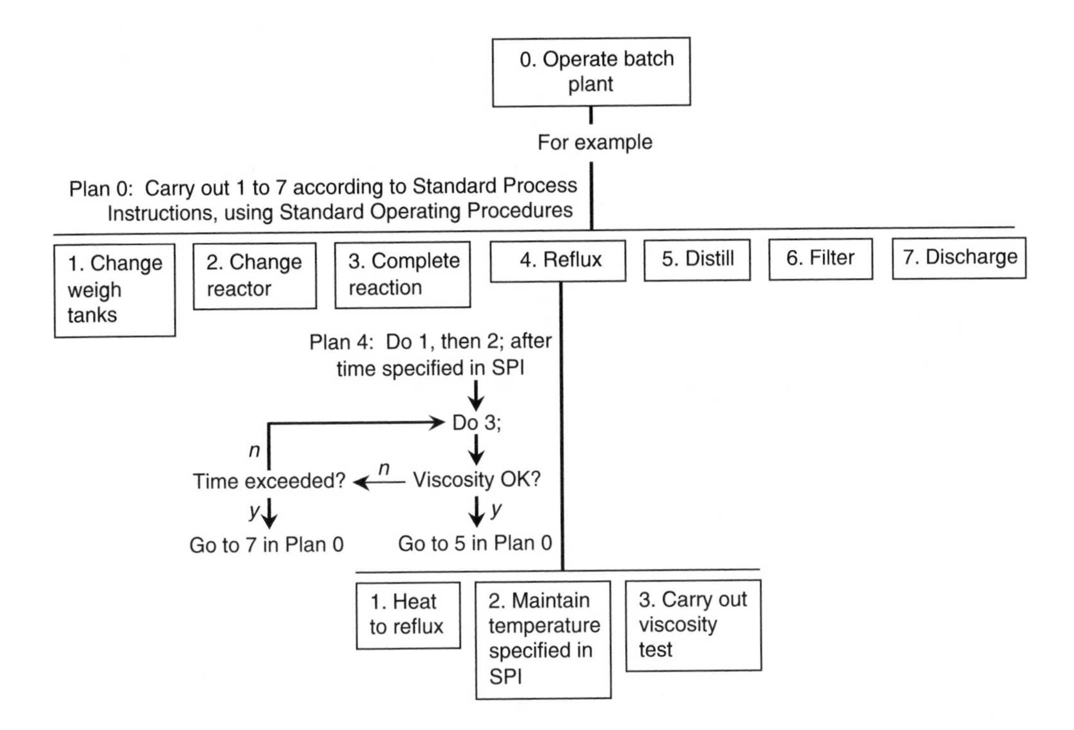

Figure 6 Example of Hierarchical Task Analysis, redrawn with some textual alterations, from Kirwan and Ainsworth (1992). This shows three levels of decomposition; and 'Plans' for executing the tasks, with decision points. Time sequencing is in the plans. The hierarchy of tasks is from top to bottom.

IDEF₀

'Integrated DEFinition' is an integrated suite of methods developed by the US Air Force 'Integrated Computer-Aided Manufacturing' programme, and has become a widely-used technique among engineers and other professions. The suite of techniques include activity modelling ($IDEF_0$), information modelling ($IDEF_1$), data modelling ($IDEF_{1x}$), dynamic modelling for simulation ($IDEF_2$), computerised process description ($IDEF_3$), object oriented design ($IDEF_4$) and ontology capture ($IDEF_5$). It is aimed at software development, and as a consequence most non-software-engineering users rely on $IDEF_0$.and pay little attention to the other techniques. The method is described fully in IDEF (1993) and IDEF (1981).

The core of the method is the 'activity-in-a-box', shown in Figure 7.

It is evident that the key aspects of any process can captured in this diagram. These units may then be aggregated to comprise a process, as shown in Figure 8. It should be noted how outputs from one activity can be inputs, controls, or resources for another activity.

Some of the advantages of $IDEF_0$ are:

- It is a widely-used technique, known among many professions.
- It emphasises the interconnectedness of activities, information flows and control flows, and makes one think about resources.
- There is a temporal flow from left to right, enabling iterations to be shown.
- The method is very clear in its structure, with rules for its construction of diagrams. Like HTA, this provides a good check on the data that have been collected, enabling the analyst to see quickly where there are omissions or insufficiencies.

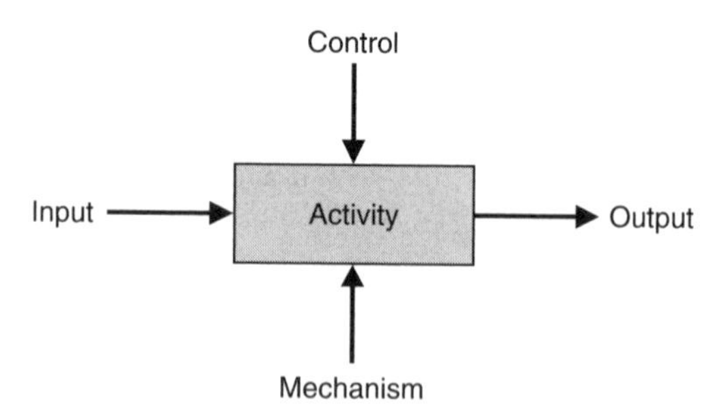

Figure 7 Basic units of $IDEF_0$ Activities are contained in boxes; the top surface is the control interface, left and right sides are for inputs and outputs, and the bottom surface is for mechanisms or resources to perform the activity.

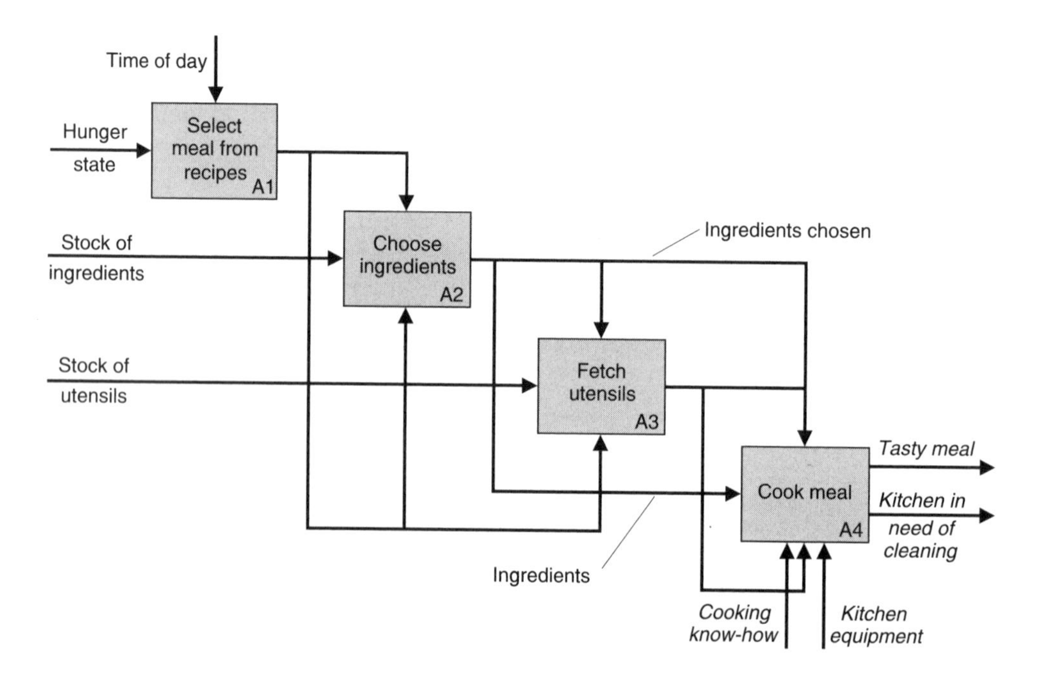

Figure 8 Example of IDEF diagramming. This is a hight-level diagram, the boxes of which could be decomposed in other IDEF diagrams. The diagonal arrangement of activities is part of the technique. Demonstrated here is the emphasis on inputs, outputs, controls and resources. Also evident are the interfaces to the rest of the world.

Some disadvantages are:

- The diagrams rapidly become cluttered with lines, and can become confusing. There are rules for how much to put on a page, but if one follows these one may have to track a process across many pages. Knowing what to aggregate and what to leave alone is an important skill to develop.
- Because of the rule-bound nature of the technique, the IDEF priesthood can always find reasons to argue with your diagrams. This can waste a lot of time.

Comparing $IDEF_0$ with HTA, the difference in emphasis will be obvious, although the kinds of data captured, and the techniques used to capture the data for these, will usually be the

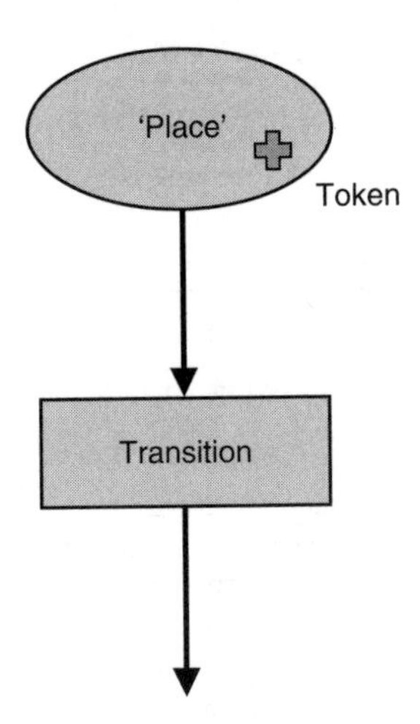

Figure 9 Basic elements of a Petri net — a circle, also known as a place, to represent the state of an element of a system; a square, to represent a transition from one state to another, and a token, which moves from circle to circle, undergoing transformations.

same. The point being made here is that you should select the representation method to fit the needs of the problem, and the needs of its stakeholders.

Petri nets

Petri nets (and their extension, coloured Petri nets), are a very flexible representation technique, developed in the software engineering domain. A good description of these will be found in Kristensen *et al.* (1998), with a more succinct version in Kirwan and Ainsworth (1992). Figure 9 shows the basic elements of Petri nets. The ellipse ('circle') represents the state of an element of a system, and the oblong ('square') represents a transition from that state to a new state (which would be in another circle). For example, the initial state might be 'Fly is alive', the transition might be 'Fly and swatter make contact', and the subsequent state would be 'Fly is dead'. This would be a terminal state, both for the Petri net and the fly.

Hence, the places (circles) model the various possible states of the system element. Depending on where you choose to place the problem boundary, they may also model the states of a set of system elements, or even the whole system, including its human operators.

The actual state of the system is determined by the tokens, which are passed from place to place, undergoing transformations as they go. Figure 10 shows an example of a Petri net, with tokens. The state of the system is at start-up, with the machine slots all empty (all tokens in the 'empty slots' place). Likewise, there are four fixtures with parts ready for processing in the buffer. The process operator is available, too. As work happens, it can be shown by the movement of the tokens from place to place. Note that for any transition to happen, the places feeding it must have suitable tokens available, and the output places must be able to accept tokens.

Coloured Petri nets are an extension to Petri nets to allow discrimination between classes of tokens, places and transitions. For example, 'blue' tokens might only undergo transformation by 'blue' transitions, and so on.

Petri nets are well-known and much used in software engineering (see, for example Kristensen *et al.* (1998). As a result, there are many applications available for modelling Petri nets, and for performing a host of subsequent analyses; validity, timings, logical integrity of the model, etc.

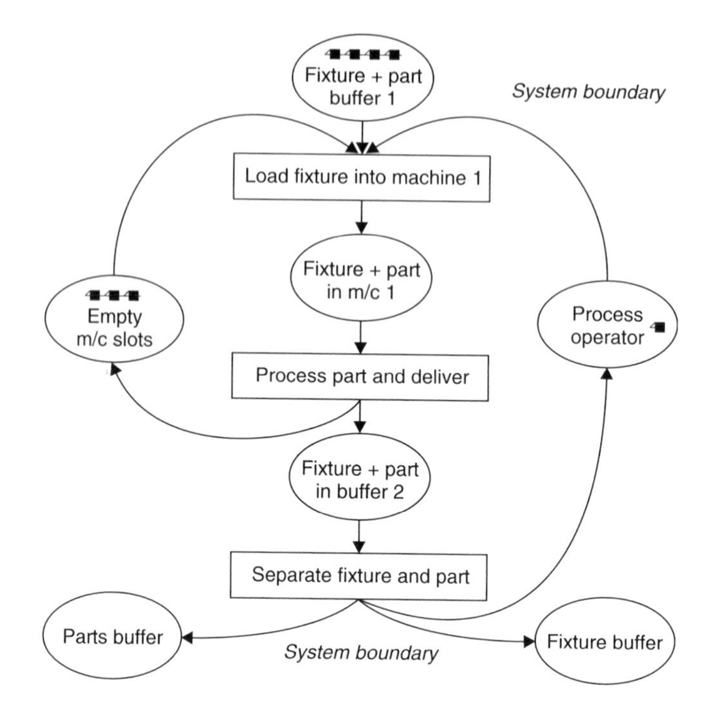

Figure 10 Example of a Petri net for a system comprising a Process Operator (PO) and a semi-automatic machine. Elsewhere, parts to be machined are affixed to fixtures, and are placed in a buffer (at top of diagram). The PO loads them into the machine when empty slots are available. When the parts emerge, the PO separates the parts from the fixtures, and places them in buffers, to go elsewhere in the process. The tokens show the state at the start of the shift; the three machine slots are empty, and four loaded fixtures are available. The PO is available to start work. Because all the places associated with the first transition have tokens, work can happen, and it duly starts.

Advantages of the technique are:

- The logic of the technique has been exhaustively elaborated and evaluated.
- It provides a means of capturing both states and transitions between them in one technique.
- It is very flexible in use.
- There is a host of well-documented analyses and tools available.
- It is widely-known, especially in the software domain.

And its disadvantages are:

- It is time-consuming to apply.
- The diagrams can become very complicated.
- While sound logical diagrams can be created, they may not match reality. This depends on the skills of the analyst.
- The diagrams are not easily interpreted by lay stakeholders.

Link analysis

This is a simple technique, of considerable value in studies of process lay-outs, efficiency and error-reduction. As the name suggests, it emphasises links between activities, using the principle that actions or activities that follow one another should be closely linked in space and/or time. The method is described in Chapanis (1965) and Kirwan and Ainsworth

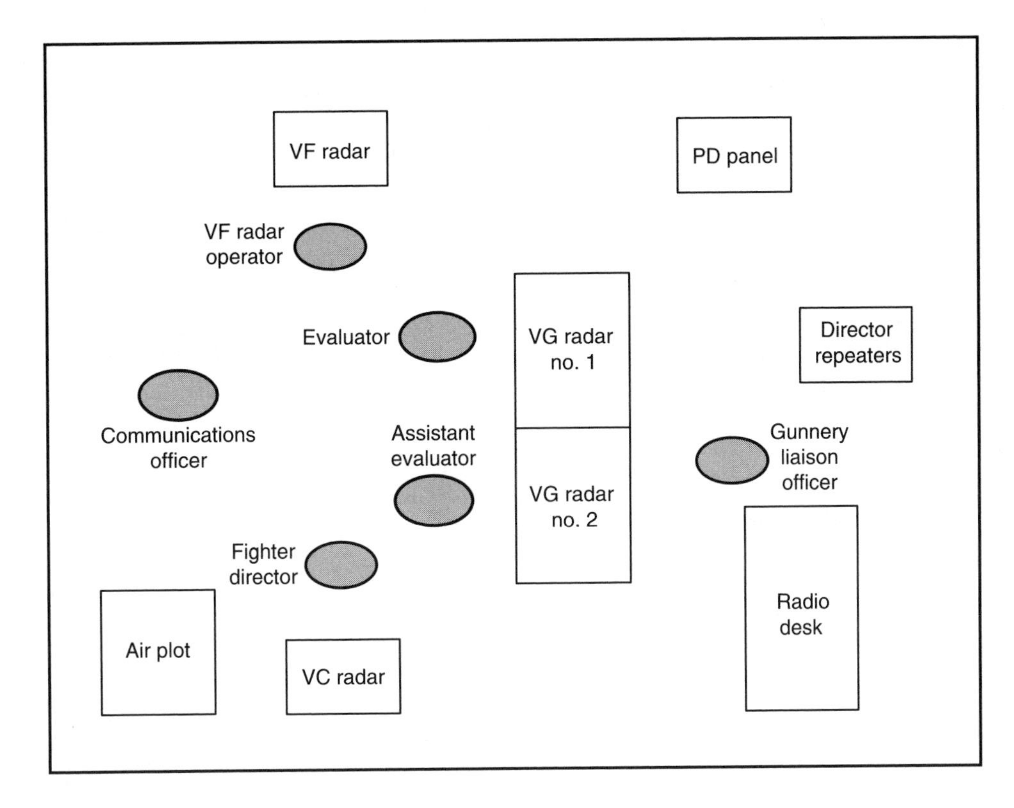

Figure 11 Plan of the CIC, showing usual positions of the decision-makers in the room.

(1992) — the latter being much more accessible. First, one must gather the data, as indicated earlier in this chapter, paying particular attention to sequential information. The question, 'and what happens after this?' is important. However, it is wise to verify this by direct observations; people are not always good at describing what they do.

The example is redrawn from a study by Chapanis for the layout of a Combat Information Centre on an aircraft carrier in the 1950s. Figure 11 shows the layout of this control room, and of the tactical decision makers within it (there were many others, operating the individual consoles).

Table 4 shows the 'Link Matrix', derived from the 'who talks to who' information that was gathered during an observation period.

From this it is possible to plot the links onto the plan, as shown in Figure 12.

It is now possible to rearrange the layout of the room, to improve the communications, as shown in Figure 13.

Influence diagrams

These are described in some detail in Kirwan and Ainsworth (1992). The role of these diagrams is to discover the influencing factors for a given behaviour within a process, and to show this graphically, as in Figure 14 below. Note that this approach is based on a philosophy of cause-and-effect for behaviour, which, while appropriate for behaviour over a short time period, is likely to be inappropriate for longer-term periods of behaviour, where feedback loops may be significant in altering behaviour. If feedback is significant, then a different approach ought to be adopted.

The technique is based on the judgements of 'experts' (i.e., professional experts; people with considerable experience of the behaviour; or people who have the best available

Table 4 A Link Matrix. Entries in the Table Indicate the Number of 'Significant' Communications between the Decision Makers, Including the Equipment Available to Them

		People					
		Comm. Officer	Evaluator	Asst. Evaluator	GLO	Fighter Director	VF Operator
People	Comm. Officer		6			6	
	Evaluator	6		5		7	
	Asst. Evaluator		5				
	GLO						
	Fighter Director	6	7				
	VF Operator						
Machines	VC Radar					8	
	Air Plot	1	9				
	Radio Desk	1					
	PD Panel				2		1
	VG Radar no. 1		9				
	VG Radar no. 2			9	7	4	
	VF Radar				3		9
	Director Repeaters				7		

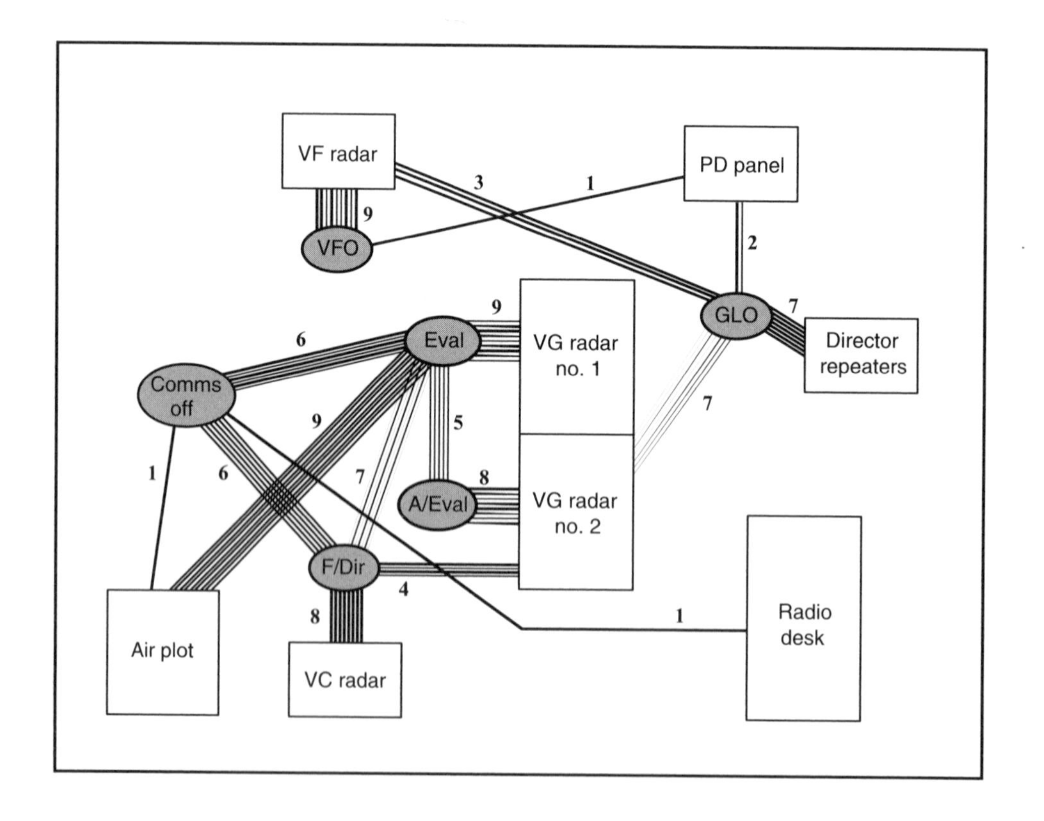

Figure 12 Plan view of CIC, with links superimposed. The plan shows that there is a lot of cross-talk and distant conversations happening in the room, with the implication that errors might occur because of this.

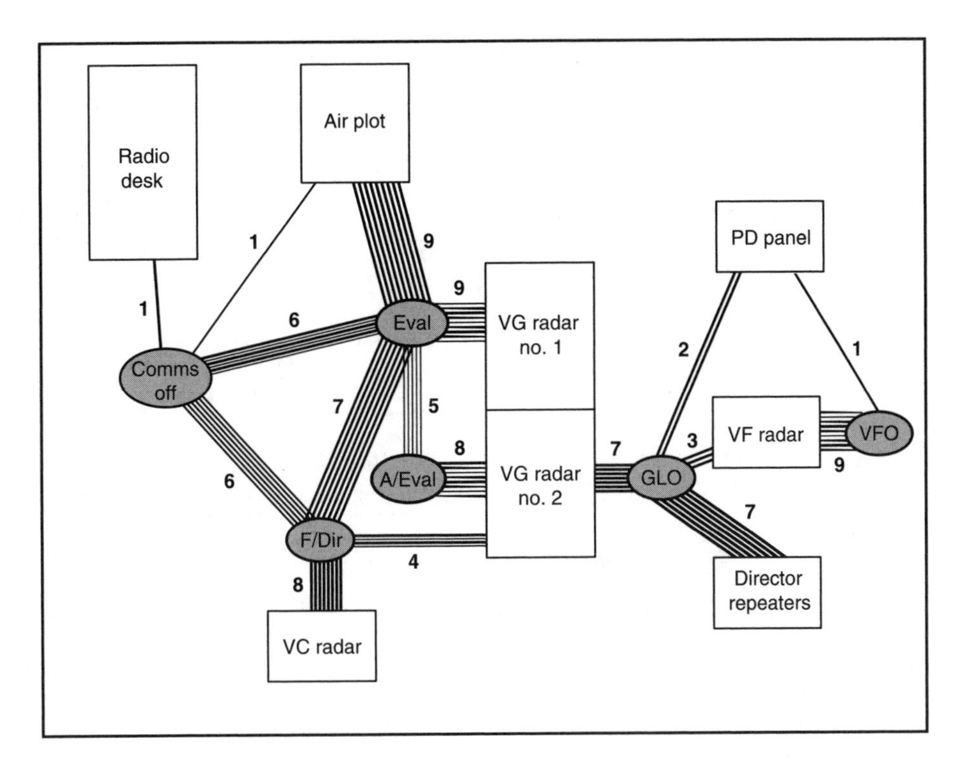

Figure 13 Plan of CIC, rearranged to improve communications among decision-makers. Comparison with Figure 10 will show shorter communications, and no cross-talk.

understanding). The first step is to make sure that the process has been running for some time, so that it is safe to assume that there is a group of process experts available. The next step is to identify a scenario of interest (for example, an operation that carries a significant risk of disaster—loading torpedoes into submarines might be one), with its boundary conditions. This is then divided into a number of 'events', each of which will be explored with this technique. Next, a group of experts is chosen from the available group. Their first task is to discuss the event, and to extract from the participants the primary influencing variables, then the secondary, and so on. For each of the influences that are identified, bipolar descriptive terms are generated (for example, for displays, there might be 'easy to read/difficult to read'; 'beside controls/far from controls'; and so on), to make sure that there is a common understanding of the influences as they are identified. A good way to do this is to use Post-Its, stuck to a wall and relabelled and moved around as the group reaches towards consensus. Once consensus is reached, a more formal diagram can be created, as in Figure 14.

The next step is to quantify the probabilities of occurrence of each of the links. This is normally performed bottom-up, from the most distant influences up towards the target event. These probabilities can then be combined to calculate the overall probability of the event occurring.

Finally, these probabilities can be used in Risk management exercises, and the like.

The advantages of this approach are:

- Any level of task decomposition can be considered.
- The structure underlying the assessment is open and can be audited.
- It makes use of the best available expertise.
- It is useful when no other source of information is available.

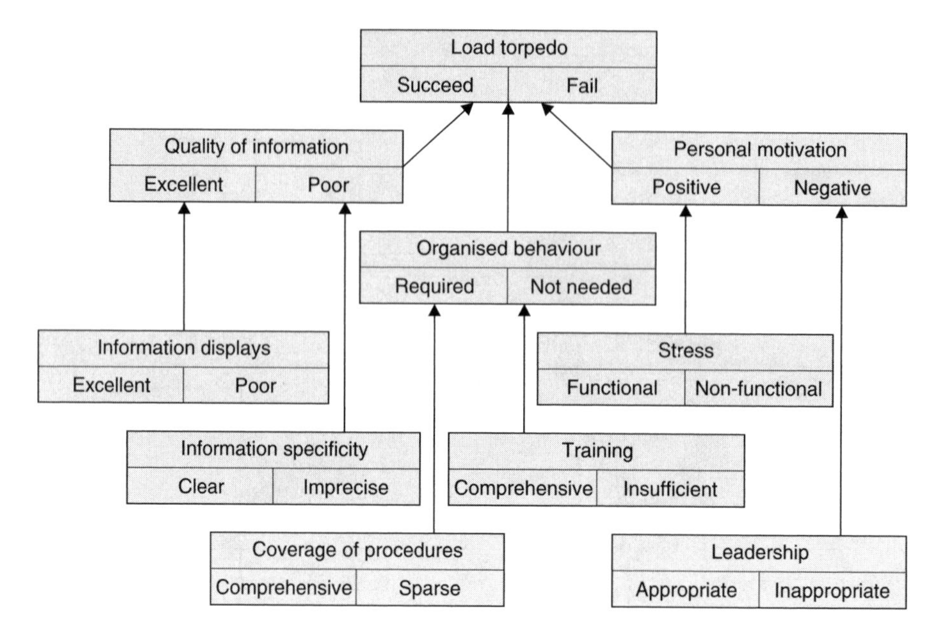

Figure 14 Example of influence diagram, after Kirwan and Ainsworth (1992). The event at the top can be influenced by the variables below by chaining. It is possible to add probabilities to the links, to enable calcualtion of the likelihood of the top event. Each box has a label and a bipolar description to ensure commonality of understanding.

- Generic templates can be used, which list the factors likely to be of importance for editing by the group. This ensures that they think of all the important factors, and reduces the group time required.

It has its disadvantages, too:

- It is costly in terms of resources, and of the administration required to get experts together for the time required.
- The elicitation of the structure and the probabilities depends heavily on the skill of the analyst.
- The 'cause-and-effect' assumptions on which the method is based are likely to hold only for short-term events.

Scenario analysis

Scenarios are brief narratives, diagrammatic or verbal or both, of the operation of some part of a process. These may be created to try to reconstruct an accident, or to explore a prototype process, or any of a range of reasons. A good source of information is the Software Engineering Institute at Carnegie-Mellon University (www.sei.cmu.edu/architecture/scenario_paper/ieee-sw3.htm); an alternative, from a human factors perspective is in Hogg *et al.* (2002). Useful tools for doing this are applications such as Flash™ and Macromind Director™, both widely available.

The first step is to identify the reason(s) for the scenario (for example, 'will two operators really be enough to perform the start-up of this process?'). These will then guide the development of the scenario and its storyline, actors, locations, etc. This should be created with the help of the process stakeholders. The storyline can then be evaluated in detail in relation to the process — 'if that is the next step in the story, then the operator would do A, and would need B to do it'. These steps are documented, together with process successes and failings. These can then be used in process redesign, or in training needs analyses, role definitions, etc.

It is wise to evaluate several different scenarios to make sure that, where they overlap in using process functions, the functions will satisfy all the different scenarios. This will help to ensure a robust process.

Advantages of this technique are:

- It establishes a relationship between the stakeholders and the analysers/designers, without the interference of jargon.
- It can compare alternative ideas at any level of analysis.
- It focuses design effort at the most important areas of the process.
- Scenarios can be made to encompass all roles in a process.
- Scenarios can explore the most unusual of circumstances for a process comparatively easily and quickly.
- The process of choosing scenarios forces analysers/designers to consider a range of future uses and possible changes to a process.

The associated disadvantages are:

- It is costly in terms of resources, and of the administration required to get people together for the time required.
- The quality of the results depends heavily on the quality of the scenarios and on the skills of the analysts involved.
- If computer simulations are deemed necessary to explore the scenarios, they can consume considerable extra time and resources.

The process of process analysis

By now, it should be apparent that some sort of process is required to knit the above into a coherent approach for actually doing process analysis. Fortunately, this thought has occurred to many people, and various methodologies have been proposed. One of the more generic methods, which has a strong socio-technical flavour to it and for which there is both recognition among practitioners and readily-available sources is Checkland's Soft Systems Methodology, which we describe briefly next.

Soft systems methodology (SSM)

This was developed by Checkland in the 1980s within a software engineering environment (Checkland, 1981, 1995; Checkland and Scholes, 1990), and gradually has become part of the philosophy of process analysis and design in almost all professions. Its strength lies in its recognition that the boundaries of systems do not lie at the edges of technical solutions, but include social aspects as well—in this way, it is a true socio-technical vision of process re-design (see Chapter 29). His methodology reflects this too.

Checkland developed his approach over many years in response to the difficulty in applying the approach of hard systems engineering from the engineering disciplines thinking to the much fuzzier problems of business. Hard systems engineering tends to emphasise:

- System boundaries that ignore human and social issues.
- Well-defined and measurable system objectives, omitting the rest.
- Top-down decomposition of systems into sub-systems through an hierarchy of modules.

Probably the most significant aspect of his approach is that it rejects the idea of developing a formal problem statement at an early stage. What most process analysts have discovered, and Checkland has stated as a principle, is that formally stating the problem too early tends to hide many significant, underlying issues. Thus, the method itself can restrict what will be elicited—conclusions will reflect starting positions. This is seldom clever, though it can be quick.

In Checkland's approach the process of process improvement is as important as the outcome. The experience of applying an SSM approach will change the organisation. This will arise from changed views about the problem and possible solutions, and because of this, although progress will be made and changes introduced, in effect the change process never ends. This is Checkland's second significant contribution.

In principle, an SSM project is carried out by the people of the organisation, with a consultant-facilitator. It covers:

- Problem situation described within its environment.
- Analysis, using checklists, models and frameworks as necessary and involving stakeholders (the key principle here it to keep the project loose and wide ranging for as long as possible).
- Conceptualisation of the problem space and of the possible solutions, again with the strong involvement of the stakeholders. Rich pictures, discussed earlier, will be useful here.
- Comparison of solutions with 'reality', perhaps involving simulations and other techniques.
- Definition of chosen solution.
- Design, development and deployment of the solution, again with the strong involvement of the stakeholders. If any process solution is to be effective, it is necessary for its stakeholders to feel that they own it, and participation in its design and development helps to do this.

During the analysis phase and later, Checkland suggests the use of the CATWOE checklist:

- *Clients*: those who benefit or suffer from the execution of the process.
- *Actors*: those involved in the execution of the process.
- *Transformations*: the activities, movements, etc. that take place in and around the process.
- *Weltanschauung or world-view*: the organisation's policies and stated values; 'the way we do things here'.
- *Owners*: those people who hold responsibility for, and authority over, the process, its sub-processes and support processes (for a discussion of process ownership and corporate governance, see Siemieniuch and Sinclair (2002).
- *Environment*: The political, legal, economic, social, demographic, technological, ethical, competitive, geographic environment in which the process is embedded.

Checkland's third significant contribution is his choice of selection criteria for evaluating candidate solutions for the process problem being addressed—his 'Five Es for Selection':

- Efficacy (does it work well?)
- Efficiency (does it do so with minimum resources?)
- Effectiveness (does it contribute to the enterprise goals?)
- Ethicality (is it moral with respect to the stakeholders in the process and the business?)
- Elegance (is it beautiful?)

Checkland's SSM has much to commend it:

- Its deep appreciation of the realities of processes and the understanding people have of them.
- Its provision of a philosophy for process analysis and development that is in tune with the zeitgeist of the decades around the year 2000.
- Its understanding that process improvement is a never-ending process.
- Its lack of rigidity; process improvement is not a 'one size fits all' arena for methods. There is a saying: 'To a consultant with a hammer, every problem is a nail'. The insufficiency of this approach can be very evident in the world of processes.

However, criticisms have been made of SSM:

- There is no real method for people to follow, allowing half-hearted attempts and faulty approaches to be adopted. This is a significant criticism.
- SSM can ignore environmental and structural determinants and questions of power. Not all organisational members have equal choice and it is naive to think that everyone can openly discuss problems.
- Openness and togetherness are implicit and explicit values of SSM. But these are not values to be found in a failing process environment, characterised by confusion, contradiction, organisational politics and dissipation of good effort.

Nevertheless, it is an approach gaining ground among many professions, and it can be discerned in the approaches used by professionals in the ergonomics community when they are working on larger projects.

Other approaches

There are other approaches which are in use, perhaps mandated by customers, which can also deliver good solutions. We outline their names and sources that describe them, in Table 5.

Selecting techniques to use in your process analysis method

If you choose to use an integrated methodology such as SSADM, you will in effect be told which techniques to use at which stage, and will be guided through the whole process. With luck, this will be appropriate to your problem and its context. If, on the other hand, you use a more generic technique such as SSM (a more popular choice, as Older *et al.* (1997) point out), you will have some choice over the techniques to use. If you are to choose, then there needs to be some criteria for you to use. Fortunately, these have been suggested by authors such as Luczak (1997) and Killich *et al.* (1999). They suggest the following:

- Does the technique provide an integrated view of activity-related control, information and object flow?
- Could the technique form part of a co-operating work system: comprises functional tools such as CAD and mediating tools such as CSCW, project management, etc.?
- Does the technique have the ability to represent roles as organisational components?
- If the process includes team-working, does the technique have the ability to represent distributed decision making?
- Can the technique accommodate weakly-structured activities, where there may be no definite ordering of activities (maintenance, for example)?

Table 5 A Listing of Other, Common Methodologies for Performing Process Analysis. There are Many More of These

Acronym	Name and comments	Source
PRINCE	*Projects in Controlled Environments* Launched in 1996, PRINCE has become widely used in both the public and private sectors and is now the UK's de facto standard for project management. PRINCE2 is said to provide a number of organisational benefits, including: a controlled and organised start, middle and end; regular reviews of progress against plan and against the Business Case; flexible decision points; and automatic management control of any deviations from the plan.	PRINCE home page: http://www.open.gov.uk/ccta/prince/prince.htm
DSDM	*Dynamic Systems Development Method* DSDM is a public domain Rapid Application Development (RAD) method which has been developed by a consortium of vendors and user organisations. It is considered to be the UK's de-facto standard for RAD. DSDM provides a generic process which must be tailored for use in a particular organisation dependent on the business and technical constraints. It consists of five main phases: feasibility study; business study; functional model iteration; design and build iteration; implementation.	DSDM Consortium Web-site: http://www.dsdm.org/
SADT	*Structure Analysis and Design Technique* The purpose of SADT is to organise, display and provide a straightforward technique for identifying and cross referencing information necessary for publication of acceptable requirements and design specifications. SADT is a tool to structure information and allows the engineers a better understanding of the problem that is to be solved. Therefore SADT is mainly used for mind-setting and structuring the analysis and design phases.	SADT homepage: ttp://panoramix.univ-paris1.fr/CRINFO/dmrg/MEE97/misop001/
SSADM	*Structured Systems Analysis and Design Methodology* One of the most widely used application development methods in the UK. The overall objective is to produce a detailed data and program specification for a computer system in a standard format by applying standard techniques and procedures. The latest version of the method (SSADM4+) describes a framework, known as the System Development Template (SDT), for the capture of user requirements, the analysis of information needs and the design and specification of information systems.	McLaren, R., Hornby, P., Robson, J., O'Brien, P., Clegg, C. and Richardson, S. (1991). Systems Design Methods — The Human Dimension. DTI
UML	*Universal Modelling Language* A generic technique sweeping through software engineering, and likely to be adopted in other disciplines. Has the advantage of being able to handle team-working and co-operative working.	Booch *et al.* (1998)

- Does the technique enable hierarchical modelling, the use of model layers and purpose-driven views
- Is the technique transparent, consistent and self-descriptive, for the benefit of multi-users?
- Is the technique software-based, for speed and accuracy?
- Is the technique easy to use?
- Do you have confidence in the technique?

None of the techniques described above (nor any not mentioned) actually satisfy all of these criteria.

The practicalities of process analysis

Whichever method you choose, and whichever of the techniques you import into the method, there are a number of practical considerations to bear in mind. We discuss these briefly.

Audit trails

It is important that you document your investigations; what you have done, who has been interviewed, notes of the content of data-gathering exercises, how you cleansed and analysed the data, and so on. This is for two reasons; firstly, we live in litigious times, and being able to show integrity in what you did, and that what you did represented 'state of the art' at that time, will be the best defences in court for both you and your employer. Second, documentation affords provenance for your conclusions and recommendations offered to the process stakeholders. A good source of information about audit trails is Miles and Huberman (1994).

The 'correct' level of detail in process analysis and process definition

One of the most significant principles in socio-technical systems theory is the 'minimum critical specification'. The meaning of this in our context is that, given a procedure to be specified, don't over-specify it. Only specify what must be specified, and leave the other parts to the person or team to decide for themselves. The first reason for this follows from Checkland's approach, outlined above; processes and procedures always happen in a context, and this may change over time. You will not be able to predict these, so leave room for manoeuvre for the people in the context who have to make the process or procedure work. Secondly, once you specify a process or procedure in detail, you create the opportunity for central control, with the immediate danger of micro-management from a distance. This is a recipe for blunders and inefficiencies, and considerable bickering between the process people and the distant management. So, although there may be a case for detailed data capture, be cautious about the level of detail you show in your formal representations.

How do you determine what must be specified and formalised? The most obvious drivers are health and safety considerations, legal requirements, fulfilling customer needs, and the need for measurement points to be able to evaluate process performance and to identify where process improvements will be required.

Characteristics of experts

Much of the data, information and knowledge required for process analysis and definition will come from interviews and other person-to-person techniques. Management of this interface is important, to ensure data quality, and we list below some of the identifiable characteristics of subject matter experts that will be of importance when trying to elicit information. It should be borne in mind here that any time you try to elicit information about a process from a stakeholder, that person is in effect an expert, however little of value that person might know. You may wish to refer to Chapter 8 in this book.

What is expertise? This derives from superior specialist knowledge, better technique, or more experience. The former is relatively easy for people to discuss; the latter two have a large content of tacit knowledge, which by definition is hard to describe and to impart to others. It is here where your skills will really be required. Furthermore, experts don't provide the information in the order and format you might desire; they will deliver it in the way that they themselves use it, and it will be for you to reorganise it into a more useful form. It is here that Rich Pictures and the like come to the fore.

An important issue that you will encounter is that experts have different models of the task and the process compared to novices, and they may differ in their models among themselves. From a process analysis point of view, this raises the problem of reconciliation of different viewpoints (we discuss this below); from a process design point of view this points to the need for careful, flexible design of the processes and their interfaces, so that they can be tailored to the needs of the individuals who will operate them.

One technique that has proved useful in tackling this issue is the development of a high level meta-model of the overall process under consideration. In essence this can be a very basic overview diagram of the process that contains enough hooks on which the various experts may hang their own perspectives. Figure 15 shows an example of such a diagram. It provides an overview of three main processes and the links between them: a capability development process, the product development process(es) into which new capability will be introduced and the set of support processes required for both of these.

Finally, it should always be borne in mind that experts can be wrong (just like you and us). Unfortunately, it is not obvious when they are wrong, and this usually can only be detected when one attempts the reconciliation process. And you cannot do reconciliation unless you have spoken to more than one expert—there is a rule lurking here.

There are some implications which stem from these considerations, known as the Gardner–McKenzie rules:

- Knowledge elicitation presupposes a high level of interpersonal understanding, common culture, and common language. Therefore, you should make sure you have spent time with the process, understanding it and the jargon people use about it.
- Clumsy elicitation may lead experts to reject the process, e.g., by giving minimalist or incomplete answers, or an open refusal to continue. In other words, you will need to cultivate a good 'bedside manner', as doctors do. This means that all the mistakes are yours; the expert leads the conversation; you treat the expert with respect; you are trustworthy and maintain any confidences you may be told; and so on.
- The opportunities for misinterpretation are frequent and subtle. In other words, unless you know the domain, you may easily be misled. The early defences against this are to obtain information from several experts (as many as your resources will allow), and to learn about the area as quickly as possible. The later defence is to be very open about what you are doing, and take your analyses back to the users for rechecking frequently.
- The range of misinterpretations is greater than you imagine. The defences against this are the same as above.
- Expert's knowledge may not be organised as you wish, it may be limited to situations, and it may be wrong. Again, it helps if you know the area when you begin. Then the use of organising techniques such as lists, Rich Pictures, etc. can be very helpful. Finally, there is the testing of your analyses with the users, as above.

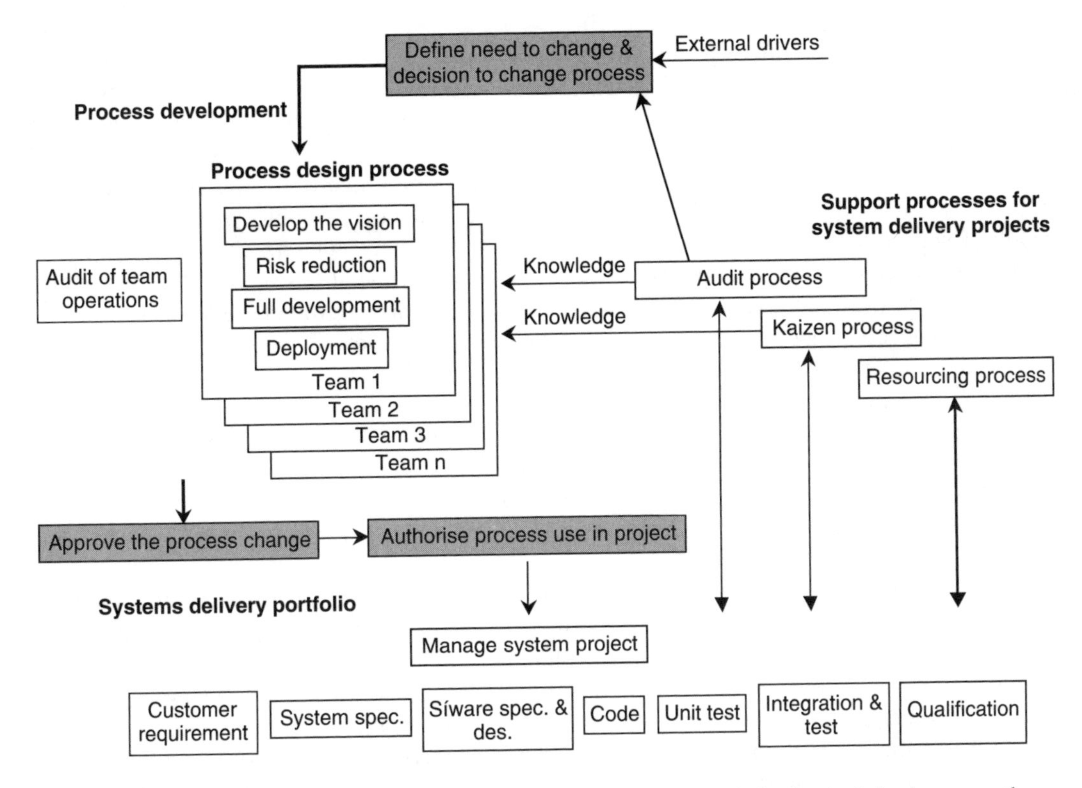

Figure 15 High level meta-model of three generic processes and the basic links between them. Generic diagrams such as Acknowledgment are useful in getting experts started in their contributions, and in helping you to understand the differences between experts' viewpoints.

- Experts may have only a partial knowledge of the extent of their knowledge; this often shows as overconfidence. This is not easy to detect in one individual, but talking to several experts will usually reveal where this has happened.
- Experts often know a great deal about a topic, but not a necessary fact. Expertise comes in many forms, and one of the most important is in knowing what not to do, followed closely by knowing what to do at an early point in time. Neither of these requires knowledge of particular facts, and this applies to many other classes of knowledge as well. If you need to know a particular fact, it may be necessary to talk to engineers, or observe for yourself what happens.
- Experts often rationalise their skills. For example, describe how you ride a bicycle. If you think you can describe this, try describing it in finer detail. Eventually, you will discover a train of thought going through you head like; 'Well, I must be doing X'. At this point, you are rationalising what you do. You may not be correct. There is not much you can do about this, because other experts may well use the same rationalisations. Bear in mind that when you reach the point where rationalisations may be occurring, you may have reached the point where you are exceeding the minimal critical specification discussed above.

The reconciliation of differing viewpoints

This can become a non-trivial exercise. In some cases, the different viewpoints may arise from errors; these will fairly rapidly be identified, tracked and corrected. It gets more difficult when there are genuine alternatives to the way a process may be executed. These may arise as a result of the minimum critical specification, where people are able to develop different ways to do the process. This situation may reveal itself by the differences between the alternatives being fairly small. However, you may well find yourself in trouble if you make the easy decision to go for the average answer. For example, assume you are looking at a process run 24 hours a day, 7 days a week (electricity generation is a case in point). This will probably require 5 shifts to enable continuous operation. These five shifts, at the beginning of their own shift, may set up the process slightly differently, because each shift has slightly different goals; one shift might want to maximise their wage packets, so they drive the process slightly harder. Another might be concerned not to have shut-downs, because they require extra effort. So they make things happen slower. These different ways of setting up the process may have evolved over a period of time, by a process of minor trials and errors. If the parameters of the process interact (as they usually do), then there will be particular combinations which succeed in delivering the goals, and others that won't. If you average the parameter settings, you may well find that the result is particularly bad. The reason is that between the performance peaks that the shifts may have discovered, there may be performance valleys separating them, and averaging may leave you in one of these.

Furthermore, if a team has learnt over a period of time a particular way to run the process that works well for them, they may be extremely resistant to any suggestions from outside.

In short, it is a balancing act, requiring skill rather than science, to decide whether there is one best way or several, whether to try to detail everything or to leave things a little vague, and to achieve acceptance of the results of your analysis. This comes with experience.

Perhaps the best comment on the problems of reconciliation and dissent that you will encounter is from Geer (1981); the real value of task analysis may be that you personally learn so much from the exercise about the process; this can be put to very good use in the future.

The last word

Congratulations. You have arrived here. Now, it is time to do some process analysis, because there is no substitute for learning how to do it. We remind you of the dictum of Henry Petroski, a wise engineer: 'Falling down is a part of growing up'. (Petroski, 1992). Forwards!

References

Allen, P.M. (1998). Evolving complexity in social science. *Systems: New Paradigms for the Human Sciences*, edited by G. Altman and W.A. Koch (New York: Walter de Gruyter).

Ashby, W.R. (1962). Introduction to cybernetics. *Principles of Self-organising Systems,* edited by H.V. Foerster and G.W. Zopf. (New York: Pergamon).

ASME (1972). *ASME Standard — Operation and Flow Process Charts.* (New York: American Society of Mechanical Engineers).

Becker, G.S. (1992). Nobel lecture: the economic way of looking at behaviour. *The Nobel Foundation,* December 9.

Booch, G., Rumbaugh, J. and Jacobson, I. (1998). *The Unified Modelling Language User Guide.* (Reading MA: Addison Wesley).

Chapanis, A. (1965). *Research Techniques in Human Engineering.* (Baltimore, MD: Johns Hopkins Press).

Checkland, P.B. (1981). *Systems Thinking, Systems Practice.* (Chichester, UK: J. Wiley & Sons).

Checkland, P.B. (1995). Soft systems methodology and its relevance to the development of information systems. *Information Systems Provision: the Contribution of Soft Systems Methodology,* edited by F.A. Stowell. (Maidenhead, McGraw-Hill), pp. 1–17.

Checkland, P.B. and Scholes, J. (1990). *Soft Systems Methodology in Action.* (Chichester, UK: J. Wiley & Sons).

Cullen, I. (1995). *Address to EPSRC 'CDP Town Meeting'.* (London: EPSRC).

Curtis, W., Hefley, W. and Millar, S., (2002). *People Capability Maturity Model* (P-CMM), Version 2. (Philadelphia: Systems Engineering Institute, Carnegie Mellon University).

Davenport, T.H. (1993). *Process Innovation.* (Boston: Havard Business School Press).

Endsley, M.C. (1998). Situation awareness, automation and decision support: designing for the future. *CSERIAC Gateway,* **9**, 11–13.

Endsley, M.R. and Kaber, D.B. (1997). The use of level of automation as a means of alleviating out-of-the-loop performance problems: a taxonomy and empirical analysis. *13th Triennial Congress of the International Ergonomics Association—IEA'97,* Tampere, Finland, June 29–July 4th 1997, Finnish Institute of Public Health.

Endsley, M.R. and Kiris, E.O. (1995). The 'out-of-the-loop' performance problem and level of control in automation. *Human Factors,* **32**, 381–394.

Geer, C. (1981). *Human Engineering Procedures Guide,* Wright-Patterson Air Force Base, OH.

Gregg, D. (1996). *Emerging Challenges in Business and Manufacturing Decision Support.* The Science of Business Process Analysis, ESRC Business Process Resource Centre, University of Warwick, Coventry, UK, ESRC Business Process Resource Centre.

Hogg, D., Sturrock, F. and Wykes, K. (2002). HMI assessment of the Nimrod MRA4 flight deck. *Contemporary Ergonomics 2002—proceedings of the Annual Conference of the Ergonomics Society,* edited by P.T. McCabe. (London: Taylor & Francis).

Humphrey, W.S. (1989). *Managing the Software Process.* (Reading, MA: Addison-Wesley).

Hunt, V.D. (1996). *Process Mapping.* (New York: J. Wiley & Sons).

IDEF (1981). *ICAM Architecture Part II—Vol IV—Functional Modelling Manual* (IDEF0). Wright-Patterson Air Force Base, OH, Materials Laboratory, Air Force Wright Aeronautical Laboratories, Air Force Systems Command.

IDEF (1993). *Integration Definition For Function Modeling (IDEF0).* Gaithersburg, MD, National Institute of Standards & Technology.

Kauffman, S.A. (1989). Principles of adaptation in complex systems. *Lectures in the Science of Complexity,* edited by D.L. Stein. (New York: Addison-Wesley), pp. 619–712.

Killich, S., Luczak, H., Schlik, C., Weissenbach, M., Weidenmaier, S. and Zeigler, J. (1999). Task modelling for cooperative work. *Behaviour and Information Technology,* **18**, 325–338.

Kirwan, B. and Ainsworth, L. (1992). *A Guide to Task Analysis.* (London: Taylor & Francis).

Kristensen, L.M., Christensen, S. and Jensen, K. (1998). The practitioner's guide to coloured Petrie nets. *International Journal of Software Tools fior Technology Transfer,* **2**, 98–132.

Luczak, H. (1997). Task analysis. *Handbook of Human Factors,* edited by G. Salvendy. (London: J. Wiley & Sons).

March, A. (1986). *The views of Deming, Juran and Crosby.* Harvard Business School.

Marion, R. (1999). *The Edge of Organisation: Chaos and Complexity Theories of Formal Social Systems* (Thousand Oaks, CA: Sage Publications).

Martin, P. (1998). Time for new trade-offs. *Financial Times.* (London: Pearson) pp. 42–43.

Miles, M.B. and Huberman, A.M. (1944). *Qualitative Data Analysis.* (London: Sage).

Morton, C. (1998). *Beyond World Class.* (Basingstoke, U.K.: Macmillan Press).

Older, M.T., Waterson, P.E. and Clegg, C.W. (1997). A critical assessment of task allocation methods and their applicability. *Ergonomics,* **40**, 151–171.

Petroski, H. (1992). *To Engineer Is Human: The Role of Failure in Successful Design.* (Random House).

Prahalad, C.K. and Hamel, G. (1990). The core competence of the corporation. *Harvard Business Review,* **68**, 79–91.

Rasmussen, J. (1997). Risk management in a dynamic society: a modelling problem. *Safety Science,* **27**, 183–213.

Rasmussen, J. (2000). Human factors in a dynamic information society: where are we heading? *Ergonomics,* **43**, 869–879.

Rycroft, R.W. and Kash, D.E. (1999). *The Complexity Challenge.* (London: Pinter).

Salvendy, G., Ed. (1997). *Handbook of Human Factors and Ergonomics.* (New York: John Wiley & Sons).

Sandom, C. (2001). Situation awareness. *People in control: Human Factors in Control Room Design,* edited by J. Noyes and M. Brandsby (London: Institute of Electrical Engineers). p. 60.

Sawhney, M. and Prandelli, E. (2001). Communities of creation: managing distributed innovation in turbulent markets. *California Management Review*, **42**, 268–302.

Senge, P. (1992). Building learning organisations. *Journal for Quality and Participation* (March).

Shepherd, A. (1986). Issues in the training of process operators. *International Journal of Industrial Ergonomics*, **1**, 49–64.

Siemieniuch, C.E. and Sinclair, M.A. (1999). Organisational aspects of knowledge lifecycle management in manufacturing. *International Journal of Human-Computer Studies*, **51**, 517–547.

Siemieniuch, C.E. and Sinclair, M.A. (2002). On complexity, process ownership and organisational learning in manufacturing organisations, from an ergonomics perspective. *Applied Ergonomics*, **33**, 449–462.

Siemieniuch, C.E. and Sinclair, M.A. (2002). The role of the Process Owner in avoiding the 'drift to disaster' in processes. *2nd International Conference on Occupational Risk Prevention*, Maspalomas, Gran Canaria.

Siemieniuch, C.E. and Sinclair, M.A. (2004). CLEVER: A process framework for knowledge lifecycle management. *International Journal of Operations and Production management*, **24**, 1104–1125.

Stewart, T.A. (1997). *Intellectual Capital: the New Wealth of Companies*. (New York: Doubleday).

Turnbull, N. (1999). Internal control: guidance for directors on the Combined Code. (London: Institute of Chartered Accountants in England and Wales), p. 18.

Waldrop, M.M. (1992). *Complexity— the emerging science at the edge of order and chaos*. (New York: Penguin).

Webb, E.J., Campbell, D.T., Schwsartz R.D. and Sechrest, L. (1966). *Unobtrusive Measures*. (Skokie, IL: Rand McNally).

Chapter 39

Conclusions

John Annett

The growth of ergonomics science

Ergonomics, or human factors engineering, has existed as a recognised field of activity for a little over half a century. Stimulated by the exigencies of the second world war, the technological developments of the cold war and post-war expansion of the chemical processing and power generation industries, by advances in civil and military aviation and space travel and the current explosion of information technology, it has become during the closing year of the 20th century a recognised academic discipline with specialist university departments offering undergraduate and postgraduate degrees. Scientific disciplines are defined as much by the methods they employ as by the objects of their study and so it is with ergonomics.

In those, still remembered, early days ergonomists came into the field armed with their own training in disciplines such as anatomy, physical anthropology, human physiology, psychology and production engineering. Each of these disciplines brought a ready-made and distinctive array of methods. The early texts such as Morgan *et al.* (1963) and Murrell (1965) dealt extensively with methods already established in their base disciplines for measuring body dimensions and metabolism, visual and auditory acuity, illumination and environmental noise and many others. But even at that early stage cross-discipline techniques were beginning to emerge from new problems. For example, although the speed and accuracy of human movement had been the subject of study since the late 19th century (Woodworth, 1899) the recognition of the role of the human operator as a servo in manual control systems (Hick and Bates, 1950) brought new ways of measuring the demands made by the dynamic properties of the controlled systems and efficiency of human manual controllers (Fitts, 1951). Chapanis (1959) presented a broad view by showing how methods derived from work study engineering, such as activity sampling, link analysis, and critical incident analysis, could be brought to bear on problems of 'human engineering'. He emphasised the importance of experimental and statistical methods employed principally by experimental psychologists studying human performance who drew on a long tradition of measuring reaction time and sensory thresholds under different conditions.

The array of methods now available to the ergonomist and reviewed in this book is considerable and continues to grow as new problems arising from new technologies come to the fore. The traditional issues of workplace design and posture, of legibility of inputs and the stresses imposed by environmental conditions are still with us but the development of information-dependent systems including significant elements of computing and the co-operation of many individuals via communication networks has brought to the fore a new range of issues such as the evaluation of team work. Although the concept of the human–machine system as a primary object of study has long been familiar the degree of complexity is ever expanding opening up new problem areas, for example the 'glass cockpit' and a demand for more sophisticated methods for both measuring and modelling system performance.

Strategies, methods and techniques

If the pioneers of ergonomics were specialists applying known techniques to problems of human work the modern ergonomics professional is more like a general medical practitioner or even an architect who may be required to draw on a variety of disciplines and methods to deal with a specific problem, be it the design of a warning system or the training of operatives. Like other 'clinicians' faced with a unique real world problem one needs not just a set of methods but a *strategy* for deciding how best to satisfy the client's needs and consequently what methods to employ. The practitioner must know where it is necessary to draw on the expertise of specialists to employ appropriate techniques in much the same way that the GP calls on the radiologist to assist a diagnosis or the architect calls on the structural engineer to calculate the stresses involved in a proposed building design.

By the term 'strategy' I mean an approach to discovering the real nature of the client's needs. For example, a client may 'present' with a demand for an improved training scheme but a careful investigation may suggest to the consultant that greater benefit might be achieved by re-designing the task to be learned. Such a strategy has to be based on thoughtful examination of the problem and its symptoms, by careful questioning of the client, including designers, managers and operatives. Many factors may have to be kept in mind including the time, physical, human and financial resources available as well as the costs associated with any likely solution. A junior ergonomist under contract to a large organisation may not have the degree of flexibility required for such an open-ended approach. He or she has perhaps been asked to produce a measure of the mental workload imposed by a given task and that is what has to be done. I well remember as a very junior clinical psychologist being required by the consultant psychiatrist to produce an IQ value for a patient knowing only that if the figure I obtained using the standardised test turned out to be 70 or less the individual would be automatically incarcerated as 'mentally defective'. Broad strategies are difficult to define in any field but they mostly depend on avoiding tunnel vision and, like Sherlock Holmes, noticing that what was strange was that the dog did *not* bark in the night! To employ a strategy to investigate a problem rather than simply to apply a technique requires both skill and experience in determining what course of action is likely to be in the best interests of the client by determining possible diagnoses and in negotiating 'cures'.

Analysis and measurement

The methods described in this volume have been classified into five broad categories, general approaches (which I take to be strategies in the sense discussed above) methods of design, the analysis of work, the analysis of the physical workplace and work systems. Each of these categories includes a variety of techniques which are often tightly defined standard procedures which must be followed if the procedure is to be valid. These are usually those techniques which, like 'intelligence' tests, are designed to place a numerical value on a particular variable whether it be rate of oxygen uptake or the mental workload imposed by a task. A distinction has been made (Annett, 2002a; Booher and Hewitt, 1990) between those methods and techniques which are aimed at understanding the problem and those which aim to measure specific qualities of the worker, the task or the workplace. The former embrace those methods designed primarily to describe, analyse and model the observable and inferred processes underlying the behaviour of the system. These methods include task and systems analysis, simulation and modelling. Understanding 'the problem' means identifying the major sources of variation in system performance much as proposed by Chapanis (1951) and this can only be done by collecting data relevant to system performance, especially inputs and outputs, but also bearing in mind the values of the organisation. Such data can normally be obtained from existing records (see Chapter 3) but system values may be harder to tie down. Obvious cases are the acceptable trade-off between quality and quantity of product or between cost and reliability in running a system. We have seen in the case of the railway system how these organisational values can change overnight as a result of a high-profile incident.

This volume describes various methods, or rather strategies, for analysing the actual or predicted performance of complex human machine systems. Over the years these have developed from basic behavioural descriptive methods to more sophisticated 'cognitive' methods such as HTA, CTA and GOMS with increasing attention to simulation and modelling as an important tool for system development. As Laughery and Meister (see Chapter 9), put it 'computer modelling and simulation is the way of the future.'

Even when using computerised tools which incorporate theoretical or historical databases, modelling is a creative activity that can be contrasted with quantification and measurement as such. I have argued (Annett, 2002a) that the concept of validity takes on a different significance as between analytic and evaluative (measurement) methods. A valid analysis or model will behave in the same manner as the system modelled in the sense that it will respond appropriately to the full range of demands placed upon it by the natural working environment. Modelling has the advantage over simple analysis in that it incorporates a dynamic response to varying situations. If task analysis describes what *should* happen, modelling provides estimates of what is *likely* to happen and thus adds an element of prediction. The primary value of a 'paper' task analysis is that it draws attention to key features of the task structure and to potential problems whereas the value of a simulation lies in how well the model predicts actual system performance. However, the adequacy of prediction depends on both the structure of the model and the quality of the data used for test runs and when predictions fail it is not always easy to say in which the problem lies.

The validity of methods whose primary purpose is to measure some attribute, such as workload, imposed by a task or a system requires a somewhat different interpretation. Here, as in classical psychometrics, validity can mean either construct validity or predictive validity. Construct validity, that the test measures a coherent theoretical entity, is a continuing problem for ergonomics. Constructs such as 'fatigue', 'comfort', 'workload' have a long history whilst constructs such as 'usability' and 'presence' are of more recent origin in the fields of human–computer interaction (HCI) and virtual reality (VR). There is a growing array of methods involving subjective judgement (Annett, 2002b). The early authorities (Morgan *et al.*, 1963), using methods based largely in the physical science disciplines warned of the dangers of subjectivity yet there seems no way of avoiding the problems associated with the measurement of such important constructs. Whether these qualities can be measured in any strict sense of the term is debatable (Annett, 2002b). There are two key problems. One is whether subjective experience can justifiably form part of an objective scientific enquiry and the second is whether constructs such as mental workload can be shown to have true quantitative structure in measurement theory (Michel, 1997; Kline, 1998). A case can be made that expressions of opinion including ratings and rankings, even though referring to entirely private events, can be legitimately regarded as public data. As such they can be used to predict other objective data. For example reports of subjective or perceived fatigue might be used to predict error rates, although curiously enough this has rarely been done. A useful approach to establishing the validity of a subjective construct is shown in the work of Åhsberg (1998) in developing the Swedish Occupational Fatigue Inventory (SOFI). The approach is based on the statistical method of structural equation modelling, sometimes called confirmatory factor analysis, in which a source group of 'manifest' or 'indicator' variables is used to predict the values of one or more 'latent' variables. Manifest variables can include both subjective ratings, such as feelings of sleepiness and objective measures of performance or physiological variables such as heart rate. The modelling process determines the weightings of the indicator variables which best predict the latent variable. With an acceptably good fit the model may then be used to predict performance variables. Whilst this multivariate statistical approach to the validation of ergonomic constructs seems right in principle it may be thought impractical to produce general measures in view of the very considerable effort in data collection and analysis which would be required. For this reason we are still without a generally applicable measure of mental workload (see Chapter 18), despite numerous studies extending over many years.

The second problem of the quantitative structure of subjective ratings pointed out by measurement theorists (Michel, 1997) may be overcome by reference to conjoint measurement which can give an assurance of additivity. This technique has been applied to the Subjective Workload Assessment Technique (SWAT) by Reid and Nygren (1988) but has not yet been widely used with ergonomics scales. Subjectivity may refer either to the respondent's internal and private feelings, such as fatigue, but also to a personal response to a public stimulus such as the urgency of a warning signal. Hellier and Edworthy (2002) suggest that in the latter case good agreement between observers (reliability) can be obtained using Stevens's free modulus magnitude estimation technique and quantitative structure can be reasonably assumed. Clearly methods of measuring ergonomically interesting variables are in urgent need of thoroughgoing basic research and not just *ad hoc* attempts to satisfy the immediate needs of clients. It can be difficult to do much-needed basic science in the context of commercially driven applied work, which probably accounts for the great majority of current activity in ergonomics, but without such a programme the question of measurable standards must remain on hold.

Conclusions

This is the third edition of this text on methods in ergonomics and it is to be hoped not the last since it is clear that the variety and sophistication of the strategies, methods and techniques available to the ergonomist are undergoing continuous development and change. This is due in part to the changing nature of human work and the complexity of the functional systems which constitute our highly technological environment and partly due to the increasing sophistication of the available methods of analysis and measurement. As in all branches of both pure and applied science the choice of an appropriate strategy in dealing with the problem as presented is something it is difficult to teach except through experience and familiarity with the many techniques which are available. I suggest that, wherever possible the strategy should be to make sure that the real problem is not something of which the client is unaware and that the techniques employed are those which are most appropriate to the real problem. Sometimes analytic methods will have to be adapted to new situations for which there is no specific precedent. It is unwise to be a slave to a detailed procedure when the primary aim is to understand but, on the other hand, in using a truly standardised measurement method strict adherence to recording and scoring procedures is paramount. There are still open questions about the validity and quantitative probity of many subjective measurement methods in current use and more fundamental work is needed. However, we should not be deluded into believing that a measure of, say, fatigue or workload is to be preferred simply because it is derived from the use of a physical instrument. Many ergonomic constructs are by their very nature multidimensional and can best be assessed by statistical analysis of a battery of measures and it is here that more basic research is needed.

References

Åhsberg, E. (1998). Perceived fatigue related to work. University of Stockholm, National Institute for Working Life Monograph series no. 19.

Annett, J. (2002a). A note on the validity and reliability of ergonomic methods. *Theoretical Issues in Ergonomics Science*, **3**, 228–232.

Annett, J. (2002b). Subjective rating scales: science or art? *Ergonomics*, **45**, 966–987.

Booher, H.R. and Hewitt, G.M. (1990). Manprint tools and techniques. In H.R. Booher, (Ed.) *Manprint: An Approach to Systems Integration.* (New York: Van Nostrand Reinhold) pp. 343–390.

Chapanis, A. (1951). Theory and methods for analyzing errors in man-machine systems. *Annals of the New York Academy of Sciences*, **51**, 1179–1203.

Chapanis, A. (1959). *Research Techniques in Human Engineering.* (Baltimore: The Johns Hopkins Press).

Fitts, P.M. (1951). Engineering psychology and equipment design. In S.S. Stevens (ed.), *Handbook of Experimental Psychology.* (New York: Wiley), pp. 1287–1340.

Hellier, E. and Edworthy, J. (2002). Subjective rating scales: scientific measures of perceived urgency? *Ergonomics*, **45**, 1011–1014.

Hick, W.E. and Bates, J.A. (1950). *The Human Operator of Control Mechanisms.* Permanent record of research and development. Monograph 17.202. (London: HMSO).

Kline, P. (1998). *The New Psychometrics: Psychology and measurement.* (London: Routledge).

Michel, J. (1997). Quantitative science and the definition of measurement in psychology. *British Journal of Psychology*, **88**, 355–383.

Morgan, C.T., Cook, J.S., Chapanis, A. and Lund, M.W. (1963). *Human Engineering Guide to Equipment Design.* (New York: McGraw-Hill).

Murrell, K.F.H. (1965). *Ergonomics: Man in His Working Environment.* (London: Chapman & Hall).

Reid, C.B. and Nygren, T.E. (1988). The subjective workload assessment technique: a scaling procedure for measuring mental workload. In P.A. Hancock and N. Meshkati (eds), *Human Mental Workload*, (Amsterdam: North Holland). pp. 185–218.

Woodworth, R.S. (1899). The accuracy of voluntary movement. *Psychological Review Monograph Supplements*, **3**, no.3.

Index

A

B

C

H

I

N

O

P

Q

R

S